Advances in Fish Science
and Technology

Advances in Fish Science and Technology

**Papers presented at the
Jubilee Conference of the Torry Research Station
Aberdeen, Scotland. 23–27 July 1979**

Edited by

**J J Connell, Director,
and staff of
Torry Research Station**
Aberdeen, Scotland

Published by

Fishing News Books Ltd
Farnham, Surrey, England

British Library CIP Data

Advances in fish science and technology.
   1. Fish-culture
   I. Connell, John Jeffrey
   664'.94   SH151

   ISBN 0 85238 108 5

ISBN 0 85238 109 3 Limp Conference Edition

Made and printed in Great Britain by
The Garden City Press Limited, Letchworth,
Hertfordshire SG6 1JS

Contents

List of Contributors

Foreword

About 50 years ago a number of countries, of which the United Kingdom was one, recognized that there was a need to expand research on the preservation of fish. This led to the establishment of a number of publicly funded national institutes devoted solely to this kind of work. The succeeding years have seen a gradual increase in the scale of fish research and development in every fishing nation in the world.

Among these institutes was Torry Research Station, which was opened in July 1929 in the district of Aberdeen, Scotland, known as Torry. From a very small beginning with a handful of staff, Torry has grown into an important and well-equipped laboratory employing over 200 people. At first concerned with chilling practices on board fishing vessels, the work of the Station and its contemporaries has expanded to include all aspects of the handling, processing, preservation, distribution and marketing of fish and fishery products. All kinds of activity are now included, from investigations of background science to applied technology and providing information and advice.

As a way of marking 50 years' existence, Torry conceived the idea of an international conference at which many of those who had contributed to fish R & D could come together both to discuss their work and to celebrate the jubilee. Essentially the meeting was to pay tribute to 50 years of co-operative international work by a large number of individual scientists and technologists. It was planned in two parts. The first half consisted of a series of retrospective and prospective accounts of 15 major areas of interest, given by acknowledged experts. These set the scene for the second half, consisting of papers describing the latest ideas, results and developments.

This book contains almost all the contributions made at the conference, although mostly for reasons of space some had to be left out. It provides the reader with a comprehensive and unique record of how the subject has developed. It sets down very fully current advances. It speculates and points to the future. Thus it is an invaluable document for all those concerned with the flourishing branch of food science and technology that is called fish science and technology.

As with other branches, fish science and technology is changing rapidly and, in a number of respects, the future is uncertain. We can be sure, however, that continued demands will be made of the scientist and technologist. Thus as stocks of unfamiliar fish and shellfish as well as unfamiliar forms of edible materials derived from them become available, the technologist has an important task in showing how these can be used to maximum effect. In recent years a great deal of attention in the UK and elsewhere has been devoted, for example, to the utilization of shoaling pelagic species and of recovered fish and shellfish flesh. Krill has been in the news. I note that several of these topics feature prominently in this book. In responding to these demands, Torry, in common with other laboratories, is directing its technical effort primarily towards advice and support for industry, but also to officials and users of fish. In line with current thinking about the formulation and management of R & D, there is a close liaison between the customers who use and need the results and the contractors in the laboratories who provide the results.

Fish is meant to be eaten. The needs of the fish eater are therefore paramount and must be kept constantly in mind. Laboratories like Torry have a dual role in suggesting ways in which these needs in respect of quality, wholesomeness and safety can be met, while at the same time helping to sell fish at a profit.

Fish technologists also have a key role to play in helping developing countries to make greater use of their fish resources and reduce the considerable waste of fish caused by spoilage or pests. Participants at the Jubilee Conference have frequently been engaged in valuable projects of this kind, often under the aegis of the international development agencies.

The fish scientist, working in the background but always looking at practical requirements, contributes through the knowledge and understanding he can bring to bear on the difficult problems underlying the properties and behaviour of fish as food. Although the book alludes to fish science and technology as apparent separate entities, they are in fact a continuum of endeavours with a common aim. Fish science and fish technology are equally important and can be equally challenging.

When I opened the Jubilee Conference at the University of Aberdeen in July 1979, it was my pleasant task to thank all those who through their work or presence had supported the occasion. I now wish to acknowledge equally those who have helped to fashion this book: especially the authors for their diligence and scholarship and the staff of Torry Research Station for their work in reviewing, editing, typing and indexing.

B D Hayes
Permanent Secretary
Ministry of Agriculture, Fisheries and Food
(April 1980)

PREFACE

One of the events held in connection with the fiftieth anniversary of the institution of Torry Research Station was an international conference on fish science and technology at Aberdeen, Scotland, from 23 July to 27 July, 1980. This book contains almost all the papers presented at the conference; a few had to be excluded for technical reasons or because of their length. The members of Torry staff who organized the conference were fortunate in attracting from all over the world, by special invitation, leading experts willing to speak and write on selected subjects. In addition, an open request was sent out to researchers to present their latest results and ideas. The outcome is an extensive compilation of valuable material which will stand, for many years, as the definitive account of the current state of advancement in fish science and technology.

Fish technology is part of the general subject of food technology. It is concerned with all those situations where fish is used as a food commodity, from the point of capture to the point of consumption. It embraces handling, machine operations, chilling, freezing, smoking, salting, drying, packaging, distribution, marketing, product innovation and development, nutritional considerations, standards and specifications, quality control. It is essentially foreshadowing the industrial application of existing or newly acquired knowledge. Fish science likewise is part of food science. It deals with understanding the background principles related to technology and is usually further removed from industrial application. It is a multidisciplinary subject embracing chemistry, biochemistry, physics, microbiology, engineering and statistics.

Particular aspects of fish science and technology have been considered at meetings in the past, notably at the series of conferences sponsored by the Food and Agriculture Organization of the United Nations, but they have never been drawn together into an integrated whole. It is also at least a decade since a comprehensive book in this subject area has been published. The organizers felt, therefore, that all the topics included in fish science and technology should be reviewed afresh in the first half of the conference. They asked their chosen specialist reviewers to describe the development of each topic over its most formative period, to appraise current trends and to look forward speculatively. The aim was not to write a formal history but rather to emphasize the ways in which developments unfold and are influenced. All the lectures in this first half were presented in plenary session in order to provide the participants with an opportunity to experience the complete span. In the book the lectures are grouped into two opening chapters dealing first with fish technology and secondly with fish science. There is no significance in the order of presentation of these groups beyond the fact that fish science in many ways serves fish technology and is dependent upon its needs.

The second half of the conference was thrown open to the latest research and views, given at over 60 lectures arranged in three simultaneous sessions. The accounts in this half provide an engrossing record of the intellectual effort that is being brought to bear at this time on the problems of fish science and technology throughout the world. The papers have been grouped into 14 chapters dealing with different aspects. Comparing all the papers with those presented at conferences of similar kind held in the past, it is interesting to note how things have changed. Thus, topics like chemical inhibitors of spoilage, freezing at sea and irradiation, that used to be prominent, are absent this time. Topics like minced fish and krill have assumed greater importance. In other topics of older vintage, important new insights have been gained, confirming the healthy, progressive and ever changing state of research and development on fish as food.

Although valuable discussion took place on most of the papers, a decision was made by the organizers not to record it for inclusion in this book. To have done so would have unduly delayed publication and added to costs.

In writing an appreciation of the papers, I do not believe I need to deal with every one: for the most part they speak for themselves. What I propose to do is to pick out highlights and in doing so I will try not to be invidious.

The starting points for the fish industry are, of course, fish resources and because of this the industry has changed markedly as exploitable resources have changed. The way in which this has happened, and is likely to happen, is reviewed by K C Lucas in an authoritative initial paper. It is clear that utilization will be increasingly constrained by a number of biological and legal factors. The fascinating origins and influence of scientific and technical ideas related to the use of fish as food are traced by G H O Burgess, whose skill as a social historian is well illustrated. I suspect that members of some of the world's best-known fish technology laboratories will learn a great deal from this paper. Methods of handling fish have in some cases developed radically over the past 50 years with the introduction of mechanization and labour-saving devices, but in others little has changed. G C Eddie, in his inimitable way, describes these changes, where we stand and what the future holds. Preservation is at the core of much else at the conference. Despite many efforts the old-established methods, including freezing, still hold sway and according to P Hansen this situation is unlikely to change much although he notes trends in their relative importance. A similar theme is explored by M A Steinberg in a useful and comprehensive statistical analysis of world utilization. Although basic preservation methods are still used extensively, the patterns of marketing and of end product presentation have been revolutionized. In parallel to this, systems of quality assurance have been introduced widely and, as E G Bligh relates, are likely to grow in importance. Chapter 2 embraces 9 reviews dealing with various aspects of fish science. J J Connell and J M

Shewan start with an examination of quality assessment methods. Then follow three papers dealing with various important groups of chemical constituents and their reactions, namely proteins, lipids and other components (organic and inorganic). Each has a vast literature which the respective authors (J Olley, Z E Sikorsii, R G Ackman, R Hardy and S Ikeda) compress admirably into readable and illuminating accounts. Fish is for eating, a fact which K J Carpenter elaborates in his lively review of nutritional aspects. The suitability of many fish for use can vary very considerably depending upon their biological state. R M Love covers this field, which, through his own research, he has made a very personal one. Microorganisms in fish and fish products have important roles in spoilage and in causing disease. The complexities of these subjects are tackled thoroughly and systematically by J Liston who also speculates boldly on possible future lines of research. Relatively little research is carried out on the physical properties and processes of fish but A C Jason and M Kent show how necessary their understanding is for many operations in the fish industry. Finally, S F Pearson and M R Hewitt take examples of such operations and of particular pieces of equipment used to process fish in order to demonstrate the value of applying the ideas of engineering science.

Turning now to the research papers, chapter 3 is concerned with the relatively new flesh-containing materials recovered by various types of machine. These materials, the problems associated with their use and their potentialities have excited a great deal of interest over the past few years. The papers add significantly to our understanding of various aspects of this new field. A good many snags remain, not least that of marketing suitably derived products, but they are being vigorously worked on. Of particular interest is the unexpected phenomenon discovered by S Svensson that is related to the stability of minces.

The next chapter draws together papers on various kinds of processing and on utilization. Despite the importance of canning it attracted relatively few papers; that of Durant is on the neglected subject of changes during storage. One of the main strands in fish technology is finding uses for species and fish materials that were hitherto considered unusable. Several examples of this, and the related subjects of reducing waste and upgrading, are represented. Belly burst is a constant problem on which A Gildberg has thrown new light. The chemistry of smoking has excited renewed interest and several interesting contributions are included in chapter 5. Chilling and freezing are such important processes in the fish industry that it was felt necessary to devote a separate chapter 6 to them, even though there was some overlap with the theme of underutilized species.

Krill is a prime example of a large resource looking for viable outlets. The submission of six papers on krill warranted formation of the separate chapter 7. Useful new background information on the storage and nutritional properties of krill is presented as well as ideas for product outlets. No krill or krill products will be sold, however, unless the price is right and the operation is profitable. In order to emphasise this point it was felt desirable to include the interesting economic analysis of krill exploitation carried out by J K McElroy. Clearly, there is some way to go before the right mixture of conditions is found for full commercial success.

By-products are often an essential element in viable fish processing industries. Their efficient production and the acceptable disposal or treatment of their effluents and emissions are all worthwhile aspects for study as shown in chapter 8.

The success of fish technology often depends upon getting it across to the right people at the right time. Lessons learnt in this direction in two different fishing nations are described in the two papers of chapter 9.

Information to supplement the earlier paper by R M Love is given in chapter 10 for the three underutilized species mackerel, blue whiting and mandi.

Chapter 11 deals with a subject of never-ending fascination: quality assessment. Both sensory and nonsensory methods are well featured with many new findings. Flavour and odour have in the past received a disproportionate amount of attention; the paper by R B Weddle on texture helps to restore the balance.

The reactions of proteins in processed fish have been investigated since the earlier days of fish science but many mysteries still remain. The large group of papers in chapter 12 is eloquent of the continued effort in this field. Several of the papers present convincing evidence for the occurrence of a number of simultaneous reactions leading to the formation of different types of noncovalent and covalent cross-links. The background paper of K Hjelmeland and J Raa nicely complements the earlier one by Gildberg. In a number of instances the effective use of the relatively new and powerful technique of gel electrophoresis is deployed.

Special subjects are covered in the final three chapters, the four papers of which all break fresh ground.

A book of this size and complexity is the product of many hands all of whom it would be tedious to list even if I could remember them. I thank them collectively here, but certain individuals need special mention. First, the authors for their splendid efforts in producing good papers to the requested deadline. Secondly staff of Torry Research Station for editorial work and time-saving consultation with authors. Especial thanks go to John Waterman, of my staff, for undertaking the Herculean task of proof-reading and indexing with amazing speed and accuracy. Fourthly, the typing and photography staff of Torry for a good deal of extra work. Finally, Vivien and Bill Redman, Directors of the publishers, Fishing News Books Ltd, who, through the expenditure of long hours on careful and systematic production, have achieved a commendable rapidity and quality of publication.

J J Connell
Aberdeen, Scotland

April 1980

1 Past, present and future of fish technology

How changes in fish resources and in the regime of the sea are affecting fisheries management, development and utilization

Kenneth C Lucas

1 Introduction

Fisheries is an integrated activity directed towards food production and changes in the availability of the resource have frequently affected catching, processing and marketing. However, in the past few years two important new influences have been in play: the establishment by a large and growing number of coastal states of exclusive economic or fishing zones, and the near-exhaustion of alternative supplies of the more familiar types of fish. With the pressure of increased demand for fish caused by population growth and rising *per capita* consumption, an urgent need and an unprecedented opportunity have arisen to devise and establish effective systems of fisheries management in order to ensure a maximal supply of food and the maintenance of a healthy fisheries industry on a world-wide basis. The traditional response of industry to changed supply situations of moving to less exploited resources is no longer a viable proposition.

The establishment of EEZs presents coastal states with greatly increased opportunities to manage and develop their fisheries in a rational manner. However, if these opportunities are to be realized, resource management must be viewed as an integral part of fisheries development, aiming at the best social and economic returns from these resources, not only for individual nations but in terms of the international food economy. The need to exploit all potential fishery resources for human use has focused increasing attention on fish stocks which up to now have been neglected or underexploited, and on development of technologies to make possible their full exploitation. In this work now as in the past, research institutions such as the Torry Research Station have an important part to play.

2 Changes in resource availability

2.1 *Stock collapses*

The most striking changes in resource availability are those which occur when a stock which has previously supported large catches collapses to a level at which little or nothing can be caught. Sometimes this can occur naturally. Examination of layers of deposits in deep water off California has allowed the history of the changes in pelagic fish stocks in this area to be traced back over a couple of thousand years or more. During this time, the dominant species – anchovy and sardine – have fluctuated, with periods of some decades between abundance and near absence. The period of the 1930s when the sardine supported what was at that time one of the world's biggest fisheries for canning and reduction,

seems, historically, to have been one of unusually high sardine abundance.

Occasionally the cause of the decline is obvious as in the case of stocks which clearly exist in conditions close to the limits of the environmental tolerance. Cod, for example, are temperate-water fish, and at West Greenland, are close to their limit. Early in this century there were few cod there. As the average water temperature rose in the 1920s, young fish from the Iceland stock established themselves off West Greenland and by the 1930s were providing annual catches of some 300–400 tons annually for large European fleets. Since then, temperatures have dropped; the stock has collapsed, and now the only cod fishing in these waters is done by locally-based Greenland boats.

More often in recent years the collapse has been due to overfishing – the classic example being that of whales. Following the development of the harpoon gun and the floating factory ship, the Atlantic whaling industry grew by the 1930s to one of the world's largest fisheries (using that term in its wider sense). As early as 1935 the whaling industry and others were expressing concern about possible overfishing. The International Whaling Commission was established in 1964, and from its beginning set controls on the catches. Unfortunately the limits were set rather too high and more important, were not adjusted quickly enough when things started to go wrong. As a result one stock after another – first the blue and humpback whales, then fin and then many sei whale stocks were fished down to a level where no significant catches could be taken.

This failure had several causes. In the early years the scientific evidence (largely due to lack of well funded and well directed research) was poor. Later when clearer advice was forthcoming, the whaling countries were slow to react, or accepted only part of the necessary cuts in catches.

A similar sequence of events has occurred in a number of important pelagic fish stocks especially of herrings and their relatives, anchovies and sardines, the management of which was reviewed at a scientific symposium held in Aberdeen in 1978. Many of these have collapsed, partially or completely, since the California sardine industry collapsed in the 1940s. In several cases, including that of the California sardine, the relative importance of natural events and heavy fishing in the collapse is obscure. In others the role of fishing is clear. In the case of the North Sea herring, events paralleled those of Antarctic whaling very closely. At first the scientific advice was conflicting, many scientists feeling that reproductive success would not be affected by fishing.

1

When scientists finally agreed that controls were needed, the reductions in catches were implemented too slowly, and the quotas were greater than had been recommended.

Taken together, the collapse of these pelagic stocks meant a very great reduction in the supply of these species, principally affecting the fish meal and canning industries. The aggregate of the peak catches of the half-dozen principal stocks is over 18 million tons. In 1977 catches from these stocks were less than 1·5 million tons. Possibly only half of this difference could be made up permanently, since the peak catches clearly could not be sustained, but the loss to the world fish supply has been large.

2.2 *Increased catching costs*
Only a minority of stocks have collapsed as a result of heavy fishing, but virtually all exploited stocks have been reduced in abundance to a greater or lesser extent. In most cases the biological reactions to this reduced abundance, *eg*, better survival of the very young fish, means that the productivity of the stock is maintained. The effect on the fishery depends on how closely catch rates (*eg*, the average catch per day or per year of a fishing vessel) are determined by the abundance of the stock. In some cases (for example some purse seine fisheries) improved searching can maintain the catch rates while the abundance falls. More typically, particularly in trawl fisheries, catch rates fall more or less in proportion to the fall in abundance. This means that the costs of landing a given quantity of fish increases as the abundance falls.

In some cases these increases can be substantial. Exploiting a stock to a level approaching maximum sustainable yield usually involves reducing the abundance by at least half. This means that costs per unit of output are doubled. If the fish being exploited is valuable, so that economics do not place a brake on rising costs, the amount of fishing can expand well past this level, with greatly increased costs. For example an ICNAF working group of biologists and economists reckoned in 1968 that, taking the North Atlantic cod fisheries as a whole, costs (at the prices then current) could be reduced by US$50–100 million without any loss of catch. These excessive costs have effects throughout the fishery: the individual fisherman must work harder, pay more for fuel, and see less return for his labour, while the price to the consumer rises. Since the total supply of fish will not change much, the processor may be less affected than most, although even he will be under pressure to find higher valued products from fish that has become more costly to catch.

2.3 *Global resource limitations*
The oceans' production of living material is enormous. The annual production of plants, and of the small animals that feed directly on them, is probably in the order of 100 000 and 10 000 million tons respectively—way beyond the range of any present or foreseeable demand for fish. The trouble is that few people want to eat diatoms or copopeds and in any case they are expensive to catch in quantity. The more convenient packages, for both eating and catching, occur further along the food chain—for example, herring or cod or tuna—and at each stage along the chain there is a great drop in the available production.

Argument about the oceans' potential food supply is therefore inherently sterile in the absence of prior agreement about what is worth eating and how much can be spent on catching. Given sufficiently broad tastes and a willingness to give unlimited subsidies to the fisheries, the resources are, by the standards of current production, essentially unlimited. If a more practical viewpoint is taken, one that considers the desirability of different types of animals, the picture is very different.

FAO has recently prepared for its Committee on Fisheries, a review of the current state of the world's fishery resources, including a consideration of the major groups of fish taken over the world as a whole. At one extreme the catches of the most valuable, the most vulnerable and the most heavily exploited group—the salmons and their relatives—can only be increased by better conservation and management. The same is true of some of the shoaling pelagic species—anchoveta, herring and some sardine stocks—which, although less valuable, seem particularly vulnerable to heavy fishing. At the other extreme are the small animals such as krill and the many species of small mesopelagic fish. These could produce large catches, but it is not certain that they could be caught at a reasonable cost and sold at an economic price. In between are many familiar groups of fish—shrimps, tunas, cods, hakes, *etc*, which, though not generally depleted, cannot offer much greater catches. If additional supplies are to be found they must come, for the most part, from smaller species, many of them more difficult and more expensive to catch, and presenting major problems of handling, processing and marketing.

2.4 *Extended fisheries jurisdiction*
The most significant event in world fisheries has been the recent extension of fisheries jurisdiction of most coastal states, usually to 200 nautical miles. The total area involved is enormous approaching that of the land area of the world—as is the quantity of fish. Before the general expansion of limits non-local fleets caught over 15 million tons of fish off the shores of other countries. While some non-local fishing is done in all oceans the main effort is concentrated in half a dozen areas, principally in the North Atlantic and North Pacific, and off northwest and southwest Africa. Thus two-thirds of the distant water catches were taken off developed countries (Canada, USA, Norway, New Zealand, *etc*) and most of the other third off west African states, nearly all semi-desert with low populations (Namibia, Mauritania, Morocco, Senegal, *etc*).

In the long run most of these states will expect to replace these foreign fleets with vessels they own and operate themselves. There will probably be exceptions: Americans may continue to find it unattractive to fish in the inhospitable waters of the Bering Sea except for high-priced species like crab or halibut. Mauritania and Namibia have such small populations that it may be difficult for them to find crews for fishing boats as well as staff for the more comfortable, and perhaps economically more rewarding, work in processing plants ashore.

Even where foreign vessels are fully replaced, the change cannot come overnight. There will be an interim period—already well under way—during which foreign vessels will operate under one arrangement or another with the coastal state. In general therefore the processing industry as a whole can expect to have roughly the same supply of fish. The difference will be that more of it will come from locally owned and based vessels, and more of the total volume of processing will be done in the countries off whose shores the catch is taken.

Other changes can be expected as a result of greater coastal state interest in fishery management and because of their increased ability to put management systems into effect. In the long run this should lead to a steadier and, on the average, greater supply of fish; and to increased ability to control the costs of catching. In the short run there may be a localized drop in catches as coastal states rebuild stocks that have been depleted in the past. For example there has been a drop in catches from several of the cod stocks in the northwest Atlantic since Canada proclaimed extended jurisdiction and took action to rebuild these stocks.

3 Traditional reactions to changes in resource availability

3.1 *Industry*
More than any other primary food producer, the fishing industry is accustomed to short-term fluctuations in raw material supplies. Fishermen, who have no way to monitor the growth of their crop, have learned to live with unpleasant surprises and to cope with the negative forces of nature. The processing industry has become so used to day-to-day or seasonal variations in the availability of raw fish that even a rather substantial change in catches from one season to another may not impel it to make significant adjustments in its production programme. There is in the industry an inherent tendency to regard fluctuations as transient phenomena with the need for adjustment perceived only when such changes follow a persistent downward trend and when the causes are expected to continue for a long period. It is therefore not astonishing that an industry in which optimism is a way of life has often been taken by surprise when catches have suddenly fallen due to the collapse of stocks. This attitude also explains the frequent opposition by the industry to the introduction of management controls, particularly when these have been imposed without adequate consultation and explanation. Even the introduction of Exclusive Economic Zones, under international discussion for several years, was at first viewed with wary scepticism in some sectors of the industry, and its full implications realized relatively late.

The history of commercial fisheries provides many examples of industry forced to face a drastic decline of raw material availability. The industry has reacted to these in various ways—sometimes after a considerable time lag. A few examples may be illustrative here.

The most visible and common response has been a shift to other, less exploited stocks except when fishing vessels lacked the necessary range or mobility to make the change. The introduction of fish preservation by ice in the last century allowed fishing vessels to move further away from their shore bases, and the widespread application of freezing at sea in the years after the Second World War made them virtually independent of shore bases. The ability to refuel in overseas ports, to tranship catches and to exchange crews, promoted the complete mobility of these fleets. The shifting of harvesting capacities from one fishing ground to another, thus allowing the industry to take advantage of optimum seasonal catching possibilities, has subjected more and more fishery resources to intensified exploitation; a development which has only recently been brought to a halt by the introduction of extended fisheries jurisdiction by coastal states. The Antarctic is virtually the only region where this option of switching capacities still exists and would be exercised in this area far more than it now is if resources such as the Antarctic krill were readily convertible into marketable products on a large scale. On the other hand, joint ventures and other agreements have allowed a considerable proportion of the world's mobile fishing capacities to continue operating within the economic zones of foreign countries.

In the initial phases of distant water fisheries development, the move from one ground to another was made easier by the availability on the new site of the same or similar species. One did not need to switch between such distinctly different animals as cod and krill. North Sea cod was supplemented with cod from Iceland and the Barents Sea and, years later, expansion of cod supplies was achieved by fishing off West Greenland and Labrador. Market acceptance problems did not occur, except in those cases where the quality of iced cod from the more distant grounds had become inferior to that of cod from nearby waters because of excessive storage time on board. Technological research has contributed much towards improving methods of on-board catch handling and preservation over the past 50 years and Torry Research Station has been prominent in this field.

Success in shifting a fishery to a different species requires careful investigation of the catching, handling, preservation and marketing ramifications. The adaptability of the industry is demonstrated by the success with which European saithe and South Atlantic hake were introduced in Europe and the United States, and Alaska pollack in the United States and in Japan.

The Japanese Alaska pollack fishery is a particularly interesting case. By 1973, catches had risen to a record level of three million metric tons, or about one-third of the total Japanese harvest. This stemmed from the development of a shredding process for pollack meat which made it possible to produce surimi from this species for use in the preparation of fish jelly (kamaboko), a highly popular product in Japan. Also, considerable market outlets were developed for pollack fillets and blocks in the United States, as a substitute for dwindling supplies of these products from other species such as cod and haddock. From 1974 to 1976, Alaska pollack stocks declined very rapidly due to pressure of fishing. By 1977 the situation had become critical, with Japan facing total exclusion from the resource, now enclosed within the new EEZs of the USSR and the USA. Two resulting problems had to be faced: the need to find alternative gainful employment for the fishing fleet if it were not to be written off; and the need to secure adequate supplies of a favourite food item in the Japanese market. The measures taken, with the support of the Japanese Government, exemplify the reactions of an industry to changes in resource availability. In negotiations with the USSR and the USA Japan obtained concessions for the phasing-out period. In addition, part of the vessel capacity was scrapped, the Government paying compensation to the owners involved, while some of the vessels were transferred to other regions, including the South Atlantic and the Antarctic. A commercially oriented exploratory fishing programme was launched in various parts of the world, some of the operations being carried out under agreements negotiated with coastal countries on the understanding that joint ventures or other agreements for commercial cooperation might be concluded in due time. To continue to meet market needs for surimi, various avenues are being explored, including substitution of pollack by other raw material, such as blue whiting, hake and mackerel; purchases of fish from USSR vessels; technological research to permit increased utilization of pollack, and diversion to surimi manufacture of pollack supplies formerly used for other purposes.

The Japanese Alaska pollack story shows that in critical situations of changing raw material availability, governments may find it necessary to intervene, not only to protect investment in a certain sector, but particularly to ensure a continued supply of fish to the domestic consumer. This latter aim may also be achieved, of course, through increased imports. This was the case with the German herring industry which, until the late fifties, got its raw material chiefly from the national fleet operating seasonally in the North Sea. In the off-season this supply was supplemented by imports. By the sixties and early seventies, the decline in the North Sea herring fisheries led to a situation in which 80–90% of the supply had to be imported, most of it from European countries. The final collapse of the North Sea herring fisheries paved the way for increasing supplies of herring from Canada and the United States.

At the same time, the industry embarked on an intensive product development programme, using mackerel, pilchard and sardines as herring substitutes. In the past year, an interesting trade in frozen pilchard fillets has been developed from South America to supply Germany and other European countries. So far, the processing sector of the German fishery industry has been able to secure, in this way, a reasonable raw material supply. For the longer term, it cannot be excluded however that the suppliers of raw material will wish to develop their processing industry themselves and thus supply the final product.

The historical development of the canned sardine industry has shown that industry adjustment is not necessarily confined to switching sources of raw material. The California sardine industry – in the thirties one of the great fisheries of the Western Hemisphere – died a lingering death after the Second World War and has today altogether vanished. After the stocks collapsed, there were limited attempts to bring raw material in from southern California where some catches could still be found, but this soon proved uneconomic. Alternative species were either available only in small quantities (mackerel) or not fished because they produced too low returns for the fishermen (jack mackerel) or because the industry was not allowed to use them for reduction purposes (anchovy). While some catching and canning capacities were converted to tuna, most vessels and pro-

cessing equipment were either taken out of operation or transferred to other production areas, such as Mexico, South Africa and Peru. The South African pilchard industry, in particular, profited from the demise of the Californian sardine industry, rapidly capturing its overseas markets for canned sardines, eg, in southeast Asia and later in Europe. The Klondike-paced expansion of the South African pilchard industry was so closely linked to the Californian sardine fishery that – a historical joke of fishery management – pilchard quotas for the Walvis Bay fishery were first set at one-third of the peak catch of the California sardine fishery.

Equipment and machinery from Californian plants also served to start up the Peruvian fish meal industry. More recently, when the pilchard fishery off South Africa collapsed, Peru began to emerge as a major supplier of canned pilchard to South Africa. Meanwhile, vessels and equipment from the Peruvian fish meal industry have contributed to developing fish meal production in Mexico and in other parts of the world. However, reduction capacities in Peru are still considerable. It should be remembered that in California and in South Africa, the tremendous absorption capacities of the fish meal industry, though not the main cause of overfishing, was certainly a contributing factor and it is to be hoped that in this respect history will not repeat itself in Peru.

While in food fish production, switches from one source of raw material to another usually necessitate adjustments in the production process and sometimes considerable marketing promotion effort, changes of this kind can be made by the fish meal industry with relative ease – provided that supplies are available at reasonable cost. The Norwegian reduction industry successively moved from herring to mackerel to capelin within a matter of a decade – and is now having a go at blue whiting. The fish meal industry should therefore be the first candidate for the utilization of unconventional resources, such as mesopelagic fish. This would allow the use of current fish meal raw material to be upgraded to products for human consumption. We will come back to this later.

3.2 Resource management

Traditionally fishery management has been dominated by biological considerations. One reason is that fisheries, offering fascinating and fertile opportunities for quantitative biological research, have attracted some of the best work in the dynamics of natural animal populations, particularly in the period just before and just after the Second World War. This happened just when the need for better management was becoming fully recognized (at least by European fishermen) as a result of the remarkable recovery of North Sea and other stocks during the war years. 'Fishery science' has therefore tended, up to now, to be more or less synonymous with 'fishery biology' with the contribution that other branches of science such as economics could make to fisheries and to fishery management, for the most part ignored. For this state of affairs, the absence of more than a handful of competent fishery economists is more to blame than any narrow-mindedness on the part of biologists. Indeed the latter have been active in recent times in calling for better fishery economics, and for a more widely based approach to fishery management.

Another reason for the dominance of biological considerations has been the international nature, in the past, of most important fisheries. Biological factors are the same for all countries. Economic or social considerations may be very different (although it should be noted that the effects of these differences on management policies can be exaggerated). Thus for example countries with such diverse conditions and economic theories as Denmark (on behalf of Greenland), Federal Republic of Germany, and the USSR can agree to act on the biologists' proposal to increase the mesh size of trawls used for catching cod off West Greenland, thus increasing the total weight caught. They can also agree, given adequate biological research, on what mesh size would be required to maximize the yield. It is much less easy for them to agree on which mesh size to use when the trawl fisheries are directed at a mix of species, with each country having its own pattern of preference—cod only for Portugal, cod or redfish for the Federal Republic of Germany, and almost anything for the USSR. Even less easy is it to reach any agreement on what measures, if any, should be taken on control of fishing effort to reach economic or social objectives.

No one can claim that management for biologically definable objectives (eg, attaining MSY) has been entirely successful—witness the collapse of the whale stocks or North Sea herring. However, in many heavily fished stocks, some attempts have been made to introduce appropriate measures and in many cases definite progress has been made. Until very recently even this modest claim could not be advanced in respect to other objectives. In particular very little attention has been paid to preventing the excessive rises in costs that occur in uncontrolled fisheries. On the contrary, biological objectives of reducing fishing mortality have been achieved with regulations that greatly reduce the efficiency of fishing and add to costs. This may be achieved somewhat indirectly, eg, through closed seasons; or it may happen directly by the prohibition of more efficient types of gear. It is not for nothing that salmon management in western North America has been referred to as an example of irrational conservation.

4 Challenges for the future

4.1 Fisheries management
A major challenge to fishery administrators is the need to fashion comprehensive systems that manage fisheries further than just fish. This will not come about merely by choosing a suitable management measure, eg, closed season for instance, or a catch quota, or publishing a regulation. The measure in question must, among other things, contribute to wider social and economic objectives (for instance more food, better living for the fisherman, improved returns on investments, improved balance of payments). The actual values set out in regulations (length of closed season or size of catch quotas, for instance) must be appropriate relative to the biological events happening in the fish stock, as well as a clear view of fishing objectives; and the regulations must be obeyed. This requires the full involvement of many people, and efficient communication between them.

The largest single group of actors who need to be involved is the staff of the fishing industry. One would imagine that this was the group for which the need to communicate was most obvious. Yet many management measures with sound theoretical bases have failed because they were not adequately explained to the industry. Without industry understanding and support, from the captain and the deckhand on the trawler (vital, for instance, in the case of size limits) to people involved in fishing enterprise management and other functions ashore, implementation and enforcement of most management measures are impossible. Other participants in the complex business of fisheries management include those who determine investment policy, both nationally, and in international investment banks. Also involved are the shapers of fiscal and tax policies: investment in a fishery is often encouraged by low-interest loans, or tax exemptions on profits reinvested long after the total investment in that fishery has grown past any desirable level. Communication is essential with and between legal authorities as well as the sections of national administrations directly and explicitly for fisheries.

The range of expertise required is wide. Biologists must determine the state of the resource—and the catches required to attain a given objective. Economists and sociologists (or at least people knowledgeable about the true needs of the fishing community) must clarify the objectives of management and the strategies for achieving them. Lawyers and others must determine how the management measures are to be implemented and enforced.

In all of these areas most developing countries need help. Few of them—and far from all developed countries—have expertise among their nationals covering all these fields. They may need help, for example, in evaluation of their resources, or in framing legislation. Sometimes the need can be met by pooling the resources, of a region—one country may, for instance, be able to provide expertise in fish population dynamics, another in the legal aspects of resource management. The regional bodies of FAO are proving to be increasingly effective in promoting this pooling of effort, as have many longer established bodies such as ICES. Even so, especially in the next few years, additional outside support will be needed to meet the urgent challenges of the new situation. Though assistance along these lines (eg, training in stock assessment) has been a feature of FAO's activity for a long time, it needs to be increased and widened to cover such matters as fisheries policy, planning and analysis, and surveillance and enforcement and other practical aspects of comprehensive fisheries management. This is one reason why the Director-General of FAO has established a major programme of assistance to developing countries for the development and management of the resources of their economic zones.

4.2 Food supplies from fisheries
Technical assistance to fisheries of developing nations will also be required in order that these may make the best use for their people of the resources within their EEZs. While fisheries can provide a valuable contribution to economic development in the form of increased employment, income and foreign exchange, the greatest challenge to be met in the context of present conditions in the developing world is the provision of increased food fish supplies for the disadvantaged sec-

tors of their population. The establishment of the EEZ will not in itself ensure that fishery products will be in sufficient supply for all those who need or want them. Fisheries development in this sense must not be seen merely as a matter of increasing production but also as a problem of achieving a more equitable distribution and use among various groups of consumers.

Average fish consumption *per capita* in developing countries (7·6kg) is only a little more than half the world average (13·1kg). Of world food fish supplies 38% are absorbed by the developed countries and another 34% are taken by centrally planned economies, leaving 28% to the developing countries (*Table I*). And yet—in relative terms—fish is much more important to the consumer in the developing world; about 60% of the population in these countries derive more than 30% of their animal protein from 'meat products' from fish and for many countries in all continents fish must be regarded as a main and indispensable item in the diet. Only in a handful of developed countries does fish play a similar role. There can, therefore, be no doubt that fisheries are called upon to continue to make a significant contribution to food supplies in developing countries.

Statistical records over the last 15 years or so unfortunately do not indicate that we are on the way to meeting this challenge: *per capita* consumption of fish in developing countries increased by 600g since the early sixties; in developed countries by 3·5kg and in eastern Europe and the USSR by some 9kg. Efforts have to be stepped up considerably in order to reverse this trend. Otherwise, it is the lower income sector of the population in developing countries that will suffer most acutely, since the demand of those who can pay for the products will be satisfied first.

Projections of future demand for fishery products indicate that approximately 110 million tons of food fish will be required annually by the year 2000, plus, perhaps, 20 to 25 million tons for animal feed purposes. This means that present food fish supplies will have to be doubled. About 38 million tons (about 35%) will be needed in the developing countries, 28 million tons (26%) in developed countries and 42 million tons (39%) in the centrally planned economies.

These projections are based on estimated population and income growth and assume constant price relationships between fish and other food products. If the latter assumption should prove incorrect and fish should become more expensive relative to other food, supplies would most likely move more to the developed countries and centrally planned economies and less would be available for those most in need of more fish as food and for whom fish is an essential component of the diet.

We must therefore ask: are there sufficient fishery resources available to satisfy this demand increase? And: can these be harvested cheaply enough in order to provide low-cost food for the needy people in developing countries? The answer to the first question can definitely be answered in the affirmative. We cannot be as certain about the second—although there are indications that a sizeable proportion of present and potential food fish supplies can be obtained at relatively low cost in comparison to other food-producing systems.

To deal first with the resources: given effective management, the potential annual harvest from traditional or conventional marine fishery resources is estimated at slightly above 100 million tons (*Table II*). However, it is unrealistic to assume that even in the best possible circumstances the maximum for a wide variety of stocks can be achieved in any single year. For all practical purposes, the overall potential is probably more in the vicinity of 80 to 90 million tons, *ie*, 20 to 30 million tons more than the production in recent years. If we add to this figure the potential from inland fisheries estimated at about 15 million tons and from both freshwater and saltwater aquaculture, which can probably be increased to 15 or 20 million tons from its present production of 6 million tons, we arrive at a potential food fish supply of 110 to 125 million tons, a quantity sufficient to satisfy human food needs but not fully the requirements of the fish meal industry. Assuming that fish meal will continue to play its part in the market for compound animal feed, additional raw material supplies will be needed to cover overall demand for fishery products. These can only come from unconventional resources, such as mesopelagic species, oceanic squid and/or Antarctic krill. We need to utilize these resources to the tune of 10 to 20 million tons if we are to meet all needs by the turn of the century. The potentials of unconventional species are enormous and there should be no problem of attaining the target if effective methods of exploitation and utilization can be developed (*Table II*).

As to the question of exploitation costs, it seems likely that most unconventional fish can only be harvested and brought home at relatively high cost. There are some indications that Antarctic krill and, perhaps, mesopelagic fish, could be economically converted into meal for animal feeding. Frozen krill is marketed in Japan at a price of US$700 to 1 000 per ton—a fairly high price

Table I
FOOD FISH CONSUMPTION IN 1972–74 (THREE-YEAR AVERAGE) AND PROJECTED DEMAND FOR 1985 AND 2000[1]

| | Fish consumption 1972/74 | | | Projected demand | | | | | |
| | | | | 1985 | | | 2000 | | |
	Million t	%	Kg per capita	Million t	%	Kg per capita	Million t	%	Kg per capita
Developed countries	19·2	38	25·7	23·1	33	28·4	27·9	26	30·9
Developing countries	14·0	28	7·6	22·4	32	9·0	38·0	35	11·0
Centrally planned economies	16·7	34	13·6	25·3	36	17·3	41·9	39	24·5
World	49·9	100	13·1	70·8	100	14·9	107·8	100	17·7

[1] The table records only 'basic' projections; 'supplementary' (higher) projections have also been made using higher populations and income growth rates
Source: FAO Commodity Projections

Table II
PRESENT CATCHES AND POTENTIALS FOR INCREASES, FROM MAJOR GROUPS OF SPECIES

Stock	Potential '000 t	Total world catch¹ '000 t 1966	1977	% of potential 1966	1977	Potential increase from natural stocks ('000 t) Total	Management¹	Increased effort	Aquaculture
Conventional marine resources									
Salmon	650	453	480	70	74	170	170	—	+
Flounder and cods	6 700	4 692	3 786	70	57	2 914	2 500	400	(+)
Herring and anchovy, etc	15 600	13 709	1 801	88	12	13 799	13 800	—	—
Shrimps and lobsters	1 670?	830	1 542	50	92	128	25	100	(+)
Tuna	2 260	1 031	1 592	46	70	668	100	550	(+)
Other demersal	37 100	9 628	17 097	26	46	20 003	4 000¹	16 000	—
Other pelagic	40 200	13 045	28 655	32	71	11 545	3 000	8 500	(+)
Unconventional marine resources									
Cephalopods	(50 000?)	833	1 165	2	2	(50 000)	—	(48 000)	—
Other molluscs	Not defined	2 115	3 011	—	—	—	—	Probably large	++
Krill	(50 000?)	—	123	—	0·2	(50 000)	—	(50 000)	—
Mesopelagic fish	(50 000?)	—	—	—	—	(50 000)	—	(50 000)	—
Total conventional marine resources	104 810	43 388	53 361	41	51	50 120	19 970	30 150	
Total marine²		48 121	62 178						

¹ Figures of catches exclude discards at sea. Potential increase by utilization of these discards included under 'management'
²Includes species groups (seaweeds, crabs, *etc*) not considered in this table
(+) Denotes aquaculture possibilities limited to fattening with specially provided food
++ or + Denotes opportunities to increase production from aquaculture using natural food sources
Source: FAO Information (Fishery Resources and Environment Division)

compared to chicken, but reasonable in relation to production costs of beef, veal and pork. Some of the potential demersal stocks are rather distant from the principal consuming centres, *eg*, those on the Patagonian shelf and in the Antarctic, and will therefore be destined primarily to the better-paying outlets in the developed countries. Others, such as blue whiting and croaker, are found near potential markets, but so far encounter problems of consumer acceptance and/or excessive processing costs due to low yields. However, the considerable R and D effort devoted to, for example, blue whiting is expected to result soon in acceptable products competitive on the markets of developed countries.

On the other hand, a number of other fish resources are known to be exploited at comparatively low cost. Notable among these are the small pelagic species and other resources providing raw material for the reduction industries. The 20 to 25 million tons annually going into fish meal are landed at prices ranging from US$30 to 70 per ton. This is very low compared to other protein food producing systems. Also, some food fish, particularly in developing countries, is brought ashore for less than US$100 per ton; and Alaska pollack is valued at US$150 to 200 per ton in Japan. In semi-intensive fish culture in Africa, tilapia can be produced at about US$350 per ton. These are just a few examples.

Contrary to widespread belief, fisheries are also in a relatively favourable position in respect of one important cost element rapidly rising: fossil fuel energy! When protein yield is compared, the energy required to harvest some seafoods, *eg*, shoaling small pelagic fish, is of the same order as that needed to grow field crops and it is only a fraction of that needed for broiler or beef production. The likely changes in fish production following the establishment of EEZs, *ie*, gradual replacement of capital-intensive, high-energy-consumption distant water fisheries by smaller-scale coastal fisheries, will reinforce this favourable position.

Overall, it would seem therefore, that fisheries are capable of providing relatively cheap raw material for food production. Certainly, more detailed comparative studies are required to determine the future place of fish in the world food economy, but the indications are that we should not fare too badly.

4.3 *Technological requirements*
It will not be easy to bring all resources into maximum utilization for human consumption. There is an enormous gap between what are loosely termed marine resources and fish products actually on the plate as part of a meal and this gap cannot be closed without substantial improvements of technology in all areas of fish catching, handling, processing, distribution and marketing. It has been shown time and time again, particularly in internationally-funded emergency feeding programmes, that people will not eat nutritious food simply because they are told it is good for them. There is a disturbing tendency for rejection or adverse reactions to be stronger in the case of people from the lower socio-economic groups who, almost by definition, are in need of better nutrition.

The fish products that will be eaten in quantity will therefore have to possess organoleptic and visual characteristics which identify them with existing products or have a possible 'up market' appeal. Above all, they will have to be palatable and attractive. The FPC Type 'A' debacle is perhaps the clearest indication of this. People do not want mysterious gritty powders which do them good. FAO acceptability and market testing of fish powder (FPC Type 'B') over the last five years has shown that there are few adverse reactions to a product, represented as fish, which has a fishy taste.

7

Rejections can often be predicted on the basis of a careful consideration of food habits and culinary practices. FPC Type 'B' is now being marketed commercially both in its own right and in combination with other ingredients in soups. A highly successful soup product 'Fish Tea' has been introduced in the Caribbean. Fish flavoured curry powder is on sale in Singapore and considerable sales of a powdered soup made from *Stolephorus* sp., boiled on board, dried and ground, are reported from Singapore and Malaysia.

FAO does not feel that the future is restricted to limited volumes of powdered products for emergency feeding and soups, but foresees a whole range of products tailored both to particular areas and species. Consideration of the rather high utilization rate and the range of species used in southeast Asia reveals a potentially promising source of product ideas for other areas. There is need for systematic evaluation of these products in order to determine whether there are species/demand combinations in other parts of the world that could be similar. Although the large food industries in developed countries can undertake these assessments and the associated product developments, it is more likely that FAO, or the international system, will have to take on the responsibility for spearheading investigations in the developing world. Past experience of the success rate of new products makes this a daunting prospect but nevertheless suggests that a broad approach involving many products is more likely to result in success than concentration on only a few. We do not see fuller utilization resulting from a crash research programme to create one particular product, but on a diversified approach in which products are transferred from one area to another, the manufacture of traditional products is increased through limited industrialization, and new products from unfamiliar species are introduced.

The difficulties of promoting fuller utilization are perhaps best exemplified by consideration of the small pelagic species. FAO is developing a strong programme of investigation of those species, which will lead to further consideration of other latent resources, and unconventional species. It is estimated that 25 to 35 million tons of small pelagic species per year could be used for direct human consumption if the technology of converting them to food were available. At present, only 11 million tons are consumed as food, the balance being made into fish meal or not caught because of lack of utilization prospects. It is not the intent of this paper to argue against the production of fish meal, which is seen as a valuable protein component, although often of limited nutritional benefit to developing countries, but to point out that there are alternative resources which could replace that part of the resource upgraded from fish meal to products for human consumption.

The difficulties of using small pelagic species start with the nature of the fisheries for them which are seasonal and characterized by high volumes landed in short periods. These are often 'glut' traditional markets, and are wasted. Ways of increasing the radius of distribution and of storing the catch, as well as extending the fishing season, are therefore required. The quality of the protein of these small fish is just as good as the larger ones but it is not surprising that only a few species are used for human consumption when one considers the following

points: the small fish cannot be eviscerated economically on a large scale; they tend to spoil rapidly; the flesh tends to be soft; the fish bony and the skin fragile. These are the problems of intrinsic raw material quality that the technologists have to tackle. Present indications and experience suggest that there is no difficulty in the actual catching operation but problems start as soon as the fish leave the water. In order to overcome rapid softening they must be chilled – an expensive process when prices paid for the fish are taken into consideration. Chilled sea water (CSW) systems are either in use or under investigation in many parts of the world. They show considerable promise in reduction of belly bursting and induction of rancidity which are the two major indicators of incipient spoilage.

Breakage of the fish during unloading from CSW systems or on transfer from boxes is the next problem. Containerization on board seemed an attractive solution but, although still promising, the iceberg effect of agglomerated lumps of ice slows cooling and in addition design concepts of the vessel must change. More work is clearly required.

Following landing, fresh fish distribution can be extended over a much greater radius if the raw material is in first-class condition. In many developing countries this can be assured if handling improvements, particularly chilling, are introduced, as most catches are made in the immediate vicinity of the landing place. Increasing fresh distribution is mainly a process of education and agreeable economic returns. The technological inputs are choice of boxes or alternative containers for insulated transport, and ensuring minimum quality loss during storage.

Assuming that technological advances can result in the landing of raw material of sufficiently good quality for processing, the range of options, or available products, is, as described previously, very large. It is clear, however, that new technology will be required in order to make full use of the raw material because present processes are generally not capable of massive throughput to produce cheap products. Mechanical evisceration or at least separation of gut material from flesh is required. Other options may be meat and bone separation, salted or acidified minces, incorporation of antioxidants, and so forth. Both traditional and new techniques can play their part. There are excellent prospects for improved dried or fermented fish products packaged in better materials. At the same time, new technologies are available for trial. A recent report from the US Office of Technology Assessment (OTA) in examining emerging food marketing technologies considered that highest priority items should include engineered or fabricated foods and the retortable pouch which would both be ideally suited to small pelagic fish utilization.

In addition to the small pelagic species there are many other real or apparent leakages from the total fish protein pool that would contribute to increased availability if they were stopped. Post-harvest losses can be reduced but not eliminated, at reasonable cost, provided methods are worked out. More importantly, advances should be carefully communicated down to the fishermen and processors who stand to benefit from increased earnings. The wastage of up to five million tons of potentially edible fish in shrimp trawler by-catch also represents a technological challenge to devise economically feasible ways of utilization.

All evidence points to there being numerous technological possibilities, many of which may prove to be economic. The obstacle is a shortage of experienced technologists to apply them to fishery resources and to put the results into practice. This shortage is particularly felt in the developing countries.

4.4 Unconventional resources

The lack of exploitation of the unconventional resources (those which have not yet been exploited for any purpose) is a function of the accessibility of such resources as the mesopelagic species, Antarctic krill and oceanic squids. Perhaps these resources have not been required to date and some at least could prove to be prohibitively expensive to exploit. From the standpoint of processing and marketing, very little is known about them mainly because of the extremely high cost of exploratory fishing.

Little is known of the mesopelagic species, except that they exist in enormous quantities. The only place where limited use has been made of them is in South Africa for fish meal, and here some problems were encountered. The mesopelagics are an example of a resource for which catching technology has not yet been worked out and when it has technology will have to be developed on the processing side. They are the obvious replacement for the small pelagics which might be switched from fish meal to human consumption. In FAO we have a programme to catch samples and submit them to technological assessment, both for fish meal and potential direct consumption. Samples should be available in August 1979 to anyone interested.

Antarctic krill resources are better researched but estimates of availability vary widely. There are no problems with catching large quantities. The impediments to development of a fishery are the lack of economically viable products and the logistic constraints of the hostile environment in which krill is found. Continued research and development of high value products for human con-

sumption and products for animal feed will clearly be required.

The oceanic squids are the least known of the unconventional species and have never been caught in any quantity. Estimates of abundance have only been made by recovery of parts of them from the stomach contents of other species about which more is known. They represent the far-distant horizon, where in the future investigations on available quantity, catching techniques and possible uses will have to be made.

5 Conclusions

The establishment of EEZs provides the opportunity to introduce rational and comprehensive systems of fisheries management and development. To install these systems, developing countries will require technical assistance of many kinds. FAO has therefore strengthened its capacity to help countries in matters of fisheries planning and in the various technical fields associated with a rational exploitation and utilization of marine fishery resources. Some donor countries and the United Nations Development Programme have agreed to support this initiative. Strengthened FAO regional fishery bodies and a new generation of regional fishery development projects will be the basis for implementation of much of this global programme.

An important aspect of technical assistance will be the development, adaptation and transfer of appropriate technology in fish catching, handling and utilization for the benefit of developing countries. Our aim is to enhance the capabilities and to increase the self-reliance of developing countries in these areas. We intend to achieve this by practical demonstration programmes carried out through regional co-operation in technology. Technological institutes in developed countries—for instance Torry Research Station—have made important contributions in the past and we count on their future collaboration so that we may benefit from a synergistic working partnership between capable institutions in industrialized and developing nations.

Response to change: the influence of fish science and technology on fish industries
G H O Burgess

1 The origins of fish science and technology

It is relatively easy to describe, at least superficially, the structure and habits of a fish, but more difficult to investigate the chemical composition of its tissues, even in an elementary way, or to explain how and why it changes with time after death. Even the simplest analysis draws on a considerable body of theory and of practical techniques and is essentially experimental in its approach. Perhaps this is why, although fish were described by Aristotle, whose observations were annotated and extended by many of the great Renaissance anatomists, it is only in the seventeenth and eighteenth centuries that occasional observations on composition are recorded. During the nineteenth century, with the increasing rate

of scientific discovery, more is published on the chemistry of fish tissues, but it was only by the end of the century that there were signs that science was beginning to be called upon to help in improving the traditional ways of handling and processing fish.

Fish science can be looked upon as the synthesis of parts of a number of scientific disciplines concerned with fish as food, notably physics, chemistry and microbiology. Fish technology is the application of fish science to the industrial art of fish handling and processing, frequently through the intervention of the engineer. It is therefore not surprising that fish science and technology developed so recently, since corpus of scientific knowledge was earlier too insubstantial to be of much value to industry.

One factor in Europe that quickened public interest in the fisheries was the series of exhibitions held in the second half of the nineteenth century. The first was in Arcachon, France, in 1866 and at least nine major exhibitions occurred between 1866 and 1883, when the Great International Fisheries Exhibition was held in London. It is instructive to examine the catalogues of some of these exhibitions, and the papers produced in connection with them, since they give a good idea of the state of development at the time. The reader is left with the strong impression that although new and improved methods of fishing were being introduced, largely as a result of the introduction of the use of steam, processing and handling methods remained little changed.

There are tantalizing glimpses into the operations of industry and indications of desire for improvement. Essays at the Edinburgh international exhibition of 1882 included topics such as curing and preserving fish, utilization of fish offal, fish supply of great cities with special reference to the best methods of delivering the fish in good condition to market and the best modes of preserving ice (Herbert, 1883). At the London exhibition, discussion went further and included not only technological questions but was also concerned with statistics collection, sociological problems of fishermen and marketing. Most papers were nevertheless disappointingly superficial and demonstrate that fish science and the fish industry had not yet met (Anon, 1884).

Some of the exhibitors' claims in the catalogue certainly could not have been substantiated. A producer of natural ice in Norway states that he can 'furnish 25 million tons block-ice . . . that will melt ten percent slower than common ice'. A London company exhibits Sanitas antiseptic and disinfecting fluid, which apparently consists of hydrogen peroxide and which can be used for preserving fish. The Antitropic company advertises Glacialine 'to preserve fish in a sound and fresh condition for a few days in the hottest weather'. On the other hand, the Hallam company of Derby advertises refrigerating machinery for a steam carrier vessel and, for a fishing smack, a refrigerator connected to a freezing chamber or cold store (Anon, 1884).

In the rest of Europe and the United States of America, the situation was similar; although many countries were giving attention to fisheries biological research, virtually nothing was being done in fish technology. The classical work by Stevenson on the preservation of fishery products for food, published at the end of the century, gives scant attention even to those scientific aspects that were understood (Stevenson, 1899). Nevertheless, the US Government had in 1871 authorized the establishment of the Fish Commission (later the Bureau of Commercial Fisheries now absorbed into the National Oceanic and Atmospheric Administration) and although its interests at first were biological (Smith, 1910), about 1879 it began to encourage Professor W O Atwater in·a series of pioneer studies of the proximate composition and nutritive value of fish (Atwater, 1892).

In fact, some chemical work on fish, particularly on nutritional aspects, was already in progress in Europe, and at one stage Atwater worked with von Voit in Munich and Kühne in Heidelberg. At least part of the impetus for this work came from the increasing amount of food legislation; there was rising concern in many countries during the second half of the nineteenth century about food adulteration and the need for accurate and reliable analytical methods resulted in rapidly improved techniques.

Most early engineering applications of concern to the fish technologist were in refrigeration (Burgess, 1965) where by 1900 systems were recognizably related to modern ones. Indeed, one gets the subjective impression that up to 1900 many major advances were made in refrigeration and that by comparison progress in equipment in the next three or four decades was relatively minor.

Knowledge of fish and fish product microbiology was vanishingly small and studies of the ways of measuring quality or of the effects of processing methods had not apparently been examined at all. Nevertheless, it can be claimed that fish science had come into existence by the end of the century. The first government fish technology institute, the Norwegian Fiskeridirectoratets Kjemisk Tekniske Forskningsinstitutt was opened in 1892: perhaps this was the first formal and official acknowledgment that scientific research, funded by government, had a part to play in developing and improving the methods used by industry in handling, processing, storing and distributing fish.

2 The years of foundation 1900–1945

Although it is convenient to consider the beginning of the twentieth century as the start of a period of development in the relationship between applied science and fish industries, there is little evidence of quickening tempo during the first decade. If fish industries drew at all on the growing body of scientific information relating to food, this is not revealed by records now available. This is perhaps surprising in view of the considerable amount of marine biological work by then being done in many countries, and the close association with it of enlightened people in the fish industries.

Much work was done on food composition and analytical methods, and fish was amongst the commodities examined. Some items of scientific interest later became important, such as the studies of the so-called extractives, the low molecular-weight, water-soluble, nitrogenous constituents of fish muscle, where American and Japanese workers made particularly valuable contributions. Attention might also be drawn to the classic paper by Anderson, Assistant Medical Officer of Health in Aberdeen, who in 1908 published an account of the characteristic changes occurring in spoiling fish (Anderson, 1908) which has since been much quoted by workers on organoleptic assessment. Nevertheless, the literature shows work on fish science and technology to have been patchy, unsystematic and not unified, and large areas were unexplored.

It was the threat of war, war itself and its aftermath that galvanized governments to allocate resources urgently to scientific research on food, including fish, and to encourage practical application of results. In all industrialized countries problems were broadly similar: towns and cities had grown enormously and food supply lines even in peacetime had become greatly extended. Improvements in transport had made such growth possible, but for some commodities existing technology was being stretched to the limit, waste and risk of waste was growing, and the quality of many foods was becoming poorer and more variable.

Steps were therefore taken by many countries to provide scientific and technical advice for industry, generally by establishing new laboratories funded by government. Urgent action was required to overcome serious and immediate shortcomings in commercial practice in the food handling and processing industries, but more often than not the scientific base on which effective advice could be established was inadequate. The new laboratories provided what practical advice they could, and this was generally sound and to the point, but much of their effort was taken up in research aimed at understanding the properties and behaviour of specific food commodities.

The early development of fish technology laboratories is hence best understood as part of a more general pattern of food research. Here, broad trends in the development of fish technology throughout the world will be mentioned, with particular reference to fish industries. Important areas will inevitably be glossed over or omitted, partly because selection is a personal matter, but mainly because emphasis is laid on industry. Development of a smoking kiln or identification of the cause of a particular type of spoilage, all call insistently for mention because, in theory at least, their industrial value is measurable. Research directed towards a deeper understanding of the behaviour of fish tissues during handling, processing and storage, is more difficult to justify in terms of specific and identifiable industrial advantage, and in consequence has sometimes been discounted, occasionally by those who should have had reason to know better. It is, however, on this essential foundation of fish science that successful technological innovation has been, and will continue to be, built. Fish science is the well-spring that allows successful technological innovation in industry; science is applied to industry through technology and it is for this reason alone that technology will receive more attention here.

The US Bureau of Fisheries gave attention for some years to such problems as the improvement of canning and freezing, but it was found unsatisfactory to try to work in commercial plants whose purpose was to supply the market for profit. Finally, the Bureau obtained funds to enable it to build a fishery products laboratory in Washington, DC (Smith, 1921). This was opened in 1918 and was probably the first purpose-built laboratory devoted entirely to fish technology. It was roughly 14m × 24m, on two floors with an attic, and contained cold stores and a wide range of equipment which would allow work to be done on most industrial processes. One investigation was to compare air and brine freezing, and this provides an early example of one way in which such laboratories throughout the world have constantly assisted their industries by introducing the testing equipment from elsewhere; the Bureau imported from Denmark the first Ottesen freezer in the US.

In the Commissioner's Report for 1922 it was stated 'Under conditions obtaining during the past few years the need for and importance of technological investigations have been felt by the industry as never before in its history'. It went on to describe by way of example the rapid growth of the Californian industry which had occurred without technological backing. 'Products, time, labor, and capital have been wasted through ignorance and for lack both of the development of standard methods that will yield high quality products and of specific information as to the best and most economical procedure.' (O'Malley, 1922). The Bureau had in fact found it necessary to open, at San Pedro in California, a temporary laboratory for the canning industry.

In Europe there were developments which also eventually led to the establishment of laboratories. Pioneer work of importance was being undertaken in 1915–16 by Plank, Ehrenbaum and Reuter on the preservation of fish by freezing; this was done under contract to the Berlin Central Marketing Board (Plank et al, 1916). In 1917 in the UK, fish was one of the first topics to be considered by the newly created Department of Scientific and Industrial Research, which itself was a result of the official realization of how little science and technology was being applied to improve industrial processes (Waterman, 1979). Indeed, it was the London cold storage industry that was pressing for research on the freezing and cold storage of fish. In France, the Institut Scientifique et Technique des Pêches Maritimes was established in 1919 and there canned fish was an early interest. During the next 15 years, many of the well-known fish technology institutes in the northern hemisphere were established. The following list is not comprehensive, but shows how widely distributed these centres were. They include the Technological Stations at Halifax (1924) and Prince Rupert (1925) in Canada, and the Institut für Fischverarbeitung, Germany (1925) later to become the Institut für Biochemie und Technologie, though in fact this institute began modestly from the initiative of Rudolf Baader in 1920 (Ludorff, 1962). The Torry Research Station was eventually opened in 1929 following an earlier and unsuccessful attempt to do research work in industry. The Fiskeriministeriets Forsøgslaboratorium, Denmark, was founded in 1930. Mention should also be made, however, of the Imperial Fisheries Experimental Station in Japan (1929) which in 1949 became the Tokai Regional Fisheries Research Laboratory. In many instances, technological investigations were done as a subsidiary activity in marine biological institutes, as for example in the Polish Maritime Station at Hel, the forerunner of the Sea Fisheries Institute at Gdynia. This list does not include non-governmental laboratories such as those associated with teaching, though mention must be made of the University of Washington, Seattle, USA, which was probably the first to institute a regular course in fish technology. Its College of Fisheries was founded in 1919.

Each institute, was, of course, concerned with the special problems of its own fisheries; for example, in the early days the US Bureau devoted much attention to canning, the German laboratory to conserves of all kinds and the Halifax laboratory to salt fish. The French laboratory was also concerned with canning and quality assessment in the 1930s. Nevertheless, there is a unifying thread running through much of the published research from 1920–1940, when war again interrupted much of the work, and it is therefore on these two decades that attention will mainly be focused. A major contribution of all the technological laboratories throughout their existence has been to interpret findings obtained within the laboratories or from outside. One is struck time and again by the way in which a finding published in one country is taken up, reinterpreted and made available to industry elsewhere.

In the earliest years there was a widespread interest in fish freezing, and this undoubtedly mirrored the concern of industry. Freezing had long been seen as the solution to many of the problems of fishing, with its periods of glut and famine, and the volume of storage space was growing in every developed country. Nevertheless, the quality of the product frequently fell far short of what was acceptable, and in consequence, frozen fish had acquired a bad reputation with consumers. It was beyond the ability of industry at that time to develop solutions to the problem of poor quality or even to identify causes, and industry in every country turned to government for help.

Work on freezing perhaps constituted the first major contribution by science and technology to improvement in the fish industries of the world and although at first sight perhaps surprising, since it might have been expected that it would be in the handling of fresh fish that the earliest developments would have occurred, it was in fact to be expected. Freezing was a new technology, new methods and equipment were being designed, storage techniques were changing and it was believed that the new systems would replace conventional stowage with ice whose limitations, it was felt, were already well understood, and which in any event did not involve costly capital investment. Industry had to avoid making expensive and damaging mistakes with frozen fish.

The earliest technological reports devoted much attention to the mechanics of the operation, giving what today seems undue attention to comparing methods but dwelling hardly at all on the more important issues of the nature of the changes brought about by freezing and cold storage, and how these might be controlled. In fact, developments in physics and engineering during the late nineteenth and early twentieth centuries had been rapid and not matched by a growth in understanding of the underlying biophysical and biochemical changes wrought by freezing. Plank *et al* (1916), whose work has already been mentioned, considered the damage caused by freezing to be almost entirely mechanical. Harden F Taylor's impressive review of the *Refrigeration of Fish* was not published until 1927 (Taylor, 1927) but much of the information in it appears in Tressler's famous compilation, first published four years earlier (Tressler, 1923). This present paper is not a review of the history of refrigeration itself, which has been recently covered in Thévenot's comprehensive book (Thévenot, 1978).

Throughout the period from 1900–1945, technologists were engaged primarily in determining the kinds of practice and process capable of giving the best results, and of identifying limitations in methods proposed or in use. It has proved difficult in making this review to identify particular laboratories or individuals as having been responsible for specific findings. Some practices, such as glazing, were already in use in industry before the turn of the century (Stevenson, 1899), perhaps 30 years before the reason for its effectiveness was fully realized and it became a widely accepted and recommended practice. Some ideas developed simultaneously and independently in different places, but mainly, in a field where many young scientists were working, it is to be expected that findings will be overlapping and complementary and the closer one approaches the present day, the greater the difficulty of identifying the 'onlie begetter' of an idea. Furthermore, many improvements arise within industry, or as a result of collaboration between industry and laboratories.

The work of Plank and his colleagues attracted much attention and the theory advanced by them, that the rate of freezing, which could be shown to affect the histological structure of the flesh, was of prime importance, had an almost seductive influence. In no country was detailed attention given to storage time and temperature (see, for example, Nuttall and Gardner, 1918; Stiles, 1922). Taylor (1927) discussed storage temperature in one paragraph, saying that it cannot be too low and that whilst $0°F$ ($-17\cdot8°C$) is excellent $-5°F$ ($-20\cdot6°C$) is better. Stiles (1922), following Plank, recommended $-7°C$. Although, therefore, there was a general view that the lower the temperature the better, there was little precise information either on the nature of changes occurring in cold storage or their relationship to time and temperature.

It appears to have been the group of workers at the Low Temperature Research Station at Cambridge, under the guidance of Sir William Hardy, who in the 1920s began to study the interrelationships of freezing rate, storage temperature and denaturation of meat protein. Reay, Torry's first Director, worked in the University of Cambridge before joining the staff of the newly established Torry Research Station, and knew people in Hardy's team. Reay applied these new ideas to fish, and by 1929 had concluded that storage temperature was of great importance. In 1930 he commented 'when brine-freezing was first introduced, the successful results obtained by freezing and immediately thawing fish, tended towards over-confidence in the "quick-freezing" theory. It has been shown that when a period of storage is introduced into the cycle of changes in temperature, even the most rapid freezing does not prevent the occurrence of great alteration in the muscle if the temperature of storage is within a certain range.' (Reay, 1931a).

In a paper for industry (Reay, 1931b) he set down the precise relationship between time and temperature of storage and also demonstrated some of the underlying changes in protein behaviour associated with cold storage deterioration (Reay, 1933, 1934). The work was quickly taken up (Finn, 1934), extended and checked by many others over the next decade. The advice to industry was quite clear: if good quality products were to be obtained it was essential to freeze quickly and store at a uniformly low temperature. Although over the period up to 1945 storage temperatures on average began to fall, the most marked results from this work did not become apparent until the 1950s and later.

It is, of course, only too easy to claim for technological research all the credit for improved performance by industry, but to do so would be false. The history of the technological laboratories in all countries known to the author has been a commendable one of collaboration with industry, where there have been many prepared to spend money to improve installations and to make their own trials. Developments by the refrigeration industry itself, as improved materials have become available and by improved design, have played a major role. Nevertheless, the justification and pressure for improvement came largely from the work of the technological laboratories which made it possible eventually to specify in very precise detail what storage conditions were necessary if the results were to be acceptable to the consumer.

The new laboratories devoted attention to the problems of the spoilage of wet fish and its control, mostly by efficient chilling. Programmes generally fell into three categories: chemical and biochemical aspects, microbiological problems, and practical studies of the best ways to chill and the best results to be obtained with efficient chilling.

There was a need to know more about the biochemistry of dying and dead tissues and the natural flora of fish. Much of the early biochemical work was not of immediate industrial concern, although it was an essential background for future applied work. The composition of fish oils and the oxidation mechanism was widely studied, no doubt because of the pharmaceutical value of some oils and the contribution of oxidation products to off flavours. Attention was given to chemical methods of measuring freshness and the usefulness of the measurement of total volatile bases and trimethylamine (TMA) for assessing freshness was debated from the early 1930s onwards. Reference should be made here to the pioneer work of the Canadians on TMA (Beatty and Gibbons, 1937). In spite of its many limitations, the measurement of TMA has been more widely used commercially as a means of assessing freshness than any other chemical technique and even today no other method of chemical analysis is regarded as anything approaching it in usefulness. Other workers in the 1930s studied problems such as glycogen content and breakdown, and the consequential changes in pH.

On the microbiological side, workers were beginning to grope their way through the as yet unmapped jungle of marine microbiology. Marine bacteria were known to have lower optimum growth temperatures than those from temperate terrestrial areas (see, for example, Hunter, 1920a, b; Harrison, Perry and Smith, 1926; Fellers, 1926), but there were formidable difficulties to be solved in the identification and taxonomy of groups not easily characterized by techniques then commonly in use. From 1930–1940 the essential groundwork was done which enabled workers after 1945 to make rapid progress in the characterization of the spoilage flora.

Some microbiological work was far in advance of its time. Hess in Canada demonstrated about 1931 the storage properties of sea water chilled with ice (Huntsman, 1931) and in 1934 described experiments in which he showed the effects of storage temperatures just below 0°C in greatly reducing the growth rate of bacteria (Hess, 1934a, b). Thirty years later industry and government technologists were assessing the possible value of chilled sea water and 'superchilling'. Although superchilling was eventually found unfavourable, stowage in chilled or refrigerated sea water has become an accepted method with many industrial applications.

One particular area of microbiological study was of considerable importance; these were the problems of microbiological spoilage of salt white fish known as 'pink' and 'dun'. Pink is caused by one of a group of halophilic bacteria and the associated dun by a halo-tolerant mould. Research was aimed at reducing or eliminating spoilage caused by them. There is an extensive literature on the subject from the technological laboratories of many countries including Germany, France, UK, US and Canada in the 1930s. It was demonstrated that the organisms responsible for both conditions occurred in solar salt. Advice on methods of control was based on the need for cleanliness and low temperatures below 5°C, at which neither type of spoilage organism will grow. Research was also done on the composition of curing salt, especially in France and Canada.

Interest in bacteria was only one facet of the more general concern with spoilage and its control. At first, workers referred to autolysis as being a major cause of spoilage, but later it was realized that proteolysis, unless gut enzymes were involved, was due to bacterial activities, underlining once again the need for cleanliness in handling and processing fish.

Although ice had been used for chilling fish for a century, and more in some fisheries, it was not always used effectively. Laboratories began to demonstrate the improvements that could be obtained by using methods already available. Scientists and technologists devoted much effort, and have continued to do so, in demonstrating techniques and offering training courses to industry. Most laboratories have regarded research, development and education as inseparable activities.

The literature of the pre-1945 period devotes little attention to problems of a physical or engineering nature. Perhaps this was partly a result of the lack of suitably trained and motivated people. Physicists and engineers were at that time likely to be trained, and have interests, in subjects far removed from food research, and it may also have been felt that physics and engineering were inherently less likely to offer solutions to the main problems as then seen than chemistry and biochemistry. Some work was, of course, published and furthermore, mention should be made of the contribution of fish working machinery manufacturers, and particularly of the company established in Lübeck in the early part of the century by Rudolf Baader. This machinery has become increasingly sophisticated and costly and in order to make the maximum use of it, companies have found it beneficial to assess the efficiency of all associated operations.

One other feature of the years of foundation calls for comment. Most technological studies were done on a small scale, often by individuals or a group of two or three, and advice was provided on the basis of small-scale laboratory investigations. There are few examples from any country of major exercises, involving collaboration between industry and government-employed scientists and technologists.

Results of work were published in scientific and technical journals or trade papers. A few special publications, written to meet the recognized needs of a relevant part of industry, were written, but were not common; two noteworthy exceptions were the two Canadian publications, the *Progress Reports* of the Pacific Coast Stations and the sister publication from the Atlantic Coast. From the beginning, these were models of their kind. They contained short informative accounts of work in progress, reviews of findings elsewhere and interpreted in terms of Canadian conditions, and discussions of matters of general technological concern. They did not aim to be contributions to the scientific literature but were readable, sometimes racy, accounts of findings of interest and value to the Canadian industry. They were eagerly read all over the world and the decision to discontinue them around 1960 was taken only because it was felt that by that time other and better means existed for communicating with industry (Hachey, 1965).

Fish industries were only slowly changing. Ice was still the main agent for controlling spoilage in unprocessed fish, and although the extent of freezing was growing, it was doing so slowly. Large quantities of fish were still preserved by traditional techniques which had changed little and although forward-looking companies were beginning to try new techniques, there was no powerful stimulus for change. This stimulus was provided by the war which, before its end, involved countries in every continent of the world.

3 The years of rapid development 1945–1970

World conflict in 1939–1945 as that of 1914–1918, caused a great leap forward in the rate of application of new technology to the industries of the world, including fish industries. It is relevant to consider briefly some general features of the increasing importance of science and technology.

By 1939 warfare was already closely involved with technological development and as the war progressed so scientists were drawn more and more into every aspect of it. All branches of the military machines of the major belligerents were served by scientific teams whose activities extended far outside indisputable military affairs, such as rockets or radar, to problems of feeding and clothing. Furthermore, all governments, whether at war or not, began to intervene in matters affecting the operation of industry, in a manner virtually unimaginable before 1939, at least in western countries.

When peace returned, attitudes had changed. Politicians accepted that science and technology could contribute enormously to the welfare of the population. They had fresh in their minds the achievements of scientists during hostilities and they required no persuading that teams of scientists, if provided with adequate facilities, could solve almost any problem. Many believed that industry should be strongly encouraged by government to apply the fruits of research. An urgent edge was given to fish research in many countries because hostilities had created severe food shortages, and governments turned to the oceans as an immediate source of protein. Unlike agriculture, where farm stocks and crops have to be nurtured and disease or bad weather may destroy the harvest, fisheries can be prosecuted at any time, provided stocks are plentiful, and many major fishing grounds had remained virtually untouched for five years.

Expenditure on fish technology research and development can hence be seen as part of a more general pattern of government expenditure on industrial research. Government resources were also frequently provided to help to re-equip fish and fishing industries or to set them up where none existed previously, and government employees with the requisite training were frequently called upon to advise on how best to apply the new technology. It is also worth recording that the Food and Agriculture Organization of the United Nations (FAO) was established in 1945.

In the period from the late 1940s to the 1960s, most established laboratories grew in size and closer links were forged with industry, often by involvement in joint developments. New laboratories were opened, mostly financed by government but some jointly with industry or even co-operatively by industry alone; a considerable number were in developing countries. It is impossible even to name them all here; for this type of information the reader should consult reference works published by FAO, such as that of Kreuzer and Tapiador (1975). Previously, little was known in any systematic way about the technology of warm water marine fish and there had been a tendency to assume that the technology developed in temperate areas was also applicable, with little modification.

Literature and advice, tailored to the needs of particular sections of the fish industry and on a far wider scale than previously, now became available. First in the field were government organizations in the US and Canada, but by the mid-1950s most of the major fish industries of the world could draw on technical literature prepared by experts working side by side with industry. The role of FAO in assisting the exchange of information between experts, and in helping governments in developing areas to establish new institutes, deserves special mention. So also does the indebtedness of fish industries and the scientific community alike to FAO for organizing a splendid series of international meetings, of which the first was held in Washington DC in 1961 (Heen and Kreuzer, 1962). Such meetings brought industry into contact with internationally known experts at a time when far-sighted companies were beginning to employ their own technologists.

For the decade immediately after 1945 most laboratories were largely occupied with short-term problems of industry, though background work was done as well. In Europe in particular, the need for fish and the shortage of vessels led fishermen on the larger boats, which went further afield, to use methods that would previously have been unacceptable. Large catches were easily made, and more fish was sometimes caught than could possibly be handled before spoilage supervened. Holds were frequently filled to the hatches, insufficient ice was used and deep stowage led to severe crushing. Gutting and washing were sometimes performed perfunctorily or not at all.

Although, however, in many laboratories attention was focused on handling and stowing at sea, it was widely recognized that not enough was understood about the general problems of the spoilage of fresh fish and over the next decade workers, building on the foundation of pre-war studies, produced an extraordinarily wide and comprehensive range of papers. Most of the practical advice on spoilage now available to industry is built on this foundation.

Bramsnaes (1965) in introducing his impressive review of the handling of fresh fish said 'Many of these results have not yet been fully utilized by the fish industry in general. On the other hand, some important findings have already had a decisive impact on the routine methods employed in getting the fish from the fishing grounds to the fishmonger's slab or to the processing plant.' No doubt these remarks are relevant to the current period.

A continuing interest in almost all laboratories has been and remains, how to measure spoilage, particularly of fresh fish. Broadly, methods fall into three categories, sensory, chemical and physical; many examples could be quoted. Some indication of the amount of work in progress in the 1950s on methods of measuring quality can be obtained from the list of contributed papers to a meeting held in the Netherlands in 1956 on the chilling of fish. (Hess and Subba Rao, 1960). In the section on subjective (sensory) methods of assessment, workers from Denmark, the UK, Canada and France contributed papers. On objective methods, workers from the UK, the US, Australia, Canada, Portugal, Japan and Sweden contributed papers. All came from research institutes funded by government.

Some methods have been subsequently adopted by industry, for quality control purposes, or have been incorporated into national inspection schemes. Every company of importance, especially if it manufactures frozen products, has its own quality control section using techniques, sometimes modified, which were first devised and tested in technological laboratories. Mention should also be made of the Codex Alimentarius Commission, a body set up in 1962 jointly by the World Health Organization and FAO. It is concerned in part with the standards of foods entering international trade and although this is not the place to discuss its work in detail, attention is drawn to the major role played by scientists and technologists in its activities, which have significantly affected the fish industry in many areas because the deliberations of Codex Committees have produced what are effectively definitions of good quality products and provide a guide to appropriate manufacturing practice.

What is quite clear is that it would have been beyond the technical capability of most companies handling and processing fish, in every part of the world except possibly North America, to have developed such methods unaided and they have become increasingly important not only for quality control but also for the setting of standards and the implementation of various inspection schemes.

Laboratories were, of course, concerned with specific national problems; for example, in the early fifties work was done in Iceland, on the causes of a yellow-brown discoloration in salt cod. Previously, heavy financial losses had been sustained by the salt fish industries not only of Iceland but also Norway, Canada and Faroes. The cause was traced to the presence of copper in the curing salt and the findings were confirmed in Canada and elsewhere. In Germany the laboratory provided much of the background information on the storage properties of products in aluminium cans, thus assisting the industry to introduce new technology. In France much work was done with the canning industry also, but on utilization of frozen tuna and sardines as raw material for canning. This involved work on freezing, storage and thawing methods.

In the UK the Torry smoking kiln, whose design was first published about 1938, began to be adopted widely by industry in the late 1940s and early 1950s; it has had a critical effect on the profitability of this part of the industry and few traditional kilns are now in use in the UK.

In the development of fish industries, however, science and technology have made such a large contribution to the application of refrigeration to the chilling and freezing of fish, especially freezing at sea, that the achievements here alone would more than justify all expenditure on research and development.

In Section 1 it was mentioned that freezing and cold storage research was an important element in the early research programmes. By 1945 much was known about the conditions necessary for preparing, freezing and storing commercial species of the northern hemisphere so that the quality of the thawed product was acceptable. In the 1950s a combination of factors led to a rapid growth in the frozen food industry in western developed countries, and fish and fish products formed the most important commodity group. No doubt the production of better designs of deep freezer, cold store, refrigerated transport and conservator cabinet assisted in this phenomenal growth; for fish, perhaps the invention of the fish finger was of major importance. Whatever the reasons, there is no doubt that scientists and technologists played a key role in this growth. Many studied specific problems, such as the microbiology of fish products or the control of rancidity by the use of appropriate packaging materials. Others examined the more general questions concerned with specification of product and process.

The development of freezing at sea deserves special mention. The idea was not new in 1945; it had first been proposed in the nineteenth century and some of the earliest practical studies were made in the 1880s, as implied by the advertisement of the Hallam Company at the 1883 exhibition. Some relatively successful commercial operations, a few of which ran for several years, were done before 1945. The reasons for their eventual demise seem to have been numerous and not necessarily technological. They include, however, in some instances a lack of knowledge of the critical factors that must be controlled if an acceptable product is to reach the consumer, a lack of adequate infrastructure on shore to deal with the product, and possibly also various social and economic factors. There is, for example, no doubt that consumer and trade prejudice played a significant part in condemning some earlier attempts.

After 1945, conditions were more favourable for the development of freezing at sea. If the more distant fishing grounds which were being fished were to continue to be fished economically, freezing at sea offered the only feasible way forward. Although these grounds were being fished by conventional trawlers stowing their catches in ice, the use of ice was being stretched to the limit and beyond, and stale fish was in consequence sometimes reaching the consumer. In many countries, including USSR, Japan, UK, Poland, Germany and the Netherlands, to mention only a few, scientists, technologists and engineers in government-financed organ-

15

izations collaborated with the fish industry, equipment manufacturers and shipbuilders in the designing, building and operation of vessels equipped to freeze-at-sea. The actual solution of problems involved a number of different approaches depending on national differences, economics, national tastes and preferences and on the species actually caught. Every approach, whether it involved the freezing of whole fish or fillets, the use of factory ships or freezer trawlers, required a very large technological and scientific input for its solution.

It is hard to envisage such a large-scale effort, involving much collaboration with industry in every country, being undertaken anywhere before 1945. Some of the problems that had to be solved required the design of equipment or investigation of the nature of refrigerant systems. Others required fundamental investigation of the *rigor mortis* of fish, how best to bleed very fresh fish, and problems of texture. Answers were found in research stations to these difficulties and were tested in industry and forthwith taken up. Much of the work was truly co-operative, and this is what led to a successful conclusion. It is also probably true that given time, freezing at sea would have developed anyway, though not necessarily in the way it did and certainly not so rapidly. Indeed, at least one company, Salvesen of Leith in Scotland, was doing investigations on its own before 1950 (Lochridge, 1950) and over the next few years built and operated three freezing-at-sea vessels; it drew, however, considerably on the existing knowledge of the requirements for good quality frozen fish. The contribution of scientists and technologists was critical in determining the spread of the development which was indeed remarkable. Within a decade of 1945, freezing vessels were being built in many countries and within two decades the oceans were all being heavily fished by freezers. In the long term, freezing-at-sea has probably been more effective in changing the fish industries of the world from traditional activities to modern operations, than any other single development.

Not all scientific work aimed at assisting industry in its operations has been ultimately successful. The reasons for lack of success are not of course the same in each instance. Mention should be made of the work of Tarr in Canada who had long been searching for an additive which could be safely added to ice to enhance its preservative properties. Tarr *et al* (1950) published results which showed that certain broad spectrum antibiotics, the tetracyclines, could increase the storage life of commercial fish in the laboratory by an extraordinary degree, sometimes almost doubling it. In industry, results were rather less spectacular but still economically attractive and antibiotic dips and ices were employed in some countries for a number of years. Here scientists advised on such problems as the maintenance of effective 'dips', how to prepare antibiotic ice and the limitations of all methods of preservation (Tarr, 1960; Castell and Dale, 1963). The practice has been discontinued everywhere, and in the more stringent atmosphere of the late 1970s it is difficult to imagine such a process even being permitted. The same comment would apply perhaps to the irradiation of fish, a procedure having certain attractions and investigated very fully in the 1950s and 1960s, particularly in the US but never so far as is known applied commercially to fish.

Although it is instructive to know why these developments were not eventually successful, this is not the intention of the present paper. The important point is that scientists and technologists independent of industry drew together all the available evidence on these and other techniques and enabled the fish industry to make judgements on the value of each of them. At the time that the use of tetracyclines was being studied, there was probably a commercial advantage in applying them. When it was no longer worth while to use them, they were abandoned.

One other example should be quoted of a relatively costly development programme, mostly in the US, to obtain a fish protein concentrate which could be made from fish not suitable for other food purposes and could be used to alleviate protein malnutrition in deprived areas of the world (Pariser *et al*, 1978). Other laboratories, for example in Norway, Canada and South Africa, also did work on this topic but the main expenditure was in the US. Perhaps the ultimate lack of commercial interest in the results was due to the pressure for the programme of work being primarily political. Industry could not make a profitable product and in any event, by the time the work was completed, other uses had been found for the raw material.

By way of contrast might be mentioned work on minced fish currently in progress in many laboratories throughout the world. Machines developed first in Japan but now manufactured elsewhere as well, can separate flesh from skin and bones by a squeezing action, resulting in a relatively coarse mince of flesh. Interest everywhere is concentrated on ways of using this material since, in some respects, its properties differ from those of normal flesh. The problems of utilization can be seen from various levels. The apparently simplest and most productive approach is 'merely' to develop new products. In practice this can be difficult because flavour, texture and colour can be influenced by mincing and in the longer term more needs to be known about how to handle and modify fish proteins if some of the problems are to be solved. The point to stress, however, is that all fish industries are concerned to know how to handle this material and to solve this problem would be to bring great benefit to industry and consumers. If any answers are found, there will be no delay in applying them!

4 Breaking new ground

During the late 1950s and 1960s in particular, fish industries in most of the developed western countries began to change in structure. Processing companies, particularly those making frozen fish products, began to grow in size, often by amalgamation, as an increasing quantity of fish went into frozen branded products. In the centrally planned economies, the size of individual enterprises has of course always been on a relatively considerable scale. Nevertheless, the impression is gained that there has everywhere been a signficant increase in the amount of scientific and technical expertise actually within fish processing organizations and this is a trend that will surely continue. There are now clear indications that in developing areas the same thing is happening, especially in those companies with export markets.

For the future, the nature of the advice that the technological laboratories will be called upon to give will change, and indeed is already changing. This will occur partly because many companies themselves can undertake some development work, although their capacity to do so may be limited. More to the point is the fact that supplies are becoming increasingly uncertain and whereas a decade ago companies could be assured of supplies of raw material that were reasonably uniform in quality, of the same species and at a stable price, this is no longer true. A firm making fish fingers, for example, could in the past draw on a number of sources of frozen cod blocks, but declining catches and the closure of some grounds has made it increasingly difficult to identify reliable and continuing sources of supply of the favoured species. On the other hand, the need to make maximum use of everything that can be caught becomes ever more urgent as world population rises.

Scientists and technologists need to solve, in partnership with industry, problems which are becoming increasingly more complex and difficult but where the rewards for success are correspondingly great. There are large resources of small fish, many virtually untapped or employed for animal feed, which could be used for human consumption if only the way was clear. Scientists from Russia, Poland, Japan and West Germany, to mention only some, have been working on ways of utilizing krill.

Some progress has been made in these fields, but what has been done so far falls well short of what is needed. What ideally is required is a means of separating flesh from small fish, retexturing it, perhaps altering its flavour and appearance, and hence producing a uniform raw material in quantity which can be successfully used in the manufacture of the type of product the consumer wants. Similarly, such technology might be applied to krill; it might, for instance, be possible to prepare relatively large pieces of crustacean meat by using appropriate methods.

Within the next few years, an increasing proportion of laboratories' research effort is likely to be spent on these difficult projects involving particularly flavour and protein structure. Perhaps some types of work which have formed the main parts of many research programmes will now begin to assume a less important role.

One other trend has become apparent in the last decade and this is the greater degree of intervention by governments to ensure that the quality of products conforms to criteria laid down elsewhere. Scientists and technologists are increasingly likely to be drawn into this type of work, perhaps advising on criteria, helping industry to meet them, developing methods of measurement or demonstrating their unsuitability for particular tasks. Whatever the precise role, it seems inevitable that more effort will be devoted to such questions as added water and product composition as the ability of industry to modify the properties of components of products becomes greater.

The next two decades are going to be searching ones. Fish industries throughout the world will continue to change as supplies become more difficult and the need continues to make use of everything that is caught. In every part of the world fish industries are becoming more technically orientated, and indeed are being forced in this direction by external pressures. The need for support from scientists and technologists has probably never been greater; the speed of change and perhaps also its direction will depend very much on the quality of the research and development now being undertaken in laboratories throughout the world.

5 References

ANDERSON, A G. On the decomposition of fish. 26th Ann. Rept. Fish-1908 ery Board Scot. for 1907, Part 3. *Sci. Invest.*, 13–39

ANON. Intern. Fish. Exhib. London 1883. Fish. Exhib. Literature, 14 1884 vols. Clowes, London

ATWATER, W O. The chemical composition and nutritive values of 1892 food fishes and aquatic invertebrates. *Ann. Rept.* US Commission Fish and Fisheries 1888, 679–868

BEATTY, S A and GIBBONS, N E. The measurement of spoilage in fish. 1937 *J. Biol. Bd. Can.*, **3**, 77–91

BRAMSNAES, F. In *Fish as Food*, Ed G Borgstrom. Academic Press, 1965 New York, Vol 4, 1–63.

BURGESS, G H O. *Developments in handling and processing fish*. Fish-1965 ing News (Books) Ltd, London. 132pp

CASTELL, C H and DALE, JACQUELINE. Antibiotic dips for preserving 1963 fish fillets. *Bull. Fish. Res. Bd. Can.*, No. 138. 70pp

FELLERS, C R. Bacteriological investigations of raw salmon spoilage. 1926 *Univ. Wash. Publs. Fisheries*, 1, 157–188

FINN, D B. The denaturation of fish muscle protein by freezing. Con-1934 trib. *Can. Biol. and Fisheries*, 8, 311–320

HACHEY, H B. *History of the Fishery Research Board of Canada*. 1965 Manuscript Rept. Series (Biol) No. 843. Fishery Research Board, Ottawa. 499pp

HARRISON, F C, PERRY, H M and SMITH, P W P. *The bacteriology of* 1926 *certain sea fish*. Natl. Res. Council Can. Tech. Rept. No. 19

HEEN, E and KREUZER, R (Eds). *Fish in Nutrition*. Fishing News 1962 (Books) Ltd, London. 445pp

HERBERT, D (Ed). *Fish and Fisheries*. A selection from the prize essays 1883 of the International Fisheries Exhibition, Edinburgh 1882. William Blackwood, Edinburgh. 352pp

HESS, E. Cultural characteristics of marine bacteria in relation to low 1934a temperature and freezing. Contrib. *Can. Biol. and Fisheries*, 8, 461–474

—— Effects of low temperatures on the growth of marine bacteria. 1934b Contrib. *Can. Biol. and Fisheries*, 8, 491–505

—— and SUBBA RAO, G N (Eds). *The chilling of fish*. FAO Interim 1960 Committee on Fish Handling and Processing. Netherlands Ministry of Agriculture, Fisheries and Food, The Hague. 276pp

HUNTER, A C. Bacterial decomposition of salmon. *J. Bacteriol.*, 5, 1920a 353–361

—— Bacterial groups in decomposing salmon. *J. Bacteriol*, 5, 543–552 1920b

HUNTSMAN, A G. The processing and handling of frozen fish, as exem-1931 plified by ice fillets. *Bull. Fish. Res. Bd. Can.*, No. 20. 52pp

KREUZER, R and TAPIADOR, D D. Directory of fish technology re-1975 search institutes in the IPFC region. Indo-Pacific Fish. Council. Regional Studies No. 6, FAO, Bangkok

LOCHRIDGE, W. 'Fairfree'—Fishing vessel and floating factory devel-1950 opment. *Trans. Inst. Engrs. Shipbuilders Scot.*, 93, 504–537

LUDORFF, W. Institut für Fischverarbeitung. *Arch. Fischereiwiss* 13 1962 Beiheft 1, 128–153

NUTTALL, G H F and GARDNER, J S. The histological changes in frozen 1918 fish and the alterations in the taste and physiological properties. *J. Hyg.*, 17, 56–62

O'MALLEY, H. Report of the US Commissioner of Fisheries 1922 1922 (Appendix IX), Government Printing Office, Washington

PARISER, E R, WALLERSTEIN, M B, CORKERY, C J and BROWN, N L. 1978 *Fish protein concentrate: panacea for protein malnutrition?* MIT Press, Cambridge, Mass. 296pp

PLANK, R, EHRENBAUM, E, and REUTER, K. Die Konservierung von 1916 Fischen durch das Gefrierfahren. *Abhandlung zur Volksernährung* (5) Berlin: Zentral-Einkaufsgesellschaft

REAY, G A. *The low-temperature preservation of the haddock*. Dept. 1931a Sci. Ind. Research Rept. Food Invest. Board, 1930, 128–134

—— Cold storage of fish. *Fish Trades Gaz.*, 49, 24 1931b

—— The influence of freezing temperatures on haddock's muscle, Part 1933 1. *J. Soc. Chem. Ind.*, 52, 265–270T

—— The influence of freezing temperatures on haddock's muscle, Part 1934 2. *J. Soc. Chem. Ind.*, 53, 413–416T

SMITH, H M. The United States Bureau of Fisheries: Its establishment, 1910 functions, organization, resources, operations and achievements. *Bull. Bureau Fish.*, 1908, 28, 1367–1411

—— Report of the US Commissioner of Fisheries 1919. Government 1921 Printing Office, Washington

STEVENSON, C H. *The preservation of fishery products for food*. 1899 Government Printing Office, Washington. 563pp

STILES, W. *Preservation of food by freezing with special reference to*
1922 *meat and fish.* Dept. Sci. Ind. Research Fd. Invest. Board, Sp.
Rept. 7. 186pp
TARR, H L A. Use of preservatives and antibiotics in the preservation
1960 of fresh fish. In *Hess and Subba Rao, 1960*, 17–23
TARR, H L A, SOUTHCOTT, B A and BISSETT, H M. Effect of several
1950 anitbiotics and food preservatives in retarding bacterial spoilage of fish. Prog. Rept. Pac. Coast Sta., *Fish. Res. Bd. Can.*,
No. 83, 35–38

TAYLOR, H F. Refrigeration of Fish. Report of the US Commissioner
1927 of Fisheries for 1926 (Appendix 8). Bureau Fish. Doc. 1016.
Government Printing Office, Washington
THEVENOT, R. *Essai pour une histoire du froid artificiel dans le monde.*
1978 Institut International du Froid. Paris. 507pp
TRESSLER, D K. *Marine Products of Commerce.* Chemical Catalog Co,
1923 New York. 762pp
WATERMAN, J J. *Torry Research Station 1929–1979. A Brief History.*
1979 Ministry of Agriculture, Fisheries and Food, London

Past, present and future of fish handling methods
G C Eddie

1 Introduction

This paper discusses the methods and equipment used in such operations as:

bringing the fish on board the catcher vessel
handling and stowage on board
transfer of catches at sea
handling of ice
unloading of catches
gutting, filleting, deboning

Processes of preservation, and the transport and distribution of fish and fish products, are discussed comprehensively in other papers to be presented to the Conference and will be mentioned only to the extent necessary to clarify particular points. Even so, the field to be covered is very wide and embraces a number of apparently disparate topics. Whole books could be written about some of the subjects listed above. Moreover, without some years of study, it would be difficult to acquire sufficient understanding of the economic and social circumstances and histories of the major fishing nations to allow the painting of a reasonably accurate picture on a world scale. Choice of methods and design of equipment of the kinds under discussion depend very much upon such factors as the ultimate product, its value and quality, and on local practices such as, for example, whether the catch is unloaded directly from the ship into the shore processing plant, or is displayed for sale by auction on the quayside. The author is therefore indebted to a number of people who provided him with information on current practice and experimental work in other parts of the world, including Stewart Roach and also Robert Payne and other colleagues in FAO, as well as to his former colleagues in the White Fish Authority and in Torry Research Station for up-to-date information on methods and equipment in the UK.

Despite the great variety of methods and equipment, common underlying trends and some general problems and broad opportunities can be discerned. An attempt will be made in the paper to identify these and discuss them. Particular machines and methods will not be described in detail: not only would this be inappropriate in the context of this Conference, but also it would mean that in a paper of this length there would be only sufficient space to provide something resembling an exhibition catalogue, and rather less useful. In any case, the details of design that make a fish pump different from a general service water pump constitute the technical capital of the designer and manufacturer, as do the finer points of the design of a filleting machine; they are seldom published, or discussed in scientific and technical conferences, and so are not readily available to authors of papers like the present, unless the development work has been carried out in the public domain.

2 Major developments

In the first half of this century, the operation of bringing the catch on board was either not mechanized – as is still the case in many subsistence and artisanal fisheries – or it was only partly mechanized. If the fishing vessel was equipped with a power-driven winch or capstan, it was simple enough to apply the power to such operations as brailing, bringing the cod-end on board and hauling of gillnets; mechanical linehaulers also made an appearance during this period.

The next big advances were the invention of the power block and later, simple net reels, in North America, and the development of the stern-ramp trawler in the United Kingdom. Later still, net drums capable of handling heavy bottom trawls were also developed in the United Kingdom. The power block made it feasible to handle very large purse seines and this led quickly to the use of submersible pumps to empty large catches of small pelagics from the purse seine into the ship. After dewatering, the fish could be conveyed below (see *Plate 1*).

Things were more complicated in the major demersal trawl fisheries. The advent of stern trawling removed the need for heavy physical effort in hauling the net, but this did not lead to immediate reductions in manpower, because the men were also needed for two other operations: mending the net, and sorting and gutting the catch. The first of these two problems was solved partly by the advent of tougher, man-made fibres, but mainly by the development of the net drum: quite small vessels can be equipped with two or even three net drums, each carrying a complete trawl. The second problem was solved by the development of efficient mechanical gutting machines. Why these did not appear until a whole generation later than the first effective filleting machines is not easy to explain, except that the problems facing the designer and development engineer are different in at least one fairly fundamental way; this is discussed later.

The use of pumps to bring the catch on board is not confined to the purse seiners. The method developed in the USSR for harvesting the *kilka* in the Caspian Sea, by light-attraction and pumping, is too well known to require description; there is also the hydraulic mollusc dredge developed by the White Fish Authority, and

18

Plate 1 Pumping from purse seine, dewatering and conveying to rsw tanks

broadly similar devices for harvesting sub-littoral seaweed. To raise the harvested material from the sea bed into the ship, there is a pump on the harvesting head, working on the injector principle, and powered from a centrifugal water pump on board the vessel. The same injector principle is employed in the latest prototype pumps for unloading salmon in the Pacific Northwest of America.

The availability of pumps of this type, and of submersible pumps with hydraulic or electric, makes it feasible to pump from the net to a height well above sea level without fear of losing suction. Pumps are now in occasional use for emptying midwater trawls into large stern trawlers without bringing the cod-end on board at all (see *Plate 2*)–for example in the krill fishery in the Southern Ocean. This avoids the problem that, when the cod-end is lifted above the water surface, the krill tend to lock together, and are difficult to spill out of the cod-end. The use of a pump to empty the cod-end can thus not only save time, but also avoids physical damage to the catch, and facilitates its direct and orderly conveyance into chilled buffer storage.

The use of large midwater trawls and purse seines tends to result in very large single hauls of fish; the difficulty of handling such large hauls by conventional methods, without physical damage and delays in initiating the chilling process, may be avoidable by wider adoption of pumps and chilled sea water stowage (see *Plate 3*). By contrast, the revival of longlining through the introduction of fully automatic systems with the catch coming on board at a more steady

Plate 2 Attaching a fish pump to the cod-end of a midwater trawl

19

Plate 3 Large haul (approximately 40 tonnes) pelagic fish in midwater trawl

rate, must be as welcome to the food technologist as it is to those biologists who would like to see the introduction of more selective methods of capture.

3 Handling on board

Meanwhile, the sorting of mixed species catches remains as much of a problem as what to do with the unwanted bycatch is after it has been separated from the commercially valuable species. There will probably never be a general mechanical solution to this problem apart from the drastic one of comminuting or digesting the whole catch; solutions will be specialized, and take various forms, according to local circumstances: special designs of trawl and purse seine; mechanical or hydraulic separating devices on board.

Size grading is much more straightforward: in the commonest types of grading machine, the fish pass over and ultimately drop through the spaces between gradually diverging vibrating bars or contra-rotating rollers (see *Plate 4*). It is interesting to note that biologists concerned with population dynamics have made use of size grading machines, and no doubt could make use of the length measuring modules of filleting machines and gutting machines if they were fitted with recording devices.

There is little to be said about shipboard machines for washing the fish. They came into use in the North Atlantic trawl fisheries in the 1950s; Waterman considered that their most useful function was as a 'delay line' and conveyor which produced a smooth flow of fish from the deck to the hold, so that the men below had time to ice and stow the catch more carefully. It therefore really did not matter too much if the research into how carefully the fish should be washed, if at all, never produced a very decisive answer. It is also still not entirely clear to the author whether and when fish should be gutted or not––except for herring where unfortunately it seems impracticable–or merely bled. Even bleeding does not always seem to be necessary to produce an acceptably white fillet as, for example, from Pacific grey cod. Perhaps the practical answer nowadays is whether the next machine or piece of plant through which the fish is to pass will accept ungutted fish or not.

Conveyor systems on board ship are of fairly conventional design although, of course, choice of general type is limited by ship motion. Flexible conveyor belts, roller conveyors, bucket elevators and worm conveyors are however large rigid objects that impose constraints on the layout of the processing space, and present a considerable cleaning and maintenance problem. Hydraulic conveying is much used in other food industries, but might be an additional safety hazard: unnoticed escape of water from piping systems serving the processing space has already endangered at least one ship. In the case of Antarctic krill, where rates of handling of 20t/h or more may be required, the use of pneumatic conveying may be attractive, especially in association with fluidized-bed freezing or similar IQF methods, followed by attrition-peeling. The giant squid, when we come to harvest them, will present quite other handling problems.

Plate 4 Grader roller type

4 Preservation on board

When preservation on board is by chilling, the choice of whether to use conventional stowage in crushed ice or stowage in chilled sea water depends largely but not entirely on which is better suited to subsequent operations such as unloading or transfer at sea; presentation for sale or temporary stowage awaiting further processing; or transport from the landing place to some remote centre. The choice between use of boxes or removable sections of hold in the first case, as against conventional pounds and shelves, and portable containers as against fixed tanks on the other, is similarly governed. At least this is so when considering the design of new vessels. In existing vessels, conversion to boxes or containers may be unacceptable because of inefficient use of hold space, but in the design – and price estimation – of new vessels, it is often forgotten that by far the cheapest part of a fishing vessel is that section containing the hold; it is also not sufficiently widely realized that provision of very large hatches is no problem, at least in single-decked vessels, provided that they are not removed at sea.

In some fisheries, where stowage is in conventional pounds (pens) and shelves, there has for many years been a reluctance to apply enough ice. For this and other reasons, there have been attempts to develop equipment that will provide a flow of crushed ice through a flexible hose (see *Plate 5*). Slinging machines for icing railcars were in use many years ago, and sub-cooled flake ice is sometimes conveyed pneumatically within the factory where it is made, but the problem of conveying crushed ice near its melting point through a pipe by means of a stream of air has never been satisfactorily solved to the author's knowledge: sooner or later, in some circumstances, blockage occurs.

Some years ago, however, a mechanical system was developed in Denmark by Baek-Olsen and Petersen for automatically mixing ice with the catch on board vessels engaged in fishing for shore-based fish meal plants on voyages of several days. The proportion of ice mixed with the fish was metered according to the temperature of the fish and the likely duration of the voyage, so that excess ice did not remain when the fish came to be processed. The use of ice improved the yield of fish meal from a given weight of fish by over 10% and also gave a higher quality product.

The use of chilled sea water was for long standard practice in some tuna fisheries and subsequently developed in the 1950s for more widespread application by the Fisheries Research Board of Canada (Roach *et al*, 1961; Roach *et al*, 1967) and afterwards in northwest Europe. It allows rapid stowage and provides rapid chilling of the catch, and lends itself to unloading of the ship, or transfer of the catch at sea, by pump.

Chilled sea water can also be used in portable containers (see *Plate 6*), which can be filled at sea (Hewitt and McDonald, Anon, 1973), lifted out of the fishing vessel by crane and transported by road or rail to a distant point, the fish remaining protected from physical damage and kept at chill temperatures throughout the journey from the fishing grounds to the processing plant. There are two difficulties: one is that the entire system of buying and selling, transport and processing must be adapted to use of containers; the other is that it is uneconomic to provide several sets of containers for one

Plate 5   Applying ice from crusher blower through a flexible hose

21

Plate 6 'Alcoa' portable container

vessel; thus the decision to implement the system depends upon almost industry-wide agreement. In this case, as in freezing at sea, innovation is facilitated by vertical integration of the industry. It must be added, however, that the smaller, specialized enterprises are often the most ready to adopt new methods.

Stowage of the catch in boxes with crushed ice offers the same advantages as the use of containers with chilled sea water: reduced handling of the fish; protection against physical damage and rises in temperature on unloading. It has other potential advantages also. Boxed fish can be transferred at sea from one trawler to another in surprisingly rough conditions, using soft fendering systems (see *Plate 7*). Because the fish is in boxes, and not being transferred by pump or in flexible floating cod-ends, it will not suffer damage even if it is in *rigor mortis*, and it will not be damaged through prolonged immersion in sea water. Extensive trials were carried out in commercial conditions some 10 or 12 years ago; the main problem turned out to be the slow rate of breaking out the stowage in the ship transferring its catch and of re-stowing it in the receiving vessel—a problem also experienced in other transfer operations, outside the fishing industry.

In distant water fisheries, where at the time of landing the catch the first-caught fish is significantly less fresh than that caught later in the voyage, boxing at sea allows much easier and much improved segregation and sorting of the different quality grades than does conventional stowage in pounds (pens) and shelves, as will become apparent later below. It allows rapid discharge of the vessel by crane, immediately on arrival, and thus facilitates rapid turn-round; the vessel can even discharge the fish at a port nearer to the fishing grounds instead of at the market or processing plant she is serving, whereas in conventional stowage in pounds and shelves, discharge has to wait until the last possible moment if the catch is to remain protected as long as possible by ice. The absence of damage from hooks, shovels and trampling results in significantly higher yields of edible material. The main practical difficulties in introducing boxing

have been to establish confidence regarding the weight of fish in the box, and to persuade the fishermen not to overfill the boxes in consequence. Incidentally, it is not easy to design a practical box that is long enough to hold sprag cod and large haddock, that can be handled by two men on a moving ship, which contains on average some convenient round number of pounds or kilogrammes of fish, which occupies less space empty than full, but still gives a stowage rate of $350kg/m^3$ or better, and which is arranged so that meltwater does not enter the box below. Such a box was designed at Torry Research Station by Waterman and used successfully in the trawler fleets.

5 Landing of the catch

In the smaller ports, and in the case of the artisanal fleets and family-owned vessels, fish intended for human consumption may be landed by the crew of the vessel. In the larger ports, and in the cases of company-owned vessels and large reduction plants, special labour and equipment are employed.

The containers and boxes already mentioned, and the portable sections of fishroom developed in France a decade or so ago, are handled by more or less conventional cranes, often mobile. The discharge of a large vessel stowing its catch in conventional pounds (pens) and shelves, and with relatively small hatches to give access to the hold, presents quite different problems: many attempts were made some 20 years ago to develop solutions that would be an improvement on filling a basket with fish, attaching it to the end of a rope, hoisting it up and swinging it ashore. At that time, in the British distant water ports, the unloading operation, based on methods of this kind, represented 10% of the total costs of production. About 2 tons of fish per shift were unloaded and laid out for auction for every man employed. Although some of these men were ex-fishermen no longer able or willing to go to sea, it was felt that costs of production and dependence on such a

Plate 7 Transfer of fish at sea

22

large labour force should be reduced. Systems based on the use of bucket elevators, and ship-to-shore conveyors that, of course, had to possess several degrees of freedom to allow for relative movement, were studied, devised and put to trial. Similar efforts were taking place in Western Germany and other countries with deep-sea trawling fleets. The general conclusion was that the most elegant method of getting fish out of a big trawler with narrow hatches, unless it boxed its catch at sea, was in a basket on the end of a rope. Since then, bucket elevators and conveyors, sometimes built into the ship, have come into service, but to a limited extent.

These experiments and innovations were mainly aimed at the saving of labour and at reducing costs. Less attention was paid to those aspects of the landing operation which affected the quality of the catch. Fish were handled two or three times, and removed from the protection of the ice. They suffered gross physical damage, as already noted, from trampling, from shovels used to clear away ice and — especially on the eastern seaboard of north America — from the use of hooks or spikes in the wrong places. In addition, although species and sometimes sizes would be segregated, fish caught at different times during the voyage would become mixed. The extent of this problem was examined in one British distant water port 20 years ago in conjunction with the proposed introduction of an improved scheme of quality control. This was based upon test panel examination of a sample withdrawn at random from each group of containers in the auction display. Each vessel landed from 100 to 200 tons, caught during some 10 days of fishing; each group of containers hold a total of $2\frac{1}{2}$ tons of fish. By code marking a proportion of the fish caught in every haul during one particular voyage, it was found that in any one $2\frac{1}{2}$-ton group of containers there were fish differing in time of catching by from two to six days; the fish from one haul selected at random, in a voyage where there were 80 hauls altogether, were scattered through about one-quarter of all the containers in which that vessel's catch was displayed. In theory, it would be possible to devise a sorting system using conveyors and controlled by a freshness meter, but it would seem to be much more practical, and have other advantages as already noted, to change the system of stowage and box the fish at sea.

Fish for human consumption are in some cases discharged by pump (Roach *et al*, 1964). This method is easiest to apply in conjunction with stowage in chilled sea water, and the state of the art is most advanced in those fisheries, such as the salmon and herring fisheries of Pacific Northwest America, where such methods of stowage are standard practice. Both chilled sea water stowage and discharge by pump are economical in labour and in that region, labour costs are high; the use of pumps to empty the catch from purse seines into the boat is familiar; the systems of harvesting moreover include transfer of catches at sea, in order to enable the catchers to spend as high a proportion of time as possible in fishing, while the fish are running; the receiving scows and carrier vessels also stow the fish in chilled sea water; the use of dewatering devices makes it easy to weigh catches if desired. Similar systems, for similar reasons, are coming into use to a growing extent in the fisheries for mackerel in the outer English Channel and Celtic

Sea. In other human consumption fisheries, adoption of discharge by pump has been limited and slow, for several reasons. One reason is that the change is easier to make where the catch is sold on contract rather than presented for sale by auction to a number of potential buyers who are unfamiliar with the slight differences in appearance, salt content and other aspects of quality of csw fish as compared to iced fish. Systems for discharge of fishing vessels by pumping — including air unloaders — are capital intensive and best suited to situations where some loss of time through queueing at the unloading point is acceptable to the vessel operators. Another reason is that many existing types of fishing vessels, and especially the larger trawlers of conventional design, may not possess sufficient reserves of lateral stability to allow the installation of chilled sea water tanks or the flooding of the hold. A further reason put forward for rejection of discharge by pump has been damage to the fish. The damage and loss caused by existing conventional methods has not always been taken into account. Nevertheless, damage can be commercially significant, as is witnessed by recent efforts to develop improved pumps for handling salmon in Canada. Seemingly it is not percentage of damaged fish that is the ruling criterion, but the absolute loss in money terms: that is to say, the higher the unit value of the materials, the lower the acceptable damage rate. The same sort of criterion seems to apply to the yield from filleting and skinning machines.

The improved pump is of the injector type with an annular jet surrounding the venturi throat. It has been found that damage can be reduced by bleeding air into the flow at a point near the throat. The mollusc dredge, referred to earlier, and which has an elevating system working on the same principle, causes damage to live mussels and is therefore of limited use for gathering them for purposes of re-laying elsewhere; it would be interesting to examine whether an air bleed would solve this problem.

Catches stowed in bulk, not in sea water, can of course be discharged by pump if a source of clean water is available. Pumping systems are regarded by some as mild sources of pollution, because of the effluent from the dewatering screens. The obvious alternative to water as the transport medium is, of course, air. The designers of air unloaders have apparently achieved the technical feat of being able to bring to rest without damage a large salmon travelling through air at a speed considerably faster than a man can run, and of dealing with fish of varying sizes and none of which are symmetrically shaped about all three main axes. Unfortunately, existing air unloaders are noisy, because they use Roots type blowers, and they tend to use more power than water pumps for a given output.

For unloading fish intended for reduction to meal and oil, the use of pumps and air unloaders is widespread but by no means universal: it is surprising how the use of mechanical grabs, with the concomitant mess and losses, has been tolerated for so long.

In some cases, the fish can be expected to flow easily towards the inlet of the pump or unloader, but some species — notably sprats, and probably also krill — are less obliging, and form a solid mass with an apparent equilibrium angle of repose approaching the vertical. Means have therefore to be provided to cause the fish to flow towards the inlet, and also to ensure that the inlet is kept covered so as not to such too much air. Manpower can not be entirely eliminated.

The use of chilled sea water systems for stowage combined with pump unloading can be expected to become more widespread; they are especially suited to fisheries where catch rates are high, where the catch is comparatively homogeneous as regards species, size and time of capture, and where the entire catch is destined for the same consignee. In other cases, container or boxing seem to have distinct advantages.

There is little to say about methods of handling fish frozen at sea. The large tunas have to be handled individually, but their value justifies such methods. In the case of the less valuable standard fillet blocks and blocks of whole frozen fish, methods of handling, stowage and discharge are similar to those used in cargo ships: in the United Kingdom, after one early application of an elevator on the paternoster principle combined with conveyors, discharge has been by standard endless looped-canvas elevator belts (see *Plate 8*). During the development stage, fears were expressed by vessel owners that blocks of whole cod frozen in vertical plate freezers, 10cm thick and 50kg in weight, would stick together in the stowage; fortunately such fears were groundless, and the unprotected blocks were mechanically strong enough to withstand stowage, discharge and transport to cold store with little or no damage. By contrast, one fleet of trawlers, catching and freezing small pelagic fish for consumption by low-income

Plate 8 Unloading a freezer trawler by 'Banana Conveyor'

groups in a developing country, produced large blocks of fish that were not sufficiently cohesive to allow stowage, transfer to reefer ships, discharge, cold storage and transport to the point of retail sale, unless they were packed into individual cartons of corrugated paper board. The cost of these cartons was 5% of the total costs of production and they occupied no less than 15% of the hold space. On the other hand, if the ships had been equipped and operated so as to produce solid, cohesive, well-frozen blocks, the operators of the retail outlets would have had considerable difficulty in breaking them up into the standard 2kg rations in which the frozen fish was sold!

Some years ago FAO convened a conference on fishing ports, harbours and markets. Among the many points made is one which reinforces the arguments against the practice of exposing the catch for auction that are put forward by those concerned with the quality of the fish reaching the consumer; namely the usual fish auction, as known in northwest Europe, utilizes a vast area of expensive concrete for only a few hours a day. In these days of boxing at sea and sophisticated telecommunications, it is possible to discharge vessels on arrival and to sell by sample; this last indeed is standard practice in some ports, and for some species, even in western Europe.

Be that as it may, few fish markets seem to have been designed with enough space to allow easy flow of fish in bulk to the traders' premises, and from there to the waiting transport. There is often considerable congestion and risk of injury from the multitude of barrows. Some 25 years ago the department of government then responsible for Torry financed a study by experts in materials handling and production engineering, who produced a proposal based on the use of overhead monorail conveyors to carry the containers of fish to the merchants' premises and the boxes of fillets to the waiting trains. There was much consultation during which freezing at sea was being developed and new large processing plants were already being built outside the fish docks; the consequence was a sharp fall in the number of traders; the scheme was too late. Of the great inland muncipal markets Billingsgate in London is destined for replacement; no doubt the new market has been thoroughly work-studied and computer simulated. Tokyo, meanwhile, remains as congested and colourful as ever although, for cutting up the larger specimens of fish, mechanization, in the form of band-saws, is beginning to take the place of swords of awe-inspiring length.

6 Fish processing machinery

Presumably the adoption of machine tools has been stimulated by the desire to replace manpower (or womanpower) that has become too expensive or is no longer available, as the very name of the Iron Chink illustrates. It has been facilitated by the concentration of processing operations in large modern factories, where the volume of material to be handled is great, instead of, as formerly, in fishermen's cottages. A much greater proportion of the catch is nowadays filleted or otherwise prepared before it leaves the port of landing.

From this it might be inferred that the development of fish-working machine tools was a series of deliberate acts in response to an economic demand; this, however, is only partly true, if at all. Several of the early machines seem to have been inventions by people from various walks of life other than the fishing industry, but possessing the right kind of mechanical flair, in response to the technical challenge rather than to an identified demand supported by market research. Indeed the inventions provided the industry with new opportunities. Only in recent years has development been more deliberate.

The first successful machines were for use in the salmon industry of Pacific Northwest America and the herring industry of northwest Europe. These machines were capable of processing fish of only a limited range of size but could be adjusted by hand before the start of a working shift. This was acceptable for salmon and herring because all the fish caught in one day would be of much the same size. The demersal species presented much more of a problem, with the variation in length, between successive fish, in the ratio of two to one or more. A successful filleting machine for cod was not developed until after the Second World War (Niema, 1946).

Some early prototypes were remarkable for their ingenuity and complexity rather than for their practical potential: in the case of one filleting machine tested in Hull in the late 1940s, it was remarked that so many operators were needed that they could have processed the fish at the same rate by hand! The availability of really practical machines in the early 1950s (Baader, 1950; Anon, 1951) helped to make feasible the factory trawler; this was one of the first applications for filleting machines because of the high throughputs required, with filleting yields a secondary consideration so long as catch rates were high. Hand filleting on board factory trawlers may have been physically possible but provision of the extra living accommodation and services on board ship for the additional crew that would have been required would have added greatly to costs. Filleting machines have come into increasing use on land as large, modern, hygienic processing plants have been built, but only for those applications where labour is scarce or dear, where the raw material is suitable and machine yields acceptable, and where the amount of material to be processed gives a high enough utilization of the machine. As remarked earlier, the level of yield that is acceptable depends partly on the value of the raw material. Through development, machines have become smaller, less complex and comparatively cheaper in terms of throughput, especially when yields are taken into account. Development continues, partly because the variety of species and ranges of sizes continue to change as new resources are exploited and existing resources come under increasingly heavy fishing pressure. The next stages in development may be machines working on new principles, as suggested later below one other development greatly to be desired is machines that are more easily cleaned or are self-cleaning.

Development is expensive and time consuming. It is not yet possible to respond to an immediate demand for a new type of machine, as was made when there was a glut of small haddocks in the North Sea in the early 1960s: development groups must try to anticipate requirements by some years. Experience and flair are still very valuable assets: whereas, no doubt, any group of engineers, given sufficient money, will produce a machine tool to carry out a required process on a given species of fish, this might well turn out to be rather complex—like the shrimp peeler that more or less simulated the actions of human fingers. Nevertheless, straightforward programmes of investigation, and the use of well-known engineering development techniques, can play a useful part, and one or two examples follow.

After complaints from the industry about the rough appearance of some of the products from one particular model of filleter, Torry and the White Fish Authority mounted a big programme of tests based on the assumption that the cause was the freshness or otherwise of the raw material, the degree of resolution of *rigor mortis* or some such similar phenomenon. The engineer in charge of the tests, however, developed a suspicion that a pair of flexible knives in the machine were sometimes vibrating, and the use of a high-speed cine camera borrowed from the Royal Aircraft Establishment confirmed this.

A programme of measurement of fish at sea, in which all fisheries research and development establishments in the United Kingdom took part, provided the statistical data on the growth and form of gadoid species, on which the rational design of a gutting machine could be based. (It should be added, however, that this did not tell the development engineers how to devise a tool to remove the upper rear part of the liver from the gut cavity.) Subsequently, a similar programme was undertaken to provide the anatomical data on which to base the development of a new concept of filleting machine. One general conclusion will be no surprise to biologists or to manufacturers of filleting machines: the shape of the skeleton is independent of size over the practical range of sizes, and it is also independent of what the biologists call the condition factor. It is therefore very simple in principle to devise a filleting machine: the machine has merely to ascertain the size of the skeleton, and also exactly where the skeleton is in relation to the bed of the machine; knowing its location, size and shape, the machine can then cut the skeleton out of the fish. The alternative approach is to use the cutting tool itself to continually sense where the skeleton is, and to control its own movements through feedback—as is partly the case in hand filleting. No development along these lines has yet come near to fruition.

The problem of devising an acceptable gutting machine for fish like cod or trout is rather different: the gut cavity must be slit as far as the anal vent but, preferably, the cut should stop there. The anal vent is in a soft structure, the dimensions and physical properties of which are very variable, and which will yield by varying amounts as a cutting tool is applied to it.

The Shetland gutting machine (see *Plate 9*), which was an invention in the sense mentioned earlier and not the result of a deliberate development programme—that came later—had two elegant features. One was that the fish proceeded through the machine in a line at right angles to their major axes. The machine was therefore very cheap and compact by comparison with any design in which the fish proceed in the direction of their own length. It is, however, much easier to locate, support and measure a fish like a cod if it is towed lengthwise through a machine. The other feature which made the machine simple and compact was that the measurement of the

Plate 9 Shetland Gutting Machine (fish coffins and revolving circular saw blade)

fish and the first gutting operation took place simultaneously, instead of in sequence at different stations; this was achieved by using the hard parts of the head as a fulcrum for the lever mechanism on which the tool was mounted. A less desirable feature of this mechanism was the extremely high accelerations to which parts of it were subjected; one version incorporated components made in carbon-fibre reinforced resin in order to reduce inertia and hence the severe' impact loads on the bearings.

Mechanical filleters and gutting machines are among the most complex machine tools of any kind ever produced. Surprisingly, almost all of the measuring devices used in them, also any associated memory stores, are mechanical, and information on anatomy and shape is stored in cams and similar devices; there has been little use of electronics – no doubt because of the adverse environment – or of fluidics.

Until recently most existing filleters and gutting machines had to be fed by hand. About 10 years ago, the stage had been reached when the operations on the fishing deck of a big stern trawler could be carried out by only three to five men, and considerable effort was therefore made to reduce the number required in the factory deck. Sorting the fish and grading it into various size ranges have already been mentioned. Also developed were machines for orienting all fish so that they were moving along a conveyor head first and all lying on their right (or all on their left) sides. As a result of the introduction of extended economic zones this work has had to be done all over again for mackerel instead of cod. With mackerel, there are problems, not yet overcome, because the scales carry protuberances that prevent the use of techniques that worked with cod, such as sliding the fish down a vibrating inclined plane.

Alternative solutions to the mechanization of gutting and filleting have been proposed or investigated: vacuum gutting in the United States, for example. It has always puzzled the author why more work has not been done on producing frozen deboned cod steaks; the problems seem simpler than filleting. Other species and other fisheries present different problems – flatfish, bivalve molluscs, univalves, and crustacea. Among the various interesting and sometimes bizarre principles put to trial can be mentioned the opening of scallops by the application of local heat to the shell in the area where the adductor muscle is attached (as a principle, very successful); attempts to peel crustaceans by explosive decompression (unsuccessful); and an investigation into the possibilities of very fine very high speed water jets, such as are used for cutting sandstone (no application found). For peeling large shrimps, prawns and crabs use was made for some years of a hand tool in the form of a hollow sharpened needle through which compressed air was injected, until it was discovered that some operators began to suffer respiratory troubles.

For peeling the smaller sizes of shrimps, various types of machine have been developed. One famous manufacturer at one stage concluded that the only way to deal with the smallest sizes was to dehydrate them and then subject them to operations analogous to the milling of grain and the winnowing of the chaff. In the case of Antarctic krill, a rather less drastic and very successful method has been developed by Karnicki, now of FAO, and his colleagues in Poland, in which the krill are first frozen and subsequently subjected to abrasive or so-called attrition peeling. The product of this process is very attractive, and retains the organoleptic properties of fresh shrimp. Krill that have been peeled by machines of conventional type, using rubber rollers, are tough and tasteless because of the leaching and other effects of the use of the large flows of water that are apparently essential to the successful operation of this type of machine.

More than once, a demand has been made for the development of a machine to carry out some processing operation formerly carried out by hand, because of the growing shortage or expense of manual labour, and on closer examination, the volume of production in the geographical region affected has been insufficient to justify development of a machine *ab initio*. If a machine already exists in some other part of the world that seems capable of adaptation, well and good; or if there seem to be possibilities of an export market if a machine were developed, then that may be the solution; if the species is a mollusc or crustacean, the solution may be to export it unprocessed to countries like France and Japan. Mixed species demersal catches present the same sort of problem several times over. A rather different problem, which may have the same general solution, is how to recover the edible portion of those unconventional species that are of such awkward size or shape, or have such difficult bones or other characteristics, that they cannot be marketed directly and for which it seems difficult or impossible to devise filleting machines.

The general solution in these cases, other than to make fish meal or FPC Type B, may be to digest the fish chemically or to recover the edible flesh by some method of mechanical comminution. In this paper, we can be concerned only with the second of these processes, using deboning machines, mincing machines and similar devices.

Deboning machines were first used to recover the edible flesh left on the skeleton after the operation of filleting; the product was used to manufacture second-grade fish fingers and fish cakes, and as a filler in sausages and similar products. Since then, a variety of products has been developed, using under-prized species and processing plant waste material, by Steinberg and his colleagues in Seattle and by others. A parellel development which has so far seen much greater commercial application is the production of surimi (Miyauchi *et al*, 1973); yet another, still in the stage of field trials, is the minced salted fish and similar products, intended for consumption in developing countries, developed by Bligh and by others; yet again, there are the heat-coagulated pastes and other products made from Antarctic krill. All of these depend upon the use of machines for mincing the flesh, or for pressing it or extruding it out of the fish. The variety of attractive products is already large, but further possibilities will be opened up if machines can be developed, perhaps working on new principles, that will produce longer fibres, or separate the oil from the muscle, or do whatever it is that the food technologist may require in order that a usable and versatile raw material is presented.

The object of such developments is to utilize, for human consumption, very large quantities of raw material at present underprized or unexploited. It will perhaps allow the fishery industry to produce a wider range of more or less standard products of which the best existing example is the fish finger, irrespective of the temporary state of the stocks of the various species. By removing the market requirement that fish must be of certain species and also of certain preferred sizes, such developments may change to some extent the objectives of resource management. At any rate, in so far as the operations of comminution are carried out at sea, they make it more difficult to recover information on the state of the resources. Moreover, a capability of making standardized products from whatever species is suitable and temporarily abundant, raises the question, at least in theory, as to whether to harvest 10 tons of capelin or 1 ton of cod, although in that particular example the option preferred by most people at present may be a foregone conclusion.

7 Role of the engineer

The justification for grouping the various topics discussed above into one paper is that the various operations and equipments mentioned are the concern of the mechanical engineer, as is much else in the fishery industry. Because of our limited knowledge, much of the development work – in common with that in fishing gear technology and in ship performance and behaviour – has to be done in the field and on the full scale. Modelling and computer simulations are of very limited use, and even when experimental work can be done on shore – as for example in developing a prototype gutting machine – raw material of known treatment and history has to be produced. All the developments mentioned in the paper have involved the engineers concerned in going to sea, observing performance and behaviour and testing prototype equipment and systems in commercial conditions. The necessity of adopting this approach has been accepted to varying extents in different parts of the world at different times. One of the main constraints to fisheries development in many countries is the lack of organized groups of engineers and others, able and willing to do such work in the field and especially at sea.

Such people are not easy to find, and even a small group is expensive to equip and maintain. Scarcely any firm in manufacturing industry or in shipbuilding choose to do so, and only a few enterprises in the fishery industry itself are big enough. Moreover, a good many of the kind of developments mentioned in the paper, and many others, some of which are mentioned in other papers at the Conference, originate in government agencies. There are very sound and respectable economic reasons why this should be so, which do not concern us here. Governments, therefore, have a special responsibility for practical development in the field of fisheries technology, and the lack of the appropriate organization is as much a constraint on development in some parts of the world as is the lack of suitable engineers. The development of practical processes and equipment to produce acceptable products from the resources at present underutilized or unexploited depends largely upon removal of these constraints, and this will also create the capability for disseminating the results throughout industry, if and when the decision is made to implement them.

Many of the operations discussed in the paper, and the design of the equipment to carry them out, are affected

27

by, or have an effect upon, the behaviour, performance and design of the fishing vessel. The same is true of methods and equipment for detecting and capturing the fish and for processing it on board. Fishing vessels are complex ships fitted with a large number of interacting engineering systems and the design of which is usually space-limited. It is therefore at least desirable that the work of the various groups concerned with design and development of each of the systems should be well-co-ordinated. Moreover, automatic long-lining systems have as much or more in common with filleting machines as they have with the design of trawls; trawls have something in common with air blast freezers and with certain methods of size-grading used on fish farms, and so on. The engineers concerned with these various branches of technology require guidance from food scientists, food technologists and fishery biologists. More and more often in the course of development work, questions will be asked, to answer which will require further scientific work. At the same time it has to be borne in mind that the methods of controlling and financing government science are not necessarily appropriate to full-scale development work in the field.

With that proviso, it is to be hoped that in the future there will develop even closer associations than at present between engineers and scientists, and also much closer links between those engaged in research and development in the field of fishing technology and those concerned with handling, processing and preservation; and especially between those concerned with research and development in the field of utilization and marketing, and those whose task it is to monitor the state of exploitation of the resources and to advise on the management of the fisheries. Otherwise each group could well present the other with serious problems that could have been avoided by a joint approach; or else opportunities will be missed, like that some years ago when it was demonstrated beyond reasonable doubt that the consumer did not require his or her fish fingers to be made from cod – species under less heavy fishing pressure were equally acceptable. Fortunately there are growing links at the consultative and decision-making levels as well as between those responsible for carrying out the practical work of research and development. It should therefore be possible to devise and carry out the programmes needed to develop practical systems for exploiting the very big opportunities facing world fisheries, and referred to in the first paper presented to the Conference, and to have these systems ready by the time they are needed or, preferably, soon enough to give us time to consider the options and test each one, and thus discern how best, and to what extent, world fisheries can and should contribute to food supplies in the future.

8 References

ANON. The new filleting machine (Baader 99) in action. *The Fish*
1951 *Trades Gaz.*, No. 3577, 14–17
ANON. Chilled sea water containers to improve herring quality. *Fish.*
1973 *News Int.*, June, 30
BAADER, R. Patent Specification 701650 – application made in Ger-
1950 many on 19 May
HEWITT, M R and McDONALD, I. Fresh fish transported in special
1972 containers. *Fish. News Int.*, December, 30
MIYAUCHI, M, KUDO, G and PATASNIK, M. 'Surimi' (A Semi-
1973 processed Wet Fish Protein). Marine Fisheries Research
Paper 1026. *Marine Fish. Review*, 35, No. 12
NIEMA, J V. Fish filleting machine cuts all commercial sizes. *Fd. In-*
1946 *dust.*, 18, 1530
ROACH, S W, HARRISON, J S M and TARR, H L A. Storage and trans-
1961 port of fish in refrigerated sea water. *J. Fish. Res. Bd. Can.*,
Bull 126
ROACH, S W, CLAGGETT, H and HARRISON, J S M. An air lift pump for
1964 elevating salmon, herring and other fish of similar size. *J. Fish.
Res. Bd. Can.*, 21, 845
ROACH, S W, TARR, H L A, TOMLINSON, M and HARRISON, J S M.
1967 Chilling and freezing salmon and tuna in refrigerated sea
water. *J. Fish. Res Bd. Can.*, Bull 160

Fish preservation methods

Poul Hansen

1 Introduction

Oil and water constitute about 80% of the weight of fresh or 'wet' fish muscle. The rest is protein, normally 16–18%, and various water-soluble constituents, such as salts, trimethylamine oxide, ammonia and carbohydrates. The water content of the muscle of wet white fish low in oil is nearly 80% while that of some fatty species may seasonally fall below 60% when the fat content reaches its maximum. All wet fish are perishable, as spoilage organisms on the skin, gills and in the intestinal tract multiply rapidly after a lag phase, to reach numbers exceeding one million per gram wet muscle. Subsequently, the bacterial enzymes convert the TMAO into TMA and decompose the amino acids and proteins forming ammonia, hydrogen sulphide and other undesirable compounds characteristic of microbial spoilage.

The principal aim of fish preservation is to delay, reduce or inhibit the microbial spoilage. In the case of fatty fish, the preservation may also aim at reducing or inhibiting oxidation and other undesirable changes in the fish oils, which are highly unsaturated and capable of going rancid at various stages of processing.

Some fish preservation methods such as canning and curing change the character of the fish substantially, while the sequence of freezing, cold storage and thawing normally aims at retaining the fresh fish character as much as possible. This is also the aim of chilling and other methods of short-term preservation of wet fish.

The following will deal with the preservation of fish as food leaving out preservation for animal feeding.

2 Short-term preservation by chilling

2.1 *Fresh or 'wet' fish*

The most important means of preservation of fresh fish in tropical and temperate climates is by chilling to about 0°C. The most common chilling media are wet ice, mixtures of ice and sea water or ice-cold sea water. The normal storage life of cold water fish chilled to 0°C immediately *post mortem* is one to two weeks in temperate climates, while fish from warm tropical waters keep somewhat longer at 0°C (Jones and Disney, 1977). The storage life at 0°C, however, depends on a number of factors as illustrated in *Fig 1*, which summarizes the results of a number of observations of the quality and storage life of small rainbow trout (150–200g), chilled to 0°C immediately *post mortem* (Hansen, 1973).

The eating quality of cooked trout was measured on a hedonic scale ranging from ideal (10), very good (8), regular (6), borderline (4) to poor (2) and very poor (0).

Curve a shows the rapid deterioration due to bile salt and gut enzyme reactions around the belly cavity of whole ungutted trout buried in wet ice. These reactions, reducing the storage life to less than one week, may be eliminated by eviscerating the trout immediately *post mortem*, which was the case with lots b, c and d.

Curve b shows the deterioration due to oxidative rancidity on the belly wall surfaces, reducing the storage life of eviscerted trout buried in wet ice to about one and a half weeks. The rancidity may be prevented by vacuum-packing the trout in airtight, plastic pouches immediately after evisceration, as was the case with lots c and d.

Curve c shows a storage life of just over two weeks at 0°C for trout eviscerated and vacuum-packed immediately *post mortem*. This lot deteriorated mainly due to bacterial spoilage leading to increases of bacterial count to several millions per gram of trout fillet.

Curve d shows a storage life at 0°C exceeding three weeks for trout eviscerated, vacuum-packed and irradiated in an electron accelerator immediately *post mortem*. The doses used, 50–200rads, greatly reduced the number of bacteria and increased the bacterial lag phase so much that no bacterial spoilage was noted within the first four weeks of chill storage. The eating quality, however, decreased slowly but steadily, probably due to reactions such as the conversion of adenosine triphosphate (ATP) into hypoxanthine in the muscle tissues.

The storage lives given above are calculated as from the time of death to the borderline quality (4). If an average score of (6) *ie* regular, is used, however, it is seen that irradiation adds little or nothing to the chill storage life of eviscerated and vacuum-packed trout. In other words, the reduction of the bacterial count by irradiation did not increase the storage life of this trout as a first class product, *ie* of more than regular quality.

These findings apply in many cases also to other wet fish species which are designed to reach the consumer as first class products, as well as to methods other than irradiation which aim to reduce bacterial growth and activity. This may explain why neither irradiation nor antibiotic treatments have been generally accepted in spite of considerable investigation and commercialization efforts to promote them on a large scale in the 1960s and the 1950s, respectively. Other reasons for failure of these preservation methods have been the increasing public resentment to such 'unnatural' or unbiological treatments of foods. The latter reservation does not apply to the CO_2 treatment of wet fish in chill storage, which is now being studied by several fish technology institutes. There is no reason to expect substantial improvements in preservation by combining chill storage with CO_2 treatments, but advantages for this combination may be found in special cases where an extension of the total storage life is essential.

The form of the curves of *Fig 1* illustrating quality changes of wet trout is by no means applicable to all other fish species. White fish, for instance, have very low oil contents and normally no oil oxidation problems. Some other fish do not suffer from gut enzyme autolysis even if chill-stored in the round. Some tropical fish keep much longer than Danish trout. Thus, a tropical fatty fish of similar size, thread herring (*Opisthonema libertate*), was recently found to keep for three to four weeks in ice water at 0°C without having been eviscerated (Hansen and Jensen, unpublished).

Quality score   *Fig 1*

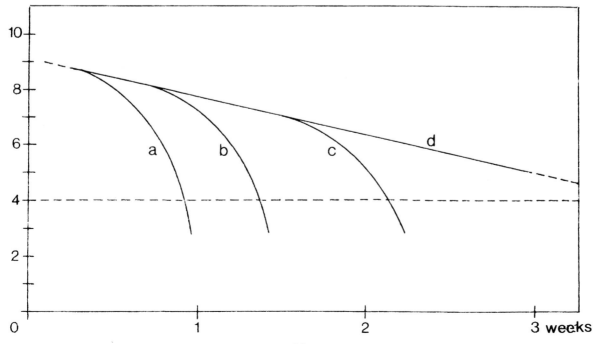

2.2 Lightly cured fish products

For culinary reasons fresh fish are sometimes lightly cured by salt, sugar or acid treatments, which may also increase chill storage life (Jarvis, 1950). The traditional Swedish 'Gravlax' are raw salmon fillets, the meat surfaces of which are covered with a layer of salt, sugar and dill in such a way that most of the salt and sugar is absorbed into the fillet within the following couple of days. The combined effect of a final salt content around 4–5% and a lactic acid fermentation, supported by the sugar, lowering the pH to between 5 and 5·5 will safeguard against production of botulinum type E toxin and secure a chill storage life of several weeks (Riemann, 1972). An improved storage life also occurs in the traditional Latin American 'Ceviche' cure, which consists of soaking cubed raw fish in sour citrus fruit juice.

There is a choice of many fish species for these light cures, which produce convenience foods suitable for watertight retail packaging and distribution from the chill store cabinets found in all modern supermarkets. Among the new fish products being distributed via such supermarket cabinets held at temperatures between 0°C and 5°C are also heat-pasteurized products consisting of minced or ground fish meat mixed with milk and flavouring agents. Some of these products are first hermetically packed in small units, which are then pasteurized by immersion into boiling or hot water until all parts of the product have been brought above 70°C. They should then be rapidly chilled to near 0°C, at which temperature they have a storage life of several weeks.

Similar products, fish hams and fish sausages, have been developed in Japan, where chemical preservatives such as furyl furamide and nitrofurazone have been used to give a long storage life at ambient temperature, eg, one month even in the summer season (Tanikawa, 1971). The traditional Japanese 'kamaboko' is another heated fish paste product which is, however, considered to be a rather perishable food (Tanikawa, 1971).

3 Preservation by freezing

3.1 Freezing of wet fish

Natural freezing must have been known for centuries as a means of long-term preservation of fish in countries with sufficiently cold winters. Records of the early commercial distribution of naturally frozen fish, however, show that the fish quality suffered drastically even if bacterial spoilage is prevented. Similar sad records come from the first attempts in the last century to use mechanical refrigeration to freeze fish for long-term preservation. In fact, such crude methods gave rise to public prejudice against frozen fish (Cutting, 1955). A great deal of research and development was required to establish a sound frozen fish technology. A main requirement was to distinguish between the initial freezing and the subsequent cold storage.

The greater part of the natural water content of fresh unsalted fish is converted into ice during the freezing process, which should be completed within a few hours. Slow freezing over days results in the formation of large ice crystals within the muscle and in irreversible damage to the fish quality. On the other hand extremely fast freezing within minutes can result in physical breakage to the muscle.

Only when the freezing process is completed should fish be stowed in the cold store, which is normally not designed for temperature changes but only for the maintenance of temperature. The control of the freezing process may be costly and difficult, particularly when fish are frozen on fishing vessels at sea. However, there are very simple tests which are sometimes quite effective in industry. For example, salmon fishermen have learnt to check on the completion of the freezing by judging the sound made when two fish are banged against each other!

There are two main kinds of quality defects which can occur during frozen storage: those affecting the fish surfaces, and those affecting the entirety of the tissues. Both are slowed down by lowering the storage temperatures, and it is now generally recommended that for long-term storage fish should be kept at temperatures of −30°C or lower.

Surface damage is often associated with evaporation or sublimation causing 'freezer burn', that is the drying of exposed surfaces, which thereby become spongy. In the case of fatty fish this is often followed by rapid oxidation of the surface oil, which becomes discoloured and rancid. Fig 2 on page 31 shows the effect of the storage temperature on the quality and frozen life of trout eviscerated and frozen immediately post mortem. The trout stored at −10°C became rancid and unpalatable within three to four months, those stored at −20°C within seven to eight months while those stored at −30°C kept well much longer (Aagaard, 1968). In other species rancidity often occurs even faster, giving only a few months of storage life at −20°C. The surfaces may be protected, however, by glazing or vacuum-packing, or by antioxidant treatment such as dipping into a weak ascorbic solution, all of which have been found to increase the storage life of trout at −20°C considerably.

White fish muscle with a low lipid content does not normally go rancid, but related oxidation processes give rise to cold store flavours reminiscent of the flavours typical of dry salt fish. The texture may also gradually deteriorate. TMAO present in some gadoid species tends to split into DMA and formaldehyde during cold storage, a reaction which may be related to the adverse texture change (Sikorski and Olley, 1980).

The thawing process completes the cycle providing wet fish for direct consumption or for processing and further preservation, which may be a second freezing. Double freezing, however, may lead to some loss of eating quality and yield. Frozen raw materials are often considered suitable for smoking and canning, but not for the production of certain herring delicatessen products such as marinated fillets.

Those freezing fish raw materials should have the subsequent cold storage and thawing in mind. It is normally recommended that raw materials should be frozen in uniform blocks of limited thickness, which can be thawed in industrial plants within a few hours or overnight. Whole crustacea and large fish such as tuna, however, are often frozen individually. Some crustacea such as large deep-sea shrimp (Pandalus borealis) are cooked at sea before freezing. This procedure gives a much higher yield of edible meat than any sequence involving a lengthy post mortem chill storage of the shrimp raw material.

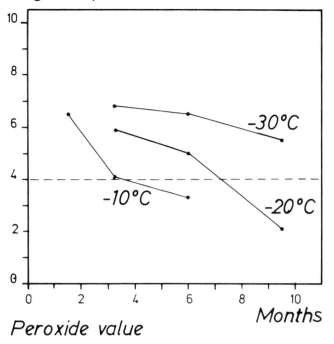

Organoleptic score (Taste)

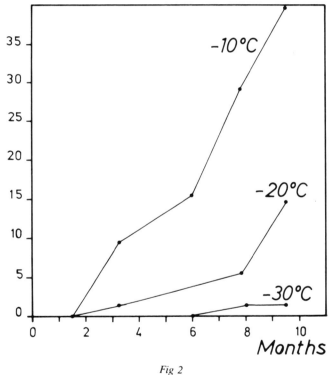

Peroxide value

Fig 2

heat-pasteurize shrimp before freezing. The best hygiene is achieved by vacuum-packing in plastic pouches maintaining a limited thickness, for instance 15mm. The product is afterwards pasteurized by immersion in hot water at 80–90°C for a few minutes. After subsequent chilling, the product is frozen. The main purpose of such pasteurization is the elimination of staphylococci and other dangerous organisms, but most other microorganisms are also destroyed giving the product a substantial chill storage life after thawing.

3.3 *Freezing of mince*
Increasing quantities of raw or cooked fish mince products are being frozen in the industrialized countries. Raw mince of some gadoid species has a very limited storage life due to rapid deteriorative texture and flavour changes. As with the fillets mentioned above these reactions may be associated with the breakdown of the TMAO to DMA and formaldehyde (Proceedings, 1976).

4 Preservation by drying

4.1 *Whole fish, split fish and fish fillets*
Large quantities of fish are preserved simply by drying, and a very good storage life may be obtained by lowering the water content of unsalted fish to below about 10%. Fish or fish muscle with low oil contents are usually more suitable for drying than fish with large contents of oil, which may rapidly oxidize and become rancid.

In the North Atlantic countries the main species which are preserved by drying only are those of the gadoid family. Since thickness is a critical factor in drying, the larger fish are often split or filleted. Even so, air drying is a lengthy process taking weeks or months. When relying on natural climatic conditions for drying, the results may be rather variable. Some slight fermentation producing cheesy flavours during drying may be acceptable or even desired by connoisseurs, but adverse weather conditions can lead to total loss because of putrefaction during the early stages of drying. Controlled atmosphere drying plants overcome these problems but are highly energy consuming, a fact which in the future may prohibit their use. The temperature during drying of intact raw fish or fish muscle from temperate waters should not exceed about 30°C to avoid the break-up of the tissues due to heat denaturation. The equivalent temperature limit for tropical water fish can be somewhat higher, around 50°C (Anon, 1977). In tropical regions large quantities of small fish are dried, often in the sun. These include some oily fish which are accepted locally in spite of pronounced rancidity and discoloration. Traditional uncontrolled fish drying in the tropics often involves great problems with insect infestation both during and after drying, resulting in great losses of fish protein. Controlled solar drying, however, may overcome this problem because it operates above 45°C at which temperature fly maggots cannot survive (Anon, 1977).

In Asia some fish are dried from the raw state while others are dried after cooking. The latter include shark fins, various species of shellfish and bony fish (Tanikawa 1971). Some products are seasoned or smoked after drying.

3.2 *Freezing of cooked or cured products*
As well as the ubiquitous fish finger increasing numbers of new frozen fish products are being introduced which have been cooked or cured in various ways before freezing. One example is frozen 'gravad' Greenland turbot, a convenience food similar to the gravlax mentioned above. Another example is cooked, peeled shrimp, which are often vacuum-packed in plastic pouches. A very high hygienic standard is required for these shrimp products since they are normally consumed after thawing without reheating. Some producers therefore

4.2 Comminuted fish

Most consumers of dry fish prefer to purchase recognizable fish or fish segments, but these dry slowly and may therefore in the future become exceedingly costly. Fish mince, on the other hand, can be dried quickly and may therefore become more economical in spite of the high cost of the industrial equipment required for fast drying. Spray drying of fish flesh suspensions was tested in Denmark some years ago. Herring in this form was spray dried without complications, while cod dried as a fibrous material which tended to stick to the inside surfaces of the drier. Roller drying, on the other hand, is unsuitable for drying fish suspensions containing more than 2% oil, but is very suitable for the drying of suspensions of lean fish like cod which comes out as fibres of light colour. The roller dried cod product is an attractive but rather expensive product. It is likely, however, that similar and nutritious fish protein products could be produced from much cheaper fish such as headed and gutted blue whiting. The economics may also be improved by partial dehydration and purification of the suspension before roller-drying, for instance, by adjusting the pH to about 4, adding a few per cent of salt and pressing, a method being presently developed in Denmark (Anon, 1978). Recent studies have shown that oil fish mince may also be roller-dried provided that some starch or other non-oily filler is added before drying. All these recent developments aim at producing attractive looking dry food products with a high content of fish protein (Nielsen, 1979).

Fish meal for human consumption, FPC type B, has been produced in quantity in Norway from whole fatty fish such as mackerel and capelin. It is a relatively cheap product containing 70–75% protein, 10–15% ash, up to 10% moisture and up to 10% oil. This product is suitable for relief feeding in some disaster areas, but marketing in many countries is hampered by its sandy and gritty texture and unattractive appearance.

5 Preservation by salting

5.1 Influence of size, oil content and salt concentration

The size and the oil content of the fish are the main factors governing the technology of salt preservation. Small fish and thin flatfish may be salted whole. Larger fish have to be eviscerated, split, filleted or opened before salting otherwise the salt does not penetrate fast enough to prevent spoilage in the centre part of thick whole fish. With oily fish contact with air must be avoided to prevent oxidative rancidity during and after salting; fish low in oil do not need such protection.

In the case of long-term preservation by salt at temperatures above zero the water phase must be nearly saturated with salt. Smaller concentrations suffice for short-term chill storage when combined with other preservation methods. In Japan, hard salting means at least 25kg of salt to 100kg of fish, while in 'slack' salting only 10–15kg of salt are used initially (Tanikawa, 1971). To the slack salted fish products, which have become more popular than the hard salted ones, chemical preservatives such as nitrofurazone are added. Chemical antioxidants such as BHT and BHA are also used in fish salting in Japan in order to retard or prevent oxidative rancidity of saltcured fatty fish (Tanikawa, 1971).

5.2 Kench salting or dry salting

The simplest way to salt fish is to layer the fish with solid salt forming a stack, from which the brine formed in the process drains away. Restacking and resalting may be required. In Greenland split cod is first kench salted for about a week, then restacked and resalted, the total curing time being about one month. The resulting saltfish contains about 24% protein, 57% of water and 18% of salt which means that the water phase is saturated with salt. Even so, the saltfish do not always keep well if stored at temperatures above 10°C. Halophile or salt-loving bacteria may thrive on saltfish causing reddish discolorations on the surface and strong 'cheesy' odours and flavours. In tropical moist climates saltfish may turn pink within a few days. Saltfish, which are somewhat hygroscopic, may also pick up moisture from the air, lose brine and ultimately spoil.

The Scandinavian saltfish of cod produced for export to warm climates are often dried to a final water content below 50% (Strøm, 1950).

5.3 Dry-salting to make brine

The fish are layered with dry salt in barrels or other containers in which the brine formed in the process remains. More salt or saturated salt brine may be added, the object being normally to have the fish completely submerged in the brine soon after initiating the salting. In some cases weights are placed on top in order to press the fish below the surface. In other cases great care is taken to have the barrels filled up completely in order to exclude air, the oxygen of which might otherwise cause rancidity. Most fish species salted in barrels or similar containers are fatty and need protection against oxidative rancidity during the long period of time between initiating salting and ultimate consumption.

As an example of the widespread former trade in this commodity, millions of barrels of Atlantic herring were salted in Iceland for subsequent export to other Nordic countries where the barrels were chill stored for several months until emptied for further processing into various delicatessen products such as 'Gaffelbidder' or marinated fillets in sauces containing vinegar, sugar, onion, herbs and, perhaps, benzoic acid or other chemical preservatives. Chemical preservatives have been also used when the raw material is initially 'slack' salted. In addition, sugar and spices can be added during the initial salt curing (Bramsnaes, 1946). The production of and trade in brine-cured herring are nowadays shadows of their former sizes.

Large pilchards such as the Latin American *Sardinops sagax*, a very large resource along the Pacific coast, may be salted and utilized for semi-conserves in ways similar to the salt curing of Icelandic herring. Headed and gutted *Sardinops sagax* which were salted in barrels in Chile, have been taken to Europe for experimental storage at +5°C during which they remained useful for filleting and marinating for more than one year (Hansen, unpublished).

While large herring and pilchard must be headed and eviscerated, or at least gilled, 'nobbed' or otherwise partially eviscerated before salting, this is not always done with smaller fish such as sprats and anchovies. Various *Engraulis* species are suitable for hard salting in barrels for long-term preservation. In fact, some species need many months of storage under brine before they 'mature', a process in which the texture of the meat softens as required for consumption. As a part of the maturation process, visceral enzymes appear to bring some proteins into solution, and strong odours and flavours appear in the product.

The resources of small pelagic fish suitable for the anchovy process, *ie* hard salting of whole ungutted fish, are enormous, and the salting process is simple and suitable for mass production. Some of these salted fish can be filleted when mature and packed in oil as a high cost delicatessen product; the market for them appears to be rather small, however. A relatively large-scale production of this kind occurs in Brazil where the market is low-income groups in the north. It is reported, however, that the quality is poor. The fish are often dark brown and rancid, affected with red bacteria or partially fermented or spoiled (Dos Santos, 1977). There seems to be an urgent need for research and development in the field of brine curing so that acceptable products can be made cheaply and consistently.

5.4 Salting with fermentation
Whereas white fish are salted for preservation only, there is usually a certain degree of fermentation involved in the salting of many of the smaller fatty species. As just pointed out enzyme action is considered an essential part of the softening or maturation of salted herring and anchovy delicatessen products. Softening is considered desirable, however, only to a certain point, and in Europe it is considered a total loss if the product liquefies. In Asia, on the other hand, some small fish are salted in order to produce liquid sauces or soft pastes for food use. In the Philippines, for instance, small fish are salted for fermentation during which the flesh disintegrates. A free liquid sauce called patis is recovered, while the residual solids may be ground, if necessary, and sold separately as bagoong (Avery, 1950). Many other products of this kind have been described (Mackie *et al*, 1971).

Experiments have been undertaken recently in southeast Asia to preserve communited fresh fish by a lactic acid fermentation controlled by small amounts of salt together with the carbohydrates necessary for acid production (Stanton and Yeoh, 1977).

5.5 Salting combined with other curing
As already mentioned, salting is often combined with air drying of white fish. This combination is not suitable in the case of fatty fish, which are instead often salted and then smoked during which the moisture content is slightly reduced. Different components of the wood smoke act as either antioxidants protecting the fish oils against oxidative rancidity or bacteriastats. There are two broad types of smoke cure: cold and hot. In the former the temperature of the smoke is low, in the latter it is high enough to cause complete or partial cooking. Detailed descriptions of the development of smoke curing as a preservation technique are available elsewhere (Cutting, 1955). In the ancient but nowadays almost vanished English 'red herring', heavy salting was combined with hard smoking of whole herring. During one week of cold smoking products of this type lose about 20% of weight. In Europe generally these types of hard-cured products are being replaced by milder cures, the products of which must be chill stored and keep for rather limited periods of time; smoke is used as a flavouring and not as a preservative.

Among the mild cures favoured are those using a combination of salt and vinegar on fillets of herring and similar fish. Danish 'marinated' herring are produced in two stages, the first being a treatment in a 12% salt brine containing 4·5% acetic acid (Christiansen, 1962). Two or more weeks later the fillets are repacked with a sauce adjusted to give a final pH value of about 4–5·5 and a salt content in the fillet of about 4%. Lower salt contents are often found, and the main preservation agent is in that case the vinegar, often supported by chemical preservatives such as benzoic and sorbic acid.

Whilst frozen raw materials are in general not suitable for the herring delicatessen industry, the final products in some cases can be frozen, but this is not yet used to any great extent. There may be an advantage in arresting the ripening process by freezing when the optimum eating quality has been obtained and any further changes would lead to a loss of quality (Christiansen, 1962).

6 Preservation by canning and bottling

Salmon and sardines have been canned in large quantities for more than 100 years. More recently tuna, mackerel and other fatty species have also been very important raw materials. Lean white fish species, on the other hand, because the products are in general inferior, are canned on a much smaller scale; canned cod roe and liver, however, have established, although limited, markets in Europe.

Some fatty species not mentioned above are canned either as 'salmon-type' which means that they are packed raw in the can with added salt, or as 'tuna type', that is precooked before being drained and enclosed hermetically in the can. Precooking reduces the natural water content of the fish and a slight drying may be applied. Mackerel, pilchards, sardines and sprat are normally precooked before canning. Slight smoking or addition of herbs or vegetables is often used to achieve special flavours in the finished product. Olive or, more often nowadays, other vegetable oils are added to some canned sardine products; tomato or other vegetable sauces are used for several different species.

As regards can closing and sterilization technology, fish and meat canning do not differ substantially. In order to soften the bones canned fish are often heated in the retorts far beyond the temperature that is required for sterilization.

Cans for the retail market are normally quite small, each holding around one or a few hundred grammes. The cost of the can itself is quite a large fraction of the price of the finished product. Even so, canned fish are often among the cheapest consumer packs of animal protein. The very long storage life at ambient temperature has helped to keep the cost of distribution low and give certain canned fish products an almost worldwide distribution. Despite a good deal of research into plastic containers, they have not as yet started to displace tinplate and aluminium cans.

Small quantities of bottled fish products such as marinated shellfish and fish spreads are produced.

7 Future developments

The world fisheries of the future are likely to include more species and a larger proportion of small fish and shellfish than in the past. A number of small and medium size fish, which up to now have been reduced to fish meal and oil, will be preserved for food in the future. This will require new and improved preservation methods both at sea and ashore.

Recent advances made in the bulk handling of small fish indicate that the chill preservation at sea may be greatly improved. More vessels will be equipped with chilled or refrigerated sea water systems and some may supplement these systems with carbon dioxide in order to extend the storage life of the catches. Systems for ice packing of the larger fish at sea are capable of improvement, particularly in the tropical areas where efficient ice packing may provide storage lives of several weeks. The fishing vessels of the future will be equipped to grade their catches and stow them in boxes, containers and tanks according to particular requirements.

The larger fish of well known species will tend to be reserved for fresh fish markets or for high-price delicatessen products. A number of such products will be developed and retail packed for distribution via supermarket sales cabinets operating near 0°C. Some of these products will combine two or more preservation methods such as salting, pH adjustment and heat pasteurization. Chemical preservatives, on the other hand, will probably be restricted even more than today.

The industries producing cheap standard products will have to rely to an increasing extent on smaller fish and shellfish of less attractive species. Machinery for heading, gutting, filleting, skinning and deboning will be developed for these species, and the edible parts will be preserved by canning, curing or freezing. Canning will remain very important for the preservation of the small fatty fishes, which could also be salt cured much more than today provided that larger markets are developed.

It is expected that mince of the less oily species will be dried on a large scale to provide cheap high protein products.

8 References

AAGAARD, J. Fryselagring af fed fisk. *Konserves,* 26, 78–84
1968

ANON. Sun-powered fish drying. *Aust. Fish.*, 36 (5), 24–25
1977

ANON. Annual Report for 1977. Technological Laboratory, Ministry
1978 of Fisheries, Lyngby, Denmark

AVERY, A C. *Fish Processing Handbook for the Philippines.* Research
1950 Report 26. US Government Printing Office, Washington

BRAMSNAES, F. *Saltbehandling af Sild.* Communication No. 58. Tech-
1946 nological Laboratory, Ministry of Fisheries, Lyngby, Denmark

CHRISTIANSEN, E. *Sildehalvkonserves.* Communication with a summ-
1962 ary in English. Technological Laboratory, Ministry of Fisheries, Lyngby, Denmark

CUTTING, C L. *Fish Saving.* Leonard Hill Ltd, London
1955

DOS SANTOS, C A M. Fish inspection and quality control in Brazil. In:
1977 *Handling, Processing and Marketing of Tropical Fish,* 73–81. Tropical Products Institute, London

HANSEN, P. Quality and storage life of iced trout. In: Annual Report
1973 for 1972, 42–50. Technological Laboratory, Ministry of Fisheries, Lyngby, Denmark

JARVIS, N D. Curing of Fishery Products. Research Report 18. US
1950 Government Printing Office, Washington

JONES, N R and DISNEY, J G. Technology in fisheries development in
1977 the tropics. In: *Handling, Processing and Marketing of Tropical Fish,* Tropical Products Institute, London

MACKIE, I M, HARDY, R and HOBBS, G. *Fermented Fish Products.*
1971 Fishery Report No. 100. FAO, Rome

NEILSEN, J. Manufacture of Edible Products by Roller-Drying Small
1979 Fatty Fishes. Report. Technological Laboratory, Ministry of Fisheries, Lyngby, Denmark

PROCEEDINGS of the Conference on Production and Utilization of
1976 Mechanically Recovered Fish Flesh (Minced Fish), Torry Research Station, Aberdeen

RIEMANN, H. Control of *Clostridium botulinum* and *Staphylococcus*
1972 *aureus* in semi-preserved meat products. *J. Milk Food Technol.*, 35, 514–523

SIKORSKI, Z E and OLLEY, J. Structure and Proteins of Fish and Shell-
1980 fish. *This volume*

STANTON, W R and YEOH, Q L. Low salt fermentation method for
1977 conserving trash fish waste under SE Asian conditions. In: *Handling, Processing and Marketing of Tropical Fish,* 277–282. Tropical Products Institute, London

STROM, J. (Ed) *Norsk Fiskeri og Fangst Handbok,* Vols I and II. Alb.
1950 Cammermeyers Forlag, Oslo

TANIKAWA, E. *Marine Products in Japan.* Koseisha-Koseikaku Com-
1971 pany, Tokyo

Past, present, and future methods of utilization   *Maynard A Steinberg*

1 Introduction

The societal evolution of man that resulted in the settlement of communities also increased the reliability of his food supplies and expanded food sources considerably over those available to him as a hunter and gatherer of food or as a travelling companion to migrating herds of animals. Most settlements were located within reach of water, if not actually on the shores of streams, lakes and oceans. This, of course, made fresh fish an important source of animal protein. In time simple technologies of food preservation developed, perhaps first in climates where the heat of the sun was sufficient to produce dried fish—carvings in Egyptian tombs built about 2500 BC show fish being split for sun-drying (Kreuzer, 1974; Cutting, 1956). This product form became common throughout the world, even in climates where the sun shone infrequently. Budding food technologists soon learned that heat from the cooking fire could do about the same thing as the sun and even make desirable changes in the flavour of the product when drying took place over a smoky fire. Perhaps about that time the preservative effects of salt were observed, and fish were brined and then dried or smoked or were preserved in dry salt alone or in concentrated brine.

Other methods of preservation not yet common in the western world were developed by the Chinese (Schafer, 1977), who noted at some time in the seventh century the advantages of preservation of whole fish by autolysis in concentrated brine. Since then fermented fish sauces have been common in southeast Asia as cooking ingredients and as nutritious condiments at the table.

The use of fish in these forms continues today almost universally; and, until Francis Bacon buried a chicken in snow to see if that would lengthen the time it would stay edible, the methods by which these products were prepared – drying, salting, with or without smoking – were the only methods of preservation known (Fyson, 1972).

The origins of these forms have to be due in part to such factors as climate, availability of salt, abundance and regularity of fishery resources, the ingenuity of those concerned with the preservation and selling of fish, *etc*; but how do we explain the unique forms in which some of the catch is used in some parts of the world, particularly in Asian countries? Regional and national cultures must also play an important role in determining how fish are used.

Culture determines the forms and methods of preparation and even whether certain things are acceptable as food. These social or religious practices are among those that we hold to most stubbornly, but even they are showing the strain of clinging to the past. For a number of years cola drinks have been accepted eagerly throughout the world. In too many countries they are regular parts of the diet regardless of their high cost and poor nutritional value. A more recent phenomenon is the readiness with which fast-food restaurants are accepted. Even strongly traditional societies such as those of the Indians and the Japanese have responded eagerly to the offerings of this type of food service. People change their food habits when the change satisfies personal needs, whatever they might be. In many industrial countries there has been a decrease in the *per capita* consumption of fresh fruits and vegetables during the past 20 or so years and an increase in the consumption of these foods in frozen, processed forms. This trend is changing the practice of daily shopping, despite severe limitations on home storage space for perishables, and is probably due to a great increase in the number of working wives and the growing demand that they make for foods that require minimum effort and time in preparation for the table.

Changes in consumer attitudes towards fast foods, eating away from home, and consumption of fresh fruits and vegetables might well be reflected in the trends of production of fishery products, too. We will examine production data over the past 25 or so years to determine if this is so and, if it is, the nature of the trends.

2 Disposition of the catch

Since the late 1940s total fishery landings have increased regularly, as has the tonnage used for human consumption (*Fig 1*). Although the amount used for food increased, the amount used for industrial products increased more rapidly, so the percentage of landings used for food has decreased from nearly 90% in 1948 to 70% in 1975.

Even though the food-fish catch is decreasing as a percentage of the total catch, the amount of fish available as food has increased at a faster rate than has the population. In the late 1940s and early 1950s the *per capita* availability of food fish was about 19lb. It had increased to about 28lb in the mid1970s. This does not mean that increased quantities of fish were uniformly available to the people of the world. It does mean that countries with large fleets of catcher and processing vessels made large increases in landings of food fish for their domestic use and for export (FAO, 1958 to 1976).

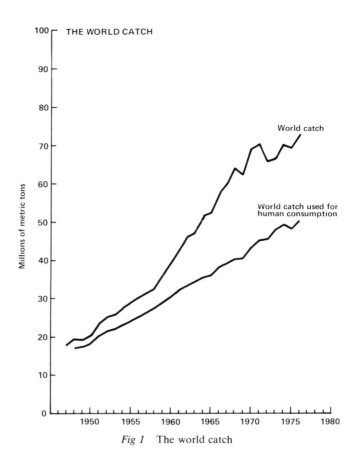

Fig 1 The world catch

As one should expect, the forces at work in changing demands for the forms in which other foods are marketed are similar to those applied to fishery products. If we are correct in the assumption that many of these changes are due to working wives, who have less time for food preparation in the kitchen, it is obvious that the increase in income more than compensates for the greater cost of eating out, the higher price of processed and preserved food, the initial cost of refrigerators and freezers, and the cost of operation of these appliances.

The questions with which we are particularly concerned are those of how catches are used for food and for industrial purposes, how these uses have changed during relatively recent years, and what might be expected in the way of changes in the near future.

More fish for food has been and continues to be marketed in the fresh form than in any other, although there are signs of levelling off since the 1960s (*Fig 2*). The amount of the catch marketed as cured has not increased for nearly 20 years and, in fact, has been exceeded by the amount going to frozen and canned products.

Another way of looking at this is as the percentage of catch used for food in these same forms (*Fig 3*); this gives us a somewhat different feel for world trends. Up to the mid-1960s, more than 50% of the catch used for food was processed into the fresh form. Since that time there has been a steady reduction in the amount of the food catch produced in that form until in the mid-70s it fell to less than 40%. This trend applied also to that part of the catch that went to cured products. Since 1948 and perhaps earlier, the proportion of the catch processed into cured products dropped from some 30% of the food catch to about 15%.

35

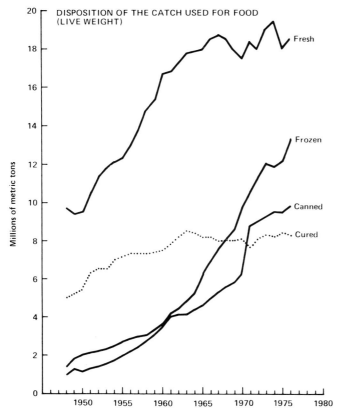

Fig 2 Disposition of the catch used for food (live weight)

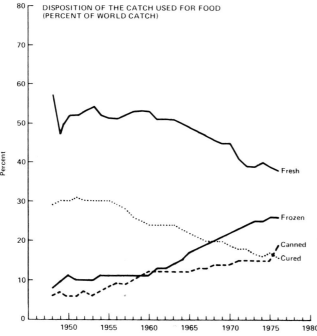

Fig 3 Disposition of the catch used for food (percentage of world catch)

Both canned and frozen products claim an increasing proportion of the catch. The frozen-product share of the food catch has increased by about four times since 1948, from about 6% in 1948 to about 26% in 1976.

Neither canning nor freezing is a high-technology method of preservation. After all, canning of food began commercially in the very early part of the nineteenth century. The first patent may have been issued by the British Patent Office in 1810 (Cruess, 1938). Canned products became popular because of the stability of the products at room temperature, the year-round availability of otherwise seasonal products, and the availability of products native to distant areas. Freezing as a method of preservation offered much the same benefits, with the additional advantage of a more nearly fresh-like product. Differently from canned foods, enjoyment of the benefits of frozen foods and the resulting growth in production of products in this form required something of the consumer, *ie*, the purchase of home refrigerators in which frozen foods could be held for several days, or for months in refrigerators fitted with freezing compartments. A general improvement in income and a desire on the part of the consumer to reduce food spoilage and improve the quality of food held at home, together with greater convenience, resulted in the reciprocal growth of the refrigeration and frozen-food industries (*Fig 4*). In 1967 world production of mechanical home refrigerators was about 23 million units. In 1976 cumulative production since 1967 was almost 312 million units (United Nations, 1976). This is an average production increase of 31 million units per year during the 10-year period.

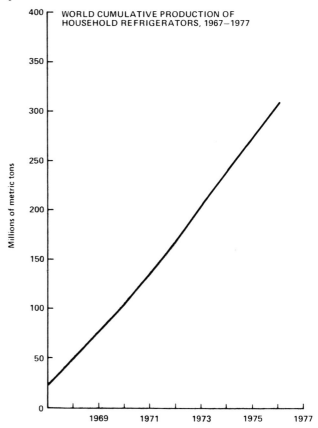

Fig 4. World cumulative production of household refrigerators, 1967–1977

3 Products, past and present

3.1 Fish

3.1.1 *Fresh fish* A more detailed examination of data on the breakdown of the catch into product forms by continents suffers absence of information from the

36

USSR, the People's Republic of China, and Taiwan, after 1971. Even so, it is clear that the amount of the catch consumed as fresh fish in Asia is considerably greater than in other parts of the world (*Fig 5*). Japan is by far the largest user of fresh fish in Asia, although this is difficult to quantify because data on disposition of the catch to fresh and frozen are combined. The use of fresh products in Europe has declined gradually over the years. In North America the tonnage used for fresh products has been relatively stable, but an increase in population during this period indicates that *per capita* use of fresh fishery products has declined. Production for the fresh market in Africa and South America has not changed a great deal. There may be a trend towards an increase in production of this form, but one suspects that this is probably a matter of the data catching up with the facts more than it is a change in production.

Although a greater live-weight equivalent of fish is consumed fresh than in any other form, we have less detailed information on fresh fish than we do on any other form. Statistics give us the total weight of fillets produced, 158 000 tons in 1976 compared with 41 000 tons in 1955, but nothing on round fish, eviscerated head-on, eviscerated head-off, or steaks. This is simply because these data are extremely difficult to collect. It is a fair assumption that most fresh fish is not processed beyond evisceration. It is also a fair assumption that considerably more fresh fish is consumed than is shown here.

3.1.2 *Frozen fish*   The first assignment of a United States patent for an artificial method of freezing fish was made in 1861 (Taylor, 1927). The process consisted of freezing in pans on an ice-salt mixture. After glazing, frozen fish were stored in insulated chambers cooled by vertical metal tubes filled with the freezing mixture. Although the idea was attractive, 90 years passed and a number of engineering improvements were made in countries throughout the world before as much as 5% of the catch used for food was processed into frozen products; by 1976, more than 25% of the catch used for food was processed for freezing.

3.1.2.1 *Fillets*   The category of frozen fish is broken down into fillets, herring, and miscellaneous products (*Fig 6*). Most frozen fillets are produced in Europe; and the largest European producers are Norway, the Federal Republic of Germany, Denmark, Poland and the United Kingdom, with production of about 100 000 tons each. Iceland is also a large European producer, roughly comparable to Canada, with production of about 75 000 to 100 000 tons of product annually. Cod is the species most commonly used. Other important species or groups of species are ocean perch (*Sebastes*), saithe (*pollock*), and the flatfishes.

The United States is a large user but a comparatively small producer of frozen fillets. Most are imported; and the most frequently used species are cod and ocean perch, followed by the group of hakes, pollock and flatfishes. Production is characterized by relatively small amounts of each of a sizeable number of species.

Fillets are the first form beyond steaks to represent a convenience-type product. There is little that can be done to that form to increase its value through greater attractiveness or convenience; consequently, whatever efforts have been made to do so have been directed to the cosmetics of an added sprig of parsley, a wedge of

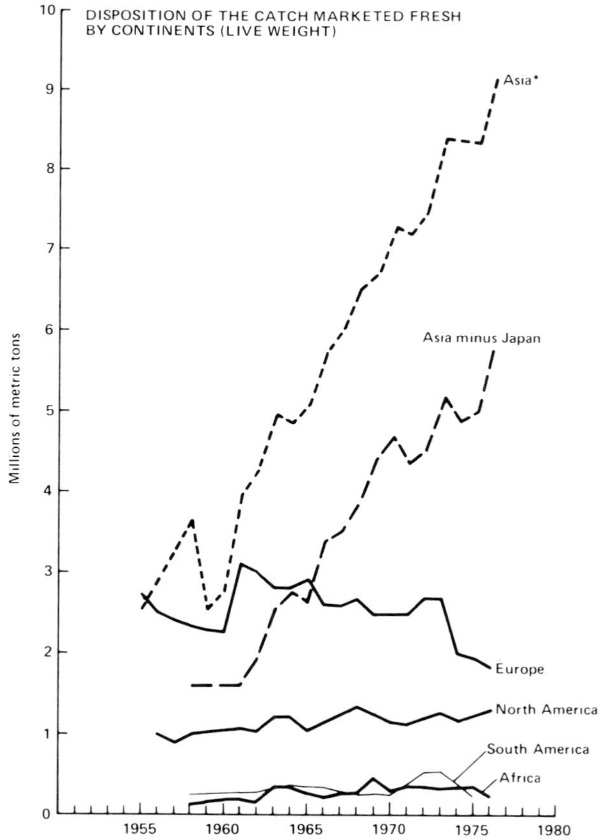

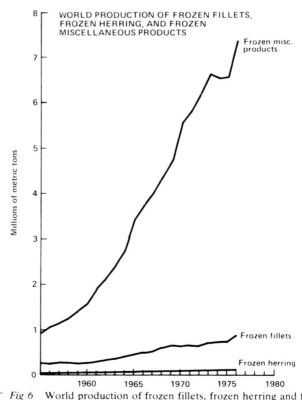

Fig 5   Disposition of the catch marketed fresh by continents (live weight)

Fig 6   World production of frozen fillets, frozen herring and frozen miscellaneous products

lemon, or a pat of butter, a few sliced almonds, and a French name; or to convenience for institutional users through packaging the fillets in layers separated by sheets of plastic film to simplify the problem of separating frozen products for large-scale cooking.

Recently, South Africa and Argentina have increased production of hake (whiting) fillets in response to the increasing demand for bottom-fish fillets and fillet blocks at a price below the premium demanded for cod.

During the last 20 years, frozen-fillet production has shown steady but unspectacular growth – about 240 000 tons in 1955 to about 800 000 tons in 1976.

3.1.2.2 *Herring* Although frozen herring is listed separately by FAO among frozen products, the amount involved is very small in comparison with some products included among 'Miscellaneous Fish Products, Frozen'.

Data on frozen herring refer only to the Atlantic herring, *Clupea harengus*. Quantities of Pacific herring, *Clupea pallasi*, have been taken by the wide-ranging fleets of Germany, Poland and the USSR; but products from these catches are not included in the category of frozen herring. During the past 25 years frozen herring has amounted to some 80 000 to 100 000 tons annually except for the years 1967 to 1970, when production fell to a low of 40 000 tons. This drop in production does not correspond closely to herring catches, which did not fall off appreciably in the northeast Atlantic until 1969.

3.1.2.3 *Miscellaneous frozen products* The group called 'Miscellaneous Frozen Products' is the largest of all frozen products. In 1970 world production was more than 7 million tons (*Fig 6*). Production by Japan and the USSR accounted for nearly 90% and the remaining 10% was almost totally European and North American. Production data for the USSR tell us only that the 3·4 million tons of products produced in 1976 included an unspecified quantity of chilled fish. The growth rate of USSR products in this group was slightly over 8% annually.

Except for unique products such as fish sticks, portions, fillet blocks, and minced-fish blocks, this group consists of a great variety of species, many of which are frozen at sea.

Fish blocks are the raw materials of fish sticks (fingers) and fish portions. Production figures for blocks are given first for the year 1957. At that time and continuing until the early 1970s these products were made from fillets exclusively. Beginning in 1971 or 1972, the process of removing the flesh from headed and gutted fish and trimmings and frames from filleting operations offered industry the opportunity of increasing yield, reducing waste, and reducing labour costs in cutting fish and forming minced blocks. The production of both fillet and minced-fish blocks is a significant part of the fish-processing industry, but there are very few statistics to show how significant. Canada and Iceland are the only countries that report block production to FAO. Iceland was turning out cod-fillet blocks (29 000MT) in 1949 (National Marine Fisheries Service, 1949), five years before the statistics show the production of finished products from these blocks. These data are reported as exported frozen fillets from major exporting countries and are therefore incomplete. The United States is the major importer, and its imports (Food and Agriculture Organization of the United Nations, 1958 and subsequent) far exceed reported exports.

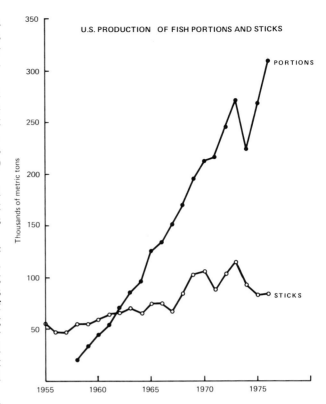

Fig 7 US production of fish portions and sticks

Figure 7 shows US production of fish sticks and portions for the years 1955 to 1976. Fish sticks, which are mostly consumed at home, may well have reached a production plateau of about 75 000 to 80 000 tons. Portions, on the other hand, are largely institutional products, first used in hospitals, factory cafeterias, and other mass-feeding situations. When franchised fast-food restaurants became established in the early 1960s, portions were logical partners to the hamburgers that were the staple product of much of that industry. As the popularity of this type of restaurant increased, so did production of fish portions. In fact, the weight of flesh in the portions and sticks produced in 1976 equalled 30% of the weight of all imported and domestically produced fillets and steaks.

Miscellaneous frozen products of Japan increased from nearly 1·5 million tons in 1965 to almost 3 million tons in 1976. During that 12-year period products from each of the four groups – tunas, mackerels, saury, and squid and cuttlefish – averaged about 10% of the total; and products from a mixed demersal-pelagic group that includes Alaska pollack, marlin, cod and herring-like fish made up about 30%, most of which is Alaska pollack. A large single group, also some 30% of the total, was itself a collection of miscellaneous and unidentified products lumped as 'other'.

Minced fish is a particularly interesting and important component of this group. This product is not unique to the Japanese, but its manner of use is, and it is an excellent example of cultural influences on the form and method of preparation of food. Although not identified, it is likely that most minced flesh is from Alaska pollack because of the abundance, availability and flesh characteristics of that resource. Minced fish is called surimi in

Japan, and the Japanese product is probably best known by that name in other countries as well. In 1965, 2 000 tons were produced. Eight years later, production was more than 400 000 tons (*Fig 8*). This enormous rate of increase is due to the function served by the product, *ie*, retention by the frozen muscle protein of the elastic gel-forming property essential to the texture of kneaded products (Miyauchi *et al*, 1973). Frozen surimi is water-washed, minced flesh to which is added sucrose, sorbitol or glucose and sodium tripolyphosphate. Its development permitted the rapid expansion of the kneaded-fish-products industry because it allowed factory-ship operations in the north Pacific Ocean and the Bering Sea as well as allowing shore plants in Hokkaido to ship more economically to kamaboko plants in Honshu because of the longer retention of quality and space-saving characteristics of blocks of frozen surimi. In 1965 only 25% of surimi production was from factory vessels. Five years later 54% was produced on factory vessels (Okada and Noguchi, 1974).

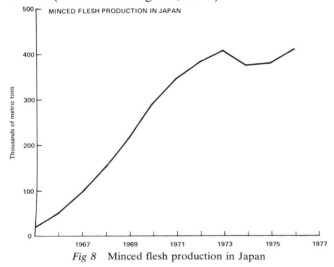

Fig 8 Minced flesh production in Japan

A family of finished products is made from frozen surimi produced on factory vessels and from wet surimi made in shore plants (Okada *et al*, 1973). They are: chikuwa, a broiled, sausage-shaped product sold fresh and frozen; kamaboko itself, prepared by steaming, boiling or frying; satsumaage, a deep-fat-fried fish cake; fish sausage; fish ham; and, more recently, fish hamburger. Production of some of those products is shown in *Fig 9*.

A brief list of miscellaneous frozen products from North America and Europe includes:

Canada—Pacific and Atlantic salmon, halibut, cod and other bottomfish, including fish sticks and portions, and blocks
United States—fish sticks and portions and unidentified unfilleted fish
France—cod and tuna
Norway—fish blocks
Germany—bottom fish, mostly cod, haddock, ocean perch and wolffish, fish blocks
Iceland—a mixture of bottom fish and blocks
Netherlands—mackerel.

Altogether, a not very informative collection of data on a group of products that is enormous in total amount and value.

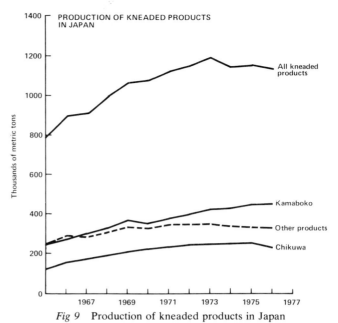

Fig 9 Production of kneaded products in Japan

3.1.3 *Cured fish* The curing of fish involves more different processes than either freezing or canning; and, of course, many cured products are ultimately frozen or canned. Curing includes drying, smoking, salting and combinations of these. It sometimes refers to pickling and fermentation as well, but for discussion here the term is applied to dried, salted or smoked fish.

Smoking is coincident with the discovery of fire, and drying may well precede that. Salting is also reported to be a development of the neolithic period, and all three methods of preservation of food take a place of importance next only to fire and stone tools (Jarvis, 1950).

Curing was critical to the survival of early man. Its importance continued at least into the Middle Ages, when cured fish provided the people of temperate climates with animal protein during the winter when meat was scarce, expensive and of poor quality. The importance of cured fish products in international trade was so great that during England's wars with Spain in the sixteenth century, the trade in dried fish between the two countries was never interrupted. According to some, England became wealthier from cured fish products in the sixteenth century than did Spain from the gold taken from South America (Jarvis, 1950).

Cured fish are largely products of Asia (*Fig 10*). Japan, Indonesia, the Philippines and Thailand account for the bulk of production from about 7 million tons of fish used for that purpose in Asia in 1976. The allocation of the catch to cured products has risen steadily and rapidly in contrast with that in other continents, where the allocation is either declining, as in Europe, or is nearly constant at a few hundred thousand tons annually, as in Africa and North and South America.

Production data on dried, salted and smoked fish show that about 85% of the product is dried or salted fish. Smoked-fish production, which was constant during the period 1953 to 1963, shows an upward trend since the mid-1960s (*Fig 11*).

The discussion of these products will include: stockfish (dried, unsalted cod and related species); salted cod,

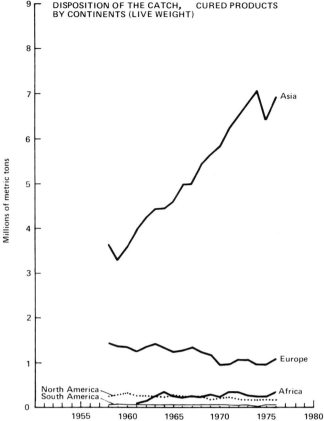

Fig 10   Disposition of the catch, cured products by continents (live weight)

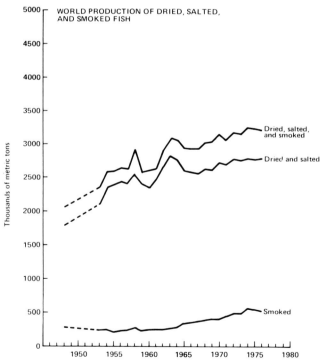

Fig 11   World production of dried salted and smoked fish

of species; herring, *Clupea harengus*, smoked, or smoked and then frozen; and a miscellaneous group of smoked or smoked-and-frozen fishery products.

Stockfish is a product of relatively small production, varying between 20 000 and 50 000 tons annually during the past 25 years (*Fig 12*). Although there is some small production of stockfish from Alaska pollack in Japan and Korea, it is primarily a product made from cod in Norway, where it has been produced for more than 1 000 years (Tressler and Lemon, 1951). The happy circumstance of abundant supplies of inshore cod early in the year permits starting the drying process with fish of excellent quality. Contrary to popular ideas, the cool, dry air of the Arctic Circle is highly suitable for the air-drying of fish.

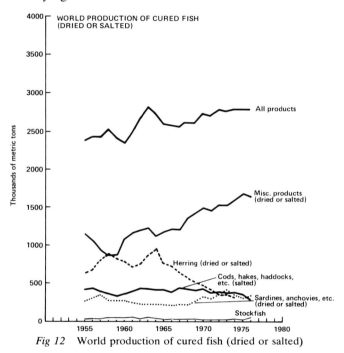

Fig 12   World production of cured fish (dried or salted)

Among dried or salted products stockfish is the most homogeneous in species composition. It is also produced in smallest volume. The product of largest volume in this group and of the greatest diversity in species composition is referred to, appropriately, as 'miscellaneous products, dried or salted'. Species used are both freshwater and marine. Marine species include the sturgeons, eels, salmonids, smelt, shads and milkfishes, flatfishes, redfishes, basses, congers, jacks, mullets, tunas, billfishes, sharks and rays, and, undoubtedly, a host of others. This group is also different from stockfish in that production has increased steadily since 1948. In 1976 it was more than 1·6 million tons.

Japan and the USSR are the largest producers of miscellaneous dried or salted products. Together they account for nearly half of world production in this group of products. Among the great variety of Japanese products we find 'fish flour', presumably a fish protein concentrate, produced since 1948 and perhaps earlier. At that time some 5 900 tons were produced. Recent production figures show that this has dwindled to some 200 tons annually since 1973. Another Japanese product of special interest is 'shaved, dried fish' or 'katsuo-bushi'. It is used in the preparation of soups and broths (Tani-

hake, haddock, *etc*; dried-salted and dried-unsalted herring, *Clupea harengus*; dried-salted and dried-unsalted sardines, anchovies, *etc*; miscellaneous dried-salted and dried-unsalted products from a great number

40

kawa, 1971). Katsuo-bushi deserves a few words of description because it is elaborate and unique. It is made from a number of species, but bonito is favoured. Fillets or strips of fillets are broiled, partially dried in air, then smoked and dried over wood fires. After the first broiling or smoking, the cracks that occur in the flesh are filled with a paste of flesh of the same species. The fillets are then broiled or smoked repeatedly, about 10 times; by this time they have become quite hard. They are then sun-dried before they are placed in a closed container for three to four days to permit equilibration of moisture. This softens the exterior, which permits them to be shaved with specially shaped knives, exposing a brilliant reddish-brown colour. The shaved fillets are sun-dried for an additional two days and then are placed in a box containing an inoculum of *Penicillium* or *Aspergillus*. The mould covers the fillets. This takes about two weeks. The fillets are removed from the box, dried in the shade and then in the sun. The mould is removed by brushing, and the fillets are again stored in a box for mould formation. This process is repeated until the surface colour has changed from grey-green to light brown. The entire process requires about three months. The yield of the finished product is 5% to 6% of the weight of the round fish.

3.1.3.1 *Dried and salted fish* Dried or salted herring (*Clupea harengus*) stands out in this group. This species, although not necessarily in these forms, has made exceedingly important contributions to the animal-protein supply of northern Europe. Its value in international trade was sufficient to bring about wars between Hanseatic cities and Baltic states over rights to fishing grounds for herring and between England and Holland for control of the herring fishery off the coast of East Anglia.

It is difficult to find another species that is used in so wide a variety of forms. It is commercially dried, dry-salted, salted and dried, marinated in chunks and fillets in a number of spiced sauces, and smoked in a number of different ways. In the northeast United States, small herring are canned in oil and in any of several sauces and are then sold as Maine sardines. Herring are made into fish meal and oil. Their scales have had value as a source of the guanine used for coating glass beads to give them the appearance of pearls.

For a brief period in the 1950s dried herring and salted herring exceeded the production of any other group of dried or salted fish. Since that time these products have shown a nearly uninterrupted decline. In the past decade major producers have been Canada, Iceland, Netherlands, Poland and, marginally, the United Kingdom. In 1976 production of dried or salted herring products was 334 000 tons, about equal to that of the group of salted cods and to that of the dried and salted sardines and anchovy group. These latter two groups have been reasonably steady in production for the past 20 years, cod, hake and haddock being about 400 000MT annually and sardines, anchovies, *etc*, about 250 000 to 300 000MT annually.

3.1.3.2 *Smoked fish* When smoking of fish was an early method of processing, it was probably a reasonably effective method of short-term preservation at ambient temperature. The salt content of the finished product was high in most cases, and the moisture content was low

from extended exposure to hot fires. In many developing countries where smoked fish is commonly used, this is still the case, as it is in industrial countries where the demand for some traditional products requires that they have the characteristics of the original hard-smoked products.

Many consumers enjoy the flavour of lightly smoked fish but not the high salt and low moisture content that tends to inhibit the growth of spoilage organisms and pathogens. Efforts to satisfy this preference have sometimes been fatal for the consumer because of the mishandling that can result from the fairly widespread belief that smoked fish is preserved.

Smoked fish is produced in a number of countries on every continent, but in most cases reported production is quite small. This is most apparent in the case of smoked Atlantic herring, which is almost exclusively a product of northern Europe. The kippered herring of the United Kingdom has usually been about 20 000 tons annually except for 1974 and 1975, when annual production exceeded 50 000 tons. This product accounts for about half the total production of all forms of smoked Atlantic herring (*Fig 13*).

All other smoked fish are lumped together as miscellaneous smoked products. They exceed smoked herring production by about ten times. In Africa, Ivory Coast and Sierra Leone are major producers, as are the Philippines and Thailand in Asia and Poland and Germany in Europe. The USSR, however, accounts for more than 50% of world production of all smoked fish. From 1955 to 1975 that country increased production of smoked fish by 400%. Poland, too, has increased production considerably, although the total cannot be compared with that of the USSR.

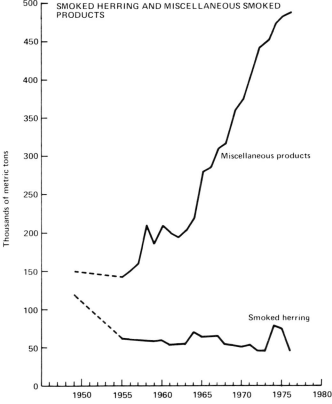

Fig 13 Smoked herring and miscellaneous smoked products

41

The origin of smoked fish is so deep in the history of man that one is tempted to speculate that the relatively small world production of half a million tons of product is explained better by the absence of reporting from a number of countries, perhaps most especially in Africa, but certainly including the United States, where production from many small smokehouses is unknown even to local regulatory authorities.

3.1.4 *Canned fish products and fish preparations* These products include marinated, fermented and 'elaborately prepared' products as well as canned (*Fig 14*). Production in 1955 was 1·7 million tons and in 1976 nearly 4·6 million tons. Canned products (*Fig 15*) are Pacific salmon, herring, sardines, anchovies, *etc*; tunas, bonitos and skipjack; and miscellaneous canned products from marine and freshwater species, *eg*, fish canned in brine, marinated non-herring species, fish balls, roe, fish in tomato sauce, *etc*.

Canned Pacific salmon is packed by the United States, Canada, Japan and the USSR. The Soviet product is last among canned miscellaneous products. The United States produces about half the pack on which we have data, and Canada and Japan produce the other half, in about equal amounts. The pack is showing a gradual decline.

Canned tuna, bonito and skipjack are largely accounted for by US production of tuna, which is about half of world production. Japan, Italy, Spain and France produce most of the rest. The US product is from several species: albacore in solid, chunk and flake packs, and unspecified packs of 'light meat' in solid, chunk and flake forms from yellowfin, which is preferred, and skipjack, which is playing an increasingly important role in the canned-tuna industry. Among the other major producing countries, Italy identifies the species processed there as bluefin in oil. Spain's pack is mostly bonito.

Among miscellaneous canned products the Japanese packs of mackerel in flakes and chunks are the largest volume of identified products. In 1976, 185 000 tons were produced, somewhat under the average of 212 000 tons for each of the previous four years. With the exception of the USSR's total production of 850 000 tons of canned fish, crustaceans, and molluscs in 1976, no other total pack of products in this group approached the Japanese pack of canned mackerel. An examination of the forms of the Soviet pack in this group is not possible because the data describe their production only as 'caviar', about 8 000 tons, and 'fishery products, all kinds, in airtight containers'.

Herrings, sardines, anchovies, *etc*, are another important group of products in this category. As is the case for dried, salted and smoked herring, this species packed in cans or jars is almost entirely a European industry, in which the Federal Republic of Germany is the largest producer.

Fish preparations not canned are made from a wide variety of species. Most, if not all, the products have their roots deep in the culture of the producing countries. The data on specific forms probably represent only a small percentage of the different and in some cases uniquely national forms into which fish are processed. During the period 1955 to 1976 production has increased from a little over 0·5 million tons to 1·8 million

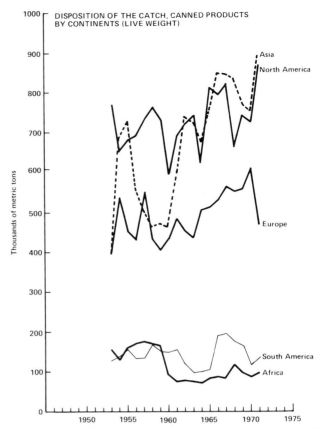

Fig 14   Disposition of the catch, canned products by continents (live weight)

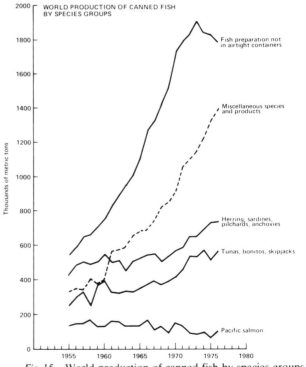

Fig 15   World production of canned fish by species groups

tons (*Fig 15*). Traditional and ethnic products comprise some 40% of all products in this group and include 78 000 tons of fish paste produced in Burma in 1976; more than 1 million tons of fish cakes, kamaboko, fish sausage and marine products in Japan; about 100 000 tons of fish paste, steamed and salted fish and fish sauces from Vietnam; and more than 200 000 tons of spiced, marinated and similar products from herring, sprats and other herring-like fish in the USSR, the United Kingdom, Poland, Norway and The Netherlands. Japan's production is about 75% of this group of products not canned. Kamaboko is by far the most significant; in 1976 production was 1 million tons.

3.1.5 *Meal and oil* Fish meal and oil are the major industrial fish products. Fish meal is part of our food supply indirectly because it is fed first to poultry, swine and fish as part of a complete ration. Fish oil, after refining and hydrogenation, is used directly as food. The unhydrogenated oil is used for a variety of nonfood purposes.

Production of meal in 1955 was about 1·7 million tons, 75% of which was derived from menhaden, herring, tuna and other oily species (*Fig 16*). The rest of the product came from lean bottom-fish species. To a large degree, bottom-fish material is waste from other processing operations. Its use has increased by only about 65% since 1955. In contrast, meal from oily species has increased by 367% in that same period.

Body- and liver-oil production in 1955 was 339 000 and 64 000 tons, respectively. Production changed in 1976 to 938 000 tons of body oil and 41 000 tons of liver oil, presumably a reflection of decreasing demand for liver oil as a source of vitamins A and D.

There are strong and conflicting attitudes towards the use of fish protein for industrial products. An opponent of this use states, 'I do not think we can justify feeding fish to animals at all . . . In many cases, poor distribution limits the availability of fishery products to coastal areas. To effectively utilize fish from coastal areas, preservation techniques must be improved' (Cutting, 1962). Others have made the point that the largest producers of fish meal are often protein-deficient countries that export their product to industrial countries, thus precluding domestic use of this source of protein (Borgström, 1962). The argument continues with the fact that exported fish meal brings hard currency to producing countries; Peru's exports of meal and oil in 1971, for example, had a value of more than $330 million. This is an impressive sum, and its potential for meeting human needs must be recognized. It should also be recognized that in the same terms, the 28 million tons of fish converted to meal and oil in 1971 would have benefited more people had it been possible to use that fish for food. And that limitation is the rub. Approximately 90% of the meal produced is from pelagic species (anchoveta, herring, menhaden, anchovy, pilchard, saury, *etc*) (*Fig 16*). These species are oily, have dark-coloured flesh, strong flavours, and become rancid easily. Their dense schooling behaviour favours low-cost capture and the economics of meal manufacture. Although many people use small, oily species for drying, salting, smoking and fermenting to fish sauces, the 21·6 million tons of fish processed into meal in 1976 represent an enormous technological challenge in handling and processing into food products that meet with wide acceptability. In some parts of the world, consideration is already being given to work towards this end. The majority of oily species used for meal could be processed into food to the benefit

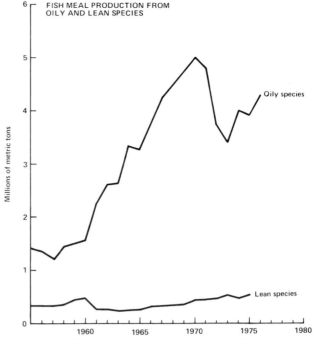

Fig 16   Fish meal production from oily and lean species

of both protein-deprived populations and manufacturers of meal and oil, who could as well be processing these species into acceptable food. At the peak of production of industrial products (1970) nearly 40% of the total catch of about 69 million tons could have been put to use directly as food, if we knew how to process, preserve, distribute and market profitably the products into which these species might be made.

3.2 *Shellfish*
3.2.1. *Crustaceans and molluscs, fresh, frozen, dried, salted,* etc. As discussed here, crustacean and molluscan products consist of all forms except live or fresh, unpeeled or unshucked animals. World production in 1955 was about 200 000 tons and in 1976 was nearly 1.1 million tons (*Fig 17*). The increases in production for the

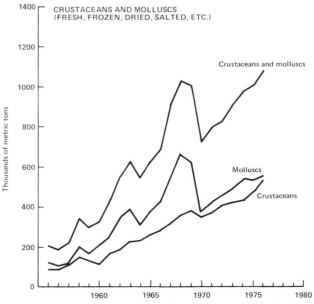

Fig 17   Crustaceans and molluscs (fresh, frozen, dried, salted, etc)

years 1967 to 1970 are due almost entirely to large catches of squid by Japan.

Crustaceans have considerably higher value per unit weight than molluscs. Shrimp, a species of universal appeal, is the most common species group among the crustaceans and is processed into a number of simple forms: head-on, dried and frozen; head-off, fresh or frozen; peeled, fresh or frozen, *etc*. Processing itself is often international in that heading and freezing may be done in an exporting country and additional processing into a final form in the importing country. In North America, Mexico and the United States are the major producers. In Asia, India, Malaysia and Thailand are large producers. Other crustacean products are lobster tails, often frozen, and a host of different species of crabs taken from Arctic, Antarctic, temperate and tropical waters.

Molluscan products were about equal in production to crustacean products, approximately 100 000 tons each, in 1955. In 1976 production was again nearly equal at about 550 000 tons of each group of products. During the intervening years molluscan products were usually greater in tonnage, in some years by nearly twice (*Fig 18*). Squid are the most prominent species group. They are processed into fresh, salted, dried, frozen and pickled products and amount to between 40 and 90% of world production of molluscan products, depending on the year. Japan is the leading producer, and frozen products are far and away the most common form, despite the ubiquitous presence of dried squid in southeast Asian countries. Korea and Thailand, the other two major producers of squid products, turn out together about 5% to 7% of Japanese production.

Netherlands and Spain produce cultured mussels in quantities second only to squid. They are processed as fresh, frozen, dried and salted products. Because our data are limited to processed products, we have no information on what may be the form in which this mollusc is most commonly marketed, *ie*, chilled, whole animals.

3.2.2 *Canned shellfish*

Most of these products are heat-processed. Most are canned. Some are simply cooked, marinated, fermented, or what is sometimes elegantly referred to as 'elaborately prepared' products. Maximum world production was in 1973, when 310 000 tons were produced. The United States produces about 75% of the crustacean products in this group, mostly as cooked crabmeat, crab sections, clams, lobster meat, and as raw and cooked breaded shrimp, cooked and peeled shrimp, and whole cooked shrimp.

Production of canned molluscs is somewhat less than that of canned crustaceans. The United States produces slightly more than half the world's products, 57 000 tons in 1976. About half of that, some 25 000 to 30 000 tons, is clam chowder, a product that may be unique to the United States and Canada. This is a soup that contains potatoes, tomatoes, onion, spices and clams. It might be called more appropriately a vegetable soup, as its clam content is not more than 7% to 10%. Other products are canned oysters, whole and minced clams, and squid. Japan produces significant amounts of canned oysters, squid and unidentified molluscan products.

3.3 *Summary of product trends, past to present*

These

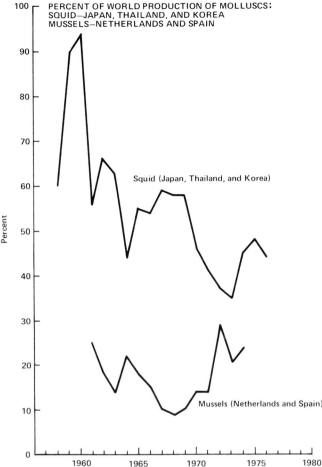

Fig 18 Percentage of world production of molluscs: squid – Japan, Thailand and Korea; mussels – Netherlands and Spain

data should first be viewed in the context of world catches and production of food and industrial products during the period 1955 to 1976 (*Table I*). In these years the total catch increased by 164%. The amount used for food decreased from 84% of the total catch to 70%. Conversion of the food catch to food products was constant at about 55%. Conversion of the industrial catch to meal and oil was inexplicably high at 36% in 1955 and dropped to a more reasonable 26% in 1976. The increase in production of food products was a respectable but modest 107% when compared with the increase of 252% in production of meal and oil.

Table I
THE CATCH AND ITS UTILIZATION

Category of catch	Year		Magnitude of change
	1955	1976	
	'000 MT	'000 MT	%
World catch	28 300	74 717	164
Catch for human consumption	23 700	51 952	119
Food products[1]	13 218	27 379	107
Industrial products	1 662	5 856	252

[1] Based on estimate of about 65% conversion of catches reported to be used for fresh products

When all these data are examined in response to the question of how utilization patterns have changed during this period, we find that: fresh fish was and still is the most common form in which fish is used for food, although frozen products are gaining rapidly (*Figs 1* and *2* and *Table II*). Cured products, which were second to fresh products in 1955, have been displaced from that position by frozen products. Despite the very large volume of fresh and cured products in 1976, 51% of all food-fish products, there was a general slowing of production of these forms and an increase in rate of production of frozen and canned products and preparations not canned.

Table II
CHANGES IN UTILIZATION OF FISHERY PRODUCTS, 1955–1976

Category of product	Year		Magnitude of change
	1955	1976	
	000 MT	000 MT	%
FISH			
Fresh or chilled[1]	7 891	11 719	49
Frozen	1 232	8 267	571
Cured	2 649	3 232	22
Canned	1 156	2 798	142
Preparations	522	1 786	242
Meal and oil	1 662	5 856	252
SHELLFISH			
Fresh, frozen and cured	213	1 087	410
Canned	77	276	258

[1] Based on estimate of about 65% conversion of catches reported to be used for fresh products

Details of the changes between 1955 and 1976 (*Table III*) show that the growth rate for production of fresh fillets exceeds that for other forms of fresh fish, although the tonnage of fish involved in fillet production is dwarfed by other uses.

Among frozen products, the convenience of fillets is reflected in a 228% increase in production in this period. Miscellaneous frozen products show phenomenal growth, due in small part to such products as fish portions and sticks, in greater part to the family of kamaboko products of Japan, and mostly to undescribed products. Frozen Atlantic herring has shown little growth.

Cured products as a group show very little increase. Dried and salted cods, hakes and haddock and salted, dried and smoked herring decreased. The decreases in herring products in all forms reflect the severe decline of the herring resource rather than a change in consumer demand.

Despite the drop in production of several cured products, there was a marked increase in miscellaneous smoked products.

The total of canned fish and fish preparations increased by 67%; but canned salmon declined, probably due to a combination of high costs and a shift in consumer preference to the frozen product. Canned tuna, bonito and skipjack production also grew in their inexorable fashion. The real growth in canned fish is in the group of miscellaneous products, just as it is in miscellaneous smoked products and frozen products.

The production of fish meal from oily species was greater in 1970 and 1971 than in later years. The decrease undoubtedly is due to a decline in the anchoveta resource and other resources rather than to a decrease in demand.

Shellfish are considerably less abundant than finfish,

Table III
CHANGES IN UTILIZATION OF FISHERY PRODUCTS, 1955–1976

Category of product	Year		Change
	1955	1976	
	000 MT	000 MT	%
FISH			
Fresh	7 891	11 719	49
Fillets	41	158	285
Other[1]	7 850	11 561	47
Frozen	1 232	8 267	571
Fillets	242	794	228
Herring	72	105	46
Miscellaneous	918	7 368	703
Cured	2 649	3 232	22
Dried and salted	2 444	2 696	10
Stockfish	39	51	31
Cods, hakes, haddock	411	297	−28
Herring	634	334	−47
Sardines, anchovies, etc	261	370	42
Miscellaneous	1 099	1 644	50
Smoked	205	536	161
Herring, smoked	61	46	−25
Miscellaneous smoked	144	490	240
Canned fish and fish preparations	1 680	4 584	173
Canned	1 156	2 798	142
Pacific salmon	138	105	−24
Herrings, sardines, anchovies, etc	434	738	70
Tunas, bonitos, skipjack	256	561	119
Miscellaneous species	328	1 394	325
Preparations not canned	524	1 786	241
Meal and oil	1 662	5 856	252
Meal (lean species)	333	549	65
Meal (oily species)	926	4 328	367
Oil, body and liver	403	979	143
SHELLFISH			
Fresh, frozen, dried, salted	213	1 087	410
Crustaceans	125	534	327
Molluscs	88	553	528
Canned	77	276	258
Crustaceans	24	164	583
Molluscs	53	112	111

[1] Based on estimate of 65% conversion of catch reported to be used for fresh products

as is shown in *Table II*. The production of shellfish products in all forms shows phenomenal growth, an increase of 410% from 1955 to 1976. Fresh and frozen crustaceans and molluscs make up the bulk of this group and parallel fish product forms in this sense. The high acceptability of this group of products and the potential of krill suggest interesting implications for the future.

4 Future utilization

The fresh fish market is of greater importance in developing countries than in industrial countries despite the fact that industrial countries have the transportation and distribution systems that permit rapid transport for considerable distances from landing sites. When the economics warrant, fresh fish and shellfish can be moved several thousand miles by air within 24 hours or less. Ironically, however, fresh fish are less important in industrial countries because of well-developed facilities for alternative methods of preservation, such as freezing, that offer a number of major advantages and only the slight disadvantage that most frozen products only look like fresh when thawed. In developing countries, the relatively limited alternatives to dried, cured and cooked products place a greater demand on the less-well-developed distribution and transport systems. The results, of course, are that fresh marine fish travel a relatively short distance inland and that, in most cases, greater use is made of freshwater fish by inland populations. In some southeast Asian countries the direction of transport is reversed because of the ability of some esteemed freshwater species to stay alive for many hours in only enough water to keep their gills wet. However, some recent developments in the use of refrigerated, gaseous environments for holding fresh fish show some promise in increasing the incentive for transporting a number of species of fish, including bottomfish, in the whole or dressed form, considerable distances, perhaps even to the retail market, because these atmospheres permit retention of quality at a high level for a significantly longer time than is now possible with ice.

Most clues as to the ways in which we will use our fishery resources come from the past; but this brief history cannot reflect yet the pressures of very recent events such as extended jurisdiction and its effect on the shape of future utilization. We already see changes in the distribution among nations of developed resources. The growing response to these changes, joint ventures and other such arrangements, will continue to supply larger user countries with products for some time, and one can predict with confidence a greater degree of cross-fertilization of technology between former processing countries and their new partner-suppliers of processed products. This inflow of different needs will take the shape of different processes and products, some of which will inevitably find their way to new markets, where acceptance by the consumer will be the incentive for major shifts in product form. As an example, some processors of kamaboko have expressed an interest in co-operative efforts to apply that technology to the production of fish blocks and breaded products.

Utilization will also be influenced strongly by the raw materials with which the industry has to work. Strong consumer demand for familiar species in recognizable forms will continue, and industry will continue to satisfy this demand to the extent that resources permit. We are probably close to that limit now, even assuming the rebuilding of stocks under improved management regimes. Other potential resources are of several kinds: there are species of enormous abundance about which some of us are learning the characteristics and processing problems and are trying to develop uses. Krill is a good example. Still other resources include those with special problems that limit their use. Several species of hake, one or two flatfishes, a high percentage of Alaska pollack and the blue whiting are in this group. Still other resources offer promise on the basis of availability, but we have little or no information on the extent of the resource, flesh characteristics or processing properties. Atka mackerel and pomfret, both of the northeast Pacific, are such resources. Another group includes familiar species that have good but often small markets as food and are used as much or more for industrial purposes. This group includes herrings, mackerel, sardines, anchovy, sprats, etc. Finally there are resources, such as the menhaden, that are used exclusively for industrial purposes because of strong flavour and small size.

The steady increase in demand for food products will also require increases in production from fully utilized resources without increases in catches. We discard as waste about 50% of the weight of fillets cut from fish and enormous quantities of fish that are too small for filleting or other processes or are a bycatch of other fisheries and are discarded at sea or landed at various stages of decomposition for conversion to poor-quality fish meal. These practices are wasteful of fuel, labour, time and resources. They will stop when we learn how to use profitably these potential sources of food.

The krill resource and its association with the whales and the Antarctic have excited the imagination of the world with its potential as an exotic and almost 'limitless' source of animal protein. However, the public view of the use of this resource for food is not consistent with reality, that is, problems of extremely rapid autolysis, physical damage when handled and resultant acceleration of lysis and loss of yield, discoloration on holding for only a few hours, discoloration from phytoplankton in feeding krill, difficulties of processing, and product acceptability. An excellent review of the utilization of this resource has been prepared by Grantham for FAO in 1977. It is recommended reading. It suffers only from an inadequate appreciation of how severely petroleum costs may limit the use of resources so distant from most countries of the world.

In this context a proposal for the novel use of krill should be mentioned, that is, the direct use as feed for salmon released annually as young fish from the southern tip of South America, from which they would move eastward around the world feeding on such crustaceans and returning to their home streams in two to four years, depending on the species (Joyner et al, 1974).

Mechanical deboning is an efficient process for obtaining fish flesh separate from skin and bones. It is likely to be used more commonly than it is now because it reduces waste through recovery of edible flesh from already processed fish frames and is an economical way of removing flesh from small fish that do not lend themselves to filleting. However, the unmodified minced flesh of most species is grainy when first formed into larger pieces and often becomes tough or rubbery during frozen storage. In some cases, advantage can be taken of this property when elasticity is an essential characteristic of the product. Plans for doing just that with small, lean, bycatch species that are now processed into meal are under way jointly between the Thailand Fisheries Laboratory in Bangkok and the International Development Research Centre of Canada. The product is the widely accepted fishball, which is a boiled product the quality of which is determined by its elasticity (personal communication, Dr W H L Allsopp, International Development Research Centre, Vancouver, Canada, July 1979). The use of deboned fish flesh has been proposed for portions and sticks of acceptable texture, based on pH control and the use of salt. Modification of texture to something resembling the flakiness of intact muscle has also been proposed through the addition of soluble alginates to minced flesh, followed by precipitation of the alginates as the calcium salt (Keay and Hardy, 1978).

Minced flesh has been shown to be a highly useful major ingredient in processed meat products (frankfurters and luncheon meats) at meat-replacement levels up to 30% (Steinberg and Dassow, 1976; Bacus, 1978). The finished products have the same acceptability as the 'all-meat' products. The fat content is lower, if lean fish is used; the protein content is higher; and ingredient cost for the meat processor is lower. The fish mince need not require any texture alteration because it is mixed uniformly with pigmented meat and other ingredients. Combining of species may be desirable and should have no adverse effects as long as strong-flavoured species are not used. However, even in these cases, it may be possible to reduce flavour to acceptable levels through washing and possibly other treatments. Of course, the costs of such treatments must be carefully studied. The possibilities of the use of fish in this way are limited more by food regulations than by technology. A logical extension of this idea is fabricated foods in which flavour, odour and texture are controlled by the selection of ingredients from among meats, vegetables, cereals and fish.

Strong points in this use for minced fish flesh are that such products are consistent with the growing demand for convenience foods, they respond to the consumer's increasing concern for the nutritional value of main-course dishes, they offer good opportunities for cost control; and, of particular importance, no product will have a precedent product against which it can be judged to determine its authenticity. Each product should stand on its own and be judged solely on the bases of palatability, nutrition, convenience and price.

It is likely that most such products will be frozen, but we have already seen a trend towards increased use of canned products, and one might anticipate the preservation of fabricated foods in this form as well. The cost of some canned products such as tuna is increasing rapidly because of the long distances vessels must travel to get to the resource and because of continually increasing competition for declining resources. Some canned products that once were staples for low-income groups are now or soon will become luxuries affordable only by the wealthy. The availability of highly palatable, reasonably priced, canned products made entirely or partly from fish or from mixtures of fish or of fish, shellfish and other ingredients may fill a void in the constellation of canned fishery products.

Products of this type are already under development at the Central Institute of Fisheries Technology, Cochin, India, where oily species that have poor acceptability because of strong flavours when preserved in traditional ways are minced and combined with cassava and other ingredients in traditional and novel forms (personal communication, Dr W H L Allsopp, International Development Research Centre, Vancouver, Canada, July 1979).

A group of highly acceptable canned products has been prepared from mixtures of raw and cooked chunks of flesh from a number of species of the northeast Pacific. The moisture content and texture are controllable based on the ratio of cooked to raw flesh as well as the species selected. The products were flavoured with liquid smoke and developed as a snack food; but it was obvious that this could be a prototype for a number of differently flavoured, canned, staple products (Tretsven, unpublished manuscript, National Marine Fisheries Service, Seattle, Washington, 1975).

Some large hake resources of North and South America as well as the arrowtooth flounder of the northeast Pacific and probably other species elsewhere are used only to a limited extent because the microscopic parasite with which they are infested secretes a proteolytic enzyme that reduces the flesh to a very objectionable paste during cooking. The incidence of development of this pasty texture is sufficient to prevent full use of these resources. Dilute solutions of hydrogen peroxide inactivate the enzyme(s) and prevent this major change in texture, thus making possible fuller use of these resources in the minced form, at least. The use as fillets or dressed fish must await the development of methods for introducing peroxide solutions into solid muscle.

The oily species used for most fish meal production are another enormous potential source of food products. Most of these are already used for food, but the limited market for the forms in which they are used leaves huge surpluses that are better used for industrial products than not at all. One obvious but perhaps controversial alternative food use is the production of fish protein concentrate. Originally thought of as an ingredient for fortifying other traditional low-protein foods, a protein concentrate is now in an industrial pilot-plant stage in Japan, where it is proposed for use as a functional, textured extender of comminuted meat products. It has been made from water-washed, alcohol-extracted deboned, headed and gutted Alaska pollack or from sardines or krill ['Continuous Manufacturing Technique for Protein Concentrate Foodstuff from Fishes' (patented), Dr Taneko Suzuki, Tokai Regional Fisheries Laboratory, Tokyo; informational circular, 'Marinbeef as a New Foodstuff', Niigata Engineering Co Ltd, Tokyo, February 1979].

In Sweden a 'functional' concentrate has been developed from whole and from deboned sardines and from cod-offal. Its use has been demonstrated in fish balls and fish sticks, and its suitability for use in meat products has been suggested (personal communication, Osterman, Astra Development AB, Södertälje, Sweden, June 1979).

Working in co-operation with FAO, Norwegian industry has prepared an FPC Type B, dark in colour, strong in fish flavour and odour, and designed for countries where there is preference for dried, salted and smoked fish.

Concentrates prepared by extraction with organic solvents usually retain only two functional properties: nutrition and water absorption. These are valuable properties, but they exploit only a few of the capabilities inherent in fish muscle protein. There are often problems with the rate of hydration in those cases where hydration is required to approach the moisture content of wet flesh. Concentrates made from whole fish also have some negative associations. The Swedish and Japanese products will be worth watching because of their massive potential.

More versatile functional protein concentrates have been prepared by relatively simple chemical modification of the muscle protein. These modified proteins have good water absorption and are excellent emulsifiers and whipping agents. Chemical treatment reduces their nutritional value slightly, but this is not considered to be significant because highly functional additives such as

47

these are intended for use as small percentages of the product to which they are added, and their contribution to the nutritional value of the product is expected to be negligible (Groninger, 1973; Groninger and Miller, 1975 and 1979; Miller and Groninger, 1976). The acceptance by the food industry of modified fish proteins such as these depends on proof of safety, reliability of supply and cost. Early animal feeding studies are strongly supportive of safety. Reliability of supply is assured on the basis that almost any species can be used. Cost is still to be determined; but rough estimates place these products in a cheaper price class than egg albumin, a widely used protein with similar functional properties.

The serious decline in the anchoveta resource has resulted in at least a temporary change in the composition of the pelagic resources off the coast of Peru and Chile, with implications for food production. The populations of jack mackerel and pilchard have increased rapidly, and Chilean companies are now planning to divert part of the catch of jack mackerel in particular to canned food products for export (Martin, 1979).

Utilization can be improved dramatically through uses already proposed, some of which can be nearly immediate in application; others will wait until additional developmental work is done on products. Still another improvement in utilization through waste reduction might be accomplished through treatments such as pasteurization of dried, salted and smoked products in tropical countries. These products are processed and handled under conditions that favour insect infestation. Consumption by insects results in losses as high as 50% (Anon, 1978; Cutting, 1962). Low levels of radiation have been shown to be effective in destroying or stopping the feeding of flesh-eating insects at all stages of their life cycle. Packaging of the product is required to prevent reinfestation. Processing costs, not including packaging, have been estimated to be between 0.6% and 1.0% of the value of the product, which, according to data from Nigeria and some countries in Asia, were approximated at US$2 per kilo in 1978. This price factors in losses, of course, and suggests very strongly that a major reduction in losses could absorb added costs of processing, packaging, and amortization of capital expenditures. An alternative procedure, still to be tested, is packaging in heat-stable flexible pouches such as are in use for boil-in-the-bag products. Evacuated pouches of product passed through steam or hot water would be free of viable insects and might also enjoy the benefits of reduced level of moulds and bacterial pathogens, depending on the amount of heat the product can take without changes in sensory properties.

5 References

ANON. Advisory Group Report on Radiation Treatment of Fish and
1978 Fishery Products, March 13–16, Manila, The Philippines. FAO, Rome

BACUS, J N. Fish and meat combinations offer economical sausage
1978 potential. *The Natl Provisioner*, 30 Sept, 18–19

BORGSTRÖM, G. In: *Fish in Nutrition*, Eds E Heen and R Kreuzer.
1962 Fishing News (Books) Ltd, London

CRUESS, W V. *Commercial Fruit and Vegetable Products*, 2nd ed.
1938 McGraw-Hill Book Co, Inc, New York and London. 798pp.

CUTTING, C L. *Fish Saving*. Philosophical Library, New York. 372pp
1956

—— The influence of drying, salting, and smoking on the nutritive
1962 value of fish. In: *Fish in Nutrition*, Eds E Heen and R Kreuzer. Fishing News (Books) Ltd, London

FAO. *Yearbook of Fishery Statistics*, Vols 7, 9, 12, 17, 19, 23, 25, 27,
1958– 29, 31, 35, 37, 39, 41 and 43. FAO, Rome
1976

FYSON, N L. *World Food*. B T Batsford Ltd
1972

GRANTHAM, G J. *The utilization of krill*. FAO, Rome
1977

GRONINGER, H. Preparation and properties of succinylated fish
1973 myofibrillar protein. *Agric. and Food Chem.*, 21(6), 987–981

GRONINGER, H and MILLER, R. Preparation and aeration properties of
1975 enzyme-modified succinylated fish proteins. *J. Food Sci.*, 40, 327–330

—— Some chemical and nutritional properties of acylated fish protein.
1979 *J. Agric. and Food Sci.* (In press)

JARVIS, N D. *Curing of Fishery Products*. Fish and Wildlife Service
1950 Research Report 18. US Government Printing Office, Washington, DC

JOYNER, T, MAHNKEN, C V W and CLARK, R C JR. Salmon–future
1974 harvest from the Antarctic Ocean? *Mar. Fish. Rev.*, 36(5), 20–28. MFR Paper 1063

KEAY, J N and HARDY, R. Fish as food–Part 1: The fisheries resource
1978 and its utilisation. *Process Biochem.*, 13(7), 20–21 and 28

KREUZER, R (Ed). In: *Fishery Products*. Fishing News (Books) Ltd,
1974 Surrey, England 462pp

MARTIN, E G. *The Wall Street J.*, July 13
1979

MILLER, R and GRONINGER, H. Functional properties of enzyme-
1976 modified acylated fish protein derivatives. *J. Food Sci.*, 41, 268–272

MIYAUCHI, D, KUDO, G, and PATASHNIK, M. Surimi–a semi-processed
1973 wet fish protein. *Mar. Fish. Rev.*, 35(12), 7–9. MFR Paper 1020

NATIONAL MARINE FISHERIES SERVICE. *Fishing Statistics of the United*
1949 *States*. US Government Printing Office, Washington, DC

OKADA, M, MIYAUCHI, D and KUDO, G. 'Kamaboko'–the giant
1973 among Japanese processed fishery products. *Mar. Fish. Rev.*, 35(12), 1–6. MFR Paper 1019

OKADA, M and NOGUCHI, E. Trends in utilization of Alaska pollock in
1974 Japan. In: *Fishery Products*, Ed R Kreuzer. Fishing News (Books) Ltd, Surrey, England. 462pp

SCHAFER, E H and 'T'ang'. In: *Food in Chinese Culture*, Ed K C Chang.
1977 Yale University Press

STEINBERG, M A and DASSOW, J A. Fish as an ingredient in processed
1976 meats. *The Natl Provisioner, 10 April, 48—49 and 54—57*

TANIKAWA, E. *Marine Products in Japan*. Hokkaido University Press,
1971 Hokkaido, Japan. 268–273

TAYLOR, H F. *Refrigeration of Fish*. US Fish and Wildlife Service. US
1927 Government Printing Office, Washington, DC

TRESSLER, D K and LEMON, J McW. *Marine Products of Commerce*.
1951 Reinhold Publishing Co, New York

UNITED NATIONS. *Yearbook of Industrial Statistics*. Rome
1976

Methods of marketing, distribution and quality assurance   E Graham Bligh

1 Introduction

Fishing is one of man's oldest industries and international trade in fishery products has been going on for centuries. The development of the world's fisheries and the traditional marketing of staple items like salted and cured fish has played a very significant role in the history of the world. I naturally think of Canada in this respect,

since it was immediately after Cabot's voyage to our waters in 1497 that British and other European ships came to harvest the bountiful cod on The Grand Banks off Newfoundland. It was gold that took men to the Canadian west, but it was salmon that enticed many of them to stay. The events are similar for numerous other countries and the challenge continues even today as we look to the oceans to provide increased quantities of protein to meet growing world demand.

In his book, *Marine Products of Commerce*, Tressler (1923) wrote that fish production was increasing rapidly both in quantity and in value. 'Fish are marvellously abundant. Thousands of species exist whose names are known only to the systematist.' At that time the important fisheries of the world were almost entirely in northern waters and were primarily associated with cod and herring. Radcliffe (1923) reported, 'On the whole, the fisheries of the tropics are unimportant and hold forth little promise of offering serious competition with those of temperate regions. Although less is known of the character and extent of the fisheries of the south temperate zone, it can be stated with assurance that they are incapable of supplying the vast quantities obtained in north temperate waters.' We now know all too well that the living resources of the oceans are not endless and that conservation and proper management to sustain populations of commercial species is imperative. In fact our northern stocks have been for the most part overfished and the continuing growth of world landings has resulted from the very substantial exploitation of southern waters. Southern species like shrimp and tuna have contributed a great deal to the economics of several developing countries and helped to improve the total value of world fisheries.

Trade in fish products has traditionally been global in nature and it is expected to become more so as a result of extension of exclusive fishing zones. Over the past 50 years fish landings have about tripled from around 25 million metric tons in the 1920s to 73·5 million metric tons in 1977. Approximately one-third of all landings in some form or another enter international markets and the proportion has not changed substantially for some time. It could change markedly, however, as many nations assume responsibility for fishing their coastal waters and are prepared to custom produce seafoods to satisfy world markets.

There are several unique features about fish that inhibit or plague trade and development. The major one is that it is very perishable and requires special attention and facilities if quality is to be maintained. This remains as a major problem in the marketing of fish even today and results in major losses throughout the industry in both developed and developing countries. It is often used as an excuse to avoid the consumption of seafoods. In many instances there is justification for complaint, and blame can be placed on improper use of available technology and the lack of knowledge in the industry and in the marketing system. Another handicap is that fish, unlike agricultural produce, must be transported from the sea to inland markets. It is recognized that marine protein could drastically reduce famine in the world today if distribution systems could be set up to transport available fish resources to where they are most needed.

The fishing industry and trade are steeped in tradition, and many fishing methods, products and eating habits have been developed and acquired with time. Under these circumstances, change and the application of science and technology is often difficult and slow. On the other hand, a variety of traditional seafoods are highly priced today; they have become gourmet items and are no longer available as a cheap source of protein. Another feature of the fishing industry is that it is based on a common resource which may fluctuate as a result of natural and environmental factors and consequently it does not, as a general rule, attract long-term investment.

Since the establishment of the Torry Research Station in Aberdeen 50 years ago, remarkable progress has taken place in the fishing industry. A vast number of technological problems associated with quality assurance, marketing and the distribution of fishery products have been investigated and much science and technology has been applied in the industry. The fact that we now catch three times as much fish and that exports have increased from $1 billion to $9 billion is a great achievement. Fisheries technology at that time was in its infancy and the trade was facing acute problems requiring improved processing methods and equipment. Many of the problems still prevail today and some of the remedies developed are no longer economic or practical. Several new developments flourished for a period and died for one reaso or another. Even though prices have increased dramatically, fish products have maintained their relative importance on the market. Certain traditional products are enjoying a resurgence in demand and like a number of other seafoods are classed as luxury items. With the development of a variety of new products, consumers have today acquired an entirely different conception of fish as food.

2 Marketing and disposition of world landings

During the past 40 to 50 years there have been marked changes in the disposition of fish landings and in the marketing methods and product forms of fishery products. These changes are shown graphically in *Fig 1*. Examination of the curves reveals that fresh fish remains as the dominant seafood with the production beginning to pick up in the 1960s and doubling by 1977. Frozen seafoods have shown the most spectacular growth from practically nothing to more than 12 million metric tons. Products cured by salting, drying and smoking as the chief means of preserving fish for distant markets have maintained their position in international trade but total production has not changed since the early 1960s. There has been steady growth in canned fish production with about a fivefold increase since 1938. Large increases in landings of industrial fish began in the 1960s in both Latin America and in Scandinavia. Actually there was an over-supply causing a decline in fish meal prices.

The percentage disposition of world fish landings is summarized in *Table I* showing the relative changes between the various product forms since 1938. Although fresh fish production doubled, the proportion of fish sold in this form showed a steady decline from 50% to 30%. This is a legacy of the marked increase in frozen seafoods which on a percentage basis increased threefold. The relative position of cured products also fell by about 50% while canned fish, including a number of luxury items, showed a slow and steady improvement. The percentage of the total catch used for non-food purposes doubled in approximately 20 years. Today we convert about 40% of world landings into processed seafoods with the remainder being divided equally between fresh and industrial fish.

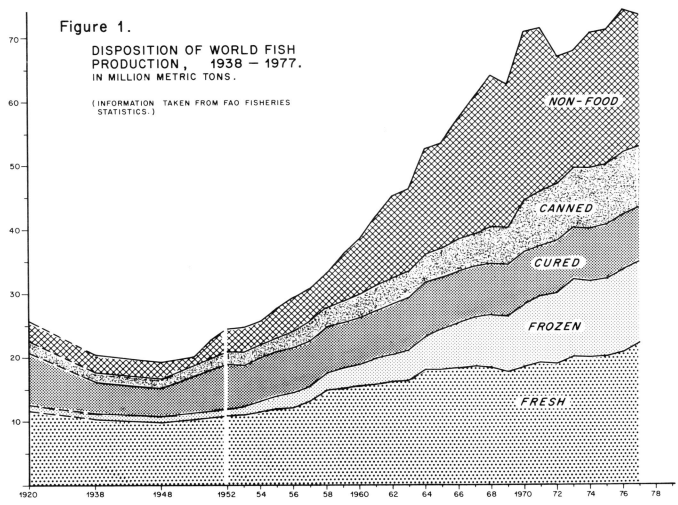

Figure 1.

DISPOSITION OF WORLD FISH
PRODUCTION , 1938 — 1977.
IN MILLION METRIC TONS.

(INFORMATION TAKEN FROM FAO FISHERIES
STATISTICS.)

NON-FOOD

CANNED

CURED

FROZEN

FRESH

Fig 1   Disposition of world fish production 1938–1977 in million tons (*FAO Fisheries Statistics*)

Table I
PERCENTAGE DISPOSITION OF WORLD FISH CATCH[1]

	1938–1948	1954–1958	1959–1962	1963–1966	1967–1970	1971–1974	1975–1977
Fresh	50	44	39	34	28	29	30
Frozen	5	7	9	11	13	16	17
Cured	24	24	18	15	13	12	11
Canned	8	9	9	9	9	12	13
Non-food	13	16	25	31	37	31	29

[1] Taken from FAO Fisheries Statistics

Market highlights over the years show that prior to 1950 the primary products were fresh and cured fish. In the 1950s herring and industrial fish were the most important in volume but salmon was the most valuable commodity. Europeans were experiencing difficulty in marketing herring and Britain mounted a project to sell high-quality frozen kippers in the US. Cured fish was the most important export item and frozen fish was showing indications of acceptance. Nevertheless, consumers were reluctant to purchase frozen fish if fresh fish was available because of unfortunate experiences regarding quality. The European trade was criticized by the Organization for European Economic Co-operation for poor retailing practices, lack of trained personnel, and often inadequate facilities. In reviewing this organization's report, Girard (1952) stated that the 'lack of sufficiently good quality was possibly the most important factor hampering fish consumption in Europe'. Similar studies in Canada had revealed that 40% of Atlantic coast fresh fish spoiled before it reached Toronto and Montreal and this led to submissions for compulsory fish inspection and greater research on quality improvement.

The fish meal market took a slump in the 1960s primarily as a result of the large landings of industrial fish by Peru and Chile. The market for luxury items like shrimp, tuna and salmon showed rapid growth. The shrimp fishery in Mexico was second only to that of the US and similar fisheries were becoming developed in Venezuela, India and Pakistan. The 1960s saw increased production of frozen fish reflecting the increase in the number of freezer trawlers and improved consumer acceptance of frozen seafoods. Facilities for handling frozen products were expanded in a number of European countries and in 1966 12 large freezer trawlers were added to the British fleet. Although the Catholic Church relaxed rules on meat consumption, the seafood trade continued to expand and export values reached record levels. As groundfish stocks started to decline, technologists gave more serious attention to non-traditional species (*eg*, hake) and fish protein concentrate.

During the early 1970s world landings continued to increase as industrial fish production remained high, particularly in Latin America. The importance of

50

greater use of fish as human food was stressed by FAO and world food organizations and extensive programmes were mounted in this regard. Peru, Chile and several other countries resolved to divert the production of some industrial fish to meet domestic protein requirements. International trade showed steady growth with the developing countries dominating in shellfish and fish meal. Several developing countries benefited from rising shrimp consumption in the US which consumed one-third of the world's total production. The tuna industry was threatened by the mercury crisis but suffered no great hardship. Prices were at record levels and there was concern that consumer resistance could be provoked against seafoods. Several sectors of the industry were hit by rising costs, the scarcity of conventional fish stocks, and the effects of extended fishing zones on the mobile factory trawler fleets. Groundfish exports increased as the popularity of frozen fish improved in Europe and the US. Alaska pollack and hake entered world markets as cod stocks continued to decline. The herring shortage in Europe stimulated herring exports from Canada and the transition from a meal to a food fishery. Research and development efforts were stepped up to create products and markets for mackerel, capelin, blue whiting, minced fish, small pelagic species, krill and edible fish meal.

Since Japan is the world's foremost producer and importer of fishery products, the current market situation in that country is worthy of mention. Prior to the extension of exclusive fishing zones, Japan took about 40% of its landings from foreign waters. The limits now being placed on distant water factory trawlers has brought about a shift in Japanese attitude from catching of fish to securing adequate fish supplies. In the last 10 years imports have increased from 3% to 11%. As an example, Canadian fish exports to Japan were valued at about $4 million in 1968 and have now increased to $142 million (75% as herring roe). The Japanese demand for squid has stimulated new fisheries in Canada, New Zealand and Argentina. A joint project is currently under way with Japan in the UK on the pilot production of surimi from blue whiting. Similar studies are under way elsewhere since the expected decline in Alaska pollack landings could seriously jeopardize the kamaboko industry.

3 Distribution and quality assurance

Methods for preserving the quality of fish from the time they are caught until they pass through the distribution network and reach the consumer have been studied by technologists the world over. Even though tremendous advances have been made, quality assurance is still the foremost problem of an industry that deals with a highly perishable commodity. Fish spoil rapidly. In contrast to animal products, the bacteria associated with fish can thrive at ice temperature, the actomyosin system of fish is more sensitive to denaturation, and the highly unsaturated fats are very susceptible to oxidative rancidity. From a technological standpoint, it would seem appropriate under the circumstances to freeze or otherwise process most fish products rather than attempt to market them in the fresh condition. The consumer, however, prefers fresh fish even though it is in many instances

impractical to supply it. Quality again has played a role in establishing this preference.

Fifty years ago the many factors responsible for fish spoilage were little understood and prevention was not always assured. Quality and product uniformity could not always be predicted nor guaranteed. Processors were faced with major difficulties since much irreparable damage would occur on the fishing vessels before the fish arrived at the plant. Spoilage patterns differed between species and from where they had come. Even for a given species, there are seasonal factors related to sexual maturity and migration, water and air temperatures, and feeding habits which drastically affect spoilage rates and the storage characteristics of seafoods. Up until the mid-1960s technologists were accustomed to working only on dead fish, and it is significant to note from the record of the FAO Symposium on the Significance of Fundamental Research in the Utilization of Fish (Anon, 1965) that 'It was recognized that the physiological condition of fish at the time of capture . . . could have a marked effect on the subsequent shelf-life of fish and fish products'.

In the sections that follow efforts have been made to give a general overview of some of the technological problems and advances related to the quality and distribution of the various classes of fishery products.

4 Fresh fish

As the starting material for all fish products, fresh fish preservation deserves special consideration. Although much quality deterioration can occur prior to landing the fish on shore, this paper is restricted to land-based operations.

The major processes responsible for fresh fish decomposition are bacterial spoilage, autolysis and oxidative rancidity, but as stated by H L A Tarr (1956a), 'There is no doubt that more edible marine products are discarded as unfit for human food purely as a direct result of bacterial spoilage than from any other known factor or combination of factors'. In essence, if proper precautions are taken to prevent bacterial spoilage, the problems of autolysis and oxidation are in many cases of no major consequence. According to Castell (1954), the key steps are: prevent physical damage; gut and thoroughly wash the fish with clean water to remove blood, digestive juices, slime and faeces; chill in layers with clean ice; through proper sanitation and hardware, avoid bacterial contamination of cut fillets; and keep chilled at all times.

There have been innumerable discussions on how to properly ice fish and on what type of ice is best but as J J Waterman (Torry Advisory Note No. 21) put it, 'ice is ice' and the important point is to use plenty of good-quality ice to chill and keep the fish at ice temperature under the prevailing circumstances.

The wholesale use of artificial ice greatly facilitated and expanded the marketing of fresh fish. As could be expected, producers soon wanted to further extend shelf-life to broaden markets and this stimulated a great deal of study on the use of chemical preservatives and antibiotics. As summarized by Tarr (1956b), the effectiveness of nitrite as a preservative for iced fish was first demonstrated in Canada and later chlortetracycline

(CTC) was shown to be very effective at retarding the bacterial spoilage of fish. Numerous other substances were tested but Canada approved two for use; if properly used (see Castell, 1963) they could be expected to double the shelf-life of fresh fish. When used as a dip for fresh groundfish fillets, these additives had a very beneficial impact on extending the markets for fresh fish in Canada and the US. However, because of their potential human health hazard, approval for both substances has now been withdrawn and Canadian workers have resumed the search for a replacement. Trials with EDTA (ethylenediaminetetraacetic acid) have shown promise but it is not as effective as CTC and health authorities have reservations on permitting its use as a fresh fish perservative.

Physical means of controlling bacterial spoilage have also been well researched by fisheries technologists the world over. Superchilling was given thorough trials, both for use ashore and at sea, but, as reported by Carlson (1969), the detrimental effects of partial freezing have to be overcome and a reliable mechanical system is needed to maintain the critical temperatures required. Superchilled temperatures can also be obtained by adding salt to iced fish. This can be effective in chilling and transporting bulk quantities of fish intended for curing or canning where salt uptake and texture changes are of no major consequence.

Mechanically refrigerated or ice-chilled sea water has been demonstrated as an excellent medium for chilling and transporting whole fresh fish. Although there has been little or no application to the retailing of fresh fish, it is being used at sea and for the road transport of bulk quantities of whole small fish and salmon. Ice-chilled sea water systems have recently been installed at several Canadian fish plants for the holding of large quantities of dressed groundfish prior to processing. Salt uptake must be monitored and thorough mixing with compressed air and/or recirculation is essential to avoid thermal stratification. The use of carbon dioxide for agitation has been shown to have no major advantage and carbonation of the fish caused excessive curd and internal can pressure in canning operations.

Irradiation as a means of preserving fresh fish has also received much attention and effort. Extensive tests have shown that seafoods sterilized by irradiation are unacceptable due to 'irradiation' odours and flavours. Pasteurization doses of irradiation extended shelf-life but the process was never shown to be economic and off odours and flavours could still be detected. Based on the fact that only centrally located installations would be feasible, this posed another transportation factor that was unattractive to the industry.

The North American groundfish industry was quite concerned that it no longer had a reliable shelf-life extender for fresh fillets. Recent Canadian trials on the hypobaric storage of fresh fish have not been too encouraging but work is continuing. In this system, iced fish were held in a mechanically refrigerated vacuum chamber at or just below 0°C wherein the atmospheric pressure was maintained at 10mm Hg. Air in the chamber was vented at a rate equivalent to two exchanges per hour while keeping the relative humidity at 95% or more.

Some may recall earlier work done at Torry which indicated that carbon dioxide could be effective in helping to preserve fresh fish. In a very recent US report, Veranth (1979) has claimed that iced salmon, sealed in an atmosphere of 60% carbon dioxide/20% oxygen/20% air, kept twice as long as regularly iced fish. These results and the Torry records possibly warrant further examination.

Although much science and technology has helped to improve and guarantee the quality of fresh fish, one of the most significant contributions in this area has been the introduction of synthetic packaging materials. Many problems were linked with the old wooden fish boxes that were often saturated with slime and bacteria and left a trail of the same mixed with melt water. These and similar containers were a great restraint on the distribution of fresh fish products and some carriers would not handle them at all. Water-tight and air-impermeable plastic boxes, trays, pouches and films have revolutionized the transportation and distribution of both fresh and frozen fish. One-way styrofoam containers provide excellent insulation for iced fish and enable dry shipment by a variety of carriers, including aircraft, without any risk of leakage or smell. Specially fabricated corrugated cartons with waterproof linings and good insulating properties can also accommodate iced fish in plastic retail packs. With modern packaging materials and containers, it is also quite simple to ship by air live shellfish, eels and a variety of processed seafoods to distant markets.

A discussion on fresh fish would not be complete without some mention of objective methods for assessing quality. This has been the subject of many conferences and the following quote by Shewan and Ehrenberg (1956) is appropriate, 'It is clear that no matter what objective method was used, it would ultimately have to be related to direct sensory assessment'. With that in mind, the TMA (trimethylamine) and TVB (total volatile bases) tests have been extensively used as indicators of bacterial spoilage. The hypoxanthine test is an objective measure of autolysis and the TBA (thiobarbituric acid) test is the most reliable for determining rancidity. The latest Torry-meter appears to be the only electronic instrument to give reproducible indications of fresh fish quality.

5 Frozen fish

Frozen fish had a stormy and difficult beginning and only recently has flourished on a world-wide basis. Many consumers, however, remember the poor quality of the past and this has created a lasting impression that all frozen fish is inferior. The fact is that frozen fish today can, in many instances, be superior in quality to much of the fresh fish currently being retailed. Nevertheless, there are still many problems which are more related to education and attitude than to improved technology. The tremendous amount of good research work done on frozen fish, particularly at Torry and at Halifax, can serve as an excellent base for improving the image and quality of frozen fish.

Looking back at the old freezing and cold storage methods and equipment, it was little wonder that the product was of inferior quality. Fish to be frozen had to be immersed in or sprayed with refrigerated brine or racked on cooling coils in cold storage plants. With the new and efficient blast and plate freezers of today, there no longer should be any problem with slow freezing. Good quality cold storage plants, home freezers and infrastructures are expanding in many countries. Storage temperatures and loading rates can stand improvement and many airports still do not have adequate facilities for handling iced and frozen goods. Furthermore, similar hardware is available for use on fishing vessels that can provide the ultimate in initial quality. The various means of transporting fresh and frozen fish were reviewed by Young (1951) and by Foley (1969) and my only comment at this time is that many of the historical problems have been overcome. Modern refrigerated railway cars (coupled with fast trains), trucks, containers and cargo vessels are equipped to handle frozen fish. In North America, most fish is transported by refrigerated truck and the major problems are encountered at the delivery end of the system.

The major quality changes in frozen fish are texture deterioration and rancidity. In the early days, it was difficult to determine what was going wrong and how to evaluate the quality of frozen fish. The chemistry of protein denaturation, the cause of inferior texture, was not understood. Taste panels were used as an assessment tool but there were differences of opinion on their validity and on how they should be conducted (Connell, 1969). Love's cell fragility test and the soluble protein nitrogen test were used primarily as objective measures for protein deterioration in frozen groundfish. It was soon discovered that all species did not react the same to frozen storage, some performed better than others, and that individual storage patterns had to be determined (Dyer and Peters, 1969). Evidence from Torry and Halifax suggested a relationship between lipid and protein deterioration but this still has not been fully elucidated. However, the studies of Dingle and Hines (1975) cleared up a great deal of the misunderstanding regarding protein denaturation in groundfish and the differences between species and tissues. They demonstrated the presence of an enzyme which transformed trimethylamine oxide to dimethylamine and formaldehyde in gadoid tissues which by promoting the formation of formaldehyde increased the rate of protein denaturation. The enzyme was found to be particularly active in gadoid kidney but it was also active in the muscle tissue of selected species which explains their sensitivity to frozen storage denaturation (eg silver hake and pollock). The authors further commented that the rapid denaturation in frozen minced flesh can be accelerated by the presence of kidney tissue or tissues from these reactive species.

Even though the chemistry of protein denaturation is not yet fully understood, a great deal of excellent work has been done on factors which affect the quality and storage life of frozen fish. These included: the effects of pre-icing and slow freezing on quality; the need for low and constant storage temperatures; the effects of periods of increased temperature as encountered in transport; and such things as the prevention of drip, dehydration and oxidation.

The introduction of freezer trawlers brought a new series of technological problems related to the freezing of fish at sea and subsequent handling and processing ashore. There were innumerable questions, such as: is it better to freeze pre-rigor or post-rigor; whole fish or fillets; does bleeding improve colour; what causes gaping; how best to thaw; and the effects of thawing on subsequent storage either fresh or frozen. Information has now been assembled on all these factors and sea-frozen pre-rigor fish generally produce products superior to those produced by traditional methods. Thawing, however, can produce problems and a great deal of effort has been devoted to the development of equipment to minimize detrimental effects. Many fish today are thawed at ambient temperatures, in flowing water, in warm-moist air (eg modified Torry kiln) and in warm water even though more exotic techniques are available such as vacuum and dielectric thawing.

Mention should also be made that freezing at sea has been a tremendous help to a number of fisheries such as those for tuna and shrimp. In addition, it has greatly facilitated the exploitation of species like squid and silver hake that do not stand up well to icing.

One of the major factors that stimulated the frozen fish trade was the introduction of the fish stick. This product alone created a whole new industry which has now been further expanded by the addition of breaded and battered portions, IQF fillets and like products which facilitate portion control, institutional feeding, the fish and chip trade, and the fast food industry.

Current technological investigations centre around non-traditional species like blue whiting, argentine, fatty species such as mackerel, and minced fish. Many recognize the potential of minced fish (see J N Keay, 1975) but it has a number of limitations, the major one being its instability in frozen storage. The most important minced fish product produced today is Japanese surimi which is a chemically stabilized frozen minced fish preparation used in the manufacture of kamaboko.

Perhaps the foremost problem in the frozen fish industry today is the fact that choice fish tend to be sold fresh and the seconds are frozen. Another problem is the adverse effect of temperature increases during the reprocessing of frozen blocks into fish sticks, etc. Freezing temperatures are still not well maintained at the end of the distribution chain, particularly in retail outlets. It is suspected that physical damage due to increased mechanical handling of fresh fish (eg during unloading) is contributing to loss of frozen quality.

6 Cured fish

These traditional products today are probably consumed primarily by ethnic groups and population sectors that have developed a taste for them as speciality items. At the same time, they continue to be sold in traditional markets where the infrastructures of handling fresh and frozen products are not available.

In the earlier years these products also had major quality and production problems requiring the attention of fisheries technologists. First, the chemistry and bacteriology of salting and smoking fish was a void. Techniques for preparing the products were vague and varied from one location to another, resulting in products varying in quality and uniformity. The drying process was laborious and subject to unfavourable weather conditions. There were also problems related to the quality of salt used which could cause discoloration and irregularity in salt penetration. In addition, there were halophilic bacteria, moulds that caused 'dun', and insects. Smoke houses also left much to be desired, being generally crude structures with little means of control and dependent primarily on the skills of the operator.

Technologists developed information and equipment which drastically altered the cured fish industry (see Beatty and Fougere, 1957, and Burgess and Bannerman). Mechanical smoke houses, driers and fish washers were developed and processing methods were altered and improved to guarantee quality and to control factors that caused deleterious effects. Similar methods and techniques were prescribed for the production and quality control of salt-cured herring and mackerel. The influence of fat content, sexual maturity and contaminating bacteria on the preparation and quality of these products was examined.

Today artificial smoke and electrostatic smoke houses are in use but they tend to be more associated with canning operations. With the growing shortage of energy, many producers are updating drying and smoking equipment to reduce energy consumption and this has stimulated interest in better use of solar energy and systems incorporating heat pumps. Plastic and fibreglass containers have facilitated the processing and distribution of cured fish products. A recent observation suggested that new entrants to this trade were experiencing difficulty in getting started because much of the literature is now outdated and there are few experienced technologists remaining in this field.

A final point worthy of mention is that the salting of minced fish has attracted attention in several countries and indications are that it has commercial potential.

7 Canned fish

Many kinds of fish produce excellent canned products and this is an ideal way of preserving a highly perishable food. Unlike most other seafoods, the long shelf-life and the ease of distribution are decided advantages and much canned fish can be classed as a luxury item. Nevertheless, the commodity has not been without its difficulties.

Many of the problems with canned fish can be related to the quality of the raw material. Since a number of species used for canning occur in glut quantities, storage of the raw material has been a problem. Refrigerated sea water, as mentioned earlier, is an excellent medium in this regard but fish can only be stored this way for a short time. Much material is, therefore, frozen for later processing. Improper holding can result in excessive salt uptake from sea water or brine, frozen storage can often lead to rancidity and dehydration, and thawing can produce detrimental effects. It has been shown that such factors can lead to undesirable colour and flavour changes, curd formation, and inferior texture in the sterilized product.

In the canning of fish, processing conditions must be rigidly controlled for each individual species and product to ensure product uniformity and quality. For example, the pre-cooking step required to reduce water content of most fish species must be varied and is more critical with some species than others. Excessive retort-

ing and improper cooling of the sterilized cans produces brown discoloration, especially in shellfish.

Today's modern canneries contain much sophisticated and automatic equipment such as continuous cookers and electrostatic smokers, automatic weigh scales, mechanical can filling, and temperature-programmed retorts. Such equipment has streamlined operations and greatly increased daily production.

Black discoloration from iron sulphide continues to plague the canning industry. Enamelled cans and parchment liners have lessened the problem considerably as will the increased use of aluminium cans. The avoidance of stale raw material and the replacement of iron hardware with stainless steel equipment are other measures of prevention.

The chemistry of struvite (magnesium ammonium phosphate) formation in canned seafoods has been well documented but contamination can still be a troublesome problem. The use of sea water in the processing of canned fish can be a source of magnesium and promote the formation of crystals, particularly with stale raw material having elevated levels of ammonia. Lowering the pH with a suitable acid is effective in preventing crystal formation if other measures fail.

More recent developments in canned fish have been easily-opened cans with rip tops and the 'box bande', and the aluminium can as already mentioned. Tests to date with retortable pouches and aluminium trays have not been particularly encouraging. Costs are relatively high and the operations are slower than with cans. However, they offer potential for delicately flavoured seafoods that could benefit from less heat treatment.

8 Fish meal and oil

The fish reduction industry has flourished in the last decade with record prices and production levels. Due to extensive studies on the nutritive value and performance of fish meals in animal feed, the fish meal industry has been able to sustain sales against stiff competition. As has often been said, more is known about the role of fish in animal nutrition than in human diets. At the same time, new and sophisticated equipment has been developed to meet the demands of this rapidly expanded industry. For example, large capacity mechanical unloaders, continuous cookers, multi-staged driers and evaporators, electronic monitors, odour abatement and effluent facilities and other equipment for the production of 'whole' meals. With the increasing concern and cost of energy, process engineers are re-examining the reduction process in an effort to reduce energy consumption. The effects of raw material quality on processing efficiency and on quality and uniformity of the finished meal and oil are also receiving increased attention. Studies continue on the problems associated with bulk storage of large quantities of whole fish with and without chemical perservatives like nitrite and formaldehyde. Bulk handling of fish meal is now common practice with the aid of anti-oxidants to prevent overheating. The high demand for fish roe has resulted in plant alterations in some areas to enable the recovery of at least part of the roe prior to rendering.

Many can remember when technologists were urged to find new uses for fish oils. New synthetic vitamins and polymers had destroyed the markets for both vitamin oils and industrial oils. Considerable effort was placed on greater use of fish oils in human nutrition and today most fish oils are consumed as margarine and shortenings (see Ackman, 1974). The current question over the potential health hazard associated with the consumption of docosenoic acids is a matter that is being seriously examined. Mention should be made that there appears to be a growing demand for fish oils for industrial purposes at competitive prices.

Although there have been continuous pleas to convert more industrial fish into human food, the results of extensive studies on fish protein concentrate and edible fish meal to date have not been very successful nor economically encouraging.

9 References

ACKMAN, R G. Marine lipids and fatty acids in human nutrition. In:
1974   *Fishery Products*, Ed R Kreuzer. Fishing News (Books) Ltd, London. 112–131

ANON. Report of the FAO Symposium on the significance of funda-
1965   mental research in the utilization of fish. *FAO Fisheries Reports*, No. 19. 52pp

BEATTY, S A and FOUGERE, H. The processing of dried salted fish. *Bull.*
1957   *Fish. Res. Bd. Can.*, No. 112. 54pp

BURGESS, G H O and BANNERMAN, A McK. *Fish Smoking* – A Torry
—   Kiln Operator's Handbook. Torry Research Station, Aberdeen. 44pp

CARLSON, C J. Superchilling fish – a review, In: *Freezing and Irradia-*
1969   *tion of Fish*, Ed R Kreuzer. Fishing News (Books) Ltd, London. 101–103

CASTELL, C H. Spoilage problems in fresh fish production. *Bull. Fish.*
1954   *Res. Bd. Can.*, No. 100. 35pp

CASTELL, C H Antibiotic dips for preserving fish fillets. *Bull. Fish. Res.*

1963   *Bd. Can.*, No. 138. 70pp

CONNELL, J J. Changes in the eating quality of frozen stored cod and
1969   associated chemical and physical changes. In: *Freezing and Irradiation of Fish*, Ed R Kreuzer. Fishing News (Books) Ltd, London. 323–338

DINGLE, J R and HINES, J A. Protein instability in minced flesh from
1975   fillets and frames of several commercial Atlantic fishes during storage at $-5°C$. *J. Fish. Res. Bd. Can.*, 32, 775–783

DYER, W J and PETERS, J. Factors influencing quality changes during
1969   frozen storage and distribution of frozen products, including glazing, coating and packaging. In: *Freezing and Irradiation of Fish*, Ed R Kreuzer. Fishing News (Books) Ltd, London. 317–322

FOLEY, M A. Transport of frozen fish. In: *Freezing and Irradiation of*
1969   *Fish*, Ed R Kreuzer. Fishing News (Books) Ltd, London. 414–421

GIRARD, M. Fish marketing in Western Europe. *FAO Fisheries*
1952   *Bulletin*, Vol 5(1), 3–15

KEAY, J N. Conference proceeding – *The Production and Utilization of*
1975   *Mechanically Recovered Fish Flesh (Minced Fish)*, 7/8 April. Torry Research Station, Aberdeen. 108pp

RADCLIFFE, L. The importance of the fisheries. In: *Marine Products of*
1923   *Commerce*, Ed D K Tressler. Reinhold Publishing Corporation, New York. 227–238

SHEWAN, J M and EHRENBERG, A C S. The sensory assessment of iced
1956   white fish by a panel technique. In: *Chilling of Fish*. Fish Processing Technologists Meeting, Rotterdam, The Netherlands, 25–29 June. 144–150

TARR, H L A. Fresh fish preservation. In: *Chilling of Fish*. Fish Pro-
1956a   cessing Technologists Meeting, Rotterdam, The Netherlands, 25–29 June. 114–116

——   Use of preservatives and antibiotics in the preservation of fresh
1956b   fish. In: *Chilling of Fish*. Fish Processing Technologists Meeting, Rotterdam, The Netherlands, 25–29 June. 19–23

TRESSLER, D K. *Marine Products of Commerce*, 18. Reinhold Publish-
1923   ing Corporation, New York. 762pp

VERANTH, M F. CO_2-enriched atmosphere keeps fish fresh more than
1979   twice as long. *Fd. Proc.*, 40, 76–79

YOUNG, O C. Transportation of fishery products. In: *Marine Products*
1951   *of Commerce*, Ed D K Tressler and J M Lemon. Reinhold Publishing Corporation, New York. 307–327

2 Past, present and future of fish science

Sensory and non-sensory assessment of fish
J J Connell and J M Shewan

1 Introduction

This paper is not intended to review comprehensively the development of these subjects over the past 50 years because the literature is voluminous and some parts of it have been adequately reviewed at various times in the recent past (Farber, 1965; Soudan, 1969; Houwing, unpublished; Gould and Peters, 1971; Spinelli, 1971; Uchiyama, 1971; Iyer, 1972; Tromsdorff, 1973; Kvale, 1973; Mills, 1975; Burt, 1976). In particular the paper by Farber, which contains nearly 550 references, should be consulted for details of all the work prior to 1965 dealing with freshness tests. Much of the work under review is concerned with chemical methods of assessing the quality of spoiling chilled fish and shellfish; this aspect has a fascination for researchers but the magnitude of their efforts gives rise to feelings of indigestion on the part of those who have to assess the field.

The purpose is to evaluate critically the status and roles of all the methods of assessment with particular reference to developments over the last decade and to future possibilities. Only a selection of the literature will be quoted. The definition of quality is a wide one because different aspects are important in different circumstances. Thus, as well as the usual tests dealing with chemical and bacterial spoilage in the chilled and frozen states references will include methods dealing with, for example, intrinsic quality, defects, composition and microbiological quality. The main use to which methods of quality assessment is put is in industrial production and commercial selling of fish and fish products. A subsidiary use is in the regulation of quality by official agencies. An important minor use is in research and development. The type of method required differs depending upon these uses, and attention is drawn to the area of application of methods.

Because of their primacy in any system of quality assessment, sensory methods will be dealt with first and then non-sensory methods.

2 Sensory methods

These are of two kinds: subjective and objective (British Standards Institute, 1975). In the former, biases in judges are not minimized and personal opinion is allowed free rein. These methods are typified by the consumer trial in which a group of the ordinary public numbering at least 20 to 50 are asked their views about or preferences for some sensory aspects of the product. In the latter, biases are deliberately minimized by the use of specially trained judges or assessors who concentrate on a particular well-defined attribute of the product and who operate as a panel of one person but usually more.

2.1 *Subjective assessments*
Relatively few investigations of this type of method

which deal specifically with fish products have been recorded. In one of these (Connell and Howgate, 1971) the preferences of consumers for frozen fish was examined in order to define changes in quality at different temperatures of storage. Quality in this case was defined by the 9-point hedonic rating scale of Peryam and Girardot (1952). This scale, or variants of it, has been used a great deal in consumer studies of foods generally and is now used frequently at Torry Research Station for a variety of studies. Within limitations the test procedure has proved useful in comparing or screening samples. For example, in the work just referred to, the relative importance to consumers of the off-flavours and poor textural characteristics of frozen fish were distinguished. In a somewhat similar investigation Rasekh *et al* (1970) were able to correlate closely consumer preference for canned tuna of different quality as assessed on a hedonic scale with a combination of quality factors, namely colour, mechanical shear value, fibre content, percentage fat and pH.

A question of contemporary interest is the consumers' perception and valuation of minced fish in products such as fish fingers. Even expert opinion differs as to the proportion of mince which can be included before changing perceived quality. Preliminary unpublished trials conducted by the Long Ashton Research Station in conjunction with Torry show that on the basis of hedonic tests, the acceptability of minced fish differs between adults and children, the latter often preferring fingers made entirely of mince, whereas the opposite is true of the former.

For over 10 years the United Kingdom White Fish Authority have been conducting consumer tests on fish products based largely on the method of measured plate-waste. In this, the fish product under test is included incognito as part of a normal meal in, say, a school restaurant, and the amount consumed by each person measured. The proportion consumed is assumed to be a measure of acceptance under near normal circumstances. The method has been applied successfully to assess, for example, the acceptability of different species of fish made up into various products. In a similar way observations have been made on the consumption behaviour of schoolchildren in developing countries when meals containing fish protein concentrate were given (Tagle *et al*, 1976).

Even simpler tests can give interesting new insights. Batches of fish fingers containing known different quantities of bones were prepared at Torry. Consumers were asked to eat the batches of cooked fingers and to report if they detected bones and if so how many. The results provide information on the level of bone content which is just detectable by the group of people involved. Such information could be used to set standards for the permissible bone content of fish products; existing standards tend not to be securely based in this respect.

In an experiment which is one of the few attempts to obtain measured information about the basis for food selection, Lamb (1975) compared the reasons for the acceptance of fish as a menu item to that of other foods like meat. In many developed countries fish is generally reckoned to be less desirable than other flesh foods when presented in competition but the fundamental reasons for this and for consumers' preferences for different kinds of fish product are not understood. To those interested in developing new fish products which must compete in the market place any information on this subject could be useful in guiding the characteristics which the products should have. Perhaps this is one of the new promising areas for research.

Subjective sensory tests of eating quality are undoubtedly those which ultimately count because they come closest to assessing those consumer responses which matter. There is, therefore, a case for using them as a basis of comparison for all other kinds of tests. If one is testing, for example, the usefulness of a new non-sensory chemical method of assessing quality, the results from it ideally should be compared to the results of subjective consumer evaluations on the same samples. Because of the difficulty and expense normally involved in arranging fully representative consumer tests, this is rarely possible and instead comparisons with the results of objective sensory tests are usually employed. In general the results of subjective and objective sensory assessments correlate well, especially where the differences between samples tends to be large. In other cases correlations may not be so obvious. Thus, trained tasters can easily differentiate between the changes in quality of fish during the early stages of spoilage, but it is less clear that the generality of ordinary consumers in some populations would show marked preferences between samples in this stage. A non-sensory method which correlates well with an objective sensory method is then considered in principle to offer a workable measure of quality.

Methods for the subjective assessment of fish are not intrinsically different from those applicable to foods in general and few radically new developments in them seem likely.

2.2 Objective assessments
A number of methods for the objective sensory analysis of foods have been described (Amerine et al, 1965; American Society for Testing Materials, 1968; Drake and Johansson, 1974) but four have been used fairly frequently for fish; paired comparison, ranking, triangle testing and scoring of attributes.

It has been pointed out that the first three types of test really give information about the discriminatory abilities of the judges rather than about the sample. They are, therefore, perhaps logically a separate category of tests. For the purpose of convenience, however, they are grouped under objective assessments.

The first method has been useful for a number of applications. For example, increases in the toughness of fish stored in the frozen state were detected by paired comparison with unfrozen controls (Love, 1966). The determination of the threshold for detection of oil taints in fish and shellfish has been achieved using ranking (Howgate et al, 1977). The ability of consumers to dif-ferentiate and recognize different species of fish or parts of fish by odour or taste has been measured using triangle and comparison tests on a number of occasions at Torry.

Scoring is, however, by far the most commonly used method particularly for assessing freshness of chilled fish. The reasoning behind those scoring systems which are in active use is rather different and bears examination. A useful comparison of the scoring systems used in different European countries for fresh fish is contained in an unpublished paper by Houwing.

The most extensive scheme, but not necessarily the best operationally, is that developed at Torry over 25 years ago. This is based on discriminatory descriptive changes in the attributes of appearance, odour, flavour and texture of the raw or cooked fish, each change being denoted by a number (Shewan et al, 1953). The use of subjective or qualifying terms such as slight, very or good is abjured as far as possible. Also the scores for the different attributes are kept separate and not amalgamated because it was believed that they changed independently and not necessarily in step. The scheme was originally designed for chilled gadoid species but with small variations has been extended to almost all the commercial species in the UK. Farber (1965) was of the opinion that the discriminatory ability of judges was poorest at the borderline between fresh and possibly spoiled, that is at the stage of incipient spoilage. This may occur with some schemes but is not a feature of the Torry scheme in the correct use of which the standard error of scores of a trained panel does not vary with the stage of spoilage. A feature of the Torry scheme, however, is the fact that over most of the scale the scores are linearly related to the number of days the fish are held in melting ice.

The basic Torry scheme is used a great deal in the UK fish industry to control the quality of wet fish both in everyday factory operations and in specifications. Recently modifications involving the pooling of attributes have been introduced into the scheme to make it more useful for industrial use. The sensory attributes of frozen fish such as off-flavour and firmness change in a progressive fashion with time of storage and these changes can be arranged into a fully descriptive scoring system (Baines et al, 1969) using intensity scales. It was found that the degrees of pre-freezing spoilage and of deterioration in the frozen state could be assessed in the same sample with reasonable accuracy.

A similar but somewhat simpler scheme for chilled fish was developed in France (Soudan et al, 1957). Here 13 different sensory attributes are combined into a single score to denote stage of spoilage. This scheme and that of Wittfogel mentioned below form the basis for the European Economic Community Scheme of grading of certain species of chilled fish at the point of first sale (EEC, 1970); it is in use in at least seven of the community partners. In this, four grades of freshness are laid down (E, A, B and C) corresponding to various stages of spoilage; E and C are freshest and unfit for human consumption, respectively. A few features of this grading system were considered to require modification for UK conditions. We have here an example of a more or less continuous scale of change being segregated into a few distinct stages for the purpose of defining grades of quality. The extent to which the EEC scheme is being applied is not fully but in view of the large quantity of fish involved it is believed that it is the single most important method of quality assessment in existence.

In the German Federal Republic systems for chilled and frozen fish (Wittfogel, 1958; Kietzmann, 1968) have been constructed which are based on the Karlsruhe 9-grade method of assessing the quality of foods in general (Paulus *et al*, 1969). A further development (Kietzmann, 1969) uses 4-, 5- and 7-point scales for odour, texture and flavour, respectively. The systems use a combination of descriptive terminology and value judgements. A Danish method of freshness uses exclusively value judgements (*eg* Herborg and Villadsen, 1975). In the Bergen Fisheries Technological Laboratory in Norway, a 5-point scale for four attributes is used.

A comparative trial in which the same samples of fish in various stages of spoilage were assessed by the different European schemes (Houwing, unpublished), gave the interesting result that all the scores appeared to be linearly related to one another. This suggests that if well-trained assessors are used the nature of the scoring system is not of overwhelming importance. It may be, however, that some schemes are more discriminatory than others.

In some circumstances the concentration of a single substance so dominates the odour or flavour spectrum that an assessment of it alone gives a sufficiently good measure of quality. This is the case with the spoilage of elasmobranchs where a sensory assessment of the intensity of ammonia is adequate (Vyncke, 1978).

All the scoring systems dealt with so far have assessed quality without direct reference to storage life. A new approach was introduced by Learson and Ronsivalli (1969) and Charm *et al*, (1972) who used the characteristics of fish stored to particular given stages of spoilage to assess the storage life of unknowns. The technique makes use of the odour of the product when held in a closed container. It is not clear to what extent some of these methods have been used outside research laboratories.

Scoring systems of a rather different kind have been introduced into the so-called defects tables of several different national (*eg* USA) and international (Codex Alimentarius) standards. Defects are quality faults which if sufficiently serious or numerous lead to all the product being declared not up to minimum standard. These include the presence of bones, parasites, skin, viscera, freezer burn and poor filling of cans. In all cases the assessment of the number or extent of faults is carried out by visual examination supplemented in the case of bones by feeling with fingers. Degree of freshness or extent of deterioration in the standards is not carried out by scoring but by a semi-subjective requirement that in order to be acceptable the product must not possess objectionable sensory features; so far it has not been found possible to reach agreement on exact definitions of what constitutes objectionable features.

Inspection for defects of contamination and for departures from standards of quality is, of course, practised very widely in industry and should be included in any discussion of objective-sensory methods. The methods used vary greatly depending upon company practice and internal specifications; they depend for their success upon the vigilance and careful scrutiny of operatives, production supervisors and quality control staff. The types of examination include superficial visual examination, colour comparisons using standard shade cards, odour and tasting testing.

In the last few years there has been some interesting research carried out on the further application of either the flavour profile method of Cairncross and Sjöström (1950) or the texture profile method of Brandt *et al* (1963). At Torry, for example, attempts have been made to unravel the complex of individual odour and flavour notes which combine to give the overall impression of flavours such as cooked haddock or badly frozen stored cod (Howgate, unpublished). As many as 10 or 20 notes have been identified and the intensity of each estimated on a short numerical scale. The objective, as described in another paper at this conference, is to aid the progress of chemical fractionation of the substances which constitute the overall flavour. A major difficulty in this work is handling statistically the large volume of data so that valid comparisons can be made. Developments in the texture profiling of fish products were initiated by Webb *et al* (1975) for shrimp and by Bruun, Ikkala and Sorensen (1976) who wanted to devise textural definitions for fish minces frozen and stored under different conditions. Howgate (1978) has also described a texture profile for fish and fish products and has applied the statistical method of factor analysis to the results in order to reduce the data. Other papers at this conference describe progress in investigations of texture profiles. It seems likely that further advances in the fundamental understanding of sensory properties using profile techniques will occur and will allow us to improve the schemes which are used in industry. For the present, however, profile techniques as just described for fish and fish products do not appear to have moved outside research.

Considering objective-sensory methods as a whole we can envisage progress in the next few years being made towards further limited international agreement on schemes and towards schemes suitable for recovered and reformed fish flesh.

3 Non-sensory methods

These by definition are always objective and are of three kinds depending upon whether they are related to (*i*) some sensory attribute of quality (for example, eating quality); (*ii*) safety and wholesomeness; (*iii*) composition or identity.

Non-sensory methods of the first kind are needed to replace or supplement sensory testing for spoilage or deterioration. In order to achieve an ideal substitution a non-sensory method must either correlate exactly under all conditions with the sensory attribute under test or be causally related to it. Few if any non-sensory methods for fish products have reached this stage of perfection, though some are sufficiently close to be of practical usefulness. The advantage of non-sensory methods is their convenience, normally lower cost and the fact that they can be standardized against reference standards.

The second kind measure the concentration in the product of potentially or actually harmful substances or micro-organisms. The former include heavy metals, radio-isotopes, pesticides and biotoxins. Although these are of obvious importance and of considerable current concern they will not be discussed here because they are of more relevance to the specialized surveillance and analysis of food in general; some reference, however, to micro-organisms will be made.

The third kind include the measurement of water, protein, lipid and weight which again because of their general nature will not be discussed in detail; aspects of colour, fish content, glaze and bone content and species identity will, however, be examined.

3.1 Spoilage and deterioration
In view of the difficulties inherent in the methodology and interpretation of sensory data it is not surprising that non-sensory methods should have been widely sought for assessing the quality factors of fresh, spoiling, frozen and other processed fish. As long ago as 1891 Eber proposed a simple chemical test for putrefaction based on volatile amines, and since then numerous chemical, biochemical, physical and microbiological methods have been investigated. The references given at the beginning of this paper deal comprehensively with the concepts behind these methods, many of which have proved of little practical use. We will confine ourselves to those which on present evidence are of utility.

3.1.1 *Chemical and biochemical methods* Of these the determination of total volatile basic nitrogen (TVBN) is perhaps the oldest established, of widespread usefulness and correlates reasonably adequately with sensory changes during spoilage or deterioration. It has the advantage of simplicity, cheapness and relative rapidity; its main disadvantages are that the sample is destroyed, the conditions of volatilization have to be standardized or specified exactly and it is not sensitive for the early stages of spoilage. In recent years the scope of the method for chilled fish has been further explored (Pearson and Musslemuddin, 1969a,b; Guardia and Haas, 1969; Vyncke, 1970; Uchiyama et al, 1970; Karnicka and Jurewicz, 1971; Amu and Disney, 1973; Sidhu et al, 1974; Ohta et al, 1975). TVBN increases only slowly during the chilled storage of most freshwater fish (for example, Nair et al, 1971) principally because of their low or negligible content of trimethylamine oxide (TMAO); it is accordingly of little general use for this group. For shrimps it has been found that the ratio TVBN/free amino acid N correlates better with sensory quality than TVBN alone (Cobb et al, 1973; Cobb and Vanderzant, 1975).

Experience with the methods seems to vary in different hands and even in cases where it might be expected to give good results it is inconsistent. One source of variability in iced fish storage appears to be leaching of volatile amines from different parts of the body (Mowlah, 1975; Karnop, 1976). Variability with TVBN, and also with trimethylamine (TMA), arises through biological variations in the concentrations of the precursors. The method has been applied to products such as dried, smoked and dried fish but little to frozen fish.

In species such as most marine fish and pike among the freshwater fish, which contain substantial amounts of TMAO, the determination of the degradation pro-

duct TMA has been used extensively as a more specific index of bacterial spoilage. Its advantages and disadvantages are similar to those of TVBN with the main difference that the substance measured can, in principle, be determined exactly; its analysis on the other hand is more complex and costly. Recent advances with the determination of TMA centre around more specific and accurate methods using gas-liquid chromatography (Miller et al, 1972; Keay and Hardy, 1972; Murray and Gibson, 1972; Gruger, 1972; Ritskes, 1975; Tromsdorff, 1975; Tokunaga et al, 1977), improved autoanalyser techniques (Kato and Uchiyama, 1973; Ruiter and Weseman, 1976), a specific gas sensor (Storey, 1975) and a specific electrode system (Chang et al, 1976). An extensive series of experiments on cod stored experimentally or sampled from markets have provided what is probably the best published statistical relationships between TMA and sensory scores (Burt et al, 1976a,b; Connell et al, 1976).

During the frozen storage of certain gadoids such as cod and hake, TMAO is enzymatically dissociated into dimethylamine (DMA) and formaldehyde (F) at a rate dependent upon the temperature. Thus, the determination of DMA concentration in a frozen stored sample could provide a measure of the amount of deterioration that had simultaneously occurred. More elegantly, determinations of TMA and DMA by a single method could provide information on both pre-freezing spoilage and deterioration on frozen storage (Castell et al, 1974). The assessment of amount of F in a frozen sample by means of a simple colour test (Connell, 1966) provides an alternative which has been found to be a useful and rapid screening test in industry.

The determination of ammonia arising from the bacterial decomposition of urea, creatine and other nitrogenous compounds is agreed to be of little use as a method for fish in general but recently good results have been obtained with elasmobranchs (Vyncke, 1970; 1978).

Bacteria produce a wide variety of substances in addition to amines through their action on substrates in fish flesh. Some of these are:

Substrate	Main products
Sulphur amino acids	hydrogen sulphide, dimethyl sulphide, methyl mercaptan
Glucose, ribose	lower fatty acids
Lipids	carbonyls
Proteins	tyrosine, indole, skatole, putrescine, cadaverine
Histidine	histamine

In the past, analyses of several of these products have been explored as methods of assessing spoilage but none has proved to be entirely satisfactory or of more than limited application. A number of more recent studies have confirmed this view with a few important exceptions. Thus, the content of fatty acids does not correlate well with quality in iced roughhead grenadier (Botta and Shaw, 1975), iced capelin (Botta et al, 1978) or herring (Reinacher, 1977). The content of indole in iced oysters is also a poor indicator (Luizzo et al, 1975). Histamine content was found in one study not to be always a good index of quality in canned tuna (Meitz and Karmas, 1977) though the opposite view has been expressed (Nerisson, 1975). High concentrations of bacteria-produced ethanol are associated with spoiled tuna which has been subsequently canned (Lerke and Huck, 1977), with spoiled canned salmon (Cosgrove, 1978) or with spoiled iced cod (Gajewska, 1977). An index based on the concentration of the higher amines which accumulate during the decomposition of rockfish, salmon, lobster and shrimp has been proposed as a method of distinguishing three grades of spoilage (Meitz and Karmas, 1978). The determination of the volatile compounds which reduce permanganate is a useful method in some circumstances and acts as a summation of several of the groups of substances listed in the table (Farber and Lerke, 1968; Luizzo et al, 1975; Witas and Olbromska, 1976).

All the methods for freshness assessment based on the measurement of the products of bacterial action are only applicable in the stages of spoilage where bacterial numbers begin to rise sharply. Before then chemical tests have to be based predominantly on the products of autolytic enzyme breakdown. The most important and useful of these has been found to be the decomposition products of adenosine triphosphate (ATP). In many species ATP is degraded and its products accumulate at a steady rate during the first five or so days of icing, that is before the first signs of bacterial decomposition occur. Within a species the concentration of ATP in the living flesh is approximately constant. Thus, a measurement of ATP decomposition products can, in principle, offer a means of estimating freshness in the early stages as well as the later stages of storage. From this point of view the most important products are inosine and hypoxanthine (Hy). The latter is particularly useful as an index in a wide range of marine and freshwater fish and of shellfish. The thorough review of Burt (1976) should be consulted for details of Hy. In species and under conditions where the accumulation of Hy is either too rapid or too slow to render it a useful index, a measurement of the so-called K-value can provide an alternative. The K-value is the ratio (expressed as a percentage) of inosine plus Hy to total amount of ATP-related compounds (Ehira et al, 1970; Ehira and Uchiyama, 1973).

The concentration of Hy correlates well in many cases with sensory values and in those cases where a direct comparison of Hy with TMA has been carried out, the former has been found to be superior in this respect. Inosine and Hy are relatively stable and therefore can be used to assess the freshness of several kinds of processed fish. Despite these advantages the analysis of these compounds entails destruction of a small sample of fish, takes about an hour at least and is somewhat complex. The extent to which inosine and Hy are used as quality indices in industry is not known but they are now often used for this purpose in research.

The concentrations of some other compounds such as sugars and sugar phosphates also change as a result of enzyme actions during the period before bacterial attack but have not been found to be of general usefulness as quality indices.

Lipid oxidation leading to rancid odours and flavours is a main type of deterioration occurring in frozen fish. Some old established chemical methods of measuring lipid oxidation are still being used. Among these are the determination of peroxide value and of thiobarbituric acid value (which measures essentially the concentration of malonaldehyde). In some cases these determinations may have some value but their general application has not been firmly established.

All the chemical or biochemical methods so far investigated suffer from the fact that they are not unequivocally or causally linked with sensory responses. More often than not the concentration of the substance measured changes in a manner which is only adventitiously similar to the change in sensory attribute. What is required for any ideal method of this kind is the identification of the substance which is sensorily important in the product under examination, the establishment of the relationship between the concentration of this substance and the magnitude of the sensory response and finally the measurement of the substance in the product. None of these requirements has been brought together in a single method. Furthermore, it is highly unlikely that a single substance will adequately measure all the sensorily important changes that occur in spoiling or deteriorating fish and thus more than one substance would in general need to be measured in order to assess quality fully. It is debatable whether research should continue on the search for new or improved chemical or biochemical methods until such causal relationships have been established. The task of doing so is formidable but in our view no significant advances in this area will be achieved until the underlying understanding of odour and flavour changes has been achieved.

Another example of a non-causally-related chemical test is the measurement of salt-soluble protein in frozen stored products. This measure correlates reasonably well with the firmness or toughness of the product, both being linked in some complex fashion through the denaturation and cross-linking of the flesh proteins.

There are indications that some substances may bear some relationship to sensory properties. TMA has a 'fishy' odour and perhaps contributes to the predominant character of bacterially spoiled fish. Hy is said to be bitter; the flavour of some spoiled fish has a bitter note. Inosine-5-monophosphate enhances sweet meaty flavours and its disappearance during spoilage may have important sensory consequences. The accumulation of low concentrations of volatile sulphur compounds during spoilage is likely to have considerable effect on the nature of the overall odour in view of their very low sensory thresholds (Herbert et al, 1975). Only a few amino acids appear to be important for flavour in the major types of fish product and their status as potential indices of quality is uncertain. A very specific association has been found between the concentration of the substance hept-cis-4-enal and the cold storage off-flavour which develops in frozen white fish (McGill et al, 1974). This appears to be a causal relationship so that in this instance the measurement of the concentration of a single substance could provide an excellent non-sensory assessment of quality. Unfortunately, the determination of the substance is by no means easy. A number of recent attempts have been made to identify by gas-liquid chromatography or mass spectrometry the very large number of compounds present in spoiled or deteriorated fish (for example, McGill et al, 1977); developments in this direction will be watched with interest.

To summarize the status of chemical and biochemical methods: the general view is that at present none can be relied upon by itself to give a wholly satisfactory measure of quality but they are useful as adjuncts to or confirmation of sensory results. It is interesting that so far no non-sensory method of assessing spoilage or deterioration has been included in any international standard.

3.1.2 *Physical methods* Those which in the past have been used for assessing spoilage include pH, buffering capacity, refractive index and opacity of eye lens or eye fluid, firmness of flesh, viscosity of flesh homogenate, fluorescence. None of these has proved particularly useful.

A promising new approach was provided by the finding that the electrical properties of the skin and underlying musculature of fish change in a regular manner during spoilage. This principle was eventually embodied in an ingenious instrument which measured the electrical impedance across the body of a fish at two different frequencies and displayed the ratio on a dial (Hennings, 1965). The design was, however, not entirely suitable for many industrial situations. Further developments along somewhat different lines have led to a more practical instrument (Jason and Richards, 1975; Cheyne, 1975) which is finding fairly widespread use in industry and for official inspection. The great advantage of such an instrument is that it gives an instantaneous reading which is obtained without destroying or affecting the sample. Single readings with it on individual fish do not give a very accurate measure of quality as determined sensorily, but it is particularly useful when assessing the average quality of a batch of 10–20 fish. Recent applications of this instrument are described later in this conference.

Several other physical tests or semi-physical methods for measuring spoilage or deterioration have been proposed over the past decade or so but none has been taken up on a lasting basis. Among these are the measurement of degree of contractility of muscle fibres on the addition of ATP as a test of protein or textural deterioration in frozen fish (Partmann, 1969); moisture loss as a criterion of freshness or texture in abalone (James and Olley, 1970) and shrimp (Shelef and Jay, 1971); firmness of cooked minced flesh as a measure of toughness in frozen fish (Main et al, 1972); intensity of peaks in the X-diffraction pattern of frozen fish as a measure of pre-freezing freshness (Niwa and Miyaki, 1971); spectrophotometric detection of rancidity in frozen fatty fish (Danopoulos and Ninni, 1972); shear values and water holding capacity as measure of texture deterioration in lobster (Dagbjartsson and Solberg, 1973) and in rockfish (Patashnik et al, 1976); viscosity of homogenates as a measure of textural deterioration in frozen fish (Nishimoto and Love, 1974); optical density measurements on homogenates of frozen fish in relation to textural deterioration (the cell-fragility method: for recent development see Whittle, 1975).

As with most chemical or biochemical methods, the physical methods so far investigated suffer from the major disadvantage that they are not causally related to sensory impressions. Some physical methods neverthe-less probably come close to being so related. Thus, the elasticity of the Japanese product kamaboko can be simply assessed by folding a disc of the material; if cracks appear, the product is of lower quality. More accurately, the gel strength measured by a penetrometer gives a good assessment of elastic quality. Firmness and waterholding capacity of fish flesh as measured instrumentally also may be related to the sensory impressions of toughness and dryness. Further detailed investigation of the relationships between rheological or physical properties and sensory impressions is needed, however, before fully satisfactory physical methods can be designed.

3.1.3 *Bacteriological methods* Since bacteria are the main agents of spoilage it would seem logical to use their numbers as an index of quality. Lerke and Farber (1969) were the most recent investigators to look into this possibility. They developed a simple technique for direct counting of bacteria and were able to differentiate with confidence between unspoiled and spoiled fish on the basis of bacterial numbers alone. Despite its promise and use by at least one purchasing agency to control the quality of lots, further experience with the technique has not been reported.

More precise estimations of bacterial numbers can be obtained by incubation techniques but these are at present too time-consuming and expensive to be of value for routine industrial use. Developments in rapid counting methods based on impedance changes in bacterial cultures may change this position (Richards et al, 1978).

Bacterial counts are, of course, not directly related to sensory properties and cannot give useful information in the early stages of spoilage. Furthermore, the sample of fish is normally destroyed at some stage in the measurement.

3.2 *Safety and wholesomeness*
There is considerable concern in the major fish processing companies to turn out products which are bacteriologically safe. Testing for standard plate count and for certain pathogens are prominent features of most industrial quality assurance programmes. Over the past few years there has been growing consultation and collaboration between practising and research bacteriologists particularly through the aegis of FAO, various Codex Alimentarius Committees and Expert Consultations. Considerable progress has been made in establishing the general principles for microbial criteria for foods, including fish products, particularly those involved in international trade, in order to protect the consumer and meet the requirements of fair practices. Such criteria as defined for Codex purposes, for example, consist of a statement of the micro-organisms of concern (bacteria, viruses, yeasts, moulds, parasites) and for their toxins, the analytical methods for their detection and quantification and statistical considerations of the number of field samples to be withdrawn, size of sample units and proportion of sample units conforming to the limits considered applicable.

Accordingly, consideration has to be given, among other things, to the microbiology of the raw materials, the effect of processing on the microbiology, including possible microbial contamination and growth during subsequent handling and storage. These criteria, all of which apply to fish and fishery products, can be grouped into three categories (FAO/WHO, 1977).

(*1*) A microbiological guideline, where no standard or code of practice exists, which should be established only when a microbiological criterion is urgently required for a product moving in international trade.

(*2*) A microbiological specification, which is attached to a code of practice, is of an advisory nature and intended to increase assurance that the hygienic provisions of the Code have been adhered to.

(*3*) A microbiological standard which is attached, for example, to a Codex Alimentarius Standard, which is mandatory and is intended primarily for use in the case of disputes.

At present so far as is known no microbiological standards are in operation for any fish or fishery products; but microbiological specifications are constantly being used by the larger food processing and distributing firms or other agencies purchasing fish or ingredients for fish products from each other or by regulatory authorities concerned with imports of fish or fishery products into areas under their jurisdiction.

Undoubtedly one of the main reasons for the production of microbiological specifications, certainly in the UK, is the occurrence of food poisoning outbreaks traceable to foods in international trade. Thus, as a result of several cases of food poisoning traceable to imported cooked shrimps and prawns into the UK, specifications were agreed upon by the importers and the public health authorities. Such specifications reinforce the tenet that there must be some justification for them such as the prevention of food poisoning. Indeed some assert it is their only justification; others maintain that additional substantial benefits can also accrue, such as a general rise in product quality; a reduction in faulty end products and of losses to industry concerned; a product of better value and of increased safety to the consumer.

All these aspects are best summarized by the concept of Good Commercial Practice (GCP) or Good Manufacturing Practice (GMP). This appears to be increasingly accepted by surveillance agencies in major importing countries such as USA and Canada. The concept involves not only raw materials, proper hygiene of the premises, factory, stores, personnel and the process itself, but ensures that the correct manufacturing process is used along with proper storage facilities and adequate control over sale and distribution. In the UK, at least, these concepts have been endorsed by some of the major food processors and distributors (Goldenberg and Elliott, 1973; Baird-Parker, 1974).

The official approach in the UK to microbiological specifications is that where adequate legislation is available to enforce standards of cleanliness and codes of good commercial practice, the need for microbiological standards or specifications of the final product can be reduced to a minimum or even be eliminated. As Sir Graham Wilson stated in 1970, when he was Director of the Public Health Service in the UK 'control of the process (and one might add knowledge of its critical control points) is of far greater importance than examination of the finished product'.

However, there has been an increasing desire, particularly among surveillance agencies, for microbiological specifications or standards for products in international trade but it must be recognized that the devising of meaningful and practical numerical values gives rise to many difficulties. In general, such specifications for fish and fishery products include Standard Plate Counts, counts for so called indicator organisms such as faecal coliforms and *Staphylococcus aureus* and the presence or absence of hazardous organisms such as *Salmonella* spp, *Clostridium botulinum* and *Vibrio parahaemolyticus*.

At present there are at least two major problems which arise in the implementation of microbial specifications, namely those of sampling and of standardizing the microbial methods used. Both of these problems are fully discussed in recent publications of the International Commission for the Microbiological Specifications for Foods (1974; 1978).

It is generally agreed that the Standard Plate Count has certain merits for indicating whether good commercial practice has been followed and that the numbers of indicator organisms (faecal coliforms; *E. coli* or *Staphylococcus aureus*) give information on the quality of factory and personnel hygiene. The presence of food poisoning pathogens such as *Salmonella* spp or *Cl. botulinum* is considered to be unacceptable, although in practice it may be virtually impossible to guarantee their complete absence and the most that can be achieved is to minimize the risks involved through adequate processing, handling and subsequent cooking.

Examples of suggested specifications for some of the more important fish and fishery products in international trade have been listed in the ICMSF publications referred to earlier.

With the growing utilization of scombroid fishes like mackerel there has been renewed interest in the problem of scombroid poisoning and the role of histamine in it (Arnold and Brown, 1978). Several new or modified methods for the determination of histamine have been suggested, but it seems that a measure of the substance alone may not give unequivocal reassurance or otherwise of safety.

3.3 *Composition and identity*

Determinations of gross chemical composition (moisture, protein, lipid and ash) are a main feature of industrial quality control operations. The purposes of the analyses are to provide information on the suitability of lots for different kinds of process and to ensure that certain compositional standards such as minimum fish content are adhered to. The main advance in this field over the years has been the invention of reliable machines and instruments which provide more rapid and cheap results on numerous samples. The ideal of a cheap instrument which measures these parameters instantaneously under industrial conditions is still elusive.

With the increase in the production of frozen fish, the practice of glazing has been growing. In order to protect the consumer against abuses arising from the addition of excessive amounts of water as glaze to retail products it has been necessary to develop improved methods for the determination of glaze. Whilst none of the current methods is entirely satisfactory one at least provides a reasonable basis for standardization (Codex Alimentarius, 1979).

The bone content of fish products has become of increasing importance with the greater use of minced or recovered fish flesh and with the growing formulation of standards. A variety of methods for measuring the total bone content of products have been proposed based on flotation (Dambergs and Regier, 1970; Patashnik *et al*, 1974), dissolution of the non-bone constituents in urea (Yamamoto and Wong, 1974) or alkali (Bon, unpublished) and by proteolysis with papain (Howgate, unpublished). There is, however, doubt as to whether within limits the total bone content as determined non-sensorily is a quality factor necessarily of industrial or consumer importance. What is more to the point is the number of bones that can be determined by sensory means.

There has been recent debate over the best method to measure the fish content of products such as frozen battered and breaded portions. The method of stripping off the coating after thawing is simple, convenient and rapid (Werren and Allhands, 1971) but appears (Analytical Methods Committee, 1978) not to be as accurate as the determination of flesh content by the old established method of Stubbs and More (1919). On the other hand the results with the latter have been claimed to be apparently superior only by chance and that the stripping method is on balance to be favoured (Aitken and Howgate, unpublished). Analysis for basic composition of the batter removed in the stripping method has been advocated as a means of adding precision (Flemming, 1971).

The large number of fish species in commercial use and some evidence that an unacceptably high proportion of cheaper varieties are sold as more expensive has led to an upsurge of interest in objective methods of identifying the nature of fish in products where gross morphological features are lacking. All of these methods are based on electrophoretic identification of the constituent muscle proteins or fragments of these. The technique is being extended to products such as canned fish and dried fish in which the protein is denatured. Investigations in the past decade includes Learson, 1970; Mackie, 1972; Mackie and Taylor, 1972; Coduri and Rand, 1972; Morel, 1977; Kokuryo and Seki, 1978. Further applications are described during the course of this conference.

Colour is an important quality attribute the non-sensory measurement of which has received relatively little attention as far as fish products are concerned. A reflectance system for the canned tuna industry has been developed (Little, 1969a,b,c) and used in an investigation of consumer preference (Rasekh *et al*, 1970). In view of the importance of colour to consumers and the growing use of a wide variety of fish species in products, the measurement of this attribute deserves more research effort.

4 References

AMERICAN SOCIETY FOR TESTING MATERIALS. *Basic Principles of Sensory*
1968   *Evaluation*. American Society for Testing Materials, Philadelphia. 105pp
AMERINE, M A, PANGBORN, R M and ROESSLER, E B. *Principles of*
1965   *sensory evaluation of food*. Academic Press Inc, London and New York. 602pp

AMU, L and DISNEY, J G. Quality changes in W. African marine fish
1973   during iced storage. *Tropical Sci.*, 15, 125–138
ANALYTICAL METHODS COMMITTEE. Determination of fish content of
1978   coated fish products. *Analyst*, 103, 971–973
ARNOLD, S H and BROWN, W D. Histamine (?) toxicity from fish pro-
1978   ducts. *Adv. Fd. Res.*, 24, 114–154
BAINES, C R, CONNELL, J J, GIBSON, D M, HOWGATE, P F, LIVING-
1969   STON, E I and SHEWAN, J M. A taste panel technique for evaluating the eating quality of frozen cod. In: *Freezing and Irradiation of Fish*, Ed R Kreuzer. Fishing News (Books) Ltd, London. 528pp
BAIRD-PARKER, A C. The production of safe foods. *Post-graduate*
1974   *Medical Journal*, 50, 644–647
BOTTA, J R and SHAW, D H. Chemical and sensory analysis of rough-
1975   head grenadiers stored in ice. *J. Fd. Sci*, 40, 1249–1252
BOTTA, J R, NOONAN, P B and LAUDER, J T. Chemical and sensory
1978   analysis of ungutted offshore capelin stored in ice. *J. Fish. Res. Bd. Can.*, 35, 976–980
BRANDT, M A, SKINNER, E Z and COLEMAN, J A. Texture profile
1963   method. *J. Fd. Sci.*, 28, 404–409
BRITISH STANDARDS INSTITUTION. *Glossary of terms relating to sensory*
1975   *analysis of foods*. BSI 5098.
BRUNN, A, IKKALA, P and SORENSEN, T. Effect of frozen storage on
1976   the functional properties of separated fish mince. In: *The production and utilisation of mechanically recovered fish flesh*, Ed J N Keay. Torry Research Station, Aberdeen. 108pp
BURT, J R. Hypoxanthine: a biochemical index of fish quality. *Process*
1976   *Biochem.*, 11, 17–19
BURT, J R, GIBSON, D M, JASON, A C and SANDERS, H R. Compari-
1976a   son of methods of freshness assessment of wet fish II. *J. Fd. Technol.*, 11, 73–89
——— III. *J. Fd. Technol.*, 11, 117–128
1976b
CAIRNCROSS, S E and SJÖSTRÖM, L B. Flavour profiles – a new
1950   approach to flavour problems. *J. Fd. Technol.*, 4, 308–311
CASTELL, C H, SMITH, B and DYER, W J. Simultaneous measurement
1974   of trimethylamine and dimethylamine in fish and their use for estimating quality of frozen stored gadoid fillets. *J. Fish. Res. Bd. Can.*, 31, 383–389
CHARM, S E, LEARSON, R J, RONSIVALLI, L J and SCHARTZ, M.
1972   Organoleptic technique predicts refrigerated shelf life of fish. *J. Fd. Technol.*, 26, 7, 65–68
CHANG, G W, CHANG, L W and LEW, K B K. Trimethyl-
1976   amine – specific electrode for fish quality control. *J. Fd. Sci.*, 41, 723–724
CHEYNE, A. How the GR Torrymeter aids quality control in the fishing
1975   industry. *Fish News Intern.*, 14, 71–76
COBB, B F, ALANIZ, I and THOMPSON, C A. Biochemical and micro-
1973   biological studies on shrimp: volatile nitrogen and amino nitrogen analysis. *J. Fd. Sci.*, 38, 431–436
COBB, B F and VANDERZANT, C. Development of chemical test for
1975   shrimp quality. *J. Fd. Sci.*, 40, 121–124
COBB, B F, THOMPSON, C A and CURTER, C S. Chemical characteris-
1973   tics, bacterial counts and potential shelf life of shrimp from various locations on the NW Gulf of Mexico. *J. Milk and Fd. Technol.*, 30, 463–468
CODEX ALIMENTARIUS. Draft standards of frozen fillets for various
1979   kinds.
CODURI, R J and RAND, A G. Vertical plate gel electrophoresis for
1972   differentiation of fish and shellfish species. *J. Assoc. Offic. Anal. Chem.*, 55, 464–466
CONNELL, J J. A simple colour test for cold storage deterioration in
1966   cod. *J. Sci. Fd. Agric.*, 17, 329–332
CONNELL, J J and HOWGATE, P F. Consumer evaluation of fresh and
1971   frozen fish. In: *Fish Inspection and Quality Control*, Ed R Kreuzer Fishing News (Books) Ltd, London. 290pp
CONNELL, J J, HOWGATE, P F, MACKIE, I M, SANDERS, H R and
1976   SMITH, G L. Comparison of methods of freshness assessment of wet fish IV. *J. Fd. Technol.*, 11, 297–308
COSGROVE, D M. Rapid method for estimating ethanol in canned sal-
1978   mon. *J. Fd. Sci.*, 43, 641–643
DAGBJARTSSON, B and SOLBERG, M. Parameters of texture change in
1973   frozen stored cooked lobster tail meat. *J. Fd. Sci.*, 38, 242–245
DAMBERGS, N and REGIER, L W. Estimation of bone material in fish
1970   protein concentrate. *J. Fish. Res. Bd. Can.*, 27, 591–595
DANOPOULOS, A A and NINNI, V L. Detection of frozen fish deteriora-
1972   tion by UV spectrophotometric method. *J. Fd. Sci.*, 37, 649–651
DRAKE, B and JOHANNSON, B. Sensory evaluation of food. Annotated
1974   bibliography supplement I and II. *SIK Rapport* No. 350
EBER, W. A chemical measure of putrefaction. *Zeit. Fleisch und*
1891   *Milchhygiene*, 1, 118–119
EEC REGULATION No. 2445/70. Determining common marketing
1970   standards for certain fresh or chilled fish
EHIRA, S and UCHIYAMA, H. Formation of inosine and hypoxanthine in
1973   fish muscle during storage in ice. *Bull. Tokai Reg. Fish. Res. Lab.*, 75, 63–73
EHIRA, S, UCHIYAMA, H, UDA, F and MATSUMIYA, H. A rapid method

1970 for determination of the acid-soluble nucleotides in fish muscle by concave gradient elution. *Bull. Jap. Soc. Sci. Fish.,* 36, 491–496

FAO/WHO. Microbiological specifications for foods. Report of 2nd
1977 Joint Expert Consultation: Geneva

FARBER, L. *Freshness Tests in Fish as Food Vol. IV.* Academic Press
1965 Inc., London and New York. 65–126

FARBER, L and LERKE, P. Colourimetric determination of volatile
1968 reducing substances. *J. Fd. Sci.,* 32, 616–617

FLEMMING, R. Towards a uniform study and evaluation of fish fingers.
1971 *Archiv. für Lebensmittel – hygiene* 22, 154–157

GAJEWSKA, R. Changes in objective indices of cod freshness during
1977 storage. *Przemysl Spozywczy,* 31, 397–399

GOLDENBERG, N and ELLIOT, D W. The value of agreed non-legal
1973 specifications. In: *The Microbiological Safety of Foods,* Eds B C Hobbs and J M B Christians, Academic Press Inc, London and New York

GOULD, E and PETERS, J A. *On Testing the Freshness of Frozen Fish.*
1971 Fishing News (Books) Ltd, London. 80pp

GRUGER, E H. Chromatographic analysis of volatile amines in marine
1972 fish. *J. Agr. Fd. Chem.,* 20, 781–785

GUARDIA, E J and HAAS, G J. Evaluation of muscle hypoxanthine and
1969 volatile bases as potential quality indices for industrial bottom fishes from the Gulf of Mexico. *Fishery Industrial Res.,* 5, 117–120

HENNINGS, C. In: *The Technology of Fish Utilization,* Ed R Kreuzer.
1965 Fishing News (Books) Ltd, London. 154

HERBERT, R D, ELLIS, J R and SHEWAN, J M. Isolation and identifica-
1975 tion of the volatile sulphides produced during chill storage of North Sea cod. *J. Sci. Fd. Agric.,* 26, 1187–1194

HERBORG, L and VILLADSEN, A. Bacterial infection/invasion in fish
1975 flesh. *J. Fd. Technol.,* 10, 507–513

HOWGATE, P F. *Aspects of Fish Texture in Sensory Properties of Foods,*
1978 Eds G C Birch, J G Brennan and K J Parker, Applied Science Publishers, London. 326pp

HOWGATE, P F, MACKIE, P R, WHITTLE, K J, FARMER, J, McINTYRE,
1977 A D and ELEFTHERIOU, A. Petroleum tainting in fish Rapp. P.-v. Reun. Cons, int. Explor. Mer., 171, 143–146

INTERNATIONAL COMMISSION ON MICROBIOLOGICAL SPECIFICATIONS FOR
1974 FOODS (ICMS). *Micro-organisms in Foods I,* Eds F S Thatcher and D S Clark, University of Toronto Press, Toronto

—— II
1978

IYER, H K. Methods for sensory evaluation of quality and application
1972 of statistical methods to sensory evaluation problems with special reference to fishery products. *Fishery Technol.,* 9, 104–108

JAMES, D G and OLLEY, J N. Moisture and pH changes as criteria of
1970 freshness in abalone and relationship to texture of the canned product. *Fd. Technol. in Australia,* 22, 350–357

JASON, A C and RICHARDS, J C S. The development of an electronic
1975 fish freshness meter. *J. Phys. E. Sci. Instr.,* 8, 826–830

KARNICKA, B and JUREWICZ, I. *Changes in content of trimethylamine-
1971 oxide, trimethylamine and total volatile base nitrogen in fresh and frozen fish.* Prace Morskiego Instit. Rybackiego B 16, 193–203

KARNOP, G. Local distribution of total volatile bases in tissues of whole
1976 fish during storage in ice. *Archiv. für Fischereiwiss,* 27, 159–169

KATO, N and UCHIYAMA, H. An automatic analysis of trimethylamine
1973 in fish muscle. *Bull. Jap. Soc. Sci. Fish.,* 39, 899–903

KEAY, J N and HARDY, R. The separation of aliphatic amines in dilute
1972 aqueous solution by gas chromatography and application of this technique to quantitative analysis of tri- and dimethyl amines in fish. *J. Sci. Fd. Agric.,* 23, 9–19

KIETZMANN, V. Low temperature preservation of sea fish. *Archiv. für
1968 Lebensmittel – hygiene* 19, 238–244

—— Evaluation of quality of frozen fish and fish portions. In: *Freezing
1969 and Irradiation of Fish,* Ed R Kreuzer. Fishing News (Books) Ltd, London. 528pp

KOKURYO, H and SEKI, N. Identification of fish species in dried fish
1978 sticks. *Bull. Jap. Soc. Sci. Fish.,* 44, 67–70

KVALE, O. Various methods for determining microbiological spoilage
1973 of fish and fish products. *Tidsskrift for Hermetik-industri,* 59, 177–179

LAMB, C W. High School student's perception of fish as a menu item.
1975 *Marine Fisheries Rev.,* 37, 25–27

LEARSON, R J. Collaborative study of a rapid electrophoretic method
1969 of fish species identification. I. *J. Assoc. Offic. Anal. Chem.,* 52, 703–704

—— II. 53, 7–9
1970

LEARSON, R J and RONSIVALLI, L J. New approach for evaluation of
1969 quality of fishery products. *Fishery Industrial Res.,* 4, 249–259

LERKE, P and FARBER, L. Direct bacterial count as a rapid freshness
1969 test for fish fillets. *Appl. Microbiol.,* 17, 197–201

LERKE, P A and HUCK, R W. Objective determination of canned tuna
1977 quality: identification of ethanol as a potentially useful index.

1977 quality: identification of ethanol as a potentially useful index. *J. Fd. Sci.,* 42, 755–758

LOVE, R M. The use of tasters for investigating cold storage deteriora-
1966 tion in frozen fish. *J. Fd.Technol.,* 1, 141–146

LITTLE, A C. Reflectance characteristics of canned tuna. I. *Fd. Tech-
1969a nol.,* 23, 1301–1304

—— II. *Fd. Technol.,* 23, 1466–1468
1969b

—— III. *Fd. Technol.,* 23, 1468–1472
1969c

LUIZZO, J A, LAGARDE, S C, GRODNER, R M and NOVAK, A F. Total
1975 reducing substances test for ascertaining oyster quality. *J. Fd. Sci.,* 40, 125–128

McGILL, A S, HARDY, R, BURT, J R and GUNSTONE, F D. Hept-cis-4-
1974 enal and its contribution to the off-flavour of cold stored cod. *J. Sci. Fd. Agric.,* 25, 1477–1489

McGILL, A S, HARDY, R and GUNSTONE, F D. Further analysis of vol-
1977 atile components of frozen cold stored cod and influence of these on flavour. *J. Sci. Fd. Agric.,* 28, 200–205

MACKIE, I M. Some improvements in the polyacrylamide disc electro-
1972 phoresis method of identifying the species of cooked fish. *J. Assoc. Public Anal.,* 10, 18–20

MACKIE, I M and TAYLOR, T. Identification of species of heat-sterilised
1972 canned fish by polyacrylamide disc electrophoresis. *Analyst,* 97, 609–611

MAIN, G, ROSS, R I and SUTTON, A H. A texturometer for measuring
1972 the toughness of cooked fish. *Lab. Practice,* 21, 185–188

MEITZ, J L and KARMAS, E. Chemical quality index of canned tuna as
1977 determined by HPLC. *J. Fd.Sci.,* 42, 155–158

—— Polyamine and histamine content of rockfish, salmon, lobster and
1978 shrimp as an indicator of decomposition. *J. Assoc. Offic. Anal. Chem.,* 61, 139–145

MILLER, A, SCANLON, R A, LEE, J S and LIBBEY, L M. Quantitative
1972 and selective gas chromatography of dimethyl and tri-methylamines in fish. *J. Agr. Fd. Chem.,* 20, 709–711

MILLS, A. Measuring changes that occur during frozen storage of fish:
1975 a review. *J. Fd. Technol.,* 10, 483–496

MOREL, M. Identification of fish species by gel electrophoresis using
1977 iso-electric focussing. *Science et Pêches,* 275. Information l'Institut Scientifique et Technique des Pêches Maritimes, 1

MOWLAH, G. Muscle volatile substances for chemical assessment of
1975 spoilage for industrial prawns of the River Dakatia during iced storage. *Bangladesh J. Biol. Sci.,* 4, 23–25

MURRAY, C K and GIBSON, D M. An investigation of the method of
1972 determining trimethylamine in fish muscle extracts by forma-tion of the picrate salt. I. *J. Fd. Technol.,* 7, 35–46

—— II. 47–51

NAIR, R B, THARAMANI, P K and LOBURG, N L. Studies on chilled stor-
1971 age of freshwater fish. *J. Fd. Sci. Technol.* (Mysore), 8, 53–56

NERISSON, P. Histamine as indicator of spoilage of fish. *Revue des
1975 Travaux,* Institute des Pêches Maritimes, 39, 471–482

NISHIMOTO, J and LOVE, R M. Study of viscosity of homogenates of
1974 thawed cod muscle in 65% sucrose solution. *Bull. Jap. Soc. Sci. Fish.,* 40, 1071–1076

NIWA, E and MIYAKI, M. Studies on frozen foods. 1: Application of
1971 X-ray diffraction method for estimating quality of frozen fish. *Bull. Jap. Soc. Sci. Fish.,* 37, 163–165

OHTA, F, KIKUCHI, H and ISHIGAMI, T. Nucleotides and volatile bases
1975 as quality indices of iced fish. *Mem. Fac. Fish. Kagoshima Univ.,* 24, 173–179

PARTMANN, W. Experiments on the determination of quality of frozen
1969 cod. *Kaltetechnik.,* Klimat. 21, 325–328

PATASHNIK, M, KUDO, G and MIYAUCHI, D. Bone particle content of
1974 some minced fish muscle products. *J. Fd. Sci.,* 39, 588–591

PATASHNIK, M, MIYAUCHI, D and KUBO, G. Objective analysis of tex-
1976 ture of minced black rockfish during frozen storage. *J. Fd. Sci.,* 41, 609–611

PAULUS, K, GUTSCHMIDT, J and FRICKER, A. Karlsruhe evaluation
1969 scheme—development, employment and modification. *Lebensmitt.-Wiss. u Technol.,* 2, 132–139

PEARSON, D and MUSLEMUDDIN, M. The accurate determination of
1969a total volatile nitrogen in meat and fish. I. *J. Assoc. Publ. Anal.,* 7, 50–54

—— II. *J. Assoc. Publ. Anal.,* 7, 73–82
1969b

PERYAM, D R and GIRARDOT, N F. Advanced taste-test method. *Fd.
1952 Engin.,* 24, 7, 58–61

RASEKH, J, KRAMER, A and FINCH, R. Objective evaluation of canned
1970 tuna sensory quality. *J. Fd. Sci.,* 35, 417–423

REINACHER, E. Quality of fish in relation to constituents which may be
1977 determined by GLC. *Information für die Fischwirtschaft,* 24, 234–236

RICHARDS, J C S, JASON, A C, HOBBS, G, GIBSON, D M and CHRISTIE,
1978 R H. Electronic measurement of bacterial growth. *J. Phys. E. Sci. Instrum.,* 11, 560–568

RITSKES, T M. Gas chromatographical determination of tri-
1975 methylamine and dimethylamine in fish, fishery products and other foodstuffs. *J. Fd. Technol.,* 10, 221–228

64

RUITER, A and WESEMAN, J M. Automated determination of volatile
1976 bases (trimethylamine, dimethylamine and ammonia) in fish
and shrimp. *J. Fd. Technol.*, 11, 59–68
SHELEF, L A and JAY, J M. Hydration capacity as an index of shrimp
1971 microbiological quality. *J. Fd. Sci.*, 36, 994–997
SHEWAN, J M, MACINTOSH, R G, TUCKER, C G and EHRENBERG,
1953 A S C. The development of a numerical scoring system for the
sensory assessment of the spoilage of wet white fish stored in
ice. *J. Sci. Fd. Agric.*, 4, 283–298
SIDHU, G S, MONTGOMERY, W A and BROWN, M A. Postmortem
1974 changes and spoilage in rock lobster muscle. II. *J. Fd. Technol.*, 9, 371–380
SOUDAN, F. Organoleptic changes in frozen fish. *Revue Génerale du
1969 Froid et des Indus. Frig.*, 60, 513–527
SOUDAN, F. Organoleptic changes in frozen fish. *Revue Génerale du
1957 Froid et des Indus. Frig.*, 60, 513–527
SOUDAN, F. Normes des fraîcheur du poisson frais. *Annales de la Nutrition
et de l'Alimentation*, 9, No. 1
SPINELLI, J. Biochemical basis of fish freshness. *Process Biochem.*, 6,
1971 36–37
STUBBS, G and MORE, A. The estimation of the approximate quantity
1919 of meat in sausages and meat pastes. *Analyst*, 44, 125–127
TAGLE, M A. *Acceptability testing of FPC type B.* Report FAO/
1976 TF/INT 120 (NOR)–Phase 1. FAO, Rome. 154pp
TAGLE, M A, VOLAND, S and JAMES, D G. Acceptability testing of
1976 FPC type B in selected developing areas. In: *Handling, processing and marketing of tropical fish.* Tropical Products Institute,
London. 511pp
TOKUNAGA, T, IIDA, H and MIWA, K. G.C. Analysis of amines in fish.
1977 *Bull. Jap. Soc. Sci. Fish.*, 43, 219–227
TROMSDORFF, H. Evaluation of fish quality. *Inform. für die Fisch-
1973 wirtschraft*, 20, 138–142
TROMSDORFF, H. Importance of volatile aroma compounds in fish.

1975 *Deutsche Lebensmittel–Rundschau*, 71, 201–207
UCHIYAMA, H. Freshness of fish: current studies and applications. *J.
1971 Fd. Hygiene Soc. Jap.*, 12, 267–276
UCHIYAMA, H, EHIRA, S, KOBAYASHI, H and SHIMIZU, W. Significance
1970 of measuring volatile bases and trimethylamine nitrogen and
nucleotides in fish muscle as indices of freshness of fish. *Bull.
Jap. Soc. Sci. Fish.*, 36, 177–187
VYNCKE, W. Determination of ammonia content of fish as an objective
1970 quality assessment method. *Medel. van de Facult. Landbouw.
Rijksimir. Gent.*, 35, 1033–1046
VYNCKE, W. Determination of ammonia in dressed thornback ray as a
1978 quality test. *J. Fd. Technol.*, 13, 37–44
WEBB, N B, HOWELL, A J, BARBOUR, B C, MONROE, R J and
1975 HAMANN, D D. Effect of additives, processing techniques and
frozen storage on the texture of shrimp. *J. Fd. Sci.*, 40,
322–326
WERREN, J C and ALLHANDS, F H. Collaborative study of methods for
1971 determining the fish content of frozen breaded fish products.
J. Assoc. Offic. Anal. Chem., 54, 640–642
WHITTLE, K J. Improvements in the Torry-Brown homogenator for
1975 the cell fragility method. *J. Fd. Technol.*, 10, 215–229
WITAS, T and OLBROMSKA, E. Volatile reducing value as a quality
1976 indicator for fish and fish products. *Przemysl Spozywczy*,
30, 146–148
WILSON G. Concluding remarks to symposium on microbiological
1970 standards for foods. *Chem. and Ind.*, 273–274
WITTFOGEL, H. Entwurs eines Punktbewertungsschemas für die Sin-
1958 nenprüfung von Frischfisch. *Arch. Lebensmittelhyg.*, 9,
279–280
YAMAMOTO, M and WONG, F. Simple chemical method for isolating
1974 bone fragments in minced fish flesh. *J. Fd. Sci.*, 39, 1259–1260

Structure and proteins of fish and shellfish. Part 1 *June N Olley*

The authors of the two papers on fish proteins were invited to write a joint paper following their previous collaboration (Sikorski, Olley and Kostuch, 1976); at that time however they were working together in Hobart. Exigencies of postal strikes and pressure of work has meant that in this instance they have only had time to exchange ideas and have thus prepared two manuscripts.

Repetition of subjects already covered in the literature is not a challenge. The structure of fish and shellfish, particularly at the electron microscope level, has been most adequately reviewed by Howgate (1979), muscle contraction and the troponin system by Ebashi (1974), SH and S–S bonds in muscle, albeit of meat, by Hofmann and Hamm (1978) and the changes in fish proteins during cold storage by the authors themselves (Sikorski *et al*, 1976). I have therefore tried to 'see' proteins through Torry's eyes by going through 50 years of Annual Reports and picking out the highlights which are related in any way to the proteins of the edible parts of seafood flesh. After tabling these relevant observations year by year (*Table I*), I followed on from the observations and noted whether the ideas were followed through at Torry or taken up by other laboratories elsewhere in the world. I also wished to find out if any interesting observations had been left lying fallow and forgotten. At first sight some of the observations listed might not seem appropriate in a paper on fish protein, but protein is the main component of fish flesh and is likely to be affected by the phenomena mentioned.

Despite the fact that Torry has been under four different Government Ministries in the last 15 years, the staff has remained remarkably constant and it is interesting in following through the annual reports to see the expertise and dedication of certain people. Things have come and things have gone, but those members of the staff who were forging new fields and disciplines have been undeterred.

Torry has always been divided into groups or sections. In the earlier days there were protein groups, lipid groups, amino acid and extractive groups, *etc*. In the later period the divisions between groups have been delineated more by the nature of the problems than by fundamental biochemistry, bacteriology, *etc*, but grouping has remained. In the late 1950s and early 1960s many seminars were held which helped to give cross fertilization between the earlier groups and in some cases these resulted in a restructuring of the work.

With hindsight, it is very easy to say that more would have come out of some of the earlier work if particular parameters had always been measured. pH is a very obvious example here. This would have made many experiments unwieldably large and would have damped the creativity of individual scientists.

The exciting and intriguing ball of 'why proteins denature and toughen' has passed from group to group. Sometimes the protein chemists seemed to hold the key; for short periods the lipid chemists appeared to hold it, now perhaps more the lipid oxidation team. Bacteriologists have had something to say and now the ball seems to be rather firmly in the biophysicists' court because they can study flesh tissue without destroying it. They are able to study individual water molecules and charge at the molecular level. Thus the biochemist who may be trying to seek a change in one S–S bond could be looking for a 0.5% difference, while the biophysicist may detect a large perturbation. Acker (1963) pointed out that in dried foods lengthy storage experiments could be replaced by measuring the sorption isotherm of the product. It may be possible to replace the storage studies in the frozen state by measurements of specific conductivity and microwave attenuation on the product.

However it is clear that all disciplines must co-ordinate their efforts. It is hoped that by playing a game of snakes and ladders up and down *Table I* that some insights will be achieved. Participants' comments on initial and later observations are given in *Table II*. For my part I would like to highlight four areas in the table and make my points with a few figures from our research in Tasmania.

Two ex-members of Torry, P E Doe and myself, are actively working on drying of abalone in Hobart. Dried abalone is a luxury product highly prized in southeast Asia. *Figure 1* shows how much we still have to learn about the physical properties of dried products.

Australia has a vast coastline, many species of fish and few fish technologists. *Figure 2* shows how we are using the data of R M Love to determine whether we can make general predictions by pulling a few fish out of the sea in one area.

Figure 3 stresses the importance of sorption isotherms in fish protein products and highlights the ease with which solvent:water extracted fish flours can be expected to grow moulds.

Table III indicates that changes during frozen storage of fish are more likely to be limited by electrical charge than by diffusion.

For a comprehensive review of the state of the 'art' in the field of protein technology the reader is referred to Part 2 of the article.

Table I
FIFTY YEARS OF PROTEIN AND RELATED STUDIES AT TORRY RESEARCH STATION
ORIGINAL OBSERVATIONS, LATER STUDIES AND CONTROVERSIAL ISSUES

Year of Annual Report	Initial observation	Ref.	Later observations and comments
F(1930)	Loss of water-holding capacity, salt solubility, and power to gel in the proteins of brine-frozen fish muscle. Loss of glossy pellicle and translucency in the smoke cure from poorly frozen fish.	Reay (1931)	Changes in elasticity, resilience, hardness, particle character, salt solubility and emulsifying capacity of fish minces described by Sorensen (1976).
F(1931)	Recommendations that cold stores for fish should be run at −20°C to −25°)C rather than at −4° C to −12°C to make fish suitable for smoking.	Broadbank (1932)	Temperatures as low as −40°C used in Scandinavia and Japan. Lowering the temperature of a cold store from −17°C to −30°C doubles energy consumption. Poulsen and Jensen (1978) suggest in view of power crisis, higher temperatures and better organization of the flow of foods
F(1931)	Globulin gels in the precipitated state readily denatured at 0°C by high concentrations of salt. Process much slower in distilled water. Globulin gels when frozen in phosphate buffer remain clear but lose salt solubility.	Reay (1932)	
F(1932)	Globulins of haddock muscle most rapidly denatured at −3°C after freezing fish in brine at −20°C, − tested by loss of glossy pellicle in smoke cure and maximum drip loss. 2% salt added to minced muscle prevents drip loss but denaturation proceeds, as in the absence of salt.	Reay (1933)	
F(1933)	Lemon sole protein much more stable than that of other species.	Reay (1934)	Connell (1961) found lemon sole myosin more stable than other fish myosins. Kim *et al* (1977) find that after 17 years' storage myofibrillar proteins are completely insolubilized to 5% salt, but do not indicate if more than one reaction rate is involved. Appreciable quantities of those amino acids present in sole (Jones, 1959) which protect actomyosin from insolubilization in salt solution (Ohnishi *et al*, 1978). Seasonal variation in amino acid composition of fish make it impossible at present to predict which species should be stabilized by their free amino acid pools. Taurine, the most prevalent amino acid in fish muscle (Mackie and Ritchie, 1974), not yet studied for its effect on actomyosin aggregation and shortening.
F(1935)	Separation of 'intracellular' and 'stroma' proteins, estimate of collagen content of fish muscle.	Reay (1936a)	
F(1935)	Poorer pellicle in smoked pre-rigor fish.	Reay (1936b)	Decreased salt solubility of protein as fish enters rigor (Deuticke effect, see Dyer and Dingle, 1961).
F(1935)	Amino acid analysis of cod muscle. 55% of the nitrogen identified.	Sharp (1936)	See Anon* (1959c).
F(1935)	Formaldehyde in the flesh of salted cod and ling, suspected source trimethylamine oxide (TMAO).	Reay (1936c)	Reay considered that high values for formaldehyde not all due to oxidation of amine during analysis. For production of formaldehyde from trimethylamine oxide in fish flesh, see Anon* (1970b), Anon* (1975a), Sikorski *et al* (1976), Kostuch and Sikorski (1977). Connell (1957; 1958; 1962a) discusses the textural properties of dried and freeze-dried fish. Production of formaldehyde not yet been considered as a source of toughening in dried cod. Lall *et al* (1975) have shown that the TMAO splitting enzyme is quite stable up to 60°C.

F(1938)	Rate of freezing of haddock not critical, slow freezing caused some gaping.	Reay (1939)	Rate of freezing once again not considered very critical after 40 years (Cutting, 1977). Slow freezing of those species which use urea as an osmoregulator can lead to ammoniation of flesh, proteins in pH range 8·7–8·8 have extensive swelling properties (Hamoir *et al*, 1973).
F(1939)	Only fish stored one day on ice before freezing will give Grade I smoked products. Prelude to freezing at sea.	Reay (1949)	Salt solubility of proteins begins to fall immediately with iced storage (Love *et al*, 1965). This is contrary to the effects noted in Anon* (1964a).
F(1940–1946)	The war years. Dehydration of minced cooked fish.	Cutting *et al* (1956)	Few later developments, see Herborg *et al* (1974), Aitken *et al* (1967) and Carpenter (1980) for nutritional value.
F(1947)	Lipids of fish muscle bound to protein.	Anon (1949)	Lipids of both mammalian and fish muscle mostly removed from proteins with weakly polar solvents such as acetone (Olley, 1961).
F(1952)	Large scatter in results for protein extractability using the Dyer *et al* (1950) method.	Anon (1953a)	See Anon* (1956a). Asghar and Yeates (1978) review the whole meat literature on the effects of rigor, pH, buffers, *etc* on the extractability of muscle proteins by salt. Some interesting lessons here. Connell (1968) reviews causes of variability in results for fish.
	Identification of species by electrophoresis of proteins extracted from fish muscle with low ionic strength salts.	Anon (1953b)	See Anon* (1967a). Libraries of species built up in South Africa (Moodie and Eva, 1973; 1974); Australia (Bremner and Vail, 1979); USA (Coduri, 1972); India (Devadasan and Nair, 1971); and Spain (Alvarez *et al*, 1974). This country uses densitometry for quantitative assay. Is time approaching for international libraries of species using isoelectric focusing (Gershman, 1979)?
	Study on the alkaline hydrolysis of fish proteins – lengths of peptides.	Anon (1953c)	Temperature rather than alkali strength important in production of ammonia (Nottingham, 1955a). Rate of peptide bond hydrolysis influenced more by alkali concentration than temperature (Nottingham, 1955b). Olley (1972) points out that despite the fact that alkali-treated herring proteins were sometimes found to be toxic, work proceeded 20 years later on alkaline-hydrolysis of fish protein concentrate.
F(1953)	Preparation of 'pure' actin and myosin from cod.	Anon (1954a)	So-called impurities in myosin molecule now found to be intimately concerned with ATPase activity. These small molecular weight components high in phenylalanine. Carp myofillin contains 85 residues per 100 000g (Laki, 1971a).
	Rehydration of air-dried and vacuum-dried cod fillets.	Anon (1954b)	
F(1954)	Amino acid composition of cod actin and myosin.	Anon (1955a)	Molecules resemble those of rabbits and molluscs. Compare Connell and Howgate (1959b) with data for various species (Laki, 1971b).
	Histology of air-dried, vacuum plate-dried and freeze-dried fillets. No biochemical evidence of cross-linking to explain differences in ability to reconstitute in water.	Anon (1955b)	Optimum pore size in the dehydrated fibres for rapid reconstitution (Connell, 1957). Evidence for thioether cross-links and production of lanthionine only under the very severe conditions of heating dehydrated fish at 80°C for three weeks with 33% moisture (Connell, 1958).
F(1955)	Method for increasing reliability of salt solubility method of measuring protein denaturation using dissected myotomes.	Anon (1956a)	Only wide myotomes around No. 12 counting from anterior end used in this estimation. 95% protein then soluble in 5% NaCl. Values as low as 72% if cod in semi-starved condition (Ironside and Love, 1956). Note that in this method pH of the salt is adjusted to pH 7–7·5, with 0·2M bicarbonate, not buffered. Poulter and Lawrie (1977) note that fish are extracted in pH range 6·5–7·0. pH and buffering capacity of fish change during spoilage (Cutting, 1953). Starved fish extracted at a much higher pH than fed fish. Solubility of myosin highest at pH 6·5 (Asghar and Yeates, 1978). Should the system be buffered or natural? Total saline extractable protein nitrogen not well correlated with toughness measurements (Gill *et al*, 1979). Would measurements of troponin or myosin light chains be better?
	Increases in collagen content of muscle when growth slows down.	Anon (1956a)	
	Factors affecting dehydrated fillets.	Anon (1956b)	
F(1956)	6–16% decrease in salt-soluble protein just by freezing and thawing. Toughening can proceed without accompanying fall in soluble protein.	Anon (1957a)	
	New protein in the expressible fluid of frozen cod, component of muscle fibrils?	Anon (1957b)	This protein only 7% of low ionic strength extracts, but 30–60% of the protein of expressible fluids (Connell, 1964a).
F(1957)	Much drip in cod frozen pre-rigor.	Anon (1958a)	

67

	Protein deterioration as measured by salt solubility, a steady process with defined minima.	Anon (1958b)	Desolubilization of cod 'actomyosin' a first order reaction (Love, 1962a). Desolubilization of myosin biphasic: first order reaction followed by a much slower one (Connell, 1962b). The activation energy of protein insolubilization, cell fragility changes and phospholipase activity (Love, 1962a) closer to that of the theoretical specific conductivity of the fish ($91 kJ mol^{-1}$) (Kent and Jason, 1975) than the much lower activation energy of water diffusion. See Anon* (1976a) for implications of diffusion.
	X-ray pictures reveal β-protein, the 'stretched' or denatured α-protein in stored dehydrated fish.	Anon (1958c)	Thermoelastic measurements related to texture of rehydrated cod muscle. Striking changes in the elasticity of molluscan muscle with extent and methods of brining and drying. See *Fig 1* and Young *et al* (1973). Molluscan muscle paramyosin and actin filaments (Olley and Thrower, 1977).
	Acquisition of analytical ultracentrifuge. Studies on molecular weight, shape and properties (aggregation, *etc*) of cod myosin and actin.	Anon (1958d)	
T(1958)	Rates of denaturation of protein in cold store depend on freezing rates and whether ice crystals intra or extra cellular.	Anon (1959a)	At a given temperature when the system is in equilibrium, the vapour pressure after different rates of freezing must be the same, but the initial formation of ice crystals could cause pools of strong salt solutions, which would solubilize actomyosin and carry it along the cell to be subsequently denatured (Love, 1958). The discontinuity in the rate of protein insolubilization below the NaCl eutectic (Love, 1958) is at variance with the calculation of an activation energy. See Anon* (1958b). See also Reay* (1939).
	Development of the 'cell fragility' method for cold storage changes.	Anon (1959a)	Method modified later to eliminate pH effects (Love and Muslemuddin, 1972) and made semi-automatic (Whittle, 1975). When cod and haddock are compared by modified method, both show same rate of reaction despite the fact that haddock does not produce formaldehyde (Mackie and Thomson, 1974). Cementing together of myofibrils as seen in the cell fragility test not caused by the denaturation of actomyosin by formaldehyde. Love and Olley (1965) noted that the slower the insolubilization of actomyosin in cold storage the greater the discrepancy between protein changes measured by salt solubility and cell fragility.
	Temperature of storage between 0–18°C does not affect rehydration of pre-cooked freeze dried cod. Rapid deterioration at 37°C. Toughness of dehydrated fish related to swelling properties and protein solubility. α structure of protein not altered by dehydration itself.	Anon (1959b)	Changes on storage indicate alteration of protein by browning reactions which can occur at negligible RH (Jones, 1962).
	Study of myosin aggregation in the frozen state. Myosin molecule split by 0·5M urea.	Anon (1959c)	Urea liberates nucleic acids, nucleotides, tropomyosin, actin and deaminase from myosin (Connell and Olcott, 1961). These components stabilize the molecule. See Anon* (1960b) and Anon* (1965b).
	Amino acid content of fish flesh proteins from four species found to be very similar.	Anon (1959c)	British food fishes analysed (Connell and Howgate, 1959a). Analysis of fish meals reveals little difference in amino acid composition of widely different species of fish (FAO, 1970).
T(1959)	Less protein denaturation when fish frozen pre-rigor.	Anon (1960a)	The difference in protein solubility which occurs on freezing and thawing pre- and post-rigor fish (Love, 1962c) is maintained during storage, *ie* no change in *rate* of denaturation (Connell, 1964b). Pre-rigor fish should have lower osmalality, so it would be expected that more ice would be formed yet Love (1962b) finds less ice and more water free for diffusion (Storey, 1970).
	Myosin from different species of fish have different stabilities.	Anon (1960b)	Work needs repeating (Connell, 1968) as aggregation rates influenced to some extent by the presence of small amounts of nucleotides and pyrophosphates (Mackie, 1966). See Anon* (1959c).
	Differences in secondary or tertiary structure from mammalian muscle as more rapid rate of attack by trypsin.	Anon (1960b)	
	No changes in sulphydryl groups on cold storage, but loss of ATPase activity.	Anon (1960b)	
T(1960)	Excessive softening of fish at −3°C.	Anon (1961a)	This statement in complete contradiction to usual finding of maximum denaturation around this temperature (Love and Elerian, 1964). Do salt solubility and cell fragility always represent toughening? Cathepsins may be more activated than denaturation at this temperature (Rehbein *et al*, 1978; Konagaya, 1978).

Sensitivity of untrained taste panel to texture changes assessed.	Anon (1961b)	Panel could detect a difference of 3½ weeks cold storage at −14°C (Love, 1966).
Haddock only keep two days before freezing or flesh softens. Lemon sole and plaice keep four days.	Anon (1961c)	See Anon* (1975b). Flatfish have lysozyme in skin which protects against bacterial attack (Murray and Fletcher, 1976).
Conversion of collagen to gelatin during canning of herring studied. Brining inhibits reaction. Skin collagen strengthens muscle.	Anon (1961d)	
Actin does not change on cold storage of fish. Attention fixed on myosin.	Anon (1961e)	
T(1961) Shrinkage of frozen pre-rigor fillets on rapid thawing during winter.	Anon (1962a)	Effects described and reviewed by Jones (1964). Effects minimized by high temperature cold storage with breakdown of ATP. Suggest that energy dissipated as heat rather than used for thaw contraction (McDonald and Jones, 1976). Lactate production and ATP breakdown are accelerated if thaw rigor is permitted (Yamanaka et al, 1978).
Threefold increase in the skin collagen of herring during winter.	Anon (1962b)	Starvation causes the myocommata and skin of cod to thicken, a weaker but thicker collagen accumulates with age (Love et al, 1976).
H-meromyosin more unstable than mammalian H-meromyosin. Actin unchanged in freeze-dried cod. Actomyosin completely insoluble. Myosin 50% extractable. Extractability of actomyosin misleading.	Anon (1962c)	Pyrophosphate + Mg ion is required to split actinmyosin bond. Hydrogen bond breaking solvents, eg urea, to split hydrogen bonds and reducing agents to split disulphide bonds (Connell, 1957; 1962b).
Attempts to allocate lipoprotein distribution amongst cell organelles; myofibrils, mitochondria and sarcoplasmic reticulum.	Anon (1962d)	Abortive study due to lipid hydrolysis during fractionation and 'debris' surrounding myofibrils as seen with electron microscope (Sheltawy and Olley, 1966).
Hydration of proteins. (Contract work with Nottingham University.)	Anon (1962e)	Speculative paper on nature of water in solid proteins. Electronic conductivity of protein can be raised a millionfold by addition of small amounts of electron accepting molecules (Eley and Leslie, 1963). Results too complex for food systems. Kent (1972) points out that multilayers of water may form before all the sites on the monolayer are filled.
T(1962) No protein denaturation by cell fragility method when fish super-cooled to −15°C. Denaturation of muscle slowed down by 10% glycerol.	Anon (1963a)	When no/less ice formed, proteins are more hydrated and bathed in weaker salt solutions, thus less susceptible to denaturation. 10% glycerol reduces area of ice by 9% (Love and Elerian, 1965). However, in the −30°C region more water could mean more diffusion and thus be deleterious.
Rigor above 17°C produces breakdown of connective structures and tough fillets on cooking.	Anon (1963b)	Rigor tensions have different response to temperature to that of beef and pork, but similar to rabbit. Compaction and toughening of muscle above 17°C related to biochemical changes (Burt et al, 1970). Contraction toughening to be distinguished from more closed form of the protein helix near the isoelectric point with lower pH?
Kinetics of myosin denaturation. Long chain fatty acids, particularly oxidized unsaturated ones, increase myosin aggregation.	Anon (1963c)	Observations confirmed in other laboratories, subject reviewed by Sikorski et al (1976). Reaction of malonaldehyde (breakdown product of lipid oxidation) with myosin does not follow simple rate law. Reaction maximum in model systems at −24°C (Poulsen and Lindelov, 1978). Does not agree with decreasing protein denaturation as temperature lowered from −3°C to −30°C. See Anon* (1958b).
T(1963) Spoiled cod denatures less rapidly on cold storage.	Anon (1964a)	Kostuch and Sikorski (1977) would attribute this to TMAO being degraded to TMA during spoilage, it can then not be a precursor of formaldehyde during frozen storage. TMAO breakdown more rapid in iced cod than other species studied (Mackie and Thomson, 1974).
Removal of bound water from proteins at lower temperatures results in more rapid denaturation at higher temperatures.	Anon (1964b)	Osmotic concentrations of the unfrozen aqueous phase at a given temperature will be the same for the fish whatever the freezing history (Kelly and Dunnett, 1969; Storey and Stainsby, 1970). However, low temperature nadirs on freezing with subsequent reactions at higher temperatures have been found to have a deleterious effect on enzymes (Fennema, 1975). Mackenzie (1963) suggests extraction of specially located water molecules from protein helix at low temperatures. Denaturation as measured by cell fragility or gentle blending of muscle cells (Childs, 1973) affected by low temperature nadirs at freezing, but not the total salt solubility of the proteins (Kelly and Dunnett, 1969), indicating as suggested by Childs (1973) that changes may be at the myofibrillar level.

	Hydrogen ion titration curve of washed cod myofibrils. No evidence for loss of titrateable groups at –14°C. Molecular weight of pure cod myosin determined.	Anon (1964c)	Very small but statistically significant increases in the available lysine content of whiting, lemon sole and skate detected after seven months at –8°C. May indicate unfolding of protein chains (Poulter and Lawrie, 1977). Yet a decrease in available lysine in stored cod. Both cod and whiting produce formaldehyde on cold storage. Recovery of added formaldehyde greater when added to previously cold stored fish (Poulter and Lawrie, 1978). Less reactive groups available after cold storage? No change in easily reactable or total SH groups in above four species stored at –8°C.
T(1964)	Colour test for deterioration of frozen cod.	Anon (1965a)	See Anon* (1970b) – colour due to formaldehyde formation. Effect of formaldehyde noted in same year in Japan, but identified there (Tokunaga, 1964).
	ATP and pyrophosphate stabilize cod myosin.	Anon (1965b)	See Anon* (1960b).
	Detergent stable cross-links not formed to any extent in frozen cod. Monothioglycol (reducing and disulphide bond splitting substance) reduces protein denaturation in frozen muscle minces.	Anon (1965c)	
	Electron microscopy reveals degradation of the sarcoplasmic reticulum on iced storage of fish and a reduction in the distance between filaments and myofibrils on cold storage.	Anon (1965d)	Work published in comprehensive review of light and electron microscopy of vertebrate and invertebrate muscle tissue (Howgate, 1979). Author makes a plea that future workers should try to relate the firmness of fish flesh to properties of the cross-striated muscle fibres and the tough and chewy nature of molluscan flesh to the smooth or oblique-striated fibres. He also asks that processing should be followed under the microscope, especially the scanning electron microscope. Japanese have studied cryoprotective effect of amino acids on actomyosin filaments with electron microscope (Ohnishi et al, 1978).
T(1965)	Marked reduction of interfibrillar space on cold storage of cod, interfilament space reduced when muscle has suffered 'freezer-burn'.	Anon (1966)	For descriptive details see Connell (1968), Aitken and Connell (1977) and Howgate (1979). Freezer burn does not change the polyacrylamide gel pattern of fish-extracted with SDS + dithiothrytol (Sikorski, 1977).
T(1966)	Polyacrylamide gel electrophoresis of proteins.	Anon (1967a)	Closely related species now differentiated by electrophoresis or sarcoplasmic proteins (Mackie and Jones, 1978).
	A myosin-like protein extracted from frozen cod stored for long periods. Protein does not combine with actin.	Anon (1967b)	
	Actomyosin splits slowly into actin and myosin at 0°C in presence of 0·5M KCl.	Anon (1967c)	
	Dipole relaxation of frozen fish muscle differs markedly from that of ice.	Anon (1967d)	
	Fish protein concentrate prepared by two novel-processes.	Anon (1967e)	The concept of solvent-extracted fish protein concentrate now somewhat discredited (Pariser et al, 1978). Contrary to popular belief fish flour not a stable product, being particularly susceptible to mould attack (Goldmintz, 1971). Any protein which has had low molecular weight extractives removed will have a high water activity. FPC Type B in which concentrated press liquors are added back (Norsildmel, 1975) would have lowered water activity and be much more stable. Fig 3.
T(1967)	First explanation for gaping in fish fillets.	Anon (1968a)	A series of papers explaining the complex interplay of factors involved, summarized by Love (1975) – effect of temperature and rigor tensions (Burt et al, 1970).
	Electrophoretic identification of cooked fish.	Anon (1968b)	Method (Mackie, 1972). Little follow-up as yet. Complaints would be at restaurant level, public unaware that fish can be checked. Japanese differentiate actomyosin light chains of cooked fish, kamobako and fish sausage with 1% SDS + 8M urea (Seki, 1976).
T(1968)	Cold storage deterioration greatest in fresh cod, no such effect with haddock, although organoleptic cold storage changes in haddock significantly faster than in cod.	Anon (1969)	See Anon* (1964a) for possible explanation. Haddock does not break down TMAO on cold storage, therefore freshness does not affect amount of substrate for formaldehyde formation.
	Myosin-like compound which develops in cold storage may result from oxidation of sulphur compounds in protein molecule.	Anon (1969)	
T(1969)	Connective tissue network in musculature of some species much thicker than in others. Skate and catfish never gape for this reason.	Anon (1970a)	Each species tends to have characteristic collagens; thermal and mechanical stability are not related (Yamaguchi et al, 1976). Shellfish collagens resemble those of lowest vertebrates in low contents of amino acids and high contents of hydroxy acids. Thermal stability correlates positively with the former and negatively with the latter (Pikkarainen et al, 1968).

	Colour test for deterioration of cod in cold store found to be due to reaction of formaldehyde from TMAO reacting with tryptophan under strongly acid conditions.	Anon (1970b)

	Correlation of sensory tests with taste panels.	Anon (1970b)	Until this date many tests were of doubtful utility. If any two parameters change with time they are bound to be correlated. In this period the first attempts were made to account for the variance in fish texture in terms of objective variables (Connell and Howgate, 1968; 1969).
	Continuity of liquid water in frozen fish may assist protein denaturation.	Anon (1970c)	See Anon* (1976a). Kent (1975) makes the assumption that diffusion can occur through unfreezable water as well as unfrozen water and equates the diffusion constants in freezing and drying. Duckworth and Smith (1963) showed with isotopes that the lowest moisture content at which solutes diffused was slightly above the calculated monolayer value.
	On cooking to 95°C 15% of water soluble fish protein is not coagulated and remains in solution.	Anon (1970d)	Thermal treatment does not cause a full denaturation of protein (Podeszewski and Swiniarska, 1977).
T(1971)	Concentration of dimethylamine closely related to the quality of frozen fish.	Anon (1972a)	Dimethylamine (DMA) the other product of the enzymic breakdown of TMAO. Unlike formaldehyde does not combine with protein, so is a better index of formaldehyde production on a molecular basis. However, ratio dimethylamine/protein insolubilized not the same for all species. Is this a pH effect? (see Kostuch and Sikorski, 1977). DMA production in hake different in summer and winter (Dingle. 1978).
	Tropomyosin unchanged by dehydration.	Anon (1972b)	
T(1972)	First use of the Baader comminuting machine.	Anon (1973a)	Symposium on comminuted fish organized by Torry in 1975. Large species variation in the yields of comminuted mince (Bremner, 1977c). Comminution facilitates water washing of flesh and removal of deleterious substances, eg TMAO, but may remove stabilizing amino acids and phosphates at the same time (Bremner, 1979). See Anon* (1965b) and Ohnishi et al (1978).
	Electrophoretic identification of canned fish using cyanogen bromide.	Anon (1973b)	Mackie and Taylor (1972). Cyanogen bromide splits methionine bonds in proteins, thus not too many fragments are obtained. Isoelectric focusing after cyanogen bromide splitting sharpens up bands (Anon, 1978c).
T(1973)	Toughness and gaping of flesh after spawning related to pH and feeding.	Anon (1974a)	See Anon* (1976a).
	Some deepwater species, eg the smooth head, found to have a junket-like texture.	Anon (1974b)	Can texture profiling (see Anon*, 1977) together with principal component analysis or factor analysis account for all quality attributes of fish? Howgate (1977) suggests a cohesiveness factor is required in the profile.
T(1974)	Comminution increases the production of dimethylamine and formaldehyde in Gadoid flesh during frozen storage. None produced in haddock.	Anon (1975a)	See Anon* (1973a). Kostuch and Sikorski (1977) and Bremner (1977a).
	Softening of blue whiting if held before freezing.	Anon (1975b)	Group II pseudomonads produce a marked softening of sterile fish muscle, suggesting proteolytic activity (Shau and Shewan, 1968), but could this be pH effect? See Kelly et al (1966). Softening of haddock during iced storage entirely attributed to pH (Connell and Howgate, 1969).
	Texture of comminuted minces vary from sloppy to rubber-like depending on mechanical mixing and additives.	Anon (1975c)	Lean species toughened by mixing and salt mackerel softened (Ravichander and Keav, 1976). Alginates and polyphosphates soften (Wylie, 1976).
T(1975)	Mobility of water in frozen products studied in detail, also effects of temperature fluctuations in cold stores.	Anon (1976a)	Microwave attenuation and diffusion of water in frozen systems (Fennema, 1975; Kent and Jason, 1975) have similar activation energies. Temperature function integration (Olley, 1978) might not apply where fluctuations to higher temperatures could carry small molecules such as formaldehyde to new substrate. Freezer burn which exacerbates lack of equilibrium in the system would also invalidate TTT studies.
	Season, grounds and species alter the pH of fish.	Anon (1976b)	Storage properties of the proteins of different species are not comparable unless studied at the same pH. Kelly and Dunnett (1969) predict that high pH fish would have more water frozen at a given temperature than low pH fish due to osmotic differences. Model systems in which the pH is held constant (Connell, 1964b) do not simulate actual circumstances. Well fed farmed fish may have too low an ultimate pH (Love and Hume, 1975). Fish in poor condition might be expected to become less tough because of high pH. Yet soluble protein decreased much more rapidly in frozen storage of spawning fish

(see Fig 2).

(Castell and Bishop, 1973). Maybe osmalality is more important than pH as spawning fish have increased quantities of low molecular weight water soluble extractives (Dambergs, 1964). Measurement of viscosity of homogenates of thawed cod in 65% sucrose solutions may be a useful index of texture, as both pH and cold storage changes affect the reading in the same direction (Nishimoto and Love, 1974).

Although fish from different grounds have different pH values, the underlying contribution of pH to texture similar for all grounds (see Fig 2).

	Enzymically prepared liquid protein hydrolysates fed to lambs and calves.	Anon (1976b)	Parallel experiments in other countries (Wessels and Atkinson, 1973; Hale and Bauersfield, 1978).
	Fish fibres formed by extrusion of alkaline elements into precipitation baths.	Anon (1976b)	Spinning 'dopes' from fish and crustacea very different. Fish proteins disperse well in alkali so only low concentrations and temperatures needed; less danger of toxicity (see Anon*, 1953c).
T(1976)	Texture profiling – juicy, firm, springy, chewy, fibrous.	Anon (1977a)	Profiling system devised at Torry (Howgate, 1977) is particularly applicable to cold stored fish as it differentiates between initial and secondary characteristics not allowed for in the general terminology of Jowitt (1974). Difference between initial and secondary characteristics lucidly described by Love (1968) and have been encountered in Australian species (Bremner, 1977b; Bremner et al, 1978). Love (1968) attributes softness in flesh to high pH (starvation), polyphosphates and long periods in ice with breaks of myofibrils at Z-band. Coconut texture attributed to cooking of pre-rigor or thaw-rigor fillets (Jones, 1966).
T(1977)	Inherent softness of blue whiting flesh – lower pH for filleting suggested.	Anon (1978a)	
	Stitch pumping of papain into mackerel for canning.	Anon (1978b)	
	Blue whiting makes a good quality surimi.	Anon (1978b)	Thorough study of this resource for hamburger-type products by the Japanese (Suzuki et al, 1978). Blue whiting for special grade 'surimi' only obtained by boxing and icing, or storage in chilled sea water (Moore, 1979). Freezing at sea produced a product of inferior quality.
	Large-scale acid ensilage of fish waste, viscera and livers.	Anon (1978c)	Quality of amino acid composition does not appear to be a problem. Some tryptophan destruction on long-term storage (Gildberg and Raa, 1977). Thiaminase activity in some fish and shellfish waste may be more of a problem (Disney et al, 1977). Enzyme not destroyed by dehydration at 38°C (Gerry et al, 1977). Thiaminase activity should not be confused with interference in thiamine assay by heme-proteins (Hilker, 1976).

F: Food Investigation Board
T: Torry Research Station
* Indicates a cross-reference within the table

Table II
PARTICIPANTS' COMMENTS ON INITIAL AND LATER OBSERVATIONS ON table I

Year of Annual Report and initial reference in Table I	Name of participant	Observation
F(1931): Broadbank (1932)	W W Foster	The temperature requirements for frozen fish set by some countries may not be economically feasible. McBride and Richardshon (1979) indicate that −18°C is not justified as a legal requirement for retail distribution.
F(1933): Reay (1934)	A C Jason	Stability of lemon sole confirmed quantitively by instrumental measurement (Jason and Lees, 1971).
F(1933): Reay (1934)	J J Matsumoto	Noguchi and Matsumoto (1971) *did* study the effect of taurine on actomyosin solubility, viscosity, ATPase activity and super precipitation. Taurine had a small positive to negative effect. Glutamate was the best protectant.
	J Olley	Taurine was tested by Noguchi and Matsumoto (1971) at concentrations about ten times those found normally in marine fish flesh while glutamate was tested at hundredfold concentrations (see Mackie and Ritchie, 1974).
F(1940–1946): Cutting et al (1956)	A C Jason	Demonstrated excellent samples of minced dried cod, smoked herring and other products and questioned why there has been no commercial development since.
F(1952): Anon (1953c)	J Raa	Lysinoalanine produced during the alkaline hydrolysis of proteins may be toxic.
	A Gildberg	Lysinoalanine has the same elution characteristics as trytophane on conventional chromatographic columns and the latter amino acid is overestimated after alkaline hydrolysis of proteins.

F(1957): Anon (1958b) See also Anon (1946b)	J J Matsumoto	The approach of tying in protein changes with the electrical properties of fish muscle would need to stand with the fact that the aggregation is sometimes accelerated by freezing down below a definite level of subzero temperature (Snow, 1950), as may the 'fusion' of myofibrils frozen down to liquid air temperatures (Love, 1967). Besides activation energy, the difference in energy level between the aggregated and dispersed state must be considered (Matsumoto, 1979). The controlling thermodynamic factor in the deconformation of proteins above 0°C is the free energy factor while the entropy factor becomes greater at low temperatures.

F(1957): Anon (1958d)

In the freeze denaturation of myofibrillar proteins, both (A) the aggregation without transconformation, and (B) the denaturation accompanying the transconformation or unfolding must occur. This is indicated by the following observations:

(A) (1) aggregation of myosin,
    (2) loss of filament forming or paracrystal,
    (3) forming capacity of myosin and LMM
(B) (1) loss of ATPase activity of myosin and HMM
    (2) loss of polymerizing capacity of actin,
    (3) loss of activity of troponin
    (4) loss of enzymic activity of LDH *etc.*

A's take place in α-helical proteins or at α-helical portions of a molecule,
B's take place in non-helical proteins or at non-helical portions of a molecule.

F(1957): Anon (1956c)	J Regenstein	Despite the different structure as shown by electron microscopy there is a great similarity and biochemical interchangeability of the myofibril proteins (Lehman and Szent-Gyorgyi, 1975).
T(1958): Anon (1959a)	A C Jason	There is no eutectic in frozen fish (Riedel, 1956).
T(1962 onwards)	J R Burt	The effects of polyphosphates on water binding capacity of proteins and on cohesion of flakes have been omitted from the table.
	W W Foster	Can phosphate brines be used without losing the fibrosity of the flesh?
T(1963): Anon (1964b)	M Kent	See 'A Comprehensive treatise on water' (Francks, 1975).
T(1963): Anon (1964a)	P Howgate	Formaldehyde is produced during iced storage of saithe in greater quantities than in frozen storage (Mackie and Thomson, 1974), yet iced fish softens rather than toughens.
	Z E Sikorski	Formaldehyde desolubilizes proteins to a greater extent at higher ionic strengths as would pertain in the frozen state (Kostuch and Sikorski, 1977).

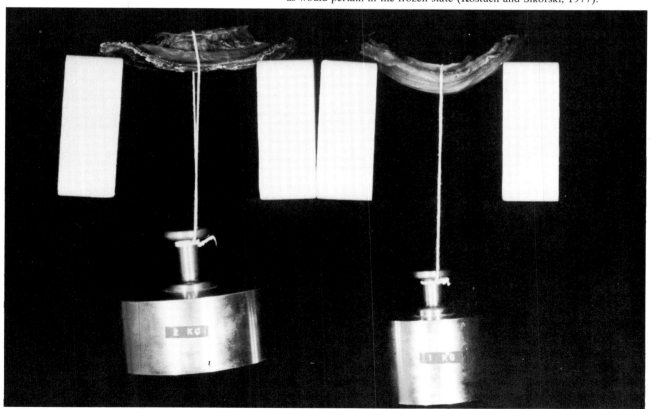

Fig 1 The effect of different pre-drying treatments and drying procedures on the texture of dried abalone (Doe and Veith, unpublished, 1979). The abalone on the left supporting a mass of 2kg was brined overnight, brought to the boil and dried in a wind tunnel at 37°C dry bulb and 22°C wet bulb temperature. The abalone on the right, barely able to support a mass of 1kg, was held overnight in saturated brine and not cooked. Drying was at 53°C dry bulb and 28°C wet bulb temperature. The abalone was successively rebrined and dried until a satisfactory consistency was obtained

73

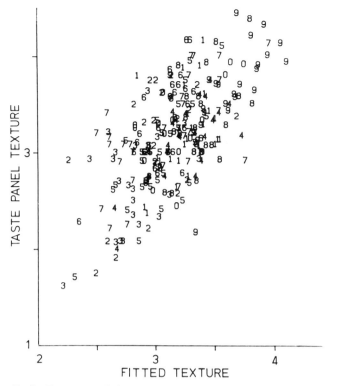

Fig 2 Texture prediction on male codfish from 10 widely different locations (Spitzbergen expedition; Love *et al*, 1974). Texture as assessed by taste panel *versus* texture as determined from regression on pH, length and water content. Numbers represent the 10 locations coded 0, 1, 2, . . . 9. The lack of clustering of the locations shows that the same regression relationship applied over all sites. pH was the most important single predictor of fitted texture (Love, Ratkowsky, Lowry and Olley, unpublished results, 1979)

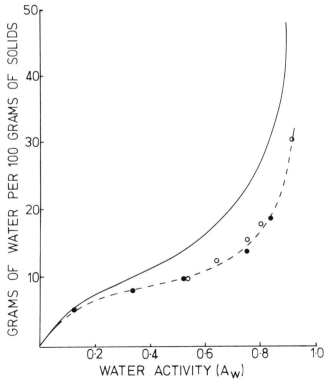

Fig 3 Water sorption isotherms for fishery products: ——— freeze-dried cod (Hashmi, 1979); – – – – – isopropanol: water extracted steam stripped fish;●, hake (Rasekh *et al*, 1971) and ○, shark (Thrower, 1979).

Table III
ACTIVATION ENERGIES IN FROZEN TISSUE

30–40 kJ/mole	90–140 kJ/mole
Water molecules	Charge activity
Diffusion[1]	Specific conductivity
Microwave attenuation[1]	
	Biological change[2]
	Free fatty acid production in
	cod[a] and peas[b]
	Cell fragility[a]
	Protein denaturation[a]

[1] Data of Kent and Jason (1975)
[2] All biological changes were assumed to be first order reactions: a, data from Love (1962a); b, data calculated from Fig 3 of Fennema (1975)

References

ACKER, L. Enzyme activity at low water contents. In: *Recent Advances in Food Science–3. Biochemistry and Biophysics in Food Research,* Eds J Muil Leitch and D N Rhodes. Butterworths, London. 239–247
1963

AITKEN, A and CONNELL, J J. The effect of frozen storage on muscle filament spacing in fish. *Int. Inst. Refrig. Commissions C1 and C2, Karlsruhe,* 187–191
1977

AITKEN, A, JASON, A C, OLLEY, J and PAYNE, P R. Effects of drying, salting and high temperature on the nutritive value of dried cod. *Fish. News Int.,* 6(9), 42–43
1967

Alvarez, E, SIMAL, J, CARRO, A and CREUS, J M. Estudio mediante electroforesis de disco de la influencia del tamano, de la congelacion, de la liofilizacion y del secado al sol en las proteinas de los pescados. *Anal Bromatol.,* 36(1), 1–42
1974

ANON. Lipro-protein complexes. G.B. Dep. Sci. Ind. Res., Rep. Food Invest. Board, Food Invest. Rep. 1948, 16
1949

—— Cold storage of frozen 'white' fish. G.B. Dep. Sci. Ind. Res., Rep. Food Invest. Board, Food Invest. Rep. 1952, 35
1953a

—— Fish muscle proteins. *Ibid,* 38
1953b

—— Alkaline hydrolysis of fish proteins. *Ibid,* 38
1953c

—— Fish muscle proteins. G.B. Dep. Sci. Ind. Res., Food Invest. Board, Food Invest. Rep. 1953, 34
1954a

—— Protein denaturation and the texture of dehydrated fish. *Ibid,* 34
1954b

—— Fish muscle proteins. G.B. Dep. Sci. Ind. Res., Food Invest. Board, Food Invest. Rep. 1954, 45
1955a

—— Protein denaturation and the texture of dehydrated fish. *Ibid,* 46
1955b

—— Deterioration during cold storage (protein denaturation and effect of freezing rate). G.B. Dep. Sci. Ind. Res., Food Invest. Board, Food Invest. Rep. 1955, 27
1956a

—— Protein denaturation and the texture of dehydrated fish. *Ibid,* 34
1956b

—— Rates of freezing and storage temperature. G.B. Dep. Sci. Ind. Res., Food Invest. Board, Food Invest. Rep. 1956, 11–12
1957a

—— Fish muscle proteins. *Ibid,* 23
1957b

—— Freezing before rigor mortis. G.B. Dep. Sci. Ind. Res., Food Invest. Board, Food Invest. Rep. 1957, 11
1958a

—— Protein denaturation. *Ibid,* 11–12
1958b

—— Texture of dehydrated fish. 'Browning' of dehydrated fish. *Ibid,* 19–20
1958c

—— Fish muscle proteins. *Ibid,* 25–26
1958d

—— Freezing and cold storage. Torry Res. St., Aberdeen, Scot., Ann. Rep. 1958, 9–10
1959a

—— Drying. *Ibid,* 17–18
1959b

—— Fish muscle proteins. *Ibid,* 25–26
1959c

—— Protein denaturation. Torry Res. St., Aberdeen, Scot., Ann. Rep. 1959, 9
1960a

—— Fish muscle proteins. *Ibid,* 23–24
1960b

—— Effect of temperature on freshness. Torry Res. St., Aberdeen, Scot., Ann. Rep. 1960, 6
1961a

—— Determination of toughness. *Ibid,* 15–16
1961b

—— Quality of sea-frozen fish. *Ibid,* 17–18
1961c

—— Chemical changes in herring during during heat processing. *Ibid,* 25–26
1961d

—— Fish muscle proteins. *Ibid,* 32
1961e

—— Cold storage of frozen fish before rigor mortis. Torry Res. St., Aberdeen, Scot., Ann. Rep. 1961, 14–15
1962a

—— A seasonal variation in the collagen content of herring. *Ibid*, 24
1962b
—— Proteins. *Ibid*, 30–31
1962c
—— Lipoproteins. *Ibid*, 32
1962d
—— Hydration and dehydration of proteins. *Ibid*, 35
1962e
—— Cold storage change at different temperatures. The use of glycerol.
1963a Torry Res. St., Aberdeen, Scot., Ann. Rep. 1962, 13–14
—— Effects of accelerating rigor mortis in cod upon the quality of
1963b frozen fillets. *Ibid*, 14–15
—— Proteins. *Ibid*, 30–31
1963c
—— Effect of initial freshness on deterioration of frozen cod. Torry
1964a Res. St., Aberdeen, Scot., Ann. Rep. 11–12
—— Studies on bound water at very low temperatures. *Ibid*, 12–13
1964b
—— Proteins. *Ibid*, 35–36
1964c
—— Chemical test for quality of frozen cod. Torry Res. St., Aberdeen,
1965a Scot., Ann. Rep. 11
—— Proteins. *Ibid*, 32–33
1965b
—— Protein reactions during cold storage of cod. *Ibid*, 33
1965c
—— Investigations at cellular level. *Ibid*, 38–39
1965d
—— Investigations at the cellular level. Torry Res. St., Aberdeen,
1966 Scot., Ann. Rep. 50–51
—— Zone electrophoresis of sarcoplasmic proteins. Torry Res. St.,
1967a Aberdeen, Scot., Ann. Rep. 1966, 53
—— Myofibrillar proteins from frozen cod muscle. *Ibid*, 53–54
1967b
—— Transformation of actomyosin into myosin in model systems. *Ibid*,
1967c 54
—— Dipole relaxation in frozen fish muscle. *Ibid*, 57
1967d
—— Fish protein concentrate. *Ibid*. 42
1967e
—— Freezing at sea. Torry Res. St., Aberdeen, Scot., Ann. Rep. 1967,
1968a 14–20
—— Cooking. *Ibid*, 34–36
1968b
—— Quality of frozen fish. Torry Res. St., Aberdeen, Scot., Ann. Rep.
1969 1968, 22–24
—— Freezing at sea. Torry Res. St., Aberdeen, Scot., Ann. Rep. 1969,
1970a 18–21
—— Quality assessment in frozen fish. *Ibid*, 23
1970b
—— Dehydration during cold storage. *Ibid*, 24–25
1970c
—— Cooking. *Ibid*, 37–38
1970d
—— Frozen fish. Torry Res. St., Aberdeen, Scot., Ann. Rep. 1971, 8–9
1972a
—— Fish proteins and quality assessment. *Ibid*, 24
1972b
—— New products Torry Res. St., Aberdeen, Scot., Ann. Rep. 1972,
1973a 14
—— Canned fish. *Ibid*, 18
1973b
—— Freezing fish at sea. Torry Res. St., Aberdeen, Scot., Ann. Rep.
1974a 1973, 4
—— Deepwater species. *Ibid*, 6–7
1974b
—— Chilling, freezing and cold storage. Torry Res. St., Aberdeen,
1975a Scot., Ann. Rep. 1974, 5–6
—— Blue whiting. *Ibid*, 7–8
1975b
—— Fish products. *Ibid*, 9–10
1975c
—— Freezing, cold storage and thawing. Torry Res. St., Aberdeen,
1976a Scot., Ann. Rep.
—— Novel processes. *Ibid*, 15–16
1976b
—— Quality assessment and quality control. Torry Res. St., Aberdeen,
1977 Scot., Ann. Rep. 1976, 7–8
—— Blue whiting. Torry Res. St., Aberdeen, Scot., Ann. Rep. 1977, 8
1978a
—— Product innovation. *Ibid*, 15
1978b
—— Analytical notes. *Ibid*, 28
1978c
ASGHAR, A and YEATES, N T M. The mechanism for the promotion of
1978 tenderness in meat during the post-mortem process: a review.
Crit. Rev. In: *Food Sci. and Nutr.*, 10, 115–145. CRC Press,
Inc

BREMNER, H A. Storage trials on the mechanically separated flesh of
1977a three Australian species. 1. Analytical tests. *Food Technol.
Aust.*, 29, 89–93
—— Storage trials on the mechanically separated flesh of three
1977b Australian mid-water fish species. 2. Taste panel evaluations.
Food Technol. Aust., 29, 183–188
—— Production and storage of mechanically separated fish flesh from
1977c Australian species. Fishexpo '76 Seminar. Report of Proceedings. Dept. Primary Industry, Aust. Govt. Publishing Service.
319–332 ·
—— Processing and freezing of the blue grenadier (*Macruronus*
1979 *novaezelandiae*). Fishexpo '79 Perth, Australia. (In press)
BREMNER, H A, LASLETT, G M and OLLEY, J. Taste panel assessment
1978 of textural properties of fish minces from Australian species. *J.
Food Technol.*, 13, 307–318
BREMNER, H A and VAIL, A. Identification of fish and fish products by
1979 electrophoresis. Division of Food Research. Report of
research 1978–79, CSIRO, Sydney, Australia. (In press)
BROADBANK, J G Fish. G.B. Dep. Sci. Ind, Res. Food Invest. Board,
1932 Rep. 1931, 9–10
BURT, J R, JONES, N R, McGILL, A S and STROUD, G D. Rigor ten-
1970 sions and gaping in cod muscle. *J. Food Technol.*, 5, 339–351
CARPENTER, K J. Fish in human and animal nutrition. *This volume*
1980
CASTELL, C H and BISHOP, D M. Effect of season on salt extractable
1973 protein in muscle from trawler caught cod and on its stability
during frozen storage. *J. Fish. Res. Bd. Can.*, 30, 157–160
CHILDS, E A. Quantitative changes in whole myofibrils and myofibril-
1973 lar proteins during frozen storage of the cod. *J. Food Sci.*, 38,
718–719
CODURI, R J. Vertical plate gel electrophoresis for the differentiation
1972 of fish and shellfish species. *J. Assoc. Off. Anal. Chem.*, 55,
464–466
CONNELL, J J. Some aspects of the texture of dehydrated fish. *J. Sci.
1957 Food Agric.*, 8, 526–537
—— The effect of drying and storage in the dried state on some
1958 properties of the proteins of food. In: *Fundamental Aspects of
the Dehydration of Foodstuffs*. London Society of Chemical
Industry. 167–177
—— The relative stabilities of the skeletal muscle myosins of some
1961 animals. *Biochem. J.*, 80, 503–509
—— The effects of freeze-drying and subsequent storage on the pro-
1962a teins of flesh foods. In: *Freeze drying of Foods*. National
Academy of Sciences, National Research Council. 50–58
—— Changes in amount of myosin extractable from cod flesh during
1962b storage at −14°. *J. Sci. Food Agric.*, 13, 607–617
—— The expressible fluid of fish fillets. VI. Electrophoretic analysis of
1964a the expressible fluid of cod muscle. *J. Sci. Food Agric.*, 13,
269–278
—— Fish muscle proteins and some effects on them of processing. In:
1964b *Symposium on Foods–Proteins and their reactions*, Eds
H W Schultz and A F Anglemier. Avi. Publ. Co, Westport,
Conn. 255–293
—— The effect of freezing and frozen storage on the proteins of fish
1968 muscle. In: *Low Temperature Biology of Foodstuffs*, Eds
J Hawthorn and E J Rolfe. Pergamon Press, Oxford.
333–355
CONNELL, J J and HOWGATE, P F. The amino acid composition of some
1959a British food fishes. *J. Sci. Food Agric.*, 10, 241–244
—— Studies on the proteins of fish skeletal muscle. 6. Amino acid
1959b composition of cod myofibrillar proteins. *Biochem. J.*, 71,
83–86
—— Sensory and objective measurements of the quality of frozen
1968 stored cod of different initial freshness. *J. Sci. Food Agric.*, 19,
342–354
—— Sensory and objective measurements of the quality of frozen
1969 stored haddock of different initial freshness. *J. Sci. Food
Agric.*, 20, 469–474
CONNELL, J J and OLCOTT, H S. The nature of the components liber-
1961 ated by treatment of cod myosin with alkali or with low concentrations of urea. *Arch. Biochem. Biophys.*, 94, 128–135
CUTTING, C L. Changes in the pH and buffering capacity of fish during
1953 spoilage. *J. Sci. Food Agric.*, 4, 597–603
—— The influence of freezing practice on the quality of meat and fish.
1977 *Aust. Refrig. Air Cond. Heat.*, 30(2), 25–41
CUTTING, C L, REAY, G A and SHEWAN, J M. Dehydration of fish.
1956 G.B. Dep. Sci. Ind. Res., Food Invest. Board, Spec. Rep. 62.
160pp
DAMBERGS, N. Extractives of fish muscle. 4. Seasonal variations in fat,
1964 water soluble protein, and water in cod (*Gadus morhua* L)
fillets. *J. Fish Res. Bd. Can.*, 21, 703–709
DEVADASAN, K and NAIR, M R. Studies on the electrophoretic pat-
1971 terns of fish muscle myogens. *Fish Technol.*, 8(1), 80–82
DINGLE, J R. Quality determination in the flesh of Merluccius species
1978 and other Gadidae caused by the formation of formaldehyde
during frozen storage. FAO Fisheries Rep. No. 203 Supplement I. Technical Consultation on the Latin American Hake
industry. 70–83

DISNEY, T G, TATTERSON, I N and OLLEY, J. Recent developments in
1977 fish silage. In: *Proc.* Conference on the handling, processing
and marketing of tropical fish. Tropical Products Institute,
London. 231–240

DUCKWORTH, R B and SMITH, G M. Diffusion of solutes at low mois-
1963 ture levels. In: *Recent Advances in Food Science*, Eds
J M Leitch and D N Rhodes. Butterworths, London. Vol. 3,
230–238

DYER, W J and DINGLE, J R. Fish proteins with special reference to
1961 freezing. In: *Fish as Food*, Ed G Borgström. Academic Press,
New York. Vol 1, 275–327

DYER, W J, FRENCH, H V and SNOW, J M. Proteins in fish muscle. I.
1950 Extraction of protein fractions in fresh fish. *J. Fish. Res. Bd.
Can.*, 7, 585–593

EBASHI, S. Regulatory mechanism of muscle contraction with special
1974 reference to the Ca-troponin-tropomyosin system. *Essays in
Biochemistry*, Eds P N Campbell and F Dickens. Academic
Press, London. 10, 1–36

ELEY, D D and LESLIE, R B. Hydration of solid proteins, with special
1963 reference to haemoglobin. In: *Recent Advances in Food Sci.*,
Eds J M Leitch and D N Rhodes. Butterworths, London. Vol
3, 215–217

FAO Available amino acid content of fish meals. *FAO Fisheries
1970 Reports*, No. 92. 66pp

FENNEMA, O. Activity of enzymes in partially frozen aqueous systems.
1975 In: *Water Relations of Foods*, Ed R B Duckworth. Academic
Press, London. 397–413

FRANCKS, F. *Water—a comprehensive treatise*. Plenum, New York
1975

GERRY, R W, O'MEARA, D C, HARRIS, P C, BRYAN, T A and-
1977 BLAMBERG, D L. Possible sources of marine protein. Technical
Report 20. Maine University, Maine Sea Grant. 10pp

GERSHMAN, L L. Report on fish and other marine products. *J. Assoc.
1979 Off. Anal. Chem.*, 62, 351–352

GILDBERG, A and RAA, J. Properties of a propionic acid/formic acid
1977 preserved silage of cod viscera. *J. Sci. Food Agric.*, 28,
647–653

GILL, T A, KEITH, R A and SMITH-LALL, B. Textural deterioration of
1979 red hake and haddock muscle in frozen storage as related to
chemical parameters and changes in the myofibrillar proteins.
J. Food. Sci., 44, 661–667

GOLDMINTZ, D. Survival of microorganisms in fish protein concentrate
1971 stored under controlled conditions. In: *Developments in industrial microbiology*. Nat. Marine Fish Service. Vol 12, 260–265

HALE, M B and BAUERSFIELD, P E. Preparation of a menhaden hy-
1978 drolysate for possible use as a milk replacer. *Mar. Fish. Rev.*,
40(8), 14–17

HAMOIR, G, PIRONT, A, GERDAY, C H and DANDO, P R. Muscle pro-
1973 teins of the coelacanth, *Latimeria chalumnae* Smith. *J. Mar.
Biol. Assoc. UK*, 53, 763–784

HERBORG, L, VILIEN, F, BRUUN, A and EGGUM, B. Roller-dried fish
1974 protein. In: *Fishery Products*, Ed R Kreuzer. Fishing News
(Books) Ltd, London. 309–311

HILKER, D M. Thiamine-modifying properties of fish and meat pro-
1976 ducts. *J. Nutr. Sci., Vitaminol.*, 22 (Suppl), 3–6

HOFMANN, K and HAMM, R. Sulfhydryl and disulphide groups in
1978 meats. *Adv. Food Res.*, Eds C O Chichester, E M Mrak and
G F Stewart. Academic Press, New York. 24, 1–111

HOWGATE, P. Aspects of fish texture. In: *Sensory Properties of Foods*,
1977 Eds G C Birch, J G Brennan and K J Parker. Applied Science
Publishers Ltd, Barking, England. 249–269

—— Fish. In: *Food Microscopy*, Ed J Vaughan. Academic Press, New
1979 York. 343–392

IRONSIDE J I and LOVE, R M. Measurement of denaturation of
1956 fish protein. *Nature (London)*, 178, 418–419

JASON, A C and LEES, A. Estimation of fish freshness by dielectric
1971 measurement. Department of Trade and Industry Report
T/71: 31

JONES, N R. The free amino acids of fish. II. Fresh skeletal muscle
1959 from lemon sole (*Pleuronectes microcephalus*). *J. Sci. Food.
Agric.*, 10, 282–286

—— Browning reactions in dried fish products. In: *Recent Advances in
1962 Food Science Vol. 2. Processing*, Eds J Hawthorn and
J M Leitch. Butterworths, London. 74–80

—— Problems associated with freezing very fresh fish. *Proc:* Meeting
1964 on Fish Technol. Fish Handling and Preservation
Scheveningen, OECD. 31–55

—— Freezing fillets at sea. In: Fish quality at sea. *Proc.* Conference on
1966 the design of fishing vessels and their equipment in relation to
the improvement of quality. Grampian Press, London. 81–88

JOWITT, R. The terminology of food texture. *J. Texture Stud.*, 5,
1974 351–358

KELLY, K, JONES, N R, LOVE, R M and OLLEY, J. Texture and pH in
1966 fish muscle related to 'cell fragility' measurements. *J. Food
Technol.*, 1, 9–15

KELLY, T R and DUNNETT, J S. Low temperature freezing of cod. *J.
1969 Food Technol.*, 4, 105–115

KENT, M. Measurement of hydration of biological molecules from
1972 dielectric measurements at centimetre wavelengths. In: *Pro-
tides of the biological fluids*, Ed H Peeters. Pergamon Press,
Oxford. 359–366

—— Fish muscle in the frozen state: time dependence of its microwave
1975 dielectric properties. *J. Food Technol.*, 10, 91–102

KENT, M and JASON, A C. Dielectric properties of foods in relation to
1975 interactions between water and the substrate. In: *Water Relations of Foods*, Ed R B Duckworth. Academic Press, London.
211–231

KIM, H K, ROBERTSON, I and LOVE, R M. Changes in the muscle of
1977 lemon sole (*Pleuronectes microcephalus*) after very long cold
storage. *J. Sci. Food Agric.*, 28, 699–700

KONAGAYA, S. Screening of peptidases in fish muscle. *Bull. Tokai Reg.
1978 Fish. Res. Lab.*, No. 94, 1–28

KOSTUCH, S and SIKORSKI, Z E. Interaction of formaldehyde with cod
1977 proteins during storage. *Int. Inst. Refrig. Comm. C1 et C2
Karlsruhe*, 199–208

LAKI, K. Size and shape of the myosin molecule. In: *Contractile pro-
1971a teins and muscle*, Ed K Laki. Marcel Dekker Inc, New York.
179–217

—— Actin. *Ibid*, 97–133
1971b

LALL, B S, MANZER, A R and HILTZ, D F. Preheat treatment for
1975 improvement of frozen storage stability at −10°C in fillets and
minced flesh of silver hake. *J. Fish. Res. Bd. Can.*, 32,
1450–1454

LEHMAN, W and SZENT-GYORGYI, A G. Regulation of muscle contrac-
1975 tion. Distribution of actin control and myosin in the animal
kingdom. *J. Gen. Physiol.*, 66, 1–30

—— The effect of critical freezing temperature on the behaviour of cod
1967 muscle proteins during subsequent storage: a histological
study of homogenates. *Bull. Jap. Soc., Sci. Fish.*, 33, 746–752

LOVE, R M. Studies on protein denaturation in frozen fish. III. The
1958 mechanism and site of denaturation at low temperatures. *J.
Sci. Food Agric.*, 9, 609–617

—— Protein denaturation in frozen fish. VI. Cold storage studies on
1962a cod using the cell fragility method. *J. Sci. Food Agric.*, 13,
269–278

—— New factors involved in the denaturation of frozen cod muscle
1962b protein. *J. Food Sci.*, 27, 544–550

—— Protein denaturation in frozen fish. VII. Effect of the onset and
1962c resolution of rigor mortis on denaturation. *J. Sci. Food Agric.*,
13, 534–545

—— The use of tasters for investigating cold storage deteriorations in
1966 frozen fish. *J. Food Technol.*, 1, 141–146

—— Histological observations on the texture of fish muscle. In: *Rheol-
1968 ogy and Texture of Foodstuffs*. Society of Chemical Industry,
London. 120–133

—— Towards a valid sampling technique. A. Chemistry and Anatomy.
1970 *The Chemical Biology of Fishes*. Academic Press, London.
1–59

—— Variability in Atlantic cod (*Gadus morhua*) from the northeast
1975 Atlantic: a review of seasonal and environmental influences on
various attributes of the flesh. *J. Fish Res. Bd. Can.*, 32,
2333–2342

LOVE, R M, AREF, M M, ELERIAN, M K, IRONSIDE, J I M, MACKAY, E M
1965 and VARELA, M G. Protein denaturation in frozen fish. X.
Changes in cod muscle in the unfrozen state, with some further
observations on the principles underlying the cell fragility
method. *J. Sci. Food Agric.*, 16, 259–267

LOVE, R M and ELERIAN, M K. Protein denaturation in frozen fish.
1964 VIII. The temperature of maximum denaturation in cod. *J. Sci.
Food Agric.*, 13, 534–545

—— Protein denaturation in frozen fish. IX. The inhibitory effect of
1965 glycerol in cod muscle. *J. Sci. Food Agric.*, 16, 65–70

LOVE, R M and HUME, A H. The quality of farmed products. *Fish
1975 Farming Int.*, 2(1), 36–37

LOVE, R M and MUSLEMUDDIN, M. Protein denaturation in frozen fish.
1972 XIII. A modified cell fragility method insensitive to the pH of
the flesh. *J. Sci. Food Agric.*, 23, 1239–1251

LOVE, R M and OLLEY, J. Cold storage deterioration in several species
1965 of fish, as measured by two methods. In: *The Technology of
Fish Utilization*, Ed R Kreuzer. Fishing News (Books) Ltd,
London. 116–118

LOVE, R M, ROBERTSON, I, LAVETY, J and SMITH, G L. Some biochemi-
1974 cal characteristics of cod (*Gadus morhua* L) from the Faroe
Bank compared with those from other fishing grounds. *Comp.
Biochem. Physiol.*, 47B, 149–161

LOVE, R M, YAMAGUCHI, K, CREAC'H, Y and LAVETY, J. The connec-
1976 tive tissues and collagens of cod during starvation. *Comp.
Biochem. Physiol.*, 55B, 487–492

LUMLEY, A, PIQUE, J and REAY, G A. Handling and stowage of fish in
1929 trawlers. G.B. Dep. Sci. Ind. Res., Food Invest. Board, Rep.
1928, 67–70

MACKENZIE, A P. Discussion in paper by Love, R M and
1963 Elerian, M K. The irreversible loosening of bound water at
very low temperatures in cod muscle. In *Proc:* 11th Int. Cong.
Refrig., Munich. 887–894

MACKIE, I M. The effect of adenosine triphosphate, inorganic
1966 pyrophosphate and inorganic tripolyphosphate on the stability
of cod myosin. *Biochem. Biophys. Acta*, 115, 160–172
—— Some improvements in the polyacrylamide disc electrophoretic
1972 method of identifying the species of cooked fish. *J. Assoc.
Public Anal.*, 10, 18–20
MACKIE, I M and JONES, B W. The use of electrophoresis of the
1978 water-soluble (Sarcoplasmic) proteins of fish muscle to differentiate the closely related species of hake (*Merluccius* sp).
Comp. Biochem. Physiol., 59B, 95–98
MACKIE, I M and RITCHIE, A H. Free amino acids of fish flesh. *Proc:*
1974 IV. Int. Congress Food Sci. and Technol. Vol I, 29–38
MACKIE, I M and TAYLOR, T. Identification of species of heat-sterilised
1972 canned fish by polyacrylamide-disc electrophoresis. *Analyst
(London)*, 97, 609–611
MACKIE, I M and THOMSON, B W. Decomposition of trimethylamine
1974 oxide during iced and frozen-storage of whole and commi-
nuted tissue of fish. *Proc:* IVth Int. Congress Food Sci. and
Technol. Vol I, 243–250
McDONALD, I and JONES, N R. Control of thaw rigor by manipulation
1976 of temperature in cold storage. *J. Food Technol.*, 11, 69–71
McBRIDE, R J and RICHARDSON, K C. The time-temperature toler-
1979 ance of frozen foods: sensory methods of assessment. *J. Food
Technol.*, 14, 57–67
MATSUMOTO, J J. Paper to joint American and Japanese Chemical
1979 Society Meeting, Honolulu
MOODIE, I M and EVA, A. Electrophoresis of fish proteins. Fishing
1973 Industry Research Institute, Cape Town. 27th Annual Report,
34–37
—— Electrophoresis of fish muscle proteins. *Ibid*, 57–61
1974
MOORE, M. Blue whiting. British trials to make 'surimi' prove disap-
1979 pointing. *Fish. News Int.*, 18(2), 30–31
MURRAY, C K and FLETCHER, T C. The immunhistochemical localiza-
1976 tion of lysozyme in plaice. (*Pleuronectes platessa* L) tissues. *J.
Fish. Biol.*, 9, 329–334
NISHIMOTO, J. and LOVE, R M. A study of the viscosity of homogenates
1974 of thawed cod muscle in 65% sucrose solution. *Bull. Jap. Soc.
Sci. Fish.*, 40, 1071–1076
NOGUCHI, S and MATSUMOTO, J J. Studies on the control of the-
1971 denaturation of fish muscle proteins during frozen storage. II.
Preventive effect of amino acids and related compounds. *Bull.
Jap. Soc. Sci. Fish.*, 37, 1115–1122
NORSILDMEL, Norse Fish Powder. Fish protein concentrate for human
1975 consumption. Norsildmel A/L, Bergen, Norway.
NOTTINGHAM, P M. The alkaline hydrolysis of haddock actomyosin. I.
1955a Ammonia formation. *J. Sci. Food Agric.*, 6, 82–86
—— The alkaline hydrolysis of haddock actomyosin. II. Peptide bond
1955b formation. *J. Sci. Food Agric.*, 6, 86–90
OHNISHI, M, TSUCHIYA, T and MATSUMOTO, J J. Electron microscopic
1978 study of the cryoprotective effect of amino acids on freeze
denaturation of carp actomyosin. *Bull. Jap. Soc. Sci. Fish.*, 44,
755–762
OLLEY, J Phospholids in fish lipoproteins. *Biochem. J.*, 81, 29–30P
1961
—— Unconventional sources of fish protein. *CSIRO Food Res. Q.*, 32,
1972 27–32
—— Current status of the theory of the application of temperature
1978 indicators, temperature integrators and temperature function
integrators to the food spoilage chain. *Int. J. Refrig.*, 1, 81–86
OLLEY, J and THROWER, S J. Abalone – an esoteric food. In: *Adv. Food
1977 Res.*, Eds C O Chichester, E M Mrak and G F Stewart.
Academic Press, New York. 23, 143–186
PARISER, E R, WALLERSTEIN, M B, CORKERY, C J and BROWN, N L.
1978 *Fish protein concentrate Panacea for Protein Malnutrition?*
MIT Press, Cambridge, Mass. 296pp
PIKKARAINEN, J, RANTANEN, J, VASTAMAKI, M, LAMPIAHO, K, KARI, A
1968 and KULONEN. E. *European J. Biochem.*, 4, 555–560
PODESZEWSKI, Z and SWINIARSKA, J. Effects of freezing upon cooked
1977 fish flesh. *Int. Inst. Refrig. Comm. C1 and C2 Karlsruhe*,
249–252
POULSEN, K P and JENSEN, S L. Quality-economy relations of frozen
1978 foods. *Inst. Int. Refrig. Comm. C2, D1, D2*. Budapest, Hun-
gary. 85–94
POULSEN, K P and LINDELOV, F. Acceleration of chemical reactions
1978 due to freezing. *ISOPOW = II* (Second International Symposium on the Properties of Water in Relation to Food Quality
and Stability), Osaka (Japan). (In press)
POULTER, R G and LAWRIE, R A. Studies on fish muscle protein.-
1977 Nutritional consequences of the changes occurring during
frozen storage. *J. Sci. Food Agric.*, 28, 701–709
—— Studies on fish muscle protein: changes in sulphydryl groups and
1978 the binding of added formaldehyde during frozen storage.
Lebensm–Wiss u–Technol., 11, 264–266

RASEKH, J G, STILLINGS, B R and DUBROW, D L. Moisture adsorption
1971 of fish protein concentrate at various relative humidities and
temperature. *J. Food Sci.*, 36, 705–707
RAVICHANDER, N and KEAY, J N. The production and properties of
1976 minced fish from several commercially important species. In:
*The production and utilisation of mechanically recovered fish
flesh (minced fish)*, Ed J N Keay. Ministry of Agriculture,
Fisheries and Food, Torry Res. St., Aberdeen, Scot. 18–24
REAY, G A. The low-temperature preservation of the haddock. G.B.
1931 Dep. Sci. Ind. Res., Food Invest. Board, Rep. 1930, 128–134
—— Freezing and cold storage of fish. G.B. Dept. Sci. Ind. Res., Food
1932 Invest. Board, Rep. 1931, 202
—— Freezing and cold storage. G.B. Dep. Sci. Ind. Res., Food Invest.
1933 Board, Rep. 1932, 181–187
—— Cold storage of different species of white fish. G.B. Dep. Sci. Ind.
1934 Res., Food Invest. Board, Rep. 1933, 171–173
—— The proteins of fish. G.B. Dep. Sci. Ind. Res., Food Invest. Board,
1936a Rep. 1935, 65
—— The freezing and cold storage of fish. *Ibid*, 67–70
1936a
—— Testing for formaldehyde in salt ling. *Ibid*, 82–83
1936c
—— The freezing and cold storage of fish. G.B. Dep. Sci. Ind. Res.,
1939 Food Invest. Board, Rep. 1938, 96–98
—— The freezing and cold storage of white fish. G.B. Dep. Sci. Ind.
1949 Res, Food Invest. Board, Food Invest. Rep. 1939, 43–47
REHBEIN, H, KRESS, G and SCHREIBER, W. An enzymic method for
1978 differentiating thawed and fresh fillets. *J. Sci. Food Agric.*, 29,
1076–1082
RIEDEL, L. Calorimetric investigations into the freezing of fish. *Käl-
1956 technik*, 8, 374–377
SEKI, N. Identification of fish species by SDS-polyacrylamide gel elec-
1976 trophoresis of the myofibrillar proteins. *Bull. Jap. Soc. Sci.
Fish.*, 42, 1169–1176
SHARP, J G. Estimation and identification of the amino-acids of the
1936 protein of fish muscle. G.B. Dep. Sci. Ind. Res., Food Invest
Board, Rep. 1935, 65–66
SHAW, B G and SHEWAN, J M. Psychrophilic spoilage bacteria of fish.
1968 *J. Appl. Bact.*, 31, 89–96
SHELTAWY, A and OLLEY, J. Lipid distribution and recovery during salt
1966 fractionation of cod muscle. *J. Sci. Food Agric.*, 17, 94–100
SIKORSKI, Z E. Protein changes in muscle foods due to freezing and
1977 frozen storage. *Inst. Int. Refrig. Comm. C1 and C2, Karlsruhe*,
25–39
SIKORSKI, Z, OLLEY, J and KOSTUCH, S. Protein changes in frozen fish.
1976 *Crit. Rev.* In: *Food Sci. and Nutr.*, 8(1), 97–129. CRC Press,
Inc.
SORENSEN, T. Effect of frozen storage on the functional properties of
1976 separated fish mince. In: *The production and utilization of
mechanically recovered fish flesh. (minced fish)*, Ed. J N Keay.
Torry Res. St., Aberdeen, Scot. 56–65
SNOW, J M. Proteins in fish muscle. III. Denaturation of myosin by
1950 freezing. *J. Fish Res. Bd. Can.*, 7, 599–609
STOREY, R M. The diffusion of water in frozen cod. *Int. Inst. Refrig.
1970 Commissions II, IV, V and VII Leningrad*, 235–238.
STOREY, R M and STAINSBY, G. The equilibrium water vapour press-
1970 ure of frozen cod. *J. Food Technol.*, 5, 157–163.
SUZUKI, T, KANNA, K, OKAZAKI, E and MORITA, N. Manufacture of
1978 meat textured fish protein concentrate from various fishes.
Bull. Jap. Soc. Sci. Fish, 44, 1275–1281
THROWER, S J. Preparation of flours of known mercury and selenium
1979 content from commercially exploited species of Australian
fish. Fishexpo '79 Perth, Australia. (In press)
TOKUNAGA, T. Studies on the development of dimethylamine and for-
1964 maldehyde in Alaska pollock muscle during frozen storage.
Rep. Hokkaido Reg. Fish. Res. Inst., 29, 108–122
WESSELS, J P H and ATKINSON, A. Soluble protein from trash fish and
1973 industrial fish. Fishing Industry Research Institute, Cape
Town, 27th Annual Report, 15–18
WHITTLE, K J. Improvement of the Torry-Brown homogeniser for the
1975 cell fragility method. *J. Food Technol.*, 10, 215–220
WYLIE, A. Alginates in the processing of minced fish. In: *The produc-
1976 tion and utilisation of mechanically recovered fish flesh (minced
fish)*, Ed J N Keay. Ministry of Agriculture, Fisheries and
Food, Torry Res. St., Aberdeen, Scot., 87–92
YAMAGUCHI, K, LAVETY, J and LOVE, R M. The connective tissues of
1976 fish. VIII. Comparative studies of hake, cod and catfish collagens. *J. Food Technol.*, 11, 389–399
YAMANAKA, H, NAKAGAWASAI, T, KIKUCHI, T and AMANO, K. Studies
1978 on the contraction of carp muscle. I. Remarkable differences
between *rigor mortis* and thaw rigor. *Bull. Jap. Soc. Sci. Fish.*,
44, 1123–1126
YOUNG, F, JAMES, D G, OLLEY, J and DOE, P E. Studies on the pro-
1973 cessing of abalone. IV. Dried abalone; Products, quality and
marketing. *Food Technol. Aust.*, 25, 142–149

Structure and proteins of fish and shellfish. Part 2

Z E Sikorski

1 Introduction

The results of basic research on composition and properties of fish proteins, initiated at the Torry Station (Reay, 1931 and 1932; Connell, 1964), and followed up also by others (see Part 1 of this paper), have been successively applied for interpreting changes occurring in the course of traditional preservation of marine foods. A better insight into alterations brought about in muscle proteins by endogenous and microbial enzymes, by acids, salts and heat, as well as by interactions with other flesh constituents and added substances is a prerequisite for rationalizing the processing parameters in order to standardize the quality and for developing new methods of manufacture of 'tailor-made' fishery products. Some general ideas of complex changes occurring in muscle proteins due to processing are given in the following figures.

2 Proteins and structure of fresh fish meat

The main processes occurring in the proteins of fish after catching under the influence of endogenous and bacterial enzymes, as well as changes in the characteristics of the flesh, are presented schematically in *Fig 1*. Considering the sequence of events in the muscle one can easily understand the reason for the dramatic change in toughness of the cooked meat during the early period *post mortem* (*Fig 2*).

3 Marinades and salted fish

Peter Biegler, the experienced German specialist in fishery technology, wrote in his book *Der Fisch*: 'Ein mild gesalzener Schinken oder mild gesalzener Kaviar oder mild gesalzener Lachs dürften schon den Göttern gut gemundet haben' (Biegler, 1960). The changes in proteins leading to the development of these highly praised sensory attributes of salted fish and marinades are shown in *Fig 3*. The results presented by Reddi *et al* (1972) indicate that the muscle cathepsins cannot play any significant role in the ripening of salted fish, as salt in considerably low concentrations brings about a large decrease in their activity towards fish proteins. On the other hand, the activity of the endopeptidases of the alimentary tract has been shown by Luijpen (1959) not to be severely affected by sodium chloride at rather high concentrations. Salt in large proportion, however, influences unfavourably the properties of muscle proteins—the products of a heavy cure do not display the attractive sensory attributes praised in the opening sentence.

4 Denaturation and functional properties

Different factors involved in fish preservation and processing not only influence the activity of proteolytic enzymes, but also bring about denaturation of the muscle proteins and their interactions with other compounds present originally in the flesh or added by the processor.

MUSCLE OF FRESHLY CAUGHT FISH

RELAXED EXTENSIBLE SARCOMERES, ACTIN AND MYOSIN MICROFIBRILS UNCOUPLED

pH about 7, high protein extractability and hydration, firm, cohesive, toughness after cooking depends on the degree of muscle shortening

ATP→ADP, release of Ca^{2+} from sarcoplasmic reticulum

FISH MEAT IN RIGOR MORTIS

PARTLY CONTRACTED SARCOMERES, ACTOMYOSIN, PARTLY RUPTURED MYOCOMMATA

pH about 6, low protein extractability and hydration, rigid, tough after cooking especially when onset of rigor takes place at about 18°C

cathepsins, Ca^{2+} activated proteinase, collagenase→←β glucuronidase, and other lysosomal enzymes

TENDER FISH MEAT

PARTLY HYDROLYSED SARCOPLASMIC PROTEINS, SLIGHTLY DISINTEGRATED SARCOMERES DUE TO HYDROLYTIC CHANGES IN TROPONINE, Z-LINE, AND M-LINE. DISRUPTED COLLAGEN STRUCTURE

pH about 7, high protein extractability and hydration, plasto-elastic, tender after cooking

endogenous enzymes→←bacterial enzymes

AUTOLYSED FISH MEAT

PARTLY HYDROLYSED PROTEINS, NONPROTEIN NITROGENOUS COMPOUNDS

pH above 7, soft, sticky

Fig 1 Changes in the muscles of fresh fish

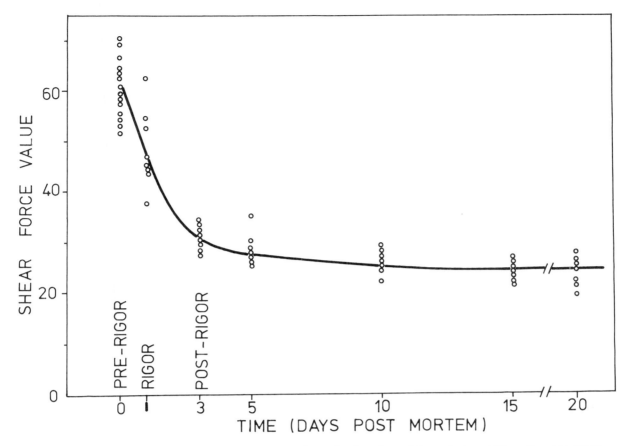

Fig 2 The influence of the state of gutted rainbow trout in ice on the shear force of the cooked meat (*From Dunajski, 1979*)

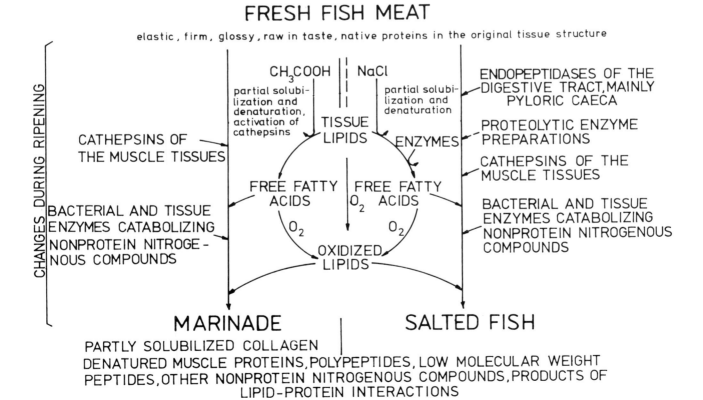

Fig 3 The ripening of marinades and salted fish

These changes are in part desirable, *eg* the increase in digestibility or formation of the appealing colour and flavour of the foods, or else are undesirable, as they are reflected in a decrease in extractability and solubility of proteins, loss in gel-forming ability, water-holding and fat-emulsifying capacity, and other functional properties, which contribute to the development of acceptable sensory quality of various processed fishery products.

Many secondary changes in proteins, following the denaturation, are caused by interactions of —SH groups, both of protein and non-protein origin. Aggregations of protein molecules can occur due to sulfhydryl-disulfide exchange, without net change in available —SH groups (Buttkus, 1970).

In fishery products a large part in the denaturation and secondary changes in proteins is played by lipids and their oxidation products (*Fig 4*). Jarenbäck and Liljemark (1975) have shown that the effectiveness of fatty acid peroxide in decreasing the extractability of cod myofibrils is several times higher than that of the corresponding fatty acid.

5 Protein changes in frozen fish

This subject has been recently reviewed (Connell, 1968; Sikorski *et al*, 1976; Partmann, 1977; Sikorski, 1978), thus only the manifold direct and indirect influence of mineral salts, not exhaustively covered elsewhere, will be shown here.

In *Fig 5* different ways of participation of salts in causing protein changes in frozen fish are presented, *ie* the direct stabilizing effect exhibited on the native con-

$$LH + R^{\cdot} \xrightarrow{O_2} L^{\cdot} + LOO^{\cdot} + LOOH + RH + S$$
$$PH + L^{\cdot} + LOO^{\cdot} + S \rightleftharpoons P^{\cdot} + LH + LOOH + POO^{\cdot} + PHS + PHSPH$$

LH – LIPID   R˙ – RADICAL
PH – PROTEIN   S – LIPID SCISSION PRODUCTS

$$
\begin{aligned}
P^{\cdot} + P^{\cdot} &\longrightarrow PP \\
P^{\cdot} + POO^{\cdot} &\longrightarrow POOP \\
P^{\cdot} + L^{\cdot} &\longrightarrow PL \\
POO^{\cdot} + L^{\cdot} &\longrightarrow POOL \\
POO^{\cdot} + LH &\longrightarrow POOL + H^{\cdot} + POOH + L^{\cdot} \\
PH + S &\longrightarrow PHS + PHSPH
\end{aligned}
$$

Fig 4   The potential reactions of proteins with lipid radicals and their oxidation products

formation of proteins, the binding to polar groups, the interference with water structures, the role in lipid changes, which lead to lipid-protein interactions, and in generating formaldehyde from trimethylamine in the tissues. Formaldehyde is known to be a potent factor in protein changes in several species of fish (Tokunaga, 1964; Castell, 1971; Sikorski *et al*, 1976; Kostuch and Sikorski, 1977) (*Fig 6*).

The relation of freezing changes in proteins to the eating quality and functional properties of muscle foods is not yet quite clear. In most species of fish a decrease in protein extractability in the thawed tissues is accompanied by corresponding deterioration in texture of the cooked meat, reflected by increased toughness, chewiness, rubberiness, stringiness or the appearance of properties described as cardboardy, crumby, dry and fibrous. Extensive toughening also takes place in the flesh of frozen stored Antarctic krill. On the other hand, in some fish no correlation could be established between the

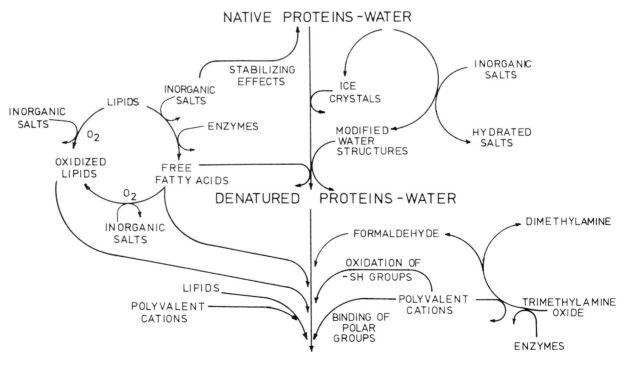

Fig 5   Protein changes in frozen stored fish meat (*From Kołodziejska and Sikorski, 1979*)

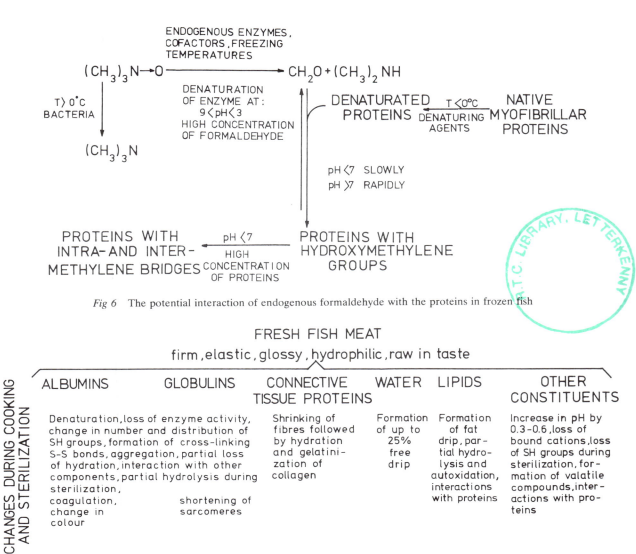

Fig 6 The potential interaction of endogenous formaldehyde with the proteins in frozen fish

FRESH FISH MEAT

firm, elastic, glossy, hydrophilic, raw in taste

CHANGES DURING COOKING AND STERILIZATION	ALBUMINS	GLOBULINS	CONNECTIVE TISSUE PROTEINS	WATER	LIPIDS	OTHER CONSTITUENTS
	Denaturation, loss of enzyme activity, change in number and distribution of SH groups, formation of cross-linking S-S bonds, aggregation, partial loss of hydration, interaction with other components, partial hydrolysis during sterilization, coagulation, change in colour	shortening of sarcomeres	Shrinking of fibres followed by hydration and gelatinization of collagen	Formation of up to 25% free drip	Formation of fat drip, partial hydrolysis and autoxidation, interactions with proteins	Increase in pH by 0.3–0.6, loss of bound cations, loss of SH groups during sterilization, formation of volatile compounds, interactions with proteins

COOKED FISH MEAT

tender, soft or tough, plasto-elastic, succulent, moist or dry

Fig 7 The influence of heating on the muscle components and on the properties of fish meat

protein extractability and eating toughness of the meat. According to Bremner (1978) in spiny flathead and ocean perch a slight softening even occurs during storage, although the protein extractability drops markedly. The published results indicate that the sensory toughness of defrosted fish depends on protein extractability, pH, and moisture of the meat (Bremner *et al*, 1978).

6 Heating and structure of proteins

A simplified sheet of changes in meat constituents due to heating is given in *Fig 7*. Of course, the rate of change in individual proteins and of their interactions with other compounds as well as the final sensory effect as perceived by the consumer depend on the parameters applied. This can be demonstrated on the examples of the influence of cooking fish to various temperatures (*Fig 8*), krill meat at different levels of pH and added salt (*Fig 9*), or fish gels containing different amounts of added gelatin (*Fig 10*), on the rheological properties of the products.

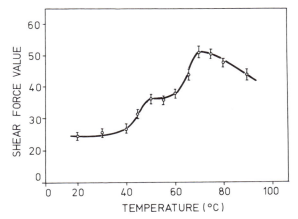

Fig 8 A typical relationship between the cooking temperature and the shear force of post rigor Pacific cod meat (*From Dunajski, 1979*)

Heat treatment plays a very important role in the manufacturing of various fish jellies and sausages. The homogeneous fish sausage 'emulsion', which forms as a

81

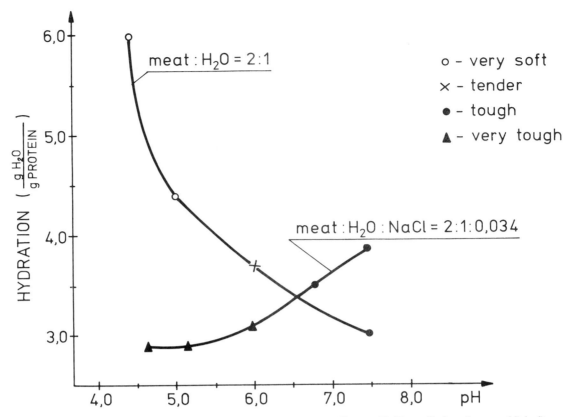

Fig 9 The hydration of cooked krill meat (100°C, 20min) at different pH (*From Grabowska, unpublished*)

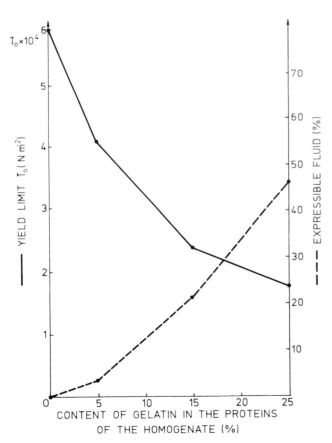

Fig 10 The rheological properties of fish gels containing gelatin, cooked at 85°C (*From Sadowska and Sikorski, 1976*)

result of mincing and mixing in a silent cutter (*Fig 11*) changes under the influence of heating into a jellied product, displaying the desirable rheological properties of a viscoelastic plastic solid (*Fig 12*). The rheological characteristics of a pure fish meat sausage formulation, as represented by the shear rate D, sepend on the content of water (W), fat (F) and protein (p), and the pH in the system:

$$\lg D = a - bpH - Clg\frac{F}{p} / lg\frac{W}{p} - a/$$

while the texture of the final cooked product, as measured by the yield limit (τ_0), is predetermined by the viscosity of the raw sausage formulation (Sadowska and Sikorski, 1976):

$$\lg \tau_0 = k - N \lg D$$

The values of τ_0 corresponding to different levels of sensory preference of the texture of fish sausages are given in *Fig 13*.

7 Manufacturing concentrated fish protein products

The quality of fish protein concentrates, isolates and preparations depends not only on the degree of removal of the volatile odoriferous compounds and lipids but also, especially as regards the technological value, on changes in proteins induced by the procedures of extraction and/or modifications (*Fig 14*). Therefore, in attempts to produce concentrated fish protein products which would find a wide application in the food industries, including sausage manufacture, many approaches

FISH MEAT

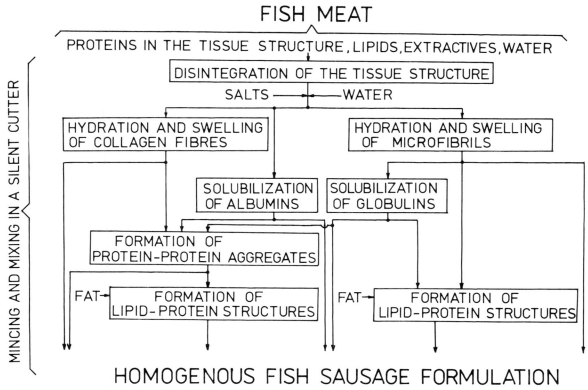

PROTEINS IN THE TISSUE STRUCTURE, LIPIDS, EXTRACTIVES, WATER

MINCING AND MIXING IN A SILENT CUTTER

DISINTEGRATION OF THE TISSUE STRUCTURE

SALTS —— WATER

HYDRATION AND SWELLING OF COLLAGEN FIBRES

HYDRATION AND SWELLING OF MICROFIBRILS

SOLUBILIZATION OF ALBUMINS

SOLUBILIZATION OF GLOBULINS

FORMATION OF PROTEIN-PROTEIN AGGREGATES

FAT → FORMATION OF LIPID-PROTEIN STRUCTURES

FAT → FORMATION OF LIPID-PROTEIN STRUCTURES

HOMOGENOUS FISH SAUSAGE FORMULATION

Water solution of proteins, salts, and other extractives, scraps of various connective tissue membranes, fragments of and intact whole myofibrils or bundles of them, aggregates of myofilaments, other particulate constituents, fat and protein aggregates

Fig 11   The formation of a fish sausage formulation

FISH SAUSAGE FORMULATION

Colloidal liquid-solid system, behaving as a non-Newtonian liquid with pseudoplastic properties, plastic body, a viscous liquid with elastic properties, or a viscoelastic solid

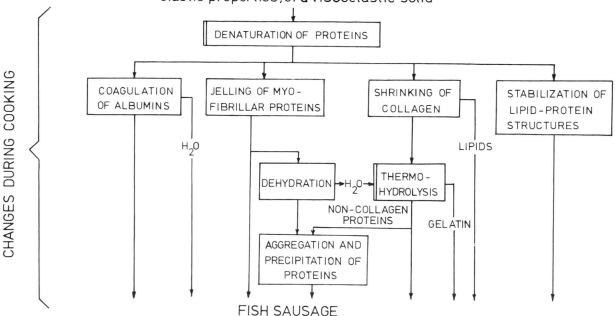

CHANGES DURING COOKING

DENATURATION OF PROTEINS

COAGULATION OF ALBUMINS

JELLING OF MYO-FIBRILLAR PROTEINS

SHRINKING OF COLLAGEN

STABILIZATION OF LIPID-PROTEIN STRUCTURES

H_2O

DEHYDRATION → H_2O → THERMO-HYDROLYSIS

LIPIDS

NON-COLLAGEN PROTEINS

GELATIN

AGGREGATION AND PRECIPITATION OF PROTEINS

FISH SAUSAGE

Jellied structure, displaying the rheological properties of a viscoelastic plastic solid

Fig 12   Changes in a sausage formulation due to heating

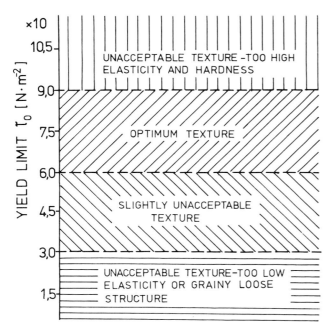

Fig 13 The texture of comminuted fish sausages (*From Sadowska, 1976*)

are being investigated with the aim of minimizing damage to the original protein structure or else to induce changes which were beneficial for the functional properties of the product (*Fig 15*) (Sikorski *et al*, in press). Thus, besides the standard isopropanol extraction which has been applied in industrial scale a range of procedures is being used experimentally in different laboratories, including among others low temperature extraction, hydrolysis, isolation of fractions and spinning into fibres (*Fig 16*), and the plastein reaction.

8 Further research needs

By following with understanding the recommendations given in textbooks and industrial codes of practice, which are the result of experience acquired over the centuries and of years of industrial research, it is possible to produce traditional fishery products of superior quality, providing that the raw material comes up to standard. In such cases there is no need for further fundamental research. Development work, however, is necessary to rationalize the process in order to save time and labour. On the other hand any attempts to produce, for example, a substitute having the quality of a mild cured matje herring, using a different raw material should be based on thorough study of the complex changes in proteins which arise due to the action of enzymes and interactions with other flesh constituents. The same is also true with respect to utilizing for human consumption the vast quantities of available unconventional marine resources, which, like the krill, are unsuitable for traditional processing (*Fig 17*). By using proper techniques they can be converted to valuable food products (Bykov, 1978; Sirkorski *et al*, 1979).

9 References

BIEGLER, P. Fischwaren—Technologie, In: *Der Fisch*, Vol 5, Ed C
1960 Baader. Der Fisch Clara Baader. Lübeck. 81pp
BOKSZCZANIN, J. Otrzymywanie strukturowanych sztucznych pro-
1978 duktów mięsnych z frakcji aktomiozynowych błękitka i
      antarktycznego kryla. Thesis, Politechnika Gdańska, Gdańsk
BREMNER, H A. Mechanically separated fish flesh from Australian
1978 species—a summary of results of storage trials. *Fd. Techn. in
      Australia*, 30, 393–401
BREMNER, H A, LASLETT, G M and OLLEY, J. Taste panel assessment
1978 of textural properties of fish minces from Australian species. *J.
      Fd. Technol.*, 13, 307–318
BUTTKUS, H. Accelerated denaturation of myosin in frozen solution, *J.
1970 Fd. Sci.*, 35, 558
BYKOV, V P. Osnovnye rezultaty tekhnologitseskikh issledovanij
1978 krilya. *Rybn. Khoz.*, 10, 60–64
CASTELL, C H. Metal-catalysed lipid oxidation and changes of proteins
1971 in fish. *J. Am. Oil Chem. Soc.*, 48, 645
CONNELL, J J. Fish muscle proteins and some effects on them of pro-
1964 cessing. In: *Proteins and their Reactions*, Eds H W Schultz and
      A F Anglemier. AVI, Westport, Conn. 255pp
—— The effect of freezing and frozen storage on the proteins of fish
1968 muscle. In: *Low Temperature Biology of Foodstuffs*, Eds J
      Hawthorn and E J Rolfe. Pergamon, Oxford. 333pp
DUNAJSKI, E. The texture of fish muscle, *J. Texture Studies*. (In press)
1979
JARENBÄCK, L and LILJEMARK, A. Ultrastructural changes during frozen storage of cod (*Gadus morhua* L.). III. Effects of linoleic
acid and linoleic acid hydroperoxides on myofibrillar proteins.
J. Fd. Technol., 10, 437

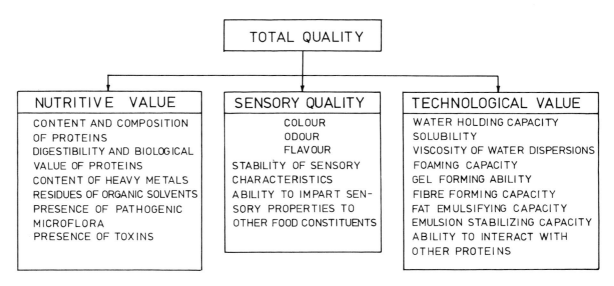

Fig 14   The attributes of quality of concentrated fish protein products

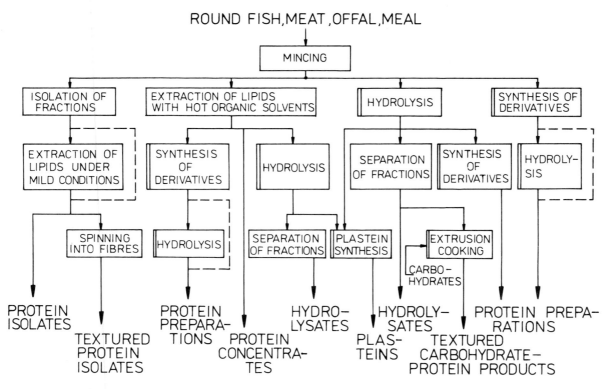

ROUND FISH, MEAT, OFFAL, MEAL

MINCING

ISOLATION OF FRACTIONS

EXTRACTION OF LIPIDS WITH HOT ORGANIC SOLVENTS

HYDROLYSIS

SYNTHESIS OF DERIVATIVES

EXTRACTION OF LIPIDS UNDER MILD CONDITIONS

SYNTHESIS OF DERIVATIVES

HYDROLYSIS

SEPARATION OF FRACTIONS

SYNTHESIS OF DERIVATIVES

HYDROLYSIS

SPINNING INTO FIBRES

HYDROLYSIS

SEPARATION OF FRACTIONS

PLASTEIN SYNTHESIS

EXTRUSION COOKING

CARBO-HYDRATES

PROTEIN ISOLATES

TEXTURED PROTEIN ISOLATES

PROTEIN PREPARA-TIONS

PROTEIN CONCENTRA-TES

HYDRO-LYSATES

HYDROLY-SATES

PLAS-TEINS

PROTEIN PREPA-RATIONS

TEXTURED CARBOHYDRATE-PROTEIN PRODUCTS

Fig 15   Main processes for manufacturing different concentrated fish protein products

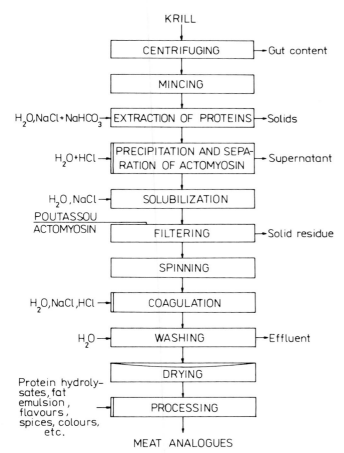

KRILL

CENTRIFUGING → Gut content

MINCING

H₂O, NaCl + NaHCO₃ → EXTRACTION OF PROTEINS → Solids

H₂O + HCl → PRECIPITATION AND SEPA-RATION OF ACTOMYOSIN → Supernatant

H₂O, NaCl → SOLUBILIZATION

POUTASSOU ACTOMYOSIN

FILTERING → Solid residue

SPINNING

H₂O, NaCl, HCl → COAGULATION

H₂O → WASHING → Effluent

DRYING

Protein hydroly-sates, fat emulsion, flavours, spices, colours, etc. → PROCESSING

MEAT ANALOGUES

Fig 16   A flow sheet of a process for preparing textured krill protein isolate (*From Bokszczanin, 1978*)

KOLODZIEJSKA, I and SIKORSKI, Z E. Inorganic salts and functional
1979   properties of fresh and frozen muscle proteins, *Proc:* 25th
European Meeting of Meat Research Workers, Budapest
KOSTUCH, S and SIKORSKI, Z E. Interaction of formaldehyde with cod
1977   proteins during frozen storage. *Inst. Int. Refrig. Comm. C1 and
C2, Karlsruhe*
LUIJPEN, A F M G. De invloed van het kaken op de rijping van
1959   gezouten Maatjesharing, Diss. Schotanus u. Jens, Utrecht,
after V. Meyer, *Arch. Fischereiwiss.*, 15/1964/, 246–251
PARTMANN, W. Some aspects of protein changes in frozen foods, *Z.*
1977   *Ernährungswiss.*, 16, 167–175
REAY, G A. The low temperature preservation of the haddock. Dep.
1931   Sci. Ind. Res., Fd. Invest. Board, 128
—— Freezing and cold storage of fish. Dep. Sci. Ind. Res., Fd. Invest.
1932   Board, 202
REDDI, P S, CONSTANTINIDES, S M and DYMSZA, H A. Catheptic ac-
1972   tivity of fish muscle. *J. Food Sci.*, 37, 643–647
SADOWSKA, M. Modelowanie reologicznych właściwości układów
1976   białkowo-lipidowo-wodnych. Thesis, Politechnika Gdańska,
Gdańsk
SADOWSKA, M and SIKORSKI, Z E. Interaction of different animal pro-
1976   teins in the formation of gels. *Lebensm. – Wiss. u. –Technol.*,
9, 207–210
SADOWSKA, M and SIKORSKI, Z E. Evaluation of technological suitabil-
1977   ity of fish meat by rheological measurements. *Lebensm.
Wiss. u. Technol.*, 10, 239–245
SIKORSKI, Z E, OLLEY, J and KOSTUCH, S. Protein changes in frozen
1976   fish. Crit. rev. In: *Food Sci. and Nutr.*, 8, 97–129
SIKORSKI, Z E. Protein changes in muscle foods due to freezing and
1978   frozen storage. *Int. J. Refrig.*, 1, 173–180
SIKORSKI, Z E, BYKOWSKI, P and KNYSZEWSKI, J. The utilization of
1979   krill for food, *Proc:* Second International Congress on
Engineering and Food – Food Process Engineering 1979, Hel-
sinki University of Technology, Espoo, Finland
SIKORSKI, Z E, NACZK, M and OLLEY, J. Modification of functional
properties of fish protein concentrates. Crit. rev. In: *Food Sci.
and Nutr.* In Press.
TOKUNAGA, T. Studies on the development of dimethylamine and for-
1964   maldehyde in Alaska pollack muscle during frozen storage. I.
Torry Research Station, Translation No. 743, Rep. Hokkaido
Reg. Fish. Res. Inst., 29, 108

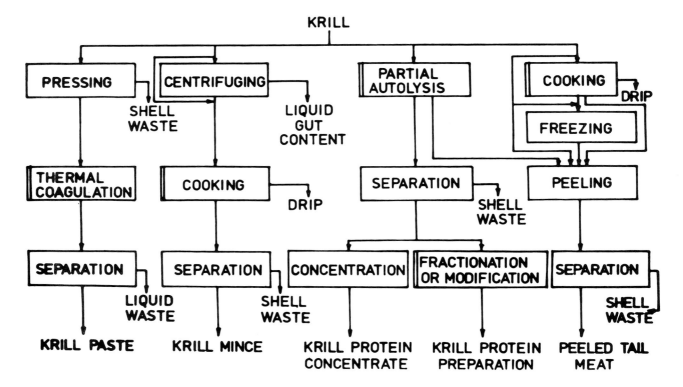

Fig 17 A general flow sheet of krill processing

Fish lipids. Part 1

$\bar{R}\ G\ Åckman$

1 Introduction

'Jack Sprat could eat no fat
His wife could eat no lean'

This classic rhyme illustrates one of the major problems of modern food science, that of catering to the varying demands of a public who are subjected to a barrage of often ill-advised directives in what and what not to eat. The producers of the popular animal meats such as chicken, beef and pork have developed highly standardized packages of protein, and with a few exceptions such as bacon prefer not to even mention fat, as well as to eschew mention of the even more notorious cholesterol. What is the situation with fish? The well-informed discriminating consumer can in fact find a wide choice of fish and fishery products providing pleasure for the palate as well as all necessary protein, and vitamins and minerals in desirable proportions (Piggott, 1976). The problem is to make available the existing knowledge on fats, and fat soluble materials, in clear and concise form.

The functionality of fat in edible fish and fish parts is often overlooked. What we call fat generally carries natural flavour components and provides and preserves others generated during cooking, pickling or other processing. A certain amount of fat and fatty acid assists in providing smoothness of texture during mastication of lean fish (Yamada, 1972; Childs, 1974). In fatty fish the influence of fat on texture is even more important. What is this fat and how does it differ from other fats?

The decade from approximately 1955 to 1965 was probably one of the most fruitful in fisheries technology.

Part of the advance stemmed from the realization that the water, fat and protein of fish were not really separate entities but interacted in the edible product to affect quality, especially during frozen storage. It is now 1979, and after some 25 years a new generation of fisheries scientists could perhaps benefit from a reminder of the seminal work of colleagues who are no longer available to guide work on the muscle lipid problems of the newer species now entering many fisheries. A selection of papers, by no means complete or even comprehensive, where lipids and fatty acids were particularly important factors includes Dyer and Morton (1956), Dyer *et al* (1958), Ironside and Love (1958), Love (1958), Love and Ironside (1958), Dyer and Fraser (1959), Fraser and Dyer (1959), Lovern *et al* (1959), Olley and Lovern (1960), Bligh (1961), Fraser *et al* (1961), Castell *et al* (1962), Lovern (1962a, b), Lovern and Olley (1962), Olcott (1962), Castell and MacLean (1964a, b), MacLean and Castell (1964), Anderson *et al* (1965), Castell *et al* (1965), Hanson and Olley (1965), Olley and Duncan (1965), Bligh and Scott (1966), Castell *et al* (1966a, b), Sheltawy and Olley (1966), Dyer (1967), Roubal (1967), Takama *et al* (1967), Castell and Spears (1968), Castell *et al* (1968), Dyer (1968), Addison *et al* (1969), Bosund and Ganrot (1969), Castell and Bishop (1969), Olley *et al* (1969), Castell (1971), and Braddock and Dugan (1972).

One key to these studies was the abandonment of the practice of studying the fats of lipids of fish muscle by first drying the tissue to chipboard, and then extracting what one could of the so-called fat by prolonged extraction with the traditional diethyl ether (Anon, 1945). In general, freeze-drying should also be avoided, as it leaves lipids susceptible to oxidation (Koizumi et al, 1978). Instead, the water-miscible solvents acetone or isopropanol (Dambergs, 1956, 1959, 1963, 1964, 1969a, b; Drozdowski and Ackman, 1969), or the more sophisticated chloroform-methanol homophase systems derived from the Folch procedure (Bligh and Dyer, 1959; Hanson and Olley, 1963a; b; Lovern, 1965; Balsingham, 1972; for special purposes see also Korn and Macedo, 1973; Hurst, 1974; Rzhavskaya et al, 1977) should be used. Such lipid extraction systems have revealed that fish flesh (and also that of shellfish) contains not just fat, but fairly complex lipid systems. Some of these systems are exemplified in *Table I*.

The second key lay in the adaption of silicic acid absorption chromatography to marine lipids, initially in columns (Bligh and Scott, 1966) and later in the form of thin-layer chromatography (TLC) (Roubal, 1967). The separations thus achieved showed that *lean* white fish muscle contained a minimum of about 0·7% of basic cellular lipid, of which 85–95% was 'polar' lipids, mostly phosphatidyl ethanolamine and phosphatidyl choline. Some details of analyses of this type are included in *Tables I* and *II*. The balance of this type of basic lipid includes sterol ester and free sterol, free fatty acid (very little in fresh samples, rather more after storage (see

Section 2)), and triglyceride. Eventually it became apparent that this basic mixture represented the structural lipid of cell walls, and that any excess of triglyceride and/or certain other non-polar lipids such as wax esters or glyceryl ethers, provided the 'fat' of fatty fish. In this particular discussion the term *lipid* will be used for total multiple-component systems, and the term *fat* for selected anatomical deposits which are patently mostly triglyceride even to the naked eye. The seasonal and anatomical variations, for which a few examples are given in *Table II*, will be discussed in more detail below.

The third key to our understanding of fish fats and lipids was the introduction of gas-liquid chromatography (GLC). By this means the details of the fatty acids of both lipids and fats of marine origin become available (*Fig 1*). It is a curious fact that as few as 10, or even eight (Lambertsen, 1978), fatty acids fairly accurately describe the composition of most fish oils. Lists of these usually start with the saturated acids; 14:0[1] (myristic or tetradecanoic), 16:0 (palmitic or hexadecanoic) and 18:0 (stearic or octadecanoic). The next group to be considered includes monoethylenic acids which can occur in several isomers and which will be discussed in

---

[1] A shorthand notation will be employed, with the chain length and number of *cis* ethylenic bonds (*eg* 14:0), followed by the position of the bond closest to the terminal methyl group counted from the terminal methyl group (the ω carbon) and denoted ωn (*eg* cis,cis-9,12-octadecadienoic acid would be 18:2ω6).

Table I

DISTRIBUTION OF SPECIFIC PHOSPHOLIPIDS AND MAJOR NEUTRAL LIPIDS IN TISSUES OF FOUR INVERTEBRATES, AND IN MUSCLE OR BODY OF FIVE FISH

Name	Oyster[1]	Abalone[2]	Crab[3]	Crayfish[4]	Cod[5]	Hake[6]	Trout[7]	Pilchard[8]	Wrasse[9]
Species	O. edulis	H. midae	C. opilio	A. pallipes*	G. morhua	M. capensis	S. iridis*	S. ocellata	T. duperreyi
Sample	total flesh	total flesh	muscle	muscle	white muscle	muscle	muscle	whole	muscle
Total lipid (g/100g sample)	1·7	1·1	0·75	1·77	0·59	1·55	5·3	5·0	0·5–1·0
Polar lipid (g/100g sample)	0·71	0·69	0·56	0·53	0·52	0·46	0·63	0·91	0·5
% phosphatidyl choline	—	37	62	54	69	63	67	53	59
% phosphatidyl ethanolamine	—	32	28	26	19	21	21	25	13
% sphingomyelin	—	2	5	5	—	4	2	6	12
% phosphatidyl serine	—	—	—	11†	5	3	4	2	10
Neutral lipid (g/100g sample)	0·69	0·30	0·19	1·24	0·07	1·0	4·68	3·9	—
% triglyceride in NL	58	—	33	91	35	~100	94	~100	~100
% sterol in NL	28	40	46	9‡	50	—	1·1	—	—

[1] Watanabe and Ackman, 1974	[6] de Koning, 1966a	* Freshwater species
[2] de Koning, 1966b	[7] Gray and MacFarlane, 1961	† Includes phosphatidyl inositol
[3] Addison et al, 1972	[8] de Koning and McMullan, 1966	‡ Cholesterol figure is 0·12g/100g sample
[4] Cossins, 1976	[9] Patton, 1975	
[5] Addison et al, 1968		

Table II

EXAMPLES OF DIFFERENCES IN LIPID COMPOSITION BETWEEN LIGHT AND DARK MUSCLE, OR WITH SEASON, IN SOME MACKEREL SPECIES

Name	Mackerel[1]		Mackerel[2]		Mackerel[3]	Mackerel[4]		Black Sea scad[5]	
Species	S. scombrus		S. scombrus		S. scombrus	S. japonicus		T. med. pont	
Sample	Dec. m.	June m.	Light m.	Dark m.	Dorsal m.	Aug. m.	Jan. m.	Light m.	Dark m.
Total lipid (g/100g sample)	24·1	9·1	10·2	14·4	2·1	10·8	15·5	4·9	10·5
Polar lipid (g/100g sample)	0·84	0·88	0·5	1·6	0·85	1·1	0·99	1·15	2·07
Neutral lipid (g/100g sample)	22·2	7·9	9·1	10·7	1·23	9·4	14·4	2·2	6·1
% triglyceride in NL	97	91	90	74	—	—	—	—	—
% sterol in NL	3	4	—	—	—	—	—	—	—

[1] Hardy and Keay, 1972	[3] Viviani et al, 1967	[5] Shchepkin et al, 1974
[2] Ackman and Eaton, 1971a	[4] Ueda, 1976	

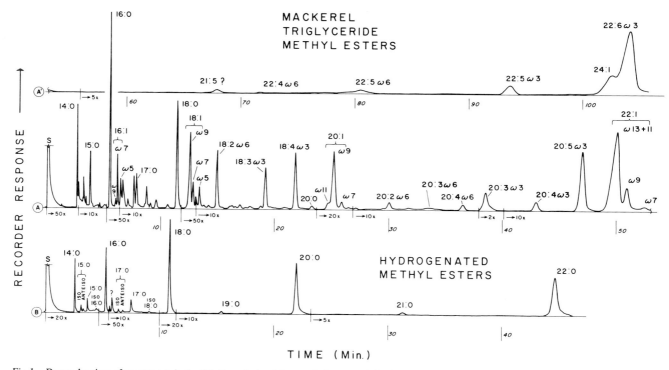

Fig 1 Reproduction of an open-tubular GLC analysis of the methyl esters of the body triglycerides of a Canadian Atlantic mackerel *Scomber scombrus*, and of the same methyl esters after total hydrogenation (*From Ackman and Eaton, 1971a; reproduced by permission of* Can. Inst. Food Sci. Technol. J.)

greater detail later. In summary, the four common monoethylenic fatty acids (and in each chain length, the major isomer) are: 16:1ω7 (palmitoleic or 7-hexadecenoic); 18:1ω9 (oleic or 9-octadecenoic); 20:1ω9 (gadoleic or 11-eicosenoic); and 22:1ω11 (cetoleic or 11-docosenoic). In practice (*Tables III and IV*) these can be summarized simply by chain length and number of bonds unless biochemical considerations based on the different monoethylenic positional isomers are under discussion. This composite monoethylenic

data is the result of gas liquid chromatographic analysis (GLC) on packed columns (Ackman, 1969). The separation of most monoethylenic fatty acid isomers can be achieved through the use of open-tubular (capillary) columns (Ackman, 1972). The polyethylenic fatty acids are a special problem as they are very susceptible to oxidation during lipid isolation and later in handling as fatty acids or methyl esters, including the final step of GLC (Ackman *et al*, 1967b). They are *not* stable in frozen fish muscle or organs (see Section 2), and

Table III

COMPARISON BETWEEN (ω/ω%) NORTH ATLANTIC PHYTOPLANKTER FATTY ACIDS AND THOSE IN SPECIFIC LIPIDS OF SELECTED INVERTEBRATES

Sample	12 Phytoplankters[1]			Oyster flesh[2]		Oc. quahaug[3]		Euphausiid[4]		Euphausiid[5]		Euphausiid[6]		Prawn, cultivated[7]	
			Aver-												
Species	Range	No.	age	*O. edulis*		*A. islandica*		*M. norvegica*		*E. superba*		*E. pacifica*		*M. rosenbergii*	
Lipids	Total lipid			PL	TG	PL	TG	PL	TG	PL	TG	PL	TG	Total (a)	Total (b)
Fatty acids															
14:0	0·5–33	12	6·8	6·0	8·1	0·4	3·5	1·2	7·7	1·8	6·9	2·2	14·0	1·5	1·1
16:0	7–37	12	22·3	31·5	42·0	12·9	32·4	18·5	19·0	23·3	17·9	19·7	19·3	14·9	15·2
18:0	0·1–5·7	12	1·2	12·7	7·5	6·3	7·9	1·5	2·4	0·7	0·9	1·3	1·7	8·0	8·3
16:1	0·3–47	12	16·9	4·0	7·6	2·0	13·5	1·7	5·1	2·7	6·0	3·9	8·3	1·0	1·3
18:1	0·3–15	12	4·1	4·0	11·3	4·8	12·4	17·9	18·8	15·0	17·1	15·5	16·2	19·7	17·3
20:1	0·1–0·6	8	0·3	4·0	3·6	2·8	7·4	2·1	9·9	0·4	0·4	0·7	1·9	—	—
22:1	0·3	1	—	1·4	0·9	—	0·4	0·1	8·0	0·2	0·4	—	0·4	—	—
18:2ω6	0·1–14	12	3·4	0·9	1·3	0·3	0·4	2·0	1·5	3·7	3·2	3·6	3·3	29·8	24·2
18:3ω3	0·1–37	10	7·9	4·9	1·1	1·3	1·3	0·8	1·0	1·2	1·3	1·6	1·9	2·5	2·0
18:4ω3	0·1–24	12	6·4	1·0	0·6	—	0·7	0·4	2·4	0·9	1·7	1·1	4·2	—	—
20:4ω6	0·1–25	7	2·4	2·6	0·5	2·1	0·2	0·9	0·3	0·3	0·4	2·4	1·2	2·4	3·2
20:5ω3	4–22	11	10·7	6·9	2·2	15·3	5·5	15·7	7·2	16·5	17·6	18·7	6·3	7·0	12·1
22:5ω6	0·4–1·1	4	0·2	—	—	0·2	—	0·4	0·1	—	—	—	—	—	—
22:5ω3	0·1–0·6	6	0·1	—	—	2·2	0·4	0·5	0·3	0·6	0·4	—	—	—	—
22:6ω3	0·5–25	10	4·9	2·3	0·9	23·9	2·8	28·6	8·0	14·4	9·2	22·9	4·4	3·2	5·0
Totals	—	—	87·6	82·2	87·6	74·5	88·8	92·3	91·7	81·7	83·4	93·6	83·1	90·0	89·7

PL = phospholipids, TG = triglycerides
[1] Ackman *et al*, 1968
[2] Watanabe and Ackman, 1974
[3] Ackman *et al*, 1974b
[4] Ackman *et al*, 1970
[5] Mori and Hikichi, 1976
[6] Takahashi and Yamada, 1976
[7] Sandifer and Joseph, 1976: the (a) group were fed a commercial diet, the (b) group the same augmented with 'ω3' fatty acids from the lipids of the shrimp *Penaeus setiferus*

freeze-drying is a very risky business. Fortunately, in this review the discussion will be based on published and unpublished fatty acid data, taken at its face value, since marine oils tend to be stabilized somewhat by their natural antioxidants such as α-tocopherol (Astrup, 1964; Ackman and Cormier, 1967; Hardy and Mackie, 1969; Chernyshov, 1972; Higashi et al, 1972; Ackman, 1974), while squalene, carotenoids and vitamins compete with polyethylenic fatty acids for oxygen. In a tank of fish oil the top few centimetres may have no dissolved (free O_2) oxygen because there are so many oxygen-labile compounds and structures available to interact with oxygen in any form (Heide-Jensen, 1965). Similarly, an apparent *decrease* in the Bunsen coefficient for oxygen in marine oils with increasing temperatures (Ke and Ackman, 1973) probably reflects the greater reactivity of oxygen and polyethylenic fatty acids at higher temperatures. Mild autoxidation is normal in fish oils, but fortunately is relatively well tolerated by animals (Lang, 1965).

Curiously enough, this susceptibility to autoxidation has not often seriously interfered with the isolation of polyethylenic fatty acids by degree of unsaturation through the use of argentation TLC (Bottino, 1971; Ackman, 1972; Bandyapadhyah and Dutta, 1975; Sen et al, 1976). It has been shown that the peroxides which readily form in diethyl ether can lead to epoxide formations in the presence of silver ion (Chen et al, 1976) but this technique is otherwise extraordinarily useful (Sen et al, 1976; Ota and Yamada, 1975). The polyethylenic fatty acids can be stabilized for isolation, separation and comparison by forming mercuric acetate or similar adducts (Ota and Yamada, 1969; Pohl et al, 1969; Wagner and Pohl, 1964; White, 1966). Urea complexing (Ackman et al, 1963; Ackman and Hooper, 1968; Ghosh et al, 1976) is useful for effecting concentrations on a large scale, but is not widely used due to lack of specificity as to type of fatty acid separated. For practical purposes, commercial fish and marine mammal 'oils' are triglycerides, the notable exception being sperm whale oil which includes wax esters. As triglycerides, marine oils freely enter many food industries (Gauglitz et al, 1974). These or any other oils can be investigated by GLC of the intact molecules (Litchfield et al, 1971, 1978; Litchfield, 1972; Takagi et al, 1977). The procedure shows wax esters as a group of peaks separate from triglycerides (Ackman et al, 1975b) and special triglyceride details can be observed which complement total fatty acid analyses (*Fig 2*). For example oils from freshwater fish which contain C_{18} acids in high proportions (Mitra and Dua, 1978) show a distribution curve with a maximum at approximately $(1 \times C_{16}) + (2 \times C_{18}) = C_{52}$ components, whereas marine fish oils show more longer-chain combinations, perhaps with a maximum at $(C_{54}–C_{56})$ (compare Litchfield, 1972 with Ikekawa et at, 1972 and Matsui et al, 1976). The arrangement of something like 10 major (5–25%) fatty acids, 25 minor (0·5–5%) fatty acids, and as many more fatty acids in trace amounts, in various combinations of threes (Bottino, 1971) is the real reason why some marine oils cannot be efficiently fractionated (eg herring oil mDrozdowski and Ackman, 1969) while others can be simply winterized (menhaden, see below) or treated with solvents such as furfural and petroleum

Table IV

COMPARISONS BETWEEN FATTY ACIDS OF POLAR LIPIDS AND TRIGLYCERIDES OF VARIOUS FISH WITH DIFFERENT FEEDING PATTERNS AND HABITATS

PL=phospholipids, TG=triglycerides, PE=phosphatidyl ethanolamine, PC=phosphatidyl choline

Sample Species	Atl. menhaden[1] B. tyrannis		Chil. anchovy[2] E. ringens		Atl. herring[3] C. harengus		Atl. herring oils[4] C. harengus	Capelin M. villosus			Cod flesh[8] G. morhua			Cod liver[8] G. morhua	Pac. herring C.h. pallasi	Pac. sablefish[5] A. fimbria	Pac. albacore tuna[9] T. alalunga	Landlocked Coho salmon[10] O. kisutch	
Lipid Fatty acid	PL	TG	PL	TG	PL	TG	Range of 12 I.V.[5]	Lean[6]	Fat[7]	TG	PE	PC	TG	Total (TG)	Comm. Oil[4]	Flesh	Total	PL	TG
14:0	1·6	8·1	4·5	11·2	1·8	5·6	4·6–8·4	9·2	6·2		0·5	1·4	4·4	4·9	5·7	6·7	2·4	3·0	3·2
16:0	24·6	25·4	25·2	20·4	21·4	12·5	10·1–15·0	8·5	8·9		8·8	23·9	14·1	12·4	16·6	11·1	11·9	15·6	10·3
18:0	9·4	4·3	5·7	6·8	3·2	1·1	0·7–2·1	1·8	0·5		4·4	2·6	2·6	1·8	1·8	1·9	3·5	4·7	3·5
16:1	3·6	8·4	7·0	7·9	4·6	13·6	6·3–12·0	15·1	13·6		1·7	4·2	12·0	11·6	7·6	6·7	8·0	8·9	10·7
18:1	18·5	17·9	15·0	12·2	13·0	15·5	9·3–21·4	7·6	7·7		14·4	12·6	22·4	22·6	22·7	29·0	15·6	25·3	21·2
20:1	0·7	1·8	2·6*	2·8*	2·4	13·7	11·0–19·9	22·7	19·9		4·0	1·7	12·1	7·6	10·7	18·1	3·0	2·3	3·4
22:1	<0·1	0·1	2·0†	2·3†	1·6	19·4	14·8–30·6	19·6	20·0		0·3	0·4	11·4	5·2	12·0	14·8	1·2	0·6	0·5
18:2ω6	1·5	1·7	3·2	3·3	0·3	1·1	0·6–2·9	0·4	1·0		0·8	0·9	0·4	1·4	0·6	0·2	2·1	3·6	5·6
18:3ω3	0·7	2·3	–*	–*	0·2	0·3	0·2–1·1	0·1	0·2		0·3	0·2	0·2	1·3	0·4	0·2	1·3	2·6	2·9
18:4ω3	0·6	3·9	1·5	2·5	1·4	1·2	1·1–2·5	0·9	2·0		0·3	0·3	1·2	1·9	1·6	0·3	1·6	0·7	0·8
20:4ω6	2·1	0·3	–†	–†	0·3	0·3	0·2–0·5	0·1	0·1		1·6	2·4	0·5	0·5	0·4	0·3	0·9	1·3	1·6
20:5ω3	7·5	10·1	7·4	10·1	12·2	6·8	3·9–8·8	6·1	9·3		10·7	19·4	7·3	12·6	8·1	1·4	9·3	3·7	4·8
22:5ω6	0·3	0·2	–†	–†	0·2		0·1–0·4	0·1	0·1						0·2	0·1		0·6	1·8
22:5ω3	0·5	0·4	0·9	1·0	0·8	0·2	0·5–1·3	0·4	0·5		1·9	1·6	0·4	1·0	0·8	0·5	0·5	4·1	4·4
22:6ω3	23·2	8·3	15·0	9·2	32·7	3·1	2·0–6·2	2·6	3·7		40·5	27·3	7·4	10·6	4·8	1·0	28·7	9·9	10·6
Totals	94·8	93·2	90·0	89·7	96·7	94·4	—	95·1	93·7		90·2	98·9	96·9	95·4	94·0	92·3	90·0	86·9	85·3

[1] Ackman et al, 1976
[2] Masson and Burgos, 1973
[3] Addison et al, 1969
[4] Ackman and Eaton, 1966a
[5] Ackman et al, 1969
[6] Ackman et al, 1967a
[7] Eaton et al, 1975
[8] Addison et al, 1968
[9] Ueda, 1967
[10] Braddock and Dugan, 1972

* Figure for combined 20:1 and 18:5ω3, probably roughly equal
† Figure for combined 22:1 and 20:4ω6, likely with 22:1>20:4

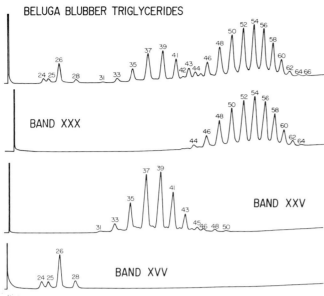

BELUGA BLUBBER TRIGLYCERIDES

BAND XXX

BAND XXV

BAND XVV

Fig 2 Reproduction of the GLC analyses (above) of intact (hydrogenated) triglycerides of blubber fat of the beluga whale *Delphinapterus leucas* and (below) of the respective TLC bands of triglycerides containing no isovaleric acid (XXX), one isovaleric acid molecule (XXV), and two isovaleric acid molecules (XVV). The numbers on the peaks refer to the total numbers of carbon atoms in the fatty acid chains, the distribution in the XXX band being similar to that of most marine fish triglycerides (*From Litchfield* et al, *1971; reproduced by permission of* Lipids)

naphtha (Chilean anchovy, Contraras *et al*, 1971) or assorted solvents (Indian oil sardine, Revankar *et al*, 1975). Once the fatty acids are split off the glycerol they, or derivative esters, can be separated (Schlenk and Sand, 1967).

Later work has been done in this direction on both laboratory (Jangaard, 1965) and semi-commercial scales (Jangaard, 1966). The simple process of winterization of menhaden oil does produce two fractions of different iodine values (*eg* 178 and 126; *Table V*), but in the case of Atlantic herring oil the solid layer usually has an iodine value only a few units lower than the liquid phase. Little progress in this field has been made in recent years and the subject is adequately covered by Bailey *et al* (1952). The commercial fish oils which do form stearine layers, *eg* menhaden (*Table V*) and (formerly) the North Pacific pilchard *Sardinops sagax* are low (ω5%) in 20:1 and 22:1 fatty acids (Ackman and Sipos, 1964), have high iodine values (165–195), and are thus set apart from the clupeid type oils of lower iodine value (100–135) which will be considered below. The work on the distribution of the fatty acids in the triglyceride molecules is limited, but does suggest that in the fish oil triglycerides there is usually a polyethylenic fatty acid such as 20:5ω3 or 22:6ω3 in the 2 position, a saturated or monoethylenic fatty acid in the 1 position, and likely a monoethylenic fatty acid in the 3 position (Malins and Wekell, 1970; Litchfield, 1972). However, none of these are hard and fast rules because of the superabundance of fatty acid types and other factors (Leger *et al*, 1977), and it should be emphasized that triglycerides of marine mammals are different, with the polyethylenic fatty acids concentrated at position 1 and 3. Wax esters, *etc* will be considered separately (see below).

2 Distribution of fat in fish

The biochemistry, transport and metabolism of fat in fish is adequately reviewed elsewhere (Sargent, 1976a). As shown in *Table II* the distribution of lipids of the cellular type in muscle is such that in lean fish dark (red, or lateral line) muscle has about twice the lipid of white muscle (*eg* skipjack tuna, Nishimoto and Takebe, 1977). The percentage of cellular lipid in the latter is normally not seriously altered by season (illustrated in depth for cod *Gadus morhua* by Jangaard *et al*, 1967; for sardine *Sardinops melanosticta* by Hayashi and Takagi, 1977; for chub mackerel *Scomber japonicus* by Ueda, 1976; 1977; and for 'horse' mackerel *Trachurus japonicus* by Toyomizu *et al*, 1976. However, the effect of season and activity on cellular lipid is a complex question beyond the scope of this section of this review (Johnston and Goldspink, 1973; Patterson *et al*, 1974; Lambertsen and Hansen, 1978; Love *et al*, 1975; 1977). Triglyceride distributed through fish muscle tends to have a homogeneous fatty acid composition (Kobayashi *et al*, 1973; Wessels and Spark, 1973; Ackman *et al*, 1975a).

The lean muscle fish generally have fatty livers (*eg* cod, Jangaard *et al*, 1967) which show seasonal effects, but in other species the muscle can show fluctuating levels of neutral fat (*ie* mostly triglyceride; *eg* sprats, Hardy and Mackie, 1969; menhaden, Dubrow *et al*, 1976; or various mackerel species, *Table II*). Ueda (1967, 1977) and Hayashi and Takagi (1977) also show in detail variations in the neutral fat in muscle. In lean muscle distribution of this muscle fat is best shown histologically (Yamada and Nakamura, 1964; Yamada, 1972; Nishimoto and Takebe, 1977) but can be considered on a gross anatomical basis if a good extraction method is used (Brandes and Dietrich, 1958; Mannan *et al*, 1961a,b; Fraser *et al*, 1961; Kobayashi *et al*, 1973; Nishimoto and Takebe, 1977).

The belly flap (*Fig 3*) is a notoriously fat section of many fish bodies (*eg* 29% lipid in belly flap of mackerel versus 18·3% lipid in dark muscle and 7·6% in light muscle; Ackman and Eaton (1971a)). More recently the role of subdermal fat has been recognized. For example, mackerel skin is 50% fat (Ke and Ackman, 1976) and in male mackerel the skin fat can total 40% of the fat in the whole fish (*Fig 4, cf* Lohne, 1976). Possibly this fat is functional in assisting water flow past the fish body (*cf* Bone, 1972). The distribution of fat in North sea capelin, herring and mackerel is thoroughly documented by a series of papers from Norway (Mohr, 1972; Flo *et al*, 1972; Mohr and Ormberg, 1977; Mohr *et al*, 1973; Lohne, 1976; Mohr *et al*, 1976a,b). The same problems apply in freshwater fish, but have not received much attention in recent years (Jafri, 1973) except for eels (Sumner and Hopkirk, 1976; Wills and Hopkirk, 1976; Polesello *et al*, 1977) and carp (Steffens, 1974). Some fat distribution data can be adduced from proximate composition tables for both freshwater and marine fish muscle (Bonnet *et al*, 1974; Sidwell *et al*, 1974; Exler *et al*, 1975; and Kinsella *et al*, 1977). Latterly special cases of fat deposition in fish in fish bone, swim bladder, lipid sacs, *etc*, have been much discussed (*eg* Butler and Pearcy, 1972; Bone, 1972; 1973; Lee *et al*, 1975; Phleger, 1975; Phleger and Grimes, 1976; Phleger *et al*, 1976; DeVries and Eastman, 1978; Phleger *et al*, 1978).

The whole problem of fat and fat distribution is inexorably intertwined with rearing of marine fish (Owen *et al*, 1972; Patterson *et al*, 1974; Kamoi *et al*, 1975; Stirling, 1976) or with aquaculture (Takeuchi, 1978) of salmonids (Robinson and Mead, 1973; Sidorov *et al*, 1977; Shabalina and Ostroumova, 1976), catfish (Gibson and Worthington, 1977) and carp (Aman and Smirnova, 1973; Farkas *et al*, 1977; 1978; Takeuchi *et al*, 1978). The use of artificial food and growth conditions make much of the literature data irrelevant to any discussion of 'normal' fat distributions or of fatty acid composition data. Most species of marine organisms try to obtain an optimal fat and fatty acid composition and their behaviour and food preferences lead towards this objective. However, simple fat-water-protein relationships are likely to be rare (Hardy and Mackie, 1969; Dabrowsky, 1978).

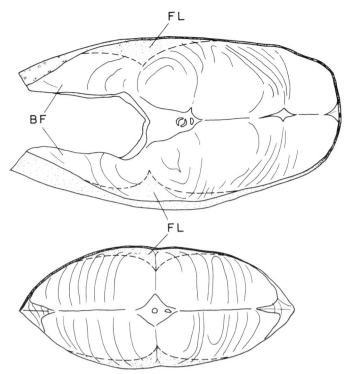

Fig 3 Cross-sections typical of fish bodies. In the forepart (above) the belly flap (BF) encompasses the visceral cavity and the flank muscle (FL, also known as the red, dark, or lateral line muscle) extends along the body through the commercially important midsection (below), often thickening towards the tail. The species shown is the halibut *Hippoglossus hippoglossus (Redrawn from Mannan* et al, *1961a; reproduced by permission of* J. Fish. Res. Bd. Can.)

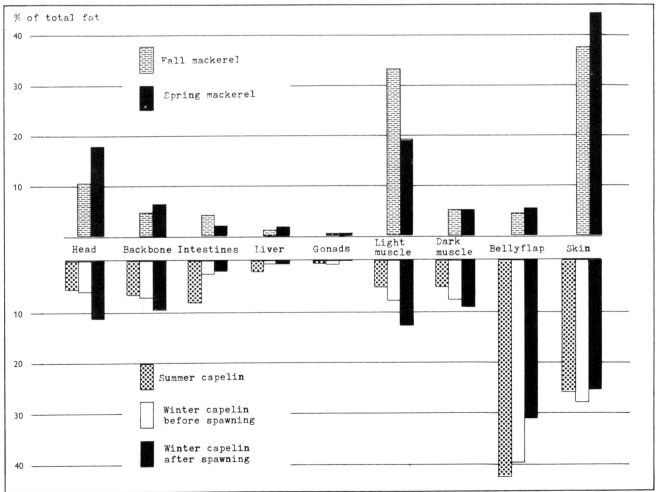

Fig 4 Distribution of total fat in various body parts and organs of mackerel *S. scombrus* (above) and capelin *Mallotus villosus* (below) of Norwegian origin. Note that the percentage of fat in a given part (*eg* mackerel dark muscle) can be higher than in a larger part such as light muscle which may contain more of the total fat (*From Lohne, 1976; Fisheries and Marine Service translation No. 4119, Ottawa; reproduced by permission of* Meld. fra S.S.F.)

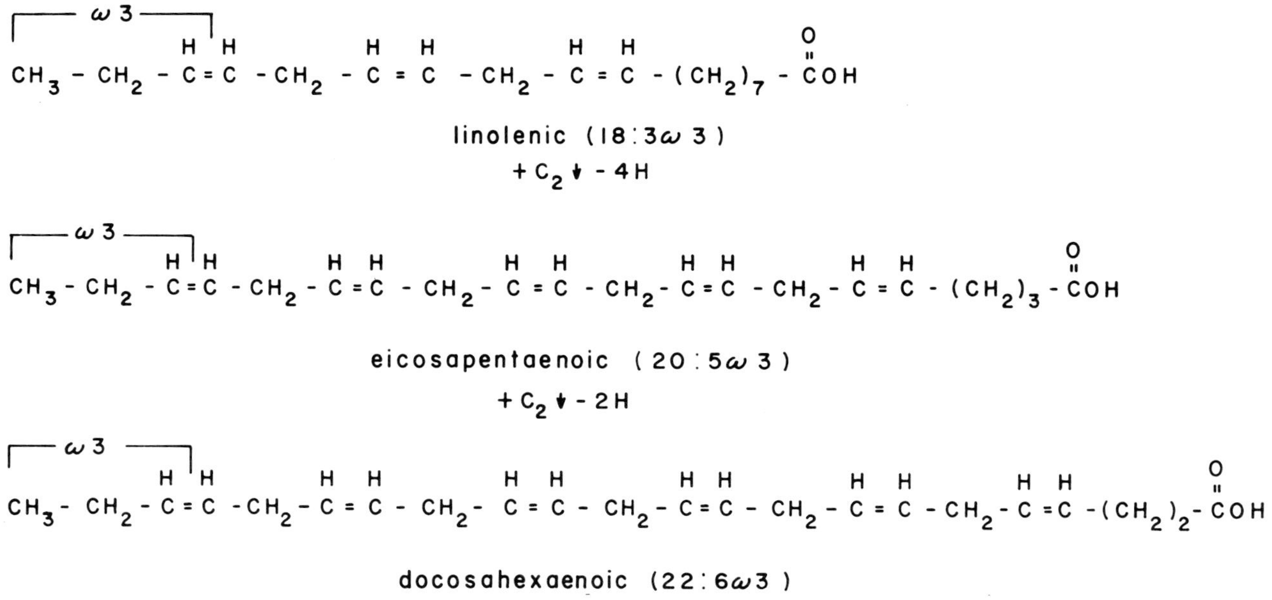

$$CH_3 - CH_2 - \overset{H}{\underset{}{C}} = \overset{H}{\underset{}{C}} - CH_2 - \overset{H}{\underset{}{C}} = \overset{H}{\underset{}{C}} - CH_2 - \overset{H}{\underset{}{C}} = \overset{H}{\underset{}{C}} - (CH_2)_7 - \overset{O}{\overset{\|}{C}}OH$$

(with $\omega 3$ bracket shown above the terminal CH₃–CH₂–C=C)

linolenic (18:3ω3)

+ C$_2$ ↓ - 4H

$$CH_3 - CH_2 - \overset{H}{\underset{}{C}} = \overset{H}{\underset{}{C}} - CH_2 - \overset{H}{\underset{}{C}} = \overset{H}{\underset{}{C}} - CH_2 - \overset{H}{\underset{}{C}} = \overset{H}{\underset{}{C}} - CH_2 - \overset{H}{\underset{}{C}} = \overset{H}{\underset{}{C}} - CH_2 - \overset{H}{\underset{}{C}} = \overset{H}{\underset{}{C}} - (CH_2)_3 - \overset{O}{\overset{\|}{C}}OH$$

eicosapentaenoic (20:5ω3)

+ C$_2$ ↓ - 2H

$$CH_3 - CH_2 - \overset{H}{\underset{}{C}} = \overset{H}{\underset{}{C}} - CH_2 - \overset{H}{\underset{}{C}} = \overset{H}{\underset{}{C}} - CH_2 - \overset{H}{\underset{}{C}} = \overset{H}{\underset{}{C}} - CH_2 - \overset{H}{\underset{}{C}} = \overset{H}{\underset{}{C}} - CH_2 - \overset{H}{\underset{}{C}} = \overset{H}{\underset{}{C}} - CH_2 - \overset{H}{\underset{}{C}} = \overset{H}{\underset{}{C}} - (CH_2)_2 - \overset{O}{\overset{\|}{C}}OH$$

docosahexaenoic (22:6ω3)

Fig 5 Condensed scheme of the elongation by marine organisms of dietary linolenic acid (18:3ω3) to eicosapentaenoic acid (20:5ω3), and then to docosahexaenoic acid (22:6ω3). Note the retention of the basis ω3, ω6 and ω ethylenic bonds which establish the familiar relationships among these fatty acids

3 Saturated fatty acids

It is a paradox that recent analyses of fish oils suggest that the higher the iodine value of a fish oil, the higher the total for the *saturated* fatty acids (*cf Tables V* and *VI*, see also Ackman and Eaton, 1966a). This may in fact be one reason why winterization is effective with high iodine value oils (*eg* menhaden, *Table V*) and not with low iodine value oils (*eg* herring). Palmitic (16:0) acid, as in most animals, is the principal saturated fatty acid (10–30% of the total), with some degree of interchangeability with myristic (14:0) acid (5–10% of the total). If the total proportion of saturated fatty acids is in the range of 20–30%, then stearic (18:0) acid is ordinarily no more than 1–3% of all fatty acids, and the 20:0, 22:0 and 24:0 acids are just detectable at levels of 0·01–0·1%. These fatty acids can all be biosynthesized by the organism, but it is also clear that they are freely absorbed from dietary fats. This has been shown, for example, by the odd-numbered straight-chain fatty acids which accumulate temporarily in the smelt *Osmerus mordax* (Addison and Ackman, 1970; Addison *et al*, 1973; Paradis and Ackman 1976a,b) and in the mullet *Mugil cephalus* (Sen and Schlenk, 1964; Deng *et al*, 1976). Laboratory feeding of single-cell protein containing odd-chain fatty acids leads to their deposition in the fish (Shimma and Nakada, 1974a,b). All animals can form these odd-chain acids if a proportionate molecule primes the two-carbon fatty acid chain extension process instead of an acetate molecule. Similarly the amino acid skeletons of leucine and isoleucine can lead respectively to the odd-carbon number *iso* and *anteiso* structures which are mostly C$_{15}$ and C$_{17}$ (*eg Table VI*). Some *iso* −14:0, −16:0 and −18:0 is usually detectable (Ackman, 1972), but the proportions are much less than the corresponding straight-chain acids. This suggests that most of the *iso* and *anteiso* odd-chain fatty acids are contributed to the lipids of the aquatic food chain by bacteria endogenous to the intestines of all living organisms (Hamid *et al*, 1978). Bacteria also contribute odd-number straight-chain fatty acids (Kunimoto *et al*, 1975). The fatty alcohols of North Atlantic copepod wax esters also have C$_{15}$ and C$_{17}$ methyl-branched and odd-carbon structures which are probably oxidized to the corresponding acids and then deposited in the fats of capelin, mackerel and herring (Ratnayake and Ackman, unpublished). These particular fatty acids have no known function in fish depot fats or in the phospholipids. In the analysis of the latter methyl esters of the *iso* acids may on GLC be confused with acetals which are derived from plasmalogens (Kaiser *et al*, 1978). This analytical problem is discussed elsewhere (Ackman, 1972).

The three isoprenoid fatty acids are derived from phytol (Lough, 1973); they are minor (total ⩽ 0·5%) but ubiquitous components of marine oils of either fish or marine mammal origin (Ackman *et al*, 1967c; Ackman and Hooper, 1968). In depot fats there is a rough correspondence between an 'effective' chain length equal to the true alkyl chain length plus one or two carbons, and the proportions of total fatty acids. For example: 4,8,12-trimethyltridecanoate, 2,6,10,14-tetramethylpentadecanoate and 3,7,11,15-tetramethylhexadecanoate may reflect the proportions of 14:0, 16:0 and 18:0 respectively. The other obvious methyl-branched fatty acid in fish oils is 7-methylhexadecanoic, which is derived from one or more unsaturated precursors (Pascal and Ackman, 1975). All are difficult to study except with capillary GLC (Ackman *et al*, 1967c).

4 Monoethylenic fatty acids

The monethylenic fatty acids of fish oil can be synthesized by fish and other marine organisms from acetate units by the same aerobic pathway used elsewhere in nature. In addition to 18:1ω9 (oleic), there is 16:1ω7 (palmitoleic acid; *Tables III–VI*), which can be chain-extended to 18:1ω7 (*cis*-vaccenic acid). The latter usually forms 10–30% of the total of 18:1 isomers (*Table V*). There has been speculation that the *total* percentage of (16:1 + 18:1) in a fish oil of a given species is controlled

Table V

FATTY ACID COMPOSITIONS (ω/ω%) AND IODINE VALUES OF COMMERCIAL MENHADEN OILS FROM DIFFERENT AREAS (AND OF SOME WINTERIZATION PRODUCTS), OF MEXICAN ANCHOVY COMMERCIAL OILS, AND OF CHILEAN ANCHOVY LABORATORY EXTRACT OILS

Oil					Menhaden oils			Anchovy oils			
Origin	Gulf of Mexico			Atlantic Ches. Bay and Mid. Atl.	Atlantic and fractions			Mexican		Chilean	
Fatty acid*	Western	Miss. delta	Eastern		Crude	Winter-ized	Stearine	1975	1978	San Vicente	Talca-huana
14:0	11·2	10·6	10·9	9·7	8·9	8·3	10·9	8·4	8·2	10·3	8·7
15:0	0·6	0·7	0·8	0·6	0·5	0·5	0·7	0·6	0·7	0·4	0·4
16:0	22·2	22·0	21·4	19·9	21·2	18·3	30·9	18·7	20·1	16·7	16·5
17:0	0·7	0·8	0·8	0·6	0·8	0·9	0·9	0·9	0·9	0·5	0·7
18:0	3·2	3·5	3·4	3·3	3·0	2·7	4·6	3·0	3·1	3·1	2·8
19:0	—	—	—	—	0·3	0·3	0·2	0·4	0·1	—	0·2
20:0	0·3	0·4	0·4	0·2	0·2	0·2	0·3	0·2	0·3	0·1	0·1
21:0	—	—	—	—	<0·1	<0·1	<0·1	<0·1	<0·1	—	—
22:0	—	—	—	—	0·1	0·1	0·1	0·1	0·1	<0·1	—
24:0	—	—	—	—	0·1	0·1	0·1	0·1	<0·1	0·1	0·2
Σ Saturates	39·1	39·3	38·3	34·9	35·7	32·1	49·3	32·9	34·7	32·1	30·5
16:1ω7	15·5	14·6	14·3	11·9	11·2	11·8	8·7	9·3	7·2	11·3	9·8
7 Methyl-hexadecenoic†	—	—	—	—	0·5	0·3	0·1	0·4	0·8	—	—
18:1ω9	6·4	6·3	6·0	9·9	10·2	9·9	8·3	9·8	9·8	7·0	7·2
18:1ω7	2·9	2·9	3·3	3·8	1·8	1·8	2·1	2·3	2·6	1·7	1·1
18:1ω5	—	—	—	—	0·1	0·1	0·2	0·2	0·1	0·3	0·1
20:1ω11	—	—	—	—	0·1	0·2	0·1	0·1	<0·1	0·2	0·2
20:1ω9	1·0	1·0	0·9	2·4	1·7	1·8	1·7	0·8	2·8	4·8	7·3
20:1ω7	0·2	0·3	0·3	0·5	0·2	0·2	0·2	0·2	0·2	0·8	0·7
20:1ω5	—	—	—	—	<0·1	0·1	<0·1	<0·1	<0·1	0·1	0·1
22:1ω13+ω11	<0·1	<0·1	<0·1	0·8	0·8	0·7	0·6	0·6	3·3	3·5	4·7
22:1ω9	0·1	0·2	0·3	0·3	0·1	0·2	0·1	0·2	0·2	0·2	0·8
22:1ω7	<0·1	<0·1	0·1	<0·1	<0·1	<0·1	<0·1	0·1	0·1	0·1	0·2
22:1ω5	—	—	—	—	ND	ND	<0·1	<0·1	<0·1	<0·1	<0·1
24:1	0·5	0·6	0·3	0·3	0·2	0·4	0·2	0·3	0·4	0·2	0·5
Σ Monoenes	27·1	26·6	26·4	30·4	27·1	27·2	22·4	24·2	28·5	30·2	32·6
16:2ω6	—	—	—	—	0·1	0·1	0·1	0·1	—	0·3	0·2
16:2ω4	1·4	1·6	1·7	1·5	1·2	1·1	0·8	0·9	—	0·9	1·4
18:2ω9	0·1	0·2	0·1	0·1	0·1	0·1	0·1	0·1	—	0·1	<0·1
18:2ω6	1·1	1·1	1·1	1·1	0·9	1·0	0·7	1·3	1·2	1·3	1·1
20:2ω9	0·2	0·4	0·3	0·5	0·2	0·2	0·2	<0·1	—	0·1	<0·1
20:2ω6	0·1	0·3	0·2	0·3	0·1	0·2	0·1	0·3	0·2	0·1	0·1
Σ Dienes	3·3	4·0	4·1	3·9	2·5	2·8	2·0	2·7	1·4	2·7	2·8
16:3ω4+ω3	2·0	2·1	2·0	1·0	1·7	1·9	1·1	1·1	0·5	1·9	1·9
18:3ω6	0·3	0·2	0·4	0·3	0·1	0·2	0·1	0·2	0·2	0·2	0·1
18:3ω3	1·0	1·0	0·9	0·8	1·2	1·3	0·9	0·8	1·1	0·3	0·4
20:3ω6	—	—	—	—	0·1	0·1	0·1	0·1	0·1	0·1	<0·1
20:3ω3	—	—	—	—	0·1	0·1	0·1	<0·1	0·1	<0·1	<0·1
Σ Trienes	3·8	3·8	3·8	2·2	3·2	3·5	2·4	2·2	2·0	2·5	2·5
16:4ω1	1·0	0·7	1·1	1·3	1·1	0·6	0·8	1·6	1·0	4·3	2·8
18:4ω3	2·3	2·0	1·9	2·1	2·8	3·0	1·9	2·3	2·4	1·9	1·7
18:4ω1	0·4	0·4	0·5	0·5	0·3	0·4	0·4	0·2	0·1	0·3	0·4
20:4ω6	1·0	1·4	1·3	0·9	0·5	0·6	0·4	0·8	0·9	0·3	0·2
20:4ω3	0·6	1·2	0·9	1·0	1·2	1·2	1·0	0·5	0·5	0·5	0·8
22:4ω6+3	—	—	—	—	0·2	0·2	0·2	0·2	<0·1	<0·1	<0·1
Σ Tetraenes	5·3	5·7	5·7	5·7	6·0	5·9	4·5	5·5	4·9	7·2	6·0
20:5ω3	12·3	12·7	12·8	11·1	13·9	15·3	10·7	18·9	13·8	18·5	14·5
21:5ω3	1·0	0·8	0·6	0·5	0·5	0·5	0·3	0·6	0·4	0·7	0·8
22:5ω6	0·4	0·5	0·4	0·2	0·3	0·3	0·2	0·3	0·2	<0·1	<0·1
22:5ω3	2·0	2·0	2·0	2·2	1·7	1·9	1·2	1·4	1·1	1·8	1·6
Σ Pentaenes	15·7	16·0	15·8	14·0	16·4	18·0	12·4	21·1	15·5	21·0	16·9
22:6ω3	5·1	4·6	6·0	7·8	9·7	10·9	7·1	11·8	12·4	4·3	8·9
Iodine Value: Calculated	145	146	147	151	166	178	126	188	167	167	167
Wijs	—	—	—	—	175	187	136	198	156	163	163

* Some minor components were not detected under analytical conditions employed
† Calculated from hydrogenated analysis

by the fish since this figure is found to be similar in oils from certain species fished on both sides of the Atlantic (eg capelin, *M. villosus*, Eaton *et al*, 1975, or sand launce *Ammodytes* sp., Ackman and Eaton, 1971b). The origin of the 20:1ω9 has never been suspect as it could always be made by chain elongation of 18:1ω9. *Tables V* and *VI* show that the proportion of the minor isomers 20:1ω11 and 20:1ω7 relative to the major isomer 20:1ω9 are similar in several fish oils, an exception being the Pacific herring oil where 20:1ω11 is present in more than the usual proportions. *Figure 1* shows the monethylenic isomers separable by capillary GLC.

Before it was understood that the dominant 22:1 isomer in marine fish oils was 22:1ω11 (*Table VI*), it was proposed that herring laid it down to help control buoyancy or as an energy reserve (Ackman and Eaton, 1970). Once the isomer structure for 22:1 was established (Ackman and Castell, 1966) it became more difficult to support this view, since, if a fish such as herring controlled the biosynthesis of 22:1, only the 22:1ω9 isomer could be produced. Recently our examination of *freshwater* fish oils (Ackman *et al*, 1979) showed that 22:1ω9 was the most important isomer in two oils out of four; in one other the ratio for important isomers was

$\omega11/\omega9 = 3/1$, and in the fourth a different pair was important, with $\omega9/\omega7 = 4/5$ (Ackman *et al*, 1979). Other work on oils from menhaden *B. tyrannis* (*cf Tables V* and *VI*) shows that even in marine fish 22:1ω11 is by no means always important *in toto* or relative to 22:1ω9.

The whole problem of the 22:1 fatty acids in fish is simplified when it is realized that their origin is the fatty alcohols of copepods. There has been abundant work on the proportions and importance of wax esters in these crustacea (recent work or reviews include Kayama *et al*, 1976; Sargent, 1976a,b; Sargent *et al*, 1976; Kayama *et al*, 1977; Sargent *et al*, 1977; 1977b; 1978) and thus it is known that most of the common North Atlantic copepods have fatty alcohols with a very high content of 22:1 structure. More detailed analyses show that this is mostly 22:1ω11 (Pascal and Ackman, 1976; Ackman *et al*, 1974c; Ratnayake and Ackman, unpublished). The transformation of the fatty alcohol of wax ester to fatty acid takes place in crossing the intestinal mucosa of the fish (Patton and Benson, 1975; Patton *et al*, 1975; Sargent *et al*, 1976; 1977; *cf* rat, Bandi and Mangold, 1973). *Table VII* compares the isomer proportions of some 22:1 fatty alcohols recovered from fish muscle and skin lipids (fatty acids are given in full in *Table VI*) with those of the copepod. In the copepod wax esters the fatty acid moiety tends to have much less 22:1 acid (*Table VI*) than the fatty alcohol moiety has 22:1 alcohol (*Table VII*), and in fact there is less 22:1 fatty acid than in most of the fish oils of *Table V*. The exception, the Pacific herring oil, has more 16:0, less 16:1ω7, and more 18:1ω9 than the other fish oils, all of which come from the western North Atlantic. This very strongly suggests that the small crustacea on which this lot of Pacific herring might have fed, if important in the diet of the fish, had little or no wax ester rich in 22:1 fatty alcohol. In fact such crustacea do exist in the area along with those containing wax esters (Lee, 1974; Sargent and Lee, 1975). Lambertsen and Hansen (1978) have recently published extensive fatty acid data for capelin, mackerel and herring which may resolve questions on these fatty acids.

The menhaden lives in somewhat limited inshore areas, and although a predator on zooplankters when young, in the adult feeds by filtration of phytoplankters (Weaver, 1974). Any acquisition of copepods is thus incidental, whereas the capelin, mackerel and herring feed by preference mostly on copepod species (Lambertsen and Mykelstad, 1972; Leim and Scott, 1966). The general conclusion on the dietary origin of the 22:1 in marine fish oils extends to 20:1ω9, which is also at least 10–20% of copepod fatty alcohols (*Table VII*: Ackman *et al*, 1974c), but it is not known why the copepod chooses to form 22:1ω11 fatty alcohol instead of 22:1ω9 fatty alcohol.

Once deposited, the 22:1ω11 enters the food web and is redistributed to other species such as the omnivorous cod. This dietary origin for 22:1ω11 extends to marine mammals, which deposit fat freely, when it is available, with very slight alterations in fatty acid composition (Ackman *et al*, 1971; 1975c,d). The greater percentage of total 22:1 in blubber of fin whales from the North Atlantic, relative to those of the extreme South Atlantic or Antarctic, has been noted before (Ackman and Eaton, 1966b; Ackman *et al*, 1970). The main food of

fin whale in the Antarctic is the krill, *E. superba*, which contains little 22:1 (*Table III*), whereas in the north Atlantic herring and capelin are eaten as well as some larger zooplanktonic crustacea such as *M. norvegica* (*Table III*). The Baltic herring provides another interesting example of exogenous origin for 22:1. Although the species is the same as the North Atlantic herring, the low salinity of the Baltic discourages the copepods common in the adjacent North Atlantic, lowering the availability of dietary 22:1 in either acid or alcohol form. Thus although some Baltic herring can have reasonably high fat contents with polar lipids similar to other marine fish already discussed (Bosund and Ganrot, 1969), their oils are fairly low in 22:1 (Bosund and Ganrot, 1969; Stodolnik and Podeszewski, 1972). In phospholipids of marine organisms (*Tables III* and *IV*) 22:1 is seldom found in more than trace amounts, which provides further evidence that it has no metabolic role in fish or species other than some copepods.

Fish oils are still discussed in terms of iodine value. In general the low iodine values (100–135) of many fish oils are due to the inclusion of high proportions (20–35%) of 22:1. When iodine values are higher, as in cod liver oil (145–165), 22:1 is usually much less important (10–20%), the fish being long-lived and accumulating 22:6ω3. In the alternative case of filter-feeders (*eg* menhaden, 165–185) very little 22:1 is present (*Table V*) and 20:5ω3 is more important. This fluctuation in 22:1 may be the underlying basis for a relationship between iodine value and polyunsaturated fatty acid content (Ackman, 1966).

Concern over possible adverse effects of docosenoic acids in the health of animals, in either unprocessed (fish fat) or partially hydrogenated form (FAO, 1978), appears to be partly founded on the excessive susceptibility of laboratory rats and is not confirmed by an extended study in non-human primates (Ackman *et al*, unpublished).

5 Polyethylenic fatty acids

These fatty acids give marine oils their most specific characteristics (Gauglitz *et al*, 1974). The polyethylenic fatty acids with five and six ethylenic bonds originate in unicellular phytoplankton or in some areas in seaweeds. As shown in *Table III*, the average fatty acid composition for 12 North Atlantic phytoplankton includes all the principal fatty acids found in the oils and lipids of the higher organisms. The two common 'plant' C_{18} fatty acids, 18:2ω6 (linoleic acid) and 18:3ω3 (linolenic acid) do not accumulate in most fish oils to the extent of more than 1% or 2% of fatty acids (*Tables IV, V* and *VI*). This is also true of marine invertebrates feeding directly on plants with lipids rich in these acids (*Table III*). It must be concluded that such organisms direct part of their intake of these two fatty acids to energy, and limit formation (*Fig 5*) of the longer chain acids such as 20:5ω3 and 22:6ω3 (de Moreno *et al*, 1976). At first glance it is the latter fatty acid which attracts attention in fish oils, since it appears to be the termination of a process which is probably 18:3ω3→20:3ω3→20:5ω3→22:6ω3. Space permits only mention of a few recent papers on the subject of this conversion (Owen *et al*, 1975; Yone and Fujii, 1975; Shabalina and Ostroumova, 1976;

Table VI

Fatty acids (ω/ω%) of copepod triglycerides and wax esters, and of total lipid (essentially depot fat triglyceride), recovered from capelin, mackerel and herring muscle and skin lipids, compared with commercial oils[1]

Fatty acid	Copepod Trigly-ceride	Copepod Wax esters	Capelin Total acids	Mackerel Total acids	Atlantic herring Total acids	Capelin Total acids	Mackerel Total acids	Atlantic herring Total acids	Pacific herring Total acids
			Fish muscle and skin lipids			Fish commercial oils			
12:0	0·32	0·62	0·55	0·15	0·17	0·16	0·11	0·10	0·20
Iso 14:0	0·07	0·25	0·16	0·02	0·07	0·06	0·03	0·03	0·03
14:0	19·84	36·42	6·60	8·60	7·18	7·85	7·81	8·77	6·81
Iso 15:0	0·31	0·59	0·41	0·18	0·27	0·14	0·20	0·22	0·01
Anteiso 15:0	0·12	0·19	0·24	0·07	0·07	0·04	0·07	0·09	0·09
15:0	1·05	0·97	0·66	0·40	0·44	0·28	0·41	0·34	0·34
Iso 16:0	0·19	0·06	0·21	0·10	0·03	0·03	0·12	0·03	0·03
16:0	28·98	11·15	11·74	17·59	13·68	8·81	15·93	14·84	22·74
Iso 17:0	0·06	0·05	0·16	0·12	0·04	0·03	0·15	0·03	0·01
Anteiso 17:0	0·07	0·05	0·13	0·04	0·01	0·03	0·05	0·01	0·07
17:0	0·83	0·42	0·27	0·54	0·43	0·17	0·46	0·33	1·29
Iso 18:0	ND	ND	ND	ND	0·01	ND	0·07	0·04	ND
18:0	1·04	0·35	0·00	2·22	0·64	0·72	1·73	0·97	2·72
19:0	0·09	0·16	0·1	0·15	0·07	0·06	0·19	0·10	0·01
20:0	0·51	0·38	0·00	0·12	0·08	0·08	0·11	0·15	0·05
22:0	0·20	ND	ND	ND	0·01	ND	0·01	ND	ND
24:0	0·09	ND	ND	ND	0·07	0·17	0·03	0·01	ND
Total saturated	53·8	52·7	22·6	30·3	23·3	18·6	27·5	26·1	34·4
Σ14:1	0·07	ND	0·32	0·08	0·07	0·06	0·11	0·07	0·22
Σ15:1	ND	ND	ND	0·06	ND	ND	ND	0·03	ND
16:1ω9	0·05	0·35	ND	0·42	0·09	0·03	0·29	ND	0·07
16:1ω7	8·89	12·33	12·89	9·57	15·74	15·42	8·20	7·22	7·53
16:1ω5	0·73	1·04	0·61	ND	0·32	0·73	0·54	0·52	0·45
Σ17:1	0·13	0·12	ND	0·16	ND	ND	0·15	0·04	ND
18:1ω9	3·14	3·46	6·15	9·47	4·36	4·40	8·61	12·27	29·72
18:1ω7	0·91	0·56	2·67	4·69	2·37	3·43	3·78	3·66	4·98
18:1ω5	0·41	0·15	0·35	0·59	0·53	0·62	0·54	0·64	0·41
Σ19:1	ND	ND	ND	0·06	ND	ND	0·04	0·07	ND
20:1ω11	0·33	0·36	0·44	0·34	1·21	1·20	0·24	0·50	1·01
20:1ω9	4·12	4·37	14·83	7·41	12·03	14·53	10·59	14·37	4·40
20:1ω7	0·34	0·55	1·44	0·93	1·72	1·84	1·13	0·94	0·37
20:1ω5	0·02	0·01	0·06	0·03	0·18	0·23	0·09	0·19	ND
22:1ω11 (+13)	5·16	4·59	18·41	9·40	15·40	17·45	12·74	20·92	3·92
22:1ω9	0·34	0·65	1·86	0·65	3·60	1·70	1·00	1·36	0·87
22:1ω7	0·11	0·11	0·42	0·09	0·43	0·42	0·19	0·33	0·12
Σ24:1	0·49	0·24	0·53	0·09	0·30	0·59	0·69	0·52	0·47
Total monoethylenic	25·2	28·9	61·0	43·9	58·4	62·7	48·9	63·7	54·5
16:2ω6	ND	0·20	0·76	0·04	0·03	0·84	ND	ND	0·75
16:2ω4	0·09	0·52	0·13	0·48	0·25	0·06	0·45	0·23	0·27
16:3ω3	0·26	0·01	0·96	0·48	0·88	0·82	0·52	0·39	0·44
16:4ω3	0·32	1·35	ND	0·49	1·30	ND	0·44	0·39	0·01
16:4ω1	ND	ND	1·52	ND	ND	1·47	ND	0·03	0·42
18:2ω6	0·97	0·83	0·92	1·04	1·11	0·78	1·28	0·78	0·67
20:2ω6	0·04	0·04	0·17	0·13	0·03	0·05	0·24	0·09	0·14
18:3ω6	0·16	0·14	0·09	0·10	0·02	0·01	0·07	0·04	0·07
18:3ω3	1·08	1·00	0·87	0·74	0·45	0·20	0·99	0·39	0·18
18:4ω3	3·23	5·63	1·57	1·96	1·60	1·36	2·47	0·93	0·67
20:4ω6	0·29	0·19	0·29	0·27	0·22	0·14	0·36	0·24	ND
20:4ω3	0·38	0·62	0·21	0·55	0·34	0·28	0·50	0·22	0·29
20:5ω3	8·38	5·81	6·11	9·39	7·29	9·35	7·64	2·85	5·31
21:4ω3	0·21	0·24	0·36	0·21	0·10	ND	0·41	0·13	0·09
22:5ω6	0·09	0·01	0·07	0·02	0·20	0·03	0·01	0·03	ND
22:5ω3	0·60	0·23	0·52	1·17	0·75	0·60	0·57	0·37	0·22
22:6ω3	4·90	0·64	1·87	8·73	3·72	2·70	7·66	2·70	1·53
Total polyethylenic	21·0	17·5	16·4	25·8	18·3	18·7	23·6	10·2	11·1

* ND not determined
[1] Ratnayake and Ackman, unpublished

Table VII

COMPARISON OF PROPORTIONS OF DOCOSENOYL ISOMERS IN COPEPOD FATTY ALCOHOLS WITH FATTY ALCOHOLS AND TOTAL FATTY ACID DOCESENOIC ACIDS OF FISH LIPIDS[1]

	Data from oxidative fission						
	Copepod	Fish muscle and skin lipids				Commercial oils	
	alcohol	Capelin		Mackerel		Capelin	
22:1 Structure		Alcohol	Acid	Alcohol	Acid	Alcohol	Acid
ω13	2·5	0·3	1·7	2·6	3·6	0·3	1·5
ω11	87·6	88·2	87·1	81·0	88·8	89·7	87·5
ω9	7·9	9·2	9·1	14·0	6·6	8·2	8·8
ω7	2·0	2·3	2·1	2·4	0·9	1·8	2·2

	Data from open-tubular GLC								
	Copepod	Fish muscle and skin lipids				Commercial oils			
	alcohol	Atlantic herring		Mackerel		Atlantic herring		Pacific herring	
22:1 Structure		Alcohol	Acid	Alcohol	Acid	Alcohol	Acid	Alcohol	Acid
ω11(+13)	89·8*	86·8	78·4	91·9	91·5	90·9	92·5	82·5	79·8
ω9	8·2	10·5	19·0	7·3	7·2	7·5	6·0	8·9	17·7
ω7	2·0	2·7	2·3	0·7	1·4	1·6	1·5	5·6	2·4
ω5	—	0·0	0·4	—	—	—	—	—	—

* 33·7% of total copepod fatty alcohols (20:1ω9 was 28·9%)
[1] Ratnayake and Ackman, unpublished

Gatesoupe *et al*, 1977; Kanazawa *et al*, 1977; Rahn *et al*, 1977; Takeuchi and Watanabe, 1977a,b,c; Takeuchi *et al*, 1978; Tinoco *et al*, 1979).

In many invertebrate lipids 20:5ω3 is the dominant polyunsaturated fatty acid. This is possibly due to its being prevalent in dietary algae (*Table III*, see also Ackman and McLachlan, 1977) and could account for 20:5ω3 being important in menhaden and anchovy oils (*Table V*).

To account for 22:6ω3 in marine organisms one has to consider the development of the vertebrate nervous system. For some reason the nervous system prefers 22:6ω3 to 20:5ω3 acids and indeed in mammals there is only a trace of 20:5ω3 in brain lipid fatty acids whereas 22:6ω3 is very important (Crawford, 1974; Lemarchal, 1978). The supreme exaltation of 22:6ω3 in this role is the recognition of tridocosahexaenoin in the reflecting spheres of the lining of part of the eyeball in several fish species (Nicol *et al*, 1972; 1975). The optic nerve is clearly a starting point for this evolutionary development. Probably all of the more mobile marine organisms, and certainly those high on the evolutionary scale, require a little dietary 22:6ω3. All food organisms contribute 20:5ω3 and at least a little 22:6ω3 to each step of the food web when eaten. If this fatty acid is regarded as a metabolic dead-end in depot fats, it is possible to see why 22:6ω3 is less important in the oils of short-lived species such as herring and capelin than in oils of long-lived species such as cod. In higher animals such as the rat it is known that the biosynthetic pathways for formation of these fatty acids can be reversed (*eg* 22:5ω3→20:5ω3). For energy purposes this reversed pathway may be total, but it is likely that in fish or marine mammals depot fat the 22:6ω3 acid, as well as 22:5ω3, could be exploited as sources of 20:5ω3, or even 18:3ω3, if required for particular biochemical processes, such as prostaglandin formation (Dyerberg *et al*, 1978).

Some of the same arguments apply to 18:2ω6. The acid is the 'essential' fatty acid of mammals as it is a precursor of 20:4ω6, which in turn is a key raw material for one or more prostaglandins (Horton, 1974; Ogata

and Nomura, 1975). In discussing marine mammals it is necessary to recall that they are terrestrial animals which have returned to the sea, but likely they retain a need for 18:2ω6 and 20:4ω6 (Ackman *et al*, 1965; Ackman and Hooper, 1974). In fish the role of 18:2ω6 is obscure. Although 18:2ω6 can be elongated to 20:4ω6 (Farkas *et al*, 1977), neither ω6 fatty acid is much used. A very abundant literature exists, especially for salmonids, to show that in marine fish the essential fatty acids are those of the linolenic (18:3ω3) family (Kayama, 1974; Takeuchi and Watanabe, 1977a; 1978). There is of course a reasonable acceptance of fish oils, including the lipids of fish meals, in feedstuffs especially for poultry and pigs (Gauglitz *et al*, 1974; Stansby 1978). The same acceptability applies to fish in the human diet, despite variable fat contents and the presence of longer-chain, highly unsaturated fatty acids (Stansby, 1973; Tinoco *et al*, 1979).

Minor fatty acids of the polyunsaturated types include a series with 'ω' values of 1, 4 or 7 (Ackman, 1964; Yamada, 1972; Kayama, 1974). Chiefly these are 16:2ω4 and 16:4ω1 (*Table VI*). All are of algal origin (Ackman, 1964; Ackman *et al*, 1968b; Ackman and McLachlan, 1977) and are metabolically inert in animals. Most fish oils include 1–2% total of fatty acid of this type, and in depot fats they sometimes can be used to indicate the feeding habits of a fish, for example in the sturgeon *Acipenser oxyrhynchus* (Ackman *et al*, 1975a). Odd chain fatty acids with several ethylenic bonds of the *cis*-monoethylenic type have been found to be seasonally important in the mullet *Mugil cephalus* (Sen and Schlenk, 1964; Deng *et al*, 1976; Schlenk, 1970). In the smelt *Osmerus mordax* a similar phenomenon is clearly dietary in origin (Paradis and Ackman, 1976a,b). These acids have 'ω' values of 2 and 5 (Schlenk, 1970) but a 21:5 acid, prevalent at 0·1–0·5% of fatty acids in most species of invertebrates and vertebrates, had the regular ω3 structure (Mayzaud and Ackman, 1978). An octadecapentaenoic acid of marine plant origin (Ackman *et al*, 1974d; Joseph, 1975) has recently been found in animal lipids (Mayzaud *et al*, 1976).

Other unusual fatty acids found in low proportions in fish oils or lipids include those with furanoid structures (Gunstone *et al*, 1976; Glass *et al*, 1977; Scrimgeour, 1977); *trans*-6-monoethylenic unsaturation (Hooper *et al*, 1973; Pearce and Stillway, 1976) and two 7-methylhexadecenoic acids (Hooper *et al*, 1973; Pascal and Ackman, 1975).

6 Wax esters and glyceryl ethers

At this point it is convenient to note that wax esters in marine organisms fall into two main classes, those rich in 16:0 and those rich in 22:1 acids (Nevenzel, 1970; Sargent *et al*, 1976). The influence of the former class on the fatty acid composition of fish oils is not well understood. The chain length total tends to be in the range C_{32} to C_{36} (note that the copepod wax ester, rich in 22:1 alcohol, has a high content of 14:0 acid, *Table VI*). Thus species with a high proportion of 16:0 fatty alcohols in their wax esters will tend to have 16:0, 16:1 or 18:1 as the main fatty acids. These three structures are such common and important fatty acids that their influence on the aquatic food web is much more difficult to assess than is the case for the 22:1 alcohol discussed above.

Some fish have wax esters as part of their body lipid. The white barracudina *Paralepis (Notolepis) rissoi kroyeri* (Ackman *et al*, 1972; 1974a) is a case in point. In addition to conversion to a defatted fish protein concentrate for possible human nutrition, barracudina meal was prepared and found satisfactory for poultry nutrition (Ackman *et al*, 1974a). In this study the partitioning of wax esters and triglycerides from isopropanol was investigated but was not very successful. The digestibility of the barracudina oil in chicks was not as high as a pure triglyceride oil, but considering that wax ester is the major lipid class an appreciable proportion was digested (*Table VIII*), thus contributing to the nutritional benefits of the fish meal.

The 'castor oil' fish *Ruvettus pretiosus* contains wax esters (Nevenzel *et al*, 1965; Sato and Tsuchiya, 1970) which are responsible for the purgative effect (Sato and Tsuchiya, 1970; Arai and Kinumaki, 1977; Kinumaki *et al*, 1977). Other adverse effects in animals are also associated with wax esters (Totani *et al*, 1975), but in general wax esters are not often included in lists of harmful substances found in fish fats (Wurziger and Dickhaut, 1977). Sato and Tsuchiya (1970) comment that 'side dishes' of marine products containing wax esters are traditional in Japan. Recent work on capelin, mackerel and herring body and skin fats has shown that herring do not usually accumulate wax esters, whereas capelin seem to have a consistent proportion in lipids whilst mackerel show more variability (Ratnayake and Ackman, unpublished). This work establishes (*Table IX*) some sort of baseline for acceptability of wax esters in fish intended for human consumption since herring, capelin and mackerel are widely accepted and valuable species (Jangaard, 1973).

Numerous proposals attempt to link the presence of wax esters in fish with depth of habitat (*eg* Hayashi and Yamada, 1975a,b,c; 1976; Kayama, 1975; Sargent *et al*, 1976; 1977; Mori *et al*, 1978) but the hypothesis is not proven. Although they have lower specific gravities than triglycerides (Lewis, 1970), other considerations may be more important.

Two related materials; I = glyceryl ethers (alkyldiacylglycerols) and II = plasmalogens (alk-1-enyldiacylglycerols) should be mentioned:

$$I \quad \begin{array}{l} CH-O-R \\ | \\ CH_2-O-R \\ | \\ CH-O-R \end{array} \qquad II \quad \begin{array}{l} CH-O-CH=CH-R^1 \\ | \\ CH_2-O-R \\ | \\ CH-O-R \end{array}$$

These are now well-defined lipids amenable to mod-

Table VIII
DIGESTIBILITY OF BARRACUDINA AND MENHADEN OILS

Oil	Test group No.	% Of test oil excreted (ether soluble)	% Of test oil excreted as soaps	% Digestibility
Barracudina	1	12·8	0·20	87·0
	2	24·6	0·85	74·6
	3	14·5	—	85·5
Menhaden	1	7·0	0·14	92·9
	2	5·2	0·40	94·4
	3	5·9	0·85	93·2

Note: There were five birds in each group and the test oils were fed at a level of 12% (basal 88 + oil 12). Work carried out by B. March, Vancouver, BC, Canada

Table IX
SAMPLE ORIGIN AND RESULTS OF EXAMINATION OF LIPIDS FOR FATTY ALCOHOLS[1]

Sample origin	Number of fish; sex	Average length cm	Average weight, g	%(ω/ω) Lipid in sample	%(ω/ω) Unsaponifiable material in lipid	%(ω/ω) Fatty alcohol in lipid	%(ω/ω) 16·0	Selected alcohol components Σ18:1	Σ20:1	Σ22:1
Capelin commercial oil (1977)	—	—	—	100	3·16	0·49	12·3	3·2	30·8	39·4
Capelin muscle and skin (1975)	20	14·4	13·4	10·8	1·96	0·19	6·8	1·6	18·7	68·4
Mackerel commercial oil (1977)	—	—	—	100	1·35	0·11	18·2	7·0	17·5	33·3
Mackerel muscle and skin	3	36	762	12·2	3·79	1·81	13·1	4·9	27·4	39·3
Pacific herring commercial oil (1973)	—	—	—	100	1·08	0·05	18·4	35·3	5·0	5·6
Atlantic herring commercial oil (1973)	—	—	—	100	1·91	0·06	9·5	4·5	29·1	43·9
Atlantic herring muscle and skin (1)	—	—	—	3·3	1·62	<0·01	—	—	—	—
Atlantic herring muscle and skin (2)	2	35·6	415	17·8	0·59	0·03	10·7	7·6	21·5	22·7
Atlantic herring muscle and skin (3a)	2, males	31·5	405	22·8	0·74	0·01	—	—	—	—
Atlantic herring muscle and skin (3b)	1, female	29·8	245	17·0	0·58	<0·01	—	—	—	—
Atlantic herring, whole extract[2]	male	—	—	—	1·06	0·03	19·3	28·0	9·8	22·2
Atlantic herring, whole extract[2]	female	—	—	—	1·00	0·17	16·3	22·2	14·9	21·1
Icelandic herring oil, commercial[2]	—	—	—	100			10·3	~5·0	~31	~52

[1] Ratnayake and Ackman, unpublished
[2] Karleskind, 1967. Additional material by private communication

97

ern analytical technology (Myher, 1978). The biochemistry of these materials is reviewed elsewhere (Malins and Wekell, 1970; Snyder, 1970; Yamada, 1972; Sargent, 1976a;). At one time dogfish (*Squalus acanthias*) livers were considered as a commercial source of glyceryl ethers (Bailey *et al*, 1952). The fish liver oils, especially from the elasmobranchs, offer a rich assortment of lipid classes (Sargent *et al*, 1973), and a few examples are given in *Table X*. The ether lipids also can lead to purgative or other effects in mammals (Totani *et al*, 1975; Kinumaki *et al*, 1977). More recently the combination of modern lipid chemistry and wider interest in mid-water or deep-sea species of fish as potential food sources has revealed that glyceryl ethers are major components of the muscle lipids of such fish (Mori *et al*, 1972; Sargent *et al*, 1973; Hayashi *et al*, 1978; Kawai *et al*, 1978; Nakayama *et al*, 1978; Mori *et al*, 1978).

Structurally related materials such as hydroxyalkyl glycerols (Hallgren and Stallberg, 1974) or methoxyglyceryl ethers (Hallgren *et al*, 1974; Hayashi and Takagi, 1978) are usually minor components in fish lipids, and with the plasmalogens, although of acute biochemical interest (Kaiser *et al*, 1978), fall outside the scope of this review. This also applies to hydrocarbons such as squalene (Ackman *et al*, 1968a; Yamada, 1972; Linko and Kaitaranta, 1976) which may be 'natural' in some fish, as distinct from pristane (Ackman, 1971) which is biogenic, but like many other hydrocarbons is not 'natural' in fish oils and lipids (Zsolnay, 1977).

7 Acknowledgements

The Zapata Haynie Corp gave permission to publish data on their menhaden and anchovy oils. The cooperation and analytical capabilities of C A Eaton, J-L Sebedio and W M N Ratnayake are gratefully acknowledged.

Table X
QUANTITATIVE ANALYSIS OF LIVER OILS FROM SOME CARTILAGINOUS FISH[1]

Lipid class	Ratfish	Dogfish	Soupfin shark	Basking shark
Ionic and other polar lipids (%)	2·4 ± 0·5	8·9 ± 0·8	4·6 ± 0·4	4·9 ± 0·6
Sterols (%)	1·7 ± 0·2	—	—	—
Triacylglycerols (%)	24·2 ± 0·7	53·6 ± 0·9	92·5 ± 0·7	47·2 ± 2·2
Alkyldiacylglycerols (%)	66·0 ± 0·8	37·5 ± 0·4	—	—
Alk-1-enyldiacylglycerols (%)	5·8 ± 0·6	—	—	—
Sterol esters and wax esters (%)	—	—	1·5 ± 0·2	—
Hydrocarbons (%)	—	—	1·4 ± 0·1	47·9 ± 2·5

* Average of four determinations by tubular thin-layer chromatography
[1] Mangold and Mukherjee, 1975

8 References

ACKMAN, R G. Structural homogeneity in unsaturated fatty acids of
1964　marine lipids. A review. *J. Fish. Res. Bd. Can*., 21, 247–254
—— Empirical relationships between iodine value and polyunsaturated fatty acid content in marine oils and lipids. *J. Amer. Oil*
1966　*Chem. Soc*., 43, 385–389
—— Gas-liquid chromatography of fatty acid and esters. In: *Methods*
1969　*of Enzymology*, Ed J M Lowenstein. Academic Press, New York. Vol 14, 329–381
—— Pristane and other hydrocarbons in some freshwater and marine
1971　fish oils. *Lipids*, 6, 520–522
—— The analysis of fatty acids and related materials by gas-liquid
1972　chromatography. In: *Progress in the Chemistry of Fats and Other Lipids*, Ed R T Holman. Pergamon Press, Oxford. Vol 12, 165–284
—— Marine lipids and fatty acids in human nutrition. In: *Fishery*
1974　*Products*, Ed R Kreuzer. Fishing News (Books) Ltd, West Byfleet, UK., 112–131
ACKMAN, R G and CASTELL, J D. Isomeric monoethylenic fatty acids
1966　in herring oil. *Lipids*, 1, 341–348
ACKMAN, R G and CORMIER, M G. α-Tocopherol in some Atlantic fish
1967　and shellfish with particular reference to live-holding without food. *J. Fish. Res. Bd. Can*., 24, 357–373
ACKMAN, R G and EATON, C A. Some commercial Atlantic herring
1966a　oils; fatty acid composition. *J. Fish. Res. Bd. Can*., 23, 991–1006
—— Lipids of the finwhale (*Balaenoptera physalus*) from North Atlantic waters. III. Occurrence of eicosenoic and docosenoic fatty
1966b　acids in the zooplankter *Meganyctiphanes norvegica* (M. Sars) and their effect on whale oil composition. *Can. J. Biochem*., 44, 1561–1566
—— Biochemical implications of seasonal trends in the iodine values
1970　and free fatty acid levels of commercially produced Atlantic Coast herring oils. *J. Fish. Res. Bd. Can*., 27, 1669–1683
—— Mackerel lipids and fatty acids. *Can. Inst. Food Sci. Technol. J*., 4,
1971a　169–174
—— Investigation of the fatty acid composition of oils and lipids from
1971b　the sand launce (*Ammodytes americanus*) from Nova Scotia waters. *J. Fish. Res. Bd. Can*., 28, 601–606
ACKMAN, R G and HOOPER, S N. Examination of isoprenoid fatty acids
1968　as distinguishing characteristics of specific marine oils with particular reference to whale oils. *Comp. Biochem. Physiol*., 24, 549–565

—— Long-chain monoethylenic and other fatty acids in heart, liver and
1974　blubber lipids of two harbour seals (*Phoca vitulina*) and one grey seal (*Halichoerus grypus*). *J. Fish. Res. Bd. Can*., 31, 333–341
ACKMAN, R G and McLACHLAN, J. Fatty acids in some Nova Scotian
1977　seaweeds: a survey for octadecapentaenoic and other biochemically novel fatty acids. *Proc. NS Inst. Sci*., 28, 47–64
ACKMAN, R G and SIPOS, J C. Pilchard oil: An analysis for component
1964　fatty acids with particular reference to the C_{24} chain length. *J. Fish. Res. Bd. Can*., 21, 841–843
ACKMAN, R G, ADDISON, R F and EATON, C A. Unusual occurrence
1968a　of squalene in a fish, the eulachon, *Thaleichthys pacificus*. *Nature*, 220, 1033–1034
ACKMAN, R G, BURGHER, R D and JANGAARD, P M. Systematic iden-
1963　tification of fatty acids in the gas-liquid chromatography of fatty acid methyl esters: a preliminary study of seal oil. *Can. J. Biochem*., 41, 1627–1641
ACKMAN, R G, EATON, C A and HINGLEY, J H. Menhaden body lipids:
1976　details of fatty acids in lipids from an untapped food resource. *J. Sci. Food Agric*., 27, 1132–1136
ACKMAN, R G, EATON, C A and JANGAARD, P M. Lipids of the fin
1965　whale (*Balaenoptera physalus*) from North Atlantic waters. II. Fatty acid composition of the liver lipids and gas-liquid chromatographic evidence for the occurrence of 5,8,11,14-nonadecatetraenoic acid. *Can. J. Biochem*., 43, 1521–1530
ACKMAN, R G, EATON, C A and KE, P J. Canadian marine oils of low
1967a　iodine value: Fatty acid composition of oils from Newfoundland turbot (Greenland halibut), certain Atlantic herring, and a sablefish. *J. Fish. Res. Bd. Can*., 24, 2563–2572
—— Utilization of barracudina as a source of wax esters and proteins.
1974a　In: *Fishery Products*, Ed R Kreuzer, Fishing News (Books) Ltd, West Byfleet, UK, 221–226
ACKMAN R G, EATON, C A and LINKE, B A. Differentiation of
1975a　freshwater characteristics of fatty acids in marine specimens of the Atlantic sturgeon, *Acipenser oxyrhynchus*. *Fish. Bull*, 73, 838–845
ACKMAN R G, EPSTEIN, S and EATON, C A. Differences in the fatty
1971　acid compositions of blubber fats from Northwestern Atlantic finwhales (*Balaenoptera physalus*) and harp seals *Pagophilus groenlandica*. *Comp. Biochem. Physiol*., 40B, 683–697
ACKMAN R G, EPSTEIN, S and KELLEHER, M. A comparison of lipids
1974b　and fatty acids of the ocean quahaug *Arctica islandica* from Nova Scotia and New Brunswick. *J. Fish. Res. Bd. Can*., 31, 1803–1811

ACKMAN, R G, LINKE, B A and HINGLEY, J. Some details of fatty acids
1974c and alcohols in the lipids of North Atlantic copepods. *J. Fish. Res. Bd. Can.*, 31, 1812–1818

ACKMAN. R G, MANZER, A and JOSEPH, J D. Tentative identification
1974d of any unusual naturally-occurring polyenoic fatty acid by calculations from precision open-tubular GLC and structural element retention data. *Chromatographia*, 7, 107–114

ACKMAN, R G, SEBEDIO, J-L and KOVACS, M I P. Role of eicosenoic
1979 and docosenoic fatty acids in freshwater and marine lipids. *Freshwater Biology*. (Submitted for publication)

ACKMAN, R G, SIPOS, J C and JANGAARD, P M. A quantitation prob-
1967b lem in the open-tubular gas chromatography of fatty acid esters from cod liver lipids. *Lipids*, 2, 251–257

ACKMAN, R G, SIPOS, J C and TOCHER, C S. Some linear, *iso, anteiso*
1967c and multiple-branched fatty acids of marine origin: analyses on packed and open-tubular gas-liquid chromatographic columns with particular reference to pristanic and phytanic acids. *J. Fish. Res. Bd. Can.*, 24, 635–650

ACKMAN, R G, TOCHER, C S and McLACHLAN, J. Marine phytoplank-
1968b ter fatty acids. *J. Fish. Res. Bd. Can.*, 25, 1603–1620

ACKMAN, R G, EATON, C A, SIPOS, J C, HOOPER, S N and CASTELL,
1970 J D. Lipids and fatty acids of two species of North Atlantic krill (*Meganyctiphanes norvegica* and *Thysanoessa inermis*) and their role in the aquatic food web. *J. Fish. Res. Bd. Can.*, 27, 513–533

ACKMAN R G, EATON, C A, KINNEMAN, J and LITCHFIELD, C. Lipids of
1975b freshwater dolphin *Sotalia fluviatilis*: Comparison of odontocete bioacoustic lipids and habitat. *Lipids*, 10, 44–49

ACKMAN, R G, HINGLEY, J H, EATON, C A, SIPOS, J C and MITCHELL
1975c E D. Blubber fat deposition in mysticeti whales. *Can. J. Zool.*, 53, 1332–1339

ACKMAN, R G, HINGLEY, J H, EATON, C A, LOGAN, V H and
1975d ODENSE, P H. Layering and tissue composition in the blubber of the northwest Atlantic sei whale (*Balaenoptera borealis*). *Can. J. Zool.*, 53, 1340–1344

ACKMAN, R G, HOOPER, S N, EPSTEIN, S and KELLEHER, M. Wax
1972 esters of barracudina lipid: A potential replacement for sperm whale oil. *J. Amer. Oil Chem. Soc.*, 49, 378–382

ACKMAN, R G, KE, P J, MacCALLUM, W A and ADAMS, D R. New-
1969 foundland capelin lipids: Fatty acid composition and alterations during frozen storage. *J. Fish. Res. Bd. Can.*, 26, 2037–2060

ADDISON, R F and ACKMAN, R G. Exceptional occurrence of odd-
1970 chain fatty acids in smelts (*Osmerus mordax*) from Jeddore Harbour, Nova Scotia. *Lipids*, 5, 554–557

ADDISON, R F, ACKMAN, R G and HINGLEY, J. Distribution of fatty
1968 acids in cod flesh lipids. *J. Fish. Res. Bd. Can.*, 25, 2083–2090

—— Free fatty acids of herring oils: Possible derivation from both
1969 phospholipids and triglycerides in fresh herring. *J. Fish. Res. Bd. Can.*, 26, 1577–1583

—— Lipid composition of the queen crab (*Chionoecetes opilio*). *J. Fish.*
1972 *Res. Bd. Can.*, 29, 407–411

—— Seasonal and local variations in odd-chain fatty acid levels in Nova
1973 Scotia rainbow smelt (*Osmerus mordax*) *J. Fish. Res. Bd. Can.*, 30, 113–115

AMAN, M A B and SMIRNOVA, G A. The effect of prolonged holding of
1973 fish in pounds on lipid metabolism. *Ryb. Khoz.*, (4), 69–70

ANDERSON, M L, STEINBERG, M A and KING, J F. Some physical
1965 effects of freezing fish muscle and their relation to protein-fatty acid interaction. In: *The Technology of Fish Utilization*, Ed R Kreuzer. Fishing News (Books) Ltd, London. 105–110

ANON. *Official and Tentative Methods of Analysis of the Association of*
1945 *Official Agricultural Chemists*. Sixth edition, Ass. Off. Agric. Chem., Washington

ARAI, K and KINUMAKI, T. Feeding Test on nutritive value of
1977 kamaboko (fish jelly product) made from fish meat containing wax. *Bull. Tokai Reg. Fish. Res. Lab.*, No. 91, 93–99

ASTRUP, H. The oxidation of a highly unsaturated herring oil. *Chem.*
1964 *and Ind.* (London), Jan. 18:107

BAILEY, B E, CARTER, N M and SWAIN, L A. Marine oils with particu-
1952 lar reference to those of Canada. *Fish. Res. Bd. Can.*, Ottawa, Bull. No. 89. 413pp

BALSINGAM, M. Extraction of oils from marine fish by using various
1972 solvent systems and the determination of the iodine value of the extracted oils. *Malaysian Agr. J.*, 48, 222–230

BANDI, Z L and MANGOLD, H K. Substrate specificity of enzymes
1973 catalysing inter-conversions of long-chain acids and alcohols in the rat. *FEBS Lett.*, 31, 97–100

BANDYAPADHYAY, G K and DUTTA, J. Separation of methyl esters of
1975 polyunsaturated fatty acids by argentation thin-layer chromatography. *J. Chromatog.*, 114, 280–282

BLIGH, E G. Lipid hydrolysis in frozen cod muscle. *J. Fish. Res. Bd.*
1961 *Can.*, 18, 143–145

BLIGH, E G and DYER, W J. A rapid method of total lipid extraction
1959 and purification. *Can. J. Biochem. Physiol.*, 37, 911–917

BLIGH, E G and SCOTT, M A. Lipids of cod muscle and the effect of
1966 frozen storage. *J. Fish. Res. Bd. Can.*, 23, 1025–1036

BONE, Q. Buoyancy and hydrodynamic functions of integument in the
1972 castor oil fish, *Ruvettus pretiosus* (Pisces: Gempylidae). *Copeia*, (1), 78–87

—— A note on the buoyancy of some lantern-fishes (Myctophoidei).
1973 *J. Mar. Biol. Ass. UK*, 53, 619–653

BONNET, J C, SIDWELL, V D and ZOOK, E G. Chemical and nutrition
1974 values of several fresh and canned finfish, crustaceans, and molluscs. Part II. Fatty acid composition. *Mar. Fish. Rev.*, 36, 8–14

BOSUND, I and GANROT, B. Lipid hydrolysis in frozen Baltic herring. *J.*
1969 *Food Sci.*, 34, 13–18

BOTTINO, N R. The composition of marine oil triglycerides as deter-
1971 mined by silver ion-thin-layer chromatography. *J. Lipid Res.*, 12, 24–30

BRADDOCK, R J and DUGAN, L R, Jr. Phospholipid changes in muscle
1972 from frozen stored Lake Michigan coho salmon. *J. Food Sci.*, 37, 426–429

BRANDES, C-H and DIETRICH, R. Betrachtungen uber die Beziehungen
1958 zwischen dem Fett- und Wassergehalt und die Fett verteilung bei Konsumfischen. *Veroff. Inst. Meeresforsch.* Bremerh. 5, 299–305

BUTLER, J L and PEARCY, W G. Swimbladder morphology and specific
1972 gravity of myctophids off Oregon. *J. Fish. Res. Bd. Can.*, 29, 1145–1150

CASTELL, C H. Metal-catalysed lipid oxidation and changes of protein
1971 in fish. *J. Amer. Oil Chem. Soc.*, 48, 645–649

CASTELL, C H and BISHOP, D M. Effect of hematin compounds on the
1969 development of rancidity in muscle of cod, flounder, scallops and lobster. *J. Fish. Res. Bd. Can.*, 26, 2299–2309

CASTELL, J H and MacLEAN, J. Rancidity in lean fish muscle. II.
1964a Anatomical and seasonal variations. *J. Fish. Res. Bd. Can.*, 21, 1361–1369

—— Rancidity in lean fish muscle. III. The inhibiting effects of bacter-
1964b ial activity. *J. Fish. Res. Bd. Can.*, 21, 1371–1377

CASTELL, C H and SPEARS, D M Heavy metal ions and the develop-
1968 ment of rancidity in blended fish muscle. *J. Fish. Res. Bd. Can.*, 25, 639–656

CASTELL, C H, BISHOP, D M and NEAL, W E. Production of
1968 trimethylamine in frozen cod muscle. *J. Fish. Res. Bd. Can.*, 25, 921–923

CASTELL, C H, DALE, J and DAMBERGS, N. Non-bacterial spoilage in
1962 Atlantic groundfish: Metal-induced rancidity. *Can. Fisherman*, 49(9), 30–33

CASTELL, C H, MacLEAN, J and MOORE, B. Rancidity in lean fish mus-
1965 cle. IV. Effect of sodium chloride and other salts. *J. Fish. Res. Bd. Can.*, 22, 929–944

CASTELL, C H, MacLEAN, J, MOORE, B and NEAL, W. Rancidity in
1966a lean fish muscle. V. The effect of amino acids. *J. Fish. Res. Bd. Can.*, 23, 27–43

CASTELL, C H, MOORE, B A, JANGAARD, P M and NEAL, W E. Oxida-
1966b tion rancidity in frozen storage cod fillets. *J. Fish. Res. Bd. Can.*, 23, 1385–1400

CHEN, S L, STEIN, R A and MEAD, J F. Expoxidation of unsaturated
1976 fatty esters in argentation chromatography. *Chem. Phys. Lip.*, 16, 161–166

CHERNYSHOV, V I. Role of lipid antioxidants in manifestation of some
1972 physiological features in fish. *Nauchnyye doklady vyssley shkoly*, 15, 40–45

CHILDS, E A. Functionality of fish muscle: Emulsification capacity. *J.*
1974 *Fish. Res. Bd. Can.*, 31, 1142–1144

CONTRARAS, R O, MIGLIARO, O A and RAFFO, R A. Continuous frac-
1971 tionation of Chilean anchovy oil with furfural. *J. Amer. Oil Chem. Soc.*, 48, 98–100

COSSINS, A R. Changes in muscle lipid composition and resistance
1976 adoption to temperature in the freshwater crayfish *Austropotamobius pallipes. Lipids*, 11, 307–316

CRAWFORD, M A. The relationship of dietary fats to the chemistry and
1974 morphological development of muscle, liver and brain. *Riv. Ital. Sost. Grasse*, LI, 302–309

DABROWSKY, K. The density and chemical composition of fish muscle.
1978 *Experientia*, 34, 1263–1265

DAMBERGS, N. Acetone-water mixtures for the extraction and rapid
1956 estimation of fats of biological materials, particularly fish products. *J. Fish. Res. Bd. Can.*, 13, 791–797

—— Extractives of fish muscle. 2. Solvent-water ratio in extraction of
1959 fat and water solubles. *Ibid*, 16, 63–71

—— 3. Amounts, sectional distribution, and variations of fat, water-
1963 solubles, protein and moisture in cod (*Gadus morhua* L.) fillets. *Ibid*, 20, 909–918

—— 4. Seasonal variations of fat, water-solubles, protein and water in
1964 cod (*Gadus morhua* L.) fillets. *Ibid*, 21, 703–709

—— Isopropanol-water mixtures for the production of fish protein
1969a concentrate from Atlantic herring (*Clupea harengus*). *J. Fish. Res. Bd. Can.*, 26, 1919–1923

—— Isopropanol-water azeotrope as solvent in the production of fish
1969b protein concentrate from herring (*Clupea harengus*). *J. Fish. Res. Bd. Can.*, 26, 1923–1926

DE KONING, A J. Phospholipids of marine origin. I. The hake (Mer-
1966a luccius capensis, Castelnau). J. Sci. Food Agric., 17, 112–117
—— Phospholipids of marine origin. IV. The abalone (Haliotis midae).
1966b J. Sci. Food Agric., 17, 460–464
DE KONING, A J and MCMULLAN, K B. Phospholipids of marine
1966 origin. III. The pilchard (Sardina ocellata, Jenyns) with particular reference to oxidation in pilchard meal manufacture. J.
Sci. Food Agric., 17, 385–388
DE MORENO, J E A, MORENO, V J and BRENNER, R R. Lipid metabolism
1976 of the yellow clam Mesodesma mactroides: 2-Polyunsaturated
fatty acid metabolism. Lipids, 11, 561–566
DENG, J-C, ORTHOEFER, F T, DENNISON, R A and WATSON, M. Lipids
1976 and fatty acids in mullet (Mugil cephalus): Seasonal and loca-
tional variations. J. Food Sci., 41, 1479–1483
DEVRIES, A L and EASTMAN, J T. Lipid sacs as a buoyancy adaptation
1978 in an Antarticc fish. Natuure, 271, 352–353
DROZDOWSKI, B and ACKMAN, R G. Isopropyl alcohol extraction of oil
1969 and lipids in the production of fish protein concentrate from
herring. J. Amer. Oil Chem. Soc., 46, 371–376
DUBROW, D, HALE, M and BIMBO, A. Seasonal variations in chemical
1976 composition and protein quality of menhaden. Mar. Fish. Rev.,
38(9), 12–16
DYER, W J. Frozen fish muscle – chemical changes and organoleptic
1967 quality. Cryobiology, 3, 297–305
—— Determination and Storage Life of Frozen Fish. In: Low Tem-
1968 perature Biology of Foodstuffs, Ed J Hawthorne. Pergamon
Press, Oxford. 429–447
DYER, W J and FRASER, D I. Proteins in fish muscle. 13. Lipid hyd-
1959 rolysis. J. Fish. Res. Bd. Can., 16, 43–52
DYER, W J and MORTON, M L. Storage of frozen plaice fillets. J. Fish.
1956 Res. Bd. Can., 13, 129–134
DYER, W J, FRASER, D I and BLIGH, E G. Fat hydrolysis in frozen fish.
1958 I. Free fatty acid formation. Prog. Rep. Atl. Coast. Sta., Fish
Res. Bd. Can., No. 71. 17–20
DYERBERG, J, BANG, H O, STOFFERSEN, E, MONCADA, S and VANE, J R.
1978 Eisosapentaenoic acid and prevention of thrombosis and
atherosclerosis? Lancet, 15 July, 117–119
EATON, C A, ACKMAN, R G, TOCHER, C S and SPENCER, K D. Cana-
1975 dian capelin 1972–1973. Fat and moisture compositions, and
fatty acids of some oils and lipid extract triglycerides. J. Fish.
Res. Bd. Can., 32, 507–513
EXLER, J, KINSELLA, J E and WATT, B K. Lipids and fatty acid of im-
1975 portant finfish; New data for nutrient tables. J. Amer. Oil
Chem. Soc., 52, 154–159
FAO Dietary fats and oils in human nutrition. FAO Food and Nutri-
1978 tion paper No. 3, Rome. 94pp
FARKAS, T, CSENGERI, I, MAJOROS, F and OLAH, J. Metabolism of fatty
1977 acids in fish. I. Development of essential fatty acid deficiency
in the carp, Cyprinus carpio Linnaeus 1758. Aquaculture, 11,
147–157
—— Metabolism of fatty acids in fish. II. Biosynthesis of fatty acids in
1978 relation to diet in the carp Cyprinus carpio Linnaeus 1758.
Aquaculture, 14, 57–65
FLO, A, HAGEN, N and MOHR, V. Fat tissues in fish. Report on pre-
1972 liminary studies during the period 1.9–31.12.1972. Res. Proj.
B.0103. 3858 Norweg. Res. Council for Tech. and Nat. Sci.
16pp
FRASER, D I and DYER, W J. Fat hydrolysis in frozen fish. 2. Relation
1959 to protein stability. 2. Prog. Rep. Atl. Coast Sta., Fish. Res.
Bd. Can., No. 72, 37–39
FRASER, D I, MANNAN, and DYER, W J. Proximate composition of
1961 Canadian Atlantic Fish. III. Sectional differences in the flesh
of a species of Chondrostei, one of Chimaerae and of some
miscellaneous teleosts. J. Fish. Res. Bd. Can., 18, 893–904
GATESOUPE, F J, LEGER, C, BOUDON, M, METAILLER, R and LUQUET, P.
1977 Alimentation lipidique du turbot (Scophthalmus maximus L.)
II. Influence de la supplémentation en esters methyliques de
l'acide linolénique et de la complémentation en acides gras de la
série ω9 sur la croissance. Ann. hydrobiol., 8, 247–254
GAUGLITZ, E J, JR, STOUT, V F and WEKELL, J C. Application of fish
1974 oils in the food industry. In: Fishery Products, Ed R Kreuzer.
Fishing News (Books) Ltd, West Byfleet, UK. 132–136
GHOSH, A, HOQUE, M and DUTTA, J. Fatty acids of boal fish oil by urea
1976 fractionation and gas-liquid chromatography. J. Sci. Food
Agric., 27, 159–164
GIBSON, T A and WORTHINGTON, R E. Lipids changes in frozen stored
1977 channel catfish grown by tank culture: effects of dietary fat,
freezing method, and storage temperature. J. Food Sci., 42,
355–358
GLASS, R L, KRICK, T P, OLSON, D L and THORSON, R L. The occurr-
1977 ence and distribution of furan fatty acids in spawning male
freshwater fish. Lipids, 12, 828–836
GRAY, G M and MACFARLANE, M G. Composition of phospholipids of
1961 rabbit, pigeon and trout muscle and various pig tissues.
Biochem. J., 81, 480–488
GUNSTONE, F D, WIJESUNDERA, R C, LOVE, R M and ROSS, D. Rela-
1976 tive enrichment of furan-containing fatty acids in the liver of a
starving cod. JCS Chem. Comm., 630–631

HALLGREN, B and STALLBERG, G. 1-0-(2-Hydroxyalkyl) glycerols iso-
1974 lated from Greenland shark liver oil. Acta Chem. Scand., 28,
1074–1076
HALLGREN, B, NIKLASSON, A, STALLBERG, G and THORIN, H. On the
1974 occurrence of 1-0-(2-methoxyalkyl) glycerols and 1-
0-phytanylglycerol in marine animals. Acta Chem. Scand., 28,
1035–1040
HAMID, A, SAKATA, T and KAKIMOTO, D. Microflora in the alimentary
1978 tract of gray mullet. II. A comparison of the mullet intestinal
microflora in fresh and sea water. Bull. Jap. Soc. Sci. Fish., 44,
53–57
HANSON, S W F and OLLEY, J. Application of the Bligh and Dyer
1963a method of lipid extraction to tissue homogenates. Biochem. J.,
89, 101–102
—— Apparent losses in lipid recovery from tissue homogenates.
1963b Biochem. J., 89, 102
—— Observations on the relationships between lipids and protein
1965 deterioration. In: The Technology of Fish Utilization, Ed R
Kreuzer. Fishing News (Books) Ltd, London. 111–115
HARDY, R and KEAY, J N. Seasonal Variations in the chemical compos-
1972 ition of Cornish mackerel, Scomber scombrus (L), with
detailed references to the lipids. J. Fd. Technol., 7, 125–137
HARDY, R and MACKIE, P. Seasonal variation in some of the lipid
1969 components of sprats (Sprattus sprattus). J. Sci. Food Agr., 20,
193–198
HAYASHI, K and TAKAGI, T. Seasonal variation in lipids and fatty acids
1977 of sardine, Sardinops melanosticta. Bull. Fac. Fish. Hokk.
Univ., 28, 83–94
—— The lipids of marine animals from various habitat depths. VIII.
1978 Occurrence of methoxy glyceryl ethers in the flesh lipids of
deep-sea teleost fish Seriollela sp. Bull. Jap. Soc. Sci. Fish., 44,
1239–1243
HAYASHI, K and YAMANDA, M. The lipids of marine animals from
1975a various habitat depths. II. On the fatty acid composition of the
neutral lipids in six species of gadiforms. Bull. Jap. Soc. Sci.
Fish., 41, 1153–1160
—— III. On the characteristics of the component fatty acids in the
1975b neutral lipids of deep-sea fishes. Ibid, 41, 1161–1175
—— IV. On the fatty acid composition of the neutral lipids in nine
1975c species of flatfishes. Bull. Fac. Fish. Hokk. Univ., 26, 265–276
—— V. Composition of wax esters and triglycerides of the gadoid fish,
1976 Podonema longipes. Ibid, 26, 356–366
HAYASHI, K, TAKAGI, T, KONDO, H and FUTAWATARI, M. The lipids of
1978 marine animals from various habitat depths. VII. Composi-
tions of diacyl glyceryl ethers in the flesh lipids of two deep-sea
teleost fish, Seriollela sp and S. punctata. Bull. Jap. Soc. Sci.
Fish., 44, 917–923
HEIDE-JENSEN, J. Investigations of oxygen pressure and oxygen
1965 absorption during storage of fish oil. In: Fat and Oil Chemistry,
Compos-Composition-Oxidation-Processing, Fourth Scand.
Symp. Fats and Oils, Turku, 31 Aug.–3 Sept. Gordon and
Breach, New York. 131–140
HIGASHI, H, TERADA, K and NAKAHIRA, T. Studies on the role of
1972 tocopherols in fish. III. Ubiquinone and tocopherol in fish
(Part I). Bitamin, 45, 113–120
HOOPER, S N, PARADIS, M and ACKMAN, R G. Distribution of trans-
1973 6-hexadecenoic acid, 7-methyl-7-hexadecenoic acid and
common fatty acids in lipids of the ocean sunfish, Mola mola.
Lipids, 8, 509–516
HORTON, E W. The prostaglandins. In: Biochemistry of Lipids (M.P.T.
1974 Int. Rev. Sci., Biochem. Ser. One), Ed T W Goodwin. Butterworths, London. Vol 1, 237–270
HURST, R E. A method of lipid extraction suitable for use with an
1974 electron capture detector. J. Fish. Res. Bd. Can., 31, 113–116
IKEKAWA, N, MATSUI, M, YOSHIDA, T and WATANABE, T. The compos-
1972 ition of triglycerides and cholesteryl esters in some fish oils of
salt, brackish and freshwater origins. Bull. Jap. Soc. Sci. Fish.,
38, 1267–1274
IRONSIDE, J I M and LOVE, R M. Studies on protein denaturation in
1958 frozen fish. I. Biological factors influencing the amounts of
soluble and insoluble protein present in the muscle of the
North Sea cod. J. Sci. Fd. Agric., 9, 597–604
JAFRI, A K. Fat and water distribution patterns in the flesh of the
1973 common cat-fish Wallago attu. Fish. Technol., X, 138–141
JANGAARD, P M. A rapid method for concentrating highly unsaturated
1965 fatty acid methyl esters in marine lipids as an aid to their
identification by GLC. J. Amer. Oil Chem. Soc., 42, 845–847
—— Pilot plant fractionation of marine oils methyl esters. J. Fish. Res.
1966 Bd. Can., 23, 681–687
—— Utilization of herring, capelin and mackerel. Can, Fisherman and
1973 Ocean Sci., 59, (Aug) 28–32, 48
JANGAARD, P M, BROCKERHOFF, H, BURGHER, R D and HOYLE, R J.
1967 Seasonal changes in general condition and lipid content of cod
from inshore waters. J. Fish. Res. Bd. Can., 24, 607–612
JOHNSTON, I A and GOLDSPINK, G. Some effects of prolonged starva-
1973 tion on the metabolism of the red and white myotomal muscles
of the plaice Pleuronectes platessa. Mar. Biol., 19, 348–353
JOSEPH, J D. Identification of 3,6,9,12,15-octadecapentaenoic acid in

1975 laboratory-cultured photosynthetic dinoflagellates. *Lipids*, 10, 395–403

KAISER, H, GROSSE-OETRINGHAUS, S and HUDALLA, B. Ionic alkoxy-
1978 lipids from the liver of elasmobranch fishes. *J. Chromatog.*, 154, 93–98

KAMOI, I, ONMIMARU, O and OBARA, T. Growth and fat content of
1975 cultivated young yellowtail (*Seriola quinqueradiata*) by fat-added feed. *Eiyo to Shokuryo*, 28, 247–255

KANAZAWA, A, TOKIWA, S, KAYAMA, M and HIRATA, M. Essential
1977 fatty acids in the diet of the prawn. I. Effects of linoleic and linolenic acids on growth. *Bull. Jap. Soc. Sci. Fish.*, 43, 1111–1114

KARLESKIND, A. Etude des alcools des insaponifiables. 1. Applications
1967 a l'étude des huiles marines. *Rev. Franc. Corps Gras*, 14, 251–258

KAWAI, N, NAKAYAMA, Y and AKEHASHI, H. Studies on the muscle
1978 lipids of deep-sea fishes (II). Identification of glyceryl ethers in the unsaponifiable matters. *Shokuhin Eiseigaku Zasshi*, 19, 73–77

KAYAMA, M. The essential fatty acids of fish. *Yushi*, 27, 110–116
1974

—— Studies on the lipids of micronektonic fishes caught in Sagami and
1975 Suruga Bays, with special reference to wax esters. *Yukagaku*, 24, 435–440

KAYAMA, M, IKEDA, Y and KOMAKI, Y. Studies on the lipids of marine
1976 zooplankton, with special reference to wax ester distribution in crustaceans and its *in vivo* formation. *Yukagaku*, 25, 329–334

KAYAMA, M, IKEDA, Y and MANKURA, M. Studies on the biosynthesis
1977 of wax ester by zooplankton, micronekton and marine micro-organism. *Yukagaku*, 26, 398–404

KE, P J and ACKMAN, R G. Bunsen coefficient for oxygen in marine
1973 oils at various temperatures determined by an exponential dilution method with a polarographic oxygen electrode. *J. Amer. Oil Chem. Soc.*, 50, 429–435

—— Metal-catalysed oxidation in mackerel skin and meat lipids. *J.*
1976 *Amer. Oil Chem. Soc.*, 53, 636–640

KINSELLA, J E, SHIMP, J L, MAI, J and WEIHRAUCH, J. Fatty acid con-
1977 tent and composition of freshwater finfish. *J. Amer. Oil Chem. Soc.*, 54, 424–429

KINUMAKI, T, ARAI, K, SUGII, K and ISEKI, S. Nutritive value of fish
1977 containing a large amount of alkoxydiglyceride of wax ester in meat. *Bull. Tokai Reg. Fish. Res. Lab.*, No. 91, 73–91

KOBAYASHI, M, FUKUSHIMA, J and NOGUCHI, S. Variations in the com-
1973 position of fatty acid in fish oils in various parts of individual fish. *Kaseigaku Zasshi*, 24, 511–515

KOIZUMI, C, IIYAMA, S, WADA, S and NONAKA, J. Lipid deteriorations
1978 of freeze-dried fish meats of different equilibrium relative humidities. *Bull. Jap. Soc. Sci. Fish.*, 44, 209–216

KORN, S and MACEDO, D. Determination of fat content in fish with a
1973 nontoxic, noninflammable solvent. *J. Fish. Res. Bd. Can.*, 30, 1880–1881

KUNIMOTO, M, ZAMA, K and IGARISHI, H. Lipids of marine bacteria. 1.
1975 Lipid composition of marine *Achromobacter* species. *Bull. Fac. Fish. Hokkaido Univ.*, 25, 332–341

LAMBERTSEN, G. Fatty acid compositions of fish fats. Comparisons
1978 based on eight fatty acids. *Fisk. Dir. Skr., Ser. Ernaering*, 1(4), 105–116

LAMBERTSEN, G and HANSEN, O. Fettsyresammensetningen i vevs-
1978 lipider fra lodde, makrell og sild. *Fiskeridirektoratets Vit-aminstitutt Rapporter og Oversikter*, No. 4, 68pp

LAMBERTSEN, S G and MYKLESTAD, H. Lipids in 'red feed' (*Calanus*
1972 *finmarchicus*), an important source of food for herring and capelin. *Proc: 6th Nordic Fat Symposium*, Grenaa, 1971, 84–91

LANG, N. Biological properties of oxidized fish oils. In: *The Technol-*
1965 *ogy of Fish Utilization*, Ed R Kreuzer. Fishing News (Books) Ltd, London. 223–224

LEE, R F. Lipids of zooplankton from Bute Inlet, British Columbia. *J.*
1974 *Fish. Res. Bd. Can.*, 31, 1577–1582

LEE, R F, PHLEGER, C F and HORN, M H. Composition of oil in fish
1975 bones: possible function in neutral buoyancy. *Comp. Biochem. Physiol.*, 50B, 13–16

LEGER, C, BERGOT, P, LUQUET, P, FLANZY, J and MEUROT, J. Specific
1977 distribution of fatty acids in the triglycerides of rainbow trout adipose tissue. Influence of temperature. *Lipids*, 12, 538–543

LEIM, A H and SCOTT, W B. Fishes of the Atlantic Coast of Canada.
1966 *Fish. Res. Bd. Can. Bull.*, No. 185, 485pp

LEMARCHAL, P. Role biologique de l'acid linolénique. *Rev. Franc.*
1978 *Corps. Gras.*, 25, 303–308

LEWIS, R W. The densities of three classes of marine lipids in relation
1970 to their possible role as hydrostatic agents. *Lipids*, 5, 151–153

LINKO, R and KAITARANTA, J. Hydrocarbons of Baltic herring lipids. *Riv.*
1976 *Ital. Sost. Grasse*, LIII, 37–39

LITCHFIELD, C. *Analysis of Triglycerides*. Academic Press, New York.
1972 104–138

LITCHFIELD, C, ACKMAN, R G, SIPOS, J C and EATON, C A. Isovaleroyl
1971 triglycerides from the blubber and melon oils of the beluga whale (*Delphinapterus leucas*). *Lipids*, 6, 674–681

LITCHFIELD, C, GREENBERG, A. J, ACKMAN, R G and EATON, C A.
1978 Distinctive medium chain wax esters, triglycerides and diacyl glyceryl ethers in the head fats of the Pacific beaked whale, *Berardius bairdi*. *Lipids*, 13, 860–886.

LOHNE, P. Fat separation – new knowledge can open new processing
1976 possibilities. *Meld. fra. S.S.F.*, No. 3, 9–14

LOUGH, A K. The chemistry and biochemistry of phytanic, pristanic
1973 and related acids. In: *Progress in the Chemistry of Fats and Other Lipids*, Ed R T Holman. Pergamon Press, Oxford. Vol 14, 1–48

LOVE, R M. Studies on protein denaturation in frozen fish. III. The
1958 mechanism and site of denaturation at low temperatures. *J. Sci. Food Agric.*, 9, 609–617

LOVE, R M and IRONSIDE, J I M. Studies on protein denaturation in
1958 frozen fish. II. Preliminary freezing experiments. *J. Sci. Food Agric.*, 9, 604–609

LOVE, R M, HARDY, R and NISHIMOTO, J. Lipids in the flesh of cod
1975 (*Gadus morhua* L.) from Faroe Bank and Aberdeen bank in early summer and autumn. Mem. Fac. Fish., Kagoshima Univ., 24, 123–126

LOVE, R M, MUNRO, L J and ROBERTSON, I. Adaption of the dark
1977 muscle of cod to swimming activity. *J. Fish. Biol.*, 11, 431–436

LOVERN, J A. The lipids of fish and changes occurring in them during
1962a processing and storage. In: *Fish in Nutrition*, Eds A Heen and R Kreuzer. Fishing News (Books) Ltd, London. 86–111

—— Autolytic changes in the lipids of fish flesh. In: *Recent Advances in*
1962b *Food Science*, Eds J Hawthorn and J M Leitch. Butterworths, London. 194–201

—— Some analytical problems in the analysis of fish and fish products.
1965 *J. Ass. Offic. Agric. Chem.*, 48, 60–68

LOVERN, J A and OLLEY, J. Inhibition and promotion of post-mortem
1962 lipid hydrolysis in the flesh of fish. *J. Food Sci.*, 27, 551–559

LOVERN, J A, OLLEY, J and WATSON, H A. Changes in the lipids of cod
1959 during storage in ice. *J. Sci. Food Agric.*, 6, 327–337

MACLEAN, J and CASTELL, C H. Rancidity in lean fish muscle. I. A
1964 proposed accelerated copper-catalysed method for evaluating the tendency of fish muscle to become rancid. *J. Fish. Res. Bd. Can.*, 21, 1345–1359

MALINS, D and WEKELL, J C. The lipid biochemistry of marine organ-
1970 isms. In: *Progress in the Chemistry of Fats and Other Lipids*, Ed R T Holman. Pergamon, Oxford. Vol 10, 337–363

MANGOLD, H K and MUKHERJEE, K D. New methods of quantitation
1975 in thin-layer chromatography: Tubular thin-layer chromatog-raphy (TTLC). *J. Chromatog. Sci.*, 13, 398–402

MANNAN, A, FRASER, D I and DYER, W J. Proximate composition of
1961a Canadian Atlantic fish. I. Variations in composition of differ-ent sections of the flesh of Atlantic halibut (*Hippoglossus hippoglossus*). *J. Fish Res. Bd. Can.*, 18, 483–493

—— Proximate composition of Canadian Atlantic fish. II. Mackerel,
1961b tuna and swordfish. *J. Fish. Res. Bd. Can.*, 18, 495–499

MASSON, L and BURGOS, M T. Fatty acid composition of the Chilean
1973 anchovy (anchoveta: *Engraulis ringens*) and of its neutral and polar fractions. *Grasas y Aceites*, 24, 327–330

MATSUI, M, WATANABE, T and KAWABATA, T. Fatty acids structures
1976 contained in several freshwater fish. *Bull. Jap. Soc. Sci. Fish.*, 42, 233–237

MAYZAUD, P and ACKMAN, R G. The 6,99,12,15,18-heneicosapenta-
1978 enoic acid of seal oil. *Lipds*, 13, 24–28

MAYZAUD, P, EATON, C A and ACKMAN, R G. The occurrence and
1976 distribution of octadecapentaenoic acid in a natural plankton population. A possible food chain index. *Lipids*, 11, 858–862

MITRA, R and DUA, R D. Studies on characterization and variation in
1978 triglyceride fatty acids from *Puntius sarana* body lipids. *J. Amer. Oil Chem. Soc.*, 55, 881–885

MOHR, V. Fatty tissue in animals. *Tidsskr. Kjemi. Bergv., Mettalurgi*,
1972 32(9), 23–24, 26, 29–30

—— Fettiviv i fisk. Samlet, avsluttende oversikt over prosjektet,
1977 Forsknings-prosjekt III 651.02, Norges Fiskeriforskning-sgrad. Institutt for teknisk biokjemi, Norges tekniske hogs-kole, Trondheim. 13pp

MOHR, V and ORMBERG, A. Fettviv i fisk. Oversikt over resultater fra
1977 prosjektets siste del. Forskningsprosjekt III 651.02, Norges Fiskeriforskningsrad. Institutt for teknisk biokjemi, Norges tekniske hogskole, Trondheim. 14pp

MOHR, V, MOLLER, M and FLO, A. Fat tissues in fish. Report I,
1973 Research Project 38C, Norweg. Fish. Res. Bd., Inst. For Tech. Biochem., Nor. Tech. Univ. Trondheim. 25pp

MOHR, V, MOLLER, M, ORMBERG, A, FLO, A, HALVORSEN, J and
1976a PADGET, E. Fettviv i fisk, Rapport II, Forskningsprosjekt 38c, Norges Fiskeriforskningsrad. Institutt for teknisk biokjemi, Norges tekniske hogskole, Trondheim. 20pp

MOHR, V, ORMBERG, A, HALVORSEN, J and PADGET, E. Fat tissues in
1976b fish. Research Project III. 651–672. Report III for period 1.1.1975 to 31.12.1975. 36pp

MORI, M and HIKICHI, S. Studies on the Antarctic Krill, *Euphausia*
1976 *superba*. III. Lipid composition. Nippon Suisan Kabushiki Kaisha Chuo Kenkyusho Hokuku, No. 11, 12–17

MORI, M, HIKICHI, S, KAMIYA, H and HASHIMOTO, Y. Three species of

1972 teleost fish having diacylglyceryl ethers in the muscle as a major lipid. *Bull. Jap. Soc. Sci. Fish.*, 38, 56–63

MORI, M, YASUDA, S and NISHIMURO, S. Two species of teleosts having
1978 wax esters or diacylglyceryl ethers in the muscle as a major lipid. *Bull. Jap. Soc. Sci. Fish.*, 44, 363–367

MYHER, J J. Separation and determination of the structure of acylg-
1978 lycerols and their ether analogues. In: *Handbook of Lipid Research Fatty Acids and Glycerides*, Ed A Kuksis. Plenum, New York. Vol 1, 123–196

NAKAYAMA, Y, KAWAI, N, MORI, T, MATSUOKA, S and AKEHASHI, H.
1978 Studies on the muscle lipids of deep-sea fishes (I). Investigation about the lipids and their unsaponifiable matters. *Shokuhin Eiseigaku Zasshi*, 19, 68–72

NEVENZEL, J C. Occurrence, function and biosynthesis of wax esters in
1970 marine organisms. *Lipids*, 5, 308–319

NEVENZEL, J C, RODEGKER, W and MEAD, J F. The lipids of *Ruvettus*
1965 *pretiosus* muscle and liver. *Biochemistry*, N.Y., 4, 1589–1594

NICOL, J A C, ARNOTT, H J, MIZUNO, G R, ELLISON, E C and
1972 CHIPAULT, J R. Occurrence of glyceryl tridocosahexaenoate in the eye of the sand trout *Cynoscion arenarius*. *Lipids*, 7, 171–177

NICOL, J A C, ZYZNAR, E S, THURSTON, E L and WANG, R T. The
1975 tapetum lucidum in the eyes of cusk-eels (Ophidiidae). *Can. J. Zool.*, 53, 1063–1079

NISHIMOTO, J and TAKEBE, M. Studies on lipid in the muscle of skipjack
1977 (*Katsuwonus pelamis*). I. Distribution of lipid in skeletal muscle. Mem. Fac. Fish., Kagoshima Univ., 26, 111–118

OGATA, H and NOMURA, T. Isolation and identification of prostaglan-
1975 din E$_2$ from the gastrointestinal tract of shark *Triakis scyllia*. *Biochem. Biophys. Acta.*, 388, 84–91

OLCOTT, H S. Oxidation of fish lipids. In: *Fish in Nutrition*, Eds A
1962 Heen and R Kreuzer. Fishing News (Books) Ltd, London. 112–116

OLLEY, J and DUNCAN, W R H. Lipids and protein denaturation in fish
1965 muscle. *J. Sci. Food Agric.*, 16, 99–104

OLLEY, J and LOVERN, J A. Phospholipid hydrolysis in cod flesh stored
1960 at various temperatures. *J. Sci. Food Agric.*, 11, 644–652

OLLEY, J, FARMER, J and STEPHAN, E. The rate of phospholipid hyd-
1969 rolysis in frozen fish. *J. Fd. Technol.*, 4, 27–37

OTA, T and YAMADA, Y. Studies on the lipids of *Hypomesus olidus*
1969 Pallas. I. Analysis of fatty acids in the triglycerides of visceral lipids by gas-liquid chromatography. *Bull. Jap. Soc. Sci. Fish.*, 35, 1138–1149

—— Fatty acids of four fresh-water fish lipids. *Bull. Fac. Fish.*, Hok-
1975 kaido Univ., 26, 277–288

OWEN, J M, ADRON, J W, MIDDLETON, D and CAWLEY, C R. Elonga-
1975 tion and desaturation of dietary fatty acids in turbot *Scophthalmus maximus* L., and rainbow trout, *Salmo gairdnerii* Rich. *Lipids*, 10, 528–531

OWEN, J M, ADRON, J W, SARGENT, J R and COWLEY, C B. Studies on
1972 the nutrition of marine flatfish. The effect of dietary fatty acids on the tissue fatty acids of the plaice *Pleuronectes platessa*. *Mar. Biol.*, 13, 160–166

PARADIS, M and ACKMAN, R G. Localization of a marine source of odd
1976a chain-length fatty acids. I. The amphipod *Pontoporeia femorata* (Kroyer). *Lipids*, 11, 863–870

—— II. Seasonal propagation of odd chain-length monoethylenic fatty
1976b acids in a marine food chain. *Lipids*, 11, 871–876

PASCAL, J-C and ACKMAN, R G. Occurrence of 7-methyl-7-hexa-
1975 decenoic acid, the corresponding alcohol, 7-methyl-6-hexadecenoic acid and 5-methyl-4-tectradecenoic acid in sperm whale oils. *Lipids*, 10, 478–482

—— Long chain monethylenic alcohol and acid isomers in lipids of
1976 copepods and capelin. *Chem. Phys. Lip.*, 16, 219–223

PATTERSON, S, JOHNSTON, I A and GOLDSPINK, G. The effect of starva-
1974 tion on the chemical composition of red and white muscles in the plaice (*Pleuronectes platessa*). *Experientia*, 30, 892–894

PATTON, J S. The effect of pressure and temperature on phospholipid
1975 and triglyceride fatty acids of fish white muscle: a comparison of deepwater and surface marine species. *Comp. Biochem. Physiol.*, 52B, 105–110

PATTON, J S and BENSON, A A. A comparative study of wax ester
1975 digestion in fish. *Comp. Biochem. Physiol.*, 52B, 111–116

PATTON, J S, NEVENZEL, J C and BENSON, A A. Specificity of digestive
1975 lipases in hydrolysis of wax esters and triglycerides studied in anchovy and other selected fish. *Lipids*, 10, 575–583

PEARCE, R E and STILLWAY, L W. *trans*-6-Hexadecenoic acid in
1976 spadefish *Chaetodipterus faber*. *Lipids*, 11, 247–249

PHLEGER, C F. Bone lipids of Kona coast reef fish: skull buoyancy in
1975 the hawkfish, *Cirrhites pinnulatus*. *Comp. Biochem. Physiol.*, 52B, 101–104

PHLEGER, C F and GRIMES, P W. Bone lipids of marine fishes. *Physiol.*
1976 *Chem. Phys.*, 8, 447–456

PHLEGER, C F, PATTON, J, GRIMES, P and LEE, R F. Fish bone oil: Per-
1976 cent total body lipid and carbon-14 uptake following feeding of 1-^{14}C-palmitic acid. *Mar. Biol.*, 35, 85–90

PHLEGER, C F, GRIMES, P W, PESELY, A and HORN, M H. Swimblad-
1978 der lipids of five species of deep benthopelagic Atlantic ocean fishes. *Bull. Mar. Sci.*, 28, 198–202

PIGGOTT, G M. New approaches to marketing fish. In: *New Protein*
1976 *Foods*, Ed. A M Altschul. Academic Press, New York. Vol 2, 1–37

POHL, P, GLASL, H and WAGNER, H. Analysis of polyene fatty acids.
1969 II. A standardized micromethod for the separation and oxidative cleavage of polyene fatty acids by thin-layer chromatography and gas chromatography. *J. Chromatog.*, 42, 75–82

POLESELLO, A, MANNINO, S and PIZZOCARO, F. Influence of breeding
1977 on the chemical composition of eels. *Riv. Ital. Sost. Grasse*, 54, 27–30

RAHN, C H, SAND, D M and SCHLENK, H. Metabolism of oleic,
1977 linoleic and linolenic acids in gourami (*Trichogaster cosby*) fry and mature females. *Comp. Biochem. Physiol.*, 58B, 17–20

REVANKAR, G D, SEN, D P, HEMAVATHY, J and MATHEW, G. Solvent
1975 winterization of sardine oil. *J. Oil Technol. Ass. India*, July/ Sept., 85–87

ROBINSON, J S and MEAD, J F. Lipid absorption and deposition in
1973 rainbow trout (*Salmo gairdnerii*). *J. Fish. Res. Bd. Can.*, 51, 1050–1058

ROUBAL, W T. Oxidative deterioration of flesh lipids of Pacific cod
1967 (*Gadus macrocephalus*). *J. Amer. Oil Chem. Soc.*, 44, 325–327

RZHAVSKAYA, F M, DUBROVSKAYA, T A, MAKAROVA, A M and
1977 PRAVDINA, L V. Study of the qualitative state and the lipid fatty acid composition in relation to their method of isolation from muscle tissues of frozen and salted fishes. Tr. *Vses. Nauchno-Issled.* Inst. Morsk. Rybn. Khoz. Okeanogr., (123), 106–119

SANDIFER, P A and JOSEPH, J. Growth responses and fatty acid com-
1976 position of juvenile prawns (*Macrobrachium rosenbergii*) fed a prepared ration augmented with shrimp oil. *Aquaculture*, 8, 129–138

SARGENT, J R. The structure, metabolism and function of lipids in
1976a marine organisms. In: *Biochemical and Biophysical Perspectives in Marine Biology*, Eds D C Malins and J R Sargent. Academic Press, London. Vol 3, 149–212

—— Waxes for survival. *Spectrum*, No. 143, 8–11
1976b

SARGENT, J R and LEE, R F. Biosynthesis of lipids of zooplankton from
1975 Saanich inlet, British Columbia, Canada. *Mar. Biol.*, 31, 15–23

SARGENT, J R, GATTEN, R R and MCINTOSH, R. The distribution of
1973 neutral lipids in shark tissues. *J. Mar. Biol. Ass. UK*, 53, 649–656

—— Wax esters in the marine environment—their occurrence, forma-
1977 tion, transformation and ultimate fates. *Mar. Chem.*, 5, 573–584

SARGENT, J R, GATTEN, R R, CORNER, E D S and KILVINGTON, C C.
1977 On the nutrition and metabolism of zooplankton. XI. Lipids in *Calanus helgolandicus* grazing *Biddulphia sinensis*. *J. Mar. Biol. Ass. UK*, 57, 525–533

SARGENT, J R, LEE, R F and NEVENZEL, J C. Marine waxes. In:
1976 *Chemistry and Biochemistry of Natural Waxes*, Ed P E Kolattukudy. Elsevier, Amsterdam. 50–90

SARGENT, J R, MORRIS, R J and MCINTOSH, R. Biosynthesis of wax
1978 esters in oceanic crustaceans. *Mar. Biol.*, 46, 315–320

SATO, Y and TSUCHIYA, Y. Studies on the lipid of *Ruvettus pretiosus*. II.
1970 The composition of the unsaponifiable matters and purgative action of the oils on mouse. *Tohoku J. Agric. Res.*, 21, 176–182

SCHLENK, H. Odd numbered polyunsaturated fatty acids. In: *Progress*
1970 *in the Chemistry of Fats and Other Lipids*, Ed R T Holman. Pergamon, Oxford. Vol IX, Part 5, 587–605

SCHLENK, H and SAND, D M. Fractionation methods. In: *Fish Oils*
1967 *—Their Chemistry, Technology, Stability, Nutritional Properties, and Uses*, Ed M E Stansby. Avi Pub. Co, Westport. 75–106

SCRIMGEOUR, C M. Quantitive analysis of furanoid fatty acids in
1977 crude and refined cod liver oil. *J. Amer. Oil Chem. Soc.*, 54, 210–211

SEN, N and SCHLENK, H. The structure of polyenoic odd- and even-
1964 numbered fatty acids from mullet (*Mugil cephalus*). *J. Amer. Oil Chem. Soc.*, 41, 241–247

SEN, P C, GHOSH, A and DUTTA, J. Fatty acids of the lipids of murrels.
1976 *J. Sci. Food Agric.*, 27, 811–818

SHABALINA, A A and OSTROUMOVA, I N. Introduction of vegetable and
1976 synthetic fat into food for trout. *Izv. Gos. Nauchno-Issled.* Inst. Ozern. Rechn. Rybn. Khoz., 72, 95–102

SHELTAWY, A and OLLEY, J. Lipid distribution and recovery during salt
1966 fractionation of cod muscle. *J. Sci. Food Agric.*, 17, 94–100

SHUSTER, C Y, FROINES, J R and OLCOTT, H S. Phospholipids of tuna
1964 white muscle. *J. Amer. Oil Chem. Soc.*, 41, 36–41

SHIMMA, Y and NAKADA, M. Utilization of petroleum yeast for food. I.
1974a Effects of supplemental oil. *Bull. Freshwater Fisheries Research Laboratory*, 24, 47–56

—— II. Effect on growth and body lipids of rainbow trout fingerlings
1974b raised in cages. *Ibid*, 24, 111–119

SIDOROV, V S, LIZENKO, E I, RIPATTI, P O and BOLGOVA, O M. Total

1977 lipid content in the organs of salmon and certain other fish. *Sravnit. Biokhimiya Ryb. i Ikh Gel'mintov Lipidy,* Fermenty, Belki, 5–56

SIDWELL, V D, FONCANNON, P R, MOORE, N S and BONNET, J C.
1974 Composition of the edible portion of raw (fresh or frozen) crustaceans, finfish and molluscs. 1. Protein, fat, moisture, ash, carbohydrate, energy value, and cholesterol. *Mar. Fish. Rev.,* 36, 21–35

SNYDER, F. The biochemistry of lipids containing ether bonds. In:
1970 *Progress in the Chemistry of Fats and Other Lipids,* Ed R T Holman. Pergamon, Oxford. Vol 10, 287–335

STANSBY, M E. Polyunsaturates and fat in fish flesh. *J. Amer. Diet.*
1973 *Ass.,* 63, 625–630

—— Development of fish oil industry in the United States. *J. Amer.*
1978 *Oil. Chem. Soc.,* 55, 238–243

STEFFENS, W. Chemical composition and nutrition value of carp flesh.
1974 *Nahrung,* 18, 789–794

STIRLING, H P. Effects of experimental feeding and starvation on the
1976 proximate composition of the European bass *Dicentrarchus labrax. Mar. Biol.,* 34, 85–91

STODOLNIK, L and PODESZEWSKI, Z. Lipids in commercial fish of the
1972 Baltic Sea, Part II. Fatty acids of lipid fractions. *Bromatologia, Chemia, Toksykologia,* 5, 295–300

SUMNER J L and HOPKIRK, J. Lipid composition on New Zealand eels.
1976 *J. Sci. Food Agric.,* 27, 933–938

TAKAGI, T and ITABASHI, Y. Random combinations of acyl and
1977 alcoholic groups through overall wax esters of sperm whale head oils. *Comp. Biochem. Physiol.,* 57B, 37–39

TAKAHASHI, H and YAMADA, M. Lipid composition of seven species of
1976 crustacean plankton. *Bull. Jap. Soc. Sci. Fish.,* 42, 769–776

TAKAMA, K. ZAMA, K and IGARISHI, H. Changes in the flesh lipids of
1967 fish during frozen storage. 1. Flesh lipids of bluefin tuna, *Thunnus orientalis. Bull. Fac. Fish., Hokk. Univ.,* 18, 240–247

TAKEUCHI, M. Effect of dietary lipid on lipid accumulation in ayu,
1978 *Plecoglossus altivelis. Bull. Tokai Reg. Lab.,* Fish. Res. Lab., No. 93, 103–109

TAKEUCHI, T and WATANABE, T. Requirement of carp for essential
1977a fatty acids. *Bull. Jap. Soc. Sci. Fish.,* 43, 541–551

—— Effect of eicosapentaenoic and docosahexaenoic acid in pollock
1977b liver oil on growth and fatty acid composition of rainbow trout. *Ibid,* 43, 947–953

—— Dietary levels of methyl laurate and essential fatty acid require-
1977c ment of rainbow trout. *Ibid,* 43, 893–898

—— Growth-enhancing effect of cuttlefish liver oil and short-necked
1978 clam oil on rainbow trout and their effective components. *Bull. Jap. Soc. Sci. Fish.,* 44, 733–738

TAKEUCHI, T, WATANABE, T and OGINO, C. Use of hydrogenated fish
1978 oil and beef tallow as a dietary energy source for carp and rainbow trout. *Bull. Jap. Soc. Sci. Fish.,* 44, 875–881

TINOCO, J, BABCOCK, R, HINCENBERGS, I, MEDWADOWSKI, B, MIL-
1979 JANICH, P and WILLIAMS, M A. Linolenic acid deficiency. *Lipids,* 14, 166–173

TOTANI, Y, TOTANI, N and MATSUO, N. Studies on the lipid in skin and
1975 liver of seborrhea-rats. *Eiyo to Shokuryo,* 28, 79–86

TOYOMIZU, M, NAKAMURA, T and SHONO, T. Fatty acid composition
1976 from horse mackerel lipid–discussion of fatty acid composition of fish lipid. *Bull. Jap. Soc. Sci. Fish.,* 42, 101–108

UEDA, T. Fatty acid composition of oils from 33 species of marine fish.
1967 *J. Shimonoseki Univ. Fish.,* 16, 1–10

—— Changes in the fatty acid composition of mackerel lipid and
1976 probably related factors. I. Influence of the season, body length and lipid content. *Bull. Jap. Soc. Sci. Fish.,* 42, 479–484

—— Variations in the fatty acid composition of fish lipids and their
1977 relation to some numerical factors. *J. Shimonoseki Univ. Fish.,* 26, 141–250

VIVIANI, R, BORGATTI, A R, MANCINI, L and CORTESI, P. Changes in
1967 the muscular lipids of the mackerel (*Scomber scombrus* L.) during frozen storage. *Atti della Soc. Ital. Sci. Vet.,* 21, 706–710

VYNCKE, W and LAGRON, F. Determination of phospholipids in fish by
1973 thin-layer chromatography. *Medad Fac. Landbouwwet Rijksuniv.* Gent., 38, 235–252

WAGNER, H and POHL, P. Zur analytik von Polyenfettsauren. I.
1964 *Biochemische Zeitschrift,* 340, 337–344

WATANABE, T and ACKMAN, R G. Lipids and fatty acids of the ameri-
1974 can (*Crassostrea virginica*) and european flat (*Ostea edulis*) oysters from a common habitat, and after one feeding with *Dicrateria inornata* and *Isochrysis galbana. J. Fish. Res. Bd. Can.,* 31, 403–409

WESSELS, J P H and SPARK, A A. The fatty acid composition of the
1973 lipids from two species of hake. *J. Sci. Food Agric.,* 24, 1359–1370

WEAVER, J E. Temporal trends in fatty acid composition of juvenile
1974 Atlantic menhaden fed brine shrimp nauplii. *Trans. Amer. Fish. Soc.,* 103, 382–386

WHITE, B Jr A complentary thin layer and gas-liquid chromatographic
1966 procedure for fatty acid analysis. *J. Chromatog.,* 21, 213–222

WILLS, R B H and HOPKIRK, G. Distribution and fatty acid composi-
1976 tion of lipids of eels (*Anguilla australis*). *Comp. Biochem. Physiol.,* 53B, 525–527

WOOD, G, HINTZ, L and SALWIN, H. Chemical alterations in fish tissue
1969 during storage at low temperatures. *J. Ass. Off. Anal. Chem.,* 52, 904–910

WURZIGER, J and DICKHAUT, G. Uber schadstoffe in fettfish-
1977 zubereitungen. *Fette Seifen. Anstrich.,* 79, 165–170

YAMADA, J. Histochemical observation of fish muscle. II Occurence
1972 and status of lipids and muscle texture of tunas. *Bull. Tokai. Reg. Fish Lab.,* 12, 35–42

YA, J and NAKAMURA, S. Histochemical studies on fish tissue. I.
1964 Distribution of depot fat in fish muscle tissue. *Bull. Tokai. Reg. Fish. Res. Lab.,* 39, 21–28

YONE, Y and FUJII, M. Studies on nutrition of red sea bream. XI.
1975 Effect of ω3 fatty acid supplement in a corn oil diet on growth rate and feed efficiency. *Bull. Jap. Soc. Sci. Fish.,* 41, 73–77

ZSOLNAY, A. Inventory of nonvolatile fatty acids and hydrocarbons in
1977 the oceans. *Mar. Chem.,* 5, 465–475

Fish lipids. Part 2   *R Hardy*

After death the lipids in fish are subject to two major changes, namely, lipolysis and autoxidation. With the possible exception of one or two fish products (Van Veen, 1965) the effects produced by these changes are considered undesirable and are often the major causes of spoilage (Fukuda, 1955; Toyama, 1956).

1 Autoxidation

Of the two processes, autoxidation is the most important, particularly in the deterioration of frozen fish products causing flavour (Banks, 1939), colour (Jones, 1962) and possibly textural changes (Sikorski *et al*, 1976). But long before the widespread use of freezing as a method of preservation it was known that in the storage of fish, especially by the use of salt, care had to be taken to exclude air, otherwise the products became rancid (Cutting, 1962; Voskresensky, 1965).

In the early part of this century it was realized that an interaction between fish oils and oxygen was involved (Marcelet, 1924), but the general mechanism of lipid autoxidation was not elucidated until the 1940s following the earlier recognition of free radical chain reactions (Bateman, 1954).

In the reaction scheme proposed three mechanisms are involved, namely,

Initiation: Production of R° or $RO_2^\bullet$ radicals rate r_1

Propagation: $R^\bullet + O_2 \rightarrow RO_2^\bullet$ rate k_2

$RO_2^\bullet + RH \rightarrow RO_2H + R^\bullet$ rate k_3

and *Termination*

$2R^\bullet \rightarrow$ non-propagating rate k_4

$R^\bullet + RO_2^\bullet \rightarrow$ products rate k_5

$2RO_2^\bullet \rightarrow$ rate k_6

The initiation stage involves the production of free radicals. One of the intriguing aspects of oxidation kinetics is the process whereby this occurs, especially in pure model systems. A number of mechanisms have been proposed such as heavy metal catalysis and bi- and ter-molecular reactions between the substrate and oxygen (Ingold, 1961). At present the termolecular mechanism is favoured (Denisov, 1964). In most natural products and fish in particular such considerations are not of practical concern because many minor constituents are present that can degrade or interact to give initiating free radicals. Once the reaction is under way there is no further dependence on such sources because the hydro-peroxides produced are labile and can degrade to give free radicals (Bateman, 1954).

The production of hydroperoxides takes place in the propagation sequence of reactions. In these the free radical reacts with oxygen to give a peroxy radical which in turn reacts with the substrate to give a hydroperoxide and another free radical which can initiate the chain of events again. Hence the term free radical chain reaction. The first reaction in the sequence is fast and the second usually much slower. The reason for this is that radical-oxygen reactions are energetically undemanding which is not the case for the hydrogen abstraction reaction where the strength of the carbon-hydrogen bond that is to be broken is a major controlling factor.

In saturated fats the bond strength is high and the overall energetics of the reaction is unfavourable and so at room temperatures this reaction is negligible. Weaker carbon-hydrogen bonds are present in unsaturated fats, particularly in methylene groups adjoining the double bonds. Hydrogen abstraction becomes easier but in addition in methylene interrupted polyenoic acids the free radical can rearrange and achieve a lower energetic state.

$$-CH_2-CH = CH-CH_2-CH = CH-CH_2-$$
$$\downarrow RO^{\cdot}_2$$
$$-CH_2-CH = CH-CH-CH = CH-CH_2-$$
$$\swarrow \quad \cdot \qquad \downarrow + ROOH$$
$$-CH_2-CH-CH = CH-CH = CH-CH_2-/$$
$$\cdot$$
$$-CH_2-CH = CH-CH = CH-CH-CH_2-$$
$$\cdot$$

There is some evidence that in the oxidation of polyunsaturated acids the free radical ends up most frequently on the allylic carbon most remote from the carboxylic acid group (Schollner and Herzschuh, 1966). Together these various mechanistic factors imply that where a molecule contains few weak bonds the propagation reaction will be slow and relatively independent of oxygen concentration. Molecules with weaker bond systems will interact in a converse manner but even here, because reaction 3 has a relatively high activation energy when compared with 2, lowering the temperature of oxidation will reduce the overall rate dependency on oxygen concentration. That is, reactions that are dependent on oxygen concentrations at high temperature will become less so at low temperatures (Bateman, 1954). Without some modification step it is clear that oxidation once initiated will proceed not only unchecked but in a constantly accelerating manner brought about by the introduction of fresh radicals through hydroperoxide breakdown and mediated only by the availability of reactants.

Although this appears to occur at first, this does not continue because radicals interact with each other, terminating the reaction chain. The termination mechanism is important not only in controlling the rate but also in deciding the nature of the primary products. A fast termination reaction will tend to reduce the amounts of hydroperoxide formed with a relative increase in other products such as peroxides, alcohols and ketones. In a slow termination reaction hydroperoxides will predominate.

Making various assumptions about the different stages (Bolland, 1949; Bateman, 1954) it is possible to derive an equation describing the rate of oxidation:

$$r = r_i^{\frac{1}{2}} k_3 k_6^{-\frac{1}{2}} [RH] \quad \frac{k_2 k_6^{-\frac{1}{2}} [O_2]}{k_3 k_4^{-\frac{1}{2}} [RH] + k_2 k_6^{-\frac{1}{2}} [O_2]}$$

where r is the overall rate.

When reaction 3 is much faster than 2 the rate becomes

$$r = r_i^{\frac{1}{2}} k_3 k_6^{-\frac{1}{2}} [RH]$$

ie a function that at any one temperature changes only with substrate concentration.

Unfortunately, in fish and fish oils the mechanism and rate equation provides only a qualitative, albeit a valuable, description of the oxidation process even in the initial stages which are less influenced by the products of the oxidation reaction. It cannot give a quantitative picture because it does not take into account the complexity of such systems.

In oils, like those of fish, containing a number of different fatty acids, some very unsaturated and highly reactive, each can exert an effect on the oxidation rate of the others (Hammond and Johnson, 1972), and even the disposition of the acid within the glyceride molecule (cf section 1) can affect the oxidation rate (Raghuveer and Hammond, 1967); unsaturated acids in the 2 position of the glyceride molecule oxidize less rapidly than in the 1 or 3 position. Free fatty acids too can have an effect, sometimes pro-oxidative (Rouchaud and Lutete, 1968; Catalano and de Felice, 1970) and sometimes anti-oxidative (Govind Rao and Achaya, 1968).

Non-lipid components also can affect the reaction, thus heavy metals can act as pro-oxidants by decomposing hydroperoxides to produce free radicals (Waters, 1971), the tocopherols can act as antioxidants (Toyama and Shimazu, 1972; Olcott and Van der Veen, 1968) and even squalene is said to have an antioxidant action (Govind Rao and Achaya, 1968).

In fish the situation is made even more complicated by the presence of proteins (Sikorski et al, 1976; Fischer and Deng, 1977), especially heme proteins (Castell and Bishop, 1969) and possibly peroxidases (Christopherson, 1968), amino acids (Marcuse, 1962; Pokorny et al, 1976) and even water (Labuza, 1971).

The disposition of the lipid within the tissue will also affect its rate. Lipids associated with the dark lateral muscle tend to oxidize more readily (Banks, 1939; Takama, 1974; Fischer and Deng, 1977; Ke et al, 1978). Where the lipid pools are large, such as on cut surfaces or in large lipid cells, the rate should approximate that of the bulk phase rate for the lipid. Smaller lipid droplets should oxidize according to emulsion reaction kinetics (Smith and Ewart, 1948) giving relatively rapid oxidations and long reaction chain lengths for a low input of free radicals.

The most unsaturated lipids in all fish are the phospholipids, but notwithstanding this they do not oxidize rapidly and this it is believed is caused by the physical disposition of the lipids making it difficult for them to participate in the oxidation chain reaction (Hardy *et al*, 1979).

In addition, access to oxygen must be taken into account; exposed lipid surfaces of fillets, minces and gutted fish will oxidize more rapidly than lipids embedded in tissues through which oxygen may diffuse only with difficulty (Banks, 1939; Bito and Kiriyama, 1973; Bligh and Regier, 1976). It is thus not possible from a simple lipid compositional knowledge to say how rapidly fish or fish oils will oxidize even though some authors believe it possible so to do (Rzhavskaya *et al*, 1978).

In fish oils oxidation will proceed quite readily at ambient temperatures (Smith *et al*, 1972). By contrast, in wet fish between 0°C and ambient temperatures oxidation does not appear to be a dominant spoilage process (Smith *et al*, 1979; 1980) even though in the latter stages of spoilage of some species such as trout (Hansen, 1972), sardines and mackerel (Madhavan *et al*, 1972). rancid flavours have been reported to affect acceptability. Oxygen is undoubtedly mobilized by fish at these temperatures but there is competition for it between micro-organisms, enzymes and lipids (Smith *et al*, 1972). The effect of this is that in whole fish the internal tissues tend to be oxygen deficient and thus unless the lipids are exposed on the surface they will oxidize only slowly. This should be quite marked in fish because with the high degree of unsaturation of their lipids one would expect, according to the theory outlined earlier, that their oxidation would be oxygen pressure dependent. But even when they are exposed and subject to autoxidation, micro-organisms are present which can interact with the oxidized lipid and possibly affect the oxidation rate (Senser and Grosch, 1973). The relative weighting of these effects coupled with those discussed earlier provides an explanation as to why certain fatty species such as trout and gutted mackerel will oxidize at temperatures above 0°C (Madhavan *et al*, 1970; Hansen, 1972; Smith *et al*, 1980) whereas others such as herring (Smith *et al*, 1980) remain relatively unaffected.

It is interesting to note in this context that irradiated chill stored fish are more prone to oxidative rancidity, presumably because microbial oxygen users are destroyed and also because radical forming precursors are introduced (Lerke *et al*, 1961; Kamat and Kumta, 1972; Baldrati *et al*, 1978). Spoilage assessment of fish, though often monitored by physical and chemical methods, must of necessity rest finally on sensory evaluation. In wet fish storage, components introduced primarily by bacterial spoilage but also by enzymatic reactions contribute more to the flavour than those derived from lipid autoxidation. For these reasons it is believed that lipid oxidation is a relatively minor spoilage process in wet fish.

Notwithstanding this there is evidence to suggest that in some fatty species oxidation in cold storage is enhanced by long-term ice storage (Shenoy and Pillai, 1971; Hardy and Smith, 1976; Bilinski *et al*, 1978; O'Keefe and Noble, 1978; Smith *et al*, 1980). Whether this is due to the introduction of free radical generating species such as hydroperoxides, the development of more reactive lipids such as free fatty acids, the activation of catalysts through the lysis of cells or the removal of an antioxidant is not known. Treatment of the fish with salt water prior to freezing causes a somewhat analogous effect but here it is believed that the enhancement of rancidity is caused by the salt increasing the amount of unfrozen water, thus allowing greater interplay between the reacting entities (Banks, 1937; Nair *et al*, 1974).

In all fish where appropriate studies have been carried out lipid oxidation occurs in frozen storage. The importance of the mechanism in the deterioration of the frozen fish is determined to a large extent by the type and disposition of the lipid in the fish. The fatty species like herring and mackerel that possess lipid reserves in the flesh are most subject to oxidation and this becomes the dominant spoilage mechanism in the low temperature storage of such fish. Fish such as cod which are deficient in such muscle tissue fats contain structural lipids associated with the membranes. These too will oxidize, although somewhat more slowly, and as such will contribute to the deterioration in cold storage.

In fatty fish, oxidation takes place primarily in the depot fats which are composed of triglycerides. The rate decreases with decreasing temperature usually by a factor of 2 to 3 for every 10°C decrease. But even within a species, different batches of fish subjected to identical conditions of storage will show differing oxidation rates. Some fatty fish with relatively saturated and low total lipid contents (after overwintering or spawning) will oxidize more rapidly than fish with high unsaturated fat contents in peak biological conditions (Agzhitova, 1969); in other fish the converse holds (Banks, 1952; Mendenhall, 1972; Rzhavskaya *et al*, 1978). The reasons for this are not known; in sprats, spring lipid depleted fish contain more tocopherol and ubiquinones as well as a lower proportion of unsaturated fatty acids in the depot lipids than their autumnal counterparts, which might in this case make the former fish less prone to oxidation (Hardy and Mackie, 1969). However, as pointed out earlier, assessments made on such simple analyses are likely to be erroneous. On the whole, small or butchered fish are more subject to oxidation, presumably because oxygen availability is not rate controlling (Bito and Kiriyama, 1973; Hardy *et al*, 1973; Cole and Keay, 1976; Hiltz *et al*, 1976).

Conditions of cold storage have important consequences; if dehydration occurs the rate of oxidation increases (Banks, 1952) whereas storage under conditions of low water loss by packaging, glazing or freezing in water has a protective influence (Tarr, 1948; Hardy *et al*, 1973; Bilinski *et al*, 1979). Two factors may be involved here. The loss of water which will permit ready access by oxygen and thus ensure its diffusion will not be rate controlling. But in addition studies on foods in which the water activity has been reduced have shown increased oxidation rates due to the activation of the metal catalysts present (Labuza, 1971). This activation is caused by hydroperoxides replacing water molecules co-ordinated with the metal, thus facilitating metal peroxide cleavage reactions, increasing radical concentration and also the production of breakdown products.

The reasons why non-fatty fish oxidize slowly are not clear. Certainly, when extracted, the lipids oxidize readily (McGill, 1976) as does the excised muscle when incubated with a variety of oxidation catalysts (Castell and Bishop, 1969).

In such fish the lipid is bound to protein often within the cell membrane matrix and it may be supposed that radical reactivity will be severely restricted so that, unless the lipids are conveniently lined up, lipid peroxy radical attack on nearby lipid molecules will not occur. Under such circumstances propagation will only ensue when the radicals are transferred to a more mobile entity such as unbound lipids or water.

Studies carried out on the frozen storage of cod and haddock show that only the phospholipids oxidize, which suggest that in any radical transfer reaction neutral lipids and free fatty acids do not participate (Hardy et al, 1979). This may be contrasted with studies on the fatty fish 'Jack Mackerel' in which marked oxidation of the free fatty acids occurred (Shono et al, 1973).

One certain effect of the oxidation in non-fatty fish like cod is the introduction of malodours, often called cold-storage flavours. The compounds most responsible for this are unsaturated carbonyls (McGill et al, 1977) and in particular hept-cis-4-enal, a by-product of the oxidation of the n-3 polyenoic acids that are present.

It is not clear whether lipid autoxidation in such fish has other side-effects. Studies on model systems containing phospholipids and proteins show that the two will react with each other during autoxidation (Labuza, 1971; Sikorski et al, 1976). It is possible, therefore, that in tissue oxidation protein denaturation will result, perhaps by co-oxidation of the sulphydryl groups or by carbonyl cross-linking of free amino groups.

From the foregoing it can be seen that nearly any processing of the fish which will make the lipids more accessible to other components in the tissues and to oxygen will affect the oxidation rate. In most instances, especially during frozen storage, the rate increases whether the products arise from simple gutting (Hansen, 1963; Hardy and Smith, 1976) or filleting, minced fish (Hiltz et al, 1976; Moledina et al, 1977) or fish meal (Karpovichyute et al, 1972; El-Lakany, 1973).

Indeed, in the latter the reaction can proceed so rapidly that the meal will burn and often in fatty fish meals little unreacted polyunsaturated acids will remain (Talabi, 1971).

Superficially, the oxidation of fish meals seem to be rather different to that encountered in meat meals. In the latter the rate is said to be less than in the flesh and this it is suggested is due to dehydration affecting the catalytic activity of the heme proteins (Labuza, 1971). However, methods of measuring oxidation, especially in the wet state, are not well advanced and the differences may be more apparent than real.

Cooking without drying has variable effects. In some instances it seems to advance oxidation (Cole and Keay, 1976; Parson et al, 1977; Ke et al, 1978), in others retardation is caused (Bosund and Ganrot, 1970; Sen and Bhandary, 1978). The latter seems to be the more general phenomenon and is contrary to the observations made on the beef, turkey and chicken where cooking is said to increase the rate of oxidation in subsequent storage (Sikorski et al, 1976).

2 Control or prevention of autoxidation

The overall reaction scheme depicted earlier indicates four areas in which the oxidation rate can be influenced, namely storage at low temperature, prevention of initiation, exclusion of oxygen and removal of free radicals. All have their application in fish technology but the most important is control by storage at low temperature and the most effective is a combination of this process coupled with removal of oxygen. The other methods are not so universally applicable because modifying the initiation mechanism or removal of free radicals requires the introduction of controlling agents to the lipid sites within the tissues and this causes problems (Banks, 1952; Crawford et al, 1972; Srikar and Hiremath, 1972).

The most common method of control is storage at low temperatures, for every 10°C reduction the oxidation rate falls by a factor of 2 to 3. With fatty fish such as herring (Banks, 1952) or mackerel (Hardy and Smith, 1976) shelf-lives in excess of six months at −30°C can be achieved. White lean fish can be stored even longer before rancid cold-storage flavours become obtrusive (Connell and Howgate, 1968). Dehydration affects both types of fish and considerably reduces the shelf-life. Glazing helps to minimize this effect as does overwrapping with waxed paper, aluminium foil or plastic films, but a more dramatic effect can be achieved by freezing in water (Bilinski et al, 1979; Smith et al, 1980). Using this method, fatty fish can be stored for up to 15 months at −30°C. The effect of the added water is believed to be twofold. First there is the prevention of dehydration and secondly the method tends to remove occluded air between the fish and also interposes a barrier for oxygen ingress. The bulk storage of whole fish by this method provides not only a cheap, simple process but also helps to reduce mechanical damage during handling operations.

The method is not so suitable for singly frozen fish, fillets or processed fish. For the first two, packaging in oxygen-impermeable material is the best practical alternative and of course in the guise of canned fish it has been practised for a long time. In many modern canning processes the oxygen content of the pack is reduced by air evacuation or steam cooking before sealing and so unless the can leaks little oxidation can occur (Lindsay, 1977).

More recently vacuum packaging has been used as a method of control. In this technique fish are sealed in flexible bags constructed of materials that have a low oxygen permeability (Ke et al, 1976; Shevchenko and Antonov, 1976; Hobbs and Hardy, 1968). From mechanistic considerations based on the reactivity of the polyenoic acids present in fish the oxidation rate should be dependent on oxygen concentrations and so even if some oxygen diffuses through into the pack the rate should be reduced (Vakhrusheva and Leonidov, 1974). This effect should be less marked, however, the lower the temperature of storage.

Control of oxidation by chemical means is somewhat variable and would seem to depend on whether the agents can be supplied to the reaction sites. Such control is exercised either by reducing or preventing initiation or by removal or lowering the concentration of free radicals

All fish contain heme protein and these will react with hydroperoxides to produce radicals and initiate autoxidation (Kochi, 1967; Marcuse, 1968; Waters, 1971). Deactivation of these proteins is difficult by chemicals acceptable in the diet (Polesello and Nani, 1972; Fischer and Deng, 1977) although some worthwhile protection seems to be obtained from the use of EDTA and phosphate which may reduce the activity of the heme proteins (Gordon, 1971; Farragut, 1972). There are however a number of substances that will react with hydroperoxides without producing free radicals (Hiatt, 1975). Some such as glutathione and various compounds containing free sulphydryl groups are endogenous in fish and may exert a modifying influence. With the possible exception of sulphur dioxide few chemical additives that function primarily in this manner (Ingold, 1961) have been assessed in fish tissue.

A wide range of compounds that can remove free radicals are natural components in fish and these together with synthetic compounds have been assessed as antioxidants. These include water-soluble components such as ascorbic acid (Tarr, 1947), sodium erythrobate (Tanaka, 1973; Bilinski et al, 1979), riboflavin derivatives (Totani et al, 1975), citric acid, glutamates (Hiremath, 1973) and a whole variety of phenolic substances (Farragut, 1972; Teets, 1975; Deng et al, 1977). The results have been variable, in some instances retardation was obtained, in others no beneficial effects were observed (Banks, 1952; Crawford et al, 1972). As stated earlier, their efficiency seemed to depend on whether the antioxidant could be delivered to the autoxidizing site. When this could be achieved in fillets (Farragut, 1972), minces (Moledina et al, 1977) and fish meal (Talabi, 1971) oxidation was reduced.

When the different methods are used in combination an effect is sometimes obtained which exceeds the expected summation effect. ie a synergistic retardation is obtained.

3 Measurement of oxidation

Measurement of oxidation in fish products presents a major problem that has not been satisfactorily resolved. The simple mechanism of oxidation described earlier shows how hydroperoxides are produced but not the secondary products of oxidation and peroxide breakdown which, in complex lipid oxidizing systems, can be numerous (Badings, 1970; Labuza, 1971; Loury, 1971; McGill, 1976). Individual determination of all these is impracticable and therefore a true quantitative assessment cannot be made. Instead, certain components or group of components are determined in the hope or belief that some simple correlation exists between their production and degree of oxidation or perceived oxidation, which we call rancidity. Because of the variable composition of fish and hence its autoxidation it would be fortuitous if such a stoichiometry existed between such components and degree of oxidation but possibly not between these components and rancidity, especially if the components determined are responsible for the rancid flavours.

Hept-cis-4-enal is believed to be the major rancid flavour in white fish and its measurement correlates well with sensory evaluation of rancidity in such fish (McGill,

1976). Other workers have shown various unsaturated carbonyls to be responsible for rancid flavour in fish and their measurement too should give a good correlation (Meijboom and Stronk, 1972; Ke et al, 1975).

This is not true of the measurement of hydroperoxides in fish although at least in the early stages of oxidation in fish oils a reasonable correlation is obtained between oxidation and hydroperoxide content (Smith et al, 1972). The reason for this is that in fish oil oxidation, in the early stages, hydroperoxide formation greatly exceeds its decomposition. Thus, so long as the chain length of the propagation reaction is long, hydroperoxide determination will correlate well with degree of oxidation. In fish tissues, especially those rich in heme proteins, hydroperoxide decomposers abound and hydroperoxide determinations are less reliable as quantitative measures of oxidation.

The reaction of thiobarbituric acid (TBA) with fish or lipid extracts is also used to estimate oxidation (Dahle et al, 1962; Vyncke, 1975) but this method too has drawbacks (Smith et al, 1972). TBA is said to react with carbonyl components, malonaldehyde in particular, which are hydroperoxide decomposition products. Again in fish oil oxidation one might expect such reactions to follow a fairly consistent pathway and thus a correlation, though a more tenuous one than hydroperoxide determination, should be obtained between production of TBA reacting substances and autoxidation. This appears to be so.

In fish tissues this is less likely and a good correlation is not obtained. In the oxidation of cod and haddock, for instance, the correlation between malonaldehyde produced and that of hept-cis-4-enal was non-extant, malonaldehyde production or retention in the tissue was not marked (McGill, 1976).

These are the two most common methods used for estimating oxidation but numerous others have been and are used such as gas liquid chromatography for the fatty acids (Keay et al, 1972; Ushkalova and Altuf'eva, 1973), infra-red spectroscopy (Smith et al, 1972), reaction with benzidine (Pokorny and Janicek, 1966), reaction with phloroglucinol (Pool and Prater, 1945), UV spectrometry (Parr and Swoboda, 1976), determination of oxirane oxygen (Rzhavaskaya et al, 1977), derivatization with 2,4 dinitrophenylhydrazone (Golovkin and Perkel, 1970; Smith et al, 1972) and volatile carbonyl compounds (Altuf'eva et al, 1970), but all for one reason or another have their failings. For instance, although in the oxidation of fish oils the increase in the molar production of carbonyls plus hydroperoxides is almost equal to oxygen uptake, in fish tissue carbonyl production from other reactions exceeded those from autoxidation and negated the value of the method.

In recent years a number of studies on the oxidation products of unsaturated esters have been made using reductive techniques to aid and simplify the analysis (Frankel et al, 1977).

The method appears to have some merit for the determination of lipid oxidation in fish but at the moment is too complex for routine analysis.

Even in the laboratory it is difficult to measure autoxidation in fish tissues. Measurement of oxygen usage is not wholly satisfactory because autoxidation is not the only pathway for oxygen mobilization, especially in wet fish where microbial and enzymatic usage can predominate (Smith *et al*, 1972).

4 Lipid autolysis

Lipid hydrolysis is a common *post mortem* feature in fish and fish products. The major products are free fatty acids (FFA) and glycerol; incomplete hydrolysis to give FFA and partially esterified glycerides appears to be rare (Lovern and Olley, 1962).

The effect of this hydrolysis on fish oils is undesirable and the FFA if present over a certain concentration are usually removed before further processing (Stansby, 1967). In fish and fish products, however, the consequences of lipolysis on acceptability are not so clear. Thus although in taste panel score sheets of chill and frozen stored fish the terms rancid and soapy are often used as descriptors, no correlation appears to have been carried out between the development of these flavours and fatty acid production. Attention has concentrated rather on whether the acids produced interact with the proteins and thus affect the textural qualities of the fish (Sikorski *et al*, 1976). It has been known for a long time that fatty acids and soaps can denature proteins and it was probably this that caused earlier workers to examine the hypothesis that lipid hydrolysis may be a factor in protein denaturation (Dyer and Morton, 1956; Dyer and Fraser, 1959; King *et al*, 1962). The hypothesis has not been verified. Protein denaturation is a complex phenomena especially in the very labile proteins of fish (Connell, 1969) and so although lipid hydrolysis may well be involved, it is only one of a number of possible causative factors. In some fish (Olley and Lovern, 1960) although by no means all (Lovern, 1962), lipid hydrolysis and protein denaturation proceed at similar rates but this of course does not constitute proof that one effect causes another (Lovern and Olley, 1962).

Some authors have reported that lipid oxidation occurs more rapidly in tissues containing FFA (Takama *et al*, 1971; Shono and Toyomizu, 1973), and this may represent a secondary effect of lipid hydrolysis. In addition, lipid oxidation has also been implicated in the denaturation of lipids (Sikorski *et al*, 1976) and thus it is possible that in addition to possible denaturation caused by FFA production, oxidation products derived from these may enhance the process.

Except in strongly acid products and oils extracted by alkaline treatment of livers (Stansby, 1967) where chemical hydrolytic mechanisms predominate, hydrolysis occurs mainly through the intervention of lipolytic enzymes. The detailed mechanism of this reaction *in vivo* is not known. It has been postulated, however, that the enzyme binds in a hydrophobic manner or through electrostatic forces to the matrix containing the lipid substrate in such a way that its hydrolytic reaction site, possibly a serine oxygen, is in close proximity to the lipid ester grouping (Brockerhoff, 1974). Brockerhoff also suggests that the specificity and rate of reaction is determined by the nucleophilicity of the attack, steric hindrance and the hydrophobicity of the ester. As with all enzyme reactions the overall rate is conditioned also by other factors such as concentration, pH and temperature (Desnuelle, 1961; Wills, 1965) and especially calcium ions (Wills, 1960) which are believed to act, at least

in part, by their ability to remove the fatty acids as insoluble calcium soaps (Wills, 1965).

The hydrolysis of both triglycerides and phospholipids is normally a stepwise process. Thus the initial reaction of lipases with triglyceride is with the primary alcohol ester group and then rather more slowly with the secondary alcohol ester group so that a mixture of glycerides, free fatty acids and glycerol is produced (Constantin *et al*, 1960).

Enzymatic hydrolysis of phospholipids is believed to be even more selective, in that hydrolysis at position 2 of the glyceride molecule appears to be essential before further hydrolysis can occur. It has been considered in the past that specific phospholipases are required for further hydrolysis (Whitaker, 1972) but more recently the suggestion has been made that non-specific esterases of the serine-histidine type (Van Den Bosch *et al*, 1973) are quite able to perform this task.

In view of these remarks it is interesting to note that in fish tissues lipolysis, especially phospholipolysis, rarely results in the production of partially hydrolyzed lipids (Lovern and Olley, 1962). Attack once initiated on the lipid molecule seems to go to completion, which suggests that the initial stages of the reaction are slow compared with the hydrolytic steps.

The lipolytic activity of fish tissue appears to be species dependent. All species contain phospholipids and although most are subject to lipolysis, those in certain elasmobranch species are remarkably stable (Lovern, 1962). In addition there is some evidence that only the choline and ethanolanine phosphatidyl esters are hydrolyzed and those of serine and inositol remain inviolate (Bligh, 1961).

Triglycerides containing tissues of many species are subject also to hydrolysis of these lipids (Botta *et al*, 1973a,b; Stodolnik *et al*, 1974) but this appears to be less common than phospholipolysis (Lovern and Olley, 1962) and further is often dependent on the tissue site (Botta *et al*, 1973a,b; Tsukuda, 1976). For instance, in herring there is evidence that hydrolysis of both phospholipid and triglyceride occurs (Addison *et al*, 1969) but in the flesh muscle tissue no hydrolysis of the glycerides could be discerned (Olley *et al*, 1962).

Temperature has a marked effect on the hydrolysis of lipids in fish tissues. At 0°C, in fish that have not been frozen, some workers have shown a marked lag phase before the maximum rate of hydrolysis is attained (Lovern and Olley, 1962). Elsewhere others studying essentially the same species of fish, cod, did not observe this effect (Dyer and Fraser, 1959).

At temperatures around the freezing point the rate of hydrolysis increases with decreasing temperature until, such as in cod, a maximum rate is obtained at −4°C (Lovern and Olley, 1962). At lower temperatures the rate falls off in a very marked fashion but is still measurable even at −30°C. Analysis made on cooked fish indicates that lipolysis occurs at elevated temperatures (Quaglia and Audisio, 1974; Strokova and Smirnova, 1978) and even after canning lipid hydrolysis still continues in the stored canned fish (Dominova and Pronichkina, 1977). Whether this is due to lipolytic enzyme action is not known, certainly some of the phospholipases are remarkably stable to heat (Whitaker, 1972) but holding at high temperatures for a period of time causes a reduction in their activity (Lovern, 1962). Drying following cooking appears to inhibit hydrolysis (Lovern, 1962). The effect however seems to be dependent not only on species but also on the extent to which the fish have been dried. Very dry fish exhibit little hydrolysis but as the water content increases so a marked increase in FFA production occurred (Koizumi et al, 1978).

There is evidence to suggest that other forms of processing such as comminution (Takama et al, 1971; Wood and Hintz, 1971) and smoking (Shevchenko and Lapshin, 1975; Shevchenko and Antonov, 1976) enhance lipolysis. Salting on the other hand does not seem to have a marked effect on the rate although, in salt cod, hydrolysis ceased when the lipid contained 50% FFA which is somewhat earlier than that found in iced and frozen cod (Lovern, 1962).

5 References

ADDISON, R F, ACKMAN, R G and HINGLEY, J. Free fatty acids of herring oils: possible derivation from both phospholipids and triglycerides in fresh herring. J. Fish. Res. Bd. Can., 26(6), 1577–1583
1969

AGZHITOVA, L A. Quality assessment of frozen fish. Rybn. Khoz., 45(4), 53–55
1969

ALTUF'EVA, K A, SOKOLOVA, O M and USHKALOVA, V N. Evaluating the quality of frozen fish according to the contents of volatile carbonyl compounds. Rybn. Khoz., 46(5), 64–66
1970

BADINGS, H T. Cold storage defects in butter and their relation to the autoxidation of unsaturated fatty acids. PhD Thesis, Landbouwhogeschool, Wageningen, Holland
1970

BALDRATI, G, PIRAZZOLI, GOLA, S and AMBROGGI, F. Use of ionizing radiation for the preservation of fresh fish: radiopasteurization of saurels (Trachurus trachurus). Ind. Conserv., 53(1), 8–10
1978

BANKS, A. Rancidity in fats. I. The effect of low temperature, sodium chloride and fish muscle on the oxidation of herring oil. J. Soc. Chem. Ind. Lond., 56, 13T–15T
1937

—— Kippers in cold storage. Food Manuf., 14, 83–85
1939

—— The freezing and cold storage of herring. DSIR Food Investigation Special Report No. 55, HMSO, London
1952

BATEMAN, L. Olefin oxidation. Q. Rev. Chem. Soc., 8, 147–167
1954

BILINSKI, E, JONAS, R E E and LAU, Y C. Chill stowage and development of rancidity in frozen Pacific herring, Clupea harengus pallasii. J. Fish. Res. Bd. Can., 35(4), 473–477
1978

—— Control of rancidity in frozen Pacific herring: use of sodium erythrobate. J. Fish. Res. Bd. Can., 36, 219–222
1979

BITO, M and KIRIYAMA, H. Effect of storage temperature on discoloration of skipjack meat during frozen storage. Bull. Tokai Reg. Fish. Res. Lab., No. 75, 87–94
1973

BLIGH, E G. Lipid hydrolysis in frozen cod muscle. J. Fish. Res. Bd. Can., 18, 143–145
1961

BLIGH, E G and REGIER, L W. The potential and limitations of minced fish. Proc: The production and utilisation of mechanically recovered fish flesh (minced flesh), 7–8 April 1976, Torry Research Station, 54–55
1976

BOLLAND, J L. Kinetics of olefin oxidation. Q. Rev. Chem. Soc., 3, 1–21
1949

BOSUND, I and GANROT, B. Effect of pre-cooking on lipid oxidation and storage life of frozen fish. Lebensmittel-Wissenschaft und Technologie, 3(4), 71–73
1970

BOTTA, J R, RICHARDS, J F and TOMLINSON, N. Thiobarbituric acid value, total long-chain free fatty acids, and flavour of Pacific halibut (Hippoglossus stenolepis) and chinook salmon (Oncorhynchus tshawytscha). J. Fish. Res. Bd. Can., (1), 63–69
1973a

—— Flesh concentration of various long-chain free fatty acids of Pacific halibut (Hippoglossus stenolepis) and chinook salmon (Oncorhynchus tshawytscha) frozen at sea. J. Fish. Res. Bd. Can., 30(1). 79–82
1973b

BROCKERHOFF, H. Lipolytic enzymes. In: Food related enzymes, Ed J R Whitaker. Advances in Chemistry Series 136, Amer. Chem. Soc.
1974

CASTELL, C H and BISHOP, D M. Effect of hematin compounds on the development of rancidity in muscle of cod, flounder, scallops and lobster. J. Fish. Res. Bd. Can., 26(9), 2299–2309
1969

CATALANO, M and DE FELICE, M. Autoxidation of fats. I. Influence of free fatty acids. Rivista Ital. Sostanze Grasse, 47(10), 484–492
1970

CHRISTOPHERSON, B O. Formation of monohydroxy-polyenic fatty acids from lipid peroxides by a glutathione peroxide. Biochim. Biophys. Acta., 164, 35–46
1968

COLE, B J and KEAY, J N. The development of rancidity in minced herring products during cold storage. Proc: The production and utilization of mechanically recovered fish flesh (minced fish), 7–8 April, Torry Research Station, 54–55
1976

CONNELL, J J. Properties of fish proteins. In: Proteins as human foods, Ed R A Lawrie. Proceedings of the 16th Easter school in agricultural science, University of Nottingham. Butterworths
1969

CONSTANTIN, M J, PASERO, L and DESNUELLE, P. Quelques remarques complémentaires sur l'hydrolyse des triglycerides par la lipase pancréatique. Biochim. Biophys. Acta., 43, 103–109
1960

CRAWFORD, D L, LAW, D K and McGILL, L S. Shelf life stability and acceptance of frozen Pacific hake (Merluccius productus) fillet portions. J. Fd. Sci., 37(5), 801–802
1972

CUTTING, C L. The influence of drying, salting and smoking on the nutritive value of fish. In: Fish in Nutrition, Eds E Heen and R Kreuzer. Fishing News (Books) Ltd, London
1962

DAHLE, L K, HILL, E G and HOLMAN, R T. The thiobarbituric acid reaction and the autoxidants of polyunsaturated fatty acid methyl esters. Arch. Biochim. Biophys., 98, 253–261
1962

DENG, J C, MATTHEWS, R F and WATSON, C M. Effect of chemical and physical treatments on rancidity development of frozen mullet (Mugil cephalus) fillets. J. Fd Sci., 42(2), 344–347
1977

DENISOV, E T. Simple reactions of formation of free radicals in liquid phase oxidation. Zh. Fiz. Khim., 38(1), 3–15
1964

DESNUELLE, P. Pancreatic lipase. In: Advanced Enzymology, 23, 129–162. Interscience
1961

DOMINOVA, S R and PRONICHKINA, A V. Variations of the lipids in canned Caspian sardines in oil during storage. Rybn. Khoz., No. 3, 58–60
1977

DYER, W J and MORTON, M L. Storage of frozen plaice fillets. J. Fish. Res. Bd. Can., 13, 129–143
1956

DYER, W J and FRASER, D I. Proteins in fish muscle. 13. Lipid hydrolysis. J. Fish. Res. Bd. Can., 16, 43–52
1959

EL-LAKANY, S M H. Storage changes in natural and model lipid-protein systems. Univ. Br. Columb. Pupls. Diss. Abstra., B 33 (12, Part 1), 5584
1973

FARRAGUT, R N. Effects of some antioxidants and EDTA on the development rancidity in Spanish mackerel (Scomberomorus maculatus) during frozen storage. NOAA (US) Technical Report NMFS SSRF 650 IV and 12pp
1972

FISCHER, J and DENG, S C. Catalysis of lipid oxidation: A study of mullet (Mugil cephalus) dark flesh and emulsion model system. J. Fd Sci., (3), 610–614
1977

FRANKEL, E N, NEFF, W E, ROHWEDDER, W K, KHAMBAY, B P S, GARWOOD, R S and WEEDON, B C L. Analysis of autoxidised fats by gas chromatography-mass spectrometry. Lipids, 12, 901–907
1977

FUKUDA, H. Application of anti-oxidant to shikokara. Bull. Jap. Soc. Scient. Fish., 21(8), 934–936
1955

GOLOVKIN, N A and PERKEL R L. Analysis of carbonyl compounds in oxidised fish fats. Trudy Vses. Nauchno-Issled., Inst. Zhirov 27, 230–239
1970

GORDON, A. Polyphosphate treatment of fish. Fd. Mf., 46(7), 57–58
1971

GOVIND RAO M K and ACHAYA, K T. Unsaturated fatty acids as synergists for antioxidants. Fette Seifen Anstr-Mittel, 70, 231–234
1968

HAMMOND, E G and JOHNSON, D C. Factors affecting the proportion of peroxide types during cooxidation of fatty ester mixtures. Proc: 11th Wld. Congr. Int. Soc. Fat Res., Gotenberg, Sweden
1972

HANSEN, P. Fat oxidation and storage life of iced trout. I. Influence of gutting. J. Sci. Fd. Agric., 11, 781–786
1963

HANSEN, P. Storage life of prepacked wet fish at 0°C. II. Trout and herring. J. Fd. Technol., 7(1), 21–26
1972

HARDY, R and MACKIE, P R. Seasonal variation in some of the lipid components of sprats (Sprattus sprattus). J. Sci. Fd. Agric., 20, 193–198
1969

HARDY, R, SMITH, J G M and YOUNG, K W. Influence of packaging in fish processing and technology. Proc: Conferences in conjunction with the 8th Intern. Exhib. for the Fd. and Allied Ind., London, 23–26 October, BPS Exhibitions Ltd
1973

HARDY, R and SMITH, J G M. The storage of mackerel. Development of histamine and rancidity. J. Sci. Fd. Agric., 27, 595–599
1976

HARDY R, McGILL, A S and GUNSTONE, F D. Lipid and autoxidative changes in cold stored cod. J. Sci. Fd. Agric. (In press)
1979

HIATT R R Hydroperoxide destroyers and how they work. CRC Crit. Rev. Fd. Sci. Nutr., 7(1), 1–12
1975

HILTZ, D F, LALL, B S, LEMON, D W and DYER, W J. Deteriorative changes during frozen storage in fillets and minced flesh of silver hake (Merluccius bilinearis) processed from round fish held in ice and refrigerated sea water. J. Fish. Res. Bd. Can., 33(11), 2560–2567
1976

HIREMATH, G G. Prevention of rancidity in frozen fatty fishes during cold storage. Indian Food Packer, 27(6), 20–24
1973

HOBBS, G and HARDY, R. Packaging of fish. Plastics and Polymers, 36, 445–448
1968

109

INGOLD, K U. Inhibition of the autoxidation of organic substances in
1961 the liquid phase. *Chem. Revs.*, 61, 563–589

JONES N R. Browning reactions in dried fish products. In: *Recent*
1962 *Advances in Food Science*, Eds J S Hawthorn and J Muilleitch.
Butterworths, London. 74–80

KAMAT, S V and KUMTA, U S. Studies on radiation of medium fatty
1972 fish. I. Control of radiation induced oxidative changes in white
pomfret (*Stromateus cinereus*) by vacuum packaging. *Fish.*
Technol., 9(1), 8–16

KARPOVICHYUTE, V V, GOLIK, M G and FESTA, N Y A. Effect of mois-
1972 ture on chemical composition of low-fat fish flour during storage. *Rybn. Khoz.*, No. 7, 72–74

KE, P J, ACKMAN, R G and LINKE, B A. Autoxidation of polyunsatu-
1975 rated fatty compounds in mackerel oil: formation of 2,4,7-
decatrienals. *J. Am. Oil Chem. Soc.*, 52(9), 349–353

KE P J, NASH, D M and ACKMAN, R G. Quality preservation in frozen
1976 mackerel. *J. Can. Inst. Fd. Sci. & Technol.*, 9(3), 135–138

KE, P J, LINKE, B A and ACKMAN, R G. Acceleration of lipid oxida-
1978 tion in frozen mackerel fillet by pretreatment with microwave
heating. *J. Fd. Sci.*, 43(1), 38–40

KEAY, J N, RATTAGOOL, P and HARDY, R. Chub Mackerel of Thailand
1972 (*Rastrelliger neglectus* Van Kampen). A short study of its
chemical composition, cold storage and canning properties. *J.*
Sci. Fd. Agric., 23, 1359–1368

KING, F J, ANDERSON, M L and STEINBERG, M A. The effect of linoleic
1962 and linolenic acids on the solubility of cod actomyosin. In: *Fish*
in Nutrition, Eds E Heen and R Kreuzer. Fishing News
(Books) Ltd, London

KOCHI, J K Mechanisms of organic oxidation and reduction by metal
1967 complexes. *Science*, 155(3761), 415–425

KOIZUMI, C, IIYAMA, S, WADA, S and NONAKA, J. Lipid deterioration
1978 of freeze-dried fish meats at different equilibrium humidities.
Bull. Jap. Soc. Scient. Fish., 44(3), 209–216

LABUZA, T P. Kinetics of lipid oxidation in foods. CRC Crit. Rev. *Fd.*
1971 *Technol.*, 2(3), 355–405

LERKE, P A, FARBER, L and HUBER, W. Preservation of fish and shell-
1961 fish by relatively low doses of beta radiation and antibiotics.
Fd. Technol., 15, 145–152

LINDSAY, R C. The effect of film packaging on oxidative quality of fish
1977 during long term frozen storage. *Fd. Prod. Dev.*, 11(8), 93–94

LOURY, M. Possible mechanisms of autoxidative rancidity. *Lipids*, 7,
1971 671–675

LOVERN, J A. The lipids of fish and the changes occurring in them
1962 during processing and storage. In: *Fish in Nutrition*, Eds
E Heen and R Kreuzer. Fishing News (Books) Ltd, London

LOVERN, J A and OLLEY, J. Inhibition and promotion of post mortem
1962 lipid hydrolysis in the flesh of fish. *J. Fd. Sci.*, 27, 551–559

MADHAVAN, P, BALACHANDRAN, K K and CHOUDHURI, D R. Suitabil-
1970 ity of ice stored mackerel and sardine for canning. *Fish. Tech-*
nol., 7(1), 67–72

MARCELET, H. *Les huiles d'animaux marins.* Published by Paris et
1924 Liége Libraire Polytechnique Beranger, Paris

MARCUSE, R. The effect of some amino acids on the oxidation of
1962 linoleic acid and its methyl ester. *J. Am. Oil Chem. Soc.*, 39,
97–103

—— (Ed). Metal catalysed lipid oxidation. SIK Rapport No. 240,
1968 171–192

McGILL, A S. The nature, origin and mechanism of development of
1976 the off flavour in frozen cod. PhD Thesis, St Andrews University, Scotland

McGILL, A S, HARDY, R and GUNSTONE, F D. Further analysis of the
1977 volatile components of frozen cold stored cod and the influence of these on flavour. *J. Sci. Fd. Agric.*, 28, 200–205

MEIJBOOM P W and STRONK J B A. 2-Trans. 4-cis, 7-cis-decatrienal,
1972 the fishy off-flavour occurring in strongly autoxidised oils containing linolenic acid or ω 3, 6, 9 etc., fatty acids. *J. Am. Chem.*
Soc., 49(10), 555–558

MENDENHALL, V T. Oxidative rancidity in raw fish fillets harvested
1972 from the Gulf of Mexico. *J. Fd. Sci.*, 37(4), 547–550

MOLEDINA, K H. REGENSTEIN, J M, BAKER, R C and STEIN-
1977 KRAUS, K H. Effect of antioxidants and chelators on the stability of frozen stored mechanically deboned flounder meat
from racks after filleting. *J. Fd. Sci.*, 42(3), 759–764

NAIR, T S U, MADHAVAN, P, BALACHANDRAN, K K and PRABHU, P V.
1974 Canning of oil sardine (*Sardinella longiceps*) natural pack.
Fish. Technol., 11(2), 151–155

O'KEEFE, T M and NOBLE, R L. Storage stability of channel catfish
1978 (*Ictalurus punctatus*) in relation to dietary level of
α-tocopherol. *J. Fish. Res. Bd. Can.*, 35(4), 457–460

OLCOTT, H S and VAN DER VEEN, J. Comparison of antioxidant
1968 activities of tocol and its methyl derivatives. *Lipids*, 3,
331–334

OLLEY, J and LOVERN, J A. Phospholipid hydrolysis in cod flesh stored
1960 at various temperatures. *J. Sci. Fd. Agric.*, 11, 644–652

OLLEY, J, PIRIE, R and WATSON, H. Lipase and phospholipase activity
1962 in fish skeletal muscle and its relationship to protein denaturation. *J. Sci. Fd. Agric.*, 13, 501–516

PARR, L J and SWOBODA, P A T. The assay of conjugal oxidation pro-

1976 ducts applied to lipid deterioration in stored foods. *J. Fd.*
Technol., 11, 1–12

PARSON, A M, LOVE, J D and SHORLAND, F B. 'Warmed-over' flavour
1977 in meat, poultry and fish. *Adv. Fd. Res.*, 23, 1–74

POKORNY, J and JANICEK, G. Modified determination of benzidene
1966 value in rancid fats. *Sb. Vys. Sk. Chem.-Technol. Praze Por-*
travin Technol., 9, 81–84

POKORNY, J, NGUYEN-THIEN, LUAN, EL-ZEANY, B A and JANICEK, G.
1976 *Die Nahrung*, 20(3)3, 273–279

POLESELLO, A and NANI, R. The protective action of nitrous oxide
1972 against rancidity in mullet and eel lipids. *Ann. Ist. Sper. Val-*
orizzazione Tecnol. Prod. Agric., 3, 67–76

POOL, M F and PRATER, A N. A modified kreis test suitable for photo-
1945 colorimetry. *Oil and Soap*, 22, 215–216

QUAGLIA, G B and AUDISIO, M. Study on the variability in fatty acid
1974 composition of various lipid fractions in some species of frozen
fish. Effect of the cooking treatments. *Proc: IV Int. Congress*
Food Sci. and Technol., 1, 682–688

RAGHUVEER, K G and HAMMOND, E G. The influence of glyceride
1967 structure on the rate of autoxidation. *J. Am. Chem. Soc.*, 44,
239–243

ROUCHAUD, J and LUTETE, B. Synthesis of acids by the liquid phase
1968 oxidation of n-hexane. *Ind. Engng. Chem. Prod. Res. Dev.*,
7(4), 266–270

RZHAVSKAYA, F M, KLIMOVA, T G and DUBROVSKAYA, T A. Charac-
1977 teristics of the oxidation process of various types of fats during
storage. *Vop. Pitan.*, No. 3, 79–84

RZHAVASKAYA, F M, MAKAROVA, A M and SOROKINA, E L. Fatty acids
1978 composition of the muscle tissue lipids of some marine fish.
Vop. Pitan., No. 1, 72–76

SCHOLLNER, R and HERZSCHUH, R. Zur autoxydation ungesättigter
1966 fettsaure-ester in gegenwart von methanol und protonen. II.
Die autoxydation von linolsaure-methylester. *Fette Seifen*
Anstr-Mittel, 68, 616–622

SEN, D P and BHANDARY, C S. Lipid oxidation in raw and cooked oil
1978 sardine (*Sardinella longiceps*) fish during refrigerated storage.
Lebensm.-Wiss. Technol., 11(3), 124–127

SENSER, F and GROSCH, W. Influence of micro-organisms isolated
1973 from herring on fat oxidation. IV. Metabolization of linoleic
acid hydroperoxides and hydroxy-acids at different temperatures. *Z. Lebensmittel unters. u. Forsch.*, 152(5), 274–279

SHENOY, A V and PILLAI, V K. Freezing characteristics of tropical
1971 fishes. I. Indian oil sardine. *Fish. Technol.*, 8(1), 37–41

SHEVCHENKO, M G and LAPSHIN, I I. The effect of smoking dried fish
1975 products on the stability of lipids to oxidation. *Ryb. Khoz.*, No.
11, 74–76

SHEVCHENKO, V V and ANTONOV, N A. The quality of cold smoked
1976 mackerel packaged by various methods. *Izu. vyssh. ucheb.*
Zaved., No. 1, 71–73

SHONO, T and TOYOMIZU, M. Lipid alteration in fish muscle during cold
1973 storage. II. Lipid alteration pattern in Jack Mackerel muscle.
Bull. Jap. Soc. Scient. Fish., 39(4), 417–421

SIKORSKI, Z, OLLEY, J and KOSTUCH, S. Protein changes in frozen fish.
1976 CRC Crit. Rev. *Fd. Sc. Nutr.*, 8(1), 97–120

SMITH, W V and EWART, R H. Kinetics of emulsion polymerisation. *J.*
1948 *Chem. Phys.*, 16, 592–599

SMITH, J G M, HARDY, R and YOUNG, K. Autoxidation in very unsatu-
1972 rated fats. *Proc: SAAFOST Symp., Foods and Their Fats*,
Pretoria, October 1972, 79–105

—— A seasonal study of the storage characteristics of mackerel storage
1980 at chill and ambient temperatures. *This volume*

SMITH, J G M., HARDY, R, McDONALD, I and TEMPLETON, J. The stor-
1979 age of herring (*Clupea harengus*) in ice, refrigerated sea water
and at ambient temperature. Chemical and sensory assessment. *J. Sci. Fd Agric.* (In press)

SMITH, J G M, McGILL, A S, THOMSON, A B and HARDY, R. Prelimi-
1980 nary investigation into the chill and frozen storage characteristics of scad (*Trachurus trachurus*) and its acceptability for
human consumption. *This volume*

SRIKAR, L N and HIREMATH, G G. Fish preservation. I. Studies on
1972 changes during frozen storage of oil sardine. *J. Fd. Sci. Tech-*
nol., 9(4), 191–193

STANSBY, M E (Ed). *Fish Oils.* 440pp
1967

STODOLNIK, L, PODESZEWSKI, Z and OTTO, B. Lipids in industrial Baltic
1974 Sea fish. III. Lipid fractions of fish muscle tissue stored under
ice. *Bromatologia i Chemia Toksykologiczna*, 7(3), 393–403

STROKOVA, L V and SMIRNOVA, G A. Study of the changes in the lipid
1978 component of muscle tissue of some fish species during cold
storage and heat treatment. *Vop. Pitan.*, 3, 56–59

TAKAMA, K, ZAMA, K and IGARASHI, H. Changes in the flesh lipids of
1971 fish during frozen storage. II. Flesh lipids of several species of
fish. *Bull. Fac. Fish. Hokkaido Univ.*, 24(4), 290–300

TAKAMA, K. Changes in the flesh lipids of fish during frozen storage.
1974 IV. Autoxidations of triglyceride and lecithin in cold storage.
Bull. Fac. Fish. Hokkaido Univ., 25(2), 154–161

TALABI, S O. Studies in the chemistry and technology of dried and
1971 stored fish. MSc Thesis, Aberdeen University, Aberdeen, UK

Tanaka, K. Natural tocopherol as an antioxidant for frozen fish.
1973 *Refrigeration*, (Reito) 48(548), 499–504

Tarr, H L A. Control of rancidity in fish flesh. I. Chemical antioxid-
1947 ants. *J. Fish. Res. Bd. Can.*, 7, 137–154

—— Control of rancidity in fish flesh. II. Physical and chemical
1948 methods. *J. Fish. Res. Bd. Can.*, 7, 237–247

Teets, S J. Effect of antioxidants and film packaging on the oxidative
1975 quality of fish during long term storage. MSc Thesis, University of Wisconsin, USA

Totani, Y, Totani, N and Matsuo, N. Effect of riboflavin-2', 3', 4',
1975 5'-tetrabutyrate on autoxidation of oils and fats. *J. Jap. Soc.
Fd. Nutr.*, 28(1), 41–44

Toyama, K. Protection of marine products from deterioration due to
1956 oxidation of the oil. *Bull. Jap. Soc. Sci. Fish.*, 22(6), 383–385

Toyama, K and Shimazu, M. Protection of marine products from
1972 deterioration due to the oxidation of oil. XIII. Applicability of
concentrated natural tocopherol mixture. *Bull. Jap. Soc. Sci.
Fish.*, 38(5), 487–495

Tsukuda, N. Changes in the lipids of frozen fish. *Bull. Tokai Reg. Fish.
1976 Res. Lab.*, 84, 31–41

Ushkalova, V N and Altuf'eva, K A. Oxidation of fat tissue of the
1973 fish *Coregonus peled*. *Pisch. Tekhnol.*, No. 6, 73–75

Vakhrusheva, M N and Leonidov, I P. The effect of vacuum on the
1974 quality of moderately salted fat herrings during storage. *Rybn.
Khoz.*, 3, 77–80

Van Den Bosch, H, Aarsman, A J, De Jong, J G N and Van
1973 Deenen, L L M. Studies on lysophospholipases. I. Purification and some properties of a lysophospholipase for beef pancreas. *Biochim. Biophys. Acta.*, 296(1), 94–104

Van Veen, A G. Fermented and dried sea-food products in South-
1965 East Asia, In: *Fish as Food*, Ed G Borgström, Academic Press,
NY. 3, 227–250

Voskresensky, N A. Salting of herring, In: *Fish as Food*, Ed G Borg-
1965 ström, Academic Press, NY. 3, 107–131

Vyncke, W. Evaluation of the direct thiobarbituric acid extraction
1975 method for determining oxidative rancidity in mackerel
(*Scomber scombrus* L). *Fette, Seiten Austrichtmitt.*, 77,
239–240

Waters, W A. The kinetics and mechanism of metal-catalysed aut-
1971 oxidation. *J. Am. Oil Chem. Soc.*, 48, 427–433

Whitaker, J R. *Principles of enzymology for the food sciences.* Marcel
1972 Dekker, NY. 488

Wills, E D. The relation of metals and —SH groups to the activity of
1960 pancreatic lipase. *Biochim. Biophys. Acta.*, 40, 481–490

—— Lipases. In: *Advances in Lipid Research*, Eds R Paoletti and D
1965 Kritcheusky. Academic Press, NY. 3, 197–240

Wood, G and Hintz, L. Lipid changes associated with the degradation
1971 of fish tissue. *J. Ass. Off. Anal. Chem.*, 54(5), 1019–1023

Other organic components and inorganic components *Shizunori Ikeda*

The chemical composition of fish and shellfish is the physiological and metabolic reflection of their living state. Individual compounds form their own pools whose sizes are greatly dependent on species, body size, season, environmental factors, nutritional status, and even on the type of muscle sample (Love, 1974). These differences in both the amount and nature of the muscle components are responsible for the difference in flavour of fresh fish as well as in the subsequent spoilage patterns (Jones, 1967). Therefore, biochemical, food chemical and technological studies must be combined for utilization of marine products to be effective.

Since there are so many kinds of compounds to be covered in this review, and several reviews have been published in the past on this subject, the present author placed the focus on limited recent studies.

1 Non-protein nitrogenous compounds

Analyses of the muscle constituents of fish and shellfish have revealed that 95% or more of the non-protein nitrogen is accounted for by the following compounds: amino acids, imidazole dipeptides, guanidine compounds, trimethylamine oxide, urea, betaines, nucleotides and compounds related to nucleotides.

1.1 *Amino acids*
There is a specific difference in the free amino acid patterns of muscle between dark- and white-fleshed fish (Suyama and Yoshizawa, 1973; Konosu *et al*, 1974); sometimes more than 1 000mg/100g of histidine occurs in the former group.

The extremely high levels of histidine in the dark-fleshed fish is often associated with so-called histamine toxicity, which causes an allergy-like illness, 'scombroid poisoning', when spoiled foods from these fish are ingested. Various aspects of the histamine problem have recently been reviewed by Arnold and Brown (1978).

Although histamine formation is due to the growth of histidine decarboxylase-positive bacteria, the major pathway of histidine degradation in fish tissues is deaminative with glutamate as the final product (Kawai and Sakaguchi, 1968). Most fish have histidine deaminase in their liver and kidney, but mackerel has this enzyme in muscle as well (Kawai *et al*, 1967; Mackie and Fernandez-Salguero, 1977).

Taurine, a sulphonic amino acid, is often a major constituent of nitrogenous extractives particularly from marine invertebrates (Allen and Garret, 1971).

Molluscan and crustacean muscle is usually rich in free amino acids whose nitrogen accounts for 20–70% of the non-protein nitrogen (Konosu, 1971). Prawns sometimes contain a level of more than 1 000mg/100g of glycine.

1.2 *Carnosine, anserine and balenine*
Eel muscle has an exceptionally high content (500–600mg/100g) of carnosine (Suyama *et al*, 1970). Dark-fleshed fish, including tuna and skipjack, also contain considerable amounts of carnosine, while many other species of fish have very small amounts (Suyama and Yoshizawa, 1973). Anserine occurs abundantly in dark-fleshed fish (Suyama *et al*, 1970; Tamaki *et al*, 1976), and in shark (Suyama and Suzuki, 1975). Balenine occurs only in very small amounts in fish, but remarkably high levels of the compound are found in whales (Suyama *et al*, 1970).

1.3 *Guanidine compounds*
Creatine is frequently phosphorylated and occurs as creatine phosphate (CrP), a phosphagen, in resting vertebrates (Watts and Watts, 1974). Creatine kinase catalyses the phosphorylation of ADP to ATP. The content of creatine in fish muscle varies greatly in different species (160–720mg/100g). The ratio of creatine nitrogen to total non-protein nitrogen is higher in white-fleshed fish than in dark-fleshed fish (Sakaguchi *et al*, 1964; Konosu *et al*, 1974). Generally, the creatine levels in dark muscle are lower than those in the ordinary muscle.

The creatinine levels in fish muscle are as low as 6–48mg/100g, apparently not correlating with the creatine levels in the muscle of a number of fish examined (Sakaguchi et al, 1964; Konosu et al, 1974). Creatinine is formed by the spontaneous cyclization of CrP under physiological conditions. Non-enzymatic conversion of creatine to creatinine increases when fish muscle is heated (Aitken and Connell, 1979).

Six phosphagens, other than CrP, are known; these are the phosphorylated forms of the guanidine bases, arginine, glycocyamine, hypotaurocyamine, taurocyamine, ophelline and lombricine, in marine invertebrates.

Octopine accumulation and the concomitant decrease of arginine were observed during cold storage of muscle of squid and scallop (Endo and Simidu, 1963; Hiltz and Dyer, 1971). Octopine biosynthesis is catalysed by octopine dehydrogenase in molluscan muscle and involves the reductive condensation of pyruvate and arginine with NADH. The substrate arginine is liberated by the hydrolysis of arginine phosphate, and the other substrates, pyruvate and NADH are produced by glycolysis (Grieshaber and Gade, 1977).

1.4 Trimethylamine and its related compounds
Trimethylamine oxide (TMAO) is widely distributed in marine species, and the highest values were reported in elasmobranch tissues (500–1 500mg/100g), followed by squid and gadoid muscle (Yamada, 1967; Harada, 1975).

Biosynthesis of TMAO in fish proceeds possibly through glycine betaine, choline and TMA (Watts and Watts, 1974). Some seaweeds have been reported to contain high levels of TMAO as well as TMA (Fujiwara-Arasaki and Mino, 1972), and there is a possibility of direct incorporation of TMAO by aquatic animals feeding on the seaweeds.

More TMA is produced from TMAO by bacterial action than by fish tissue enzymes. TMA is clearly an important factor of fishy odour (Jones, 1967; Kikuchi et al, 1976). In the bacterial reduction of TMAO to TMA, the participation of cytochromes as electron carriers together with the enzyme TMAO reductase was suggested (Sakaguchi and Kawai, 1978).

TMA levels are usually much lower than TMAO levels in fresh fish muscle. Enzymatic reduction of TMAO to TMA in the dark muscle of several dark-fleshed fish has been observed (Amano and Yamada, 1965; Tokunaga, 1970). Non-enzymatic reduction of TMAO to TMA occurs in the presence of certain catalysts or by heating during the processing of various sea foods (Tokunaga, 1975; Uchiyama et al, 1976). The accumulation of dimethylamine (DMA) and the simultaneous occurrence of formaldehyde (FA) were observed during storage of fish muscle (Amano and Yamada, 1965; Yamada, 1968; Tokunaga, 1970; Babbit et al, 1972; Castell et al, 1973; Dingle et al, 1977). Gadoid species generally possess high levels of the enzyme responsible for the dissociation of TMAO into DMA and FA in ordinary muscle.

DMA as well as TMA and TMAO in the presence of nitrite can be involved in the formation of the carcinogenic nitrosodimethylamine (NDMA) (Malins et al, 1970; Tozawa and Sato, 1974; Iyengar et al, 1976;

Kunisaki et al, 1977; Ohshima and Kawabata, 1978). Keefer and Roller (1973) observed that certain aldehydes catalyse the nitrosation of some amines.

A close relationship between the strength of greening of canned tuna meat and the concentration of TMAO in raw fish was demonstrated (Koizumi, 1967).

1.5 Urea
All tissues of embryo and adult marine elasmobranchs are particularly rich in urea (Read, 1968). Urea is formed in marine elasmobranchs principally by the ornithine-urea cycle (Watts and Watts, 1974), thus allowing ammonia to be detoxicated. Urea formed by this cycle is much more than that from purine metabolism.

Muscle lactate dehydrogenase of some marine elasmobranchs requires the presence of urea to establish optimal kinetic properties (Yancey and Somero, 1978).

The production of large amounts of ammonia during storage of elasmobranch muscle is reported to be largely due to bacterial urease action; it gives an offensive smell together with TMA (Simidu, 1961).

1.6 Betaines
Glycine betaine is one of the naturally-occurring betaines and occurs generally in high concentrations in comparison with other betaines (Beers, 1967). The amounts in muscle of marine invertebrates, such as molluscs and crustaceans, are 200–1 400mg/100g, and they exceed by far those of fish muscle (Konosu and Hayashi, 1975). β-Alanine betaine was found in some marine invertebrates and fishes (Konosu and Hayashi, 1975; Konosu et al, 1978).

Homarine is distributed in marine invertebrates among which molluscs and crustaceans retain 30–250mg/100g in muscle (Hirano, 1975). Homarine is known to be a metabolite of tryptophan. It was suggested, however, that homarine in marine invertebrates is derived from their diet, marine red algae (Yabe et al, 1966).

Carnitine is also encountered often in various invertebrates (Beers, 1967; Hayashi and Konosu, 1977).

1.7 Nucleotides and related compounds (see 2.1)

1.8 Role of these components in taste
Glutamic acid gives a meaty taste to an extract of 'katsuobushi' (dried bonito) with the synergistic action of inosine 5'-phosphate (IMP), while histidine hardly contributes to its taste. Histidine, however, is still considered to participate in taste production possibly through its buffering activity (Konosu, 1973). Anserine is said to contribute to 'mouth satisfaction' (Jones, 1967). Glutamic acid and adenosine 5'-phosphate (AMP) are the main substances in producing the meaty taste of abalone (Hashimoto, 1965). Glycine and betaine (glycine betaine), both present in large amounts in abalone meat, give sweetness to the meat. Free amino acids play a key role in characterizing 'Uni' (unripe gonads of sea urchin) flavour; glycine, valine, alanine, glutamic acid, and especially methionine are important (Hashimoto, 1965). The removal of IMP and guanosine 5'-phosphate (GMP) results in lesser flavour in 'Uni'.

Recently, Hayashi et al (1978) reported that only

seven nitrogenous compounds, glycine, arginine, alanine, glutamic acid, AMP, cytidine 5′-phosphate, and GMP, together with four kinds of inorganic ions, Na^+, K^+, Cl^-, and PO_4^{3-} were essential for producing the characteristic taste of boiled crab extracts.

In relation to the recent increasing trend of fish culture, some trials were undertaken to clarify the difference in relative contents of muscle components which should reflect the flavour difference between cultured and wild fish. No significant difference between the cultured and wild fish in yellowtail (Endo et al, 1974), red sea-bream (Konosu and Watanabe, 1976), and ayu, *Plecoglosus altivelis* (Suyama et al, 1977) was observed in the distribution patterns of principal nitrogenous components including free amino acids, creatine and TMAO. The cultured fishes, however, were found to be less palatable than the wild fishes.

1.9 *Fish muscle proteases*
It has not been demonstrated conclusively that fish muscle proteases participate in the autolysis of fish meat.

Cathepsin D, or haemoglobin-splitting enzyme, is an enzyme that has been studied most frequently among the proteases of fish muscle (Groninger, 1964; Siebert and Schmitt, 1965; Makinodan and Ikeda, 1969b; Wojtowicz and Odense, 1972).

Cathepsins A, B and C (Makinodan and Ikeda, 1976), carnosinase (Partmann, 1976), anserinase (Jones, 1956), and various dipeptidases (Siebert and Schmitt, 1965; Konagaya, 1978) have been detected in fish muscle.

Makinodan et al (1969a) found a unique alkaline proteinase active optimally at pH 8·0 and at 60–65°C in fish muscle. Later, the existence of a fish muscle alkaline protease was confirmed by several workers (Iwata et al, 1973), and some properties have been clarified by using the partially purified enzyme (Makinodan and Ikeda, 1977; Iwata et al, 1979).

Recently, a new neutral proteinase (tentatively called subendopeptidase) was found in carp muscle (Makinodan et al, 1979). The enzyme showed a relatively broad pH-dependence, with optimum pH at 7·2 for haemoglobin.

The object of further studies is to clarify the relationship between these muscle proteases and autolysis in fish meat.

2 Nucleotides, carbohydrates and their related compounds

This section will emphasize the *post mortem* biochemical changes of nucleotides, glycogen, sugars and sugar phosphates.

2.1 *Nucleotides and related compounds*
In fish muscle, adenine nucleotides (5–8 μmol/g) occupy more than 90% of total nucleotides, and the main component is adenosine 5′-triphosphate (ATP).

Rigor mortis is induced with a decrease in the ATP level of muscle (Partmann, 1965). The intensity of *rigor mortis* in skipjack was dependent upon the amounts of ATP decomposed per unit time (Yamanaka et al, 1978). In scallop meat arginine phosphate and ATP decreased rapidly between $-6°C$ and $-8°C$ and disappeared

almost completely in about two hours (Hiltz et al, 1974). This is probably because adenosinetri phosphatase was activated by Ca^{2+} released as a result of cellular destruction caused by ice crystal formation (Heber et al, 1973).

The principal course of ATP decomposition in fish is as follows (Tarr, 1966; Eskin et al, 1971):

$$ATP \rightarrow ADP \rightarrow AMP \rightarrow IMP \rightarrow Inosine \nearrow \begin{array}{l} Hypoxanthine \\ + Ribose \end{array} \searrow \begin{array}{l} Hypoxanthine \\ + Ribose\text{-}l\text{-}P \end{array}$$

In general, the reactions 'ATP→IMP' take place in a relatively early stage after death, or before the muscular pH reaches a constant level. In this period the reactions 'IMP→Inosine (HxR)→Hypoxanthine (Hx)' occur more slowly than those in the sequence 'ATP→IMP'. Accordingly IMP is readily accumulated in fish meat in the early stage after death. HxR and Hx concentrations increase with lapse of time, or with decrease in freshness. The decomposition of HxR proceeds by the two pathways as shown above. The process with the production of ribose is assumed to be the principal one, since free ribose is found more abundantly than ribose-l-P (Tarr, 1966).

Many papers have been published on the changes of ATP-related compounds during storage, since not only IMP attracts attention for its action in enhancing flavour but the amounts of HxR and Hx reflect the freshness of fish (Dingle and Hines, 1971; Ehira and Uchiyama, 1973; 1979).

Arai (1966) determined nucleotides in the muscle of 12 species of marine invertebrates. The total amounts of nucleotides were remarkably large in squid and prawn (8·9–10·3μ mol/g; ATP comprising 70–80% of the total. According to Arai (1966), the following pathway is common for the decomposition of ATP in marine invertebrates: ATP→ADP→AMP→Adenosine→HxR→Hx. However, accumulation of IMP was also noticed in lobster (Dingle et al, 1968), shrimp (Suryanarayama et al, 1969), and crab (Sasano and Hirata, 1973). A review on nucleotides and other chemical components of marine invertebrates, especially abalone, has been published by Olley and Thrower (1977).

The total amount of nicotinamide-adenine dinucleotide (NAD) and reduced NAD in ordinary muscle of various fish ranged from 4-38mg/100g, and that of nicotinamide-adenine dinucleotide phosphate (NADP) and reduced NADP from 0·3–11mg/100g (Shimizu et al, 1969). Either of these amounts in dark muscle was about twice as large as that of ordinary muscle. At least four hydrolytic enzymes are known to participate in the decomposition of NAD, although the enzyme activities in fish muscle are very low (Abe et al, 1976). The disappearance of NAD in fish muscle affects the rates of various oxidation-reduction reactions in which NAD participates. For example, the browning of fishery products (Yamanaka, 1975c), the ultimate *post mortem* pH (Bito, 1978), and the maintenance of fish meat colour (Shimizu et al, 1969) are affected by the reactions involving this coenzyme. Therefore, attention should be paid to the technological significance of NAD and NADP in fish muscle.

2.2 Glycogen, sugars and sugar phosphates

Dark-fleshed fish generally contain much more glycogen (about 1% in the muscle) than white-fleshed ones. A large amount of glycogen (1–8%) is also deposited in the body of molluscs.

The *post mortem* decrease of pH in fish meat depends upon the amount of lactic acid produced by the decomposition of glycogen. However, there was not always a parallel relationship between the decrease in glycogen and the production of lactic acid (Manohar, 1970).

Glycogen of fish muscle is decomposed by two pathways, the glycolytic and amylolytic enzyme systems (Tarr, 1966; Eskin et al, 1971). The principal pathway is presumed to be the latter in fish (Burt, 1966; Nagayama, 1966), while it is the former in mammals. Tarr (1968) pointed out that glycogen was decomposed by the two above-mentioned pathways and that the gluconeogenetic system, glycogen synthetic system, and hexose monophosphate cycle functioned at the same time.

In fish meat free glucose is produced by *post mortem* glycolysis (Tarr, 1968). On the other hand, a minute amount of glucose (0.1–3.9μ mol/g) is present originally in the muscle of living fish (Hoffman et al, 1970). In general, the hexose phosphate contents decrease after death (Burt and Stroud, 1966), while the pentose phosphate contents increase transitionally after death and then decrease (Tarr, 1966).

The final product of glycolysis in mollusc muscle is not lactic acid but pyruvic acid (Shibata, 1977). Molluscan muscle shows a low activity of lactic dehydrogenase, and possibly a mechanism different from that of fish muscle is operative for controlling sugar metabolism. Succinic and propionic acids and alanine are also glycolytic products of the adductor muscle of mussels (Kluytmans et al, 1977).

2.3 Browning of canned skipjack meat

The mechanisms of browning of processed marine products differ depending on the kind of raw material and the methods of processing and storage. Yamanaka et al (1973) found that glucose 6-phosphate (G6P) and fructose 6-phosphate (F6P) are mainly responsible for an orange discoloration of canned skipjack meat. When the sugar phosphates accumulated in the early stage of glycolysis and heating, the Maillard-type browning reaction took place at a high temperature producing orange-discoloured meat (Hasegawa et al, 1976). Yamanaka (1975a,b,c) elucidated the cause of accumulation of G6P and F6P as follows: in the meat which gave a strong orange colour on cooking, ATP had decreased rapidly during thawing and disappeared completely after thawing. In normal meat, however, ATP decreased only slowly. In the former meat glycogen decreased extensively, but lactic acid increased only slightly with a small fall of pH. The NAD content was 60–80mg/100g in normal meat, while it was 1–2mg/100g in the meat liable to turn orange. Anserine, creatine and histidine were the main amino components participating in this amino-sugar reaction.

3 Organic acids

Lactic acid is produced from glycogen by glycolysis in fish meat (see 2.2). Since the glycogen content of muscle is large in dark-fleshed fish, the lactic acid content is also large in the commercial fresh meat of these fish, exceeding 1 000mg/100g in some species.

Succinic acid is one of the important intermediates in the tricarboxylic acid cycle and a constant component of the muscle of fish and shellfish. In addition to this acid, such organic acids as pyruvic, fumaric, citric, malic and acetic acids are known to be contained in muscle, although they are low in level. Studies in the early period after death revealed that succinic acid was abundant especially in shellfish, reaching a level of 150-650mg/100g and presumed to be related to the taste of shellfish. However, some authors are dubious about the role of succinic acid as a flavour component (Takagi and Shimizu, 1962; Konosu et al, 1967). The succinic acid level was always less than 40mg/100g in clams immediately after catch, and reached about 200mg/100g after three days of storage in air in summer (Konosu et al, 1967). The contents of succinic and fumaric acids in prawns were so small (7-22mg/100g) that these acids had no significant effect on the taste of prawns, as demonstrated by the omission test (Take et al, 1966). No organic acids were essential for the characteristic taste of boiled crab extracts (Hayashi et al, 1978). However, it is possible that some organic acids may participate in producing the characteristic taste of shellfish, since at least the natural form of fumaric acid has a refreshing acidic taste.

4 Volatile compounds

4.1 Volatile sulphur compounds

Hydrogen sulphide (H_2S) and methanethiol (CH_3SH) were detected in fresh tissues of fish (McLay, 1967; Wong et al, 1967; Ooyama, 1973). Dimethyl sulphide (DMS) and dimethyl-β-propiothetin (DMPT) were detected in fresh tissues of marine invertebrates (Ronald and Thomson, 1964; Ackman and Hingley, 1968; Brooke et al, 1968; Tokunaga et al, 1977).

During storage of fish and shellfish, a number of volatile sulphur compounds are produced by bacterial action (Ronald and Thomson, 1964; Wong et al, 1967; Miller et al, 1972; Angelini et al, 1975; Herbert et al, 1975). In these cases, the following compounds are identified: H_2S, CH_3SH, ethanethiol, 2-methyl-2-propanethiol, 2-butanethiol, 1-butanethiol, DMS, dimethyl disulphide, carbon disulphide. On the other hand, volatile sulphur compounds were not detected in stored samples of sterile muscle of fish (Miller et al, 1973; Herbert and Shewan, 1975).

Volatile sulphur compounds are also produced during heat-processing of fish and shellfish (Hughes, 1964; Mendelsohn and Brooke, 1968; Khayat, 1977; McGill et al, 1977; Tokunaga et al, 1977).

4.2 Volatile non-nitrogenous compounds

4.2.1 *Carbonyl compounds* Volatile carbonyl compounds are produced during heat-processing of fish and shellfish (Hughes, 1963; Gadbois et al, 1967; Miwa et al, 1976). Carbonyl compounds have also been found in processed fish (Yurkowski and Bordeleau, 1965; Kasahara and Nishibori, 1975a) and in stored fish (Matsuto et al, 1967; Wong et al, 1967; Miller et al, 1972; McGill et al, 1977). In processed and stored fish, the following compounds have been identified: alkanals (C_1–C_{11}, C_{14}, C_{16}, C_{18}), alkenals (C_3–C_{10}), alkadienals (c_6, C_9, C_{10}), 2-alkanones (C_3–C_9, C_{11}), 3-alkanones (C_5–C_7), and 3-alken-2-ones (C_3, C_5, C_8), etc. Development of these compounds is considered to be mainly due to oxidation of oil in fish (Crawford et al, 1976).

4.2.2 *Fatty acids*   Formic, acetic and propionic acids were detected in the fresh tissues of various species of fish; higher acids were also detected in some species (Ooyama, 1973; Miwa et al, 1976). However, the acid content was very small in each case. Volatile fatty acids increased during spoilage (Kasahara and Nishibori, 1975b; Miwa et al, 1976), whereas no significant change occurred in the quantities of acids during cooking (Ooyama, 1975).

4.2.3 *Miscellaneous compounds*   Volatile alcohols and hydrocarbons were identified in fish and shellfish (Wong et al, 1967; Mendelsohn and Brooke, 1968; Shimomura et al, 1971; Miller et al, 1972; Angelini et al, 1975; McGill et al, 1977). In most cases, only a relatively small number of compounds among the following compounds was found in each species: methanol, ethanol, 1-propanol, 3-methyl-1-butanol, 1-penten-3-ol, butene, 1-pentene, 1-hexene, 1-octene, benzene and toluene. In an exceptional case, seven alcohols and 14 hydrocarbons were detected in cod (McGill et al, 1977).

As other miscellaneous compounds, diethyl ether, phenols, furans, chloroform, methylene chloride and bromobenzene were detected in chill-stored cod (Wong et al, 1967; McGill et al, 1977).

Most of these compounds which were identified in fish are considered to be formed by oxidation of oil during storage and heating (Crawford et al, 1976).

4.3 *Role of volatile components in odour*
Jones (1967) reviewed the flavour of fish, and discussed how volatile components are related to odour.

Generally the odour of fresh fish and shellfish is not pronounced except for some special cases and studies on the odour of fish and shellfish in relation to their components are mainly concerned with storage or cooking. Characteristic aroma and unpleasant odour depend on various factors such as the odour and flavour thresholds of the volatile components, their concentrations and temperature. The odour thresholds of fatty acids and volatile bases are of the order of 0.1–100ppm, while those of sulphur compounds and aldehydes are of the order of 0.1–100ppb (Guadagni et al, 1963; McGill et al, 1974; Kikuchi et al, 1976). Therefore, odour of the food is strongly influenced by trace quantities of these latter compounds. DMS at low concentration is known to be partly responsible for the characteristic aroma of fresh oyster (Ronald and Thomson, 1964), soft-shell clam (Brooke et al, 1968), Antarctic krill, and some kinds of shrimp (Tokunaga et al, 1977). On the other hand, DMS at a high concentration is known to be a component of certain offensive odours such as 'petroleum odour' in some lots of canned salmon (Motohiro, 1962), 'blackberry odour' in Labrador cod (Sipos and Ackman, 1964) and 'petroleum refinery odour' in Nova Scotia mackerel (Ackman et al, 1972). Ronald and Thomson (1964) showed methanethiol and dimethyl disulphide presumably to be associated with a rotten odour in air-stored

oyster. Herbert and Shewan (1975) demonstrated that the production of H_2S, methanethiol and DMS in amounts above their odour threshold levels contributed to the off-odour of spoiled cod fillets. McGill et al (1974, 1977) identified cis-4-heptenal as the causative component of the cold storage flavour of cod. Kasahara and Nishibori (1975b) identified 3-methylbutanal (or 2-butanone) as one of the components of unpleasant odour of dried Alaska pollack of inferior quality. Volatile fatty acids are related to the 'fishy and rancid odour' and among them C_4- and C_5-acids have a strongly rancid odour. It is recognized that although the amounts of lower fatty acids increase with the progress of spoilage, the unpleasant odour in spoiled fish is due not only to the acids but also to other volatile compounds (Kikuchi et al, 1976; Miwa et al, 1976).

Flavour change and flavour enhancement are known to occur through interactions between different volatile compounds. Kikuchi et al (1976) found that when acetic acid is added to a TMA solution, the TMA odour changes to an unpleasant boiled or canned fishy odour, and when butyric acid or valeric acid is added to acetic acid the odour is intensified.

Several reports by Japanese workers are available on the role of volatile compounds in flavour of cooked fish (Shimomura et al, 1971; Kasahara and Nishibori, 1975a) and in flavour of soup stocks of 'Niboshi' (cooked and dried small fish) and 'Katsuoboshi' (dried bonito) (Honma et al, 1974; Nishibori and Kasahara, 1978).

Recently, the subject of flavour components in heated fish was reviewed by Aitken and Connell (1979), and off-flavours in meat and fish by Reineccius (1979).

4.4 *Browning caused by oxidation of lipid*
Many studies have been made on the mechanism of the discoloration, or so-called rusting, caused by the oxidation of lipids of marine products. This mechanism is so complicated that there has been no agreement on it. However, it is widely accepted that carbonyl compounds from oxidized lipids are involved in the mechanism (Matsuto et al, 1967). Accordingly, rusting may be regarded as the browning induced by a type of aminocarbonyl reaction. Little has still been elucidated, however, on how the nitrogenous compounds participate in the development of rusting. Fujimoto and Kaneda (1973) presumed that aldol condensation in the presence of volatile bases as catalysts might serve as the early stage of the mechanism. On the other hand, Nakamura et al (1973) reported that the non-volatile basic nitrogenous compounds made a greater contribution to the browning than the volatile ones.

5 Vitamins

Much of the earlier work on the vitamin contents of fish and shellfish has been reviewed by several authors (Jacquot, 1961; Braekkan, 1962; Higashi, 1962). Recently, remarkable progress has been made in physiological studies on vitamin requirements of fish in relation to pisciculture (Halver, 1972). A few studies, however, have been published in recent years on vitamins from the viewpoint of utilization of fish meat (Kennedy and Ley, 1971).

As is clear from the physiological role, the vitamin contents are markedly higher in the internal organs than in the muscle. In general, dark muscle contains larger amounts of many vitamins than ordinary muscle. The vitamin contents (by rough calculation from referring to a number of reports) in the edible portion of important fish and shellfish are as follows:

Vitamin A (IU/100g): lamprey and eel 2 500–23 400; dogfish, 200–300; other fish, 100–200; Antarctic krill (Yanase, 1971), 380. Vitamin D (IU/100g): lean fish, 0–40; fatty fish, 300–1 700; eel 4 700. α-Tocopherol ($\mu g/g$): fish, 1–50; lobster, 10–20; mussels, 1–11; lipid of fish muscle (Ackman and Cormier, 1967), 210–330$\mu g/g$ lipid; lipid of lobster, 90–1 700$\mu g/g$ lipid. Thiamine ($\mu g/g$): fish, 0.4–1.8. Riboflavin ($\mu g/g$): fish, 0.3–7.2. Vitamin B_6 ($\mu g/g$): fish, 1–16, molluscs, 0.6–4.0. Vitamin B_{12} ($\mu g/100g$): fish, 0.03–12.7. Folic acid ($\mu g/100g$): lamprey and eel, 6–29; other fish, 1–11; Antarctic krill, 66. Pantothenic acid ($\mu g/g$): fish, 1–19; shellfish, 3–30. Nicotinic acid ($\mu g/g$): fish 50–160; molluscs, 12–26; Antarctic krill, 70. Biotin ($\mu g/100g$): molluscs (Miyake and Noda, 1962), 7.4–16.4. Choline (mg/g): fish, 0.4–0.6, squid, 0.8–1.1. Inositol ($\mu g/g$): fish (Sato et al, 1967), 73–680; molluscs (Miyake and Noda, 1962), 120–330. Ascorbic acid ($\mu g/g$): fish (Ikeda et al, 1963), trace-40.

6 Inorganic compounds

Earlier work on mineral constituents of fish was reviewed by Causeret (1962), and the functions of inorganic compounds in animals by Bowen (1966).

The concentrations of heavy metals in the edible portion of fish and shellfish attract attention from the food-hygienic point of view (Ikebe et al, 1977; Kumagai and Saeki, 1978). It is well known that mercury and cadmium are the causative agents of 'Minamata disease' and 'Itai-itai disease', respectively. The total mercury concentration in the body of skipjack is higher in muscle than in visceral organs, and methyl-mercury comprises about 90% of the total mercury (Katuki et al, 1975). Heavy metals other than mercury are generally more abundant in visceral organs than in muscle. Cadmium and copper accumulate particularly in the liver (Yamamoto et al, 1978). Of the mercury contained in skipjack muscle, 57–68% was in the myofibrillar protein fraction and 14–28% in the sarcoplasmatic protein fraction (Arima and Umemoto, 1976). The sulphhydryl group of myofibrillar protein shows a higher affinity to mercury than does that of sarcoplasmic protein.

Mercury contained in fish meat is removed effectively when bleaching during the production of kamaboko is performed in an acidic cysteine solution (Suzuki, 1974). In red sea bream cultured by feeding a diet supplemented by cysteine and pectin, which have a high affinity for mercury, mercury was found to be released efficiently from the muscle and internal organs (Kikuchi et al, 1978).

In fish meat magnesium is more readily released than calcium during storage (Taguchi et al, 1969). This appears to explain why the crystals of struvite ($Mg\cdot NH_4\cdot PO_4\cdot 6H_2O$) are readily produced when less fresh raw materials are used for canning. Haemocyanin, which contains copper, participates in the development of the blue meat of some canned crab (Boon, 1975).

7 Acknowledgements

The author wishes to acknowledge the assistance of M Sakaguchi, R Yoshinaka, and M Sato who collected references and read all or part of the manuscript, and of K Hayama who made a fair copy of the tables.

Table IV
CONTENTS OF CREATINE AND CREATININE IN FISH MUSCLE

Species	Creatine (mg%)	Creatinine (mg%)	Ratio[4] (%)
Shark *Mustelus kanekonis*[1]	485	48	12
Shark *Lamna cornubica*[1]	507	33	12
Yellowfin tuna *Neothunnus albacora*[2]	599	24	27
Yellow-tail *Seriola quinqueradiata*[2]	322	34	20
Mackerel *Scomber japonicus*[2]	230	25	19
Red sea bream *Chrysophrys major*[3]	718	17	60
Flounder *Paralichthys olivaceus*[3]	464	11	46
Puffer *Fugu vermiculare porphyreum*[3]	561	21	54

[1] Suyama and Suzuki (1975); [2] Sakaguchi et al (1964); [3] Konosu et al (1974); [4] Non-protein N accounted for by creatine-N + creatinine-N

Table I
FREE AMINO ACID CONTENTS OF MUSCLE OF SOME FISHES

	Flounder[1] (mg%)	Puffer[2] (mg%)	Jack[3] mackerel (mg%)	Mackerel[4] (mg%)	Yellowfin[5] tuna (mg%)	Shark[6] (mg%)
Asp	±	1	1	—	1	7
Thr	4	10	15	11	3	7
Ser	3	4	3	6	2	10
Glu	6	4	13	18	3	12
Pro	1	13	6	26	2	7
Gly	5	20	10	7	3	21
Ala	13	22	21	26	7	19
Val	1	2	6	16	7	7
Met	1	±	1	2	3	6
Ile	1	2	1	7	3	5
Leu	1	3	5	14	7	8
Tyr	1	2	1	7	2	5
Phe	1	1	1	4	2	4
Try	—	—	—	—		—
Lys	17	128	54	93	35	3
His	1	1	289	676	1 220	8
Arg	3	20	3	11	1	6
Tau	171	123	75	84	26	44

[1-4] Konosu et al (1974); [5] Suyama and Yoshizawa (1973); [6] Suyama and Suzuki (1975)

Table II
FREE AMINO ACID CONTENTS OF MUSCLE OF SOME MOLLUSCS AND CRUSTACEANS

	Abalone[1] (mg%)	Short-necked clam[2] (mg%)	Sepia[3] (mg%)	Crab[4] (mg%)	Prawn[5] (mg%)	Krill[6] (mg%)
Asp	9	21	—	15	—	52
Thr	82	13	9	30	13	54
Ser	95	24	27	158	133	43
Glu	109	103	3	62	34	35
Pro	83	16	749	239	203	217
Gly	174	329	831	190	1 222	116
Ala	98	130	181	173	43	106
Val	37	14	3	20	17	63
Met	13	11	7	9	12	34
Ile	18	10	6	17	9	48
Leu	24	20	12	33	13	86
Tyr	57	16	8	20	20	48
Phe	26	20	2	13	7	53
Try	20	—	15	—	—	—
Lys	76	25	29	18	52	145
His	23	9	48	16	16	17
Arg	299	94	983	397	902	266
Tau	946	664	160	50	150	206

[1] Hashimoto (1965); [2] Konosu et al (1965); [3] Endo and Sinidu (1963); [4] Severin et al (1972); [5] Hujita (1961); [6] Suyama et al (1965)

Table III
CONCENTRATION OF HISTIDINE AND IMIDAZOLE DIPEPTIDES IN FISH MUSCLE

Species	mg/100g Non-protein nitrogen	Histidine	Carnosine	Anserine	Balenine	Imidazole-N /NPN (%)
Southern bluefin[1]	562	667	±	767	0	64·1
Yellowfin tuna[1]	614	1 220	54·8	234	0	65·0
Skipjack[1]	802	1 340	252	559	0	69·3
Swordfish[1]	564	831	130	370	0	61·0
Shark[2]						
Mustelus mitsukurii	1 410	11·2	0	24·0	0	0·6
Lamna cornubica	1 450	8·0	0	1 060	0	17·2
Eel[3]	386	2·8	542	±	±	35·0
Horse mackerel[3]	375	323	0	2·6	0	20·9
Sea perch[3]	345	14·2	0·1	±	0·4	1·0
Flat fish[3]	303	8·4	0	±	0	0·7

[1] Suyama and Yoshizawa (1973); [2] Suyama and Suzuki (1975); [3] Suyama et al (1970)

Table V
TMAO AND TMA CONTENTS OF MUSCLE AND VISCERA OF SOME AQUATIC ANIMALS

Species	TMAO (mg%) Muscle	Viscera	TMA (mg%) Muscle	Viscera
Elasmobranch				
Mustelus manazo[1]	1 460	489 (liver)	0·8	15·8 (liver)
Raja hollandi[1]	1 360	612 (liver)	0·6	12·3 (liver)
Teleost				
Scomber japonicus[2]	27·9	22·2 (liver)	0·8	2·1 (liver)
Pagrosmus major[2]	39·7	22·3 (liver)	3·1	6·9 (liver)
Theraga charcogramma[3]	657	53·2 (liver)	0·7	2·9 (liver)
Mollusc				
Nordotis discus[4]	0·5	1·6	7·4	8·7
Batillus cornutus[4]	1·6	0·5	16·1	13·5
Mytilus edulis[4]	2·7	2·1	4·8	4·4
Pecten albicans[4]	269	12·3	24·8	7·8
Sepioteuthis lessoniana[4]	750 (mantle)	32·1 (mgg)	2·1 (mantle)	3·1 (mgg)
Crustacea				
Metapenaeopsis barbata[4]	311	87·3	9·1	22·2
Pachygrapsus crassipes[4]	213	19·8	10·5	7·8

[1] Suyama et al (1960); [2] Takada and Nishimoto (1958); [3] Tokunaga (1970); [4] Harada et al (1968, 1970, 1971. 1972)
mgg = mid-gut gland

Table VI
CONCENTRATION OF UREA IN MUSCLE OF FISH AND SHELLFISH

Species	Urea (mg%)	Species	Urea (mg%)
Marine elasmobranchs[1,2]		Marine teleosts[3]	
Mustelus griseus	1 740	Melanogrammus aeglefinus	1·98
Mustelus kanekonis	1 820	Gadus callarias	2·3
Prionace glauca	1 600	Salmo salar	1·3
Isuropsis glauca	1 410	Clupea harengus	2·9
Lamna cornubica	1 520	Germo germo	2·3
Squalus mutsukurii	1 530	Sparus sp.	2·6
Mustelus manazo	1 718	Sardina melanosticta	0·45
Raja hollandi	2 167	Limanda sp.	9·91
Fresh water elasmobranchs[3]		Shellfish[3]	
Prissit microdon	650	Homarus americanus	0·07
		Portunus trituberculatus	0
Coelacanth[4]			
Latimeria chalumnae	1 860		

[1] Suyama and Suzuki (1975); [2] Suyama and Tokuhiro (1954); [3] Shewan (1951); [4] Lutz and Robertson (1971)

Table VII
CONTENTS OF GLYCINE BETAINE IN MUSCLE OF SOME FISH AND SHELLFISH

Fish[1] species	(mg%)	Shellfish[2] species	(mg%)
Dogfish (adult)	70	Abalone	668
Dogfish (embryos)	120	Fan-mussel	964
Rays	210	Scallop	211
Sharks	260	Oyster	805
Cod	102	Hard clam	727
Witch	75	Squid	733
Carp	10	Octopus	1 434
Pike	14	Krill	365
River lamprey	54	Prawn	539

[1] Shewan (1951); [2] Konosu and Hayashi (1975)
Fish

Table VIII
COMPARISON OF THE PROPERTIES OF FISH MUSCLE PROTEINASES (CATHEPSIN) IN D)

Property \ Origin	Albacore[1]	Cod[2]	Carp[3]
Optimum pH	2·4–2·5	4·3	2·8–3·0
Optimum temperature	42°C (15min)		50°C (60min)
Heat stability	82% of the activity lost after 10min at 60°C		Stable at 37°C. Loses all of the activity after 10min at 55°C
Substrate specificity	Active against haemoglobin	Active against haemoglobin	Active against haemoglobin casein and endogenous protein
Activator			GSH, 2-Mercapto ethanol
Inhibitor	p-CMB, Iodoacetamide, N-Ethyl maleimide		

[1] Groninger (1964); [2] Siebert and Schmitt (1965); [3] Makinodan and Ikeda (1969b)

Table IX
VOLATILE SULPHUR COMPOUNDS

Compound	Fresh materials of
Hydrogen sulphide	Squid[1], mackerel[1], cod[2], herring[3]
Methanethiol (Methyl mercaptan)	Mackerel[1], herring[3]
Ethanethiol (Ethyl mercaptan)	
Dimethyl sulphide	Pacific oyster[4], soft-shell clam[5], Antarctic krill[6], shrimp[6], cod[2,7] Atlantic mackerel[8], herring[3]
Carbon disulphide	Cod[2]

Compound	Stored materials of
Hydrogen sulphide	Pacific oyster[4], cod[9,10]
Methanethiol	Pacific oyster[4], cod[9,10]
Ethanethiol	Pacific oyster[4], haddock[11]
Dimethyl sulphide	Pacific oyster[4], haddock[11], cod[2,9,10], Canary rockfish[12]
1-Propanethiol	Pacific oyster[4]
2-Methyl-2-propanethiol	Pacific oyster[4]
1-Butanethiol	Pacific oyster[4]
2-Butanethiol	Pacific oyster[4]
Carbon disulphide	Cod[2]
Dimethyl disulphide	Pacific oyster[4], haddock[11]

Compound	Heated materials of
Hydrogen sulphide	Soft-shell clam[14], herring[3,15], cod[16] Yellowfin tuna[17], Albacore tuna[17]
Methanethiol	Soft-shell clam[14], herring[3,15], cod[16] Chum salmon[18]
Dimethyl sulphide	Soft-shell clam[14], Antarctic krill[6] herring[3,15], cod[16], Chum salmon[18]
Diethyl sulphide	Soft-shell clam[14]
Ethyl-n-butyl sulphide	Soft-shell clam[14]
Dimethyl disulphide, 2,5-dimethylthiophene, propylthiophene, 2-acetylthiazole, 2,4-dimethylthiazole, 4,5-dimethylthiazole, 2,4,5-trimethylthiazole, 2-ethyl-4,5-dimethylthiazole, 5-ethyl-2,4-dimethylthiazole	Cod[16]

[1] Ooyama (1973); [2] Wong et al (1967); [3] McLay (1967); [4] Ronald and Thomson (1964); [5] Brooke et al (1968); [6] Tokunaga et al (1977); [7] Sipos and Ackman (1964); [8] Ackman et al (1972); [9] Herbert et al (1975); [10] Herbert and Shewan (1975; 1976); [11] Angelini et al 1975); [12] Miller et al (1972); [14] Mendelsohn and Brooke (1968); [15] Hughes (1964); [16] McGill et al (1977); [17] Khayat (1977); [18] Motohiro (1962)

Table XIV
VITAMIN CONTENTS OF THE EDIBLE PORTION OF FISH AND SHELLFISH[1–3]

Vitamin A (IU/100g): lamprey and eel, 2 500–23 400; dogfish, 200–300; other fish, 100–200; Antarctic krill[4] 380
Vitamin D (IU/100g): lean fish, 0–40; fatty fish, 300–1 700; eel, 4 700
α-Tocopherol (μg/g): fish, 1–50; lobster, 10–20; mussel, 1–11
Thiamine (μg/g): fish, 0·4–1·8
Riboflavin (μg/g): fish, 0·3–7·2
Vitamin B_6 (μg/g): fish, 1–16; mollusca, 0·6–4·0
Vitamin B_{12} (μg/100g): fish, 0·03–12·7
Folic acid (μg/100g): lamprey and eel, 6–29; other fish, 1–11; Antarctic krill, 66
Pantothenic acid (μg/g): fish, 1–19; shellfish, 3–30
Nicotinic acid (μg/g): fish, 50–160; molluss, 12–26; Antarctic krill, 70
Biotin (μg/100g): mollusca[5], 7·4–16·4
Choline (mg/g): fish, 0·4–0·6; squid, 0·8–1·1
Inositol (μg/g): fish[6,7], 73–680, mollusca[5] 120–330
Ascorbic acid (μg/g): fish[8], trace–40

[1] Jacquot, 1961; [2] Higashi, 1962; [3] Braekkan, 1962; [4] Yanase, 1971; [5] Miyake and Noda, 1962; [6] Chang et al, 1960; [7] Sato et al, 1967; [8] Ikeda et al, 1963

118

Table X
VOLATILE CARBONYL COMPOUNDS

A: Volatile fraction from materials heated at temperature higher than 90°C
B: Volatile fraction from raw or uncooked materials
○*: Type of isomer was not described

Type of storage or processing	Fresh			Freeze-stored			Chill-stored			Salted	Dried
Species	Mackerel[1]	Herring[2,3]	Horse mackerel[4]	Cod[5]	Halibut[6]	Cod[7]	Haddock[8]	Canary rockfish[9]	Soft-shell clam[10,11]	Cod[12]	Alaska pollack[13]
Sample analysed (Compound)	A	A	A	A	B	B	B	B	A,B	A	A
Formaldehyde		○		○	○					○	
Ethanal	○	○	○	○	○	○	○	○	○	○	○
Propanal	○	○	○	○	○	○			○	○	○
2-Methylpropanal	○	○					○		○	○	
Butanal	○	○				○	○		○	○	○
2-Methylbutanal		○		○			○				
3-Methylbutanal	○	○	○	○				○	○		○
Pentanal		○		○	○					○	○
2-Methylpentanal				○							
Hexanal			○	○				○	○		○
Heptanal				○	○				○		
Octanal				○						○	
Nonanal		○		○	○						
Decanal										○	
Undecanal		○									
Tetradecanal				○							
Hexadecanal				○							
Octadecanal				○							
Propenal			○	○				○			
trans-2-Butenal		○		○						○*	
cis-2-Pentenal		○*		○						○*	
trans-2-Hexenal		○*		○						○*	
cis-3-Hexenal				○							
trans-2-Heptenal		○*		○						○*	
cis-4-Heptenal				○							
trans-4-Heptenal				○							
cis-2-Octenal				○						○*	
trans-2-Octenal				○							
trans-2-Nonenal				○							
trans-2-Decenal				○							
trans-2, cis-4-Heptadienal				○							
trans-2, trans-4-Heptadienal				○							
trans-2, cis-6-Nonadienal				○							
trans-2, cis-4-Decadienal				○							
trans-2, trans-4-Decadienal				○							
2-Furaldehyde				○							
Benzaldehyde				○							
o-Tolualdehyde				○							
Terephthalaldehyde				○							
Phenylacetaldehyde				○							
Acetone		○	○	○	○	○	○		○	○	
2-Butanone		○		○	○	○			○	○	
3-Hydroxy-2-butanone									○		
3-Methyl-2-butanone										○	
2-Pentanone				○	○	○			○	○	
3-Pentanone						○			○		
2-Hexanone		○		○	○						
3-Hexanone				○							
2-Heptanone	○	○		○					○		
3-Heptanone				○					○		
4-Heptanone									○		
2-Octanone		○		○							
2-Nonanone		○		○							
5-Nonanone				○							
2-Undecanone		○		○							
3-Buten-2-one							○		○		
3-Hexen-2-one				○							
3-Octen-2-one				○							
1-Octen-3-one				○							
3,5-Octadien-2-one				○							
Acetophenone				○							
Cyclohexanone				○							
2,3-Butanedione							○		○	○	

[1] Miwa et al (1976); [2] Hughes (1961); [3] Hughes (1963); [4] Shimomura et al (1971); [5] McGill et al (1977); [6] Matsuto et al (1967); [7] Wong et al (1967); [8] Angelini et al (1975); [9] Miller III et al (1972); [10] Mendelsohn and Brooke (1968); [11] Gadbois et al (1967); [12] Yurkowski and Bordeleau (1965); [13] Kasahara and Nishibori (1975b)

119

Table XI
VOLATILE FATTY ACIDS

A: Volatile fraction from materials heated at temperature higher than 90°C
B: Volatile fraction or extracts from raw or uncooked materials

Type of storage or processing	Fresh									Freeze-stored	Dried	Canned				
Species	Squid[1]	Mackerel[1]	Skipjack[1]	Seabream[1]	Alaska pollack[2]	Mackerel[2]	Herring[3]	Tuna[4]	Mackerel[4]	Cod[5]	Alaska pollack[6]	Squid[7]	Mackerel[7]	Skipjack[7]	Seabream[7]	Herring[4]
Sample analysed → Compound	B	B	B	B	B	B	B	A	A	A	A	A	A	A	A	B
Formic acid	O	O	O	O			O	O	O			O	O	O	O	O
Acetic acid	O	O	O	O	O	O	O	O	O	O	O	O	O	O	O	O
Propionic acid	O	O	O	O	O	O	O		O		O	O	O	O	O	O
Butyric acid					O	O	O	O	O		O		O	O	O	
Isobutyric acid					O	O		O	O		O					
Valeric acid						O					O					
Isovaleric acid					O	O					O					
Caproic acid											O					
Isocaproic acid						O										
Palmitic acid										O						

[1] Ooyama (1973); [2] Miwa *et al* (1976); [3] Hughes (1960); [4] Miyahara (1961); [5] McGill *et al* (1977); [6] Kasahara and Nishibori (1975b); [7] Ooyama (1975)

Table XII
VOLATILE ALCOHOLS

Type of storage	Fresh	Freeze-stored	Chill-stored		
Species	Horse mackerel[1]	Cod[2]	Cod[3]	Haddock[4]	Canary rockfish[5]
Sample analysed → Compound	A	A	B	B	B
Methanol				O	
Ethanol	O	O	O	O	O
1-Propanol					O
3-Methyl-1-butanol					O
1-Hexanol		O			
1-Heptanol		O			
3-Pentanol		O			
1-Penten-3-ol					O
1-Octen-3-ol		O			
Methylbenzyl alcohol		O			
2-Phenylethanol		O			

[1] Shimomura *et al* (1971); [2] McGill *et al* (1977); [3] Wong *et al* (1967); [4] Angelini *et al* (1975); [5] Miller III *et al* (1972)

Table XIII
VOLATILE HYDROCARBONS

Type of storage	Fresh	Freeze-stored	Chill-stored		
Species	Horse mackerel[1]	Cod[2]	Cod[3]	Haddock[4]	Soft-shell clam[5]
Sample analysed → Compound	A	A	B	B	A,B
Heptan		O			
C9-branched paraffin		O			
Butene					O
1-Pentene					O
1-Hexene	O				
1-Octene	O				
Benzene			O	O	
Toluene		O	O	O	O
		O ↑			

o, m and p−xylene, ethylbenzene, propylbenzene, isopropylbenzene, butylbenzene, p-cymene, cyclohexane, styrene, methylstyrene

[1] Shimomura *et al* (1971); [2] McGill *et al* (1977); [3] Wong *et al* (1967); [4] Angelini *et al* (1975); [5] Mendelsohn and Brooke (1968)

8 References

ABE, H, SHIMIZU, C and MATSUURA, F. Occurrence and distribution of
1976 NAD(P) splitting enzymes in fish tissues. *Bull. Jap. Soc. Sci. Fish.*, 42, 703–711

ACKMAN, R G and CORMIER, M G. α-Tocopherol in some Atlantic fish
1967 and shellfish with particular reference to live-holding without food. *J. Fish. Res. Bd. Can.*, 24, 357–373

ACKMAN, R G and HINGLEY, H J. The occurrence and retention of
1968 dimethyl-β-propiothetin in some filter-feeding organisms. *J. Fish. Res. Bd. Can.*, 25, 267–284

ACKMAN, R G, HINGLEY, J and McKAY, K T. Dimethyl sulfide as an
1972 odor component in Nova Scotia fall mackerel. *J. Fish. Res. Bd. Can.*, 29, 1085–1088

AITKEN, A and CONNELL, J J. *Chemical and physical effects of heating*
1979 *on foodstuffs: fish*, Ed R J Priestley. Applied Science Publishers Ltd, London. 417

ALLEN, A and GARRETT, M R. Taurine in marine invertebrates. *Adv.*
1971 *Mar. Biol.*, 9, 205–253

AMANO, K and YAMADA, K. The biochemical formation of formal-
1965 dehyde in cod flesh. In: *The Technology of Fish Utilization*, Ed R Kreuzer. Fishing News (Books) Ltd, London. 73–78

ANGELINI, P, MERRITT, C Jr, MENDELSOHN, J M and KING, F J. Effect
1975 of irradiation on volatile constituents of stored haddock flesh. *J. Fd. Sci.*, 40, 197–199

ARAI, K. Nucleotides in the muscle of marine invertebrates—A
1966 review. *Bull. Jap. Soc. Sci. Fish.*, 32, 174–179

ARIMA, S and UMEMOTO, S. Mercury in aquatic organism. 2. Mercury
1976 distribution in muscles of tunas and swordfish. *Bull. Jap. Soc. Sci. Fish.*, 42, 931–937

ARNOLD, S H and BROWN, W D. Histamine (?) toxicity from fish pro-
1978 ducts. *Adv. Food Res.*, 24, 113–154

BABBIT, J K, CRAWFORD, D L and LAW, K D. Decomposition of
1972 trimethylamine oxide and changes in protein extractability during frozen storage of minced and intact hake (*Merluccius productus*) muscle. *J. Agric. Fd. Chem.*, 20, 1052–1054

BEERS, J R. The species distribution of some naturally-occurring
1967 quaternary ammonium compounds. *Comp. Biochem. Physiol.*, 21, 11–21

BITO, M. Changes in NAD and ATP levels and pH in frozen-stored
1978 skipjack meat, in relation to amount of drip. *Bull. Jap. Soc. Sci. Fish.*, 44, 897–902

BOON, D D. Discoloration in processed crab meat—A review. *J. Fd.*
1975 *Sci.*, 40, 756–761

BOWEN, H J M. *Trace Elements in Biochemistry*. Academic Press,
1966 London. 234pp

BRAEKKAN, O R. B-Vitamins in fish and shellfish. In: Eds E Heen and
1962 R Kreuzer. *Fish in Nutrition*, Fishing New (Books) Ltd, London. 132–145

BROOKE, R O, MENDELSOHN, J M and KING, F J. Significance of
1968 dimethyl sulfide to the odor of soft-shell clams. *J. Fish. Res. Bd. Can.*, 25, 2453–2460

BURT, J R. Glycolytic enzymes of cod (*Gadus callarias*) muscle. *J. Fish.*
1966 *Res. Bd. Can.*, 23, 527–538

BURT, J R and STROUD, G D. The metabolism of sugar phosphates in
1966 cod muscle. *Bull. Jap. Soc. Sci. Fish.*, 32, 204–212

CASTELL, C H, NEAL, W E and DALE, J. Comparison of changes in
1973 trimethylamine, dimethylamine, and extractable protein in iced and frozen gadoid fillets. *J. Fish. Res. Bd. Can.*, 30, 1246–1248

CAUSERET, J. Fish as source of mineral nutrition. *Fish as Food*.
1962 Academic Press, New York and Londdon. Vol 2, 205–234

CHANG, V M, TSUYUKI, H and IDLER, D R. Distribution of phos-
1960 phorus compounds, creatine and inositol in tissues. *J. Fish. Res. Bd. Can.*, 17, 565–582

CRAWFORD, L, KRETSH, M J and GUADAGNI, D. Identification of vol-
1976 atiles from extracted commercial tuna oil with a high docosahexaenoic acid content. *J. Sci. Fd. Agric.*, 27, 531–535

DINGLE, R R and HINES, J A. Degradation of iosine 5'-
1971 monophosphate in the skeletal muscle of several North Atlantic fishes. *J. Fish. Res. Bd. Can.*, 28, 1125–1131

DINGLE, J R, HINES, J A and FRASER, D I. Postmortem degradation of
1968 adenine nucleotides in muscle of the lobster, *Homarus americanus*. *J. Fd. Sci.*, 33, 100–103

DINGLE, J R, KEITH, R A and LALL, B. Protein instability in frozen
1977 storage induced in minced muscle of red hake. *Can. Inst. Fd. Sci. Technol. J.*, 10, 143–146

EHIRA, S and UCHIYAMA, H. Formation of inosine and hypoxanthine in
1973 fish muscle during ice storage. *Bull. Tokai Reg. Lab.*, No. 75, 63–73

—— Denaturation of myofibrillar protein of iced fish in relation to its
1979 lowering of freshness—Changes in Ca²⁺–ATPase activity and extractability during the period from death to spoilage. *Bull. Jap. Soc. Sci. Fish.*, 45, 121–127

ENDO, K, KISHIMOTO, R, YAMAMOTO, Y and SHIMIZU, Y. Seasonal
1974 variations in chemical constituents of yellowtail muscle. 2. Nitrogenous extractives. *Bull. Jap. Soc. Sci. Fish.*, 40, 67–72

ENDO, K and SIMIDU, W. Studies on muscle of aquatic animals. 37.
1963 Octopine in squid muscle. *Bull. Jap. Soc. Sci. Fish.*, 29, 362–365

ESKIN, N A M, HENDERSON, H M and TOWNSEND, R J. *Biochemistry*
1971 *of Food*. Academic Press, New York and London. 1–29

FUJIMOTO, K and Kaneda, T. Studies on the brown discoloration of fish
1973 products. 5. Reaction mechanism in the early stage. *Bull. Jap. Soc. Sci. Fish.*, 39, 185–190

FUJIWARA-ARASAKI, T and MINO, N. The distribution of
1972 trimethylamine and trimethylamine oxide in marine alga. *Proc: 7th Int. Seaweed Sym.* (1971), Japan. Univ. of Tokyo Press. 506–510

GADBOIS, D F, MENDELSOHN, J M and RONSIVALLI, L J. Effects of
1967 radiation, heating and storage on volatile coabonyl compounds in clam meats. *J. Fd. Sci.*, 32, 511–515

GRIESHABER, M and GADE, G. Energy supply and the formation of
1977 octopine in the adductor muscle of the scallop, *Pecten jacobaeus*. *Comp. Biochem. Physiol.*, 58B, 249–252

GRONINGER, H S Jr. Partial purification and some properties of a pro-
1964 teinase from albacore (*Germo alalunga*) muscle. *Arch. Biochem. Biophys.*, 108, 175–182

GUADAGNI, D G, BUTTERY, R G and OKANO, S. Odour thresholds of
1963 some organic compounds associated with food flavours. *J. Sci. Fd. Agric.*, 14, 761–765

HALVER, J E. The vitamins. *Fish Nutrition*, Academic Press, New York
1972 and London, 29–103

HARADA, K. Studies of enzyme catalyzing formation of formaldehyde
1975 and dimethylamine in fishes and shells. *J. Shimonoseki Univ. Fish.*, 23, 165–241

HARADA, K, DERIHA, T and YAMADA, K. Distribution of
1972 trimethylamine oxide in fishes and other aquatic animals. IV. Arthropods, chinoderms and other invertebrates. *J. Shimonoseki Univ. Fish.*, 20(3), 115–130

HARADA, K, FUJIMOTO, T and YAMADA, K. Distribution of
1968 trimethylamine oxide in fishes and other aquatic animals. I. Decapodan mollusca. *J. Shimonoseki Univ. Fish.*, 17(2), 87–95

HARADA, K, TAKEDA, J and YAMADA, K. Distribution of
1970 trimethylamine oxide in fishes and other aquatic animals. II. Bivalvian mollusca. *J. Shimonoseki Univ. Fish.*, 18(3), 11–19

HARADA, K, YAMAMOTO, Y and YAMADA, K. Distribution of
1971 trimethylamine oxide in fishes and other aquatic animals. III. Gastropodan mollusca. *J. Shimonoseki Univ. Fish.*, 19(2–3), 105–114

HASEGAWA, K, WADA, T, SHIMODA, Y SAWADA, T, FUJII, Y, NAKAM-
1976 URA, K and ISHIKAWA, S. Studies on the prevention of the browning of canned skipjack and albacore (orange discoloration). 7. Effect of the treatment for raw materials and conditions of storage on the orange discoloration of canned meat (2). *Bull. Jap. Soc. Sci. Fish.*, 42, 187–195

HASHIMOTO, Y. Taste-producing substances in marine products. In:
1965 *The Technology of Fish Utilization*, Ed R. Kreuzer. Fishing News (Books) Ltd, London. 57–61

HAYASHI, T and KONOSU, S. Quaternary ammonium bases in the
1977 adductor muscle of fan-mussel. *Bull. Jap. Soc. Sci. Fish.*, 43, 343–348

HAYASHI, T, YAMAGUCHI, K and KONOSU, S. Contribution of extrac-
1978 tive components to the taste of boiled crabs. Abst. 5th Int. Cong. Food Sci. Technol., 159

HEBER, U, TYANKOVA, L and SANTARIUS, K A. Effects of freezing on
1973 biological membranes *in vivo* and *in vitro*. *Biochim. Biophys. Acta*, 291, 23–37

HERBERT, R A, ELLIS, J R and SHEWAN, J M. Isolation and identifica-
1975 tion of the voltaile sulphides produced during chill-storage of North Sea cod (*Gadus morhua*). *J. Sci. Fd. Agric.*, 26, 1187–1194

HERBERT, R A and SHEWAN, J M. Precursors of the volatile sulphides
1975 in spoiling North Sea cod (*Gadus morhua*). *J. Sci. Fd. Agric.*, 26, 1195–1202

HERBERT, R A and SHEWAN, J M. Roles played by bacterial and auto-
1976 lytic enzymes in the production of volatile sulphides in spoiling North Sea cod (*Gadus morhua*). *J. Sci. Fd. Agric.*, 27(1), 89–94

HIGASHI, H. Relationship between processing techniques and the
1962 amount of vitamins and minerals in processed fish. In: *Fish in Nutrition*, E Heen and R Kreuzer. Fishing News (Books) Ltd, London. 125–131

HILTZ, D F, BISHOP, L J and DYER, W J. Accelerated nucleotide
1974 degradation and glycolysis during warming to and subsequent storage at −5°C of prerigor, quick-frozen adductor muscle of the sea scallop (*Placopecten magellanicus*). *J. Fish. Res. Bd. Can.*, 31, 1181–1187

HILTZ, D F and DYER W J. Octopine in postmortem adductor muscle
1971 of the sea scallop (*Placopecten magellanicus*). *J. Fish. Res. Bd. Can.*, 28, 869–874

HIRANO, T. On the distribution and seasonal variation of homarine in
1975 some marine invertebrates. *Bull. Jap. Soc. Sci. Fish.*, 41, 1047–1051

HOFFMAN, A, DISNEY, J G, GRIMWOOD, B E and JONES, N R. Glucose
1970   levels in fresh *Tilapia* muscle. *J. Fish. Res. Bd. Can.*, 27, 801–3
HONMA, N, SHIOZAKI, K, SHIBUYA, U and ISHIHARA, K. Volatile car-
1974   bonyl compounds in unheated and heated 'Niboshi-soup'.
        Kaseigaku Zasshi, 25, 362–369
HUGHES, R B. Chemical studies on the herring. III. The lower fatty
1960   acids. *J. Sci. Fd. Agric.*, 11(1), 47–53
—— Chemical studies on the herring. VI. Carbonyl compounds formed
1961   during heat processing of herring. *J. Sci. Fd. Agric.*, 12(12),
        822–826
HUGHES, R B. Chemical studies on the herring (*Clupea harengus*). 7.
1963   Further observations on the production of carbonyls in heat-
        processed herring. *J. Sci. Fd. Agric.*, 14, 893–903
—— Chemical studies on the herring (*Clupea harengus*). 9. Preliminary
1964   gas-chromatographic study of volatile sulphur compounds
        produced during the cooking of herring. *J. Sci. Fd. Agric.*, 15,
        290–292
IKEBE, K, TANAKA, Y, TANAKA, R and KUNITA, N. Contents of heavy
1977   metals in food. 6. Contents of heavy metals in fishes, shell-
        fishes, meats, poultries and whales. *Shokuhin-Eiseigaku
        Kaishi*, 18, 86–97
IKEDA, S, SATO, M and KIMURA, R. Biochemical studies on L-ascorbic
1963   acid in aquatic animals. 2. Distribution in various parts of fish.
        Bull. Jap. Soc. Sci. Fish., 29, 765–770
IWATA, K, KOBAYASHI, K and HASE, J. Studies on muscle alkaline pro-
1973   tease. 1. Isolation, purification and some physicochemical
        properties of an alkaline protease from carp muscle. *Bull. Jap.
        Soc. Sci. Fish.*, 39, 1325–1337
—— Studies on muscle alkaline protease. 7. Effect of the muscular
1979   alkaline protease and protein fractions purified from white
        croaker and horse mackerel on the 'Himodori' phenomenon
        during kamaboko production. *Bull. Jap. Soc. Sci. Fish.*, 45,
        157–161
IYENGAR, J R, PANALAKS, T, MILES, W F and SEN, N P. A survey of
1976   fish products for volatile *N*-nitrosamines. *J. Sci. Fd. Agric.*, 27,
        527–530
JACQUOT, R. Organic constituents of fish and other aquatic animal
1961   food. In: *Fish as Food*, Ed G Borgstrom. Academic Press, New
        York. Vol 1, 145–192
JONES, N R. Anserinase and other peptidase activity in skeletal muscle
1956   of codling (*Gadus callarias*). *Biochem. J.*, 64, 20pp
—— Fish flavors. In: *Symposium on Food: The Chemistry and Physiol-
1967   ogy of Flavors*, Eds H W Schultz, E A Day and L M Libbey.
        The AVI Publishing Co, Westport, Connecticut. 267–295
KASAHARA K and NISHIBORI, K. Flavoring volatiles of roasted fish
1975a   meat. *Bull. Jap. Soc. Sci. Fish.*, 41, 43–49
—— Unpleasant odors of Alaska pollack. 1. Acidic, basic and carbonyl
1975b   compounds. *Bull. Jap. Soc. Sci. Fish.*, 41, 1009–1013
KATUKI, Y, YASUDA, K, UEDA, K and KIMURA, Y. Studies on amounts
1975   of trace elements in marine fishes. 2. Distribution of heavy
        metals in bonito tissue. Ann. Rep. Tokyo Metr. Res. Lab.
        P.H., 26, 196–199
KAWAI, A and SAKAGUCHI, M. Histidine metabolism in fish. 2. Forma-
1968   tion of urocanic, formiminoglutamic, and glutamic acids from
        histidine in the livers of carp and mackerel. *Bull. Jap. Soc. Sci.
        Fish.*, 34, 507–511
KAWAI, A, SAKAGUCHI, M and KIMATA, M. Histidine metabolism in
1967   fish. 1. Histidine deaminase and urocanase activities in fish
        muscle and liver. Mem. Res. Inst. Food Sci., Kyoto Univ., 28,
        18–25
KEEFER, L K and ROLLER, P P. *N*-nitrosation by nitrite ion in neutral
1973   and basic medium. *Science*, 181, 1245–1247
KENNEDY, T S and LEY, F J. Studies on the combined effect of gamma
1971   radiation and cooking on the nutritional value of fish. *J. Sci.
        Fd. Agric.*, 22, 146–148
KHAYAT, A. Hydrogen sulfide production by heating tuna meat. *J. Fd.
1977   Sci.*, 42, 601–609
KIKUCHI, T, HONDA, H, ISHIKAWA, M, YAMANAKA, H and AMANO, K.
1978   Excretion of mercury from fish. *Bull. Jap. Soc. Sci. Fish.*, 44,
        217–222
KIKUCHI, T, WADA, S and SUZUKI, H. Significance of volatile bases and
1976   volatile acids in the development of off-flavor of fish meat. *J.
        Jap. Soc. Fd. Nutr.*, 29, 147–152
KLUYTMANS, J H, DE BONT, A M T, JANUS, J and WIJSMAN, T C M.
1977   Time dependent changes and tissue specificities in the accumu-
        lation of anaerobic fermentation products in the sea mussel
        Mytilis edulis. *Comp. Biochem. Physiol.*, 58B, 81–87
KOIZUMI, C. Blue or green meat of tuna fish—A review. *Bull. Jap. Soc.
1967   Sci. Fish.*, 33, 883–887
KONAGAYA, S. Screening of peptidase in fish muscle. *Bull. Tokai Reg.
1978   Fish. Res. Lab.*, 94, 1–28
KONOSU, S, FUJIMOTO, K, TAKASHIMA, Y, MATSUSHITA, T and
1965   HASHIMOTO, Y. Constituents of the extracts and the amino acid
        composition of the protein of short-necked clam. *Bull. Jap.
        Soc. Sci. Fish.*, 31(9), 680–686
KONOSU, S. Distribution of nitrogenous constituents in the muscle
1971   extracts of aquatic animals—A review. *Bull. Jap. Soc. Sci.
        Fish.*, 37, 763–770

—— Taste of fish and shellfish with special reference to taste-
1973   producing substances—A review. *J. Jap. Soc. Food Sci. Tech-
        nol.*, 20, 432–439
KONOSU, S and HAYASHI, T. Determination of β-alanine betaine and
1975   glycine betaine in some marine invertebrates. *Bull. Jap. Soc.
        Sci. Fish.*, 41, 743–746
KONOSU, S, MURAKAMI, M, HAYASHI, T and FUKE, S. Occurrence of
1978   β-alanine betaine in the muscle of New Zealand whiptail and
        southern blue whiting. *Bull. Jap. Soc. Sci. Fish.*, 44,
        1165–1166
KONOSU, S, SHIBATA, M and HASHIMOTO, Y. Concentration of organic
1967   acids in shellfish with particular reference to succinic acid. *J.
        Jap. Soc. Fd. Nutr.*, 20, 186–189
KONOSU, S and WATANABE, K. Comparison of nitrogenous extractives
1976   of cultured and wild red sea breams. *Bull. Jap. Soc. Sci. Fish.*,
        42, 1263–1266
KONOSU, S, WATANABE, K and SIMIZU, T. Distribution of nitrogenous
1974   constituents in the muscle extracts of eight species of fish. *Bull.
        Jap. Soc. Sci. Fish.*, 40, 909–915
KUMAGAI, H and SAEKI, K. Patterns of increasing mercury content
1978   with growth for wild and cultured red sea breams, *Chrysophrys
        major*. *Bull. Jap. Soc. Sci. Fish.*, 44, 269–272
KUNISAKI, N, MATSUURA, H and HAYASHI, M. A food-hygienical study
1977   on the formation of *N*-nitrosodimethylamine from
        trimethylamine-*N*-oxide and nitrite. *Bull. Jap. Soc. Sci. Fish.*,
        43, 1287–1292
LOVE, R M. *The Chemical Biology of Fishes*. Academic Press, London
1974   and New York. 265pp
LUTZ, P L and ROBERTSON, J D. Osmotic constituents of the
1971   coelacanth *Latimeria chalumnae*. *Biol. Bull.*, 141(3), 553–560
MACKIE, I M and FERNANDEZ-SALGUERO, J. Histidine metabolism in
1977   fish. Urocanic acid in mackerel (*Scomber scombrus*). *J. Sci. Fd.
        Agric.*, 28, 935–940
MAKINODAN, Y, HIROTSUKA, M and IKEDA, S. Neutral proteinase of
1979   carp muscle. *J. Food Sci.*, 44. (In press)
MAKINODAN, Y and IKEDA, S. Studies on fish muscle protease. 2.
1969a   Purification and properties of a proteinase active in slightly
        alkaline pH range. *Bull. Jap. Soc. Sci. Fish.*, 35, 749–757
—— Studies on fish muscle protease. 3. Purification and properties of a
1969b   proteinase active in acid pH range. *Bull. Jap. Soc. Sci. Fish.*,
        35, 758–766
—— Studies on fish muscle protease. 6. Separation of carp muscle
1976   cathepsins A and D, and some properties of carp muscle
        cathepsin A. *Bull. Jap. Soc. Sci. Fish.*, 37, 239–247
—— Alkaline proteases of carp muscle: Effects of some protein
1977   denaturing agents of the activity. *J. Fd. Sci.*, 42, 1026–1033
MAKINODAN, Y, YAMAMOTO, M and SHIMIZU, W. Studies on muscle of
1963   aquatic animals. 39. Protease in fish muscle. *Bull. Japan. Soc.
        Sci. Fish.*, 29, 776–780
MALINS, D C, ROUBAL, W T and ROBISCH, P A. The possible nitrosa-
1970   tion of amines in smoked chub. *J. Agric. Fd. Chem.*, 18,
        740–741
MANOHAR, S V. Postmortem glycolytic and other biochemical changes
1970   in white muscle of white sucker (*Catostomus commersoni*) and
        Northern pike (*Esox lucius*) at 0°C. *J. Fish. Res. Bd. Can.*, 27,
        1997–2002
MATSUTO, S, NAGAYAMA, F and ONO, T. Volatile monocarbonyls in
1967   frozen halibut. *Bull. Jap. Soc. Sci. Fish.*, 33, 586–590
McGILL, A S, HARDY, R, BURT, J R and GUNSTONE, F D. Hept-*cis*-
1974   4-enal and its contribution to the off-flavour in cold stored cod.
        J. Sci. Fd. Agric., 25, 1477–1489
McGILL, A S, HARDY, R and GUNSTONE, F D. Further analysis of the
1977   volatile components of frozen cold stored cod and the influ-
        ence of these on flavour. *J. Sci. Fd. Agric.*, 28, 200–215
McLAY, R. Chemical studies on the herring (*Clupea harengus*). XI.
1967   Quantitative estimation of volatile sulphur compounds pro-
        duced during the cooking of herring. *J. Sci. Fd. Agric.*, 18(12),
        605–607
MENDELSOHN, J M and BROOKE, R O. Radiation, processing and stor-
1968   age effects on the head gas components in clam meats. *Fd.
        Technol.*, 22, 1162–1166
MILLER, A III, SCANLAN, R A, LEE, J S and LIBBEY, L M. Volatile
1972   compounds produced in ground muscle tissue of canary
        rockfish (*Sebastes pinniger*) stored on ice. *J. Fish. Res. Bd.
        Can.*, 29, 1125–1129
—— Volatile compounds produced in sterile fish muscle (*Sebastes
1973   melanops*) by *Pseudomonas putrefaciens*, *Pseudomonas
        fluorescens*, and *Achromobacter* species. *Appl. Microbiol.*, 26,
        18–21
MIWA, K, TOKUNAGA, T and IIDA, H. Studies on protecting methods of
1976   occurrence of bad odors and their removing methods in
        fisheries processing factories. 2. Cooking odor and drying odor
        of fish. *Bull. Tokai Reg. Fish. Res., Lab.*, No. 86, 7–27
MIYAHARA, S. Gas-liquid chromatographic separation and determi-
1961   nation of volatile fatty acids in fish meat during spoilage. *Bull.
        Jap. Soc. Sci. Fish.*, 27(1), 42–47
MIYAKE, M and NODA, H. Vitamin B group in the extracts of mollusca.
1962   2. On Vitamin B_2, inositol, pantothenic acid, biotin and niacin.

Bull. Jap. Soc. Sci. Fish., 28, 597–601

Motohiro, T. Studies on the petroleum odour in canned chum salmon. 1962 *Mem. Fac. Fish., Hokkaido Univ.*, 10, 1–65

Nagayama, F. Mechanism of breakdown and synthesis of glycogen in 1966 the tissue of marine animals—A review. *Bull. Jap. Soc. Sci. Fish.*, 32, 188–192

Nakamura, T, Yoshitake, K and Toyomizu, M. The discoloration of 1973 autoxidized lipid by the reaction with VBN or Non-VBN fraction from fish muscle. *Bull. Jap. Soc. Sci. Fish.*, 39, 791–796

Nishibori, K and Kasahara, K. Studies of the flavor of 'Katsuobushi'. 1978 7. Identification of non-carbonyl neutrals. *Bull. Jap. Soc. Sci. Fish.*, 44, 389–391

Oshima, H and Kawabata, T. Mechanism of the *N*-nitro-1978 sodimethylamine formation from trimethylamine and trimethylamine oxide. *Bull. Jap. Soc. Sci. Fish.*, 44, 77–81

Olley, J and Thrower, S J. Abalone—an esoteric food. *Adv. Food* 1977 *Res.*, 23, 143–186

Ooyama, S. Studies on the odor of canned fish. 1. Volatile components 1973 from the muscle of fresh fish. *Kaseigaku Zasshi*, 24, 694–698

—— Studies on the odor of canned fish. 3. Volatile components of 1975 steam-boiled fish. *Kaseigaku Zasshi*, 26, 470–473

Partmann, W. Changes in proteins, nucleotides and carbohydrates 1965 during rigor mortis. In: *The Technology of Fish Utilization*, Ed R Kreuzer. Fishing News (Books) Ltd, London. 4–13

—— Eine Carnosin spartende Enzymaktivität im Skelettmuskel des 1976 Aales. *Arch. Fisch Wiss.*, 27, 55–62

Read, L J. Urea and trimethylamine oxide levels in elasmobranch 1968 embryos. *Biol. Bull.*, 135, 537–547

Reineccius, G A. Off-flavors in meat and fish: A review. *J. Fd. Sci.*, 1979 44, 12–21, 24

Ronald, A P and Thomson, W A B. The volatile sulphur compounds 1964 of oysters. *J. Fish. Res. Bd. Can.*, 21, 1481–1487

Sakaguchi, M, Hujita, M and Simidu, W. Studies on muscle of aqua-1964 tic animals. 43. Creatine and creatinine contents in fish muscle extractives. *Bull. Jap. Soc. Sci. Fish.*, 30, 999–1002

Sakaguchi, M and Kawai, A. Presence of b- and c-type cytochromes 1978 in the membrane of *Escherichia coli* induced by trimethylamine *N*-oxide. *Bull. Jap. Soc. Sci. Fish.*, 44, 999–1002

Sasano, Y and Hirata, F. Studies on freezing storage of tanner crabs: 1973 Relation between the quality of meat and the nucleotides content. *Bull. Jap. Soc. Sci. Fish.*, 39, 951–954

Sato, M, Fukusaka, Y, Ishiguro, K and Ikeda, S. Distribution of 1967 myo-inositol in fishes. *Abs. Ann. Meet. Jap. Soc. Sci. Fish.*, April, 96pp

Severin, S E, Boldyrev, A A and Lebedev, A V. Nitrogenous 1972 extractive compounds of muscle tissue in invertebrates. *Comp. Biochem. Physiol.*, B43(2), 369–381

Shewan, J I M. The chemistry and metabolism of the nitrogenous 1951 extractives in fish. *Biochem. Soc. Symposia*, No. 6, 28–48. Biochemical Society, London

Shibata, T. Enzymological studies on the glycolytic system in the 1977 muscles of aquatic animals. *Mem. Fac. Fish., Hokkaido Univ.*, 24, 1–80

Shimizu, C, Abe, K and Matsuura, F. Levels of oxidized and reduced 1969 nicotinamide adenine dinucleotides in fish tissues. *Bull. Jap. Soc. Sci. Fish.*, 35, 1034–1040

Shimomura, M, Yoshimatsu, F and Matsumoto, F. Studies on 1971 cooked fish—odor of cooked horse mackerel. *Kaseigaku Zasshi*, 22, 106–112

Siebert, G and Schmitt, A. Fish tissue enzymes and their role in the 1965 deteriorative changes in fish. In: *The Technology of Fish Utilization*, Ed R Kreuzer. Fishing News (Books) Ltd, London. 47–52

Simidu, W. Nonprotein nitrogenous compounds. In: *Fish as Food*, Ed 1961 G Borgstrom. Academic Press, New York and London. Vol 1, 353–384

Sipos, J C and Ackman, R G. Association of dimethyl sulphide with 1964 the 'blackberry' problem in cod from the Labrador area. *J. Fish. Res. Bd. Can.*, 21, 423–425

Suryanarayama, S V, Rangaswamy, Rao, J R and Lahiry, N L. 1969 Nucleotides and related compounds in canned shrimp. *J. Fish. Res. Bd. Can.*, 26, 704–706

Suyama, M and Tokuhiro, Y. Urea content and ammonia formation 1954 in the muscle of cartilaginous fishes. *Bull. Jap. Soc. Sci. Fish.*, 19, 935–938

Suyama, M, Koike, J and Suzuki, K. Studies on glycolysis and forma-1960 tion of ammonia in muscle and blood of elasmobranchs. *J. Tokyo Univ. Fisheries*, 46, 51–65

Suyama, M, Nakajima, K and Nonaka, J. Nitrogenous constituents of 1965 Euphasia. *Bull. Jap. Soc. Sci. Fish*, 31(4), 302–306

Suyama, M, Hirano, T, Okada, N and Shibuya, T. Quality of wild 1977 and cultured ayu. 1. On the proximate composition, free amino acids and the related compounds. *Bull. Jap. Soc. Sci. Fish.*, 43, 535–540

Suyama, M and Suzuki, H. Nitrogenous constituents in the muscle 1975 extracts of marine elasmobranchs. *Bull. Jap. Soc. Sci. Fish.*, 41, 787–790

Suyama, M, Suzuki, T, Maruyama, M and Sato, K. Determination of 1970 carnosine, anserine, and balenine in the muscle of animal. *Bull. Jap. Soc. Sci. Fish.*, 36, 1048–1053

Suyama, M and Yoshizawa, Y. Free amino acid composition of the 1973 skeletal muscle of migratory fish. *Bull. Jap. Soc. Sci. Fish.*, 39, 1339–1343

Suzuki, T. Developing a new food material from fish flesh. 3. 1974 Removal of mercury from fish flesh. *Bull. Tokai Reg. Res. Lab.*, No. 78, 67–72

Taguchi, T, Suzuki, K and Osakabe, I. Magnesium and calcium 1969 contents of fish and squid tissues. *Bull. Japan. Soc. Sci. Fish.*, 35, 405–409

Takada, K and Nishimoto, J. Studies in the choline in fish. 1. Content 1958 of choline and the similar substances in fishes. *Bull. Jap. Soc. Sci. Fish.*, 24(8), 632–635

Takagi, I and Shimizu, W. Constituents and extractives nitrogens in a 1962 few species of shell-fish. *Bull. Jap. Soc. Sci. Fish.*, 28, 1192–1198

Take, T, Honda, R and Otsuka, H. On the tasty substances of prawn 1966 and shrimp. *J. Jap. Soc. Fd. Nutr.*, 17, 268–274

Tamaki, N, Ishizumi, H, Masumitu, N, Kubota, A and Hama, T. 1976 Species specificity on the contents of anserine and carnosine. *Yakugaku Zasshi*, 96, 1481–1486

Tarr, H L A. Post-mortem changes in glycogen, nucleotides, 1966 sugar phosphates, and sugars in fish muscle. *J. Fd. Sci.*, 31, 846–854

——Postmortem degradation of glycogen and starch in fish muscle. *J.* 1968 *Fish. Res. Bd. Can.*, 25, 1539–1554

Tokunaga, T. Trimethylamine oxide and its decomposition in the 1970 bloody muscle of fish. 2. Formation of DMA and TMA during storage. *Bull. Jap. Soc. Sci. Fish.*, 36, 510–515

—— On the thermal decomposition of trimethylamine oxide in muscle 1975 of some marine animals. *Bull. Jap. Soc. Sci. Fish.*, 41, 535–546

Tokunaga, T, Iida, H and Nakamura, K. Formation of dimethyl sul-1977 fide in Antarctic krill, *Euphausia superba*. *Bull. Jap. Soc. Sci. Fish.*, 43, 1209–1217

Tozawa, H and Sato, M. Formation of dimethylnitrosamine 1974 (DMNA) in sea foods. 1. Effects of hemoglobin and ascorbate on DMNA formation in Alaska pollock roe. *Bull. Jap. Soc. Sci. Fish.*, 40, 425–430

Uchiyama, S, Kondo, T and Uchiyama, M. Studies on the formation 1976 of volatile amines from trimethylamine oxide by γ-irradiation. 2. Influence of co-existing metal ions and protein. *Shokuhin-Eiseigaku Zasshi*, 17, 352–356

Watts, R L and Watts, D C. Nitrogen metabolism in fishes. *Chemi-1974 cal Zoology*. Academic Press, London and New York. Vol 8, 369–446

Wojtowicz, M B and Odense, P H. Comparative study of the muscle 1972 catheptic activity of some marine species. *J. Fish. Res. Bd. Can.*, 29, 85–90

Wong, N P, Damico, J N and Salwin, H. Investigation of volatile 1967 compounds in cod fish by gas chromatography and mass spectrometry. *J. Assoc. Offic. Anal. Chemists*, 50, 8–15

Yabe, K, Tsujino, I and Saito, T. Studies on the compounds specific 1966 for each group of marine algae. 4. Occurrence of homarine and trigonelline in red algae, *Tichocarpus crinitus*. *Bull. Fac. Fish. Hokkaido Univ.*, 16, 273–277

Yamada, K. Occurrence and origin of trimethylamine oxide in fishes 1967 and marine invertebrates. *Bull. Jap. Soc. Sci. Fish.*, 33, 591–603

—— Post-mortem breakdown of trimethylamine oxide in fishes and 1968 marine invertebrates. *Bull. Jap. Soc. Sci. Fish.*, 34, 541–551

Yamamoto, Y, Ishii, T and Ikeda, S. Studies on copper metabolism in 1978 fishes. 3. Existence of metallothionine-like protein in carp hepatopancreas. *Bull. Jap. Soc. Sci. Fish.*, 44, 149–153

Yamanaka, H. Orange discolored meat by canned skipjack. 4. Causes 1975a on the accumulation of G6P and F6P(1). *Bull. Jap. Soc. Sci. Fish.*, 41, 217–223

—— Orange discolored meat by canned skipjack. 5. Amino com-1975b pounds responsible for orange discoloration. *Bull. Jap. Soc. Sci. Fish.*, 41, 357–363

—— Orange discolored meat by canned skipjack. 6. Causes on the 1975c accumulation of G6P and F6P(2). *Bull. Jap. Soc. Sci. Fish.*, 41, 573–578

Yamanaka, H, Bito, M and Yokoseki, M. Orange discolored meat of 1973 canned skipjack. 2. The main compound responsible for orange discoloration. *Bull. Jap. Soc. Sci. Fish.*, 39, 1299–1308

Yamanaka, H, Nakagawasai, T, Kikuchi, T and Amano, K. Studies 1978 on the contraction of carp muscle. 1. Remarkable differences between *rigor mortis* and thaw *rigor*. *Bull. Jap. Soc. Sci. Fish.*, 44. 1123–1126

Yanase, M. Chemical composition of *Euphausia superba* and its util-1971 ization as condensed solubles for human food. *Bull. Tokai Reg. Fish. Res. Lab.*, No. 65, 59–66

Yancey, P H and Somero, G N. Urea-requiring lactate dehydro-1978 genases of marine elasmobranch fishes. *J. Comp. Physiol.*, 125. 135–141

123

YURKOWSKI, M and BORDELEAU, M A.
1965   Carbonyl compounds in salted cod. 2. Separation and indentification of volatile monocar-bonyl compounds from heavily salted cod. *J. Fish. Res. Bd. Can.*, 22, 27–32

Fish in human and animal nutrition

Kenneth J Carpenter

1 Intensive animal production

This is a huge subject. It is easier to begin with animal nutrition because modern animal production has rationalized the formulation of diets. To take the extreme case, the broiler chicken is killed and marketed at about eight weeks of age. In general, the fastest possible growth is also the most economic. And 'fast' means doubling its initial body weight five to six times in that short period. To sustain that rate of growth the chick (or any animal) needs a highly concentrated diet, with at least 20% of its energy coming from protein and special provision of vitamins and minerals. For egg production pullets are kept indoors and will lay five times their own weight of eggs (*ie* 250–300) in a year, and are then usually killed.

Complete, dry mix diets can be prepared in stable form and in any season of the year to meet the needs of these animals. Most of the continuing studies are not to produce better diets, but to produce equally good ones at a lower cost. The major cost items are the cereals used as the staple source of energy and the protein concentrates, mostly from oil seeds plus some animal products, and possibly a synthetic amino acid. Topping up the mixture with a standard pre-mix of extra vitamins and minerals is relatively cheap and done as a routine. Since the relative abundance of individual products varies from one harvest to another, and market prices are continually changing, compounders use flexible formulae produced by 'least-cost' computer calculations. It is the professional nutritionist's job to programme the computer, specifying minimum nutrient levels for the final mix and corresponding analytical data for each possible ingredient, with an upper limit for its use in some cases. The specifications for 'protein' are usually expressed as so much total nitrogen and lysine, 'methionine and cystine' and possibly threonine (the essential amino acids most commonly limiting in mixed diets).

Some 35% of the world's fish catch is used for reduction to fish meal (and oil) and the great majority of the meal is fed to poultry. It is the high protein content (60–75%) and its good quality that makes the meal a valued ingredient – usually commanding a higher price than any other protein concentrate except milk powder. In elasmobranchs a significant proportion of the crude protein ($N \times 6.25$) is urea, but they are not used to a significant extent for fish meal.

The protein quality factor comes partly from fish meal protein having a higher lysine content than most vegetable proteins; and the cereal proteins are particularly deficient (FAO, 1970). It is the muscle protein in fish that is richest in lysine; connective tissue is poorer (Arnesen, 1969). Whole fish, which provide most of the world's fish meal, have of course a predominance of muscle protein, but even filleter's offal, a second source of meal, still has a large proportion of muscle (Carpenter *et al*, 1957). Since the lysine in protein is subject to

heat-damage through its ε-NH_2 group, there has obviously been concern as to whether the heat used in drying fish meal was seriously damaging. 'Flame-drying' sounds particularly horrific, but as long as a 'crumb' of material is still moist, whatever the temperature of the hot air around it, the crumb will be cooled by the evaporation of its water (Delort-Laval and Zelter, 1963). Where meal was, in past years, found to have a low available lysine content, this was due, it is now realized, to storage conditions where the temperature rose as a result of spontaneous exothermic oxidation of its oil (*cf* review by Carpenter and Booth, 1973).

Fish oil is a valuable raw material for hydrogenation and use in direct human consumption as margarine, and most of the world's fish meal is made from pelagic species rich in oil. The processors aim therefore to remove as much oil as possible by steam-cooking the fish, pressing and centrifuging. This recovers some 80% of the total oil, but the dried fish meal still contains some 10% oil on analysis. With this oil now being exposed to contact with oxygen, the most unsaturated fatty acids will begin to oxidize. Traditionally, piles of newly made meal were turned daily for it to 'cure' (*ie* oxidize) without over-heating. The oxidized fat was largely undigested by poultry, but harmless (Lea *et al*, 1966). With the advent of very large scale manufacture in Peru using anchovy with its particularly reactive oil, the practice of adding an anti-oxidant, usually 'ethoxyquin', to the meal was developed. This avoided the need for 'curing', and the risk of overheating and damage to lysine, and the oil was deposited in poultry tissue to a greater extent.

The upper limit to the use of fish meal in a poultry diet depends on the risk of 'fishy' or 'kippery' flavour appearing in meat in which the characteristically fishy fatty acids have been laid down and oxidize during cooking (Wessels *et al*, 1973). The common upper limit is an amount of meal contributing 1% fish oil to the diet (*ie* about 10% meal). In countries importing meal this usually has no practical effect since economics impose a lower level of use, in combination with vegetable proteins. But in some producing countries such as South Africa and Peru, with policies of restricting imports of vegetable protein concentrates, higher levels are used and people become accustomed, it seems, to a kippery taste to their chicken meat.

It is argued by some that the whole business of using fish for feeding to animals rather than directly to humans is immoral, especially as much of the fish is caught where people are poor, and the meal is exported to wealthier countries. Being scientists gives us no exemption from trying to promote the general good, but as with so many moral issues it is not obvious where that lies, nor is the issue of the inefficient cycling of potential food through animals confined to fish. If western man were to consume more of his calories directly as grain, and less as meat, there would be a large saving of grain. Some could be distributed to poorer people and/or the total crop acreage could be reduced, sparing the world's reserves of fuel and fertilizers. Of course, neither 'fish' nor 'grain' is a homogeneous item, uniformly suitable as human food.

2 The contribution of fish in the UK diet

This leads us on to the role of seafood in human diets. There have been two long-standing ideas in the public mind. White fish is recommended, especially for invalids, as being more easily digested. This really means 'leaving the stomach more rapidly', as would be expected with a food low in fat, and having muscle fibres that separate easily.

Another belief is that fish, and especially oysters, are aphrodisiac. Again, this could possibly be explained by the low fat and total calorie content of oysters, and their not being cooked in fat or eaten with other things; so that post-prandial 'digestion' does not have to take its usual priority. Certainly the idea has a long history. Brillat-Savarin (1884), the famous French gourmet and writer on transcendental gastronomy, was a believer in this characteristic of fish in general, and quotes an early experiment. The Sultan Saladin (famous for fighting the Crusaders) was puzzled by religious devotees pledging themselves to asceticism. He seemed to regard it as a pathological effect of malnutrition. To test his theory he brought two men in from their caves in the desert, fed them well on meat for two weeks, then had two of his charming concubines put in their room. To his surprise, they resisted the ladies' blandishments. For the next two weeks he fed them on fish and challenged them again; this time they were unable to resist. It is an interesting experimental design, and shows some degree of replication, though it clearly should have been better balanced to avoid confounding time and diet. No better experiment seems to have been reported.

The more conventional, modern work has concentrated on the nutrients contributed by different groups of foods in relation to our needs. As for animals, estimates have been made of our requirements for the individual nutrients and these, with some safety margin added, are expressed as Recommended Daily Allowances (RDA). Even in early childhood, the human is extremely slow growing, gaining perhaps 5g per day when weighing 16kg (compared to a gain of 400g per day by a piglet of the same weight) and taking years rather than weeks to double its weight. Maximum gain, which comes in adolescence, is 15g per day. Since tissues contain approximately 20% protein, these correspond to gains of 1g and 3g protein per day, respectively. As a consequence children do not need such a high ratio of individual nutrients to total calories in their diet as do young growing animals. Adults need even less and, with more precise knowledge, recommended minimum intakes, particularly for protein, have been reduced. Past and present recommendations for children are summarized in *Table I*.

Again, in order to calculate the adequacy of diets, analytical values are needed for each food item. These must include the trace nutrients since possible deficiencies are not made up with vitamin-mineral mixes as in

Table I

CHANGES IN THE AMOUNTS OF PROTEIN AND ENERGY CONSIDERED
SATISFACTORY FOR YOUNG CHILDREN (1–3 YEARS OLD)[1]

Year	Protein g/kg body weight/day	Energy kcal/kg body weight/day	Percentage of calories from protein
1948	3·3	100	13·2%
1963	2·5	100	10·0%
1965	1·5	—	—
1968	2·1	92	9·1%
1969	1·7	105	6·5%
1973	1·7	102	6·8%

[1] The data are taken from Whitehead (1973) except for the final estimates which were those of FAO/WHO (1973)

animal feeding. Analysing materials for food tables is an enormous, tedious job, but *The Chemistry of Flesh Foods and their Losses on Cooking* (McCance and Shipp, 1933) is a classic and fascinating account of people applying the scientific method of investigation to this humdrum subject. That the work is never finished is illustrated by the recent finding of Dyer *et al* (1977) that frozen 'fish sticks' may contain 500mg Na/100g, compared with the '60mg' expected. This was explained by processors having added sodium polyphosphate to reduce drip.

There are always new questions being raised, improved methods of analysis for micro-nutrients and so on. Thus Gordon *et al* (1979) have just published a large study on the B-vitamin levels in Pacific Coast fish of different sizes, and on the effect of frozen storage. There have been other recent studies of the quality of the protein in fish prepared in different ways, including smoking and salting (Del Valle *et al*, 1976; Vervack *et al*, 1977; Bodwell and Womack, 1978) and using shredded tissue machine-picked from filleters' offal for fish sticks (Crawford *et al*, 1972). In general the differences have proved to be small, though Hoffman *et al* (1977) have found significant damage to the lysine in some heavily smoked and dried tropical fish.

Nearly every method of preparing fish results in the liver and viscera being discarded; nor are the bones eaten. In this way the value of the food as a source of vitamins (particularly A and D) and of calcium and phosphorus is much reduced (Higashi, 1962). The fat content of the flesh varies greatly with species, from less than 1% to over 25%, with the percentage of total calories coming from protein ranging inversely from 95 to 24 (Geiger and Borgström, 1962). At one time the protein of fish flesh had been thought to be inferior to that of carcass meat, because it was lower in tryptophan, but this has not been confirmed either by later analyses or in feeding trials (Konosu and Matsuura, 1962; Miller *et al*, 1965).

To place the contribution of nutrients from fish into some perspective, we must consider their part in a total diet and how far that diet meets man's requirements. Records for the UK are collected in systematic, large-scale surveys by a Government Department (National Food Survey Committee, 1978). I will consider these first and then return later to contrasting situations from the developing countries. Data taken from the most recent survey are set out in *Table II*. These values refer, of course, to hypothetical 'average' people. They are eating about 15g fish/day which make up 1% of their calories, and by chance this percentage corresponds to

	Recommended minimum daily intake[1]	% of recommendations supplied by:			
		(a) Total food	(b) Fish products	(c) Dairy products	(d) Meat, poultry etc
Energy (kcal)	2 400k cal	94[2]	1·0	14·8	16·2
Total protein, g	39·3g	184[3]	3·5	23·9	31·1
Calcium, mg	550mg	183	1·1	60·4	2·4
Iron	11mg	98	1·6	2·8	24·1
Thiamine	0·95mg	129	0·8	13·0	13·5
Riboflavin	1·41mg	128	1·0	38·5	20·0
Nicotine acid equivalents	15·7mg	185	3·3	15·2	35·2
Vitamin C	29·2mg	178	0·1	8·3	1·5
Vitamin A (retinol equivalents)	700μg	210	0·1	15·4	37·1
Vitamin D	3·19μg	83[4]	13·9	9·7	1·3

[1] These data are calculated to represent the requirement of the averags mix of men, women and children in a household
[2] Significant numbers of calories are consumed outside the household in sweets, ice cream, beverages, etc
[3] There is also a higher protein standard equivalent to 60g/head/day for use as a basis in menu planning. The supply corresponds to 120% of this
[4] The vitamin D standard makes no allowance for the considerable amount of the vitamin synthesized under the skin receiving direct sunlight. It is thought to be acceptable to have only 60% of the RDA provided by the diet

Source: National Food Survey Committee, 1978

the world average also. Fish provides 3% of the protein but both protein and calories are already present in the diet at above RDA levels. Nor do any of the other nutrients in fish seem to have a greater significance here. We cannot therefore make a case for the necessity of the fish in 'nutrient' terms, unlike the milk group with its riboflavin and meat with its iron. It does, of course, continue to provide pleasure and variety.

That sounds almost shocking – we belong to a basically puritan culture and like to feel that as scientists we are concerned with a desperate struggle to produce the bare necessities of life, rather than its frivolities. But let us not be hypocrites. As Burgess (1977) has said elsewhere, fish is no longer the cheap food it was in 1929 when people were eating three times as much. Some relative costs of buying protein from different foods is set out in *Table III*. In any case, commercial fishing has always had a cost; first a human cost and then everything concerned with their harvest and marketing uses fuel; to quote one estimate, 18 calories for each nutritional calorie of cod fillet (Rawitscher and Mayer, 1977). Many animal protein foods cost even more in fuel, and it cannot be argued that they are all necessities, but we take the pleasures that we feel we can afford – food, drink, travel or whatever.

Table III
PROTEIN CONTAINED IN THE AMOUNTS OF DIFFERENT FOODS BOUGHT FOR ONE PENNY

	g		g
Liver	1·8	White bread	2·9
Cheese	1·8	Canned beans	1·5
Milk	1·7	Breakfast cereal	1·3
Eggs	1·5	Potatoes	1·0
Poultry	1·4		
Fat fish	1·0		
Beef	1·0		
Pork	0·9		
White fish	0·8		

Source: National Food Survey Committee, 1978

3 Problem with the affluent diet

So far we have only considered the UK diet in terms of whether or not it meets RDAs. Are these the only criteria to be met for a diet to be a healthy one? In the past, nutritional advances have mostly depended on short-term experiments using young animals, with rapidity of weight gain as the measure of response – and implicitly as the ideal. The recognized results of this work include the identification of the vitamins and thus the rational prevention of deficiency diseases in man. This was a great achievement to which many fine workers dedicated their lives, but we must not, for that reason, regard it as the 'whole truth' of the subject.

The very successes generated the application of the '100-yard dash' mentality to human nutrition; the attitude of increasing nutrient intake so as to provide the greatest possible safety margin against any risk of deficiency, when really man is in a different event, more analogous to the 50-mile walk.

I remember – it was in 1955 – being shown round the Harvard medical students' common room which had a constant, free supply of full-cream milk – provided from a benefactor's will – so that whenever a student was thirsty he would drink milk rather than water, and so increase his calcium and vitamin intake. No thought was given to the extra butter fat and other calories consumed in response to thirst rather than hunger. And a large surplus of body fat (from excess calorie intake) is very difficult to shed and, statistically, an impediment to a long and vigorous life.

This is not really in contradiction to the experience gained from feeding animals. It is not the laboratory rat reared on a high-protein diet, or kept indefinitely on *ad libitum* feeding, that has the longest life span. Nor does the knowledgeable farmer continue to provide a concentrated diet for the pigs that he decides to keep on for breeding. The dry sow, *ie* between lactation periods, is commonly given a small, grain-based ration and for the rest, left to satisfy her appetite with grass or green leaves (which yield more bulk than calories).

Public health statistics indicate that the more affluent and educated, with their diet virtually unrestricted by lack of purchasing power, are not on average living longer. Although fatal infectious disease is now only a minor problem, there is in each age range a higher incidence of cancer and coronary heart disease (CHD). The major, known factor contributing to these statistics is cigarette smoking, but there are strong indications that diet is also a factor (*eg* Department of Health and Social Security, 1974). Investigators can only hunt for any sort of clue as to what detrimental factors may be at work now that were not present to the same extent 50 years ago, and how to make the diet safer.

Most of the modern investigations involving the consumption of fish products by man have turned on their possible helpful effect for people at risk from CHD. The basic cause of the disease has been an area of great controversy, but it is agreed that obstruction of flow in an artery feeding the wall of the heart starts from the deposition in the vessel of cholesterol-containing plaques. Further, there is consistent evidence that having a high level of cholesterol circulating in one's serum is one important factor increasing the risk of the disease (review by Truswell, 1978). Although the body can synthesize cholesterol, it is generally thought prudent for those in vulnerable groups to limit their intake of cholesterol-rich foods such as eggs. Some doctors have also forbidden shellfish to their patients since Connor et al (1963) reported that eating them raised the levels of serum cholesterol in rabbits; however, Schulze and Truswell (1976) have done further analyses and suggest that such a restriction is not really necessary. A major factor increasing cholesterol levels in the blood is a high level of saturated fat in the diet; in practice this comes mainly from ruminant animal products.

Rats will only show hypercholesterolaemia when stressed with high levels of cholesterol and bile salts added to their diet. Under these conditions, adult rats with their cholesterol levels already raised, showed a return to normal when 10–25% freeze-dried whole fish was added to their diet at the expense of casein and beef tallow (Peifer et al, 1962). The same result was obtained with the lipid fraction obtained from the dried fish by methanol-hexane extraction, whereas the residue was inactive. This was true for all four species used–menhaden, salmon, mullet and perch. Further experiment indicated that it was the fatty acid fraction of the oils, rather than the 'unsaponifiables', that was responsible for the effect (Peifer et al, 1965). Further, from using fractions of menhaden oil with different unsaturation (as measured by iodine value), it seemed that fish oil was more potent than plant oils of corresponding IV.

Stansby (1969) has reviewed these and other papers and concluded that the cholesterol-lowering effect has been shown with oil from every species of fish tested, and each species of experimental animal. This also extends to man (eg Ahrens et al, 1959; Kingsbury et al, 1961). In a recent human study (von Lossonczy et al, 1978), using monks and nuns who ate 200g mackerel (contributing 54g lipid) per day for three weeks, their serum cholesterol levels had fallen significantly as compared with values obtained after a similar period with cheese as the dietary supplement. The average difference was 16mg/100ml (ie about an 8% fall). Also the high-density lipoprotein fraction actually rose by about 6%; and an increase in this material is now thought to be favourable, and to be associated with a reduced risk of CHD (Gordon et al, 1977).

One medical practitioner who specializes in the dietary treatment of patients who already have coronary heart disease, and whose regimen centres on the replacement of meat by seafood, has presented data indicating almost a doubling of average survival (from five to eleven years) in the 80 patients who followed the treatment, as compared with the 76 who did not diet (Nelson, 1972). Obviously any such study is difficult to assess when subjects are not allocated to treatment at random.

The other approach, with equal problems, is the epidemiological one. One group of people with low serum cholesterol values are the Greenland Eskimos, even compared with Eskimos who have emigrated to Denmark (Dyerberg et al, 1977). CHD is uncommon in this community. Analysis of their diet (high in seal and whole meat) shows a fatty acid pattern similar to that of a high-fish diet (Bang et al, 1976). This is presumably because aquatic food chains are passing on a largely common pattern of fatty acids. Obviously these studies show us that a high intake of fat of a 'fishy' type can be consistent with a low risk of CHD, but we cannot say that it is the cause of this relative immunity because of the many other special features of life in Greenland.

According to Borgström (1962), the highest recorded consumption of fish was amongst the community on the island of Tristan da Cunha in the South Atlantic, when studied by a Norwegian expedition in 1937 (Henriksen and Oeding, 1946). At that time they were still living largely on their own resources and life was hard. The colonies of seals and of seabirds that used to provide much of the people's diet had been almost wiped out, and they were reduced at some seasons to a diet of fish and potatoes, with the potatoes as the greater luxury. Their estimated consumption of fish, including shellfish, was 300g/head/day. Calorie intakes were low and the people were generally quite lean.

The doctors in the expedition were impressed by the good state of the people's health. Teeth were sound though mottled as we would now expect with little sugar to be had, and a high fluorine intake. Numbers were small and, of course, there was little opportunity for spread of infection, but the general impression was of healthy old age with some evidence of tumours in their records, but not much. Certainly there was no indication that so much fish had been harmful. Taylor et al (1966) also review the earlier records, and point out that the islanders' diet changed entirely after 1940.

4 Investigations of possible toxicity problems

Within the last few years, as Ackman (1980) will already have recounted, a different line of investigation has led to a query as to the safety of partially hydrogenated fish oils (PHFO) in relation to heart disease. The history behind this has been summarized by FAO (1978). In short, rapeseed oil, fed at high levels to young rats, has been found damaging to their hearts; this showed up particularly under cold stress in one experiment (Darnerud et al, 1978) and is, at least largely, due to its high content of erucic acid. This acid survives the partial hydrogenation process used for incorporating rapeseed oil into margarine. As a consequence of this, plant breeders have developed new seeds of negligible erucic acid. The only other food fats rich in C22 acids are some of the fish oils; and with partial hydrogenation for their common use in margarine and mayonnaises, a large proportion of these are converted to a C22:1 acid, 'cetoleic' with its cis double bond at the n-11 position, as compared with the n-9 position in erucic acid.

A particularly interesting recent study (Loew et al, 1978) compares the results of feeding monkeys for four months on diets containing 25% fat−either rapeseed or partially hydrogenated herring. Each oil contained 24% C22:1 fatty acids. A control treatment contained a 3:1 mix of lard and corn oil. During the trial, serum from the 'rapeseed oil' animals showed elevated levels of cholesterol, triglycerides and glutamic oxaloacetic transaminase (an enzyme produced within cells so that leakage into serum is used as an indicator of cell damage in some conditions); with the fish oil the levels were all normal. All the monkeys survived the trial in apparently normal health. However, post mortem study of both heart and skeletal muscle showed severe lipidosis in the 'fish oil' animals (compared with mild-to-moderate lipidosis in the controls). The rapeseed oil tissue showed the most severe lipidosis; there was also approximately twice as much (33%) of C22:1 in the heart triglycerides of the 'rapeseed' group as of the 'fish oil' group.

There is not even a hint from any epidemiological studies that populations with a relatively high intake of C22:1 acids have more myocardial lesions. So what can people with responsibilities for public health and food regulations reasonably conclude about fish oils from this and other data? FAO (1978) recommends urgent studies of heart tissues from different populations and suggests that until the picture becomes clearer it would be prudent for fish oils to be blended with other oils in the manufacture of margarine. I do not see what else they could do, but it is hard to be at the receiving end of a 'presumed guilty until proven innocent' attitude.

With the proliferation of research institutes, and scientific periodicals, one begins to feel that every possible claim will eventually be published. Certainly a Russian group has recently reported that rats receiving canned fish had an exceptionally high incidence of tumours (Neiman et al, 1978) and another group has found that certain fish products, when included in a high-salt diet, can raise the blood pressure of rats (Chakkapak and Lichton, 1970). Any food high in salt is contraindicated for someone with a tendency to elevated blood pressure, and it has been suggested that the high incidence of this condition in Japan may be explained by the popularity of salted foods there.

Last in the series of possible troubles is the presence of toxic materials that come from contamination of water and the inability of fish to re-excrete them, so that they are concentrated in going from one trophic level to another in the sea. The same materials−methyl mercury, cadmium, lead, chlorinated organic pesticides−will accumulate in human tissues too, over long periods. In the past it was thought safe for toxic materials to run to waste into the sea or even a large lake because of the almost limitless dilution there. However, aquatic life has extraordinary powers of concentrating solutes−the chalk cliffs of Dover are a monument to this ability. The clearest example of such poisoning from fish is 'Minamata disease' in Japan, found amongst people eating fish from an estuary with waste water running into it from a factory using a process that involved mercury (review by Clarkson et al, 1976). It was established that the 150 or so deaths were from methyl mercury poisoning.

Following this tragedy many more analyses for heavy metals have been carried out on fish (eg Zook et al, 1976). Also, acceptable daily intakes for them and for chlorinated organic compounds have been reconsidered (review by Munro and Charbonneau, 1978), and some fishing from some stocks has ceased as a result. Further, there is always a possibility that a chemical fairly harmless in itself, for example sodium nitrite added as a preservative, may combine with another constituent of the food−in this example trimethylamine−during processing to produce a carcinogen, dimethylnitrosamine, as in some batches of fish meal when the fish were preserved with nitrite (Koppang and Helegebostad, 1966). Fortunately it was only mink and sheep that died before the possibility of this happening in a food was appreciated.

Nutritionists, when they were only concerned to 'top up' the public's diet, with a good breakfast to start the day, a vitamin pill after lunch and so on, were fairly popular fellows. That they are now looking critically at possible dangers in all sorts of tasty things from fried eggs (for their cholesterol) to kippers (for their salt and dye), makes them targets of popular ridicule for fussing over the insignificant. Probably any particular 'fuss' will turn out to be nothing, but somewhere there is something to be found and common sense will not serve as a guide. The power of micro-organisms invisible to the eye to bring death, and of a vitamin in parts per million of the diet to bring life, seemed equally nonsensical when first proposed.

5 Fish in the developing countries

What the protein requirement of a population is thought to be has a profound effect on what kind of food it is thought to be most in need of for its nutritional standards to be improved. Twenty years ago it was repeated at meeting after meeting that the number one priority was the 'protein problem' or the 'impending world protein crisis'. To give just one quotation of a complete paragraph:

> 'About 60% of the world's population receive fewer than 2 200 calories per day, and 80% must be content with less than 30g of animal protein per day. Over 60% have a daily ration of less than 15g animal protein. These are disquieting figures, and it is no secret that millions of people die from malnutrition owing to shortage of animal protein even today.'

This is taken from the Washington conference on 'Fish in Nutrition' (Meseck, 1962) and the speaker draws as his conclusion: 'In short, I re-emphasize that a comparatively small effort put into fisheries development will quickly result in improved standards of living and nutrition'.

Certainly all the statistics still show that, whereas people in the more affluent countries of the world eat about 90g protein per day, of which half is 'animal', the typical member of an underdeveloped country may be eating no more than 45g total protein of which only 8g is 'animal'. From there we come into an area where statements of 'what is', and value judgements as to what people 'should do', become intertwined. The background is discussed elsewhere (Porter and Rolls, 1973); nutritionists now believe that in a typical impoverished community with cereal as its staple food and with little fat available, the population four years old and over would be expected to do very well if they could just buy enough of their ordinary pattern of diet to satisfy their appetite together, in some areas, with supplementation with a trace nutrient, vitamin A, B$_1$ and/or iodine. This is illustrated by a classic experiment in which malnourished children receiving only 8·8g animal protein per day grew at a 'catch-up' rate when allowed to eat 85% extraction wheaten bread *ad libitum*. (Widdowson and McCance, 1954).

For the children under four years of age there is often a problem of their diet (*eg* maize gruels) being too bulky for them to eat sufficiently at two or three family meals per day. For them some means of concentrating or enriching the diet by boiling the cereal with milk rather than water for example, would make the difference. The toddlers should therefore have priority for what milk is available. For the community as a whole, if the overall food supply is increased, the protein will take care of itself.

How does this relate to the role of fish in the nutrition of poor people in the developing countries? Their situations are, of course, very diverse. At one extreme we may imagine coastal villagers who can only paddle canoes out to the nearby fishing ground in the calmest season. With an outboard motor they could fish all year, increase their food supply and repay the cost by selling fish surplus to their need to neighbours a few miles inland. This sounds the sort of thing FAO is organizing, encouraged by nutritionists and economists alike so long as the resource is not likely to become depleted.

But millions of poor are living in inland villages, quite unable to afford the high cost of bringing fresh fish from the nearest source of supply. It was for such people that there have been many attempts to organize the production and supply of protein concentrates—from leaf material, single-celled organisms and, of course, fish. There is a very interesting review (Pariser *et al*, 1978) of the history of projects to make and market fish flour or protein concentrate (FPC), with emphasis on the initial enthusiasm followed by gradual disillusionment of those concerned with the big US Government project in the 1960s. In essence, each project has been prompted by the feeling that, since fish meal can be made cheaply enough to be fed as a high quality protein supplement to chickens, it could and should be available, in up-graded form, for children and others suffering from malnutrition. This raises several quite different issues, *all* of which have to be satisfied for the idea to work.

It has to be remembered that the chicken has important differences from humans that are relevant here. It seems insensitive to the taste of its food, does not masticate in its mouth and welcomes grit in its food, and is more tolerant of food-borne pathogens. In practice the cost of a product meeting standards set for human food has proved to be several times that of ordinary fish meal. However, to be thinking only about these problems represents, as Pariser *et al* (1978) have said, a kind of technological myopia:

'The production of FPC is a high-technology enterprise, requiring massive inputs of capital, energy, and expertise. All of these inputs are characteristically lacking in the developing world. The attempt to transfer FPC technology demonstrates that technological solutions that seemplausible in a developed context are less advantageous and usually inappropriate when transferred to a developing context.'

Perhaps at another level it reflects the willingness of scientists in the past to turn their backs to the economic facts of life and to have little sympathy or wish to understand the customs and values of people living in quite different societies from our own. The first generation of work on this problem has been almost entirely fruitless. The challenge to fisheries to make a greater contribution to the food needs of the poor remains, but it is obviously not an easy one.

6 References

ACKMAN, R G. Fish lipids. In: *This volume*
1980
AHRENS, E H, INSULL, W, HIRSCH, J, STOFFEL, W, PETERSON,
1959 M, FARQUAR, J W, MILLER, T and THOMASSEN, H J. The effect on human serum-lipids of a dietary fat, highly unsaturated but poor in essential fatty acids. *Lancet*, 115
ARNESEN, G. Total and free amino acids in fish meals and vacuum-
1969 dried codfish organs, flesh, bones, skin and stomach contents. *J. Sci. Fd. Agric.*, 20, 218
BANG, H O, DYERBERG, J and HJØRNE, N. The composition of food
1976 consumed by Greenland Eskimos. *Acta. Med. Scand.*, 200, 69
BODWELL, C E and WOMACK, M. Effects of heating method on protein
1978 nutritional value of five fresh or frozen food products. *J. Fd. Sci.*, 43, 1543
BORGSTRÖM, G. Fish in World Nutrition. In: *Fish as Food*, Ed G
1962 Borgström. Academic Press, London. Vol 2, 267
BRILLAT-SAVARIN, J A. A Handbook of Gastronomy (English transla-
1884 tion of 'Physiologie du Goût'). Nimmo and Bain, London. 119
BURGESS, G H O. Making the most of fish supplies. *Proc. Nutr. Soc.*,
1977 36, 285
CARPENTER, K J and BOOTH, V H. Damage to lysine in food process-
1973 ing: its measurement and its significance. *Nutr. Abstr. Rev.*, 43, 423
CARPENTER, K J, ELLINGER, G M, MUNRO, M I and ROLFE, E J. Fish
1957 products as protein supplements to cereals. *Brit. J. Nutr.*, 11, 162
CHAKKAPAK, M S and LICHTON, I J. Elevation of systolic blood
1970 pressure in rats fed diets containing fish and soybean proteins. *J. Nutr.*, 100, 1081
CLARKSON, T W, AMIN-ZAKI, L and AL-TIKRITI, S K. An outbreak of
1976 methyl-mercury poisoning due to consumption of contaminated grain. *Feder. Proc.*, 35, 2395
CONNOR, W E, ROHWEDDER, J J and MEAK, J C. Production of hyper-
1963 cholesterolaemia and antherosclerosis by a diet rich in shellfish. *J. Nutr.*, 79, 443.
CRAWFORD, D L, LAW, D K and BABBITT, J K. Nutritional character-
1972 istics of marine food fish carcass waste and machine-separated flesh. *J. Agr. Fd. Chem.*, 20, 1048
DARNERUD, P O, OLSEN, M and WAHLSTROM, B. Effects of cold stress
1978 on rats fed different levels of docosenoic acids. *Lipids*, 13, 459
DELORT-LAVAL, J and ZELTER, S Z. Effect of the method of flame-
1963 drying on the biological value of herring meals. *Ann. Zootech.*, 12, 193
DEL VALLE, F R, BOURGES, H, HASS, R and GAONA, H. Proximate
1976 analysis protein quality and microbial counts of quick-salted, freshly made and stored fish cakes. *J. Fd. Sci.*, 41, 975
DEPARTMENT OF HEALTH AND SOCIAL SECURITY. Diet and coronary
1974 heart disease. *Report on Health and Social Subjects* 7. HMSO, London
DYER, W J, HILTZ, D F, HAYES, E R and MUNRO, V G. Retail frozen
1977 fishery products—proximate and mineral composition of the edible portion. *J. Inst. Can. Sci. Technol. Aliment.*, 10, 185
DYERBERG, J., BANG, H O and HJØRNE, N. Plasma cholesterol concen-
1977 tration in Caucasian Danes and Greenland Eskimos. *Danish Med. Bull.*, 24, 52
FAO. Amino-acid content of foods and biological data on proteins.
1970 *FAO Nutritional Studies* No. 24. FAO, Rome
—— Dietary fats and oils in human nutrition. *FAO Food and Nutrition*
1978 *Paper* No. 3. FAO, Rome
FAO/WHO. Energy and protein requirements. *FAO Nutr. Meeting*
1973 *Series*, No. 52. FAO, Rome
GEIGER, E and BORGSTRÖM, G. Fish protein-nutritive aspects. In: *Fish*
1962 *as Food*, Ed G Borgström. Academic Press, New York. Vol 2, 29

GORDON, D T, ROBERTS, G L and HEINTZ, D M. Thiamine, riboflavin
1979 and niacin content and stability in Pacific Coast seafoods. *J. Agric. Fd. Chem.*, 27, 483

GORDON, T, CASTELLI, W P, HJORTLAND, M C, KANNEL, W B and
1977 DAWBER, T R. High density lipoprotein as a protective factor against coronary heat disease. *Amer. J. Med.*, 62, 707

HENRIKSEN, S D and OEDING, P. Medical survey of Tristan da Cunha.
1946 In: *Results of the Norwegian Scientific Expedition to Tristan da Cunha.* Jacob Dybwad, Oslo. Vol I, No. 5

HIGASHI, H. Vitamins in fish – with special reference to edible parts.
1962 In: *Fish as Food*, Ed G Borgström. Academic Press, New York. Vol 1, 411

HOFFMAN, A, BARRANCO, A, FRANCIS, B J and DISNEY, J G. The
1977 effect of processing and storage upon the nutritive value of smoked fish from Africa. *Tropical Sci.*, 19, 41

KINGSBURY, K J, MORGAN, D M, AYLOTT, C and EMMERSON, R.
1961 Effects of ethyl arachidonate, cod liver oil and corn oil on the plasma cholesterol level. *Lancet*, 739

KONOSU, S and MATSUURA, F. Tryptophan content of fish meat with
1962 special reference to the protein score. In: *Fish in Nutrition*, Eds E Heen and R Kreuzer. Fishing News (Books) Ltd, London. 261

KOPPANG, N and HELGEBOSTAD, A. Toxic hepatosis in fur animals. III.
1966 Conditions affecting the formation of the toxic factor in herring meal. *Nord. Ved.-Med.*, 18, 26

LEA, C H, PARR, L J, L'ESTRANGE, J L and CARPENTER, K J. Nutri-
1966 tional effects of autoxidised fats in animal diets. 3. *Brit. J. Nutr.*, 20, 123

LOEW, F M, SCHIEFER, B, LATDAL, V A, PRASAD, K, FORSYTH, G W,
1978 ACKMAN, R G, OLFERT, E D and BELL, J M. Effects of plant and animal lipids rich in docosenoic acids on the myocardium of Cynomolgus monkeys. *Nutr. Metab.*, 22, 201

McCANCE, R A and SHIPP, H L. The chemistry of flesh foods and their
1933 losses on cooking. *Med. Res. Council, Spec. Rept. Ser.* No. 187

MESECK, G. Importance of fisheries production and utilization in the
1962 food economy. In: *Fish in Nutrition*, Eds E Heen and R Kreuzer. Fishing News (Books) Ltd, London. 23

MILLER, E L, CARPENTER, K J and MILNER, C K. Chemical and
1965 nutritional changes in heated cod muscle. *Brit. J. Nutr.*, 19, 547

MUNRO, I C and CHARBONNEAU, S M. Principal hazards in food safety
1978 and their assessment: environmental contaminants *Feder. Proc.*, 37, 2582

NATIONAL FOOD SURVEY COMMITTEE. Household Food Consumption
1978 and Expenditure: 1977. HMSO, London

NELSON, A E. Diet therapy in coronary disease: effect on mortality of
1972 high-protein, high sea-food, fat-controlled diet. *Geriatrics*, December, 103

NEIMAN, I M, ANDRIANOVA, M M, BELOSHAPKO, A A, GORTALUM, G M,
1978 KOLOSNITSYNA, N W and FINOGENOVA, M A. An experimental study of the after-effects of feeding rats on some types of canned fish products. *Voprosy Pitaniya*, 5, 60. [Cited in N.A. & R., Ser. A, 49, 2969 (1979)]

PARISER, E R, WALTERSTEIN, M B, CORKERY, C J and BROWN, N L. *Fish*
1978 *Protein Concentrate: Panacea for Protein Malnutrition?* The MIT Press, Cambridge, Mass.

PEIFER, J J, JANSSEN, F, MUESING, R and LUNDBERG, W O. The lipid
1962 depressant activities of whole fish and their component oils. *J. Amer. Oil Chem. Soc.*, 39, 292

PEIFER, J J, LUNDBERG, W O, ISHIO, S and WARMANN, E. Studies of
1965 the distribution of lipids in hypocholesterolemic rats. 3. Changes in hypercholesterolemia and tissue fatty acids induced by dietary fats and marine oil fractions. *Arch. Biochem. Biophys.*, 110,270

PORTER, J W G and ROLLS, B A. *Proteins in Human Nutrition.* Acade-
1973 mic Press, London

RAWITSCHER, M and MAYER, J. Nutritional outputs and energy inputs
1977 in seafoods. *Science*, 198, 261

SCHULZE, A and TRUSWELL, A S. Sterols in British shellfish. *Proc. Nutr. Soc.*, 36, 25A

STANSBY, M E. Nutritional properties of fish oils. *World Rev. Nutr.*
1969 *Diet.*, 11, 46

TAYLOR, E C, HOLLINGSWORTH, D F and CHAMBERS, M A. The diet of
1966 the Tristan da Cunha Islanders. *Brit. J. Nutr.*, 20, 393

TRUSWELL, A S. Diet and Plasma lipids – a reappraisal. *Amer. J. Clin.*
1978 *Nutr.*, 31, 977

VERVACK, W, VANBELLE, M and FOULOU, M. The amino acid composi-
1977 tion of some fish products. *Revue des Fermentations et des Industries Alimentaires*, 32, 171

VON LOSSONCZY, T O, RUITER, A, BRONSGEEST-SCHONTE, H C, VAN GENT,
1978 C M and HERMUS, R J J. The effect of a fish diet on serum lipids in healthy human subjects. *Amer. J. Clin. Nutr.*, 31, 1340

WESSELS, J P H, ATKINSON, A, VAN DER MERWE, R P and
1973 DE JONGH, J H. Flavour studies with fish meals and with fish oil fractions in broiler diets. *J. Sci. Fd. Agric.*, 24, 451

WHITEHEAD, R G. The protein needs of malnourished children. In:
1973 *Proteins in Human Nutrition*, Eds J W G Porter and B A Rolls. Academic Press, London. 103

WIDDOWSON, E M and McCANCE, R A. Studies on the nutritive value
1954 of bread and on the effect of variations in the extraction rate on the growth of undernourished children. *Med. Res. Council, Spec. Rept Ser.* No. 287. HMSO, London

ZOOK, E G, POWELL, J J, HACKLEY, B M, EMERSON, J A, BROOKER, J R
1976 and KNOBL, G M. National Marine Fisheries Service prelimin- ary survey of selected seafoods for mercury, lead, cadmium, chromium and arsenic content. *J. Agr. Fd. Chem.*, 24, 47

Biological factors affecting processing and utilization *R Malcolm Love*

1 Introduction

In some ways the people who process chickens are to be envied. The moment a little chick hatches the date of its impending death is immediately noted, and for its short and intensive life the food, light, temperature and walking space are strictly controlled so that plastic bundles of almost identical foodstuff for man can be lined up on the shelf of a shop.

Not so for the fish processor. Fish can be handled and processed in a multitude of ways, after which the pro- duct in our mouths may show quite a range of flavours and textures. However, some of the greatest variations are present in the fish at the time it was caught. Frozen fish toughen with storage until they are no longer acceptable as food, but a fish caught on a particular ground on a certain date may be almost as tough even though it has never been frozen.

K J Whittle (unpublished, quoted by Love, 1975) thawed, cooked and examined some cod (*Gadus morhua*) which had been caught on five different grounds of the North Atlantic, frozen on the ship and stored at −30°C for just three months. Under these

circumstances, they should have been just about ideal, indistinguishable from very fresh fish. The samples from four of the grounds were indeed like this, but those from the Faroe Bank were so tough to eat as to be unaccept- able, and the cold-store flavour and cold-store odour were very much stronger than in the fish from the other grounds.

A processor of my acquaintance needed cod with a very white flesh to make into fish fingers, and was delighted with a ship-load of cod from a certain ground. He was asked by telephone whether he would like a load caught by the sister-ship in the same area just a few days later, and accepted with alacrity – only to find that when the fish actually arrived the flesh was so dark as to be useless for his special purpose.

Even the rates at which fish spoil when they are packed in melting ice vary according to the place and time of catching. A former director of Torry, Dr G A Reay, reported in 1955: 'Large differences are often observed in the spoilage behaviour of different catches which have been similarly treated. These differ- ences cannot be fully accounted for by any of the factors such as temperature, size, sex or maturity. An explanation must therefore be sought in intrinsic differences, possibly of a physiological or bacteriological character in the fish themselves'.

The following year it was reported (Reay, 1956) that fish caught in April, May and June seemed to spoil more rapidly than at other times. The differences in spoilage rate could apparently only sometimes be correlated with heavier initial bacterial loads, and unknown factors were again mentioned.

It was then reported (Reay, 1957) that factors not directly linked with the season were involved. Cod caught on the North Cape of Norway spoiled the most rapidly, those from the North Sea less rapidly and those from the Faroe Bank least of all, becoming inedible some six days later than North Cape cod. By this time, 'intrinsic quality' seemed to be acquiring a mystique of its own.

Now when bacteria invade fish muscle after death and make it unacceptable as a foodstuff, they break down the structural proteins to simpler substances, some of which have a bad taste. However, the main factor by which a fresh, seaweedy-smelling material acquires a 'fishy' or ammonia-like smell is the reduction by the bacteria of the tasteless trimethylamine oxide to the fishy-tasting trimethylamine. For some reason there is much more of the precursor substance trimethylamine oxide in various species of fish caught in the Atlantic than in those caught in the North Sea (Shewan, 1951), and there is also a marked seasonal variation in this substance, which was shown by Ronold and Jakobsen (1947) to range in the flesh of herrings (*Clupea harengus*) from a minimum in July and August to values three times as great in February. The cause of the variation was however unknown, as it still is today. Hughes (1959a) noticed a similar change in the same species, but values were minimal in June and July in herring caught off Peterhead (northeast Scotland) and minimal in September in those from near Inverness.

Ronold and Jackobsen (1947) thought that the presence of trimethylamine oxide would retard the desirable maturing process of herring in a can during post-canning storage, by inhibiting the formation of a reducing atmosphere. Hughes (1959a) on the other hand found that herrings with a high trimethylamine oxide content gave a good canned product. Trimethylamine oxide does, however, pose problems in the canning of tuna flesh, since when heated anaerobically it combines with met-myoglobin and cysteine to form an undesirable green colour (Koizumi and Nonaka, 1970). The significance of trimethylamine oxide in the reaction was demonstrated by Koizumi and Hashimoto (1965), and Koizumi *et al* (1967) found that at least 8mg% of trimethylamine oxide nitrogen needed to be present to cause discoloration in *Thunnus alalunga*. The situation was complicated by the uneven distribution of trimethylamine oxide along the body of the fish—this is biological variation of a different kind. The proportion of met-myoglobin increases in the muscle of tuna during frozen storage at the higher range of temperature (Bito and Honma, 1967), but the reaction is very slow below $-30°C$.

Hughes (1959b: *clupea harengus*) showed not only seasonal variations in individual free amino acids of the flesh, but also variations in the total free amino acids according to where the fish had been caught. The same author (Hughes, 1964) also demonstrated an apparent variation between the free glucose contents of herrings caught on different grounds, in addition to the usual seasonal variation.

Demonstrating variations of this kind in both the free amino acids and glucose suggests possibilities of variable production of orange or brown colours during heat-processing. The reaction (the 'Maillard' reaction) between amino acids and glucose intermediates has in fact been observed at 100°C and over in skipjack (*katsuwonus vagans*) by Yamanaka (1975), and while seasonal and ground-to-ground variations in the actual discoloration of any species of fish has not so far been reported, the possibility clearly exists.

The fact that 'nutritional' substances show enormous apparently random variations is also of significance when we are considering fish as a foodstuff. Lovern (1933) observed 'sudden large fluctuations' in the concentration of vitamin A (retinol) in the liver oil of the halibut (*hippoglossus hippoglossus*) and could find no correlation with the fishing ground or stomach contents. Love *et al* (1959) indeed quote vitamin A values in various fish livers as ranging from 179 to 9 819 000 International Units per 100g, which almost implies that the concentration is unimportant to the fish, or that the vitamin is present fortuitously.

2 'Condition'

All of the phenomena we have described so far relate to the fish as a food, and no explanations have up to now been offered for any of them. However, some variations in fish are widely assumed to result from seasonal fluctuations in the diet or the effect of spawning. It is sometimes very obvious that newly-caught fish are much more lively than at other times, and the musculature can have a firm or 'springy' feel as distinct from a soft or putty-like feel, when the dent made by poking the flank remains after one's finger has been removed. Local fishermen are as knowledgeable as anyone about these effects, and their views are interesting. They believe that fish are in better 'condition' (firmer, more lively) if the water in which they are swimming is tidal, that is, flowing, as distinct from still or deep pools. We appear to have seen this effect in cod kept in an experimental flume in which they were made to swim continuously: such fish seem healthier, their eyes seem brighter, their skins seem to be more glossy, than the corresponding attributes of fish kept in still water. The fishermen believe that the exercise is good for them. They also believe that fish from a hard, stony ground is in better condition, the flesh being firmer and the keeping quality better in melting ice than fish from a muddy or sandy bottom, and the fishermen will go to considerable trouble to obtain such fish, even at the risk of damaging their nets.

To complete the account of unexplained variations in fish as a foodstuff, it is interesting to note that there is a rhythmic annual cycle of the smell of the raw cod which is observed by trained panels of people, and a similar cycle in the cooked flavour as eaten (J M Shewan, quoted by Jason and Lees, 1971). There are also fairly reproducible seasonal variations in the dielectric properties of the skin of the same species as measured by the 'Torrymeter' (Jason and Lees, 1971).

3 Recent findings

3.1 *The approach to the problem*

So much of the historical review given in the first part of this account is unexplained that it is quite a problem to know where to begin a scientific assessment of the situation, as distinct from an account of 'folklore'.

The approach made in this laboratory with cod (*gadus morhua*) has been twofold. First, fish were caught with the experimental trawler *Sir William Hardy* from a widely-spaced series of fishing grounds ranging from Aberdeen to Spitzbergen and to West Greenland. Each fish was packed in melting ice for a day until it were fully in *rigor mortis*, and then examined in detail in the ship's laboratory – sex, weight, length, fullness of stomach, weight of liver, colour of bile and size of pancreas. Its age was assessed by counting the rings on an otolith. Samples of urine were taken for nitrogen determination and samples of white muscle were dissected and frozen for subsequent determination of *post mortem* pH and water content. Dark muscle was also dissected out and frozen for measuring its total haem pigment content. The liver was frozen for lipid and glycogen determinations. The remainder of the fish was wrapped in aluminium foil and frozen entire (minus the head). Back at the laboratory it was thawed as soon as possible, and the complete fillet removed. 'Gaping', the appearance of holes or slits (*Fig 1*) was assessed on a subjective scale, using graded photographs of gaping fish for comparison. The remaining musculature was steam-cooked without seasoning and the texture (the firmness or softness in the mouth) was assessed, again on a subjective scale, by a group of people accustomed to this type of assessment. Cold storage at −30°C never exceeded three months, so cold-storage deterioration should not have been an important influence. On some occasions the flavour was assessed in a similar way.

These steps have been described in some detail to show the immense amount of work involved – over 1 000 fish were examined following several voyages. The attributes of a fish as food were then statistically examined in relation to all the other parameters. The basic objective was to see how much of the variation was caused by nutritional factors or reproduction, and how much the cod from different grounds differed intrinsically from each other.

The second stage of the work, carried out in parallel with the first, involved holding living cod under controlled conditions in aquaria at the laboratory. In this way the fish could be starved, re-fed on different diets or made to swim continually in imitation of migratory stocks.

We shall now consider the attributes of fish as a foodstuff one by one. Astonishingly, they are all influenced by 'biological variation'.

3.2 *Texture*

Cooked fish should crumble easily in the mouth and disintegrate completely. Some samples of fresh fish feel very sloppy, and do not give a pleasant sensation. Others can be tough (though never as tough as some tough cooked beef muscle, for example), and since excessive cold storage engenders further toughening, the sample would then be unfit to eat. Tough, cold-stored fish feel succulent during the first bite, but nearly all the fluid is then squeezed out and one is left with a small pad of 'dry' fibrous material that is difficult to swallow, and, in my case at least, remains between the teeth as tough fibres which are difficult to dislodge. In cattle, the connective tissue is a very important factor in toughness, and as the beast grows older the collagen becomes ever more insoluble and less softened by cooking, so that eventually it is impossible to chew. Fish are different, however, in that the collagen is very unstable and hydrolysed at a relatively low temperature, and on cooking it all appears

Fig 1 A severely gaping cod fillet compared with a normal fillet

to have been converted to gelatin. Collagen does not appear to be a factor in the texture of fish muscle.

Figure 2 shows that the age of a fish influences the texture very slightly, older ones being tougher, but the effect is small. Female fish seem to show the effect more strongly than males but some September-caught, well-fed fish do not appear to show the effect at all or even show it in reverse.

The main factor to influence the texture of cooked fish is the *post mortem* pH (*Fig 3*).

After the death of a fish, some of the residual carbohydrate in the muscle is converted anaerobically to lactic acid which, together with any lactic acid already present from struggling before death, causes the pH to fall. The fall in pH is proportional to the amount of lactic acid present (MacCallum *et al*, 1968; Kida and Tamoto, 1969) but not necessarily to the initial amount of

glycogen present in the muscle (MacCallum *et al*, 1967) some of which is converted hydrolytically to glucose, which does not influence the pH.

Broadly speaking therefore, studies on the texture of cooked fish muscle reflect the nutritional state of the fish before death. However, as we shall see later, the changes in pH are not a straightforward reflection of the quantity of food consumed, although the pH of the muscle of starving fish is always high.

Since, as already mentioned, freezing toughens fish muscle further, it has been suggested that fish should not be frozen if the *post mortem* pH is lower than 6·6 (T R Kelly, 1969) or even 6·7 (K O Kelly, 1969).

3.3 *Gaping*
In *Fig 1* is seen a picture of a gaping fish. Until recently, the causes were quite unknown, although discussion

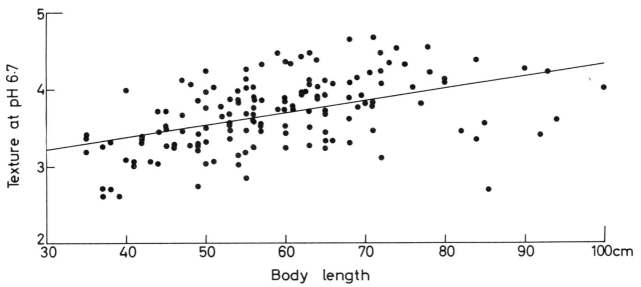

Fig 2   The influence of body length on the texture of cooked cod muscle as eaten, pH values being adjusted to an arbitrary value of 6·7 (Love *et al*, 1947b)

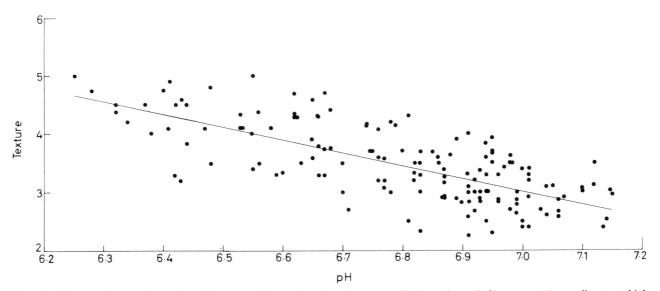

Fig 3   The influence of *post mortem* pH on the texture of cooked cod muscle as eaten. On the scale used, '3' represents 'normal' texture, higher values being tough or firm, lower values soft or sloppy (Love *et al*, 1974b)

133

with people in the fishing industry elicited the opinion that fish gape at around the spawning season, when the flesh has become soft.

What holds a fish together? A fillet can be seen to consist of very many little blocks of muscle which are usually joined in all directions by gleaming sheets of connective tissue, and which fall apart when the fish is cooked. Histological examination shows that connective tissue from any myocomma (the connective tissue sheet) comes out at right angles and passes between the individual muscle cells, perhaps proceeding without interruption to the next myocomma of the fillet (*Fig 4*).

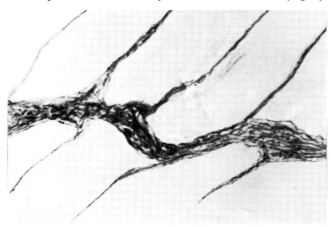

Fig 4 Histological section of cod muscle showing the junction between contractile muscle cells (light areas) and a myocomma (thick dark stripe). The thin dark processes are part of the myocomma and pass between and around each muscle cell, probably connecting with the next myocomma. When these processes break at the myocomma face, the fillet will gape

When these hollow 'cylinders' of connective tissue break at the tips of the muscle cells, the fish gapes.

During starvation or spawning when fish muscle becomes soft, the water content of the muscle rises as protein is removed and extracellular space increases. If the conclusions reached by people in the fishing industry had been correct, the gaping should therefore have increased with the water content, but in fact the opposite proved to be the case. *Figure 5* shows that it is the fish in best condition, that is, with the lowest water content, which gape, while starving fish do not gape at all under the same conditions.

The strength of muscle contraction in *rigor mortis* may be weaker in starving fish, and this could help preserve the connective tissue and reduce gaping. However, Love *et al* (1972) showed that isolated strips of cod connective tissue are extraordinarily sensitive to small changes in the pH: the mechanical strength at pH 7·1 is four times as great as at pH 6·2. The effect is reversible, so weak connective tissue at low pH becomes stronger in a neutral buffer.

Now the pH of cod muscle is close to neutrality while the fish is still alive, but after death, as already mentioned, some of the muscle glycogen breaks down anaerobically to lactic acid, and the pH falls. In fish in poor condition there is very little glycogen in the muscle and the pH remains close to neutrality after death. In well-nourished fish, however, the larger quantity of muscle glycogen gives rise to considerable quantities of lactic acid, and in sea-caught fish a pH as low as 6·0 is occasionally seen. It takes about 15 hours at 0°C after death for the pH to reach its minimum value (Mac-Callum *et al*, 1967).

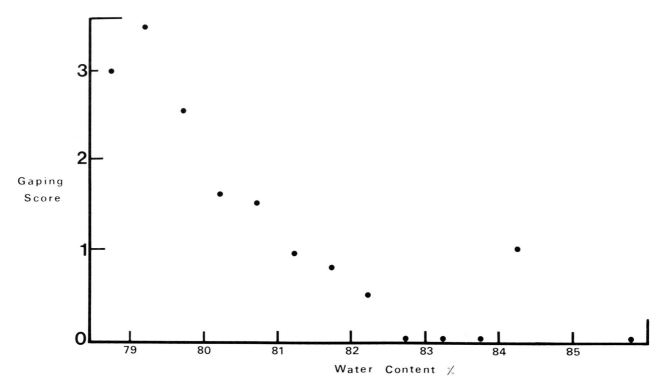

Fig 5 The amount of gaping, assessed on a subjective scale, in frozen-thawed cod of different water contents. The water content of the muscle rises during starvation, and this figure shows that well-fed fish gape more than starved ones. Severely starved fish do not gape at all

134

Figure 6 shows the results of actual gaping measurements: the relationship with pH is very close. Freezing always makes the gaping worse.

From these two sections we see that a low *post mortem* pH will make fish so firm that they should not be frozen, or, if they are frozen whole, thawed and filleted, the gaping may be so bad that the appearance is very poor and the fillets cannot be skinned mechanically. pH is absolutely central to fish technology, therefore its seasonal and ground-to-ground variation will be considered later.

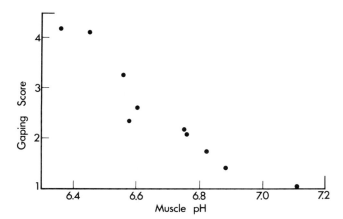

Fig 6 The influence of the *post mortem* pH on the gaping of cod fillets made from frozen-thawed 'round' fish

3.4 Flavour

3.4.1 In fresh fish

Variations in the taste of a fish may result from small quantities of material being absorbed directly from the surrounding water or from the diet.

As an example of the former, there are certain Russian lakes in which the mineral composition of the water varies according to that of the surrounding earth, and the same differences appear in the fish tissues (Anon, 1969) affecting the manganese, iron, copper and zinc—and perhaps other minerals—in the muscle and liver. There is no record of the flavour being affected, but clearly this kind of influence is likely to produce small differences. Fish with a particular fondness for molluscs acquire higher concentrations of zinc (Vinogradov, 1953), so there is scope for ground-to-ground variation here also.

Changes in the taste originating in the diet of the fish have, however, in fact been described. The flavour of the cooked muscle is pleasant if herrings (*Clupea harengus*) have been feeding on crustaceans of the copepod family, for example *Calanus finmarchicus*, but a bitter taste results if they fed on certain larvae such as those of *Mytilus* (Liicke, 1954, quoted by Rakow, 1963). Cod (*Gadus morhua*) acquire an unpleasant smell if they have been eating 'blackberry'—a marine invertebrate—(Sipos and Ackman, 1964) the compound responsible being identified as dimethylsulphide. This has also been found in the mackerel, *Scomber scombrus*, originating in the pteropod *Spiratella retroversa* (Ackman *et al*, 1972). Its precursor, β-dimethylpropiothetin, probably occurs in certain algae from brackish water (Granroth and Hattula, 1976).

Fish from various saline lakes of west Canada have

been found to contain geosmin (Yurkowsky and Tabachek, 1974) originating in metabolites of the bacterium *Streptomyces*. The compound imparts a 'muddy' flavour to the flesh. An 'earthy' taint has been found to result from consuming an odoriferous species of *Actinomyces* (Kesteven, 1942, quoted by Connell, 1974) and a 'muddy-earthy' taint has been noted in the flesh of *Salmo gairdneri* captured in earth pounds (Iredale and York, 1976).

Since these taints seem to originate in foreign substances introduced either by mouth or through the body surface, there is scope for seasonal or ground-to-ground variation, depending on the source substance. Such variations have indeed been noted in the fishing industry—periodic 'weedy' smells in cod from Bear Island, and a smell resembling iodine in cod from Betty's Bay (South Spitzbergen) have been described. The origins of these particular contaminants are not known at present.

3.4.2 In frozen fish

A clear indication of ground-to-ground variation in the cold-storage flavour of cod was mentioned briefly in the introduction to this paper. K J Whittle (unpublished 1970: quoted by Love, 1975) cold stored batches of cod from various grounds in the North Atlantic for three months at −30°C, then thawed, cooked and tasted them.

The cold-store flavour was estimated on a scale of 0 (absent) to 5 (very strong), with the upper taint of commercial acceptability about 3. Under the well-nigh perfect storage conditions used, the fish being wrapped in aluminium foil there should have been almost no flavour detectable and in most batches this proved to be the case: the average score for fish from Aberdeen Bank was 1·7, Faroe Plateau 1·5, southeast Iceland 1·7 and northwest Iceland 1·4: all very low. However, that of fish from the Faroe Bank was just over 3, an unacceptably strong flavour, and the cold-store odour was similarly strong.

To explain this phenomenon, we need to know the origin of the cold-storage flavour. This has been identified by McGill (1974) and McGill *et al* (1974) as *cis*-4-heptenal, and it arises principally from the atmosphere oxidation of those unsaturated fatty acids which are located primarily in the phospholipids (McGill, personal communication). When Love *et al* (1975) compared the concentrations of white-muscle lipids of Faroe Bank cod with those from Aberdeen Bank cod, they found that whatever the nutritional state, they were richer in the cod from the Faroe Bank. The muscle of this species maintains almost no stores of triglyceride—only about 1% of the total lipid (Ross and Love, 1979). Consequently most of it is ('structural') phospholipid which can be as high as 88% of the total (same author). Any increases or decreases in the total lipid of cod resulting from feeding or starvation are therefore in fact increases or decreases in phospholipid (Love *et al*, 1975), and therefore changes in the potential of the tissue for producing *cis*-4-heptenal.

Finally, Ross and Love (1979) stored a batch of fish in an aquarium for a few weeks, killed, filleted and froze them and stored the fillets alongside fillets from fed controls. After five weeks both groups were thawed, cooked and tasted, and the taste-panel score for cold-store flavour was 3·4 in the fed controls but only 1·3 in the stored group—a very big difference. Concentrations of *cis*-4-heptenal in the two groups were 23nm/kg of wet muscle in the fed group and only 3·5 in the starved group.

3.5 *Colour*

The flesh of 'white' fish is surprisingly red if the fillet is removed immediately after death. However, the fish are often bled by cutting the 'throat' before gutting, and even if they are simply gutted and packed in ice, most of the blood drains out provided the fish were kept chilled and were gutted without too much delay. The flesh of fish frozen at sea tends to be darker than those gutted and packed in melting ice, because the blood did not have time to drain completely before freezing commenced.

However, along the lank of a white fish just under the skin lies a band of naturally-dark muscle, and no amount of washing removes this particular reddish-brown colour, which is mostly myoglobin and serves to transport oxygen to the mitochondria.

Consumers of fish in Great Britain tend to be conservative in their choice of seafood, and they also expect white fish to look white and not vary in their degree of pigmentation. A considerable seasonal variation in the colour of dark cod muscle was reported therefore with alarm by people in the fishing industry.

In a large survey of cod, Love *et al* (1974a) found that while the proportion of dark muscle did not vary from ground-to-ground, the intensity of its pigmentation showed some marked variations. The colours in cod from various grounds off Aberdeen, Faroe, Iceland and Norway were very similar, but that in cod from two grounds off Spitzbergen was about twice as intense. The colour in Bear Island cod lay between the two extremes. Now the Spitzbergen cod migrates further than any of the other stocks (Trout, 1957) swimming past Bear Island and down the coast of Norway to spawn at the Lofoten grounds every February, then returning to Spitzbergen to feed. A visit by the writer to Tromsø in the month of February confirmed that the colour of the dark muscle of the migratory Spitzbergen stock of cod was a dark chocolate-brown, confirming its unusual nature at both ends of the spawning run. Since the myoglobin content of mammalian muscle has been shown to increase with its physical activity (Lawrie, 1950, working with horses, hens and pigs), it appears likely that the dark muscle of cod which is used for steady swimming (Love, 1970, among others) also darkens with increased swimming (see *Figs* 7 and 8).

As a final check, Love *et al* (1977) made a group of cod swim in a circular tank by chasing them with a travelling gantry—the apparatus was designed by Hawkins and has been fully described by him (Hawkins, 1971). After 27 days of enforced swimming the intensity of pigmentation was nearly double that of a similar control group, which was kept quietly in a circular tank with a similar light intensity and temperature to the experimental group was paler than that of newly-caught fish, so had faded appreciably during the enforced inactivity.

Pigment in the dark flesh of fish is therefore in a dynamic state, influenced by recent activity. Distribution diagrams of pH variations, both seasonally and

from ground to ground, would be of very great value to those hoping to freeze their catch at sea.

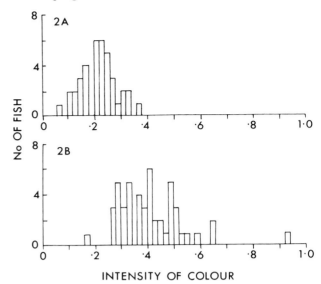

Fig 7 The distribution of intensity of pigmentation of the dark muscle among batches of cod caught on (*a*) Faroe Bank (*b*) West Spitzbergen. Faroe Bank fish are a 'stationary' stock, while those sometimes present at Spitzbergen are strongly migratory. Note the uneven distribution of pigmentation among individual Spitzbergen fish, in addition to their darker colour

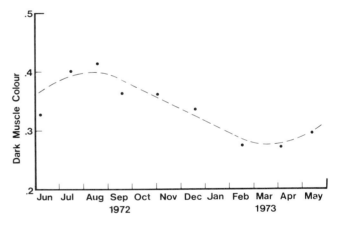

Fig 8 Seasonal variation in the intensity of colour in the dark muscle of cod caught off Aberdeen (average values from 20 fish). The peak may be related to maximum swimming activity in catching prey

The pH of the fish while still alive is close to neutrality. The *post mortem* conversion of glycogen to lactic acid is generally complete in about 12 hours (MacCallum *et al*, 1967), and in our work we waited until the day following capture before measuring pH. It might be thought that the values would be uniformly high when the fish were starving during the winter (because little lactic acid would be formed) and uniformly low during the summer feeding, when the muscle glycogen would increase. The real situation is not as straightforward as this.

A survey over the years 1969–1979 in which fish were caught monthly near Aberdeen throughout the year showed that almost always the pH of the muscle of the majority of fish in a haul was greater than 6·6. However, for a certain short period some very low values were observed, and most of a haul would have an unacceptably low pH for the purposes of freezing technology.

Figure 9 shows that the pH values fall in June, and again perhaps in January–February. Readings taken in other years show that the summer fall can occur at any time from June to August, inclusive, and that the fall in January does not often occur. However, when the pH falls in summer, it is for a shorter period than that for which food is available, and sometimes seems to be only for three weeks or so, as in *Fig 9*. Further work on fish from other grounds confirmed that the phenomenon was a general one, and that the period of low pH was always short, with the possible exception of cod from west Scotland.

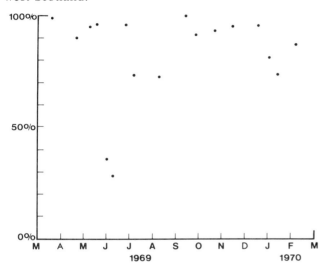

Fig 9 Seasonal variation in the *post mortem* pH of cod caught off Aberdeen. Note the sharp but short-lived drop in June. Results are expressed as the percentage of a batch of 30 fish with pH of 6·6 or more

Experiments with cod kept in an aquarium at the laboratory shed light on the phenomenon. Fish were starved for about three months and re-fed, using the special 're-feeding' diet devised by Bouche *et al* (1971) for the first two weeks, and herring or squid muscle thereafter. The results (Love, 1979) showed that, after re-feeding for about three months, the pH fell markedly, in one fish a value as low as 5·87 being recorded, the lowest value ever observed by us in this species.

Further work in this laboratory (Love, unpublished) has shown that while re-feeding starving fish causes the liver fat and liver glycogen to rise sharply about 70 days after the commencement of re-feeding, and a fall in muscle pH corresponding with a rise in muscle glycogen, the situation is sharply reversed soon after, spontaneously, and with the fish still being fed daily under identical temperature and lighting conditions. It is as though the fish over-compensated for lack of food once feeding was restored, and then 'came to its senses', as it were, and returned the values to normal. This observation, now made in three independent experiments, suggests that the well-known annual peak-and-decline in liver fat seen in the summer may well not be a simple response to summer feeding and later food shortage, but partly to a built-in mechanism of over-compensation and spontaneous redistribution of energy reserves. The topic of over-compensation is discussed more fully by Love (1980).

4 Conclusions

We may therefore summarize our thoughts arising from this work as follows:

'Biological variation' affects all known attributes of fish as food for man: the surface appearance, colour of flesh, colour of skin, texture of the cooked product in the mouth, cooked flavour of the fresh product and cooked flavour of the cold-stored product all vary according to the ground where the fish were caught and some of them vary with the season. There is as yet no indication of genetic differences between fish from the different grounds, although the difference in skin colour, while governed mostly by the colour of the ground, might contain a genetic factor (Love 1974). However, skin colour does not influence the acceptability of the fish as a food. The principal factor affecting the texture and the extent of gaping is the *post mortem* pH of the flesh, a low pH giving in each case the undesirable effect. A low pH occurs as a result of re-feeding after the winter starvation, but the effect is short-lived and in fish in the wild from one ground lasts in general only about three weeks. The recovery of 'desirably' high pH values appears to be spontaneous and not related to food shortage. Undesirable flavours in fresh fish relate to certain waters containing contaminants or to the consumption by the fish of certain organisms or algae which contain strong-tasting compounds.

The premature development of a strong cold-store flavour (and odour) results in cod from grounds with an abundant food supply, which raises their flesh phospholipid concentration somewhat. This gives rise under frozen conditions to higher concentrations of *cis*-4-heptenal, the cold-store flavour compound. Starvation of cod reduces the formation of this compound, so there will almost certainly be a rhythmic seasonal variation in this attribute. Variations in the colour of the white muscle of cod result from inadequate bleeding at death, or freezing too soon after death. Variations in the intensity of the colour of the dark muscle result from variations in swimming activity. Migratory stocks of cod, principally the Spitzbergen–Bear Island–Lofoten Stock, therefore have darker dark muscle than have fish from other grounds, and a seasonal cycle in dark muscle colour seems to follow the activity exhibited in chasing food.

The outlook for future work is promising. It encompasses among other things the discovery of new techniques with which to specify the biological condition of a fish, and so to be able to state how such a fish would behave during subsequent processing as food. New parameters currently under examination in this laboratory are the concentration of protein in the endolymph, the colour of the bile, the sizes of the pancreas and spleen, and the heart weight. The studies of muscle pH and water contents and of liver fat and glycogen have been in progress for some years.

5 References

ACKMAN, R G, HINGLEY, J and MACKAY, K T. Dimethyl sulphide as
1972   an odour component in Nova Scotia fall mackerel. *J. Fish. Res. Bd. Can.*, 29, 1085–1088
ANON. Geochemical ecology of freshwater fish. Vop. Biol., Mater.
1969   Konf., 159–164, Ed H Maurina. Izd. 'Zinatne', Riga, USSR. From *Chem. Abstr.*, 72, 52385K (1970)

Bito, M and Honma, S. Studies on the retention of meat colour of
1967 frozen tuna, IV. Acceleration of discolouration of tuna meat
by freezing and its relation to storage temperatures. *Bull. Jap. Soc. Sci. Fish.*, 33,33–40.

Bouche, G, Murat, J C and Parent, J P. Study of the influence of
1971 synthetic (dietary) regimes on protein synthesis and the carbo-
hydrate and lipid reserves in the liver of starving carp. *Crit. rev. Seanc. Soc. Biol.*, 165, 2202–2205

Connell, D W. A kerosene-like taint in the sea mullet, *Mugil cepha-
1974 lus* (Linnaeus). 1. Composition and environmental occurrence
of the tainting substance. *Aust. J. Mar. Freshwat. Res.*, 25, 7–24

Granroth, B and Hattula, T. Formation of dimethyl sulphide by
1976 brackish water algae and its possible implication for the
flavour of Baltic herring. *Finn. Chem. Lett*,, 148–150.

Hawkins, A D. Europe's first fish behaviour tank. *World Fish.*, 20,
1971 8–10

Hughes, R B. Chemical studies on the herring *(Clupea harengus)*. I.
1959a Trimethylamine oxide and volatile amines in fresh, spoiling
and cooked herring flesh. *J. Sci. Fd. Agric.*, 10, 431–436

—— Chemical studies on the herring *(Clupea harengus)*. II. The free
1959b amino-acids of herring flesh and their behaviour during
post-mortem spoilage. *J. Sci. Fd. Agric.*, 10, 558–564

—— Chemical studies on the herring *(Clupea harengus)*. X. Histidine
1964 and free sugars in herring flesh. *J. Sci. Fd. Agric.*, 15, 293–299

Iredale, D G and York, R K. Purging a muddy-earthy flavour taint
1976 from rainbow trout *(Salmo gairdneri)* by transferring to artificial and natural holding environments. *J. Fish. Res. Board Can.*, 33, 160–166

Jason, A C and Lees, A. Estimation of fish freshness by dielectric
1971 measurement. *Spec. Rep.*, Torry Res. Sta. 31pp

Kelly, K O. Factors affecting the texture of frozen fish. In: *Freezing
1969 and Irradiation of Fish*, Ed R Kreuzer. Fishing News (Books)
Ltd, London. 339–342

Kelly, T R. Quality in frozen cod and limiting factors on its shelf life.
1969 *J. Fd. Technol.*, 4, 95–103

Kida, K and Tamoto, K. Studies on the muscle of aquatic animals. IV.
1969 On the relation between various pH values and organic acids
of Nagazuka muscle *(Stichaeus grigorjewi* Herzenstein). *Sci. Rep. Hokkaido Fish. Exp. Sta.*, 11, 41–51

Koizumi, C and Hashimoto, Y. Studies on 'green' tuna. I. The signifi-
1965 cance of trimethylamine oxide. *Bull. Jap. Soc. Sci. Fish.*, 31, 157–160

Koizumi, C, Kawakami, H and Nonaka, J. Studies on 'green' tuna.
1967 III. Relation between 'greening' and trimethylamine oxide
concentration in albacore meat. *Bull. Jap. Soc. Sci. Fish.*, 33, 131–135

Koizumi, C and Nonaka, J. A green pigment produced from tuna
1970 metmyoglobin. *Bull. Jap. Soc. Sci. Fish.*, 36, 1258 only

Lawrie, R A. Some observations on factors affecting myoglobin
1950 concentrations in muscle. *J. Agric. Sci., Camb.*, 40, 356–366

Love, R M. *The Chemical Biology of Fishes*. Academic Press, London
1970 and New York. 547pp

—— Colour stability in cod *(Gadus morhua* L.) from different grounds.
1974 *J. Cons. perm. int. Explor. Mer.*, 35, 207–209

—— Variability in Atlantic cod *(Gadus morhua)* from the northeast
1975 Atlantic: a review of seasonal and environmental influences on
various attributes of the flesh. *J. Fish. Res. Bd. Can.*, 32, 2333–2342

—— The post-mortem pH of cod and haddock muscle and its seasonal
1979 variation. *J. Sci. Food Agric.*, 30, 433–438

—— *The Chemical Biology of Fishes*. Academic Press, London and
1980 New York. Vol 2

Love, R M, Hardy, R and Nishimoto, J. Lipids in the flesh of cod
1975 *(Gadus morhua* L.) from Faroe Bank and Aberdeen Bank in
early summer and autumn. *Mem. Fac. Fish., Kagoshima Univ.*, 24, 123–126

Love, R M, Lavety, J and Garcia, N G. The connective tissues of
1972 fish. VI. Mechanical studies on isolated myocommata. *J. Fd. Technol.*, 7, 291–301

Love, R M, Lovern, J A and Jones, N R. The chemical composition
1959 of fish tissues. Dep. Sci. Indust. Res., Spec. Rep. 69. HMSO, London

Love, R M, Munro, L J and Robertson, I. Adaptation of the dark
1977 muscle of cod to swimming activity. *J. Fish Biol.*, 11, 431–436

Love, R M, Robertson, I, Lavéty, J and Smith, G L. Some biochemi-
1974a cal characteristics of cod *(Gadus morhua* L.) from the Faroe
Bank compared with those from other fishing grounds. *Comp. Biochem. Physiol.*, 47B, 149–161

Love, R M, Robertson, I, Smith, G L and Whittle, K J. The texture
1974b of cod muscle. *J. Texture Stud.*, 5, 201–212

Lovern, J A. Vitamin A in fish oils. Food Invest. Board Ann. Rep.
1933 1932. HMSO, London. 198–199

MacCallum, W A, Jaffray, J I, Churchill, D N, Idler, D R and
1967 Odense, P H. Postmortem physiochemical changes in
unfrozen Newfoundland trap-caught cod. *J. Fish. Res. Bd. Can.*, 24, 651–676

MacCallum, W A, Jaffray, J I, Churchill, D N, and Idler, D R.
1968 Condition of Newfoundland trap-caught cod and its influence
on quality after single and double freezing. *J. Fish. Res. Bd. Can.*, 25, 733–755

McGill, A S. *An investigation into the chemical composition of the
1974 cold storage flavour components of cod*. IFST mini-symposium
on freezing, Institute of Food Science and Technology, GB. 24–26

McGill, A S, Hardy, R, Burt, J R and Gunstone, F D.
1974 Hept-*cis*-4-enal and its contribution to the off-flavour in
cold-stored cod. *J. Sci. Food Agric.*, 25, 1477–1489

Rakow, D. On the feeding habits of fishes and their effect on the
1963 quality of the fish flesh. *Archiv. Lebensmittelhyg.*, 14, 261–263

Reay, G A. Objective assessment of quality. Dep. Sci. Indust. Res.,
1955 Food Invest. 1954. 34 only

—— Factors affecting initial and keeping quality. Dep. Sci. Indust.
1956 Res., Food Invest. 1955. 24 only

—— Factors affecting initial and keeping quality. Dep. Sci. Indust.
1957 Res. Food Invest. 1957. 3 only

Ronold, O A and Jakobsen, F. Trimethylamine oxide in marine
1947 products. *J. Soc. Chem. Ind.*, Lond., 66, 160–166

Ross, D A and Love, R M. Decrease in the cold store flavour deve-
1979 loped by frozen fillets of starved cod *(Gadus morhua* L.). *J. Fd. Technol.*, 14, 115–122

Shewan, J M. The chemistry and metabolism of the nitrogenous
1951 extractives in fish. *Biochem. Soc. Symp.*, Lond., 6, 28–48

Sipos, H C and Ackman, R G. Association of dimethyl sulphide with
1964 the 'blackberry' problem in cod from the Labrador area. *J. Fish. Res. Bd. Can.*, 21, 423–425

Trout, G C. The Bear Island cod: migrations and movements. *Fishery
1957 Invest.* Lond., Ser II, 21, (6), 51pp

Vinogradov, A P. *The elementary composition of marine organisms
1953 *(translators J. Efron and J K Setlow). Sears Foundation, New Haven. 647pp

Yamanaka, H. Orange discoloured meat of canned skipjack. V.
1975 Amino acids responsible for orange discolouration. *Bull. Jap. Soc. Sci. Fish.*, 41, 357–363

Yurkowski, M and Tabachek, J A L. Identification, analysis and
1974 removal of geosmin from muddy-flavoured trout. *J. Fish. Res. Bd. Can.*, 31, 1851–1858

Microbiology in fishery science

J Liston

1 Introduction

Fishing is a very ancient activity of man and the arts
which we now describe as technology were certainly well
developed by the time of the Pharaohs (Kreuzer, 1974).
Fishery science, however, is a recent discipline developed by necessity from ichthyology, chemistry,
engineering and other disciplines under the practical
threat of overfishing which resulted from the industrial
technology of the late nineteenth and early twentieth
centuries. It is certainly no older and probably younger
than the science of microbiology which really began with
the discoveries of Louis Pasteur and his contemporaries
in the 1850s and 1860s. Scientific interest in bacteria in
the oceans resulted in isolations of strains from various
marine sources including sediments and fish (Certes,
1884; Russel, 1892) but the general scientific interest of
earlier investigators did not result in much information
of practical importance to fishery science. The primary
attention of bacteriologists around the turn of the century tended to be focused on human and animal disease.
Recognition of the involvement of shellfish (Cameron,
1880; Conn, 1894; Foote, 1895) as carriers of human

enteric disease led to an extensive investigation of the natural occurrence of enteropathogenic bacteria in fish and other aquatic animals. When it was shown that fish and marine invertebrates living in unpolluted water did not carry the kind of bacteria commonly found in human and mammalian animals' or birds' intestines interest in fish by clinical or health-related bacteriologists waned though attention continued to be paid to the problems of mussels and oysters (Johnstone, 1907; 1915; Jordan, 1925; Hunter and Harrison, 1928; Dodyson, 1928; Arcisz and Kelly, 1955). It was shown quite early (Stiles, 1911; Bodin and Cherrel, 1913; Wells, 1916) that bivalve molluscs would cleanse themselves of undesirable (ie 'enteric') bacteria if held in clean flowing sea water and this subsequently led to the depuration processes widely used today. The perishability of fish so obviously reduces the value of this commodity and limits the size of the market for the fresh product that there has probably always been high interest in fish spoilage among fish peddlers and fish consumers. The phrase 'stinking fish' implies corruption and it was well recognized as a result of Pasteur's researches that corruption is associated with bacteria. As early as 1907, Anderson clearly showed that spoilage was dominantly caused by bacteria and through the work of a number of investigators in different countries it was established that the bacteria most apparently involved were those found normally on fresh fish (Hunter, 1920a; 1922; Fellers, 1926; Harrison, 1929; Reed and Spence, 1929). This sparked a whole series of studies on the bacteriology of fresh and spoiling fish with attention being concentrated on those species of fish or shellfish of commerical significance (Lumley et al, 1929; Sanborn, 1930; Stewart, 1932; Griffith, 1937).

For the period from the 1920s to 1945, the results of studies on fish bacteria and their involvement in spoilage is rather completely covered in the review by Reay and Shewan (1949). The early researchers had shown that fish muscle tissues are sterile in healthy animals while more or less large populations of bacteria are associated with the external surfaces, gills and intestine of fish. Estimates were quoted of 10^2–10^6 bacteria/cm^2 skin surface, similar or slightly higher gill counts and intestinal counts of very few to in excess of 10^8/g depending on whether or not the fish were feeding. During spoilage skin counts increased to 10^7 organisms/cm^2 or more and gill counts increased comparably.

The composition of the microfloras of the different species of fish tested were mostly reported to be dominated by Gram-negative bacteria usually identified as *Achromobacter*, *Flavobacterium*, Pseudomonas or less frequently *Vibrio* or enterobacterial genera. However, there were a few reports of large numbers of Gram-positive bacteria. Since in many early studies fish were obtained from commercial sources the heterogeneous nature of the microflora and the presence of organisms associated with animals including micrococcus is not unexpected. However, the results are representative of many different species of fish including salmon (Hunter, 1922; Fellers, 1926; Snow and Beard, 1939), North American bottom fish species (Reed and Spence, 1929; Gibbons, 1934; Dyer, 1947), North Sea bottom fish (Stewart, 1932; Thjøtte and Sømme, 1938; Shewan, 1946) and pelagic species (Aschehoug and Vesterhus,

1943; Reay and Shewan, 1949). Some representative early data are shown in *Table I*. The more recent and carefully controlled spoilage studies clearly indicated that Gram-negative bacteria of the genera *Achromobacter*, *Flavobacterium* and *Pseudomonas* became dominant after a few days' ice storage (Shewan, 1946). When fish were held at high temperatures a mixed flora resulted with some Gram-positive survivors (Fellers, 1926).

It was reported by Suwa[1] as early as 1909 that trimethylamine oxide which occurs in dogfish flesh is reduced to the amine by bacteria. However, it was not until 1939 that the productive research of (mostly) Canadian scientists (Beatty, Watson, Tarr and Collins)[1] clearly established that fish spoilage bacteria bring about the reduction of TMAO to trimethylamine by a coupled reaction involving oxidation probably of lactate to acetic acid, CO_2 and H_2O through an activation step involving a 'triamine-oxidase'.

The importance of temperature and pH was recognized in the review. It was noted by reference to the works of Hess (1934a,b), Bedford (1933), Kiser (1944) and others that fish bacteria are mostly psychrophilic, growing between 0°C and about 30°C with some strains growing as low as −7·5°C. The very large effect on growth rate of temperature changes near 0°C as contrasted to comparable changes at higher temperature ranges was noted: eg, *P. fluorescens* showed a Q_{10} of 3·7 in the range 5–20°C but 8·4 in the range 0·5°C (Hess, 1934b). Biochemical activities were shown to persist at the lower range of growth explaining why spoilage proceeded near 0°C. Fish bacteria were shown to be sensitive to low pH. Most would not grow below pH 6·0 and it was suggested that this is one reason for the stable population during *rigor mortis* when pH of flesh was reported to be in the range 6·2 to 6·5.

Thus there was quite good information available by the late 1940s concerning the numbers and types of bacteria likely to occur on fish, the bacterial changes occurring during spoilage and the relationship between major chemical changes in the fish and bacterial activity.

If we now jump 28 years to another review by Shewan (1977) we will see the advances which have occurred in the understanding of fish bacteriology and the spoilage process during this time. Advances in bacteriological techniques and classification systems arising in part out of the developments in molecular biology in the 1960s enable the microbiologist to identify the spoilage bacteria with greater assurance than before. It can now be seen that the organisms which dominate the spoilage microflora are actually *Pseudomonas* and *Alteromonas*, and their characteristics can be related to the spoilage process. Information on fish and bacteria is now available from many parts of the world and differences are now apparent. Important new species have been identified and biochemical processes are better understood.

2 Microbiology of fish

2.1 Quantitative

The numerous reports in the literature since 1950 of fish counts largely confirm the values discussed by Reay and

[1] See Reay and Shewan, 1949

Shewan (1949) for temperate or cold water fish. Surface counts mostly range from 10^2–10^5 organisms/cm^2 of skin, 10^3–10^7/g for gills and 10–10^8/g for intestine. In our experience fish taken from cold ($<10°C$) unpolluted water may have counts as low as 10/cm^2 and when temperatures are high they may occasionally surpass 10^6/cm^2. These values are quoted from counts in which plates were incubated at 20–25°C. Lower counts, often one-tenth of the 20–25°C count, may be obtained by incubation at 35–37°C and sometimes somewhat higher counts may be obtained at 0–5°C. This emphasizes the psychotrophic nature of the fish microflora. Our studies indicate little difference in the count whether or not sea water is used in the media so long as there is sufficient salt (usually 0·5% NaCl) to accommodate the low halophilism of many of the bacteria from marine fish (*Table II*).

Shewan (1977) quotes data which indicate that fish from warm waters frequently carry greater numbers of bacteria than cold-water fish and yield higher counts when incubation temperatures are 35–37°C. This suggests that the population on fish in warm waters is more mesophilic.

There are similar variations seen in the microbial populations of shrimp which are fished in all oceans of the world and taken from both cold and warm waters. Cold-water shrimp not infrequently have counts of 10–10^3/g though the total range is large and up to 10^7/g has been reported on occasion. Warm-water shrimp and prawns frequently show counts of 10^6/g on capture (Cann, 1974; Vanderzant *et al*, 1970). Counts of shrimp and other bottom-dwelling creatures may be difficult to evaluate because they tend to be contaminated with sediment material.

Some of the more recent data available on bacterial population numbers of fish and other marine creatures are shown in *Table III*.

2.2 *Qualitative*

In his review Shewan (1977) lists nine cases out of a total of 17 in which the microflora of fish skin showed over 80% Gram-negative rod bacteria. All nine cases were cold-water fish. Among the others two showed a lesser dominance (65–67%) of Gram-negatives and a third showed over 40% Gram-negatives: in these three cases it appears there may be some doubt concerning water

Table I

FLORA OF SEA FISH ACCORDING TO VARIOUS AUTHORS, AS SHOWN BY PERCENTAGE OF ISOLATIONS IN THE VARIOUS GENERIC GROUPS

Author	Micro.	Achro.	Flavo.	Pseudo.	Bacillus	Misc.	
Reed and Spence (1929) Haddock	4	23[1]	8	22	24	18	
Stewart (1932) Haddock	22	57[1]	11	5	—	5	
Bedford (1933) Halibut	16	34[1]	30	—	—	20	
Shewan (1938b) Herring	24	43[1]	13	11	—	9	
Thjøtte and Sømme (1938) Cod	14	48[1]	25	5	—	8	
Thjøtte and Sømme (1943) Misc. species	3	21[1]	6	4[2]	2	64[2]	(chiefly *Vibrio*)
Wood (1940) Misc. species	48	19	17	7	9	—	
Aschehoug and Vesterhus (1943) Herring	17	25[1]	18	40	—	1	
Snow and Beard (1939) Salmon	13	54[1]	5	8	2	19	
Dyer (1947) Cod	73	5	3	4	—	16	
Gianelli (1957) Porgy	53	21	7	6	—	13	
Liston (1957) Skate	3	19	9	65	—	4	(including *Coryne*)
Liston (1957) Lemon sole	1	22	5	69	—	3	(including *Coryne*)
Georgala (1958) Cod	1	32	6	50	—	11	(including *Coryne*)

[1] Probably includes organisms classed as *Pseudomonas* by current methods
[2] Note that most of the bacteria in the miscellaneous classification should be added to *Pseudomonas*

Table II

BACTERIAL COUNTS ON ICED PACIFIC OCEAN FISH

Days in ice	Hake (Merluccius productus) TSA[1]	Hake (Merluccius productus) TSA–S[2]	Rockfish (Sebastodes spp.) TSA	Rockfish (Sebastodes spp.) TSA–S	English sole (Parophrys vetulus) TSA	English sole (Parophrys vetulus) TSA–S
	\multicolumn{6}{c}{(count expressed as $\log_{10}$ per cm^2 skin)}					
0	1·18	1·04	2·08	2·15	1·80	1·38
3	2·49	3·16	1·38	1·80	2·30	2·11
7	6·33	6·96	5·54	5·16	4·97	4·77
10	6·27	6·04	4·92	5·30	6·26	6·62
14	7·27	7·76	7·16	6·96	7·18	7·13

[1] TSA = Trypticase soy agar made with distilled water
[2] TSA–S = Trypticase soy agar made with aged sea water
(All counts were incubated at 10°C for 21 days)

temperatures. Fish taken from truly warm waters had a major population of Gram-positive bacteria. This appears to represent a true distinction between the microflora of fish from warm and cold waters and supports Shewan's hypothesis that fish bacterial populations are more affected by the environment than by (fish) species.

Cold-water marine fish carry mainly *Moraxella*,[1] *Acinetobacter*,[1] *Pseudomonas*, *Flavobacterium* and *Vibrio* while warm-water fish carry mostly micrococci, coryneforms and bacillus (*Table IV*).

Table III
BACTERIAL COUNTS ON FISH RECENTLY REPORTED IN THE LITERATURE

	Count × 10⁻³		Count × 10⁻³
Cod	47	North Sea fish	0·1–100
Pollock	52	Indian sardine	10–10 000
Whiting	77	Flatfish	10
Channel catfish	69–198 000	Mullet	10
Hake	0·15	Shrimp (cold water)	0·53–169
Rockfish	0·14	Shrimp (tropical)	1–10 000
English sole	0·06	Shrimp (pond)	1·5–13
		Crab	6–10

The bacteria in fish intestines vary somewhat depending on the food being consumed but normally contain *Vibrio*, *Achromobacter*, *Pseudomonas* and *Aeromonas* in addition to smaller numbers of Gram-positive bacteria including *Clostridium*.

Freshwater fish may show slightly lower skin and gill

[1] *Moraxella* and *Acinetobacter* include most of the bacteria formerly identified as *Achromobacter*

counts than marine fish but the difference is hardly significant. The kinds of bacteria on freshwater fish are quite variable and are undoubtedly influenced by the microflora of the water. This was shown by Horsley (1973) in his study of the changes in skin bacteria of Atlantic salmon during an upstream migration. A somewhat more extensive study by Yoshimizu and Kimura (1976) showed differences in the intestinal flora of Pacific salmon species during out-migration, ocean residence and as returned spawners. Some of their data are shown in *Table V*. Note the high *Vibrio* content in ocean fish which has been reported by others in marine fish from time to time (Thjøtte and Sømme, 1943; Liston, 1957; Simidu *et al*, 1969) and the shift to *Aeromonas* in fresh water. Trust's (1975) data on the microflora of the gills of Pacific salmon taken in fresh waters of British Columbia and including both wild and hatchery fish confirm the dominance of *Pseudomonas* and (surprisingly) *Cytophaga*, and also the occurrence of *Aeromonas* and the coryneform group. Warm-water freshwater fish often carry high numbers of coryneform bacteria and sometimes Enterobacteriaceae even including *Salmonella*. These undoubtedly reflect the microflora of the water in which the animals live and it is interesting in this respect to note the consistent occurrence of *C. botulinum* in fish captured where sediments are contaminated with this organism (Huss *et al*, 1974). One may speculate on the significance of this in aquaculture systems throughout the world, particularly where natural earth ponds are used (Bach *et al*, 1971; Cann *et al*, 1975).

Table IV
BACTERIA IN FISH AND SHELLFISH, EXPRESSED AS A PERCENTAGE OF THE TOTAL FLORA FOR VARIOUS GENERIC GROUPS

Fish type	Pseudomonas	Vibrio	Achromobacter[1]	Coryneform	Others[2]	Reference author
N Sea fish						
(1932)	5	—	56	—	39	Shewan (1971)
(1960)	16	1	23	18	42	Shewan (1971)
(1970)	22	1	41	18	19	Shewan (1971)
Haddock (N Atlantic)	26	2	45	4	23	Laycock and Regier (1970)
Flatfish (Japan)	20	13	30	17	20	Simidu *et al* (1969)
'Pescada' (Brazil)	32	—	35	4	29	Watanabe (1965)
Shrimp (N Pacific)	10	—	47	3	40	Harrison and Lee (1969)
Shrimp (Gulf-Ocean)	22	2	15	40	21	Vanderzant *et al* (1970)
Shrimp (Pond culture)	—	5	2	83	13	Vanderzant *et al* (1971)
Scampi (UK)	3	—	11	81	5	Walker *et al* (1970)
Mullet (Queensland)	18	—	9	12	61[3]	Gillespie and Macrae (1975)

[1] Includes *Moraxella* and *Acinetobacter*
[2] Includes *Flavobacterium, Micrococcus, Bacillus*, Coliforms, and Yeast
[3] Micrococci constitute 49%

Table V
MICROBIOLOGY OF PACIFIC SALMON INTESTINE
(Yoshimizu and Kimura, 1976)

	Freshwater (immature)	Transplant (50% sea water)	Ocean	Freshwater (adult-spawners)
Terrestrial bacteria	100%	49%	—	34%
Marine/halophiles	—	51%	100%	66%
Bacterial genera	Aeromonas Enteric Bacteria	Aeromonas Pseudomonas Vibrio	Vibrio Achromobacter Aeromonas Pseudomonas	Vibrio Aeromonas Pseudomonas

The microflora of crustaceans, like that of fish, seems to vary with the temperature of the water in which the animals live. Gram-negative bacteria belonging to *Achromobacter* and *Pseudomonas* (and sometimes *Flavobacterium*) dominate the microflora of most crustaceans from cold waters while warm-water species carry mostly Gram-positive bacteria belonging to the coryneform group or micrococcus (Cann, 1977). Sometimes, however, temperate zone crustaceans also carry a high proportion of coryneform bacteria (Hobbs *et al*, 1971; Lee and Pfeifer, 1975). Some illustrative data are shown in *Table VI*.

Freshwater fish are subject to more exposure to fungus contamination. Infections with *Saprolegnia* species are not uncommon, particularly with dying fish (*eg* spawned out Pacific salmon).

Fungi (Wood, 1965; Johnson, 1968) are common in plankton and it is remarkable that they occur so seldom on fresh-caught fish.

Apart from passive transfer of enteric viruses by polluted shellfish there is no human significance to the viruses of fish. All are concerned with fish disease.

Table VI
MICROBIOLOGY OF EDIBLE INVERTEBRATES

	Pseudomonas	Achromobacter	Coryneform	Micrococcus	Other
			% incidence		
Scallops (Queens)	16	22	17	22	23
Shrimp (*Metapenaeus*)	1	5	55	24	15
Shrimp (*Parapeneopsis*)	17	26	7	39	11
Shrimp (*Pandalus*)	18	38	17	—	27
Crab (*Cancer*)	18	60	3	2	17
Antarctic krill	25	21	40	—	14

2.3 Yeasts, moulds and other micro-organisms

Though yeasts are often reported to be present in small numbers on fish there is only limited information on them available in the literature (Phaff *et al*, 1952; Ross and Morris, 1965; and Koburger *et al*, 1975. *Rhodotorula*, *candida* and *torulopsis* seem to be most abundant. *Rhodotorula* were also reported to be the most abundant genus (70%) isolated from a variety of freshwater fish in a recent study by Cantoni *et al* (1976). A listing of some genera reported to have been isolated from fish and shellfish is shown in *Table VII*.

It is certainly common experience to encounter pink yeast colonies of the rhodotorula type in small numbers on plates inoculated from fresh fish and shellfish but they never occur in high numbers. In irradiated fish, on the other hand, yeasts sometimes seem to become a dominant part of the post-irradiation storage flora.

Table VII
YEASTS ON MARINE FISH AND SHELLFISH

Candida	5 or more species	Fish and shrimp
Cryptococcus	2 species	Fish and shrimp
Debaromyces	2 species	Fish
Hansenula	1 species	Shrimp
Hanseniaspora	1 species	Fish
Pichia	1 species	Fish and shrimp
Rhodotorula	8 or more species	Fish and shrimp
Sporobolomyces	1 species	Shrimp
Torula	1 species	Oyster (frozen)
Torulopsis	7 or more species	Fish and shrimp
Trichosporon	3 species	Shrimp

Moulds are rarely reported associated with marine animals except where a diseased condition exists. Moulds of the genus *Pulluleria* (*Aureobasidium*) (Phaff *et al*, 1952; Koburger *et al*, 1975) have been isolated from fresh shrimp and chitinoclastic fungi are frequently observed growing on crab shells. Indeed, a widespread disease condition appears to exist in certain stocks of the Alaska tanner crab in which a chitinoclastic fungus seems to penetrate the shell and invade the underlying tissue (A K Sparks, pers. comm.).

3 Spoilage

There is no question that spoilage of fish is due primarily to bacterial action. The Torry group has shown in repeated experiments (Shewan, 1971) that aseptically excised fish muscle tissue held under sterile conditions for several weeks at refrigerator temperatures does not develop objectionable odour or appearance. Endogeneous biochemical changes which occur in fish *post mortem* are important in conditioning the substrate for bacterial action but autolysis as such is insignificant as a spoilage mechanism except perhaps in the belly-burn phenomenon which occurs in some heavily feeding fish (*eg* herring) when stored ungutted or in fish held at high temperatures conducive to rapid enzyme activity. Another possible exception is the rapid texture change which may occur in hake and sometimes in other species infected with *Myxosporida*.

The typical pattern of bacterial growth during spoilage is shown in *Fig 1*. It can be seen that this has the form of a classical population curve and the phases can be correlated with endogenous and bacterial enzyme-caused changes in the chemistry of fish muscle. The well-known curve from Reay and Shewan (1949) shown in *Fig 2* is still a satisfactory and true summary of some of the gross changes occurring at least in gadoid fish. The rate of increase of bacteria is a function of the temperature of storage and also the species of fish involved. Fish spoil at different rates and this seems to be related at least partly to the rate of increase of bacteria on them (*Table II*).

The apparent lag period of bacterial growth occurs during the development of *rigor mortis* and growth does not really commence until rigor is resolved. The accumulation of lactic acid in the muscle drops the pH in most fishes to the range 6·2 to 6·4, which is marginal for the growth of most spoilage bacteria but not absolutely inhibitory. Flatfish including particularly halibut are known to undergo a much greater pH drop to below 6·0 and it is interesting that these fish typically show more extended storage life than other relatively high pH types (Shewan, 1977). It is not clear whether pH is a factor in

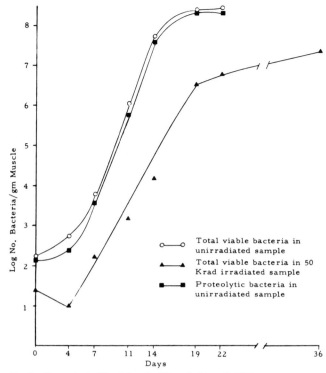

Fig 1 Bacteria in English sole fillets held at 0–2°C

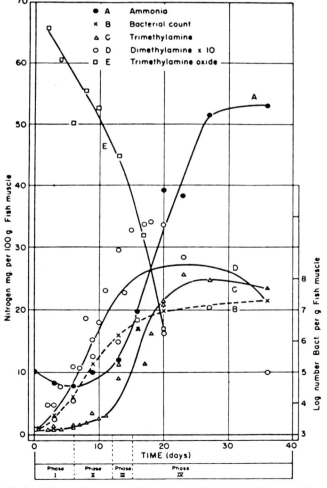

Fig 2 Effect of storage period on bacterial count and decomposition of fish muscle (Reay and Shewan, 1949)

the greatly extended shelf-life reported for some warm-water fish, though it seems more likely that this is a consequence of the kind of bacteria present rather than fish chemistry.

Bacteria seem to be confined to the surface layers until spoilage is advanced (Shewan, 1971). Penetration into the tissues or along blood vessels has been reported by earlier investigators (*eg* Beatty and Gibbons, 1937) but most recent research suggests that this is not the case in whole or eviscerated fish held at low temperature (Shewan, 1977). In fillets, penetration is more rapid but most activity still occurs at the surfaces. This is a function of the oxidative nature of the bacteria involved and many of the processes leading to spoilage changes appear to be oxidative in nature.

There is a shift in bacterial types during the storage period. *Pseudomonas* rapidly assume a dominant position in the microflora. *Moraxella*/Acinetobacter (*Achromobacter*) and *Flavobacterium* persist but at a decreasing relative level. This is illustrated by the data in *Table VIII*. Interestingly, a similar pattern is reported

Table VIII
PERCENTAGE COMPOSITION OF MICROFLORA OF HADDOCK FILLETS DURING STORAGE AT 3°C
(from Laycock and Regier, (1970)

Days in storage	Pseudo-monas	Achromo-bacter	Flavo-bacterium	Coryne-form	Other
0	26	45	15	4	8
4	70	19	1	8	2
8	87	12	0	0	1
13	65	29	0	1	4

for meat and poultry held under refrigeration, suggesting that common factors are involved. The *Pseudomonas* which become dominant include a number of different types which however share some common properties. They grow rapidly at low incubation temperatures (*Table IX* and *Fig 3*) and they utilize a variety of compounds of the type in fish muscle juice quickly

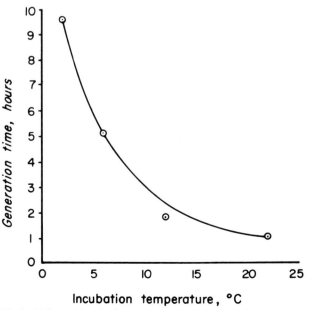

Fig 3 Effect of incubation temperature on generation time of bacteria from iced fish

143

Table IX
GENERATION TIMES AND LAG PHASES FOR BACTERIA ISOLATED FROM SPOILING FISH

Incubation temperature	Days of isolation	Number of organisms	Generation time, hours		Lag, hours	
			Mean	Range	Mean	Range
22°C	16	145	1·1	0·7–4·0	2·6	0–11
	9	120	1·3	0·8–5·0	2·2	0–6
	3	39	4·5	1·2–18	2·4	0–6
12°C	16	156	2·1	0·8–7·5	7·7	0–18
	9	125	2·3	1·0–11·0	7·2	0–18
	3	38	5·8	2·1–13	21	0–76
2°C	16	164	9·6	3·9–20·5	24	2–72
	9	35	11·6	8·7–16·0	15	4–45
	3	3	21	16·6–43		

and efficiently (*Fig 4*). They may be distinguished from the other psychrotrophic bacteria by the rate at which they utilize substrate. In *Fig 5* are shown rates of oxidation of alanine (a representative substrate) and all the spoilage *Pseudomonas* were found to fall into group 1.

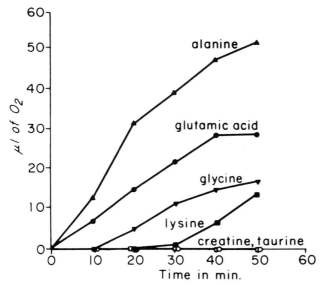

Fig 4 Utilization of fish NPN components by *Pseudomonas*

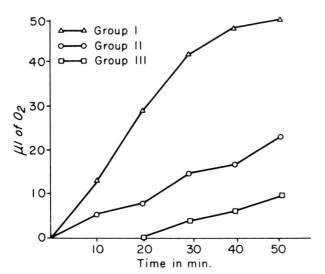

Fig 5 Mean value of oxygen uptake by bacteria from three-day iced fish grouped by rate

While a variety of *Pseudomonas* types can be found in the flora of spoiling fish and shellfish, certain types seem more important than others. From a broad chemical standpoint this may be because of ability to effect certain chemical changes. However, there is an uneven distribution of these properties as can be seen in *Tables X* and *XI*.

Table X
BIOCHEMICAL ACTIVITIES OF BACTERIA ISOLATED FROM FILLETS STORED IN ICE

Days ice storage	No. of isolates	TMAO reduction	Gelatin liquef.	Lipase	Volatile base
		Percentage positive			
0	56	13	29	18	32
4	94	5	12	20	28
8	98	15	32	37	52
12	77	18	32	31	34

Table XI
PERCENTAGE ORGANISMS FROM ICED FISH FILLETS PRODUCING TMA AND TVB

Days in ice	TMA Pseudomonas	TMA Achromobacter	TVB Pseudomonas	TVB Achromobacter
0	12	14	75	43
4	8	5	32	11
8	22	15	45	45
12	21	18	42	29
[18	48	0	95	100]

[ ] Separate experiment

Recent studies have identified those bacteria which produce spoilage odours when grown in pure culture on sterile fish muscle blocks or in press juice as being perhaps most important in spoilage (Castell *et al*, 1957; 1959; Lerke *et al*, 1965; Shaw and Shewan, 1968; Herbert *et al*, 1971; Miller *et al*, 1973; Shewan, 1974). Compounds probably responsible for the spoilage odours have been partly identified by gas chromatography and mass spectrometry and a list is shown in *Table XII*. The S-containing compounds are perhaps most important as spoilage odour components and *Pseudomonas fluorescens*, *P. perolens*, *P. putida* and *P. putrefaciens* have been shown to produce them. It has been concluded from the work of a number of investigators (Chai *et al*, 1968; Herbert *et al*, 1971) that the low GC *Pseudomonas*, called *Alteromonas* by some authors, are of primary importance since they typically produce S compounds from S-containing amino acids. These types have recently also been found to be of importance in spoilage of chicken and meat (Gill and Newton, 1978;

McMeekin *et al*, 1978). Certainly in fish spoilage the appearance of H₂S producers at levels in excess of 10^6/cm² seems to signal spoilage (*Table XIII*).

Table XII
PARTIAL LIST OF VOLATILE COMPOUNDS PRESENT IN SPOILING FISH MUSCLE IN SMALL AMOUNTS

Ethyl mercaptan	Ethanol
Methyl mercaptan	Methanol
Dimethyl sulphide	Acetone
Dimethyl disulphide	Acetoin
Hydrogen sulphide	Butanal
Diacetyl	Ethanal
Acetaldehyde	Methyl butanal
Propionaldehyde	

Table XIII
CHANGES[1] IN TMAO REDUCING AND H₂S PRODUCING BACTERIA ON STORED FISH

			Days Hake Rockfish English sole			
stored	TMAO[2]	H₂S_E	TMAO	H₂S	TMAO	H₂S
0	1·18	1·28	1·20	0	0	0
3	2·46	1·53	0·63	0	0·81	0·81
7	6·40	6·15	4·60	4·30	5·48	5·18
10	5·69	5·37	4·67	4·61	5·08	0
14	6·98	6·66	6·28	6·15	6·24	6·06
18	—	—	6·67	0	7·06	7·06
22	—	—	7·97	0	8·02	0

[1] Calculated by extrapolation from properties of isolates
[2] Estimated log count/g of organisms able TMAO→TMA
[3] Estimated log count/g of organisms producing H₂S

A most informative paper was recently published by Van Spreekens in which the properties of a relatively small but representative sample of fish spoilage bacteria were presented. Her data indicate quite clearly the range of properties of the active spoilage bacteria and some of her results are shown in *Table XIV*. The massive production of ammonia and volatile acids in the later stages of spoilage is no doubt due to oxidation of amino acids and other NPN (non-protein nitrogen) compounds. Studies during the 1960s by Dr Jong Chung (1968) in our Institute seemed clearly to establish that proteinase production by spoilage bacteria is repressed in the earlier stages of spoilage but is derepressed in the later stages of spoilage and this causes increased protein hydrolysis and leads to a rapid replenishment of the amino acid pool (*Figs 6* and *7*). Presumably this influences production of all end products derived from amino acids including H₂S from cysteine and (CH₃)₂S and CH₃SH from methionine. These substances were noted by Shewan (1977) to be produced rapidly after eight and 10 days' storage, respectively, which is close to Chung's observed time of derepression.

Thus the general bacteriology of spoilage in most

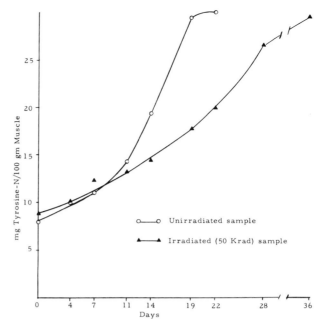

Fig 6   Tyrosine-*N* in English sole fillets held at 0–2°C

cases may be summarized according to the scheme in *Fig 8*. *Pseudomonas* types having the shortest generation times (and lag periods of growth) at temperatures in the 0–5°C range and with an enhanced capability to utilize NPN components of the muscle fluids as growth substrates rapidly outgrow the other bacteria present and become the dominant organisms in a population which increases to over 10^6 bacteria per cm² (or g). This population oxidizes amino acids and lactic acid, reduces trimethylamine and in later stages produces proteinases and a variety of S-containing compounds mostly derived from cysteine and methionine as well as other compounds variously identified which together produce the objectionable odours perceived as spoilage. Dominant in the production of the spoilage odours are low GC pseudomonads, probably *Alteromonas*, though other *Pseudomonas* may also be involved (Van Spreekens, 1977).

However, there are situations in which this neat pattern does not quite fit the facts. Walker *et al* (1970) have reported spoilage patterns for shrimp in which *Achromobacter* (*Moraxella*) and Corynebacterium strains dominate (*Table XV*). Crabmeat has been reported to spoil through action of Gram-positive bacteria. When fish are irradiated spoilage seems to occur through growth and activity of *Moraxella* types or yeast and sometimes *Aeromonas* or even *Lactobacillus* if airtight pouches are used. There are thus some unre-

Table XIV
PROPERTIES OF FISH SPOILAGE BACTERIA
(Van Spreekens, 1977)

Test	P. putrefaciens	Group Alteromonas[1]	Alteromonas[2]	Pseudomonas[3]
Fish spoilage	Strong odours	Strong odours (D)[4]	Odours (D)	Strong odours
Shrimp spoilage	Mostly strong odours	Strong odours (D)	Strong odours (R)[4]	Strong odours
H₂S cysteine	+	∓(2/10)	±	−
Gelatin liquefied	+	+	+	+
TMAO reduced	+	+	−	−
Hypoxanthine from IMP	+	+	+	−

[1] Non-defined spoilers   [2] Pseudomonad-like shrimp spoilers   [3] Pseudomonas Groups I and II   [4] D = delayed, R = rapid

145

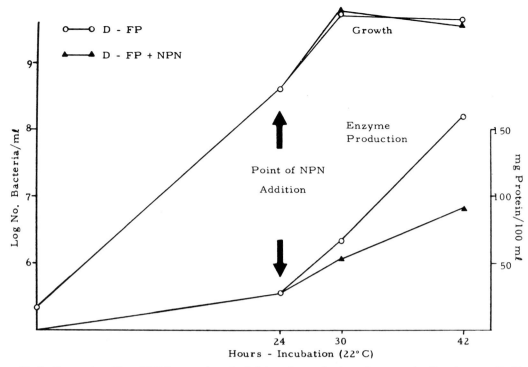

Fig 7 Repressive effect of NPN extract from fresh fish on the production of protease by *Pseudomonas* D–62

solved questions in the bacteriology of spoilage which require further investigation.

Fig 8 Sequence of events during spoilage

1. Spoilage bacteria naturally present on fish
2. Amino acid and other NPN substrate pool present
3. Selective growth of organisms (mostly *Pseudomonas*) which actively oxidatively deaminate amino acids
4. Repression of proteinase production derepressed by selective use of amino acids by *Pseudomonas* bacteria.
5. Amino acid recruitment to substrate pool by bacterial hydrolysis of protein
6. Ammonia and volatile fatty acid production sharply increases due to 5
7. Specific 'spoiler' types of bacteria produce S-containing and other odorous compounds

Table XV
SPOILAGE FLORA OF SCAMPI (I)
(Walker et al, 1970)

Days in ice	Pseudo-monas	Achromo-bacter	Flavo-bacterium	Coryne-form
2	25	32	2	34
4	3	9	1	79
6	5	58	3	33
8	0	72	1	13
10	8	70	1	19

4 Micro-organisms of public helath significance on fish

There are only two species of pathogenic bacteria which might truly be said to occur naturally on fish. These are *Clostridium botulinum* type E and *Vibrio parahaemolyticus*.

C. botulinum type E has been called the fish botulism organism and has been shown to occur in marine and lake sediments and in fish intestine. It is unusual in being able to grow and produce toxin at low temperatures (down to 3·3°C in suitable media) (Hobbs, 1976). However, it does not seem to grow and produce toxin in living fish but is carried passively. From a practical standpoint therefore it only becomes a hazard in mishandled processed products. Nevertheless, this organism and perhaps strains designated as type F and non-proteolytic B should be expected to occur in small numbers on fish.

V. parahaemolyticus is perhaps more truly an indigenous organism because it is halophilic and is known to grow in the marine environment and on fish and shellfish when temperatures are high enough (in practice probably above 15°C). The organism is distributed widely in inshore marine areas and can be readily isolated all year round from marine animals in warm-water areas (Liston, 1976) but is only found on animals during summer months in temperate zones. When *V. parahaemolyticus* is ingested at a level of 10^6 cells or more it frequently causes a characteristic food poisoning syndrome. However, not all *V. parahaemolyticus* are pathogenic and since the level of natural occurrence only rarely approaches infective numbers food poisoning from this organism again normally involves mishandling of seafood products. It is very sensitive to heat above 48°C and to cold, particularly 0–5°C, and is killed by exposure to fresh water. Most outbreaks in Japan derive from consumption of raw fish (Okabe, 1974) and elsewhere from eating shrimp and crab recontaminated after cooking and held at temperatures permitting rapid growth (Barker *et al*, 1974). Characteristically *V. parahaemolyticus* grows very rapidly when conditions are right (*Tables XVI and XVII*).

Other potentially pathogenic bacteria occasionally associated with fish and shellfish include *C. perfringens, Staphylococcus, Erysipelothrix, Edwardsiella, Salmonella, Shigella, Franciscella, Vibrio cholerae* and other Vibrios.

All of these organisms are probably derived by contamination from terrestrial sources (Shewan, 1971) with

146

Table XVI
OUTGROWTH OF *V. parahaemolyticus* IN SHRIMP AT 20°C[1]

Hours at 20°C	Raw	Precooked	Cooked
	Count × 10³/g		
0	3	3·5	5·4
1	11·0	8·5	13·0
3	120·0	320·0	120·0

[1] Average of seven experiments: 48 samples

Table XVII
OUTGROWTH OF *V. parahaemolyticus* IN CRABMEAT
HELD AT 20°C AND 30°C[1]

Hours stored	20°C	30°C
	Count × 10³/g	
0	1·4	1·4
1	3·0	6·2
3	140·0	130·0

[1] Average of four experiments: 24 samples

the possible exception of NAG Vibrios, lactose-positive Vibrios and possibly *V. cholerae*.

Vibrio cholerae can be very persistent in inshore marine environments as evidenced by the recent incidents involving crab in Louisiana, USA, while the dangers associated with the other Vibrios are still partly conjectural. *Salmonella* may also persist in fish for long periods. A recent study by Lewis (1975) showed that *S. typhimurium* persisted in warm-water seafish and freshwater species for 30 days and some of the fish even showed signs of actual Salmonellosis. Andrews *et al* (1977) recently reported isolating *Salmonella* from 4·5% of samples of fresh and 1·5% of frozen channel catfish from retail markets in USA. *Edwardsiella tardi* was isolated from 0·6% of fresh samples only. This suggests indigenous contamination of catfish both in USA and Brazil. *C. perfringens* has been quite widely reported on fish (Taniguti, 1971; Sohn *et al*, 1973; Burow, 1974; Matches *et al*, 1974). Studies by Dr Matches and his group in our Institute suggest that sewage may be the ultimate source of this organism, at least in Puget Sound fish.

There have been reports of *Yersinia* on fish (Zen-Yoji *et al*, 1975) but so far no hazard from this organism has been established.

Bacteria occur on fresh tuna which can cause a public health problem. These organisms actively decarboxylate histidine to histamine and this is associated with scombroid poisoning, a condition which affects people consuming tuna, mackerel or related fishes containing more than 100mg histamine/100g tissue. Other amines produced by the bacteria may also be involved since histamine itself ingested at the level in toxic fish does not apparently cause typical symptoms (Kim and Bjeldanes,

1979). Bacteria identified as being involved include *Proteus morgani*, *Klebsiella pneumoniae* and *Hafnia alvei* (Omura *et al*, 1978).

The occurrence of these various pathogenic bacteria is important to the fish processor who must design processing and handling procedures to exclude, eliminate or inhibit them. Equally it is important to the processor and retailer to know what the occurrence on fish is of organisms commonly used by regulatory agencies as indices of hazardous conditions. Typically such organisms include coliforms, fecal coli or *E. coli*, *Staphylococcus* and sometimes enterococci. It has already been noted that such organisms, with the possible and puzzling exception of staphylococci, should not be present on fresh caught fish. They are nevertheless rapidly added to fish when they are handled and as this happens at the point of capture it is not surprising that low levels of these organisms are found on iced fish as they are unloaded from the boats (Spencer and Georgala, 1958). But levels on properly handled fresh fish should be low. It was reported in *Microorganisms in Foods 2* (ICMSF, 1974) that fresh and frozen fish showed fewer than 100 fecal coli/100g and less than 100 coagulase-positive staphylococci/g even though total counts were often up to 10^6/g. The values for fish and other fresh or frozen seafoods are shown in *Table XVIII*. Changes in the proportions and numbers of these indicator organisms are valuable in assessing the bacteriological effectiveness of processing and handling procedures though there is often considerable doubt concerning their reliability as measures of potential hazard.

5 Processing

5.1 *Butchering, filleting and wet fish handling*

The numbers and types of bacteria on fish are affected by simple primary processing operations. Since bacteria are mainly confined to the skin, gills and intestine of freshly caught fish it might be expected that evisceration, beheading, filleting and skinning would greatly reduce the bacterial count on the final product. This depends on avoidance of cross-contamination and the addition of extraneous bacteria from the environment. In practice, even with the use of chlorinated process water, machine processing, antiseptic dips for hands and knives and other precautions, fillets, steaks and other products from fresh fish processors usually carry counts of 10^3/g–10^5/g of fillet flesh though occasionally lower counts are achieved and at other times much higher counts occur.

The major change qualitatively from primary processing is usually an increase in the relative proportion of Gram-positive bacteria present and a general appearance of bacteria associated with humans including some

Table XVIII
COUNTS ON RAW COMMERCIAL PRODUCT (GIVEN AS PERCENTAGE OF TOTAL TESTED)
(ICMSF, 1974)

	SPC at 25°C			Coagulase + Staph/g			Fecal coliforms/100g		
	<10⁵	10⁵–10⁶	>10⁶	<10²	10²–10³	>10³	<100	100–360	>360
Fish gutted and frozen at sea	64	30	6	99	<1	0	100	0	0
Fish fillets or frozen-blocks	52	43	5	99	1	<1	84	15	1
Frozen minced fish	31	28	41	96	4	0	67	7	26
Raw frozen scallops	68	27	5	–NA–			91	5	4
Raw frozen lobster tails	27	38	35	91	8	1	82	8	10
Raw frozen shrimp	24	11	65	95	4	1	83	8	9

staphylococci and enteric bacteria. This is apparent from the results in *Table XIX*.

Table XIX
EFFECT OF PROCESSING ON FISH MICROFLORA
(from Gillespie and Macrae, 1975)

	Newly caught fish	Auction[1]	Fillets	Retail fillets[2]
Bacterial count	8.4×10^3/cm²	7×10^4/cm²	7×10^5/g	8.6×10^5
		Percentage distribution		
Pseudomonas	18	15	6	11
Moraxella	8	15	6	7
Other G− bacteria	8	6	2	12
Micrococcus	49	41	76	45
Coryneforms	12	19	7	22
Other G+ bacteria	4	2	—	—
Others				

[1] An additional lot showed *Pseudomonas* 45%, Micrococcus 22%
[2] An additional lot showed counts of 10^8, *Pseudomonas* 57%, *Moraxella* 26%, Micrococcus 65%

A number of authors have studied the processing of shrimp. Quite variable results have been reported as can be seen from the data summarized in *Table XX*. Cooked products usually show a lower count than uncooked but subsequent operations cause an increase in count. Shrimp on world markets typically are reported to have counts which average 10^6/g (ICMSF, 1974). Processing may or may not bring about major changes in composition of the microflora of shrimp (*Table XXI*).

Table XX
EFFECT OF PROCESSING ON BACTERIAL COUNTS IN SHRIMP
(Ridley and Slabyj, 1978; Lee and Pfeifer, 1977)

	Maine, USA				Oregon, USA		
	Hand peeled	Machine peeled				Tropical	Pandalus
Initial	10^4	10^3	10^5	10^6	10^5	10^7	10^5
Brining	10^4						
Storage		10^{8a}	10^5	10^6			
Blanch		10^2			10^3	10^5	10^3
Peeling	10^3	10^3	10^3	10^4	10^5	10^5	10^5
Cook		<10					
Brining			10^2	10^4			10^5
Grading or packing	10^4	10^4	10^3	10^4	10^5	10^6	10^6

[a] From a storage hopper

Table XXI
EFFECT OF PROCESSING ON MICROFLORA OF PANDALUS SHRIMP
(Lee and Pfeifer, 1977)

	Initial	Final	Initial	Final
Pseudomonas	30	44	17	19
Moraxella	20	3	16	5
Acinetobacter	16	13	19	29
Flavobacterium	13	21	11	19
Arthrobacter	14	6	17	14
Other G+ Rods	6	5	7	4
Micrococcus	0	6	3	5
Other	0	2	10	5

Crab are normally cooked, cooled and picked. Cooking usually causes a sharp reduction in count but since the batch processes widely used do not ensure uniform temperatures, some undercooking usually takes place and so all heat-sensitive bacteria may not be destroyed (Cann, 1977). The count rises as a result of the picking operation and even though brining (for shell fragment separation) and washing cause another reduction, counts rise again during final operations. The results of a recent study by Lee and Pfeifer (1975) in *Table XXII*

clearly show this and also illustrate the effects of human contact and, no doubt, the brine which result in a shift from a dominantly Gram-negative microflora to one dominated by Gram-positives, particularly micrococcus and *Arthrobacter*. However, it should be noted that Gram-negatives regained their dominance during draining and final packing operations. It is common practice to bulk pasteurize picked crabmeat at 85°C in eastern USA. The pasteurized crabmeat is then held under refrigeration and may be kept for up to three months. Ward *et al* (1977) have reported recently on the bacteriology of crab treated in this way. They found the process reduced the aerobic count from 10^5 to about 10/g but the anaerobic count was reduced only from 10^6/g–10^7/g to 10^5/g. After three months the total count was 10^8/g–10^9/g of both aerobes and anaerobes. The final anaerobe population was mainly *Lactobacillus* with a few unidentified cocci also present, while aerobes appeared to consist of *Pseudomonas, Proteus* and *Brevibacterium. Clostridia* were not present. Crabmeat may have unusual effects on bacteria, perhaps due to the combined effects of heating and brine treatment. Earlier Slabyj *et al* (1965) found that staphylococci would not grow competitively on crabmeat while *Salmonella* will.

Table XXII
CHANGES IN THE BACTERIAL COUNT DURING CRAB PROCESSING (*Cancer*)
(from Lee and Pfeifer, 1975)

	Bacteria/g	Dominant species (%)
Raw crab	2.6×10^4	*Moraxella* (45), *Pseudomonas* (18), *Acinetobacter* (15)
Cooked crab	4.3×10^4	*Moraxella* (65), *Arthrobacter* (10)
Picked meat	1.3×10^5	*Moraxella* (31), *Arthrobacter* (28), Micrococcus (16)
Mat brined and washed	3.3×10^3	Micrococcus (30), *Moraxella* (20), *Acinetobacter* (15)
Finished meat	3.4×10^4	*Moraxella* (27), Micrococcus (19), *Acinetobacter* (18)

Mincing by use of deboner machines is being used more and more widely as the value of fish raw materials increases. It is a valuable procedure for recovery of fillet frame meat, flesh recovery from small fish and treatment of unappealing little-used species. Mostly, minced fish is converted into frozen products either as prepared foods or as complete fish blocks or, more recently, as fish block fillers or binders. Bacteriologically, minced fish by analogy with minced beef (hamburger) would be expected to be of poor quality. However, this is not the case in the Japanese industry where minced fish produced under closely controlled high sanitation conditions is used in the manufacture of fish sausage. Some years ago the bacteriological quality of minced fish produced in North America was very poor (see *Table XIX*) and this was found to be due to bad practices such as accumulating fillet waste over a day's run before deboning it. Recent reports on minced fish blocks have shown a better situation. Licciardello and Hill (1978) reported values for blocks well within acceptable ranges (*Table XXIII*). In an interesting study Raccach and Baker (1978) showed that minced products could be produced from cod, pollock and whiting with only an approximate one log increase in count but noted that products spoiled rapidly when thawed and held at 3°C.

It seems probable that minced fish applications will increase because of the highly economic yields obtained

Table XXIII
MICROBIAL QUALITY OF MINCED FISH BLOCKS (1975/1976)

	APC (Number per gram)	Coagulase + Staph.	Coliforms
Alaska pollack	$10-10^5$	0–16	0–3[c]
Cod frames	10^4-10^{7a}	0–16	0–15
Cod	$10^{3.5}-10^{6b}$	0–24	0–43
Other N Atlantic fish	10^3-10^6	0–24	0–43

[a] 13% above 10^6
[b] 95% below 10^6
[c] Fecal coliforms detected in only four blocks

using this processing method. The extreme susceptibility of this product to bacterial contamination should be clearly recognized.

5.2 Freezing

The effect of freezing on the bacterial population on fish is erratic and difficult to predict. There is generally some reduction in count and the numbers will continue, in most cases, to fall during storage in the frozen state (*Table XXIV*). In general, Gram-negative bacteria are more sensitive to freezing than Gram-positive bacteria and bacterial spores are highly resistant. *Salmonella* and other members of the Enterobacterioceae are among the more sensitive bacteria (Raj and Liston, 1961) but there are great variations in the response of these organisms when present on fish. Indeed, the variability of response has cast some doubt on the practice of freezing portions of foods for later bacteriological analysis (DiGirolamo *et al*, 1970). Even in the case of cold-sensitive micro-organisms such as *V. parahaemolyticus* there may be some survival after freezing (*Table XXV*).

Unfortunately from a practical standpoint, the conditions of freezing and subsequent cold storage most desirable for quality, *ie* rapid freezing, low non-fluctuating temperature during storage, are most protective of the bacteria present. It is thus safest to assume that freezing simply preserves the bacterial *status quo* of the product.

Freezing is of course an effective method of halting bacterial action. Though a few micro-organisms have been reported to be able to grow at $-7.5°C$, in practice there is no significant bacterial activity below $-5°C$ in seafoods. When frozen seafoods are defrosted and stored at refrigerator temperatures they seem to undergo bacterial spoilage similar to that of unfrozen products. There is no convincing evidence that they spoil either more quickly or more slowly.

5.2.1 Prepared frozen products

There is an increasing number of prepared frozen seafood products in the world markets. These include both raw and precooked seafoods and range from relatively simple products such as fish-sticks and breaded fish portions to compounded shaped or extruded materials incorporating potatoes or other starch sources, stabilizers and flavouring agents as well as the ubiquitous breading. Some data are shown in *Tables XXVI* and *XXVII* which illustrate the bacteriology of breading and batter operations (Raj, 1968). Breading and batter introduce different micro-organisms to the seafood including particularly Gram-positive bacteria and moulds. Because egg is often used in batter formulations there has been concern over the possible introduction of *Salmonella*. The precooking operation normally involves deep-fat frying but this does not necessarily

Table XXIV
EFFECT OF FREEZING ON MICROFLORA OF FISH (OCEAN PERCH) (from Lee et al, 1967)

	Fresh	Frozen
Bacteria count	$4.9 \times 10^5/g$	$8.3 \times 10^4/g$
Coliforms	$0.6/g$	$0/g$
Pseudomonas	22.9%	26.9%
Achromobacter	28.4%	22%
Flavobacterium	16.0%	12.1%
Gram and rods	27.2%	24.8%
Micrococcus	4.3%	1.5%

Table XXVI
BACTERIAL COUNTS IN SEAFOOD BATTER AND BREADING (Raj, 1970)

	Plate count 20°C	Plate count 35°C	Coliforms MPN/g	E. coli MPN/g	Staphylococci positive
Dehydrated mix	2.1×10^4	1.5×10^4	80	30	7
Reconstituted batter (Vat)	52.4×10^4	26.1×10^4	3 700	125	70
Batter – on line	1.71×10^6	1.45×10^6	6 800	382	100
Breading	1.6×10^4	1.5×10^4	44	0	0

Note: No *Salmonella* were isolated from 49 samples tested

Table XXV
SURVIVAL OF V. parahaemolyticus (STRAIN 17802SM) IN REFRIGERATED AND FROZEN OYSTERS[a]

Days of storage	+11°C	+8°C	+5°C	+1°C[d]		−15°C[d]		−30°C[d]	
0	58.0[b]	58.0	58.0	58.0	320.0	58.0	320.0	58.0	320.0
1	—[c]	—	—	—	190.0	—	54.0	—	57.0
2.5	9.7	6.9	5.3	6.8	—	16.0	—	8.7	—
8	10.0	4.2	4.4	1.6	19.0	3.5	21.0	9.5	—
14	—	0.18	1.6	—	16.0	—	9.1	—	49.0
	+11°, +8°, and +5°C Completed								
17				2.2	—	2.0	—	2.8	—
21				0.39	—	2.2	—	2.4	—
24				—	5.6	—	13.0	—	85.0
27				0.05	—	1.3	—	6.7	—
30				—	0.03	—	19.0	—	75.0
42				0.002	0.005	0.45	2.7	4.1	28.0
59				Completed		0.73	—	5.1	—
66						—	0.75	—	29.0
82						—	1.7	—	13.0

[a] Depurated oysters contaminated by natural uptake
[b] Counts given in thousand organisms/g as obtained on TSASS at 37°C
[c] Not done
[d] Duplicate experiments

149

Table XXVII
EFFECT OF PROCESSING ON BACTERIA OF PUBLIC HEALTH SIGNIFICANCE IN
PRECOOKED FROZEN SEAFOODS
(Raj, 1969)

	Plate count 20°C	Plate count 35°C	Coli- forms MPN/g	E. coli MPN/g	Staphy- locci % positive
Frozen blocks	10^4–10^5/g	10^4/g	<10–10^2	<10	64
Cut, battered and breaded	10^5–10^6/g	10^5/g	10^1–10^3	<10–10^2	73
Precooked	10^4–10^5/g	10^3–10^4/g	<10–10^1	· 0	52

Note: No Salmonella were isolated from 293 samples tested

raise the temperature of the entire product to a steriliz-
ing level.

Nevertheless, in general the bacteriological quality of
frozen prepared seafoods is good and has indeed shown
an improvement over the years as processors become
more conscious of the need for a high level of hygiene in
their preparation (see Tables XIX and XXVIII).

Table XXVIII
MICROBIAL QUALITY OF FROZEN BREADED SEAFOOD PRODUCTS

	Percentage of samples containing Plate count <10^6/g US[1]	Plate count <10^6/g Canada[2]	E. coli <3/g US	E. coli <3/g Canada	S. aureus <100/g US	S. aureus <100/g Canada
Fish sticks	99·8		96·1		100	
'Other breaded'		100		100		100
Fish cakes	99·4		90		99·7	
Scallops	100		98·6		100	
Fish portion	99·2		97·9		100	
Breaded raw shrimp	82·2	7·3	98·8	99	99·7	83·3
'Precooked'	—	97·4		100		—

[1] From Baer et al (1976)
[2] Import product values from Neufeld (1971)

Table XXIX
CANNED FISH PRODUCTS CANADA 1954–1968
(from Neufeld, 1971)

Product	Sterility test results (%) 35°C Swells	35°C Th(+)[1]	50°C Swells	50°C TH(+)
Fish spreads	0	6·7	0	0
Salmon	0	0·3	0·4	<0·1
Domestic clams	0	0·2	0·8	28·2[2]
Imported clams	0·5	0·8	3·3	30·8[2]
Imported shrimp	0·1	2·4	2·8	9·7

[1] Th+ indicates positive in thioglycollate
[2] Due to thermophile sporeformers which resist 90min at 121°C

6 Canned seafoods

Canned seafoods fall into two categories from a bacteriological point of view: these are (a) fully processed
commercially sterile products and (b) semi-conserved
products. The fully processed products would include
canned tuna, salmon, shrimp, crab, sardines and other
fish, fish balls, etc. The heating process applied to these
products is designed to destroy all pathogenic bacteria
(12D botulinum cook) and normal numbers of other
organisms. Problems arise mainly from spore formers
present in the unprocessed material in excessive numbers or, more commonly, gaining entry to the can after
processing due to improper seaming or contaminated
cooling water. Flat sour spoilage due to thermophiles
such as Bacillus stearothermophilus which survive pro-

cessing and multiply during slow cooling or storage at
excessively high temperatures (45°C or above) can be a
problem in seafoods. Swollen or 'blown' cans usually
contain Clostridia (often C. sporogenes) which have
survived an improper processing cycle or have infected a
'leaker'. The most common cause is a deficient double
seam. Faulty seaming is compounded by the use of
contaminated cooling water because of the vulnerability
of seams due to the stresses which occur during cooling.

It has been reported that some canned fish products
are not always sterile (Neufeld, 1971). This may be due
to the oil pack used since oil will protect bacterial spores
against heat to some extent. Another specific seafood
problem is the periodic difficulty encountered in canning clams due to a heat-resistant spore former which is
probably derived from sediment (Table XXIX).

Semi-conserved products are usually preserved by a
combination of treatments including pickling in salt or
acid or by fermentation and a mild heat treatment. In
modern practice most semi-conserved products are held
under refrigeration and this is probably the reason for
the improved record of these products as compared with
some years ago (Shewan and Liston, 1955). They are
not, of course, sterile and depend on a delicate balance
of factors to maintain stability. The factors may include
low pH, salt, a_w control, specific acids or other preservatives, anaerobic conditions and of course refrigeration.
Failure can occur as a result of a shift in any of these and
this can permit growth of acidophilic bacteria or yeasts
or moulds and even sometimes of dangerous bacteria.
Since these delicatessen items are rarely given additional processing before eating it is extremely important
that they be prepared under hygienic conditions and
handled with great care.

Of course the most feared consequence of faulty
processing of canned foods is botulism. Botulism due to
canned fish products is fortunately rare. Recent incidents involving commercially packed products have
been due to a failure of can integrity, permitting contamination and growth after processing (Hobbs, 1976).
There have been incidents due to improper processing in
home canned seafoods. One recent type A incident in
Seattle was due to insufficient venting of the home
pressure cooker. An unusual incident involving staphylococcal food poisoning from improperly canned salmon
was recently reported in Canada.

7 Salting and drying

Preservation of fish and shrimp by drying or by salting or
by a combination of these processes is still widely
practised throughout the world though this method is
less important in industrialized countries. The principal
effect on micro-organisms is due to the lowering of a_w,
though NaCl itself in higher concentrations may be
lethal for some bacteria and yeasts due to osmotic
effects. Surprisingly, even sensitive bacteria such as
Salmonella which contaminate a dried product may
persist on it for some time.

Microbiological changes occur mainly during the salting and drying process with growth sometimes taking
place in the early stages while the a_w level still permits
this ($a_w > 0.90$) (Dusseault, 1958). The final population
on salted and dried fish is generally dominantly com-

150

posed of micrococci and Gram-positive rods (Liston and Shewan, 1958).

Dried fish, shrimp or other seafood products are readily contaminated by mould spores which germinate and grow if the product becomes slightly moist. Mycotoxigenic moulds have been identified on such products.

The largest single dried fish commodity accounting for more than one-third of world fish production is of course fish meal. Microbiological concerns of the fish meal industry have mainly been directed to preservation of the raw materials and contamination of the finished product. The first concern has really not been resolved but a variety of compounds including formaldehyde, organic acids and antibiotics have been tested with varying degrees of success. Of course, preservation is not simply a bacteriological problem, for the small pelagic fish caught in such enormous quantities by purse seine are usually feeding so that gut enzymes are major factors in the rapid liquefaction which can occur in masses of fish held unrefrigerated in vessel holds or in outside holding bins at tropical temperatures. The major problem of product contamination has been *Salmonella*. Fish meal among other feed components has been widely blamed for the apparent dissemination of *Salmonella* serotypes throughout the world. Certainly there are some obvious cases of *Salmonella* contamination of fish meal held unprotected in the open by birds and rodents. One benchmark study of the menhaden industry showed that raw fish was contaminated in the boats and in holding areas of the plants (Morris *et al*, 1970). More importantly, it was observed that only the first 30–45min of each day's production yielded *Salmonella* and this was shown to be due to growth of the organisms in the moist fish material left in presses and conveyers overnight when the plant was shut down. Once the equipment in the processing area was fully heated up *Salmonellae* were destroyed and did not appear in the final product. Thus Morris *et al* recommended that chlorinated water should be used for pumping fish, boats and machinery should be completely cleaned and sanitized after each use period and the first portion of each day's meal should be reprocessed.

Spoilage of salt fish may occur due to growth of halophilic bacteria or halotolerant moulds. The halophilic bacteria are frequently derived from contaminated solar salts (Bain *et al*, 1958). They include rod-shaped *Halobacterium* and coccal *Halococcus* forms (Gibbons, 1974) and are most troublesome at the wet stack stage in salting and drying processes when the a_w is still above 0·75. They may grow on fully dried salt fish which is locally dampened. These bacteria cause a pink or red discoloration and produce areas of softness and decay. The most troublesome spoilage moulds are those causing dun and brown discoloration associated with localized decay. Anti-mould substances such as propionic and sorbic acids are useful as treatments in inhibiting mould growth (Phillips and Wallbridge, 1977).

8 Smoking

Quite large amounts of fish are still treated by smoking processes. In the industrialized countries this process is primarily designed to produce a product of desirable appearance, odour and flavour, but in much of the world smoking is used as a preservative process. Preservation is achieved essentially by drying.

Smoking may be done as hot smoking or barbecuing or cold smoking. In hot smoking the internal temperature of the product will normally exceed 60°C while cold smoked products rarely exceed 35°C.

Many smoking procedures call for a preliminary brining for texture and flavour and in some cases fish may be held in brine and freshened prior to smoking (*eg* mild cured salmon, herring). The brining will bring about an initial change in the microflora as noted earlier.

Generally, smoking extends the shift from a Gram-negative to a Gram-positive microflora. Coryneform bacteria and micrococci and bacillus are frequently the dominant forms present (Lee and Pfeifer, 1973). However, in cold-smoked fish a typical pseudomonad spoilage flora develops during subsequent storage. In the case of hot-smoked products the internal temperature reached in the fish during smoking is important. It has been reported that chub smoked at 140°F for 30min had a microflora dominated by cocci (61%) and non-sporing rods (34%). When smoked at 160°F the product showed 81% spore-forming rods with cocci and non-spore-forming rods composing only 9% and 10% respectively of the microflora.

As a result of botulism outbreaks caused by hot-smoked Great Lakes white fish the dangers of botulism due to growth of *C. botulinum* type E in smoked fish are now very widely recognized. The US Food & Drug Authority has promulgated regulations for processes which, if followed, would virtually exclude any danger of botulism from this source. These require that the fish be brought to an internal temperature of 82°C for 30min if brined to contain 3·5% water phase salt or not less than 65°C for 30min if salt content is 5·0% in the water phase. However, cases of botulism from smoked fish are traceable to improper holding practices. Unless products are dried to an a_w below 0·93 during the smoking or by heavy salting they should be considered potentially capable of supporting growth of *C. botulinum* and should therefore be held under refrigeration or consumed shortly after preparation. The combined use of nitrite and salt is quite effective in preventing growth of *C. botulinum* but, of course, nitrite is currently looked on with disfavour because of suspected carcinogenicity.

Hard-smoked and essentially dried-fish products are normally spoiled by moulds. Some of the moulds reported from fish are listed in *Table XXX*. In a recent study of smoked and dried fish from tropical regions Phillips and Wallbridge (1977) reported isolating *Aspergillus* (six species), *Wallemia* (*Spirendonema*), *Penicillium* (five species), *Acremonium* and *Rhizopus*. Most of these fungi are capable of growth if the relative humidity rises above 70%.

Table XXX
FUNGI ON SMOKED AND SALTED FISH

Genus	Source	Author
Aspergillus sp.	Smoked fish	Graikoski (1973)
Oospora Nikitinski	Salted fish	Malevich (1936)
Oospora spp.	'Dun' of salted fish	Frank and Hess (1941)
Penicillium sp.	Smoked fish	Graikoski (1973)
Sporendonema epizoum	Salted fish	Van Klavern and Lequendre (1965)
Sporendonema spp.	'Dun' of salted fish	Frank and Hess (1941)

Production of mycotoxins, particularly aflatoxin, by moulds growing on foods has recently become a matter of increased concern because of the demonstrated carcinogenicity of these compounds. Wu and Salunkhe (1978) have recently reported on the isolation of 114 strains of fungi from dried shrimp—a common commercial product in many parts of the world. While 27 of the strains were toxigenic by a chick embryo test, two were confirmed to be aflatoxin producing strains of *A. flavus* and were shown to be capable of producing the toxin on dried shrimp. They also noted that dry shrimp showing mould growth had a moisture content of 28–31%, except for one sample with 10%, while shrimp showing no overt mould growth had moisture contents in the range 9–12·8%. The importance of protecting such products against rehydration during storage or transport is apparent (*Table XXXI*).

Table XXXI
FUNGI ON DRY SHRIMP
(Wu and Salunkhe, 1978)

Aspergillus[1]	11 species[2]	54 strains (16 toxic)
Penicillium	9 species[2]	52 strains (9 toxic)
Rhizopus	2 species	4 strains
Cladosporium	1 species	1 strain (toxic)
Alternaria	1 species	1 strain (toxic)
Tricothecium	1 species	1 strain

[1] Includes two aflatoxin-producing *A. flavus*. Most common species *A. ruber* 11%)
[2] Most common species *P. expansum* (13%)

9 Fermentation

Fermentation is practised as a means of preserving or altering the flavour of fish products to a greater degree in the Orient than in Europe or North and South America. There are many different fermented products and most are at least mentioned by Van Veen in his 1953 review and by Tanikawa (1965). Of the major products, fish sauces, which are produced by allowing fish, small crustaceans or squid to digest in a brine formed by adding 20–25% salt to the raw fish, seem to be mainly the product of fish enzyme action. Bacteria rapidly diminish in number during the fermentation period (*Table XXXII*). Some studies suggest that bacteria are involved in flavour development (Saisithi *et al*, 1966). Specifically, micrococci, clostridia and pediococcus have been separately indicated as flavour producers. However, Orejana (1978) in a recent investigation could find no evidence of microbial involvement.

Table XXXII
BACTERIAL CHANGES IN PHILIPPINE FISH SAUCE
(Orejana, 1978)

Time fermented	Experimental (anchovy) log count/ml		Commercial (scad) log count/ml
0 days	7·23		—
15 days	2·95		—
40 days[1]	1·	(1 month)	2·8
7 months	—		2·2
8 months	—		1·4
12 months	—		1·3

[1] Count remained unchanged through 140 days

Mixed fermentations of fish or shellfish and vegetable or cereals are also used to prepare products such as

i-sushi (Tanikawa, 1965). In these cases both lactic acid bacteria and moulds may be involved. It has been suggested that the lactic acid bacteria may produce antibacterial substances which stabilize the final product.

Mould fermented seafood products are popular in Japan. The category known as koji-zuki (Tanikawa, 1965) is prepared by adding a koji to salted fish or roe. The koji is prepared by growing *Aspergillus oryzae* on steamed rice. The mould contributes proteinases and other enzymes which produce desirable flavour and texture changes. Another favourite Japanese product which involves moulds is katsuobushi. This is prepared by boiling, smoking, sun drying and then deliberate moulding of bonito fillets to produce a hard stick-like product which is used to prepare soup stock. The moulds which grow on the partly (sun) dried product are mostly *Aspergillus* and *Penicillium*, and they assist final drying, reduce fat content and improve flavour.

There are many other fermentation processes used for fish and shellfish products in southeast Asia and some use yeasts and moulds as well as bacteria. In most cases virtually nothing is known about the micro-organisms involved though it is clear that in many cases lactic acid bacteria are important. Recently it has been suggested that the stability of some products is due to the production of antibacterial compounds, probably by the lactobacilli (Lindgren and Clestrom, 1978a,b).

The modern practice of fish ensilage which is carried out in Europe and is being introduced to other parts of the world does involve a lactic fermentation. Lactic starter cultures including *Lactobacillus plantarum*, *Pediococcus* and others are added to the mixture of fish and carbohydrate source such as cereal meals, cassava, molasses, *etc*, and a controlled digestion proceeds (Disney *et al*, 1977). This microbial process shows great promise as a means of utilizing fish which might otherwise be wasted to produce a high-quality animal food. A controlled mould based fermentation using 'ragi' (a mixed culture of amylolytic moulds and yeast in rice flour—similar to the koji mentioned earlier) and a mixed lactic starter has been shown to be effective (Stanton and Yeoh, 1977).

10 Novel processes

10.1 *Fish protein concentrate (FPC)*
FPCs are distinguished from fish meals by the fact that in their preparation a greater or lesser proportion of the lipids are removed so that the final product has a protein content in excess of 70%. They are also prepared under hygienic conditions so that the final bacterial count is low and potentially pathogenic bacteria are absent (Finch, 1970). Processes involving the use of organic solvents to remove fats usually result in FPCs which have few bacteria since the solvents themselves are toxic to micro-organisms (Goldmintz and Hull, 1970). However, even the milder protein precipitation (Finch, 1970; Chu and Pigott, 1973) and biological processes (Finch, 1970; Tarky, 1971; Pigott *et al*, 1978) produce low-count FPCs because processing conditions are quite bactericidal.

10.2 *Irradiation*
Studies on the use of high-energy radiation (particularly

γ-radiation and high-energy electron beams) to destroy bacteria and thereby extend the shelf-life of fish and other seafoods, began in the 1940s and have continued intermittently ever since (Hayner and Proctor, 1953; Hannan, 1956). Dose levels of 100 krad or greater are highly effective in destroying common spoilage bacteria on fish (Liston and Matches, 1968) (*Table XXXIII*). However, below 1 000 krad the effect is limited to an extension of shelf-life under refrigerated storage. Higher dose levels in the region of 3 500 to 5 000 krad (3·5–5 megarad) effectively sterilize food products but the labile lipids of fish make them particularly susceptible to undesirable flavour changes at such high radiation levels.

It has already been noted that the most remarkable effect of radiation at the lower pasteurizing level (100–200 krad) is to bring about a shift in the apparent spoilage flora from *Pseudomonas* to *Achromobacter* (probably *Moraxella*) and Gram-positive bacteria (micrococcus, lactobacillus, cornyneforms) and yeasts. At slightly higher (> 300 krad) dose levels yeasts and Gram-positives dominate (Pelroy and Eklund, 1966). Spoilage in aerobic packs is generally due to *Achromobacter* though yeasts sometimes dominate (Pelroy *et al*, 1967). Much higher levels of count seem to be necessary before spoilage is apparent in irradiated and stored fish than in non-irradiated fish similarly treated (> 10⁸/g as compared with about 10⁶/g). In our laboratory, *Aeromonas* species have been found to predominate in vacuum-packaged low-dose irradiated fish though other workers have also noted dominance of lactobacillus (Pelroy and Eklund, 1966).

Table XXXIII
BACTERIAL COUNTS ON IRRADIATED FILLETS (ENGLISH SOLE) PREPARED FROM ONE-DAY ICED FISH

| Dose, krad | Days after catch | | | | |
| | 1 | 6 | 11 | 15 | 22 |
	(Bacterial count in log No. Cells/g)				
0	2·46	2·78	5·31	7·9	8·05
25	2·14	2·40	4·64	5·9	5·81
50	1·1	1·59	2·79	3·7	5·65
100	1·1	0·88	2·19	4·23	5·29
150	0	0·40	1·46	1·8	3·99

Fillets packaged and stored at 1°C

Research on radiation processing of food is once again being actively pursued in countries other than the US and it is to be hoped that new information will cast more light on the sometimes peculiar microbiology of irradiated fish.

One major problem of radiation-pasteurized fish is *C. botulinum* growth. This is well reviewed in a recent article by Hobbs (1977).

11 Public health microbiology

With the notable exception of molluscs, fish and shellfish taken in cold or temperate waters are rarely a major cause of human disease (*Table XXXIV*). Molluscs are recognized as a potential health hazard because they are frequently harvested from estuarine and intertidal areas which may be subject to pollution from human or land animal sources. In feeding, bivalve molluscs pass large quantities of water over a mucous system which selects out microscopic particulates and carries them into the alimentary canal. The net effect is to concentrate bacteria and viruses present in the water. There are two hazards to human consumers in this system: if pathogenic bacteria or viruses are in the water they will be concentrated by the molluscs and consumed with them, or if toxic dinoflagellates such as *Gonyaulax catanella* are present they will be digested by the mollusc and toxin may be stored in the tissue. The former situation can lead to outbreaks of gastroenteritis or virus disease such as infectious hepatitis while the latter can cause the much more lethal paralytic shellfish poisoning. Primary control of both situations depends on regular surveillance and strict observance of harvesting closures when a dangerous situation is revealed by laboratory testing of water or animals.

Another relatively rare public health problem can arise when pathogenic micro-organisms are passively transferred by fish. Recent examples are the transmission of cholera by crabs in Louisiana, USA, and the *Salmonella* outbreak due to smoked eels in Hanover, Germany (Center for Disease Control, 1978; Hurms and Kruse, 1976). A potential hazard in the case of catfish was noted by Andrews *et al* (1977).

Fish and other foods of aquatic origin are also subject to the same problems of contamination and pathogen outgrowth due to mishandling or inadequate processing after landing as are other foods. *Staphylococci* and *Salmonella* will grow readily on fish if conditions are suitable (Matches and Liston, 1968). However, the only particular problems are due to *Vibrio parahaemolyticus* and *C. botulinum* type E and possibly type F (Craig and Pilcher, 1966). It has been noted earlier that sickness due to these organisms is almost always a result of errors in processing or handling (a possible exception was the outbreak diagnosed *ex post facto* in Washington State in which oysters and clams with an unexpectedly high natural load of *V. parahaemolyticus* were consumed (Barker *et al*, 1970).

Unfortunately in tropical regions there is reason to suppose that fish and other marine and freshwater products play a larger role in human disease. The figures for Japan are quite conclusive (Okabe, 1974). The vast bulk of gastroenteritis cases in that country can be traced to seafoods. Of course, seafoods constitute the major animal protein source in the Japanese diet and this in itself may explain the high incidence. Nevertheless, the uniquely fish-borne pathogens *V. parahaemolyticus* and *C. botulinum* type E are prominent as a cause of sickness. This is true, at least so far as *V. parahaemolyticus* is concerned, in other southeast Asian countries. So

Table XXXIV
PERCENTAGE OF OUTBREAKS DUE TO FISH AND SHELLFISH (FS), MEAT AND POULTRY (MP) AND UNKNOWN FOODS (U) IN FIVE COUNTRIES (adapted from Todd, 1978)

	FS	MP	U
Canada[1]	5·8	40·3	17·6
USA[1]	9·3	27·3	28·3
England and Wales[1]	0·5	12·0	82·9
Australia[2]	12·5	43·7	31·3
Japan[3]	35·4	–NA–	33·6

[1] 1973–1975
[2] 1967–1971
[3] 1968–1972

far, few major problems have resulted from the increased importation of fish and crustacea from tropical areas to Europe and USA and this may be a consequence of the rather careful handling of these commodities. But it is important that importing countries maintain a consistent surveillance on seafood products coming in from areas in which dangerous bacteria are known to be endemic.

A short but useful review of disease from fish was written by Brown and Dorn (1977) and a more comprehensive earlier discussion is provided by Bryan (1973).

12 Fish disease

Nobody knows the extent of disease incidence among wild stocks of fish and other aquatic animals. The sea is too vast and still largely inaccessible to human observation on a continuing basis. It is only after calamitous epidemics occur resulting in fish kills of enormous proportions that evidence of disease is easily available, but even then it is not always possible to identify the causal agents.

Diseases of fish in rivers or small freshwater lakes are more easily seen, though even in such small systems the recycling efficiency of the food chain ensures rapid disappearance of cadavers.

Most work has been done on those species which are raised in hatcheries or feeding stations. Of course, while most of the aquaculture is in the Orient, most of the science capability is in the West and so it is no surprise that most is probably known about diseases of salmonid fish though something is known about carp, since they are raised for food in Europe and Israel. Japanese scientists have been identifying some of the diseases which affect those species of marine animals which are extensively cultured (mariculture) in that country.

The field is currently under such intense scrutiny by a range of scientists in different parts of the world that it would be impossible to adequately discuss the present status. The interested reader is referred to reviews by Sniezsko (1973), Roberts (1976), Egusa (1976), Fryer and Amend (1977).

However, it can be noted that the range of pathogens is quite wide as can be seen from the fragmentary listing in *Table XXXV*.

Table XXXV
SOME BACTERIA PATHOGENIC FOR FISH

Aeromonas salmonicada	Salmon furunculosis
Corynebacterium sp.	Kidney disease of salmonids
Cytophaga psychrophilia	Cold water disease of salmonids
Flexibacter columnaris	Cotton wool disease (gill disease)
Haemophilus piscium	Ulcer disease of trout
Mycobacterium spp.	Fish tuberculosis
Myxobacterium sp.	Bacterial infections hatchery fish
Pasteurella piscida	Kidney and liver disease of sea fish
Pseudomonas spp.	Localized lesions and septicaemia
Vibrio anguillarum	Vibrio disease in fresh and seawater fish
Vibrio spp.	Local lesions and generalized disease
V. parahaemolyticus	Invasive disease of shrimp, crab and oysters

One recent achievement worthy of note is the development of effective systems for the vaccination of fish against particular diseases. This technique is now widely applied in the Pacific northwest to protect young salmon against Vibrio disease (Fryer and Amend, 1977).

With the possible exception of *V. parahaemolyticus* which has been shown to cause disease in shrimp, crabs and oysters, none of the fish disease organisms is known to be pathogenic to man. Of course, the interesting observation by Lewis (1975) on the apparent salmonellosis in mullet and other fish under experimental conditions raises questions for this genus also.

13 Future directions in research

Predictions of where research will go in the future are usually doomed to failure because research by its nature creates its own directions as new ideas are generated and new leads followed. Nevertheless, it is certainly possible to point to current areas of research interest and to identify some research needs.

Microbial research on spoilage is sure to continue. The relatively recent findings concerning the Alteromonas group and their role in the decay of meat, fish and poultry are still being digested. Meat scientists are actively researching *post mortem* change and some particularly interesting work is being published by Gill and his colleagues in New Zealand. One sees the thrust of the work here being concentrated on biochemical interactions between tissue components and microbial enzymes and histological work to determine textural effects of bacterial spoilage activities.

Coupled with the bacteriological studies one would expect to see further exploration of tests for quality related to spoilage. The chemical 5-5'dithiobis-2-nitrobenzoic acid (DTNB) used by Gillespie to detect volatile sulphide compounds from spoilage bacteria shows real promise in preliminary studies in our laboratories. But no doubt there are even more novel and ingenious methods being worked on elsewhere.

A simple technology to suppress or retard spoilage which does not depend on ice or refrigeration would be enormously valuable to artisanal fishermen in developing countries. This is a microbiological problem which might conceivably be resolved using sugar and lactic starters or other competitive 'non-spoiling' bacteria.

Fish fermentations which involve bacteria, yeasts and moulds present a largely unexplored field of opportunity to the fish microbiologist. Some of the traditional systems might well offer prospects for a more modern industry. Clearly, fish ensilage under tropical conditions and using indigenous carbohydrate sources also is a potentially fruitful field for research.

If, as seems likely, the trend towards shelf-stable food products is maintained, it is likely that an increased number of dehydrated or intermediate moisture fish products will be developed. This will bring forward a need for knowledge of the yeasts and moulds which can grow at low a_w levels. We are already aware of the potential public health problems which are posed by the growth of moulds on some foods due to possible mycotoxigenesis. However, because most dried-fish products currently fall into a low value category, work in this area is not being vigorously pushed.

There is still considerable waste from fisheries operations and while much of the waste material can be made into fish meal, this is not necessarily its best use. Biological systems for converting fish waste into a stable human food have not advanced to any great extent. Yet the potential of such systems is great since they provide the opportunity to combine waste agricultural products with fish waste. It seems likely that a more detailed study of Japanese and other southeast Asian mixed vegetable fish fermentations might yield applicable results here.

The human risk factors associated with the food supply are under constant scrutiny. While much current research is directed towards cancer and microconstituents of foods, it is well to note that bacterial food poisoning is still the most troublesome of food-borne illnesses. There is need for continuing research in this field. We still do not yet understand the mechanism of intoxication by *V. parahemolyticus* nor can we identify with certainty the pathogenic forms of this organism in nature. As commercial fisheries are extended into areas and stocks which have been taken previously only on a small local scale it seems probable that new problems of a microbiological character will arise. Certainly we have seen in recent years an apparent increase in the number of human intoxications from fish in the USA. Parasites such as anisakis are of increasing concern. These are subject areas for increased research effort.

The enormous interest in aquaculture and particularly in mariculture has made obvious the depth of our ignorance concerning microbial diseases of fish and other aquatic animals other than salmonid fishes. The diseases of wild marine animals is a particularly difficult field of study because diseased animals are quickly disposed of by predators and therefore are only rarely taken by commercial fishermen. But when such animals are raised in the confined and crowded conditions of hatcheries and ponds or channels disease may become all too evident. It has been stated repeatedly that the three basic science needs of the aquaculturalist are genetics, nutrition and disease studies and certainly in the last of these there is a major role for the microbiologist.

Finally, we might usefully consider research in more basic areas of microbiology. The sea and the animals it contains constitute a unique set of microenvironments different in many respects from land areas and terrestrial animals. There are many fascinating microbial associations within the vast marine world which have had only cursory examination. One might cite here the luminous organs of fish, many of which depend on associations of luminous bacteria in particular organ structures on the animal. The relationship of chitin-digesting bacteria to crustacea and to fish which feed on crustacea needs to be investigated further. The role of bacteria – whose total biomass in the oceans probably exceeds that of all other creatures – as food for marine animals is obviously of importance to fisheries since the food of many benthic organisms is probably small animals which feed on bacteria. Some commercially important animals are themselves excellent systems for microbiological research: this is the case with the oyster which can be used to study microbial interactions *in situ* and has already provided useful information on genetic exchange processes and bacteriophage reactions. The marine Vibrios which are often associated with fish and shellfish are a particularly interesting group of micro-organisms which show an unexampled capacity for genetic variation which may indicate a hitherto unsuspected system for dealing with short-term environmental changes. Psychrophilism is a way of life for most marine bacteria and there is much

still to be learned about this phenomenon both at the cellular and biochemical levels. And so on and so on!

There are indeed an enormous number of interesting research projects to be tackled by microbiologists with access to fish and water. Some of these could have an immediate importance to the fishery scientist while others are perhaps of more fundamental or at least long-range significance. It may seem a far cry from the genetics of Vibrio to stinking fish, but in the microcosmic world of bacteria they may be intimately connected and even interdependent. The present climate for research support unfortunately is so keyed to the need for immediately applicable results that these types of interdependencies may be overlooked. Yet it is on such basic concepts that our understanding of practical issues such as fish spoilage is based. Fortunately the world is full of intelligent and pertinaciously curious young scientists and through their efforts our knowledge and understanding of microbiology in fisheries will continue to grow.

14 References

ANDERSON, A G On the decomposition of fish. Ann. Rept. Fish. Bd.
1907   (Scot.) Part III, *Scient. Invest.*, 26, 13–39

ANDREWS, W, WILSON, C R, POEBNA, P and ROMERO, A. Bacteriologi-
1977   cal survey of the channel catfish (*Ictalurus punctatus*) at the retail level. *J. Food Sci.*, 42, 359–363

ARCISZ, W and KELLY, C B. Self purification of the soft clam *Mya*
1955   *arenaria. Pub. Health Repts.*, 70, 605–614

ASCHEHOUG, V and VESTERHUS, R. Investigations of the bacterial flora
1943   of fresh herring. *Z. fur Bakt. Parsit. Infect.*, Abt III, 106, 5–27

BACH, R, WENTZEL, S, MULLER-PRASUHN, G and GLASKER, M.
1971   Farmed trout as a carrier of *Clostridium botulinum* and a cause of botulism. III. Evidence of *Clostridium botulinum* type E on a fish farm with processing station and in fresh and smoked trout from different sources. *Archiv. Lebensmittel Hyg.*, 22, 107

BAER, E F, DURAN, A D, LEININGOR, H V, READ, R B, SCHWAB, A H
1976   and SWARTZENTRUBER, A. Microbiological quality of frozen breaded fish and shellfish products. *Appl. Environ. Microbiol.*, 31 (3), 337–341

BAIN, N, HODGKISS, W and SHEWAN, J M. The bacteriology of salt
1958   used in fish curing. In: *The Microbiology of Fish and Meat Curing Brines.* Ed B P Eddy. HMSO, London. 1—11

BARKER, W, HOOPER, D and BAROSS, J. Shellfish related gastroen-
1970   teritis. *New Eng. J. Med.*, 283, 319

BARKER, W H, WEAVER, R E, MORRIS, G K and MARTIN, W T.
1974   Epidemiology of *Vibrio parahaemolyticus* infections in humans. In: *Microbiology 1974*, Ed D Schlessinger. USA American Society for Microbiology, Washington DC. 257–262

BEATTY, S A and GIBBONS, N E. The measurement of spoilage in fish.
1937   *J. Biol. Bd. Can.*, 3, 77–91

BEDFORD, R H. Marine bacteria of the North Pacific Ocean. The
1933   temperature range of growth. Contrib. *Can Biol. and Fisheries N S.*, 7, 433–438

BODIN, E and CHERREL, F. Bacterial purification of oysters in filtered
1913   sea water. *Compt. rend. Acad. Sci. Paris*, 156, 342–345

BROWN, L D and DORN, C R. Fish, shellfish and human health. *J.*
1977   *Food Prot.*, 40, 712–717

BRYAN, F L. Activities of the Center for Disease Control in Public
1973   Health Problems related to the consumption of fish and fishery products. In: *Microbial Safety of Fishery Products*, Eds C O Chichester and H D Graham. Academic Press, New York

BUROW, H. Untersuchungen zum *Clostridium perfringens* – Befall bei
1974   Jekuhlten Seefischen and Meismuscheln in der Turkii. *Arch. fur Lebensmit.*, 25, 39–52

CAMERON, C A. On sewage in oysters. *Brit. Med. J.*, 2, 471
1880

CANN, D C. Bacteriological aspects of tropical shrimp. In: *Fishery*
1974   *Products*, Ed R Kreuzer. Fishing News (Books) Ltd, West Byfleet, Surrey, England, 338–344

—— Bacteriology of shellfish with reference to international trade. In:
1977   *Handling Processing and Marketing of Tropical Fish.* Tropical Products Institute, London. 377–394

CANN, D C., TAYLOR, L Y and HOBBS, G. The incidence of *C.*
1975   *botulinum* in farmed trout raised in Great Britain. *J. Appl. Bact.*, 39, 331–336

CANTONI, C, SIANO, S and CALCINARDI, C. Yeasts in freshwater fish.
1976   *Arch. Veterin. Italiano*, 27, 64–65
CASTELL, C H, GREENHOUGH, M F and JENKIN, N L. The action of
1957   Pseudomonas on fish muscle. 2. Musty and potato like odors.
    J. Fish. Res. Bd. Can., 14, 775–782
CASTELL, C H., GREENHOUGH, M F and DALE, G. The action of Pseu-
1959   domonas on muscle. 3. Identification of organisms producing
    fruity and oniony odors. *J. Fish. Res. Bd. Can.*, 16, 13–19
CENTER FOR DISEASE CONTROL *V. cholerae* Louisiana. Morbidity and
1978   mortality. *Weekly Report*, 27, 341
CERTES, A. De l'action des hautes pressions sur les phénomènes de la
1884   putréfaction et sur la vitalité des micro-organisms d'eau douce
    et d'eau de mer. *Compt. rend. Acad. Sci.*, 99, 385–388
CHAI, T, CHEN, C, ROSEN, A and LEVIN, R E. Detection and incidence
1968   of specific species of spoilage bacteria on fish. II. Relative
    incidence of *Pseudomonas putrefaciens* and fluorescent pseu-
    domonas on haddock fillets. *Appl. Microbiol.*, 16, 1738–1741
CHU, C L, and PIGOTT, G M. Acidified brine extraction of fish. *Trans-*
1973   *actions ASAE* American Society of Agricultural Engineers, St
    Joseph, Mich, USA, 16, 949–952
CHUNG, J. PhD. Thesis, University of Washington, Seattle, WA
1968
CONN, H W. The 'oyster epidemic' of typhoid fever at Wesleyan
1894   University. *Med. Rec.*, 46, 743–746
CRAIG, J and PILCHER, K. *Clostridium botulinum* type F: Isolation
1966   salmon from the Columbia river. *Science*, 153, 311–312
DIGIROLAMO, R, LISTON, J, MATCHES, J R. The effects of freezing on
1970   the survival of Salmonella and *E. coli* in Pacific oysters. *J. Fd.*
    Sci., 35, 13–16
DISNEY, J G, TATTERSON, I N and OLLEY, J. Recent developments in
1977   fish silage. In: *Handling, Processing and Marketing Tropical*
    Fish. Tropical Products Institute, London. 231–240
DODYSON, R W. Report on mussel purification. *Min. Agric. Fish.*
1928   *Invest. Ser.* II. 10, 1–498
DUSSEAULT, H P. The salt tolerance of bacteria from lightly salted fish.
1958   In: *The Microbiology of Fish and Meat Curing Brines*, Ed B P
    Eddy. HMSO, London. 61–66
DYER, F E. The microorganisms from Atlantic Cod. *J. Fish. Res. Bd.*
1947   *Can.*, 7, 128–136
EGUSA, E (ed). Recent advances in fish pathology: proceedings of an
1976   international seminar on fish diseases. *Fish. Path.*, 10
FELLERS, C R. Bacteriological investigations on raw salmon spoilage.
1926   *Univ. Wash. Publ. Fish.*, 1, 157–188
FINCH, R. Fish protein for human foods In: *CRC Critical Reviews in*
1970   *Food Technology*, Ed T E Furia. The Chimcal Rubber Co,
    Cleveland, Ohio. Vol 1(4), 519–579
FOOTE, C J. A bacteriological study of oysters with special reference to
1895   them as a source of typhoid infection. *Med. News.*, 66,
    320–324
FRANK, M and HESS, E. Studies on salt fish. V1. Halophilic brown
1941   moulds of the genus *Sporendonema* emend. (Ciferri and
    Redaelli). *J. Fish. Res. Bd. Can.*, 5, 287–292
FRYER, J L and AMEND, D A (eds). *International symposium on diseases*
1977   *of cultured salmonids*. Tavolek Laboratories, Seattle, Wash.
GEORGALA, D C. The bacterial flora of the skin of North Sea cod. *J.*
1958   *Gen. Microbiol.*, 18, 84–91
GIBBONS, N E. The slime and intestinal flora of some marine fishes.
1934   Contrib. *Can. Biol. Fish*, 8, 275–290
——Halobacteriaceae In: *Bergey's Manual of Determinative Bacteri-*
1974   *ology*, Eds R E Buchanan and N E Gibbons. Williams and
    Wilkins Co, Baltimore, USA. 269–273
GILL, C O and NEWTON, K G. The ecology of bacterial spoilage of
1978   fresh meat at chill temperatures. *Meat Sci.*, 2, 207–217
GILLESPIE, N C and MACRAE, I C. The bacterial flora of some Queens-
1975   land fish and its ability to cause spoilage. *J. Appl. Bact.*, 39,
    91–100
GOLDMINTZ, D and HULL, J C. Bacteriological aspects of fish protein
1970   concentrate production *Dev. Ind. Microbiol.*, 11, 335–337
GRIFFITH, A S. A review of the bacteriology of fresh marine-fishery
1937   products. *Food Res.*, 2, 121–134
HANNAN, R S. *Science and technology of food preservation by ionizing*
1956   *radiations*. Chemical Pub. Co, New York
HARRISON, F C. The discoloration of halibut. *Can. J. Res.*, 1, 214–239
1929
HARRISON, J M and LEE, J S. Microbiological evaluation of Pacific
1969   shrimp processing. *Appl. Microbiol.*, 18, 188–192
HAYNER, G A and PROCTOR, B E. Investigations relating to the poss-
1953   ible use of atomic fission products for food sterilization. *Food*
    Technol., 7, 6–10
HEALTH AND WELFARE CANADA. Staphylococcal enterotoxemia from
1978   canned salmon – Quebec. *Canada Diseases Weekly Report*, 4,
    153
HERBERT, R A, HENDRIE, M S, GIBSON, D M and SHEWAN, J M. Bac-
1971   teria active in the spoilage of certain sea foods. *J. Appl.*
    Bacteriol., 34, 41–50
HESS. Cultural characteristics of marine bacteria in relation to low
1934a temperatures and freezing. Contrib. *Can. Biol. Fish.*, 8,
    461–474
—— Effect of low temperatures on the growth of marine bacteria.

1934b Contrib. *Can. Biol. Fish.*, 8, 491–505
HOBBS, G. *Clostridium botulinum* and its importance in fishery pro-
1976   ducts. In: *Advances in Food Res.*, 22, 135–185
—— *Clostridium botulinum* in irradiated fish. *Fd. Irrad. Inf.*, 7, 39–54
1977
HOBBS, G, CANN D C, WILSON, B B and HORSLEY, R W. The bacteri-
1971   ology of 'scampi' (*Nephrops norvegicus*). III. Effects of
    processing. *J. Fd. Technol.*, 6, 233–251
HORSLEY, R W. The bacterial flora of Atlantic salmon (*Salmo salar*) in
1973   relation to its environment. *J. Appl. Bact.*, 36, 377–386
HUNTER, A C. Bacterial decomposition of salmon. *J. Bacteriol.*, 5,
1920a 353–361
—— A pink yeast causing spoilage in oysters. *US Dept. Agric. Bull.*,
1920b 819, 1–24
—— The sources and characteristics of the bacteria in decomposing
1922   salmon. *J. Bacteriol.*, 7, 85–109
HUNTER, A C and HARRISON, C W. Bacteriology and chemistry of oys-
1928   ters, with special reference to regulatory control of production,
    handling and shipment. *US Dept. Agr. Bull.*, 64, 1–75
HURMS, F and KRUSE, K P. Food poisoning caused by smoked eel.
1976   *Archiv. f. Lebensmit.*, 27, 88–91
HUSS, H H, PEDERSON, A and CANN, D C. The incidence of *Clos-*
1974   *tridium botulinum* in Danish trout farms. I. Distribution in fish
    and this environment. *J. Fd. Technol.*, 9, 445
ICMSF. *Microorganisms in Foods 2*. University of Toronto Press,
1974   Toronto, Canada. 92—104
JOHNSON, T W. Saprobic marine fungi. In: *The Fungi*. Academic
1968   Press, New York. Vol. III
JOHNSTONE, J. Report on various bacteriological analyses of mussels
1907   from Lancashire and Wales. *Proc. Trans. Liverpool Biol. Soc.*,
    21, 328–370
—— The methods of cleansing living mussels from ingested sewage
1915   bacteria. *Rept. Lancashire Sea-fish Lab.*, 23, 57–108
JORDAN, E O. Viability of typhoid bacilli in shell oysters. *J. Amer.*
1925   *Med. Assn.*, 84, 1402–1403
KELLY, M D, LUKASCHEWSKY, S and ANDERSON, C G. The bacterial
1978   flora of Antarctic krill (*Euphausia superba*) and some of their
    enzymatic properties. *J. Fd. Sci.*, 43, 1196–1197
KIM, I and BJELDANES, L F. Amine content of toxic and wholesome
1979   canned tuna fish. *J. Fd. Sci.*, 44, 922–923
KISER, J S. Effect of temperatures approximating 0°C upon growth
1944   and biochemical activities of bacteria isolated from mackerel.
    Food Research, 9, 257–267
KOBURGER, J A, NORDEN, A R and KEMPLER, G M. The microbial
1975   flora of rock shrimp – *Sicyonia brevirostrus. J. Milk Fd. Tech-*
    nol., 38, 747–749
KREUZER, R. Fish and its place in culture. In: *Fishery Products*, Ed
1974   R Kreuzer. Fishing News Books (Ltd), West Byfleet, Surrey,
    England. 22–47
LAYCOCK, R A and REGIER, L W. Pseudomonads and achromobacters
1970   in the spoilage of irradiated haddock of different pre-
    irradiation quality. *Appl. Microbiol.*, 20(3), 333–341
LEE, J S and PFEIFER, D K. Aerobic microbial flora of smoked salmon.
1973   *J. Milk Food Technol.*, 33, 237–239
—— Microbial characteristics of Dungeness crab (*Cancer magister*).
1975   *Appl. Microbiol.*, 30, 72–78
—— Microbiological characteristics of Pacific shrimp (*Pandalus jorda-*
1977   *nii*). *Appl. Environ. Microbiol.*, 33, 853–859
LERKE, P, ADAMS, R and FARBER, L. Bacteriology of spoilage of fish
1965   muscle. III. Characterization of spoilers. *Appl. Microbiol.*, 13,
    625–630
LEWIS, D H. Retention of *S. typhimurium* by certain species of fish
1975   and shrimp. *J. Amer. Vet. Med. Assoc.*, 167, 551–552
LICCIARDELLO, J and HILL, W S. Microbiological quality of commer-
1978   cial frozen minced fish blocks. *J. Food Prot.*, 41, 948–952
LINDGREN, S and CLEVSTROM, G. Antibacterial activity of lactic acid
1978a bacteria. 1. Activity of fish silage, a cereal starter and isolated
    organisms. *Swedish J. Agric. Res.*, 8, 61–66
—— Antibacterial activity of lactic acid bacteria. 2. Activity in
1978b vegetable silages, Indonesian fermented foods and starter
    cultures. *Swedish J. Agric. Res.*, 8, 67–73
LISTON, J. The occurrence and distribution of bacterial types of
1957   flatfish. *J. Gen. Microbiol.*, 16, 205–216
—— *Vibrio parahaemolyticus*. In: *Food Microbiology: Public Health*
1976   *and Spoilage Aspects*, Eds M P Defigueiredo and D F Splitt-
    stoesser. Avi. Publ. Co, Westport, Connecticut, USA
LISTON, J and MATCHES, J R. Single and multiple doses in the radiation
1968   pasteurization of seafoods. *Fd. Technol.*, 22, 893–896
LISTON, J and SHEWAN, J M. Bacteria brought into brines on fish. In:
1958   *The Microbiology of Fish and Meat Curing Brines*, Ed
    B P Eddy. HMSO, London. 35–41
LUMLEY, A, PIQUE, J J and REAY, G A. The handling and stowage of
1929   white fish at sea. *Food Invest. Board Special Report No. 37.*
    HMSO, London
McMEEKIN, T A, GIBBS, P A and PATTERSON, J T. Detection of vola-
1978   tile sulfide producing bacteria isolated from poultry processing
    plants. *Appl. Environ. Microbiol.*, 35, 1216–1218
MALEVICH, O A. A new species of halophilic mould isolated from
1936   salted fish (*Oospora nikitinskii*). *Mikrobiologiya*, 5, 813–817

MATCHES, J R and LISTON, J. Growth of salmonella in irradiated and
1968 non-irradiated seafoods. *J. Fd. Sci.*, 33. 406–510
MATCHES, J R, LISTON, J and CURRAN, D. *Clostridium perfringens* in
1974 the environment. *Appl. Microbiol.*, 28, 655–660
MILLER III, A, SCANLAN R A, LEE, J S and LIBBEY, L M. Volatile
1973 compounds produced in sterile fish muscle (*Sebastes
melanops*) by *Pseudomonas putrefaciens*, *Pseudomonas
fluorescens* and an *Achromobacter* species. *Appl. Microbiol.*,
26, 18–21
MORRIS, G K, MARTIN, W T, SHELTON, W H, WELLS, J G and
1970 BRACHMAN, P S. Salmonella in fish meal plants: relative
amounts of contamination at various stages of processing and a
method of control. *Appl. Microbiol.*, 19, 401–408
NEUFELD, N. Influence of bacteriological standards on the quality of
1971 inspected fisheries products. In: *Fish Inspection and Quality
Control*, Ed R Kreuzer. Fishing News (Books) Ltd, London.
234–240
OKABE, S. Statistical review of food poisoning in Japan especially by
1974 *Vibrio parahaemolyticus* In: *International symposium on
vibrio parahaemolyticus*, Eds T Fujino, G Sakaguchi, R Saka-
zaki and Y Takeda. Saikon Publ. Co, Tokyo, Japan
OMURA, J, PRICE, R J and OLCOTT, H S. Histamine forming bacteria
1978 isolated from spoiled skipjack tuna and jack mackerel. *J. Fd.
Sci.*, 43, 1779–1781
OREJANA. F. Proteolysis and control mechanisms in fish sauce fermen-
1978 tation. PhD Thesis, University of Washington, Seattle, USA
PHAFF, H J, MRAK, E M and WILLIAMS, O B. Yeasts isolated from
1952 shrimp. *Mycologia*, 44, 431–451
PELROY, G A and EKLUND, M W. Changes in the microflora of
1966 vacuum-packaged irradiated Petrale sole (*Eopsetta jordani*)
fillets stored anaerobically at 0·5°C. *Appl. Microbiol.*, 15,
92–96
PELROY, G A, SEMAN, J P, EKLUND, M W, Changes in the microflora
1967 of irradiated Petrale sole (*Eopsetta jordani*) fillets stored
aerobically at 0·5°C. *Appl. Microbiol.*, 15, 92–96
PHILLIPS, S and WALLBRIDGE, A. The mycoflora associated with dry
1977 salted tropical fish in *Handling, Processing and Marketing of
Tropical Fish*. Tropical Products Institute, London. 353–356
PIGOTT, G M, BUCOVE, G O and OSTRANDER, J G. Engineering a plant
1978 for enzymatic production of supplemental fish proteins.
RACCACH, M and BAKER, R. Microbial properties of mechanically
1978 deboned fish flesh. *J. Fd. Sci.*, 43, 1675–1677
RAJ, H. Public health bacteriology of processed foods. I. Dehydrated
1968 and reconstituted batter. In: *Microbiology of Dried Foods*,
Eds E H Kampelmacher, M Ingram and D A A Mossel.
Grafische Industrial, Haarlem, The Netherlands. 371–384
—— Public health bacteriology of processed foods. *Lab. Practice*,
1970 19(4), 374
RAJ, H and LISTON, J. Survival of bacteria of public health importance
1961 in frozen seafoods. *Fd. Technol.*, 15, 429–434
REAY, G A and SHEWAN, J M. The spoilage of fish and its preservation
1949 by chilling. *Advances in Food Res.*, 11, 343–398
REED, G M and SPENCE, C M. The intestinal and slime flora of the
1929 haddock. A preliminary note. Contrib. *Can. Biol. and Fisheries*, NS 4, 259–264
RIDLEY, S C and SLABYJ, B M. Microbiological evaluation of shrimp
1978 (*Pandalus borealis*) processing. *J. Fd. Prot.*, 41, 40–43
ROBERTS, R J. Bacterial diseases of farmed fishes. In: *Microbiology in
1976 Agriculture, Fisheries and Food*, Eds F A Skinner and J G Carr.
Academic Press, London
ROSS, S S and MORRIS, E O. An investigation of the yeast flora of
1965 marine fish from Scottish coastal waters and a fishing ground
off Iceland. *J. Appl. Bacteriol.*, 28, 224–234
RUSSEL, H L. Bacterial investigations of the sea and its floor. *Bot.
1892 Gaz.*, 17, 312–315
SAISITHI, P, KASEMSARN, B O, LISTON, J and DOLLAR, A M. Mic-
1966 robiology and chemistry of fermented fish. *J. Fd. Sci.*, 31,
105–110
SANBORN, J R. Certain relationships of marine bacteria to the decom-
1930 position of fish. *J. Bacteriol.*, 19, 357–382
SHAW, B G and SHEWAN, J M. Psychrophilic spoilage bacteria of fish.
1968 *J. Appl. Bacteriol.*, 31, 89–96
SHEWAN, J M. The salt curing of herring. In: *Report of the Director of
1938b Food Investigation for the year 1938*. HMSO, London.
115–117
—— Bacteriology of fish. Am. Rept. Food Investig. Bd. (Gt. Britain)
1946 1940–1946, 28
—— The microbiology of fish and fishery products–a progress report.
1971 *J. Appl. Bacteriol*, 34, 299–315
—— The biodeterioration of certain proteinaceous foodstuffs at chill
1974 temperatures. In: *Industrial Aspects of Biochemistry*. Federa-
tion of European Biochemical Societies. 475–490

—— The bacteriology of fresh and spoiling fish and the biochemical
1977 changes induced by bacterial action. In: *Handling, Processing
and Marketing of Tropical Fish*. Tropical Products Institute,
London. 51–66
SHEWAN, J M and LISTON, J. A review of food poisoning caused by fish
1955 and fishery products. *J. Appl. Bacteriol.*, 18, 522–534
SIMIDU, U, KANEKO, E and AISO, K. Microflora of fresh and stored
1969 flatfish (*Kareius bicoloratus*). *Bull. Jap. Soc. Sci. Fish.*, 35,
77–82
SLABYJ, B M, DOLLAR, A M and LISTON, J. Post irradiation survival of
1965 *Staphylococcus aureus* in seafoods. *J. Fd. Sci.*, 30, 344–350
SNIEZSKO, S F. Recent advances in scientific knowledge and develop-
1973 ments pertaining to diseases of fish. *Adv. Vet. Sci. Compar.
Med.*, 17, 291–314
SNOW, J E and BEARD, P J. Studies on the bacterial flora of North
1939 Pacific Salmon. *Fd. Res.*, 4, 563–585
SOHN, J Y, RYEOM, K, KIM, Y, LEE, M, ON, O and RYN, J K. A study
1973 on the distribution of *Clostridium welchii* in fish and shellfish in
Korea. Repts. Korean National Institute of Health, 10, 79–88
SPENCER, R and GEORGALA, D L. Bacteria of public health significance
1958 on white fish prior to brining. In: *The Microbiology of Fish and
Meat Curing Brines*, Ed B P Eddy. HMSO, London. 299–301
STANTON, W R and YEOH, Q L. Low salt fermentation merial flora of
1977 the slime and intestinal contents of the haddock (*Gadus
aeglefinus*) *J. Marine Biol. Assoc. Clostridium perfringens* in
seafoods. *Bull. Fac. Fish.*, Nagasaki Univ., No. 31, 1–67
STEWART, M M. The bacterial flora of the slime and intestinal contents
1932 of the haddock (*Gadus aeglefinus*). *J. Marine Biol. Assoc. UK*,
18, 35–50
STILES, G W. Shellfish contamination from sewage polluted waters
1911 and from other sources. *US Dept. Agric. Bur. Chem. Bull.*, 136
TANIGUTI, Z. Studies on *Clostridium perfringens* in seafoods. *Bull. Fac.
1971 Fish.*, Nagasaki Univ., No. 31, 1–67
TANIKAWA, E. *Marine Products in Japan*. Hokkaido University, Hok-
1965 kaido, Japan
TARKY, W. Recovery of proteins from fish waste by enzyme. MS
1971 Thesis, University of Washington, Seattle, USA
THJØTTE, TH and SØMME, O M. The bacteriological flora of normal
1938 fish. A preliminary report. *Acta Path. Microbial. Scand.
Suppl.*, 37, 514–526
—— The bacteriological flora of normal fish. Norsk. Videnskap Akad.
1943 Oslo I Nat. Naturu. Klasse
TODD, E C D. Food borne disease in six countries–a comparison. *J.
1978 Fd. Prot.*, 41, 559–565
TRUST, T J. Bacteria associated with the gills of salmonid fishes in
1975 freshwater. *J. Appl. Bacteriol.*, 38, 225–233
UNKLES, S E. Bacterial flora of the sea urchin, *Echinus esculentus*.
1977 *Appl. Environ. Microbiol.*, 34, 347–350
VANDERZANT, C, MROZ, E and NICKELSON, R. Microbial flora of Gulf
1970 of Mexico and pond shrimp. *J. Milk Fd. Technol.*, 33, 346–350
VANDERZANT, C and NICKELSON, R. Comparison of extract release
1971 volume, pH and agar plate count of shrimp. *J. Milk Fd.
Technol.*, 34, 133
VAN KLAVEREN, F W and LEGENDRE, R. Salted cod. In: *Fish as Food*,
1965 Ed G Börgstrom. Academic Press, NY and London. Vol 3,
133
VAN SPREEKENS, K J A. Characterization of some fish and shrimp
1977 spoiling bacteria. Anton. van Leewenhoek. 43, 283–303
VAN VEEN, A G. Fish preservation in Southeast Asia. *Advances in
1953 Food Res.*, 4, 209–231
WALKER, P, CANN, D and SHEWAN, J M. The bacteriology of 'scampi'
1970 (*Nephrops norvegicus*). I. Preliminary bacteriological, chemical and sensory studies. *J. Fd. Technol.*, 5, 357–385
WARD, D, PIERSON, K and VAN TASSELL, K R. The microflora of
1977 unpasteurized and pasteurized crabmeat. *J. Fd. Sci.*, 42,
597–600
WELLS, W F. Artificial purification of oysters. *Pub. Health Rept.*, 31,
1916 1848–1852
WOOD, E J F. Studies on the marketing of fresh fish in Eastern
1940 Australia. Part II. The bacteriology of spoiling marine fish.
Aus. Comm. Coun. Sci. Ind. Res. Div. Fish. Rept. No. 3
—— *Marine Microbiol. Ecology*. Chapman and Hall, London
1965
WU, M T and SALUNKHE. Mycotoxin potential of fungi associated with
1978 dry shrimps. *J. Appl. Bacteriol.*, 45, 231–238
YOSHIMIZU, M and KIMURA, T. Study on the intestinal flora of sal-
1976 monids. *Fish. Pathol.*, 10, 243–259
ZEN-YOJI, H, SAKAI, S, MARUYAMA, T and YANAGAWA, Y. Annual
1975 Report. Tokyo Metropolitan Research Lab. of Public Health,
25, 1–7

Physical properties and processes

A C Jason and M Kent

1 Introduction

Physics provides both our understanding of processes in the fish industry and the possibility of controlling them. The discipline covers vast areas and its exploitation has barely commenced. In order to give an idea of the potential, *Table I* lists various areas of physics and some fields of application in which a sound understanding of the basic physical properties can profitably serve the needs of both industry and consumer.

Many of the detailed physical properties of technological importance vary from fish to fish, from species to species; they vary with composition and with time and conditions of storage. We find that little is known of these basic properties in detail or the manner in which they vary. We propose to examine here the more important areas and to point out some gaps in our knowledge of the physical behaviour of fish, gaps that could profitably be filled.

Table I
PHYSICS IN FISH TECHNOLOGY

Area	Applications
Thermal behaviour	Freezing, thawing, tempering, cooking, canning, cold storage, chilling, refrigerated display
Heat transfer	
Mass transfer	Drying, brining, smoke curing, cold storage, drip loss
Electrical characteristics	Thawing, freshness measurement, moisture determination, fat determination, electrostatic smoking, cooking
Mechanical and rheological behaviour	Machine forming, extrusion, cutting, conveying, texture measurement
Sorption	Drying, smoke-curing, cold storage, hygrometry
Optics	Smoke-curing, cold storage
Irradiation	Pasteurization

2 Thermal behaviour

2.1 *Specific heat, conductivity and diffusivity*

Almost all operations in handling and processing fish involve change of temperature which, in the case of freezing and thawing, is accompanied by change of phase. To understand such thermal behaviour and to design equipment for bringing about thermal changes in a desired manner there are two essential requirements; the first is the availability of accurate and detailed data on thermal properties, the second is the ability to calculate heat fluxes and temperature profiles for typical geometrical configurations and any given set of boundary conditions. Phase change below the temperature at which water begins to crystallize is a complicated process which does not occur completely at a fixed temperature but takes place continuously as the temperature is progressively reduced. It is this behaviour that leads to much difficulty in setting down a quantitative description of thermal processes in the freezing and thawing operations.

In order to solve the Fourier heat flow equation

$$\rho c \frac{\mathrm{d}\theta}{\mathrm{d}t} = \mathrm{div} \ (k \ \mathrm{grad} \ \theta)$$

(where ρ is the density, c is the specific heat, k is the thermal conductivity and θ is the temperature at any point in the system at time t), we require a knowledge of the temperature dependent properties ρ, c and k. The thermal diffusivity, $k/\rho c$, and its temperature variation are thus implicit in all solutions.

Very little accurate data exists concerning thermal properties of fish products (Jason and Jowitt, 1969; Adam, 1969) particularly at temperatures immediately below the freezing point where a considerable proportion of the water changes phase over a very small temperature range. This change of phase continues down to about –40°C below which some 8% of the total water is apparently unfreezable (Riedel, 1956). Whether the unfreezable water is 'bound' to protein or structured in some other way so that freezing cannot take place is yet to be decided. It is certainly known from the X-ray diffraction work of Aitken (1966) that only ice I exists in frozen fish; thus any other 'structuring' if not direct bonding must be of a vitreous nature. So called 'glass-transitions' have been observed in beef muscle by Simatos *et al* (1975) using the techniques of differential thermal analysis (dta) and differential scanning calorimetry (dsc). No eutectic points are observed due to the interference of other solutes, a direct consequence of Gibb's phase rule: the solidification of the 'unfreezable' water appears to take place in the form of a glass. Under certain conditions, rewarming can result in further ice formation taking place. Much depends on the thermal history of the sample but differences in estimates of unfreezable water content most likely arise from different experimental treatments. Also it has been shown (Kent, 1975b) that there is a time dependence of the ice fraction which is probably diffusion controlled. Under these circumstances, as has been noted by Duckworth (1971), Meryman (1966) and Krasnitskaya *et al* (1976), equilibrium may not be reached at low temperatures in a reasonable experimental time.

Long (1954, 1955), on the other hand, interpreted thermal behaviour during freezing on the basis of an analysis of the eutectic characteristics of the sodium chloride solution present in the muscle tissues, deriving theoretically the temperature dependence of ρ, c and k from the individual values and concentrations of constituent elements present in the matrix of water, ice and non-aqueous material. Extensive measurements of thermal conductivity were made on cod (*Gadus morhua*) over a wide range of temperature, including the region immediately below the freezing point of water. This data was found to be closely in accord with the theoretical predictions. Apart from the observation of a eutectic point, which is doubtful, Long's data remains the definitive work to this day, albeit for one species only.

The corresponding definitive data for specific heat capacities were derived by Riedel (1956) from measurements he made on enthalpy as a function of temperature. Although these data are more extensive than those obtained by others, either before (Chipman and Langstroth, 1929; Lobsin, 1939; Short and Staph, 1951; Jason and Long, 1955) or since Kachmanukyan (1972), they relate only to a single sample taken from each of three species of lean fish. Riedel (1978) has now finalized his work with a theoretical expression for calculating enthalpies of eight different food systems.

More detailed information is required both for a satisfactory understanding of the physics of freezing and thawing and for the calculation of process times and temperatures. This entails the need to determine c and k with sufficient accuracy over the range of variation of composition normally encountered in each of the principal species. Considering that such data is needed for the design of freezers and thawers it is surprising that so little effort has been made in this direction. The effects of composition alone are considerable. The difference in water content between lean cod (~81%) and very fatty herring (~60%) is reflected in the enthalpy and in the relative freezing and thawing behaviour of these two species. It is of course a difficult problem. Thermal diffusivity is a function of thermal conductivity, specific heat capacity and density and is extremely temperature dependent. Several orders of magnitude change in its value from $-30°C$ to $-1°C$ can account for the difference observed in freezing and thawing times under similar conditions of surface heat transfer. The larger part of the difference is accounted for by the enormous values of the specific heat capacity at temperatures just below the initial freezing point.

Figure 1 shows the variation of k, ρ and c with temperature in the temperature range of interest for cod (81% moisture content) and illustrates the complexity of the problem of having to solve the Fourier equation for any set of boundary conditions. Albasiny (1956) obtained numerical solutions and, using data provided by Long (1954), computed temperature profiles produced during the freezing of slabs of cod fillets. Sanders (1980) has greatly improved the method and his results illustrating the temperature profiles in uniform slabs of lean fish at various times during the freezing and thawing processes can be seen in these proceedings.

2.2 Surface heat transfer

It is evident that the rates of freezing and thawing are conditioned by the intrinsic thermal properties and by the rate of surface heat transfer. As an example of the latter, consider the mechanism of surface heat transfer in a humidified air blast thawer (Merritt and Banks, 1964). Although much of the heat is conveyed to the cold surface by the well-understood mechanism of convective heat transfer, a significant part of the total thermal flux is derived from the latent heat of condensation of the water vapour conveyed in the moist air as it condenses on the cold surface. The rate of heat transfer dH/dt attributable to vapour condensation is related to the thickness of the boundary layer Δz, to the diffusion coefficient D of water vapour in air and to its heat of liquefaction L. To a certain extent dH/dt is enhanced by increasing the air velocity so as to reduce Δz. Practical considerations, however, limit this expedient. The rate of condensation heat transfer on a unit surface area is then given by

$$dH/dt = D\ L\ \Delta\rho/\Delta z$$

where $\Delta\rho$ is the density difference between the water vapour in the air and that at the surface of the product. dH/dt may also be increased by reducing the value of Δz.

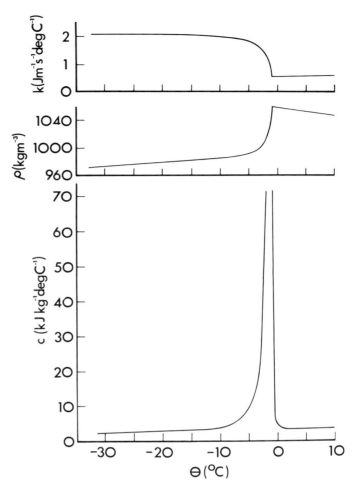

k, ρ and c as functions of θ. Cod muscle.

Fig 1 Variation with temperature of thermal conductivity k, density ρ and specific heat capacity c for cod (*Gadus morhua*) muscle

Removal of the non-condensable gas constituents of air (*ie* nitrogen, oxygen, *etc*) and using water vapour as the sole heat transfer medium (Jason, 1972) will achieve this end. This is effected simply by enclosing the frozen product in a chamber from which air is rapidly extracted by means of a liquid ring pump (in which the working fluid is water) and introducing water vapour into the chamber by the process of evaporation from a constant temperature water bath within. Because the temperature of the water vapour cannot exceed that of the bath, no overheating of the product is possible.

Effusion of dissolved gases in the fish tissues necessitates continuous pumping to minimize the build-up of non-condensable gases which, it can be shown, would accumulate at the product surfaces. Notwithstanding such precautions, their very presence can significantly impede heat transfer particularly at opposing contiguous surfaces. Factors conditioning optimization of product throughput have been investigated (Christie and Jason, 1975), but there is much scope for improving the process by the application of several concepts in the field of vapour dynamics.

2.3 Canning

In connection with condensation heat transfer, mention must be made of its relevance in the canning process which does not appear to have been systematically studied, particularly in relation to fish canning. The subject is of special interest here because it is widely acknowledged that in order to achieve the highest levels of product sterility and wide margins of safety, fish products are invariably overprocessed and in many cases their eating qualities impaired. There are two contributory physical factors giving rise to such overprocessing, both of them related to thermal behaviour. The first has already received some mention in relation to vapour phase or vacuum thawing: heat transfer rates at can surfaces within a retort are limited in an uncertain manner by the presence of non-condensable gases. If the surface heat transfer coefficient could be increased so as to make the boundary condition non-limiting (say by removing the non-condensable gases more efficiently by venting at the bottom of the retort, instead of the top), then this aspect of the problem would be removed. There then remains but one more obstacle to rapid processing: that of heat transfer within the material inside the can. To a certain extent, unless the can contains a significant proportion of freely flowing liquid in addition, as for example in soups, little improvement can be achieved. Although the problem of solving the Fourier equation to predict the integrated lethality at all locations and at all times is much simpler in this case than for say freezing or thawing, nothing is known in detail of the high temperature thermal diffusivity values for any fish product. The problem becomes extremely complex if convective heat transfer occurs within the can when there is some liquid flow (Hiddink, 1975). Further progress in these aspects of canning awaits the initiation of an extensive programme to determine the high temperature thermal properties of fish products of all types. Advances in the theory of combined convective and conductive heat transfer are also likely to be of practical relevance.

An alternative approach to the problem is the establishment of a universal empirical model for temperature distribution within a can. Such a model could offer computational simplification but, more significantly, would overcome difficulties arising from discrepancies between theory and observed behaviour (Skinner and Jowitt, 1977; Skinner, 1979).

2.4 Radiative heat transfer in refrigerated cabinets
Open refrigerated display cabinets in shops represent a critical source of product deterioration in the frozen food chain due to radiative heat transfer between the surface packages and the surroundings (Lorentzen, 1963). Various methods of reducing this heat gain have been suggested such as low emissivity packaging materials (Eskilson, 1967) or heat reflecting screens above the cabinets (Lorentzen, 1968). A thorough study of these proposals was published in 1973 by Hawkins, Pearson and Raynor and although not a study of frozen fish, the authors choosing peas as a subject material, the results are extremely relevant to the frozen fish industry.

First, these workers measured the emissivities of various packaging materials, some 200 in number, such as aluminium foil with various layers of lacquer, silvered plastics and waxed paper. The emissivity is also numerically equal to the absorptivity and represents the fraction of incident energy that is absorbed. Values obtained ranged from 0·05 for 0·025mm aluminium foil up to 0·90 for waxed paper. In actual tests with these materials, the former maintained the surface package temperatures some 15°C to 16°C lower than the latter under identical conditions. Although plain aluminium foil is unlikely to be suitable as a packaging material, emissivity values of 0·3 are obtainable for useful materials. An expression was derived which relates various parameters to the temperature T_p of the packages at the load line in a typical display cabinet. Since heat radiated to the packages from the surroundings is balanced by heat transferred by conduction from the cold wall evaporator and by convection from a separate convective evaporator, we have

$$F A s e_p e_r (T_r^4 - T_p^4) = K_w (T_p - T_w) + K_c (T_p - T_c)$$

where F is a geometrical factor related to the field of view of a radiator or package at each position, A is the area of the package, e_p and e_r are the thermal emissivities of the wrapping material and room surfaces respectively; s is Stefan's radiation constant; T_p is the wrapper top surface temperature in °K and T_r is the radiator (room surface) temperature; K_w and K_c are thermal coefficients for conductive and convective heat transfer respectively; T_w is the operating temperature of the cooled wall evaporator and T_c is that of the finned convective evaporator.

Hawkins et al (1973) found that product temperatures could be reduced by about 5°C by using an assembly of corner cube reflectors placed at a height of 2m above a cabinet to reflect radiation from the top surface of the packages back to the packages themselves. Provided that the reflector also screens the packages from outside radiation, temperature gradients in the product are minimal because both radiative and convective gains are negligible.

Of equal importance, perhaps, is the fact that refrigerated food cabinets consume large amounts of electrical energy and this has given cause for concern in the operation of supermarkets where high fuel costs reduce narrow profit margins. In some countries heat from compressors imposes an additional load on air-conditioning plant and exacerbates the energy problem. Much of the load could be conserved by the widespread use of reflective night blinds: Kindleysides and Hale (1979) have shown that a reduction of up to 35% is possible under practical operating conditions.

3 Electrical properties

3.1 General discussion
Probably more widely studied and understood are the electrical properties of fish muscle and other foodstuffs. These include conductivity, dielectric and piezoelectric properties with the heaviest emphasis, for practical reasons, being placed on dielectric phenomena. The great interest in dielectric properties stems from the need to be able to design equipment properly such as microwave and radio-frequency heaters to act as thawers, temperers or cookers. Consequently it is found that a large body of data exists in the RF region (Ede and Haddow, 1951; Bengtsson et al, 1963) and more at the microwave oven frequencies of 915MHz and 2 350MHz (Bengtsson and Risman, 1971; Ohlsson and Bengtsson,

1975). We ourselves have made studies over the range from dc to 35GHz thus acquiring a very broad knowledge of dielectric behaviour in fish products (Kent, 1977; Kent and Jason, 1975).

Apart from these dielectric heating applications, knowledge of dielectric behaviour can be employed in a variety of useful tasks, principally because they reflect transient changes in the structure and physical behaviour of many foodstuffs. Examples of the application of such knowledge are the estimation of fish freshness by means of the Torrymeter (Jason and Richards, 1975) and the estimation of moisture content in fish meal (Kent, 1972a). Microwave attenuation measurements also provide a powerful method of following phase changes which take place in water in fish tissues during freezing, thawing and frozen cold storage (Kent, 1975a,b).

3.2 Radio frequency and microwave thawing

In the frequency band 27–70MHz thawing can be effected by placing the material to be heated in slab form between two plates of a simple capacitor. No contact is made with the electrodes, the useful part of the power being dissipated through the mechanism of dielectric losses. This dissipated power may be written in SI units

$$P = \epsilon'' f V E^2 / 1 \cdot 8 \times 10^{10} d^2 \ \mathrm{W}$$

where ϵ'' is the loss factor of the material at frequency f, ie the imaginary component of the complex dielectric permittivity, E is the rms potential developed across the opposing surfaces of the slab, V is its volume and d its thickness.

Problems of thermal runaway can be encountered because the loss factor ϵ'' at such a frequency has contributing to it a component of the dc conductivity that varies inversely with frequency and directly with temperature. It has been shown that the temperature dependence of this conductivity (Kent and Jason, 1975) is in part due to the eutectic-like behaviour of frozen flesh and in part due to the thermal activation of charge carriers by the normal processes expected in such conducting systems.

Temperature gradients induced by the conditions of thawing (such as when the material is continuously conveyed between electrodes) or through microwave applicators are enhanced by further electromagnetic heating and, as before, conditions of thermal instability are exacerbated by greater power dissipation in the warmer portions (Jason, 1974).

Upon increasing the frequency to the microwave region (915MHz or 2 340MHz are the permitted frequencies for this kind of application) the problems are by no means diminished, in fact as far as thawing is concerned new problems are introduced. The penetration of electromagnetic energy into a material such as frozen cod decreases with increasing frequency and increasing temperature (Fig 2). Such poor penetration, especially as the material approaches the thawed state, means once again a high degree of local heating at the surface. One attempt to solve this problem recently (Priou et al, 1977; 1978) has incorporated a small air refrigeration unit in the microwave thawer to keep the surface frozen and thus make more uniform the dissipation of power within the material. This works reasonably

well but can still become unstable at higher temperatures. For this reason most manufacturers of microwave units sell them as tempering units and in many applications this limited function is all that is required prior to mechanical operations such as extrusion or cutting. It is also cheaper to operate them this way since, to raise the temperature of a frozen block to $-5°C$ from $-20°C$, only a quarter of the total heat needed for complete thawing is required.

Further understanding of the mechanisms of instability in dielectric heating requires a detailed knowledge of the rf and microwave dielectric properties of fish at all temperatures and of the electromagnetic field configuration.

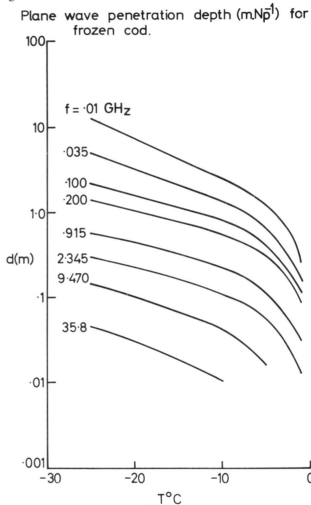

Fig 2 Penetration depth of electromagnetic radiation at various frequencies into a slab of cod at various temperatures

Various models exist to explain the observed temperature effects in unfrozen material (Mudgett et al, 1978), relating these properties closely to those of pure liquid water. Strangely, it appears that, at these frequencies, the dielectric properties of foods such as meat and fish are almost entirely those of the 'free' water present and dissolved salts: there is little dielectric evidence of any water structuring in whole intact muscle. When frozen or dried muscle is studied, however, a different picture emerges. With most of the 'free' water frozen out, the dielectric behaviour in the rf and microwave fields is basically that of water molecules in strongly hindered rotational states; that is to say, whereas pure water has a well characterized relaxation frequency (20GHz at 25°C), unfrozen water at subzero temperatures and 'bound' water at all temperatures have no clearly defined relaxation behaviour, the relaxation being broadly distributed over the rf and microwave frequency bands (*Fig 3*). Thus, the dielectric properties of frozen products, like their thermal properties, are almost entirely attributable to the unfrozen water content.

3.3 *Moisture determination*

The loss behaviour of unfrozen fish exhibits a similar pattern of frequency dependence to that of the frozen fish except that the magnitude of the loss factor at a given frequency is then dependent on the somewhat different concentrations of 'free' and 'bound' water and on the ionic concentration. The contribution from the 'free' water is reduced at all frequencies when the moisture content is reduced, as is the effect of ionic conductivity at the lower end of the spectrum. The influence of bound water remains relatively unaffected until most of the 'free' water has been removed. Clearly, such a dependence can be used to advantage in situations where the moisture content is the variable to be monitored. This is the position in much processing which involves drying, in particular in fish meal production. Laboratory tests show a marked dependence of microwave attenuation on moisture content. The attenuation α is given by

$$\alpha = (2\pi/\lambda)\ \{[\epsilon'/2]\ [\sqrt{|(1 + \tan^2 \delta)} - 1]\}^{\frac{1}{2}}$$

where λ is the wavelength of the radiation in free space,

ϵ' = relative permittivity, ϵ'' = loss factor, $\tan \delta = \epsilon''/\epsilon'$ and ϵ' and ϵ'' are the real and imaginary components of the complex permittivity.

If a suitable sampling head can be devised this method offers potential as a means of controlling the output from fish meal dryers. It has already found use in other food and agriculture applications such as grain and rice dryers.

It should be noted that it is not only the loss factor ϵ'' that is moisture dependent. ϵ', the relative permittivity, exhibits a similar, though less strong, dependence on water content (*Fig 4*). This is still valuable, however, and leads directly to a useful moisture-sensing device based on microwave strip line or microstrip (Kent, 1972a; 1973). This form of transmission line is usually considered to propagate TEM waves (as in free space or coaxial line). Its asymmetry allows it to be immersed in any medium when it is found that the apparent attenuation, or insertion loss, of the line depends on the permittivity of that medium. Hence it becomes an obvious means of measuring water content, especially of fluids and other medium to high moisture content materials. It has even been used for estimating, non-destructively and non-invasively, the water content of whole fish (*Fig 5*) (Steele and Kent, 1979).

All measurements so made appear to have some temperature-dependence but in the case of powder-like material a further problem is caused by variations in density. This problem is not easily eliminated, an independent means of measuring density being required.

Some workers (Stuchly and Kraszewski, 1965; Kalinski, 1978) have used the fact that the loss factor and permittivity have different relationships to moisture and

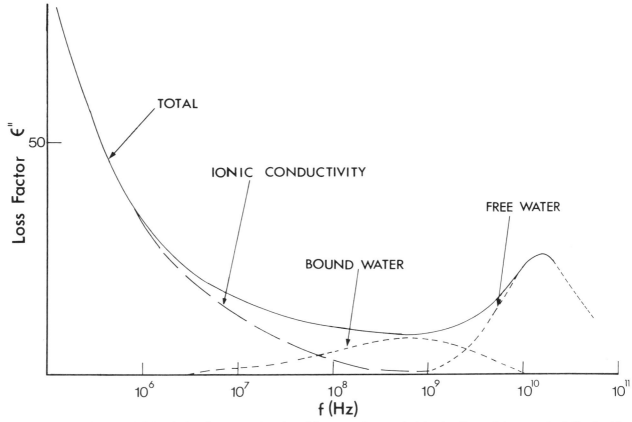

Fig 3   Schematic diagram of dielectric loss factor as a function of frequency for a typical foodstuff containing water and dissolved ions

162

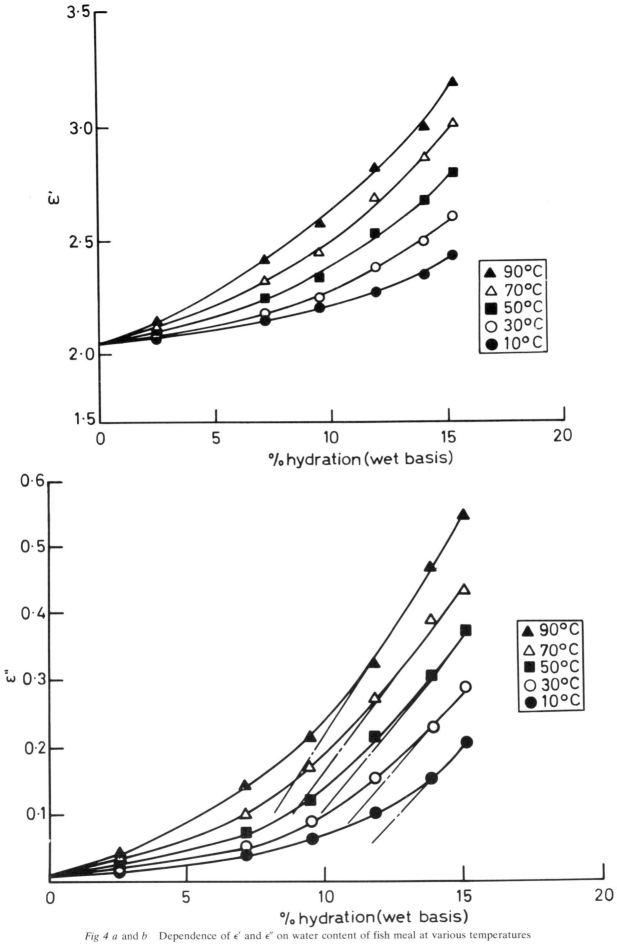

Fig 4 a and *b*   Dependence of ϵ' and ϵ'' on water content of fish meal at various temperatures

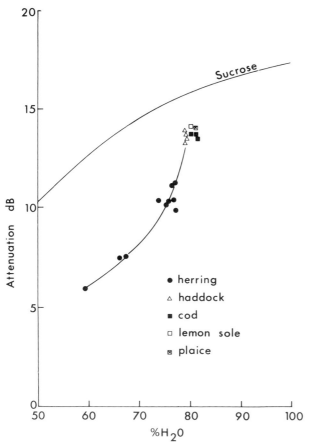

Insertion loss versus moisture content for various fish species. Also shown is the calibration for a solution of sucrose

Fig 5 Stripline measurement of water content of whole fish

density. Simultaneous measurements of both, or of some other properties dominated by each of the dielectric parameters (*eg* attenuation and phase shift), yield pairs of simultaneous equations from which the density can be eliminated.

Other workers have resorted to a totally different technique for density correction such as γ-ray absorption (Mládek, 1973) (see 6.3). Instruments based on this type of measurement exist but are very expensive. In the case of fish meal, density is a problem that is difficult to study due to changes in powder properties with temperature and with time. Even control of density is difficult since the product changes according to the type of fish used or the composition of the raw material (*eg* fish frames or whole fish).

If circumstances permit direct contact between electrodes and the material, then electrical conductivity is a strong contender for the measurement of moisture. It is found that the low-frequency conductivity of fish meal varies exponentially with moisture content, whereas it varies linearly with bulk density, so that relatively small density variations need present no problem. Temperature is easily monitored and its effects corrected. However, the method has some disadvantages because solid material accretes on the surfaces of the electrodes and water condenses on them when they are cooler than the dew point temperature. Nevertheless, it is understood

that this method has been applied successfully to the measurement of moisture in fish meal.

3.4 *Fish freshness measurement*

The electrical properties of fish muscle are by no means constant with time and they are quite sensitive to biological condition and to handling. Following the work of Osterhout in 1922 on plant tissues, Makarov (1952) and Hennings (1963) showed that certain of these properties are related to the age-in-ice of fish. During the last decade we have taken this work further at Torry and have shown that a particular type of measurement, the logarithm of electrical power factor (tan Φ_0) measured at a frequency in the audio range, is functionally related to objective sensory assessment scores, irrespective of age-in-ice (Jason and Lees, 1971). This functional relationship shown in *Fig 6* forms the basis of the GR Torrymeter used for the assessment of fish freshness (Jason and Richards, 1975).

The meter as designed is suitable only for wet unfrozen fish. Freezing fish, whilst preserving the quality of appearance, *etc*, disrupts the tissues sufficiently to give false low values on the freshness meter. However, it can be used in this way to determine whether or not fish that looks fresh has been frozen or not.

4 Mass transfer

4.1 *Drying*

The advance of modern techniques in the frozen food industry, including distribution and domestic storage, has resulted in a rapid decline in the demand for dried or partially dried fish products during the past half century.

When Torry Research Station was first set up in 1929, the production and consumption of stockfish, saltfish, heavily smoked and well dried kippers, and other dried products still represented a significant part of the activities of the fish industry in Europe. During the war years 1939–1945, when the need arose to safeguard food supplies, a number of novel comminuted dried fish products were developed at Torry (Cutting *et al*, 1956) which were more acceptable to the British palate than were the existing traditional dried or salt-cured products. Semi-industrial scale production continued for a short while, but ceased with the conclusion of the war. Subsequently, the earlier work of Ede (1949) and Gane (1951) on freeze-dried foodstuffs led to large-scale experimental production of many products, including those of fish, at the then Ministry of Food Experimental Factory, adjoining Torry Research Station. The method finally evolved, known as Accelerated Freeze Drying (AFD), was based on a method-enhanced heat and mass transfer, originally proposed by Falk *et al* (1919) but independently developed by Dalgleish and Thomson (1958). Freeze-dried fish, which reconstitutes rapidly in water and is acceptable during a limited period of storage, has not succeeded commercially (except in a minor way when mixed with other dried packaged foods in the form of curries, *etc*) largely because it is expensive to produce and to store, and because taste and texture deteriorate more rapidly than does the normal frozen product (Anon, 1957; 1958) unless very great care is taken to reduce the moisture content to very low levels (Matheson and Penny, 1961).

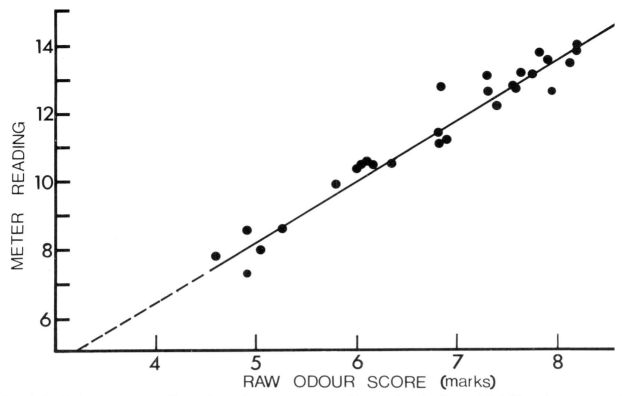

Fig 6 Relationship between meter readings and raw odour assessment for cod inspected at Aberdeen and Hull fish markets (meter reading $F = 10 \log_{10} (\Phi/2\pi f)$, $f = 2\text{kHz}$)

The decline in the need to dry fish intended for human consumption, even partially, in order to confer a small extension on the shelf life of smoked products, has continued and such products today suffer very little weight loss during processing. Nevertheless, it is desirable to achieve whatever small loss is thought to be necessary for reasons of texture, in the shortest possible time. In this area drying is still of contemporary importance.

The amount of fish dried for animal consumption utilizes a major fraction of the world's resources (FAO, 1977). The process is slow, being limited both by low rates of heat transfer in the early stages of drying and by low rates of mass transfer in the latter stages. It is also energetically more wasteful than it need be. The detailed mechanisms of fish meal drying are complicated and not well understood, though Myklestad (1973) has empirically investigated some physical aspects of the process.

It is evident, therefore, that the need persists for an adequate knowledge of the way in which water interacts with the non-aqueous substrate and of the factors that affect its mobility both within a product and at its surface.

The gel-like structure of fresh fish muscle is such that most of the water molecules are only slightly more strongly bonded to the substrate than to each other. This helps to explain the small difference between the dielectric properties of water and fish muscle. Removal of water is achieved by breaking the bonds, the strength of which remain substantially constant until only a few layers of water molecules surround those of the substrate. At any given water content there exists a relationship between the Gibbs isosteric heat of binding ΔH and the equilibrium vapour pressure p given by the Clausius-Clapeyron equation

$$\frac{d(\ln p)}{dT} = \frac{\Delta H}{RT^2}$$

where R is the universal gas constant, p is the vapour pressure and T is the absolute temperature.

The value of ΔH may be expected to increase when the concentration of solutes in the water significantly reduces the vapour pressure as indicated by Raoult's Law (see *eg* Karel, 1975).

In the still drier state the interaction between water and the substrate is very well described by the multi-molecular adsorption theory of Brunauer *et al* (1938) (BET). The BET theory relates the amount of water, x, adsorbed at pressure p and temperature T to the amount that would be adsorbed, x_m, in a single monomolecular layer. This leads to the equation

$$\frac{p}{x (p_0 - p)} = \frac{1 + (p/p_0) \{\exp l (E_m - E_1)/RT) - 1\}}{x_m \exp [(E_m - E_1)/RT]}$$

p_0 being the saturation vapour pressure of water at temperature T and E_m and E_L are the heats of adsorption of the first monomolecular layer and the other multilayers, respectively. This equation has also been derived by a statistical mechanical treatment by Hill (1946). Its importance in relation to the previously discussed dielectric properties of fish meal and other low-moisture materials lies in the fact that the occupancy of each layer can be expressed separately as

$$x_m = x (1 - p/p_0)$$

for monolayer water and

$$x_{mult} = x.p/p_0$$

for multilayer water.

165

Since $p/p_0 = \exp[(E_L - \Delta H)/RT]$ and the value of ΔH increases from E_L to E_m as p decreases from p_0 to 0, the occupancy of the multilayer becomes increasingly temperature-dependent as p/p_0 approaches zero. This behaviour is the principal factor determining the temperature-dependence of the microwave dielectric properties of a number of materials, including fish meal (Kent, 1972b).

The BET parameters for cod muscle at 30°C obtained by Jason (1958) are: $\bar{x}_m = 0 \cdot 088 \pm 0 \cdot 009$ kg H_2O/kg substrate and $E_m - E_L = 5 \cdot 11 \pm 0 \cdot 71$ kJ mol^{-1}. The BET relationship adequately describes the sorption behaviour for values of x $0 \cdot 6$ kg/kg substrate, above which the relationship cannot be adequately tested owing to additional influences such as that of solvent-solute interaction.

The above observations have some technological relevance, especially to the drying of fish meal. During this process, which takes place at elevated temperatures, much of the monolayer water is thermally 'pumped' into the multilayer state from which some of it is evaporated. At the end of the process, as the material cools, some of the remaining multilayer molecules fall back into the monolayer and the additional energy that had been used to activate them is wasted unless it can be recovered in some way. Any such improvement in the process clearly requires more data relating to sorption, particularly at temperatures above about 50°C. (In another connection – that of cold storage behaviour – data are also required at sub-zero temperatures).

The conventional drying process might well be improved by avoiding energetically wasteful overdrying, as such an improvement would carry with it the advantage of higher rates of throughput. The problem is one of both measurement and control of the moisture content of the product as it emerges from the dryer and of measurement of moisture content as it enters as raw material. The solution is far from simple: microwave methods of measurement, as discussed already, hold much promise but, as we have indicated, there are difficulties associated with variations of density, temperature and composition. Control presents difficulties of a different nature largely resulting from transient disturbances to the system produced by massive fluctuations in the nature, volume and moisture content of the input material. These problems are not necessarily insuperable as the advent of microprocessors could much simplify application of the feed-back and feed-forward control system that is required.

In the early stages of drying, when the thermal diffusivity and the water diffusion coefficients of the meal are both high, conditions are favourable for rapid weight loss. As the meal becomes progressively drier, the converse increasingly applies. Further improvements in fish meal drying could therefore be brought about by novel ways of achieving rapid heat transfer in the latter stages such as, for example, fluidized bed and microwave heating. As always, economic considerations apply severe constraints but a feasibility study coupled with some essential experimentation would undoubtedly help to clarify the situation.

4.2 Diffusion of water during drying

The prospect of increasing the diffusion-limited drying rate is meagre because of the very nature of the diffusion process. As we have seen, water molecules are bound to the substrate at two levels corresponding to the heats of adsorption E_L and E_m. The mechanism of diffusion (Jason, 1958) is one of thermal activation in which molecules hop from one site to an adjacent site over a saddle point on a potential surface. When they reach a surface exposed to the atmosphere, multilayer and monolayer molecules will evaporate if the energy received from lattice vibrations exceeds E_L and E_m respectively. Movement to the surface is characterized by a coefficient of diffusion D which is the sum of two components, D_1 for multilayer water and D_2 for monolayer water. Each component varies with temperature in accordance with Boltzmann statistics, so that

$$D = D_1 + D_2$$
$$= D_0 \exp(-E_1/RT) + D_0 \exp(-E_2/RT)$$

D_0 is related to the lattice frequency and has the approximate value

$$D_0 = d^2 kT/3h$$

where d is the distance between adjacent sites, k is Boltzmann's constant and h is Planck's constant. $E_1 = E_L - E_s$ and $E_2 = E_m - E_s$, E_s being the energy required to remove a water molecule from the saddle point.

Under conditions of atmospheric pressure T can only exceed 273K by 100K so that there is no possibility of increasing D beyond the value obtaining in current industrial practice. The rate of drying dm/dt during the diffusion-limited or 'falling rate' period of drying is a function of both the geometry of the material in its comminuted state and of the diffusion coefficient, such that, after time t in the dryer

$$\mathrm{d}m/\mathrm{d}t = (6\,m_0/\pi^2) \exp(-Dt/a^2)$$

where m_0 is the initial moisture content and a represents some sort of mean particle radius. The drying rate is therefore sensitively dependent on the radius a and suggests that further comminution of pre-dried meal would amply repay the extra effort involved. It should be noted however that many other considerations of a physical nature apply to the drying of fish meal (Myklestad, 1973), such as bed depth, meal structure and particle surface area.

The process of extracting oil from fatty fish before drying is carried out both for the commercial value of the oil itself and because drying would otherwise be an extremely slow operation. Oil distributed throughout the muscle surrounds the individual fibres and impedes the movement of water by virtue of the very low diffusion coefficient of water in oil (7×10^{-12}m^{2}s^{-1} at 30°C) compared with that in muscle ($3 \cdot 4 \times 10^{-10}$m^{2}s^{-1}). The resultant diffusion coefficient D has been shown by Jason (1965) to be related to the fat content by the following expression

$$1/D = 1/D_N + \beta F$$

where D_N is the diffusion coefficient of water in fat free tissue and β is a constant.

Table II shows that D increases rapidly with decreasing fat content so that the drying rate dm/dt increases correspondingly. Optimization of oil extraction is therefore extremely important in relation to both energy and cost.

Table II
VALUES OF DIFFUSION COEFFICIENT OF WATER IN FATTY FISH TISSUE

Fat content %	Diffusion coefficient D (10^{-10} m^{2}s^{-1})	
0	3.42 ± 0.34	
5	1.2	
10	0.74	
15	0.53	s.d. ± 20%
20	0.41	
25	0.34	
30	0.28	

4.3 Drying behaviour in smoking kilns

Smoke curing is a process in which simultaneous drying and smoking produces products of characteristic taste and texture. The process is dominated by the need to dry within specified ranges of weight loss and drying rate which are best achieved under controlled conditions. During the past 25 years the Torry kiln (Cutting, 1950) has largely displaced the traditional vertical smoking kiln, mainly because of economy in operation and the facility it offers for closely controlling the smoking and drying processes.

During the smoke curing of lean fish the drying rate is almost entirely conditioned by the rates of convective heat and mass transfer, ie during the 'constant' rate period. Fatty fish dry successfully under both constant and falling rate conditions. Conservation of total enthalpy in an incremental volume of a kiln requires that

$$\delta h_A + \delta h_F = 0$$

where δh_A is the enthalpy change of the moist air flowing in the kiln and δh_F is that of the wet fish, from which Doe (1969) has derived the differential equation for any set of conditions. Basically

$$\delta h_A = \left(\frac{\partial m_a}{\partial t} \cdot \frac{\partial h_a}{\partial x} + \frac{\partial m_v}{\partial t} \cdot \frac{\partial h_v}{\partial x} + h_v \frac{\partial^2 m_v}{\partial x \partial t} \right) dx \, dt$$

where $\partial m_a / \partial t$, $\partial m_v / \partial t$ are the mass flow rates of dry air and water vapour respectively; x is the distance along the kiln and t is time. The equation is further developed by incorporating expressions describing evaporative cooling, mass transfer rates under constant and falling rate conditions and the time of transition between them; when fish dimensions, packing density and dimensions of the kiln are introduced, together with initial and boundary conditions, a very complicated expression results. Doe (1969) obtained sets of solutions using standard finite difference approximations chosen to proceed in a marching fashion. The solutions enabled him to derive the variations with time and distance along the kiln of fish weight, dry bulb temperature and partial pressure of water vapour (Fig 7), and to investigate the effects on drying rate of wet bulb depression, kiln loading, air temperature and fish fillet thickness.

4.4 Smoke deposition

The smoking process starts with the pyrolysis of wood under conditions of restricted oxygen supply. The physics and chemistry of the production of smoke are complicated and have not been extensively investigated. We

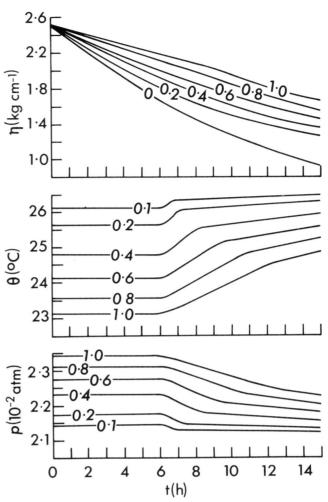

Fig 7 Variation with time t at various distances X/L along kiln of:
(a) fish weight per unit kiln length
(b) dry bulb temperature
(c) partial pressure of water vapour p
Figures on curves indicate values of X/L (After Doe, 1969)

are, however, not here concerned with this aspect of the process. Suffice it to say that pyrolysis produces two discrete phases, a disperse liquid phase and smoke vapour. Individual smoke constituents are distributed between them according to Nernst's partition law (Foster et al, 1961) but, although the mass concentration in the particle phase is always greater than that in the vapour phase, both the phase distribution and the particle size distribution change with the age of the smoke and with removal of constituents by deposition. The physical properties of wood smoke and the physical factors involved in its deposition on water and fish surface have been most extensively studied by Foster (1957; 1960) and further investigated by Foster and Simpson (1961) and by Foster et al (1961). Wood smoke particles are in general smaller than the wavelength of visible light. Their size in any given sample varies over a wide range, as can be seen in Fig 8, and follows a log normal distribution (Foster, 1957; 1960). Their average radius, however, varies over a comparatively small range −0.08 μm to 0.14 μm under practical conditions−and thus their light-scattering properties do not change markedly with the ageing of the smoke. Such comparative size stability gives rise to a similar relation-

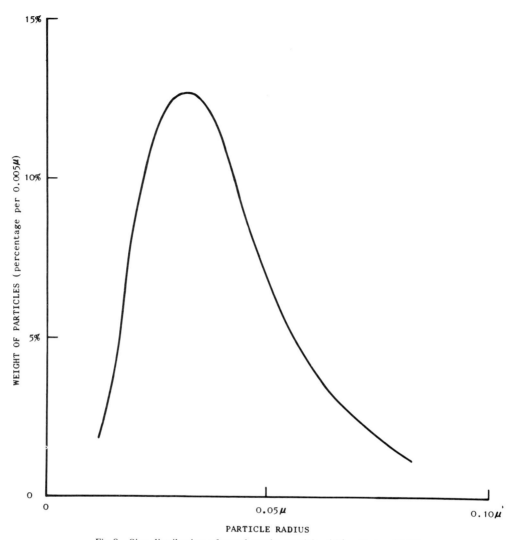

Fig 8 Size distribution of wood smoke particles (*After Foster, 1960*)

ship between gravimetric density of the particles and optical density for all size distributions in the range encountered. The relationship follows the Beer-Lambert law for the intensity I of a parallel beam of light transmitted through a column of smoke of length l, viz.

$$I = I_0 \exp\left(- k\rho l\right)$$

where I_0 is the intensity of the incident light, k is a constant and ρ is gravimetric density. The optical density per unit length $(1/l) \log (I_0/I)$ thus varies linearly with gravimetric density. In commercial practice an optical density per unit length is typically about 0.3m^{-1} for a corresponding gravimetric density of 0.15kg m^{-3}.

The smoke particles are mostly sufficiently small to remain in suspension so that they are not themselves likely to be deposited on to fish under normal smoking conditions. They do, however, serve as a reservoir for volatile components from the vapour. The rates of deposition of smoke constituents on water and wet fish surfaces are identical and are considerably greater than on dry surfaces. Water readily absorbs the vapours but the contribution of the particles to the smoking process is negligible. Foster and Simpson (1961) demonstrated that the rate of deposition of smoke is approximately proportional to the rate of evaporation of water, an observation which suggests that the mechanism of deposition is similar to that of condensation mass transfer of water vapour such as occurs in vacuum thawing, that is, by diffusing through the stagnant layer of air at the surface.

Enhancement of rates of smoking by electrostatic deposition of the particles is not as great as might be supposed—Foster (1957) found this to be only a factor of two. Generally speaking electrostatic smoking appears to confer only a marginal advantage at the expense of considerable complication and cost. When the particles are propelled on to fish arrayed in two dimensions on a grounded conveyor under the action of an imposed electrostatic field, the rate of deposition, although locally very high, does not greatly exceed the integrated rate taking place in three dimensions in a normal kiln. The field-free method (Foster and Jason, 1959) overcomes the dimensional limitation of conventional electrostatic smoking by allowing deposition to occur in a Torry kiln under the influence of the mutual repulsion of pre-charged particles which pass through an ionizing screen. However, despite this simplification, the gains remain marginal and the method has not been adopted commercially.

The complex mixture of smoke vapour constituents diffuses into the fish tissues after deposition but it is not known how transport takes place in physical terms. Some indication of this behaviour may be obtained by time-lapse photography of sections of previously smoked kippers illuminated in ultraviolet light. *Figure 9* shows that the presence of fluorescent constituents of the smoke are apparent at a depth of a few millimetres below the cut surfaces an hour or so after smoking is complete and that the process continues for six days at least. The fact that no fluorescing substances can be observed to diffuse through the skin might be attributed to the lack of smoke deposition on its surface (which dries more rapidly than the cut surfaces) but is more likely to result from the impermeability of the skin to these constituents. If transport is diffusion controlled, then a very rough estimate of the diffusion coefficient suggests that its value lies between 10^{-9} and $10^{-10} \text{m}^2\text{s}^{-1}$ at ambient temperature.

4.5 Salt diffusion

Most smoked-cured products are salted or brined prior to smoking and, although the human palate is not particularly sensitive to variations of salt content in fish, it is generally considered desirable to exercise some control of the process in industrial practice to ensure that its concentration does not exceed a value that would lead to a significant rejection rate.

Provided that an adequate supply of saturated brine is in contact with the surfaces of fish, sodium chloride diffuses rapidly into muscle tissue. Peters' investigation of the process (1971) showed that:

'The diffusion of sodium chloride from a saturated brine solution into cod muscle is characterized by a constant diffusion coefficient of $1 \cdot 4 \times 10^{-9} \text{m}^2\text{s}^{-1}$ (at 23°C). The diffusion is thermally activated with an activation energy of 17kJ $\text{K}^{-1} \text{mol}^{-1}$. Both the diffusion coefficient and the activation energy correspond closely to those of NaCl in water.'

The concentration C_s of NaCl at a distance x from one surface of a slab of fish of thickness L at time t after immersion in the brine is

$$\frac{C_s - C_{s0}}{C_{s1} - C_{s0}} = 1 - \frac{4}{\pi^2} \exp\left[-\frac{D\pi^2 t}{L^2}\right] \sin\frac{\pi x}{L}$$

where C_{s0} is the initial concentration of NaCl in the muscle and C_{s1} is its equilibrium value, and D_s is the diffusion coefficient. *Figure 10* is an example of how well the theoretical equation closely fits experimental values of a salt concentration profile.

Early experiments by Reay (1935), Fougere (1952) and Chipman (1930) showed that in dilute solutions fish gains water during brining, while in concentrated solutions the net water flux is outward. Crean (1961) observed that there is a linear relationship between the overall salt and water contents of fish. Peters (1971) repeated this work carefully under closely controlled conditions and with well-defined sample geometry. He argued that since fish muscle behaves as though it were simply water in respect to the diffusion of NaCl, D_w and D_s, the diffusion coefficients of water and salt, respectively, are identical. Thus we may write, when $C_{s0} = 0$,

$$c_w = \left(\frac{C_{w1} - C_{w0}}{C_{s1}}\right) C_s + C_{w0}$$

which, with C_{w0}, C_{w1} and C_{s1} constant for a given set of conditions, shows that the salt content is a linear function of the water content, thus giving theoretical support for Crean's empirical observation. *Figure 10* shows concentration profiles of water and salt in a slab of cod muscle immersed in concentrated brine. Note that the flux of both water and salt is outwards. When the brine concentration is less than about $0 \cdot 2$kg NaCl/kg H_2O, there is a net increase in the water content of the muscle which takes place against solvent counterflow (resulting from the inward flux of solute) and the resulting transient behaviour becomes very complicated. Peters found the complexity of mathematical analysis of such behaviour daunting.

4.6 Salt and the diffusion of water

The presence of salt in fish muscle has virtually no influence on the rate of drying of brined fish in the range of weight loss normally obtaining in the smoke curing process. Smoke cured fish, such as kippers, smoked haddock, smoked mackerel, *etc*, usually contain about $2 \pm 1\%$ sodium chloride in the aqueous phase after brining, and in consequence the initial water vapour pressure is reduced only marginally. Although subsequent drying tends progressively to saturate the surface water with salt, and so to reduce the vapour pressure difference across the stagnant boundary layer by as much as 25%, saturation conditions are rarely produced because the flux of water to the surface is adequate to prevent such severe desiccation.

Peters (1971), in his definitive work on the subject of the mass transfer of solvent and solutes in fish muscle, has demonstrated that water in excess of that necessary to dissolve the salt present diffuses in exactly the same manner as it does in the absence of salt. He showed that a cluster of nine water molecules surrounds each of the sodium ions in a shell of hydration and that the water so grouped diffuses at a much slower rate. He was thus able to explain the two-stage diffusion behaviour which had earlier been noted by Del Valle and Nickerson (1968). In salted fish, water therefore exists in the form of two 'species'. The first constitutes a fraction $(M_0 - 2 \cdot 77 M_s)/(M_0 - M_e)$ of the total water available for diffusion and the second $(2 \cdot 77 M_s - M_e)/(M_0 - M_e)$ where M_0, M_e represent the initial and equilibrium masses of water present, respectively, and M_s is the mass of dissolved sodium chloride. The factor $2 \cdot 77$ represents the mass ratio between 9 moles of H_2O and 1 mole of NaCl, viz. $(9 \times 18):58 \cdot 5$. Diffusion coefficients D_1 and D_2 associated with the first and second species, respectively, differ considerably, the former being considerably greater than the latter.

Sorption isotherms for muscle containing NaCl display a somewhat complicated behaviour. During initial desorption the value of equilibrium moisture content C_{we} measured on a salt-free solids (SFS) basis depends on C_s (SFS) at any given relative humidity. After partial rehydration in the adsorption process there is only a slight dependence of C_{we} on C_s and the isotherm differs little from that of unsalted muscle. Peters (1971) has shown that the desorption behaviour is consistent with the

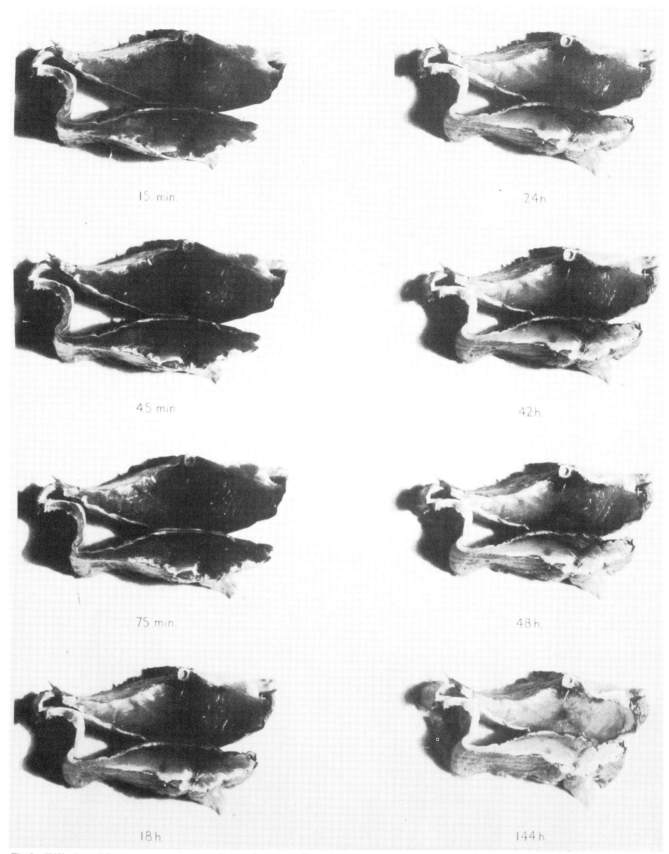

Fig 9 Diffusion of fluorescent smoke constituents into a section of smoked herring. The section has been folded with the skin inside for convenience of display. The times indicated below each photograph indicate the period elapsed after removing the fish from the smoking kiln

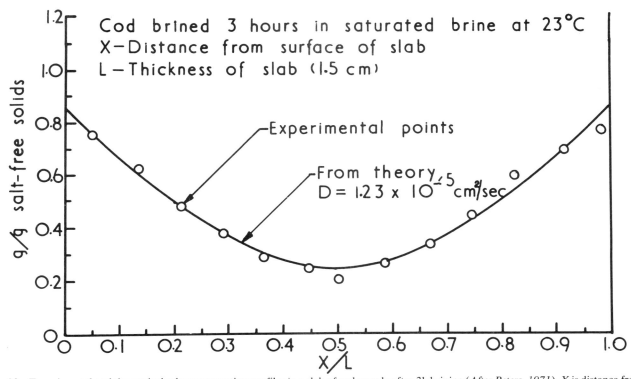

Fig 10 Experimental and theoretical salt concentration profiles in a slab of cod muscle after 3h brining (*After Peters, 1971*). X is distance from surface, L is thickness of slab (15mm)

water-clustering model and attributes the differing adsorption behaviour to the breaking down of the cluster structure during drying prior to rehydration.

Theory can serve to predict drying behaviour in industrial processes and points to some procedures that could be advantageously applied. We have seen that, during the smoke-curing process, drying rate is unaffected by the presence of salt. This knowledge enables us to calculate, without further complication, the dynamic interaction between the mechanism of water migration within the fish and the characteristics of the smoking kiln in which it is drying. While conditions of convective heat and mass transfer dominate this type of drying, they are less important in the drying of fish meal where bulk thermal properties and intra-particle diffusion dominate. In the practice of salt curing, attention must be paid to some seemingly paradoxical behaviour during drying of both heavily salted and lightly salted fish. The existence of a negative temperature coefficient for D_1 in the first phase of diffusion has already been noted, but it has been demonstrated also that drying rates decrease with decreasing relative humidity both in heavily salted fish (Linton and Wood, 1945) and in lightly salted fish (Legendre, 1955)—an effect attributed to the progressive formation of a thick crust of salt at the surface of heavily salted fish but not necessarily so when it is lightly salted. A possible explanation for a reversal of the expected behaviour in both instances may reside in the production of saturated conditions immediately beneath the surface leading to cluster formation and to the early dominance of D_2. Linton and Wood (1945) found that heavily salted cod dries optimally at 26°C when the relative humidity lies between 45% and 55%, and Legendre (1955) that lightly salted fish should be dried at 27°C and between 50% and 55% RH in the first stage

and between 60% and 65% RH subsequently. Peters (1971) points out the benefit of breaking the process at the instant that the transition moisture content is reached (in order to obtain maximum benefit from the dryer) and then subjecting the fish to normal press-piling.

The value of D_2 (about $0.3 \times 10^{-10} \mathrm{m^2 s^{-1}}$) was found to be invariant with temperature in the range 15° to 30°C while D_1 varied from $2.6 \times 10^{-10} \mathrm{m^2 s^{-1}}$ to $1.8 \times 10^{-10} \mathrm{m^2 s^{-1}}$ in the same range. When values of log D_1 were plotted as a function of $1/T$ the slope of the regression line was unexpectedly positive, implying that E was negative, which is insupportable on the basis of a simple statistical mechanical model. Peters speculated that, since the diffusion coefficients were obtained from drying curves of slabs of salted muscle, the negative effect of temperature rise may possibly be ascribed to collision or associations between inward moving ions and outward diffusing water molecules. These considerations were shown by Peters (1971) to account for the very slow rate of drying of salt fish. Initially water is removed at a rate dependent on the value of D_1 until it falls below a certain moisture content which depends on the initial salt concentration. After this point is reached it migrates at a much lower rate dictated by the value of D_2 (*Fig 11*). Transition moisture contents C_{wB} were determined over the entire range of initial salt concentration C_s in 80 separate experiments and in each case the ratio C_{wB}/C_s was found to be close to 2.78 as expected.

4.7 *Freezer burn*
The type of drying that occurs in cold stores, known as freezer burn, leads to quality deterioration and to economically undesirable weight losses. For some years

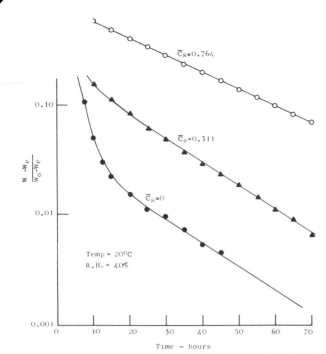

Fig 11 Effect of salt content on drying characteristics of cod muscle. C_s values are those of initial salt concentration (kg/kg SFS) (*After Peters, 1971*)

attempts have been made to minimize the extent of such desiccation by constructing cold stores of suitable design. However, such designs have been entirely empirical because, until recently (Storey, 1975), no reliable data had been systematically gathered and a basic physical description of the process was not available. For the most part the mechanisms controlling this type of drying differ from those already described in this section but, since they are discussed in detail elsewhere (Storey, 1980), we here make only brief reference to them.

Drying under such conditions is essentially a slow isothermal process which differs from the more rapid freeze-drying process taking place *in vacuo* under an externally imposed temperature gradient. The rate of sublimation of ice through the surface of the frozen product is initially determined by the ambient conditions and the thickness of the stagnant boundary layer in the surrounding atmosphere. The ice front recedes through the non-aqueous material and vacates a porous layer of increasing thickness. As it grows, its permeability to water vapour is reduced and a further impedance is added to that of the boundary layer. When the thickness increases to such a value that the permeability becomes limiting, the sublimation rate falls progressively from the initial convective rate and then varies inversely with the thickness of the porous layer. Eventually, all the ice is removed and a second falling rate period commences during which the moisture content of the porous layer slowly equilibrates with the environment. Such a condition would be disastrous in commercial products and is rarely allowed to occur.

5 Mechanical and rheological behaviour

The structure of fish muscle is exceedingly complex at all levels, molecular to macroscopic, and varies from species to species. While possessing some of the attributes of a gel it nevertheless has structural anisotropy which varies according to position. In vertebrate fish, for example, the myotomes (muscle 'flakes') are arranged in the form of roughly conical layers tapering tailward and nesting coaxially along, and symmetrically disposed about, the major axis. Within the conical structure, the muscle cells are arranged in bundles inclined at various angles to the axis. The connective tissue that 'cements' the myotomes together (myocommata) provides additional structural elements.

Biological condition, chill storage and cold storage, which influence the texture of fish (Love, 1980), also affect the mechanical properties. *Rigor mortis*, and rough handling, too, has pronounced rheological consequences.

When fish are allowed to enter *rigor mortis* at temperatures in excess of 0°C then 'gaping' results, that is the individual myotomes are seen to separate giving the flesh a very poor appearance. Using various techniques, Burt *et al* (1970) measured the tension developed in wet muscle in rigor as a function of temperature and also the breaking stress. In addition, the actual physical positions of the breaks were noted. Their results showed that rigor tension increases with increasing temperature and the breaking stress decreases. A point is thus reached (at about 25°C) where the rigor tension exceeds the breaking stress and the muscle pulls itself apart. Also, significantly, the breakage occurs more frequently at the myotome – myocommata interface as the temperature increases and thus explains the phenomenon of high-temperature gaping. Unfortunately no relationships could be found between rigor tension or temperature and biochemical factors such as lactate and ATP concentration or even changes in these factors. Nevertheless, these physical measurements have resulted in a clearer understanding of the gaping process.

The effect of cooking, a process which is most difficult to characterize physically, undoubtedly alters both elastic and yield properties of the muscle components and hence of the matrix. To this must be added the effects of such treatments as salting, frying, smoking, canning and comminution.

In the face of such complexity of macroscopic structure and of physical changes at all levels, it is hardly surprising that nothing is known in physical terms of the mechanical and rheological properties of fish. The basic difficulty is first to characterize these properties and then to measure them, neither of which is easy. Yet even elementary physical descriptions of some aspects would be of pragmatic value in relation to the sensory assessment of 'texture'.

6 Radiation

6.1 *Radiation pasteurization*
Radiation pasteurization, until recently, received a great deal of attention, especially in the USA. Considerable resources were devoted in an attempt to exploit the lethal effects of high-energy radiation which was earlier recognized to be capable of destroying the bacteria responsible for spoilage. Initially results looked exceedingly promising. Using a shipboard irradiator based on a

Co^{60} source of γ-rays, the American Bureau of Fisheries irradiated surf clams (*Spisula solidissima*), haddock (*Melanogrammus aeglefinus*), herring smelt (*Argentina silus*), butterfish (*Poronotus triacanthus*), skate (*Raja naevus*), angler (*Lophius americanus*) and squid (*Loligo pealei*) (Carver *et al*, 1968). The conclusion of this work was that radio pasteurization extended the shelf life of all of these species, in the case of herring smelt, for example, by up to 21 days.

Similarly, hopeful results were obtained by Ehlermann and Munzner (1969) using the far safer and more convenient method of an on-board X-ray facility designed for the handling of fish. This type of radiation source is more convenient since it requires only 10% of the lead screening required by Co^{60} and, thus, leads to a great reduction in weight. When not in use such a source also emits no hazardous radiation. The possibility of collision at sea or sinking thus creates no additional problem of radioactive pollution. The research for this ambitious seagoing project was made possible by an ingeniously designed high power X-ray tube, the so-called pot anode tube, containing a large hollow cylindrical anode through which fish were conveyed. The combined effects of the convergence of the radial radiation field and its attenuation within the target fish, resulted in a tendency towards a desired uniform radiation dose. The large-scale machine for handling several tonnes per hour employed more conventional X-ray tubes arranged in tandem to irradiate both upper and lower sides of the fish. A greater throughput was achieved than was possible with the pot anode. However, despite earlier promising results, a recent report by this same team indicates that no useful extension of shelf life occurs, certainly for haddock and redfish. The spoilage retardation becomes effective only after some 21 days' storage in ice, when this rather poor quality is maintained (Ehlermann and Reinacher, 1978). These results were shown to be entirely consistent with earlier American findings which had been perhaps too enthusiastically interpreted. The final conclusions of Ehlermann and Reinacher was that 'onboard irradiation of whole, unpackaged or packaged ocean fish is not worthwhile'. This did, however, leave open the possibility for irradiation of fillets or high cost rapidly spoiling products such as shrimp meat.

Slow acceptance of the method in the late sixties and early seventies now seems in hindsight to have been extremely fortunate. The conditions prevailing at that time excluded the technique purely on the grounds of cost: a shipboard X-ray irradiator would at that time have nearly doubled the price of a ship. With the current contraction in distant-water fisheries future application also seems to be unlikely. Economic considerations in 1968 dictated voyage extension of at least eight days to make the system profitable but today the achievement of such an advantage would be less significant.

Further comment is unnecessary but it should be stressed that many publications have appeared on this matter other than those quoted. For further information the recent paper of Ehlermann and Reinacher contains some 50 references. Irradiation of food in general is a major topic covered by a large quantity of literature and the deliberations of expert committees (FAO/IAEA/WHO) and this short comment should not be

regarded as being a definitive review of the subject.

6.2 *Bone detection*
A further potential use of X-rays was investigated by Moran *et al* (1965). This was the use of the radiation in its more well-known mode to visualize and thus detect bones in fish products. The system consisted of a low-energy X-ray source and a fluorescent disc detector. The sample to be scanned was placed on a turntable interposed between the source and detector. The fluorescent disc some 13mm in diameter was connected to a photomultiplier by a light pipe. Interposed between the disc and the light pipe, however, was a rapidly rotating disc with a small hole of 0·4mm^2 in area. The combination of this 'flying hole' and the rotation of the sample turntable produced a scanning effect not unlike that in the well-known James Logie Baird prototype television system.

Results with this bone detector were promising. In fish fingers (sticks) 20mm thick, 100% of all bones larger than 0·45mm were detected, 80% of those between 0·3mm and 0·45mm and 50% in the range 0·2mm to 0·3mm. For thinner samples the results were even better.

In fillets the results depended on the position of the bone in the fillet, *ie* the thickness of flesh at that point. In fillets thicker than 19mm no bones smaller than 0·45mm could be detected reliably. Bones some 0·6mm in diameter would be reliably detected subject to this fillet thickness limitation.

The failure of this technique to find general application is probably indicative of unfavourable economies possibly coupled with a more general public acceptance of the occasional bone than was believed originally.

6.3 *Density measurement*
As was discussed in the section on electrical measurements, moisture measurement in fish meal and other powders is hampered to a certain extent by fluctuations in bulk density which occur in dryers. A practical solution to this problem has been obtained by Mládek (1973) who showed that the density of various foodstuffs can be monitored by measuring the absorption of low energy γ-rays with an apparatus producing an electrical signal the magnitude of which is proportional to the density of the material. Precautions need to be taken to eliminate Compton scattering by selecting a source of suitable energy and collimating the radiation.

Microwave systems incorporating γ-ray density correction are now available commercially but, unfortunately, the cost of the correction is an order of magnitude larger than that of the microwave section. Consequently, the overall cost becomes comparable with the cost of the dryer itself.

7 Conclusions

We have discussed various aspects of physics, mostly classical, that have direct relevance to the fish industry. There have been modest achievements in their application but much work remains to be done and there are large gaps in our knowledge. However, as new species come to be utilized and new processes developed, additional problems will emerge that will call for us to undertake the necessary applied physical research. There is a great need to build up a large body of background knowledge relating to the underlying physical processes relevant to fish technology. Much of this knowledge will have immediate relevance; while some, including the results of speculative research, will help to provide 'seed corn' for the future. Additionally, the gathering of basic data on physical properties of fish and fish products is a task that we must set ourselves in the coming years. Many of these properties remain to be determined over the wide range of conditions encountered in numerous processes.

Although it was not intended that this paper should be a comprehensive review of all work so far published in the field, our aim has been to give a broad view with occasional detail where we have felt ourselves competent to do so. Because the subject of physics is so extensive it is inevitable that some significant work has been overlooked. If this is the case we apologize both to the authors concerned and to the readers of this paper. Nevertheless, we hope that we have succeeded in our main purpose and that this paper will stimulate further work and new ideas in the application of physics to fish utilization.

8 References

ADAM, M. *Bibliography of Physical Properties of Foodstuffs*. CAE,
1969   Prague. 375pp
AITKEN, A. Nature and orientation of ice in frozen fish. *J. Fd. Technol.*,
1966   1, 17–24
ALBASINY, E L. The solution of non-linear heat-conduction problems
1956   on the Pilot Ace. IEE Convention on Digital-Computer Techniques. *Proc. Inst. Elec. Eng.*, 103, B Suppl. No. 1, 158–162
ANON. GB Dept. Sci. Ind. Res., Food Invest. Rept., 19
1957
—— GB Dept. Sci. Ind. Res., 17, Food Invest. Rept.
1958
BENGTSSON, N E, MELIN, J, REMI, K and SODERLIND, S. Measure-
1963   ments of the dielectric properties of frozen and defrosted meat and fish in the frequency range 10–200 MHz. *J. Sci. Fd. Agric.*, 14, 592–604
BENGTSSON, N E and RISMAN, P O. Dielectric properties of foods at 3
1971   GHz as determined by a cavity perturbation technique. *J. Microwave Power*, 6, 107–123
BRUNAUER, S, EMMETT, R H and TELLER, E. Adsorption of gases in
1938   multimolecular layers. *J. Am. Chem. Soc.*, 60, 309–319
BURT, J R, JONES, N R, McGILL, A S and STROUD, G D. Rigor ten-
1970   sions and gaping in cod muscle. *J. Fd. Technol.*, 5, 339–351
CARVER, J H, CONNORS, T J, RONSIVALLI, L J and HOLSTON, J A.
1968   Shipboard irradiator studies. Final report TID 24332. US Atomic Energy Commission, Division of Technical Information, Oak Ridge, Tennessee
CHIPMAN, H R. Ms. Rept Biol. Bd. of Can. No. 32
1930
CHIPMAN, H R and LANGSTROTH, G O. Some measurements of the
1929   heat capacity of fish muscle. *Trans. Nova Scotia Inst. Sci.*, 17, 175–184
CHRISTIE, R H and JASON, A C. Vacuum thawing of foodstuffs. In:
1975   *Food Engineering and Food Quality*. Proceedings of the 6th European Symposium, Cambridge, 8–10 Sept. Society of Chemical Industry, London. 153–174
CREAN, P. Light pickle salting of cod. *J. Fish. Res. Bd. Can.*, 18,
1961   833–844
CUTTING, C L. The Torry Research Station controlled fish-smoking
1950   kiln. *Dept. Sci. Ind. Res. Food Investigation Leaflet* No. 10. HMSO, London
CUTTING, C L, REAY, G A and SHEWAN, J M. Dehydration of fish:
1956   British experiments (1939–1945) and the development of a warm air drying process. *Food Investigation Special Report* No. 62. HMSO, London
DALGLEISH, J M and THOMSON, H P. Improvements in and relating to
1958   the preservation of foodstuffs. Provisional UK Pat. No. 27044/58
DEL VALLE, F R and NICKERSON, J T R. Salting and drying fish. 3.
1968   Diffusion of water. *J. Fd. Sci.*, 33, 499–503
DOE, P E. A mathematical model of the Torry fish drying kiln. *J. Fd.*
1969   *Technol.*, 4, 319–338
DUCKWORTH, R B. Differential thermal analysis of frozen food sys-

1971   tems. I. Determination of unfreezable water. *J. Fd. Technol.*,
        6, 317–327
EDE, A J. Physics of the low temperature vacuum drying process. *J.*
1949   *Soc. Chem. Ind.*, 68, 3302, 33–40
EDE, A J and HADDOW, R R. The electrical properties of food at high
1951   frequencies. *Food Manufacture*, 26, 156–160
EHLERMANN, D A E and MUNZER, R. Zur Strahlenkonservierung
1969   von Fischproduckten Ein Fertiggericht aus Kabeljau, *Kältetech. Klim.*, 21, 331–332
EHLERMANN, D A E and REINACHER, E. Some conclusion from ship-
1978   board experiments on the radurization of whole fish in the Federal Republic of Germany. In: *Food Preservation by Irradiation*. International Atomic Energy Agency. Vol 1, 321–331
ESKILSON, P. Strålningsvarmets inflytande på varu temperaturer i
1967   moderna frysdiskr. *Kylteknisk tidskrift*, 26, 69–71
FALK, K G, FRANKEL, E M and McKEE, R H. Low temperature vac-
1919   uum food dehydration. *J. Ind. Eng. Chem.*, 11, 1036–1040
FAO. *Yearbook of Fishery Statistics: Fishery Commodities*. HMSO,
1977   London. Vol 45
FOSTER, W W. Some of the physical factors involved in the deposition
1957   of wood smoke on surfaces with ultimate reference to the process of smoke curing. PhD. thesis, University of Aberdeen
—— The size of wood smoke particles. In: *Aerodynamic capture of*
1960   *particles*. Pergamon Press, Oxford. 89–96
FOSTER, W W and JASON, A C. Improvements in and relating to the
1959   smoking of foodstuffs. UK Pat. No. 814,121
FOSTER, W W, SIMPSON, T H and CAMPBELL, D. Studies of the smok-
1961   ing process for foods. II. The role of smoke particles. *J. Sci. Fd. Agric.*, 12, 635–644
FOSTER, W W and SIMPSON, T H. Studies of the smoking process for
1961   foods. I. The importance of vapours. *J. Sci. Fd. Agric.*, 12, 363–374
FOUGERE, H. The water transfer in codfish muscle immersed in sodium
1952   chloride solutions. *J. Fish. Res. Bd. Can.*, 9(8), 388–392
GANE, A J. Freeze-drying of foodstuffs. In: *Freezing and Drying*, Ed
1951   R J C Harris. Inst. Biol.
HAWKINS, A E, PEARSON, C A and RAYNOR, D. Advantages of low
1973   emissivity materials to products in commercial refrigerated open display cabinets. *Proc. Inst. Refrig.*, 69, 54–64
HENNINGS, C. Ein neues elektronisches Schnell-verfahren zur Ermit-
1963   tlung der Frische von see-fischen. *Lebensmittelunters. u. Forsch*, 119, 461–477
HIDDINK, J. *Natural convection of liquids with respect to sterilization of*
1975   *canned foods*. Centre for Agriculture, Publishing and Documentation, Wageningen, Holland
HILL, T L. Statistical mechanics of multimolecular adsorption. *J.*
1946   *Chem. Phys.*, 14, 263–267
JASON, A C. A study of evaporation and diffusion processes in the
1958   drying of fish muscle. In: *Fundamental Aspects of the Dehydration of Foods*. Society of Chemical Industry. 103–135
—— Effects of fat content on diffusion of water in fish muscle. *J. Sci.*
1965   *Fd. Agric.*, 16, 281–288
—— Improvements in or relating to thawing frozen foodstuffs. UK Pat.
1972   No. 1272 396
—— Microwave thawing—some basic considerations. In: *Meat*
1974   *Freezing—Why and How?* Meat Research Institute Symposium No. 3, Bristol. 4331–4337
JASON, A C and JOWITT, R. Physical properties of foodstuffs in relation
1969   to engineering design. *Dechema-Monographien*, 63, 21–72
JASON, A C and LEES, A. Estimation of fish freshness by dielectric
1971   measurement. *Department of Trade and Industry Report* No. 71/7, Torry Research Station, Aberdeen
JASON, A C and LONG, R A K. The specific heat and thermal conduc-
1955   tivity of fish muscle. *Proc: IXth Int. Con. of Refrig. Inst. Int. du Froid*
JASON, A C and RICHARDS, J C S. The development of an electronic
1975   fish freshness meter. *J. Phys. E: Sci. Instrum.*, 8, 826–830
KALINSKI, J. An industrial microwave attenuation monitor (MAM)
1978   and its application for continuous moisture content measurements. *J. Microwave Power*, 13, 275–281
KAREL, M. Physico-chemical modification of the state of water in
1975   foods—a speculative survey. In: *Water Relation of Food*, Ed R B Duckworth. Academic Press, London. 639–657
KENT, M. The use of strip-line configuration in microwave moisture
1972a   measurement. *J. Microwave Power*, 7, 185–193
—— Complex permittivity of protein powders at 9·4 GHz as a function
1972b   of temperature and hydration. *J. Phys. D: Appl. Phys.*, 5, 394–409
—— The use of strip-line configurations in microwave moisture
1973   measurements. II. *J. Microwave Power*, 8, 189–194
—— Time domain measurements of the dielectric properties of frozen
1975a   fish. *J. Microwave Power*, 10, 37–48
—— Fish muscle in the frozen state: time dependence of its dielectric
1975b   properties. *J. Fd. Technol.*, 10, 37–48
—— Microwave attenuation by frozen fish. *J. Microwave Power*, 12,
1977   101–107

KENT, M and JASON, A C. Dielectric properties of foods in relation to
1975 interactions between water and the substrate. In: *Water Relations in Foods*, Ed R B Duckworth. Academic Press, London. 211

KHACHMANUKYAN, S G. Experimental determination of enthalpy and
1972 heat capacity of fish in the temperature range between −20°C and −196°C. *Trudy VNIRO*, 88, 165–170

KINDLEYSIDES, L and HALE, M J. Energy conservation in the oper-
1979 ation of refrigerated food display cabinets. *Proc:* IFST, 12, 31–38

KRASTNITSKAYA, A A, JONOV, A G and BOGOLYUBSKIJ, O K. Freezing
1976 and storage of fish culinary products. *Ryb. Khoz.* No. 11, 55–59

LEGENDRE, R. The artificial drying of highly salted codfish. *J. Fish. Res.*
1955 *Bd. Can.*, 12(1), 68–74

LINTON, E P and WOOD, A L. Drying of heavily salted fish. *J. Fish.*
1945 *Res. Bd. Can.*, 6, 380–391

LOBSIN, P P. Physical properties of fish. *Trans. Inst. Marine Fisheries*
1939 *and Oceanography, USSR,* 13, 5–50

LONG, R A K. The cooling and freezing of fish: some thermodynami-
1954 cal aspects. PhD thesis, University of Aberdeen. 113pp

—— Some thermodynamic properties of fish and their effect on the
1955 rate of freezing. *J. Sci. Fd. Agric.*, 6, 621–633

LORENTZEN, G F. The freezer chain – problems and possibilities. *Proc.*
1963 *Inst. Refrig.*, 59, 109–154

—— Forsøk med reduksjon av innstrålingen i åpne frysedisker
1968 kjøleteknikk og fryserinaering, 20, 35–39

LOVE, R M. Biological factors affecting processing and utilization.
1980 *This volume*

MAKAROV, T I. *Proceedings of the All Union Institute of Sea Fishing and*
1952 *Oceanography*, 20, 21–27

MATHESON, N A and PENNY, I F. Storage of dehydrated cod. Part II.
1961 *Food Process Packaging*, 30, 123–127

MERRITT, J H and BANKS, A. Thawing of frozen cod fillets. *J. Refrigera-*
1964 *tion*, 8, 115–119 and 150–152

MERYMAN, H T. *Cryobiology*. Academic Press, New York. 11
1966

MLÁDEK, J. Determination of the moisture content of loose materials
1973 by the microwave method. *Zem. Tedinika*, 19, 453–458

MORAN, J M, WISE, D P, TETRAULT, R and CARVER, J H. Prototype
1965 automatic bone detector. *Fd. Technol.*, 19, 728–733

MUDGETT, R E, MUDGETT, D R, GOLDBLITH, S A, WANE, D I C and
1978 WESTPHAL, W B. Dielectric properties of frozen meats. *Proc:*
13th Microwave Power Symposium, Ottawa, Canada

MYKLESTAD, O. Physical aspects of the drying of fish meals. *J. Sci. Fd.*
1973 *Agric.*, 24, 1209–1215

OHLSSON, T and BENGTSSON, N E. Dielectric food data for microwave

1975 sterilisation processing. *J. Microwave Power*, 10, 93–108

OSTERHOUT, W J V. *Injury, recovery and death in relation to conductiv-*
1922 *ity and permeability*. J B Lippincott Co, London

PETERS, G R. Diffusion in a medium containing a solvent and solutes,
1971 with particular reference to fish muscle. PhD thesis, University of Aberdeen. 102pp

PRIOU, A, FOURNET, C, GAILLARDIN, J C, DEFICIS, A and GIMONET, E.
1977 Thawing of beef slabs by microwave energy. *Proc:* 12th Microwave Power symposium, Minneapolis, USA

PRIOU, A, FOURNET-FAYAS, C, DEFICIS, A and GIMONET, E. Micro-
1978 wave thawing of large pieces of beef. *Proc:* 13th Microwave Power Symposium, Ottawa, Canada

REAY, G A. Report of UK Food Investigation Board, 70–75
1935

RIEDEL, L. Kalorimetrische Untersuchungen über das Gefrieren von
1956 Seefischen. *Kältetechnik*, 8, 347–377

—— Eine Formel zur Berechnung der Enthalpie fettarmer Lebensmit-
1978 tel in Abhängigkeit von Wassergehalt und Temperatur. *Chem. Mikrobiol. Technol. Lebensm.*,5, 129–133

SANDERS, H R. A computer program for the numerical calculations of
1980 heating and cooling processes in blocks of fish. *This volume*

SHORT, B E and STAPH, H E. The energy content of foods. *Ice Refrig.*,
1951 121, 23–26

SIMATOS, D, FAURE, M, BONJOUR, E and COUACH, M. Differential
1975 thermal analysis and differential scanning calorimetry in the study of water in foods. In: *Water Relations of Foods*, Ed R B Duckworth. Academic Press, London. 193–209

SKINNER, R H. The second order linear system and an empirical can
1979 centre temperature by story model. *Proc:* Int. Meeting of Food Microbiology and Technology. Medicina Viva, Parma, Italy

SKINNER, R H and JOWITT, R. A reappraisal of available temperature
1977 history models for heat sterilization processes. *Proc:* BEFCE Mini Symposium in Mathematical Modelling. Lund Inst. Technol., Sweden

STEELE, D J and KENT, M. Microwave stripline techniques applied to
1979 moisture measurement in food materials (paper presented at the International Microwave Power Institute Symposium, Ottawa, 1978). *BFMIRA technical circular* No. 673

STOREY, R M. The migration of water in frozen cod muscle. MPhil
1975 thesis, University of Leeds. 110pp

—— Modes of dehydration of frozen fish flesh. *This volume*
1980

STUCHLY, S and KRASZEWSKI, A. Method for determination of water
1965 content in solids, liquids and gases by means of microwaves and arrangements for application of this method. Polish Patent 51.731

The application of engineering science to fish preservation. Part 1 *M R Hewitt*

1 The beginnings of engineering science

The definition of engineering science is rather difficult and it always appears to be anomalous. Engineering is always described as an art, but without some scientific training it would be impossible for the engineer to bridge the gap between pure scientific thought or discoveries and their practical embodiment by the craftsman.

The first example of bridging this gap followed the development of the beam engine. In 1719 Newcomen, who was a millwright with no scientific training, introduced his engine (Rolt, 1967) which he had developed by application of the craftsman's art. Many of these engines were in service before Smeaton, who was a well-educated man, was able to improve the efficiency by the application of scientific methods and thus married science to the art of engineering.

Obviously there were considerable economic benefits to be gained by improvements in the efficiency of prime movers and most attention was centred on such developments. In order to attain any marked improvement in performance the application of scientific principles was necessary.

The scientific approach naturally spread to other fields but it is not surprising that, in the field of fish processing, where the overall benefit is fairly limited, engineering science was not applied to machinery developments until well into the twentieth century.

2 Growth in the fish processing industry

In the field of fish processing the growth of the application of engineering science in the early days was relatively slow. Machinery development tended to be carried out using limited and tried empirical methods and, until fairly recently, has depended more on the skill and art of the craftsman than on scientific methods. This did not necessarily inhibit development, since the handling of biological material such as fish does not lend itself to ready analysis and usually precludes the use of the short-cuts which can be made to an experimental programme by the use of scientific analysis.

Even as late as 1963, Eddie (1964) indicated that 'the total number of engineers in the British Fishing Industry with qualifications equivalent to graduate membership of a senior engineering institution is perhaps 10 or 12 and the Torry Research Station of the Department of

Scientific and Industrial Research and the Industrial Development Unit of the White Fish Authority account for most of this number'. Indeed, of these 10 or 12, perhaps only half could be described as working in the field of fish process engineering.

Allowing for the fact that this is only one example of one developed nation the indications are that the application of engineering science in fish processing throughout the world is not very broad. Probable exceptions are where engineering developments have taken place in other sectors of the food industry and can be applied to fish. Handling machinery associated with canning and packaging are particular examples.

There are other reasons for this limited use of engineering science. Because priority in any fish industry has been given to catching it follows that priority has been given to development of vessels, engines and fishing equipment and it is only during this century when distribution has become widespread and voyages have become longer that there has been the need to industrialize and improve the simple methods of preservation and processing formerly employed.

Despite the limited field, the recent growth of the application of engineering science to processing has been fairly rapid. The increase in requirements for machines and systems in the fish processing industry has been world-wide as the cost of labour and the demand for more sophisticated products has risen and longer voyages have increased the likelihood of spoilage.

Forty to 50 years ago there was only one machine (Drews, 1974) available for splitting herrings for kippering. Now a whole host of machines is available for handling, filleting and processing many different species.

One factor which has inevitably retarded development has been the necessity to develop machinery for a specific product or process, this because of the wide differences in the species to be handled and in the tastes and requirements of consumers. In these circumstances there will be a limited return on any investment in research and development because of the limited market for the final product. Development costs often inhibit the development of machines and systems for a given process, since these costs may render a process uneconomic even where large savings in labour can be made. This is particularly so in the case of a small manufacturer. The result is that a large proportion of the development work is inevitably carried out either by government departments or by universities and not upon strict commercial terms.

3 The pattern of development

Where developments have taken place in research institutes and universities they are normally well documented. Where major developments have been made the only likely record will be a patent application, so that many smaller inventions pass unrecorded. Many developments are still carried out using engineering art rather than science and these are often very successful. It is difficult, therefore, to build up a comprehensive picture of the history and development of engineering science in the field of fish process engineering. There are, however, several interesting examples which serve to illustrate the application of scientific methods.

Surprisingly, many of the applications have stemmed from scientists working on the spoilage characteristics of fish. These workers have identified the needs of industry or have recognized the possibilities of improvement to existing processes by the application of scientific principles. Most engineers or scientists have found that application of mechanical handling and processing methods to biological materials is not necessarily very simple and some problems remain intractable. These difficulties have been highlighted by several authors (Drews, 1974; Hewitt, 1975). The major problems are the possibility of contamination during processing and the possibilities of mechanical damage. The rates of bacterial and biochemical spoilage during handling and processing are also important and serve as constraints to the engineer. Therefore it is not surprising that many of the earlier ideas for development came from biologically oriented scientists rather than from engineers. An example is the development of the mechanical smoking kiln in the late 1930s and early 1940s.

It is apparent, however, that, with the current availability of biochemical, biophysical and physical data for many species of fish, the pattern of development has changed, with the emphasis now being placed on engineers and physicists. These, of course, are still dependent upon the support of biologists and biochemists, which is necessary when analysing the effect of any new handling and processing method.

There are still wide gaps in the availability of some physical data which is almost inevitable with the wide range of species of fish which have to be processed throughout the world. It cannot be emphasized enough that although empirical development programmes can now be designed more systematically, it is impossible to obtain the full benefits of mathematical modelling and the full advantage of modern techniques, such as the use of digital computers, without certain basic data. The emphasis, therefore, will still be on empirical methods until the necessary data is available. It must be pointed out, however, that whenever models are used they must be related to real situations by appropriate experiments (Doe, 1975).

The aim of a scientific approach will always be to minimize the number of measurements that have to be made and to obtain a general solution with widespread application. This has not, in general, been easy with such a variable raw material as fish but some of the difficulties may be reduced in the future by use of microprocessors where a large amount of data can be stored and be available to allow fine adjustments in machine control.

4 Application

The application of engineering science in fish processing has been diverse and it would be impossible in a paper of this length to review all the developments. A few examples have therefore been selected, with which it is hoped to indicate the steps that have been taken and the advantages that are to be gained by the application of scientific methods.

4.1 The Torry smoking kiln
One of the earlier applications of scientific methods in engineering developments was undoubtedly the Torry

smoking kiln. Although this was developed in collaboration with engineers, it was principally the work of Cutting (1942; 1950) around 1940 which led to the improvement of the method of preservation which had been used for centuries, that is the smoking and drying of fish.

It was well known that fish could be smoked and dried in a heated current of air and the traditional type of kiln, which was in use at that time, is still in use today. This kiln, however, cannot readily be controlled and is subject to the vagaries of weather and the variability in raw material. In addition there appeared to be no available design parameters on which to base a system of control, thus making it difficult to obtain a uniform product.

Traditional products had been developed over the years but it became possible, from systematic analysis of biochemical properties and spoilage characteristics, to define product requirements closely. Having defined the product it is then possible by the use of scientific principles and by the necessary experiments to specify the design parameters for a kiln.

The examination of these parameters was obviously very detailed and the breadth of the study is interesting. Not only was it appreciated that certain parameters such as temperature, air velocity and humidity would have to be controlled but the limitations imposed by fish properties such as conductivity and rate of diffusion of water on the drying rate were also considered. It became possible to specify the complete process and lay down values of air velocity, temperature and humidity.

In collaboration with Hardy, Cutting (1942) was able to develop a mechanical kiln which enabled the process to be controlled in most conditions of temperature and humidity. The design incorporated many features which indicate the emphasis placed on a systematic and scientific approach and the attention paid to the important details. The layout of the kiln is shown schematically in *Fig 1*.

Every attempt was made to ensure a uniform velocity distribution in the curing chambers and thus to ensure uniform drying conditions. To eliminate the swirl component imported by the fan, aerofoil sections in the bends of the ducts and a diffuser wall were incorporated. For the control of humidity, a makeup and recirculation system are important features. In addition the design incorporated smoke producers with multiple fires so that the quantity of smoke could be readily controlled.

To provide variable heating throughout the drying cycle, the system was carefully balanced. The heat of the fires producing the smoke is included in the overall requirement and reheating of the air after it passed through part of the working section in order to reduce its relative humidity was also incorporated.

A further interesting aspect is the essentially scientific approach in recognition that, although the design of all kilns may be standard, there will be differences between models. In consequence the design includes adjustable slots in the diffuser walls and Cutting (1950) recommended that measurements of airflow, humidity, smoke density and temperature should be made when setting up the kiln.

This development, which was a major step forward in

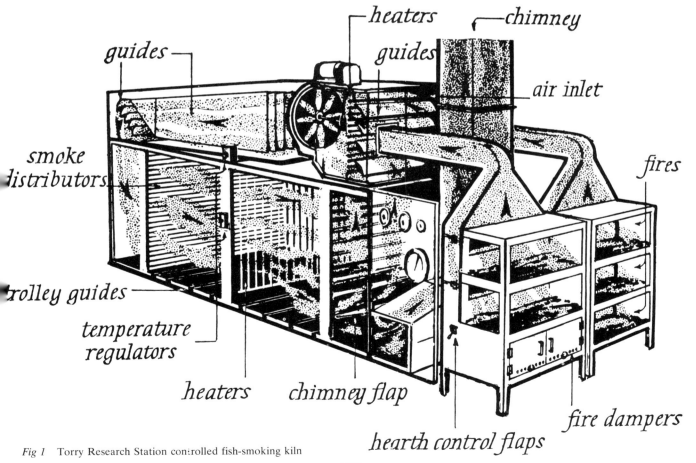

Fig 1   Torry Research Station con:rolled fish-smoking kiln

177

the field of fish processing, incorporated many novel features and it would not have been possible without collaboration between scientists and engineers.

One interesting point made by Cutting, which perhaps illustrates the limitation to any development, is that no metal parts should be used in the smoke producer. This remark was obviously perfectly valid from experience at the time, where the smoke producers closely copied traditional fires in their method of operation, but does not apply today in the design of modern smoke producers where the method of operation is different.

4.2 Cutting machinery

One of the most significant areas of engineering development has been fish splitting and filleting machinery. Probably first was a machine for producing split herrings for the manufacture of kippers and this work was started about 50 years ago (Drews, 1974). The machine developed, which is still produced today by Fisadco, is basically oriented to one product and is therefore mainly in use in the UK. It is an important development, however, because it represents the replacement of a number of relatively complicated and laborious manual processes by one machine.

Possibly the most significant breakthrough in the design of this machine was the use of circular knives, which enabled a smooth cut to be made, and circular brushes to clean out the gut cavity. The use of circular knives is almost standard practice in filleting machines marketed today by leading manufacturers such as Baader, Arenco, etc.

Appearance plays a large part in the sale of a food product and circular knives have enabled good appearance of cut flesh to be maintained. This being so it is not surprising that a considerable amount of work has gone into optimizing the design and velocity of cutting blades. Several theoretical works have been published (Zhilin, 1966; Kawka, 1974) and the results have been reviewed by Johnston (unpublished) in an attempt to rationalize all the findings. Most of the emphasis has been placed on the determination of blade forces and hence the cutting power. Johnston experienced some difficulty in obtaining a common factor in the models for blade forces which, of course, depend on the geometry of the machine used and on the feed rate. One difficulty is obtaining representative figures for the cutting force, since this will depend on the material being cut. For instance, Zhilin (1966) made some observations on the difference in resistance between the skin and fish flesh, and the installed power will obviously be even greater still where it is necessary to cut through bone.

Kawka (1974) indicated that there is a minimum speed of 12m/s for a clean cut, which is important for the purposes for presentation. The relationship of cutting force with cutting speed (tip speed) for beheading cod and hake with different blade profiles is shown in Fig 2.

It is probably true to say that most cutting machine development is carried out empirically with data being collected systematically. However, using predictions for cutting speed (Kawka, 1974) it is possible for blades to be designed for each application and for the power requirements to be determined more accurately. It is likely, however, that for the foreseeable future, although modern materials and techniques will be used, the development of cutting machinery will involve a combination of engineering science with a considerable amount of painstaking work involving 'trial and error'.

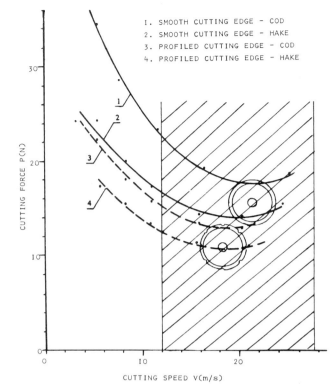

Fig 2 Correlation of cutting forces for beheading with cutting speed (Kawka, 1974)

4.3 Size grading

The size grading of fish is often important for both processing and marketing. Several mechanical systems have been developed or are in the process of being developed. Most mechanical graders, either proposed or in commercial use, select fish on the basis of their thickness and a typical grader is shown in Fig 3. There is therefore reliance on the correlation between either thickness and length or thickness and weight. These correlations have been shown (Ionas, 1977) to have a variance of up to 15% so that there will inevitably be some scatter in the length of fish which will fall into a given grade size.

A second limitation, which is readily apparent, is illustrated in Fig 4. Whenever it is desired to divide the size range, one has the probability that 50% of the fish of the size at the divider will fall into either the larger or smaller range. There will also be a certain quantity of larger or heavier fish in the smaller grade and vice versa, due to a number of factors such as biological variation, carry-over due to misalignment and similar mechanical problems.

In full knowledge of these limitations, it has been possible to give some prediction of the accuracy of such methods of grading and Johnston (unpublished) has devised a set of parameters which can be used to design variable gap graders within predictable limits, knowing the size distribution and variability in a given species.

4.4 Thawing

Scientific work on the thawing of frozen fish really

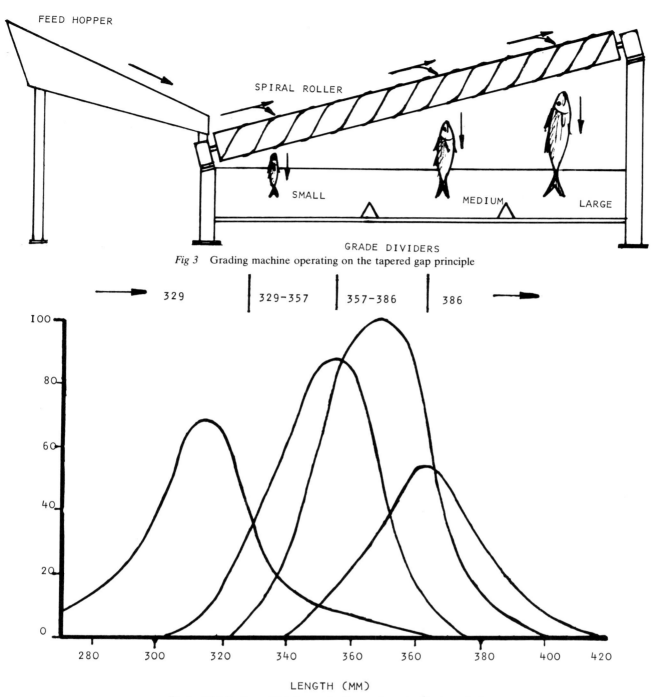

Fig 3 Grading machine operating on the tapered gap principle

Fig 4 Distribution of fish by grades (tapered gap haddock grader)

expanded in the 1950s when frozen fish became a raw material for commercial processing on a large scale. The area where a significant contribution was made was in the thawing of blocks of frozen fish. In 1967 Merritt (1969) reviewed the state of development of thawing machines and further review was carried out by Burgess *et al* (1974). Many different methods had been proposed and tested and this is another area where engineers have not necessarily taken the lead. For example Jason and Sanders (1962) have made a significant contribution in electrical methods and neither would describe them-

selves as engineers. A similar case has arisen with the development of vacuum thawing (Jason, 1973).

A wide range of methods has been explored by various workers, including still air, humid air blast (Merritt and Banks, 1964), warm water (Hewitt, 1969), dielectric, electric resistance (Jason and Sanders, 1962), and vacuum thawing (Everington and Cooper, 1972; Jason, 1973). One factor common to all methods is that they have been studied systematically and that the development has been carried out using scientific principles followed by experimental verification and operating within the quality constraints dictated by the raw material.

The early determination of the limitations imposed by a given method is important. The raw material imposes a limitation on thawing medium temperature if spoilage, in every sense, is to be minimized. Thus the limitations in thawing time can be predicted from known material properties which were determined during earlier investigation on freezing by Plank *et al* (1916) and Riedel (1956). The application of numerical methods and digital computers, of course, is of great assistance in the determination of thawing times.

One example of the scientific approach is the recognition of the role of condensation as a major contributor to heat transfer in humid air blast thawing. Merritt and Banks (1964) have estimated that condensation accounts for 50% of the heat input to the fish and this has led to a design incorporating saturated air.

Early elimination of some methods has been possible where the transition from theoretical considerations to a practical engineering solution could not be made.

4.5 *Geometrical relationships*
The design of fish cutting and handling machinery is dependent upon the establishment of the physical parameters of the fish. To enable mechanisms to be designed to adjust to the wide range of sizes normally experienced, without incorporating complicated measuring systems, the development of correlations between the length, thickness, breadth and weight are essential. Some of the correlations have been published for a limited number of species and one leading machinery manufacturer, Baader (1978), include some data in their catalogue.

Ionas (1977) has reviewed the Russian work on the establishment of relationships between thickness, weight and depth of various species and has attempted to establish accurate relationships. The earliest work reviewed was published by Kislevich (1914–1915) but, apart from the work of Torin (1927), little appears to have been published until the 1950s.

It is interesting to note that the relationships based on geometric similarity such as:

$$\text{Surface area } \alpha \text{ length}^2$$
$$\text{Weight } \alpha \text{ length}^3$$

appear to give the best correlations rather than some of the polynomial solutions, which bears out a point made by Doe (1975) that the mathematical model should be tailored to fit the real situation as closely as possible. One of the relationships is illustrated in *Fig 5* where differences due to biological variation are of the order of 15%.

The geometric and weight information currently being amassed for different species will be of continuing importance for the development of fish processing and handling machines. It is hoped that in the near future generalized correlations will be available to enable prediction of the various other parameters required from a single measurement of the fish.

This concept has been considered in the development of machinery by several workers in the field, notably the White Fish Authority, Torry Research Station and the Institute of Fishery Technology Research in Tromsø (Anon, 1976). In all cases the proposal is to use programmable micro-computers to control machine operations rather than using mechanical linkages as in existing machines. The potential for storing information and providing well-controlled operations is enormous, once sufficient primary data has been accumulated.

4.6 *Modelling of refrigerated sea water and chilled sea water systems*
Although the development of refrigeration falls within the scope of the first part of this paper, the analysis of refrigerated sea water systems for freezer trawlers and

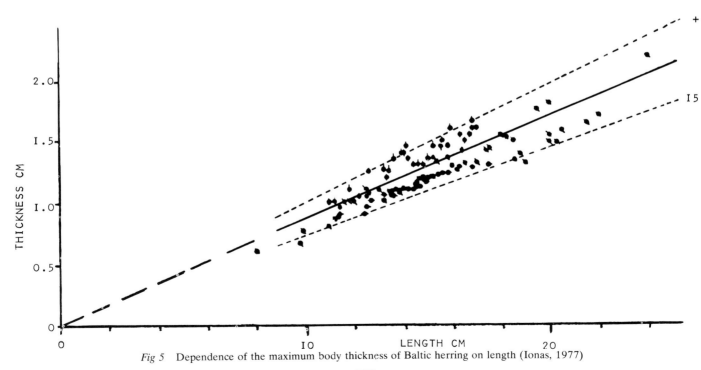

Fig 5 Dependence of the maximum body thickness of Baltic herring on length (Ionas, 1977)

the estimation of heat leak in purse seiners exhibit the application of modelling techniques to engineering design.

The physical requirements for different stages in the operation of a freezer trawler working in warm waters have been evaluated and a system proposed (McDonald *et al*, 1974). The sizing of the equipment required was determined from the analysis of a number of trawl logs from existing situations, whereas other workers (Chaplin and Heywood, 1968) have used simulated logs for similar predictions. A computer programme was also developed (Fraser and Hewitt, 1964) in order to predict the design requirement for different components. The operations that were considered are shown in *Fig 6* and are based on the type of layout shown in *Fig 7*.

In the development of the systems a combination of analytical and empirical approaches produced a practical engineering solution. This was subsequently supported by a computer programme which will permit the effect of alterations in the variables to be determined rapidly, without extensive experimentation.

The selection and use of either refrigerated or chilled sea water to cool large quantities of pelagic species aboard purse seiner vessels depends upon commercial considerations but there are some technical limitations, particularly with chilled sea water. With refrigerated sea water, the heat transfer into the tanks from the surroundings will be small compared with the total capacity of the refrigeration plant, and will not normally be a limitation except when large quantities of fish have just been added to the tank and the temperature is being reduced. With sea water cooled by adding ice, however, there will normally be a limitation on the quantity of ice that can be carried. An accurate estimate of heat gain is therefore more important in this case so that appropriate designs can be made.

It is possible to construct mathematical models for this type of installation and the complexity of these can depend on the accuracy required. In general, simple models can be used if an extrapolation from the ideal to the practical case can be made.

Magnusson (1972) adopted the latter approach and predicted that, for a particular method of construction and installation of the tanks, the heat gain will be 15 times that for ideal insulation. The heat gain for the ideal situation can be readily calculated using the following equation:

$$\frac{1}{U} = \frac{1}{h_0} + \frac{x_1}{k_1} + \frac{x_2}{k_2} + \frac{x_3}{k_1} + \frac{1}{h_i}$$

where U = overall heat transfer coefficient (W/m²°C); h_0 = outside heat transfer coefficient (W/m²°C); x_1 = thickness of steel plate, ships side (m); k_1 = conductivity of steel (Wm/m²°C); x_2 = thickness of insulation (m); k_2 = conductivity of insulation (Wm/m²°C); x_3 = thickness of tank lining (m); and h_i = inside heat transfer coefficient (W/m²°C).

The individual heat transfer coefficients can be readily obtained (Threlbeld, 1962; Anon, 1972), enabling the steady state heat transfer into the tanks to be calculated. It is then possible to predict the heat gain in any installation where a similar tank construction is used by applying the Magnusson coefficient.

For most practical purposes this would be sufficiently accurate but it cannot be universally applied to different methods of tank construction and insulation. A more refined model, which took account of different construction methods and of transient conditions, could be developed and this would permit more accurate predictions to be made.

One of the problems with modelling on such a large scale is that empirical verification is often difficult and this is particularly the case when work on board commercial fishing vessels is necessary. Some measurements on RSW/CSW tanks have been reported (Magnusson, 1972; Hewitt *et al*, 1977; 1978) and the difficulties have been highlighted. It is in a situation such as this, where confirmation by experiments is difficult, that considerable advantage is to be gained by the application of theoretical analysis.

5 Engineering science and future developments

Some examples have been given in the preceeding sections which illustrate the application of engineering science to fish processing machinery or systems. It is not possible to give a comprehensive picture but it is hoped that the examples selected have given a broad view of the sort of development that has taken place within the short space of 50 years.

Even today, due to the variable nature of the basic raw materials, empirical methods are still much in evidence and will continue to be so for some time to come. Some systems, however, lend themselves to a certain degree of analysis and scientific methods may be used to determine the engineering parameters necessary for either machinery or system design.

The main difficulty had been to obtain mathematical models which fit the physical situations accurately. Inevitably the models selected have only been appropriate for part of the range of application and would need considerable refinement to provide accurate and all-embracing solutions to the complex problems.

It is difficult to predict what the future developments in engineering science as applied to fish processing machinery will be, particularly in the light of the world fisheries situation today. Most of the advanced fishing nations are suffering from cutbacks in fish supplies and this is hardly the climate in which to expect rapid developments, since budgets for research and development will normally be reduced. However, inventions which increase efficiency and reduce waste are likely to come to the forefront (Drews, 1975) and therefore attract money for development. Greater filleting yields will be increasingly important and this may be brought about by improvements to the control of mechanisms which may be possible with the use of microprocessors (Nicholson, 1977) as long as the base data is comprehensive and accurate correlations can be obtained. Machines for handling and processing shredded fish flesh or presently unusable parts of the fish will be given high priority.

It is unlikely that the development of new machinery will be possible without a great deal of work involving trial and error but it is certain that engineering information currently available will assist in these developments. Not every invention will be successful but, considering the number of engineers and scientists involved

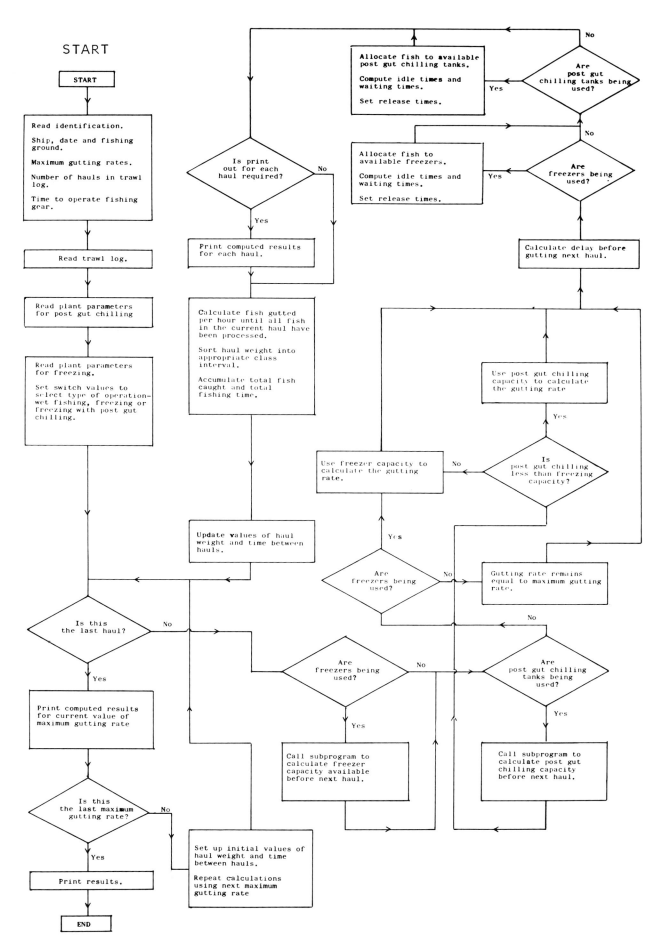

START

END

Fig 6 Computer simulation flow diagram for freezer trawler factory operation

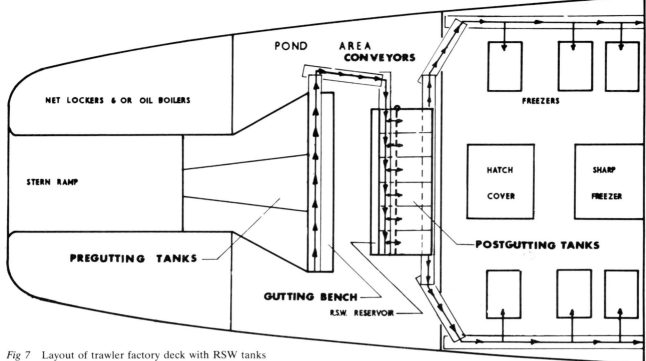

Fig 7 Layout of trawler factory deck with RSW tanks

in developments world-wide, the range of machines available today and the difficulty in handling delicate raw materials, the success rate will be as high as in most industries.

6 References

ANON. *Handbook of Fundamentals.* ASHRAE. Chapter 20
1972

ANON. Annual Report, Fiskeriteknologisk Forskingsinstitutt, Tromsø
1976

BAADER, R. Machinery for the fishing industry. Catalogue
1978

BURGESS, C H O, HEWITT, M R and JASON, A C. A review of current
1974 methods of thawing fish and future prospects. *Proc. Inst. of Refrig.*

CHAPLIN, P D and HAYWOOD, K H. Operational research applied to
1968 stern freezer trawler design. *Proc:* Joint Meeting of I. Mar. E. and Grimsby Inst. Engnrs. and Shipbuilders

CUTTING, C L. Engineering problems of the smoke-curing of fish.
1942 *Chem. Ind.*, 61(35), 365–368

—— The Torry Research Station controlled fish-smoking kiln. *Food*
1950 *Investigation Leaflet* No. 10, HMSO

DOE, P E. Mathematical modelling of food operations and its implica-
1975 tion for food quality. *Proc:* 6th European Symposium–Food engineering and food quality

DREWS, J. Mechanised fish processing aboard ship and ashore. *Fish.*
1974 *News Int.*, 13, No. 11

—— J. Development of fish boning machines. *Proc:* 6th European
1975 Symposium–Food engineering and food quality

EDDIE, G C. Some recent developments in mechanical engineering in
1963– the deep-sea fishing industry. *Proc. I. Mech. E.*, 178, Part 1,
1964 No. 27, 743–778

EVERINGTON, D W and COOPER, A. Vacuum heat thawing of frozen
1972 foodstuffs. Annex 2, *Bull. Int. Inst. Refrig.*

FRASER, A G and HEWITT, M R. A computer programme to assist in
1964 the design of fish handling and freezing requirements for trawlers. *Int. Inst. Refrig. Annex 1*, 45–49

HEWITT, M R. Thawing of frozen fish in water. In: *Freezing and Irradi-*
1969 *ation of Fish*, Ed R Kreuzer. Fishing News (Books) Ltd, London. 201

—— Quality and mechanical aspects of food processing. *Proc:* 6th
1975 European Symposium–Food engineering and food quality

HEWITT, M R, KELMAN, J H and McDONALD, I. Chilled sea water sys-
1977– tems for the preservation of fish. Various papers presented at
1978 meetings of the Institute of Refrigeration

IONAS, C P. Geometrical similarity in body shape of fish. *Trudy*,
1977 VNIRO, 78, 6–21

JASON, A C and SANDERS, H R. *Fd. Technol.*, Champaign., 16, 101
1962

JASON, A C. *Proc. Inst. Fd. Sc., Techn.*, 7, 146
1973

KAWKA, T. Analysis on determination of highest working parameter in
1974 circular cutters in mechanical processing of fish. *Acta Icth. Pis.* Szczecin, 4, 39–51

KISLEVICH, K A. *Caspian-Volga Herring*, Tr. Astrakhanskoi
1914– nauchno-promyslovoi ekspeditsu T.S. (3)
1915

MAGNUSSON, O M. Transport of herring and mackerel in tanks. *Scand.*
1972 *Refrig.*, 1(5), 156–161

McDONALD, I, HEWITT, M R, KELMAN, J H and MERRITT, J H. A
1974 refrigerated sea-water system for UK freezer trawlers. *Int. Inst. Refrig. Annex 1*, 169–174

MERRITT, J H and BANKS, A. Thawing of frozen cod in air and water.
1964 Annex 1, *Bull. Int. Inst. Refrig.*

MERRITT, J H. Evaluation of techniques and equipment for thawing
1969 frozen fish. In: *Freezing and Irradiation of Fish*. Fishing News (Books) Ltd, London. 196–200

NICHOLSON, F J. *Electronics in Grading and Orientation of Round Fish.*
1977 Institution of Electronic and Radio Engineers, Royal Institution, London

PLANK, R, EHRENBAUM, E and REUTER, K. *Z. ges Kälteindustr.*, 23, 37
1916

RIEDEL, L. Kalorimetrische Untersuchungen über das Gefrieren von
1956 Seefische. *Kaltetechnik*, 9(22), 38–42

ROLT, L T C. The mechanicals. *Progress of a Profession.* Heinemann
1967

THRELKELD, J. *Thermal Environmental Engineering.* Prentice Hall
1962 New York. 24

TORIN, P V. *Relationship between length of fish and its weight*, Tr.
1927 Sibirskoi ikhtiologioscheskoi laboratorii 2(3)

ZHILIN, N I. Some rules governing fish cutting procedure. *Ryb. Khoz.*,
1966 No. 12, 65–67

The application of engineering science to fish preservation. Part 2

S Forbes Pearson

1 Introduction

The story of man's struggle for food and survival can be considered to be a story of competition. The primary competition has always been to obtain the food supply itself and this competition has been waged with opponents who at first sight are better and more specifically adapted for the task of obtaining the food. The flexibility and reasoning power of man has enabled him to obtain the position of dominant predator on the earth despite his apparent physical disadvantages. It is sometimes thought that the dominant position of man is of very recent origin stemming from the scientific and industrial revolutions of the eighteenth and nineteenth centuries. The psalmist writing nearly 3 000 years ago wrote: 'Thou hast made him a little lower than the angels, and hast crowned him with glory and honour . . . Thou hast put all things under his feet' (Anon, 1000 BC). This displays a clear understanding of the dominant position of mankind relative to the other predators and food gatherers. In general, man has been successful in the primary competition for food and the numbers of mankind have tended to increase throughout history.

Success in the primary competition for foodstuffs did not always mean that primitive man was free to enjoy the fruits of his labour without further competition. The acquired foodstuffs became the subject of a secondary competition with forces of spoilage and decay which were even more difficult to combat than the primary competition for the food itself. The most baffling competition undoubtedly came from the bacteria which could reduce wholesome food to a dangerous and inedible mess within a few days.

The earliest form of preservation was probably the drying of strips of meat to produce a concentrated and sustaining food which was not susceptible to decay. It is interesting to note that this early method of preservation operated by removing the water from the food to an extent which rendered the food inaccessible to bacteria. The majority of more sophisticated methods of preservation operate on the same principle. The preservation of food by salting, pickling, smoking and fermentation developed before the invention of writing. These methods of preservation produce significant changes in the flavour and texture of the food so that what results is essentially different from the fresh product and may be preferred to the fresh product by an educated palate.

Table I compares the main characteristics of 12 different methods of food preservation.

The only method which gives long-term storage with minimal deterioration from the fresh condition is freezing followed by low temperature cold storage. Freezing and low temperature cold storage has become the dominant method of long-term fish preservation and other methods of fish preservation, such as smoking or canning, frequently make use of freezing either prior to the other preservation process or subsequent to it.

Early methods of fish preservation owed little to engineering science and were almost entirely empirical in their development. A notable exception to this was the development of the Torry kiln for the mechanical smoking of fish. The Torry kiln, which is described in literature produced by the Research Station (Cutting, 1950) allows the smoke curing of fish to take place under much more controlled conditions than were possible with the old natural-draught kilns. However, attempts to increase the independence of the curing process by adding mechanically refrigerated dehumidification plants were not commercially successful because the number of days per year when the additional plant was of significant use did not pay for its installation. Other science-based attempts to introduce accelerated mechanical drying of fish were also not commercially successful because there is a declining demand for dried fish when frozen fish can be made available.

Fish canning is an important means of preservation which is likely to be in widespread use for many years. The percentage of the fish catch which is canned, however, is unlikely to increase because canned fish is significantly different from the fresh product and the total energy requirements for a canning system must be high. This is especially so if the product being canned has been chilled or frozen beforehand.

Freezing and low-temperature cold storage is the most important method of long-term fish preservation at present and it is also the method which appears to give most scope for the application of engineering science. Although it is possible that new methods of food and fish preservation will be evolved, there are not, at present, any new methods which seem likely to be serious competitors of freezing and low-temperature storage.

We have long been accustomed to synthetic flavours and preservatives in the food industry. There is a possi-

Table I

	Allowable storage duration	Physical change	Effect on weight	Special packing needed	Special storage needed	Suspicion of health hazard	Cost
Drying	Months	Great	Loss	No	No	No	Low
Salting	Months	Great	Gain	Yes	No	No	Low
Smoking	Days	Moderate	Loss	No	No	Yes	Moderate
Pickling	Months	Great	Gain	Yes	No	No	Moderate
Fermenting	Years	Great	Small	Yes	No	No	Moderate
Canning	Years	Great	Gain	Yes	No	No	Moderate
Sugaring	Months	Great	Gain	No	No	No	High
Chilling	Days	Slight	Small	No	Yes	No	Moderate
Freezing	Years	Slight	Small	No	Yes	No	Moderate
Freeze Drying	Years	Slight	Loss	Yes	No	No	High
Antibiotics	Days	Slight	Nil	No	No	Yes	—
Irradiation	Days	Slight	Nil	No	No	Yes	Moderate

bility that the major threat to freezing and low-temperature storage as the dominant method of food preservation will come not directly from other methods of food preservation but from the development of synthetic foods themselves which do not require sophisticated methods of preservation. A step in this direction has taken place with the introduction of textured vegetable protein as a meat substitute. The most usual source of vegetable protein is the soya bean. Those who have had the opportunity of comparing the textured vegetable protein with steak diane cannot lood forward to its widespread use with any relish. Another possible source of protein in the future arises from the conversion of crude oil into edible protein by the action of certain yeasts. At present, it is reported that the protein is not suitable for sustained human consumption because of the relatively large content of RNA. As RNA is apparently implicated in the production of gout in the human animal, it would probably have a most unfortunate effect on customer relations if this type of protein were promoted too enthusiastically for direct human consumption. Presumably, however, the protein could be used as feeding stuff for other animals fortunate enough, or unfortunate enough, not to suffer from gout.

Although a future can be envisaged in which 'natural' foods are thought of as primitive and disgusting, it is not a future which twentieth-century man can look forward to with any pleasurable anticipation. We shall therefore turn our attention to mechanical refrigeration as the best hope of preserving the flavour and texture of foods as we know them for the foreseeable future.

2 Mechanical refrigeration

Mechanical refrigeration came into widespread use in the final quarter of the nineteenth century. The industrial revolution was introduced almost without benefit of engineering science, the steam engine having been developed by practical engineers with little theoretical knowledge. As the steam engine became established, it became increasingly important to understand the thermodynamic basis of heat engines and the science of thermodynamics developed fairly rapidly. The steam engine operates on the Rankine cycle which was described by Professor Rankine of Glasgow University. Except in the case of very large installations, the Rankine cycle has been superseded as a source of power but to this day the vast majority of mechanical refrigeration plants operate on the reversed Rankine cycle. The reversed Rankine cycle did not always hold this preponderant position. In the early days much refrigeration was carried out by the use of cold-air machines which used essentially the same thermodynamic cycle as a modern gas turbine. The disadvantage of the cold-air machine is that the thermodynamic efficiency is very low compared to the vapour compression system and for this reason cold-air machines have dropped out of favour except for the occasional application in aircraft cabin cooling. The most thermodynamically suitable of the early classical refrigerants was anhydrous ammonia. The very unpleasant properties of ammonia impelled engineers to seek a substitute and for many years carbon dioxide was especially favoured for marine applications where it was felt that the hazards of ammonia in the confined spaces

of a ship were verging on the unacceptable. Unfortunately, the thermodynamic efficiency of carbon dioxide is very poor relative to the efficiency of ammonia becasue the critical temperature of carbon dioxide is low. The implication of this is that under extreme conditions it becomes impossible to use the reversed Rankine cycle with carbon dioxide as a working fluid. For this reason and also because of the high pressure involved, carbon dioxide has gone completely out of use as a refrigerant. Ammonia is still one of the most widely used refrigerants, especially for large installations, but intensive work was carried out in the 1930s in the USA to develop a safe non-toxic refrigerant. Eventually a whole family of refrigerants known as halocarbons was developed. These compounds are, in general, stable, non-corrosive and non-toxic though some are much more stable than others. Since the introduction of halocarbons there has been a steady increase in their usage and area of application. The use of ammonia is now confined to relatively large plants.

It is ironical that the halocarbons which were developed in the interests of safety have recently come under attack for two reasons. It has been suggested that the most common of the refrigerants, R12, which is a very stable compound, may be accumulating in the upper layers of the atmosphere where it could interfere with the protection afforded from ionizing radiation by ozone. The other doubt which arises is over the safety of R22, which in high concentration has been observed as an apparent cause of an increased rate of mutation of certain strains of bacteria. This effect of R22 suggests that it might have a long-term effect on higher forms of animal life. Further evidence is being sought about the behaviour of each refrigerant and, in the meantime, the use of these substances for freezing by direct contact and as propellants for aerosols must be discouraged. Although the classical refrigerant, ammonia, is acutely toxic, it is a substance readily produced and broken down by natural methods and thus cannot have any long-term environmental effects.

3 Insulation

The efficient use of mechanical refrigeration requires that power requirements should be kept to a minimum by the separation of refrigerated areas from areas at normal temperature by insulating substances. Very many natural and synthetic substances are now available but in the early days refrigerating engineers and cold store constructors had to use such materials as were readily available. The major problem with cold store insulation is that moisture tends to permeate through it towards the low temperature. If the rate of moisture ingress to the insulation is not restricted then serious build-up of moisture or ice within the insulation may take place. Materials used for insulation in the nineteenth century included hair felt, sawdust, ashes and cork. Hair felt is still used for the lagging of water pipes but hair felt, sawdust and ashes can be understood to have proved very unsatisfactory for low-temperature insulation on account of the difficulty of vapour-sealing them. Cork board set in bitumen proved to be a highly satisfactory insulant and it was only the rise in price of cork coupled with the appearance of cheap synthetic insulants which rendered cork board obsolete except for certain loadbearing functions. The need for a vapour seal on the warm surface of the insulation was understood at a very early stage and reference to this is made in manufacturers' catalogues dated from the 1890s (De La Vergne, 1897). In those days, tarred or waxed paper was often used as the vapour seal. A light and relatively effective insulation was sometimes provided by arranging a grid of light wood battens with tarred paper sandwiched between in such a way that narrow, air-filled cavities were produced within the structure. The main insulating value was therefore provided by the relatively stagnant air within the cavities. These cavities were generally less than one inch wide but even at this width convection effects must have been significant.

The insulating effect of cork is also provided by air cells within the material but in this case the dimensions of the air cells are such that convection is not important. The main disadvantages of cork are its inflammability, its high water vapour transmission and its limited availability and consequent high price. The first synthetic insulation to challenge the position of cork was rigid, expanded rubber sold under the trade name of Onazote. This material had excellent resistance to the transmission of water vapour, was rot- and vermin-proof and had very high insulation value. The cellular pores of the material were filled with hydrogen sulphide gas which gave a characteristic odour when Onazote was freshly cut or broken. The major disadvantages of Onazote were cost, liability to damage by aromatic solvents and the possibility of corrosion caused by aqueous solutions of the hydrogen sulphide. Onazote held its position till the advent of expanded polystyrene which is a much cheaper and lighter material with lower compression strength and higher water vapour transmission. The low cost of expanded polystyrene has kept it in a competitive position till the present day, despite its disadvantages of inflammability and very low resistance to solvents. The water vapour transmission of expanded polystyrene is much reduced in an extruded form, which can also have high compressive strength. However, even the extruded form is still very liable to damage from solvents. The next synthetic insulant to come to prominence was foamed rigid polyurethane. The rigid polyurethane had the advantage that it could be foamed *in situ* and would adhere tenaciously to many surfaces, thus producing a structure of considerable strength when adhered to sheet metal. Foamed polyurethane had a conductivity similar to the other insulants which were available at the time. The development of foamed polyurethane using a halocarbon blowing agent, usually R11, to enhance the foaming effect of the reaction, produced a rigid foam of markedly lower conductivity than had previously been experienced. The reason for this was a reduction in internal convection because of the lack of mobility of the large R11 molecule within the cellular pores of the foamed material. There is a progressive loss of insulating value with time as the R11 diffuses out of the pores of the insulation. However, in most applications where the polyurethane foam is contained within a metal casing, the loss of insulating value is not of major significance over a period of some years. The lower insulating value to which the polyurethane tends with time is equivalent to the best values obtained with other insulants. The

major disadvantage of foamed polyurethane is that it is inflammable and that when burned it produces copious quantities of smoke and toxic products. There have been many fatalities as a result of smoke and vapours produced from flexible polyurethane foam in burning houses and fire authorities are increasingly reluctant to permit the use of polyurethanes where fires are a possibility.

The water vapour transmission of polyurethane, even where there is a closed cell structure, is significantly higher than for expanded polystyrene and especial care is required in vapour-sealing this material.

The undoubted technical advantages of rigid polyurethane foam as an insulant and a structural material have led to a search for materials with similar properties but with reduced fire risk. An improved material with similar properties is rigid polyisocyanurate foam. This material is able to withstand temperatures up to 150°C and has an improved resistance to burning and spread of flame when compared to polyurethane. The isocyanurate insulation is not in itself a fire hazard as it is self-extinguishing but in the presence of fire, isocyanurate will produce dense smoke and toxic vapours so that though less dangerous than polyurethane, it could add to the hazards of any fire which took place.

The reaction against polyurethane has led to the increasing use of foams derived from phenol formaldehyde resin. Such foams have poorer insulation value than polyurethanes and much higher water vapour transmission. They have, however, been used for many years as an insulant, for example, in house cavity wall insulation where the water vapour transmission is not a disadvantage, the water vapour pressure gradient being in the opposite direction to the gradient in a cold store. Rigid phenolic foam insulation, despite its inferior mechanical and thermal properties compared to polyurethane, is finding increasing use as an insulating material for pipelines and large vessels, especially in locations where fire would be serious or where the local authority opposes the use of polyurethane.

It is probable that the use of rigid polyurethane foam will be progressively restricted to applications where the foam is bonded in place between sheets of metal.

Table II shows some of the significant properties of some modern insulating materials. The figures are taken from manufacturers' data.

4 Machinery

The mechanical refrigeration industry grew out of the age of steam at a time when electric motor drives were not available. It is natural therefore that the first refrigeration compressors were large, slow-running machines directly coupled to a steam engine as the source of power. Though bulky and expensive to build, these machines were very efficient and very reliable. The advent of electric motor drives brought the introduction of higher speed compressors with enclosed crankcases, the seal between the refrigerant and the atmosphere being transferred from a stuffing box on the piston rod to a gland seal on the rotating crankshaft. A typical speed for a first-generation refrigeration compressor would be of the order of 60 rev/min. The second-generation refrigeration compressors with enclosed crankcases ran

Table II

	Density lb/ft³	Compressive strength lb/in²	Conductivity Btu/ ft² hr °F/in	Moisture vapour transmission 'perm inches'	Fire hazard
Cork board	6	40	0·30	120	Yes
Extruded polystyrene	2	30	0·24	1	Yes
Polyurethane	2	30	0·16	3	Yes
Polyisocyanurate	2	25	0·16	4	Slight
Phenol formaldehyde foam	2	30	0·22	10	Nil

at the daring speed of 300 rev/min. These second-generation machines were also very robust, reliable and efficient. Many of them are running to this day. To save space and to reduce costs still further, it was necessary to design higher speed machines. Most manufacturers introduced ranges of V and VW machines which, being capable of force and couple balance, could be arranged to run at speeds up to 1 500 rev/min. It is generally recognized that there has been some sacrifice of reliability and efficiency compared to slower running machines but recent developments have proved that high-speed compressors can be very reliable if they are protected from the effects of wet suction vapour. Reciprocating compressors are now available which run at speeds of up to 3 600 rev/min though these speeds are restricted to small hermetic compressors running on 60Hz supply. The speed at which modern reciprocating compressors run is a function of their size. The largest standard reciprocating machines are designed for speeds of 750 rev/min or 1 000 rev/min. Other machines are generally designed for four-pole motor speed, that is 1 440 rev/min on 50Hz supply and 1 760 rev/min on 60Hz supply. There is little benefit to be gained from running industrial reciprocating compressors at two-pole speeds because gas velocity effects through the automatic valves become dominant and the compressor cannot therefore be much reduced in size. Two-pole squirrel cage motors also are not significantly cheaper than four-pole motors of the same horsepower so that there is little benefit in two-pole speeds for industrial compressors.

The oil-injected screw compressor has been developed as an alternative to reciprocating compressors for large refrigerating duties. The screw compressor is a positive displacement machine with pure rotary motion and fixed volume ratios. Because of its inherent balance, this type of machine is also widely used for portable air compressors. There is a tendency to develop smaller screw compressors as an alternative to the medium-sized reciprocating compressor. Screw compressors generally run at 2 900 rev/min. The efficiency of the screw compressor is not as high as the efficiency of the best reciprocating compressors, especially at part load. This is not of over-riding importance in very large freezing plants where the compactness of the screw compressor makes it a sensible choice. But the low efficiency is likely to restrict the development of smaller screw compressors and is also likely to militate against the widespread use of screw compressors for low-temperature cold storage.

The original design of screw compressor has two meshing rotors. A new type of screw compressor has been developed in which there is a single screw sealing against a pair of wheels which rotate about axes at right angles to the axis of the main rotor. Efficiencies of the single-screw machine appear to be similar to efficiencies of the twin-screw compressor but it is claimed that the twin-screw machine is quieter. Twin-screw machines are also designed to run at 2 900 rev/min and this means that the electric motor will be noisy unless precautions are taken.

5 Applications

The earliest use of refrigeration to preserve fish was by the application of ice. This method had been known since ancient times but it was not till the nineteenth century that ice began to be used on a large scale. At first natural ice was used, originally from the local winter's supply which had been stored in ice houses. Latterly natural ice was imported from Norway. The remains of ice houses can be seen in many places round the coast and in the country. Ice houses were essentially a roofed over pit in which natural ice could be stored, insulated by straw or other material. The importation of natural ice became a thriving trade. In 1872, Grimsby imported 22 000 tons of natural ice from Norway (Waterman, 1963). Fishing ports like Grimsby had been developed by the enterprise of the railway companies who foresaw the possibility of a traffic between fishing ports and centres of population. It was not long before the technology which made the railways possible was also applied to the local production of ice for fish preservation both on board trawlers and during the journey from the fishing port to the main inland markets. The first ice factory was opened in Lowestoft in 1875 and the other fishing ports followed suit within a few years. The largest of the ice factories was probably the Hull Ice Co's factory at St Andrew's Dock, Hull. The fishing industry has changed so much in the past few years that this factory, once capable of producing about 1 500 tons of ice per day, is now out of production.

The successful use of refrigeration on land to produce ice tempted the more enterprising trawler owners to install refrigerating plant for use on board trawlers at sea. One of the earliest attempts to use mechanical refrigeration on a fishing vessel is believed to have been the installation of a cold-air machine on the *St Clements* of Aberdeen in 1883. Little is known about this attempt but the equipment and techniques available at that time were not adequate for commercial success, especially when abundant fish was available within easy sailing distance of the fishing ports.

It is not generally realized that the most effective method of keeping chilled fish is by surrounding it with melting ice. Melting is the operative word. If refrigeration is applied to prevent the melting of the ice or to eliminate the need for ice altogether, then the fish will suffer loss of moisture and become a less marketable article. The most common form of refrigeration supplied to wet fishing trawlers was the deck-head grid system operated from a small refrigerating plant usually in the fo'c'sle. The deck-head grid system was effective in preventing the melting of ice on the way to the fishing grounds. It could also be used with some advantage while fish was being stowed in ice. On the return trip, however, it was important that the ice should be allowed to refrigerate the fish by melting and the most effective way of ensuring this was to turn the refrigerating plant off. This fact was not generally understood in the fishing industry and there was sometimes disappointment at the quality of fish produced from these refrigerated trawlers. The deck-head cooling system was not essential to the operation of the trawler and was therefore frequently neglected by the owners and their engineers. This often resulted in ineffective operation which allowed the ice to melt on the homeward journey and thus preserved the fish quality. By 1960 about one-third of the British distant-water trawlers carried cooling plant of this type.

Scientists and engineers who understood the dual function of the ice in intercepting heat leakage from outside the vessel and in cooling the freshly caught fish by melting were aware that better stowage rates could be achieved if the wall of ice, which ideally should have been placed between the fish and the ship's side, were replaced by a narrow jacket containing refrigerated air. The first attempt to use refrigeration in this way was probably on the American trawler *Storm* in 1936. The Canadians, who had also experimented with jacketed cold stores on land, built two trawlers using this method in 1951. From about 1960, the *Wilton Queen* of Lowestoft and a series of sister ships were built with jacketed fish-rooms. Though mechanically successful, these plants gave disappointing results because it was not generally understood that the best that the refrigerating system could accomplish was to economize on the use of ice. Quality improvement over what could be achieved with a sufficient quantity of melting ice was not possible at temperatures above freezing.

5.1 Superchilling

By the late fifties all the fishing nations of the western seaboard of Europe were facing the same problems of diminution of the local catch due to over-fishing and consequent need to range ever further afield in search of good fishing. White fish well stowed in crushed ice does not retain good quality after about 15 days' storage. The distant-water fishing grounds are sufficiently remote to make it probable that this storage life would be exceeded by a significant proportion of the catch. The initial reaction of trawler owners and naval architects was to make the trawlers faster. This is a self-defeating policy for two reasons. To be faster, the ships have to be longer and this results in a larger fish-room, which takes longer to fill. Secondly, the power required to propel a surface vessel varies not directly with the speed but as some power of the speed. This means that vessels of the proportions of distant-water trawlers of the 1950s required a disproportionate amount of additional power to produce increases in speed which would have had a significant effect on the quality of the catch. The obvious answer to

the problem of quality from the distant waters was freezing at sea, but this method was not readily applicable to existing trawlers. Portuguese vessels with mechanical refrigeration for chilling began to operate their fish-rooms at temperatures as low as −3°C. It was found that by the use of these temperatures, cod fish could be kept in an edible condition for up to 32 days. In 1965 experiments were carried out into 'superchilling', as this method of storage had come to be called by the White Fish Authority on the Hull trawler *Boston Phantom* (White Fish Authority, 1965). On this vessel the fish were cooled in ice in aluminium boxes by the circulation of air at about −3°C. Improvements in the acceptable life of wet fish were achieved but the benefits were not sufficient to make this method attractive to trawler owners. In particular, the difficulty of unloading fish which had become frozen into the crushed ice and the appearance of the fish which had been treated in this way made it unprofitable to handle such fish on the open markets, which were the only method of sale at that time. If the eating quality of cooked cod fish had been the sole criterion, it is probable that the method would have been worthwhile, especially if introduced earlier.

5.2 Chilled sea water

Chilled sea water is used extensively in North America, mainly for fatty fish like herring or salmon. Chilled sea water is also used extensively by the Scandinavians for herring and mackerel. This method appears to be acceptable where the catching rate is such that the fish could not be chilled in any other way or where the fish is to be subsequently cooked and canned. The method has not found favour for white fish. At one stage there was a possibility that commercial interests might introduce the use of antibiotics in chilled sea water or crushed ice. Fortunately this came to nothing.

5.3 Preservation by freezing

The use of mechanical refrigeration to freeze fish dates back to the end of the nineteenth century. Results were not particularly good because the temperatures used for freezing and cold storage in those days were not low enough to prevent fairly rapid deterioration of the product. Freezing in cold air was not particularly effective, not only because the air was not of low enough temperature but also because extended surface coolers to give compact air to refrigerant heat exchangers were not available. The most successful freezing methods in those early days appeared to involve the use of sodium chloride brine. The patented Ottesen process envisaged the use of a eutectic solution of sodium chloride brine on the assumption that there would be less penetration of salt from a eutectic solution into the fish. Whether or not this process really resulted in a reduction of salt penetration is not clear, but a considerable number of Ottesen type plants were put into operation with some technical success. The only major fishery still using direct immersion brine freezing is the tuna fishery. Traditionally, the tuna are frozen in holds fitted with special grids being first chilled and then frozen in a progressively concentrated sodium chloride solution. When the tuna is thoroughly frozen, the brine solution is pumped overboard and the tuna maintained at low temperature in the refrigerated hold. It is significant that practically all the tuna frozen in this way is destined for subsequent cooking and canning before sale.

In 1926 Hellyer Bros of Hull converted a meat-carrying vessel for the brine immersion freezing of halibut. This vessel was renamed *Arctic Prince* and subsequently another vessel, renamed *Arctic Queen*, was also converted for freezing halibut. The halibut was caught off the coast of Greenland using long lines and dories. The reasons why this enterprise did not continue are not clear but it has been said that intensive fishing resulted in a decline in the rate of catch. Most prewar freezing was by brine immersion or sprayed brine mist. It appears that none of these prewar freezing plants was sufficiently successful to result in the widespread adoption of the brine freezing method for white fish.

After the war, the emphasis changed completely from brine immersion freezing to air blast and plate contact freezing. This was partly due to the availability of improved equipment such as standardized finned evaporators and horizontal plate freezers which by that time were in common use in the USA. But much of the credit for the abrupt change in direction must go to Christian Salvesen, who built on their whaling expertise to develop new forms of trawling, mechanized fish handling and freezing. In 1947 a former naval vessel was converted to a stern trawler with factory deck and freezing equipment. Much of the inspiration and the detailed planning for this work came from Sir Denis Burney and W Lochridge. The fishing trials were so successful and the quality of the fish so high that subsequent fullscale commercial exploitation was assured.

The first commercial exploitation of the work done on the experimental trawler *Fairfree* was the conversion of the French trawler *Mabrouk*, operated from Casablanca (Lochridge, 1956). The *Mabrouk* operated successfully for many years freezing gutted, headless fish. A compound ammonia refrigerating plant was used to cool the secondary refrigerant which was calcium chloride brine.

5.4 'Fairtry'
The lessons learned on the experimental trawler *Fairfree* were put into practice by Christian Salvesen in the novel factory trawler *Fairtry*. The *Fairtry* was a much larger ship than a conventional trawler, being 245ft BP × 44ft beam with a gross tonnage of 2 600. Freezing was by means of horizontal plate freezers and especially designed air-blast freezers which had been evolved for the *Fairfree*. Calcium chloride brine was used as the secondary refrigerant and the primary refrigerant was R12 compressed by single-stage vertical compressors of L Sterne and Co design and cooling the calcium chloride brine in a flooded shell and tube evaporator. So successful was this vessel of 1954 that Eddie was able to state in 1971 that there were by that time about 900 freezer trawlers and factory trawlers of over 1 000 tons operating in the oceans of the world (Eddie, 1971). The majority of these vessels were based on the *Fairtry* design. The *Fairtry I*, which was built at John Lewis, Aberdeen, was followed by *Fairtry II* and *Fairtry III*, which were built by Simons Lobnitz of Renfrew. Developments in filleting machinery, freezing equipment and refrigerating machinery led to some changes. Practically all freezing was now by horizontal plate freezers. Though calcium chloride brine and refrigerant 12 were retained, the

compressors now installed were Sternes 6VQC compressors running at 600 rev/min. The VQC compressor, as its name implies, was a compound machine with cylinders arranged in a V configuration. The *Fairtry*s were technically very successful and became the pattern for the fishing fleets of the world. Unfortunately, conditions and developments in Britain were not so favourable for commercial success in the long term. Vessels of this size and complexity required high catching rates to be profitable. The decline in catching rates due to overfishing and the foreseeable exclusion of British trawlers from many traditional fishing grounds led to the eventual withdrawal of all the *Fairtry*s from service.

At the same time as the *Fairtry*s were being put into operation, Torry Research Station was working on the development of freezing plant which could be applied to the existing British trawler fleet. These vessels were much smaller, simpler and cheaper than factory trawlers. Scientists at Torry Research Station had already established the requirements for the long-term preservation of frozen fish (Reay, 1951). It remained to the engineers to develop methods of applying these techniques on board fishing vessels. Filleting at sea and the loading of horizontal plate freezers were considered to be impracticable on the traditional distant-water trawler. A vertical plate freezer was therefore designed and tested on shore (Yule and Eddie, 1953). Sea trials were then carried out on the motor fishing vessel *Keelby*. The vertical plate freezer proved practicable to operate at sea and the quality of the whole gutted fish produced by the vessel was extremely good. With the objective of proving that existing distant-water trawlers could be made more profitable by conversion of part of the wet-fish-room for freezing and cold storage, a large-scale experiment was carried out by the White Fish Authority using the trawler *Northern Wave* (White Fish Authority, 1957). A single-stage direct expansion R12 plant was installed in this vessel by J and E Hall Ltd, who also developed a variant of the vertical plate freezer originally designed at Torry Research Station. The vertical plate freezer employed pumped circulation of refrigerant 12 and the blocks of frozen fish were unloaded through the bottom of the freezer directly into the low-temperature hold. The experiment was a technical success and encouraged trawler owners to proceed with other vessels. The concept of freezing the first caught part of the catch only and thus prolonging the useful life of the trawler fleet was not, however, taken up. The only trawlers converted were the old *Junella*, which had a small freezing plant and low-temperature cold store installed by J Marr and Sons as a commercial experiment and the *Ross Fighter* which was later converted to 100% freezing by Ross Trawlers in order to get into freezing at sea without waiting for specially built trawlers. The first and only British part-freezing trawler designed and built as such was the *Lord Nelson*, which was built in Germany in 1961. The *Lord Nelson* used direct expansion R12 refrigeration with single-stage J and E Hall compressors and pump circulation of the refrigerant using glandless pumps. Eventually the freezing plant operated well and reliably, demonstrating the basic soundness of the design, but protracted troubles with leakage of the relatively expensive refrigerant led to subsequent vessels going back to the use of a secondary refrigerant. The next British stern trawler was the *Junella*, built in Aberdeen for J Marr and Son of Hull in 1962. The *Junella* took the name of the previous side trawler which had been the first freezing trawler to operate commercially from Hull. The *Junella* had no provision for the storage of fish in ice. The ship was totally dependent on the freezing plant which consisted of two L Sterne and Co Ltd 4VQC compressors running on the more expensive refrigerant R22. The R22 was confined to a short and simple refrigerating circuit which cooled the secondary refrigerant, trichloroethylene, down to temperatures in the region of −40°F. The cold store was refrigerated by pumping trichloroethylene through a grid of galvanized steel pipes and the freezing was carried out in side unloading vertical plate freezers of L Sterne and Co design. The *Junella* served as a pattern for a large number of commercially successful British freezing trawlers. Direct expansion was not used again on British freezing trawlers till the Ranger Fishing Co used it successfully in a series of vessels designed to freeze fillets.

6 Recent developments

Recent developments have had a catastrophic effect on the British fishing industry and have altered the basic parameters of all fishing. The multiplication of the price of fuel oil and the loss of traditional fishing grounds, due to extended national limits, have rendered most of the British distant-water fleet uneconomic. The effects of the oil price rise and the extension of fishing limits are felt throughout the world and they are bound to cause changes in fishing methods. In addition, the increasing gulf between amenities and working conditions on shore in western-type economies and the working and living conditions which must obtain on board a large freezing trawler is making it progressively more difficult to obtain competent crews.

In the short term, the attention of the British fishing industry has been turned to the exploitation of different fish stocks, such as mackerel and blue whiting. The exploitation of blue whiting for human consumption appears to be very difficult owing to the nature of the fish but the exploitation of the mackerel stock by freezing trawlers has been very significant and has resulted in a new export trade with West Africa. Whether the mackerel stocks can sustain the present and projected rate of fishing remains to be seen. The catching rates of powerful stern trawlers can be very high when mid-water trawling for mackerel. The majority of distant-water freezing trawlers were designed to freeze up to about 30 tons/day of white fish. Under present catching conditions, much larger throughputs of mackerel would be desirable. In 1978 the stern trawlers *Conqueror* and *Defiance* were modified to process an increased throughput of mackerel. These vessels had originally been designed to operate both the freezers and the cold stores on low-temperature trichloroethylene refrigerated by a two-stage R22 plant. It was desired to increase the throughput of frozen fish as much as possible without adding to the generating capacity or increasing the refrigerating machinery. This was done by changing the freezing side of the plant to direct expansion of R22 while the cold store remained on trichloroethylene. There was neither space for large refrigerant surge drums nor power avail-

able for massive circulating pumps because it was expected that the refrigerating compressors would, as a result of the direct expansion operation, be absorbing significantly more power. The existing freezers on the trawlers were removed and replaced by ten-off 20-station vertical plate freezers of APV design and manufacture on each trawler. The problem of circulating refrigerant to the freezers without pumps was solved by using a specially designed low-pressure receiver which allowed an over-feed of refrigerant to each freezer and re-evaporated the excess refrigerant before returning it to the compressor as vapour. Unfortunately the *Conqueror* ran aground on her first voyage after conversion and became a total loss. The *Defiance*, however, continues to fish and freeze successfully, having daily throughputs in the region of 55 tons. Although the converted plant is somewhat of a compromise and therefore not ideal, the crew have had little difficulty with the plant and it has performed reliably despite the increased load upon the refrigerating machinery. Novel forms of refrigerant control and hot gas defrosting had to be devised for this vessel. These methods have since been used successfully on land cold stores also. It remains to be seen whether other trawlers will be converted to use the low-pressure receiver but it is a matter of some regret to the author that the low-pressure receiver was not available at the time when major investments in new freezing trawlers were being made.

7 Air-blast freezers

The preferred method of freezing on British distant-water trawlers has been the plate freezer. On land, however, and on the larger factory vessels of Eastern European nations, air-blast freezers have been extensively used. The major disadvantage of the air-blast freezer is the space which it takes up and the high fan power which is often required. Significant reductions in space have been obtained by the use of spiral belting in air-blast freezers and some air-blast freezers which use spiral belting have also achieved economies in fan power by directing the air in counterflow with the product. Designs of continuous air-blast freezers which use a smaller proportion of their total power requirement for fan power than many cold stores have also been developed (Pearson, 1977). Some of these freezers are installed and operating in Aberdeen. Torry Research Station has also developed a continuous freezer using a smooth, stainless steel belt, which produces a frozen fillet of very fine appearance. At the present stage of development, the Torry freezer has a rather high power requirement but there seems to be no technical reason why the air flow methods used in high-efficiency air-blast freezers should not be applied to freezers with a continuous stainless steel belt.

8 The future

The author does not foresee any rival to freezing and low-temperature storage for the preservation of fish in the immediate future. But the ways in which fish are caught and preserved by freezing are bound to change in response to the changing world situation. There is already a tendency for fish landings to depart from the traditional distant-water and middle-water ports and to return to a larger number of small ports around the coast. This tendency is linked to the provision of freezing and cold storage facilities for freshly caught fish at these ports. A major factor in the shift to smaller ports, smaller vessels and shorter trips is undoubtedly the pleasanter, though no less hard, environment on the smaller vessel. This is an indication that the future should be studied, 'As if people mattered' (Shumacher, 1973).

The rising cost of fuel means that fishing methods which consume less energy will become more competitive. It matters little whether the energy is spent dragging a large net through the water by brute force or in quartering the distant oceans in search of fish. It is therefore to be expected that Danish seining and purse seining will, to some extent, supersede trawling as the favoured method of marine fishing. Consideration must also be given to fish trapping and fish farming, both of which require far less energy than any form of fish hunting.

Present trends would indicate an increasing number of fish-processing plants and landing points on the coast close to fishing grounds. It is to be expected that catching rate and fishing effort for white fish will be controlled so that a stabilized and economic yield is obtained. In the initial stages it is expected that quality will be maintained by short trips and copious use of ice. In the longer term, it is likely that quality will be further improved by the adoption of some form of whole-fishing freezing on smaller, coastal fishing vessels.

The situation qith regard to pelagic fish such as herring and mackerel is different. Assuming that the fishing effort is regulated to maintain stocks at an economic value, these fish will always be caught in large quantities. It is therefore reasonable to predict that the pelagic fishing vessel of the future will be a large purse seiner with continuous air-blast freezing and low-temperature cold storage facilities on board. An alternative might be purse seiners with chilled sea water storage and motherships for the freezing operation. It is assumed that catching of immature fish for meal and oil will be outlawed and that high-quality fish for human consumption will be the objective of the fishing effort of the future. Experience has shown that the highest quality can be achieved by freezing as soon as possible after catching and after as little handling as possible. For this reason, it seems likely that after an initial phase involving the landing of high-quality wet fish and its processing at many points round the coast, the fishing pattern in British waters will revert to the freezing of whole fish at sea for subsequent thawing, processing and possibly refreezing on shore.

Thanks to years of devoted research by scientists at Torry Research Station and other institutions, the requirements which must be met to preserve fish in the highest quality condition are well understood and the appropriate techniques have been worked out. It remains to be seen whether the conomic and political climate will make it worth while for the fishing industry to make use of this knowledge and experience.

9 References

ANON. Psalm 8, V5 and 6
1000 BC

CUTTING, C L. The Torry Research Station controlled fish-smoking
1950 kiln. *Food Investigation Leaflet* No. 10, DSIR

DE LA VERGNE. *Mechanical Refrigeration and Ice Making.* The
1897 Knickerbocker Press, New York. 5th ed., 96

EDDIE, G C. The Expansion of the Fisheries. *Proc:* I. Mech. E., 185,
1971 42/71

LOCHRIDGE, W. Mechanisation in fishing vessels. *Trans. I.E.S. 1955–*
1956 56, paper 1210, 99, 511

PEARSON, S F. Performance of a high efficiency air blast freezer. *Proc:*
1977 Inst. R., 73, 49

REAY, G A. The preservation of white fish at sea. *Mod. Refrig.* 54,
1951 376–379

SCHUMACHER, E F. *Small is Beautiful.* Blond and Briggs Ltd
1973

WATERMAN, J J. Development of equipment for trawlers. *Eng.*, 542
1963

WHITE FISH AUTHORITY. Report on an experiment into the freezing of
1957 fish at sea. White Fish Authority, June.

——Commercial trial of superchilling on M T *Boston Phantom.* Indus-
1965 trial Development Unit. *Technical Memorandum* TM31

YULE, P A A and EDDIE, G C. A vertical plate freezer for whole fish.
1953 *Mod. Refrig.*, 56, 441

3 Minced fish

The Cornell experience with minced fish

J M Regenstein

In a world that is short of food, and especially short of high-quality protein, the misuse and underexploitation of our aquatic resources must be stopped. Unfortunately, the fish delivery system can still be represented as a very leaky pipeline (*Fig 1*): the fishing effort concentrates on the most valuable species with the by-catch ('trash' fish) often being discarded. These underutilized species (which should not be called 'trash fish' if we want to use them for food!) often die before being thrown back overboard. In the Great Lakes, fishermen estimate that as much as 90% of the catch is discarded; coastal fishermen off Long Island, New York, often cite figures as high as 70%. Even allowing for the infamous exaggeration of a fisherman, this represents a potential loss of valuable human food. After the catch is landed there is also waste in fish transportation: offal, spoilage and filleting wastes occur at various points between the sea and consumer. The most important human food is the meat adhering to the bones, which themselves might constitute a potential human food. Can we improve the system?

Many of the underutilized species are not used because of consumer unfamiliarity, boniness, bad names and unpleasing looks as whole fish, though many of them have excellent white flesh. Filleting waste after the heads and viscera are removed from the racks (skeletons) may be over 50% meat. In major fish markets, such as New York City's Fulton Fish Market, fish may be left to waste at the end of the day. And all of these problems have a potential solution. In all these cases, minced fish can be used to create nutritious food from waste, or if you will, to literally create something out of nothing. We can use either the type of deboner made by firms such as Bibun, Yanigiya or Baader which yields a fairly nice quality flesh (using drums having 3, 5, 7mm holes) or the higher pressure type of equipment which offers a pastier product but higher yields. Possibly a sequential combination of the two types of equipment might give the best overall result.

Once the fish flesh has been minced, we must decide what to do with it and how best to handle it. Given the history of deboning in the USA, particularly within the

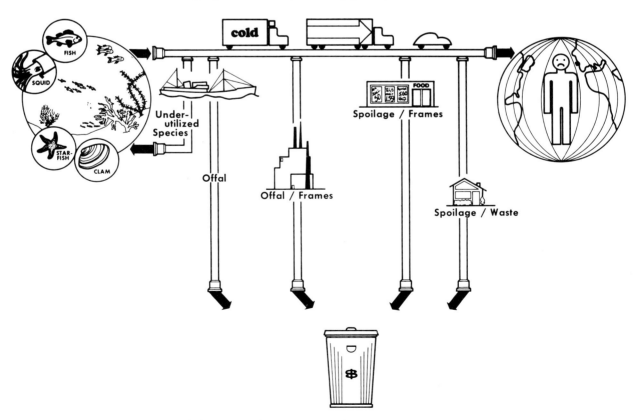

FOOD FROM THE SEA - 1

Fig 1 The seafood pipeline: a very leaky system with lots of waste

US Department of Agriculture, the fishing industry is fortunate that the product has always been identified as 'minced' fish. Most of this paper will deal with the product development and marketing aspects of minced fish; we will then turn briefly to some technical considerations. Throughout, we will note how other fishery by-products have been incorporated into our new products.

The first way to use minced fish is to use it to make traditional fish products, thus demonstrating how minced fish products compare with the more traditional forms. Unfortunately, from a marketing point of view, these products may simply represent the replacement of one fish product of higher economic value with one of lower economic value. Although this may allow for some new market development, in general it will not extend fish to new consumers; and because many of these new products are breaded-battered and/or deep fried (often desired by the consumer), they do not fully exploit the nutritional benefits of fish. Our fish crispy is a product that was designed for this market. It is a battered-breaded product that can be prepared either in a deep fryer or in the oven. This is because we have learned that many institutions, particularly schools, do not have deep fryers. Flavour and texture can be improved by adding minced clam, which also permits us to call the product 'seafood crispies' rather than 'fish crispies'. In co-operation with Drs Hood and Zall at Cornell University, we have been using scallop mantles obtained as a by-product from a scallop shucking operation, instead of the more expensive minced clams. Mantles are not usually used in this country, though we are told they are eaten regularly in other countries, eg France. In fact, disposal of this material has presented a problem to the processor. In both this product and in seafood chowders the mantles perform quite well. The formulation for the seafood crispies is shown in Table I.

Table I
FORMULA FOR SEAFOOD CRISPIES
(modified from Baker et al, 1976b)

Unbreaded crispies	
Ingredient	%
Minced fish	63.0
Chopped scallops	17.0
Vegetable oil	4.0
Textured soy protein (Promate III, Griffith Labs)	4.0
Cracker meal	4.0
Onion (fresh, diced)	3.0
Fibrous soy protein (SPF 200, Ralston Foods)	3.0
Seasoning mix*	2.0

* Seasoning mix: salt 17.5g, onion powder 0.6g, garlic powder 0.4g, pepper (black) 0.5g, celery seed 0.2g, MSG 0.8g.
Procedure: Mix all ingredients together until sticky. Allow to hydrate, refrigerated, at least 1h. Shape into 10g balls (about ⅓oz), batter (#4310, Modern Maid Food Products) and bread (#6071, Modern Maid Food Products).
For cooking by frying: Package and freeze.
For cooking by oven heating: Prefry at 190°C (375°F) for 45s, cool, package, freeze.
Cooking instructions: For frying: fry at 190°C (375°F) for 1½–2min if thawed, 3–4min if frozen. For oven heating: heat on a cookie sheet in a preheated oven at 205°C (400°F) for 15min if thawed, 20min if frozen

Another product line is that of chowders. Again, we can create various forms, such as Manhattan (tomato) or New England (milk) seafood chowders and/or fish chowders. Some products must be prepared without

seafood so they can be used by people with allergies to shellfish, a surprisingly common occurrence. In addition to using scallop mantles in this product, we have also used a clam broth prepared from the water used to wash clams just after they were minced (Hood et al, 1976). Thus, the condensed, canned seafood chowders contained over 60% underutilized material. An important aspect of our 'chowder' product was to use high levels of fish (about 28%) and mantles (10%), which means that a reasonable portion of this soup could make a significant contribution to a consumer's daily protein requirement. Under the United States Department of Agriculture's School Lunch Program, this product could contribute significantly to the protein requirement of schoolchildren, something soups usually cannot accomplish. It has been calculated by a commercial concern that a 28% minced-fish chowder could compete on price with clam chowders containing around 7% minced clams. The final product as produced for the retail trade was of even higher quality than a school lunch chowder. Formulas are shown in Tables II and III.

Table II
FORMULA FOR MANHATTAN-STYLE SEAFOOD CHOWDER
(modified from Baker et al, 1976a)

Ingredient	%
Minced fish	24.0
Tomatoes (canned)	18.0
Clam broth	11.0
Water	10.0
Scallops	10.0
Tomato paste	6.0
Carrots (diced)	5.0
Celery (chopped)	4.0
Modified tapioca starch (#3435, Stein Hall Co, Inc)	3.0
Potatoes (dehydrated, ¼in diced)	2.5
Green peppers (fresh, diced)	2.0
Vegetable oil	2.0
Salt	1.04
Onions (dehydrated, diced)	0.50
Sugar	0.40
Parsley (dehydrated)	0.30
Hydrolyzed plant protein (Seafood Base 85, Nestle Co)	0.20
Thyme (leaf)	0.04
Pepper	0.02

Procedure: Slurry starch with about one-half of the water. Heat the remaining ingredients together, stirring to break up the minced fish. When nearly boiling, add the starch slurry and heat until thickened. Fill cans and process. The chowder was diluted with an equal quantity of water and reheated prior to serving

Table III
FORMULA FOR NEW-ENGLAND-STYLE SEAFOOD CHOWDER
(modified from Baker et al, 1976a)

Ingredient	%
Minced fish	27.0
Clam broth	25.0
Water	24.6
Chopped scallops	10.0
Potatoes (dehydrated, diced)	4.8
Modified tapioca starch (#3435, Stein Hall Co, Inc)	3.0
Vegetable oil	2.0
Vee Creme (Nestle Co)	1.5
Salt	1.1
Hydrolyzed plant protein (Seafood Base 85, Nestle Co)	0.6
Onions (dehydrated, diced)	0.3
Thyme (leaf)	0.03
Pepper	0.03

Procedure: Reserve part of the water to slurry the starch. Put the remainder of the ingredients in the kettle, bring to a boil, stirring to break up the fish. Add starch slurry, heat until thickened. Fill cans and process. The chowder is diluted with an equal quantity of milk and reheated prior to serving

A word about the species used might be appropriate. Most of our research has been done with minced white sucker (*Catostomus commersoni*) from the Great Lakes which is sold in the Midwest of the USA as mullet (not to be confused with the salt-water form from the Gulf of Mexico). It seems to be an excellent fish for mincing, and with 0·05% polyphosphate added it has a shelf-life under commercial conditions that approaches one year. However, the supply of this fish is seasonal, freshwater fishing effort in the US is limited, and problems of commercial fishing near the shorelines exist. Recreational fishermen have been known to shoot at commercial boats suspected of interfering with game fishing.

At the time of our chowder market tests, the Great Lakes were frozen so that we had to import minced cod from Canada. These minced blocks were 'dry' and had some of the preliminary signs of the textural change attributable to gadoids which we will discuss later. We are currently trying to identify potential salt-water species of fish for further development of a US minced fish industry.

The final product produced for market testing was a 1lb package of frozen minced fish sold directly to the consumer (Baker *et al*, 1977). The package contained a recipe book which explained how to use the new product (see *Fig 2*). In many ways, this is the most exciting minced fish product concept tested. The versatility of the product, its high nutritional value (both high-quality protein and low levels of polyunsaturated fat), and its reasonable price made it an excellent buy and gave the consumer the option of using it interchangeably with ground beef. In fact, the fat content is often so low

(1–2%) that many consumers prefer to add additional vegetable oil (also polyunsaturated) for the best mouth-feel and juiciness.

Either as a retail or commercial product, minced fish can be used in a number of forms and, therefore, our work has been both at the product concept level and at the recipe development level. In the latter case, some recipes are clearly only for home and restaurant use.

In the USA we find that fish often stands out as the animal protein source of choice for the increasing number of people who are weight conscious and who consider health important. The US Senate's Select Committee on Nutrition has supported an increase in fish consumption in the US diet. Unfortunately, the consumer is often wary of fish and often does not know how to cook it properly. Proper broiling and poaching to ensure a juicy, moist product is a skill that Americans must learn.

Many of the products we are proposing are not dependent on this type of cooking, but rather draw on cooking skills analogous to those used with ground beef. I might add that there are many products available in the American market that call for the addition of 1lb of ground beef. Among these are mixes for products such as chili, tacos, hamburger, meatloaf, meatballs, Italian sausage, Chinese food, stews, *etc*. All these can be made using minced fish instead of ground beef. In America, about 40% to 45% of all beef ends up as ground beef. Thus even a small dent in this market could be significant for the fish industry.

Of course, consumers can also prepare these products directly from scratch using their favourite recipes and

Fig 2 Pictures of various minced fish products packaged for market testing under the Cayuga Brand label (Cayuga Brand is Cornell University's Department of Poultry Science's brand name. It is used during market testing to keep products from being immediately identified as Cornell's. The back of the product, however, does indicate that the university is the producer)

replacing the 1lb of ground beef with 1lb of fish that they mince themselves; our product, however, has greater attractions.

Examination of institutions (particularly schools) has been fruitful as we have found that many minced fish products such as chili, pizza sauce, meatballs for spaghetti, sloppy jonah and lasagna oceana are acceptable to them. Names like 'sloppy jonah' and 'lasagna oceana' give some romance to the products and at the same time clearly indicate the non-beef origin of the meat. We also feel that introducing exciting, appetizing fish products into schools may help develop a more favourable attitude among children towards fish; the same children who would probably not even try a product called 'fish sloppy joe' or 'fish lasagna'.

Minced fish can be used in many ethnic recipes (Italian, French and Chinese) which call for fish or ground beef. The use of caramel colouring, and/or home gravy-making products such as the US products Gravymaster or Kitchen Bouquet gives minced fish a meat-like colour and reduces concern about the colour of the deboned flesh. In many of these products, any fish flavour is masked by the spices just as a beef flavour would be.

Minced fish can also be used for many gourmet fish dishes. Recipes here often call for ground fish and have been traditionally prepared by grinding or chopping fresh fish. Minced fish is often more convenient and cheaper. Among the products developed in this category are fish newburg, a crepe filling, gefilte fish (a Jewish delicacy) (*Fig 3*), and a fish mould salad made of minced fish, mayonnaise, dill weed, celery and sweet relish all

Fig 3 Gefilte fish made from minced fish. Centre well contains beet-coloured and flavoured horseradish sauce that is often used as a dip

set with gelatine. This salad is particularly attractive if formed in a fish-shaped mould (*Fig 4*). Unfortunately, except for a few special products such as those in *Figs 3* and *4*, pictures of our products look as though they could have been made from meat or poultry. The important point is that they taste like meat in the meat replacement products while tasting like fish in the fish products.

Our work with gefilte fish was particularly interesting. This product is generally made from freshwater fish, but we found that acceptable products could be made from salt-water species. If we used flounder rack meat prepared in a Beehive deboner, the meat turned grey on

storage, but had an acceptable off-white colour and no flavour problems in the final boiled product, although the texture was a little soft. Some gefilte fish is being marketed in the USA using Canadian minced freshwater fish.

It should be noted that people of the Jewish faith who observe the kosher dietary laws do not mix meat and milk products (Regenstein and Regenstein, 1979). Because fish is considered neutral, however, it can be used in many recipes which call for dairy and meat together such as cheeseburgers and tacos served with cheese. The use of minced fish in this capacity could therefore help increase its market wherever there are significant Jewish populations, *eg* New York State.

A special product that we have developed is a puffed snack cracker made from a mixture of canned fish and commercial starch that is retorted, sliced and deep fried (*Fig 5*). Other groups in the USA are working on extruded products for the snack market. These are valuable markets in the short run but do not provide the kind of long-term health-giving foods that we should be striving to provide.

Another product idea that we have explored is the formation of a canned tuna fish substitute; the texture is clearly different but the uses would be similar. Surprisingly, our taste panel work has shown that this product is often more acceptable in flavour and overall acceptability than tuna. (Dr Mendelsohn at the NMFS Laboratory in Gloucester, Massachusetts, has also worked on this product).

We have also found a use for the fish discarded during mincing, in this case a fish broth. In the USA, the only broth generally available for chowders, newburghs, *etc*, is chicken soup. Using deboning waste and fish heads we can make a good fish broth. Dr Hood is currently investigating the formation of a broth directly from flounder racks without mincing.

Now let us turn to Dr Goodrich's marketing effort on some earlier products (Goodrich and Whitaker, 1977; 1978; 1979) prepared either completely or partially at the university. The packaging design was worked out by our own media service group. The economics of production could not be controlled and the products were priced on an estimate of costs and mark-ups or competitively with comparable products. As might be expected, our actual costs tended to be significantly higher than our commercial projections. Unfortunately, various constraints prevented us from testing the importance of nutritional labels or product promotion based on nutritional or weight-control benefits.

Arrangements for market testing were made with a fish store in Rochester, New York, and/or supermarkets in Rochester or Ithaca. Two days a week (Thursdays and Fridays from 9am to 5pm demonstrators were present in the stores giving out samples of our products or food prepared from minced fish. The demonstrators thus only exposed a limited number of consumers to the product. The product being tested was, of course, available at that time but was then kept in the store for several weeks after the demonstrators were withdrawn. Some point-of-purchase sales information was available and some local publicity was obtained. The co-operating stores supplied data on other sales and customer traffic so that comparative measures of sales could be obtained. Suc-

Fig 4   Moulded fish salad

Fig 5   Fish chips made from minced fish, modified amylopectin (corn) starch and powdered instant rice which was dried after retorting (Moledina, 1975)

cess was defined as the attainment of a greater market share than the average share for the product area. All products were successful by this standard (*Figs 6–8*).

Clearly the amount of promotional support was minimal and the customer exposure limited; also, the price was somewhat arbitrarily set. In the case of the chowders, the Manhattan cost more to make than the New England, but, based on competitive products, the price of the New England was higher. We have not tested price elasticity in any way; however, in the light of rising beef prices, a test of this kind might be quite interesting.

Another problem, particularly with the retail 1lb frozen minced fish, is the question of positioning in the store. It is contended that placing the minced fish nearer to the meat counter and promoting it as interchangeable with meat might increase sales. In spite of these problems and limitations, it is still felt that there is a market for minced fish.

With all these limitations, why do we market test? Experience has taught us that the US industry is not interested in university-sponsored research on new products until some practical testing has been performed. Also, the products were formulated as best we could but we did not try all the possible variants. Experience has also shown us that industry wants to be able to give a product its own identity. An overly-perfected product from a university which is available to all, leaves no opportunity for the entrepreneur to use an individual brand name. Thus incentive for risk-taking is decreased significantly.

As we started to get favourable responses to our activity, we became more and more aware of the problems of supply. In addition, when working with the Canadian minced cod, we became aware of the problem associated with trimethylamine oxide in many gadoids.

Trimethylamine is formed from TMAO in fresh fish whereas dimethylamine and formaldehyde are formed in frozen fish. We currently have projects under way studying the effects on this problem of food additives and of handling minced fish; also a bacteriological project examining *P. putrefaciens*, the major spoilage organism of fresh fish; and a project examining texture, chemical changes in the amines and changes in organoleptic properties of the fish. We have not yet solved these problems, but we do feel that some of the food-approved additives might at least slow down the undesirable changes.

We have found that immediate heating of the minced fish after deboning destroys the enzymes involved and produces a stable frozen material.

We have also begun to study the effect of modified atmosphere storage on fresh gutted hake. Eventually, we hope that this line of research might lead to the possible sale of minced fish directly to the consumer in the fresh form.

Thus we feel that we have started to solve the problem of wastage by finding ways to use fish more efficiently. With a continued effort in this area, we hope to see more people eating fish, even in the stubborn United States.

Acknowledgements

I wish to acknowledge the support of the New York Sea Grant Institute and its director Donald Squires; the work and discussions with colleagues Dana Goodrich, Lamartine Hood and Robert Zall. Special thanks go to Robert C Baker, my collaborator and colleague, who has introduced me to many aspects of fisheries and of product development.

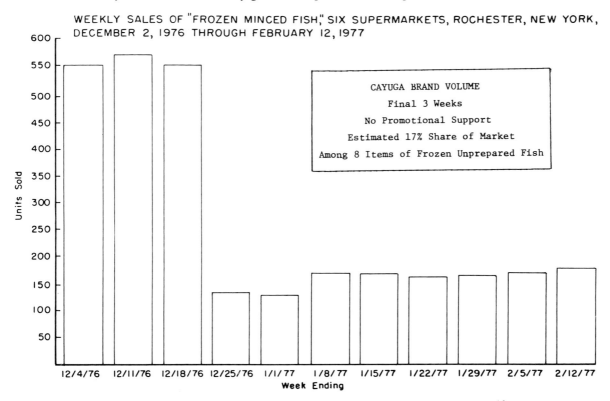

Fig 6   Sales data for 1lb packages of frozen minced fish (Goodrich and Whitaker, 1977)

197

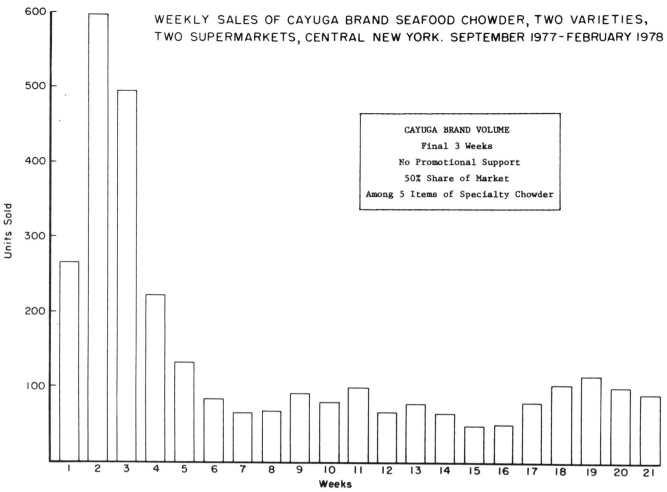

Fig 7 Sales data for Manhattan and New England Seafood Chowders (Goodrich and Whitaker, 1978)

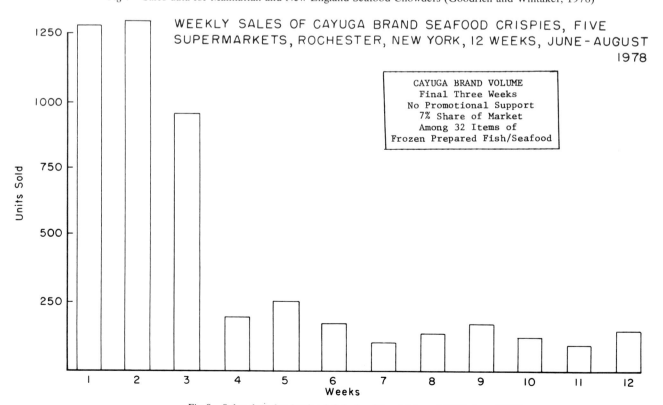

Fig 8 Sales data for Seafood Crispies (Goodrich and Whitaker, 1979)

198

References

BAKER, R C, REGENSTEIN, J M and DARFLER, J M. Development of
1976a products from minced fish: 1. Seafood chowders. *N.Y. Sea
Grant Institute Bulletin*
—— Development of products from minced fish: 2. Seafood crispies.
1976b *N.Y. Sea Grant Institute Bulletin*
—— Development of products from minced fish: 3. Frozen minced
1977 fish, *N.Y. Sea Grant Institute Bulletin*
GOODRICH, D C, JR and WHITAKER, D B. Retail market tests of frozen
1977 minced fish. *A.E. Res.* Dept. of Agricultural Economics,
Cornell University. 77–6
—— Retail market tests of minced seafood chowders. *A.E. Res.* Dept.
1978 of Agricultural Economics, Cornell University. 78–4
—— Retail market tests of minced seafood crispies. *A.E. Res.* Dept. of
1979 Agricultural Economics, Cornell University. 79–2

HOOD, L F, ZALL, R R and CONWAY, R L. Conversion of minced clam
1976 wash water into clam juice: waste handling or product development. *Fd. Prod. Dev.*, 10(11), 86, 88
MOLEDINA, K H. Effect of some treatments on quality of frozen stored
1975 mechanically deboned flounder meat (MDFM); and use of
MDFM for preparation of dehydrated salted fish-sov 'cakes.'
MS thesis. Cornell University
REGENSTEIN, J M and REGENSTEIN, C E. An introduction to the kosher
1979 dietary laws for food scientists and food processors. *Fd. Tech.*,
33(1), 89–99

Effect of time, temperature, raw material type, processing and use of cryoprotective agents on mince quality

G Rodger, R B Weddle
and P Craig

1 Introduction

Fish mince is a raw material which has attracted considerable attention from food manufacturers throughout
the world. The main reasons are twofold, the first being
the potential abundance of minces on the market, either
from off-cuts, frames or small whole fish like blue
whiting, and the second that they are only 40–50% of the
price of fillets. However, the widespread use of minces
as a replacement for fillet material in products is limited
by the following factors:

(*i*) The minces have lost the 'integrity of shape'
associated with fillets and are therefore difficult to use in
steak-type products at any significant level.

(*ii*) Fish minces lose 'quality' very quickly during
frozen storage, resulting in dry, 'bitty' products, the rate
of quality loss being dependent on storage temperature
for a given batch of raw material.

However, previous work has shown that in products
where the visual appearance of intact fillets is not a
prerequisite, *eg* fish fingers, freshly generated minces
can be used at high fillet-substitution levels with little
difference in texture and product quality. As a result, it
is obvious that it is the loss of mince quality during
storage above all else which prevents its widespread use
in products.

The Japanese have been aware of this problem of
quality loss for a long time, in the manufacture of the fish
sausage product kamaboko, and in the past few years
have studied the use of non-protein additives to help
preserve the required functional properties of the mince
(Noguchi, 1971; 1975a, b, c). They have met with
considerable success, although they still do not understand fully why particular additives have the desired
effect. However, the properties they require for kamaboko are not necessarily those which we seek in our
home products. Other researchers have implied that the
loss of fish mince quality during cold storage is accelerated by water-soluble enzymes and substrates, and that
their removal by washing prior to freezing helps improve
the storage performance of the mince (Miyauchi *et al*,
1975).

As well as these problems, the fish technologist is
faced with the additional one of assessing the quality of
minces and being able to state a particular product
situation, and, for example, the maximum fillet replacement possible in a product for a particular batch of
mince, using quick and simple techniques. The aims and
objectives of this work were therefore:

(*i*) To study, using the existing quality assessment
technique of salt solubility of protein (SSP), the rate of
quality loss of several different mince samples at various
storage temperatures.

(*ii*) To compare this assessment with those derived
from differential scanning calorimetry (DSC) and texturometer studies, thereby estimating the value of SSP
as a quality assessment technique for fish minces.

(*iii*) To investigate the possibility of using washing,
and non-protein additives, to prevent quality loss with
V-cut and frame minces. The additives used were lactose, monosodium glutamate (MSG) and sodium citrate
at 3% w/w.

2 Experimental

2.1 Preparation, treatment and frozen storage of minces
Freshly caught inshore cod (*Gadus morhua*) were gutted
and stored on ice for 5 and 21 days. The fish were then
filleted, the V-cuts and frames retained (the kidney
tissue having been excised from the frames) and passed
through a Baader 694 deboner, from which the V-cut
and frame minces were obtained. These minces were
then treated as outlined in the flow diagram in *Fig 1*
packed in polythene bags (100g/bag), sealed, blast
frozen, and stored at the required temperature.

2.1.1 *Washing* One part of mince to two parts iced
water were stirred for five minutes in a Hobart mixer set
at the slowest speed. Dewatering was carried out by
spinning the mixture in a basket centrifuge till as much
water as had been added was spun off. The effect of
washing was visibly most noticeable with frame minces
from five-days-on-ice (doi) fish. Before washing they

199

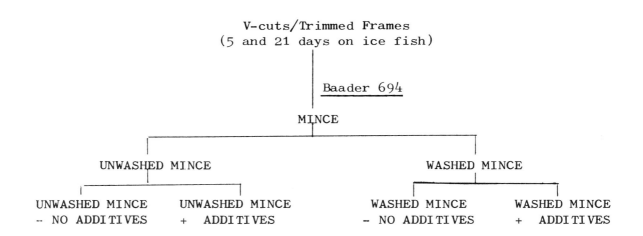

ADDITIVES: Lactose, Monosodium Glutamate (MSG),
and Sodium Citrate, at 3% w/w level

Fig 1a Flow diagram for the preparation of mince samples

Since so many samples were involved in this study, an easy to understand code was developed, and this is outlined below:

1st Symbol : 5 or 21   - Number of days fish was on ice prior to generation of mince

2nd Symbol : V or F   - Type of mince   V= V-cut
                                             F= Frame

3rd Symbol : U or W   - If mince was washed or not U = unwashed
                                                          W = washed

4th Symbol : X, L, M or N   - Type of additive   X - no additive
                                                 L - Lactose
                                                 M - Monosodium Glutamate
                                                 N - Sodium Citrate

5th Symbol : 7, 14, 29   - Temperature of storage ($-^{o}$C)

6th Symbol : 1, 4, 8, 12, 20 - Time (weeks) of storage

		V-cuts		No additives		Storage time (weeks)
e.g.		5 - V -	U -	X -	29 -	8
		Days on ice		unwashed		Storage temp.

Fig 1b Explanation of sample code

were red in colour, as a result of blood contamination. After washing and dewatering, however, they were as white as the corresponding V-cut minces.

In contrast to this, it was noted that the frame minces from 21 doi fish were very light—there was no visible pink/red colour. This could mean that storage on ice leads to some 'bleaching' of haemoglobin, or drainage of blood from the muscle tissue.

200

2.1.2 *Addition of additives* The additives used in this study were lactose, monosodium glutamate, and sodium citrate. They were added in the dry state to the minces, to give a final concentration of 3% w/w. They were mixed very thoroughly to ensure an even distribution throughout the mince samples.

2.1.3 *Packing and freezing of samples* One hundred-gram samples of the various minces were packed in polythene bags (12 × 15cm) and blast frozen at −40°C. They were then removed and stored in freezers set at the required temperatures, −7°C, −14°C and −29°C.

2.2 *Analytical procedures*
Samples were removed from the freezer cabinets after various times and allowed to thaw. Salt solubility measurements, SDS gel electrophoresis, pH and texture measurements were carried out on all the samples in a particular batch on the same day. Due to a lack of time, only certain selected samples were investigated by differential scanning calorimetry.

2.2.1 *pH measurement* Since it has been shown in the past that the pH of fish muscle influences the texture (Kelly, 1967), the pH was monitored during frozen storage to determine whether any change in texture was resulting simply from pH changes, or actual changes in the protein chemistry of the system. 10g of mince were homogenized with 20g water for 30s, and the pH of the resulting suspension measured.

2.2.2 *Salt solubility measurements* Five grams of mince were homogenized in 195ml 5% NaCl, 0·02M NaHCO₃ (pH 7·6) for 30s using a Silverson heavy-duty laboratory mixer set at maximum speed. After centrifugation of 10ml of the homogenate at 18 000rpm, an aliquot (up to 1ml) which gave a final A575 not greater than 0·15 was removed and used in the Biuret reaction.

2.2.3 *SDS gel electrophoresis* SDS gel electrophoresis was performed on all the salt extracts of the mince samples to give a picture of which proteins were becoming inextractable during cold storage. The salt extracts (after removal of the aliquot for the Biuret reaction) were made up to 4% SDS and 1% mercaptoethanol. (The solution had to be heated before the SDS would dissolve).

The preparation of the gels, buffers, *etc* was as described elsewhere (Weber and Osborn, 1969). The gels were run at 5mA/tube with a loading of 30λ of the relevant salt extract. For studies of whole mince, 0·5g mince were dissolved in 4% SDS (+ or − Mercaptoethanol − 1%) and 10λ loaded on to the gels and run under the same conditions.

2.2.4 *Texturometer measurements* The Aberdeen texturometer was used to determine textural changes in both raw and cooked mince samples. Raw samples were allowed to thaw and rise to ambient. The cooked samples were steamed for 10min and allowed to cool to room temperature. Sample application to the instrument was as previously described (Main *et al*, 1972). Measurements were carried out in triplicate for all samples.

2.2.5 *Differential scanning calorimetry* DSC studies were carried out on a Perkin-Elmer scanning calorimeter. Not all the samples generated for this trial could be accommodated for DSC analysis in the time available and in general only the control samples (*ie* without additives) were scanned.

The thermograms of samples after various times and temperatures of storage were compared to determine if there were any differences in thermal transition energies of the main protein species.

3 Results

3.1 *Salt solubility studies*
The results for the salt solubility experiments are given in *Figs 2–9*. They show the following points of interest:

(*i*) The major part of the solubility loss occurring over the 20-week storage period happens during the first week, and that the subsequent solubility loss is very slow.

(*ii*) Washing the minces prior to freezing has no great effect other than to reduce the initial and final solubilities by an amount approximately equal to the water-soluble protein content (20–30mg g⁻¹).

(*iii*) The frame minces behave similarly to V-cuts, in the general *shape* of the solubility curves, but the absolute SSP values are lower, and the difference in storage performance is less pronounced between −29°C, and −14°C and −7°C.

(*iv*) Storage of whole fish on ice for 21 days prior to the generation of the minces has a significant effect on the solubility. The maximum rate of solubility loss is again in the first week, but the initial solubility prior to freezing is only 50% of that for the 5 doi fish.

(*v*) The additives used all have some effect in preserving the solubility of the fish mince proteins. Lactose performs best, but for all the additives used, the effect is most pronouced at the storage temperature −29°C. The effect at −7°C and −14°C is slight, relative to the control.

3.2 *Discussion*
3.2.1 *Variation of solubility with time* A simple explanation for the loss of solubility during the first week could be that the act of freezing was causing the effect. However, if this were the only reason, then all the samples should be affected to the same degree. The results show that this is not so. It could be argued that the loss of solubility was caused by the changes in pH, ionic strength, or enzyme/substrate concentrations brought about by the freezing out of water. If this were so, however, a linear relationship between rate of solubility loss and temperature may be expected. That is, if this effect occurs at −7°C then it should be magnified at −29°C, since more water is frozen out. However, this argument precludes the possibility of there being a temperature optimum for these effects, and a eutectic point in the system.

3.2.2 *Effect of washing on solubility profile* Washing has been advanced as a technique for preserving the functional properties of fish minces. It is thought that the substrate TMAO (trimethylamine oxide) for the formaldehyde-forming reaction, and the enzymes involved, are

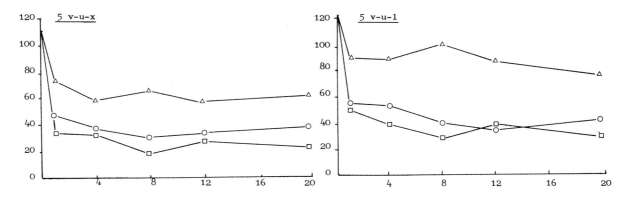

EFFECT OF TIME, TEMPERATURE AND CRYOPROTECTIVE AGENT ON SOLUBILITY OF FISH MINCE

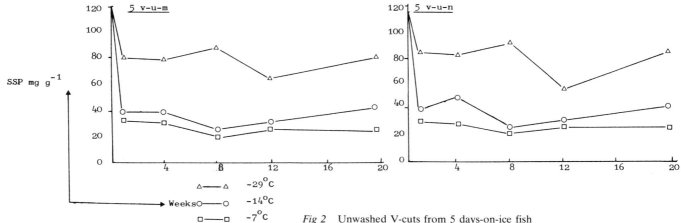

SSP mg g^{-1}

Weeks

△———△ -29°C
○———○ -14°C
□———□ -7°C

Fig 2 Unwashed V-cuts from 5 days-on-ice fish

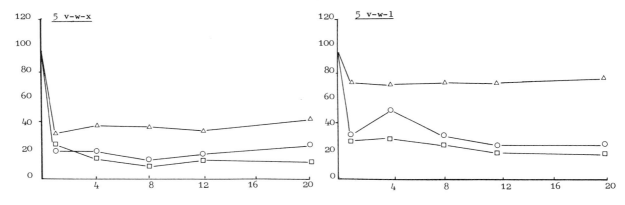

EFFECT OF TIME, TEMPERATURE AND CRYOPROTECTIVE AGENT ON SOLUBILITY OF FISH MINCE

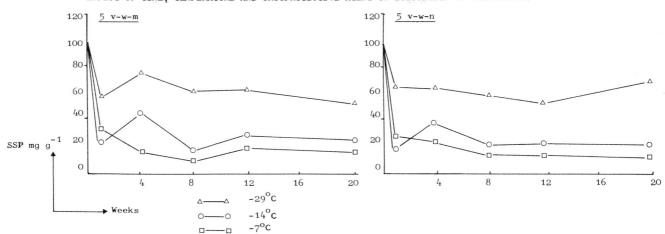

SSP mg g^{-1}

Weeks

△———△ -29°C
○———○ -14°C
□———□ -7°C

Fig 3 Washed V-cuts from 5 days-on-ice fish

202

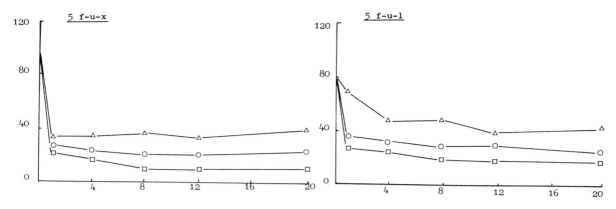

EFFECT OF TIME, TEMPERATURE, AND CRYOPROTECTIVE AGENT ON SOLUBILITY OF FISH MINCE

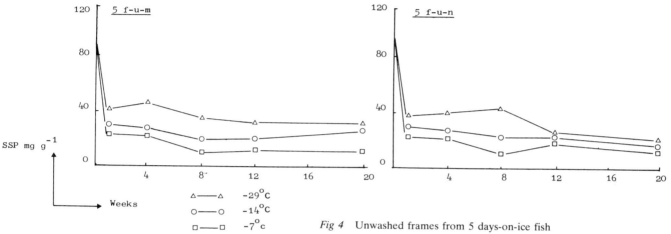

SSP mg g^{-1}

Weeks

△——△ -29°C
○——○ -14°C
□——□ -7°c

Fig 4   Unwashed frames from 5 days-on-ice fish

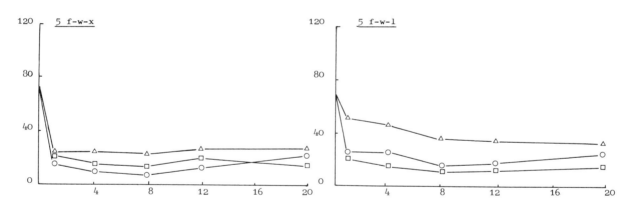

EFFECT OF TIME, TEMPERATURE, AND CRYOPROTECTIVE AGENT ON SOLUBILITY OF FISH MINCE

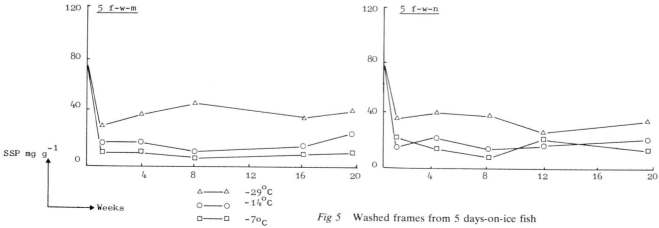

SSP mg g^{-1}

Weeks

△——△ -29°C
○——○ -14°C
□——□ -7°c

Fig 5   Washed frames from 5 days-on-ice fish

203

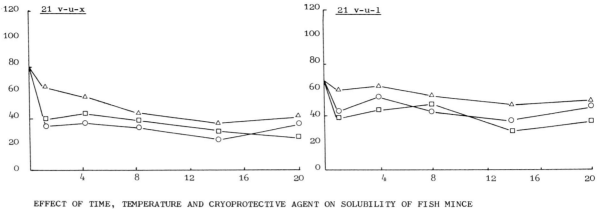

EFFECT OF TIME, TEMPERATURE AND CRYOPROTECTIVE AGENT ON SOLUBILITY OF FISH MINCE

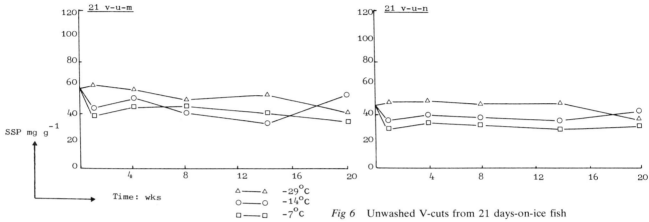

SSP mg g^{-1}

Time: wks

△———△ -29°C
○———○ -14°C
□———□ -7°C

Fig 6 Unwashed V-cuts from 21 days-on-ice fish

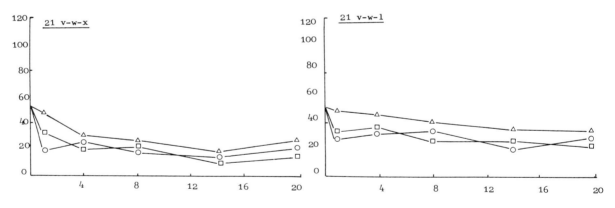

EFFECT OF TIME, TEMPERATURE AND CRYOPROTECTIVE AGENT ON SOLUBILITY OF FISH MINCE

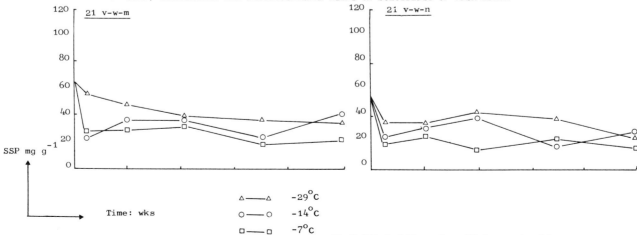

SSP mg g^{-1}

Time: wks

△———△ -29°C
○———○ -14°C
□———□ -7°C

Fig 7 Washed V-cuts from 21 days-on-ice fish

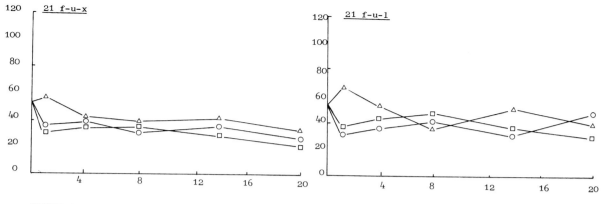

EFFECT OF TIME, TEMPERATURE AND CRYOPROTECTIVE AGENT ON SOLUBILITY OF FISH MINCE

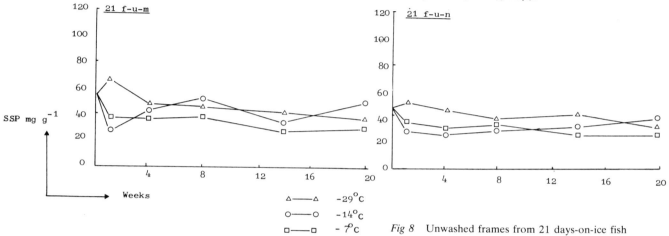

SSP mg g^{-1}

Weeks

△——△ -29°C
○——○ -14°C
□——□ - 7°C

Fig 8 Unwashed frames from 21 days-on-ice fish

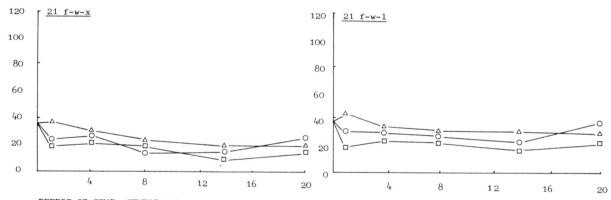

EFFECT OF TIME, TEMPERATURE AND CRYOPROTECTIVE AGENT ON SOLUBILITY OF FISH MINCE

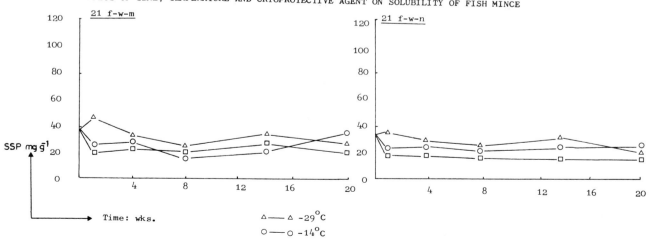

SSP mg g^{1}

Time: wks.

△——△ -29°C
○——○ -14°C
□——□ -7°C

Fig 9 Washed frames from 21 days-on-ice fish

205

removed (Sikorski *et al*, 1976), hence improving the storage performance. However, it has also been suggested that naturally occurring cryoprotective agents may be washed out resulting in increased sensitivity to cold-store deterioration of the actomyosin fraction (Umemoto and Muraki, 1969).

The results obtained in this study show neither effect has occurred. The only effects of washing were to reduce the measurable salt-soluble protein content, by removing the water-soluble protein component, and in the case of frame minces, from 5 doi fish, to dramatically improve the colour (red→white).

3.2.3 *Effect of mince type on solubility profiles*
The frame minces used in this study do not differ significantly in the shapes of their solubility curves from the V-cut minces, except that there is not the pronounced difference in solubility between the $-7°C$ and $-14°C$ stored samples and those at $-29°C$, *ie* the solubility at $-29°C$ is not preserved as well as in 5 doi fish. Why this should be so is not understood. If it were the result of blood pigments catalysing free fatty acid formation, leading to loss of protein solubility, then washing should show a beneficial rather than a neutral effect. If there were increased activity of the formaldehyde-forming enzyme system, then this too should be reduced by washing.

However, there is no evidence that washing has been effective in this way. The reason is not understood at present.

3.2.4 *Effect of ice storage period of whole fish on solubility profiles of minces*
That storage on ice for 21 days led to such a large decrease in the initial (unfrozen) solubility of the mince protein was surprising, since *post mortem* it might be expected that the action of cathepsins and other proteolytic enzymes could lead to an increase in protein extractability. The present observation could be explained if there were a consolidation of the interactions between actin and myosin in the myofibrils. However, there is conflicting evidence in the literature that this indeed happens. Some workers have found that there is no change (Dyer and Dingle, 1961) or a continual decrease in the extractability (Suzubi and Migata, 1962; Tomlinson *et al*, 1965) of myofibrillar proteins during *post mortem* storage on ice. The results of this study support the latter workers.

If the level of salt-soluble protein is found to be a vital property in determining the product applicability of fish mince, then a problem arises—storage on ice has been proposed as a method of avoiding the formation of formaldehyde after freezing, since wet storage results in bacterial decomposition of TMAO→TMA + oxygen. During frozen storage, however, TMAO→DMA + formaldehyde (FA). Thus if TMAO is used up prior to freezing, no FA should form.

The problem could arise in getting the correct balance between length of time on ice to prevent subsequent FA formation, and the loss of protein solubility occurring during ice storage (as well as flavour changes).

3.2.5 *Effect of additives on solubility profiles*
The additives in this study helped retain protein solubility relative to the control, with lactose performing best. The main effect, however, was seen at $-29°C$ and only slightly at $-7°C$ and $-14°C$. This could be explained if we consider the protective effect of the additives to be directed against a reaction which happens only very slowly at $-29°C$. Hence it may be easily controlled, whereas at $-7°C$ and $-14°C$ it may proceed very quickly, the additives having little or no effect on the rate.

In general, however, the cryoprotective effect of the additives used did not match the performance obtained by the Japanese in their studies on the preservation of surimi. A possible factor influencing the extent of cryoprotection may well be the degree of muscle tissue disintegration prior to additive addition. Surimi is a homogeneous 'paste'—hence the additives will be much better distributed throughout the matrix than in fish 'mince', where there are still some intact/partially intact cells, into which the additives must diffuse before becoming effective.

4 Texture measurements

4.1 *Results (cooked minces)*
(*i*) The textures of stored samples show a relationship to the temperature. Those stored at the lower temperature neither change texture at the same rate nor reach the same high texture value (= 'hardness') as those at the higher storage temperatures. (*Figs 10–17*).

(*ii*) No general trend in texture change is evident as a result of washing the minces, except that the final values for unwashed samples are usually lower than those for washed samples (*Figs 10–13*).

(*iii*) Frame minces tend to be slightly softer than V-cut minces (*ie* have a lower mV/volt value). This applies to the minces from 5 doi and 21 doi fish. This could be the result of the slightly higher water content of the frame minces (*Figs 14–17*).

(*iv*) The rate and extent of change from the initial texture values is greater in minces from 5 doi fish than from 21 doi fish.

(*v*) The additives used in these experiments have little effect on the texture of the minces except in the case of MSG, which may increase the 'hardness' of a mince by 50–100% relative to the control.

(*iv*) One of the most important observations to be made is that there are still textural changes occurring in the minces long after their salt solubilities have reached constant levels, especially for minces from 5 doi fish.

4.2 *Texture measurements—discussion*
The Aberdeen texturometer is a machine which compresses the sample in a sample cup by a fixed amount. What is measured is mainly the 'hardness' of the sample. This change in texture can only result from changes in the protein in the mince, for example as a result of aggregation, unfolding, *etc*. Therefore we would expect, from the solubility results, that the texture would change most slowly in the sample stored at the lowest temperature. This did indeed happen, but the relationship between solubility level and texture value is not a simple one. We have seen that the texture can still change even when the solubility is constant. If we consider the changes occurring in the proteins to be aggregation, *eg* as simplistically: Actomyosin

S—Soluble $\bullet + \bullet \longrightarrow \bullet\text{-}\bullet + \bullet \longrightarrow \bullet\text{-}\bullet\text{-}\bullet$
I—Insoluble   *S*   *I*   *S*   *I*

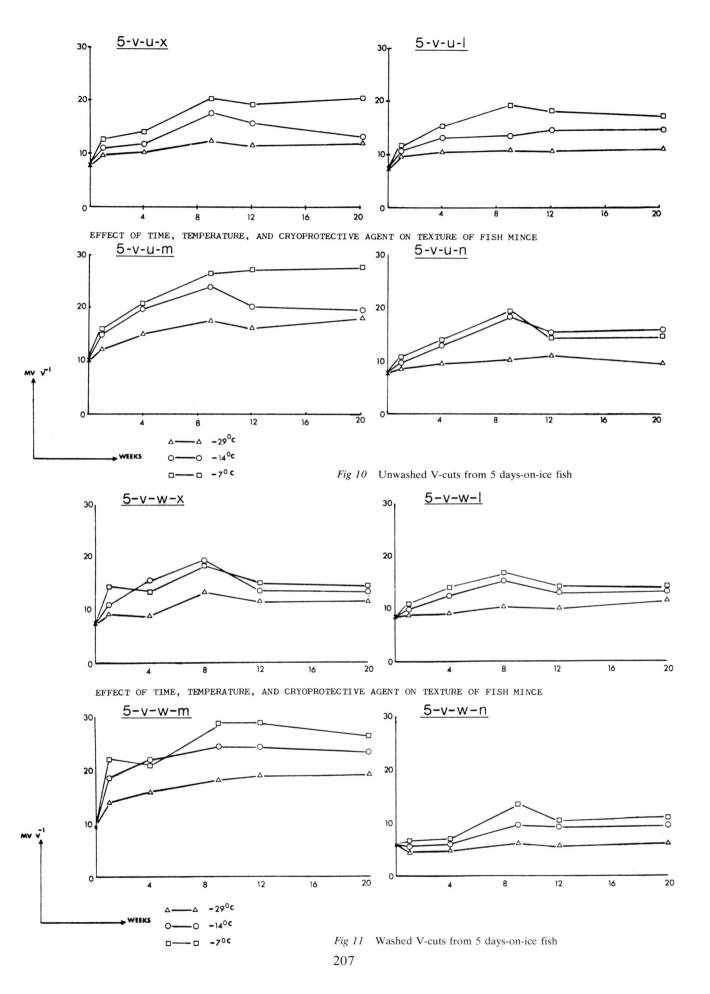

EFFECT OF TIME, TEMPERATURE, AND CRYOPROTECTIVE AGENT ON TEXTURE OF FISH MINCE

△——△ -29^{0}c
○——○ -14^{0}c
□——□ -7^{0}c

Fig 10   Unwashed V-cuts from 5 days-on-ice fish

EFFECT OF TIME, TEMPERATURE, AND CRYOPROTECTIVE AGENT ON TEXTURE OF FISH MINCE

△——△ -29^{0}c
○——○ -14^{0}c
□——□ -7^{0}c

Fig 11   Washed V-cuts from 5 days-on-ice fish

207

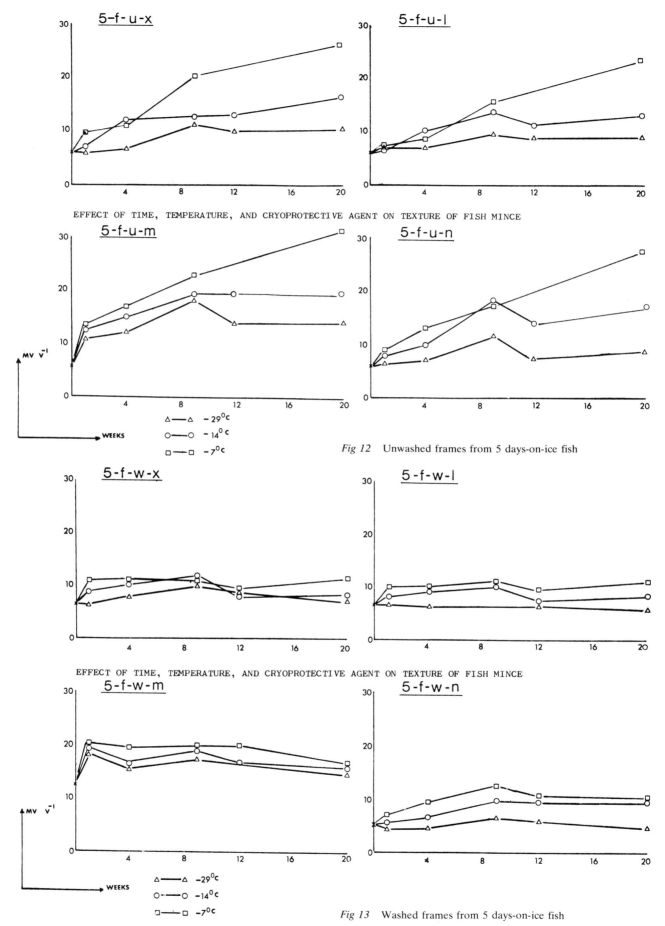

EFFECT OF TIME, TEMPERATURE, AND CRYOPROTECTIVE AGENT ON TEXTURE OF FISH MINCE

Fig 12 Unwashed frames from 5 days-on-ice fish

EFFECT OF TIME, TEMPERATURE, AND CRYOPROTECTIVE AGENT ON TEXTURE OF FISH MINCE

Fig 13 Washed frames from 5 days-on-ice fish

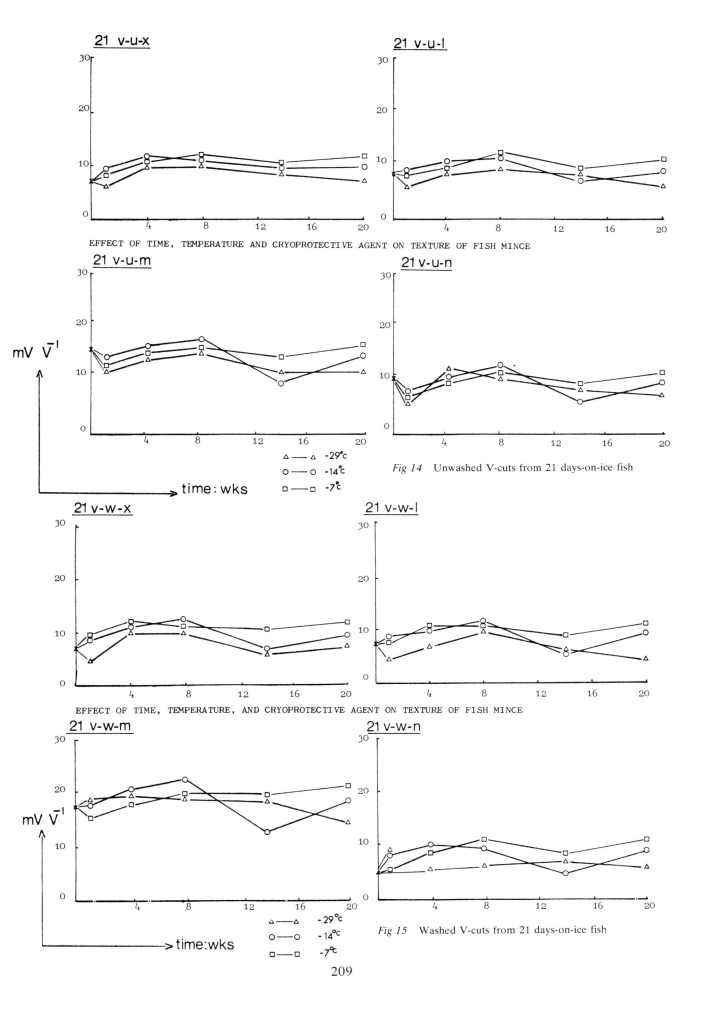

21 v-u-x

21 v-u-l

EFFECT OF TIME, TEMPERATURE AND CRYOPROTECTIVE AGENT ON TEXTURE OF FISH MINCE

21 v-u-m

21 v-u-n

mV V⁻¹

△——△ -29°c
○——○ -14°c
□——□ -7°c

time: wks

Fig 14 Unwashed V-cuts from 21 days-on-ice fish

21 v-w-x

21 v-w-l

EFFECT OF TIME, TEMPERATURE, AND CRYOPROTECTIVE AGENT ON TEXTURE OF FISH MINCE

21 v-w-m

21 v-w-n

mV V⁻¹

△——△ -29°C
○——○ -14°c
□——□ -7°c

time:wks

Fig 15 Washed V-cuts from 21 days-on-ice fish

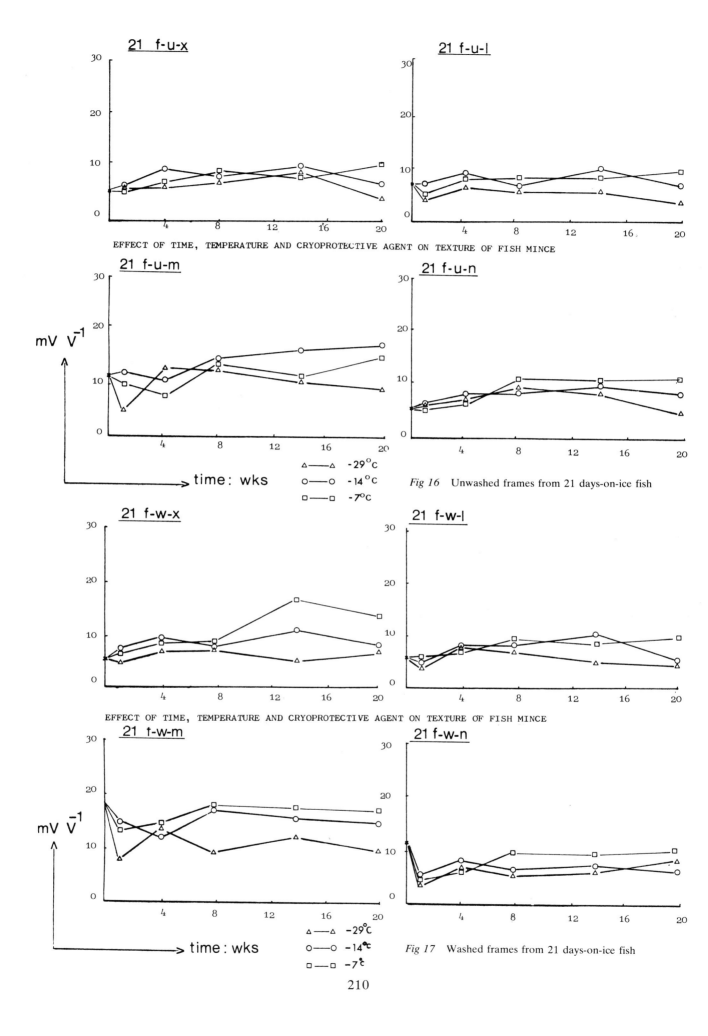

EFFECT OF TIME, TEMPERATURE AND CRYOPROTECTIVE AGENT ON TEXTURE OF FISH MINCE

Fig 16 Unwashed frames from 21 days-on-ice fish

EFFECT OF TIME, TEMPERATURE AND CRYOPROTECTIVE AGENT ON TEXTURE OF FISH MINCE

Fig 17 Washed frames from 21 days-on-ice fish

210

then we can postulate that in, say, the 'monomeric' form 'actomyosin' is soluble. On the formation of a 'dimer' or higher polymer, this solubility is lost. However, the effects of the dimeric, trimeric, *etc*, states of the protein on the texturometer reading need not be the same, which could explain the observed effect.

Neither the washing regime nor the length of iced storage had any great effect on the initial measured textures of the samples. This is surprising for the following reasons:

(*i*) The sarcoplasmic proteins contribute little to the gelling of muscle systems. Hence when they are removed by washing, and the mince is dewatered to give the original protein concentration, it should be more resilient, elastic, *etc*. This is not evident from the texturometer results. However, the texturometer does not really measure the elastic component of the system, and it would therefore be presumptuous to say that there is no effect on overall texture.

(*ii*) It was noted in the previous section that storage of fish on ice for 21 days resulted in the mince having a much reduced initial solubility relative to the 5 doi samples. However, the texturometer results show that in general there is no corresponding large change in the texturometer readings at this time, whereas the earlier results for the 5 doi fish indicated that this loss of solubility should be accompanied by a significant textural change. The reason for this is not apparent. A simple explanation could be that the protein changes in wet fish leading to loss of solubility are entirely different from those in frozen mince, and have no effect on the response to the texturometer. If formaldehyde is considered the causative agent of hardening texture, this assumption is supported by the fact that in wet fish there is no formaldehyde present since TMAO→TMA + O$_2$; nor should there be any change in ionic strength, pH or disruption of protein hydration shells which can result from freezing.

This therefore is more evidence against taking salt solubility as the *sole* criterion of fish quality, since it appears that how proteins lose their solubility is as important as the fact that they do, where the effect on texture is concerned.

The results obtained from the samples including additives showed that MSG had a very significant effect in increasing the texture scores. This is an anomalous effect which cannot be explained at present. If a large increase in texture score is taken as an undesirable trait, then the use of MSG as an additive would need to be considered carefully.

5 pH measurements

5.1 *Results (in some cases, pH values were taken only up to week 12)*

The initial pH values of the fish minces prior to freezing were significantly higher than those which are normally associated with cod (average pH 7·0 for 5 doi fish as opposed to normal pH 6·7–6·8).

The pH of the mince samples was measured after various times at the three storage temperatures. The changes in pH with time are shown in *Figs 18–25*.

(*i*) General trends: The general trend is for the pH to fall sharply during the first week of storage at all temperatures, followed by a slight rise till week 12, and then a slight fall. This trend is most obvious for those samples stored at $-29°C$ and less so for those stored at $-7°C$ and $-14°C$.

(*ii*) Effect of washing: Washing had the effect of reducing the pH of the unfrozen 5 doi samples (*ie* t = 0) but not those from the 21 doi fish. However, after freezing the pH profiles for both the washed and unwashed minces were very similar.

(*iii*) Effect of iced storage: Minces from 21 doi fish have higher pH values than the corresponding minces from 5 doi fish.

(*iv*) Effect of mince type: Frame minces in general have higher pH values than V-cut minces.

(*v*) Effect of additives: The additives used in this study have an effect on pH, which follows a general pattern. The pH of samples + Na citrate > + lactose > control > + monosodium glutamate throughout the storage period.

5.2 *pH—discussion*

The initial fall in pH during the first week of frozen storage is quite marked in most of the samples. This should not be caused by the production of lactic acid from glycogen since the glycogen store would be depleted by the end of the ice-storage period.

An explanation for the rise in pH during frozen storage and its subsequent fall is perhaps easier to find if the reaction TMAO $\longrightarrow$ FA + DMA is considered. The production of DMA will lead to a pH rise since it is a basic compound. However, there will be a finite amount of TMAO which can decompose, and if we consider that DMA may diffuse from the tissue, at some point the loss of DMA will exceed its production and hence the observed pH will fall.

Similarly, the formation of TMA from TMAO during the extended ice-storage period may account for the differences in pH between the minces from the 21 and 5 doi fish. However, in the former case, there are probably many other metabolites present as a result of bacterial spoilage which are affecting the pH.

In addition, since it is thought that blood, *etc* may accelerate TMAO decomposition, then the higher pH of frame minces would be expected. Washing out these basic materials would lead to the initial pH being lowered (time 0). However, the subsequent pH profiles would be very similar to the unwashed samples since in the latter case the basic compounds would soon reach the lower levels of the former as a result of diffusion.

6 Differential scanning calorimetry

The major result obtained from the DSC studies was that changes in protein solubility are not mirrored by similar changes in the thermal transition energies (*Fig 26*). Thermograms of samples whose solubilities are vastly different are shown in *Fig 26*. It can be seen that there is no great difference between them. However, a general trend to emerge from these experiments was the gradual broadening of the myosin transition peak, which can be suggestive of some aggregation of the protein species. This study illustrates that in this particular case, although fish muscle may be 'denatured' so far as solubility measurements are concerned, the component

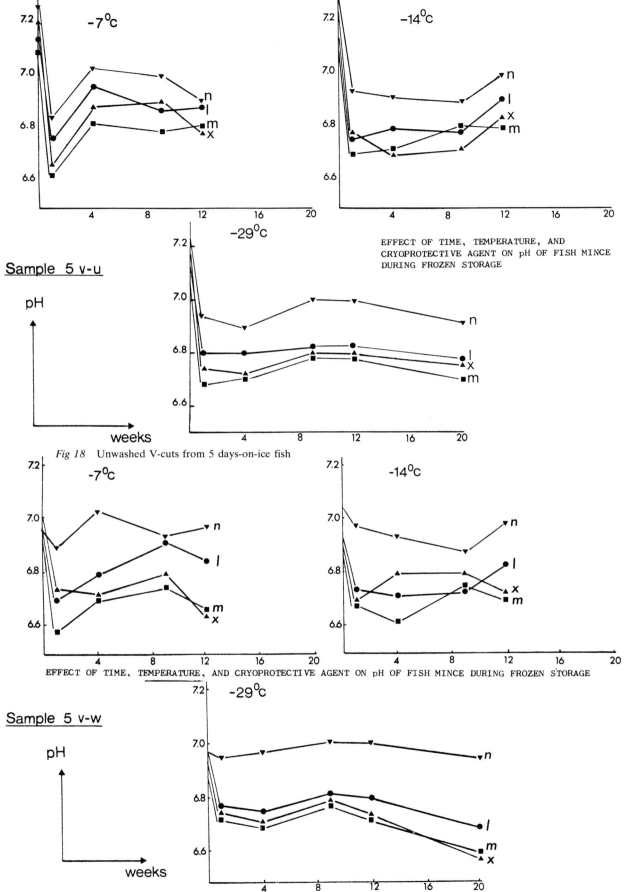

EFFECT OF TIME, TEMPERATURE, AND CRYOPROTECTIVE AGENT ON pH OF FISH MINCE DURING FROZEN STORAGE

Sample 5 v-u

Fig 18 Unwashed V-cuts from 5 days-on-ice fish

EFFECT OF TIME, TEMPERATURE, AND CRYOPROTECTIVE AGENT ON pH OF FISH MINCE DURING FROZEN STORAGE

Sample 5 v-w

Fig 19 Washed V-cuts from 5 days-on-ice fish

212

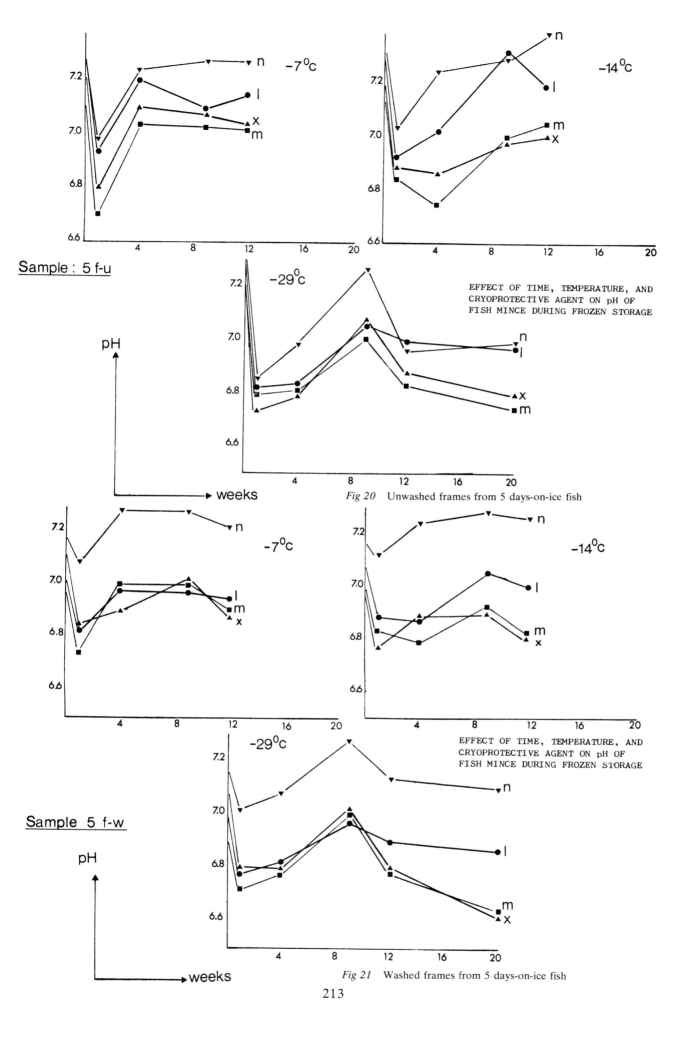

Sample : 5 f-u

pH

weeks

-7 °C

-14 °C

-29 °C

EFFECT OF TIME, TEMPERATURE, AND CRYOPROTECTIVE AGENT ON pH OF FISH MINCE DURING FROZEN STORAGE

Fig 20   Unwashed frames from 5 days-on-ice fish

Sample 5 f-w

pH

weeks

-7 °C

-14 °C

-29 °C

EFFECT OF TIME, TEMPERATURE, AND CRYOPROTECTIVE AGENT ON pH OF FISH MINCE DURING FROZEN STORAGE

Fig 21   Washed frames from 5 days-on-ice fish

213

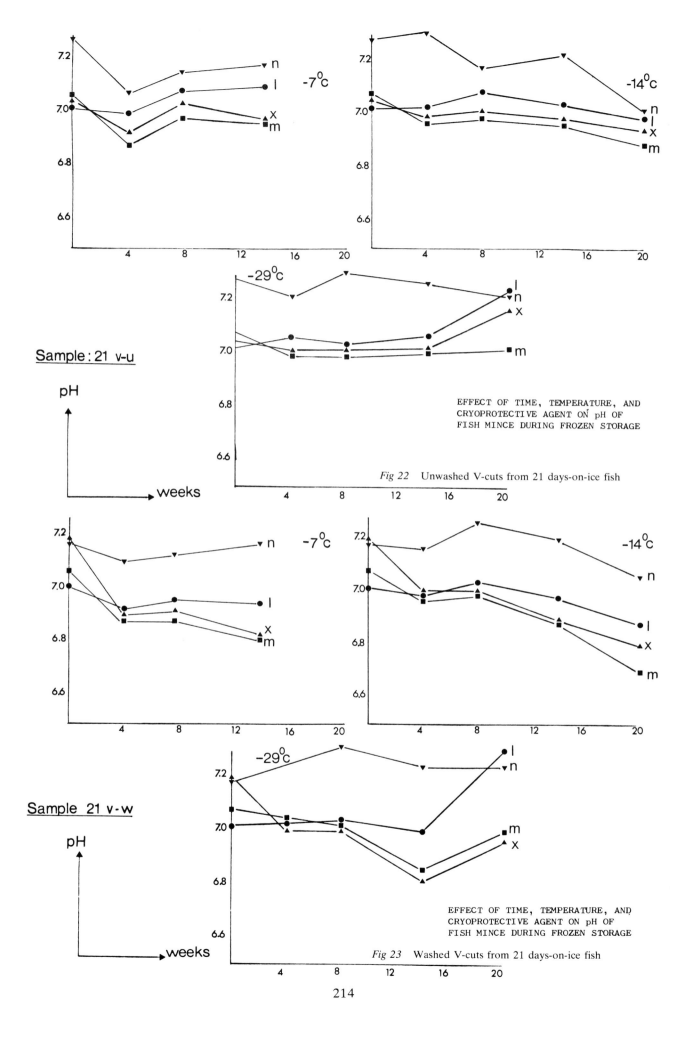

Sample : 21 v-u

pH

weeks

-7°C

-14°C

-29°C

EFFECT OF TIME, TEMPERATURE, AND
CRYOPROTECTIVE AGENT ON pH OF
FISH MINCE DURING FROZEN STORAGE

Fig 22 Unwashed V-cuts from 21 days-on-ice fish

Sample 21 v-w

pH

weeks

-7°C

-14°C

-29°C

EFFECT OF TIME, TEMPERATURE, AND
CRYOPROTECTIVE AGENT ON pH OF
FISH MINCE DURING FROZEN STORAGE

Fig 23 Washed V-cuts from 21 days-on-ice fish

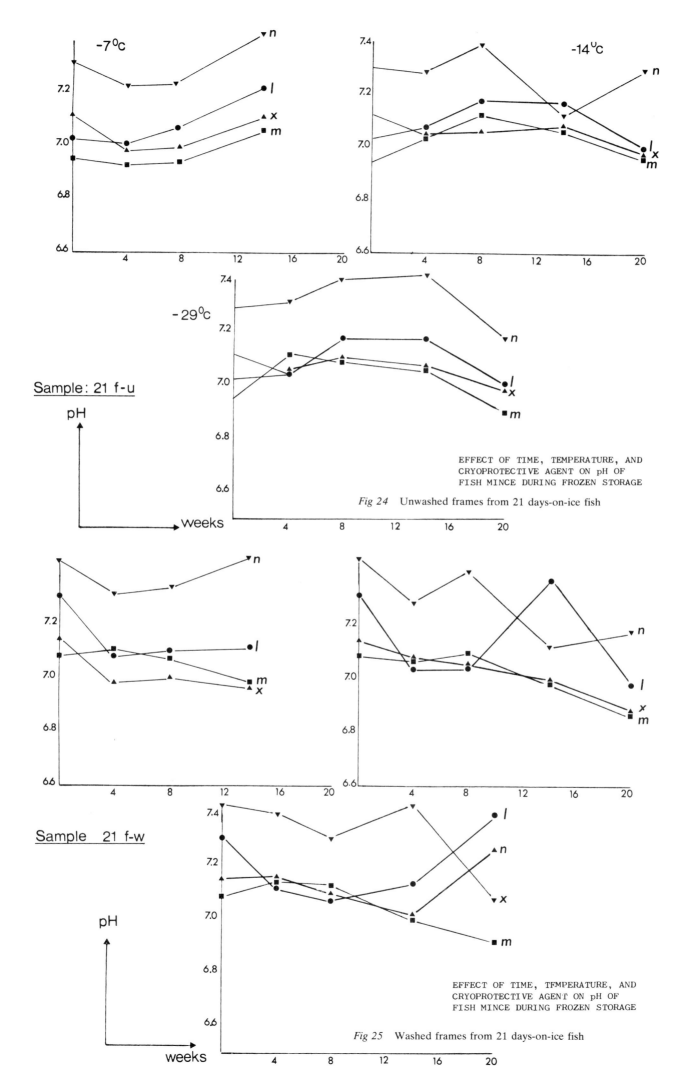

-7°C

-14°C

-29°C

Sample: 21 f-u

pH

weeks

EFFECT OF TIME, TEMPERATURE, AND
CRYOPROTECTIVE AGENT ON pH OF
FISH MINCE DURING FROZEN STORAGE

Fig 24   Unwashed frames from 21 days-on-ice fish

Sample 21 f-w

pH

weeks

EFFECT OF TIME, TFMPERATURE, AND
CRYOPROTECTIVE AGENT ON pH OF
FISH MINCE DURING FROZEN STORAGE

Fig 25   Washed frames from 21 days-on-ice fish

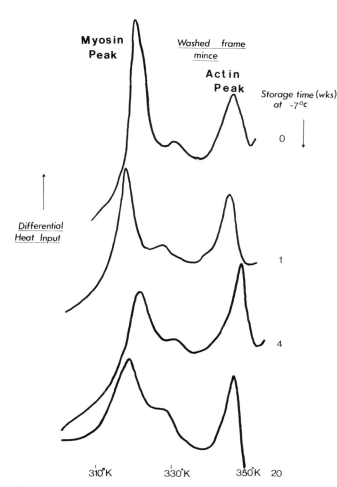

Fig 26 Thermogram of washed frame minces after different storage times at 17°C

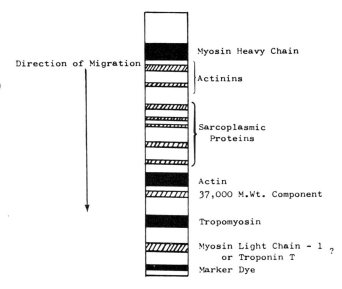

Fig 27 Typical SDS electrophoretogram (8.75% gel) of salt extract of fresh cod mince.

proteins still possess the ability to undergo thermal transitions, and hence appear native to the DSC. This is important in explaining what is happening during cold storage. Whatever is occurring, be it aggregation by ionic, hydrophobic or covalent bonding, is not leading to an absolutely fixed, rigid structure, since the molecules can absorb heat energy and 'unfold' in some way. Not enough work has been done using this technique, however, to answer all the questions raised by these results. An informative piece of work would be to 'denature' fish muscle by a variety of means, eg pH change, ionic strength variations, various cross-linking agents such as free fatty acids and formaldehyde, and compare the thermograms with those of freeze denatured muscle.

7 Gel electrophoresis

Fig 27 shows a typical electrophoretogram of a salt extract of fresh cold mince.

7.1 Variation of band patterns as a function of storage temperature

Samples for PAGE were prepared by adding SDS (to 4%) and mercaptoethanol (to 1%) to the salt extracts obtained during solubility studies. Plate 1 shows the band patterns obtained from salt extracts of the 4 types of mince used in this study—ie washed and unwashed frame and V-cut minces. The figures all refer to minces from 5 doi fish at time zero, but those from 21 doi fish

were essentially identical, except that the staining intensity was slightly less, which can be explained by the lower initial solubility of these minces.

Plates 2, 3 and 4 show the same samples after eight weeks at −29°C, −14°C and −7°C. What is clearly visible is the progressive disappearance, with increasing temperature, of the extractible myosin and actin. In the case of the gels obtained from salt extracts of minces containing the various additives, the patterns were again virtually identical.

Where the additives had helped preserve solubility, this was shown up by slightly more intense myosin heavy chains and the actin component.

7.2 The relative effects of mercaptoethanol and SDS on gel patterns

All the samples run on the gels in Plates 1 to 4 had SDS (4%) and mercaptoethanol (1%) present. As well as these, some samples containing SDS only were prepared and run, and three such runs are shown in Plate 5. The main point of interest here is the effect which mercaptoethanol has on the high molecular weight protein fraction which does not penetrate the gel in its absence. This fraction is dissociated into myosin heavy chains, leaving no material which does not penetrate the gel. Since the intensity of the actin band is relatively unchanged within each pair of samples, this suggests that the high molecular weight fraction comprises solely myosin sub-units, stabilized by disulphide linkages. Any actin extracted by salt is either in the form which migrates as shown in the plates, or in an associated state which is disrupted by SDS addition to give this form.

8 General discussion

This work has shown that the behaviour of fish minces during frozen storage does not always follow the expected pattern, as evidenced by the dramatic fall in solubility during the first week of storage, and the seemingly non-beneficial effect of washing on the technological properties. The use of additives in preserving the functional properties of the minces gave only a

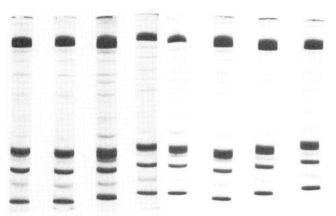

marginal improvement over the controls, under the conditions employed in this study. Therefore unless minces are comminuted/ground to a surimi-type paste, it is unlikely that the application of this technology will solve the problems facing us today.

There is still a great deal to learn about what factors influence mince keepability, and how the organoleptic properties relate to some measurable chemical/physical properties. Until we can determine this, mince will remain a very attractive raw material with, however, seemingly intractable problems.

Plate 1   SDS gels on salt extracts from four types of fresh fish mince. Samples are (left to right)
V-cut mince
washed V-cut mince
frame mince
washed frame mince
The direction of migration in this and subsequent plates was from top to bottom

Plate 2   SDS gels of salt extracts from four types of fish mince stored at $-29°C$ for eight weeks. Samples are as in Plate 1

9 References

DYER, W J and DINGLE, J R. *Fish as food*. Academic Press, New York
1961   and London. Vol 1, 275
KELLY, K O. Factors affecting the texture of frozen fish. In: *Freezing
1967   and Irradiation of Fish*, Ed R Kreuzer. Fishing News (Books), London
MAIN, G, *et al*. A texturometer for measuring the toughness of cooked
1972   fish. *Lab. Prac.*, 21(3), 185
MIYAUCHI, D, *et al* Frozen storage keeping quality of minced black
1975   rockfish (*Sebastes spp*) improved by cold water washing and by use of fish binder. *J. Fd. Sci.*, 40, 592
NOGUCHI, S, *et al* Studies on the control of denaturation of fish muscle
1971   protein during frozen storage. II. Preventive effect of amino acids and related compounds. *Bull. Jap. Soc. Sci. Fish.*, 37, 1115
—— Studies on the control of denaturation of fish muscle proteins
1975a   during frozen storage. III. Preventive effect of some amino acids, peptides, acetylamino acids and sulphur compounds. *Bull. Jap. Soc. Sci. Fish.*, 41, 243
—— Studies on the control of denaturation of fish muscle proteins
1975b   during frozen storage. IV. Preventive effect of carboxylic acids. *Bull. Jap. Soc. Sci. Fish.*, 41, 329
—— Studies on the control of denaturation of fish muscle proteins
1975c   during frozen storage. V. Technological application of cryo-protective substances on frozen minced fish meat. *Bull. Jap. Soc. Sci. Fish.*, 41, 779
SIKORSKI, Z, *et al* Protein changes in frozen fish. Crit. Revs. *Fd. Sci.*
1976   *and Nutr.*, 8(1), 97
SUZUKI, T and MIGATA, M. Post mortem changes of fish myosins. 1.
1962   Some physico-chemical changes with special reference to species and lethal conditions of fish. *Bull. Jap. Soc. Sci. Fish.*, 28, 61
TOMLINSON, N *et al* Chalkiness in halibut in relation to muscle pH and
1965   protein denaturation. *J. Fish. Res. Bd. Can.*, 22, 643
UMEMOTO, S and MURAKI, Y. Insolubilisation of fish actomyosin by
1969   frozen storage. *Bull. Tokai Reg. Fish. Res. Lab.*, 60, 191
WEBER, K and OSBORN, M. The reliability of molecular weight deter-
1969   minations by SDS polyacrylamide gel electrophoresis. *J. Biol. Chem.*, 244(16), 4406

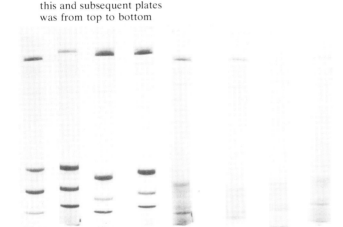

Plate 3   SDS gels of salt extracts from four types of fish mince stored at $-14°C$ for eight weeks. Samples are as in Plate 1

Plate 4   SDS gels of salt extracts from four types of fish mince stored at $-7°C$ for eight weeks. Samples are as in Plate 1

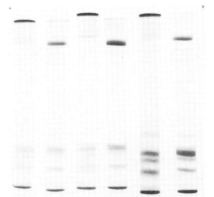

Plate 5   SDS gels of salt extracts of fish muscle proteins in the presence and absence of mercaptoethanol. Samples are (left to right)
V-cut mince (no mercaptoethanol)
V-cut mince (plus mercaptoethanol)
washed V-cut mince (no mercaptoethanol)
washed V-cut mince (plus mercaptoethanol)
frame mince (no mercaptoethanol)
frame mince (plus mercaptoethanol)

The use of response surface methodology to determine the effects of salt, tripolyphosphate and sodium alginate on the quality of fish patties prepared from minced fish, croaker

J C Deng and F B Tomaszewski

1 Introduction

The retail prices of conventional fish species have increased all over the world and fish has ceased to be an inexpensive item. As a consequence of demand exceeding supply, fish processors will soon be faced with a tightening supply of conventional fish species. The new 200-mile fishing jurisdiction will probably increase the supply of non-conventional resources in the United States, and an effort must be made to develop the technology necessary to put these new resources into the channels of commerce.

Recently, the under utilized scraps and frames of common fish and some of the non or partially utilized species (such as cod, sole, hake and croaker) have been exploited to process minced fish products in the United States, Japan, Denmark and other countries (Tanikawa, 1963; Carver and King, 1971; Herborg, 1976; Marlin, 1976; Daley *et al*, 1978). Filleting finfish leaves a considerable quantity of edible flesh remaining on the frame; however, the left-over flesh may be recovered by using a meat-bone separator (Drews, 1976). Deboning machines permit the recovery of flesh from filleting wastes and have made possible the economical processing of previously unused species having a low flesh yield (Ravichander and Keay, 1976). Fish which are under utilized due to their undesirable characteristics such as poor texture and flavour may be improved by chemical or physical treatment in the processing of the minced flesh (Drews, 1976; Marlin, 1976).

There is wide variation in the quality of minced fish products. Factors that may affect the quality are the deboning process, the washing technique, the chemical treatments (bleaching), the addition of salt, polyphosphates, alginate and/or spices, the cooking time and temperature, the storage conditions and the species (Carver and King, 1971; Drews, 1976; Lee and Toledo, 1976; Ravichander and Keay, 1976).

In Japan, fish and ham sausage have become popular. The sausage typically contains minced fish flesh with 10% pork fat, 10% starch, 2.5% salt, seasoning and preservatives (Tanikawa, 1963). In the United States, Carver and King (1971) developed fish frankfurters containing 76% fish plus added starch, oil and seasoning. Daley *et al* (1978) developed the 'Sea Dog', a sausage-type product containing minced mullet, textured soy flour, tripolyphosphate and seasonings. Thus, minced fish represents a potential source of inexpensive protein for human food if it can be incorporated into acceptable products. The purpose of this study was to develop a fish patty from under utilized species used in combination with other ingredients and to determine possible optimum combinations of these ingredients. Atlantic croaker (*Micropogon undulatus*) was used as the primary raw material for the development of the patty.

2 Materials and methods

2.1 Materials

Croaker (*Micropogon undulatus*) were purchased fresh from Jacksonville, Fla., and transported on ice to Gainesville. The fish were packaged in 5kg plastic bags and frozen at −34°C until used. Food grade sodium chloride (NaCl) was purchased from the Morton Salt Company, Chicago, Ill., food grade sodium tripolyphosphate (TPP) was obtained from the FMC Corporation, Philadelphia, Pa., and the sodium alginate, Keltone (NaAlg) was obtained from the Kelco Company (Clark, NJ). The batter and breading mixes were purchased from North America Food Service, Chicago, Ill. The soybean oil used for the deep fat frying was purchased from a wholesale store in Gainesville, Fla.

2.2 Minced flesh preparation

The whole croaker were thawed at 2°C for 18–20h. They were then headed, gutted, split, rinsed, packed on ice and stored overnight at 2°C. On the following day, the cleaned fish were passed through a deboning machine (Baader 694). The deboned flesh was washed (1 part flesh: 20 parts water) for 10min at 35°C with constant stirring (Chao and Deng, 1979). The washed flesh was packaged in plastic bags and frozen at −34°C.

2.3 Patty preparation

The washed minced flesh was thawed at 2°C for 18–20h prior to use. It was mixed (Univex Mixer, Model 122) for 10min with the proportions of NaCl, TPP and NaAlg indicated by the experimental design. The mixtures were weighed into 90g portions and shaped into a patty form with a Petri dish. Each patty was dipped into a 0.50% CaCl$_2$ solution (the interaction between NaAlg and CaCl$_2$ forms a film on the patty surface which would prevent the patty from sticking to a conveyor and facilitate commercial processing), drained, battered and breaded, then deep-fat fried for 30s at 202°C. The patties were cooled, placed in individual coded bags and frozen at −18°C until evaluation by objective and sensory methods. A schematic diagram indicating the procedure is presented in *Fig 1*.

2.4 Objective analyses

The frozen patties were cooked in an electric oven at 202°C for 30min and then cooled for 90min before evaluation. Textural properties of the patties were evaluated in terms of breaking force. The Instron Universal Testing Machine, Type TM 1900 (Canton, Mass.) equipped with load cell CCTM and plunger attachment No. 13 (1.9cm diameter) was used. Chart speed was 5cm/min, crosshead speed was 2cm/min and the clearance from the load cell was 2mm. Breaking force by compression was evaluated as the first peak height of the force curve, which is the amount of force (g) required for the plunger to break the sample. Two replicate patties were used for each treatment with three measurements on each patty.

A test was conducted to determine the flow properties of the 17 uncooked formulations. A Brookfield Synchro-Lectric Viscometer, Model RVT (Brookfield,

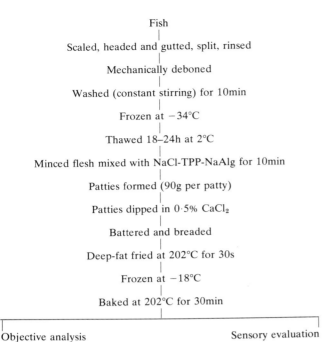

Fish
|
Scaled, headed and gutted, split, rinsed
|
Mechanically deboned
|
Washed (constant stirring) for 10min
|
Frozen at −34°C
|
Thawed 18–24h at 2°C
|
Minced flesh mixed with NaCl-TPP-NaAlg for 10min
|
Patties formed (90g per patty)
|
Patties dipped in 0·5% CaCl₂
|
Battered and breaded
|
Deep-fat fried at 202°C for 30s
|
Frozen at −18°C
|
Baked at 202°C for 30min
|

Objective analysis   Sensory evaluation

Fig 1 Flow chart for preparation of minced fish patties

Mass.) equipped with cylindrical spindle No. 4 was used to measure the apparent viscosity of the different combinations. The minced mullet was thawed as previously indicated. The appropriate proportions of NaCl-TPP-NaAlg were added to 150g of the minced flesh and mixed (by hand) for 2min. Approximately 50g of the mixture was poured into a $2\cdot9\text{cm} \times 13\cdot4\text{cm}$ cylinder and placed in a $4\pm 1°C$ water bath. The dial readings were recorded after 60s at each revolution. Each combination was measured in triplicate. Apparent viscosity was calculated from the recorded readings (Garcia-Borras, 1965).

2.5 *Sensory evaluation*
In the sensory evaluation study, eight experienced panellists were presented with four or five samples and asked to evaluate them for texture and flavour as compared to a reference. Before evaluation the frozen patties were cooked for 30min at 202°C (electric oven); the samples were cut, cooled for 15min, numerically coded and served in random order to judges in individual booths. Categories judged were texture-firmness, texture-preference and flavour. The data were evaluated using a nine-point rating scale where 5 = similar to reference, 1 = extremely poorer than reference and 9 = extremely better than reference. For the texture-firmness evaluation, 1 = extremely softer and 9 = extremely firmer than reference.

2.6 *Statistical design and analysis*
A response surface analysis (RSA) (Cochran and Cox, 1957) was performed on the data. Three factors (NaCl, TPP and NaAlg) were chosen for their effect on the various responses, and the function was expressed in terms of a quadratic polynomial equation:

$$\hat{y} = \beta_0 + \beta_1 x_1 + \beta_2 x_2 + \beta_3 x_3 + \beta_{11} x_1^2 + \beta_{22} x_2^2 + \beta_{33} x_3^2 + \beta_{12} x_1 x_2 + \beta_{13} x_1 x_3 + \beta_{23} x_2 x_3$$

which measures the linear effects, the quadratic effects and the interaction effects. Contours of a constant value were calculated by fixing one factor at a constant value and solving the equation for combinations of the other two factors. This type of statistical analysis enables one to predict responses and observe trends for combinations of the factors not necessarily included in the actual experiment. By using the APL System, contours of a constant estimated response value were calculated by fixing one value and solving the equation for combinations of the other two factors (Gilman and Rose, 1970). The effect of the third factor may be estimated by changing the value of the fixed factor and solving the equation at the new level.

Five levels of three factors (*Table I*) were chosen for study. Seventeen combinations (including three replicates of the centre point) were chosen in random order according to a central composite rotatable design configuration for three factors (Cochran and Cox, 1957). The data were analysed using the statistical analyses system (SAS) programme package for the analysis of variance calculations (Barr *et al*, 1976).

Table 1
DEFINITIONS AND LEVELS OF INDEPENDENT VARIABLES

Independent variable	Symbol	Levels of study, %				
		−1·682	−1	0	1	1·682
NaCl	x_1	0	0·21	0·50	0·79	1·0
TPP	x_2	0	0·10	0·25	0·40	0·5
NaAlg	x_3	0·1	0·28	0·55	0·82	1·0

3 Results and discussion

3.1 *Statistical analyses*
The coefficient of determination or R^2 value is a measure of how well the empirical model fits the actual data. The closer the R^2 is to unity, the better the empirical model fits the actual data (Mendenhall, 1975). The smaller the value of R^2, the less influence the independent variables in the model have in explaining the behaviour variation. The R^2 values for both the BF ($R^2 = 0·911$) and the texture-firmness ($R^2 = 0·902$) were highly significant ($\alpha = 0·01$), therefore, predictions concerning the response were made with confidence. An R^2 value of $0·814$ was observed for the flavour panel evaluation and the F-value was significant at the $0·063$ level; the model was considered approximate and used for trend analysis but not the prediction of optimum values. The F-value for texture preference had a $p > 0·01$ and $R^2(0·745)$ is low; therefore the model was only considered for trend analysis.

The R^2 values and coefficient estimate (B_i) for four regression models and the results of significance tests on the coefficients are presented in *Table II*. According to the significance tests on the estimates, NaAlg was probably the primary factor influencing the texture of the fish patties since x_3 was highly significant ($\alpha = 0·01$) for both the breaking force and the sensory evaluation of texture. The NaAlg also had an effect ($\alpha = 0·05$) on the flavour response. The negative values suggest that as the NaAlg was increased, the firmness and flavour quality of the patties decreased. The TPP (x_2) level had a significant effect on both the breaking force and the panellists' evaluation of texture. The NaCl (x_1) had a significant effect on flavour; the positive value suggesting an

219

Table II
REGRESSION COEFFICIENTS AND R² VALUES FOR FOUR DEPENDENT VARIABLES

Model term	Regression coefficient (B_i)			
		Texture		
	Breaking force	Firmness	Preference	Flavour
R^2	0.911[c]	0.902[c]	0.745	0.814
constant	559.561[c]	4.674[c]	4.772[c]	5.502[c]
x_1[a]	14.295	0.119	0.122	0.241[b]
x_2[a]	19.805[b]	0.483[b]	0.351	0.199
x_3[a]	−229.714[c]	−1.029[c]	0.861[c]	−0.288[b]
$x_1 \cdot x_1$	29.658	0.025	0.033	−0.182
$x_2 \cdot x_2$	41.087	0.180	0.254	−0.093
$x_3 \cdot x_3$	49.718	0.136	0.011	0.017
$x_1 \cdot x_2$	−119.646[b]	−0.219	0.219	−0.203
$x_1 \cdot x_3$	51.521	−0.094	0.094	−0.078
$x_2 \cdot x_3$	78.188	0.469	0.531	0.172

[a] Independent variable defined in *Table I*
[b] Significant at $\alpha = 0.05$ level
[c] Significant at $\alpha = 0.01$ level

improvement in flavour quality with an increase in NaCl. An interaction effect between NaCl-TPP ($x_1 \cdot x_2$) was noted for the breaking force.

3.2 Breaking force

Contours of a constant value were generated by fixing the level of NaAlg and plotting the response as determined by altering the levels of NaCl and TPP. Two main trends were noted (*Figs 2a–e*). One set of contours first appeared in the upper quadrants and as the NaAlg level was increased, these contours shifted upwards and towards the left. With this set of contours, as the NaCl level was increased, it was necessary to increase the TPP level in order to maintain the same breaking force. At a fixed NaAlg level, if the TPP concentration remained constant, then as the NaCl level was increased the breaking force decreased.

A second set of contours originated in the lower quadrants; as the NaAlg was increased, these contours shifted towards the right of the central axis and increased NaCl levels. At first the shift was towards increasing TPP levels, but as the NaAlg was increased, then the contours began to shift towards lower TPP levels. At a fixed NaAlg level, if the TPP concentration remained constant, then as the NaCl level was increased the breaking force increased. This set of contours was not apparent at high (1.0%) NaAlg levels.

Generally, as the NaAlg was increased, the breaking force decreased. At the lower NaAlg levels, the patties with lower breaking force were obtained at low NaCl-TPP levels.

3.3 Flow properties

The flow behaviour of the 17 uncooked formulations was measured. This information might be of value to food processors who would use extruders in the formation of their product. Generally, a pseudo plastic flow behaviour was noted ($n < 1$). A representative flow behaviour curve is presented in *Fig 3*.

3.4 Sensory evaluation

In the sensory evaluation study, the panellists were asked to compare the patties with a reference sample. The general trend for the flavour response at NaAlg level $\geq$

Fig 2 Response surface contours for TPP and NaCl effect on texture quality response (breaking force) of minced croaker at various alginate levels (a = 0.10%, b = 0.28%, c = 0.55%, d = 0.82% and e = 1.0%)

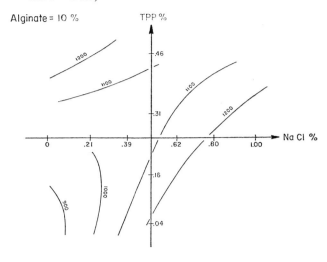

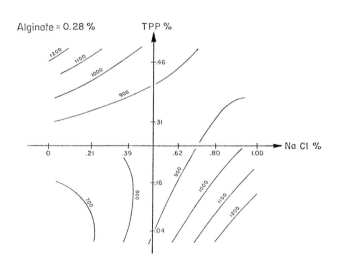

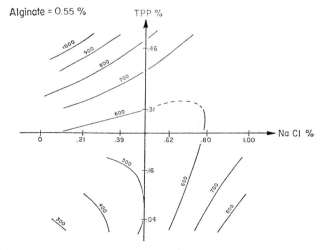

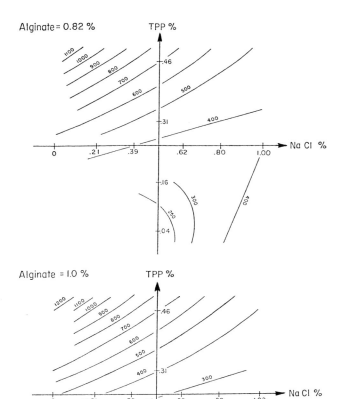

Alginate = 0.82 %

Alginate = 1.0 %

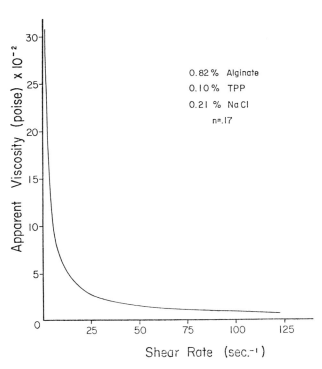

0.82 % Alginate
0.10 % TPP
0.21 % NaCl
n = 17

Fig 3 Effect of shear rate on the apparent viscosity of minced croaker at 0.82% alginate, 0.10% TPP and 0.21% NaCl

0.55% is indicated in *Fig 4*. When considering only one surface, the contour line means the same response at all points in the line. The inner contour line has a higher flavour acceptance (5.4); therefore, the maximum acceptability was in the centre. At alginate level 0.55%, the acceptability increased as the NaCl was increased to approximately 0.67%; however, above this point the acceptance decreased. Similarly, as the TPP was increased to 0.34%, the acceptability increased, but above this point the flavour acceptability decreased.

When several surface areas representing different NaAlg levels are superimposed on each other, then more of a three-dimensional view is presented. As the NaAlg was increased, the combination points between TPP and NaCl for higher acceptance decreased. Also as the NaAlg was increased, the general trend for the maximum point of flavour acceptability shifted from lower to higher TPP and from higher to lower NaCl.

The R^2 for texture-firmness was high ($R^2 = 0.902$) so that predictions concerning the response were made with confidence. Two general trends became apparent as the NaAlg level was increased (*Figs 5a–e*). One set of contours appeared in the upper quadrants and as the NaAlg level was increased to 0.685% these contours shifted upwards and towards the left. Generally, at a fixed NaAlg level, if the NaCl level was increased, then it was necessary to increase the TPP level in order to obtain the same response. Also, the greater the NaAlg concentration, the greater the amount of TPP required to maintain the similarity.

The second set of contours were visible at concentrations of TPP < 0.34%. At a fixed NaAlg level and constant TPP level, the firmness score increased as the NaCl level increased.

A simple correlation analysis between breaking force and sensory texture firmness indicated an $r = 0.939$ which was highly significant ($\alpha = 0.01$). The positive value indicates that as the breaking force was increased, the sensory evaluation of texture increased (firmer). This high correlation enables one to use the objective measurement of breaking force to predict the sensory response.

4 Summary and conclusions

In this study, it appears that NaAlg was the primary factor affecting the texture quality of fish patties prepared from minced croaker. It was found to be a significant factor ($\alpha = 0.01$) affecting both breaking force and the panel evaluation of firmness. As the NaAlg level increased, the breaking force or firmness of the patties decreased. The level of TPP also had a significant effect ($\alpha = 0.05$) on flavour. Generally, the flavour score increased as the NaCl level increased; however, above a certain point the flavour score began to decrease.

There was a high positive correlation ($r = 0.939$) between the breaking force and the panellists' evaluation of texture (firmness). This significant correlation ($\alpha = 0.01$) will enable one to use the breaking force as a tool to predict the sensory response to firmness. When analysing the texture preference data in the experimental range, it was determined that the greater preference scores were found to have a breaking force between

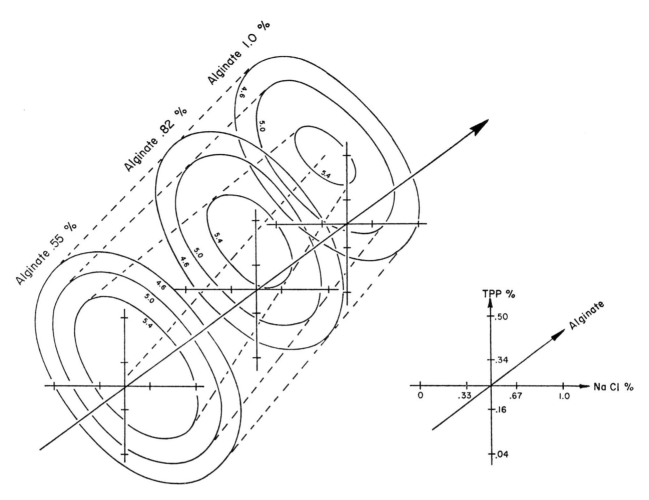

Fig 4   Response surface contours for TPP and NaCl effect on flavour
sensory response of minced croaker at various alginate levels

Fig 5   Response surface contours for TPP and NaCl effect on texture
response-firmness of minced croaker at various alginate levels
(a = 0·28%,   b = 0·415%,   c = 0·55%,   d = 0·685%   and
e = 0·82%)

Alginate = 0.415 %

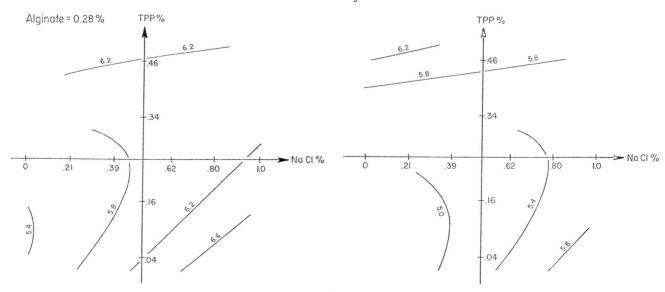

Alginate = 0.55%

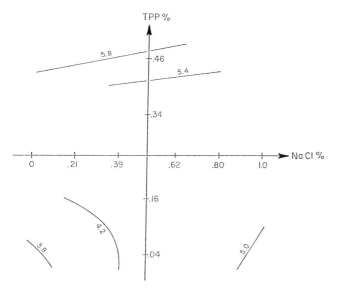

Alginate = 0.82%

TPP %

5.4
.46
5.0
.34

0   .21   .39   .62   .80   1.0   Na Cl %

.16
3.4
.04

Alginate = 0.685 %

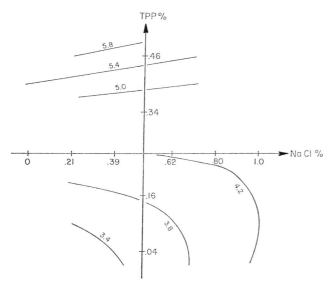

1 000g and 1 200g. These higher scores were found at low NaAlg levels (≤0·415%). Thus, since the breaking force and texture-firmness scores were correlated, the breaking force parameter may be used to determine the texture preference.

5 References

BARR, A J, GOODNIGHT, J H, SALL, J P and HELWIG, J T. *A User's Guide to SAS 76*. Institute Inc, Raleigh, NC
1976

CARVER, J H and KING, F J. Fish scrap offers high quality protein. *Food Eng.*, 43, 75–76
1971

CHAO, L and DENG, J C. *Effect of washing on the quality of minced fish flesh*. 76th Meeting of Southern Association of Agricultural Scientists, New Orleans, La
1979

COCHRAN, W G and COX, G M. *Experimental Design*. John Wiley and Sons, New York, 2nd edn
1957

DALEY, L H, DENG, J C and CORNELL, J A. Development of a sausage-type product from minced mullet using response surface methodology. *J. Fd. Sci.*, 43, 1501
1978

DREWS, J. Development of fish bonding machines. *The production and utilization of mechanically recovered fish flesh (minced fish)*, Ed J N Keay. Proc: Min. Ag., Fish. and Food. Torry Research Station, Aberdeen
1976

GARCIA-BORRAS, F. Calibrated rotational viscometer for non-Newtonian fluids. *Chem. Eng. Jan.*, 18, 176
1965

GILMAN, L and ROSE, H J. *APL/360, An Interative Approach*. John Wiley and Sons Inc., New York
1970

HERBORG, L. Production of separated fish mince for traditional and new products in Denmark. *The production and utilization of mechanically recovered fish flesh (minced fish)*, Ed J N Keay. Proc: Min. of Ag., Fish. and Food. Torry Research Station, Aberdeen
1976

LEE, C M and TOLEDO, R T. Factors affecting textural characteristics of cooked comminuted fish muscle. *J. Fd. Sci.*, 41, 391
1976

MARLIN, R E. Mechanically deboned fish flesh. *Fd. Tech.*, 30, 64
1976

MENDENHALL, W. *Introduction to Probability and Statistics*. Dunbury Press, North Scituate, Mass. 4th edn
1975

RAVICHANDER, N and KEAY, J N. The production and properties of minced fish from several commercially important species. *The production and utilization of mechanically recovered fish flesh (minced fish)*. Ed J N Keay. Proc: Min. of Ag., Fish. and Food. Torry Research Station, Aberdeen.
1976

TANIKAWA, E. Fish sausage and ham industry in Japan. In: *Advances in Food Research*. Eds C O Chichester, E M Mrak and G F Stewart. Academic Press, New York. Vol 12
1963

Spice minced fish from tilapia

Asiah M Zain

1 Introduction

In Malaysia, marine fish is the most important source of animal protein. However, with the increasing price of marine fish, coupled with the problem of depletion due to overfishing, pollution and seasonal fluctuation, fresh-water fish is the next best alternative protein food supply.

According to Ong (1977) freshwater fish culture in Peninsular Malaysia has expanded from 146ha in 1957 to 1 500ha in 1967 and to 50 000ha in 1977. At present freshwater fish contribute about 10% of the overall fish production in the country. This figure might double in the near future, based on the present rate of expansion and development of aquaculture.

In Malaysia, several species of freshwater fish are widely cultivated and are popular among the local consumers. These species fetch a good price on the local market and some are exported to neighbouring countries.

Tilapia (*Tilapia mossambica*), one of the freshwater species, is abundantly available but it has no commercial value as a fresh fish. Preliminary work done in the Department of Food Science and Technology, at University of Agriculture, Malaysia, has shown that several high protein foods such as salted-dried tilapia, tilapia canned in tomato sauce, curry sauce or chilli sauce, can be successfully prepared and well accepted by the taste panellists. Production and utilization of tilapia can be easily increased if a variety of products made from tilapia is organoleptically acceptable to local consumers.

This paper deals with trials on the production of spice minced fish from tilapia. The purpose is to utilize Malaysian tilapia, a low-cost freshwater fish, through the development of acceptable high-protein food. This work is an attempt to produce high-protein low-cost food from under utilized fish available locally.

Tilapia mossambica has been selected for the preparation of spice minced fish for several reasons as follows:

(*1*) Tilapia grows and breeds well in local conditions.

(*2*) It is abundantly available in many lakes and ponds.

(*3*) It is found to be mainly herbivorous hence requiring little management.

(*4*) It has no commercial value as a fresh fish.

At present tilapia is considered as a poor man's food or a low class food. A number of factors contribute to the present low acceptability of this fish which include price, appearance of the fish and attitudes of the consumers. Tilapia is cheap compared to other freshwater fish hence does not get into the daily diet of the higher-income group. In most developing countries anything that is cheap is frequently considered as belonging to a low class, a social attitude which is difficult to change. This has happened before to a number of local products such as 'belachan' (shrimp paste), 'ikan bilis' (dried anchovies) and 'krupuk' (fish or prawn crackers or crisps) to name a few. Interestingly enough, these products are now being recognized by food scientists and general consumers as high-protein food. They are now being served in most reputable restaurants and consumers belonging to a higher social strata have included such items in their menu. One of the reasons for the recognition and acceptance of these products is attributed to the increasing price of these items which suggest that they are not a low-class food (Zain, 1978).

2 Proximate composition of tilapia

The proximate analysis (*Table I*) shows that tilapia is low in fat. This is important in product development because the problem of rancidity can be reduced. The results for proximate analysis confirm the work done by Wood-Tsuen (1972).

Table I
PROXIMATE COMPOSITION OF *Tilapia mossambica*

Composition	% Zain (1979)	Wood-Tseun (1972)
Moisture	78·20	77·40
Ash	1·04	1·00
Oil	2·20	2·80
Protein	18·60	18·80

Table II gives the proximate composition of three different sizes and two different portions of tilapia (head and tail).

Table II
PROXIMATE COMPOSITION OF DIFFERENT SIZES AND PORTIONS OF *T. mossambica*

Composition	Portion	Group Size I	II	III
% Moisture	Head	81·03	80·10	79·10
	Tail	79·74	79·03	77·22
% Ash	Head	1·31	0·90	0·85
	Tail	1·34	0·98	0·91
% Oil	Head	1·36	2·10	2·82
	Tail	1·59	2·31	2·97
% Protein	Head	15·98	17·27	18·63
	Tail	16·80	18·03	19·39
Average length (cm)		16·80	20·07	28·00
Average width (cm)		7·20	8·50	10·20
Average fresh weight (gm)		102·35	267·65	350·25

Table III shows the amino acids composition of fresh fillets of tilapia.

Table III
AMINO ACIDS COMPOSITION OF TILAPIA

Amino acids	mg/gm of protein
Aspartic acid	115·66
Threonine	38·80
Serine	36·41
Glutamic acid	162·33
Proline	41·09
Glycine	72·33
Alanine	67·98
Cystine	5·41
Valine	48·19
Methionine	32·54
Isoleucine	43·82
Leucine	77·41
Tyrosine	33·13
Phenylalanine	43·52
Histidine	38·58
Lysine	126·46
Arginine	59·99
NH_3	60·63

The results show that variation occurs in the composition of different sizes and parts of tilapia. The oil content ranges from 1·36% in group one fish to 2·97% in group three fish. There is an inverse relationship between the moisture and oil content of the fish. The amino acid composition shows that the fish has a high nutritional value.

3 Spice minced fish

Spice minced fish or dried fish flakes is one of the

224

traditional food products of the Malaysian consumers. It is normally consumed together with rice, bread, pastries and porridge. Traditionally, the minced fish is prepared by subjecting heavily spiced or curried fish to prolonged heating until the moisture content is greatly reduced and the fish fibre disintegrates.

Preparation of this spice minced fish is carried out in the home, hence the amount and the types of ingredients used are left to the discretion of the individual household. The product is prepared from a variety of marine fish and the ingredients used include vegetable oil, salt, sugar, turmeric, coriander, cloves, cumin, fennel, garlic, ginger, pepper, onion, anise and many other natural seasonings and flavourings. The preparation of spiced minced fish is an example of a traditional method of food preservation.

In the Department of Food Science and Technology, University of Agriculture, Malaysia, tilapia has been introduced in the preparation of spice minced fish. The steps involved in the production of spice minced fish from this species is given in *Table IV*. The main features of this improved method are as follows:

Table IV
PRODUCTION OF SPICE MINCED FISH

Tilapia
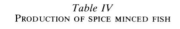
Nobbing
Cleaning
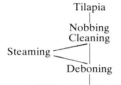
Steaming
Deboning

Addition of salt and spices

Mixing

Pressing

Drying

Grinding

Packaging

Storage

(*1*) The fish is not subjected to prolonged cooking.
(*2*) Addition of oil is omitted.
(*3*) Fewer ingredients are used.
(*4*) The process is simple, inexpensive and appropriate to the need of the developing countries.
(*5*) Desalting and rehydration of the final product is not necessary.

The description of each step involved in the improved method is:

Nobbing and cleaning: The fish is descaled, the head and the intestinal parts are removed; followed by washing.

Steaming: The fish is steamed for 3min which facilitates the separation of bones from flesh.

Deboning: The flesh is separated from the bone by slicing and scraping with a knife.

Addition of salt and spices: 5% salt and 10% other ingredients are added to the fish flesh. The ingredients used are shown in *Table V*.

Mixing: The fish flesh, salt and spices are well mixed and set aside for 30min.

Pressing: The mixture is pressed to remove the liquid which contains dissolved salt and spices.

Drying: The pressed cake is sundried.

Grinding: Grinding of the dried flakes is necessary to produce fine flakes.

Packing: The final product is packed in plastic film.

Table V
INGREDIENTS USED IN THE PREPARATION OF SPICE MINCED TILAPIA

Ingredients	%
Red chilli (pepper)	3·0
White pepper	1·0
Tamarind	1·0
Sugar	3·0
Coriander	1·0
Curry powder	1·0
Salt	5·0

The results of the analysis of protein, moisture and salt content are shown on *Table VI*.

Table VI
PROTEIN, MOISTURE AND SALT CONTENTS OF SPICE MINCED TILAPIA

Analysis	After pressing %	After drying %
Protein	21·8	53·7
Moisture	55·3	10·8
Salt	4·7	10·7

When 5% salt is added to the fresh fish flesh the salt content of flesh after pressing and drying is found to be 4·7% and 10·7%, respectively. Increasing the amount of salt added to the flesh of tilapia will render the product unacceptable due to salty taste. Poulter and Disney (1978) found that in uncooked minced fish, with salt content of 15% or above, the proteins start to denature and lose their water-holding capacity, hence the minced fish dry rapidly. With partially cooked fish, the protein is already denatured and therefore less salt is needed to obtain the optimum textural characteristics. The moisture contents of the spiced minced fish after pressing and drying were found to be 55·3% and 10·8%, respectively. At the latter moisture content the product is stable up to five months' storage at room temperature.

4 Acceptability of spice minced tilapia

The finished product has a good spice flavour which is not alien to the Malaysian taste. However, the product is found to be too spicy for the western taste. Lightly spiced minced fish can be prepared by reducing the amount of spices used. Extra-hot flavour can be obtained by increasing the amount of chilli powder. The final colour of the product depends on the amount of spices used. This product has no strong fishy odour, which is an important feature for products in some developing countries. The spice minced fish can be kept for a few months before any noticeable rancidity flavour develops, since it is partly preserved by the spices used. A mild degree of rancid flavour is masked by the spice flavour.

According to Bligh (1977) salt minced fish produced at the Halifax Laboratory, Canada, must be soaked in water to reduce the salt content prior to consumption. In the case of spice minced fish, desalting and rehydration are not necessary.

Spice minced fish is consumed directly by mixing with rice, porridge, sweet potatoes, cassava or as topping and spread for local dishes. It can also be used as ingredients in the preparation of fish sausages, fish ball, fish crackers or as condiment. Steinberg et al (1976) had successfully used minced fish as partial replacement for lean beef in sausages. Obileye and Spinelli (1978) managed to produce stable and palatable smoked tilapia from minced tilapia flesh.

5 Conclusion

The technology used in the preparation of spice minced fish shows that tilapia, an under utilized freshwater fish, can be turned into high-protein food at low cost. The method outlined in this paper is most appropriate to the needs of the developing countries. Acceptance of this product is less likely to face problems since fish and spices are common in our daily diet.

6 References

BLIGH, G. A note on salt minced fish. *Proc:* Conf. Handling, process-
1977   ing and marketing of tropical fish, TPI, London. 291
OBILEYE, T and SPINELLI, J. A smoked minced tilapia product with
1978   enhanced keeping qualities. Symp. Fish utilization technology
        and marketing in the IPFC region, Manila, Philippines
ONG, K S. The role of aquaculture in food production. *Proc:* Confer-
1977   ence on food and agriculture, Malaysia 2000. 8
POULTER and DISNEY. Preparation of protein concentrates from waste
1978   fish. Symp. Fish utilization technology and marketing in the
        IPFC region, Manila, Philippines
STEINBERG, M A, SPINELLI, J and MIYAUCHI, D. Minced fish as an
1976   ingredient in food combinations. *Proc:* Conf. Handling,
        processing and marketing of tropical fish, TPI, London.
        245–248
WOOD-TSUEN, W L. *Food composition table for use in East Asia.* FAO.
1972   113–143
ZAIN, A. Acceptability of Malaysian fishery products. Symp. Fish
1978   utilization technology and marketing in the IPFC region,
        Manila, Philippines

Stabilization of fish mince from gadoid species by pre-treatment of the fish

S Svensson

1 Introduction

Trimethylamine oxide (TMAO) is a natural component in muscles and many organs of marine fish and invertebrates but is completely lacking or is present in very small amounts in freshwater fish (Dyer, 1952; Harada and Yamada, 1974). Since many micro-organisms reduce TMAO to the 'fishy' smelling component trimethylamine (TMA), the TMA content of chilled fish is often used as an indicator of staleness (Yamada, 1968; Laycock and Regier, 1971).

Species belonging to the commercially important Gadidae family have been shown to contain an enzyme system which can catalyse the degradation of TMAO to formaldehyde (FA) and dimethylamine (DMA) (Amano and Yamada, 1964a; Yamada et al, 1969; Tokunaga, 1970; Dyer and Hiltz, 1974; Mackie and Thomson, 1974). Such enzymes have been found in dark muscle as well as in some organs of the viscera, eg pyloric caeca and the kidney and the liver tissues (Amano and Yamada, 1964b; Castell, 1971; Castell et al, 1971; Tokunaga, 1974; Dingle and Hines, 1975).

During frozen storage of fish muscle, the functional properties of the proteins are changed due to the action of eg FA produced, or of liberated fatty acids and their oxidation products (Sörensen, 1976; Hiltz et al, 1976; Sikorski et al, 1976; Dingle et al, 1977; Sikorski, 1978; Gill et al, 1979; Crawford et al, 1979). The presence of an enzyme system capable of degrading TMAO to FA and DMA in the muscle can thus cause undesirable quality changes in frozen stored fish products. This is especially pronounced for deboner-separated mince which is often contaminated with small amounts of kidney tissue rapidly making the mince unsuitable for further processing.

This paper covers work on the thermostability of the TMAO-degrading enzyme from cod and blue whiting kidney and on the storage stability of minces produced from fish pretreated in different ways.

2 Experimental

2.1 Raw materials
The experiments were performed with beheaded and gutted cod (*Gadus morhua*) (size 40–60cm) and with gutted blue whiting (*Micromesistius poutassou*), the latter caught in April, west of the Hebrides. Individual fish were packed in aluminium foil and plastic bags and stored at −40°C for up to six months before use.

2.2 Fish minces
Two main types of fish minces were prepared: fillet mince and deboner-separated mince.

Fillets of cod and blue whiting were cut from the frozen fish and minces were prepared by grinding the thawed fillets in a meat mincer through 3mm holes. The blue whiting mince was divided into three lots. Kidney tissue was added to two of these as listed in *Table I*.

Three types of deboner-separated blue whiting were prepared using a Baader 694 flesh-bone separator equipped with a drum having 5mm perforations. Before passage through the separator the fish were thawed in cold water, beheaded and washed. The kidney tissues of some of them were removed with a special tool, or they were heat treated (temperature > 80°C) with a specially designed steam jet in order to inactivate the TMAO-degrading enzyme (*Table I*).

Samples, 100g, of the different minces were packed in aluminium boxes, put into plastic bags, frozen at −35°C in a blast freezer overnight and then stored at −20°C until analysed.

2.3 Fish ball model product
The usefulness of the fish mince for further processing

226

Table I
MINCES OF BLUE WHITING

Mince	Dry substance % fresh weight	Protein (6·25 × N) %	Lipid %	Ash %
F Fillet mince	20·0	85	9	6
FK Fillet mince containing 5g kidney tissue per kg mince	19·8	85	10	5
FHK Fillet mince containing 5g heat-treated (80°C, 90s) kidney tissue per kg mince	19·8	85	9	6
D Deboner-separated mince of fish from which the kidney tissue had been carefully removed	18·1	86	7	7
DK Deboner-separated mince of fish with kidney tissue	18·3	87	7	6
DHK Deboner-separated mince of fish with heat-treated (>80°C) kidney tissue	17·8	85	7	8

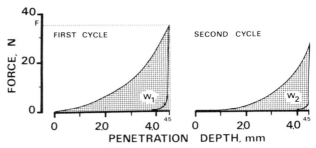

Fig 1 Typical Instron force/deformation curves for two consecutive compression cycles
F = maximal force; first cycle
W_1 = work of deformation; first cycle
W_2 = work of deformation; second cycle

was tested by measuring the texture properties of a fish ball model product as described by Andersson (1975) and Jarenbäck (1976). The formula for the model product is given in *Table II*.

Table II
FORMULA FOR THE FISH BALL MODEL PRODUCT

Ingredient	Amount %
Fish mince	56
Water + ice	32
Salt (NaCl)	2
Potato flour	7
Soya oil	3

The partly thawed ($-5°C$) fish mince was homogenized for 30s with salt and ice water in a specially designed homogenizer. After addition of potato flour and soya oil the mixture was emulsified for 1min at high speed, its temperature being kept below $+16°C$. The paste was then poured into aluminium boxes which were heated in 1% NaCl solution at 95°C for 12min. The gelled fish cakes were stored in the salt solution at 20°C overnight.

From each fish cake five discs (25mm diameter, 8mm thickness) were cut. These discs were tested in an Instron Universal Testing Machine, model 1122, by letting a punch with a diameter of 60mm compress the disc to a thickness of 3·5mm with a rate of 10mm/min. The punch was then raised and the compression cycle repeated. The maximal deformation force (F) in the first cycle, and the work of deformation (W_1 and W_2) in the two cycles were measured as is shown in *Fig 1* and the ratio W_1/W_2 was calculated.

2.4 Preparation of crude enzyme
The TMAO-degrading enzyme was extracted from the kidney tissue of cod and blue whiting using a method outlined by Yamada et al (1969). A mixture of 10g $(NH_4)_2SO_4$ and 100g homogenized kidney tissue was centrifuged at 20 000g for 20min; $(NH_4)_2SO_4$ (1·5g per 10ml) was added to the decanted supernatant in order to precipitate the enzyme. The suspension was centrifuged and supernatant and precipitate collected separately.

Another portion of $(NH_4)_2SO_4$ (0·5g per 10ml) was added to the supernatant and the mixture centrifuged. This solution was used as the 'cofactor-solution' in the enzyme assay. The precipitate with the enzyme was resuspended in 10 volumes of 25% $(NH_4)_2SO_4$ solution and spun down. The purified enzyme was dissolved in a 0·1M potassium phosphate buffer the day before use.

2.5 Enzyme assay
The enzyme activity was measured at 25°C by a procedure outlined by Harada and Yamada (1971) and Tomioka et al (1974).

The substrate consisted of 3·25ml 0·1M potassium phosphate, pH 6·1, 0·50ml 0·1M TMAO, 0·25ml 1mM methylene blue and 0·50ml of the 'cofactor solution' obtained in the enzyme preparation step. The reaction was started by adding 0·50ml of the enzyme solution and stopped after a settled time, usually 10min, by adding 5·0ml of a 20% trichloroacetic acid (TCA) solution. The DMA produced was determined by the method of Dyer and Mounsey (1945).

2.6 Chemical analysis of fish minces
The fish minces were analysed as summarized in *Table III*. For the determination of DMA, TMA and TMAO, 10g of fish mince was homogenized in 10ml distilled water, after which 10ml 10% TCA solution was added. The homogenization step was repeated and the suspension filtered.

Table III
CHEMICAL ANALYSIS OF FISH MINCE

Component	Method
Dry substance	Gravimetrically after drying at 105°C for 24h
Fat	Soxhlet technique
Protein	Kjeldahl technique
Ash	Gravimetrically after drying at 525°C to constant weight
DMA	According to Dyer and Mounsey (1945)
TMA	According to Tozawa et al (1970)
TMAO	As TMA using the technique outlined by Dyer et al (1952)

2.7 Heat treatment of crude enzyme and kidney tissue
Samples, 0·70ml, of the crude enzyme buffer solutions were heat treated in thin-walled, rotating-glass ampoules using the technique of Eriksson et al (1971).

Homogenized kidney tissue in thin-walled glass tubes (o.d. 7·5mm; i.d. 6·5mm) was heat treated by immersing the tubes in a thermostated water bath at 80°C for 90s.

The temperature in the centre of the kidney homogenate reached 78°C within 50s.

For convenience, enzyme activities obtained after heat treatment were expressed as percentage of the activity obtained after equivalent treatment at 25°C. The detection level after heat treatment was approximately 1% of the original enzyme activity.

2.8 Fractionation of cod muscle
Cod muscle was separated into three fractions:

(*i*) components insoluble in water
(*ii*) water-soluble components with molecular weight > 2 000 daltons, and
(*iii*) water-soluble components with molecular weight < 2 000 daltons

as shown in *Fig 2*.

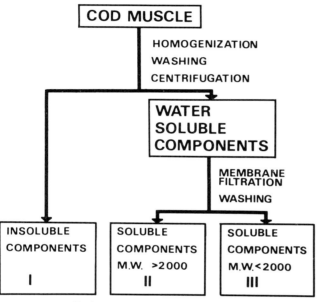

Fig 2   Fractionation of fish muscle

Comminuted cod muscle, 200g, was homogenized in 200ml distilled water during 3min and the mixture centrifuged at 40 000g for 20min. The supernatant was decanted and the precipitate resuspended in 200ml distilled water. After centrifugation and decanting the washing process was repeated twice. All supernatants were pooled and membrane filtered through an Amicon Diaflo UM2 membrane, cut-off 2 000. The concentrate (approximately 50ml) was 'washed' three times with equal portions of distilled water in order to 'wash out' low molecular weight components.

The three cod muscle fractions were freeze-dried.

3 Results

3.1 Thermostability of the TMAO-degrading enzymes
Solutions of the TMAO-degrading enzyme in 0·1M potassium phosphate, pH 7·1, were heat treated in the temperature range 60–75°C for 15–200s. The thermal inactivation of enzymes from both cod and blue whiting followed first-order kinetics, as is shown in *Figs 3* and *4*. From the slope of the lines obtained by the method of

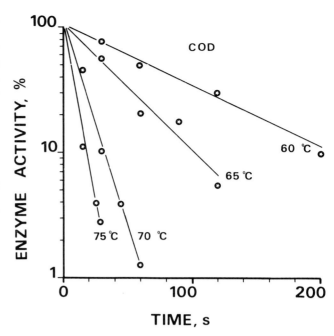

Fig 3   Thermal inactivation of TMAO-degrading enzyme from kidney tissue of cod as a function of temperature and time

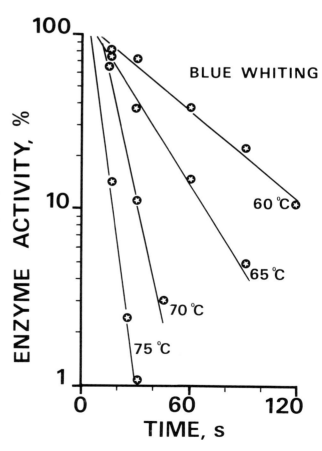

Fig 4   Thermal inactivation of TMAO-degrading enzyme from kidney tissue of blue whiting as a function of temperature and time

228

least squares, the D-values[1] for different heat treatment temperatures were determined. By plotting the logarithm of the D-values against heat treatment temperature (*Fig 5*) the z-values[2] were calculated as 13°C and 15°C for the inactivation of enzymes from cod and blue whiting, respectively.

As is shown, the TMAO-degrading enzymes in the kidney tissue of cod and blue whiting are relatively thermolabile enzymes which are rapidly inactivated by heat treatment at temperatures above 75°C; the $D_{75°C}$ values being 0·26min and 0·21min for cod and blue whiting, respectively. Thus there is no appreciable difference between the TMAO-degrading enzymes from the two fish species in their thermal inactivation behaviour.

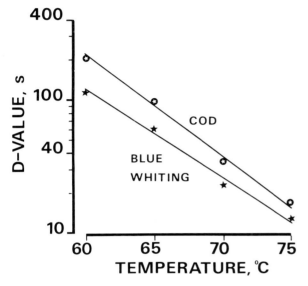

Fig 5 D-value as a function of heat treatment temperature for TMAO-degrading enzyme from kidney tissues of cod and blue whiting. Calculated z-values: cod 13°C; blue whiting 15°C

3.2 *Storage stability of frozen minces*

In order to determine the influence of heat treated respective untreated kidney tissue on the properties of blue whiting minces stored at −20°C, fillet minces with as well as without added kidney tissue, and deboner-separated mince from beheaded and gutted fish were prepared. The minces and their composition (dry substance, protein, lipid and ash) are listed in *Table I*. The amounts of TMAO, TMA and DMA in the frozen stored minces were analysed periodically during a period of 11 weeks. The texture properties of a fish ball model product prepared from frozen stored fish mince were also measured, using an Instron Universal Testing Machine.

The TMAO content was considerably higher in fillet mince than in deboner-separated mince; 96mg and 67mg TMAO-N per 100g, respectively. Probably some of the TMAO in the fish muscle had been extracted by the water during thawing and washing before mince production. Changes in TMAO content during frozen storage were noticed only for minces containing kidney

tissue (*Table IV*). The TMA level was very low and constant, 0·4mg–0·8mg TMA-N per 100g mince, in all minces during frozen storage.

Table IV
TRIMETHYLAMINE OXIDE (TMAO) CONTENT IN FROZEN STORED BLUE WHITING MINCES

Mince[1]	TMAO–N, mg/100g mince		
	Storage time, weeks		
	0	7	11
Fillet mince	96	96	98
Fillet mince + kidney tissue	96	92	86
Fillet mince + heat treated kidney tissue	97	90	88
Deboner-separated mince *without* kidney tissue	67	68	67
Deboner-separated mince *with* kidney tissue	65	68	62
Deboner-separated mince *with* heat treated kidney tissue	65	60	62

[1] For details see *Table I*

The DMA concentration in some of the minces changed drastically during frozen storage (*Fig 6*). The initial DMA level varied between 1·1 and 2·4mg DMA–N per 100g mince. As is shown in *Fig 6* the increase in DMA was negligible in minces completely free from kidney tissue (minces F and D in *Fig 6*), while the DMA production was high in fillet minces containing kidney tissue (minces FK and FHK). The increase in DMA content was independent of whether the kidney tissue had been heat treated or not. No difference was found between deboner-separated mince from fish with heat-treated kidney tissues and from untreated fish

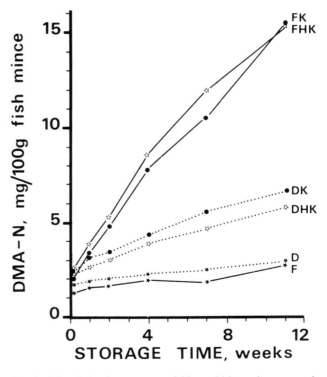

Fig 6 Dimethylamine content of blue whiting minces stored at −20°C. F, fillet mince; FK, fillet mince with added kidney tissue; FHK, fillet mince with added heat-treated kidney tissue. D, deboner-separated mince from fish without kidney tissue; DK deboner-separated mince from fish with kidney tissue; DHK, deboner-separated mince from fish with heat treated kidney tissue. (For further details see *Table 1*)

---

[1] D-value = heat treatment time required to reduce the enzyme activity to 10% of the original value at a certain temperature
[2] z-value = the temperature increment required to reduce the D-value to 10% of the original one

(minces DHK respective DK). For some reason, the heat treatment of the kidney tissue, which was aimed at a complete inactivation of the TMAO-degrading enzyme, did not reduce the DMA production in the frozen stored fish mince.

The functional properties of some of the minces, measured as texture quality of a fish ball model product, changed considerably during frozen storage. The addition of kidney tissue to fillet mince resulted in nearly the same change in force of deformation and in W_1/W_2 during frozen storage, whether the kidney tissue had been heat treated or not (minces FK and FHK in *Fig 7*). The two deboner-separated minces containing kidney tissue (minces DK and DHK) also showed similar texture properties whether the kidney part had been heat treated or not. During frozen storage, the minces with kidney tissue soon lost their binding properties so much that an acceptable model product could not be made.

Minces without kidney tissue (minces D and F), however, could be used for model products, with excellent properties, even after 11 weeks at $-20°C$.

Repeated experiments with minces made from cod instead of blue whiting confirmed the results, indicating that the thermally inactivated TMAO-degrading enzyme is reactivated when mixed with muscle tissue.

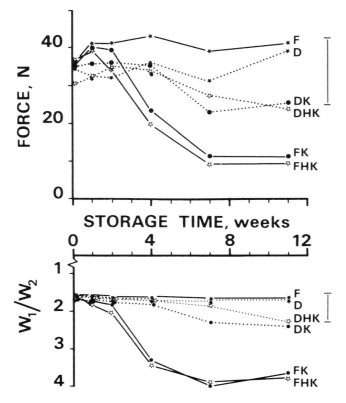

Fig 7 Texture properties of a fish ball model product produced from frozen stored ($-20°C$) mince. Force of deformation and ratio between work of deformation for two consecutive compression cycles. F, fillet mince; FK fillet mince with added kidney tissue; FHK, fillet mince with added heat treated kidney tissue; D, deboner-separated mince from fish without kidney tissue; DK, deboner-separated mince from fish with kidney tissue; DHK, deboner-separated mince from fish with heat treated kidney tissue. Fish minces producing an acceptable commercial fish ball product show a force of deformation and a W_1/W_2 ratio within the marked ranges(———). (For further details see *Table I*).

3.3 *Influence of fish muscle components on the reactivation of TMAO-degrading enzymes*

To elucidate the influence of fish muscle on the reactivation of thermally inactivated TMAO-degrading enzyme, different amounts of minced cod muscle were added to heat-treated (80°C, 90s) homogenized kidney tissue from cod. The mixtures were then frozen and stored for up to 57 days at $-20°C$. The activity of the TMAO-degrading enzyme was determined regularly. As is shown in *Fig 8*, neither the heat-treated kidney tissue nor the fish muscle itself showed any TMAO degrading activity. However, mixtures of heat-treated kidney-tissue and fish muscle with a muscle/kidney ratio of approximately 2, could catalyse the TMAO degradation. It was also noticed that the addition of minced fish muscle to untreated kidney tissue activated the TMAO-degrading enzyme system. It thus seems clear that the fish muscle contains some substance(s) which can activate the TMAO-degrading enzyme and in addition reactivate thermally inactivated enzyme molecules.

The nature of the reactivating substance was further investigated by separating the cod muscle into three fractions; (*i*) components insoluble in water, (*ii*) water-soluble components with molecular weight $> 2\,000$ daltons, and (*iii*) water-soluble components with molecular weight $< 2\,000$ daltons (*Fig 2*). After freeze-drying the fractions were added to samples of heat-treated kidney tissue and the mixtures stored for one week at $-20°C$ and then tested for TMAO-degrading activity. The presence of fraction (*i*) or (*ii*) in heat-treated kidney tissue had no effect on the inactivated enzyme while fraction (*iii*) could restore approximately 80% of the original TMAO-degrading activity, when $0.2g$ (corresponding to 10g cod muscle) was added to 1g of heat-treated kidney tissue.

It has thus been shown that cod muscle contains a water-soluble, low molecular weight component capable of reactivating the thermally inactivated TMAO-degrading enzyme from the kidney tissue. In contrast to the 'cofactor' (Yamada *et al*, 1969), the component can be 'inactivated' by heat treatment at temperatures above 80°C, as indicated by preliminary experiments.

4 Discussion

Deboner-separated fish mince from beheaded and gutted gadoid fish, as well as that from frames after filleting, will inevitably be contaminated by small amounts of kidney tissue. Usually such minces have a considerably lower storage stability than fillet mince, due to the action of a TMAO-degrading kidney enzyme on TMAO which occurs naturally in fish muscle. The storage stability of deboner-separated fish mince could be increased if the kidney tissue were eliminated or if the TMAO-degrading enzyme thermally inactivated before the fish pass through the deboner machine.

The TMAO-degrading enzyme from the kidney tissue of cod and blue whiting is a relatively thermolabile enzyme, rapidly inactivated when heat treated in a buffer solution, pH 7.1, at temperatures above 75°C (*Figs 3–5*). No measurable enzyme activity could be detected in homogenized kidney tissue heat treated at 80°C for 90s. Yamada and Amano (1965) observed that the TMAO-degrading enzyme from pyloric caeca of

Alaska pollack (*Theragra chalcogramma*) was completely inactivated when heat treated at 100°C for less than 5min, while Tomioka *et al* (1974) reported a rapid enzyme inactivation when the same enzyme in buffer solution, pH 5·0, was heat treated above 60°C. Thus, according to the thermal stability data a selective heat treatment (> 80°C) of the kidney part of the fish before their passage through the deboner should result in fish minces contaminated with an inactive enzyme and having an increased storage stability.

Frozen storage of deboner-separated minces of blue whiting, however, resulted in identical quality changes (DMA production and loss of protein functionality) whether the mince was obtained from fish with heat-treated (80°C) or with untreated kidney tissues (*Figs 6* and 7). The same results were obtained for fillet minces to which heat-treated (80°C, 90s) respective untreated kidney tissue had been added. Thus, in spite of the complete enzyme inactivation before mixing with the minced fish muscle, the DMA production and the changes in protein functionality were identical for minces containing heat-treated and untreated kidney tissue. These results indicate that the inactivated TMAO-degrading enzyme will regain its activity when frozen stored together with minced fish muscle.

Castell *et al* (1971) had previously found that the DMA production in frozen stored (−5°C) hake (*Merluccius bilinearis*) fillets was reduced but not eliminated by heat treating the fillets before freezing. The TMAO-degrading enzyme in the dark muscle was considered to be very thermostable. Neither was the DMA production completely inhibited when minces of hake were heat treated at 80°C (Lall *et al*, 1975). In addition, heat treatment in itself causes TMAO degradation (Tokunaga, 1975).

The addition of minced cod muscle to heat-treated kidney tissue resulted in an extensive reactivation of the TMAO-degrading enzyme when the fish muscle/kidney ratio was approximately two (*Fig 8*). The presence of minced fish muscle together with untreated kidney tissue also increased the activity of the TMAO-degrading enzyme, thus showing that the fish muscle normally contains a substance which can interfere with the enzyme. By separating the fish muscle into fractions according to their water solubility and molecular size properties, it could be shown that the substance activating the enzyme is water soluble and has a molecular weight less than 2 000. Work on chemical identification of the substance is now in progress.

As is shown in this investigation, the storage stability of deboner-separated minces from beheaded and gutted cod and blue whiting can be considerably increased if the kidney tissue is eliminated before the fish pass through the deboner. On the other hand, selective heat treatment of the kidney tissue in order to inactivate the TMAO-degrading enzyme had no stabilizing effect since the thermally inactivated enzyme was reactivated when mixed with minced fish muscle.

5 Acknowledgements
The skilled technical assistance of Mrs I Gangby, Mrs M Jarenbäck, Miss E Olsson and Mrs G Schallin and the helpful discussions with Dr L Jarenbäck and Mr A Liljemark are gratefully acknowledged.

6 References
AMANO, K and YAMADA, K. A biological formation of formaldehyde in 1964a the muscle tissue of gadoid fish. *Bull. Jap. Soc. Fish.*, 30, 430–435

—— Formaldehyde formation from trimethylamine oxide by the action 1964b of pyloric caeca of cod. *Bull. Jap. Soc. Sci. Fish.*, 30, 639–645.

ANDERSSON, Y. Instrumental measurements of texture/consistency of 1975 patties containing textured soy protein. *SIK – −The Swedish Food Institute Rapport* No. 376, 37pp

CASTELL, C H. Metal-catalyzed lipid oxidation and changes of proteins 1971 in fish. *J. Amer. Oil Chem. Soc.*, 48, 645–649

CASTELL, c h, SMITH, B and NEAL, W. Production of dimethylamine in 1971 muscle of several species of gadoid fish during frozen storage, especially in relation to presence of dark muscle. *J. Fish. Res. Bd. Can.*, 28, 1–5

CRAWFORD, D L, LAW, D K, BABBIT, J K and McGILL, L A. Com- 1979 parative stability and desirability of frozen Pacific hake fillet and minced flesh blocks. *J. Fd. Sci.*, 44, 363–367

DINGLE, J R and HINES, J A. Protein instability in minced flesh from 1975 fillets and frames of several commercial Atlantic fishes during storate at −5°C. *J. Fish. Res. Bd. Can.*, 32, 775–783

DINGLE, J R, KEITH, R A and LALL, B. Protein instability in frozen 1977 storage induced in minced muscle of flatfishes by mixture with muscle of red hake. *J. Inst. Can. Sci. Technol. Aliment.*, 10, 143–146

DYER, W J. Amines in fish muscle. VI. Trimethylamine oxide content 1952 of fish and marine invertebrates. *J. Fish. Res. Bd. Can.*, 8, 314–324

DYER, W J and MOUNSEY, Y A. Amines in fish muscle. II. Develop- 1945 ment of trimethylamine and other amines. *J. Fish. Res. Bd. Can.*, 6, 359–367

DYER, J W and HILTZ, D F. Sensitivity of hake muscle to frozen stor- 1974 age. Halifax Lab., *Fish. Mar. Serv. New Ser. Circ.*, 45, 4pp

DYER, W J, DYER, F E and SNOW, J M. Amines in fish muscle. V. 1952 Trimethylamine oxide estimation. *J. Fish. Res. Bd. Can.*, 8, 309–313

ERIKSSON C E, OLSSON, P A and SVENSSON, S G. Denatured hemop- 1971 roteins as catalysts in lipid oxidation. *J. Amer. Oil Chem. Soc.*, 48, 442–447

GILL, T A, KEITH, R A and LALL, B. Textural deterioration of red 1979 hake and haddock muscle in frozen storage as related to chemical parameters and changes in the myofibrillar proteins. *J. Fd. Sci.*, 44, 661–667

HARADA, K and YAMADA, K. Some properties of a formaldehyde and 1971 dimethylamine-forming enzyme obtained from *Barbatia virescens. J. Shimonoseki Univ. Fish.*, 19, 95–103

—— Distribution of trimethylamine oxide in fishes and other aquatic 1974 animals. V. Teleosts and elasmobranchs. *J. Shimonoseki Univ. Fish.*, 22, 77–94

HILTZ, D F, LALL, B. LEMON, D W and DYER, W J. Deteriorative 1976 changes during frozen storage in fillets and minced flesh of silver hake (*Merluccius bilinearis*) processed from round fish held in ice and refrigerated sea water. *J. Fish. Res. Bd. Can.*, 33, 2560–2567

JARENBÄCK, L. Frozen storage of fish mince. I. Quality changes in 1976 minced flesh from different parts of cod during frozen storage. *SIK−The Swedish Food Institute Rapport* No. 409, 41pp

LALL, B. MANZER, A R and HILTZ, D F. Preheat treatment for 1975 improvement of frozen storage stability at −10°C in fillets and minced flesh of silver hake (*Merluccius bilinearis*). *J. Fish. Res. Bd. Can.*, 32, 1450–1454

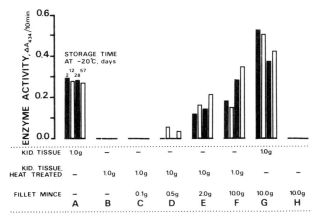

Fig 8 Influence of added minced fish muscle on reactivation of the TMAO-degrading enzyme during frozen storage (−20°C) of heat-treated (80°C, 90s) kidney tissue

LAYCOCK, R A and REIGER, L W. Trimethylamine-producing bacteria
1971 on haddock (*Melanogrammus aeglefinus*) fillets during refrigerated storage. *J. Fish. Res. Bd. Can.*, 28, 305–309
MACKIE, I M and THOMSON, B W. Decomposition of trimethylamine
1974 oxide during iced and frozen-storage of whole and comminuted tissue of fish. *Proc*: IV Int. Ccongress Food Sci. and Technol. Vol. I, 243–250
SIKORSKI, Z E. Protein changes in muscle foods due to freezing and
1978 frozen storage. *Int. J. Refrig.*, 1, 173–180
SIKORSKI, Z E, OLLEY, J and KOSTUCH, S. Protein changes in frozen
1976 fish. Crit. Rev. Fd. Sci. Nutr., 8(1), 97–129
SÖRENSEN, T. Effect of frozen storage on the functional properties of
1976 separated fish mince. In: *The production and utilization of mechanically recovered fish flesh (minced fish)*, Ed J N Keay, Torry Research Station, 108pp
TOKUNAGA, T. Trimethylamine oxide and its decomposition in the
1970 bloody muscle of fish. I. TMAO, TMA and DMA contents in ordinary and bloody muscles. *Bull. Jap. Soc. Sci. Fish.*, 36, 502–509
—— The effect of decomposed products of trimethylamine oxide on
1974 quality of frozen Alaska pollack fillet. *Bull. Jap. Soc. Sci. Fish.*,

40, 167–174
—— On the thermal decomposition of trimethylamine oxide in muscle
1975 of some marine animals. *Bull. Jap. Soc. Sci. Fish.*, 41, 535–546
TOMIOKA, K, ŌGUSHI, J and ENDO, K. Studies on dimethylamine in
1974 foods. II. Enzymic formation of dimethylamine from trimethylamine oxide. *Bull. Jap. Soc. Sci. Fish.*, 40, 1021–1026
TOZAWA, H, ENOKIBARA, K and AMANO, K. Effect of dimethylamine
1970 on the value of trimethylamine determined by the Dyer's method. *Bull. Jap. Soc. Sci. Fish.*, 36, 606–611
YAMADA, K. Post-mortem breakdown of trimethylamine oxide in
1968 fishes and marine invertebrates. *Bull. Jap. Soc. Sci. Fish.*, 34, 541–551
YAMADA, K and AMANO, K. Studies on the biological formation of
1965 formaldehyde and dimethylamine in fish and shellfish. V. On the enzymatic formation in the pyloric caeca of Alaska pollack. *Bull. Jap. Soc. Sci. Fish.*, 31, 60–64
YAMADA, K, HARADA, K and AMANO, K. Biological formation of for-
1969 maldehyde and dimethylamine in fish and shellfish. VIII. Requirement of cofactor in the enzyme system. *Bull. Jap. Soc. Sci. Fish.*, 35, 227–321

Studies on a minced squid product. Effect of raw material and ingredients on the texture of the product

Bonnie Sun Pan, D J Lee and L P Lin

1 Introduction

The harvest of squid in Taiwan was 36 335 tons in 1976, and is ever increasing. It is only reasonable to make use of such abundant resource for domestic consumption and for export. However, squid is considered a 'devil fish' by some western consumers because of its appearance. A change of image would be necessary in order to increase its acceptability. A minced squid product, similar to the Jewish gefilte fish or the Japanese kamaboko, would be a potential way of utilizing it.

As with all other minced products, rheological property is of major concern for the minced squid product. Since actomyosin is the basic structural protein of minced fish products, the unique biochemical properties of the squid protein (Matsumoto, 1959; Tsuchiya *et al*, 1978) could give rise to particular textural properties.

It is the objective of this paper to study the effects of the raw material and ingredients on the texture of a minced squid product.

2 Material and method

2.1 Material
Squid of 12cm mantle length of the Sepidae family, mainly *Sepia exculenta*, was obtained from the fish market in Keelung.

2.2 Preparation of minced squid product
The skinned squid mantle was ground in a meat grinder (holes 1·6mm in diameter).

The ground squid was mixed without salt for 10min in an electrical mortar at 67rpm, then with salt for 10min and finally with cassava starch for 15min.

The squid mince was filled into Saran casing (3·8cm diameter) and allowed to set at room temperature for various lengths of time and then heated in water at 90–95°C for 30min.

2.3 Texture evaluation
2.3.1 Folding test
The textural quality of minced squid product was graded according to the flexibility and elasticity of the product sliced in 3mm thickness after folding into halves and quarters as follows:

Grade AA: no cracks on folding into quarters
    A: no cracks on folding in half
    B: cracks on folding in half
    C: breaks into pieces on folding in half
    D: breaks into fragments with finger pressure.

2.3.2 Instrumental measurement
A Yamamoto Food Checker was used to determine the breaking force in grams, and deformation in millimetres. The gel strength in g-mm was calculated by multiplication of the two parameters.

2.4 Scanning electron microscopy (SEM)
Specimens of minced squid product were fixed with 2·5% glutaldehyde at 5°C for 2h and dehydrated rapidly in a graded series of ethanol solutions followed by critical point drying with CO_2. The specimens were then coated using an ion coater IB–2 and examined with an Hitachi S–550 SEM.

2.5 Water retention
The chopped sample was wrapped in Whatman No. 2 filter paper and centrifuged at 3 000 × g for 20min. The moisture content was determined as g/g solid.

2.6 Ammonia content
An Ionalyzer 407A (Orion Research) with ammonia electrode was used.

2.7 Actomyosin extractability
Squid actomyosin was extracted with 0·6M KCl, precipitated at 0·2M KCl and the nitrogen content of the precipitate determined by the Kjeldahl Method and expressed as a percentage of total nitrogen (Matsumoto, 1958).

3 Results and discussion

3.1 *Effect of moisture*

The content of extractable myofibrillar protein and the formation of a network structure affect product texture. Since moisture affects the concentration and hydration of myofibrillar protein, it also has an effect on product texture. *Figure 1* shows that gel strength measured as g breaking force was an inverse function of moisture content in the range of 75% to 82%. Every increase of 1% moisture caused a reduction of 31g in breaking force. Therefore, it is possible to predict and adjust the texture of the end product by manipulating the moisture content of the mince. When a more elastic product is desired, a dewatering step is suggested.

3.2 *Effect of freshness of squid*

Squid mince of different degrees of freshness was obtained by storing the mince at 5°C for various lengths of time. Actomyosin extractability, pH and NH_3 were chosen as indices of freshness. Changes in pH were slow, from 6·3 to 6·6, within the first three days as shown in *Fig 2*. On the fourth day, the pH of squid mince increased to 7·6 and then became constant. Changes in NH_3 content were similar, increasing dramatically on the fourth day from 34·5 to 196·0mg%. Actomyosin extractability changed only slightly throughout the storage period.

Breaking force was maximal when squid mince was stored for two days. Its pH was 6·4 and NH_3 content

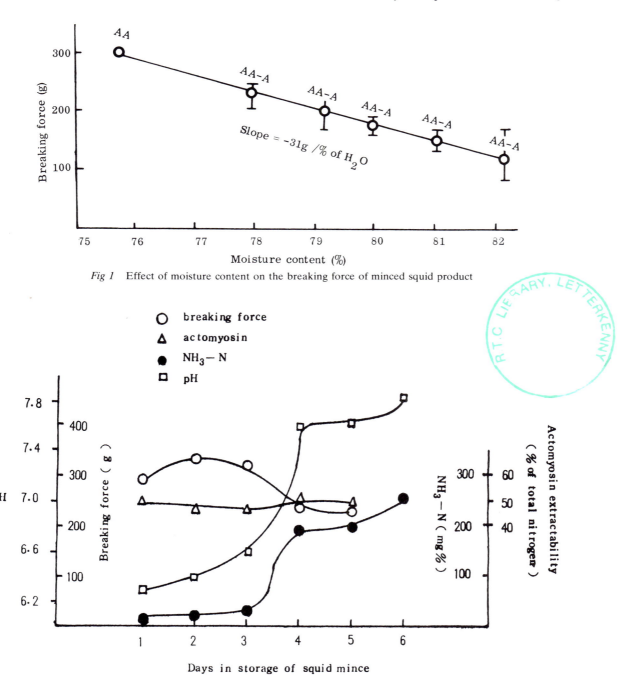

Fig 1　Effect of moisture content on the breaking force of minced squid product

Fig 2　Change in freshness of squid mince stored at 5°C as indicated by changes in pH, NH_3, actomyosin extractability on breaking force of its product

233

21·76mg%. When the squid mince deteriorated on the fourth day the breaking force of the product was much less. Therefore, if a more elastic product is desired, very fresh raw material is desirable.

Because the pronounced increases in pH and NH_3 coincided with a fall in breaking force, the individual effects of pH and NH_3 on the texture of product were investigated.

3.3 Effect of pH
The pH of squid mince was adjusted with phosphates according to *Table I*.

Table I

AMOUNT OF PHOSPHATES ADDED AND THE RESULTING pH OF THE SQUID MINCE

Phosphates added (%w/w) K_2HPO_4	KH_2PO_4	pH
4·96	14·65	6·0
5·63	12·97	6·2
7·10	9·52	6·3
8·17	7·11	6·5
9·11	4·81	6·7
9·92	3·03	6·9
10·45	1·77	7·2
10·72	1·05	7·4

At pH 6·7 the breaking force of the product was maximal. Below pH 6·7, the breaking force and water retention both increased with pH (*Fig 3*). Above pH 6·7, a decrease in breaking force but an increase in water retention were observed. Therefore, pH 6·7 and water retention of 1·72 to 1·80g/g solid give maximal gel strength (*Table II*).

The optimal pH to obtain maximal breaking force observed by pH adjustment was different from that obtained by autodegradation, namely pH 6·7 and pH

Table II

EFFECT OF pH ON TEXTURE OF MINCED SQUID PRODUCT

pH	Moisture %	Water retention (g–H_2O/ g–solid)	Breaking force (g)	Breaking force[1] (g)	Deforma- tion (mm)	Folding test
6·0	70·0	1·46	225	225	10·3	C
6·2	70·7	1·46	230	251	10·9	C
6·3	70·5	1·69	297	313	7·9	B
6·5	71·6	1·70	263	313	7·1	B
6·7	72·3	1·73	348	418	8·3	B
6·9	71·9	1·84	190	249	7·2	B
7·2	73·1	1·92	148	244·	7·6	B
7·4	72·8	1·90	190	277	7·7	B

[1] Breaking force corrected to that at 70% moisture

6·4, respectively. The difference may arise from differences in ionic strength in the two materials.

3.4 Effect of NH_3
Squid mince stored at 5°C was sampled daily for NH_3 content (*Table III*). The samples were then buffered to pH 6·7. The effect of NH_3 on the texture of the product

Table III

RELATIONSHIP BETWEEN ACTOMYOSIN EXTRACTABILITY, BREAKING FORCE AND NH_3 DEVELOPED IN CHILLED STORAGE[1] OF MINCED SQUID PRODUCT AT CONSTANT pH 6·7

Days in storage	NH_3–N (mg%)	Moisture (%)	Actomyosin (%)[2]	Breaking force (g)	Corrected breaking force (g)[3]
2	21·76	72·8	47·1	286	373
3	34·50	73·2	46·3	264	363
4	196·00	72·5	51·0	212	290
5	200·57	72·9	48·0	189	279

[1] At 6°C
[2] % total nitrogen
[3] Breaking force corrected to that at 70% moisture

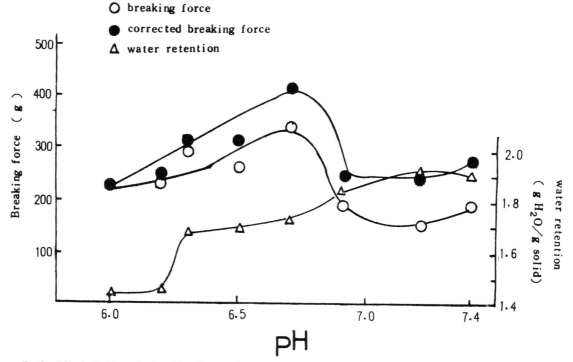

Fig 3 Effect of pH on the breaking force and water retention of minced squid product (buffered with phosphate)

at constant pH is shown in *Fig 4*. The dependence of breaking force on NH_3 was calculated as $0.54g/mg\%$ NH_3. It seems, therefore, that pH is a more dominant factor than NH_3 on the texture of the minced squid product.

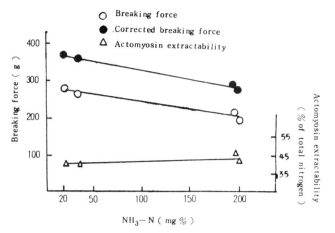

Fig 4 Relationship between actomyosin extractability, breaking force and NH_3 of minced squid product (at pH 6·7)

3.5 *Effect of salt*
The effect of salt in the range of 0% to 5% on the textural property of the minced squid product is shown in *Table IV*. The maximal breaking force was found at a concentration of 3% to 4% salt. However, the product was too salty. 2·5% salt is the maximum concentration to give maximal elasticity.

The SEM micrograph shown in *Plate 1* demonstrates the non-structured mince before salt addition (A). After mixing with salt for 10min, a structure of threadlike bundles is apparent (B).

Prolonged mixing resulted in a cross-linking network (C).

Table IV

EFFECT OF SALT CONTENT ON TEXTURE OF MINCED SQUID PRODUCT[1]

NaCl	Breaking force	Folding test
(%)	(g)	
1	203	AA
2	262	AA
3	271	AA
4	288	AA
5	223	AA

[1] Moisture content 75%, pH = 6·4

3.6 *Effect of starch*
Addition of cassava starch (5%) increased the breaking force but slightly reduced the deformation of the minced squid product (*Table V*). This suggests that starch is not involved in network formation. The increase in breaking force indicates that starch exerts a 'filler reinforcement' effect as postulated by Takagi and Simidu (1972).

The SEM micrographs (*Plate 2*) confirmed that cassava starch disperses in the myofibrillar protein matrix as partially gelatinized granules of 13μ to 17μ diameter.

3.7 *Microstructure and texture of minced fish product*
Minced squid product of high breaking force showed a uniform and compact protein matrix in SEM (*Plate*

Plate 1 SEM of minced squid products (600 ×)
(A) without salt
(B) with salt added, mixed for 10min prior to and after salt addition
(C) with salt added, mixed for 10min prior to salt addition and 20min afterwards

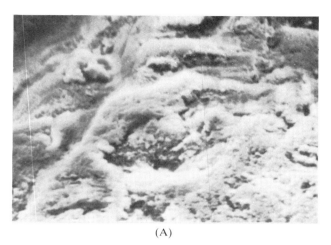

(A)

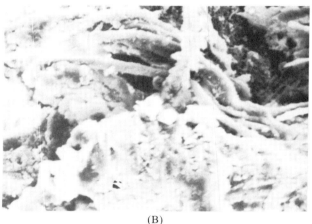

(B)

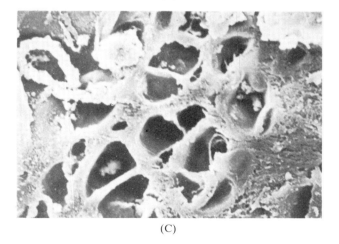

(C)

Table V
EFFECT OF CASSAVA STARCH ON TEXTURAL PROPERTIES OF MINCED SQUID PRODUCT[1]

Starch	Breaking force	Deformation	Gel strength	Folding test
(%)	(g)	(mm)	(g–mm)	
0	276	11·55	3 188	BA
5	392	10·12	3 967	BA

[1] pH = 6·45 ± 0·5; moisture content = 80 ± 1%

235

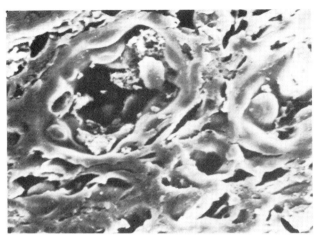

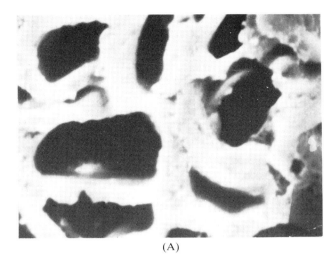

(A)

Plate 2 SEM micrograph of protein matrix of minced squid product containing cassava starch granule (400 ×)

Plate 3 (right) SEM micrographs of minced fish products
(A) compact and uniform protein matrix of minced squid product having high breaking force (1250 ×)
(B) amorphous protein matrix of minced fish product having low breaking force (880 ×)

3–A). Those of low breaking force revealed an amorphous matrix (*Plate 3–B*). It is suggested therefore that uniformity and compactness of the protein matrix play an important role in the textural quality of the minced fish product.

4. References

MATSUMOTO, J J. *Bull. Jap. Soc. Sci. Fish.*, 24, 125–132
1958
—— *Bull. Tokai Reg. Fish. Res. Lab.*, 21, 51–63
1959
TAKAGI, T and SIMIDU, W. *Bull. Jap. Soc. Sci. Fish.*, 38, 471–474
1972
TSUCHIYA, T, YAMADA, N and MATSUMOTO, J J. *Bull. Jap. Soc. Sci.*
1978   *Fish.*, 44, 181–184

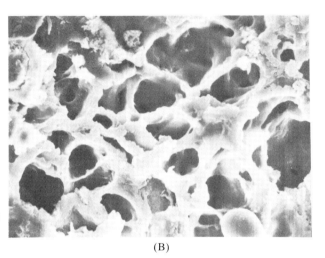

(B)

4 Process and new product investigations

A study of canned fish behaviour during storage

H Durand and Y Thibaud

1 Introduction

In order to find if a packing type is suited to the requirements of a particular canned product we studied the behaviour of several types of canned fish conditioned in two packing-cans of different types.

For these tests we made fish preserves traditionally consumed in France from two species: sardine and mackerel.

2 Canning type

The preparations were as follows:

> Sardines in oil
> Sardines in oil with lemon
> Sardines in tomato sauce
> Mackerel in tomato sauce
> Mackerel pickled in white wine.

After deheading, evisceration and cooking in steam at 103°C, the fish were put in cans and covered with sauce. The tomato and marinade sauces were prepared according to the French standards of the Sea Product Processing Industry Confederation, namely:

> White wine picklings: an aqueous sauce containing 25% of vinegar (6·5° acetic acid) and 10% of white wine;
> Tomato sauce: sauce containing 8% of dry extract and an oil content of 10%.

3 Conditioning

The products were conditioned in tin plate and aluminium cans, in 1/6 P 25 (125cc) formats for the sardine and 1/6 Cl 30 (125cc) sizes for the mackerel packs.

The aluminium containers were pressed metal cans made of Al-Mg alloy, whose inside varnish is covered with a polypropylene layer.

The tin plate cans were the non-sticking type with one lid soldered. For the mackerel they were made of MR steel (with a little metalloid content), K grade (tinning type whose middle layer structure offers better protection), E4/E2 tinning ($11·2g/m^2$ inside and $5·6g/m^2$ outside).

4 Storage

The manufactured lots were stored at 20°C and 37°C, the latter accelerating the ageing process. The storage duration was two years. The tin cans were stocked with the soldered lid turned downward.

The sample coding is shown in *Table I*.

Table I

Canning type	Storage at 20°C		Storage at 37°C	
	Aluminium	Tin	Aluminium	Tin
Sardines in oil	SHA 20	SH F 20	SH A 37	SH F 37
Sardines in oil with lemon	SCA 20	SC F 20	SC A 37	SC F 37
Sardines in tomato sauce	ST A 20	ST F 20	ST A 37	ST F 37
Mackerel in tomato sauce	MT A 20	MT F 20	MT A 37	MT F 37
Mackerel pickled in white wine sauce	MM A 20	MM F 20	MM A 37	MM F 37

5 Methods

The examinations were effected after six months, one year and two years of storage, on five cans of each type of product.

At each of these periods, the following were examined:

> Sensory attributes
> Inside and outside appearance of the can
> pH, or the oil acidity for the sardines in oil, and in oil with lemon (expressed in mg of oleic acid per 100g of oil)

After two years, the lead content of the homogenized contents was measured. These determinations were effected by atomic absorption spectrophotometry at 217·6nm, after mineralization and extraction of lead by methylisobutylketone + sodium dimethydithiocarbamate. Trimethylamine (TMA) and trimethylamine oxide (TMAO) were also determined.

6 Results

6.1 *Changes in organoleptic character*

Regarding the sardines in oil, the odour, the texture and colour remain normal during the storage period, whatever the storage and the container.

After a year of storage, a weakened fish flavour in the aluminium cans, at 20°C and 37°C, was observed. Thus it seems that the specific fish flavour is best preserved in tin cans.

The sardines with lemon remained normal and had unchanged texture and covering colour. Also, the odour and flavour were slightly reduced in aluminium cans after two years. The lemon showed a few grey-blackish spots, owing to the local de-tinning, as well as a slight metallic savour.

The sardines with tomato sauce all had a normal colour and texture. The odour and flavour in the product in aluminium cans stored two years at 37°C were both reduced. The tomato sauces were considerably changed, being much more fluid with a tendency to syneresis in aluminium, whatever the storage temperature.

Similar changes were observed with mackerel in tomato sauce. The sauces contained in aluminium were in fact totally separated in an oily phase and a thick tomato mash. In addition, metallic flavours, due to de-tinning, appeared in the tin cans as early as six months at 37°C.

Finally, in the pickled mackerel packs all the organoleptic characters were normal, except for feebly metallic flavours in tin stored for one year at 37°C.

Thus as a general rule, it appears that the flavours are slightly affected by contact with aluminium, as is the appearance of the tomato sauces. On the other hand, in tin metallic flavours appear to be caused by de-tinning of the can.

6.2 Internal and external appearance of the can

The aluminium containers were always apparently intact, inside and outside, as were the varnished bottoms of the tin cans.

With regard to the non-varnished parts of the tin cans. the following observations were made.

Externally the cans of sardines in oil were all normal. Rarely, cans stored at 20°C showed corrosion pits; those stored at 37°C showed a slight de-tinning, estimated at about 5% of the surface.

Cans of sardines with lemons stored at ambient temperature showed a slight internal corrosion adjacent to the lemon slice; on the other hand cans stored at 37°C

were very de-tinned after two years. Furthermore, in the latter case we observed swellings or leaks at the soldered bottom.

Sardines in tomato sauce displayed a homogeneous de-tinning that can attain 30% after storage at 20°C and up to 80% after storage at 37°C. The latter also showed swelling.

Similar observations were made with mackerel in tomato sauce.

As regards the pickled mackerel, the pitting attacks were homogeneous but remain slight. On the other hand, weakening of the seals in the cans stored at 37°C occurred, giving rise on occasions to leaks.

We have tried to estimate the degree of internal corrosion of the tin plate cans. We used the following scale:

Nil or very slight corrision	0
Slight corrosion with little extent	1
Slight corrosion with medium extent	2
Slight corrosion with important extent	3
Much corrosion with little extent	4
Much corrosion with medium extent	5
Much corrosion with important extent	6

The results are reported in *Fig 1*. This shows that K tin sheet offers great advantages from the de-tinning point of view. The marinades, although they are much more corrosive than the tomato sauce, have less corrosion in this type of tin. On the other hand, the seals remained weakened.

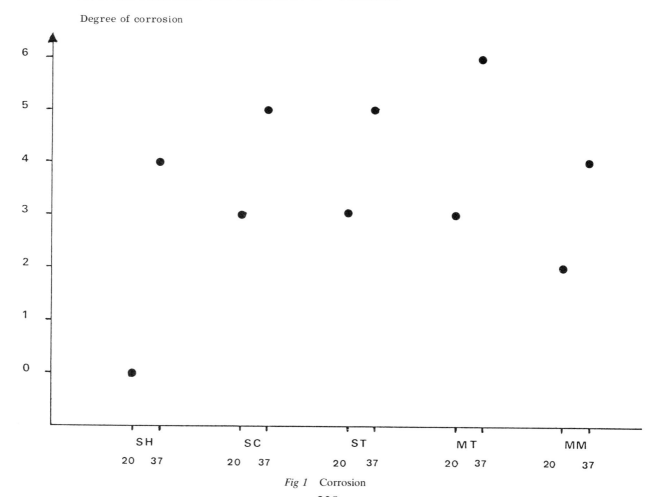

Fig 1 Corrosion

238

6.3 *Measures of pH and oil acidity*

These measures are regrouped in *Table II*. The figures are mean values, the dispersion being almost nil. Regarding the pH, we did not observe significant variations from one container to the other for the preparations in tomato sauce. For the marinades, on the other hand, the aluminium cans are distinguishable from the tin containers.

For oil acidity the preponderant factor is not the nature of the can but the storage temperature (*Fig 2*).

6.4 *Lead content*

The results are depicted in *Fig 3* and are the average values of five cans.

As expected the lead contents are very low in the aluminium cans. They correspond to the lead existing in the raw materials together with that introduced during preparation.

For the tin cans the quantities found are much larger and arise particularly from the weld. The aqueous sauces (tomato and vinegar) appear to favour dissolution of

Table II
pH AND ACIDITY OF THE OIL DEPENDING UPON THE STORAGE DURATION

Packing type and storage temperature	Duration of storage (years)	pH of the homogenized content			Acidity of the oil in g of oleic acid for 100g of oil	
		Sardines in tomato sauce	Mackerel in tomato sauce	Pickled mackerel	Sardines in oil	Sardines in oil with lemon
Tin 37°C	0·5	5·5	6	4·7	0·36	0·49
	1	5·4	5·9	4·5	0·59	0·67
	2	5·6	6	4·9	1·19	1·52
Tin 20°C	0·5	5·4	6·1	4·6	0·24	0·46
	1	5·4	5·9	4·6	0·34	0·39
	2	5·7	6·2	4·8	0·47	0·59
Aluminium 37°C	0·5	5·4	6	5	0·31	0·44
	1	5·4	6	5	0·56	0·56
	2	5·7	6·1	5·1	1·36	1·20
Aluminium 20°C	0·5	5·6	6·1	5	0·28	0·42
	1	5·6	5·9	5	0·36	0·42
	2	5·8	6·25	5	0·51	0·59

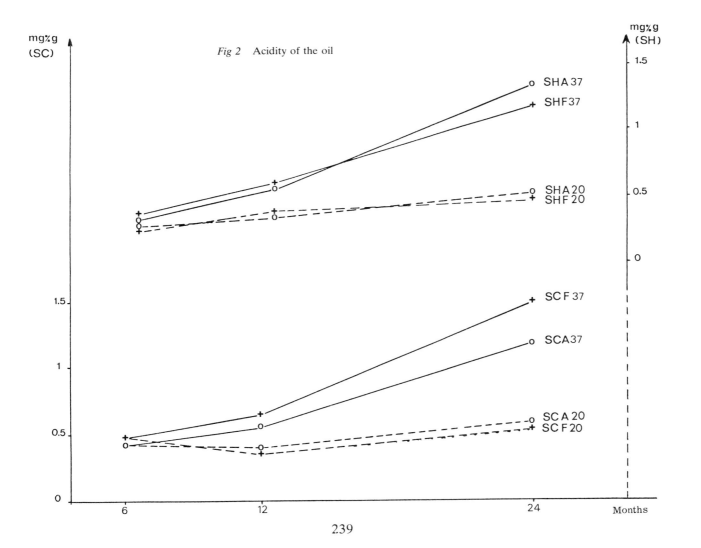

Fig 2   Acidity of the oil

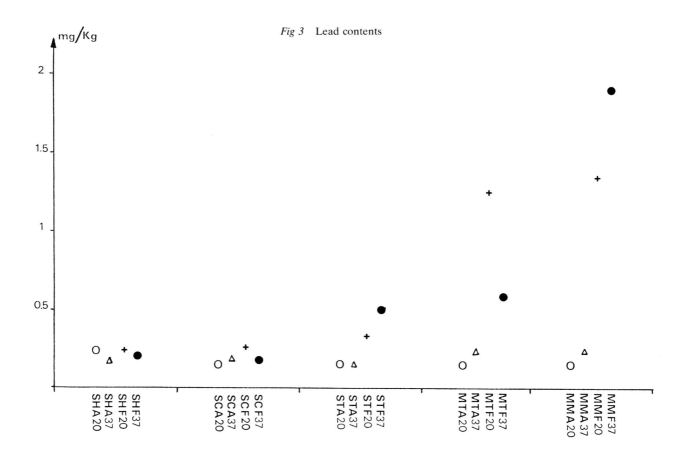

Fig 3   Lead contents

lead particularly in acid conditions, maximum values being obtained in the products with vinegar (average pH = 4·8). In the products with oil, not only the contents are low but moreover the differences of acidity observed between the products stored at 20°C and 37°C do not correspond to the corresponding lead contents. Larger quantities of lead were found in the mackerel pack as compared to the sardine pack in tomato sauce.

Results for TMAO and TMA are shown in *Table III*.

Table III

	ST F 20	*MT F 20*	*ST F 37*	*MT F 37*
TMAO	0	0	0	0
TMA	7·7	13·7	6·6	16·1

No TMAO remains in any of the packs. High lead content corresponds to high TMA content.

This illustrates the importance of the species because we know that fresh mackerel is very rich in TMAO. It appears that simultaneous reduction of TMAO to TMA and oxidation of the Pb to lead salt Pb++ is occurring.

The difference is amplified when the storage temperature is high.

7 Conclusions

These observations and measurements show that the particular flavours of fish and those of tomato sauce become less prominent in aluminium than in tin cans. On the other hand, with regard to de-tinning and consequently the advent of metallic flavours, aluminium is undeniably superior.

The measured lead contents are all relatively low, especially for aluminium; they also remain constant with storage time and temperature. Only the mackerel in marinade sauce contains lead in more important quantities, albeit under severe and unusual conditions (two years at 37°C). Thus, in order to reduce lead contamination, aluminium is only justified in the case of specially aggressive products stored at rather high temperatures.

The technology of Spanish canned mussels (*Mytilus edulis* L.) from raft cultivation

F López Capont

1 Introduction

In this report we will only consider technological questions related to the canning of raft cultivated mussels: wider issues related to cultivation will not be considered.

Mussels (with three genera *Mytilus*, *Modiolus* and *Perna*) have a wide distribution around the world. Many species exist; for example, in Mytilidae there are more than 150 species (some in freshwater).

240

2 Species suitable for canning

In general, any mussel species in the world is suitable for canning, but local variations and conditions can affect utilization. For example. *M. algossus* in Chile has an abnormal brown side; 'pearls' (produced by a parasite *Gymnophallum*) occur in some species; contamination (paralytic shellfish poison, heavy metals, pathogens) may be present.

Mussels have three principal parts;

(*i*) Meat: in the cultivated Spanish mussel 30% is a good industrial yield; occasionally this can rise to 50% in certain fresh specimens.

(*ii*) Water: occurs in the mantle cavity in variable amounts of 20–40%. Like the meat, the amount depends on the season and age of the specimen. The amount is very important for survival after landing.

(*iii*) Shell: amounts to 30%. It has very poor possibilities for economic utilization; it is very important in canning because it is the highest proportion of waste.

The biological factor of most importance from the point of view of canning is the reproductive cycle. From a practical point of view, we need to consider only four stages:

Stage O – Inactive (including also the virgin animal or less than one year old). Without gonads, increasing carbohydrates (as reserve in the mantle). Very low industrial yield.

Stage I – In cultivation, two to three months old. Gametes beginning to form. Little interest for canning.

Stage II – Fully ripe. Optimum for industrialization. Proteins, lipids and yield are at their highest values; carbohydrate beginning to fall.

Stage III – Spawning; decrease of quality and yield because about 20% are gonads.

3 Technical aspects of mussel culture in Spain

Mussel culture in Spain started around 1942 in the northwest area of Galicia. The floating or raft method was and still is used. In 1963, with 100 000 tonnes harvested, Galicia was the largest producer in the world. Today Spanish production is 175 000–200 000 tonnes per year, broken down as follows:

30–40% destined for canning (95% prepared in 'escabeche' sauce). Exported all over the world.

35–40% to national markets, fresh (depurated). Can be increased.

15% exported frozen (previously cooked). Potential for increase.

10% exported to Europe as fresh (depurated). A declining market.

Cultivation takes place on ropes suspended from floating square frames of 16m side. The frame is made usually of wood or steel and is equipped with two or four floats made of wood cubes covered with tar, cement or fibreglass. Occasionally frames equipped with semispherical plastic floats are employed because these can be stacked easily for transport. Frames made of polyurethane are being introduced and nowadays make up about 5% of the total employed; they have good resistance to weathering and marine boring organisms. Plastic frames are more expensive than those made of wood or steel but maintenance costs are less. Each frame is securely anchored to the sea bed.

Each frame has 600 ropes each about 10m in length, only about 70% of which is usable for cultivation. In Galicia 8m water depth is the maximum possible for good spawning, and illumination. For collecting spat, the number of ropes is reduced to about 500.

Today, ropes 2–2.5cm in diameter made of synthetic fibres are usual. For production the colour is black; for collecting spat it is white. Every 60cm small wooden pegs covered with coloured plastic tubes are inserted into the ropes to prevent the accumulated mussels from sliding down. Even so 1–2% are lost in this way. Under good cultivation 450–500 mussels are attached per metre of rope. About 10–20% of the animals on the rope are not harvestable mussels. The yield per metre of rope is usually 100–125kg/year and from one raft about 50 tonnes/year.

Collecting spat is easy in Galicia because natural spawning is plentiful, particularly in autumn. During one year the mussels can increase up to a good commercial size of 10–12cm, though 8–9cm is more usual. Every four to six months the ropes with their growing mussels are divided into two in order to improve the yield.

In Galicia there are about 4 000 rafts owned by about 1 100 individuals organized into groups or co-operatives. Altogether about 15 000 people are employed in cultivation and associated activities. The current first hand price for fresh mussels is 12–18 pesetas/kg (£0.08–0.12/kg). The cost of depuration is 4–5 pesetas/kg (about £0.03/kg).

Cultivated mussels stored in ice out of direct contact can be kept in good condition for up to four days.

4 Technology of canning mussels

4.1 *Raw material*
The best quality is obtained with mussels one year or 18 months old, harvested between September and December, 7–9cm long (average 35g); at sexual stage II and three to four weeks before or after spawning. They should have completely closed shells, be roughly cleaned with sea water and quickly transported (maximum 12h) to the factory. Direct exposure to the sun can induce spawning.

According to Spanish regulations, it is not necessary to depurate before canning.

Sometimes plankton can contaminate the mussels resulting in an objectionable green colour in the canned product. Little control over this problem is possible.

4.2 *Steaming*
Prior to canning the mussels are steamed for 2–4min. This does not fully cook the flesh but inactivates enzymes and opens the shell. The mussels are spread thinly on perforated trays or baskets and steamed in a closed cabinet. In a newer system the steamed mussels in a perforated container are turned over to release the water resting in the mantle cavity.

4.3 *Deshelling*
This is achieved by vibrating by hand on special continuous tables; after steaming about 30% of the muscle meat is attached to the shell, the remainder being free.

4.4 Frying

Olive oil or vegetable oil (less than 0·2% acidity) is used in a continuous frier. Frying is carried out for 1–2min after which the meat contains about 67% water.

4.5 Packing in the can

After selection of the dried mussels by hand they are hand packed directly into the cans (usually 125ml capacity). Best quality mussels are considered to be females just before spawning because they have a high concentration of carotene and an appealing reddish-yellow colour. Quality is also determined by size: fried mussels of 7·5g, 5·5g and 3g to 5g are considered as first, second and 'popular' quality, respectively.

4.6 Sauces

The cans after packing with mussels are filled with sauce. According to Spanish regulations a maximum of 30% of the net contents of the can should be sauce, but most commercial samples contain 25%.

More than 95% of Spanish production uses a pickling sauce known as 'escabeche'. The remainder is in brine or in oil. In the latter case frying is omitted.

Formulations for escabeche sauce vary from factory to factory, sometimes according to 'secret' recipes. Normally, however, they consist of roughly equal parts of vegetable oil, water and wine vinegar (up to 7% acetic acid). Spices and salt (2·5–3·5% of the total weight) are included. The traditional red colour is very important. This is imparted by paprika supplemented by xanthophyl concentrate which since it is soluble in both oil and water gives a good uniform colour.

For the correct taste the optimum concentration of acetic acid in the whole contents is about 0·8%. If the concentration rises much above 1% corrosion problems and softening of texture may arise. During storage little variation in acidity occurs.

Addition of other organic acids such as synthetic acetic acid, citric acid or tartaric acid reduce the quality.

A measurement of the concentration of solubles (mostly derived from autolysis products) in the water phase of the sauce provides a good measure of the quality of the mussels. With stale mussels the quantity is higher than with fresh.

5 Selected bibliography

ANDREU, B. Sobre el cultivo del mejillón en Galicia: biologia, cre-
1958 cimiento y produccion. *Ind. Pesqueras* (Vigo), 745–6, 44–7
—— El mejillón como primera materia para la conserva. *Informacion*
1963 *Conservera* (Valencia), 119–120, Nov., 8pp
—— Pesqueria y cultivo de mejillónes y ostras en España. *Pub-*
1968 *licaciones Técnicas de la Junta de Estudios de Pesca* (Madrid), 7, 303–320
FIDALGO, F. A. Estudio de los escabeches en conservas de mejillón.
1976 Nuevas preparaciones. Tesina de Grado. *Ciencias Biologicas.* Universidad de Santiago de Compostela, Spain
FIGUERAS, A. Desarrollo actual del cultivo del mejillón (*Mytilus edulis*
1976 L.) y posibilidades de expansion. *Advances in Aquaculture*, Eds T V R Pillay and W A Dill. Fishing News (Books) Ltd, Farnham, Surrey
FRAGA, F. Relacion entre peso y talla y composición química en el
1958 mejillón de la Ria de Vigo. *Invest. Pesq.*, 14, 25–32
—— L'utilisation possible de l'eau de cuisson des moules. *Doc. Tech.*
1963 *Cons. Pêches Medit.* FAO, 32, 3pp
FRAGA, F and LOPEZ CAPONT, F. Oligosacaridos en el mejillón. Factor
1955 de proteinas. *Invest. Pesq.*, 11, 39–52
FRAGA, F and VIVES, F. Retencion de particulas organicas por el mejil-
1958 lón en los viveros flotantes. *Reun. Produc y Pesquerias*, Barcelona, 4, 71–73
LOPEZ-CAPONT, F. Muestreo de conservas de mejillón en escabeche.
1973 *Rev. Agroq. Techn. Alimentos* (Valencia), 13(2), 233–240
—— S.O.S. . . . mejillón de Galicia. *La Voz de Galicia*, Suplemento
1976 domicial, 19 Sept
—— Las toxinas del mejillón y otros moluscos. Su problematica e
1978 importancia para España. *Rev. Agroq. Techn. Alimentos* (Valencia), 18(1), 47–63
LOPEZ-CAPONT, F and FIDALGO, A. Salsas y escabeches como factores
1977 de calidad del mejillón cultivado y en conserva. *Ibid*, 17(4), 427–434
LOPEZ-CAPONT, F and PASCUAL LOPEZ, M. Factores ambientales e
1979 incidencia del copepodo *Mytilicola intestinalis* en el mejillón de Galicia. *Bol. Int. Oceanografia*. Univ. Sao Paulo (Brasil). (In press)
PAZ-ANDRADE, A and WAUGH, G D. Raft cultivation of mussels is big
1968 business in Spain. *World Fish.*, 17(3), 50–52
RODRIGUEZ MOLINS, L and BESADA RIAL, J R. Estudios quimicos
1957 sobre el mejillón de la Ria de Vigo. *Bol. Inst. Esp. Oceang.*, 87, 29pp
RYAN, P. A raft for five mariculture nets. Techn. Report No. 544. Fish.
1975 and Marine Serv. (Can.)

The use of dilute solutions of hydrogen peroxide to whiten fish flesh

K W Young, S L Neumann, A S McGill and R Hardy

1 Introduction

Modern western methods of processing fish with its insistence on producing bone- and skin-free pieces is wasteful not only because butchering in this manner leaves behind a considerable proportion of the flesh but also because small fish cannot be processed readily by such methods and thus are not utilized to any great extent in human nutrition.

Nowadays, however, it is possible by the use of flesh-stripping machines to extract the flesh from fish efficiently, leaving behind only the skin and bones and thus making optimum use of our fish resources. The material thus obtained ranges in texture from a coarse mince to a slurry and in colour from near white to dark brown. People are accustomed to eating terrestrial animal products in this form but not fish and there is considerable resistance to doing so. There are two major

reasons for this; the first is concerned with texture and the second with colour. On the whole people, at present, prefer to eat fish as a discernible piece such as in the form of a fillet or stick (fish finger). Fortunately the deboned material (or mince) can be retextured or reformed using comparatively simple techniques to give the appearance of a fish finger or piece (Keay and Hardy, 1974a,b).

The colour of the minces presents a more intractable problem. With certain exceptions our experiments indicate that reformed white minces are accepted readily whereas coloured ones are not. This would not be of any great concern if it were possible to remove the pigments in some physical manner such as washing but, except with some white fish discoloured with a little blood, this is not possible because the pigments are complex and are difficult to remove by physical methods. As a result, attempts have been made to mask the colour with coating agents such as titanium dioxide and starches or by encapsulation in batter, but these methods have no' been very successful. Reinforcement of the colour with brown dyes and the addition of smoke flavours to give a smoked product has met with consumer approval but this technique is of limited application (Bannerman *et al*, 1973).

It has been known for many years that colours can be removed from foodstuffs by chemical oxidizing agents (Chesner and Dickinson, 1963; Sims *et al*, 1975) and indeed we attempted to bleach fish with hydrogen peroxide, in some earlier experiments, at neutral pH. A worthwhile bleaching effect was only obtained using relatively strong hydrogen peroxide solution which seemed unacceptable for food processing. This method was therefore abandoned until it was pointed out that much milder conditions could be used at high pH values to achieve bleaching (James and McCrudden, 1976). Initial experiments showed that this was so and the following is a brief account of the work carried out at Torry Research Station on the bleaching of fish with hydrogen peroxide.

2 Experimental

2.1 *Materials*

Cod (*Gadus morhua*), saithe (*Pollachius virens*) and mackerel (*Scomber scombrus*) were obtained commercially, the former two species used fresh and the latter after six months' frozen storage at $-30°C$.

Citric acid, sodium carbonate, disodium orthophosphate, boric acid, potassium chloride and hydrogen peroxide (100vol) were supplied by BDH Ltd. All except the disodium orthophosphate were of Analar grade. Sodium tripolyphosphate was classified 'food grade' and was obtained from Albright and Wilson.

Reference amino acids were purchased from Sigman Chemical Co Ltd.

Cyanogen bromide and heptafluorobutyric anhydride were supplied by the Aldrich Chemical Co.

All solvents which were pronalys grade from May and Baker Ltd were redistilled before use.

2.2 *Methods*

2.2.1 *Preparation of fish* Skinned fillets and two sizes of cube segments (1cm and 0·5cm) were used. For minces single fillets from nobbed fish and the resultant fish wastes were separately comminuted in a Baader 694 deboning machine (drum perforation size 5mm). The minces were air-blast frozen as blocks (2cm × 20cm × 40cm), wrapped in polythene, and held at –30°C for up to six days before whitening.

2.2.2 *Buffer solutions* A range of buffer solutions of pH 2 to 8, 9 and 10·5 was prepared using citric acid–disodium orthophosphate, boric acid–potassium chloride and sodium carbonate–sodium tripolyphosphate, respectively (Dawson *et al*, 1969; James and McCrudden, 1976).

2.2.3 *Whitening process* Two chemical treatments were applied. In the first, the fish in one of the forms described above (2.2.1) was added to the appropriate buffer solution in the ratio of 1 part fish to 4 parts buffer,

thoroughly mixed by manual stirring and then left standing undisturbed at ambient temperature. The second treatment was identical except that hydrogen peroxide (0·75% of buffer) was added prior to stirring.

To produce finer particle sizes, minces treated as above were homogenized in an Ultra-Turrex mixer before leaving to react.

Samples were reacted for time periods of from ¼h up to 7½h, whereupon an excess of sodium sulphite was added to reduce residual hydrogen peroxide. After adjusting the pH to 7·0 ± 0·2 with either N sodium hydroxide or hydrochloric acid, the mince was filtered and then washed with water in an MSE 300 centrifuge, the basket of which was lined with linen.

The whitened product was stored at $-30°C$ either in blocks (2cm × 10cm × 15cm) wrapped in polythene, or as fish sticks (2cm × 2cm × 9cm) coated in batter and breadcrumbs.

2.2.4 *Analysis* The peroxide value (PV) (Banks, 1937) and the epoxide value (Durbetaki, 1956) were determined on the lipid extracted from the mackerel mince using chloroform-methanol (Bligh and Dyer, 1959).

Residual hydrogen peroxide was estimated by taking 0·5g of lipid and shaking well with sulphuric acid (1:17) (2·5ml), potassium titanium oxalate (5%) (2ml) and water (5ml) in a conical flask (25ml), centrifuging and measuring the absorption of the aqueous layer at 400nm. In the case of the filtered liquors an aliquot (5ml) was titrated with potassium permanganate (Vogel, 1961).

The protein solubility was estimated by measuring the protein which could be extracted by a cold neutral 5% sodium chloride solution (Cowie and Mackie, 1968).

The percentage solids were determined by drying a fish sample (100g) overnight to constant weight in a vacuum oven at 90°C.

To estimate the amino acid content, freeze-dried mince (0·25g) was hydrolysed by refluxing under nitrogen for 24h using 50% hydrochloric acid (62·5ml) and diluting to 250ml with distilled water. The heptafluorobutyric *n*-propyl (HFB *n*-propyl) derivative (March, 1975) of the released amino acids was prepared using the filtered hydrolysate (1ml) and analysed by gas liquid chromatography (GLC). Prior to this analysis the GLC response to each amino acid derivative was determined. The quantitative GLC of the HFB *n*-propyl derivatives was carried out on a Pye 104 gas chromatograph fitted with a glass column (4m × 2mm id) packed with 3% SP 2100 on Supelcoport 80–100 mesh using a nitrogen flow of 30ml/min, programming isothermally for 15min at 90°C then 3°C/min to 240°C and measuring the peaks manually by triangulation.

The methionine content of the minces was also determined by GLC (Ellinger and Duncan, 1976) using the cyanogen bromide method.

3 Results and discussion

In all the experiments in which the fish were immersed in buffer solution a reduction in colour was observed. This, in part, was caused by dissolution of some of the pigments in the aqueous phase and was most noticeable in those fish treated at alkaline pH. In acid pH solutions the dissolution effect was less marked but here changes occurred at the protein surface, altering its translucency, making it more opaque, reflective and white. This is illustrated in *Plate 1* and in this photograph the effect of hydrogen peroxide treatment on mackerel fillet mince especially at acid pH can be seen clearly. On isolation and neutralization of the fish tissue it was apparent that maximum whitening effect was attained in the pH range 2–5 and 9–11. The latter effect is masked in *Plate 1* by pigment dissolution in the aqueous phase.

At quite an early stage in the experimental work it was realized that the whitening effect due to the hydrogen peroxide occurred mainly at the surface. This can be seen in *Plate 2* where a fillet has been treated under the conditions described by Laporte (James and McCrudden, 1976).

With reduction of fish particle size from small cubes to minces, to the finely divided particles produced in a homogenizer, the effect became more homogeneous and a whiter product was obtained with decreasing particle size. This is shown in part in *Plate 3*, the homogenized material having a whiter appearance than the mince. These results indicate that either hydrogen peroxide does not diffuse rapidly into the flesh or perhaps more likely that the reaction between the fish and hydrogen peroxide is rapid and the hydrogen peroxide concentration is markedly reduced before it can diffuse to any great extent. The measurement made on changes in hydrogen peroxide concentration in the beakers after adding to the fish would seem to support this latter view. Within minutes the level of hydrogen peroxide fell to less than 5% of the original level. Nevertheless, not all

the hydrogen peroxide is reduced because whitening increased with time although visually the maximum rate occurred immediately after mixing. As is to be expected and can be seen in *Plate 4* the whitening effect was enhanced with increasing hydrogen peroxide concentration.

The effects of treating cod, saithe and mackerel at alkaline pH 10·5 and hydrogen peroxide concentration of 0·75% for 15min are shown in *Plates 3, 5* and *6*. Longer immersion or increasing peroxide concentration has a more marked effect but it is possible by using very mild conditions to convert the saithe flesh into a remarkably white material. Mackerel flesh does not whiten completely under such conditions but elsewhere experiments have been carried out to show that white material can be obtained by somewhat harsher treatments using stronger hydrogen peroxide solution and higher temperatures (Unilever Research – Snoflake functional fish protein, 1975).

Treatment with hydrogen peroxide does therefore give a bleaching effect which may be advantageous in the marketing of the more coloured fish fleshes.

Chemical measurements were carried out primarily on the minced and homogenized flesh of cod but also to a lesser extent on mackerel flesh that had been treated with hydrogen peroxide.

The effects of the treatment on the oxidation state of the lipids of the fish as measured by peroxide and epoxide values appear to be small even with the mackerel which had lipid contents ranging from 5% to 22%.

No epoxide could be detected in the cod or mackerel after treatment at both acid and alkaline pH with 0·75% hydrogen peroxide solutions for up to 4h. Hydro-

Plate 1   Mackerel fillet mince treated in a range of buffers from pH 2–10 for 7½h. A: buffer only   B: buffer + 0·75% hydrogen peroxide

244

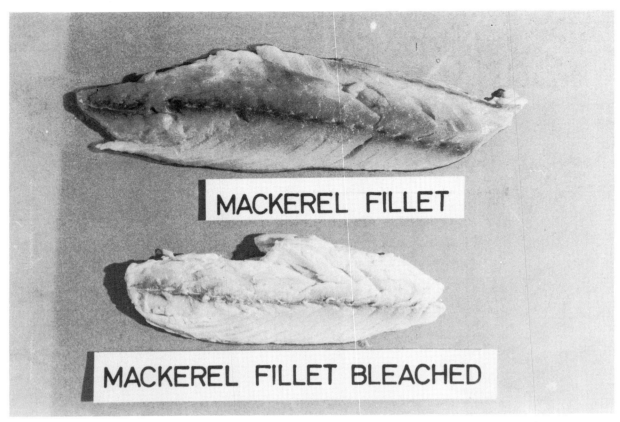

Plate 2 Mackerel fillet treated with 0·75% hydrogen peroxide in a pH 10·5 buffer solution for ¼h

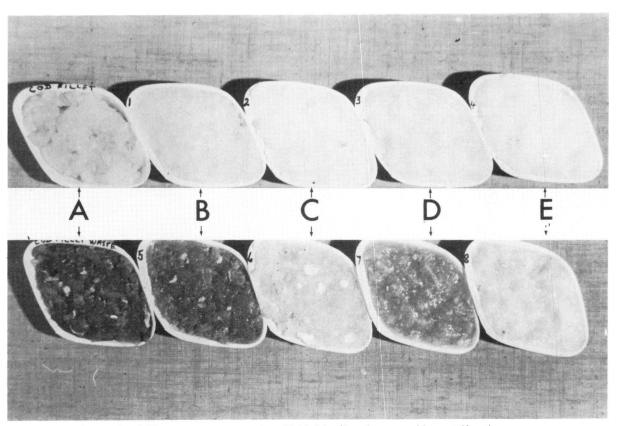

Plate 3 Cod fillet mince and cod fillet waste mince treated at pH 10·5 for ¼h and recovered by centrifugation
Top row: fillet mince   Bottom row: fillet waste mince
A: untreated   B: buffer only   C: buffer + 0·75% hydrogen peroxide   D: homogenized with buffer   E: homogenized with buffer + 0·75% hydrogen peroxide

245

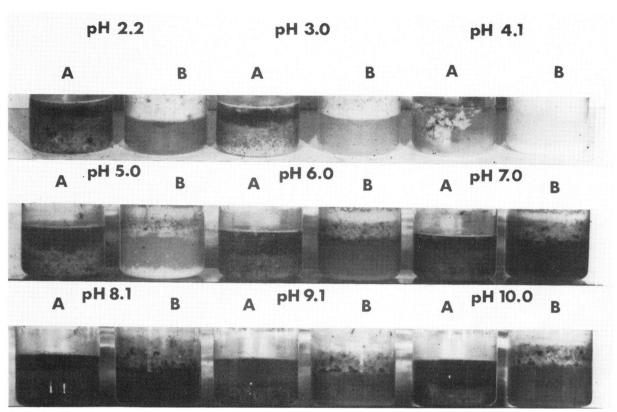

Plate 4 Saithe fillet mince treated in a range of buffers from pH 3–10 with different concentrations of hydrogen peroxide
Top row: 0·3% hydrogen peroxide in buffer   Middle row: 0·75% hydrogen peroxide in buffer   Bottom row: 1·5% hydrogen peroxide in buffer

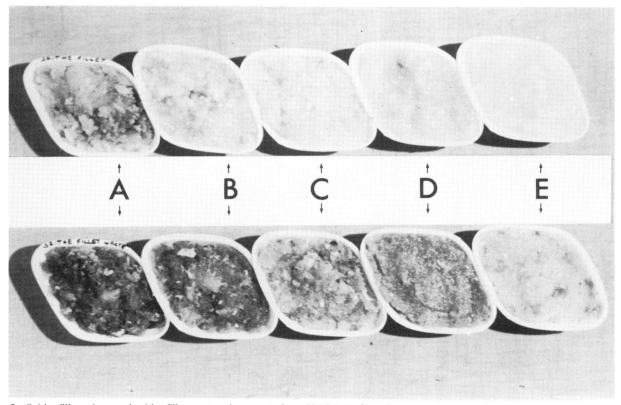

Plate 5 Saithe fillet mince and saithe fillet waste mince treated at pH 10·5 for ¼h and recovered by centrifugation
Top row: fillet mince   Bottom row: fillet waste mince
A: untreated   B: buffer only   C: buffer + 0·75% hydrogen peroxide   D: homogenized with buffer   E: homogenized with buffer + 0·75% hydrogen peroxide

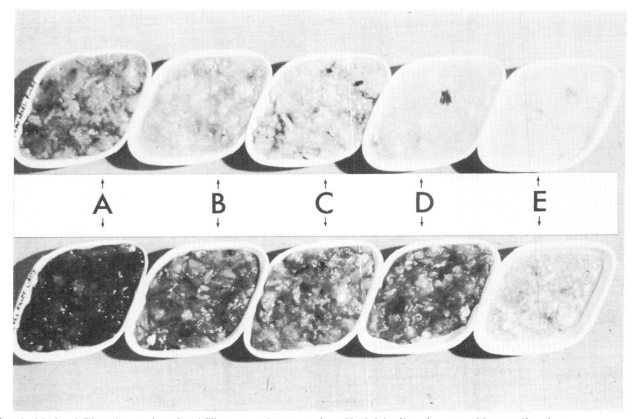

Plate 6 Mackerel fillet mince and mackerel fillet waste mince treated at pH 10·5 for ¼h and recovered by centrifugation
Top row: fillet mince   Bottom row: fillet waste mince
A: untreated   B: buffer only   C: buffer + 0·75% hydrogen peroxide   D: homogenized with buffer   E: homogenized with buffer + 0·75% hydrogen peroxide

peroxide contents did increase in the mackerel flesh lipids from 3μ to 12μ moles/g of lipid but this occurred both in the presence and absence of added hydrogen peroxide.

Measurement of the loss of protein due to dissolution was made in these experiments and it was shown that the losses incurred by treating cod mince were between 8·4% for the pH 10·5 buffered solutions and 12·2% for the pH 4 buffered solutions and were independent of the presence of hydrogen peroxide or immersion time. The solution properties of the protein were affected by the various treatments, all of which result in a loss of soluble protein and conversely an increase in the insoluble protein of up to 200%.

Studies elsewhere, *eg* Manyua (1975), on the treatment of proteins with hydrogen peroxide have shown that the sulphur amino acids, particularly methionine, are readily oxidized. Using the cyanogen bromide method the effect on protein methionine content of treating cod with hydrogen peroxide at both low and high pH was examined. Treatment at pH 4 in the presence or absence of hydrogen peroxide had no effect on the methionine content up to 15min but thereafter a decrease was noted in the solutions containing hydrogen peroxide. After 4h the methionine content had decreased to 30% of that present in the untreated fish. In solutions of pH 10·5 changes in methionine concentration in the presence and absence of hydrogen peroxide were noted after 15min, the concentration falling to approximately 90% of that in the untreated fish. After 4h no methionine could be detected in the hydrogen

peroxide-treated mince. Some difficulty was experienced in the course of these experiments in obtaining reproducible results and this, coupled with the desire to see whether the process was affecting the other amino acids, caused us to abandon the method in favour of a gas chromatographic procedure capable of analysing 20 of these acids.

Before carrying out the analyses on the fish a series of analyses were made on aqueous mixtures of known concentrations of the various amino acids. The concentrations of these mixtures were chosen to span the range of amounts that were likely to be obtained on acid hydrolysis of the fish. After concentration and derivatization a known quantity of the HFB *n*-propyl derivative of norleucine was added as an internal standard.

At first considerable difficulty was experienced in preparing derivatives of the following amino acids, methionine, histidine, arginine, cysteine and tryptophan. However, when special care was exercised to remove all traces of water from the samples before derivatization, and butylated hydroxy toluene was added to the reaction mixture prior to butylation, the derivatization of these amino acids was greatly improved.

The results of the analysis of the standards are given in *Table I* in which the coefficient of variation (the standard deviation expressed as a percentage of the mean) is quoted for each of the amino acid derivatives relative to the norleucine derivative over the given weight range placed on the column. The coefficient of variation was not constant, in some instances it was relatively low, *eg*

247

Table I
ANALYSIS OF AMINO ACID STANDARDS (28 SAMPLES ANALYSED)

Amino acid	Relative response compared with norleucine	Coefficient of variation	Weight range on column g	
Alanine	1·05	13	0·88	2·94
Glycine	1·08	9	0·98	3·26
Valine	1·01	7	0·64	2·13
Threonine	0·99	9	0·19	1·93
Serine	0·96	13	0·21	2·13
Leucine	0·95	13	0·99	3·29
Isoleucine	0·97	8	0·55	1·84
Proline	0·96	7	0·76	2·54
Cysteine	2·45	120	0·05	0·50
Hydroxyproline	0·31	34	0·02	0·04
Methionine	1·24	17	0·15	0·75
Aspartic acid	1·09	9	1·77	5·91
Phenylalanine	0·45	6	0·65	2·17
Glutamic acid	1·49	4	1·99	6·63
Lysine	1·45	9	1·27	4·23
Tyrosine	1·09	11	0·41	1·37
Arginine	1·33	7	0·27	2·69
Histidine	2·13	17	0·12	1·15
Tryptophan	3·67	29	0·18	0·45
Cystine	0·58	81	0·08	0·08

glycine, valine, threonine, isoleucine, proline, aspartic acid, phenylalanine, glutamic acid, lysine and arginine and quite close to what one would expect in normal GLC analysis, indicating perhaps that derivatization was essentially the same from sample to sample.

With the remaining amino acids derivatization appeared to be more variable and also the possibility arises that in some cases, *eg* cysteine, hydroxyproline and cystine insufficient material was placed on the column to obtain a large enough response.

To date amino acid analyses have been carried out on cod and mackerel minces and homogenates treated in buffer solution at pH 5 and 10·5 for 15min and 4h with and without the addition of 0·75% hydrogen peroxide in the buffer. In addition three and four analyses were carried out on separate portions of unprocessed cod and mackerel minces, respectively.

Several effects were discernible in the cod analysis. First of all, as can be seen in *Fig 1*, the hydrogen peroxide-treated material had reduced amino acid content and this was most noticeable in the homogenate treatment. *Table II* shows that treatment with hydrogen peroxide also resulted in a general lowering of the concentrations of nearly all the amino acids and the total amino acid content, when compared with the buffer only and untreated samples. Methionine and cystine were affected most, those disappearing almost completely in the peroxide treatment and although losses of cystine appear to be related to the degree of comminution and length of treatment it was not possible to determine any such relationship for methionine.

The measured coefficient of variation for a number of the acids, *eg* glycine, valine, aspartic acid, phenylalanine, glutamic acid and lysine in the non-peroxide-treated samples approximate those observed in the analysis of the standards. This indicated that not only did little change occur in the various buffer treatments but also that the protein hydrolysis stage was reproducible and did not cause further errors in the analysis. The remaining amino acids showed a greater variation between analyses. There is a trend towards a greater coefficient of

Fig 1   Total amino acids in cod

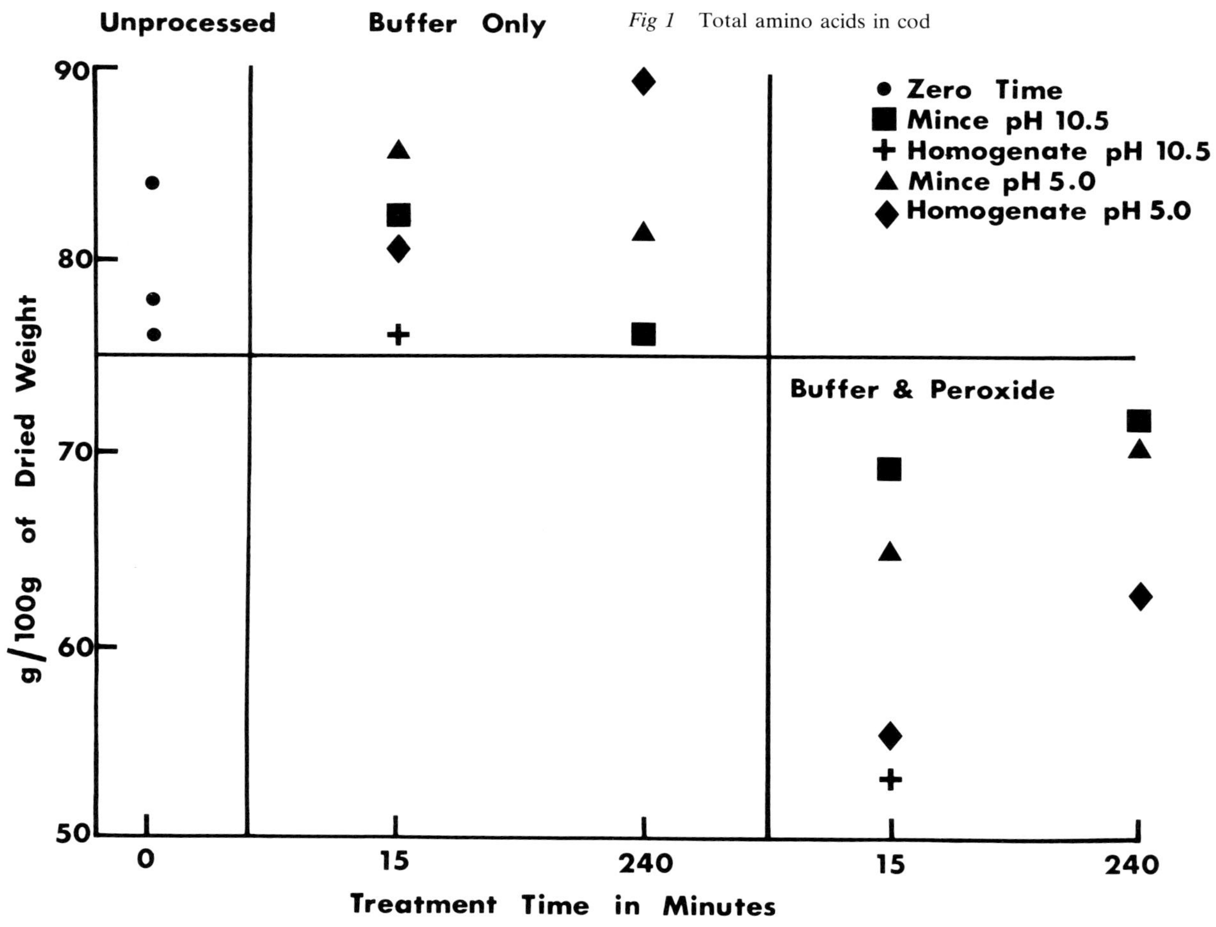

variation in the peroxide-treated samples than in either the standards or the non-peroxide-treated material. It is assumed that this was caused by a somewhat erratic attack by hydrogen peroxide on the protein substrate so that in some instances little change took place whereas in others, especially in homogenized treated material, major changes occurred.

The amino acid analyses of the mackerel which are set out in *Table III* were not so reproducible as those of the cod even in the untreated samples. It seems probable that the lipids present in these fish, 22% of the tissue weight, adversely affected the reproducibility of the method. The amino acid content of the dried material was significantly lower than that of the cod, 30–40g as against 70–80g per 100g dry weight, and this is accounted for by the presence of lipids.

Even though the mackerel amino acids showed a greater variation than those of the cod it was noticeable that the hydrogen peroxide treatment lowered the total concentration. With one or two exceptions methionine and cystine were lost in the majority of the samples regardless of pH and whether they were treated with hydrogen peroxide or not. A possible explanation for this is that during the drying and derivatization pro-

Table II
ANALYSIS OF AMINO ACIDS IN COD

Amino acid	Untreated samples			Buffer only treated samples			Buffer + peroxide treated samples		
	Number of analyses	Mean of conc g/100g dry tissue	Coefficient of variation	Number of analyses	Mean of conc g/100g dry tissue	Coefficient of variation	Number of analyses	Mean of conc g/100g dry tissue	Coefficient of variation
Alanine	3	5·58	3	7	5·00	6	7	4·35	25
Glycine	3	4·54	7	7	3·91	8	7	3·22	13
Valine	3	3·84	14	7	4·16	8	7	3·57	14
Threonine	3	3·27	10	7	3·51	12	7	1·86	31
Serine	3	2·60	7	7	3·18	11	7	1·37	41
Leucine	3	6·54	3	7	6·91	9	7	5·82	12
Isoleucine	3	3·26	22	7	3·79	5	7	3·08	15
Proline	3	2·68	3	7	3·00	14	7	2·35	11
Cysteine	3	0·71	87	7	0·78	46	7	0·34	44
Hydroxyproline	3	0·07	83	6	0·05	60	7	0·004	750
Methionine	3	0·93	140	7	0·98	109	6	0·04	100
Aspartic acid	3	8·25	1	7	8·43	8	7	6·57	19
Phenylalanine	3	1·65	1	7	1·58	7	7	1·34	11
Glutamic acid	3	15·80	4	7	16·74	10	7	14·21	11
Lysine	3	9·73	6	7	10·46	8	7	8·58	12
Tyrosine	3	3·27	19	6	3·12	17	7	2·00	28
Arginine	3	4·41	9	6	4·85	17	7	3·01	40
Histidine	3	0·87	61	6	0·83	75	7	0·93	37
Tryptophan		(a)			(a)			(a)	
Cystine	3	(b)		6	(b)		7	(b)	
Total		78·00			81·26			62·60	

(a) Destroyed by acid hydrolysis
(b) Very low and variable values

Table III
ANALYSIS OF AMINO ACIDS IN MACKEREL

Amino acid	Untreated samples			Buffer only treated samples			Buffer + peroxide treated samples		
	Number of analyses	Mean of conc g/100g dry tissue	Coefficient of variation	Number of analyses	Mean of conc g/100g dry tissue	Coefficient of variation	Number of analyses	Mean of conc g/100g dry tissue	Coefficient of variation
Alanine	4	2·79	19	8	2·42	38	8	2·27	19
Glycine	4	1·99	11	8	1·54	18	8	1·42	23
Valine	4	2·12	23	8	1·94	19	8	1·76	17
Threonine	4	1·02	69	8	0·68	85	7	0·66	50
Serine	4	0·72	74	8	0·48	112	8	0·38	66
Leucine	4	2·57	52	8	2·71	29	8	2·22	40
Isoleucine	4	1·83	31	8	1·69	16	8	1·49	18
Proline	4	1·32	14	8	1·31	18	8	1·09	18
Cysteine	4	(b)		8	(b)		8	(b)	
Hydroxyproline	4	(c)		8	(c)		8	(c)	
Methionine	4	(b)		8	(b)		8	(b)	
Aspartic acid	4	3·00	41	8	2·82	34	8	2·51	28
Phenylalanine	4	0·76	24	8	0·73	22	8	0·56	23
Glutamic acid	4	6·13	19	8	7·84	13	8	5·46	42
Lysine	4	4·67	22	8	4·20	29	8	3·58	28
Tyrosine	4	1·29	39	8	1·37	26	8	1·02	28
Arginine	4	1·51	52	8	1·78	38	8	1·69	37
Histidine	4	1·44	45	8	0·94	56	8	0·52	86
Tryptophan		(a)			(a)			(a)	
Cystine	4	(c)		8	(c)		8	(c)	
Total		33·16			32·45			26·63	

(a) Destroyed by acid hydrolysis
(b) Low and variable values
(c) Very low values

cedures co-oxidation of these acids took place with the lipids.

Overall the protein analyses show that even under relatively mild treatment of fish minces and homogenates, changes occurred which resulted in insolubilization of protein and loss of amino acids, particularly the sulphur amino acids. Similar observations, although not with fish proteins (Srinivasan, 1972; Snider and Cotterill, 1972), have been noted elsewhere but it should be pointed out that other workers (Koning and Rooijen, 1972; Schreiber *et al*, 1976; Orlova *et al*, 1977) using milk and fish protein substrates and a range of hydrogen peroxide concentrations found no change in the amino acid content.

4 Acknowledgements

The authors would like to thank J F March (John Innes Institute, Norwich) for some very useful discussion about the amino acid derivatization procedure.

5 References

BANKS, A. Rancidity in fats. I. The effect of low temperature, sodium
1937   chloride and fish muscle on the oxidation of herring oil. *J. Soc. Chem. Ind.* London, 56, 13T–15T

BANNERMAN, A, KEAY, J N, SMITH, J G M and HARDY, R. Kipper
1973   fingers—a new fish product. *Fd. Manuf.*, 48(10), 63 and 64

BLIGH, E G and DYER, W J. A rapid method of total lipid extraction
1959   and purification. *Can. J. Biochem. Physiol.*, 37, 911–917

CHESNER, L and DICKINSON, K. Processing of tripe. *Fd. Manuf.*, 38,
1963   606–608

COWIE, W P and MACKIE, I M. Examination of the protein extracta-
1968   bility method of determining cold-storage protein denaturation in cod. *J. Sci. Fd. Agric.*, 19, 696–700

DAWSON, R M C, ELLIOTT, D C, ELLIOTT, W H and JONES, K M
1969   (Eds). *Data for Biochemical Research*. Clarendon Press, Oxford. 2nd edn

DURBETAKI, A J. Direct titration of oxirane oxygen with hydrogen
1956   bromide in acetic acid. *Anal. Chem.*, 28(12), 2000–2001

ELLINGER, G M and DUNCAN, A. The determination of methionine in
1976   protein by gas liquid chromatography. *Biochem. J.*, 155, 615–621

JAMES, A L and McCRUDDEN, J E. Whitening of fish with hydrogen
1976   peroxide. *Proc: Conf.* The production and utlisation of mechanically recovered fish flesh (minced fish), Torry Research Station. 54–55

KEAY, J N and HARDY, R. Application of bone separation techniques
1974a   to pelagic and underutilised species and the preparation of derived products. In: *Second technical seminar on mechanical recovery and utilisation of fish flesh*. Ed R E Martin. Boston, Mass. 88–150

——— The application of extrusion techniques to the utilisation of com-
1974b   minuted fish flesh. In: *Second technical seminar on mechanical recovery and utilisation of fish flesh*, Ed R E Martin. Boston, Mass. 101–108

KONING, P J de and ROOIJEN, P J van. Effect of hydrogen peroxide on
1972   the amino acid composition of the proteins from cheese whey and evaporated milk. *Ned. Melk-en Zuiveltijdschr.*, 26(1), 41–44

MANYUA, J K. Hydrogen peroxide alteration of whey protein in skim
1975   milk and whey protein solutions. *Milchvissenschaft*, 30(12), 730–734

MARCH, J F. A modified technique for the quantitative analysis of
1975   amino acids by gas liquid chromatography using heptafluorobutyric n-propyl derivatives. *Anal. Biochem.*, 69, 420–442

ORLOVA, T A, KURANOVA, L K and NELICHIK, N N. Effect of hydrogen
1977   peroxide on quality of protein matter. *Trudy Polyar. nauchno-issled.* Inst. morsk. ryb. Khoy Okeanogr., 39, 125–129

SCHREIBER, W, ANTONACOPOULUS, N and KRESS, G. Amino-
1976   saeuregehalt von heringsmarinaden bei wasserstoffsuperoxyd. *Infn Fischw. Auslds*, 23(3), 92–94

SIMS, G G, COSHAM, C E and ANDERSON, W E. Hydrogen peroxide
1975   bleaching of marinated herring. *J. Fd. Technol.*, 10, 497–505

SNIDER, D W and COTTERILL, O J. Hydrogen peroxide oxidation and
1972   coagulation of egg white. *J. Fd. Sci.*, 37(4), 558–561

SRINIVASAN, A. Effect of hydrogen peroxide on buffalo milk proteins.
1972   *Biochem. J.*, 128(1), 50pp

UNILEVER RESEARCH. Snoflake functional fish protein—Users Manual.
1975   Unilever Research, Aberdeen, Scotland. 16pp

VOGEL, A I. *A textbook of quantitative inorganic analysis*. Longmans.
1961   3rd edn

Alternatives to the production of fish portions from frozen fillet blocks

K J Whittle, I Robertson and J N Keay

1 Introduction

The feasibility and advantages of freezing fresh fish as a means of preservation and of extending distribution lines were given commercial impetus 50 years ago by Clarence Birdseye. However, freezing in processing and retailing fish did not become firmly established in the United Kingdom until the mid-sixties when frozen fish fingers (fish sticks), some 10 years after being launched, had become meal-time favourites and were especially popular with children. This success must have contributed significantly to the rapid growth of freezing-at-sea between 1962 and 1972 and to the heyday of the British freezer trawler fleet in the early seventies since, by 1970, much of the frozen-at-sea cod landed in the UK was processed into frozen fillet blocks for subsequent fish finger or fish portion production either at home or abroad. The production of frozen fillet blocks had become the largest single operation in frozen fish processing in the UK and the frozen cod block had become a new commodity in the international trading market, its value being influenced markedly by the Boston market in the USA.

The successful introduction of frozen white fish portions in attractive sauces beginning about 1969; the later introduction of crispy, puffy batters as an alternative enrobing to batter and crumb; and the increased usage of mechanically recovered, deboned fish flesh in products, either as an added proportion or entirely in its own right; have each served to establish more firmly the now substantial contribution of the frozen finger and portion to the total consumer market in fish. In terms of landed weight, fish fingers alone account for about 10% of the total supplies of fish used for human consumption. They represent about 30% of the fish used in quick-frozen production and the retail market value is greater than £65 million with annual sales now of about 37 000 tonnes product.

Recent years, of course, have seen a dramatic reduction in the fortunes of the UK distant-water freezer fleet. However, fish finger and portion production has been sustained here by increased imports of fresh and frozen fish from a variety of sources; by increasing the range of species offered; and by introduction of so-called 'economy' or lower priced fish fingers in which use of mechanically recovered, deboned fish flesh has played a large part. Recovered fish flesh will be referred to as 'mince' throughout the following paper.

During this long development period spanning more than 25 years, the manufacture of frozen fillet or mince blocks has been an essential prerequisite to finger and portion production. The fundamentals of the manufacture of the frozen fillet block (also referred to as laminated blocks and industrial blocks) and subsequent processing to portions have remained more or less the same and are outlined briefly below.

2 Portion production by frozen fillet block manufacture and processing

Predetermined weights of deboned fillet, mince or mixtures of the two, perhaps including additives such as polyphosphate or salt, are packed by hand into a rectangular mould or metal frame into which is fitted a waxed card carton. The dimensions of the mould vary considerably according to source or market intended. A common size in world trade is about 7·4kg, and is sometimes referred to as an international block. The packed moulds usually are frozen in horizontal plate freezers and the rectangular, frozen blocks protected by the surrounding waxed carton are subsequently ejected for storage. The aim of the entire process is to form a smooth, regularly shaped, frozen block of standard dimensions and weight, free from internal and external air spaces or voids, from blemishes such as skin and membranes and from bones. Apart from the important aspects of overall quality and consumer acceptability, tight control of the physical properties of the block, such as shape, dimensions, weight, voids and angles, is essential for efficient and economical production of portions subsequently and, quite possibly, remote from where the blocks were made. Adherence to the specification is particularly important when considering the small sizes (less than 20g) required for some derived products since manufacture involves sawing, slicing, chopping or some combination of these, all of which produce some saw or trim waste. The waste increases if underweight or overweight portions, the result of irregularities in the original block, have to be recycled or discarded.

The formation of frozen blocks as described above is a labour-intensive process and considerable effort and control has to be exercised to produce blocks of impeccable geometry and specification. Attention to detail, standardization, strict control as well as application of common sense throughout, enables perfectly acceptable blocks to be made from a wide variety of species, using relatively unskilled labour. However, various patents have been described over the years with the aims of making it easier to produce blocks with good physical characteristics and of reducing the labour-intensive nature of production. Some years ago, a freezer company introduced and tested an automatic, double contact, horizontal plate freezer designed to freeze fillet blocks continuously. It included a conveyor system of moulds which three operators filled by hand to achieve a throughput of about 2½ tonnes/h. A disadvantage was that the blocks were formed with a slightly tapered edge to facilitate removal from the moulds and this had to be trimmed off to obtain a precise rectangular shape. None

of these proposals or developments appear to have been exploited to any significant extent.

The most efficient combination of cutting methods for portion production uses a bandsaw for the initial cuts and a slicer, chopper or circular saw for the final division. 'Sawdust' losses in fish finger production should be no more than 6·5% under absolutely ideal conditions of throughput whereas, with the bandsaw alone, losses may even be higher than 15%. However, to attain high efficiency with slicing and chopping it is necessary to raise the temperature of the frozen blocks (temper) from about −30°C to within the range −7°C to −15°C (Bezanson et al, 1973; Sanders, 1967) in order to prevent fracturing and reduce cutting waste. Depending upon the type and speed of operation, the optimum temperature seems to be −8°C to −10°C.

In the final stage of processing, the cut portions are usually enrobed or coated in batter and crumb or puffy batter alone and flash-fried. Irrespective of how they are processed, they still have to be cooled again to the storage temperature of −30°C. Subdivision of the large frozen block has the further disadvantage that portion shape is limited to a number of geometric forms which can be cut with as little waste as possible. It is true, of course, that novelty-shape portion cutters have been developed, but waste from offcuts is a critical factor with unusual shapes and the portions need to be heavily enrobed or puffy batters used if the regular outline and precise edges of the portion are to be hidden by 'rounding off'.

3 Portion production by direct forming or moulding

The idea of forming portions directly from fillet or mince is not new. Patty formers have been available for years, notably in hamburger production.

Various other ideas for moulding wet and frozen fish have been put forward and tested. Indeed, minced fish can be formed, moulded or extruded successfully but the fact remains that these various ideas have not been taken up commercially on any large scale for high-volume fish products like fingers and portions.

Recently, using a Guylew Food Former, we have examined the forming of portions directly from thawed and unfrozen fillet, fillet pieces (chunks) and mince, sometimes with added polyphosphate, using a variety of white fish species (haddock, whiting, saithe, blue whiting and cod). In all, 13 different treatments were investigated. Six varied shapes were produced as shown by the haddock fillet products in *Fig 1*, ranging in weight from about 30g to 75g. The single emergent line of portions from the former was automatically converted to five lines and transferred to the moving stainless steel belt of the Torry Continuous Air Blast Freezer. After freezing, inspection and weighing, the portions were coated with batter and crumb, flash-fried, cooled to −30°C, weighed again and stored.

The forming action of the machine was most efficient when operated at its higher rates of production from 6 000 to 8 000 portions/h, representing a throughput of 340kg/h to 450kg/h. This does not represent the upper limit of production by this type of forming process but merely the operating parameters of the machine actually used. At the start and finish of a run, the mould was not

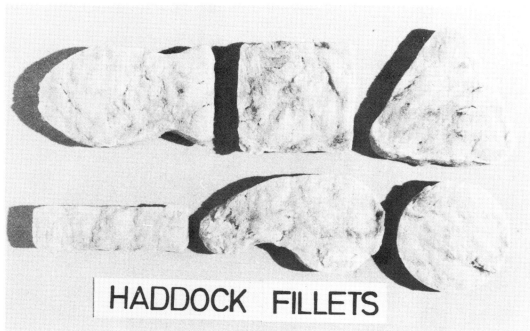

Fig 1 A variety of shapes made in the Guylew Food Former using whole haddock fillets untreated with polyphosphate. Average portion sizes reading from left to right are
Top row: 72·2g, 60·6g, 62·4g
bottom row: 31·9g, 61·4g, 47·3g

entirely filled resulting in underweight portions. These can be collected and recycled. The residual material in the former and hopper at the end of a production run was constant irrespective of throughput and in our experience was about 1kg. It too can be recovered easily and recycled. Reproducibility of weight among the whole portions in the bulk of the production run was excellent as the histogram of portion weight in *Fig 2* shows for finger shapes (32g) formed from untreated blue whiting fillets. Apart from some minor aberrations, the distribution of portion weights about the mean did not depart significantly from a normal distribution and so the coefficient of variation (standard deviation as a percentage of the mean) could be used as a measure of reproducibility. *Table I* summarizes the weight data of the uncoated portions by comparing coefficients of variation. The mean values of the pooled data for each shape varied only between 1·0 and 1·2, indicating that the variation about the mean weight for each shape represented a relatively constant proportion of the mean weight and that reproducibility in the different shapes was comparatively good. Further analysis showed that the portions from mince, chunk and fillet, respectively, ranked in order of decreasing reproducibility and that inclusion of polyphosphate had no effect on reproducibility from chunk and fillet but, somewhat surprisingly, actually reduced reproducibility when added to mince. These detailed observations within the small overall variation do not alter the conclusion that the former produced portions with a very high degree of weight reproducibility from mince, chunk and fillet, as *Fig 2* shows for the latter. The portions compared very favourably with fish finger blanks (approximately 19g) cut by hand bandsaw for which the coefficient of variation ranged from 3·9 to 5·3 (Bannerman and Smith, unpublished document) and, as far as we know, with current commercial practice.

Unfortunately, the experimental production line used in the tests above was not ideal in that the various conveyor belt speeds could not be matched exactly, the full capacity of the freezer tunnel could not be used and the maximum possible freezing time was not quite sufficient to allow continuous operation. Nevertheless, it

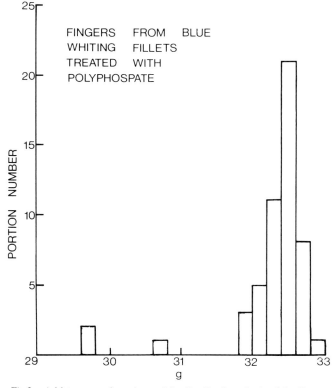

Fig 2 A histogram of portion weight distribution obtained for finger-type shapes produced in the Guylew Food Former using whole blue whiting fillets treated with polyphosphate

252

Table I
UNCOATED WET FORMED PORTION REPRODUCIBILITY BY COMPARISON OF
COEFFICIENTS OF VARIATIONS[1] (CV) FOR POOLED DATA

A. By portion shape and weight

Shape	▭	□	△	○	⊲	◠
Mean weight g (all treatments)	31·9	60·6	62·4	47·3	72·2	61·4
CV	1·0	1·0	1·1	1·2	1·0	1·0

B. By raw material and treatment

Raw material	Fillet		Chunks		Mince	
Treatment[2]	UT	T	UT	T	UT	T
CV (pooled data all shapes)	1·05	1·10	0·7	0·65	0·55	1·65

[1] With only minor aberrations, distributions about the mean portion size did not show significant skewness or departure from normal distribution so coefficient of variation (standard deviation as % of mean) could be used as a measure of reproducibility
[2] UT = not treated; T = treated with polyphosphate

was possible to demonstrate the feasibility of the approach with continuous runs up to 25kg and to confirm that whole fillet material can be mechanically formed reproducibly into various cohesive shapes.

4 Comparison of the traditional and wet forming processes for portion production

The production of portions of the desired shape and size directly from fillet or mince, in theory at least, offers a number of advantages over the traditional methods in current and widespread use. The alternative process is relatively simple in comparison with high-pressure moulding or extrusion of frozen fish and with extrusion methods which rely on the formation of pastes and which require additives or lubricants.

Direct forming overcomes the need to produce the precise, geometric frozen block of fillet or mince unless portion manufacture is to be far removed from the supplies of fish. In this case, the high costs of transporting whole fish over large distances may weigh heavily in favour of transporting only the usable portion, that is the fillet or mince. However, many processing centres remain relatively close to the ports where fish are landed. Also, we have seen recently how the greatly increased exploitation for human consumption of mackerel caught in UK waters has basically been confined to bulk chilling at the ports of landing followed by refrigerated transport by road to continental Europe. In cases where the frozen block is still considered necessary for transport or storage, the need for stringent control of the physical characteristics of the block is no longer important, except in terms of stated weight, if wet forming is to be used subsequently because the thawed block will be satisfactory for portion production. Without the constraints of a demanding physical specification, it seems possible to consider simpler means of completely automatic production of blocks of specified weight coupled directly to a machine output of skinned fillets or a mechanical separator producing mince. Alternative methods of packaging the block could also be considered.

The waste produced during wet forming is practically negligible since misshapes and residual fish can easily be recycled during the same run or a succeeding one. More importantly, perhaps, the amount to be recycled is independent of the volume of a production run. In our tests

the residual material in the former and hopper was only about 1kg per entire production run compared with a minimum loss in cutting waste or sawdust of about 7% of throughput for the most efficient combination of cutting methods for fish fingers from 7·4kg blocks. Assuming optimistically that all cutting losses actually averaged only 5% of the weight of blocks processed, an estimate can be made of the potential saving in raw material and its value in terms of lost product sales if the wet-forming process entirely replaced the traditional process for the production of fish fingers in the UK for instance. The most complete data on which to base this estimate relate to 1976. Some 37 000 tonnes of fish fingers were consumed containing about 26 000 tonnes of fish, which cost the consumer about £65 million. About 1 400 tonnes of frozen block would have been wasted in manufacture, equivalent to retail sales worth some £3·4 million. The actual percentage loss in manufacture is difficult to determine because a variety of companies are involved and detailed production information is understandably confidential. In practice, 5% cutting loss for UK production as a whole is likely to be an underestimate and it is also probably unrealistic to expect that all production would be achieved by wet-forming. However, the estimate does give a guide to potential savings on cutting losses but does not take account of additional savings in raw material which could be made by recouping the losses in production of the frozen blocks themselves (0·5–2% raw material is lost in producing the fillet block itself).

A variety of shapes and sizes of portions from 10g to 100g or more can be made with different forming heads. There is wide versatility of shape since attention can be given to shaping all the surfaces, in contrast to the so-called novelty portion cutter. If necessary, some advantage can also be taken of the tendency of the formed portions to flow when subjected to a sudden change in momentum. The wet portions are relatively fragile and must be handled as little as possible until the shape is set. Shapes in which one dimension is very much greater than any of the others as in a fish finger or shapes with small appendages are susceptible to deformation and damage in handling. This is particularly so if they are subjected to sudden forces which might twist them out of shape or to collisions with guides or other parts of a conveyor system which might result in small pieces being removed from one portion and added to another. To retain the desired shape the portions were frozen immediately in a continuous tunnel freezer followed by enrobing, flash-frying and cooling to −30°C for storage as in the traditional process. It is not necessary to freeze newly formed portions to −30°C; the temperature required would be the lowest necessary to retain shape during enrobing, flash-frying and the associated handling by machine. Recent results have shown that the portions need only be frozen to an equilibrated temperature of −2°C for shape to be retained perfectly but some work still needs to be done to determine whether chilling alone and gentle handling would suffice. At chill temperatures, coating the wet portion would probably be more successful than with the frozen portion; also the prior freezing stage would be eliminated.

The heat content changes at various stages of the traditional and alternative processes, under the con-

253

ditions outlined in *Fig 3*, are compared in *Fig 4* (Nicholson, F J, personal communication) in order to estimate what the difference in energy requirement might be solely for the heat to be added and removed in the two processes. It is assumed that (*i*) both processes begin with wet fillets at 6°C and conclude by coating with batter and crumb, flash-frying and cooling to −30°C for storage, (*ii*) the fillet blocks are removed from storage at −30°C, tempered to −7°C or alternatively −15°C and

processed with only 5% loss in cutting, (*iii*) the wet-formed fingers are immediately frozen to −2°C. The estimates of change in heat content are made in Kjoules per kg end-product and it is assumed that equilibrated temperatures obtain throughout. It was not possible to estimate the changes during flash-frying adequately but these are the same for each case. In the traditional process with tempering to −7°C about 30% more energy has to be transferred as heat added or removed.

1. TRADITIONAL PROCESS

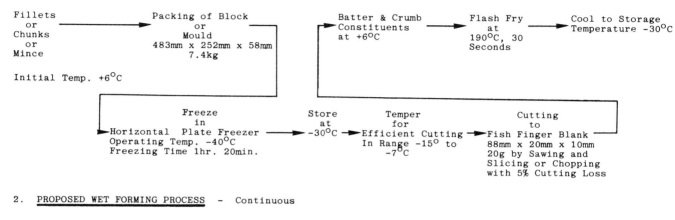

2. PROPOSED WET FORMING PROCESS − Continuous

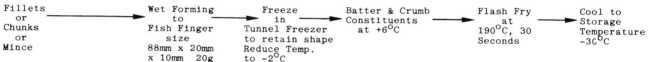

Fig 3 Comparison of fish finger product processes. A production line flow chart comparing the traditional process and the proposed wet-forming process for making portions of fish finger size. The conditions given were used in the calculation of heat content changes shown in *Fig 4*

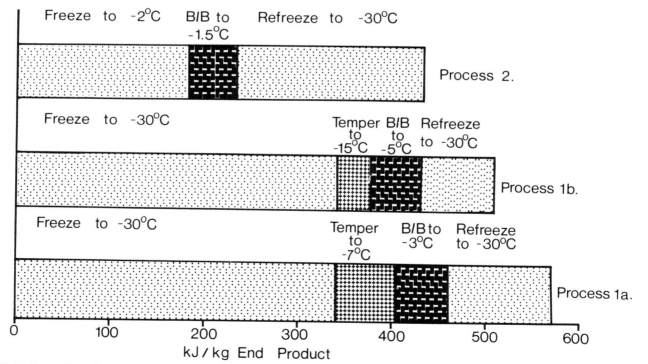

Fig 4 Comparison of the energy requirements for the traditional and proposed processes for making portions of fish finger size. The comparison is made in terms of heat content changes for the various stages of the processes expressed as kJoules per kg end-product. The conditions used are those outlined in *Fig 3*

One of the important features of the forming method is that it is not restricted to minces. Whole fillet or chunk can be formed successfully as long as the shape remains cohesive. Thus, products or portions stated to contain fillet can be produced as such. It is interesting in this regard that consumer preferences have been tested on products of the fish finger type made entirely from fillet and compared with those made entirely from mince derived from the same source. So far, the results show that there is no clear consumer preference for the product made entirely from fillet. There are products, of course, which are entirely desirable to make up from fillet, such as the larger portions often prepared in puffy batter. One of these is a frozen portion of approximately 100g for use for example in the fish-frying trade fish and chip meal and there is potential for a battered product for retail sales as well. Fillets much smaller than 100g can be composited and formed into an attractive frozen fillet shape to produce standard size portions close to 100g giving good portion control and enabling small white fish to be utilized. The frozen portions can be battered and then deep-fat fried successfully. Unfortunately, there is some resistance to the use of frozen portions generally among a large proportion of fish fryers. Thus, even although the final product from the composite portions is excellent, the fish fryers still have to be won over to the use of frozen portions. For this product there seems to be the additional requirement that the fillets should be arranged or orientated along the long axis of the fillet shape so that the randomizing moulding or forming head discussed earlier is unsuitable. An alternative approach is currently being developed in Torry Research Station.

In conclusion, wet-forming in comparison to traditional methods, offers the advantages of reduction in machine operations and in waste, savings in energy and labour, versatility of shape and size, close portion control, and a simplified production line. In some circumstances it may be economically more advantageous to store and transport a frozen block rather than the end-product. If the frozen block is still required and is followed by wet-forming, the specification of the physical characteristics of the block will be less demanding.

Consistent shortage of traditional white fish supplies suggests that examination of the merits of the wet-forming process is timely and may encourage commercial exploitation of alternative species.

5 Acknowledgements

The calculation of the estimated heat content changes was carried out by F J Nicholson of Torry Research Station. G L Smith carried out some of the statistical analysis.

Some of the background information on fish finger consumption was provided by the Public Relations Office, Birds Eye Foods Ltd.

J W Preston, Guylew Manufacturing Co Ltd, arranged the loan of the Food Former and Flow Convertor used in this work and advised on their performance.

6 References

BEZANSON, A, LEARSON, R and TEICH, W. Defrosting shrimp with
1973   microwaves. In: *Proc. Gulf and Caribbean Fisheries Institute*,
       25th Annual Session, 44–55
SANDERS, H R. Note on the tempering of fish finger blocks. *J. Fd.*
1967   *Technol.*, 2, 183–187

Tissue degradation and belly bursting in capelin

A Gildberg and J Raa

1 Introduction

Sprats, herring, mackerel and capelin (*Mallotus villosus*) caught during periods of heavy feeding are very susceptible to autolytic tissue degradation. The usual explanation has been that fish with much stomach and gut content also have high activities of digestive enzymes, which leak out *post mortem* and degrade the surrounding tissues.

Certain catches of capelin liquefy very fast. This is most pronounced in the summer season (August–September), whereas the spawning winter capelin usually is rather stable. In apparent accordance with this, the level of proteolytic enzymes, on average, is higher in catches of summer capelin (Gildberg, 1978). It has, nevertheless, been hard to find a good correlation between proteolytic activity and the tendency of belly bursting. A high level of proteases in the fish does not necessarily imply a rapid tissue degradation, and the belly bursting may sometimes occur even in fish with relatively low level of proteases (Marvik, 1974).

It is obvious that there are several other factors that may influence the tissue degradation *post mortem*. The

state of the connective tissue certainly is of some importance. It is well known that heavy feeding fish have a weaker connective tissue than starving fish (Lavéty and Love, 1972; Love, 1975).

The connective tissue is relatively weak if the *post mortem* tissue pH is low (Love *et al*, 1972b; Gildberg and Raa, 1979). Some observations indicate that the *post mortem* tissue pH is lowest when fish are caught during heavy feeding periods (Love *et al*, 1972a; Gildberg, 1978).

Quick degradation of fish protein tissue has sometimes been observed at slightly acid pH, where the digestive proteases have low activity on a model substrate like haemoglobin (Hjelmeland, 1978; Hjelmeland and Raa, 1980; Nicolaysen, unpublished). To explain these observations it is necessary to get more knowledge about enzymic action on the different tissue components.

Our studies were initially designed to reveal simple assays which could be used to predict the risk of belly bursting in catches of capelin. Although this goal has not been reached, we hope that the work gives some information that may contribute to a better understanding of autolytic degradation of fish tissue.

2 Materials and methods

2.1 Species used
Capelin (*Mallotus villosus*) caught by trawl in the Barents Sea in August and September were used.

Immediately after catching the fish were wrapped individually in aluminium foil, packed in sealed plastic bags and stored at −20°C until use. Fresh non-frozen material was used in some experiments.

2.2 Method of determining protease activity
Protease activity was determined according to Barret (1972). The incubation mixtures consisted of 0·5ml Johnson/Lindsay-buffer (Johnson and Lindsay, 1939), 0·25ml enzyme sample and 0·25ml 8% haemoglobin. The concentration of Folin positive material in the supernatant after adding TCA (5% end concentration) was determined by Lowry's method (Lowry *et al*, 1951) using tyrosine as a reference. There is one exception to this procedure. That is in *Fig 4*, where the substrates added were 4% haemoglobin and 0·8% structural glycoprotein. The glycoprotein was extracted from capelin skin by urea according to a procedure described by Robert and Comte (1968).

2.3 Soluble protease
Protease activity was determined in extracts of homogenized digestive tracts. Twenty capelin from each catch were gutted. The digestive tracts were homogenized in nine volumes of distilled water (1min in a Waring blender).

The homogenates were centrifuged (20min, 20 000g) and the protease activities in the supernatant solutions were determined.

2.4 Frequency of belly bursting
Twenty non-frozen newly caught fish were placed at 4°C in a covered plastic tray, and the number of fish with burst bellies was counted after three days. One or more holes in the belly side between the gills and the anal opening was taken as belly bursting.

2.5 pH in stomach contents
Immediately after capture the stomach contents were collected from 20 fish, and pooled. The pH in this mixture was recorded as the representative stomach pH of that particular catch.

2.6 Solubilization of skin and muscle
Fifty milligrams of lyophilized tissue were incubated for 20h at 8°C with an enzyme/buffer mixture prepared as follows: digestive tracts were homogenized in an equal volume of distilled water, and the supernatant (20 000g, 20min) was mixed with 3 vol of Johnson/Lindsay buffers with different pH values. At the end of the incubation time, the samples were centrifuged for 20min at 20 000g at about 4°C. The Folin positive material in the supernatant (= soluble protein) and degraded protein soluble in 5% TCA (= peptides/amino acids) was determined (Lowry *et al*, 1951). Bovine serum albumin (BSA) was used as the standard protein in this analytical method.

Total protein in muscle and skin substrates was determined by the Kjeldahl procedure (by multiplying total nitrogen by the factor 6·25). Reference incubations without substrate were run at otherwise identical conditions. The pH values shown are the average of pH measured at the start and at the end of the incubations.

2.7 Autolysis
Ten grams of minced capelin were mixed with 30ml Johnson/Lindsay buffer and placed on a water bath (25°C) for 18h. At the end of the incubation, the samples were centrifuged (20min, 20 000g). Soluble protein in the supernatant and peptides/amino acids soluble in 5% TCA were estimated by the Lowry method (Lowry *et al*, 1951) using BSA as a standard.

3 Results

Figure 1 shows the activities of soluble proteases in homogenates of digestive tracts with different amounts of feed content. The activity of soluble proteases, both at acid and alkaline pH, decreases with increasing amount of feed content in the digestive tract. Even the total amount of soluble acid protease activity (activity/g whole fish) decreases significantly with increasing feed content.

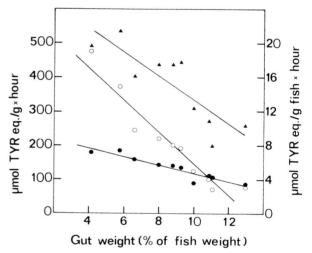

Fig 1 Activities of soluble proteases in homogenates of digestive tracts from different catches of capelin as a function of gut weight. Activities were determined at pH 4·4 (○) and 9·8 (●) after incubation for 1–2h at 23°C. Soluble activity at pH 4·4 is also given as total activity per g whole fish (▲). Each point represents different catches, and measurements were performed on batches of 20 fish from each catch

Figure 2 shows the activity of soluble protease (activity/g whole fish) at pH 6·8, and the corresponding tendency of belly bursting in the different catches. The tendency of belly bursting was usually highest in catches where the fish had levels of soluble protease in the digestive tract, but the correlation was poor.

Figure 3 shows the pH in the stomach content of capelin from different catches. There is no apparent correlation between the stomach pH and the amount of feed in the digestive tract.

Figure 4 shows the activity of digestive proteases at different pH with haemoglobin and structural glycoprotein (from capelin skin) as substrates. With haemoglobin as the substrate, the enzymes have optimal activities at pH about 3 and 9, while the glycoprotein is readily degraded only at neutral and alkaline conditions.

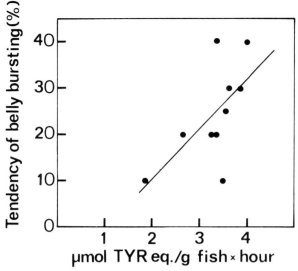

Fig 2 Frequency of belly bursting in different catches of capelin as a function of activity of soluble digestive proteases per g whole fish. Protease activities were determined after incubation for 2h at 23°C and pH 6·8. The belly bursting tendency was determined after incubating 20 fish from each catch for three days at 4°C

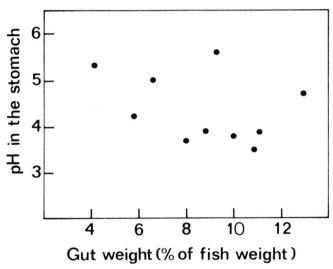

Fig 3 pH in the stomach content of capelin from different catches as a function of gut weight related to fish weight. Each point is representing batches of 20 fish

Figure 5 shows the effect of pH on the chemical/enzymatic solubilization and degradation of protein from capelin skin and muscle at low temperature. Skin was resistant to enzymatic decomposition at pH values above 6, but was rapidly solubilized at pH below 5. The solubilization of muscle protein was highest at acid and neutral pH and had a minimum between pH 5 and 6.

Figure 6 shows the autolysis of minced capelin at different pH. At slightly acid and neutral conditions the protein degradation is low, but the release of high molecular protein is highest in this interval.

4 Discussion

It is well established that fish which are caught in their feeding periods are very liable to *post mortem* tissue degradation (Love *et al*, 1972a; Love, 1975). This might

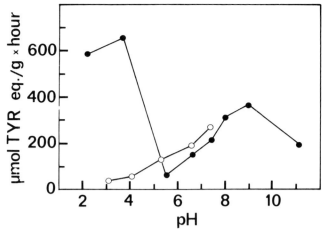

Fig 4 Protease activity in homogenized digestive tracts of capelin as a function of pH. Activities were determined after incubation for 1h at 25°C. The substrates were haemoglobin (●) and structural glycoprotein from capelin skin (○)

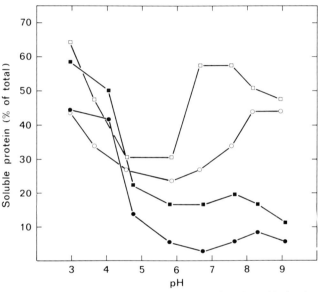

Fig 5 Soluble protein of muscle (□) and skin (■) and peptides/amino acids of muscle (○) and skin (●) after incubation for 20h at 8°C in the presence of digestive enzymes at different pH (Gildberg and Raa, 1979)

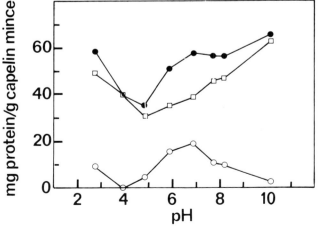

Fig 6 Autolysis of minced capelin as a function of pH: water-soluble protein (●), Folin positive material soluble in 5% TCA (□) and the difference between water-soluble and TCA-soluble protein material (○). The autolysis went on for 18h at 25°C (Gildberg, 1978)

be because the feeding fish produces high levels of digestive enzymes which leak out and degrade the surrounding tissues. Capelin with the highest stomach and gut content contained, however, least extractable proteases. Since it is unlikely that this fish actually had produced less digestive proteases after feed intake, it may be that the proteases are so firmly attached to their substrates in the digestive tract that they are not released in the enzyme extract, and thus not accounted for in the enzyme assay. The bound enzymes are, on the other hand, not diffusible and can accordingly hardly be involved in belly bursting, before advanced stages during feed decomposition. This may explain why fish sometimes become more liable to belly bursting when they are emptying the gut during captivity in seines (Marvik, 1974), and that fish with a high gut and stomach content are not necessarily the first to suffer belly bursting in laboratory tests of short duration. In accordance with this line of argument, the activity of extractable proteases should be a useful indication of the risk of belly bursting.

Although the belly bursting seemed to be most frequent in catches with high protease activity, the correlation was so poor that it cannot be used in practice to predict belly bursting.

The type of feed ingested, the condition of the connective tissue and the *post mortem* pH are additional factors which certainly also determine the rate and pattern of tissue degradation.

The pH of minced capelin caught in the summer season is about 6·5 (Gildberg, 1978), and is in the range where the protease activity of digestive enzymes is low. Slight pH changes may effect the enzyme activity. Diffusion of the stomach acid may speed up the rate of solubilization because the connective tissue is physically weakened and rendered susceptible to the digestive enzymes.

Some observations have indicated that the type of feed ingested influences the rate of acid production in the stomach. We have found (unpublished) that the pH of stomach content which mainly consisted of 'red feed' (*Calanus*) was as low as 3·2, while the stomach content of fish that had ingested mainly a small amphiepod, 'Seafly' (*Hyperia*), had neutral pH.

Nicolaysen at this meeting has showed that the rate of

autolysis at pH 6·5 was highest in sardines which were rejected by the sorter as a raw material for canning. He showed that autolysis at this pH could be significantly higher than at alkaline pH, where the protease activity was close to its optimum.

This has been demonstrated also with capelin (Hjelmeland, 1978). These observations suggest that a solubilizing principle is active at neutral/slightly acid pH, and that its activity does not coincide with the protease activity which was assayed with a soluble protein substrate.

In the present paper we show that solubilized protein accumulates at neutral pH. This may be because a separate tissue solubilizing principle is active in this pH range where the rate of further degradation to peptides and amino acids is low. Alternatively, it may be that the pH optimum of protease activity with an intact tissue substrate differs from that with a soluble protein.

5 References

BARRET, A J. Lysosomal enzymes. In: *Lysosomes — a Laboratory*
1972 *Handbook,* Ed J T Dingle. North-Holland, Amsterdam. 46–135
GILDBERG, A. Proteolytic activity and the frequency of burst bellies in
1978 capelin. *J. Fd. Technol.,* 13, 409–416
GILDBERG, A and RAA, J. Solubility and enzymatic solubilization of
1979 muscle and skin of capelin (*Mallotus villosus*) at different pH and temperature. *Comp. Biochem. Physiol.,* 63B, 309–314
HJELMELAND, K. Buksprenging hos lodde — Undersøkelse av
1978 fordøyelsesproteasenes rolle i vevsdegraderingen. Thesis, University of Tromsø, Norway. 115pp
HJELMELAND, K and RAA, J. Fish tissue degradation by trypsin type
1980 enzymes. *This volume*
JOHNSON, W C and LINDSAY, A J. An improved universal buffer.
1939 *Analyst.,* Lond., 64, 490–492
LAVÉTY, J and LOVE, R M. The strengthening of cod connective tissue
1972 during starvation. *Comp. Biochem. Physiol.,* 41, 39–42
LOVE, R M. Variability in Atlantic cod (*Gadus morhua*) from the
1975 north-east Atlantic: A review of seasonal and environmental influences on various attributes of the flesh. *J. Fish. Res. Bd, Can.,* 32, 2333–2342
LOVE, R M, HAQ, M A and SMITH, G L. The connective tissues of fish.
1972a V. Gaping in cod of different sizes as influenced by a seasonal variation in the ultimate pH. *J. Fd. Technol.,* 7, 281–290
LOVE, R M, LAVÉTY, J and GARCIA, N C. The connective tissues of fish.
1972b VI. Mechanical studies on isolated myocommata. *J. Fd. Technol.,* 7, 291–301
LOWRY, O H, ROSEBROUGH, N J, FARR, A L and RANDALL, R J. Protein
1951 tein measurement with the Folin phenol reagent. *J. Biol. Chem.,* 193, 265–275
MARVIK, S. Stengforsøk. Report to: The Research Laboratory of the
1974 Norwegian Canning Industry, Stavanger, Norway
ROBERT, L and COMTE, P. Aminoacid composition of structural
1968 glycoproteins. *Life Sci.,* 7, 493–497

Studies on the auto-oxidation of sardine oil

F Pizzocaro, E Fedeli and A Gasparoli

1 Introduction

Technological utilization of the sardine (*Sardina pilchardus*), which represents 30% of the entire Italian catch of fish, is economically important.

This fish is of low cost with respect to both production and consumption, but of great biological-food value.

Nevertheless, it is known that the sardine is a difficult fish to preserve in either the fresh or frozen state, because it rapidly becomes rancid and this leads to decay of its organoleptic characteristics (Scolari *et al*, 1955). A recent study carried out has shown (Pizzocaro *et al*, 1979a,b) that whole sardines in ice can be maintained in

an acceptable condition up to the eighth day of storage, even if slight rancidity is apparent as early as the third day; on the other hand, the frozen whole sardine can be kept for six months at a temperature not exceeding $-28°C$. As is known, rancidity is closely linked to oxidation of the unsaturated fatty acids present in the lipid fraction (Gruger, 1967). This oxidation is known as auto-oxidation because of the predominant autocatalytic nature of the reactions, and takes place through a complex chemical mechanism in which the unsaturated fatty acids tend to oxidize at the reactive allyl position, forming volatile and bad-smelling acids, alcohols and aldehydes.

2 Materials and methods

In order to obtain further knowledge about the auto-oxidation process of sardine oil at low temperatures, studies were carried out on oils from sardines caught in two Italian seas (Ligurian Sea and Adriatic Sea), frozen, and stored at $-20°C$ and $-30°C$. The tests were run at various intervals of time (0, 60, 120, 180 and 240 days). On arrival in the laboratory a few hours after being caught and then frozen, the samples of fish were immediately analysed. For both samples, the seasonal fishing period is the summer. It should be pointed out that while the sardines caught in the Adriatic Sea were in the adult stage of development (20–22cm long), those caught in the Ligurian Sea were, by contrast, in a young stage of development (10–11cm long). The following are the analytical techniques used: moisture content determined in an oven at 90°C; oil content (Bligh and Dyer, 1959); high pressure liquid chromatography (HPLC) (Gasparoli and Fedeli, 1979); determination of the peroxide number (Lea, 1952); and determination of the acidic composition by means of gas-chromatography (Polesello et al, 1977).

The HPLC analysis was effected by means of a Varian 8500 chromatograph, utilizing a Micropak Si 10 column (length 25cm, internal diameter 2·2mm), mobile phase n-hexane–ethyl ether (3:1), flow-rate 70ml/h, pressure 80atm, chart rate 50cm/h. A Perkin Elmer spectrophotometer model LC 55 was used as detector. The analyses were carried out at 230nm, a wavelength at which the peak representative of the state of oxidation of a fat presents maximum sensitivity.

The acid composition was determined by qualitative and quantitative identification of the methylesters obtained by transmethylation in ampoules with methanol–sulphuric acid mixture (95:5) (Kinsella et al, 1977), by means of gas-chromatographic analysis under the following operative conditions: column 15% BDS supported on Chromosorb W, 80–100 mesh; column temperature 210°C; evaporator temperature 250°C; carrier gas, nitrogen at a flow-rate of 60ml/min.

3 Results and discussion

Table I gives the analytical data relative to the oil content, moisture content and peroxide number found in the various determinations run up to the 240th day of storage at $-20°C$ and $-30°C$. In particular, the different oil and moisture content of the two samples stand out: 3% of oil and 74% moisture for sardines caught in the Ligurian Sea (young fish), against 13% of oil and 64% moisture for those fished in the Adriatic Sea (adult fish). The results relative to the peroxide number clearly show the aptitude of sardine oil for auto-oxidation, with an initial value of 16·0meq O_2/kg for Ligurian sardines and 60·0meq O_2/kg for Adriatic sardines, which rapidly increases in both cases, up to 240 days. A fact of immediate observation is the greater protective effect of the lower temperature ($-30°C$) with regard to auto-oxidation, compared to the temperature of $-20°C$ (Figs 1 and 2). In particular, the increase of the peroxide number for the samples originating from the Ligurian Sea shows an irregular course, a phenomenon already encountered in a previous study on sardines preserved at low temperatures (Pizzocaro et al, 1979b). This sample showed a fall in the peroxide number from 185 to 135meq/kg from day 180 to day 240 at the temperature of $-20°C$, and from 135 to 87meq/kg between day 60 and day 120 at the temperature of $-30°C$.

In contrast, this irregularity does not appear in sardine caught in the Adriatic Sea.[1] Furthermore, it is well known that measurement of the state of oxidation of an oil by means of its peroxide number is not always valid because of the instability of the lipid-oxygen dynamic equilibrium after a first relatively stable oxidation stage. This difficulty appears even greater in the case of sardine oil, whose complex composition of long chain fatty acids with a high degree of unsaturation is well known.

High pressure liquid chromatographic analysis has afforded a valuable contribution to the investigation regarding the state of oxidation of the two oils. In recent work, Gasparoli and Fedeli (1979) have shown that by means of HPLC analysis it is possible to evaluate the oxidizability of some oils by measuring a characteristic peak that can be observed at a wavelength of 230nm. The area of this peak has been found to be directly proportional to the degree of oxidation (Fig 3). The course of oxidation of the two samples of sardine oil analysed by HPLC is revealed by the increase in area of

---

[1] However, the peroxide value of Ligurian sardine after 240 days of storage is 50% of the value of the Adriatic sardine oil.

Table I

Samples	Moisture		Oil		Oil basis	
					peroxide value meq/kg	
days of storage	%		%			
	$-20°C$	$-30°C$	$-20°C$	$-30°C$	$-20°C$	$-30°C$
(a) sardine of Ligurian Sea						
Starting time	74·6		3·8		16·0	
After 60 days	73·8	74·2	3·4	3·1	138·0	135·0
After 120 days	74·2	74·2	3·3	3·3	140·9	87·0
After 180 days	73·8	73·7	3·2	4·1	185·8	112·6
After 240 days	73·8	—	3·4	—	135·5	—
(b) sardine of Adriatic Sea						
Starting time	65·0		12·9		60·0	
After 60 days	65·2	65·2	12·0	11·9	89·0	78·0
After 120 days	63·4	64·8	14·7	13·2	190·2	115·0
After 180 days	64·6	64·1	13·0	13·2	238·7	140·5
After 240 days	64·8	66·1	13·1	11·9	311·1	231·2

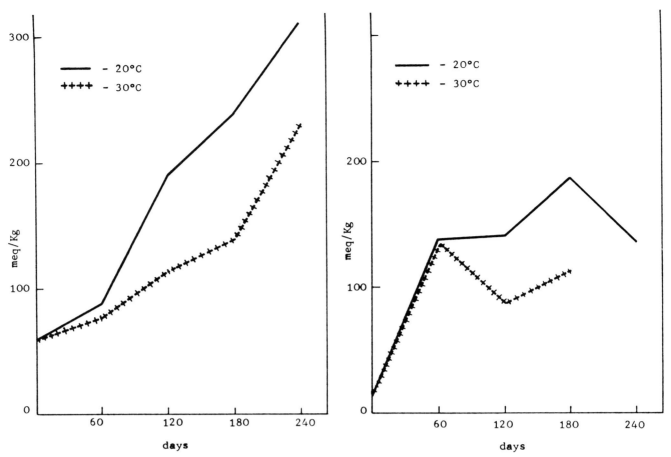

Fig 1 Evolution of peroxide number in Adriatic sardine oil at $-20°C$ and $-30°C$

Fig 2 Evolution of peroxide number in Ligurian sardine oil at $-20°C$ and $-30°C$

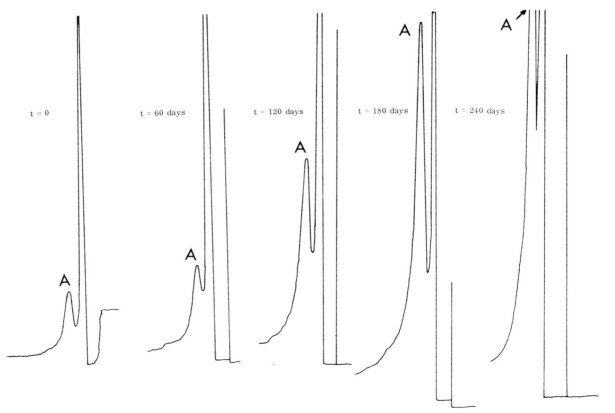

Fig 3 HPLC analysis of Adriatic sardine oil at $-20°C$. A = peak of oxidation

260

the peak (see *Table II*) which passes from an initial value of 1·7 sq cm for Adriatic sardines to respectively 12·7 sq cm at −20°C and 11·8 sq cm at −30°C after 240 days' preservation. This confirms that the temperature of −30°C is more effective in slowing down oxidation (*Figs 4* and *5*).

Table II
HPLC ANALYSIS (AREA) IN THE ADRIATIC AND LIGURIAN SARDINE OILS AT −20°C AND −30°C

Samples days of storage	Area cm²	
	− 20°C	− 30°C
(a) sardine of Adriatic Sea		
Starting time	1·70	
After 60 days	2·8	2·1
After 120 days	5·7	2·8
After 180 days	8·7	6·3
After 240 days	12·7	11·8
(b) sardine of Ligurian Sea		
Starting time	not measurable	
After 60 days	3·2	3·4
After 120 days	3·3	3·4
After 180 days	4·5 (4·2¹)	3·9
After 240 days	2·6 (1·7¹)	—

¹ Complex peaks

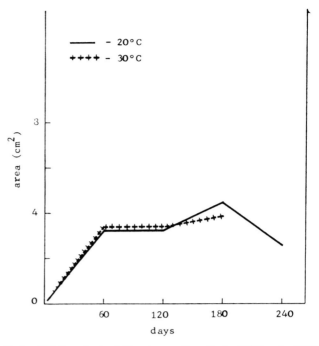

Fig 5 HPLC analysis of Ligurian sardine oil at −20°C and −30°C

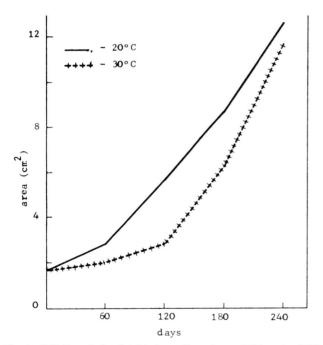

Fig 4 HPLC analysis of Adriatic sardine oil at −20°C and −30°C

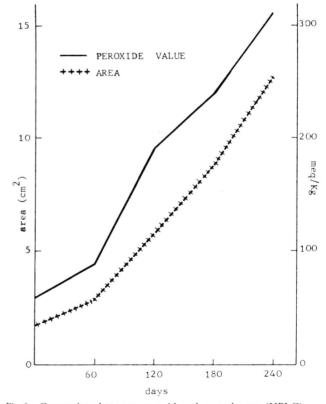

Fig 6 Comparison between peroxide values and areas (HPLC) *versus* time in Adriatic sardine oil at −20°C

It can be seen, in the case of Ligurian sardines, that the behaviour of the HPLC analysis substantially follows the abnormal course of the peroxide number, already encountered, with a reduction of the area of the peak representative of oxidation in the interval between 180 and 240 days at −20°C. *Figures 6, 7, 8* and *9* bring out the correspondence of the results of the two analytical methods graphically illustrating the areas (in sq cm) obtained by HPLC and the peroxide number versus time. Examination of the chromatogram of Ligurian sardine oil at 180 and 240 days at −20°C (*Fig 10*) also reveals that the above-mentioned reduction of the area of the characteristic peak corresponds to a complex of successive peaks which could be considered significant of a more advanced state of oxidation than that of hydroperoxides.

In parallel with the increase of the already mentioned parameters, the composition as determined by gas-chromatography shows a percentage fall of polyunsaturated acids from an initial 36% to 16% after 240 days with regard to Adriatic sardines kept at −20°C, and

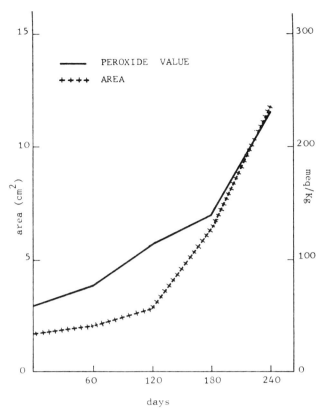

Fig 7 Comparison between peroxide values and areas (HPLC) *versus* time, in Adriatic sardine oil at −30°C

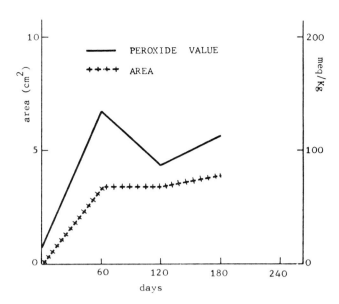

Fig 9 Comparison between peroxide values and areas (HPLC) *versus* time, in the Ligurian sardine oil at −30°C

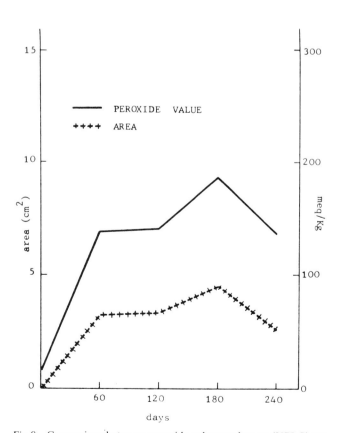

Fig 8 Comparison between peroxide values and areas (HPLC) *versus* time, in Ligurian sardine oil at −20°C

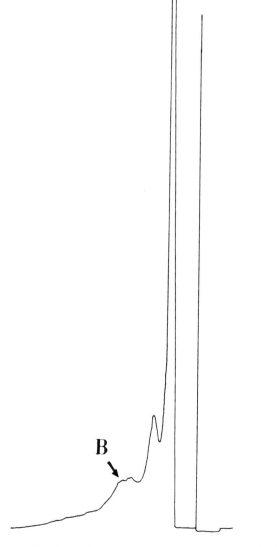

Fig 10 HPLC analysis of Ligurian sardine oil (180 days at −20°C). B = complex peaks

262

from 36% to 22% in those kept at −30°C. Ligurian sardines behave similarly, their initial percentage of polyunsaturated fatty acids (36·7%) falling to 6·5% in the sample kept at −20°C and to 4·4% in that at −30°C, after 240 days. It can be deduced from this that the great perishability of sardine oil is a direct function of the long chain polyunsaturated acids content.

4 Conclusions

While our research confirms the great tendency of sardine oil to auto-oxidation even at low temperatures, it can also establish the usefulness of high pressure liquid chromatographic analysis for identifying and measuring the degree of oxidation of sardine oil kept at −20°C and −30°C.

The sensitivity and regularity limits of conventional analytical methods (peroxide number) could be overcome in part by this new method after demonstration that the origin of the complex of peaks successive to the main one appearing in the chromatogram of oil of Ligurian sardines after 180 and 240 days' storage at −20°C is definitely related to more advanced oxidation products than hydroperoxides.

The behaviour of the parameters considered here appears to be linked to the type of sardine, for example, its race, development stage, area of capture and fishing season as may be deduced from the regularity of the oxidation of oil from Adriatic sardines compared to the irregularity of that from Ligurian sardines.

5 References

BLIGH, E G and DYER, W J. A rapid method of total lipid extraction
1959   and purification. *Can. J. Biochem. Physiol.*, 37, 911
GASPAROLI, A and FEDELI, E. Valutazione della stabilità di oli comme
1979   stibili mediante HPLC. *Riv. Ital. Sostanze Grasse*, 56, 2
GRUGER, E H. In: *Fish oils*, Ed M E Stansby. AVI Publ Co, Westport,
1967   Conn. 3
KINSELLA, J E, SHIMP, J L, MAI, J and WEIHRAUCH, J. Fatty acid con-
1977   tent and composition of freshwater finfish. *J. Am. Oil Chem.
         Soc.*, 54, 424
LEA, C H. Methods for determining peroxides in lipids. *J. Sci. Fd.*
1952   *Agr.*, 3, 586
POLESELLO, A, MANNINO, S and PIZZOCARO, F. Influenza del tipo di
1977   allevamento sulla composizione chimica dell'anguilla. *Riv.
         Ital. Sostanze Grasse*, 54, 27
PIZZOCARO, F, TROVATI, M and POLESELLO, A. Stability tests on frozen
1979a   sardines (*Sardina pilchardus* W.). Atti XV Int. Cong. of
         Refrig., Venezia
PIZZOCARO, F, SENESI, E, TROVATI, M and POLESELLO, A. Experimen-
1979b   tation sur la conservation des filets congélés de sardine. (*Sardi-
         na pilchardus* W.) soumise à divers prétraitements. Atti XV
         Int. Cong. of Refrig., Venezia
SCOLARI, C, BRUSS, O and MARSILI, S. Osservazioni sullo stato di con-
1955   servazione delle sarde congelate di produzione naziuonale.
         Atti IV Conv. Naz. del Freddo, Padova. 403

A computer programme for the numerical calculation of heating and cooling processes in blocks of fish

H R Sanders

1 Introduction

In the design and evaluation of equipment for the thermal processing of food a knowledge of the temperature distribution in the material is required. The times when the whole mass of the material has passed a given temperature, and when the average temperature has reached a required value, must be established, particularly during thawing and freezing. Other quantities of interest include the instantaneous rate of energy transfer and the cumulative amount of energy transferred at any stage in the process.

A means of calculating these parameters is required which, while not replacing experimental data, can be used to study the process.

A number of methods varying in complexity have been devised for this purpose; they are classified by Hayakawa (1977). A method suitable for implementation on a minicomputer is in regular use at Torry Research Station and is described in this paper together with some of the results obtained thereby.

2 Symbols

c	specific heat capacity	J/(kg °C)
d	block dimension (half-thickness in symmetrical case, thickness in non-symmetrical)	m
h	surface heat transfer coefficient	W/(m² °C)
n	number of subdivisions of slab (half-slab in symmetrical case)	—
t	time	s
t'	t/d^2	s/m²
x	distance across slab (from centre in symmetrical case)	m
x'	x/d	—
H	enthalpy per unit volume	J/m³
L	dh/L	—
α	base-line temperature for integration of enthalpy	°C
β	base-line temperature for integration of thermal conductivity	°C
γ	(see equation (24))	—
θ	temperature	°C
λ	thermal conductivity	W/(m °C)
ρ	density	kg/m³
Φ	(defined in equation (3))	W/m

Subscripts

av	average
e	external
i	nodal
in	initial
m	(see equation (24))
s	surface

3 Method

The most important configuration to be considered in fish processing is a regular slab with parallel faces. A block of fish with known initial temperature distribution is subjected to external heating or cooling and heat transfer is assumed to be only through the two faces. If the slab is sufficiently large in relation to its thickness so that edge effects can be neglected and if the conditions over the whole of a face are the same (although they may be different on the two faces) the problem is reduced to the one-dimensional case. The programme deals only with one-dimensional heat transfer. In addition to the slab, this includes the cylinder and the sphere with radial symmetry. However, these versions of the programme are not further discussed here.

The thickness of the slab is specified and the initial temperature is either constant throughout the block or is known at the surfaces and at regularly spaced intervals within the block. The slab is then placed in surroundings of known temperature, which may vary with time but usually remains constant. If the conditions are always the same at both faces, only half the slab, from the centre to one surface, needs to be considered for reasons of symmetry.

The one-dimensional heat equation to be solved is

$$\rho c \, \frac{\partial \theta}{\partial t} = \frac{\partial}{\partial x} \left(\lambda \, \frac{\partial \theta}{\partial x} \right) \qquad (1)$$

where ρ, c and λ may vary with temperature. The method of computation is due to Albasiny (1956).

The thermal and physical properties are replaced by two new variables, H and Φ, both functions of temperature.

H, the enthalpy per unit volume, is defined as

$$H = \int_{\alpha}^{\theta} \rho \, c \, \mathrm{d} \theta \qquad (2)$$

and Φ is defined as

$$\Phi = \int_{\beta}^{\theta} \lambda \, \mathrm{d} \theta \qquad (3)$$

α and β are suitable baseline temperatures.

With these new variables, equation (1) reduced to

$$\frac{\partial H}{\partial t} = \frac{\partial^2 \Phi}{\partial x^2} \qquad (4)$$

Making the further transformations of

$$x' = x/d \qquad (5)$$

where x' is a dimensionless measure of length with $0 \leqslant x' \leqslant 1$, and

$$t' = t/d^2 \qquad (6)$$

the equation to be solved becomes

$$\frac{\partial H}{\partial t'} = \frac{\partial^2 \Phi}{\partial x'^2} \qquad (7)$$

With these transformations the solution of the equations in terms of t' is independent of the thickness of the block.

Let the block be divided into n slices of equal thickness $\delta x'$, then the condition over any given sectional plane is constant. The sections are represented one-dimensionally by the $(n + 1)$ nodal points $x' = i/n$, $i = 0$, $1 \ldots n$.

If the values of H and Φ at the nodal points are denoted by H_i, and Φ_i equation (7) becomes

$$\frac{\mathrm{d} H_i}{\mathrm{d} t'} = \left(\frac{\partial^2 \Phi}{\partial x'^2} \right)_i \qquad (8)$$

A finite difference approximation is used to replace the second differential; in the programme an approximation neglecting fourth differences is used:

$$\left(\frac{\partial^2 \Phi}{\partial x'^2} \right)_i = (\Phi_{i-1} - 2\Phi_i + \Phi_{i+1})/(\delta x')^2 \qquad (9)$$

The thickness of the section $\delta x' = 1/n$ and equation (8) becomes

$$\frac{\mathrm{d} H_i}{\mathrm{d} t'} = n^2 \, (\Phi_{i-1} - 2\Phi_i + \Phi_{i+1}) \qquad (10)$$

where both H and Φ are functions of θ and hence Φ can be regarded as a function of H.

Equation (10) can be applied direct only when $1 \leqslant i \leqslant (n - 1)$ and special methods must be used at $i = 0$ and $i = n$.

In the symmetrical case, $i = 0$, $x = 0$ represents the centre of the block and equation (10) becomes

$$\frac{\mathrm{d} H_0}{\mathrm{d} t'} = n^2 \, (\Phi_{-1} - 2\Phi_0 + \Phi_1) \qquad (11)$$

By symmetry $\Phi_{-1} = \Phi_1$ and hence

$$\frac{\mathrm{d} H_0}{\mathrm{d} t'} = 2n^2 \, (\Phi_1 - \Phi_0) \qquad (12)$$

At the surface $i = n$, $x' = 1$ one of several conditions may obtain. The three kinds of boundary condition to be considered at the surface are

(1) A prescribed surface temperature

$$\theta_s = f(t) \qquad (13)$$

in the limiting case $f(t) = $ constant

(2) No heat gain or loss at the surface

$$\frac{\partial \theta}{\partial x} = 0 \qquad (14)$$

(3) The heat flux across the surface is proportional to the temperature difference between the surface and the external medium

$$- \lambda \, \frac{\partial \theta}{\partial x} = h \, (\theta_s - \theta_e) \qquad (15)$$

The third condition applies both to forced convection and to radiation for small temperature differences.

With the definitions for Φ and x', the third condition takes the form

$$- \frac{\partial \Phi}{\partial x'} = hd \, (\theta_s - \theta_e) \qquad (16)$$

Using the finite difference approximation

$$\frac{\partial \Phi}{\partial x'} = (\Phi_{n+1} - \Phi_{n-1})/(2\delta x') \qquad (17)$$

equation (16) becomes

$$- \frac{n}{2} \, (\Phi_{n+1} - \Phi_{n-1}) = hd \, (\theta_s - \theta_e) \qquad (18)$$

Equating the fictitious Φ_{n+1} from equations (10) and (18) gives

$$\frac{\mathrm{d} H_n}{\mathrm{d} t'} = 2n \, [n \, (\Phi_{n-1} - \Phi_n) - hd \, (\theta_s - \theta_e)] \qquad (19)$$

In the non-symmetrical case a similar equation obtains at the other surface i = 0, $x' = 0$ to replace equation (12).

The other boundary conditions can be treated as special cases within the above scheme. In the case of a prescribed surface temperature θ_s, the external temperature θ_e is given the value required for the surface, and h is made very large. By putting $h = 0$ the second condition is obeyed.

The problem has thus been reduced to the solution of a set of $n + 1$ simultaneous ordinary differential equations (10), (12) and (19) which may be solved by any suitable method.

4 Thermo-physical properties

In equation (1) three thermo-physical properties are involved, all of which are temperature-dependent in fish:

ρ density
c specific heat capacity
λ thermal conductivity.

When heat transfer from or to an external medium is considered, a coefficient of heat transfer is also involved (equation (15)). This, however, is a property of the system rather than of the material only.

4.1 Density

In common with most work in this field, the variation of density ρ with temperature has been neglected. A fixed value of $1\,000\text{kg/m}^3$ has been assumed in the calculations.

Any variation in density also has an effect on the thickness and hence on the inter-nodal distance and any solution would have to deal with variable co-ordinates with all its attendant complications. As Eyres *et al* (1946) have pointed out, to include the effect of thermal expansion on density only is a worse approximation than to neglect it altogether.

Experimental values for fish muscle (Lobzin, 1939) show an increase of density with temperature from 990kg/m^3 at $-12°C$ to $1\,060\text{kg/m}^3$ at $+15°C$. As ρd^3 is constant, and density enters into the equation as ρd^2, the effect is proportional to only $\rho^{1/3}$ so that for a change in density of 7% the effect is less than 2%.

4.2 Specific heat capacity

Specific heat capacity is not used as such in the computations, but enters the variable H, defined in equation (2).

For constant ρ this becomes

$$H = \rho \int_\alpha^\theta c\,d\theta = \rho H' \qquad (20)$$

where H' is enthalpy per unit mass.

In much experimental work it is this quantity which is determined, the specific heat capacity being calculated from it. For materials like fish, where a phase change takes place over a range of temperatures, the enthalpy and specific heat capacity are taken to include the latent heat of phase change.

An advantage of using H instead of c is that H is a monotonic function of temperature and that its changes with temperature are less violent than those of c.

Experimental values of enthalpy of fish over the temperature range of interest are provided by D G Ryutov—reported by Skvarchenko (1950)—and Riedel (1956); the latter's values for cod have been used in the present work.

In the course of computation frequent conversions from enthalpy to temperatures are required. This can be done either by a look-up table with a suitable interpolation formula or by determining a functional relationship. A number of empirical and semi-empirical representations of H or c as functions of temperature have been proposed, the latest being those of Riedel (1978). In the present work the experimental data were approximated by a piecewise polynomial in the form of a cubic B-spline. In this representation a number of values, called knots, of the independent variable (in this case temperature) are chosen, and in each interval between knots a cubic curve is fitted. At a knot the continuity of the curve and its first and second differential are usually maintained. If the data do not allow such continuity, multiple knots can be used.

Inspection and preliminary calculations indicated the need for a triple knot at $-0.89°C$, giving a change of first differential at that point. Fitting was carried out by a programme adapted from NPL Algorithm E2/03/F (Cox, 1972). A representation was achieved where the largest difference between experimental and calculated values was 0.2kJ/kg.

Table I shows the chosen knots and the parameters obtained, and *Fig 1* shows the experimental values and the fitted curve in the range $-40°C$ to $+20°C$.

Table I
B-SPLINE REPRESENTATION OF ENTHALPY AS A FUNCTION OF TEMPERATURE

Knot/°C	Spline coefficient
	0·0
−40	15·6
−15	34·5
−8	75·1
−5	98·9
−3	125·3
−2	164·9
−0·89	242·1
−0·89	319·9
−0·89	345·0
+20	370·8
	396·3

When experimental values are not available at all, or only in the frozen and thawed state, estimates of the specific heat capacities or enthalpies can be obtained by calculation. Bonacini *et al* (1974) suggest that the specific heat relationship with temperature can be much simplified without a considerable effect on any calculated temperature distribution. The properties in the unfrozen state are calculated from those of water and dry material, those in the frozen state from those of ice and dry material. The required amount of latent heat is assumed to be evolved or absorbed in a narrow temperature range (*Fig 2a*). The water content must be known accurately and suitable values are chosen for the initial, peak and final temperatures of the region of maximum change. This approach was tried in the present work, but instead of calculated values, the experimental values at $-40°C$ and at $+20°C$ were used. Enthalpies were calculated by computation of areas, and the total enthalpy

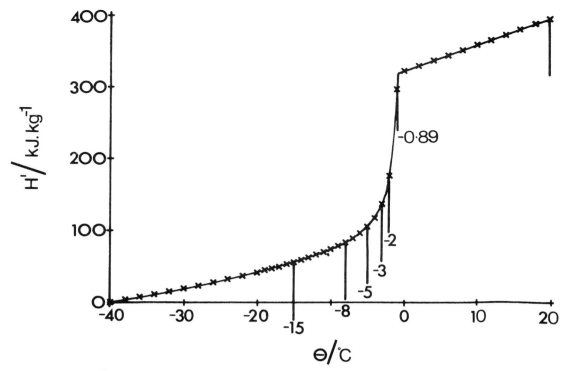

Fig 1 Enthalpy per unit mass (*H'*). Experimental values and spline fit with knots

change was made to agree with its experimental value. *Table II* compares the experimental and calculated enthalpy changes within each temperature region. *Fig 2b* shows the course of the enthalpy curve over the temperature range.

Table II
ENTHALPY CHANGE

Range $\theta/°C$	Experimental $H'/(kJ.kg^{-1})$	Calculated $H'/(kJ.kg^{-1})$
−40 to −8	83·7	59·0
−8 to −0·89	236·2	260·7
−0·89 to +20	76·4	76·6
−40 to +20	396·3	396·3

4.3 Thermal conductivity

Very few experimental data encompassing the whole freezing range are available. Lobzin (1939) gives results for pike-perch, unfrozen and down to −5°C. Khachaturov (1957) drew a mean curve through the results of other workers, but his results are weighted towards fatty fish. The most detailed results are those of Long (1955) who measured thermal conductivities of cod from −26°C to +2°C. Unfortunately his experimental results have not been tabulated and had to be read off the published graphs. In thawed fish, there is no evidence of any variation with temperature and a constant value of 0·55W/m °C has been taken. As in the case of enthalpy, the data, extrapolated to −40°C, were approximated by a spline fit. *Table III* shows the knots and parameters and *Fig 3* the experimental points and the fitted curve.

In the programme thermal conductivity is used in the integrated form of equation (3). Values of Φ are readily computed from λ, because spline fits, being polynomials, can be easily integrated.

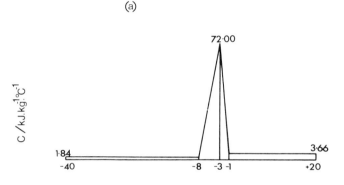

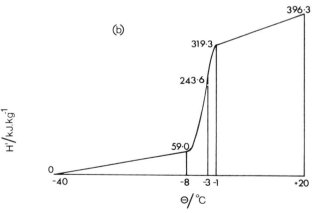

Fig 2 Calculated values of: (*a*) specific heat capacity (*c*); (*b*) enthalpy per unit mass (*H'*)

In the alternative approach, when a simplified calculated relationship is used, the experimental value for λ was used above −1°C, a linear relationship in the region of maximum change from −1°C to −8°C, and a constant value below −8°C (*Fig 4a*). This value was chosen to ensure that the calculated value for Φ at 20°C agreed

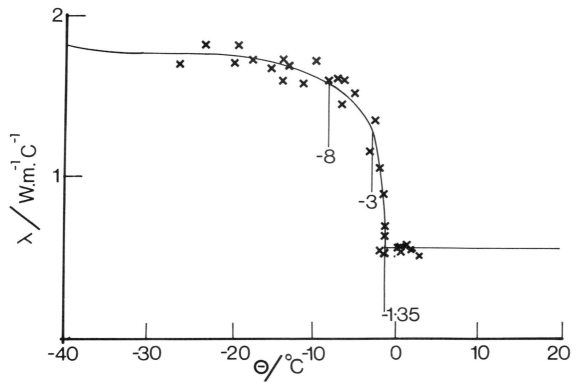

Fig 3 Thermal conductivity (λ). Experimental values and spline fit with knots

Table III
B-SPLINE REPRESENTATION OF THERMAL CONDUCTIVITY AS FUNCTION OF TEMPERATURE

Knot/°C	Spline coefficient
	1·810
−40	1·705
−8	1·897
−3	1·467
−1·35	1·182
−1·35	0·591
−1·35	0·495
+2·4	0·584
	0·547

with that obtained from the integration of the spline fit. *Figure 4b* shows the calculated Φ as a function of temperature.

4.4 *Surface heat transfer coefficient*

Reliable values of heat transfer coefficients are available for only a few conditions. In many published computations heat transfer coefficients are chosen to match the calculated with the experimental temperature distribution and are not verified independently.

For plate freezing Templeton and Nicholson (1972) have determined a range of values for the internal surface heat transfer coefficient of 358–1 550W/m² °C depending on type and flow rate of refrigerant. Typical values for air and liquid are shown in *Table IV*. There are no figures available for condensation heat transfer.

5 Programme organization

The programme, written in FORTRAN, is run on an IBM 1130 computer. The conversion routines express-

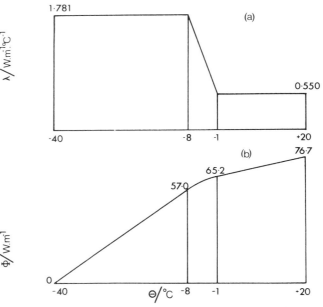

Fig 4 Calculated values of: (*a*) thermal conductivity (λ); (*b*) $\int_{-40}^{\theta} \lambda.d\theta$ (Φ)

Table IV
HEAT TRANSFER COEFFICIENT h AT FLUID VELOCITY v AND TEMPERATURE DROP $\triangle t$

	$v/(m\ s^{-1})$	$\triangle t/°C$	$h/(W\ m^{-2}\ °C^{-1})$
Air	0	—	9
	2	—	28
	6	—	62
Brine	0	6	140
	0	14	220
	0·15	3	340
	0·15	8	375

After Fleming (1967)

ing the relationship between H, Φ, and θ are incorporated into the programme. Data to be supplied for each computation include the number of subdivisions of the block n, $n + 1$ nodal temperatures representing the initial temperature distribution, the block dimension d, the external temperature θ_e, the heat transfer coefficient h, and instructions for output.

The equations are solved by an IBM subroutine using 4th order Runge-Kutta formulae (Romanelli, 1960). Automatic control of step size is included.

Enthalpies and temperatures are calculated for each node at the requested time intervals. The enthalpy values are integrated to determine the average enthalpy of the block and hence the equilibrium temperature, *ie* the constant temperature which the block would finally attain if removed from the plant and stored under adiabatic conditions. Other parameters calculated for output are the cumulative value of the energy transferred per unit area of surface (energy density) and the instantaneous rate of heat transfer (heat flux density).

The number of subdivisions used is 10 in the symmetrical case. The running time is highly dependent on this number which is therefore kept as small as practicable. Running time is also affected by the type of conversion routine. A typical time for computing a complete freezing or thawing process is 75min when functional fits to the experimental thermal properties are used, and 15min for the simplified calculated values.

There are a number of variants of the programme. One deals with the non-symmetrical case, when conditions on the two surfaces are different; 20 subdivisions are then used. There is also provision for limiting the heat flux density, because in practice the design value of the external temperature cannot always be maintained in the initial stages of heating or cooling due to the limitations of the plant.

6 Results

6.1 Freezing and thawing
The output of two typical computations is shown in *Fig 5* for freezing and *Fig 6* for thawing. Results are given at constant time intervals. The plots show the calculated nodal temperatures and a smooth curve joining them.

The differences between the freezing and thawing processes are clearly demonstrated. Heat transfer to or from the block takes place at its surface; the part of the block between the surface and the region of maximum enthalpy change has a lower specific heat capacity and a higher thermal conductivity for freezing than for thawing, causing heat to be supplied to the block more slowly during thawing than it is removed during freezing under similar conditions. This results in flatter centres of the temperature profiles during thawing and longer thawing times.

6.2 Analytical results
The heat transfer equation (1) has no general analytical solution when thermal properties vary with temperature.

For constant thermal properties equation (1) may be written

$$\rho c \, \frac{\partial \theta}{\partial t} = \lambda \, \frac{\partial^2 \theta}{\partial x^2} \qquad (21)$$

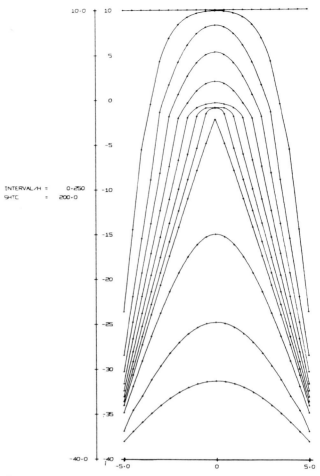

INTERVAL/H = 0·250
SHTC = 200·0

Fig 5 Temperature profiles at 0·25h intervals during freezing – experimental thermal properties – $d = 5$cm, $h = 200$W/(m² °C), $\theta_{in} = 10$°C, $\theta_e = -40$°C

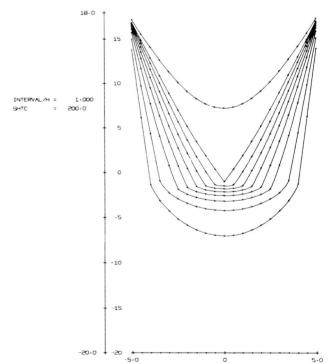

INTERVAL/H = 1·000
SHTC = 200·0

Fig 6 Temperature profiles at 1·0h intervals during thawing – experimental thermal properties – $d = 5$cm, $h = 200$W/(m² °C), $\theta_{in} = -20$°C, $\theta_e = 18$°C

268

or, introducing the thermal diffusivity $a = \lambda/\rho c$

$$\frac{1}{a}\frac{\partial\theta}{\partial t} = \frac{\partial^2\theta}{\partial x^2} \qquad (22)$$

Carslaw and Jaeger (1959) give the solution for a slab with constant initial temperature radiating at its surface into a medium at zero. Defining a dimensionless parameter $L = dh/\lambda$, the temperature θ at a point distant x' from the centre is

$$\theta = \theta_{in}\sum_{m=1}^{\infty}\frac{2L\cos(\gamma_m x')\sec\gamma_m\exp(-\gamma_m^2 at')}{L(L+1)+\gamma_m^2} \qquad (23)$$

where

$$\gamma_m \text{ is the } m^{th} \text{ root of } \gamma\tan\gamma = L \qquad (24)$$

The average temperature θ_{av} is given by

$$\theta_{av} = \theta_{in}\sum_{m=1}^{\infty}\frac{2L^2\exp(-\gamma_m^2 at')}{\gamma_m^2[L(L+1)+\gamma_m^2]} \qquad (25)$$

In the conversion subroutines c and λ are constant at temperatures above 0°C with values of $c = 3.66\text{kJ/°C kg}$ and $\lambda = 0.55\text{W/m°C}$ (ρ is constant throughout at $1\,000\text{kg/m}^3$). A direct comparison between the analytical and numerical solutions can therefore be made if the external temperatures $\theta_e = 0°\text{C}$ and the initial block temperature $\theta_{in} > 0°\text{C}$. The calculations were carried out for $\theta_{in} = 20°\text{C}$, $d = 10\text{cm}$, and $h = 200\text{W/m}^2\text{ °C}$.

The sudden initial change in surface conditions in addition to the crude discretization introduces an error in the numerical solution. The maximum error is initially at the surface but gradually moves towards the centre, decreasing quickly in value (*Table V*). After one hour the maximum difference between the two solutions is everywhere less than 0.05°C.

Comparing the computed equilibrium temperatures with the analytical solution for average temperatures shows a difference of 0.06°C at 0.1h decreasing to 0.01°C at 1h.

It can be concluded that under these conditions the numerical solution is an adequate representation of the analytical solution.

Table V
COMPARISON OF NUMERICAL AND ANALYTICAL SOLUTIONS FOR A BLOCK WITH RADIATION AT ITS SURFACE

| Time/h | Maximum difference | |
	Position 1 centre 11 surface	Value/°C
0.1	11	0.37
0.2	11	0.16
0.3	8	0.11
0.4	7	0.08
0.6	6	0.06
0.8	5	0.05
1.0	4–5	0.04
2.0	1–3	0.03
3.0	1	0.02

$d = 10\text{cm}$, $\theta_{in} = 10°\text{C}$, $\theta_e = 0°\text{C}$, $h = 200\text{W/m}^2\text{ °C}$

6.3 Effect of thermal properties

The freezing time of a block may be defined either with respect to the equilibrium temperature or to the highest temperature anywhere in the block, which for a homogeneous block cooled symmetrically from both sides is at its centre. *Table VI* shows the freezing times to

an equilibrium temperature of $-30°\text{C}$ and to a centre temperature of $-20°\text{C}$ for different surface heat transfer coefficients (h).

For high values of h, heat is removed more rapidly from the surface than for low h, but due to the resistance to heat flow within the block a larger temperature difference is established between the surface and the centre and hence between the equilibrium temperature and the centre temperature. *Table VI* also shows the equilibrium temperatures when the centre temperature reaches $-20°\text{C}$.

When the simplified calculated thermal properties are used instead of the functional relationships based on experimental values, little effect is observed on the freezing times; equilibrium freezing times are about 1% shorter, centre freezing times about 1% longer.

Table VI
FREEZING TIME t, EQUILIBRIUM TEMPERATURE θ_{equ} FOR DIFFERENT SURFACE HEAT TRANSFER COEFFICIENTS h

$h/Wm^{-2}\,°C^{-1}$	Equilibrium $-30°C$ t/h	Centre $-20°C$ t/h	$\theta_{equ}/°C$
1 000	2.0	1.9	−27.2
500	2.1	2.0	−26.6
200	2.6	2.4	−25.5
100	3.3	3.0	−24.5
50	4.7	4.2	−23.0

$d = 5\text{cm}$, $\theta_{in} = 10°\text{C}$, $\theta_e = -40°\text{C}$
experimental thermal properties

The differences are greater for thawing. *Table VII* shows the thawing times for calculated and experimental thermal properties and their differences. Times are given for the centre temperature and for the equilibrium temperature to pass the upper boundary of the latent heat region. Thawing times are underestimated by 3–4% for equilibrium and 6–8% for the centre when the calculated thermal properties are used.

Table VII clearly demonstrates the advantage of determining thawing time by equilibrium instead of centre temperatures, saving about 2h for a block of 5cm half-thickness. The consequent energy saving is about 10%.

Table VII
THAWING TIME t FOR DIFFERENT SURFACE HEAT TRANSFER COEFFICIENTS h

$h/(Wm^{-2}\,°C^{-1})$	Equilibrium thermal properties exp. t/h	calc. t/h	diff./%	Centre thermal properties exp. t/h	calc. t/h	diff./%
1 000	5.8	5.6	3	7.6	7.2	6
500	5.9	5.8	3	7.9	7.3	7
200	6.4	6.2	3	8.4	7.8	7
100	7.2	7.0	3	9.0	8.5	6
50	9.0	8.6	4	10.9	10.1	8

$d = 5\text{cm}$, $\theta_{in} = -20°\text{C}$, $\theta_e = 18°\text{C}$

Figures 7 and *8* show the temperature profiles obtained by using the simplified calculated thermal properties under the conditions of *Figs 5* and *6*. The effect of using different relationships of thermal properties with temperature is readily apparent, the experimental properties resulting in curves that are flatter towards the centre and in closer bunching of the profiles in the region of maximum enthalpy change. This effect is also demon-

269

strated by *Figs 9* and *10*, which show the progress of centre temperatures during freezing and thawing.

The use of the simplified calculated thermal properties thus gives reasonably accurate total freezing and, to a lesser extent, thawing times, but if temperature distributions during processing are required, a closer approximation to the experimental thermal properties must be used.

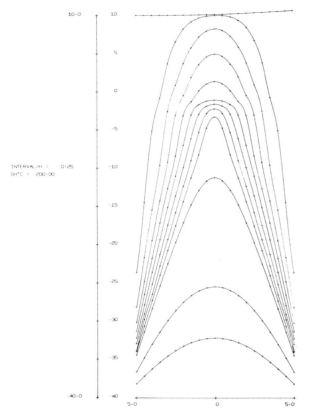

Fig 7 Temperature profiles at 0·25h intervals during freezing – calculated thermal properties – $d = 5$cm, $h = 200$W/(m² °C), $\theta_{in} = 10$°C, $\theta_e = -40$°C

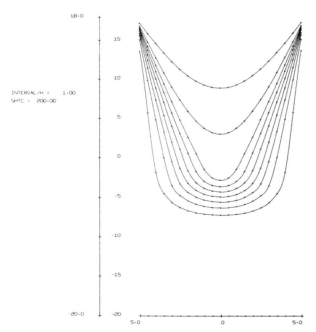

Fig 8 Temperature profiles at 1·0h intervals during thawing – calculated thermal properties – $d = 5$cm, $h = 200$W/(m² °C), $\theta_{in} = -20$°C, $\theta_e = 18$°C

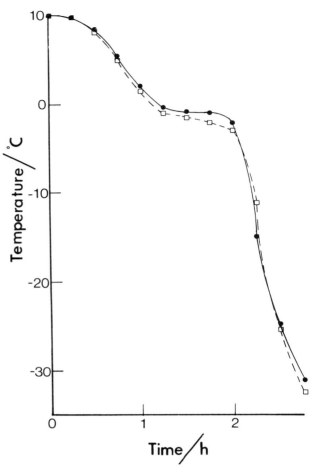

Fig 9 Centre temperatures during freezing – $d = 5$cm, $h = 200$W/(m² °C), $\theta_{in} = 10$°C, $\theta_e = -40$°C
experimental thermal properties ———•———; calculated thermal properties – – –□– – –

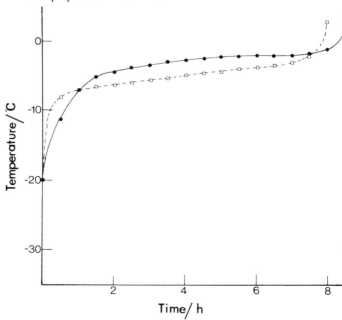

Fig 10 Centre temperatures during thawing – $d = 5$cm, $h = 200$W/(m² °C), $\theta_{in} = 10$°C, $\theta_e = 18$°C
experimental thermal properties ———•———; calculated thermal properties – – –□– – –

6.4 *Comparison with experiment*

For a comparison of computed with experimental results the conditions of processing must be known accurately. Hewitt *et al* (1974) give freezing times (time for centre to reach −20°C) for different refrigerants and contact area ratios, *ie* the proportion of the fish block in contact with the freezer plate. Two of the blocks have a contact area ratio of 1·0 and are thus directly comparable with computed results. The actual conditions under which the two blocks were frozen were supplied by the authors and are shown in *Table VIII* with the experimental and computed freezing times. A good estimate of the actual freezing time is given by the programme, which is also able to distinguish between the different conditions of treatment.

Table VIII
OBSERVED AND COMPUTED FREEZING TIME AT CENTRE FOR BLOCKS WITH
CONTACT AREA RATIO 1·0

d/cm	$\theta_e/°C$	$\theta_{in}/°C$	$h/(Wm^{-2} °C^{-1})$	observed t/h	computed t/h
5·05	−24	4	390	3·8	3·8
5·05	−22	0	310	6·6	6·7

In the same paper the effect of contact area on freezing time is discussed. Three other blocks were frozen under similar conditions but with a low contact area ratio. (Block 5 is wrongly assigned to Refrigerant 22 in the paper). These three blocks have an average freezing time of 6·6h. If it is assumed that heat transfer from the freezer plates to those portions of the block not in contact is negligible, an effective heat transfer coefficient may be obtained as the product of the actual surface heat transfer coefficient and the contact area ratio. The computed freezing time under these conditions is 6·7h, suggesting that this approach gives acceptable results.

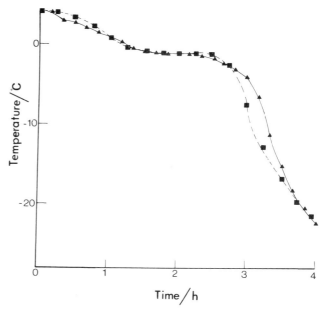

Fig 11   Centre temperatures during freezing, contact area ratio 1·0 observed ——— ▲ ———;computed – – – ■ – – –

The computed and observed centre temperatures of a block with contact area ratio 1·0 are shown in *Fig 11* and with 0·2 in *Fig 12*. Agreement in both cases is good.

7 Conclusion

Results from the programme give a satisfactory representation of thermal processes in plane blocks, heated or cooled from their surfaces. When accurate temperature distributions are required, the thermal properties of the material must be known accurately as a function of temperature. A simplified approximation to the properties is sufficient if freezing times only are required.

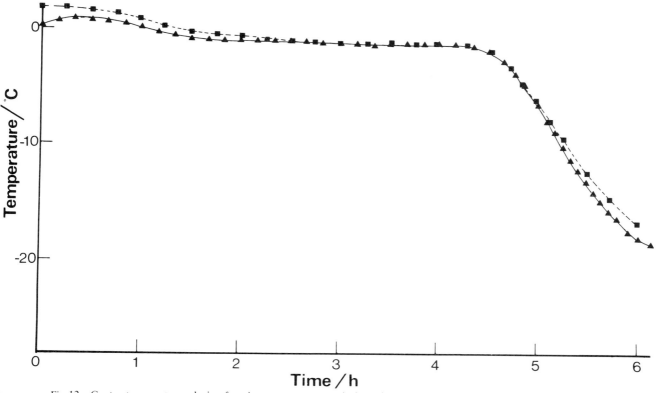

Fig 12   Centre temperatures during freezing, contact area ratio 0·2; observed ——— ▲ ———; computed – – – ■ – – –

8 References

ALBASINY, E L. The solution of non-linear heat conduction on the
1956   pilot ACE. *Proc:* Instn Elect. Engrs., 103(B), 158–162

BONACINA, C, COMINI, C, FASANO, A and PRIMICERIO, M. On the esti-
1974   mation of thermophysical properties in non-linear heat-
conduction problems. *Int. J. Heat Mass Transfer*, 17, 861–867

CARSLAW, H S and JAEGER, J C. *Conduction of Heat in Solids.* Claren-
1959   don Press, Oxford. 2nd edn, 122

COX, M G. The numerical evaluation of B-splines. *J. Inst. Math. and its*
1972   *Appl.*, 10, 134–149

EYRES, N R, HARTREE, D R, INGHAM, J, JACKSON, R, SARJANT, R J and
1946   WAGSTAFF, J B. The calculation of variable heat flow in solids.
Phil. Trans. R. Soc., A 240, 1–57

HAYAKAWA, K K. Estimation of heat transfer during freezing or
1977   defrosting of food. *Bull. Int. Inst. Refrig.*, Annexe 1977–1,
293–301

HEWITT, M R, NICHOLSON, F J, HILL, G P and SMITH, G L. Freezing
1974   times for blocks of fish in vertical plate freezers. The effect of
contact area and block density. *Bull. Int. Inst. Refrig.*, Annexe
1974–3, 39–43

KHACHATUROV, A. Thermal processes during air blast freezing of fish.
1957   *Kholod. Tekh.*, 34(3), 66–71

LOBZIN, P P. Physical properties of fish. *Trudy vses. nauchno-issled.*
1939   Inst. morsk. ryb. Khoz. Okeanogr., 13, 5–50

LONG, R A K. Some thermodynamic properties of fish and their effect
1955   on the rate of freezing. *J. Sci. Fd. Agric.*, 6, 621–633

RIEDEL, L. Kalorimetrische Untersuchungen über das Gefrieren von
1956   Seefischen. *Kältetechnik*, 8, 374–377

—— Eine Formel zur Berechnung der Enthalpie fettarmer Lebensmit-
1978   tel in Abhängigkeit von Wassergehalt und Temperatur. *Chem.
Mikrobiolog. Technol. Lebensm.*, 5, 129–133

ROMANELLI, M J. Runge-Kutta methods for the solution of ordinary
1960   differential equations. In: *Mathematical Methods for Digital
Computers*, Eds A Ralston and H S Wilf. John Wiley and Sons,
New York. Vol 1, 110–120

SKVARCHENKO, R. Consultation. *Kholod. Tekh.*, 27(4), 69–70
1950

TEMPLETON, J and NICHOLSON, F J. Heat transfer and flow characteris-
1972   tics of freezer plates refrigerated with trichlorethylene, calcium chloride brine and pump-circulated Refrigerant 22. *Bull.
Int. Inst. Refrig.*, 52, 1083–1097

Belt selection for a continuous air-blast freezer
J Graham and S Mair

1 Introduction

An increase in the demand for individually quick frozen
(IQF) food products has resulted in a need for continuous freezers. The conveyor used in a continuous freezer
has an influence on many aspects of the freezing process
and its selection is therefore important. Some conveyor
belts used in air-blast freezers impose restrictions on the
type of product that can be frozen. Others give rise to
unhygienic operating conditions and are difficult to
clean. In view of these and other difficulties experienced
with belts particular attention was given to the selection
of a belt for the continuous air-blast freezer recently
designed and developed by Torry Research Station.

1.1 *Requirements of a conveyor belt*
Apart from the requirement for a belt material suitable
for use with foods, the belt in a continuous freezer must
have a number of attributes to enable the freezer to be
operated effectively with a wide range of products.

The belt should not mark, distort or damage the product either during freezing or on release after freezing.

The belt should not significantly affect the freezing
rate of the product.

The belt should be hygienic in operation and should
be easily cleaned when freezing is finished.

The design of the belt should not give rise to an
accumulation of debris within the freezer.

The belt should be robust and of a simple design so
that there are few breakdowns and little maintenance.

1.2 *Types of belt*
There are three types of belt currently used in continuous freezers:

(*a*) Wire mesh or link belts made from stainless steel.

(*b*) Link belts made from plastic segments.

(*c*) Belts made from stainless steel sheet.

These and other possible types were considered during the course of development of the Torry continuous
air-blast freezer.

It seems that, in order to comply with the food-
handling requirements, the choice of materials used in
the construction of a belt is limited to stainless steel and
food-quality plastic. The material used must also be
suitable for operation at −30°C or below and it should
be able to withstand continuous cycling between this low
temperature and ambient. Stainless steel and some plastic materials are suitable in this respect.

1.3 *Damage and marking of product*
All open-mesh or link belts present a surface which can
leave marks or indentations on the undersurface of the
frozen product. This may not always be acceptable,
therefore a belt made from a flat sheet material is pref-
erable. Even when indentations are acceptable open-
mesh and link belts have limitations due to difficulties
encountered when releasing some frozen products. Soft,
wet products such as skin-off fillets and scampi meats
sink into the belt and form a strong bond with the irregular surface thus giving rise to product damage on release
after freezing. Often operators have to resort to the
labour intensive and inefficient practice of first loading
the product on plastic or metal sheets.

Plastic belts made from linked segments give rise to a
particular difficulty during release of the frozen product. A product such as a fish fillet is readily released
from the plastic surface as the belt bends around the
return roller but, depending on the position of the fillet
in relation to the segments of the belt, the break-off
may be from the rear of the fillet and not the leading
edge (see *Plate 1*). Since it is not possible to operate a
scraper in direct contact with this segmented type of belt,
fillets are carried under the scraper and either damaged
or returned to the freezing space.

Products frozen on a flat belt made from sheet material are more easily released but there are some difficult products. Frozen fish products such as mackerel
fillets which have a high oil content (25% to 30%)
remain structurally weak even when frozen to a temperature of −30°C. The result is that the bond between
the fish and the belt is stronger than the fillet itself and
considerable breakage can occur when they are being
unloaded. If, however, the belt is heated at the exit of the
freezer even the most fragile products can be released
from the belt without damage. The plain stainless steel
belt incorporated in the Torry freezer allows this to be
achieved without affecting the fish, whereas with open-
pattern belts it would be difficult.

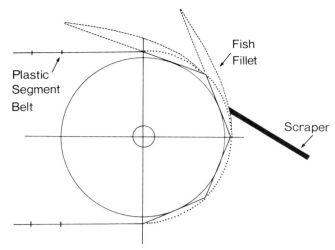

Plate 1   Diagram showing fish fillet adhering to plastic segment belt

Plain flat belts can be used to enhance the appearance of the frozen product in some instances. Fillets which show signs of gaping and fillets with a rough, uneven surface are better in appearance after freezing on a flat belt than on an open belt (see *Plate 2*).

1.4 *Hygiene*

During operation open belts accumulate particles of the product, frozen drip water and frost. In addition, the open pattern of the belt allows some of these deposits to drop off inside the freezer and accumulate beneath the belt. At the end of a freezing session it can take one hour or more to clean an open-patterned belt. Since copious quantities of water are often used, this sometimes results in a build-up of ice within the freezer space. Operating with open-mesh belts therefore gives rise to situations that are not conducive to good hygiene and a considerable effort is required in cleaning the belt and keeping the freezer free from debris.

Belts made from flat sheet material can be kept clean during operation (see *Plate 3*) and only a few minutes are required to clean the belt thoroughly at the end of freezing.

The apparent disadvantage of having to turn a flat stainless steel belt at each end of the freezer over pulleys of about 1m diameter can be used to good effect. The Torry freezer using this type of belt is designed to return the belt outside the freezing space and it is therefore a simple matter to wash and wipe the belt clean continuously or at any time. The flat belt also does not allow particles to drop off below the belt into the freezer. The

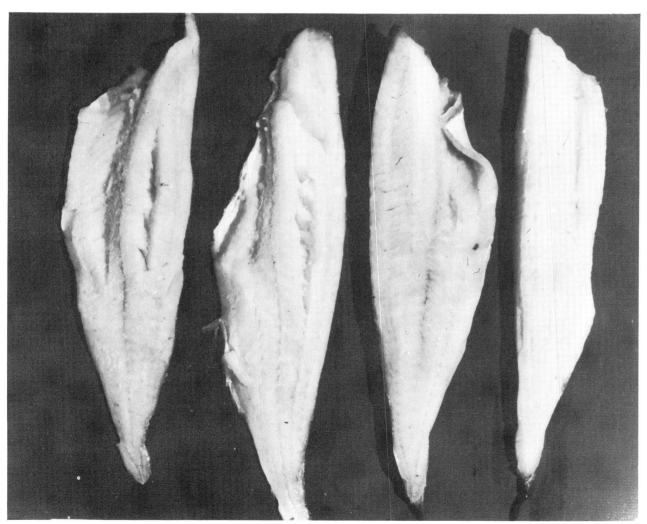

Plate 2   Improved appearance of fish fillets (RH pair) frozen flesh down on sheet stainless steel belt

273

Plate 3 Input end of freezer showing clean stainless steel belt

only place that may require an occasional clean is a plenum space downstream of the belt which is readily accessible.

Cleaning the belt at the end of freezing consists of a quick wipe over with a suitable cleaning agent and this takes only a few minutes with the belt at fast speed.

1.5 *The effect of the belt on freezing times*
It is particularly important that the freezing time of produce in a continuous freezer is kept short. Any adverse affect on freezing time results in a larger freezer and the extra cost and increased dimensions may be critical. In order to reduce the floor space required, some continuous freezers have been built in the form of spirals or have multipass belt arrangements. There are, however, advantages in having a once-through, in-line freezer and in this case every attempt should be made to reduce freezing times. Freezing times can be affected by the choice of belt since it may introduce a high resistance to the transfer of heat from the surface of the product in contact with it.

For other reasons already mentioned a belt made from a flat sheet material looks attractive and a plastic belt of this type would be cheaper than one made from stainless steel. A number of plastic materials were considered and two that looked promising were subjected to tests to determine their durability under simulated freezer conditions. The belts were run at high speed over rollers at a temperature of $-30°C$, corresponding to many years of operation at normal speed in a comparatively short time. These tests indicated that the plastic belts were durable and would give a useful service life. However, further static freezing tests showed that the insulation properties of plastic belts would significantly extend the freezing time. With a product such as a 100g fillet, for instance, a freezer using a flat plastic belt would be 50% longer than a freezer using a stainless steel belt.

Belts constructed from perforated plastic links also extend the freezing time by about 10% compared with freezing on sheet metal belts. The tests on perforated plastic links were made with a clean belt and as the spaces fill with frost and ice and restrict air circulation freezing times will be extended further in practice.

1.6 *Air flow*
One obvious difference between belts is that air can only flow over the surface of a belt made from sheet material whereas an open-pattern construction allows air to flow through the belt. Freezers using flat stainless steel belts must therefore be designed with a crossflow arrangement to enable air to flow both above the belt and over the product and below the belt to assist freezing through the metal from the underside of the product. The stainless steel sheet metal belt used in the Torry freezer is 1mm thick and imposes little resistance to heat transfer. A uniform air flow above and below the belt therefore results in equal rates of freezing from the top and bottom surfaces of the product.

Open-mesh belts relying on air to flow through the belt to the underside give rise to variable freezer operating conditions. The accumulation of frost and particles of produce restrict the air flow as freezing progresses. The crossflow arrangement that must be used with a flat sheet belt therefore results in more stable operating conditions.

1.7 *Belt durability and maintenance*
Belts made from flat sheet material will require less

274

1.7 *Belt durability and maintenance*

Belts made from flat sheet material will require less maintenance than open-mesh belts. Mesh belts allow drip water to freeze between the links and this sometimes results in high stresses due to their decreased flexibility. Also, like a chain, the strength of a belt is no better than the strength of the weakest link and the more links and joints a belt has the more susceptible it is to damage. Flat sheet belts need only have one joint and in practice they require the minimum of maintenance.

Continuous air-blast freezers have used this type of belt for a relatively short time but they have been used extensively in brine-spray freezers and in other applications not involving a low-temperature operation.

2 Conclusions

Due to its many advantages over mesh belts, a stainless steel sheet belt was selected for use in the Torry continuous air-blast freezer and the experience of two years of commercial practice has shown it to be entirely suitable for this application (see *Plate 4*).

Plate 4   Torry freezer

Aspects of optimal utilization of the food fish resource through product innovation

J N Keay

1 Introduction

Approximately 1 million tonnes per annum of fish is landed in the UK. Of this, for one reason or another, only about 300 000 tonnes, or one-third, reaches the table. That is to say about 600 000 tonnes goes to waste or to the fish meal factory. Probably half of this could be used for human consumption; so we have an immediate prospect of doubling the present food fish supply.

But the arithmetic does not end there. If to this are added the very large quantities (say 2 million tonnes) which are caught or are presently catchable in EEC waters by British and foreign vessels, it can be deduced that the UK is in a very fortunate position if there is the will to exploit it.

Succinctly, then, what is necessary is optimal utilization of the fish resource. From a fish technology point of view how can this best be achieved? In some cases all that is necessary is to apply well-established processing techniques such as smoke-curing, freezing and canning to an under-exploited resource in order to produce high-quality products, coupled with a vigorous marketing programme to inform and interest the customer and the consumer. A case in point is that of mackerel. It is now several years since it was recognized at Torry that

this species was a likely candidate for such treatment. In the laboratory and in the field all aspects of the processing of mackerel were studied and at the same time the opportunity was taken to stimulate interest among fishermen and processors. It was subsequently possible to publish an advisory leaflet showing how mackerel can be converted to excellent fresh, frozen, smoked and canned products. There was then a long and rather discouraging gestation period during which industry did not seem to be moving on this development. In recent years, however, there has been a steadily increasing commercial interest in mackerel utilization. In 1977 UK landings were of the order of 150 000 tonnes, with multiples of this quantity apparently being taken by foreign vessels. Indeed it seems that our perhaps labile mackerel stock may be under severe pressure. In this connection there is a need to look at horse mackerel or scad. Preliminary work indicates that this species may prove to be an excellent substitute for mackerel and herring.

But not all under-exploited species can be tackled in this direct way with no essential departure from the established processes of the industry. This is where product development or, as it is more appropriately called, product innovation enters the field.

2 Product innovation

The distinction between product development and product innovation is more than a mere semantic quibble. Product development in industry involves filling what is called a 'gap in the market'. This is achieved with a product suited to a company's sphere of operations. The notional product is developed—it may be anything from soap powder to salad cream—it is packaged, advertised, test-marketed and hopefully launched.

Torry, on the other hand, as a Government laboratory, is concerned to look at the fisheries resource situation as it is expertly presented by colleagues in the fisheries laboratories, to take full account of the fish/food industry and social scene and add to this expertise in fish technology or food science as applied to fish processing. From this total approach what is derived are product concepts rather than products *per se*. It is hoped that industry will adopt these concepts with individual companies exploiting them in their very individual ways. If they subsequently encounter technological problems, then the Laboratory can provide the answers or, failing this, carry out further investigations.

The main purpose of this paper is to give an account of such a programme of product innovation which has been pursued at Torry Research Station during the past few years. In addition to the intrinsic interest of individual items, the narrative will show how this relatively new venture for a fish technology laboratory has proceeded and developed over a number of years.

An early example of such successful product innovation was in the field of cook/freeze fish products. These are composite food products with a protein food base, incorporating a sauce or other ingredients, which are cooked and then frozen at the point of preparation and simply require heating (or regeneration as it is called) before serving. They can be produced in small portion sizes for domestic consumption or in large multi-portion trays. They find outlets in institutional and restaurant catering and for domestic use through the deep freeze cabinets of the supermarkets. Already well developed for meat products, cook/freeze was a neglected field as far as fish was concerned, although it offered the prospect of new and wider markets for fish and of increasing the appeal of lesser-known species. The processing variables were studied and it was found that, since cooked fish has even better frozen storage properties than raw fish, excellent products were obtained. An advisory leaflet on the subject and lectures and demonstrations stimulated commercial interest. Thus it was possible to indicate to the industry how it might raise the value of the catch and ease its supply problems while at the same time attractively widening its product range.

It is at this stage appropriate to report some innovative work carried out which, while not directly concerned with maximal utilization of the fish resource, is connected with it and could help to make the processing industry more efficient and economical with prospects of benefit to the consumer in terms of price. Substantial quantities of smoked fish are eaten in (and exported from) the UK each year. Delicious as the products are, food smoking is essentially a primitive process even when carried out in modern mechanical kilns. It is labour-intensive, can be unhygienic and, in an environ-mentally conscious age, one cannot disregard the attendant atmospheric pollution or the hazards of ingesting certain smoke components. The possibility of replacing smoke-curing with a continuous, clean and otherwise unexceptionable process by the use of effective but safer smoke flavours is attractive to fish processors and could be very much in the public interest. Earlier efforts to effect this change had, up to the time of our starting our work, come to nought because the products were unacceptable to the consumer. However, more recent experiments have shown that it is possible to obtain good results using new smoke-flavour formulations and a spray method of application of the flavour in place of dipping. A demonstration to a group of production and development managers drawn from all the processing companies was arranged. In the opinion of most of these experts the flavoured products were at least as acceptable as the traditional ones and to some they were indistinguishable. This result and consequent wider industrial interest has caused one flavour firm, in co-operation with an engineering company which specializes in high-pressure electrostatic spraying, to devise a total process which has been demonstrated on tonnage quantities in several large fish-processing factories. Subsequent consumer trials have indicated high levels of acceptability. Much has been achieved but it remains for a process to be perfected which fully meets the requirements of industry while being consistent with developing legislation. It seems likely that in the next few years, as existing plant and equipment require to be replaced, there will be a move from traditional smoke curing to the use of smoke flavours.

At the beginning of this paper mention was made of the lack of proper utilization or wastage, both of traditionally prized species and of unused resources. Filleting is the main primary method of processing fish. But when, for example, cod or haddock are filleted the yield of fillet is about 40–42% of the head-on gutted weight while the total amount of flesh on the carcase is about 65–70%. There is thus a gap to be closed. Fortunately in recent years machines, bone separators as they are called, have become available which are capable of removing all the flesh from the carcase, albeit in the form of a coarse mince. The technique presents several difficulties in handling and storage but principally one of converting the mince to marketable products. With experimentation it was realized that the texture, flavour and colour of the mince are, to a large degree, controllable, and that this could be exploited to produce not only a variety of new products but simulation of some of the more valuable existing ones. For example, recovered white fish mince can, by mechanical mixing in the presence of additives, including shellfish fish flavour and carotenoid pigment, followed by machine forming, be converted to 'prawn' or 'scampi' which, to even the most discerning taste, is indistinguishable from the now prized natural product. This is an example of how interchangeability of the species is well within our technological competence.

Recovered minced fish can, of course, be used in products such as fish fingers in place of, or as a supplement to, the conventional fillet material. But it had always seemed that a minced fish finger was probably less acceptable to the consumer than one prepared from fillet material where all the flakiness of the whole fish

276

was retained. In the light of consumer research since carried out, which will be discussed later, ideas on this matter have had to be revised. However, one can still say that for adults, at least, there appears to be some bonus in flaky texture. How can this be restored to minced fish? The key to doing this proved to be that interesting group of food additives, the alginates. The sodium and potassium salts of alginic acid are soluble in water but the calcium salts are not. By addition of calcium ions to an alginate solution precipitation of gelatinous, hydrolysed calcium alginate takes place. The viscosity of the resultant jelly is stable and highly controllable. The properties do not alter on heating as would happen, for example, with gelatin, and the gel is also freeze/thaw stable. Alginate layers are non-adhesive. That is to say layers of gel when pressed together do not adhere or fuse and can be readily separated after prolonged contact or after either freezing or cooking.

It was this last property particularly which prompted interest in the alginates to achieve layering and hence produce flakiness in minced fish portions. A small percentage of soluble alginate was incorporated into minced fish which was then spread in layers of suitable thickness and the alginate gelled by the addition of calcium ions. Within a short time the layers were mechanically strong enough to be washed under running water to remove excess calcium salt. Stacks of these layers were prepared, frozen, cut into finger-size portions, enrobed in batter and crumb and refrozen. By this means fish fingers closely resembling those prepared from fillet were obtained.

Now, this may sound like an over-elaborate means of producing flaky fish from mince but the process is very similar to that used in paper making and, by analogy, capable of large-scale industrial application.

The use of bone separation is not restricted to white fish. Bony, fatty fish such as mackerel, horse mackerel, sprat and herring can be processed in this way to yield bone-free portions, thus for the first time bringing these highly nutritious species within the range of acceptability of a much wider population of consumers, most notably children. Following the marketing example of the dairy industry with yoghurt and cottage cheese, product variety can be increased and consumer interest maintained by the incorporation into these products of sauces of different flavours. When the development work on products from fatty fish minces was being carried out it was anticipated that the products might have poor storage properties and in particular be highly susceptible to the development of oxidative rancidity. It is now known that, provided high quality starting material is used and good cold-storage practice followed, minced fatty fish products are no more susceptible to deterioration than the whole fish material (Cole and Keay, 1976).

However, these are merely examples of what can be done. Moreover, minced fish technology has now gone far beyond merely being a means of saving filleting waste. For example, it may prove to be the method of choice of using the greater part of that vast resource of blue whiting which appears annually in our western approaches. The broad concept is that some fish-processing factories of the future may be run very much on the lines of a modern bakery with various types of fish mince entering a continuous production system to be quality controlled, blended, textured, flavoured, coloured, shaped and packaged to yield a whole variety of marketable new fish products with minimum waste. These will almost certainly be frozen products and in this connection, it should be noted, predictions indicate that by 1985 home freezer ownership will be as prevalent as car ownership is today.

There is one additional and particular development in the minced fish field which merits discussion, namely, the production of kamaboko. Kamaboko is a Japanese fish product with an elastic texture which is eaten in large quantities in that country (up to $\frac{1}{2}$ million tonnes/annum). It fetches a good price and for the best grades a very high price. With new fishing limits seriously affecting supplies of raw material (mainly Alaska pollack) for its production the Japanese are seeking alternative sources of fish supply. Not all fish species are capable of conversion to kamaboko but one would, on theoretical grounds, expect blue whiting to be so – but this had to be demonstrated. A series of experiments conducted at Torry resulted in a successful laboratory-scale method of preparing surimi, an intermediate product, and in turn high-quality kamaboko from blue whiting. Interest by the White Fish Authority after they had been apprised of this work resulted in very successful pilot scale trials with Japanese participation in Stornoway in the spring of 1978. Unfortunately the short spring fishery season for blue whiting operates against the economic viability of a UK surimi production industry. If frozen stocks of blue whiting built up during the short season could be drawn upon for the preparation of surimi, thus lengthening the production period by even three months, then the economics of the process would be much more favourable. It is claimed that frozen fish is unsuitable for the production of surimi and kamaboko but this remains to be proven. Accordingly, Torry and the WFA are presently involved in a collaborative experiment which should provide a soundly based answer to this important question. At this interim stage it does seem that an acceptable product can be produced from the frozen fish.

The last development to be mentioned is a good example of how in R and D the work can take an unexpected turn. Early last year during experiments with mechanical food formers, as part of a programme of work on the conversion of minced fish to a variety of new products in continuous production lines, it was found that such machines are also capable of forming regularly shaped portions of highly reproducible weight from small fillet material. An immediate observation on this finding is that it may provide a method of producing fish fingers without recourse to the preparation and sawing of frozen fillet blocks.

However, there is another, perhaps even more interesting, aspect of this surprising result with the food former. One of the problems in the utilization of blue whiting is that, while good quality fillets can now be produced by machine these fillets are very small – about 30g in weight – too small to be marketed directly. Using the food former such fillets can be coalesced to yield composite portions. Moreover, these composite portions need not be of a regular, geometrical and artificial shape. They can, through the use of suitably shaped forming heads, simulate a true single or double fish fillet.

The result is a good-sized fillet of blue whiting which, when enrobed with puff batter, seems ideal for the fried fish trade. This trade, which is being hard-pressed in its efforts to obtain reasonably priced supplies, uses one-fifth of all the fish that is landed in this country. It is important to bear in mind that use of the technique is not restricted to blue whiting; it can be applied to all small fillet material.

3 Consumer testing

A product innovation programme of the type described above must be convincing to the industry and thus give commercial management the confidence to invest in new ideas. It must also fulfil a real national need. These criteria require that the programme should be steered by sound information on consumer attitudes to fish and fish products generally and to the new concepts being generated. In other words, the product innovation programme should be tied to a consumer testing programme.

This presents a problem at Torry as indeed it does at many research establishments. Consumer research can be expensive and time-consuming and few research laboratories have the in-house facilities to carry out this work. On the other hand, too great a separation of the technological research from the consumer research can be detrimental to both. We at Torry have had to look to other organizations for collaboration. Most notably the Long Ashton Research Station in Bristol has a unit, the Home Food Storage and Preservation Section, which is the authoritative scientific reference centre of the Ministry of Agriculture, Fisheries and Food on home food matters and we have initiated a joint programme of work with this unit. We have also had the assistance of the Market Development Unit of the White Fish Authority and of the Food Products Intelligence Centre of Imperial Foods.

This consumer testing work is at a very early stage but the sort of topics being studied and the results obtained can be outlined.

In developing minced fish technology it had, as previously indicated, been assumed that when minced fish was used in place of fillet material for the preparation of portioned products such as fish fingers, the products would have inferior eating properties. This was the basis of our work on fish flake formation using alginates. When carrying out consumer trials however, it was found that this belief was not fully supported. When fish fingers were prepared from prime quality cod fillet and prime quality cod mince, adult testers scored the fillet product higher than the minced product, but scored the minced product slightly higher than a commercial sample of fillet fish fingers. However, in the case of child testers, the fillet versus mince scoring pattern was reversed and the minced fish fingers were preferred. These results were obtained from carefully conducted tests carried out with small groups of subjects in an 'in-depth interview' situation. They require verification with larger groups since they could be very important in view of the fact that children are major, if not the major, consumers of fish fingers.

Earlier in this paper reference was made to the possibilities which minced fish technology presented as a means of expanding the market for bony, fatty fish and notably the incorporation of sauces into these products. In consumer studies with minced fatty fish fingers it has again been found that children display a different and more encouraging pattern of preference from those of their elders. In general the children score a range of fatty fish fingers as highly as cod fillet fingers but not as highly as cod mince fingers! Exceptionally, adults respond more favourably to colouring the flavoured product with tomato sauce.

In discussions among fish technologists and consumer research workers in the UK on methods of marketing fish mince, the possibility of producing suitably sized frozen blocks (0·5–1kg) to be sold through retail outlets for home preparation of fish dishes has emerged. A recent report from Cornell University (Goodrich and Whitaker, 1977) indicates that American workers have been thinking on the same lines and the published results of their consumer research work with mince from white mullet are very encouraging. Consumer studies carried out by the Long Ashton Research Station on blocks of minced cod, herring and blue whiting show similar promise. The products were generally well received and the work revealed a preference by the housewife to use recipes of her own devising rather than those supplied by the consumer research team. A key factor in the marketability of this type of product would appear to be cost. If good quality frozen fish mince, suitably packed, is available at a reasonable price it seems likely that the housewife will use it.

Finally some account of very recent trials that have been carried out in collaboration with the White Fish Authority and fish fryers on the composite blue whiting fillets described earlier. In the area where the trial was conducted (in the north of England) there was not much awareness of blue whiting; fewer than half of the fryers who took part had heard of the species. They had strongly held attitudes about the desirability of cod, notwithstanding great concern about its increasing price. In order to achieve a change towards a relatively unknown species such as blue whiting it was felt that a discount in price, as compared with cod, of the order of 25% would be required. Little concern was expressed about the fact that the portions were composite in form. A few did not even notice the difference! There were complaints about the variability in colour of the samples, but the biggest single problem appears to be the unwillingness to fry the portions in the frozen state, *ie* without prior thawing. On the basis of this trial this factor is felt to be the major constraint on market potential. One must bear in mind, however, that within the UK there are wide regional differences in the attitudes and practices of fish fryers and, for example, in recent years substantial amounts of frozen, cut portions have been used in the trade.

4 References

COLE, B J and KEAY, J N. The development of rancidity in minced
1976   herring products during cold storage. Ed J N Keay. *Proc: Conference on the production and utilization of mechanically recovered fish flesh (minced fish), Torry Research Station, Aberdeen. 66–70

GOODRICH, D C JR and WHITAKER, D B. Retail market tests of frozen
1977   minced fish. *Paper A. E. Res.*, 77–6. Department of Agricultural Economics, Cornell University, Ithaca, New York

5 Smoking

Inhibition of the trimethylamine oxide degrading enzyme in frozen smoked cod

S Moini and R M Storey

1 Introduction

It is well known that when certain gadoids, including cod, are cold stored, dimethylamine (DMA) and formaldehyde (FA) are formed during storage. Amano and Yamada (1964) suggested that these compounds arise from the action of an enzyme system on trimethylamine oxide (TMAO) present in the tissue. Under given conditions of cold storage the amounts of DMA and FA produced depend on the quantity of TMAO initially present (Kostuch and Sikorski, 1977) and minced fish flesh is known to produce several times more DMA and FA than intact fillets, in both the unfrozen and frozen states (Babbitt *et al*, 1972; Mackie and Thompson, 1974). The greater rate of breakdown of TMAO in minced flesh is probably due to a more intimate mixing of substrate and enzyme and particularly to a more even distribution of fish tissue (*eg* dark meat) with a higher TMAO splitting activity (Tokunaga, 1970; 1974).

The experiments to be described formed part of a research project investigating the fate of some smoke constituents, formaldehyde and steam-volatile ether-soluble phenols, during the cold storage of smoked fish. Much fish is still smoked by traditional methods in which the fish is exposed to wood smoke generated by heating sawdust to an appropriate high temperature. Fish is smoked today primarily to impart a desirable flavour rather than for its preservative effect and a large proportion of smoked fish is frozen shortly after smoking, cold storage being the major preservative. During cold storage deteriorative changes occur, at rates which are dependent on the temperature of storage, which result in the development of undesirable flavours. The research was directed towards a study of the role of the selected smoke constituents in the development of undesirable flavours. Minced cod fillet was chosen as the experimental material in order to reduce the variation expected between whole fillets, even though the cold-storage properties of minced fish are different from those of whole fillets.

2 Experimental materials and methods

2.1 Fish

Inshore cod, one day old in ice, was filleted, skinned, brined in sodium chloride brine without dye until the average salt content was 2% and then passed through a Baader 694 flesh stripper with 5mm diameter holes in the drum. The fillet mince was well mixed and placed in 250 aluminium foil dishes, each $160 \times 100 \times 25$mm. The surface of the minced fish was smoothed over to be as level as possible with the rim of the dish and then placed in a Torry kiln and smoked for 3h at a temperature of 28°C on one surface only. The smoke velocity

was $1 \cdot 5$m sec^{-1}, the relative humidity 72% and the smoke density was maintained at a level slightly higher than that used in normal commercial practice at $0 \cdot 98$m^{-1}. Identical dishes of fish were dried in a wind tunnel under similar conditions of temperature and humidity to provide smoke-free blanks for comparison.

Immediately after smoking the dishes of fish were placed in a small plate freezer and frozen to a temperature of -30°C in under 2h, after which time they were wrapped in aluminium foil, placed in polyethylene bags, sealed and stored for a short time at -30°C. The dishes were divided at random, to reduce any variation due to position in the smoking kiln, into lots of five dishes, the lots grouped into four batches of 10 lots (*ie* 50 dishes). Each batch was stored at one of the following temperatures, -4°C, -10°C, -15°C and -30°C, and after appropriate periods of storage samples were removed for analysis.

2.2 Analytical methods

2.2.1 *Measurement of FA* Each dish of minced fish was sliced, whilst still frozen, into 3mm layers. The comparable layers from each of the five dishes in a lot were combined to produce six samples each of five slices. These were rapidly mixed and 100g homogenized with 200ml of water, 10ml of 12M orthophosphoric acid, $0 \cdot 5$ml of silicone DC antifoam, a few glass beads and steam distilled. The first 600ml of distillate was collected and made up to $1l$ with $0 \cdot 02$M phosphoric acid. A 10ml aliquot was transferred to a boiling tube with the addition of 1ml 1% phenylhydrazine hydrochloride solution and 1ml of 2% potassium ferricyanide solution. After exactly 4min 3ml of 10M HCl were added and the intensity of the colour developed was measured at 520nm. The colour tends to fade and measurements were made as quickly as possible after the addition of the HCl.

The method was found to be highly specific for FA; acids, other carbonyls present in smoke, the phenolic fraction of wood smoke and the amines produced by the fish during storage did not interfere.

It is well known that FA rapidly binds to protein (Castell *et al*, 1973; Dingle *et al*, 1974; Connell, 1975; Kostuch and Sikorski, 1977; Partmann, 1977; Connell *et al*, 1978) and the recovery of FA added to protein can be low. It was found that the recovery of added FA was dependent on the state of the fish protein, *ie* whether cold stored or non-frozen, whether smoked or unsmoked and on the amount of FA added. *Table I* summarizes these effects.

From *Table I* it would appear that for concentrations of FA present in the sample of fish, due to production during cold storage, to smoking, or to addition of more than about 6mg/100g, the recovery of added FA is 90%

or over. The lowest recoveries were from fresh fish and well cold stored fish, with a minimum observed recovery of 75%. These recoveries were deemed to be sufficiently high for an adequate precision.

Table I
RECOVERY OF ADDED FORMALDEHYDE

Samples (all cod)	FA added[1] mg/100g	FA found[2] mg/100g	% Recovery
1 One day in ice fish, filleted,	0	0·04	—
minced and analysed immediately	2	1·60	78·4
	5	4·00	79·4
	10	8·20	81·7
2 As sample 1 but frozen and stored	0	0·20	—
at −20°C for 10 days	1·4	1·20	75·0
	3·5	3·14	84·9
3 One day in ice fish, filleted, stored	0	9·25	—
six months at −10°C and then	2	10·60	94·2
minced and analysed	5	13·60	95·4
	10	18·60	96·6
4 One day in ice fish filleted,	0	5·80	—
smoked, minced and analysed	2	7·00	89·7
immediately	5	10·00	92·6
	10	15·00	95·0
5 As sample 4 but stored as minced	0	12·00	—
smoked fillet for six months at	2	13·00	92·9
−10°C	5	15·80	92·9
	10	21·00	95·5

[1] FA added immediately prior to distillation
[2] Averages of five experiments

2.2.2 Measurement of phenols

Fifty grams of fish from five comparable layers, as described in 2.2.1 above, were weighed into a 2l round-bottom flask to which was added 170ml of water, 5ml of 10M HCl and 60g of lithium chloride. The contents of the flasks were refluxed for 3h via a Clevenger support containing 30ml of diethyl ether. This method of extraction is a modification of that published by Baltes and Bange (1976). After refluxing, the ether was separated from the water layer and dried over anhydrous sodium sulphate, filtered and the volume reduced to 1ml.

Samples (1 μl) of the smoke extract were analysed for their phenol content in a Pye 104 gas liquid chromatograph, with a 1·5m 0·4cm id column containing 5% Carbowax 20M–TPA on 100/120 mesh Chromosorb W. The column was run at 50°C for 15min, followed by 5°C min^{-1} rise to 250°C and finally 25min at 250°C. The detector and injector temperatures were 270°C and 280°C, respectively, and the carrier gas, oxygen-free nitrogen, was adjusted to a flow rate of 30ml min^{-1}.

The chromotograph was calibrated in the usual way with standard phenols and peak areas measured for the calculation of concentrations. Confirmation of the identification of the peaks was obtained by mass spectral analysis.

A major peak was found to be furfuraldehyde, which was present in the unsmoked fish as well as in the smoked fish and in smoked water samples. The presence of furfuraldehyde in unsmoked fish is thought to be due to the breakdown of carbohydrates in the fish during refluxing. The recovery from an added phenol mixture identical to that extracted from smoked fish was found to be 78 ± 3%.

3 Results

3.1 Production of FA in unsmoked samples

Figure 1 illustrates the production of FA in unsmoked minced, brined and dried samples stored at −4°C, −10°C, −15°C and −30°C up to 100 days. The anomalous behaviour at −4°C, previously noted at −5°C by Tokunaga (1974) is clearly seen. It may also be possible that some bacterial activity at this temperature reduced the amount of TMAO available for the production of FA.

3.2 Production of FA in smoked samples

The pattern of concentration of formaldehyde found in smoked fish stored for up to 180 days at −15°C is shown in Fig 2. Similar patterns were found at −4°C and −10°C, but, as expected, there was little FA production at −30°C. The FA concentration in the outer 3mm layer changes little with storage time, decreasing slightly as formaldehyde diffuses out into the next layer (the 3mm to 6mm layer) or becomes irreversibly bound. The FA concentration in the 3mm to 6mm layer increases by a factor of approximately 2·5 times over the maximum storage period of 180 days; the concentration in subsequent layers, more remote from the smoked surface, increasing to an even greater extent.

The rates of production of FA at −15°C in the 6mm to 9mm, 9mm to 12mm, 12mm to 15mm and 15mm to 18mm layers, as well as in the unsmoked blank are shown in Fig 3. It can be seen that there is a steady increase in the concentration of FA; the further the layer is away from the smoked surface the greater the increase found.

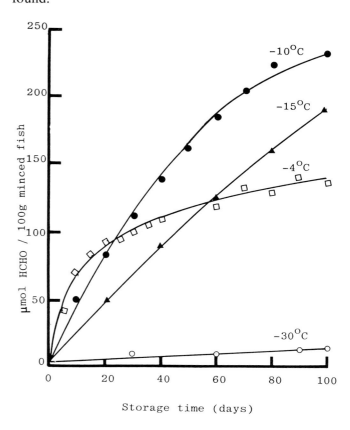

Fig 1   Production of formaldehyde in unsmoked fish

280

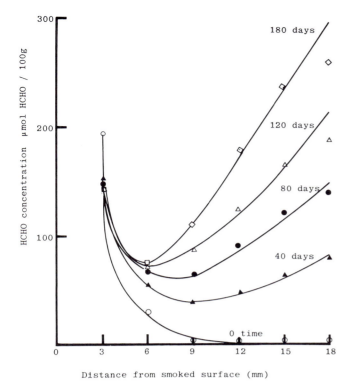

Fig 2   Concentration of formaldehyde in different layers of smoked minced cod stored at −15°C

3.3 *Distribution of smoke constituents*

The depth of penetration of the measured phenols immediately after smoking is illustrated in *Fig 4*. Some 86% of the total phenols present was initially in the outer 3mm layer and none was detected in the 12mm to 15mm and 15mm to 18mm layers. Diffusion of the phenols occurs, during subsequent cold storage, at a rate which is dependent on temperature (Moini, 1979). The rate of diffusion is not the same for all phenols, so that after storage the phenol profile is different from that immediately after smoking. Because of this, and the fact that any enzyme inhibitory activity cannot, from this work, be assigned to any specific compound or compounds, the phenol concentration is expressed as total phenols, in μ moles per 100g, to give a measure of smoke penetration. The increase in total phenol concentration with time in each layer is shown in *Fig 5*. As expected the increase in the 3mm to 6mm layer is greatest and that in the 15mm to 18mm layer the least.

4 Discussion and conclusion

It is quite clear from *Figs 3* and *5* that after storage the layers of fish which contain the most FA also contain the least phenols. It is also clear from *Fig 2* that increases in FA concentration in the layers remote from the smoked surface cannot be explained by the diffusion of formal-

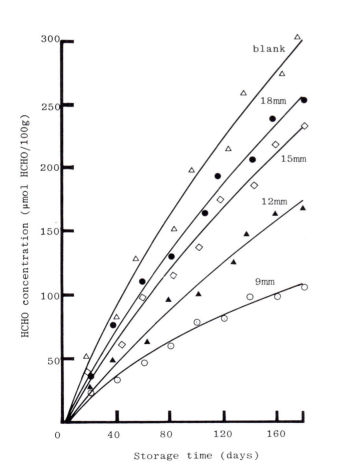

Fig 3   Rate of production of formaldehyde in different layers of smoked fish stored at −15°C

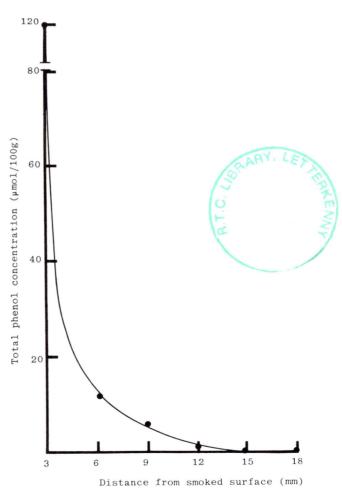

Fig 4   Distribution of phenols in smoked minced cod immediately after smoking

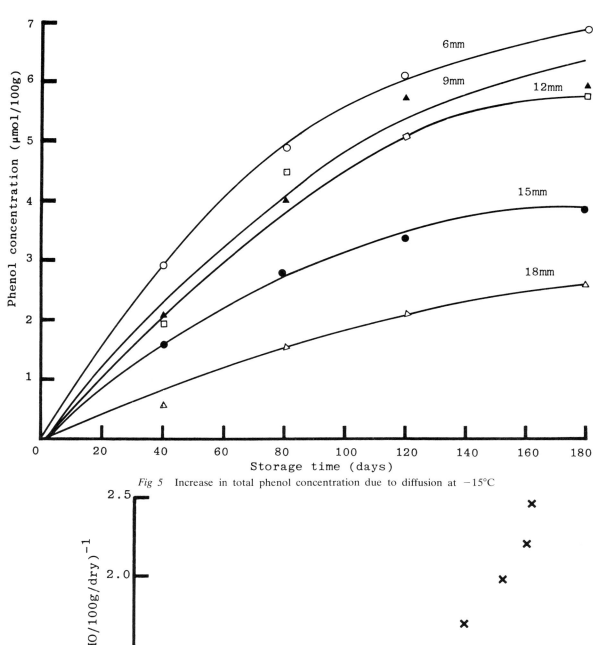

Fig 5   Increase in total phenol concentration due to diffusion at −15°C

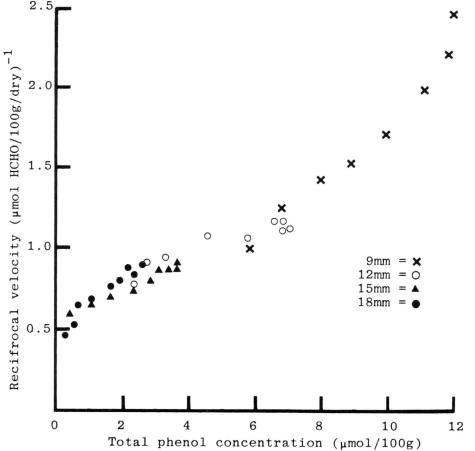

Fig 6   Inhibition of formaldehyde production in different layers of smoked fish stored at −15°C

dehyde arising from the smoke. Kostuch and Sikorski (1977) demonstrated that a concentration of FA about 10 times the maximum concentration found in the current experiments had to be added to cod to significantly reduce the breakdown of TMAO during frozen storage at −18°C. The latter observation tends to confirm the hypothesis that the inhibition of the production of FA during cold storage is not by FA arising from the smoke.

A common method of studying the relationship between the reduction of enzyme activity and the concentration of inhibitor for a constant substrate concentration is the Dixon plot (Bergmeyer, 1974). This predicts a linear relation between the reciprocal of the reaction rate and inhibitor concentration. *Figure 6* is a plot of the inverse of the rate of production of FA for each layer against total phenols concentration for the whole of the storage period at −15°C. It can be seen that there is a strong relationship between the inverse of the rate of production of FA and phenol concentration.

By varying the substrate concentration it is possible to identify the type of inhibition occurring in an isolated system. Variation of the TMAO concentration could be achieved, as by Kostuch and Sikorski, by allowing minced cod to spoil at above zero temperatures and allowing bacterial action to reduce the TMAO concentration by its reduction to TMA before smoking. However it is not possible to determine from the data obtained in the current experiments whether the reduction of FA formation is due to the reaction of the inhibitor with the substrate, TMAO, or with the enzyme or with a combination of both.

The observed range of concentration of phenols can only be considered as an indication of the penetration of smoke constituents. It is possible that non-phenolic constituents which were not measured may also be of importance, although the reactivity of phenols with protein is well known, and thus there is a strong possibility that phenols from the smoke react with the enzyme.

The practical significance of these observations is probably small, since fish is usually smoked whole or filleted and not minced, and therefore would be expected to produce formaldehyde, even in the absence of smoke constituents, at much lower rates than those reported here. Although the middle layers of smoked mince after cold storage were found by a taste panel to have changed marginally less in texture than the unsmoked blanks as well as the outer more smoky layers and the innermost layer examined, fish is not eaten layer by layer and any differences in eating quality, due to different amounts of smoke, when averaged over the whole fish are likely to be even smaller than those observed.

5 References

AMANO, K and YAMADA, K. Formaldehyde formation from trimethylamine oxide by the action of the pyloric caeca of cod. *Bull. Jap. Soc. Sci. Fish.*, 30, 639–645
1964

BABBITT, J K, CRAWFORD, D L and LAW, D K. Decomposition of trimethylamine oxide and changes in protein extractability during frozen storage of minced and intact hake. *Agric. and Fd. Chem.*, 20, 1052–1054
1972

BALTES, W and BANGE, J. The identification of liquid smoke products in foods. *Proc.*: IUFOST Sympos Advances in smoking of foods, Warsaw
1976

BERGMEYER, H W. *Methods of enzymatic analysis*. Academic Press, New York, 2nd edn, Vol 1, 565pp
1974

CASTELL, C H, SMITH, B and DYER, W J. Effects of formaldehyde on salt extractable proteins of Gadoid muscle. *J. Fish. Res. Bd. Dan.*, 30, 1205
1973

CONNELL, J J. The role of formaldehyde as a protein crosslinking agent acting during the frozen storage of cod. *J. Sci. Fd. Agric.*, 26, 1925–1929
1975

CONNELL, J J, LAIRD, W M, MACKIE, I M and RICHIE, A. Change in proteins during frozen storage of cod as detected by SDS electrophoresis. *Proc.*: 5th Int. Congress Food Sci. and Technol. (In press)
1978

DINGLE, J R, HINES, J A and ROBSON, W. Frozen storage stability of minced fish. *New Series Circular* No. 48, Halifax Laboratory, Fisheries and Marine Service, Canada
1974

KOSTUCH, S and SIKORSKI, Z E. Interaction of formaldehyde with cod proteins during frozen storage. IIR symposium on freezing, frozen storage and freeze drying, Karlsruhe. 199
1977

MACKIE, I M and THOMPSON, B W. Decomposition of trimethylamine oxide during iced and frozen-storage of whole and comminuted tissue of fish. *Proc.*: IV Int. Congress Food Sci. and Technol., 1, 243–250
1974

MOINI, S. PhD Thesis, Department of Food Science, University of Reading. (In preparation)
1979

PARTMANN, W. Some aspects of protein changes in frozen foods. *Z. Ernahrungswiss*, 16, 167–175
1977

TOKUNAGA, T. Trimethylamine oxide and its decomposition in the bloody muscle of fish. *Bull. Jap. Soc. Sci. Fish.*, 36, 510–575
1970

—— The effect of decomposed products of trimethylamine oxide on quality of frozen Alaska pollock fillet. *Bull. Jap. Soc. Sci. Fish.*, 40, 167–174
1974

Recovery of phenols from fish using a solvent extraction technique

A S McGill, J Murray and E Parsons

1 Introduction

As a method for food preservation the smoking process no longer plays a significant role in the modern food industry. It is true that the process does confer limited antibacterial and antioxidant activity (Barylko-Pikielna, 1977) but its importance lies more in the desirable flavours and odours that it provides. In recent years, however, concern has been growing over the possible hazard to health associated with the eating of smoked foods (Kumar and Ramachandran, 1973; Dalgat *et al*, 1974; Berg, 1975; Hajdu, 1976; Engst and Fritz, 1977) and this has stimulated radical changes in smoking technique. Using the results of fundamental studies (Tilgner, 1977) into the mechanism of smoke generation it has been found that under strictly controlled conditions the composition of the smoke can be altered to reduce the concentration of potentially dangerous components such as polynuclear aromatic hydrocarbons (Tilgner, 1977) and nitrosamines (Rusz and Miler, 1977).

Another development has been the use of smoke solutions. This has accelerated the development of new forms of smoking in which the food is 'smoked' by a variety of methods (Hardy and McGill, in press). Unfortunately it is frequently reported that smoked foods produced using these innovations may often lack the full flavour associated with those smoked in the traditional way. Perhaps because of this, novel methods are not used to any great extent to make smoked fish products, even though there are good economic and technical reasons for doing so.

The chemical composition of wood smoke is extremely complex (Hamm, 1977) consisting of acids, terpenes, carbonyls, phenols and polynuclear aromatic hydrocarbons but it is the phenols that characterize the major desirable odours and flavours of smoked foods (Barylko-Pikielna, 1977). The relative importance of individual phenols and their interactions with other components of the smoke or with the food have not yet been determined either from the flavour aspect or with regard to their possible toxicological or mutagenic importance. Before this information can be acquired, accurate quantitative data must be obtained.

It is apparent from the literature that difficulties still arise (Issenberg *et al*, 1971; Potthast, 1976; Luten *et al*, 1979) in obtaining quantitative recoveries of phenols from smoked foods. Reports of their isolation by distillation methods describe either long extraction times (4h) (Baltes and Bange, 1977) or the use of elevated temperatures (Potthast, 1976) whereby the food matrix and the component phenols are subjected to the possibility of further reactions. In the latter method poor recoveries were reported for aldehydic phenols. Others have used solvent extraction techniques in conjunction with alkali (Issenberg *et al*, 1971; Luten *et al*, 1979) but marked losses of phenol itself have been recorded.

In this paper we describe our experiments using a simple solvent extraction technique whereby good recoveries of phenol and aldehydic phenols are attained.

2 Experimental

2.1 *Chemicals*
[U−^{14}C] labelled phenol (specific activity 35mCi/mmole) was obtained from the Radiochemical Centre, Amersham, and the solution diluted 25 times with aqueous methanol to give an activity of approximately 2μCi/ml. Vanillin and syringaldehyde were purchased from Koch Light Laboratories and made up as a 0·1% stock solution in acetone. Aliquots (1ml) of these stock solutions were used for individual recovery experiments. All solvents were redistilled before use. Acetonitrile was supplied by Fisons and the other solvents by May and Baker Ltd.

2.2 *Gas liquid chromatography (GLC)*
Quantitative GLC was carried out on a Pye 104 gas chromatograph using a glass column (2m × 2mm id) packed with 4% OV17 on Chromosorb G AW–DMCS 80–100 mesh. Peak areas were obtained using an Infotronics Integrator.

2.3 *Radioactivity measurements*
Radioactivity in samples was estimated after dispersion in NE 260 scintillant and counting at ambient temperature in a Nuclear Chicago Isocap spectrometer. Counting efficiency was determined by the sample channels ratio technique using a set of commercially available quenched standards.

2.4 *Distribution experiment*
Phenol was added to méthanol (40ml), chloroform (40ml) and water (36ml) held in a separating funnel. After the mixture was thoroughly shaken and allowed to settle, the two layers were separated and analysed for their radioactivity.

The experiment was repeated using vanillin and syringaldehyde but the separated phases were analysed as follows:
the methanol/water layer was extracted with ether (2 × 50ml) and the pooled ether extracts concentrated to ~1ml on a rotary evaporator at 20°C. The volume was made up accurately to 2ml prior to GLC analysis. Analysis of the chloroform layer was carried out by GLC after concentration to a final volume of 2ml as before.

2.5 *Recovery of phenols from fish flesh*
2.5.1 *Phenol* Herring flesh (20g) was comminuted and placed in a homogenizer flask. The phenol solution was carefully added from a pipette by dropping the solution over as much exposed tissue as possible and allowing it to permeate for 30min. A standard Bligh and Dyer (Bligh and Dyer, 1959) type extraction was carried out using an MSE top drive homogenizer as follows: methanol (40ml) plus chloroform (20ml) was homogenized with the fish for 1min. More chloroform (20ml) was then added and homogenization continued for an additional minute. Finally water (20ml) was added and homogenization continued for a further minute. The extract was separated from the protein by pouring the slurry into a sinter funnel (18cm × 7·5cm id), filtering under a slight positive nitrogen pressure and washing the tissue with methanol/chloroform 1:1 (20ml). After extraction of the methanol/water layer with ether (2 × 50ml), the ether extract was pooled with the chloroform layer and concentrated to about 5ml using a rotary evaporator at 20°C. The sample was made up to 10ml with methanol and its radioactivity measured.

The tissue was extracted for a second time by homogenizing with methanol (40ml) and chloroform (20ml) for 1min then adding more chloroform (20ml) and homogenizing for another minute. The slurry was filtered, washed and the monophasic extract concentrated and analysed as before.

Finally the residue from the above extract was extracted for a third time using the same conditions as in the second extraction.

2.5.2 *Recovery of vanillin and syringaldehyde from fish flesh* Vanillin and syringaldehyde were added to herring flesh (20g) and extracted as in the first two extractions in the previous experiment. The extracts were pooled, the chloroform and methanol removed on the rotary evaporator at 20°C[1] and the resulting aqueous solution made slightly acid with hydrochloric acid then ether extracted (2 × 100ml). The ether was removed, the sample taken up in acetonitrile (3 × 20ml), concentrated and made up accurately to 2ml for GLC analysis.

---

[1] At this stage during the extraction of a smoked fish sample, the solution is made alkaline (pH 12) and extracted with ether (2 × 100ml) to remove some lipid and neutral components before acidification and further work-up as described above

2.5.3 Recovery of phenols using an alkaline Bligh and Dyer extraction Herring flesh (20g) treated with the phenols was extracted using the standard Bligh and Dyer method but replacing the water by N/10 NaOH (20ml) in the final homogenization step. Filtration and washing were carried out as described previously and the tissue re-extracted under alkaline conditions.

After the pooled solution was centrifuged and the phases separated[2] the chloroform phase was extracted[2] with N/10 NaOH (2 × 100ml) and this extract was added to the methanol/water phase. The methanol was removed using a rotary evaporator and the alkaline solution extracted[2] with ether (2 × 100ml) and then acidified with NHCl. Finally the acidified solution was extracted[2] with ether (2 × 100ml) and concentrated for analysis as before.

2.5.4 Recovery of phenols using Celite 545 Herring (20g) to which all three phenols had been added, was extracted twice as above (2.5.1) but no water was added in the first extraction procedure. The pooled extracts were slurried with Celite 545 (20g) and anhydrous sodium sulphate (30g) and the solvents removed on a rotary evaporator at 20°C. The Celite was transferred to a glass column (26cm × 2·5cm id), eluted with acetonitrile (300ml), and the resulting solution concentrated to give a final volume of 2ml prior to GLC and determination of radioactivity.

3 Results and discussion

The Bligh and Dyer solvent extraction technique is an acknowledged method for the quantitative recovery of lipids from fish flesh and is used extensively in our laboratory. It seemed highly probable that this solvent system would be equally efficient in extracting those phenols likely to be strongly adsorbed on smoked fish protein.

Three model substances known to be present in smoke were selected for the recovery experiments, phenol because of its volatility and vanillin and syringaldehyde because of their adsorptive nature.

Our first experiment was to determine how these phenols are distributed between the two phases of chloroform and methanol/water which are formed when the Bligh and Dyer technique is used. It was hoped that they would be exclusively held in one of the phases but the actual distribution is shown in *Table I*. However it is apparent that losses are minimal during the work-up procedure.

Table I
DISTRIBUTION OF PHENOLS IN BLIGH AND DYER EXTRACTION SOLVENTS

| Component | Recovery | Phase | |
		Methanol/water	Chloroform
	%	%	%
Phenol	99·2	36·4	62·8
Vanillin	94·5	25·8	68·7
Syringaldehyde	99·4	12·2	87·2

[2] Centrifugation necessary

The recovery of radioactive phenol from fish protein was then determined using the standard Bligh and Dyer method and the work-up procedure described earlier. It is our practice when carrying out lipid extractions of fish to filter off the protein using a Buchner funnel but its use here resulted in a loss of phenol of at least 25% by evaporation. It was decided therefore to use a sinter funnel and filter under nitrogen pressure; under these conditions no loss of the highly volatile phenol was observed. Recoveries of the phenol were checked using radiochemical assay techniques and the results indicate that two extractions are sufficient to give near quantitative recoveries (*Table II*). On this basis the recoveries of vanillin and syringaldehyde were checked (*Table II*). The removal of neutral components by the inclusion of an additional step in which the aqueous phase is ether extracted under alkaline conditions does not adversely affect recoveries (*Table II*). This experiment carried out with lean fish tissue produced very similar results.

Table II
RECOVERY OF PHENOLS FROM FISH PROTEIN

	Phenol	Vanillin	Syringaldehyde
	%	%	%
1st extraction	90·64		
2nd extraction	8·86		
3rd extraction	—		
Total after two extractions	99·50	93·0	95·5
Total after two extractions[1]	94·1	92·0	96·6

[1] Recovery after ether extraction under alkaline conditions to remove neutrals

A variation of the method, in which an alkaline Bligh and Dyer extraction was attempted, also resulted in good recoveries of the three phenols (*Table III*). However, the manipulations were complicated by the formation of emulsions which necessitated centrifugation at every stage of the work-up procedure making the process very time-consuming.

Table III
RECOVERY OF PHENOLS FROM HERRING FLESH USING BLIGH AND DYER UNDER ALKALINE CONDITIONS

Component	%
Phenol	92·5
Vanillin	96·5
Syringaldehyde	103·0

In the method of extraction and work-up used, the lipids present are also extracted and concentrated, prohibiting GLC analysis on the sample as it stands. To overcome this, the lipid extracts are themselves extracted with acetonitrile in which the phenols but not the lipids are soluble. This poses few problems in practice; however, others (Potthast and Eigner, 1975; Lehrian *et al*, 1978) have suggested an alternative technique by which the interfering lipid can be removed. Using this method the fish extracts were added to a Celite 545/sodium sulphate mixture, the solvents removed and the phenols eluted in the manner described earlier. Unfortunately the Celite did not effectively retain the lipid and the recovery of vanillin and syringaldehyde was relatively poor (*Table IV*). This loss was traced to adsorption on the sodium sulphate; a similar loss was reported by Issenberg *et al* (1971) for phenol. The results in *Table II* clearly show that when the modified Bligh and Dyer extraction technique is used, as outlined in experimental section 2.5.2, the recovery of all three phenols from fish flesh is very good. Those of the more polar aldehydic phenols vanillin (93%) and syringaldehyde (95·5%) compare very favourably with those quoted for meat homogenates, 63·5% and 16·7%, respectively (Potthast, 1976). Potthast interpreted these low values as being a consequence of carbonyl/protein amino group reactions at the high distillation temperature.

A limited analysis of phenols in smoked fish using the solvent extraction technique has been carried out (*Table V*). The major phenols are present at concentrations approximating those used in the extraction experiments for vanillin and syringaldehyde (50mg/kg fish) and significantly higher than that used in the case of radioactive phenol (0·3mg/kg fish), indicating that the method is appropriate for the concentrations found in smoked fish. This has not always been so in the work of others; Issenberg, using 50% aqueous ethanol as an extracting solvent, reported a recovery of 84% for phenol from pork containing 1 000mg/kg (20 times the level normally found in fish) so that at the lower concentrations found in fish the recoveries are likely to be lower. This contrasts with the near quantitative recovery of phenol obtained in this present work which is most probably a consequence of a more efficient lipid extraction by the Bligh and Dyer solvent system. Further losses were noted by Issenberg *et al* (1971) after separation of the neutral fraction and concentration to smaller volumes. It has been our experience that if the evaporation temperature is kept below 20°C, then no losses are observed.

To summarize, we have confirmed that characteristic wood smoke phenols can be efficiently extracted from fish flesh by using a room temperature solvent extraction procedure based upon that of Bligh and Dyer. There is no reason to expect that the recoveries quoted here for phenols from fish protein would be radically different for other flesh foods.

Table IV
RECOVERY OF PHENOLS FROM HERRING FLESH USING CELITE COLUMN

Component	%
Phenol	96·5
Vanillin	82·0
Syringaldehyde	72·0

Table V
SOME PHENOLIC COMPONENTS IN A RANGE OF SMOKED FISH SPECIES

Phenols	Concentration (mg/kg)
Guaiacol	0–25·6
Phenol	0–52·0
m-and p-xylenol	0–20·0
o-xylenol	0–10·2
Eugenol	0–21·6
Iso-eugenol	0–19·2
Vanillin	1–20·4
Syringaldehyde	23–90·7
Pyrocatechol	0–52·0

4 References

BALTES, W and BANGE, J. The detection of liquid smoke flavourings in 1977 foodstuffs. *Acta Aliment. Pol.*, 3, 325–333

BARYLKO-PIKIELNA, N. Contribution of smoke compounds to sensory, 1977 bacteriostatic and antioxidative effects on smoked foods. *Pure and Appl. Chem.*, 49, 1667–1671

BERG, J W. *Persons at high risk of cancer—an approach to cancer* 1975 *etiology and control*. Academic Press, New York

BLIGH, E G and DYER, W J. A rapid method of total lipid extraction 1959 and purification. *Can. J. Biochem. Physiol.*, 37, 911–917

DALGAT, D M, ALIEV, R G, GIREEV, G I and ABDULGAMIDOV, M M. 1974 Epidemiology of stomach cancer in Dagestan. *Vopr. Onkol.*, 20, 76–81

ENGST, R and FRITZ, W. Food hygienic toxicological evaluation of 1977 the occurrence of carcinogenic hydrocarbons in smoked products. *Acta Aliment. Pol.*, 3, 255–263

HAJDU, G. Epidemiological problems of gastric cancer in Felsoszol-1976 nok. *Magy Onkol.*, 20, 157–162

HAMM, R. Analysis of smoke and smoked foods. *Fleischwirtschaft*, 57, 1977 92–96

HARDY, R and McGILL, A S. Smoking of foods. *Process. Biochem.* 1979 (In press)

ISSENBERG, P, KORNREICH, M R and LUSTRE, A O. Recovery of 1971 phenolic wood smoke components from smoked foods and model systems. *J. Fd. Sci.*, 36, 107–109

KUMAR, K M and RAMACHANDRAN, P. Carcinoma oesophagus in 1973 North Kerala. *Indian J. Cancer*, 10, 183–187

LEHRIAN, D W, KEENEY, P G and LOPEZ, A S. Method for the 1978 measurement of phenols associated with the smoky/hammy flavour defect of cocoa beans and chocolate liquor. *J. Fd. Sci.*, 43, 734–735

LUTEN, J B, RITSKES, J M and WESEMAN, J M. Determination of 1979 phenol, guaiacol and 4-methyl guaiacol in wood smoke and smoked fish products by gas-liquid chromatography. *Z. Lebensm. Unters. Forsch*, 168, 289–292

POTTHAST, K and EIGNER, G. A new method for the rapid isolation of 1975 polycyclic aromatic hydrocarbons from smoked meat products. *J. Chromatogr.*, 103, 173–176

POTTHAST, K. Determination of phenols in smoked meat products. 1976 IUPAC/IUFOST *Proc:* International Joint Symposium on Advances in smoking of food, Warsaw, 39–44

RUSZ, J and MILER, K B M. Physical and chemical processes involved 1977 in the production and application of smoke. *Pure and Appl. Chem.*, 49, 1639–1654

TILGNER, D J. The phenomena of quality in the smoke curing process. 1977 *Pure and Appl. Chem.*, 49, 1629–1638

The development of analytical methods for investigating chemical changes during fish smoking

M N Clifford, S L Tang and A A Eyo

1 Introduction

There is much literature concerning the composition of wood smoke and smoked foods in general (*eg* reviews by Hamm, 1977; Gilbert and Knowles, 1975) but relatively little information about the chemical composition and nutritive value of smoked fish (Yuditskaya and Lebedeva, 1960; Hoffman *et al*, 1977). The present study is an attempt to obtain basic data in this neglected area; it reports an evaluation of two rapid methods that have not previously been applied to smoked fish.

2 The monitoring of lysine and total basic amino acid residues by dyebinding

The basic technique of dyebinding is well documented and this investigation followed closely the methods of Lakin and of Jones (Lakin, 1973; 1975; Jones, 1974) for the determination of optimum reaction conditions for smoked and unsmoked fish and ultimately the analysis of the total basic amino acids content. The lysine content was determined by the dyebinding difference (DBD) method of Sandler and Warren (1974) and used methyl chloroformate as the lysine-blocking agent. The blocked samples were treated with a lower concentration of dye in accordance with the recommendations of Jones (1974). The data so obtained were compared with values for FDNB-available-lysine obtained by Carpenter's Method as modified by Booth (Carpenter, 1960; Booth, 1971). A detailed account of these investigations is given by Tang (1979).

2.1 Experimental

2.1.1 *Preparation of fish samples* Fresh coley (*Pollachius virens*) were purchased from Grimsby Fish Docks as necessary at intervals between September 1977 and May 1979. The fish were filleted and the fillets retained in pairs identified by coloured beads that were attached with wire. One fillet from each pair was placed on a wire-mesh tray in a Torry mini kiln where smoke production from one firebox was in progress and the temperature was approximately 50°C. The smoking temperature was raised to 115°C and the second and third fireboxes lit as required to maintain a dense smoke. The air flow rate in the middle of the kiln was in the range 0·7 to 0·8m s⁻¹. Fillets were removed 3h and 5h after commencement of smoking.

From the upper surface of each fillet two successive layers, each of approximately 0·5cm thickness, were removed (the upper and middle layer, respectively). These layers were individually blast-frozen, freeze-dried and ground in a hammer mill to pass a 0·25mm sieve. Each sample was defatted twice (8h in total) using 100cm³ portions of diethyl ether, then packed in high-density polyethene bags and stored at 4°C in the dark for a period not exceeding one week. Each investigation employed a minimum of 11 fish.

2.1.2 The dyebinding procedure

Sample: Ground defatted coley, 150mg from 2.1.1.

Dye: Buffered Acid Orange 12 (Foss Electric, York, England) at a concentration of 4·0meq l⁻¹ for determining total basic amino acids and 3·2meq l⁻¹ for determining lysine by DBD. In all cases 40cm³ was used.

Equilibration: Sample and dye were shaken with three glass beads in a stoppered polyethene centrifuge tube (100cm³ capacity) for 1h. The supernatant was cleared by centrifugation.

Measurement of supernatant optical density: Readings were taken at 482nm after dilution if necessary with 0·2M citric acid.

Calculation of dyebinding capacity (DBC): By reference to a calibration curve the change in optical density associated with dyebinding is converted to meq of dye per mg of sample.

Calculation of dyebinding difference: This is the difference in dyebinding capacity associated with the use of methyl chloroformate, *ie*

$$DBC = \begin{array}{c}\text{DBC of untreated} \\ \text{sample (meq g}^{-1})\end{array} - \begin{array}{c}\text{DBC of methyl-} \\ \text{chloroformate} \\ \text{blocked sample} \\ \text{(meq g}^{-1})\end{array}$$

2.1.3 *Blocking of lysine with methyl chloroformate* – see Sandler and Warren (1974) and Jones (1974).

2.1.4 *FDNB-available-lysine* – see Booth (1971).

3 Results and discussion

When using Carpenter's Method it is known that ϵ-DNP-lysine is partially destroyed during hydrolysis and that the recovery rate must be determined so that a correction can be applied. The mean recovery of ϵ-DNP-lysine for unsmoked coley was 96%, comparing favourably with literature values of 96% for freeze-dried chicken meat and 95% for fish meals (Booth, 1971). The mean recovery from smoked coley was only 91% and thus correction factors of 1·043 and 1·096 were used for unsmoked and smoked coley, respectively. The greater loss in smoked samples was traced to the smoke components, particularly phenolic compounds as illustrated in *Table I*.

Table I

THE EFFECT OF KNOWN SMOKE CONSTITUENTS[1] AND CRUDE SMOKE DEPOSIT UPON THE RECOVERY OF ϵ-DNP-LYSINE IN CARPENTER'S METHOD

Compound	% Destruction of ϵ-DNP-lysine
Vanillin	6·7
Acetovanillone	6·6
Guaiacol	2·2
Syringol	3·7
Methanal	4·2
Crude smoke deposit from kiln wall	21·7
Unsmoked coley	4·0
Smoked coley	9·0

[1] Pure compounds were used with ϵ-DNP-lysine on a 1:1 molar basis

The lysine contents of the unsmoked coley fillets were 97·6 ± 1·4mg g⁻¹ and 98·3 ± 1·1mg g⁻¹ by the DBD and by Carpenter's Methods, respectively. These values are similar to the 103·7mg g⁻¹ for fresh and 94·5mg g⁻¹ for frozen coley fillets calculated from the data of Tooley and Lawrie (1974) using an assumed 80% moisture content. Statistical analysis of the present data for all control and smoked samples ($N = 72$) indicates that the DBD-lysine values are indistinguishable from the FDNB-lysine values ($t = 1·42$, $F = 1·38$, $p = 0·01$). The relationship was linear ($r_{xy} = 0·99$) over the range 74mg g⁻¹ to 100mg g⁻¹ with a regression equation $y = 1·003x - 0·2550$ (x = DBD-lysine mg g⁻¹, y = FDNB-lysine mg g⁻¹).

The lysine content of the surface layer fell progressively (see *Table II*) to some 75% of its original value over a 5h smoking period and to some 89% in the middle layer over the same period of time. These results are similar to those of Hoffman *et al* who reported that the FDNB-lysine contents of whole *Tilapia lidole* smoked under various conditions fell to 91–94% of the original values, and in the case of heavily smoked *Sardinella aurita* to 71% of the original value.

Table II
THE EFFECTS OF HOT SMOKING ON THE DBD-LYSINE AND FDNB-LYSINE
CONTENTS OF COLEY

	DBD-lysine mg g⁻¹		FDNB-lysine mg g⁻¹	
	mean[1] ± S.D.	Loss %	mean[1] ± S.D.	Loss %
Unsmoked	97·56 ± 1·38	—	98·33 ± 1·33	—
Smoked 3 hours				
Upper layer	87·17 ± 1·07	10·6	86·16 ± 1·34	12·4
Middle layer	94·73 ± 1·51	2·9	93·76 ± 1·10	4·6
Smoked 5 hours				
Upper layer	73·31 ± 2·26	24·9	75·07 ± 1·19	23·3
Middle layer	85·71 ± 2·36	12·1	86·26 ± 2·38	12·3

[1] For the unsmoked samples data from 3h and 5h have been pooled since they were not significantly different and $N = 12$. For the other values $N = 6$

A comparison of the data (see *Table III*) for DBC and DBD indicates that lysine destruction as shown by DBD is primarily responsible for the loss of dye-binding capacity such that a 25% destruction of lysine is accompanied by a 7% fall in the other basic amino acids, *ie* histidine, arginine and N terminal residues collectively.

Table III
RELATIVE DESTRUCTION OF LYSINE AND OTHER BASIC AMINO ACIDS DURING
THE HOT SMOKING OF COLEY

	DBD meq g⁻¹ (ie lysine)		DBC−DBD meq g⁻¹ (ie basic amino acids other than lysine)	
	Mean[1] ± S.D.	Loss %	Mean[1] ± S.D.	Loss %
Unsmoked	0·667 ± 0·009	—	0·362 ± 0·011	—
Smoked 3 hours				
Upper	0·596 ± 0·007	10·6	0·347 ± 0·016	4·1
Middle	0·648 ± 0·010	2·8	0·369 ± 0·013	n.d.
Smoked 5 hours				
Upper	0·506 ± 0·015	24·1	0·338 ± 0·030	6·6
Middle	0·586 ± 0·016	12·1	0·372 ± 0·021	n.d.

n.d. = not detectable – observed change substantially less than total standard deviation of mean
[1] See footnote to *Table II*

4 The use of the metaperiodate reagent to monitor the phenol content of hot smoked coley

Knowles *et al* (1975) have reported that some 60% to 71% by weight of the phenolic fraction recovered from smoked bacon consists of guaiacol (2-methoxyphenol), syringol (2,6-dimethoxyphenol) and a range of 4-substituted guaiacols and syringols. Several groups of workers have reported that this fraction may form between some 75% and 96% of the phenolic fraction in liquid smokes or the smoke deposits recoverable from various smoked meats (Lustre and Issenberg, 1970; Knowles *et al*, 1975; Potthast, 1977). Any analytical procedure intended to measure total phenols must include these guaiacols and syringols but the Emerson reagent and the Gibbs reagent that have previously been employed for such analyses do not react with the 4-substituted derivatives that are found in smoke (Lustre and Issenberg, 1970; Knowles, *et al*, 1975; Svobodova and Gasparic, 1971; Bratzler *et al*, 1969). It has recently been reported that aqueous metaperiodate produces a yellow-orange colour with guaiacol, syringol

and twelve 4-substituted derivatives of them (Clifford, 1973). This reagent has since been found well suited to the routine quantitative determination of chlorogenic acids which include relatively complex 4-substituted derivatives of guaiacol (Clifford and Wight, 1976; Clifford and Staniforth, 1979) suggesting that metaperiodate might be a good reagent for measuring smoke phenols.

The commercially available smoke phenols were treated with metaperiodate as described previously for chlorogenic acids (Clifford and Wight, 1976) to determine the optimum reaction conditions. This method was applied to synthetic mixtures of smoke phenols that simulated the composition of smoke deposits and extracts as reported by other workers (Lustre and Issenberg, 1970; Knowles *et al*, 1975). Only quantitatively major components (over 10%) were included in the synthetic smokes, *ie* guaiacol, syringol, eugenol and phenol. These mixtures were analysed immediately after preparation, after steam distillation by the method of Potthast (1977) and after extraction by the method of Issenberg *et al* (1971) to ascertain the better method for recovering phenols from smoked fish.

4.1 Experimental
4.1.1 *Colour production with metaperiodate* Phenolic extract or standard (1cm³ containing up to 7·5 μM) in 70% 2-propanol is added to 2cm³ 0·2% aqueous sodium metaperiodate at 20°C in a stoppered test tube. The components are mixed thoroughly by shaking and allowed to react for 10min at 20°C and the colour read promptly at 390nm and 468nm. The blank is prepared by adding 1cm³ of extract or standard to 2cm³ of distilled water.

4.1.2 *Solvent extraction of phenols* – see Issenberg *et al* (1971).

4.1.3 *Steam distillation of phenols* – see Potthast (1977).

5 Results and discussion

The major smoke phenols guaiacol and syringol have a linear response to 0·2% metaperiodate up to at least 7·5 μM of added standard ($r_{xy} = 0·999$ and 0·995, respectively). Both yield yellow-orange colours with an absorption maximum at 390nm. Eugenol (4-allylguaiacol) and isoeugenol (4-propenylguaiacol) behave similarly to guaiacol but vanillin (4-formylguaiacol) and acetovanillone (4-methyl ketone guaiacol) yield little colour at 390nm. In contrast syringaldehyde (4-formylsyringol) and acetosyringone (4-methyl ketone syringol) yield moderate colour and 4-allylsyringol yields intense colour at 390nm. Unfortunately neither 4-alkyl guaiacols nor 4-alkyl syringols were available but the similar behaviour of guaiacol, syringol, catechol (2-hydroxyphenol), 4-methyl catechol and 4-*t*-butylcatechol imply that such compounds would yield substantial colour at 390nm and that they will be measured in smoked-fish extracts.

The analysis of various smoke extracts and deposits produced substantial absorption at 390nm consistent with guaiacol and syringol being major components and so guaiacol was adopted as a standard. The content of total metaperiodate-sensitive phenols (TPSP) is calculated from the absorption at 390nm and reported as guaiacol-equivalents. This interpretation is somewhat arbitrary because not only do the molar extinction coefficients of individual compounds vary substantially but the literature indicates that the relative proportions of individual phenols also vary substantially from sample to sample.

Of the compounds treated with metaperiodate, syringol alone showed a secondary peak at 468nm. Repeated recrystallization of syringol did not eliminate this second peak and since smoke extracts also showed strong absorption at this wavelength it was assumed that this second peak was characteristic of syringol, and possibly of hard wood being used for smoke generation. The absorption at this wavelength has been reported as 'syringol' using a molar extinction coefficient of 922. From the two molar extinction coefficients for syringol it is possible to calculate the proportion of the absorption at 390nm which is not due to 'syringol' and this has been reported as 'guaiacols' (*Table IV*).

Table IV

THE MOLAR EXTINCTION COEFFICIENTS OF VARIOUS PHENOLS TREATED WITH METAPERIODATE

Compound	λ max nm	Molar extinction coefficient at λ max mean $\pm$ S.D.	at 390nm
Guaiacol	390	–	1 028 $\pm$ 20
Syringol 1	390	–	1 835 $\pm$ 31
2	468	922 $\pm$ 20	–
Eugenol	390	–	1 220 $\pm$ 46
Isoeugenol	390	–	1 750 $\pm$ 50
Vanillin	350	1 250 $\pm$ 50	–
Acetovanillone	370	1 000 $\pm$ 50	–
Syringaldehyde	398	700 $\pm$ 25	680 $\pm$ 10
Acetosyringone	402	535 $\pm$ 5	480 $\pm$ 5
4-Allylsyringol	369	8 650 $\pm$ 60	5 070 $\pm$ 110
Catechol	382		1 088 $\pm$ 20
4-Methylcatechol	395		990 $\pm$ 20
4-t-Butylcatechol	395		1 060 $\pm$ 20

The results presented in *Table V* for smokes 1 and 2 suggest that the solvent-extraction procedure is better for recovering phenols than the cumbersome distillation method, in contrast to the findings of Potthast (1977). This different behaviour may reflect the lower fat content in coley compared with the meat examined by Potthast. Accordingly the solvent extraction procedure was adopted for the examination of hot smoked coley.[1] The results obtained for smoke 3 are of interest since they imply that when certain phenol mixtures are treated with metaperiodate the colour produced is not simply the sum of the colours that would have been produced with the compounds treated individually. It is possibly significant that smoke 3 was the only mixture to contain eugenol but it is clear that further investigation of this effect is required.

The data presented in *Table VI* show the contents of TPSP, 'guaiacols' and 'syringol' in the upper, middle and lower layers of hot smoked coley fillets. The values are of a magnitude similar to those reported by Lustre and Issenberg (1970) for smoked sausage and pork belly and

Table V

RECOVERY OF PHENOLS FROM SYNTHETIC SMOKES

Actual composition g l^{-1}	Guaiacol equivalents present[1]	Guaiacol equivalents recovered %		
		As made	After extraction	After distillation
Smoke 1 simulating pork belly deposit of Lustre and Issenberg (1970)				
Guaiacol 0·24	0·24			
Phenol 0·20[2]	0			
Eugenol –	–			
Syringol 0·88	1·41			
	1·65	100	97	79
Smoke 2 simulating sausage deposit of Lustre and Issenberg (1970)				
Guaiacol –	–			
Phenol 0·23	0			
Eugenol –	–			
Syringol 1·54	2·46			
	2·46	100	100	83
Smoke 3 simulating bacon deposit 1 of Knowles et al (1975)				
Guaiacol 0·80	0·80			
Phenol 0·40[2]	0			
Eugenol 0·07	0·084			
Syringol 0·14	0·224			
	1·108	120	129	87

[1] Actual content corrected for molar extinction coefficient of compound where this differs from guaiacol
[2] Does not react with metaperiodate

those reported by Bratzler *et al* (1969) for bologna but lower than those reported by Potthast. It is interesting to note that 'syringol' does not penetrate the fillet to the same extent as the 'guaiacols'.

Table VI

THE CONTENT OF PHENOLS IN HOT SMOKED COLEY FILLETS

	Total metaperiodate-sensitive phenols μg g^{-1} guaiacol equivalent fwb mean $\pm$ S.D.	'Syringol' content μg g^{-1} fwb mean $\pm$ S.D.	'Guaiacol' content μg g^{-1} fwb mean $\pm$ S.D.
Upper layer	33·07 $\pm$ 0·75	9·34 $\pm$ 1·39	9·53 $\pm$ 3·13
Middle layer	10·33 $\pm$ 0·75	2·00 $\pm$ 0·15	5·13 $\pm$ 0·71
Lower layer	22·23 $\pm$ 0·75	5·54 $\pm$ 2·15	5·43 $\pm$ 0·29

Total metaperiodate sensitive phenols calculated from E_{390} using guaiacol as standard
'Syringol' – calculated from E_{468} using syringol as standard
'Guaiacol' – calculated from E_{390} after correcting for syringol contribution at this wavelength
fwb = fresh weight basis

6 Concluding remarks

The dyebinding procedure reported in this paper is well suited to the routine measurement of lysine contents in smoked fish. The metaperiodate reagent shows considerable promise as a simple means of monitoring phenols in smoked fish and smoke deposits. It is desirable that more standard smoke phenols and smoke mixtures are analysed and that metaperiodate be used in conjunction with other functional group specific reagents.

7 References

BOOTH, V H. Problems in the determination of FDNB available lysine. 1971 *J. Sci. Fd. Agric.*, 22, 658–665
BRATZLER, L J, SPOONER, M E, WEATHERSPOON, J B and MAXEY, J A. Smoke flavour as related to phenol, carbonyl and acid content of bologna, *J. Fd. Sci.*, 34, 146–148

[1] The recovery of phenols from model systems and smoked fish using a solvent extraction procedure is the subject of another paper at this symposium (McGill *et al*, 1980).

CARPENTER, K J. The estimation of available lysine in animal protein
1960   foods, *Biochem, J.*, 16, 451–464
CLIFFORD, M N. Metaperiodate – a new structure-specific locating
1973   reagent for phenolic compounds. *J. Chromatog.*, 86, 222–224
CLIFFORD, M N and WIGHT, J. The measurement of caffeoylquinic
1976   acids and feruloylquinic acids in coffee beans. Development of
the technique and its preliminary application to green coffee
beans. *J. Sci. Fd. Agric.*, 27, 73–84.
CLIFFORD, M N and STANIFORTH, P S. A critical comparison of six spec-
1979   trophotometric methods for measuring chlorogenic acids in
green coffee beans. Huitième Colloque International sur la
Chemie des Cafés Verts, Torréfiés et leurs Dérives, Association International du Café, Nogent-sur-Marne, Franc. (In
press)
GILBERT, J and KNOWLES, M E. The chemistry of smoked foods – a
1975   review. *J. Fd. Technol.*, 10, 245–261
HAMM, R. Analysis of smoke and smoked foods. *Pure and Applied*
1977   *Chem.*, 49, 1655–1666
HOFFMAN, A, BARRANCO, A, FRANCIS, B J and DISNEY, J G. The effect
1977   of processing and storage upon the nutritive value of smoked
fish from Africa. *Tropical Science*, 19, 41–53
ISSENBERG, P, KORNREICH, M R and LUSTRE, A O. Recovery of
1971   phenolic woodsmoke components from smoked foods and
model system. *J. Fd. Sci.*, 36, 107–109
JONES, G P. The use of dyebinding procedures for the evaluation of
1974   protein quality. PhD Thesis, University of Reading, England

KNOWLES, M E, GILBERT, J and MCWEENY, D J. Phenols in smoked
1975   cured meats. Phenolic composition of commercial liquid
smoke preparations and derived bacon. *J. Sci. Fd. Agric.*, 26,
189–196
LAKIN, A L. The estimation of protein quality by dyebinding-
1973   procedures, *IFST Proceedings*, 6, 80–83
——— The estimation of protein by dyebinding principles and par-
1975   ameters. *J. Sci. Fd. Agric.*, 26, 549
LUSTRE, A O and ISSENBERG, P. Phenolic components of smoked meat
1970   products. *J. Agr. Fd. Chem.*, 18, 1056–1060
MCGILL, A S, MURRAY, J and PARSONS, E. Recovery of phenols from
1980   fish using a solvent extraction technique, *This volume*
POTTHAST, K. Determination of phenols in smoked meat products.
1977   *Acta Alimentaria*, 3, 189–193
SANDLER, L and WARREN, F L. Effect of ethyl chloroformate on the
1974   dyebinding capacity of protein. *Analyt. Chem.*, 46, 1870–1872
SVOBODOVA, D and GASPARIC, J. Investigation of the colour reaction of
1971   phenols with 4-aminoantipyrine. *Mikrochimica Acta* (Wein),
384–390
TANG, S L. A chemical method for the investigation of chemical
1979   changes in fish during smoking. MSc Thesis, Loughborough
University of Technology, England
TOOLEY, P J and LAWRIE, R A. Effect of deep fat frying on the availa-
1974   bility of lysine in fish fillets, *J. Fd. Technol.*, 9, 247–254
YUDITSKAYA, A I and LEBEDEVA, T M. The phenol composition of
1960   smoke cured fish (Sostav fenolov kopchenoi ryby) *Ryb. Khoz.*,
9, 69–73

The utilization of Mediterranean sardines by means of smoking

E Senesi, G Bertolo, D Torreggiani, L Di Cesare and G Caserio

1 Introduction

Italian sardine production (*Sardina pilchardus* W.), which was about 40 000 tons in 1978, does not find a satisfactory outlet on the domestic market, so that part of it is exported while the remainder is made into animal feeding stuffs. This form of utilization is unsatisfactory due to the pressing national demand for animal proteins for direct human consumption.

With the aim of utilizing this large catch more rationally an attempt was made to diversify use of the sardine, which is at present limited to the production of canned sardines in oil, so as to reach wider consumer markets.

A series of studies have been carried out for this purpose. The first one was aimed at obtaining a frozen product ready for consumption by means of filleting and frying pretreatment (Pizzocaro *et al*, 1979), the second one at achieving a basic paste for the preparation of precooked dishes (Crivelli *et al*, 1979), while the purpose of the third one was to test the behaviour of the sardine on smoking. The present paper deals with the third approach.

Two types of products have been developed: sardine fillets smoked and preserved by either cold storage or canning.

2 Materials and methods

2.1 Smoking
Smoking was carried out using an AFOS–Torry mini kiln pilot plant, charged with shavings of beech wood and beech or poplar sawdust. The speed of the air flow within the smoking chamber was 3m/s. Ventilation was continuous in the case of fillets to be preserved by cold storage and intermittent for those to be canned. The quantity of smoke entering the smoking chamber was regulated by lighting one or two sawdust burners.

2.2 Sardines
The fish used in these experiments were caught in the northern part of the Adriatic Sea 12h to 18h before their arrival in the laboratory. During this period they were iced.

The size was 25 fish per kg. The fish were prepared and filleted by hand before being smoked.

2.3 Cold-stored smoked sardine fillets
Before smoking the fillets were dipped in a 26% (w/w) salt brine for 5min at a temperature of 20°C. The percentage of salt and the optimum dipping time were chosen after a series of preliminary tests. At the end of the dipping time the drained fillets were placed on trays and kept in a cold room at 0°C for 24h. The trays were then placed in the kiln in which smoke and air at 30°C were allowed to circulate.

After some preliminary tests the first cold smoking stage was set at 75min; after this time, the temperature of the air was raised to 82°C for 30min, so as to achieve hot smoking, with a total amount of heat sufficient to pasteurize the fillets (Deng *et al*, 1974).

The smoked fillets were packed under vacuum in plastic film of very low permeability to oxygen (BB1 supplied by W R Grace Co). The fillets were stored for six months both frozen at −20°C and refrigerated at 0–2°C. Colorimetric, organoleptic, bacteriological and chemical tests were done at regular intervals during storage.

2.4 Canned smoked sardine fillets
In this case, the fillets were dipped in brine of varying NaCl content (10%, 12% and 15%) for 5min, followed by a brief standing in the cold room at 0°C to allow the surface water to drip away. Smoking was carried out for 60min at 30°C, with dense smoke, adjusting the air circulation in such a way as to make the fillets lose a

certain amount of their initial weight. The smoked fillets were then cooked for about 1min with free-flowing steam, so as to reach 70°C inside the muscle. This was followed by a drying stage with circulation of air always in the presence of smoke, for the time necessary to lose a further amount of moisture. Finally, the fillets were arranged in layers in coated aluminium cans (105 × 76 × 21mm) which were then filled with olive oil. After closure, the cans were heated at 102°C for 30min and then rapidly cooled. The processing parameters considered for these tests were the loss of weight during the entire process and the NaCl concentration of the brine. Organoleptic and chemical analyses, the latter to determine the salt and smoke content, commercial sterility and biological sterility tests were carried out.

2.5 Organoleptic tests
The organoleptic evaluations were performed according to the test panel method, using a score from 1 (very poor) to 9 (very good) and considering appearance, flavour and texture (Larmond, 1977). The results were then subjected to analysis of variance, using the multiple comparison test.

2.6 Colorimetric analyses
A Hunter-Lab model D25 colorimeter was used and the test was based on five readings per sample. Each sample consisted of five fillets which were placed in a container 10cm in diameter with a glass base, the skin-side being downwards first, followed by the flesh side. The container was emptied after each set of determinations and the fillets arranged differently.

The following values were calculated on the basis of Hunter's L, and b values.

(1) $(a^2 + b^2)^{\frac{1}{2}}$, which expresses colour saturation;

(2) ΔE which, according to the Hunter-Scofield formula $[(L)^2 + (a)^2 + (b)^2]^{\frac{1}{2}}$ expresses the colour difference in NBS units as reported by Judd and Wyszecki (1967).

2.7 Chemical analyses
The NaCl content of the flesh was evaluated according to the AOAC method, while the phenolic components content (smoke) was determined according to Tucker's method modified by Di Cesare (1979).

2.8 Bacteriological tests
Counts of mesophiles, coliforms, sulphite-reducing clostridia and psychrophiles were run on the cold stored fillets.

A search for mesophiles was carried out on Petri dishes of tryptose agar plus 3% salt, incubated at 30°C for 48h. Coliforms were detected by means of brilliant green-bile medium, sulphite-reducing clostridia in tubes of reinforced clostridial medium plus sodium sulphite and psychrophiles on Petri dishes of tryptose agar incubated at 2–4°C for 15 days.

Commercial and biological sterility tests were carried out on canned fillets.

The commercial sterility tests were performed by incubating samples at 30°C and 55°C for 20 and 7 days, respectively.

After this period, the samples were subjected to sterility tests, innoculating suitable amounts of the product into test tubes of brain heart infusion and reinforced clostridial medium, in order to determine the presence of surviving spores of aerobes and anaerobes, respectively.

3 Results and discussion

3.1 Refrigerated fillets
Examination of the organoleptic results (Table I) shows that smoked sardine fillets stored at a temperature of 0–2°C had excellent stability (appearance and flavour) up to three months. A reduction in flavour score, statistically significant at the 5% level, was observed only after four months' storage at 0–2°C. This reduction became statistically significant at the 1% level after six months' storage. However, the reduction never rendered the product unsatisfactory.

Table I
ORGANOLEPTIC TESTS OF COLD-STORED SMOKED FILLETS

	Refrigerated (days storage at 0–2°C)						Frozen (days storage at −30°C)		
	0	15	60	90	120	180	60	90	120
Appearance	7·1	7·1	7·5	7·2	6·8	6·5	6·8	7·2	5·2[2]
Taste	7·6	7·3	7·1	7·1	6·5[1]	6·1[2]	6·8	5·8[1]	4·4[2]

Mean of the scores given by panellists

[1] Significant difference at the 5% level
[2] Significant difference at the 1% level

The smoke content, as measured by phenol concentration, remained almost constant until the end of the storage period; similarly for salt content and water activity (Table II).

Table II
CHEMICAL ANALYSES OF COLD-STORED SMOKED FILLETS

	Refrigerated (days storage at 0–2°C)						Frozen (days storage at −30°C)		
	0	15	60	90	120	180	60	90	120
Phenol (mg/100g)	6·6	6·4	—	6·7	—	6·6	—	6·5	6·4
Salt (g/100g)	3·5	3·4	—	3·5	—	3·5	—	3·4	3·4
A_w	0·87	0·87	—	0·87	—	0·87	—	0·87	0·86

The colour of the skin (Table III) revealed marked changes after only 15 days' storage; however, these modifications did not appear to have any negative effect on the judgement of the panel. In contrast, the flesh retained its colour up to two months. After this it darkened until it became stable (fourth to sixth month). Darkening was probably due to surface non-enzymatic auto-oxidation reactions.

Bacteriological tests revealed (Table IV) a very low microbial content, the values of which fell during storage. Furthermore, the microflora was found to consist of common, metabolically inactive micro-organisms. Bacteria indicative of faecal contamination (coliforms) and potentially pathogenic for man (sulphite-reducing clostridia) were absent.

Table III
COLORIMETRIC ANALYSES OF COLD-STORED SMOKED FILLETS

	Refrigerated (days storage at 0–2°C)						Frozen (days storage at −30°C)		
	0	15	60	90	120	180	60	90	120
L^1 skin	46·34	40·07	41·10	37·69	41·80	40·42	39·02	41·10	43·39
flesh	40·80	40·18	37·14	39·66	35·19	38·17	42·51	41·08	42·41
a^1 skin	−1·37	−0·89	−1·29	−0·83	−1·09	−0·46	−1·31	−1·10	−0·87
flesh	4·87	5·03	6·02	5·23	6·33	5·85	5·28	5·79	5·48
b^1 skin	8·49	6·81	6·63	6·36	6·41	7·78	6·61	6·13	6·66
flesh	16·53	16·29	15·58	14·39	14·47	14·63	17·81	16·78	16·47
$(a^2 + b^2)^{\frac{1}{2}}$ skin	8·59	6·86	6·75	6·41	6·50	7·79	6·73	6·22	6·71
flesh	17·23	17·04	16·70	15·31	15·79	15·76	18·57	17·75	17·35
ΔE^2 skin		6·50	5·55	8·92	5·00	6·03	7·35	5·75	3·50
flesh		0·67	3·95	2·44	6·15	3·38	2·17	0·98	1·72

[1] Mean of 5 readings
[2] Difference with regard to 0 days

Table IV
BACTERIOLOGICAL ANALYSES OF COLD-STORED SMOKED FILLETS

Micro-organisms (cells/g)	Refrigerated (days storage at 0—2°C)						Frozen (days storage at −30°C)		
	0	15	60	90	120	180	60	90	120
Total count of mesophiles	1 750	150	500	250	200	75	400	250	150
Coliforms	<10	<1	<1	<1	<1	<1			
Sulphite-reducing clostridia	none in 5g	idem	idem	idem	idem	idem	idem	idem	idem
Psychrophiles	300	—	—	200	900	10	—	250	500

The psychrophiles were found not to be in an active state. The very good hygienic condition of smoked sardine fillets is favoured by the fact that the water activity (A_w) was at values such as to limit growth of micro-organisms, which were also hindered by the high salt content in the flesh.

3.2 Frozen fillets
Physical and bacteriological examinations gave results very similar to those for cold-stored fillets and the comments made in the previous section also apply here.

However, some differences were observed as regards the colorimetric and organoleptic results.

Thus, the colour of the frozen flesh (*Table IV*) appeared to remain similar to that seen immediately after smoking, and showed good stability; colour differences were slight, even after four months' storage.

The organoleptic tests (*Table I*) showed that the flavour had deteriorated markedly after only three months' storage; after four months, the appearance as well was judged negatively, to the extent that the products could no longer be considered edible. The differences between the results of the organoleptic tests on the sample stored for four months, compared with the fresh ones, were statistically significant at the 1% level. In the case of the three-month sample, compared to the fresh one, a statistically significant difference at the 5% level is seen only as regards flavour.

3.3 Canned fillets
For this type of product, preliminary tests had revealed that the main defect to be avoided was the onset of a certain degree of toughness of the flesh, probably caused by the action of heat in combination with the presence of phenolic substances derived from the smoke, the salt content of the flesh after brining and the loss of moisture.

Organoleptic tests, during which the panellists were asked to judge also the texture, made it possible to determine the processing parameters (concentration of the brine, loss of weight during the entire process) most suitable for maintaining a good final texture of the muscle.

In fact, all the panellists perceived the salt content differences but the scores attributed to flavour and texture did not give statistically significant differences (*Table V*). In contrast, the tests performed on samples dipped in brine with the same salt content (15%), but which had lost a different amount of water (22% and 30% respectively) during the process, revealed a statistically significant difference at the 1% level as regards texture, while the scores attributed to flavour and appearance remained almost the same. This would seem to demonstrate that changes in texture, with the onset of toughness, may be ascribed to the loss of water during the process rather than to the initial salt content of the flesh, and that they occur when the fillets undergo an approximately 30% weight loss. The commercial sterility tests gave negative results: no swellings were seen at the end of the incubation period, and thus the presence of gas due to microbial activity could be excluded. However, bacteriological sterility tests revealed in some cases the survival of a few bacterial spores of the genus *Bacillus* but these were incapable of reproducing themselves inside the product.

This fact could be of great importance because food regulations in many countries, including Italy, prescribe that canned fish must be sterile, otherwise it is classified as a semi-conserve.

However the tendency of European Community legislators is to admit also a category of non-sterile canned goods but stable at room temperature, the ones known, for instance, in France, as 'canned food of commercial sterility' (Buttiaux, 1956).

Table V
SYNOPTIC TABLE OF CANNED SMOKED FILLETS

Salt content of brine %	Salt content of flesh %	Weight loss %	Appearance	Flavour	Texture	Commercial sterility 30°C 20 days	Commercial sterility 55°C 7 days	Biological sterility
10	1·5	22	7·2	6·3	6·4	−Ve	−Ve	few spores
12	3·0	16·5	7·0	6·8	6·4	−Ve	−Ve	sterile
15	4·3	22	6·3	6·5	6·3	−Ve	−Ve	sterile
15	4·3	30	6·4	5·9	4·3	−Ve	−Ve	few spores

4 Conclusions

It has been demonstrated that in the case of sardines, smoking is feasible as a means of ensuring the long-term preservation of this species of fish. In fact, smoked sardine fillets made in the way described can be stored for four months at temperatures of 0–2°C in vacuum packs, without any decay of the organoleptic characteristics. In contrast, storage of sardine fillets in the frozen state, again in vacuum packs, affords products which retain good quality for two months at most.

Canned smoked sardine fillets are commercially stable, although the processing conditions do not always give a sterile product. The tests carried out have shown that the main defect to be avoided is the onset of a certain degree of toughness which could probably be ascribed to loss of moisture during processing rather than to the initial salt content of the flesh.

5 References

BUTTIAUX, R. Definition et modalités d'examen bactériologique des
1956 divers variétés de conserves alimentaires. *Proc:* III Congrès international de la conserve, Rome

CRIVELLI, G, SENESI, E, MAESTRELLI, A and BERTOLO, G. Production
1979 de plats précuisinés à base de pâte de sardine. Comm. au XV Congrès international du froid, Venice

DENG, J, TOLEDO, R T and LILLARD, D A. Effect of smoking tempera-
1974 tures on acceptability and storage stability of smoked spanish mackerel. *J. Fd. Sci.*, 39, 596–601

DI CESARE, L F. Un metodo rapido per la determinazione dei com-
1979 posti fenolici nel pesce affumicato. *Tecnologie Alimentari.* (In press)

JUDD, D B and WYSZECKI, G. Color in business, science and industry.
1967 John Wiley, New York. 2nd edn., 296

LARMOND, E. Laboratory methods for sensory evaluation of food.
1977 Publication 1637. Res. Branch Can. Dep. Agri.

PIZZOCARO, F, SENESI, E, TROVATI, M and POLESELLO, A. Expérimen-
1979 tation sur la conservation des filets congelés de sardine soumis à divers prétraitements. Comm. au XV Congrès international du froid, Venice

6 Chilled and frozen storage

Storage of chilled cod under vacuum and at various concentrations of carbon dioxide

M H Jensen, A Petersen, E H Røge and A Jepsen

1 Introduction

It has long been known (Kolbe, 1882) that beef stored in an atmosphere of carbon dioxide (CO_2) has an extended shelf-life. However, it was not until the thirties that the practical use of carbon dioxide as a preservative was realized in shipments of whole chilled carcasses from Australia and New Zealand to Great Britain (Lawrie, 1974).

Today it is well known that packaging in a carbon dioxide atmosphere at low temperatures effectively extends the shelf-life of red meat and poultry (Thomson and Risse, 1971; Silliker *et al*, 1977; Rosset, 1978; Sander and Soo, 1978; Bailey *et al*, 1979). This method has not been utilized commercially, except for retail packaging of red meat cuts.

At the Technical Laboratory of the Ministry of Fisheries, Denmark, investigations have shown that packaging in a 100% CO_2 atmosphere extends the shelf-life of industrial fish by nine days at 0°C, whereas at 12°C, the effect is not significant (Olsen, personal communication, 1977). Similar results have been obtained by Norwegian investigators (Mjelde, 1974; 1975).

Considering fish for human consumption, Cann (personal communication, 1978), found, in comparison with iced controls, a shelf-life extension of seven days at 0·5°C, using CO_2 in combination with mechanically refrigerated sea water (rsw). As regards retail packaging of fish, however, only limited information is available.

The present work was carried out to investigate the effect of vacuum and various concentrations of carbon dioxide on microbiological, and thereby chemical, changes in chilled fish. The associated changes in organoleptic properties have also been evaluated and the practical significance of these results is discussed.

2 Experimental

2.1 *Preparation of fish samples*
Cod were caught by trawl in the Kattegat and brought to the laboratory. After two days in ice, pieces of washed and gutted fish (about 20cm in length) were packed at random in (*1*) air, (*2*) vacuum, (*3*) 40% CO_2, 60% N_2, (*4*) 60% CO_2, 40% N_2, and (*5*) 100% CO_2, and stored at 2°C. A Multivac R 7000/S packaging machine and polyacrylonitrile/polyethylene film were used for (*2*)–(*5*), and polyethylene film for (*1*). There was one piece of fish in each bag. At intervals during storage five bags from each of the treatments were removed for examination.

2.2 *Analysis*
2.2.1 *Chemical tests* The gas composition in the bags with carbon dioxide atmosphere was measured in quadruplicate on a Hewlet-Packard chromatograph, model 5700 A. The method of Conway and Byrne (1933) was used in analysing trimethylamine oxide (TMAO), trimethylamine (TMA) and total volatile bases (TVB). These tests were carried out in duplicate.

2.2.2 *Microbiological tests* Total viable count of bacteria (incubated at 20°C and 37°C, respectively) was estimated by plating serial dilutions of fish samples prepared by homogenizing 10g of fish meat in 90ml of sterile physiological saline solution containing 0·1% peptone. Surface plate counts were employed where colonies were to be selected for identification; pour plate counts were used elsewhere. The agar used was peptone-iron-agar containing 0·3% beef extract, 0·3% yeast extract, 2% peptone, 0·5% sodium chloride, 0·03% ferric citrate and 0·03% sodium thiosulphate (pH 7·4). H_2S-producing organisms were counted on iron-plate-agar as black colonies. It was shown by Levin (1968) that from fish these are specific spoilage bacteria *Alteromonas putrefaciens*. Lactic acid bacteria were counted on a selective agar (Davidson and Cronin, 1973) containing 10·0g/l peptone, 10·0g/l peptonized milk, 10·0g/l yeast extract, 7·5g/l glucose, 2·5g/l beef extract, 0·575g/l $MgSO_4$, 7 H_2O, 0·05g/l $MnSO_4$, 4 H_2O, 1·0g/l Tween 80, 15g/l agar (pH 5·5). Prior to use, 1·0ml of each of (*1*) freshly prepared sodium nitrite solution (6·0% wt/vol), (*2*) acitidone solution (0·1% wt/vol) and (*3*) polymyxin B solution (0·03% wt/vol) were added to each 97ml of the basal medium. All counts were calculated as the geometric mean of five individual samples taken from each of the five treatments and the values tested for significance (one side t-test). Each of the dominating colony types on surface plating from the fresh and the spoiled fish were counted separately. Five colonies of each type were picked out and transferred to a meat extract broth. After purification and isolation by repeated platings on iron-plate-agar, and/or blood agar, the bacteria were classified according to Cowan and Steel (1966). The isolated bacteria were tested for their ability to reduce TMAO to TMA, using an agar containing 0·3% beef extract, 0·3% yeast extract, 2% peptone, 0·4% NaCl, 0·4% KH_2PO_4, 0·575% K_2HPO_4, 0·05% $MgSO_4$, 0·4% agar, and 2 Resazurin tablets (Eh indicator) (pH 6·8). Prior to use, 1·0ml of each of (*1*) filter-sterilized cysteine solution (4% wt/vol), and (*2*) filter-sterilized TMAO solution (50% wt/vol), were added to each 100ml of the basal medium.

2.2.3 *Organoleptic properties* Samples from each of the treatments were cooked in a casserole by immersion for 15min in warm water at 80°C containing 2% salt. The fish were tested in two groups (two and three samples respectively) on the same day under code by a panel of eight members. Numerical scores for odour, flavour and texture were given on a scale where 10 represents the ideal fresh product and 0 the lowest quality, while 4 represents the limit of acceptable quality. The values were tested for significance, using two-dimensional tests and Duncan's multiple range test.

3 Results

3.1 Microbiological

Total viable counts at 20°C (psychrophilic flora) and the growth of H_2S-producing organisms are shown in Figs 1 and 2, respectively. In both cases the growth was indeed limited by the absence of air (vacuum treatment) and increasing levels of CO_2 in bags with controlled atmosphere.

The presence of CO_2 delayed the onset of the logarithmic growth phase for the H_2S-producing organisms up to four to six days (Fig 2). The absence of an initial lag phase in air- and vacuum-treated packs resulted in growth from the first day of storage (Fig 2). The stationary phase of growth was reached when H_2S-producing organisms and total viable counts at 20°C reached 10^7 and $10^8/g$, respectively, in air-pack treat-

ment, whereas in vacuum and in the CO_2 atmospheres the counts were 100–1 000 times lower on the last day of storage.

During 12 days of storage, the total count at 37°C (mesophile flora) in the air-packed fish reached $10^4/g$. In the four other groups the counts remained at the initial level (10^2–$10^3/g$). No lactic acid bacteria were detected.

In all groups and, in particular, in vacuum-treated fish there was increasing predominance of H_2S-producing organisms during storage (Fig 3). However, this tendency was not very marked in fish packed in 100% CO_2. In the fresh fish there was a predominance of Alcaligenes strains, while mainly Pseudomonas, Flavobacterium and Coryneform strains made up the rest of the flora. Regardless of the packaging method, however, TMAO-reducing bacteria (including Pseudomonas spp. and H_2S-producing organisms) dominated in the spoiled fish.

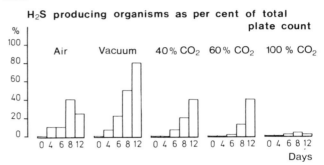

Fig 3 H_2S-producing organisms as percentage of total plate count (20°C)

3.2 Chemical

In Fig 4 it is seen that the development of TVB in fish packed in air, vacuum and 40% CO_2 was practically the same, reaching a higher level than in fish packed in 60% and 100% CO_2. The TMA development followed the same pattern, but at a lower level (Fig 5), however, the differences in TMA contents were almost eliminated on the last day of storage. The decrease in TMAO corresponded to the increase in TMA. Table I shows that the head space gases in bags with carbon dioxide atmospheres undergo appreciable changes during the first five days of storage, after which they remain practically constant.

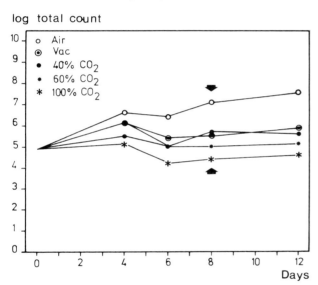

Fig 1 Log total plate count (20°C) per g in cod stored at 2°C

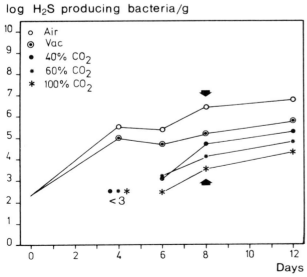

Fig 2 Log H_2S-producing organisms per g in cod stored at 2°C

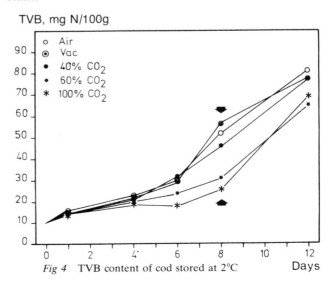

Fig 4 TVB content of cod stored at 2°C

295

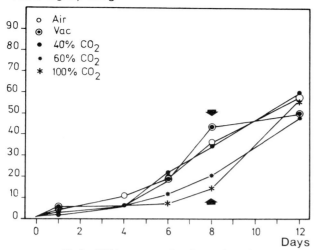

Fig 5   TMA content of cod stored at 2°C

Table I
CHANGES IN MEAN GAS COMPOSITION (IN PER CENT) DURING STORAGE AT 2°C

Days at 2°C Code	0			5			12		
	CO_2	N_2	O_2	CO_2	N_2	O_2	CO_2	N_2	O_2
40%	43·7	55·3	1·0	20·3	79·4	0·1	20·0	79·9	0·1
60%	63·0	36·3	0·9	30·5	67·5	0·4	28·3	71·6	0·1
100%	96·5	2·7	0·8	69·1	20·3	3·1	71·7	27·2	1·0

3.3 Organoleptic properties

During six days of storage, organoleptic quality scores gradually decreased (*Fig 6*). Between the sixth and the eighth day, the air-packed fish became unacceptable. At the same time the vacuum-packed fish were just acceptable, whereas the fish in carbon dioxide atmosphere were scored at about 6. At this important point the difference was significant ($P < 0.05$). So was the difference in total psychrophilic flora and H_2S-producing flora between the air-packed fish and the fish packed in 60% and 100% CO_2.

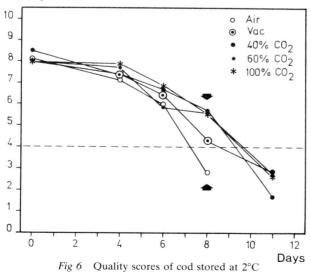

Fig 6   Quality scores of cod stored at 2°C

4 Discussion

Packaging of cod in vacuum (low oxygen tension) inhibits the growth of the total flora as previously recorded by Huss (1972). Moreover, the presence of increasing levels of CO_2 leads to an additionally inhibitory effect; this agrees with previous findings (King and Nagel, 1967; Rask, 1977a,b; Sutherland *et al*, 1977). However, during the first four days of storage, vacuum packaging has no influence on the growth of H_2S-producing organisms. This resulted in an increase in the proportion of this type of organism which was probably due to the ability of H_2S-producing bacteria to adapt, using TMAO as hydrogen acceptor (Sakaguchi and Kawai, 1977a). Consistent with these suggestions are the similarity in TMA development in the air- and vacuum-packed fish. On the other hand, our results do not support the theory that oxygen has an inhibitory effect on TMAO reductase formation and activity (Sakaguchi and Kawai, 1977b). However, their suggestions are supported by Debevere and Voets (1971) and Huss (1972), who found a higher TVB development in vacuum-treated haddock compared to the air-packed control, which, on the other hand, had a higher total plate count.

The slow-down of the growth of H_2S-producing organisms after four to six days of storage of vacuum-packed fish is probably caused by bacterial formation of CO_2.

During storage of air-packed fish TMAO-reducing *Pseudomonas* species and H_2S-producing bacteria are favoured, which is in accordance with earlier investigations (Castell *et al*, 1949; Shewan *et al*, 1960; 1971; Herbert *et al*, 1971). As mentioned above, only H_2S-producing organisms are favoured at low oxygen tension. With increasing levels of CO_2, however, these too are inhibited so allowing the *Pseudomonas* strains to predominate. Hence, regardless of the packaging method, the growth of bacteria is accompanied qualitatively by an emergence of TMAO-reducing bacteria more or less endowed with H_2S-producing properties. The limited shelf-life extension with packaging cod in vacuum or in the presence of CO_2, compared with the extended shelf-life of red meat and poultry in a carbon dioxide atmosphere, must be due to the presence of TMAO in the fish meat. This compound both supports the growth of specific spoilage bacteria and at the same time is reduced to TMA, the chemical substance most commonly associated with fish spoilage. Moreover, the unavoidable high bacterial content at the time of packaging may also reduce the effect of carbon dioxide on the shelf-life. Flavour developed by *Pseudomonas* species, not able to reduce TMAO, may also be of significance in spoilage of chilled fish in the presence of air (Castell *et al*, 1957; Castell and Greenough, 1957; 1959). However, the taste panel's comments on the differently treated fish were practically the same except for the acid taste sometimes noted in the fish packed in vacuum and in carbon dioxide. Considering the lack of growth of lactic acid bacteria, this acid taste may be due to dissolution of CO_2 in the fish water phase.

It would appear from these observations that most of the odours developing during the spoilage of the chilled cod are caused by various species of TMAO-reducing bacteria. How far the development of TMA alone is critical for the keeping quality, is, however, a matter for discussion.

In the present investigation, the relationship between TMA production and the organoleptic quality is not very clear. However, it is interesting to notice how the growth of H_2S-producing organisms was affected by the packaging method. Unfortunately, we did not measure the H_2S concentration chemically and thereby demonstrate the significance of sulphide-like odours, which, according to Shewan (1965), are characteristic of an anaerobic type of spoilage.

In spite of the quantitative and qualitative differences in microbial growth and in TMA and TVB development, there were only slight differences in the spoilage progress and a very little extension of shelf-life. It is doubtful if the gain of two days in storage time with carbon dioxide would be of any commercial advantage whereas the high costs involved and the pronounced shrinkage of the packs due to the carbon dioxide dissolving in the water phase of the fish are serious disadvantages.

5 References

BAILEY, J S, REAGAN, J O, CARPENTER, J A and SCHULER, G A. Mi-
1979 crobiological condition of broilers as influenced by vacuum and carbon dioxide in bulk shipping packs. *J. Fd. Sci.*, 44(1), 134–137

CASTELL, C H and GREENOUGH, M F. The action of *Pseudomonas* on
1957 fish muscle. 1. Organisms responsible for odours produced during incipient spoilage of chilled fish muscle. *J. Fish. Res. Bd. Can.*, 14(4), 617–625

—— The action of *Pseudomonas* on fish muscle. 4. Relation between
1959 substrate and the development of odours by *Pseudomonas fragi. J. Fish. Res. Bd. Can.*, 16(1), 21–31

CASTELL, C H, GREENOUGH, M F and JENKIN, N L. The action of
1957 *Pseudomonas* on fish muscle. 2. Musty and potato-like odours. *J. Fish. Res. Bd. Can.*, 14, 775–782

CASTELL, C H, RICHARDS, J F and WILMET, I. *Pseudomonas putrefa-*
1949 *ciens* from cod fillets. *J. Fish. Res. Bd. Can.*, 7(7), 430–431

CONWAY, E J and BYRNE, A. An absorption apparatus for the micro-
1933 determination of certain volatile substances. 1. The micro-determination of ammonia. *Biochem. J.*, 27, 419–429

COWAN, S T and STEEL, K J. *Manual for the Identification of Medical*
1966 *Bacteria*. Cambridge University Press

DAVIDSON, C M and CRONIN, F. Medium for the selective enumeration
1973 of lactic acid bacteria from foods. *Appl. Microbiol.*, 26(3), 439–440

DEBEVERE, J M and VOETS, J P. Microbiological changes in prepacked
1971 cod fillets in relation to the oxygen permeability of the film. *J. Appl. Bact.*, 34(3), 507–513

HERBERT, R A, HENDRIE, M S, GIBSON, P M and SHEWAN, J M. Bac-
1971 teria active in the spoilage of certain sea foods. *J. Appl. Bacteria*, 34(1), 41–50

HUSS, H H. Storage life of prepacked wet fish at 0°C. I. Plaice and
1972 haddock. *J. Fd. Techn.*, 7, 13–19

KING, A D and NAGEL, C W. Growth inhibition of a *Pseudomonas* by
1967 carbon dioxide. *J. Fd. Sci.*, 32, 575–579

KOLBE, H. Antiseptische eigenschaften der kohlensäure. *J. Prakt.*
1882 *Chem.*, 26, 249–255

LAWRIE, R A. *Meat Science*, Pergamon Press Ltd, Oxford. 2nd edn.,
1974 196–197

LEVIN, R E. Detection and incidence of specific species of spoilage
1968 bacteria on fish. I. Methodology. *Appl. Microbiol.*, 16, 1734–1937

MJELDE, A. Bruk av kjøling og kulsyra for å forlænge holdbarheten av
1974 vinterlodda. *Meld. Fra SSF*, nr. 2, 15–17

—— Konservering av vinterlodda med is og kulsyre. *Meld. Fra SSF*, nr.
1975 4, 9–11

RASK, K. Kontrolleret atmosfæres indvirkning på mikroorganismer.
1977a *Review*. Emballageinstituttet, København

—— Kontrolleret atmosfæres indvirkning på mikroorganismer. *For-*
1977b *søgsrapport*. Emballageinstituttet, København

ROSSET, R. Storage of meat under controlled atmosphere. *Annales de*
1978 *la Nutrition et de l'Alimentation*, 32, 545

SAKAGUCHI, M and KAWAI, A. Electron donors and carriers for the
1977a reduction of trimethylamine N-oxide in *Escherichia coli. Bull. Jap. Soc. Sci. Fish.*, 43(4), 437–442

—— Confirmation of a sequence of the events associated with
1977b trimethylamine formation by *Escherichia coli. Bull, Jap. Soc. Sci. Fish.*, 43(5), 611

SANDER, E H and SOO, M M Increasing shelf life by carbon dioxide
1978 treatment and low temperature storage of bulk pack fresh chickens packaged in nylon/surlyn film. *J. Fd. Sci.*, 43(5), 1519–1527

SHEWAN, J M. Bacteriology of fish stored in chilled sea water. In: *Proc:*
1965 *Meeting on fish technology, fish handling and preservation*, Scheveningen, OECD, Paris. 95pp

—— The microbiology of fish and fishery products – a progress report.
1971 *J. Appl. Bacteriol.*, 34(2), 299–315

—— The bacteriology of fresh and spoiling fish and the biochemical
1977 changes induced by bacterial action. In: *Handling, Processing and Marketing of Tropical Fish*. Tropical Products Institute, London. 51.

SHEWAN, J M, HOBBS, G and HODGKISS, W. The *Pseudomonas* and
1960 *Achromobacter* groups of bacteria in the spoilage of marine white fish. *J. Appl. Bact.*, 23(3), 463–468

SILLIKER, J H, WOODRUFF, R E, LUGG, J R, WOLFE, S K and
1977 BROWN, W D. Preservation of refrigerated meats with controlled atmospheres: treatment and post-treatment effects of carbon dioxide on pork and beef. *Meat Science*, 1(3), 195–204

SUTHERLAND, J P, PATTERSON, L T, GIBBS, P A and MURRAY, G J. The
1977 effect of several gaseous environments on the multiplication of organisms isolated from vacuum packaged beef. *J. Fd. Tech.*, 13(3), 249–255

THOMSON, J E and RISSE, L A. Dry ice in various shipping boxes for
1971 chilled poultry: Effect on microbiological and organoleptical quality. *J. Fd. Sci.*, 36, 74

Correlation of microbiological status with organoleptic quality of chilled Cape hake

C K Simmonds and E C Lamprecht

1 Introduction

Specifications for frozen fish and frozen fish products often include, among other microbiological requirements, a maximum tolerance for a mesophilic count, for example total aerobic count at 37°C (South African Bureau of Standards, 1977). It has been noted by the authors that mesophilic counts are frequently used by commercial fish processors for quality control purposes. While such counts may give a good indication of the hygienic standard of the product, the practice of using them to assess the quality *per se* of fish is questionable. It has long been known that psychrophilic bacteria are largely responsible for the putrefaction of fish (Shewan, 1962). It was therefore considered important to determine the relationship between the total mesophilic and psychrophilic counts, and the quality as assessed by sensory testing, of Cape hake, the major demersal fish species processed in South Africa.

2 Materials and methods

Fresh headed and gutted Cape hake were obtained from commercial trawlers, from the last catch of the voyage. The handling conditions and exact time elapsed since catching were not known, and the storage times for the tests were therefore reckoned from the arrival of the fish at the laboratory. The fish were transported from the fishing vessel to the laboratory in ice, and stored in melting ice in containers kept in a chill room at 1°C. Three to five fish were removed initially and at intervals of a few days, placed on a flamed tray, and samples taken aseptically for microbiological examination. The fish were thereafter inspected by a trained panel and the freshness assessed using a raw odour scale on which 10 points represents sea-freshness, 0 points putridity, and 5 points the borderline at which the fish is just no longer acceptable (Rowan, 1956).

Separate counts were performed on skin and flesh. Skin samples ($9.625 cm^2$) were cut on the lateral line near the middle of the fish with a sterile metal cork borer and the adhering flesh removed with a scalpel. Flesh samples (10g) were removed from below the area from which the skin had been removed. Roll tube counts were performed at 20°C and 37°C using a sea-water based medium (G) for the lower temperature and one based on distilled water (L) for the higher temperature. Incubation times were four to five days and two to three days, respectively. In two of the six trials carried out, counts were made using both media at both temperatures.

Suitable dilutions from each fish sample were also set up as plates, poured with the same medium as used for the roll tubes. From these plates colonies were picked off for identification and were streaked on G or L medium. Isolates obtained in this way were plated on appropriate media to obtain pure cultures. Ten skin and 10 flesh isolates were streaked per fish. Single colonies were also cultured in liquid medium, eg peptone water. Identification was based on morphology, motility, pigmentation, Gram reaction, type of growth in liquid culture, as well as biochemical tests such as reaction in litmus milk, liquefaction of gelatin, etc (Buchanan and Gibbons, 1974)

3 Results and discussion

3.1 Effect of incubation temperatures
Table 1 summarizes the counts and odour ratings for iced hake in six trials at 20°C and 37°C. The data have been grouped into:

(a) initial counts at the time of arrival at the laboratory

(b) counts on fish stored in ice for four to eight days, whose quality was still acceptable, and

(c) counts on fish stored in ice for eight to thirteen days, which were borderline or unacceptable

Only mean figures for all six tests combined are shown. The increase of 20°C counts with increasing time and decreasing freshness are clearly shown.

Correlation coefficients and regression lines were calculated for all six tests of the logarithms of the bacterial counts versus days in ice and odour rating. The correlation coefficients obtained for 20°C counts were all highly significant both for time ($r \geqslant 0.854$; $p < 0.01$) and for freshness ($r \geqslant -0.860$; $p < 0.01$) the numbers of bacteria increasing logarithmically with increasing time and decreasing odour rating. Initial mean skin counts ($\log_{10}$ no/cm²) calculated from regression lines ranged from 3.976 to 5.080, and slopes from 0.218/day to 0.330/day. Equivalent flesh counts ($\log_{10}$ no/g) varied initially from 3.596 to 4.635 with slopes from 0.162/day to 0.313/day. The slopes of $\log_{10}$ count versus odour rating were similarly variable viz 0.552/unit to 0.768/unit for skin counts and 0.403/unit to 0.678/unit for flesh counts. The regression lines calculated for any one test were significantly different ($p < 0.05$) from the equivalent lines in all other trials. The threshold counts, ie the $\log_{10}$ counts at the borderline value of 5 odour rating points ranged from 6.32 to 7.43 (average 7.00) for skin and 5.99 to 6.45 (average 6.23) for flesh counts. The latter value showed somewhat less variability than counts at higher and lower odour ratings.

There was no consistent relation between 37°C counts and either storage time or odour rating. Correlation coefficients for time in ice versus $\log_{10}$ count varied from -0.550 to $+0.809$ for skin and -0.740 to $+0.761$ for flesh samples. Since in many cases a negative correlation coefficient for skin counts was associated with a positive one for flesh counts or vice versa, it was considered that where the figures obtained showed a significant relation this was fortuitous.

3.2 Effect of recovery media
In trials 5 and 6 two recovery media were compared, a sea-water based medium and a medium prepared in distilled water.

There was no significant difference between counts obtained using either medium at a given temperature. Similarly the distribution pattern of bacteria did not differ significantly for the two media.

3.3 Identification of spoilage bacteria
Table II summarizes the distribution of genera found in trials 1 to 6.

Regardless of culture medium isolates from 20°C counts showed Achromobacter and Pseudomonas predominating with the latter increasing as the fish became stale. Micrococcus, initially present, decreased with time. Corynebacterium, Brevibacterium and Flavobacterium, all normally associated with the microflora of hake, were also isolated.

No significant difference was found between the distribution patterns for skin and flesh samples from the same fish, nor for different fish of the same batch in one test. Mean skin plus flesh percentages per batch were therefore used for calculating correlation coefficients

Table I
TOTAL COUNTS AT 20°C (MEDIUM G) AND 37°C (MEDIUM L) FOR ICED HAKE

| Days in ice | Odour rating | Log₁₀ Total count (20°C) | | Log₁₀ Total count (37°C) | |
		Skin (/cm²)	Flesh (/g)	Skin (/cm²)	Flesh (/g)
0	8.3	4.858	4.321	3.343	3.060
4–8	6.5	6.168	5.512	3.577	3.291
8–13	4.1	7.540	6.709	3.533	3.367

Table II
DISTRIBUTION OF GENERA AT 20°C AND 37°C USING TWO MEDIA

Trial	Days in ice	Incubation temperature	Medium	% Genera			
				Achromobacter	Pseudomonas	Micrococcus	Other
1–6	0	20	G	28·7	58·0	9·6	3·7
1–6	0	37	L	2·8	15·6	72·5	9·1
1–6	4–8	20	G	55·8	39·6	2·4	2·2
1–6	4–8	37	L	1·0	14·3	76·9	7·8
1–6	8–13	20	G	36·7	62·1	0·5	0·7
1–6	8–13	37	L	1·1	29·6	60·5	8·8
5&6	0	20	L	38·5	51·2	7·7	2·6
5&6	0	37	G	3·5	52·7	34·2	9·6
5&6	4–8	20	L	37·1	52·1	3·7	7·1
5&6	4–8	37	G	3·7	17·5	70·6	8·2
5&6	8–13	20	L	28·6	67·6	1·7	2·1
5&6	8–13	37	G	3·5	26·6	61·1	8·8

and regression lines for *Pseudomonas* and *Achromobacter* logarithmic counts *versus* days in ice and freshness. The results obtained were on average similar to those from total counts, and no significantly better prediction of fish quality could be made from data on either genus.

Isolates from 37°C counts showed *Micrococcus* predominating followed by *Pseudomonas*, with very few *Achromobacter*. As well as those mentioned above *Bacillus, Staphylococcus, Sarcina* and *Streptococcus* were also isolated.

4 Conclusion

Total counts at 20°C on either a sea-water or distilled-water based medium bear a close relationship to the freshness of raw chilled Cape hake as assessed by odour. It is unnecessary for purposes of quality assessment to classify the bacterial genera, as none of them gives significantly more accurate results than those obtained from total counts. Although significant variations in these counts exist between different batches of fish, they are adequate for the classification of raw material into classes such as sea-fresh, good, borderline and unacceptable.

Total counts at 37°C bear no significant relation to quality of chilled hake. It is thought likely, however, that they may give indications of exposure to undesirably high temperatures during processing and storage. Investigations have been commenced on the growth rates of bacteria, as measured by the two types of counts, at higher storage temperatures, but insufficient data have as yet been accumulated to draw even tentative conclusions.

5 References

ANON. *Standard specification for the production of frozen fish, frozen*
1977 *marine molluscs, and frozen fish and frozen marine mollusc products*. South African Bureau of Standards SABS 585
BUCHANAN, R E and GIBBONS, N E (co-Ed). *Bergey's Manual of*
1974 *Determinative Bacteriology*. The Williams and Wilkins Company, Baltimore. 8th edn
ROWAN, A N. The assessment of the freshness of fish by odour. FIRI
1956 Annual Report 10, 8–11
SHEWAN, J M. The bacteriology of fresh and spoiling fish and some
1962 related chemical changes. *Recent Advances in Food Science*, Eds J Hawthorn and J Mail Leitch. Butterworths, London. Vol 1, 167–193

Some observations on the ambient and chill storage of blue whiting (*Micromesistius poutassou*)

J G M Smith, R Hardy,
A B Thomson, K W Young and E Parsons

1 Introduction

The estimated viable yearly catch of blue whiting has been put at 1 million tons with catch rates of over 50t/h being not uncommon. Freezer trawlers can only freeze part of a catch this size at one time, the remainder being held at ambient temperature or if possible at chill temperatures before being frozen. Non-freezer trawlers would of course store their fish in ice or chill them by other means, *eg* in chilled or refrigerated sea water. In order to cover all possibilities we have monitored spoilage of this species during storage at ambient temperature, in ice and chilled sea water (csw) by chemical and sensory methods. Three trials were carried out during early February, March and April 1978 when the majority of the fish caught were pre-spawning, in-spawning and post-spawning respectively, to ascertain if there was any seasonal effect in storage quality.

2 Experimental

2.1 Fish

The blue whiting were caught by the research vessel *G A Reay*, during daylight hours using a pelagic trawl. The positions fished for the three trips were:

February 62°16′N 03°52′W east of Faroe Islands
March 60°20′N 06°00′W south of Faroe Islands
April 57°50′N 09°30′W west of St Kilda

The biological condition of the fish from the five trials carried out in 1978 are given by Whittle *et al* (1980) in their paper 'Seasonal variability in blue whiting and its influence on processing'. *Table I* gives details of the early February, March and April cruises and has been abstracted from their paper. The February fish could be described as being all pre-spawned fish with developing gonads and liver weight at a maximum, the water content being normal at around 80%. The March fish were mainly in-spawning with some spent fish; the gonads were maximal, liver size reducing and water content normal. The April fish had spent gonads, with livers minimal and water content higher at 83% to 84%.

Table I
VARIATION IN GROSS PERCENTAGE COMPOSITION OF BLUE WHITING CAUGHT IN FEBRUARY, MARCH AND APRIL 1978

	% Ungutted weight		
	February	*March*	*April*
Liver	6·9	5·0	2·1
Gonad (male)	3·2	8·4	1·0
Gonad (female)	2·0	5·5	1·5
Guts total	12·1	15·3	5·5

2.2 *Types of storage*
Whole fish were stored at ambient temperature and whole and gutted fish stored in ice, csw, and in csw where the water was changed every 24h.

2.2.1 *Ambient*
Whole fish were stored in 50kg capacity plastic fish boxes on the factory deck. Ambient temperatures on the factory deck were fairly constant during the February, March and April trials, being 9–13°C, 10–14°C and 10–14°C, respectively.

2.2.2 *Chill*
Approximately 30–35kg of fish were stored well iced in plastic boxes and also in csw where the ice and water were first loaded into the large plastic baths prior to loading the fish. In the csw trials about 70kg of fish were used with equal quantities of ice and sea water. Some of the plastic baths were fitted with a drainage hole to permit daily changes of sea water. During the February trial only whole fish were used in the storage experiment where the sea water was changed daily. All chilled fish were stored in the ship's chill-room at around +1°C and little re-icing was required during the trials.

2.3 *Methods*
At each sampling time three fish were filleted, skinned and minced and portions of the well-mixed mince were used to prepare extracts for the following analyses.

2.3.1 *Hypoxanthine (HX)*
A portion of the minced flesh, 30g, was homogenized with 90ml chilled 0·6 N perchloric acid and filtered. 5ml filtrate was then neutralized with 5ml 0·43 N potassium hydroxide containing 0·2 M potassium dihydrogen phosphate reagent and analysed by the method of Jones at *et al* (1964).

2.3.2 *Di- and trimethylamine (DMA, TMA)*
The remainder of the perchloric acid extract was used for the estimation of DMA and TMA by a gas-liquid chromatographic method based on that of Keay and Hardy (1972).

As tasting the fish could not be carried out on board approximately 12 fish were frozen at each sampling time. These fish were well glazed and stored in a polythene bag at −30°C until required for sensory assessment.

2.3.3 *Sensory assessment*
From the 12 fish three were picked at random, thawed, filleted and the skinned fillets steamed in casseroles for 25min and presented to four members of the staff trained in tasting white fish. The same four members of staff were used in all three trials. Besides describing the odour, texture and flavour of the fillets the panels were asked to give an overall acceptability score based on the following hedonic scale:

7 like much
6 like
5 like slightly
4 neither like or dislike
3 dislike slightly
2 dislike
1 dislike much

3 Results and discussion
The plots of TMA concentration over 11 days' storage for the three trials can be seen in *Fig 1, 2* and *3*. They show curves typical of white fish where there is an initial lag phase followed by a more rapid increase in TMA concentration. With the iced fish in each trial this phase lasts about six to seven days, thereafter the increase in concentration becomes greater as the fish change from a pre-spawned to a post-spawned state. A level of 13·6mg TMA–N/100g fish was found for whole blue whiting stored in ice for 11 days (Dagbjartsson, 1975), a figure which agrees well with our findings for the April post-spawned fish. The fish stored in csw have the highest rates of increase in TMA concentration but again the February whole and gutted fish have lower values after 11 days' storage than the March and April csw fish. Norwegian studies carried out by Karsti (1974) on blue whiting stored in refrigerated sea water (rsw) found a level of 6mg TMA–N/100g fish after four days's storage, a concentration much higher than that found for blue whiting stored in csw in this present study. The same author also found that blue whiting stored for seven days in ice were still acceptable. The result of changing the sea water can be clearly seen in the March and April trips where, especially in the March trip, concentrations of TMA in both the whole and gutted fish remain very low. Gutting the fish has little effect on the production of TMA over the first six to seven days. Thereafter results tend to vary depending upon the mode and month of storage. Changing the sea water, however, results in lower TMA concentrations in the gutted fish in all three trials.

The results of the DMA determinations are given in *Tables II, III* and *IV*. With the exception of the iced fish in the March trial gutting the fish marginally reduced the production of DMA. Lower concentrations are again determined in the pre-spawning fish caught in February, increasing as the fish change through the in-spawning stage to a post-spawning state. This increase is more apparent in the iced fish which, although having lower values than the csw fish in February, have the higher values in April. Changing the sea water results in lower concentrations of DMA in both whole and gutted fish. Zero time DMA concentrations increase from 0·6 in February to 2·0mg DMA–N/100g fish in April.

The HX concentrations of chill stored blue whiting are given in *Tables V, VI* and *VII*. On plotting these figures one finds a fairly wide scatter of results which makes

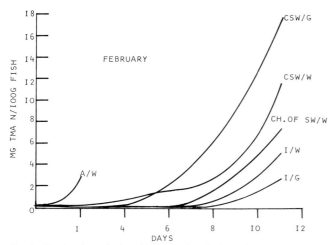

Fig 1 Production of trimethylamine in whole and gutted February blue whiting stored at chill and ambient temperatures

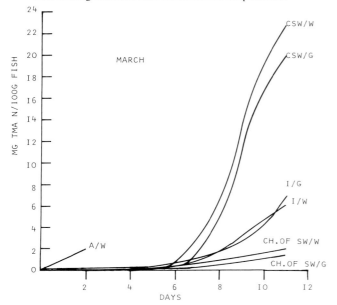

Fig 2 Production of trimethylamine in whole and gutted March blue whiting stored at chill and ambient temperatures

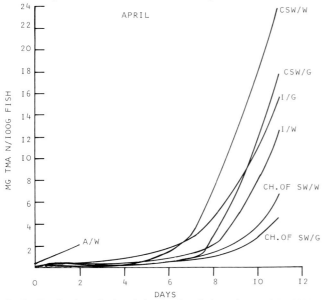

Fig 3 Production of trimethylamine in whole and gutted April blue whiting stored at chill and ambient temperatures

Table II

DIMETHYLAMINE CONTENT OF FEBRUARY BLUE WHITING STORED AT CHILL TEMPERATURE (MG DMA–N/100G FISH)

Days storage	Ice Whole	Ice Gutted	csw Whole	csw Gutted	Change of sea water Whole
0	0·58	0·58	0·58	0·58	0·58
1	0·24	0·40	0·28	0·35	0·25
2	0·38	0·55	0·37	0·29	0·20
3	0·35	0·49	0·28	0·26	0·39
4	0·73	0·45	0·60	0·62	,0·45
5	1·37	0·67	1·15	0·40	0·30
6	1·11	0·95	1·79	0·88	1·47
7	1·24	0·87	1·87	0·97	2·40
8	1·43	0·97	2·28	1·99	2·09
9	1·58	1·02	2·05	2·20	3·23
10	1·78	0·98	3·26	2·38	4·91
11	3·00	1·12	3·32	3·18	2·92

Table III

DIMETHYLAMINE CONTENT OF MARCH BLUE WHITING STORED AT CHILL TEMPERATURE (MG DMA–N/100G FISH)

Days storage	Ice Whole	Ice Gutted	csw Whole	csw Gutted	Change of sea water Whole	Change of sea water Gutted
0	0·92	0·92	0·92	0·92	0·92	0·92
1	1·16	1·31	1·15	1·66	1·15	1·66
2	1·50	1·14	1·71	1·31	1·02	1·03
3	2·57	1·87	2·07	1·23	1·46	1·20
4	1·91	1·95	3·40	1·36	1·67	1·40
5	2·22	4·55	3·41	1·96	2·21	1·94
6	2·31	4·01	2·61	2·28	2·14	2·10
7	2·02	3·18	2·92	2·49	3·99	2·06
8	2·75	3·65	5·88	2·99	3·91	2·64
9	4·12	3·89	5·97	3·12	3·28	2·77
10	4·63	4·95	7·30	3·65	3·59	2·98
11	3·68	5·65	6·02	3·78	3·55	2·08

Table IV

DIMETHYLAMINE CONTENT OF APRIL BLUE WHITING STORED AT CHILL TEMPERATURE (MG DMA–N/100G FISH)

Days storage	Ice Whole	Ice Gutted	csw Whole	csw Gutted	Change of sea water Whole	Change of sea water Gutted
0	1·86	2·08	1·86	2·08	1·86	2·08
1	2·22	1·92	2·94	2·17	1·77	2·21
2	2·12	2·33	2·59	2·05	1·57	2·39
3	2·48	2·75	3·04	2·50	1·82	2·10
4	2·32	3·42	2·47	3·05	2·21	2·17
5	3·01	3·24	3·30	2·68	2·50	2·37
6	3·36	3·95	5·29	3·59	2·25	2·89
7	5·70	5·00	4·49	3·98	3·12	2·71
8	4·13	5·64	4·76	4·43	3·84	3·28
9	7·08	5·65	5·56	4·10	3·95	3·54
10	9·05	6·33	7·49	5·29	6·66	4·49
11	11·58	8·42	7·01	6·63	6·41	5·16

drawing a curve very difficult and possibly inaccurate. The concentration scatter is more apparent in the fish caught in March and this may be due to the fact that although these fish were mainly in-spawning there was a proportion which had spawned. The fish taken for HX determinations were picked at random, no selection being made between spawned and unspawned fish and this may well have affected the results. The HX concentrations found in 11-day iced whole blue whiting are something similar to those found in haddock (*Melanogrammus aeglefinus*), cod (*Gadus morhua*) and whiting (*Merlangus merlangus*) (Jones *et al*, 1964) but are slightly higher than those found in mackerel (*Scomber scombrus*) and scad (*Trachurus trachurus*) (Smith *et al*, 1980a,b) stored for the same period in ice. Studying the tables more closely it is again evident that the concen-

Table V
HYPOXANTHINE CONTENT OF FEBRUARY BLUE WHITING STORED AT CHILL TEMPERATURE (μMOLES HX/G FISH)

Days storage	Ice Whole	Ice Gutted	csw Whole	csw Gutted	Change of sea water Whole
0	0·48	0·48	0·48	0·48	0·48
1	0·41	0·32	0·35	0·46	0·40
2	0·55	0·53	0·54	0·45	0·38
3	0·48	0·35	0·28	0·52	0·42
4	0·60	0·67	0·46	0·49	0·44
5	0·92	0·88	0·63	0·68	0·43
6	0·82	0·92	0·81	0·73	0·51
7	1·79	0·95	0·80	1·10	0·53
8	1·09	1·07	0·94	1·60	0·58
9	1·34	1·47	1·89	1·92	0·83
10	1·65	1·21	1·92	1·94	0·94
11	1·40	1·85	2·46	2·91	1·45

Table VI
HYPOXANTHINE CONTENT OF MARCH BLUE WHITING STORED AT CHILL TEMPERATURE (μMOLES HX/G FISH)

Days storage	Ice Whole	Ice Gutted	csw Whole	csw Gutted	Change of sea water Whole	Change of sea water Gutted
0	0·83	0·83	0·83	0·83	0·83	0·83
1	0·68	0·88	0·85	0·78	0·85	0·78
2	1·29	0·40	0·35	0·47	0·22	0·20
3	0·99	1·21	0·85	0·86	0·25	0·33
4	0·63	0·60	0·43	0·47	0·35	0·39
5	0·95	1·39	0·83	0·71	0·58	0·65
6	2·11	0·61	1·03	0·93	0·36	0·30
7	1·39	0·67	0·71	0·55	0·89	0·68
8	1·70	1·98	1·26	1·22	1·04	0·62
9	1·03	1·41	1·15	0·98	0·42	0·39
10	2·08	1·82	1·07	1·35	0·95	0·66
11	1·68	1·95	1·71	1·52	0·84	0·46

Table VII
HYPOXANTHINE CONTENT OF APRIL BLUE WHITING STORED AT CHILL TEMPERATURE (μMOLES HX/G FISH)

Days storage	Ice Whole	Ice Gutted	csw Whole	csw Gutted	Change of sea water Whole	Change of sea water Gutted
0	0·58	0·58	0·58	0·58	0·58	0·58
1	0·65	0·84	0·49	0·59	0·69	0·66
2	0·69	0·79	0·54	0·68	0·52	0·59
3	0·87	0·79	0·48	0·78	0·51	0·52
4	0·93	0·96	0·66	0·68	0·58	0·47
5	1·00	1·06	0·63	0·64	0·51	0·40
6	1·23	1·26	0·69	0·85	0·58	0·47
7	1·03	1·34	0·61	0·86	0·67	0·61
8	1·40	1·56	1·02	0·99	0·62	0·59
9	1·60	1·35	1·08	1·21	0·64	0·91
10	1·62	2·32	1·23	1·49	0·67	0·60
11	2·10	2·59	1·56	1·84	1·23	0·82

trations of HX in the fish stored in the daily changed sea water remain low. The csw stored fish caught in February have higher concentrations of HX during the latter stages of storage compared to the other two months and although they have higher values than the iced fish caught in January the reverse occurs in March and April; the iced fish having the higher concentrations. This reversal is similar to that occurring in DMA determinations.

Ambient temperature on board the factory deck varied little during the three trips and the results of the HX, TMA and DMA determinations are given in *Table VIII*. The HX concentrations increase slowly over the first 24h at ambient temperature after which time there is an apparent rise, this being more marked in the February fish after 48h.

The TMA concentrations for the fish stored at ambient temperature remain low to begin with followed by a more rapid increase in formation of TMA. This increase occurs after 24h, 20h and 16h approximately in the February, March and April fish, respectively.

The DMA concentrations increase somewhat erratically with time at ambient temperature, the lowest increase over 48h being that of the February fish.

Table VIII
HYPOXANTHINE, TRIMETHYLAMINE AND DIMETHYLAMINE CONTENTS OF FEBRUARY, MARCH AND APRIL BLUE WHITING STORED AT AMBIENT TEMPERATURE

Hours storage	February (9–13°C) HX	TMA	DMA	March (10–14°C) HX	TMA	DMA	April (10–14°C) HX	TMA	DMA
0	0·48	0·35	0·58	0·83	0·06	0·92	0·58	0·57	1·86
4	0·22	0·23	0·26	0·25	0·12	1·04	0·43	0·72	2·49
8	0·30	0·18	0·37	0·26	0·48	1·42	0·60	0·72	2·21
12	0·68	0·29	2·23	0·33	0·50	1·28	0·66	0·67	2·43
16	—	—	—	0·82	0·26	1·06	0·96	0·93	3·83
18	0·55	0·30	1·73	—	—	—	—	—	—
20	—	—	—	1·06	0·51	1·54	1·02	1·51	3·57
24	0·58	0·63	2·52	0·93	1·14	3·08	0·65	1·50	3·16
36	—	—	—	—	—	—	1·13	1·53	3·90
48	3·12	2·91	2·79	1·41	1·98	3·93	1·24	1·80	4·76

HX = μmoles hypoxanthine/g fish
TMA = mg trimethylamine-nitrogen/100g fish
DMA = mg dimethylamine-nitrogen/100g fish

The results of the sensory assessments are given in *Table IX*. The figures quoted in this table give the number of days the fish are considered to be still acceptable, *ie* had an average score of 4 or over followed by two consecutive scores of below 4. It is obvious from this table that the whole and gutted fish caught in April, *ie* the post-spawning fish, had a shorter chilled and ambient storage life in all media than those caught in February and March. Gutting before chilling in ice or csw improved the storage life of these April fish. In two cases after there had been at least two consecutive scores lower than 4, there followed two consecutive scores of 4 or over. This only occurred in the March whole and gutted fish stored in csw and the days when the fish were considered acceptable again are given in brackets in the March column of *Table IX*. A closer examination of the flavour, odour and texture descriptions given by the panel did not reveal any specific reason for this. One possible answer may again lie in the selection of these March fish where either pre- or post-spawned fish or a mixture of both were selected for tasting. Dagbjartsson (1975) found that blue whiting that had just been feeding on zooplankton could be stored for less than two days, but those with empty stomachs remained unspoiled for at least one week in ice. However, our taste panel found that the fish caught in April which had the lowest percentage gut weight, *Table 1* (Whittle *et al*, 1980) were only acceptable for up to one day storage in ice. Gutting apparently had a marked influence on the storage time of the February fish stored in csw, the gutted fish having twice the chilled storage life of ungutted fish. On the other hand gutting the fish held in csw changed every 24h during March had an adverse effect by almost halving the storage life from seven days for whole fish to four days for gutted fish. The April fish stored at ambient temperature were only acceptable for up to 8h compared to 24h for the February and March fish. The order of preference based on the overall acceptability scores is March, February and April.

Table IX
ACCEPTABLE CHILLED AND AMBIENT STORAGE LIFE (DAYS) OF BLUE WHITING

Storage media	Storage life (days)		
	February	March	April
I/W	7	9	1
I/G	6	9	5
csw/W	4	5 (9, 10)	1
csw/G	8	5 (8, 9)	3
Ch of SW/W	4	7	1
Ch of SW/G	—	4	1
A/W	1	1	0.33

Comparing the results from the three trials it is evident that there is a greater degree of seasonal variability in blue whiting than in other gadoid species, the overall acceptability dropping dramatically once the fish have spawned. TMA concentrations under all conditions remain low over the first five to six days, thereafter the fish in csw show the higher rates of increase. A similar result has been found in mackerel, scad and herring (Smith et al, 1979; 1980b). It was noticeable that the baths holding the csw fish had developed strong odours after 10 days' storage, but this was almost totally absent from the baths where the csw was changed daily. It is also interesting to note that whole and gutted April fish stored for 11 days in ice showed larger increases in HX and DMA, which are both produced enzymatically, than similarly treated fish in February. As was to be expected increases in HX, TMA and DMA concentrations were far more rapid at ambient temperature than at chill temperatures. In conclusion it can be said that in order to use any of the chemical indices to assess the storage life of blue whiting it would be necessary to know the month the fish were caught and the method of storage.

4 References

DAGBJARTSSON, B. Utilization of blue whiting, Micromesistius poutas-
1975 sou, for human consumption. J. Fish. Res. Bd, Can., 32(6), 747–751

JONES, N R, MURRAY, J, LIVINGSTON, E I and MURRAY, C K. Rapid
1964 estimations of hypoxanthine concentrations as indices of the freshness of chill stored fish. J. Sci. Fd. Agric., 15, 763–774

KARSTI, O. Annual Report 1973, Fiskeridirektoratets, Teknisk-
1974 Kemiske Forsöksinstitut, Bergen

KEAY, J N and HARDY, R. The separation of aliphatic amines in dilute
1972 aqueous solution by gas chromatography and application of this technique to the quantitative analyses of tri- and dimethylamine in fish. J. Sci. Fd. Agric., 23, 9–19

SMITH, J G M, HARDY, R and YOUNG, K W. A seasonal study of the
1980a storage characteristics of mackerel. Storage at chill and ambient temperatures. This volume

SMITH, J G M, McGILL, A S, THOMSON, A and HARDY, R. Preliminary
1980b investigations into the chill and frozen storage characteristics of scad (Trachurus trachurus) and its acceptability for human consumption. This volume

SMITH, J G M, HARDY, R, McDONALD, I and TEMPLETON, J. The stor-
1979 age of herring (Clupea harengus) in ice, refrigerated sea water and at ambient temperature. Chemical and sensory assessment. (In press)

WHITTLE, K J, ROBERTSON, I and McDONALD, I. Seasonal variability in
1980 blue whiting (Micromesistius poutassou) and its influence on processing. This volume

---

Preliminary investigation into the chill and frozen storage characteristics of scad (Trachurus trachurus) and its acceptability for human consumption

J G M Smith, A S McGill, A B Thomson and R Hardy

1 Introduction

During the last few years the quantities of the more popular white and fatty fish landed at UK ports have decreased considerably and the need to utilize the less accepted species for direct human consumption has become of growing importance. For example landings of mackerel (Scomber scombrus) have increased tenfold since 1973 and, although this species was not highly prized, a number of products are now to be found throughout the UK market.

Another species which to date is mainly caught for fish meal and oil production in Britain is scad or horse mackerel (Trachurus trachurus). Research vessels of the UK Ministry of Agriculture, Fisheries and Food (MAFF) have shown that the main concentrations of scad are to be found during the winter months in the English Channel, and catch rates of 8–15 tons per hour have been recorded off Plymouth, Start Point and Beachy Head (Lockwood and Johnson, 1977). Our experience confirms this MAFF report and also that it is best to fish a high head line bottom trawl during the day and a pelagic trawl at night.

Scad are used for human consumption in Mediterranean countries, the USSR, Japan and South Africa, and it is possible that they could be so utilized on the UK market. It is feasible that an export trade from the UK could be established.

It was therefore decided to study the ambient chill and frozen storage characteristics of scad caught throughout the year and this report concentrates mainly on the results of a voyage carried out during October 1978.

2 Experimental

2.1 Fish
The scad were caught by the research vessel G A Reay using a pelagic trawl off Start Point, southwest England during October 1978.

2.2 Storage procedures
2.2.1 Ambient temperature Whole fish and gutted fish were stored in plastic fish boxes of 50kg capacity on the factory deck at ambient temperature (10–15°C).

2.2.2 Chill temperature Whole fish and gutted fish (35–40kg) were stored well iced in plastic boxes and also in chilled sea water (csw) contained in large plastic baths some of which were fitted with a drainage hole to permit daily changes of the sea water. The ratio of fish to ice to water was maintained at 1:1:1, using approximately 70kg of fish in each bath.

All chilled fish were kept in the ship's chill-room at around +1°C.

2.2.3 *Frozen* Whole fish were frozen immediately after capture, both singly in the air-blast freezer and in polyethylene-lined paper sacks in the vertical-plate freezer (Cruickshank and Hewitt, 1973). In the latter case sea water was added to the sacks to above the level of the fish prior to freezing. The single fish were stored unglazed in heavy-duty polyethylene bags. This process was repeated with fish that had been held for three and six days in ice.

Further samples of whole fish stored in ice were withdrawn at intervals up to ten days, filleted and skinned. Two dozen fillets were frozen singly, the remainder being minced and frozen in small blocks 300mm × 100mm × 20mm. The fillets and blocks of mince were stored in polyethylene bags, the blocks being prewrapped in polyethylene sheets to prevent adhesion.

All frozen samples were stored on board ship and at Torry Research Station at −30°C.

2.3 *Analytical methods*
At approximately daily intervals for up to 12·5 days three fish from each of the chill storage experiments were filleted, skinned, minced and portions of the mince used to prepare extracts for hypoxanthine and di- and trimethylamine analysis. A similar procedure was carried out with the ambient stored fish at intervals up to 48h. All extracts not analysed immediately were stored at −30°C. A further 12 fish from each storage experiment were frozen and stored in polyethylene bags at −30°C to be used for sensory assessment.

2.3.1 *Hypoxanthine (HX)* Minced flesh (30g) was homogenized with chilled 0·6 N perchloric acid (90ml) and filtered. The filtrate (5ml) was neutralized with potassium hydroxide-phosphate buffer reagent (5ml) prepared as follows. 27·22g potassium dihydrogen orthophosphate is dissolved in 250ml water and to this solution is added 171ml N sodium hydroxide. The pH of this mixture is adjusted to 7·6 if necessary. To this mixture is then added 474ml N potassium hydroxide and the reagent made up to 1l with water. The 10ml mixture, after filtering, is then analysed by the method of Jones *et al* (1964).

2.3.2 *Di- and trimethylamine (DMA, TMA)* The remainder of the perchloric acid filtrate from the HX extract was used for the estimation of DMA and TMA by a gas chromatographic technique based on that of Keay and Hardy (1972).

2.3.3 *Total lipid* Fish from two size ranges, small (20–25cm) and large (28–33cm), were selected for this analysis after 0, 3 and 6 days in ice. Three fish from each range were filleted and skinned. For small fish both fillets were used but for large fish only one fillet was required (approximately 50g). In addition the lipid content of three large whole fish was determined and in this case the entire fish was minced and a portion (100g) of the well-mixed mince extracted. The extraction method used was that of Bligh and Dyer (1959) incorporating 2,6-ditertiary-butyl-*p*-cresol (BHT) (0·001%) in the chloroform.

2.3.4 *Peroxide value (PV)* This was determined on the lipid extracted as in 2.3.3 above from the single and block frozen fish and from the fillets and minces prepared and frozen as described in 2.2.3. The PV of the extracted lipid was determined using the method of Banks (1937).

2.3.5 *Sensory assessment* Six members of the staff normally used in white fish taste panels were asked to describe the odour, texture and flavour of the steamed casseroled fillets prepared from thawed fish and in addition to give an overall acceptability score on the following hedonic scale:

    7 like much
    6 like
    5 like slightly
    4 neither like or dislike
    3 dislike slightly
    2 dislike
    1 dislike much

3 Results and discussion

3.1 *Chill and ambient storage*
The increase in TMA values is practically identical for all three chilled storage media for up to seven days rising from 1·4mg to 2·1mg TMA nitrogen per 100g fish (*Table I*). Thereafter the greatest increase in TMA is noted for fish stored in unchanged csw. When the csw is changed every 24h the TMA values approximate those for fish stored in ice. The higher values in the fish stored in csw can be explained possibly by the fact that in iced storage the ice on melting removes a part of the decomposition and bacterial metabolic products whereas in chilled sea water this does not occur. Similarly, these

Table I
TRIMETHYLAMINE CONTENT OF SCAD STORED AT AMBIENT AND CHILL TEMPERATURES (MG TMA NITROGEN/100G FISH)

Days Storage	Ice Whole	Ice Gutted	csw Whole	csw Gutted	Changed csw Whole	Changed csw Gutted	Ambient Whole	Ambient Gutted
0	1·4	1·4	1·4	1·4	1·4	1·4	1·4	1·4
0·5	1·5	1·5	1·5	1·4	1·5	1·4	1·5	1·4
1	1·4	1·4	1·4	1·4	1·4	1·4	1·6	1·8
2	1·5	1·4	1·5	1·5	1·5	1·5	13·1	12·5
3·5	1·6	1·5	1·5	1·5	1·4	1·4		
4·5	1·6	1·5	1·6	1·6	1·5	1·5		
5·5	1·5	1·6	1·7	1·6	1·6	1·5		
6·5	1·5	1·9	1·7	1·6	1·7	1·6		
7·5	1·8	2·1	2·1	1·9	1·8	1·7		
8·5	2·5	2·0	3·1	2·2	1·9	1·9		
10·5	2·7	2·3	6·7	3·5	2·8	2·7		
12·5	3·4	2·8	7·9	4·3	3·9	3·6		

decomposition products will be removed when the csw is changed every 24h. The advantage of gutting the fish is mainly evident in the case of csw storage and only after seven days (*Fig 1*). It would appear therefore that over the first six to seven days the TMA concentration cannot be used as a measure of the time the fish have been stored in the chilled media.

In the case of ambient storage (*Table I*) the TMA values remain low for up to 24h increasing to 13mg TMA nitrogen per 100g fish after 48h.

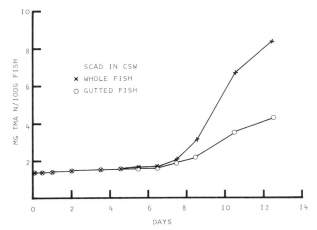

Fig 1 Scad stored in chilled sea water

The DMA values (*Table II*) of chilled stored fish increase only marginally over 12·5 days from 1·1 to a maximum of 2·0mg DMA nitrogen per 100g fish. There is no apparent increase in DMA in the fish stored at ambient temperature.

In contrast to the TMA values there is a gradual increase in HX concentration from zero time onwards (*Table III*). The rate of production of HX in chill storage varies with the treatment; thus the gutted fish stored in csw changed every 24h showed a rate of increase of 0·038 μM HX/g tissue/day whereas whole fish in ice have an increase of 0·059 μM HX/g tissue/day (*Fig 2*). Within each chill storage treatment the HX content increases linearly and from this experiment measurement of its concentration appears to be a useful indicator of storage time.

The taste panel found whole fish stored in unchanged and changed csw to be acceptable for up to 10·5 days' storage (*Table IV*), the presence of salt appearing to enhance the flavour. However, the gutted fish, especially those in changed csw, were described after 5·5 days' storage as being very salty and this possibly accounts for the lowering of the acceptability scores. Comparing the acceptability period of scad stored in ice and unchanged csw with that of mackerel (Smith *et al*, 1980b) it can be seen that scad has the longer chilled shelf-life by one to four days.

As was to be expected the fish stored at ambient temperature spoiled more rapidly and became unacceptable after 12h storage.

3.2 Frozen storage

To date October fish stored at $-30°C$ have been examined once after four months. The PV results indicate (*Table V*) that in the single and block frozen fish there has been little oxidation. The former have slightly higher values and both show a tendency to increase with time in ice prior to freezing. On the other hand the

Table II
DIMETHYLAMINE CONTENT OF SCAD STORED AT AMBIENT AND CHILL TEMPERATURES (MG DMA NITROGEN/100G FISH)

Days Storage	Ice		csw		Changed csw		Ambient	
	Whole	Gutted	Whole	Gutted	Whole	Gutted	Whole	Gutted
0	1·1	1·1	1·1	1·1	1·1	1·1	1·1	1·1
0·5	1·1	0·9	1·0	1·0	1·0	1·0	1·0	1·0
1	1·0	0·9	1·0	1·1	1·0	1·1	1·1	1·1
2	1·1	1·0	1·0	1·0	0·9	0·9	1·2	1·1
3·5	1·2	1·0	1·1	1·0	1·0	0·9		
4·5	1·3	1·1	1·2	1·0	1·0	1·1		
5·5	1·2	1·0	1·2	1·0	1·0	1·0		
6·5	1·2	1·1	1·2	1·1	1·1	1·0		
7·5	1·3	1·2	1·3	1·1	1·1	1·0		
8·5	1·3	1·2	1·4	1·2	1·3	1·2		
10·5	1·4	1·3	1·7	1·5	1·3	1·3		
12·5	2·0	1·5	1·8	1·6	1·6	1·6		

Table III
HYPOXANTHINE CONTENT OF SCAD STORED AT AMBIENT AND CHILL TEMPERATURES (μMOLES HYPOXANTHINE/G FISH)

Days Storage	Ice		csw		Changed csw		Ambient	
	Whole	Gutted	Whole	Gutted	Whole	Gutted	Whole	Gutted
0	0·01	0·01	0·01	0·01	0·01	0·01	0·01	0·01
0·5	0·07	0·14	0·26	0·15	0·26	0·15	0·01	0·13
1	0·13	0·16	0·07	0·09	0·07	0·09	0·25	0·21
2	0·11	0·10	0·19	0·14	0·12	0·12	7·09	6·98
3·5	0·22	0·18	0·15	0·20	0·18	0·21		
4·5	0·15	0·30	0·19	0·18	0·64	0·24		
5·5	0·17	0·26	0·24	0·18	0·18	0·17		
6·5	0·27	0·29	0·21	0·23	0·30	0·18		
7·5	0·46	0·52	0·39	0·25	0·28	0·26		
8·5	0·47	0·40	0·32	0·37	0·33	0·33		
10·5	0·80	0·61	0·62	0·47	0·37	0·52		
12·5	0·82	0·72	0·61	0·74	0·77	0·42		

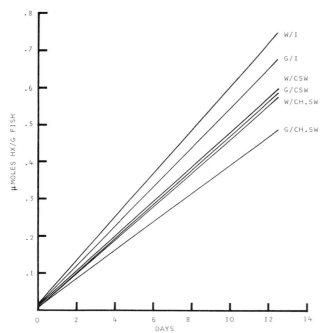

Fig 2 Regression lines for hypoxanthine against time constrained to an initial value of 0·01 μm HX/g fish
W/I: whole iced
G/I: gutted iced
W/CSW: whole chilled sea water
G/CSW: gutted chilled sea water
W/Ch. SW: whole changed chilled sea water
G/Ch. SW: gutted changed chilled sea water

minces and fillets show that oxidative deterioration is more rapid at −30°C in those products prepared from fish held for longer periods in ice prior to processing and freezing. The higher PVs in the fillets and minces are no doubt due to the greater exposure of the lipid to atmospheric oxygen, the whole fish being protected by a somewhat tougher skin. One would have expected the minces to have higher PVs than the fillets but possibly overwrapping the mince blocks in polyethylene sheeting may have helped to reduce oxidation. No rancidity was detected by the taste panel in the single or block frozen fish, but slight rancid flavours were detected in those fillets stored for six, seven and ten days in ice before freezing. It is noteworthy that fish caught in April 1978 and stored under similar conditions showed that single and block frozen fish had lower PVs than the October fish stored for four months.

Table V
PEROXIDE VALUES OF SCAD FROZEN AND STORED AT −30°C FOR 4 MONTHS, AS WHOLE FISH, FILLETS AND MINCE (ML N/500 SODIUM THIOSULPHATE PER G LIPID)

Days in ice prior to freezing	Whole fish Single	Blocks	Fillets (skin off)	Mince
0	1·2	0·5	2·5	1·4
1			2·5	2·4
3	1·3	1·0	3·9	2·6
5			3·2	5·1
6	2·8	1·9	5·3	6·4
7			6·1	4·8
10			6·5	5·5

3.3 Comparison with other species

The TMA concentration in scad, although initially higher than other fish species, eg blue whiting (Smith et al, 1980a), mackerel (Smith et al, 1980b) and cod (Mackie and Thomson, 1974) increases only from 1·4mg to 3·4mg TMA nitrogen/100g fish after 12·5 days in ice (Fig 3). It is also evident that the rate of TMA increase for scad parallels that of mackerel over 10 days' storage contrasting to those of cod and blue whiting. As TMA production is believed to be a consequence of microbial action this suggests that there was either a deficiency of trimethylamine oxide (TMAO) degrading bacteria or that some inhibiting mechanism was operating.

The DMA content of scad increases by only 1mg nitrogen per 100g of fish over 12·5 days' iced storage whereas cod and blue whiting increase by 3mg and 4mg over 13 and 14 days, respectively (Fig 4). DMA arises from the enzymatic degradation of TMAO so that it appears that this enzyme is less active or is present at a lower concentration in scad.

It is of interest to note that the rate of increase of HX, which is also produced enzymatically, was, as in mackerel, less than that observed in blue whiting and cod (Fig 5).

The lipid contents of the October fish are shown in Table VI. The considerable spread in the figures may be due to biological variations, and a larger number of samples will have to be analysed in order that a clearer picture may be formed. It is also possible that during the skinning process a portion of the subcutaneous lipid was removed with the skin, although care was taken to prevent this.

The protection against oxidative rancidity afforded by freezing the fish in bags filled with water is comparable to that observed with herring (Hardy et al, 1973) and

Table IV
OVERALL ACCEPTABILITY SCORES FOR SCAD STORED AT CHILL AND AMBIENT TEMPERATURES

Days storage	Ice Whole	Ice Gutted	csw Whole	csw Gutted	Changed csw Whole	Changed csw Gutted	Hours storage	Ambient Whole	Ambient Gutted
0	5·2	5·2	5·2	5·2	5·2	5·2	0	5·2	5·2
0·5	4·2	4·8	5·2	4·8	5·2	4·8	4	4·2	4·8
1	4·0	3·7	4·0	4·0	4·0	4·0	8	4·3	3·2
2	4·7	4·8	4·0	4·5	4·2	5·1	12	3·3	2·9
3·5	4·5	4·2	4·7	4·2	3·8	4·2	16	3·5	3·5
4·5	4·0	4·5	4·8	4·8	4·5	4·4	24	3·5	3·7
5·5	3·5	3·5	3·8	3·3	4·7	4·7	48	1·0	1·0
6·5	4·5	4·8	4·8	5·0	4·5	3·6			
7·5	3·5	4·2	4·8	4·4	3·6	2·8			
8·5	2·8	3·2	5·3	5·2	4·0	4·0			
10·5	3·2	3·3	4·7	3·7	4·5	3·3			
12·5	3·5	3·5	2·8	1·8	3·0	2·8			

The italic figures denote the point up to which the fish were considered to be acceptable by the taste panel

306

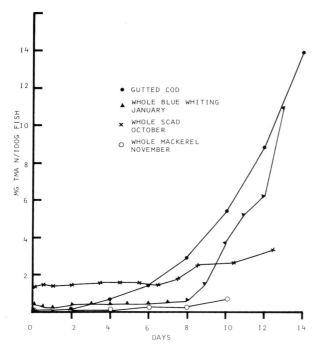

Fig 3 Trimethylamine values for various species of fish stored in ice

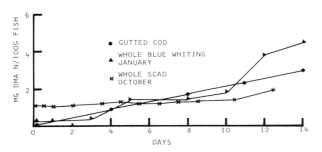

Fig 4 Dimethylamine values for various species of fish stored in ice

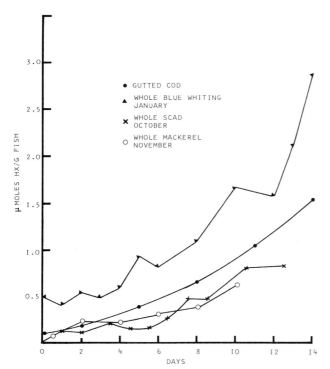

Fig 5 Hypoxanthine values for various species of fish stored in ice

Table VI
LIPID CONTENT OF WHOLE LARGE FISH AND SKINNED FILLETS FROM SMALL
AND LARGE FISH STORED IN ICE FOR 0, 3 AND 6 DAYS

Days in ice	Whole Large fish	Skinned fillets	
		Large fish	Small fish
0	16·5	13·9	6·5
	11·6	9·3	12·1
	12·2	9·1	5·7
3	13·4	7·8	6·7
	17·9	6·7	5·9
	11·5	9·3	8·7
6	15·6	7·2	6·9
	10·6	6·3	7·6
	11·8	7·7	9·0
Average	13·5	8·6	7·7

mackerel (Smith *et al*, 1980b). It is our opinion that this method of freezing not only prevents dehydration and exposure of fish surfaces to oxygen but also minimizes mechanical damage during handling of the blocks.

In conclusion the results obtained in this preliminary study indicate that up to six days the chilled storage medium does not influence the acceptability of the fish. Thereafter the whole fish stored in chilled sea water were considered by the taste panel to be the most palatable. The concentrations of TMA, DMA and HX do not accurately predict the limits of acceptability of the fish and with the possible exception of HX they do not give a satisfactory picture of storage history. The chemical indices indicate that scad has chilled and frozen storage characteristics similar to those of mackerel, but the taste panel found that the flavour deterioration in chilled storage, at least, is less marked in scad.

4 References

BANKS, A. Rancidity in fats. I. The effect of low temperature, sodium
1937 chloride and fish muscle on the oxidation of herring oil. *J. Soc. Chem. Ind.* Lond., 56, 13T–15T

BLIGH, E G and DYER, W G. A rapid method of total lipid extraction
1959 and purification. *Can. J. Biochem. Physiol.*, 37, 911–917

CRUICKSHANK, A and HEWITT, M R. Wrappers for herring frozen at
1973 sea in vertical plate freezers. *Fish Marketing Processing Packaging*, No. 13, 3–6

HARDY, R, SMITH, J G M and YOUNG, K W. Influence of packaging in
1973 fish processing and technology. Conference in conjunction with the 8th International Exhibition for the Food and Allied Industries, London, BPS Exhibitions Ltd

JONES, N R, MURRAY, J, LIVINGSTON, E I and MURRAY, C K. Rapid
1964 estimations of hypoxanthine concentrations as indices of the freshness of chill stored fish. *J. Sci. Fd. Agric.*, 15, 763–774

KEAY, J N and HARDY, R. The separation of aliphatic amines in dilute
1972 aqueous solution by gas chromatography and application of this technique to the quantitative analysis of tri- and dimethylamine in fish. *J. Sci. Fd. Agric.*, 23, 9–19

LOCKWOOD, S J and JOHNSON, P O. Horse mackerel (*Trachurus*
1977 *trachurus*). Laboratory Leaflet, MAFF, Direct. Fish. Res. Lowestoft, 38, 18pp

MACKIE, I M and THOMSON, B W. Decomposition of trimethylamine
1974 oxide during iced and frozen storage of whole and comminuted tissues of fish. *Proc: 4th Intern. Con. of Food Sci. and Tech.*, 1, 243–250

SMITH, J G M, HARDY, R, THOMSON, A B, YOUNG, K W and PARSONS,
1980a E. Some observations on the ambient and chill storage of blue whiting (*Micromesistius poutassou*). *This volume*

SMITH, J G M, HARDY, R and YOUNG, K W. A seasonal study of the
1980b storage characteristics of mackerel. *This volume*

7 Krill

Preliminary estimates on the quality and shelf-life of krill meat

P Bykowski, W Kołodziejski, J Pielichowski and Z Karnicki

1 Introduction

The technology of obtaining peeled krill meat is being worked out on a semi-technical scale at the Sea Fisheries Institute at Gdynia. The final aim of the technology is to obtain a product which will retain the consumer values of fresh krill meat which has a flavour and texture similar to shrimp meat.

Investigations carried out in 1977 (2nd Antarctic Expedition) showed that the shelf-life of a product freeze-stored in the shape of loose, frozen granules of meat in cartons or plastic bags, with free access of air, was very short. The main reason for this was that the product became rancid very rapidly. The aim of the work conducted during the 3rd Antarctic Expedition was therefore the improvement of the quality and shelf-life of the product.

2 Assumptions and methods of investigations

2.1 *Technological assumptions and procedure*
It was assumed that the main factor affecting the quality and shelf-life of krill during freezing storage is its susceptibility to oxidation. For the purpose of these investigations, several samples of peeled krill prepared according to different technological variants shown in *Table I* were carried out during the 3rd Antarctic Expedition on the RV *Professor Bogucki*.

All samples were stored at an average temperature of 248K ($-25°$C).

Table I
VARIANTS OF PEELED KRILL MEAT SAMPLES

Method of packing and storing	Boiling method	
	In salt water	In fresh water
Vacuum packed in polyethylene bags	AI	BI
In polyethylene bags, not vacuum packed	AII	BII
In cardboard boxes, not vacuum packed	AIII	BIII
Frozen in the shape of blocks of ice	AIV	BIV

2.2 *Methods of evaluation*
The evaluation of quality of samples and changes which take place during long-term freezing storage was carried out ashore. The basic method of evaluating the quality and shelf-life of peeled krill samples and comparative samples of whole krill was by periodic sensory testing. This consisted in the evaluation of flavour as the main determinant of quality. The testing was carried out in groups, applying a five-grade verbal-description scale with an additional description of the samples' charac-teristics. Chemical tests including the following were also applied:

(*1*) the ratio of non-protein to total nitrogen (Nnp/Nt),
(*2*) the content of volatile ammonium bases (VAB),
(*3*) acid, iodine and peroxide values of the lipids,
(*4*) the amino acid composition of the protein,
(*5*) the content of certain metals.

3 Results of evaluations

3.1 *Organoleptic tests*
3.1.1 *The effect on the quality and shelf-life of the product of boiling krill in salt and fresh water*   Quality changes in the various krill meat samples (the material being treated immediately after catching) are presented in *Fig 1*.

Over the whole period of evaluation, there were distinct differences in results (0·5 to 1 point according to the 5-grade scale) in favour of krill meat boiled in fresh water with the exception of those frozen in ice.

In the initial period (60 days' storage), the lower evaluation of the products obtained from boiling in salt water resulted mainly from the acquisition of a sharp, bitter, salty flavour by the meat. After 120 days' storage, the vacuum-packed krill meat boiled in fresh water showed no signs of rancidity, whereas that boiled in salt water (sea water) was distinctly rancid. In the case of packing which was not air-proof (plastic bags without vacuum packing and cardboard cartons) the differences in the rancidity of the meat were even more distinct, occurring after 90 days' storage in the variant boiled in sea water.

These differences did not occur in the two variants of samples frozen in the form of ice blocks. This might be explained by the 'de-salting' of products submerged in substantial quantities of fresh water prior to freezing which removes the flavour substances characteristic of crustaceans out of the meat on the one hand, and diminishes its consumer value, but on the other hand rinses out the salts absorbed by the products during heat treatment, slows down the fat oxidation rate and reduces the undesirable bitter flavour.

Altogether, the shelf-life of krill meat boiled in fresh water was much higher than that boiled in salt water.

3.1.2 *The effect of packing on the quality and shelf-life of a product*   Figure 2 presents the results of organoleptic tests on the effect of the four methods of packing on the shelf-life of the product.

An analysis of meat quality deterioration in all types of packing enabled us to list the methods in order of

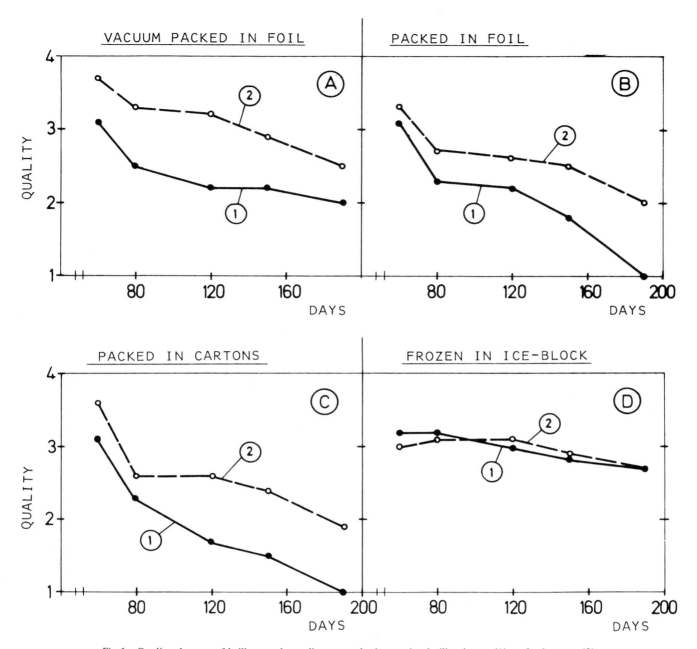

Fig 1 Quality changes of krill meat depending upon the immersion boiling in sea (1) or fresh water (2)

efficiency of safeguarding the product against unfavourable changes. This function was best fulfilled by freezing in the form of ice blocks, followed by vacuum packing in plastic bags, plastic bags without vacuum and cardboard cartons. On the other hand, as mentioned, the freezing in ice blocks diminished the consumer attraction of the meat due to the extraction of the flavours characteristic of marine crustaceans.

The diminishing of quality of the products during storage was primarily due to the appearance of signs of rancidity and its gradual intensification with time.

Vacuum packing in plastic bags proved to be the best of the methods evaluated, as the krill meat could be stored for two to three months whilst still retaining its good quality and for up to about five months whilst retaining average quality providing that very fresh meat (up to two hours' storage after catching) boiled in fresh water was utilized.

3.1.3 *Comparison of the quality and shelf-life of peeled and whole krill* At the beginning of the evaluations (the sixtieth day of storage), the level of the overall sensory evaluations of whole boiled krill (hand peeled immediately prior to evaluation) was slightly higher or almost the same as the corresponding variants for the meat of krill mechanically peeled on board ship. The rate at which the quality of whole krill diminished was decidedly slower than that of peeled krill; thus in the final stage, the difference in the levels of quality evaluations of the meat from the two products (peeled and whole krill) increased distinctly in favour of whole krill. This result suggests that in order to obtain better quality peeled krill, the peeling should be carried out ashore using krill boiled in fresh water and frozen on board ship.

3.2 *Chemical evaluations*
3.2.1 *The chemical characteristics of samples* Table II

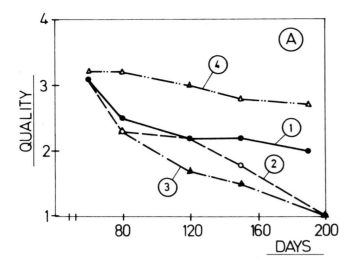

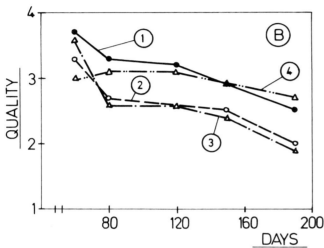

Fig 2   Results of organoleptic test on the effect of the four methods of packing on the shelf life of the krill meat
A: boiling in seawater
B: boiling in freshwater
(1) vacuum packed in polyethylene bags
(2) in polyethylene bags, not vacuum packed
(3) in cardboard boxes
(4) frozen in the shape of blocks of ice

presents the chemical evaluations of peeled krill compared to the corresponding samples of whole krill. The results show substantial differences in the chemical composition of the raw material and that there was a relatively large quantity of lipid (about 5% on average). This was also reflected in the high initial difference in taste.

The amino acid composition of the protein in selected

Table II
CHEMICAL CHARACTERISTICS OF PEELED KRILL MEAT

Sample symbol	Dry mass %	Nt × 6·25 %	Nnp/ Nt	Lipids %	Shell residue %	VAB mgN/ 100g
AI	22·9	15·9	0·195	5·4		22·02
AII	25·9	17·2	0·194	5·6	0·97	21·09
AIII	19·7	14·3	0·205	4·4		21·56
BI	23·6	16·8	0·133	5·0		13·25
BII	21·3	15·6	0·144	5·6	0·61	12·32
BIII	23·1	16·0	0·145	6·4		15·62

samples of peeled krill meat is presented in *Table III*. No distinct and systematic differences in the amino acid composition of the protein, depending upon the type of samples, were confirmed. The differences observed were most probably the result of the natural variability of samples and of analyses.

Table Table III
AMINO ACID COMPOSITION OF PROTEINS IN PEELED KRILL MEAT (IN%)

Amino acid	Type of sample	
	AI	BI
ASP	11·25	12·04
THR	4·47	4·80
SER	4·41	4·76
GLU	16·51	12·78
PRO	5·89	5·05
GLY	7·80	8·01
ALA	8·27	7·99
VAL	3·81	4·05
MET	1·50	1·12
ILE	4·08	6·94
LEU	8·79	9·30
TYR	4·34	2·81
PHE	3·51	4·34
HIS	1·86	1·90
LYS	8·09	8·58
ARG	5·43	5·54

[1] See *Table I*

The metal content in selected samples of peeled krill meat is given in *Table IV*. It can be seen that the quantity of sodium, potassium, calcium and magnesium found in samples of krill meat boiled in salt water is greater than in that boiled in fresh water. This would indicate marked absorption of salts from the sea water during the heat treatment (*Table IV*), which has a negative effect on the quality and shelf-life of the product.

Table IV
METAL CONTENT IN PEELED KRILL MEAT DEPENDING UPON THE METHOD OF BOILING THE RAW MATERIAL

		Method of boiling	
		In salt water	In fresh water
Na	µg/100g	678·6	238·8
K	µg/100g	179·2	135·3
Ca	µg/100g	196·9	152·6
Mg	µg/100g	119·7	51·7
Fe	µg/100g	0·73	1·45
Cd	µg/1000g	0·49	0·16
Cu	µg/1000g	19·97	23·22
Pb	µg/1000g	0·13	0·36
Zn	µg/1000g	11·33	13·93
Hg	µg/1000g	0·007	0·005
As	µg/1000g		0·048
NaCl%		1·81	0·53

3.2.2 *Changes of chemical indices*   The ratio of non-protein nitrogen to the overall nitrogen and the nitrogen content of volatile ammonium bases of selected samples was determined during storage. No systematic changes were observed in the values of these indices in relation to the method of packing during storage.

In view of the fact that peeled krill meat becomes rancid very rapidly, the investigation of lipid changes is important. Of the parameters studied (acid, iodine, peroxide values), only the latter was found to correlate with rancidity (*Table V*).

Table V

LIPID PEROXIDE NUMBER OF PEELED KRILL MEAT DEPENDING UPON THE METHOD OF PREPARATION OF SAMPLES, PACKING AND FREEZE STORAGE TIME AT −25°C

Type of sample[1]	Storage time (days)		
	60	120	190
AI	12·8	9·1	4·0
AII	13·2	18·9	17·1
AIII	11·1	16·8	16·8
BI	3·1	9·2	7·2
BII	6·9	16·8	15·6
BIII	11·6	15·8	16·2

[1] See Table I

4 Conclusions

(a) The heat treatment of the raw material in salt water applied during the technological processing had a very detrimental effect on the quality and shelf-life of peeled krill, irrespective of the type of packing used in these experiments.

(b) The packing and storing of peeled krill in the shape of loose granules (in plastic bags or cardboard cartons) is decidedly unsuitable for this product.

(c) Peeled krill meat obtained from raw material boiled in freshwater and vacuum packed or block frozen retains a good quality for two months and is of a satisfactory standard for six months if stored at −25°C.

Composition and properties of krill fingers

H Rehbein

1 Introduction

The efforts to utilize the Antarctic krill (*Euphausia superba Dana*) for direct human consumption resulted in the development of various new products (Grantham, 1977). On the 2nd West German Antarctic Expedition 1977–1978, krill meat (*ie* the tail muscle) was prepared from fresh krill by means of a roller peeler. In this work the suitability of krill meat for the production of krill fingers is described.

A similar product was made in Chile (Anon, 1977).

2 Materials and methods

2.1 Preparation of krill meat

The krill meat described in this work was obtained by peeling fresh raw krill with a roller peeler originally designed for the shrimp industry (supplier: Skrmetta Machinery Co Ltd, New Orleans, La, USA) (Schreiber, 1978). Peeling was carried out on board the commercial freezer trawler *Julius Fock* from January to March 1978.

After separation from the shells, the meat was washed with freshwater and in part supplied with additives. Because of the high water content of krill meat, powders of polyphosphates or solid salts of organic acids were thoroughly mixed with the meat.

Blocks of meat were frozen in a plate freezer and smaller samples in a freezer chest; both were stored at −30°C.

2.2 Production of krill fingers

In July 1978 krill fingers were prepared from blocks of frozen krill meat in an industrial plant which specializes in the production of fish fingers. The frozen meat was cut into portions of about 18g; the coating was the same as used for commercial fish fingers; the batter consisted of a mixture of wheaten flour, starch, salt and water. After addition of the breadings the fingers were fried for 20s in hot (180°C) soybean oil. Immediately afterwards they were frozen (−5°C) in a freezing tunnel. After wrapping in polyethylene bags the product was stored at −23°C to −26°C in a freezer chest.

The krill fingers had a mean weight of 30g; the fraction of krill meat was about 60%.

2.3 Chemical analysis

Dry matter, ash, salt, fat and crude protein were estimated as described by Roschke and Schreiber (1977). Chitin was determined as follows (Yanase, 1975): thawed drained raw krill (15g) or krill meat (5g) was mixed with 100ml 2 M NaOH and heated for 2h in a boiling water bath. After cooling the suspension was centrifuged (for 20min at room temperature at $1\,900 \times g$). Boiling (in 10ml 2 M NaOH) and centrifugation were repeated. The particles (shells) were washed with 100ml distilled water. The filter cake (shells) was incubated overnight in 10ml 2M HCl at room temperature. After filtration the residue was twice washed with 100ml distilled water. Finally, dry matter was estimated and taken as the chitin content of the sample.

TCA-soluble and TCA-precipitated nitrogen content was determined by adding 10ml 5% (w/v) TCA to samples of 400mg thawed raw krill or krill meat. After thorough mixing for 1min (Reax mixer, Heidolph) and incubation for 30min at 7°C in a refrigerator the suspensions were centrifuged for 30min at room temperature at $1\,900 \times g$). Then the nitrogen content was estimated in the sediment and supernatant (Roschke and Schreiber, 1977). Taurine was determined by ion-exchange chromatography using the Beckman-Multichrom-B-amino acid analyser. TMAO and DMA were estimated according to Antonacopoulos (1973).

2.4 Physical methods

For estimation of thawing drip, frozen krill meat was weighed, sealed into polyethylene bags and thawed with running tap water. Then the meat was put into a mesh covered with parafilm, which was placed on a beaker. After 30min of draining, the difference in weight against the frozen sample was determined and taken as drip loss.

The waterbinding-capacity of krill meat was defined as the amount of water in 100g meat that could not be pressed out by centrifugation. $2 \times 10g$ of drained meat were centrifuged for 30min at 5°C at $17\,000 \times g$. Then the amount of press juice was measured by weighing and subtracted from the water content.

The consistency of raw and cooked krill meat was determined with the Wolodkewitsch-Consistency-

Tester μ (Gruenewald, 1957). The samples were packed into a case and pressed by a piston through a narrow circular slit (1 mm). The force which had to be applied in this process depended on the elasticity and firmness of the krill meat: softer and more springy samples needed lower forces.

In some experiments krill meat was cooked for 15 s before being tested.

The loss of juice resulting from the cooking of krill meat was determined as follows: two samples of meat (25 g in each petri dish) were heated for 30 s in a microwave oven (Siemens HF 0520). Then the free juice was weighed.

The pH was measured on the homogenized meat.

2.5 Taste panel assessments

A taste panel of six members evaluated the organoleptic properties of krill meat or fingers using a 9-point scale (Reinacher et al, 1977). Appearance, odour/flavour, taste, texture and overall acceptability were scored.

Krill fingers were fried for 7 minutes in hot fat (165°C). Krill meat was cooked 15–30 s before examination; longer cooking resulted in a tough and stringy product.

3 Results and discussion

3.1 Composition of krill meat

Krill meat obtained by roller peeling consists of krill tail muscle, which has been washed thoroughly during the process. Therefore it had lost most of the soluble proteins and metabolites, as can be seen from the low content of taurine and the small percentage of TCA-soluble nitrogen (Table I). Due to its relatively low content of fat, heavy metals (Stoeppler and Brandt, 1979) and TMAO/DMA (Tables I and II), krill meat should be well suited for deep frozen storage.

Compared to raw krill, the content of crude protein of the meat had increased from 65% to 81% of dry matter (Table I). The fluctuations in the salt content of the meat can be explained by the variable degree of freshwater washing after peeling. Apart from residual shells, krill meat resembles cod flesh (Table II).

Table I
COMPARISON OF THE COMPOSITION OF RAW KRILL AND KRILL MEAT

	Raw krill		Krill meat	
	Sample 1	Sample 2	From sample 1	From sample 2
	% wet weight			
Dry matter	19·5	24·0	14·2	15·7
Ash	2·59	2·72	1·04	1·95[2]
Salt (NaCl)	1·94	1·31	0·53	0·58
Shells (Chitin)	n.d.[1]	n.d.	0·10	0·20
Fat	2·14	3·24	0·71	0·65
Crude Protein (N × 6·25)	12·6	15·7	11·7	12·6
	% total N			
TCA-soluble-N	63	n.d.	28	n.d.
TCA-precipitated-N	37	n.d.	72	n.d.
Taurine-N	2·17	n.d.	0·19	n.d.

n.d. = not determined
[1] On the average, raw krill contained 0·83% shells
[2] Sample 2 contained 1% polyphosphate

Table II
COMPOSITION OF THE MEAT CORE OF KRILL—AND FISH FINGERS

	Krill meat		Cod flesh
	Poor freshwater washing $n = 2$	Intensive freshwater washing[1] $n = 2$	$n = 1$
	% wet weight		
Dry matter	16·7 ± 0·1[2]	16·0 ± 0·7	17·0
Ash	1·70 ± 0·0	1·55 ± 0·15	0·96
Salt (NaCl)	1·48 ± 0·015	0·96 ± 0·045	0·36
Shells (Chitin)	0·051 ± 0·01	0·20 ± 0·02	n.d.
Fat	0·99 ± 0·0	1·12 ± 0·015	1·01
Crude Protein (N × 6·25)	11·1 ± 0·11	11·6 ± 0·26	13·9
TMAO-N	50·0 ± 1·49	25·2 ± 0·83	n.d.
DMA-N	< 1·0	< 1·0	n.d.

n.d. = not determined
[1] After washing, 0·5% (w/w) polyphosphate was added
[2] Standard deviation; n is the number of samples

Washing procedures: poorly washed krill tails only had been sprinkled with freshwater, whereas intensive washing was dipping 20 kg of krill tails into 30 litres of freshwater for 30 min. The content of TMAO-N in raw krill was 144 mg%

3.2 Waterbinding capacity and consistency of the meat

Before freezing aboard, the meat had been mixed with various polyphosphates and sodium tartrate or citrate to reduce the drip and to improve the consistency of the thawed meat.

Sodium tetrapolyphosphate and sodium pyrophosphate were most effective in reducing the drip (Table III). Furthermore, all additives except tartrate caused an increase in waterbinding capacity compared to the control. The consistency (ie the springiness measured with the Wolodkewitsch-Consistency-Tester) of the raw meat was also improved in the samples containing polyphosphates.

Table III
PHYSICAL PROPERTIES OF RAW KRILL MEAT

Additives	pH of treated meat	Thawing drip % (w/w)	Waterbinding capacity %	Elasticity (kp)
None (control)	6·79	30·6	47·4	2·67
Sodium pyrophosphate	7·07	20·7	53·3	1·80
Potassium pyrophosphate	7·08	28·3	52·4	1·90
Disodium dihydrogen pyrophosphate	5·98	29·9	39·5	5·10
Sodium tripolyphosphate	6·84	27·3	53·1	1·35
Sodium tetrapolyphosphate	6·73	20·9	50·7	1·65
Disodium tartrate	6·80	28·7	43·4	2·27
Trisodium citrate	6·98	29·6	52·7	1·80

Additives were supplied as powders to the krill meat in a concentration of 0·5% (w/w). The samples had been wrapped in polyethylene bags; they were stored at −30°C for 14 months

In an additional experiment the influence of additives on the properties of cooked krill meat was examined. Table IV shows that only sodium tetrapolyphosphate and potassium pyrophosphate markedly inhibited the loss of juice in the course of cooking the meat. In general the springiness of the raw samples was better than described above, presumably because of the shorter period of frozen storage. The results obtained with the Consistency-Tester indicated that the samples contain-

Table IV
CONSISTENCY OF COOKED KRILL MEAT

Additives	pH of treated meat	Waterbinding capacity of raw meat %	Loss of juice after cooking % (w/w)	Elasticity of meat (kp)		Consistency scores of the taste panel for cooked meat
				raw	cooked	
None (control)	7·04	55·0	21·6	1·65	5·5	7·3 ± 0·75[1]
Potassium pyrophosphate	7·47	58·3	16·6	1·55	5·0	6·7 ± 0·47
Sodium tripolyphosphate	7·07	54·5	21·8	1·45	5·0	7·2 ± 0·69
Sodium tetrapolyphosphate	6·95	52·5	14·0	1·43	6·5	7·2 ± 0·37
Disodium tartrate	7·02	51·5	20·6	1·70	5·3	6·3 ± 0·47
Trisodium citrate	7·24	57·3	20·8	1·45	4·8	6·3 ± 0·47

[1] Standard deviation

Preparation of samples as described in *Table 3*; samples were stored at −30°C for eight months

ing polyphosphates or citrate were softer than the control, either raw or cooked. After cooking, the krill meat had lost much of its springiness. However, the samples containing polyphosphates as well as the control received about 7 points (*ie* they were denoted as 'good') in the taste panel assessments.

3.3 *Evaluation of krill fingers*
3.3.1 *Taste panel assessments* The deep frozen krill fingers were stored up to 11 months; at regular intervals their organoleptic properties were scored by a taste panel.

The krill fingers were characterized by a firm, but not tough or dry, consistency. They had a neutral to slightly aromatic taste, not resembling fish. Their flavour was mainly determined by the breading.

In general, krill fingers were well accepted: on a 9-point scale (9: optimal, 6: satisfactory, 3: unaccept-

able) the fingers received 6–7 points in overall acceptability (*Fig 1C*). Even after a storage time of 11 months no decrease in quality could be detected. On the contrary, a process of 'maturing' was observed during the first months of storage, which presumably resulted from an interaction between meat and coating.

The addition of polyphosphates had no significant influence on the quality of the krill fingers (*Fig 1*), although at the end of storage the consistency of samples containing polyphosphate was softer and more sticky than that of the controls.

Comparing the consistency of the whole krill fingers against that of the meat alone (*ie* after removal of the coating), the latter was considered worse (*Fig 1A* and *B*); especially the samples of meat without polyphosphate were described as being slightly dry and tough. These properties, however, had only a minor influence on the consistency of the whole fingers, because they

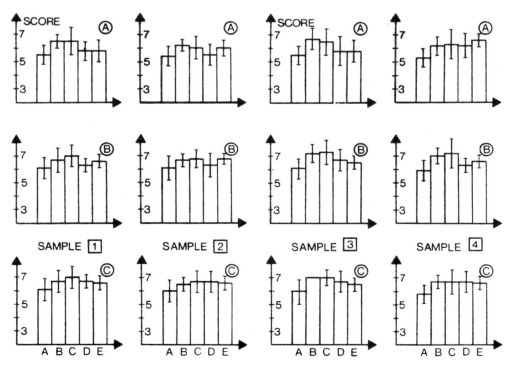

Fig 1 Taste panel assessments of krill fingers. Krill fingers had been prepared from four different blocks of krill meat: samples 1 and 2 contained no additives, whereas samples 3 and 4 had been supplied with 0·5% (w/w) sodium tripolyphosphate or sodium pyrophosphate, respectively. At the top (A), the consistency scores for the meat core are presented, followed by the scores for the consistency (B) and overall acceptability (C) of the whole krill fingers. The vertical lines indicate the standard deviation. The letters on the bottom of the figure indicate the time of frozen storage after production of the fingers. A: 3 weeks, B: 19 weeks, C: 30 weeks, D: 41 weeks, E: 48 weeks

were masked by the firm and crispy coating. On the other hand, the consistency of cod flesh in the fingers often was considered to be too soft in relationship to the coating, resulting in an unbalanced product.

3.3.2 Physical examination of krill meat from fingers
In addition to taste panel scoring of fried fingers the waterbinding capacity and consistency of meat from thawed, unheated fingers were determined. In the beginning of the deep-frozen storage trial the waterbinding capacity and elasticity of all samples increased (*Figs 2 and 3*). These observations corresponded to the process of 'maturing' mentioned above. After 20–30 weeks the textural properties of the meat got worse: the waterbinding capacity decreased and the springiness of the meat was reduced. In part, the results of the physical measurements are reflected by the taste panel assessments (*Fig 1A*).

Originally, the addition of sodium tripolyphosphate resulted in an increase of waterbinding capacity and elasticity compared to the other samples. At the end of the storage trial, however, these positive effects were reduced again.

Figs 2 and 3   Changes in the waterbinding capacity and consistency of krill meat from fingers in the course of deep-frozen storage. Parallel to the taste panel assessments described in *Fig 1*, the water-binding capacity of the meat core of krill fingers was determined (sample 1: — ● —, sample 2: — ○ —, sample 3: — ■ —, sample 4: — □ —). After removal of the coating the meat was thawed, drained and examined.

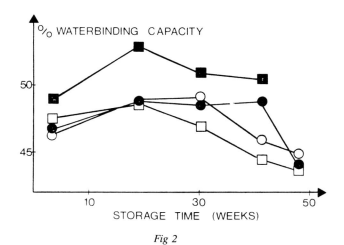

Fig 2

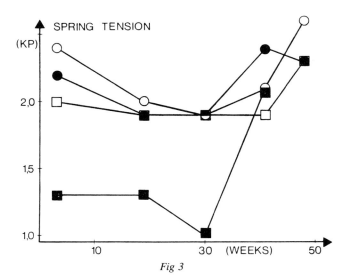

Fig 3

4 References

ANON. Venta de krill en barras apandados. *El Mercurio*, Chile 27 majo
1977   27–8
ANTONACOPOULOS, N. In: *Fische und Fischerzeugnisse*, Eds N Ludorff
1973   and V Meyer. Paul Parey Verlag, Berlin, Hamburg. 2nd edn,
226–228
GRANTHAM, G J. *The utilization of krill*. FAO, Rome
1977
GRUENEWALD, T. Ein Festigkeitsprüfgerät für Lebensmittel nach N.
1957   Wolodkewitsch. *Z. Lebensmittel–Unters. u.–Forsch.*, 105,
1–12
REINACHER, E, ANTONACOPOULOS, N and EHLERMANN, D. Die Prob-
1977   lematik der Bewertung neuer Technologien am Beispiel der
Radurisierung ('Strahlenpasteurisierung') von Frischfisch.
Deutsche Lebensmittel-Rundschau, 73, 361–363
ROSCHKE, N and SCHREIBER, W. Analytik von Krill, Krillprodukten
1977   und antarktischer Fischen. *Arch. Fisch Wiss.*, 28, 135–141
SCHREIBER, W. Krill-Rohstoff für neue Nahrungsmittel. *Gordian*, 78,
1978   101–105
STOEPPLER, M and BRANDT, K. Comparative studies on trace metal
1979   levels in marine biota. II. Trace metals in krill, krill products
and fishes from the Antarctic Scotia Sea. *Z. Lebensmittel–
Unters. u.–Forsch.* (In press)
YANASE, M. Chemical composition of the exoskeleton of Antarctic
1975   krill. *Bull. Tokai Reg. Fish. Res. Lab.*, 83, 1–6

Economic evaluations of krill fishing systems

J K McElroy

1 Introduction

Expeditions to the Southern Ocean that took an interest in krill as a fisheries resource *per se* began with the Russian vessel *Muksun* in the 1961–1962 season; 4 tonnes of krill were caught. Russian efforts on krill have grown steadily since that time.

Japanese interest in this fishery began in the 1972–1973 summer season, with the arrival in the Antarctic of the *Chiyoda Maru*, under charter to the Japanese Marine Resources Centre (Jamarc). Since then, research expeditions consisting of research vessels and chartered commercial vessels from various other nations including Poland, West Germany and Chile, have fished

for krill. Based on the knowledge gained in these first years, exploitation advances were made in the technology employed in the fishery, mainly in net design. By 1975–1976 season, vessels were now achieving very high rates of catch. The following season, 1976–1977, saw an intensification and expansion of the effort on Antarctic krill due in part to these improvements in catch rate and the growing need to re-deploy a part of the distant-water fleet of nations affected by extensions of national fisheries limits. Catches of Antarctic krill in this year topped 100 000 tonnes (FAO, 1978a).

Eddie (1977), in his forward-looking review on the harvesting of krill, suggested that catch rates were now sufficiently high that it would be possible to define economic systems of harvesting, once suitable products and processes were developed. At that time, many of the major developed fishing nations of the world, either through individuals or research teams, were engaged on work mainly aimed at developing products for human consumption. The key to successful development of products which would be acceptable on world markets apparently lies in finding ways of producing stable, intermediate (or final) products on board the catching vessel.

Of the range of types of fishing systems that have either been tried or contemplated, the single vessel stern trawler with its aimed mid-water trawls has been the most successful to date. Such a vessel may either supply a processing factory, at sea or ashore, or process its catch on board. Onboard processing may involve simply cooking and freezing. Alternatively, where such vessels have a factory deck, further processing to produce stable intermediate or final products may be carried out.

The major products that can be produced from krill, either on board a catching vessel or ashore, are whole frozen krill, tailmeats, mince or meal.

1.1 The potential market for krill products

Another study by the author (in preparation, a) gave some indication of the market potential for the main products listed above. The results of this study are summarized in *Fig 1*.

The indicative value of krill tailmeats has been taken at between $1 600–2 500/tonne. An upper limit for the market for this product on the basis of its possessing a similar texture and flavour to shrimp has been put at about 60 000 tonnes/annum product weight.

This figure would represent some 10% of the current shrimp market, in terms of product weight (edible portion) (FAO 1977; 1978b).

Potentially, the largest single outlet for krill-based products is the protein meal market. The value of krill meat will be determined mainly by its protein content. At small volumes (of less than 100 000 tonnes, say) its value will be primarily linked to that of fish meal, on the basis of its protein content. As the supply of krill meal increases, its value per unit of protein would tend to migrate away from that of fish meal down towards that of soya. On the basis of the prices prevailing for fish meal and soya in 1977, the value for krill meal (with between 48–55% protein) has been estimated to lie in the region of $300–400/tonne for volumes of krill meal of less than 500 000 tonnes/year.

Markets for the other two main products *per se* are likely to be specialized and are not dealt with further here.

The current situation, then, in respect to interest in the krill resource can be briefly summarized as follows:

(*1*) A vast resource exists which can be utilized. What is the best way to exploit this resource economically, biologically?

(*2*) There is a proven interest in the krill resource which has been heightened with recent advances in the technologies associated with catching, processing and product development.

(*3*) The market for krill-based products is relatively unknown in terms of both its size and the price the market can bear. However, there is considered to be a potentially fruitful market for certain krill products.

1.2 The area covered

This paper assesses some of the options for a commercial

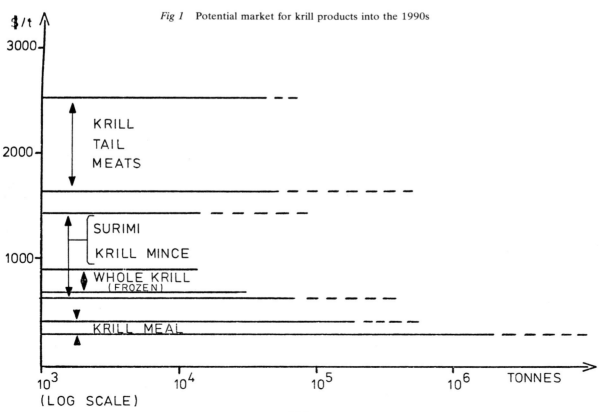

Fig 1   Potential market for krill products into the 1990s

fishery harvesting krill. The analysis concentrates on identifying the most important variables affecting its exploitation, demonstrating the relative importance of each in the systems assessed.

Figure 2 illustrates the range of potential systems that may be considered when employing a mid-water trawling vessel in this fishery. Although only the freezer trawler and fish meal factory trawler will be discussed here in detail, the method of assessment employed is applicable to the others.

2 Material and methods

2.1 *Material*

Information on costs and the technology of the systems under consideration was collected from the following sources: the literature; visits to commercial companies, Government bodies and international agencies in the UK, West Germany, Poland, Spain and Italy; letters to reefer transport companies, processing equipment manufacturers, mealing factory vessel operators, *etc*.

2.2 *Methodology*

Basically western European cost structures have been used for the year 1977. Cost adjustments to take account of krill fishing operations being based upon a South American port were made (see *Appendix I*).

Such adjustments included extra payments to crew (to take account of share money portion of wages, hardship allowances), higher fuel costs, higher insurance charges and costs of extra machinery depreciation where appropriate; these factors gave rise to a 20% increase in operating costs.

In order to carry out an economic analysis the following is required:

(*1*) Vessel characteristics and operational details (*eg* hold size, shaft horse power, processing capacity).

(*2*) Catch rate curves (*ie* distribution of catches over time).

Vessel characteristics are generally available from design data. Operational details covering:

(*i*) Catching (*eg* fishing cycle times, net capacities)

(*ii*) Processing (*eg* processing rates, throughputs, yields, product quality)

allow an assessment of the costs and earnings of a specific vessel to be made. For this analysis, calculations are based on best current practice and performance of gears, machines, vessels and crews.

Information on catch rates suitable for evaluation of sustained commercial-type operations, where available, is not generally well founded. Consequently, much of the analysis is given in terms of the levels of catch per day required for a certain operation to be viable.

The analysis assumes that prospective entrants to this fishery are risk neutral. Some of the evaluations would change depending upon the attitude to risk held by a prospective entrant.[1] Other assumptions used in the analysis are given in *Appendix II*.

3 Results

3.1 *The physical constraints on the system*

3.1.1 *Trawler design* Vessel size is an important parameter in terms of both costs and potential earnings. In general, the larger the vessel the higher are its costs, and thus the greater must be its earnings. Owing to weather

[1] Thus, where commitment to this fishery represented the major part of an entrant's holdings, he would generally be considered to be risk averse. However, if such commitment represented only a small part of an entrant's overall activities, his attitude may be taken as risk neutral. The latter position might apply to multinational food companies, large nations or centrally planned economies

SOME KRILL EXPLOITATION SYSTEM COMBINATIONS

	PROCESSING LOCATION			POSSIBLE PRODUCT RANGE			
VESSEL TYPE	ON BOARD	AT SEA	ON SHORE	WHOLE	TAIL	MINCE	MEAL
FRESH FISH TRAWLERS		●	●				●
FREEZER TRAWLERS	●	●	●	● ●	●	●	●
FACTORY TRAWLERS	●			●	●	●	●
FISHMEAL FACTORY TRAWLERS	●						●

Fig 2 Some krill exploitation system combinations

conditions, minimum vessel sizes considered to be suitable for fishing in the Southern Ocean are 35m length overall (loa) for catchers and freezers; although independent vessels operating without a mothership would probably be 45–50m or above.

A decisive parameter in terms of the economics of a trawling vessel relates to its catching power (a function of its shaft horse power). Taken alone, the ideal is a very powerful but small trawler. However, the catch has either to be preserved on board or transferred to another ship within certain time limits. (In a krill fishery, depending upon use, such limits range from about 4h to 72h (Grantham, 1977). The shorter the time after landing before the catch spoils, the greater must be the capacity to preserve the catch. In conventional demersal and pelagic fisheries, catches fluctuate widely on a day to day and hour by hour basis, from zero to over 10 times the mean catch rate for the season. When, therefore, should a processing/freezing capacity be installed to cater for catches above the average catch rate? Buffer storage (eg rsw tanks) helps dampen the wide fluctuations in catch available to the processor, thus reducing the number of processing units required to process a given proportion of the catch. Nevertheless, there is a tendency in fisheries where the annual average catch rate justifies it, to increase the size of the vessel, by incorporating an improved combination of the four basic design parameters[2]:

(1) fishing (or catching) power
(2) buffer storage capacity
(3) processing capacity
(4) storage hold capacity

Ceteris paribus, increasing the storage hold capacity of a vessel allows a greater proportion of time to be spent on the grounds actually fishing, thereby further increasing the earning capacity of the vessel if the catch rates justify it. This trend towards increasing vessel size will continue until any additional increment will produce no further increase in net benefit.[3]

Put more generally, the fundamental limiting factor on vessel size concerns the relationship between the average value of a day's catch and the cost of fishing per day. In conventional large trawlers, the average value of a day's catch can be approximately doubled by filleting, fish finger block forming, mealing the offal and trash fish catch and recovering the oil on board. This reduces the freezer plant requirement, and the capacity of the hold required is reduced by more than one-half compared with freezing the same catch as whole fish.

The increased cost of processing on board (processing machines, additional crew, etc) may or may not be off-set by the added value of the catch after processing. Furthermore, larger vessels with greater endurance are able to operate further afield, where catch rates may be substantially higher, thus compensating for the generally higher

costs. However, any piece of processing machinery is usually restricted in its ability to process different species and different ranges of any given species. Increasing size of the factory vessel, then, generally leads to increasing specialization. In an ever-changing environment in terms of catch rates, species fished, areas fished, etc, a most important requirement of a fishing vessel is its adaptability to changing circumstances. Adaptability incurs its own costs. However, the costs associated with 'non-adaptability' are those incurred in lost opportunities.

Clearly, then, the question of selecting an appropriate vessel type and design is quite complex. However, an initial screening to select the most appropriate systems has been carried out at the University of Stirling. Furthermore, the importance of the various physical constraints for different sizes of each type of vessel – the fresh fish, freezer, and factory trawlers – engaged on fishing for krill will emerge below.

3.2 The influence of increased rates of catch on the percentage utilization of a vessel

3.2.1 The effect of processing capacity The ability of a vessel to deal with a gradually increasing catch rate increases proportionately towards the processing capacity limit.

At higher potential catch rates, the processing capacity limits the proportion of the catch that is processed. Thus, though higher catch rates may be possible, the effective catch rate will be limited by the processing capacity of the vessel (Fig 3). Over an extended period of time, the hold capacity will also limit the effective catch rate in the season.

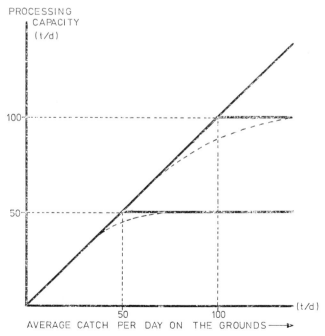

Fig 3   Relationship of processing capacity versus catch rate

3.2.2 The effect of storage hold capacity Assuming that processing capacity for a vessel is not limiting, then as the average catch rate per day on the grounds increases, the influence of the size of a vessel's hold capacity on the

---

[2] Two general points should be noted here:
(1) Any of the above parameters may be freed, but at a cost.
(2) These parameters are inter-related. In attempting to optimize for any one, the overall system may no longer be optimal.
This paper does not attempt to optimize the total system but merely indicate appropriate values for some of its parts

[3] Essentially this problem reduces to one of maximizing net revenue, where net revenue = revenue – costs. Thus, maximizing the expected value of net profits is the assumed objective function

ability of that vessel to take advantage of increased catch rates increases. This is because, once the vessel's hold is full, the trawler must return to base (port) and off-load the catch before returning to the grounds to resume fishing. Assuming the time to revert to fishing takes d days, as the time to fill the hold with increased rates of catch shortens, the relative significance of this variable (d) increases.

Thus, as the daily catch rate, c, increases, the total season's catch changes according to the following formula:

the percentage utilization
of the available catch is given by $\frac{(100 - Nd)c}{100}$

where $N = \frac{(100)}{[(k/c) + d]}$ = no. of complete cycles in a season

d = no. of days non-fishing in a cycle
so Ndʼd = no. of days non-fishing in a season
k = hold capacity (tonnes)
and c = catch rate per day (tonnes/day) on the grounds.

Figure 4 shows the influence of increasing hold size on the proportional utilization of the available catches made by a vessel during the course of a fishing season.

With a constant value for d of 4 days, and with an average season's catch rate on the grounds equal to 120t/d, vessels with hold capacities in the range of 500–800 tonnes differ in their percentage utilization of the available catch by as much as 22% (equivalent to a difference of 2 640 tonnes for the season).

4 Implications for different krill fishing systems

4.1 *The example of whole boiled frozen krill*
The operating and total costs for a UK type freezer trawler based in South America with a throughput capacity of 50 tonnes/day and a hold capacity of 600 tonnes is given in *Appendix Ia*. Such a vessel, obtaining $500/tonne for whole cooked frozen krill ex-vessel, requires catches of 16·14t/d. to cover total costs.

The costs of a larger freezer trawler (100t/d throughput capacity; 800t hold capacity) are given in *Appendix Ib*.

Figure 5 shows the relationship between the estimated average cost per tonne landed, and catch rate. The cost per tonne landed declines more and more slowly as the catch rate is increased due in part to the decreasing percentage utilization of the fishing vessel. The comparison of vessel sizes is influenced by the difference in vessel costs.

No allowance has been made for the increasing significance of randomly occurring large hauls at higher average catch rates (as indicated by dotted lines in *Fig 3*). Nonetheless, the greater proportion of fishing time spent on the grounds by the larger vessel (purely on account of its larger hold size) reduces the average cost per tonne landed to a level approaching that of the smaller vessel as the latter reaches its maximum processing rate.

With the vessel costs used here, the cost per tonne landed is higher at a given catch rate below 50t/d for the larger vessel. The graph shows how much the catch rate must be increased to achieve the same cost per tonne as the smaller vessel. Clearly, though, as average catch rates achievable by both vessels exceed 50t/d on the grounds, the advantages of a larger vessel become apparent. The possibility of a freezer trawler supplying a processing factory with whole frozen krill is considered below. Before moving on, however, two particular variants of the non-design parameters in the basic case are considered.

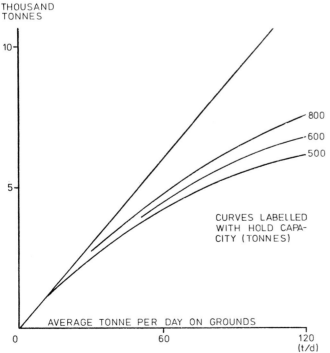

Fig 4 Landings from 100 'fishing' days *versus* catch rate

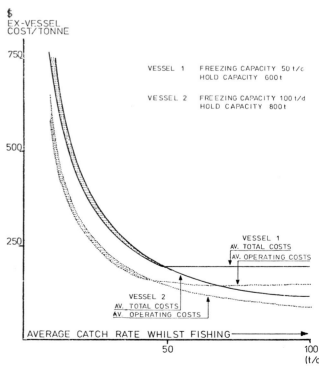

Fig 5 Cost of catching krill *versus* catch rate for vessels of different size

318

4.1.1 *A United Kingdom* versus *a South American based vessel* (The effect of distance from the fishery.) Most of the major fishing nations using large trawling vessels are situated in the northern hemisphere. The use of such vessels in the krill fishery would involve an additional round trip from a South American port of about 16 000 miles. Such a trip would add some $175 500 to operating costs.

At a price of $500/tonne for whole frozen krill, an additional catch of 351 tonne of krill is required in order to cover costs.

Figure 6 shows how much the catch rate must be increased for a UK based vessel to achieve the same cost per tonne as a South American based vessel.

As ex-vessel prices fall, the difference in catch rate required by the UK operation to break even increases dramatically. More generally, when the value of a product is low, the effect of a reduced cost structure for the same size of vessel can be substantial.

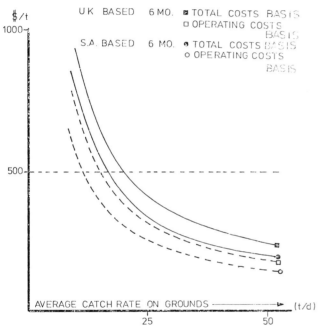

Fig 6 Comparison of the cost of catching krill for a UK based operation *versus* South American operation

4.1.2 *A six-month* versus *an eight-month krill fishery* Extending an operation from six to eight months imposes increased operating and capital related costs. Providing the catch rate at the extremes of the season is not substantially different from the effective catch rate during the normal season (a point on which there is considerable doubt), the percentage utilization of the vessel (and thus the cost per tonne required to cover costs) may be maintained.

Not surprisingly, employing a vessel for as long as possible in a fishery (and thereby increasing its percentage utilization) may produce the least cost per tonne solution for that fishery whilst still maintaining viability. However, such a strategy does not necessarily, and in fact is probably unlikely to, produce the maximum expected value of net profits for a vessel over a year. The outcome will depend upon the opportunity cost (maximum net profit expected) of employing a vessel for a

greater period in one fishery if it could be operating in another. However, in the case where a factory vessel is designed only to process krill (*ie* effectively increasing its processing capacity and thus its percentage utilization in this fishery per day on the grounds), all other things being equal, the vessel will stay on the grounds fishing up to the point where an additional day's revenue will no longer cover an additional day's operating costs.

The decision to build a factory vessel that could only effectively process krill would depend upon demonstrating that such a vessel could be profitably employed in this fishery alone (*ie* total costs for the year could be covered).

4.2 *Processing products from whole frozen krill*
4.2.1 *The location decision. 1. A southern hemisphere* versus *northern hemisphere processing factory base* The idea of using whole sea-frozen krill for land or sea based processing has been entertained by many nations. Two important cost factors affect the decision: (*1*) transport costs and (*2*) relative processing costs.

Processing costs will vary with location depending upon the local cost of inputs such as labour, power, *etc*, and the percentage utilization of the factory (and/or its specific krill processing machinery) in the course of a day and throughout the year.

Where the cost of inputs and the utilization of the factory throughout the year are the same, the processing cost will be the same. Employing realistic assumptions, the cost of processing krill tailmeats from whole krill has been calculated at between $300/tonne and $400/tonne production (McElroy, in preparation, b)

Transport costs vary with the size of the reefer vessel and the distance to be covered. However, the process yield can have a considerable effect on the choice between locating a processing plant near the western European market or in the vicinity of the Southern Ocean (*eg* in South America). *Table I* shows the effect such a decision would have on the price that could be paid for whole krill ex-vessel.

Clearly, when product yields are low, the significance of transport costs (a difference in this case of $1 000 − 150 = $850/tonne product) to the final product cost is of over-riding importance to the decision on process plant location.

4.2.2 *The location decision. II. Processing on-board* versus *processing ashore. The case of fish meal* The use of catcher vessels fishing for krill and supplying a mealing factory, whether at sea or ashore, has been investigated elsewhere (McElroy, in preparation, b). The rapid spoilage rate of krill coupled with the large proportion of time lost in steaming to and from the grounds and transferring the catch, reduces the proportional utilization of the vessel to well below any viable level.

Vessel utilization is considerably improved by processing the meal on board the catcher vessel.

Figure 7 shows how the cost per tonne of producing krill meal falls as the average production per day on the grounds rises. Two vessel sizes are considered. The dotted zones represent the boundary lines covering this capacity range. The cost of krill meal ex-vessel is unlikely to rise above $250/tonne (assuming a 20%

yield, and 54% protein content); transport costs to western Europe at $120/tonne).

At this price, and for the cost structures given here, the operating costs of the smaller vessel can be met only at full capacity utilization, but the total cost of this vessel cannot.

Employing a larger vessel in this fishery allows a lower unit cost of meal production but only at somewhat higher catch rates. The operating cost cross-over point

Table I

'THE EFFECT OF LOCATION UPON FINAL PRODUCT COST. THE EXAMPLE OF KRILL TAILMEAT PRODUCTION

Assuming a wholesale value for krill tailmeats of $2 000/t and a processing cost of about $400/t product, the maximum preprocessed value of the quantity of frozen krill required to produce 1 tonne of product is $1 600. Assume a meat yield of 15% of frozen whole krill. Transport costs to western Europe from South America are taken at $150/t

Part A
(1) Processing plant in South America; market in western Europe.
In this case only the product is shipped, thus
$(1 600 − 150)/t × 0·15 = $217·5/t raw material landed price South America.
(2) Processing plant in western Europe, near to the market.
In this case the raw material is shipped, thus
$(1 600 × 0·15)/t − $150/t = $90/t raw material landed price South America.

Part B
In order to determine the ex-vessel price for raw krill in the Southern Ocean, the cost of transhipment from the Southern Ocean to South America must be subtracted. At a distance of 1 500 miles, the 'round trip' cost of a reefer vessel on such a journey has been calculated at about $50/tonne (McElroy, in preparation, b). (The use of a reefer is generally preferable to the use of a catcher to transport the catch if the one-way distance to be covered exceeds about 500 miles (McElroy in preparation, b).)
Depending upon the processing location used, the ex-vessel price for krill would be either $160–170/t or about $40/t. At the higher value of $160–170/t, a daily rate of catch on the grounds of about 40t/d would be required just to cover operating costs for both a 50t/d and 100t/d throughput capacity freezer trawler during a six-month season (*Fig 5*). Only the larger vessel could also cover total costs, when a catch rate of some 64t/d would be required (at this price/tonne).
The significance of transport cost is reduced at higher product yields.

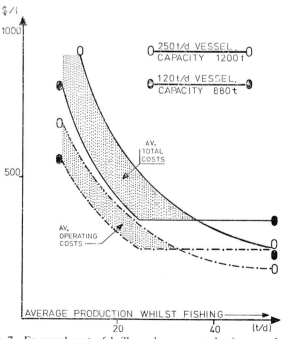

Fig 7 Ex-vessel cost of krill meal *versus* production per day for vessels of different sizes

represents a production of about 33 tonnes/day on the grounds. For the yield rates given above, the 33 tonnes represents a catch of about 165 tonnes of krill per day on the grounds.

With current nets, the Poles, West Germans and Japanese consider that catches should be limited to about 12 tonnes a haul (maximum 20 tonnes/haul) to avoid excessive clogging of the fine net meshes, the risk of net rupture, and the resulting excessive unloading times. At an average catch rate of 12 tonnes/hour (experienced while working on commercial fishing concentrations of krill), assuming hauling and shooting times about 30min, and with a fishing day effectively limited by the availability of such concentrations to about 16h, daily catches much in excess of 150t/d are unlikely (based on Bogdanov and Lyubimova, 1978; Lestev, 1978, for example). This is covered in *Table II*.

Table II

TECHNICAL CONSTRAINTS ON MAXIMUM DAILY CATCH RATE

Assumptions: 30 minutes hauling, off-loading, shooting; catch rate 12t/h trawling; 16h fishing day;
(1) Assuming maximum capacity of net is 12t/haul without undue delays.
Each fishing cycle takes 1½h. 10·67 cycles/day yield a catch of 128t/d.
(2) Assuming maximum capacity of net is 20t/haul without undue delays.
Each fishing cycle takes 2h 10min. 7·38 cycles/day yield a catch of 148t/d.
Apart from the increased risk of serious clogging of the fine meshes, increasing pressure with increased haul tonnages reduces the weight and increases the spoilage rate of krill on account of the press juice which is lost. This may amount to some 20–25% of the undeformed weight of krill.

Therefore, mealing factory trawlers with throughput capacities in the range of 120t/d to about 250t/d are currently barely capable of covering operating costs whilst fishing for krill.

None the less, they could be used in this fishery for part of the year. However, with these cost structures, the construction of new vessels for this fishery is not justified at this stage in the development of the technology of catching, at the currently indicated price of krill meal.

Future prospects for such a fishery will depend upon movements in relative costs and prices.

No attempt has been made here to include a detailed evaluation of a conventional factory trawler catching krill.

Considerations concerning the allocation of catch to different processes onboard and the use of the vessel in the off-season go beyond the scope of the present paper. Some preliminary work on this subject is presented in McElroy (in preparation, b).

5 Conclusions

5.1 *The most sensitive parameters*
The outcome of the fishing systems evaluated here are more sensitive to changes in some parameter values than to changes in others. Generally, proportional changes in parameter values, where these parameters contribute a substantial part of the total cost or value of the operation, will have the greatest effect.

In this instance the outcomes are most sensitive to

changes in the following parameters:

(*1*) The price of the product concerned

(*2*) The percentage utilization of the vessel

This may be brought about in two main ways:

(*i*) Through a change in the proportion of days actually spent on the grounds.

This may be achieved by increasing the hold capacity.

(*ii*) Through a change in the effective catch rate on the grounds.

This may be achieved by increasing the processing capacity.

(*3*) The processing plant

A change in the throughput rate of raw material, or in the yield of raw material to product weight, can be influenced by the size and type of processing machinery used.

(*4*) The cost of fishing

Maintaining or reducing the cost of fishing generally has low priority in the short-term. It is generally considered easier to affect the efficiency of a vessel in the long run than to reduce the cost of fishing. Where vessel operators can affect costs (by a change in operating base for example), then cost reduction becomes an important function.

5.2 *Summary of results involving single product systems*

(*1*) The employment of trawlers fishing for krill to supply the Japanese market with whole boiled frozen krill is an attractive though limited proposition. Minimum catch rates of 20t/d on the grounds for a hundred-day fishing season are required.

(*2*) The employment of large freezer trawlers to supply a processing factory (*eg* for tailmeat production) based in South America is attractive if large vessels are used (say 100t/d freezing capacity; 800 tonnes hold capacity), providing minimum average catch rates on the grounds of 65t/day are possible. With average catch rates between 40 t/d and 65 t/d on the grounds for such a vessel, the operation is viable only in the short term.

(*3*) The analysis showed that employing a mealing factory trawler of between 120 t/d and 250 t/day could be economic only in the short term (*ie* covering operating costs) provided high catch rates of between 120 t/day and 170 t/day on the grounds are possible and could be sustained throughout the season. At higher potential catch rates, current net designs are limiting.

5.3 *Multiple processing of krill on board a factory trawler has not been analysed in this paper*

However, there are indications that such a system might be economically attractive if the degree of compatability between its krill processing activities and its fish processing activities is sufficiently high (*ie* mincing, canning, mealing capabilities).

Alternatively, processing factories based in the vicinity of the Southern Ocean which make maximum use of the raw material (*eg* chitosan and astaxanthin) may contribute to increasing the value of the raw material.

This could be used, particularly in an integrated operation, to help reduce the burden of high catch rates required for the freezer trawlers to be viable.

The development of this fishery ultimately depends upon movements in relative factor costs, and prices, in conjunction with the indentification of suitably large markets for krill products to justify the large investments required.

Appendix Ia
FREEZER TRAWLER VESSELS AND COSTS

(*1*) Vessel characteristics and operating patterns.

Freezer trawler	
Length overall (metres)	65·75
Hold capacity (tonnes)	600
Throughput capacity (tonnes/day)	50
Speed	15 knots

(*2*) Vessel costs (est 1977)	US$
Operating costs/day at sea	3 250
Capital charges for six months (depreciation and interest) as a cost/day at sea	1 560
Insurance cost/day at sea at 4% of capital cost for six months	650
	5 460
Rounded up to	$5 500/day at sea

Note: Costs based on UK freezer trawler for 1977 adjusted for operation in Antarctica.

Appendix Ib

Large freezer trawler	
Length overall (metres)	80–85
Hold capacity (tonnes)	800
Throughput capacity (tonnes/day)	100
Speed	15 knots

Cost differences to *Ia*	US $/day at sea
Operating costs/day at sea (increased fuel, labour mainly)	$300
Capital costs/day at sea (increased processing and hold capacity)	$150
Total operating costs	$4 200/day at sea
Total costs	$5 900/day at sea

Appendix II
ASSUMPTIONS ACCEPTED FOR THE BASE CASE

(*1*) Season's length varies from year to year. The normal harvesting season lasts from mid-November to mid-April (six months). We shall **assume** that the fishing season's length is 150 days although fixed costs are allocated on a half-year's basis. The number of days spent at sea is taken at 135. Part of this time will be given up due to bad weather, breakdowns, crew sickness, searching, idle time on grounds, *etc*, and the number of days available for fishing per season is taken at 100.

(*2*) Catch rates, though variable on a haul by haul basis, have been **assumed** not to vary over the season's length, and so have been taken on a 'mean daily catch rate' basis.

(*3*) Processing capacity varies dependent upon the product to be produced, vessel size, *etc*. It is generally **assumed** that processing capacity can be fully used.

(*4*) It is **assumed** that the first season of operation is typical over the full life of an investment.

(*5*) All revenues and costs are given in US dollars. Costs and prices for 1977 have generally been used. Where these were not available the following exchange rates were employed:

£1.00 = US $1.75 for 1977
£1.00 = US $2.00 for 1979
US $1.00 = 240 Yen for 1977

Inflation index 1977 = 100; 1979 = 122.

(*6*) The optimum size of vessels and processing plant has not been determined. Those sizes on which there is information available to the author have been used. The effect of different vessel capacities upon the economies of an operation has been included in the assessment.

6 References

BOGDANOV, A S, and LYUBIMOVA, T C. Soviet studies of krill in the
1978   Antarctic Ocean. *Rybn. Khoz., Mosk.*, 10(10), 6–9 (in Russian); translated in JPRS No. 72381 on USSR resources No. 841

EDDIE, G C. The harvesting of krill. *FAO Rep.*, GLO/SO/77/2.
1977   Southern Ocean Fisheries Survey Programme, Rome

FAO.   The international market for shrimps. South China Seas
1977   Fisheries Development and Co-ordinated Programme, FAO, Manila, *SCS/DEV/76/12.*

FAO. Southern Ocean: nominal catches by countries and species, 1978a 1971–77. *FAO Fish Circ.*, (648), 23pp

FAO. Fishery products: supply demand and trade projections, 1985. 1978b *FAO. ESC: PROJ/78/*

GRANTHAM, G J. The utilisation of krill. *FAO Rep.*, GLO/SO/3. 1977 Southern Ocean Fisheries Survey Programme, Rome

LESTEV, A V. Krill fishing technique. *Rybn Khoz. Mosk.*, 10(10), 1978 52–55 (in Russian); translated in JPRS No. 72381 on USSR resources No. 841

McELROY, J K. Potential krill products and their markets. For publicain prep. a tion by IIED/IUCN

McELROY, J K. Economics of krill harvesting. For publication by in prep. b IIEC/IUCN

Development of matured krill sausage

Otto Christians

In order to utilize the abundant availability in Antarctic waters of krill (*Euphausia superba Dana*), we have searched for ways and means of separating its flesh from the shell.

Our first endeavours were aimed at separating the flesh by means of a Baader bone separator, and at establishing a method of processing the minced krill meat into food tasty enough to comply with modern eating habits. Easily perishable protein products like milk and meat can be converted by bacterial fermentation into products of various tastes, preserved without freezing or sterilization. We thought, therefore, that it would be a good idea to treat krill meat, which also is highly perishable, with a similar fermentation. During the maturation process of hard sausages, the sugar content is fermented by lactic acid bacteria, by creating acids which mainly consist of lactic acid. In this way krill mixture would be acidified and the development of putrefactive bacteria inhibited. At the same time in the acidified milieu a change from sol to gel state occurs, and by the action of bacteria on protein, fat and carbohydrate special flavour compounds are created which are responsible for the characteristic flavour. By losing moisture content during the process of maturation water activity is diminished to a degree where the development of undesirable microorganisms becomes inhibited.

These processes, which only occur when using raw meat, enable the highly perishable protein to be converted into a product with a long shelf-life. Analogous experiments with minced raw krill were not successful because of the liquefying effect of the still active proteases in the minced raw krill meat. Due to this effect we decided to use minced cooked krill meat because cooking inactivates the proteases. However, cooked meat possesses little soluble protein, because they are denaturated. Soluble protein has to be provided by another native protein; for this purpose fresh fish seems to be adequate.

The following mixture of krill meat was the basis of a series of experiments to discover conditions of maturation:

minced cooked krill meat mixture	52·0%
unsmoked bacon	26·4%
minced fresh fish	20·3%
salt	1·0%
spices	0·3%

To control the maturation process, sausages stuffed with this mixture in casings of calibre 40 were checked daily with respect to their pH value, and periodically for volatile bases, lactic acid and acetic acid content. First it had to be ascertained whether due to the addition of sugar at a temperature of 20°C and at a humidity of 85%, the krill mixture showed a reduction in pH value in a desirable way. *Figure 1* shows that acidification occurs at a low glucose concentration.

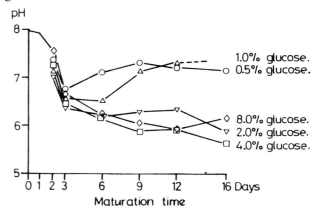

Fig 1 Influence of different concentrations of glucose on pH values during the maturation of krill sausage

After three days of maturation, however, pH increases. A stabilization to about pH6 over a longer period occurs only with a concentration of 2% and upwards. Under these conditions not even a high glucose concentration is able to reduce the pH to the required value of 5·2, as is found in the case of hard sausages.

The acidification of the mixture occurs also by adding lactose (*Fig 2*). Whereas acidification occurs in the first three days when glucose is added, addition of lactose reduces the pH value only slowly, and never to a value which inhibits putrefactive bacteria.

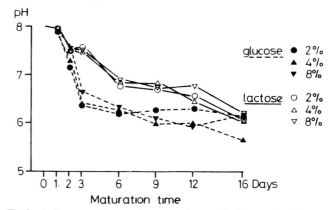

Fig 2 Influence of lactose in comparison with glucose in different concentrations on the pH values during the maturation of krill sausages

These experiments proved that the addition of sugar is necessary to cause acidification. Certain sugar combinations applied by the meat industry are useful for maturation since they cause rapid acidification and inhibition, thus we used one of these in the further experiments.

The reason that even highly concentrated sugar did not cause acidification up to pH 5·2 may be the inadequate lactic acid bacteria and/or a strong competition of some other bacterial strains in the krill mixture. Another decisive factor is the high pH value of 8 in the mixture at the beginning, which is favourable for the development of undesirable bacterial strains since they can proliferate. It was the purpose of the following experiments to find out whether mixtures containing a starter culture were able to create suitable lactic acid bacteria flora. *Figure 3* demonstrates the favourable influence of the addition of a starter culture upon the maturation of, for example, cold-smoked krill sausage.

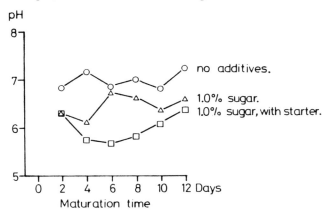

Fig 3   Influence of starter on the pH values during the maturation of cold-smoked krill sausages

When the added lactic acid bacteria become active they reduce the pH values continuously until the sugar is used up, then, due to the lack of carbohydrates the pH values start to increase slowly and evenly. It was noticed in a control test without additives that the lactic acid bacteria in the mixture did not compete with the indigenous microflora. As a consequence the pH values decrease only slowly and unevenly, and an adequate process of maturation cannot be guaranteed any more.

The addition of 1% of sugar is not sufficient; a concentration of sugar of 1·5% lowers the pH value to the range required (*Fig 4*).

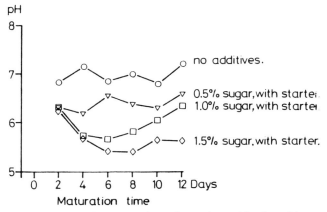

Fig 4   Influence of concentrations of sugar in combination with starter on the pH values during maturation of cold-smoked krill sausage

Sufficiently low pH can be reached only by using the cold-smoking method. Experiments with the same mixture where the mixture was matured either by hot smoke or without smoke, did not result in the desired pH level.

Figure 5 shows well the influence of three different maturation processes on the change of pH values. The pH value in hot-smoked sausages decreases less, then changes little over a period of several days of maturation, then increases again. The reason for this behaviour is that the acidifying bacteria cannot grow at the high temperature of smoking and indeed are killed by it. This has been shown by measurement of the bacteria count: samples of cold-smoked sausages contain lactic acid bacteria with a count of log 7·10, whereas after the hot-smoking process the count is log 5·60.

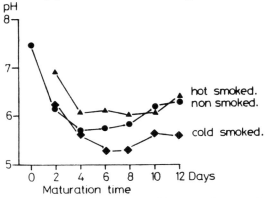

Fig 5   Influence of various smoking procedures on the pH values during the maturation of krill sausages with starter and 1·5% sugar

Samples of unsmoked sausages show initially a similar tendency of pH change to that of cold-smoked ones. However, the pH change is smaller, and after the eighth day of maturation pH values start increasing. Therefore, the cold-smoking process gives the best results as far as pH values are concerned as long as sufficient sugar and a dominant initial quantity of lactic acid bacteria are present. Examination of total volatile bases, bacterial flora and sensory results confirm the beneficial effect of cold smoking on the maturation of krill sausages. The krill sausage matured with this technique has an aroma which is similar to that of hard sausage, though an undesirable burning, prickling aftertaste can be noticed.

The latter can be explained by the high pH value of the mixture during the first phase of maturation which is favourable for the proliferation of putrefactive bacteria.

In order to attain a slightly acidified stage at the beginning, a combination of organic acids as used by the meat industry and glucono-delta-lactone (GdL) were added. In order to avoid the risk of an unsuitable acidification due to the lack of sugar, 4% was added. *Figure 6* shows that the addition of only 1% of organic acid reduces the initial pH value of the mixture to the acid side.

Stronger acidification is obtained by the addition of GdL, 1% reducing the pH to 6·46. In the course of maturation the pH values of the different mixtures tend to equalize to 4·8. This very low value reached on the twelfth day of maturation can be explained by the addition of a high concentration of sugar. The samples also tasted acid.

After suitable conditions for development of a lactic acid bacterial flora and maturation had been deter-

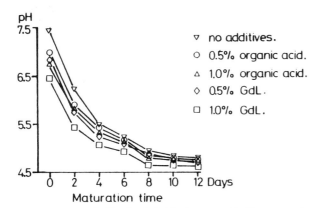

Fig 6 Decrease of pH value in krill mixture and krill sausages during maturation by addition of organic acids and glucono-delta-lactone (GdL). Krill mixture contains 4% sugar and starter

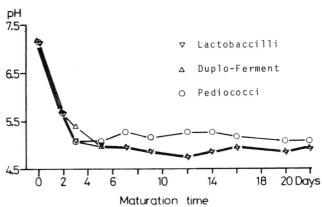

Fig 7 Influence of different starters on the pH values of cold-smoked krill sausages during maturation

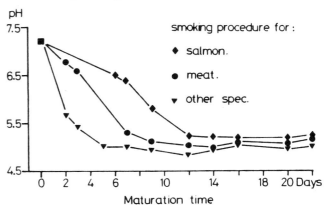

Fig 8 Influence of different cold-smoking procedures on the pH values of krill sausages during maturation; lactobacilli starter

mined, we wanted to find out in which way the different starter cultures would influence the development of flavour during maturation.

The importance of smoking on the maturation process suggested that several cold-smoking procedures should be investigated.

The three starters we used were a strain of lactobacilli, a combination of lactobacilli and micrococci, called duplo ferment, and a strain of pediococci which is used in the meat industry.

The added quantity of starters guaranteed a lactic acid bacteria level of log 6·0 in the mixtures. The cold-smoking procedures were carried out in the smoking unit of the Institut für Biochemie und Technologie, Hamburg, research station and in two commercial smoking plants, one used to smoke meat and the other to smoke salmon.

It should be mentioned that the krill sausages smoked in the meat-smoking unit were kept for five days at a temperature of 0°C before smoking, due to a breakdown of the plant. The long pre-storage time in these cases greatly influenced the maturation.

The pH values, the most important indicators for the maturation process, were taken every day. In addition, bacterial counts, total volatile basic nitrogen (TVB–N), acetic acid and total acid were measured. TVB–N and acid values are important criteria of krill sausages maturation.

Figure 7 shows changes in pH due to different starters in krill sausages smoked in the research station unit. In every case the pH values fall very quickly to the desired pH range of approx 5·0–5·2 during the first three days of maturation. Thereafter the sample with the mildly acidifying pediococci starter maintains a pH of about 5·2, whereas the samples with the two stronger acidifying starters reach pH 4·9.

The special influence of different smoking procedures on the pH values is shown in *Figure 8* where lactobacilli starter was used. Whereas the krill sausages cold smoked in the research station reached the desired acidification of 5·4 quickly within the first three days, the other samples reached this value only after seven or 12 days. It appears that the unusually long pre-storage period of five days at 0°C is responsible for the slowly decreasing pH of the samples processed in the meat-smoking plant.

The slow pH decrease observed on the salmon-smoking unit could be explained by several factors, *eg*

the temperature, humidity and the kind and time of smoke addition.

The flavour of the maturated krill sausages is determined by the various smoking procedures as well as by the starter cultures. The method of the smoking procedures has a stronger influence on the development of the flavour than the specific starter culture. As shown, the sensory values of the samples processed in the smoking unit of the research station show that all the starters seem to provide optimal conditions of flavour development.

Table I
SENSORY VALUES (FLAVOUR)
(9 = very good, 1 = very bad)

Starter culture	Smoking procedures as for		
	Special	Meat	Salmon
Lactobacilli	7	3	5
Duplo-ferment	6	5	5
Pediococci	7	5	6

The aroma of these sausages is similar to those of German hard sausages which are smoked in the special manner called 'eel smoking'.

The smoking conditions greatly influenced the formation of flavour with low organoleptic values being given by the commercial smoking units.

Therefore it is very important to maintain exactly well defined smoking procedures with regard to temperature, humidity, smoke addition, as well as minimal pre-storage times in order to give the starter cultures a chance of developing a typical flavour.

Dietary influences on the feeding of various fish meals and a soya-based meal to chickens and laying hens

J G M Smith, R Hardy and A A Woodham

1 Introduction

During the last decade there has been an increasing interest in the possible use of krill, *Euphausia superba*, as a source of protein for humans and animals. Estimates of viable krill stock in the Antarctic range from 50 to 150 million tons/year with catch rates of about 40 tons/hour being quoted. There are many problems associated with the handling of krill if it is to be used for human consumption and possibly a simpler method of processing meantime would be to convert the krill to a meal to be incorporated in animal feeds. However, when using fish-type meals there is always the possibility that tainting of the flesh may occur and with this in mind feeding trials were carried out on chickens and laying hens using four cereal-based diets incorporating a soya bean meal (SB), a commercial white fish meal (WF), a commercially dried krill meal (CDK) and a freeze-dried krill meal (FDK). Sensory assessment and chemical analysis were used to determine whether any fishy flavours or colouring had been transferred to the eggs and flesh of the hens and broilers.

An evaluation of incorporating krill meals into diets was carried out on the broilers (12 birds/diet) by measuring the total protein efficiency (TPE) and food conversion efficiency (FCE) factors. The relationship between protein intake and number of eggs produced by the laying hens (4 birds/diet) was also measured.

2 Experimental

2.1 Broiler growth and egg laying trials

The TPE growth tests were carried out as described by Woodham (1968) using groups of 12, Ross 1 male chicks to each diet. The food intake and weight gain of the four groups of birds were measured over a period of nine weeks.

The egg-laying trials were carried out by the method of Woodham (unpublished) using groups of four Honegger Blonde birds to each diet. The number and weight of eggs produced along with the weight of food eaten were recorded over a period of eight weeks.

2.2 Diets

The broiler and egg-laying diets are given in *Tables I* and *II*. The whole krill and krill meal were obtained from the first West German Expedition to the Antarctic 1975–1976 (Christians *et al*, 1978). The krill meal was prepared on board ship in a conventional fish meal plant by heating whole krill to 90–95°C followed by pressing, drying and grinding. The stickwater was not added back to the meal and most of the lipid remained in the press cake giving the meal a high lipid content of about 15% and crude protein (CP) 59·6%. As a comparison a freeze-dried krill meal was prepared from whole krill which had been stored at −30°C. The sample was left to dry in an evacuated chamber, 0·05 Torr at −20°C, for 24h, producing a meal with 65%CP.

The diets in the egg-laying trials were made isonitrogenous (15·5% CP) by the addition of maize meal, barley meal PQ 90, weatings PQ 84 and grass meal and isoenergetic, 2866kcal metabolizable energy (ME)/kg by adding lard and starch. Each of the four diets had a calories to protein ration (C/P) of 83 to 84. (In this calculation the protein is conventionally expressed in lbs and not in kg.) Similarly, in the broiler trials by the addition of barley meal, weatings, starch and lard the diets were made isonitrogenous (18% CP) and isoenergetic 3080kcal ME/kg, the CP ratio being 77.6.

2.3 Sampling

One week after commencement of the egg-laying trial one egg from each group was received almost daily. Lipid analysis and sensory assessment were carried out at approximately weekly intervals.

At the end of the growth trials the birds were slaughtered, plucked and eviscerated, the hearts and livers being retained for lipid analysis.

3 Methods

3.1 Lipid analysis

The lipid content of the eggs was determined on the total fluid content by the extraction method of Bligh and Dyer

Table I
BROILER DIETS

	CDK			FDK			WF 72			SB 91		
	% of diet	% CP³	ME²	% of diet	% CP	ME	% of diet	% CP	ME	% of diet	% CP	ME
Barley PQ90	30·99	3	845	30·99	3	845	30·99	3	845	30·99	3	845
Weatings PQ 84	18·19	3	440	18·19	3	440	18·19	3	440	18·19	3	440
Bone meal	3·00			3·00			3·00			3·00		
Salt	0·19			0·19			0·19			0·19		
Vitamin premix	0·31			0·31			0·31			0·31		
Lard							3·46		304	6·62		583
Starch	27·20		1 050	28·87		1 114	25·63		989	15·61		603
CDK	20·12	12	744									
FDK				18·45	12	683						
WF 72							18·23	12	501			
SB 91										25·09	12	610
Total	100	18	3 079	100	18	3 082	100	18	3 079	100	18	3 081
C/P Ratio¹			77·6			77·6			77·6			77·6

¹ C/P = calories/protein where P is conventionally expressed in lbs
² ME = kcal/kg
³ CP = crude protein

Table II
EGG-LAYING DIETS

	CDK			FDK			WF 72			SB 91		
	% of diet	% CP[3]	ME[2]	% of diet	% CP	ME	% of diet	% CP	ME	% of diet	% CP	ME
Maize meal	24·75	2·5	827	24·75	2·5	827	24·75	2·5	827	24·75	2·5	827
Barley PQ 90	25·83	2·5	715	25·83	2·5	715	25·83	2·5	715	25·83	2·5	715
Weatings PQ 84	15·16	2·5	356	15·16	2·5	356	15·16	2·5	356	15·16	2·5	356
Grass meal	2·50	0·5	30	2·50	0·5	30	2·50	0·5	30	2·50	0·5	30
Vitamin premix	1·00			1·00			1·00			1·00		
Salt	0·50			0·50			0·50			0·50		
Limestone flour	5·50			5·50			5·50			6·25		
Di calcium phosphate										2·50		
Lard							2·21		194	5·83		513
Starch	12·18		470	13·23		511	11·16		431			
CDK	12·58	7·5	465									
FDK				11·53	7·5	427						
WF 72							11·39	7·5	313			
SB 91										15·68	7·5	381
Total	100	15·5	2 863	100	15·5	2 860	100	15·5	2 866	100	15·5	2 822
C/P Ratio[1]			84			84			84			83

[1] C/P = calories/protein where P is conventionally expressed in lbs
[2] ME = kcal/kg
[3] CP = crude protein

(1959) and calculated as a percentage of the whole egg weight. The entire livers and hearts and 70g of the minced flesh were also extracted in a similar manner. The minced flesh, liver and heart were taken from the broiler which was used for sensory assessment. To prevent oxidation during extraction 0.001% of the antioxidant 2,6-ditertiary-butyl-*p*-cresol (BHT) was added to the chloroform.

The lipids were esterified by a method based on that of Seino *et al* (1972) using a sample size as recommended by Solomon *et al* (1974). 7ml 0.5N sodium hydroxide in 95% methyl alcohol, 3·5ml benzene and two crystals BHT were added to 350mg lipid, the mixture being refluxed for 30min. 7ml 14% boron trifluoride in methyl alcohol was then added slowly via the condenser and refluxing continued for a further 30min. After cooling and diluting with 20ml water, the contents were passed through two separating funnels containing 35ml hexane, followed by a wash with 35ml water, shaking vigorously for 1min at each step. The hexane layers were combined, dried over a small amount of anhydrous sodium sulphate and the solvent removed on a rotary film evaporator. All lipid and ester extracts were stored at $-30°C$ prior to analysis.

Gas chromatography (GC) of the methyl esters was carried out isothermally at 180°C using a 25m glass capillary column coated with EG SSX fitted to a modified Pye 104 gas chromatograph as described by McGill *et al* (1977). Quantitation of the GC peaks was accomplished using an Infotronics integrator.

The absorption spectra of the lipids were obtained scanning between 200nm and 800nm using a Cecil CE 505 Double Beam Ultraviolet Spectrophotometer, using *n*-hexane and chloroform as solvents.

3.2 Sensory assessment

The eggs were hard-boiled for 4min, sliced and presented to the taste panel under red light illumination. The chickens were halved, one half being roasted in tin foil and pieces of breast and leg from the other half being casseroled in a steam bath. The panel were asked to describe the flavour of the eggs and flesh.

4 Results and discussion

The lipid content of the eggs from the hens fed on the four diets is given in *Table III*. It can be seen that the highest lipid content is in the eggs from hens fed on the CDK diet closely followed by those from hens fed on the WF, FDK and SB diets, the difference between all four being only 1%. *Table IV* gives the lipid content of the broiler and egg-laying diets and also that of the flesh, hearts and livers. In both the egg-laying and broiler diets those containing SB meal have the highest lipid content and it is interesting to note that while the broiler fed on the SB diet has a low lipid heart content of 2.8%, those fed on CDK and FDK diets have two and three times that value respectively. Liver lipid contents show little variation and the highest flesh lipid concentration is found in the broiler fed on the SB diet.

Table III
PERCENTAGE LIPID IN EGGS FROM HENS FED ON FOUR DIFFERENT DIETS

	Diets			
No of days on diet	WF	SB	FDK	CDK
1	—	7·9	7·9	8·9
2	9·0	6·8	9·3	8·4
5	7·5	7·9	8·3	8·8
7	8·7	8·4	8·6	8·9
13	8·3	7·7	8·4	8·1
20	8·8	7·9	8·6	8·4
27	8·5	8·0	8·8	9·3
32	8·7	9·3	8·3	8·7
39	9·0	8·1	8·1	9·5
46	8·4	7·4	8·4	9·4
53	9·8	8·0	8·5	9·5
Average %	8·7	7·9	8·5	8·9

Table IV
PERCENTAGE LIPID IN THE EGG-LAYING AND BROILER DIETS AND FLESH, HEARTS AND LIVERS FROM CHICKENS FED ON FOUR DIFFERENT DIETS

	Diet		Broiler		
	Egg laying	Broiler	Flesh	Heart	Liver
WF	3·9	5·0	10·5	4·6	4·0
SB	7·0	6·3	14·0	2·8	3·5
FDK	3·4	3·8	12·3	8·5	3·3
CDK	3·2	3·5	11·5	5·3	3·7

326

Tables V to VIII show the fatty acid composition of the diets, eggs, flesh, heart and livers from the egg-laying and broiler growth trials. The percentage of saturated, monoenoic, dienoic and polyenoic fatty acids is given at the foot of each table. The fatty acid composition of all eight diets varied, sometimes quite markedly. Those containing krill meals had relatively high concentrations of 14:0, 16:1, 20:5 and 22:6 fatty acids and conversely the other two diets had higher concentrations of 18:0 and 18:1. The egg-laying diets contained significantly more 18:2 than the broiler diets, probably introduced through the incorporation of maize meal in the diet. As is to be expected, the relative concentrations of saturated and the various unsaturated acids varied also, with the krill meals containing the highest polyenoic content and the other two meals having relatively higher saturated and monoenoic acids, probably caused by the introduction of lard. With the possible exception of the 22:6 acid which tended to increase in the eggs with increasing levels in the diet, the fatty acid composition of the eggs did not reflect that of the diet. A small but significant amount of 20:4 was present in all of the eggs but was absent from the diets, and where 22:6 was absent in the diet (SB) some of this acid was found in the eggs. Marked variations of other component acids in the diets were not reflected in the eggs.

Like the eggs, the fatty acids of the broiler flesh did not reflect that of the diets but in this case polyenoic acids in the diets did not appear to be deposited to any great extent in the flesh. The composition of the flesh fatty acids was not constant but this did not appear to be related to the diet composition and it is assumed that this was due to normal biological variations.

The high content of 20:5 in the krill diets appears to raise the content of this acid in the hearts and livers of the broilers although it is only present in the flesh of the

Table V

PERCENTAGE COMPOSITION OF FATTY ACIDS IN THE DIETS, EGGS, FLESH, HEART AND LIVER OF LAYING HENS AND BROILERS FED ON DIETS CONTAINING CDK

| Fatty Acid | CDK diet | Eggs from laying hens | | | CDK diet | Broilers | | |
| | | Days on diet | | | | | | |
		7	32	60		Flesh	Heart	Liver
14:0	6·6	0·4	0·8	0·9	8·3	1·5	2·4	1·4
16:0	21·5	28·9	31·1	29·9	24·3	26·6	25·9	31·8
16:1	4·5	5·9	5·6	6·3	7·2	7·2	8·8	4·1
18:0	1·6	6·7	7·7	6·4	2·4	6·2	7·6	14·1
18:1	20·8	46·1	44·2	45·1	21·6	44·6	34·3	28·1
18:2	28·2	8·2	7·0	7·5	12·4	10·8	8·6	5·1
18:3	2·8	0·3	0·3	0·3	1·6	1·2	0·4	T
18:4	0·6	—	—	—	1·1	—	0·2	—
20:1	0·5	0·1	0·1	0·1	0·9	0·4	0·4	—
20:4	—	0·5	0·5	0·4	—	0·3	1·1	0·5
20:5	8·3	0·2	0·5	0·2	12·1	—	4·7	6·0
22:1	—	—	—	—	1·0	—	—	—
22:5	—	—	—	0·1	T	—	1·7	1·8
22:6	3·2	2·6	2·1	2·6	4·5	T	1·3	6·4
U	1·5	0·1	0·1	0·1	2·5	1·2	2·7	0·7
Saturates	29·7	36·0	39·6	37·2	35·0	34·3	35·9	47·3
Monoenes	25·8	52·1	49·9	51·5	30·7	52·2	43·5	32·2
Dienes	28·2	8·2	7·0	7·5	12·4	10·8	8·6	5·1
Polyenes	14·9	3·6	3·4	3·6	19·3	1·5	9·4	14·7

T = trace; U = unknown

Table VI

PERCENTAGE COMPOSITION OF FATTY ACIDS IN THE DIETS, EGGS, FLESH, HEART AND LIVERS OF LAYING HENS AND BROILERS FED ON DIETS CONTAINING FDK

| Fatty acid | FDK diet | Eggs from laying hens | | | FDK diet | Broilers | | |
| | | Days on diet | | | | | | |
		7	32	60		Flesh	Heart	Liver
14:0	8·5	0·6	0·8	1·0	10·5	3·4	3·9	1·9
16:0	20·3	29·7	29·9	30·2	23·6	30·1	25·2	30·2
16:1	4·7	4·7	5·9	5·0	6·9	10·6	10·4	4·0
18:0	1·4	6·3	6·3	6·3	1·9	5·8	7·0	15·8
18:1	21·1	43·6	40·2	44·9	20·6	37·1	33·6	21·7
18:2	22·7	12·0	10·2	9·0	10·3	6·5	7·4	5·6
18:3	2·3	0·4	0·5	0·3	1·8	0·8	0·7	0·2
18:4	1·9	—	—	—	2·5	0·6	0·6	—
20:1	0·4	—	0·1	—	0·6	0·3	0·4	—
20:4	T	0·5	0·5	0·3	—	—	0·9	0·5
20:5	10·7	T	0·3	0·2	13·4	1·8	5·0	6·7
22:1	T	—	—	—	1·1	—	—	—
22:5	T	—	0·9	—	T	0·6	1·6	3·5
22:6	4·2	1·8	3·4	2·5	4·9	1·0	1·6	9·1
U	1·9	0·3	0·9	0·2	1·8	1·4	1·9	0·9
Saturates	30·2	36·6	37·0	37·5	36·0	39·3	36·1	47·9
Monoenes	26·2	48·3	46·2	49·9	29·2	48·0	44·4	25·7
Dienes	22·7	12·0	10·2	9·0	10·3	6·5	7·4	5·6
Polyenes	19·1	2·7	5·6	3·3	22·6	4·8	10·4	20·0

T = trace; U = unknown

Table VII
PERCENTAGE COMPOSITION OF FATTY ACIDS IN THE DIETS, EGGS, FLESH, HEART AND LIVERS OF LAYING HENS AND BROILERS FED ON DIETS CONTAINING WF

| Fatty acid | WF diet | Eggs from laying hens | | | WF diet | Broilers | | |
| | | Days on diet | | | | | | |
		7	32	60		Flesh	Heart	Liver
14:0	1·4	0·6	1·0	0·6	2·4	1·5	1·4	1·2
16:0	25·7	29·6	29·8	25·5	24·7	28·6	24·4	29·6
16:1	2·0	5·6	4·8	3·5	2·7	7·5	6·1	4·1
18:0	10·9	6·0	7·5	8·7	19·8	7·7	10·5	13·4
18:1	31·4	45·5	47·0	52·8	34·0	44·7	36·0	31·5
18:2	23·1	9·8	6·9	6·3	7·9	7·4	10·8	8·6
18:3	2·1	0·4	0·3	0·2	1·4	0·7	0·5	0·5
18:4	0·2	—	—	—	T	—	—	—
20:1	1·0	0·1	0·1	0·2	0·6	0·8	0·5	0·4
20:4	—	0·7	0·7	0·7	0·1	—	4·3	1·8
20:5	0·4	—	0·1	—	1·1	—	1·7	2·6
22:1	—	—	—	—	0·9	—	—	—
22:5	T	—	—	—	—	—	T	—
22:6	1·5	1·5	1·8	1·2	1·5	—	1·0	5·3
U	0·5	0·3	0·1	0·1	2·9	1·2	2·7	1·0
Saturates	38·0	36·2	38·3	34·8	46·9	37·8	36·3	44·2
Monoenes	34·4	51·2	51·9	56·5	38·2	53·0	42·6	36·0
Dienes	23·1	9·8	6·9	6·3	7·9	7·4	10·8	8·6
Polyenes	4·2	2·6	2·9	2·1	4·1	0·7	7·5	10·2

T = trace; U = unknown

Table VIII
PERCENTAGE COMPOSITION OF FATTY ACIDS IN THE DIETS, EGGS, FLESH, HEARTS AND LIVERS OF LAYING HENS AND BROILERS FED ON DIETS CONTAINING SB

| Fatty acid | SB diet | Eggs from laying hens | | | SB diet | Broilers | | |
| | | Days on diet | | | | | | |
		7	32	60		Flesh	Heart	Liver
14:0	1·5	0·5	0·6	0·4	2·2	3·5	0·8	1·0
16:0	26·4	24·7	25·9	23·0	25·9	27·7	22·6	25·3
16:1	2·6	3·6	4·0	3·5	2·2	11·2	3·0	3·5
18:0	15·1	8·0	6·9	8·3	21·6	5·4	13·2	16·8
18:1	34·7	48·4	48·4	50·7	33·4	40·7	25·4	28·7
18:2	17·1	10·9	11·4	10·4	10·7	6·3	15·1	13·1
18:3	1·7	0·4	0·4	0·2	1·9	0·7	0·7	0·9
18:4	—	—	—	—	—	0·3	—	—
20:1	0·5	0·1	0·1	0·1	0·3	0·4	0·2	T
20:4	—	1·3	1·1	2·0	—	—	12·3	5·4
20:5	—	—	—	—	—	1·6	—	0·6
22:1	—	—	—	—	—	—	—	—
22:5	—	—	—	—	—	0·2	1·0	T
22:6	—	1·7	0·9	1·1	—	0·2	1·0	2·9
U	0·5	0·3	0·2	0·3	1·9	1·7	4·9	2·0
Saturates	43·0	33·2	33·4	31·7	49·7	36·6	36·6	43·1
Monoenes	37·8	52·1	52·5	54·3	35·9	52·3	28·6	32·2
Dienes	17·1	10·9	11·4	10·4	10·7	6·3	15·1	13·1
Polyenes	1·7	3·4	2·4	3·3	1·9	3·0	15·0	9·8

T = trace; U = unknown

broiler fed on the FDK diet. The fatty acid 22:6 also appears in high concentrations in the livers of the broilers fed on the two krill and white fish diets and although it is not present in the SB diet it is present in the liver of the broiler fed on that diet. The 18:0 content in the WF and SB diets was considerably higher than in the other two but this does not appear to affect the concentration of this acid in the flesh or livers of the broilers. It does however appear to increase the concentration of this acid in the hearts of the birds fed on these diets. As in the egg lipids the fatty acid 20:4 appears in the hearts and livers of the broilers although it is only present in very low concentration in one of the diets (WF). In this work the marked variations in the fatty acid dietary composition were not reflected to any great extent in those of the eggs or flesh. Other workers have found a more marked correlation. Navarro et al (1972) found that changes produced in egg yolk fat from hens fed on diets with increasing fish meal content showed a fatty acid pattern similar to those of the diets themselves. Opstvedt (1976), in a review on the effect of dietary fat on fatty acid composition of poultry products, found that the fatty acid composition of the product tended to be similar to that of the fat in the feed.

From the initial time of receiving eggs there was a distinct difference in the colour of the yolks. Those from hens fed on a FDK diet were of a dark red/orange colour while those from hens fed on a CDK diet, although still red/orange, were of a lighter hue. The yolks from the hens fed on the WF and SB diets were a normal yellowish colour. There was no apparent colouring of the flesh in any of the broilers fed on the two krill diets. Thin layer chromatography and UV analysis indicated the presence of several carotenoid pigments in the egg lipids, some of which appeared to be identical to those present in krill, probably astacene and astaxanthin. The difference in colour was detectable however in the lipid extracts from the flesh, liver, hearts and, of course, the eggs, where it was apparent that carotenoid had been carried over to these organs. The absorption spectra shown in *Fig 1* clearly shows that the carotenoid was more prominent in the flesh, liver, heart and egg lipids of the broilers and hens fed on the FDK diet. *Table IX* shows the E, 1% 1cm of the egg lipid carotenoid peak at 450nm and it can be seen that, although there was a variation in intensities, there was no visible increase in concentration of the carotenoid with time of feeding. The highest values were however in the FDK eggs followed by the CDK eggs. The spectra of the egg-laying diet lipids were similar in pattern; that of the FDK diet is shown in *Fig 2*, the two most prominent peaks at 670nm and 410nm being typical of a chlorophyll type pigment.

Table IX
$E^{1\%}_{1cm}$ OF EGG LIPIDS AT 450NM

	Days on diet			
Diet	7	25	53	60
FDK	0·304	0·268	0·265	0·282
CDK	0·274	0·300	0·228	0·183
WF	0·213	0·196	0·183	0·143
SB	0·195	0·268	0·265	0·282

quantity of the FDK diet was eaten, 3·8kg CP compared to 5·8kg CP for the other three diets. On determining the egg weight per g of protein eaten, the CDK diet produces the highest figure of 2.59, the lowest being the SB diet at 1·84. The FDK diet however requires only 24·23g protein to produce one egg and this compares reasonably well with 23·37g and 24·38g for the CDK and WF diets, respectively. The SB diet figure 31·36g protein/egg is significantly higher. It would seem therefore that in this egg-laying trial the FDK and CDK diets compare favourably with a typical white fish diet, all three producing better results than a SB diet.

Throughout the whole of the broiler growth trial the lowest weight gains were those of the broilers fed on the two krill diets (*Table XI*) and they also had the lowest food and protein intakes (*Table XII*), possibly due to low palatability. However, on determining TPE (*Table XIII*) for each diet it can be seen that after six and nine weeks' feeding the highest TPE figure is that of the broilers fed on the FDK diet although in all diets the TPE values had dropped after nine weeks. Although the SB diet produced the greatest weight gains the TPE value for this diet was the lowest. The FCE values (*Table XIII*) also show that the two krill diets compare favourably with the WF diet. It would appear therefore from this trial that krill could be added as a supplement to broiler diets

Table X
EGG, FOOD AND PROTEIN RELATIONSHIP FROM EGG-LAYING TRIAL

Diet	No. of birds	No. of eggs	Wt of eggs (g)	Wt of food eaten (g)	Crude protein eaten (g)	Egg wt per g protein	Protein g to produce 1 egg
WF 72	4	232	14 245	37 240	5 772	2·47	24·38
SB 91	4	185	10 688	37 430	5 802	1·84	31·36
FDK	4	158	9 005	24 702	3 829	2·35	24·23
CDK	4	248	15 012	37 392	5 796	2·59	23·37

As the colour of the egg yolks could have influenced taste panel members all tasting was carried out under red light illumination which negated the difference in the yolk colours. Most panel members were able to detect fishy flavours and slight fishy odours in the FDK eggs from the initial time of tasting, *ie* seven days after feeding commenced. The intensity of this flavour appeared to vary from day to day and although some members could not detect it they described the eggs as having a strong after-taste. This fishy flavour, although not discernible in the CDK eggs during the earlier stages of the trial, became more apparent during the last four weeks. Some members found the level of fishy flavour to be objectionable. No such flavours were detected in the WF and SB eggs during any time of the feeding trial. Half the panel members could detect slight fishy flavours in the flesh of the roasted and casseroled broilers fed on the FDK diet but only one member could detect them in the roast broiler fed on the CDK diet; in neither case were the fishy flavours at an objectionable level.

The results of the egg-laying trial are shown in *Table X*.

The four hens fed on the CDK diet produced most eggs closely followed by those fed on the WF diet. The hens fed on the FDK diet produced the lowest number of eggs where their average weight was 57·0g as compared to 61·4g, 60·5g and 57·8g for the WF, CDK and SB diets, respectively. It is, also evident that a far lower

Table XI
AVERAGE WEIGHT GAIN (G) OF 12 BROILERS DURING THE FEEDING TRIAL

	Number of days on diet			
Diet	13	42	56	64
CDK	312·2	1 477·7	2 194·3	2 561·0
FDK	277·0	1 270·0	1 728·2	2 045·0
WF	345·4	1 630·3	2 297·0	2 647·0
SB	366·2	1 829·1	2 254·1	2 670·8

Table XII
AVERAGE FOOD (G) AND CP INTAKE (G) OF 12 BROILERS DURING THE FEEDING TRIAL

	Food intake		CP intake	
Diet	42 days	64 days	42 days	64 days
CDK	4 883·7	8 933·7	870	1 572
FDK	3 822·3	6 914·0	688	1 245
WF	5 149·2	9 899·2	927	1 782
SB	6 057·9	10 546·2	1 090	1 902

producing results as good as that of a white fish supplemented diet.

The effect on flavour of chickens for diets containing fish meal, fish oil and stabilized fish meals was reviewed by Mostert *et al* (1969). Miller and Robisch (1969), Dreosti *et al* (1970), Atkinson *et al* (1972) and Opstvedt

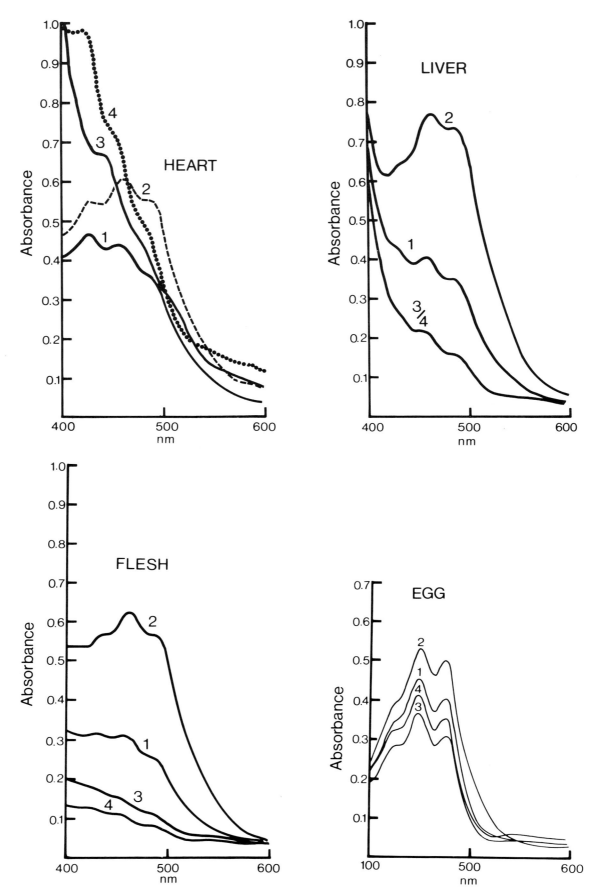

Fig 1 Absorption spectra of egg, flesh, heart and liver lipids (1%, 30%, 15% and 7·5% solutions in hexane, respectively) from hens and broilers fed on diets containing: (1) commercially dried krill meal; (2) freeze-dried krill meal; (3) white fish meal; (4) soya bean meal

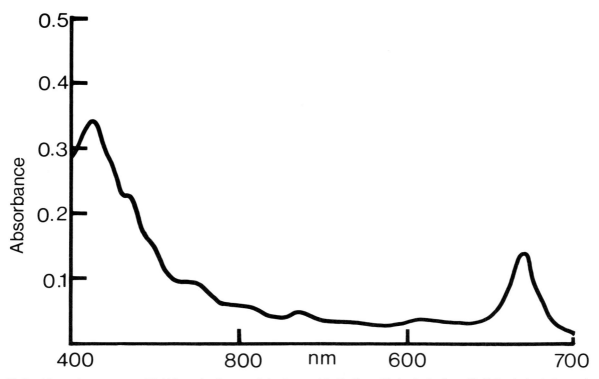

Fig 2   Absorption spectrum of lipid from the diet containing freeze-dried krill meal fed to laying hens (0·61% solution in hexane)

Table XIII
TOTAL PROTEIN EFFICIENCY (TPE) AND FOOD CONVERSION EFFICIENCY (FCE) OF 12 BROILERS DURING FEEDING TRIAL

Diet	TPE		FCE	
	42 days	64 days	42 days	64 days
CDK	1·70	1·63	3·30	3·41
FDK	1·85	1·64	3·01	3·38
WF	1·76	1·49	3·16	3·74
SB	1·68	1·40	3·31	3·96

TPE = weight gain ÷ intake
FCE = food intake ÷ weight gain

(1974) have illustrated a relationship between certain long chain fatty acids and taint in chicken flesh. Opstvedt (1974) set limits of these fatty acids in the carcass fat at which fishy flavours could be detected. In this trial fishy flavours were detected in the flesh of the broiler fed the FDK diet where the polyenoic acid level (20:5 + 22:5 + 22:6) was 3·4% and also in the broiler fed the CDK diet where only a trace of these acids was detectable. The concentration of the polyenoics in the eggs from hens fed both the FDK and CDK diets was also of the same level (2·7%) throughout the egg-laying trial; however the fishy flavour was considered far stronger in the eggs than in the broiler flesh. The manner of cooking could have an effect here. Loss of volatiles is presumably minimized by the protection of the egg shell during cooking whereas during casseroling and roasting these would be lost to the atmosphere. It is interesting to note also that the diet prepared under ideal conditions against oxidation, *ie* the FDK diet, imparted flavour to the eggs far earlier in the trial than did the CDK diet. This would be explained by the fact that fishy flavour components were retained during preparation of the freeze-dried krill meal but were lost during the heated preparation of the commercially dried krill meal.

It would seem therefore that a freeze-dried or commercially dried krill meal could be used as a replacement for a white fish meal in supplemented broiler diets but the FDK diet would not be suitable in an egg-laying diet because of flavour transfer to the egg yolk. Colour of the yolk from both krill diets could also influence acceptability.

5 References

ATKINSON, A, SWART, L G, VAN DER MERWE, R P and WESSELS, 1972   J P H. Flavour studies with different levels and times of fish meal feeding and some flavour imparting additives in broiler meats. *Agroanimalia*, 4, 53–62

BLIGH, E G and DYER, W J. A rapid method of total lipid extraction 1959   and purification. *Can. J. Biochem. Physiol.*, 37, 911–917

CHRISTIANS, O, FLECHTENMACHER, W, PAPAJEWSKI, H, ROSCHKE, N 1978   and SCHREIBER, W. Antarctic Expedition 1975/76 of the Federal Republic of Germany. *Arch. Fisch Wiss.*, 29, Beih 1, 80–91

DREOSTI, G M, VAN DER MERWE, R P, ATKINSON, A and SWART, L G. 1970   Excess tocopherol in poultry feed *vs* flavour of chickens. Fishing Ind. Res. Inst. Prog. Rep. 124

McGILL, A S, PARSONS, E and SMITH, A. Connection of glass capillary 1977   columns to a Pye series 104 chromatograph. *Chem. and Ind.*, 4 June, 456–457

MILLER, D and ROBISCH, P. Comparative effect of herring, menhaden 1969   and sunflower oils on broiler tissues fatty acid composition and flavour. *Poult. Sci.*, 48(6), 2146–2157

MOSTERT, G C, DREOSTI, G M, ATKINSON, A, SWART, L G and VAN 1969   ZYL, E. Flavour studies with high levels of unoxidized, oxidized and fat extracted fish meals in broiler tissues. *Agroanimalia*, 1, 123–128

NAVARRO, J G, SAAVEDRA, J C, BORIE, F B and CAIOZZI, M M. Influ-1972   ence of dietary fish meal on fatty acid composition. *J. Sci. Fd. Agric.*, 23, 1287–1292

OPSTVEDT, J. Influence of residual lipids on the nutritive value of fish 1974   meal. 6. Effects of fat addition to diets high in fish meal on fatty acid composition and flavour quality of broiler meat. *Acta Agric. Scand.*, 24, 61–75

—— Effect of dietary fat on the fatty acid composition of poultry 1976   products. *Meld. Norg. LandbrHoegsk.*, 55, 15, 21pp

SEINO, H, NAKASATO, S, SANGAI, T, MURUI, T and YOSHIDA, H. Prep-1972   aration of methyl esters from fats and oils containing short chain fatty acids using boron trifluoride-methanol reagent. *Yukagaku*, 26, 405–410

SOLOMON, H L, HUBBARD, W D, PROSSER, A R and SHEPPARD, A J. 1974 Sample size influence on boron trifluoride-methanol procedure for preparing fatty acid methyl esters. *JAOCS,* 51, 424–425

WOODHAM, A A. A chick growth test for the evaluation of protein 1968 quality in cereal based diets. 1. Development of the method. *Br. Poult. Sci.,* 9(1), 53–63

Nutritional experiments with krill

G Siebert, J Kühl and R Hannover

1 Introduction

The general usefulness of krill (*Euphausia superba*) for nutritive purposes is well established over many millenia by its role as one of the main components of whale food. With the whales now being almost extinct, the best possible future use of krill is sought in the field of human nutrition, more specifically as a protein source. While growth and maintenance of whales apparently did not pose any problems related to fresh krill, it is probably wise to check more carefully if products made from krill fulfil all requirements of food safety. The present paper is therefore concerned with possible effects of technological measures on krill protein as well as with some possible uses of krill shells which at present are a waste product.

2 Materials and methods

Krill shells were obtained in a pilot procedure, during the 1975–1976 West German Antarctic Expedition, by mechanical removal of shells which, after boiling, were freeze-dried. Krill meat was prepared according to Flechtenmacher *et al* (1976) and Sahrhage *et al* (1978) during the 1977–1978 expedition; after mechanical removal of shells by improved machinery, unheated or steam-heated crushed meat was spray-dried aboard the research vessel, with 200°C for incoming air, 90°C for outgoing air, and about 65°C for the meat to be dried. Animal rations were analysed by published procedures (Anon, 1976); chitin was determined as described by Roschke and Schreiber (1977). Amino acid analyses were made on acid hydrolysates, with a special programme to effectively remove glucosamine and ornithine; methionine and cystine were measured after oxidation. Tryptophan was determined with a modification of the procedure described by Concon (1975). Protein was determined according to Lowry *et al* (1951). Lactate dehydrogenase and cathepsin and trypsin inhibition were measured by published procedures (Anon, 1974a; Bergmeyer and Bernt, 1970).

Biological value, NPU and other parameters of *Table III* were measured according to Müller (1964) as described by (Kühl *et al*, 1978). The 90 days' feeding experiment was performed according to international recommendations (Anon, 1974b), with synthetic diets containing 20% protein plus 0·38% DL-methionine as described elsewhere (Grupp and Siebert, 1978). Chemical score (Block and Mitchell, 1946) and essential amino acid index (EAA) (Oser, 1951) were calculated as described.

3 Results and discussion

3.1 Analytical data

It is evident from the data in *Table I* that the samples of krill meat are of similar composition, in terms of standard food chemistry, to other sources of animal protein like fish, poultry, *etc*. The small amount of chitin in the meat samples indicates effective removal of the shells.

On the other hand, krill shells, with visibly adherent remnants of krill meat, do not analyse as more or less pure chitin but rather as a mixture of chitin and protein. If the Kjeldahl-N of krill shells (8% in *Table I*) is used to subtract protein-N (33% pure protein in *Table II*), 3·7 out of the 8% N are maximally left for chitin-N. At a monomer weight of 202 for N-acetylglucosamine and 6·9% N in chitin, at the most 53% chitin could be calculated for krill shells if all non-protein-N of the shells is assumed to come from chitin. The assumption of about 50% chitin in the shell samples which were investigated here appears to be reasonable.

Table I
ANALYTICAL COMPOSITION OF KRILL FRACTIONS (PERCENTAGE)

	Raw meat	Heated meat	Shells
Dry matter	96·2	96·1	98·3
Ash	11·9	13·3	26·2
Total lipid	2·7	6·0	6·6
Total nitrogen	12·5	12·3	8·0
crude protein = $N \times 6·25$	78·4	77·0	50·0
pure protein from amino acid analysis (Table II)	67·9	61·3	33·0
Chitin	0·77	2·1	not determined

Table II
AMINO ACID COMPOSITION OF KRILL MEAT AND SHELLS (AS PERCENTAGE OF PURE PROTEIN)

	Raw meat	Heated meat	Shells
Asx	10·2	10·9	11·0
Thr	4·23	4·58	4·67
Ser	4·10	4·37	4·42
Glx	13·7	13·9	14·3
Pro	5·17	4·99	4·91
Gly	7·14	7·47	5·00
Ala	5·92	6·26	5·58
Val	5·09	5·25	5·52
Cys	1·66	1·78	1·55
Met	2·54	2·72	2·52
Ile	4·82	4·97	5·27
Leu	7·35	7·48	8·06
Tyr	3·88	4·03	4·64
Phe	4·25	4·68	5·15
Lys	7·33	7·87	7·70
Trp	0·27[1]	0·18[1]	0·85
His	2·15	2·46	2·91
Arg	8·17	8·56	7·27
NH₃	2·07	1·96	not determined

[1] These data, although obtained consistently, are very probably too low; Dr W Schreiber (personal communication) reports 0·9 and 1·0%, respectively

3.2 *Amino acids in krill fractions*

Whereas krill meat samples were collected in the 1977–1978 trip, krill shells stem from the 1975–1976 journey. It is rather surprising that there is an almost complete agreement in amino acid patterns, once correction is made for the differences in content of pure protein (*Table II*). This agreement renders statements on the amino acid composition rather firm. Deviations of possible significance concern the glycine content (5% in shells, above 7% in meat samples), and perhaps also histidine and arginine. However, in the light of all the information available, the tryptophan data of *Table II* are very probably too low and should be assumed to be around 0·9–1·0%.

3.3 Nutritional evaluation of krill fractions

The data of *Table III* indicate the high biological value of krill meat. It seems likely that a low digestibility would be shown by krill shells whose biological value is as high as that of commercial fish meal.

Chemical scores of 43 and 59, and EAA's of 69 and 73 have been calculated for krill shells and fish meal, respectively. As is seen from the data in *Table III*, the biological value and net protein utilization are above the calculated values; in consequence, the nutritional quality of krill shells is not limited by any factor impeding digestibility. Less obvious are the reasons why krill meat samples display a diminished biological value in comparison with krill shells (*Table III*) Again, impaired digestibility does not seem to be the reason, and it has been impossible to find antitryptic activity in these meat samples. Nearly identical NPU values for krill meat and fish meal must be viewed in the light of the fact that the thermal pretreatment of heated krill did not suffice to inactivate enzymes completely (*Table IV*) whereas spray-drying as such, performed as described under section 2, apparently did not do much harm to the enzymes in the raw meat. Catheptic activity of heated krill meat amounts to about 10% of the value of raw krill meat which, on the other hand, has a lactate dehydrogenase activity sufficient to produce lactate at a rate of one-sixth the meat weight per hour. It remains to be seen if such enzymatic data could be indicative of the shelf-life of spray-dried krill meat.

3.4 Long-term feeding of krill meat samples

In the light of the existing experience (Gilberg, 1971; Christians *et al*, 1977), and since protein quality (section 3.3) apparently was not impaired by the technological steps applied to the krill meat, a first survey on nutritional safety in a 90 days' feeding experiment is recorded in *Table V*. When compared with casein plus methionine as control there is no difference whatsoever in growth and food efficiency with the samples of krill meat. However, two points deserve some closer scrutiny: liver weights of female rats−not of male animals−were significantly higher than in the control whereas kidney weights were normal (*Table V*). Histological examination does not reveal any data in the livers of rats of both sexes which could be connected with the increased liver weights. In addition, the questionable tryptophan data of *Table II* apparently did not limit growth performance of the rats and do not lend themselves as explanation of the increased liver weight.

Table V
90 DAYS' GROWTH EXPERIMENT WITH KRILL MEAT ON LABORATORY RATS

	Raw meat		Heated meat		Casein plus methionine (control)	
	♀	♂	♀	♂	♀	♂
Weight increase (g)	158	223	156	228	160	228
Food consumption (kg)	1·07	1·07	1·07	1·07	1·02	1·02
Weight increase per unit food (g/100g)	15·2	21·4	15·1	21·9	14·9	22·0
Protein efficiency ratio (PER) (g/g protein)	0·76	1·07	0·75	1·09	0·74	1·10
Fresh weight, liver (g)	7·89[1]	10·4	7·76[1]	10·4	6·8	9·1
same, g per 100g body weight	3·91	3·87	3·93	3·86	3·45	3·44
Fresh weight, kidneys (g)	1·70	2·24	1·65	2·18	1·59	2·13
Fresh weight, total caecum (g/100g body weight)[2]	1·17	1·10	1·35	1·13	1·36	1·19

[1] $p < 0.01$; no other difference is statistically significant
[2] Relative caecum weight (g/100g) after feeding of krill shells was 2.2 ± 0.24g and with $p < 0.001$ highly significantly elevated versus controls (both sexes combined)

Table III
BIOLOGICAL VALUE OF PROTEINS IN KRILL FRACTIONS

Parameter	Casein plus methionine (standard)	Fish meal (reference)	Raw krill meat	Heated krill meat	Krill shells
Biological value	98·6 ± 1·9	87·8 ± 2·8	81·3 ± 1·9	82·7 ± 2·1	88·6 ± 2·1
Apparent digestibility	89·7 ± 1·5	84·0 ± 1·8	91·6 ± 0·5	91·7 ± 0·5	75·3 ± 1·0
True digestibility	100·3 ± 1·5	93·8 ± 2·0	101·3 ± 0·6	101·4 ± 0·8	85·4 ± 1·2
Net protein utilization (NPU)	98·9 ± 2·1	82·4 ± 3·8	82·4 ± 1·8	83·9 ± 2·6	75·6 ± 2·5
Animal weight (g)	66·6 ± 5·2	53·1 ± 6·3	52·9 ± 4·8	53·8 ± 5·5	51·1 ± 3·8
N uptake (mg)	182 ± 8·6	121·3 ± 21·9	117·8 ± 10·1	121·4 ± 17·2	102 ± 8·0
Urinary N (mg)	26·4 ± 2·6	28·1 ± 2·9	36·8 ± 2·9	35·8 ± 3·6	25·4 ± 0·83
Faecal N (mg)	17·4 ± 2·3	19·3 ± 3·9	9·8 ± 0·9	10·0 ± 1·4	25·2 ± 1·1
Endogenous urinary N (mg)	19·1 ± 1·7	14·6 ± 2·1	14·5 ± 1·6	14·8 ± 1·9	14·0 ± 1·2
Intestinal loss N (mg)	15·3 ± 0·8	11·9 ± 2·6	11·5 ± 1·3	11·8 ± 2·2	10·3 ± 1·0

Table IV
ENZYMATIC PROPERTIES OF KRILL MEAT SAMPLES

	Raw meat	Heated meat
Soluble protein (% dry weight)	23	4·8
Lactate dehydrogenase (μmol/min × g dry weight)	30	5·6
Cathepsin (mg tyrosine equivalents/ hour × g dry weight)	162	16·5
Trypsin inhibitor	not detectable in either sample	

Furthermore, it is seen from *Table V* that feeding krill shells causes a highly significant increase of caecal weights, indicative of microbial participation in the utilization of some food constituents. Since krill does not cause increased caecal weight−as expected−some constituents of krill shells must have caused microbial augmentation. The probable very best candidate is chitin since many other substances including hexitols, arab

333

gum and some kinds of starch behave in the same manner. Since on feeding of krill shells only 70% of nonprotein nitrogen is recovered from the faeces (Kühl *et al*, 1978), and, since there is some difference between biological value and NPU (*Table III*) for krill shells, one has to conclude that part of the chitin-*N*—through microbial activity in the caecum—is actually made available for the nitrogen metabolism of the rat. Microscopical inspection of the rat faeces after feeding of krill shells suggests indeed a partial degradation of chitin containing structures from krill shells. Such a conclusion is justified in the light of existing knowledge on degradation and digestion of chitin (Araki *et al*, 1971; Anon, 1977; Muzzarelli, 1977). In conclusion, krill meat is a highly valuable and interesting material whose nutritional safety may require some additional studies.

4 Acknowledgements

These studies were aided by financial help from Ernährungswissenschaftlicher Beirat der Deutschen Fischwirtschaft, Bremerhaven, and Bundesministerium für Forschung und Technologie, Bonn. The authors are grateful to Dr Schreiber and his colleagues, Bundesforschungsanstalt für Fischerei, Hamburg, for making available krill samples, and for discussions and advice. Also Dr W Schneider and Dipl. agr. biol. H Cornelius, Dept. of Animal Nutrition, Dr G Fischer, Dept. of Histology and Embryology, and Dr J Nittinger from the Dept. of Biological Chemistry, University of Hohenheim and others who contributed experimentally. In the animal experiments, the technical help of Miss H Gerlach, A Brinks and M Mack is gratefully acknowledged.

5 References

ANON. Abstracts First Int. Conf. Chitin/Chitosan, Boston
1977
ANON. Biochemica Merck, Enzyme, 52–53
1974a
ANON. *Handb. Landw. Versuchs- und Untersuchungsmethodik.*
1976 Neumann-Neudamm Publishers, Melsungen. Vol. III
ANON. Wld. Hlth. Org. Techn. Rep. Ser., 539. FAO Nutr. Meet. Rep.
1974b Ser., 53, Geneva. 8pp
ARAKI, Y, FUHUOKA, S, OBA, S and ITO, E. *Biochem. Biophys. Res.*
1971 *Commun.*, 45, 751–758
BERGMEYER, H U and BERNT, E. *Meth. enzymat. Anal.*, Ed H U Bergmeyer. Verlag Chemie, Weinheim. 2nd edn, 534–538
1970
BLOCK, R J and MITCHELL, H H. The correlation of the amino acid
1946 composition of proteins with their nutritive value. *Nutr. Abstr. Rev.*, 16, 249–278
CHRISTIANS, O, FLECHTENMACHER, W and SCHREIBER, W. Neue
1977 Nahrungsmittel aus der Antarktis. *Ernähr. Umschau*, 24, 141–143
CONCON, J M. Rapid and simple method for the determination of
1975 tryptophan in cereal grains. *Anal. Biochem.*, 67, 206–219
FLECHTENMACHER, W, SCHREIBER, W, CHRISTIANS, O and ROSCHKE, N.
1976 Die Verarbeitung von Krill. *Inf. Fischw.*, 23(6), 188–196
GILBERG, Y C. Krill and its possible place in human nutrition. *Food*
1971 *Technol. in New Zealand*, 13–17
GRUPP, U and SIEBERT, G. Metabolism of hydrogenated palatinose, an
1978 equimolar mixture of α-D-glucopyranosido-1,6-sorbitol and α-D-glucopyranosido-1,6-mannitol. *Res. Exp. Med.* (Berl.), 173, 261–278
KÜHL, J, NITTINGER, J and SIEBERT, G. Verwertung von Krillschalen in
1978 Fütterungsversuchen an der Ratte. *Arch. Fisch Wiss.*, 29, 99–103
LOWRY, O H, ROSEBROUGH, N J, FARR, A'L and RANDALL, R J. Protein
1951 measurement with the folin phenol reagent. *J. Biol. Chem.*, 193, 265–275
MÜLLER, R. Vorschrift zur Proteinbewertung in Versuchen an wach-
1964 senden Ratten. *Z. Tierphysiol. Tierernähr. Futtermittelkde.*, 19, 305–308
MUZZARELLI, R A A. *Chitin.* Pergamon Press, Oxford
1977
OSER, B L. Method for integrating essential amino acid content in the
1951 nutrition evaluation of protein. *J. Amer. Diet. Assoc.*, 27, 396–402
ROSCHKE, N and SCHREIBER, W. *Arch. Fisch Wiss.*, 28, 135–145
1977
SAHRHAGE, D, SCHREIBER, W, STEINBERG, R and HEMPEL, G.
1978 Antarktis-Expedition 1975–1976 der Bundesrepublik Deutschland. *Arch. Fisch Wiss.*, 29 (Beiheft 1), 1–96

8 By-products

Utilization of 'big-eye' fish (*Brachydeuterus auritus*) for fish meal and fish protein concentrate production. A preliminary biochemical and nutritional evaluation

S O Talabi, B L Fetuga and A Ologhobo

1 Introduction

The provision of proteins of adequate quality and quantity is a major concern of most developing countries, since malnutrition is regarded as the most serious health problem in these areas.

In the search for suitable sources of protein, the sea has become particularly important. Some of the living marine resources are, however, difficult to utilize for varying technoeconomic reasons. In the Eastern Central Atlantic Fisheries region (ECAF) the big-eye, *Brachydeuterus auritus* is one of such resources. A recent estimate (Teutscher, 1979) has put the stock at 50 000 tonnes in the subtropical ECAF. Further south, the big-eye is a shrimp by-catch as well as an important component of the complex species mix that often characterizes the fisheries of the Gulf of Guinea.

The big-eye is small-sized, scaly and bony. Although estimates show that it can support a fishery of its own, no attempt has been made to develop it, mainly because of the lack of economically viable utilization outlets. Potentially these include fish meal and FPC Type B, and mince production. In Nigeria, in particular, the demand for some of the foregoing products is very high.

In the case of fish meal, because all of it is imported, the price has always been high and this has limited the expansion of the livestock industry in Nigeria. The Norwegian FPC has recorded some measure of acceptability in parts of Nigeria while canned fish is a well established item of food.

There is, however, little or no published work on the chemistry and utilization possibilities of the big-eye and other fishery resources of the ECAF. This study was therefore initiated to obtain some preliminary information on its chemistry and to examine by means of nutritional and technological studies the potential of this resource for the production of fish meal and FPC Type B.

2 Materials and methods

2.1 *Source of raw materials and processing for fish meal and fish protein concentrate*

Big-eye (*Brachydeuterus auritus*) were caught using Granton trawl net by the RV *Kiara* in the Lagos fishing grounds. They were immediately iced before *rigor mortis* set in and transported to the technology laboratories of the Nigerian Institute for Oceanography and Marine Research for analysis. Within 24h of catch, analysis of the fresh ungutted samples had commenced. This was done at least twice a month over the period from January to October 1976. Later, approximately 300kg of fresh ungutted specimens were fed into a stainless steel fish meal plant (model T100 supplied by Chemical Research Organisation, Esbjerg, Denmark). Production of fish meal was completed within 8h. The product was allowed to cool down and then milled in a hammer mill. For the production of FPC, 300kg of fresh ungutted fish were thoroughly washed in freshwater at 30°C and then fed into a Baader 694 meat bone separator. Separated meat was then fed directly into the fish meal plant. The finished product was then milled in a hammer mill for subsequent analysis.

For the raw material analysis, the catches were first sorted out into two size classes A and B, namely: those above 16cm body length and those below 16cm body length, respectively. Approximately 10kg of the whole ungutted fish in each size group was thoroughly minced in a Kenwood mincer with the bowl immersed in melting ice. The minced materials were then analysed for the following components: lipids by the procedure of Bligh and Dyer (1959); moisture; fatty acid value and iodine value (Wij's method) was by standard AOAC (1970) procedures. Total nitrogen contents were determined by means of the Kjel-Floss automatic nitrogen analyser model 16200. Samples were analysed monthly over a 12-month period.

The fish meal and fish protein concentrate samples described earlier were analysed for their proximate constituents by standard AOAC (1970) procedures. Lipid analyses were by Bligh and Dyer (1959).

Analysis of the lipid fatty acids was carried out by transesterification using BF^3/methanol followed by gas chromatography using a packed column (Supelco SP2330) operated at 208°C.

Total amino acid analysis was carried out on the samples after acid hydrolysis using a Technicon TSM amino acid analyser. Tryptophan was chemically determined by the method of Miller (1967). Available lysine content of samples was determined as described by Booth (1971); available methionine was determined by pre-digestion of samples using a combination of pepsin pancreato-peptidase E as described by Pieneazek *et al* (1975); methionine in the hydrolysate was then measured according to the method of McCarthy and Sullivan (1941).

2.2 *Nutritive quality*

Three experiments were conducted with weanling or growing albino rats to evaluate the nutritive quality of the pilot plant prepared 'big-eye' fish meal, as compared to white fish meal and herring fish meal. In a fourth experiment the protein quality of big-eye fish protein concentrate was determined in comparison to nutritional casein (British Drug Houses, Poole, England) and laboratory prepared defatted whole hen's egg (Fetuga *et al*, 1973). In experiments I, II and IV, protein quality indices were measured as protein efficiency ratio (PER), net protein utilization (NPU) and true digestibility (TD) using techniques described by the National Academy of Sciences–National Research Council (NAS/NRC, 1963). The basal diet used contained starch, 65%; glucose, 5·0%; sucrose, 10·0%; non-nutritive cellulose, 5·0%; groundnut oil, 10·0%; mineral supplement (Miller, 1963), 4·0%; and vitamin mixture, (Miller, 1963), 1·0%. In experiments I and IV the protein sources were included in the basal diet at the expense of maize starch such that they provided 10% crude protein in the final diet, while in experiment II, big-eye fish meal was included to provide 10%, 15% and 20% protein, respectively. In experiment III the replacement value of big-eye fish meal for white fish meal, in an otherwise adequate diet, was measured. The objective was to assess quality in association with the normal components of diets.

All experimental data were subjected to analysis of variance and, where significance was indicated, means were compared using Duncan's multiple range test.

3 Results and discussion

Results of the seasonal analysis of the fish are presented in *Table I*. Lipid content of the bigger fish is always higher than that of the smaller. They both show the same seasonal variations, increasing in a relatively smooth pattern from January to June and then decreasing. The iodine value, a measure of lipid unsaturation, did not vary markedly although there was a slight trend for the values to peak at the lower lipid contents. This effect has been reported in some other species of fish (Lovern, 1938a).

Table I
SEASONAL VARIATION OF LIPIDS, LIPID CHARACTERISTICS, WATER AND PROTEIN CONTENT OF *Brachydeuterus auritus*

Month	% Lipid content		% Water content		Wij's iodine value	
	Group A	Group B	Group A	Group B	Group A	Group B
January	1·3	1·0	74·0	76·3	153·5	140·5
February	1·5	1·3	74·5	74·5	140·0	142·0
March	1·8	1·5	72·6	74·5	140·5	145·0
April	2·2	1·5	74·6	77·2	141·3	143·5
May	2·5	1·7	74·2	76·5	142·3	143·5
June	3·8	2·2	74·0	78·0	135·2	139·0
August	3·5	1·8	75·8	73·9	131·0	134·0
October	2·8	1·7	70·8	72·6	146·4	155·0

Month	% Free fatty acids (on total fat basis)		% Total protein (N² × 6·25)		Non-protein N²	
	Group A	Group B	Group A	Group B	Group A	Group B
January	0·4	0·6	—	—	—	—
February	0·4	0·6	—	—	—	—
March	0·5	0·5	—	—	—	—
April	0·6	0·7	—	—	—	—
May	0·7	0·9	18·3	16·2	0·2	0·1
June	0·8	1·0	18·0	16·8	0·1	0·1
August	1·3	1·7	19·6	18·5	0·2	0·1
October	1·5	1·5	19·5	—	0·1	0·1

Free fatty acid level was low throughout the study period and it appears to be independent of the amount of lipid present.

Variations in the water content occurred, but these did not appear to be associated with changes in the lipid contents as it is sometimes observed with species containing high levels of lipids (Love, 1970).

Some variations in the total protein content were observed but these may be a function of the method used rather than actual changes. In the smaller-sized fish, protein content tended to be lower when the water content was high, but this was not the case in the larger fish.

Fatty acid analysis of the whole fish mince and the FPC prepared from the mince is presented in *Table II*. Some differences were observed, particularly in the C14, C16 and C18 fatty acids, and it is presumed that these are caused by processing variations. The analysis of the FPC produced indicates that marked losses in the polyunsaturated fatty acids did not occur during the drying process.

Table II
PERCENTAGE FATTY ACID COMPOSITION OF *Brachydeuterus auritus* (WHOLE FISH, BAADER SEPARATED MEAT AND FPC PREPARED FROM SEPARATED MEAT)

Fatty acids	Fresh whole Total lipids	Separated meat Total lipids	FPC Total lipids
C14:0	3·7	18·0	2·9
C16:0	31·5	23·9	31·9
C16:1	3·2	5·2	4·8
C18:0	10·1	6·8	9·2
C18:1	10·7	8·4	12·2
C18:2	1·0	0·7	1·0
C20:1	1·3	1·1	1·3
C18:4	0·4	0·2	0·2
C22:4	1·6	1·0	1·4
C22:1	3·9	3·6	4·0
C20:4	0·7	0·3	0·8
C20:5	4·2	4·3	3·3
C22:5	3·1	3·1	2·6
C22:6	21·2	20·1	21·5
Minor components	3·1	3·1	2·6
% Saturates	45·3	48·7	44·0
% Monounsaturates	19·1	18·3	22·3
% Polyunsaturates	31·2	29·0	29·8

Total proteins of the FPC and fish meal were relatively high (cf Table III). This is as expected, being more marked in the FPC produced from the mince.

Table III
PROXIMATE COMPOSITION OF THREE FISH MEALS AND BIG-EYE FPC (g/100g DRY MATTER)

	Dry matter	Crude protein	Crude fibre	Total lipid	Total ash	Nitrogen-free extract
Big-eye fish meal	92·6	77·6	1·04	5·3	14·7	1·4
White fish meal	94·5	68·5	0·10	8·7	14·8	7·5
Herring fish meal	96·3	71·4	0·42	7·6	10·4	8·1
Big-eye FPC	98·4	88·8	0·36	3·2	6·5	1·1

Table IV shows the close similarities in the amino acid profiles of the 'big-eye' fish meal, the white fish meal and the herring fish meal, except in the slightly higher methionine value obtained for herring fish meal and its lower tryptophan level. The data in respect of big-eye FPC compare favourably with freeze-dried hen's egg except for the obviously lower levels of methionine, cystine and tryptophan and the higher lysine, arginine and glycine. Whole egg protein has a high nutritional quality, so much so that it is used as a yardstick in measurements of protein quality; some amino acids are present in whole egg protein in relatively greater concentration than appears necessary to satisfy the patterns of human requirement. The spectrum of amino acids in big-eye fish meal and FPC is indicative of a high nutritive value. Both the lysine and methionine in all the samples studied were highly available in values ranging between 90% to 98% (*Table V*). This again indicates that processing or storage did not have any adverse effects on these amino acids.

Table IV
COMPARATIVE AMINO ACID COMPOSITION OF BIG-EYE FISH MEAL, BIG-EYE FISH PROTEIN CONCENTRATE, WHITE FISH MEAL, HERRING FISH MEAL AND WHOLE HEN'S EGG (g/16g N)

Amino acid	Big-eye fish meal	Big-eye fish protein concentrate	White fish meal	Herring fish meal	Whole hen's egg
Lysine	8·72	8·94	7·98	8·90	6·98
Histidine	3·79	3·82	3·34	3·54	2·43
Arginine	10·38	10·04	5·87	6·40	6·10
Aspartic acid	11·20	10·23	10·84	11·36	9·02
Threonine	4·49	4·80	5·00	4·50	5·12
Serine	5·14	5·46	4·78	4·77	7·65
Glutamic acid	13·96	15·55	14·94	12·84	12·74
Proline	6·61	5·81	8·20	7·46	4·16
Glycine	6·04	6·29	6·78	6·96	3·31
Alanine	6·67	5·89	5·95	6·74	5·92
Cystine	1·09	1·16	1·01	1·24	2·43
Valine	5·99	6·12	4·73	5·40	6·85
Methionine	2·49	2·89	2·81	3·00	3·36
Isoleucine	6·28	5·37	4·78	4·80	6·29
Leucine	7·52	7·78	7·94	8·20	8·82
Tyrosine	3·84	3·86	3·14	3·08	4·16
Phenylalanine	5·55	6·10	5·18	4·98	5·63
Tryptophan	0·94	0·96	1·12	0·88	1·62

Table VI
YIELD DATA ON PILOT PLANT PRODUCTION OF MINCE AND FPC FROM *Brachydeuterus auritus*

Material	Weight (kg)	% Yield
Raw fish	300	—
Offal after meat bone separation	179·4	59·8
Separated meat	120·3	40·1
FPC from separated meat	30·1	10·03
Fish meal from whole fish	66	22
Fish meal from the offal (after meat bone separation)	34·8	11·6

NB: Fish meal employed in the study was from whole fish but yield from offal has been included in the table for comparison

Table VII
COMPARATIVE NUTRITIVE VALUE OF BIG-EYE FISH MEAL, WHITE FISH MEAL AND HERRING FISH MEAL FED TO WEANLING ALBINO RATS

	Big-eye fish meal	White fish meal	Herring fish meal	SE of mean
Weight gain (g)	18·50a	32·16	28·29b	1·36†
Protein intake (g)	7·06	8·66	8·29	2·14
PER	2·62a	3·71b	3·41b	0·32†
NPU	65·43a	76·05b	73·65b	0·84†
TD%	90·44	88·96	87·94	0·52 (NS)

† Highly significant differences (P < 0·01)
(NS) = No significant differences
a,b—Mean followed by the same letters are not significantly different

Weight gain, PER and NPU were significantly lower for big-eye fish meal than for the other two fish meals. The digestibility of big-eye fish meal was not affected by protein concentration in the dietary range 10% to 20% crude protein. Weight gain and protein intake increased with increasing dietary concentration of big-eye fish meal but NPU and PER decreased slightly (see *Table VIII*). The fact that higher levels of incorporation of the big-eye fish meal resulted in increased growth response would tend to suggest the absence of any toxic factors. It

Table V
AVAILABLE LYSINE AND METHIONINE IN BIG-EYE FISH MEAL, COMPARED TO WHITE FISH MEAL, HERRING FISH MEAL AND EGG

Source	Total lysine (g/16g N²)	Available lysine (g/16g N²)	%Availability	Total methionine (g/16g N²)	Available methionine (g/16g N²)	% Availability
Big-eye fish meal	8·72	7·84	89·94	2·49	2·38	95·64
Big-eye fish protein concentrate	8·94	8·59	96·10	2·89	2·84	98·21
White fish meal	7·98	7·57	94·87	2·81	2·52	89·64
Herring fish meal	8·90	8·21	92·26	3·00	2·76	92·04
Freeze-dried hen's egg	6·98	6·86	98·21	3·36	3·30	98·14

In conventional FPC Type B production, whole fish is normally reduced in food grade fish meal plants under hygienic conditions. Published compositional data often show high values for ash which provides minerals far in excess of normal metabolic requirements. In this work the process of meat bone separation before reduction facilitated the lowering of the mineral content from 14·7% to 6·5%, thus enabling a much higher level of total protein to be obtained. While the nutritional significance of the modification remains to be shown conclusively, it can be seen that a better quality product is possible. The final yields of fish meal and FPC from the raw material are presented in *Table VI*.

3.1 Animal feeding tests
The nutritive values of big-eye fish meal and the other fish meals given at a concentration to provide 10% crude protein in the diet of albino rats are shown in *Table VII*.

is strange that in tests with rats, the nutritive quality of the big-eye fish meal was inferior to that of herring meal and white fish meal, expecially as the amino acid profiles of the samples were similar and the availability of the two critical amino acids, lysine and methionine, were the same in the big-eye meal as in the commercial fish meals.

When fed to growing rats as dietary components in which other protein sources were incorporated and in which the big-eye meal replaced part of white fish meal (*Table IX*), then gains and feed gain ratio were similar and not significantly different from controls, suggesting that both these fish meal types would have comparable values in pratical type diets. Laksesvala and Aga (1965) obtained similar results in their studies on herring meals where an inferior quality was obtained in some samples when fed as the sole source of dietary protein but not when fed as supplements to other protein sources.

Table VIII
THE NUTRITIVE VALUE OF BIG-EYE FISH MEAL FED AT VARYING LEVELS OF
PROTEIN

	Protein level (%)			
	10	15	20	SE of mean
Weight gain (g)	17·32a	28·76b	39·42c	2·34‡
Protein intake (g)	8·45a	14·17b	20·11c	0·89‡
PER	2·05	2·03	1·96	0·42 (NS)
NPU	65·82a	50·75b	49·08b	1·34†
TD (%)	93·30a	86·34ab	85·19b	0·82*

In this and subsequent tables: * significant differences (P < 0·05);
† highly significant differences (P < 0·01); ‡ very highly significant
differences (P < 0·001)
(NS) = No significant differences
a,b,c—Mean followed by the same letters are not significantly different

Table IX
THE REPLACEMENT VALUE OF BIG-EYE FISH MEAL FOR WHITE FISH MEAL IN
DIETS FOR RATS

	Treatments					
	I	II	III	IV	V	SE of means
Av. daily gain (g)	3·97	3·16	3·15	3·09	3·58	0·43 (NS)
Av. daily feed intake (g)	12·91	10·81	11·56	11·56	13·07	0·52 (NS)
Feed: gain ratio	3·25	3·42	3·67	3·74	3·65	0·45 (NS)
Total protein efficiency ratio (TPER)	1·47	1·44	1·36	1·34	1·43	0·15 (NS)
Apparent digestibility (%)	87·28	88·13	88·69	90·20	90·20	1·25 (NS)

The basal diet used in this study was composed as follows (in percentages): maize meal, 70·0; white fishmeal, 10·0; blood meal, 2·0;
groundnut meal, 10·0; rice bran, 2·0; groundnut oil, 2·0; dicalcium
phosphate, 2·5; oyster shell, 0·5; vitamin—trace mineral mixture, 0·5
and sodium chloride, 0·5. Treatment I comprised the basal diet, while
in treatments II, III, IV and V big-eye fish meal replaced 25%, 50%,
75% and 100% of the white fish meal in the basal diet. The diets were
formulated to contain 20% crude protein on a dry matter basis

Protein quality of big-eye fish protein concentrate in
comparison to casein and whole hen's egg is given in
Table X. From the results presented, it is evident that the
FPC is of high nutritive quality, being comparable to
casein but decidedly inferior to hen's egg. The potential-
ity of this modified FPC process would have to be further
investigated. However, it is obvious from this study that
dried products of very high protein content and quality
can be produced from big-eye without the use of additives. This implies that its future role as an animal protein source for humans should be investigated. In this
regard, the variable response of populations to FPC

Type B should not discourage future research
endeavours (Tagle, 1976).

Table X
COMPARISON OF THE PROTEIN QUALITY OF BIG-EYE FISH PROTEIN WITH
CASEIN AND WHOLE HEN'S EGG

Diets	Average weight gained by rats	Average protein intake (g)	PER
Casein (10%)	28·6	9·90a	2·89ab
FPC big-eye (10%)	14·8	5·67a	2·61b
FPC big-eye (20%)	36·4	17·84b	2·04b
Defatted whole hen's egg (10%)	32·7	8·30a	3·94a
SE of means	2·42*	2·66†	0·29†

The values in parentheses represent the protein levels at which the
sources were tested

4 Acknowledgement

This work was supported partly by a research grant from
the Nigerian Institute for Oceanography and Marine
Research, Lagos, under the 3rd National Plan pro-
gramme and by the regular research grant of the
Department of Animal Science, University of Ibadan,
Nigeria, for which the authors are very grateful.

5 References

ASSOCIATION OF OFFICIAL ANALYTICAL CHEMISTS. *Official methods of*
1970 *analysis.* Washington DC. 11th edn
BLIGH, E G and DYER, W J. A rapid method of total lipid extraction
1959 and purification. *Can. J. Biochem. and Physiology*, 37,
        911–917
BOOTH, V H. Problems in the determination of FDNB-available
1971 lysine. *J. Sci. Fd. Agric.*, 22, 658
FETUGA B L, BABATUNDE, G M and OYENUGA, V A. *J. Sci. Fd. Agric.*,
1973 24, 1515–1523
LAKSESVALA, B and AGA, A T. Nutritive value of new and up to 12
1965 years old herring meals. *J. Sci. Fd. Agric.*, 16, 743–749
LOVE, R M. In: *Chemical Biology of Fish*. Academic Press, New York
1979
LOVERN, J A. Fat metabolism in fishes. XII. Seasonal changes in the
1938a composition of herring fat. *Biochem. J.*, 32, 676–680
MILLER, D S. In: *Evaluation of protein quality*. NRC, Washington DC.
1963 No. 1100, 34
MILLER, E L. Determination of tryptophan content of feedstuffs with
1967 particular reference to cereals. *J. Sci. Fd. Agric.*, 18, 381–385
MCCARTHY, T E and SULLIVAN, M. A new and highly specific col-
1941 orimetric test for methionine. *J. Biol. Chem.*, 141, 871–876
NATIONAL ACADEMY OF SCIENCES/NATIONAL RESEARCH COUNCIL.
1963 *Evaluation of protein quality*. NRC, Washington DC. No. 1100
PIENEAZEK, D, RAKOWSKA, M, SZKILLADZIOWA, W and GRABAREK, Z.
1975 Estimation of available methionine and cystine in proteins of
        food products by *in vivo* and *in vitro* methods. *Brit. J. Nutr.*, 34,
        175–177
TAGLE, M A. Acceptability testing of FPC Type B. *FAO Report*
1976 FI:TF/INT 120(NOR), Rome
TEUTSCHER, F. A review of potential catches and utilisation of small
1979 pelagic fish. FAO TF/INT 298 (DEN), Rome

Preliminary studies of two techniques for the removal of oil from fish silage using commercial equipment

D Potter, I Tatterson
and J Wignall

1 Introduction

The production of fish silage from fish or fish offal by the
addition of acid is a well-established process. When low
oil content raw material such as white fish offal is used
there are no problems in utilizing the finished product to
replace fish meal in the diet of pigs. There is no risk of

taint in the carcase provided the fish oil content of the
feed is not greater than 1% on a dry matter basis (dm)
(Adamson and Smith, 1976). At a 10% dm level this
corresponds to a level of about 2% oil in the silage. The
oil content of pelagic fish ranges from around 5% to 30%
depending on species and season and it is obvious that

338

many of the silages made from such material will be of little value in many instances for feeding unless the oil content is reduced. Good quality fish oil has a number of uses and it is perhaps paradoxical that separately, both oil and de-oiled silage are of high commercial value, whilst together they constitute a product of only limited use and value.

The amount of low oil content material such as white fish offal available in the UK has decreased dramatically over recent years and almost all of it is used for the manufacture of fish meal. Furthermore, unless a species such as blue whiting is to be exploited for industrial purposes, quantities of low oil content material are likely to become even smaller in the future. So far as fish meal production is concerned a short fall in supplies of raw material has so far been countered by landings of industrial species such as sand eels and sprats, together with whatever mackerel are not needed for human consumption. These same materials are also available for the production of fish silage but whereas the separation of oil during fish meal manufacture is an established technique its removal from fish silage is more difficult and there is no published work on this subject.

Because of this and also because of the potential increase in the quantity of industrial fish being landed in the UK and its availability for production of fish silage, there is a need in the medium-term to examine the parameters controlling oil removal. However, there was a need to find an immediate solution to the problem of de-oiling silage. Consequently a series of experiments which are described in this paper was started. The approach taken was to examine the performance of commercially available de-oiling equipment with various types of silages. Because commercial equipment was used, each experiment required relatively large quantities of material and up to 1 tonne per batch was normally prepared. This necessarily limited the variables examined and when experimental difficulties occurred it was not always possible to repeat the work. Nevertheless, the trials did give an indication of what is possible although not what may ultimately be achieved.

2 Equipment

Two basic de-oiling techniques were examined:

(1) a two-step system comprising a two-phase decanter and a stacked disc centrifuge manufactured by Westfalia Separator Ltd; and
(2) a single-stage, three-phase decanter manufactured by the Alfa-Laval Company.

In the first system the decanter was needed to reduce the quantities of suspended solids to achieve an efficient operation of the centrifuge. The decanter, model CA 220, was designed for continuous operation on liquors containing a high percentage of solids and it had an operating speed of 5 100rpm. Its maximum capacity was 8 000lh^{-1}. The centrifugal separator, model SA 20 03076, was designed for continuous operation for separation of heavy and light liquid components with automatic discharge of solids at predetermined intervals. Its operational speed was 7 500rpm and quoted hydraulic capacity 10 000lh^{-6}. The layout of the devised system which consisted of two 2t capacity tanks, a clip-on tank stirrer, a variable speed monopump, and a shell and tube heat exchanger heated by hot water is shown in Fig 1. All the pipework and valves were constructed from PVC.

For the second system, the decanter and centrifuge were replaced by an Alfa-Laval decanter, model N x

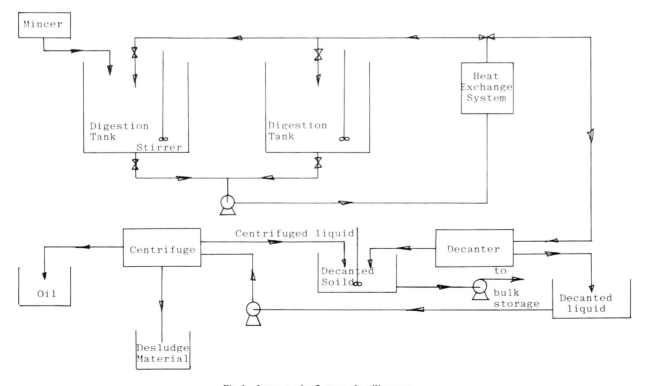

Fig 1 Layout of a 2-stage de-oiling system

3095/11 which was designed to separate continuously the three-component mixture into a solid, and a light and a heavy liquid phase. The decanter has an operational speed of 5 300rpm and hydraulic capacity of 10 000lh⁻¹. In this system two means of producing silage, the conventional mincing and blending technique and a chopper pump method (BP Nutrition, 1979), were included in the layout. In addition, facilities for two heating methods were incorporated, (*a*) direct steam injection and (*b*) a steam-heated scraped surface heat exchanger manufactured by Alfa-Laval (Contherm, Model 6 x 2). The layout, including a 1t capacity insulated plastic digestion tank, is shown in *Fig 2*.

3 Trials, results and comments

3.1 *De-oiling using a decanter and centrifuge*
This work was carried out on industrial premises using herring processing offal. Preliminary work had shown that it was essential to reduce the quantity of suspended solids present in the silage and also that temperatures of at least 60°C were necessary to achieve effective separation in the centrifuge. It was also important for practical reasons to limit, if possible, the digestion period prior to de-oiling to a maximum of 24h and, for this work, all the silages were de-oiled within this time.

The raw material, herring offal, was taken directly off the filleting line, minced and pumped to one of the two digestion tanks, 3·5% by weight of 85% formic acid was added and batches were, in the main, stirred continuously for about 24h in order to blend the acid with the fish and to aid the hydrolysis. In other experiments batches were stirred intermittently but this action did not affect the results. Silage temperatures reached about 20°C during the digestion period. The finished silages were pumped through a water-heated shell and tube heat exchanger to raise the temperature to between 65°C and 70°C before being fed to the decanter and centrifuge. It was necessary to control the feed rate to about 1t h⁻¹, about 10% of the quoted hydraulic capacity because of the limitations of the centrifuge. Steady feed conditions were also necessary to achieve a satisfactory performance of the centrifuge.

The silage was split into the fractions as shown in *Fig 1*.

In all cases the decanter solids, which had a low oil content, were subsequently added back to the aqueous phase from the centrifuge (de-oiled silage) to form the final product. Because the decanted liquors still had a relatively high percentage of suspended solids it was necessary to introduce a fairly stringent cleaning schedule for the centrifuge and, for all the tests, the centrifuge was partially desludged every 5min and fully desludged every 30min. This schedule naturally resulted in the production of a large quantity of desludge solids which were relatively oily. Whether desludge solids should be incorporated into the final product will depend upon the individual circumstances.

3.2 *Composition and mass of the fractions*
A number of runs were carried out and the compositions and the mass balance shown in *Table I* are typical of the findings.

The figures show that the aqueous phase from the centrifuge was the main component of the final product and that the oil content of this phase was reduced to 0·8%. If the decanter solids had been added to this fraction, the oil content would rise to 0·9% and, further, if the desludge solids had also been incorporated into the de-oiled silage, the final oil content would be 1·9%. An acceptable oil content depends upon the ultimate use. If silage replaces fish meal included at a 10% level in a diet then, based on a fish silage to fish meal conversion of 5:1 and given that the oil content in a total diet should not exceed 1% on a dry matter basis, an oil content of 2% is acceptable in the finished silage. Thus the system as tested was able to de-oil a silage digested for 24h satisfactorily although at only a relatively low rate considering the capacity of the centrifuge.

Because of the costs entailed in the removal of oil from fish silage, the quantity and quality of the oil are important. In this case the oil recovery was 87%, yielding 140kg of oil per 1 000kg of silage. Apart from being a much more saleable commodity, a better price can be obtained for a good quality oil, that is an oil of low FFA content and with a good colour. Previous work (Tatter-

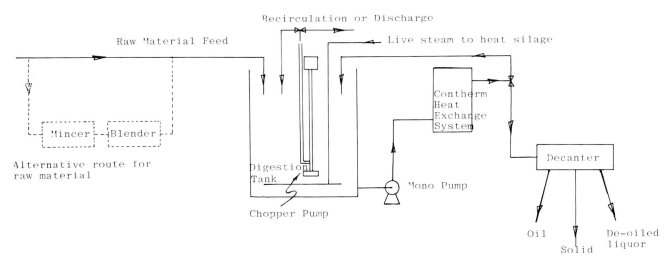

Fig 2 Layout of the de-oiling system using a 3-phase decanter

340

son, 1976) has shown the importance of early oil removal for the maintenance of quality, and the possibility of further deterioration during storage was studied. Extracted oil was stored at a temperature of 20°C for 109 days when the FFA increased by 0.3% and the iodine value fell by 1·4 units. There did not appear to be any additional deterioration in the colour of the oil during the storage period.

Table I
COMPOSITION AND MASS OF FRACTIONS – HERRING OFFAL SILAGE

| Fraction | Composition (%) | | | | Mass[3] |
	Protein	Oil	Moisture	Ash	kg
Whole silage	12·9	15·8	66·4	2·7	1 000
Decanter solids	23·0[1]	1·4	58·7	13·0	85
Decanter liquor	13·0	16·2	67·6	2·5	—
Centrifuge liquor	15·1	0·8	79·5	2·9	700
Desludge solids	15·3	7·3	73·8	2·6	150
Oil	—	—	—	—	140
Decanter solids[2] + centrifuge liquor	—	0·9	—	—	785
Decanter solids[2] + centrifuge liquor + desludge solids	—	1·9	—	—	935

[1] A factor of 6·25 was used to calculate the protein and this is not strictly applicable for this material
[2] Calculated from mass balance and analyses
[3] The weight of the various fractions exceeded the original weight because of water added during the desludging operation

3.3 *Oil separation using a single-stage, three-phase decanter*

The trials with the single-stage three-phase decanter were laboratory based and consequently it was possible to examine a wider range of conditions than in the former trials.

Because of the restricted machine time available, however, it was decided to obtain an overall view of what is possible rather than make an exhaustive study of one type of material or method of manufacture. In fact, detailed laboratory studies of some of the factors raised in the trials are in progress. The following conditions were examined in the trials:

(*1*) de-oiling of silages made from herring offal, sprats, scad and mackerel;
(*2*) methods of manufacture including two methods of heating;
(*3*) age of the silage; and
(*4*) effect of temperature and feed rate of the silage to the decanter.

3.4 *The trials*

The same basic equipment (*Fig 2*) was used in each trial and the variations used with the system were as follows:

(*a*) The fish was minced and blended with the acid by use of a heavy-duty slow-moving blender or by hand or by pulsed intermittent use of the chopper pump. This exercise on blending of the acid was necessary because the possibility of emulsions being formed appeared as an additional complication as the trials progressed.

(*b*) Whole fish and acid were blended by use of the chopper pump. The use of this pump, which gradually breaks the fish into small pieces and at the same time circulates free liquor, is central to a system of silage manufacture patented by BP Nutrition (UK) and is cur-

rently in use in industry for non-oily raw material.

(*c*) In all the trials, the silages were made by adding 3·5% by weight of 85% formic acid and then heating to 30°C by addition of steam. Digestion time was varied, ranging from 3h to several weeks before de-oiling.

(*d*) Silages were heated either by direct injection of steam or by a steam-heated scraped surface heat exchanger (Contherm) to the temperature required for de-oiling; this varied from about 50°C to 90°C.

The operating conditions of the decanter, after a satisfactory setting had been established, were not varied.

Samples of each silage were taken immediately after blending with the acid, and before and after heating to the de-oiling temperature. De-oiled liquor, oil and solids, where produced, were also sampled after the decanter. Trichloroacetic acid extracts of the whole silages, the de-oiled liquor and the solids were prepared immediately for the measurement of the degree of hydrolysis. All the samples and extracts were stored at −30°C for subsequent analysis.

4 Results and discussions

4.1 *Sprat silage*

In this series of trials, the effects of the extent of hydrolysis, the method of manufacture, and the de-oiling temperature and feed rate were examined. Each trial was carried out with silages made from the same batch of sprats; the oil content of the raw material averaged 18·3%.

4.1.1 *The effect of maturity of the silage and feed rate to the decanter*
In this trial, three ages of silages were examined; seven weeks, one day and three hours. Each silage was made by mincing and then blending the acid by a heavy duty blender followed by minimal use of mechanical agitation. For the seven-week-old silage, the mix was held at 20°C for the digestion period. The temperature of the other two was raised to and held at 30°C by live steam. Two feed rates to the decanter were used: 0·5t h⁻¹ and 1t h⁻¹, and there were two de-oiling temperatures, 70° and 80°C.

The results given in *Table II* show that an advanced degree of hydrolysis, as reached in the seven-week silage, led to improved oil removal. At a feed rate of 0·5t h⁻¹ the oil level for the most mature silage was 2·6% whilst for the others it was 3·7% and 3·9%. It is not thought that the change in the de-oiling temperature from 70°C to 80°C made a significant difference.

Table II
EFFECT OF THE DEGREE OF HYDROLYSIS AND FEED RATE—SPRAT SILAGE

Age of silage	Degree of hydrolysis %	De-oiling temperature °C	Feed rate t h⁻¹	Amount of oil in aqueous phase %
3 hours	42·3	70	0·5	3·7
			1	6·3
1 day	58·3	70	0·5	3·9
			1	6·6
7 weeks	69·0	80	0·5	2·6
			1	2·7

Feed rate to the decanter clearly had an effect; for the less mature silages doubling the feed rate from 0·5t h⁻¹ to 1t h⁻¹ raised the oil level from 3·9% to 6·6% but for the well matured silage the feed rate did not affect the efficiency of de-oiling.

In these experiments no solid phase was produced and how this affected the oil level is not known.

From these results it is concluded that the decanter did not perform as well as the two-stage operation but with the mature silage the oil content was approaching a satisfactory level.

4.1.2 *The effect of agitation*

The use of a powerful mixing action during the manufacture of the sprat silages seemed to give rise to abnormally high oil levels in the de-oiled silages, possibly due to the formation of emulsions. Consequently the effect of agitation during the digestion period was examined by preparing two silages, one by the chopper pump technique in which whole fish were chopped and blended with acid, and the liquor, even after the fish had liquefied, was continuously recirculated up to the time of de-oiling. The other silage was made by the more conventional method of mincing and mechanical blending with negligible stirring. Both silages were allowed to digest for 48h at 30°C prior to de-oiling at a temperature of 90°C.

The results showed that the stirred silage, at feed rates of $0.5t\ h^{-1}$ and $1t\ h^{-1}$ had de-oiled liquors containing about 6% and 10% oil, respectively. These were the worst results obtained with a reasonably well hydrolysed sprat silage. The other silage, made with minimum stirring, yielded liquors with oil contents of approximately 4% for both the $0.5t\ h^{-1}$ and $1t\ h^{-1}$ feed rates. The adverse effect of agitation was further demonstrated by stirring this particular silage for 6h and de-oiling immediately; the oil content of the de-oiled silage increased by 2%. It is concluded that emulsification did occur and laboratory investigations are necessary to quantify the effect.

It should be noted that problems with emulsions either did not arise or were overcome satisfactorily by the two-step de-oiling system which involved continual agitation during digestion.

4.1.3 *Effect of temperature*

Though previous laboratory work had indicated that an effective de-oiling operation could be achieved at 70°C, a few trials were carried out on the same silage, at 90°C as well as at 70°C. It was not possible to carry out a comprehensive programme on this aspect of the work at this stage but the preliminary results indicate that further reductions in the oil level, of the order of 0.5%, can be achieved by working at 90°C instead of 70°C. Whether this degree of improvement justifies working at a higher temperature depends upon the circumstances. In our case, specialized equipment and a steam supply were available; in many instances hot water may be the only heating medium and a temperature of 90°C could be difficult to achieve.

4.2 *Scad silage*

In this series, the work was limited to the study of the effect of the degree of hydrolysis and the feed rate on the performance of the decanter. All the silages were prepared from scad with an oil content of 12.8%, using a mincer and mechanical blender with intermittent agitation during the digestion period. In all cases the de-oiling temperature was 65°C, achieved by either direct injection of steam or by contherm.

The results given in *Table III* show that the removal of oil is affected by the degree of hydrolysis and by the feed rate. Thus at $0.5t\ h^{-1}$ for the mature and the one-day-old silages oil levels were reduced to about 3% whilst in the fresh silage, although the degree of hydrolysis was 30% compared with 33% for the one-day material, the oil content was only reduced to 6%. At the higher feed rate de-oiling was less efficient.

Solids were separated in each silage examined and, for the two less mature silages, constituted up to 27% of the decanter output. However, in all cases the oil content of the solids was only 1.4% and consequently could if added back to the de-oiled phase reduce the oil level in the finished product.

Table III
DE-OILING OF SCAD SILAGE OF VARYING AGES

Age of silage	Degree of hydrolysis %	Feed rate $t\ h^{-1}$	Oil content of aqueous phase %	Solids % of product
3 hours	30.2	0.5	6.0	27.2
1 day	33.4	0.5	2.9	9.8
		1	4.9	15.4
8 weeks	52.4	0.5	3.0	traces
		1	3.7	traces

4.3 *Herring offal silage*

A similar series of experiments was repeated with herring offal though these were more curtailed due to the lack of raw material. However, it was possible to carry out three trials with three materials with oil contents varying from 11.9% to 14.0%. In each experiment the silages were prepared by mincing and mechanical blending of the acid and digesting at 25°C. Three digestion periods were examined: three hours, three days and eighteen days. The results given in *Table IV* show that the oil levels were reduced to at least 2.8% at feed rates of $0.5t\ h^{-1}$ and $1t\ h^{-1}$.

Table IV
DE-OILING OF HERRING OFFAL SILAGES OF VARIOUS AGES

Age of silage	Degree of hydrolysis %	Feed rate $t\ h^{-1}$	Oil content aqueous phase %	Solids % of product
3 hours	34.9	0.5	2.8	17.4
3 days	46.9	0.5	2.6	5.0
18 days	70.9	1	2.8	0

It is interesting to note that, as with scad silage, a degree of hydrolysis of the order of 35% was adequate for oil removal. Increasing the degree of hydrolysis to 47% and 53% for herring offal and scad silages, respectively, did not achieve, at a feed rate of $0.5t\ h^{-1}$, worthwhile reductions. As seen from the scad silage results, a higher degree of hydrolysis is necessary for effective oil removal at higher feed rates. For herring offal silage with a degree of hydrolysis of 71%, it was possible to obtain an oil content of 2.8% at a feed rate of $1t\ h^{-1}$.

Solids were produced in these trials and, as expected, the amount was related to the age of the silage; the maximum was 17.8% for the three hour silage and trace quantities for the eighteen-day silage. In each case the oil level in the solids was 1.3%.

These results are the only ones in which a direct comparison may be made with the decanter/centrifuge system and it is clear that the performance of the three-phase decanter fell short of the more expensive system.

4.4 Mackerel silage

An additional series of trials was carried out with mackerel but, in this case, the investigations were concerned with examining a possible industrial system combining the optimum use of the chopper pump, the contherm and the decanter.

Whole mackerel with an oil content of 13·1% were heated to 30°C with live steam and comminuted and blended with the acid by use of the chopper pump. Since emulsification appears to be of importance, the running time of the pump was particularly noted – this was for a period of 3·75h spread, but not evenly, over a 24h period. The results which are given in *Table V* show that after a 24h digestion period, a liquor containing 3·7% oil was produced at a feed rate and temperature of 0·5t h⁻¹ and 90°C. After a further 24h digestion, during which time the hydrolysis increased from 54% to 60%, no further reduction in the oil level was achieved at the same feed rate and temperature. Increasing the feed rate to 1t h⁻¹ resulted in a 0·9% increase in the oil content of the liquor. These results confirm previous findings that no improvements in oil removal at the lower feed rate of 0·5t h⁻¹ is achieved by increasing the digestion period beyond one day.

In these trials solids were not produced.

It is perhaps dangerous to compare these results with those obtained in the partially hydrolysed sprat silage experiment but since both untreated silages contained about the same amount of oil and the degree of hydrolysis was similar a valid comparison might be claimed. Similar results led to the conclusion that, provided the chopper pump is used with some discretion, the formation of emulsions can be avoided. Controlled intermittent use is acceptable since prolonged running does not appear to be necessary to bring about initial chopping and liquefaction of the fish. In addition this mode of operation would result in a considerable saving of power. Use of the pump offers advantages in that it removes the need for a mincer and a mixer and, apart from a saving in capital costs, constitutes the basis of what is essentially a much simpler plant, requiring the minimum of supervision.

Table V
DE-OILING OF MACKEREL SILAGES OF VARIOUS AGES

Age of silage days	Degree of hydrolysis %	Feed rate t h⁻¹	Temp.¹ °C	Oil content aqueous phase %
1	53·5	0·5	90	3·7
2	60·3	0·5	93	3·7
2	60·3	1	90	4·6

¹ All silages heated with the contherm

5 Conclusions

The trials, as intended, gave some indications of the performance of two possible commercial scale techniques using existing machinery for de-oiling silage. With a two-step system it was possible to obtain liquors with less than 2% oil from herring offal silage after a digestion period of 24h up to a capacity of 1t h⁻¹. A single stage operation did not perform quite as well but it was possible to reduce oil levels below 3% with a throughput of 1t h⁻¹ over a wide range of material. Modifications to the machine are expected to improve on this performance.

Capital costs are relatively high for de-oiling equipment and it is possible to replace the decanter and centrifuge used in the first series of experiments with a single decanter which incorporates oil removal facilities. The use of this machine, if the performance meets specification, would achieve a substantial reduction in capital costs.

A number of questions were raised in these trials concerning the effect of various factors such as temperature, throughput and particularly the degree of hydrolysis on the de-oiling operation. Further studies are necessary before these questions can be resolved.

6 References

ADAMSON, A H and SMITH, P. Pig feeding trials with white fish and
1976   herring liquid protein (fish silage). *Proc:* Torry Research
        Station Symposium on Fish Silage, Aberdeen
B. P .NUTRITION. BPN Liquid fish protein plant LFP 300. BP Nutrition
1979   (UK) Ltd
TATTERSON, I N. The preparation and storage of fish silage. *Proc:*
1976   Torry Research Station Symposium on Fish Silage, Aberdeen

Propionic acid as a preservative for industrial fish

I Tatterson, S Pollitt
and J Wignall

1 Introduction

Industrial fish can be defined as species which are caught specifically for conversion into animal feeds. In 1977 up to 180 000t of such fish (sand eels, summer sprat and scad, together with whatever winter sprat and mackerel were not required for human consumption) were converted into fish meal in the UK (MAFF, 1977). During the same year world landings of industrial fish were approximately 20 000 000t (IAFMM, 1978).

Most industrial fisheries are seasonal and landings may fluctuate, with glut following scarcity, and this can lead to production planning problems. Also, since industrial fish are not normally preserved, raw material is often in poor condition and this may result in processing difficulties. The main problems caused by the spoilage of industrial fish can be summarized as follows:

(a) Loss of 'drip' liquor containing protein and oil reduces the yield of both fish meal and oil. Liquor losses can be as high as 25% by weight for species which rapidly liquefy during storage.

(b) Since unpreserved industrial fish cannot normally be stored for more than a few days the processing plant needs to be large enough to deal with the highest level of landings. This results in under utilization of capacity for much of the remainder of the year.

(c) Toxic gases such as hydrogen sulphide are produced by spoiling fish and can build up to high concentrations in the ship's hold. On occasions this has led to fatalities.

(d) The odours emitted during the production of fish meal are much worse when spoiled fish rather than fresh raw material is processed. The smell from deteriorating raw material itself can also be considerable.

Chilling is one of the most effective ways of preserving industrial fish but since they have a relatively low value compared with fish caught for human consumption the use of ice or refrigeration techniques is not normally economic. In some industrial fisheries sodium nitrite and formaldehyde are used but problems of toxicity and loss of protein qualities can arise if their level of addition is not carefully controlled. A number of other chemical preservatives has been tried and Windsor and Thoma (1974) described work in which sprats were preserved with a solution containing propionic acid (PA), ascorbic acid and ethylene diamine tetra-acetic acid. The most effective component in the mixture appeared to be PA, particularly when fish were stored in an atmosphere of carbon dioxide. In that work fish were dipped for 10min in either a solution containing 0·7% v/v PA together with the other two components or in one containing 1% v/v of the acid alone. The results were sufficiently encouraging to consider further work and it was decided to extend the study to include other species. Some of the results to date are reported in this paper.

Treatments were limited to solutions of PA alone and, to be more commercially realistic, shorter dipping times and a range of stronger dip solutions were used. Any advantage to be gained by excluding air was examined by storing some samples under both sealed and unsealed conditions. Since the work by Windsor and Thoma had dealt with fish stored only at 11°C, storage at a number of different temperatures was employed. Results obtained with thawed blue whiting, treated and stored under laboratory conditions, are given in some detail but, to check their validity, a study of blue whiting and a number of other species has been made at sea using fresh, unfrozen fish. Those results, also, are reported.

2 Experimental

2.1 Laboratory studies
Two principal experiments were carried out with thawed blue whiting. In the first, the effect of PA uptake on storage life at 15°C was investigated; fish were treated with a range of PA solutions for either 0·5min, 1min or 5min. Some of the solutions used were much stronger than would be advised for commercial practice but were included so that high PA contents in the fish could be achieved. The complete set of treatments is shown in Table I.

In the second experiment the temperature effect was studied but only the weaker solutions were used to produce acid uptakes more in line with what might be commercially practicable. Storage was carried out at five different temperatures between 0°C and 20°C. Table II shows the treatments in detail.

For both experiments the fish had been frozen and cold stored at −30°C for several months. After air thawing, batches of material were dipped in the PA solution (1·5kg fish to 2·1 dip) and then removed and drained for 10min. Lots (0·5kg) were taken from each batch, sealed, where required, in laminated 'Synthene' bags from which as much air as possible was removed, and stored at the selected temperature.

For each dip, the total weight of fish treated and the volume and strength of the solution before and after use were measured. From those data the amount of acid used, assumed to have been taken up by the fish, was calculated. Although this may not have been the best way of measuring acid uptake it was used because a suitable rapid method was not available at the time of the experiments. Currently in the work carried out at sea, PA is being measured by a gas solid chromatographic technique.

Table I
EXPERIMENT 1 – TREATMENTS, ACID UPTAKES AND STORAGE LIVES OF THAWED BLUE WHITING SAMPLES

Dip strength (%PA v/v)	Dip time (minutes)	Storage life (days)	Apparent acid content (g PA/ 100g fish)
Control (open)	—	4	—
Control (sealed)	—	5	—
5	0·5	13	0·36
	1	22	0·49
	5	27	0·70
10	0·5	24	0·57
	1	34	0·82
	5	47	1·19
15	0·5	51	0·93
	1	54	1·26
	5	56	1·46
20	0·5	73	1·45
	1	78	1·76
	5	91	2·34

Storage was carried out at 15°C

2.2 Work at sea
In the work carried out at sea, treatments were for 0·5min and 1min in 5% solution and for 0·5min in 10% solution with storage temperatures of 0°C, 10°C and 20°C. Storage was carried out in both sealed and unsealed bags, although in this paper only the former results are reported. Controls in both sealed and unsealed bags were again included.

2.3 Quality assessment
In each experiment, in the laboratory or at sea, raw odours were assessed at predetermined intervals by a panel of three to eight people using, for sealed storage, a freshly opened bag of fish on each sampling occasion. The standard cod/haddock score sheet (Shewan et al, 1953) was used generally, although sprats and sand eels were assessed by the system developed by Windsor and Thoma (1974). Average scores were plotted against storage time and the time taken for the score to fall to 2·0, arbitrarily defined as storage life, taken from each graph. A typical plot is shown in Fig 1.

3 Results and discussion

3.1 Blue whiting (laboratory studies)
Tables I and II show the results of the work carried out in

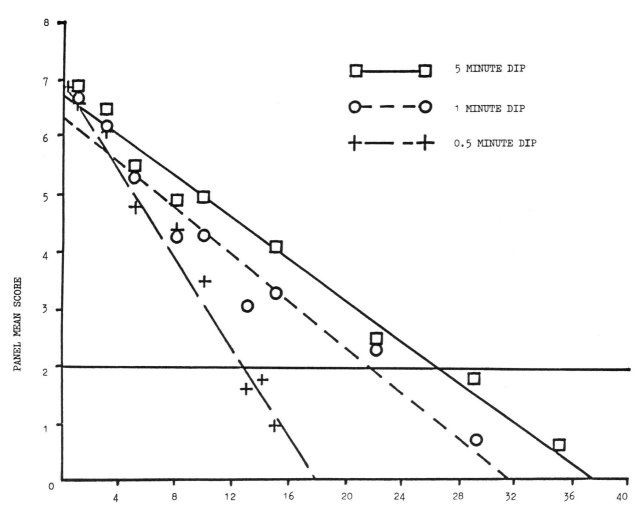

Fig 1 Panel mean score *versus* storage time for thawed blue whiting treated with 5% PA 15°C storage

the laboratory with thawed fish. To obtain some measure of the reproducibility obtained, direct comparisons can be made between the sets of results for 0·5min and 1min dips in 5% solution and that for a 0·5min dip in 10% solution, and untreated controls with storage in each case at 15°C. There is some measure of agreement between the two sets of figures although, apart from the 5% PA/1min dip treatment where there was a considerable difference between the acid uptakes, the fish in experiment 2 lasted longer than those in experiment 1. This may have been due to a difference in initial quality of the material.

For experiment 1 (*Table I*) the results show that control fish kept for about five days at 15°C. Depending upon the dip time and hence the acid uptake, storage lives of between 13 days and 27 days were achieved with 5% PA and between 24 days and 47 days with 10% PA. The amount of acid absorbed by the fish to achieve these storage lives varied between 0·36 and 1·19g/100 fish. The apparent PA uptake for each dip and treatment is plotted against dip time in *Fig 2*. It can be seen that for each dip at least 50% of the PA present after 5min was absorbed within 0·5min and that there is little advantage to be gained, particularly with 5%, 10% and 15% solutions, by dipping for periods over 1min. To obtain a storage life of three weeks at 15°C requires an acid uptake of 0·5% (*Fig 3*) and this is absorbed within approximately 5s, 16s, 36s and 60s, respectively, from 20%, 15%, 10% and 5% dips.

In *Fig 3*, storage life at 15°C has been plotted against

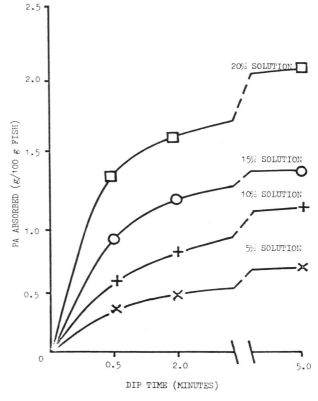

Fig 2 Apparent uptake of PA with varying dip time (thawed blue whiting)

345

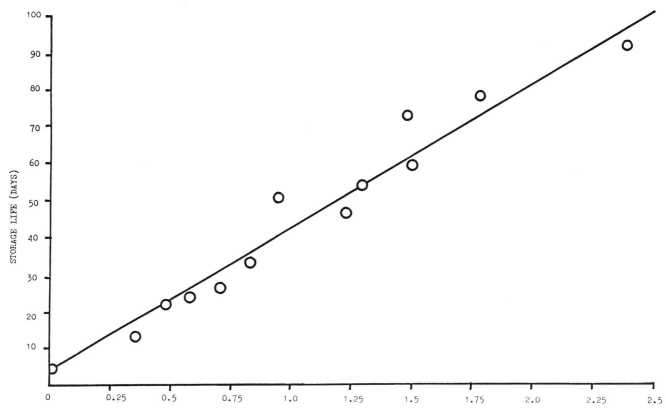

Fig 3 Storage life at 15°C *versus* apparent PA uptake (thawed blue whiting)

PA uptake and there is seen to be a good linear relationship between the two.

From *Figs 2* and *3* dipping conditions can be selected to achieve any required storage life up to about 90 days at, in this case, a storage temperature of 15°C. In practice, conditions at sea may make the use of solutions stronger than 10% undesirable from the safety point of view and this would, in effect, cut practicable storage times to something of the order of 50 days.

The results for experiment 2 (*Table II*) illustrate the effect of temperature and in *Fig 4* the storage life of fish treated in each of the three different ways together with untreated controls is plotted against storage temperature. The increased preservative effect of successively stronger treatments is apparent but it can be seen also that for both treated and untreated fish the rate of increase of storage life becomes greater as storage temperatures fall.

Apart from one instance the results for control fish in both experiments showed that material in sealed bags kept better than that in open bags. At the best the former kept over 1·5 times longer than the latter and this may have been due to the build-up of gases such as carbon dioxide over the fish together with the disappearance of oxygen.

3.2 *Blue whiting (studies at sea)*

The results of the experiment carried out at sea with blue whiting are given in *Table III*; for comparison, the results with thawed fish for the same treatment are included. Apart from those relating to 5% solution with a 0·5min dip, and storage at 10°C, there is reasonably good agreement between storage lives. Since the acid levels in the fish, although measured by different

Table II

EXPERIMENT 2 – TREATMENTS, ACID UPTAKES AND STORAGE LIVES OF THAWED BLUE WHITING SAMPLES

Dip strength (% PA v/v)	Dip time (minutes)	Storage temperature (°C)	Storage life (days)	Apparent acid content (g PA/100g fish)
Control (open)	—		30	—
Control (sealed)	—		47	—
5	0·5	0	84	0·27
5	1		93	0·28
10	0·5		86	0·58
Control (open)	—		17	—
Control (sealed)	—		25	—
5	0·5	5	57	0·27
5	1		73	0·28
10	0·5		77	0·58
Control (open)	—		9	—
Control (sealed)	—		10	—
5	0·5	10	35	0·27
5	1		31	0·28
10	0·5		44	0·58
Control (open)	—		8	—
Control (sealed)	—		9	—
5	0·5	15	17	0·27
5	1		19	0·28
10	0·5		28	0·58
Control (open)	—		4	—
Control (sealed)	—		4	—
5	0·5	20	10	0·27
5	1		11	0·28
10	0·5		12	0·58

methods, had been found to be very similar, the results agreed with our expectation. Consequently, the findings for thawed fish, which had been studied in some detail, are regarded as giving a reliable indication of the likely behaviour of fresh, acid-treated material.

346

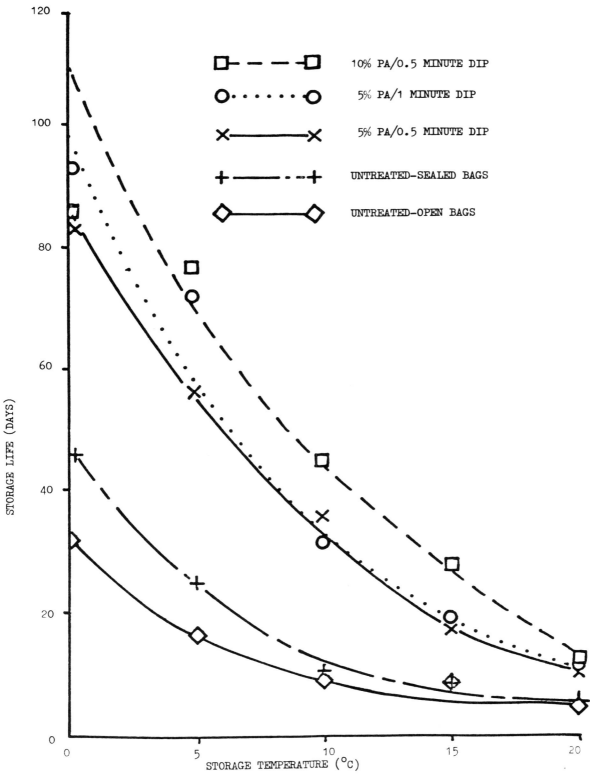

Fig 4 Storage life (sensory) *versus* temperature (thawed blue whiting)

3.3 *Mould formation*

In both experiments carried out with thawed fish the presence of moulds was observed on some of the treated samples. These were identified as *Penicillin chryso-genum* (Panes, personal communication) and the spores were probably picked up during initial cold storage. Moulds did not grow on untreated samples at any temperature nor on any sample at all at 0°C and they gener-ally appeared on fish containing relatively small amounts of PA.

3.4 *Other species (studies at sea)*

The work on fresh blue whiting was extended and the following species were studied at sea: summer sprat, summer mackerel, Norway pout, scad and sand eels; these results are given in *Tables IV* and *V*. In *Table IV*,

347

Table III
STORAGE LIVES AND ACID CONTENTS OF THAWED BLUE WHITING AND SEA FRESH BLUE WHITING

Treatment			Storage life (days)		Acid content (g PA/100g fish)	
Dip strength (% PA v/v)	Dip time (minutes)	Storage temperature (°C)	Thawed fish	Sea fresh	Thawed fish	Sea fresh
10	0·5	0	86	92		
		10	44	57	0·57	0·59
		20	12	8		
5	1	0	93	76		
		10	31	28	0·38	0·35
		20	11	10		
5	0·5	0	84	88		
		10	35	17	0·31	0·29
		20	10	6		
Control	—	0	47	41	nil	nil
	—	10	10	10	nil	nil
	—	20	4	2	nil	nil

which shows the acid contents, the fish are listed in descending order of size, although there was little difference between the size of the last three species—summer sprat, Norway pout and sand eels. Acid contents increased as the size of the fish decreased, although Norway pout contained rather more acid than either sand eels or summer sprat. This may have been due to the fact that it contained very little oil which made it easier for the acid to penetrate the flesh. The blue whiting already discussed were about the same length as the scad, but much thinner and, consequently, contained slightly more acid than the latter.

Table IV
PROPIONIC ACID CONTENT OF FISH TREATED AT SEA

Treatment		Acid uptake (g PA/100g fish)				
Dip strength (% PA v/v)	Dip time (minutes)	Mackerel	Scad	Sprat	Norway pout	Sand eels
10	0·5	0·37	0·49	0·75	0·92	0·76
5	1	0·27	0·31	0·50	0·58	0·46
5	0·5	0·24	0·27	0·46	0·59	0·44

The storage lives are listed in *Table V*. It is clear that PA can be an effective preservative for all the species examined although in some instances the effect during storage at high temperatures is only small. Problems during subsequent processing would probably arise with summer sprat where acid preservation may be unsuitable, and occasionally with sand eels. These are species which tend to autolyse and liquefy rapidly and the process was accelerated by the presence of the acid. When liquefaction did occur, a product resembling fish silage resulted and preservation was usually maintained.

Norway pout kept exceptionally well with storage lives of 207 days, 44 days and 24 days at 0°C, 10°C and 20°C, respectively, for a 0·5min dip in 10% PA solution. At the same temperatures control fish kept for 53 days, 4 days and 1 day. Sand eels absorbed slightly less acid than Norway pout, with similar treatments, but did not keep nearly so well; storage lives were sometimes only about half those of the latter species. The results for scad and mackerel were in reverse order to what might have been

expected. For the same dips and dip times, scad contained more acid than mackerel but in every instance storage lives were shorter than those of mackerel. The same behaviour but to a lesser extent was apparent in control fish.

For all the species examined storage temperature was important and should temperatures approach 20°C it is likely that acid contents of between 0·5% and 1% would be needed to obtain worthwhile storage periods. This is particularly apparent with scad and mackerel. It is perhaps worth recording that, during the work carried out on the FRV *G A Reay*, the temperature on the processing deck varied between 5°C and 17°C, depending on the time of year.

This work has not yet been completed and a number of aspects still have to be examined. Little is known of any processing problems that might arise from the presence of the acid nor is it known whether large quantities of stored treated material will behave in a similar way to the relatively small amounts of fish stored in these experiments. Both factors could have an important effect on the success of the technique.

Table V
STORAGE LIVES OF FISH TREATED AT SEA WITH PA, AND CONTROLS – STOR CARRIED OUT IN SEALED BAGS

Treatment			Storage life (days)				
Dip strength (% PA v/v)	Dip time (minutes)	Storage temperature (°C)	Mackerel	Scad	Summer sprat	Norway pout	Sand eels
10	0·5	0	116	108	50[1]	207	98
		10	16	13	37	44	26
		20	5	4	34	24	14
5	1	0	107	60	50[1]	124	63
		10	10	9	35[1]	27	18
		20	5	3	9	10	8
5	0·5	0	96	73	50[1]	104	67
		10	15	9	26	23	12
		20	4	3	6	10	6
Control		0	34	31	50[1]	53	21
		10	6	6	9	4	8
		20	4	3	2	1	4

[1] Samples had liquefied or become very soft and were no longer processable Storage was also carried out in open bags but shorter storage lives were generally found

4 Costs

At this stage it is impossible to estimate the capital outlay needed to establish such a system of preservation and difficult even to hazard a guess at running costs. If, however, it is assumed in the first place that the fish should not contain more than 0·5% PA and that application is carried out by dipping in 5% solution, the following might apply. One tonne of fish required 400kg of dip for total submersion and at a cost of £400/t for PA, the cost of the acid in the dip is £8 (20kg). During dipping the fish would remove 5kg of acid from the solution and therefore 'topping up' the dip before re-use would cost £2. If the dip were to be used five times before discarding, the total acid costs for 5t of fish would be £16 or just over £3/t of fish treated. Using the dip 10 times would reduce acid costs to around £2.50/t of fish.

5 Conclusions

(1) The storage life of blue whiting and other industrial species can be extended considerably by the application of propionic acid.

(2) There appears to a good linear relationship between storage life and PA uptake.

(3) The PA uptake by the fish can be controlled by selecting suitable dip solutions and times.

(4) Storage lives increase markedly with decreasing storage temperature and below 5°C or so considerable storage periods appear possible even when the fish contain only a small amount of PA.

(5) The exclusion of air by carrying out storage in sealed bags generally leads to improved storage lives.

6 References

IAFMM. Digest of world statistics: fish meals. *Proc:* 18th Annual
1978 Conference, Intern. Assoc. of Fish Meal Manufacturers, Potters Bar

MAFF. *Sea Fisheries Statistical Tables.* Min. of Ag., Fish. and Food.
1977 HMSO, London

SHEWAN, J M, MACINTOSH, R G, TUCKER, C G and EHRENBERG,
1953 A S C. The development of a numerical scoring system for the sensory assessment of the spoilage of wet white fish stored in ice. *J. Sci. Fd. Agric.*, 4, 283–298

WINDSOR, M L and THOMA, T. Chemical preservation of industrial
1974 fish: new preservation mixtures. *J. Sci. Fd. Agric.*, 25, 993–1005

Nutritional value of ensilaged by-catch fish from Indonesian shrimp trawlers

I Putu Kompiang, R Arifudin and J Raa

1 Introduction

The aim of the present work was to assess the prospects of producing fish silage as a method of preserving and utilizing the by-catch fish on shrimp trawlers operating in Indonesian waters. The by-catch is usually thrown overboard because its market value is too low to justify preservation in ice. Fish meal production is not a realistic alternative when the landings are scattered and irregular, and even for conventional fish meal the fish must be preserved on board to avoid serious putrefaction during the two to five days' journey back from the fishing grounds. Production of fish silage, by adding acid to the chopped fish on board the vessel, thus seemed suitable for storing the fish as a feed component. However, the data on the nutritional value of fish silage are still conflicting (Disney and Hoffman, 1976). To rule out the possibility that partial spoilage prior to ensiling could affect the nutritional value, silage was prepared from fresh fish on board the vessel and its nutritional value was compared with that of fish meal produced from the same raw material.

2 Materials and methods

2.1 Fish

The raw material for fish silage, and the reference fish meal, was the by-catch fish from shrimp trawlers operating north of West Java, south of the Biliton Islands. The by-catch contained usually more than 20 different species, but *Upeneus* (goatfish), *Saurida* (lizard fish), *Dorosoma*, *Nemipterus* and *Piracanthus* predominated. *Upeneus* and *Saurida* alone made up usually more that 60% by weight of the catch. *Caranx* sp. were always present in the catch, but was not included in the silage, because the fishermen salted it for marketing as human food. The toxic puffer fish occurred at low frequency in the by-catch, but it was not included in the silages, nor the fish meal. The two predominating species, *Upeneus* and *Saurida*, were medium fat and lean, respectively. Slight variations in the relative frequency of these species may therefore significantly affect the lipid content of the raw materials. The by-catch was thus not always of exactly the same composition.

2.2 Icing/freezing

The by-catch fish were removed from the trawl and immediately cooled in refrigerated sea water on board the vessel. After draining off this sea water, the fish was mixed with three times its weight of crushed ice and stored for two days in an isopore-insulated glass-fibre tank until the vessel had returned to Jakarta. The ice was then melted with running tap water and the fish packed in troughs and immediately frozen at $-20°C$.

2.3 Preparation of fish silage

For the first feeding experiment (*Table I*), the silage was made on board. The by-catch fish were removed from the trawl and each fish was cut into three pieces and mixed immediately with either 3% (v/w) of 90% formic acid (silage A) or 3% (v/w of 50/50 (vol) mixture of 90% formic acid and 95% propionic acid (silage B). The silages were kept in glass-fibre tanks on the deck until back on shore (three days) and then at ambient temperature indoors (c 30°C in the silage) for a total of 10 days before the diets were prepared. For the other feeding experiments, fish silage was prepared in the laboratory from the frozen fish after it was melted with running water. After 10 days, the silage was mixed with corn (1:1) and sun-dried.

Table I
GROWTH AND FEED CONSUMPTION OF CHICKENS FED THE DIETS DESCRIBED IN TABLE II

			Body weight gain (g/head/3 weeks)	Feed consumption
Soy bean meal			454b	785
Meal	of fresh fish	whole	424b	652
		press-cake	464ab	667
	of spoiled fish	whole	241c	451
		press-cake	497a	779
Fish silage		A	307c	556
		B	277c	493

a, b and c indicate significant differences ($P > 0.05$)

2.4 Fish meal

Fish meal produced from the same raw material as the silage was used as reference fish meal diets. The frozen fish was plunged into a pan where it was heated and finally boiled for 15 min under continuous stirring. The boiled fish was pressed in a small scale (20kg) fish meal press. The press-cake was sun-dried after grinding in a meat grinder. A fraction of the boiled fish was sun-dried after grinding, without pressing (= fish meal of whole fish). Corresponding fish meals were prepared also from fish which were left in the troughs at ambient temperatures (c 30°C) for 24h after being removed from the freezer.

2.5 Chemical analysis
Protein was estimated by multiplying total nitrogen, determined by the Kjeldahl procedure, by the factor 6·25.

Lipid was determined by the method of Bligh and Dyer (1959).

Amino acids were determined in an acid hydrolysate (6N HCl for 24h at 110°C) of dried sample, using an automatic amino acid analyser (Jeol. JLC–6AH).

Tryptophan was determined according to Eggum (1968). Four samples were hydrolysed in parallel, three of them containing increasing amounts of added tryptophan. The analytical result was considered to be acceptable if the analysis of the sample without added tryptophan was identical to the value obtained by extrapolating the three parallel analyses to zero tryptophan addition.

2.6 Feeding experiments
All diets contained the same amount of fish protein and were isonitrogenous. Each diet was fed to 40 one-week-old broiler chickens, divided into four pens with five males and five females. Growth and feed consumption was measured each week for three weeks.

3 Results

3.1 The chemical and microbial stability of silage
Introductory experiments showed that 1·5% (v/w) of a 50:50 mixture of formic acid and propionic acid, as prescribed by Gildberg and Raa (1977), was not sufficient to produce a stable silage of minced by-catch fish. Stable silages with a fresh acidic smell were always obtained when 3% (v/w) of propionic acid/formic acid (1:1) was added. This was usually so also with 3% (v/w) of formic acid alone, but such silage sometimes became infected by moulds, in particular *Aspergillus flavus*. This has been reported also by Gaiger (1978). The advantage

of using the propionic acid/formic acid mixture was evident when the silage was mixed with a carbohydrate carrier. Moist mixtures of 1 unit weight of silage and 1 unit weight of dry cassava, or corn meal, remained free of mould growth for at least three months at 30°C. A corresponding moist mixture of the formic acid silage spoiled, usually within a few weeks. The action of propionic acid on *A. flavus* has been further described by Strøm et al (1980).

Silage B was sterile for 21 days. This was often so also with silage A, unless it had visible mould growth at the surface. There was a slow release of ammonia. The amount of ammonia released after 21 days corresponded to less than 1·8% of the total amino nitrogen of silage A and less than 1·3% of that of silage B. This ammonia production might represent a significant degradation of the nutritional values of the protein if it derived from one, or a few essential amino acids. However, storage for 21 days caused no significant change in the amino acid composition of an acid hydrolysate of either of the two silages, except for degradation of tryptophan in the aqueous phase of the silage.

3.2 Nutritional value
The nutritional value of the fish silage diets was poor compared to diets containing fish meal or soy bean meal as the main protein source (*Table I*). Although the press-cake of 24h spoiled fish had a good nutritional value, the spoiling generated substances in the press-liquid which markedly reduced the growth performance of the chickens. The composition of the diets used in this experiment is shown in *Table II*, and the amino acid composition of some of these diets shown in *Table III*.

Boiling improved the nutritional value of the silage (*Table IV*); the chickens grew faster on the same quantity of feed, but their performance was still less than on the fish meal reference diets.

Thiamine supplementation did not improve the nutritional value of either the boiled or the non-boiled silage (*Table V*).

Removal of lipid from the non-boiled silage by ether extraction improved its nutritional value markedly (*Table IV*). The feed consumption was the same as for the fish meal diets, but the growth was slightly less. There was no significant difference in amino acid composition of the ether extracted silage and that of whole fish meal, except for a slightly lower tryptophan content (*Table VI*).

Table II

COMPOSITION AND CHEMICAL ANALYSIS OF THE DIETS USED IN THE FEEDING EXPERIMENTS PRESENTED IN TABLE I

| | | | Weight (%) composition in diets | | | Premix (vitamin/ trace mineral) | Chemical analysis | | | |
| | | | | | | | (% of dry weight) | | μg/g | pH in |
Protein source			Corn meal	Salt	Santoquin[1]		Protein	Lipid	NH[3]	suspension
Soy bean meal[2]		35	61·5	0·20	0·02	0·50	22·0	2·6	—	—
Fresh	whole	22·5	76·7	0·20	0·02	0·50	22·4	9·0	17	5·7
fish	press-cake	23·3	76·0	0·20	0·02	0·50	22·1	9·5	14	6·2
Spoiled	whole	23·5	75·8	0·20	0·02	0·50	21·2	9·6	238	5·9
fish	press-cake	23·5	75·8	0·20	0·02	0·50	21·9	11·5	182	6·7
Fish	A	99·3[3]		0·20	0·02	0·50	21·5	8·9	47	4·3
silage	B	99·3[3]		0·20	0·02	0·50	21·6	8·9	27	4·6

[1] Contained 66% ethoxyquin
[2] Was supplemented with 1·6% tricalcium phosphate and 1·1% CaCO[3]
[3] Sun-dried in corn

Table III
AMINO ACID COMPOSITION (MOLE %) OF SOME OF THE DIETS USED IN FEEDING EXPERIMENTS

Amino acid	Fish meal diet (whole fresh fish)	Fish meal diet (whole spoiled fish)	Fish silage B diet
Lysine	5·4	5·7	5·6
Histidine	1·8	1·5	2·0
Arginine	3·9	3·1	3·5
HO-proline	1·2	0·7	0·9
Aspartic acid	8·5	8·5	8·8
Threonine	4·9	4·8	5·4
Serine	5·9	5·9	6·0
Glutamic acid	14·7	14·7	14·8
Proline	4·4	5·0	6·1
Glycine	11·3	13·6	9·9
Alanine	11·8	12·3	10·4
Cystine/2	0·8	1·3	0·8
Valine	4·9	4·1	4·7
Methionine	2·2	2·2	2·2
Isoleucine	2·8	3·5	4·0
Leucine	9·0	8·3	9·6
Tyrosine	1·8	1·7	1·3
Phenylalanine	2·8	1·8	3·1
Tryptophan	0·85	0·80	0·65

Table IV
EFFECT OF BOILING ON NUTRITIVE VALUE OF FISH SILAGE

Protein source of diet	Body weight gain (g/head/4 weeks)	Feed consumption
Whole fish meal	906a	1558
Boiled silage[1]	677b	1310
Boiled silage[2]	637b	1255
Silage	487c	1208

a, b and c indicate significant difference (P > 0·05)
[1] Fish was boiled prior to making silage
[2] Silage was boiled after 10 days of storage and prior to mixing with corn

Table V
GROWTH AND FEED CONSUMPTION OF CHICKENS FED FISH SILAGE B SUPPLEMENTED WITH THIAMINE

Protein source	Thiamine (mg/kg feed)	Body weight gain (g/head/3 weeks)	Feed consumption
Whole fish meal	—	576a	908
Boiled silage	—	256b	576
Boiled silage	20	265b	598
Boiled silage	200	254b	584
Silage	—	209c	486
Silage	20	221c	485
Silage	200	219c	497

a, b and c indicate significant difference (P > 0·05)

Table VI
THE NUTRITIVE VALUE OF SILAGE AFTER REMOVAL OF LIPID BY ETHER EXTRACTION

Protein source of diet	Body weight gain (g/head/3 weeks)	Feed consumption	Tryptophan (mg/100 mg protein)
Whole fish meal	561a	964	0·85
Silage ether extracted	477b	952	0·53
Silage	387c	739	0·65

a, b and c indicate significant difference (P > 0·05)
The ether-extracted silage had 74% protein and 0·75% lipid. The diets were prepared isonitrogenous, but the extracted lipid was not replaced

4 Discussion

A series of feeding experiments with chicken showed that growth was poor on silage diets compared to diets containing fish meal of fresh fish, or press-cake meal of spoiled fish. The slow growth was in part due to low feed intake, but also the efficiency of feed conversion was low. Relatively poor weight gain has been obtained also with pigs fed on fish silage absorbed in barley (Sumner, 1978), but fish silage has been shown also to be a good pig diet (Disney and Hoffman, 1976). The experiments were designed to examine whether the poor growth was due to (a) substances generated during enzymatic autolysis, (b) thiamine deficiency, (c) degradation of amino acid, (d) toxic products in the lipid fraction.

The feed conversion efficiency of the silage diets was significantly improved by boiling the silage prior to mixing with corn. There was no difference in the performance of the chickens on the silage diet that was boiled immediately after acid addition and that boiled after 10 days of autolysis, thus suggesting that enzymatic autolysis is of minor importance in the deterioration of the nutritional value of the silage. It appears that boiling either removes a component in the raw material that impairs the chickens' ability to utilize the silage diets, or generates inhibitors of chemical transformations which yield toxic products in the dried feed.

Since some fish species contain high activities of thiaminase, animals fed raw fish diets may get thiamine deficiency. This can, however, not be the reason for the poor nutritional value of the silage diet, since it was not improved by thiamine supplementation to both boiled and non-boiled silage.

We found no significant decrease in any of the essential amino acids during storage at 30°C for 21 days, except for tryptophan; this is in accordance with Tarky et al (1973). The decrease was too small to account for the markedly lower feed conversion efficiency of the non-boiled fish silage diets than the fish meal diets. It may, however, become the limiting amino acid of the silage protein during prolonged storage, and it was certainly limiting in the diet prepared of ether extracted silage (Table VI), but in this case some tryptophan may have been extracted in ether.

An apparent explanation for the low acceptability of the silage feeds is the residual acid, since this is the only difference, except the storage period, between the fish meal and the silage that was boiled prior to acid addition. However, Disney and Hoffman (1976) reported that chicks in fact preferred a commercial dry diet to which 1% formic acid was added, and this has recently been confirmed by us with propionic acid. It seems accordingly that the factors which cause palatability problems, and probably also reduced performance of the chickens, are generated during storage of the acidified fish or during drying of the silage/corn mixture, and that these factors are not the products of enzymatically catalyzed reactions. They are, however, associated with the lipid component of the silage, since ether extraction of non-boiled silage prior to mixing with corn, yielded a feed that was equally well accepted by the chickens as the fish meal diets. However, the feed conversion efficiency was somewhat lower with the ether extracted silage diet, probably due to limiting amount of tryptophan.

In conclusion; to produce a high-quality chicken feed with a high proportion of fish silage, oil has to be removed. Loss of tryptophan can probably be compensated by other feed components. The fact that the press-cake of spoiled by-catch fish gave a good feed, might seem to favour using this fish for fish meal production. However, the toxic press-liquid after 24h of spoilage may contain 30% to 40% of the total protein, and moreover, if the fish spoiled for more than 30h it was technologically impossible to separate press-cake and press-liquid.

5 Acknowledgements

The advice and help of Dr D Cresswell, Dr L F Cook and Director S Ilyas was very much appreciated. We acknowledge FAO for economic and professional support and P T Pumar, Jakarta, for kind co-operation in obtaining by-catch fish.

6 References

BLIGH, E G and DYER, W J. A rapid method of total lipid extraction
1959   and purification. *Can. J. Biochem. and Physiol.*, 37(8), 911–917

DISNEY, J G and HOFFMAN, A. A dried fish silage product. In: *Proc:*
1976   Torry Research Station Symposium on Fish Silage, Aberdeen

EGGUM, B O. Determination of tryptophan. *Acta Agric. Scand.*, 18,
1968   127–131

GAIGER, P J. Fish silage trials in Hong Kong. IPFC Symposium on fish
1978   utilization technology and marketing in the IPFC region; Manila, Philippines. FAO–print

GILDBERG, A and RAA, J. Properties of a propionic acid/formic acid
1977   preserved silage of cod viscera. *J. Sci. Fd. Agric.*, 28, 647–653

STRØM, T, GILDBERG, A, STORMO, B and RAA, J. Fish silage: why not
1980   use propionic acid and formic acid? *This volume*

SUMNER, J. Performance of pigs fed on diets containing fish silage:
1978   evaluation in the commercial situation. IPFC Symposium on fish utilization technology and marketing in the IPFC region; Manila, Philippines. FAO–print

TARKY, W, AGARWALA, O P and PIGOTT, G M. Protein hydrolysate
1973   from fish waste. *J. Fd. Science*, 38, 917–918

Fish silage: why not use propionic and formic acid?

T Strøm, A Gildberg, B Stormo and J Raa

1 Introduction

Fish viscera represents an environmental problem to the Norwegian fishing industry. It has been estimated that 20 000–25 000 tons of fish viscera, representing approximately 50% of the total amounts of viscera from cod, saithe and haddock, are being dumped in close proximity to the factories (Stormo and Strøm, 1978).

From the chemical analysis of fish viscera, Gildberg and Raa (1977) concluded that this raw material could be an important feed resource depending on the development of a practical method for preservation of the viscera and a process for the production of a feedstuff. These authors, following their experiments in the laboratory, suggested that the fish viscera could be preserved by propionic acid at pH 4·3. Such a silage remained sterile for months at ambient temperature, even after mixing with a dry carbohydrate feedstuff, *eg* barley straw meal. The growth of moulds was not observed in such a mixture. Furthermore, by increasing the temperature in the silage the fish tissue is easily autolysed, solubilizing more than 90% of the total protein after two to three days at approximately 30°C. Following this autolysis, most of the lipids could be removed by centrifugation, giving a silage containing 11%–13% (w/w) of protein in 0·1% (w/w) or less, of lipid. Fish silage is produced in some countries: for reviews see Tatterson and Windsor (1974) and Jensen and Schmidtsdorff (1977). However, lipids are usually not separated from the silage. In Norway, the fat content of fish viscera varies greatly depending on the manual removal of liver (*ie* from cod) or not (*ie* from saithe) from the guts. Therefore, in order to produce a feed from different fish viscera with the same lipid content, a method for lipid removal was necessary.

This report describes some additional properties of the fish silage using propionic and formic acid, and outlines the industrial production of animal feed from fish viscera which has recently started in northern Norway.

2 Prevention of growth of *Aspergillus flavus*

Working with silage in Indonesia, Kompiang *et al* (1980)

noticed the occurrence of fungus on silage made with formic acid. The fungus was identified as *Aspergillus flavus*. The growth of this fungus and the following production of aflatoxins represent a serious health hazard and will greatly reduce the application of fish silage, especially in tropical areas. In order to screen the fungicide activity of some acids used for silage preparation, the growth of *A. flavus* was tested at 30°C in a mineral-medium with added glucose as carbon source (Skare *et al*, 1975) in the presence of formic, propionic and sulphuric acids. In the first experiment, formic and propionic acid, respectively, were added separately at concentrations varying from 0·05% (w/v) to 2% (w/v), with pH adjusted with HCl or NaOH to pH 4·5 before inoculation with conidia of *A. flavus*. *Table I* shows that *A. flavus* grew in all media containing formic acid, whereas propionic acid inhibited growth of the fungus at concentration of 0·2% (w/v) or more.

In the second experiment, using the same standard media, formic, propionic and sulphuric acids were added separately to different media to a concentration of 1% (w/v) and pH was varied from pH 2·5 to pH 5·5, using HCl or NaOH, respectively, for pH adjustment. Propionic acid inhibited growth at pH 5·5 and lower values, whereas formic acid inhibited growth at pH below 4·0 (*Table II*). The fungus grew well even at pH 2·5 in the presence of sulphuric acid. These two experiments clearly demonstrate the fungicidal activity of propionic acid and the advantage of using this organic acid when preparing a silage.

3 Industrial production of fish viscera silage

Most of the fish viscera that creates problems for the fish factories are from the gadoid species (cod, haddock and saithe). These species are landed along the Norwegian coast in quantities ranging from 500 tons to 20 000 tons at different factories. For the utilization of fish viscera, it is of importance to preserve the guts as soon as possible after degutting, and the fish viscera silage can then be stored for a certain period before transportation to a central plant for further processing. There are a number of animal farms close to some major fishing districts. Due to the reduced transportation cost by using the silage in the same district, processing plants for fish viscera and production of animal feed could be economically operated (Stormo and Strøm, unpublished). The total process includes three steps: (*1*) silaging, (*2*) silage processing for lipid separation and (*3*) mixing of the feed. In order to demonstrate the process, two areas were chosen for a trial for a period of one year; *ie* Vesterålen region and East-Finnmark region in northern Norway. The central silage-processing plants (capacity 1 000–2 000 tons/year) were built at Myre in Vesterålen and Vardø whereas the silaging equipment was installed at local fish factories in these districts.

Spores of *A. flavus* were inoculated in mineral medium containing glucose and varying concentrations of propionic and formic acids at pH 4·5. The flasks were incubated at 30°C and the growth (+) or no growth (0) of the fungus recorded visually

	Concentration % (w/v)				
	0·05	0·1	0·2	0·5	2·0
Propionic	+	+	0	0	0
Formic	+	+	+	+	+

Table II
GROWTH OF *Aspergillus flavus* IN THE PRESENCE OF 1% (w/v) SULPHURIC, PROPIONIC AND FORMIC ACIDS AT DIFFERENT pH

Spores of *A. flavus* were inoculated in mineral medium containing glucose and 0·1% (w/v) sulphuric propionic and formic acids, respectively, at pH varying from pH 2·5 to pH 6·0. The flasks were incubated at 30°C and growth (+) or no growth (0) were recorded visually

	pH							
	2·5	3·0	3·5	4·0	4·5	5·0	5·5	6·0
Sulphuric	+	+	+	+	+	+	+	+
Formic	0	0	0	+	+	+	+	+
Propionic	0	0	0	0	0	0	0	+

1.	Silaging	At each fish factory (1·5% formic and propionic acid 1:1)
2.	Transportation	By car
3.	Silage processing	At the fish silage processing plant
	a) Storage	10°C or lower
	b) Autolysis	30°C for 2–3 days
	c) Lipid separation	Decanter centrifuge and lipid separator (95°C)
4.	Feed production	At the fish silage processing plant
5.	Transportation	By car, in special tanks or plastic containers (30 litres)
6.	Animal feeding	At the farms. To cattle, sheep, pigs and poultry

Fig 1 Industrial utilization of fish viscera. The different process steps used in the industrial production of an animal feed from fish viscera

3.1 *Production of fish viscera silage*
In order to prepare a good silage, the fish viscera has to be minced, acid must be added to pH approximately 4·5, and the acids must be evenly mixed with the minced fish viscera prior to storage. This can be achieved by adding the acids automatically to the fish viscera before mincing in a Mutrator mincing unit (Mono Mutrator No. 100, Mutrator UK). The amount of acids is regulated by presetting manually the acid dispensing pump according to the mean capacity of the Mutrator. This is an accept-

able procedure (Stormo and Strøm, 1978), although more sophisticated regulatory principles, *eg* pH regulation, may be used. The capacity of the silaging unit is approximately 5 tons viscera silage/day, which is sufficient for the fish factories.

3.2 *Processing of the fish viscera silage*
The utilization of fish viscera locally requires the operation of a plant for further processing of the silage with a capacity of 1 000–2 000 tons/year. The equipment needed in this plant includes storage tanks for the silage, 10m³ tanks for the autolysis of the silage using steam (8 bar, 170°C) for the heating of the silage, decanter centrifuge (Alfa Laval NX 210) for the removal of heavy solids, lipid separator (Westfalia SA 12–00–006) for sludge and lipid removal, and tanks for the storage of lipids and the processed silage. Preliminary experiments showed that maximum lipid removal was achieved by heating the silage to approximately 95°C before centrifugation (Stormo and Strøm, 1979). The lipid content in the processed silage was then reduced to 0·1–0·3% (of wet weight). By processing viscera from saithe guts without prior removal of the liver, the residual lipid in the processed silage sometimes increased to more than 0·3%. Therefore an acceptable upper limit of lipids in the processed silage is set to 0·5% lipid (of wet weight) for commercial production. Raa and Gildberg (1976) reported a residual sediment in the silage after autolysis and centrifugation. Under the conditions used in the industrial centrifugation of the silage, less than 0·1% (w/w) of sludge was removed by the decanter centrifuge and the solid separator together. Even so, the operation of the decanter centrifuge was found to be necessary for smooth production of the processed silage. The processed silage was stored at 15°C or lower with periodic stirring of the silage by using air added at the bottom of the tanks.

In certain regions there are gluts of silage, and transportation to fish meal factories or animal feed producers is necessary. In order to reduce transportation costs, the processed silage can be evaporated to approximately 50% dry matter in conventional evaporators (Stormo and Strøm, 1979).

3.3 *Preparation of the silage feed*
A mixture of 55% (w/w) processed silage, 20% (w/w) minced fish heads, 15% (w/w) barley meal and 10% (w/w) grass meal was tried in preliminary experiments and found acceptable to the farmers (Dahl and Slettbakk, personal communications). This feed is a protein-rich feed (*Table III*) with approximately 35% (w/w) protein (of dry matter). The silage feed has a dry matter content of approximately 33% and is of a thick consistency which makes it easy to handle during feeding. The feed is mixed at the silage-processing plant in batches of 3–5 tons, using a Wolfking mixer and Wolfking homogenizer (7850/2). Afterwards it is transported to the farmers in trucks using special tanks for transportation of wet animal feed, or delivered in 30-litre plastic containers. The farmers keep a stock sufficient for one week's supply.

3.4 *The cost of the silage processing*
In *Table IV* we have listed the investment costs and

production cost for ensiling the viscera, for further processing of the silage and the production of the silage feed. The total costs, including transportation, amount to 0·65N kr/kg silage or approximately 3·80N kr/kg dry matter. This cost is higher than the price of fish meal on the world market (approximately 2 to 2·50N kr/kg fish meal) and should therefore make the production of the silage uneconomic. However, due to special regulations of the prices of the animal feed within Norway issued by the Norwegian government, the production can operate without any direct subsidies. Thus, the price of the silage feed, 3·15N kr/kg dry matters, compares well with a competing commercial feed of 3·30N kr/kg that is made from barley and soy meal with equivalent energy and protein content.

Table III
COMPOSITION OF THE SILAGE FEED
(The silage feed is made up of several components, giving the chemical composition of the feed listed below)

Raw materials		*Chemical composition*	
Fish viscera silage (15% dry matter)	55% (w/w)	Protein	11% (w/w)
Minced fish heads	20% (w/w)	Lipid (soxhlet)	2% (w/w)
Barley	15% (w/w)	Ash	3% (w/w)
Grass meal	10% (w/w)	Dry matter	33% (w/w)

Table IV
INVESTMENT AND PRODUCTION COSTS[1] FOR THE SILAGE PROCESS

	Silaging	Silage[2] processing plant	Silage feed production	Transportation
Equipment	Mutrator	Storage tanks	Mixer	
	Acid dispenser tanks	Autolysis tanks	Homogenizer	
	Pumps, *etc*	Decanter	Miscellanous	
		Separator		
		Steam generator		
		Storage tank		
		Buildings		
Invest. cost, N kr[3]	100 000	2–3 million	300 000	
Capacity, tons/year[4]	500	2 000	2 000	
Prod. cost[5]	0·25	0·30	0·05	0·10–0·15

[1] All prices from 1978
[2] The investment cost is greatly influenced by a possible investment in new buildings
[3] 10N kr equal approximately £1
[4] Basis for the calculation of production cost
[5] Five years' depreciation and 10% interest rate of invested capital

3.5 *Utilization of the silage feed*

Both feeding trials and feeding experiments with cattle, sheep, pigs, poultry and mink were carried out. The result from the feeding trials lasting more than two years indicated a practical daily intake of 4kg, 1kg and 4kg to cattle, sheep and adult pigs, respectively. In total quantity, due to a very high proportion of cattle in northern Norway, most of the silage feed is utilized by these animals. Results from the feeding experiments of the Norwegian University of Agriculture will be published.

4 Discussion

Fish silage has been applied as a feedstuff for many years in Scandinavian countries (Jensen and Schmidtsdorff, 1977). This demonstrates the usefulness of this method for preserving fish and preparing a feed without prior drying of the fish. Our findings have shown the advantages of using formic and propionic acids in preparing a good silage. Bacterial growth is inhibited as well as growth of moulds, including *Aspergillus flavus*. This is of particular importance when preparing silage in tropical regions. The presence of bacterial toxins in the silage is not a problem when preparing silage from fresh raw material. However, there is a risk of toxins, *eg* toxin from *Clostridium botulinum*, being present when ensiling spoiled raw material. The possible thermal destruction of both toxins and spores from *C. botulinum* under the conditions used for lipid removal (95°C) is under investigation. In addition to microbial toxins the presence of toxins from poisonous fish like puffers (*Sphaeroides maculatus*) may be a special problem in tropical waters when preparing silage from shrimp trawlers' by-catch. As far as we know there is no information regarding the stability of such toxins in fish silage.

Upon storage of fish silage there is an increase of ammonia and free fatty acids. The rate of ammonia formation is very low, amounting to 8% of protein nitrogen after 220 days at 27°C (Gildberg and Raa, 1977). The possible loss of tryptophan has been reported to be a problem (Jensen and Schmidtsdorff, 1977; Kompiang *et al*, 1980); however there is no significant decrease of the tryptophan level during storage of fish viscera silage for 60 days at 15°C (Strøm and Eggum, 1979). The level of free fatty acids in the separated lipids from fish viscera is high, due to an initial high level as well as rapid lipid hydrolysis (Gildberg and Raa, 1977). However, this resulting poor quality of the separated lipids is not a general problem with fish silage. In sprat silage the quality of the separated lipid is good (Wignall and Tatterson, 1976).

At present we are engaged in ensiling capelin (*Mallotus villosus*) to be used as fish feed and one of us (Raa) is engaged by FAO to study the application of the silage method in tropical regions like Indonesia. For these raw materials as well, it is of advantage to use formic and propionic acids when preparing the silage, the main advantages being:

(*a*) Fish or fish waste is preserved against bacterial and fungal growth, even in the presence of added carbohydrate sources.

(*b*) Growth by *Aspergillus flavus* is inhibited.

(*c*) By heating the silage to approximately 30°C and storing for two to three days, 85–90% of the total protein is solubilized.

(*d*) Lipids are easily removed from the silage after autolysis, using conventional equipment, resulting in silage containing less than 0·5% (of wet weight) lipid.

(*e*) Neutralization of the silage feed prior to feeding is not necessary.

(*f*) A balanced stable feed can be prepared without drying of the feed or the feed ingredients.

5 References

GILDBERG, A and RAA, J. Properties of a propionic acid/formic acid
1977   preserved silage of cod viscera. *J. Sci. Fd. Agric.*, 28, 647–653

JENSEN, J and SCHMIDTSDORFF, S. Fish silage; low fat and soluble fish
1977   protein products. IAFMM symposium, Szczecin, Poland
KOMPIANG, I P, ARIFUDIN, R and RAA, J. Nutritional value of ensil-
1980   aged by-catch fish from Indonesian shrimp trawlers. *This vol-
        ume*
RAA, J and GILDBERG, A. Autolysis and proteolytic activity of cod
1976   viscera. *J. Fd. Technol.*, 11, 619–628
SKARE, N H, PAUS, F and RAA, J. Production of pectinase cellulase by
1975   *Cladosporium cucumerinum* with dissolved carbohydrates and
        isolated cell walls of cucumber as carbon sources. *Physiol.
        Plant.*, 33, 229–233

STORMO, B and STRØM, T. Ensilering av fiskeslo. Report from FTFI,
1978   Tromsø) Norway
—— Videreforedling av ensilert fiskeslo. Report from FTFI, Tromsø,
1979   Norway. (In press)
STRØM, T and ÉGGUM, B. (In press)
1979
TATTERSON, I N and WINDSOR, M L. Fish silage. *J. Sci. Fd. Agric.*, 25,
1974   369–379
WIGNALL, J and TATTERSON, I N. Fish silage. *Process Biochem.*, 11,
1976   No. 12, 17–19

Recovery, utilization and treatment of seafood processing wastes

L F Hood and R R Zall

1 Introduction

Shellfish and finfish processors are faced with increasing problems of waste handling and disposal, plant-sanitation, raw material availability and cost, production efficiency, and escalating labour and energy costs. All of these factors significantly increase processing and product costs. Processors are continually looking for opportunities to increase production efficiency and profitability. Conversion of unused waste materials into marketable products not only provides such opportunities but reduces waste disposal problems.

The objectives of our work have been threefold: (*1*) recovery of protein, other nutrients and flavour materials from fish-processing wastes, (*2*) conversion of the recovered materials into food ingredients or marketable food products, and (*3*) development of procedures for (pre)treating the non-recoverable solids. Our goal has been to attain total utilization of seafood processing wastes for food or feed.

2 Preparation of clam juice from washwater

Surf clams (*Spisula solidissima*) are widely utilized as a source of minced and chopped clams, clam juice (broth), and clam strips. After the clams are shucked, the meat is washed, minced and packaged for distribution (*Fig 1*). It is necessary to wash the minced clams to remove the sand embedded in the tissue during dredging. The resulting washwater only contains about 0.5–1% solids. However, it has a distinct clam-like odour and flavour. We have developed a method for converting this washwater into a marketable food product (Hood *et al*, 1976). The process is being applied commercially and the resulting product is being marketed as clam juice (*Fig 1*).

Although the process for converting the washwater to clam juice is not a complicated one, it does include several important steps that are critical to the high quality of the finished product. The minced clams are washed in a rotating washer. After 2h, the water in the washer is transferred to a steam-jacketed kettle and boiled. The boiling step is essential to prohibit the subsequent development of fishy-like flavours. It also serves to concentrate the liquid. The duration of boiling is 10–60min, depending on the desired solids concentration in the finished product. Following boiling, the concentrated clam washwater is canned, retorted and subsequently marketed as clam juice.

In developing this process, several methods were evaluated for concentrating the washwater. These included boiling, vacuum evaporation and ultrafiltra-

tion. All of these methods were effective in concentrating the washwater to two to three times its original solids content. Products were judged by a five-member taste panel, trained to judge flavour, aroma and colour characteristics peculiar to clam juice. Six to eight samples were evaluated during each panel session. Panellist fatigue resulted if more samples were included. The processed (concentrated) washwater was compared to commercial clam juice. Two types of evaluation forms were utilized (*Figs 2* and *3*). The seven-point hedonic scale was used to record judgements on the processed washwater relative to canned clam juice. The second form asked panellists to use descriptive words to characterize sample flavour. After each tasting session, panellists discussed individual impressions and usually came to a consensus on which sample had the best clam flavour or most closely resembled commercial clam juice.

The processing methods evaluated did not yield

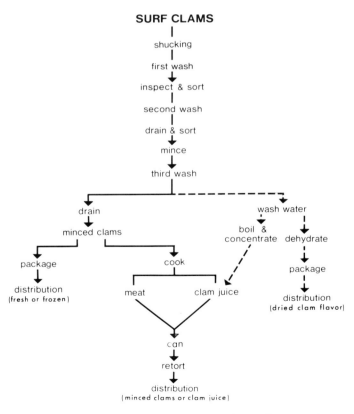

Fig 1   Flow diagram of surf clam processing. Schemes for converting washwater to clam juice or to dried clam flavour are shown on right

Tasting Session I

FLAVOR EVALUATION OF CLAM WASH WATER

Directions: You have before you 10 samples of processed clam wash water and a sample of canned clam juice "R".

Taste the reference sample "R" first and evaluate samples 1-10 against it. Samples 1-6 are unsalted so taste these first and then proceed to 7-10, the salted samples.

Flavor attribute	Intensity rating									
	Much less -5	-4	Mod. less -3	-2	Slightly less -1	Refer- ence 0	Slightly more 1	Mod. more 2	4	Much more 5
Clam flavor										
1										
2										
10										
Fish flavor										
1										
2										
9										
10										

Fig 2 Taste panel form for comparing washwater with commercial clam juice

Tasting Session II

FLAVOR EVALUATION OF CLAM WASH WATER

Directions: As you taste the following 10 samples, please jot down a few descriptive words which characterize the flavor of these samples.

Please base your comments* on the sample with respect to "R".

Sample	Description
1	
2	
3	
10	

* Feel free to make comparisons such as, "#3 is more clammy than #4".

Fig 3 Taste panel worksheet for developing flavour profile of clam washwater

equivalent products. In general, washwaters concentrated by vacuum evaporation or ultrafiltration were more fishy than those concentrated by boiling. Boiling at 95–100°C apparently removed most of the volatile flavours responsible for the undesirable fishy flavour. Lower temperature boiling at 50°C or 80°C (*ie* vacuum evaporation) removed some of the volatiles but did not yield as good a clam-flavoured product as the washwater boiled at atmospheric pressure. The condensate from the vacuum-evaporated samples tasted fishier than the corresponding concentrate. By comparing the results of these three processing techniques, it was apparent that the compounds responsible for fishy flavour in clam juice were volatile. Obviously the flavour, odour and acceptability of clam juice and other clam products are dependent upon their chemical composition and the processing treatments that they are subjected to. We are currently investigating the relationship between processing conditions and the composition of flavour constituents.

Retorting is critical to the development of optimum clam flavour in clam juice. It yields a sweeter, clam-flavoured product than was evident in the non-retorted juice. Obviously, the process by which juices are canned and retorted for safety reasons also leads to a more highly flavoured product than if the product was not retorted.

Storage stability studies have indicated that canned and retorted concentrated washwater maintains a sweet clam flavour when stored at room temperature for six months. At the end of that time, concentrated washwater was judged equivalent in flavour to commercial clam juice.

The process for converting clam washwater into clam juice is now being applied commercially. In addition to the direct economic benefits from marketing a material that was heretofore discarded, the process has resulted in a reduction in the BOD of the plant effluent and has increased the capacity for manufacturing clam juice without utilizing clams specifically for that purpose. With the price of surf clams increasing rapidly, this conversion of a waste material into a marketable food product has and will be of significant economic and pollution control benefit to the seafood processing industry.

3 Dehydrated clam flavour

Other uses for the clam washwater have been explored. One that is promising and should lead to substantial economic benefits for clam processors is the dehydration of the washwater to form a clam-flavour ingredient that could be used in formulated foods such as soups, dips and snacks (*Fig 1*). The dried clam flavour has several advantages over the clam juice as a food ingredient. These include lower storage and distribution costs, and greater versatility. In addition, dehydrated flavours are in a different product category than the clam juice and therefore would command a higher price as a food ingredient.

Several dehydration methods were evaluated for converting the clam washwater into a dried powder (Joh, 1978; Joh and Hood, 1979). These included drum-, spray-, and freeze-drying. A trained taste panel compared the rehydrated dried washwater with commercial clam juice. All of the dehydration methods yielded a sticky, hygroscopic product. Therefore a low DE dextrin was blended with the washwater before drying. All pro-

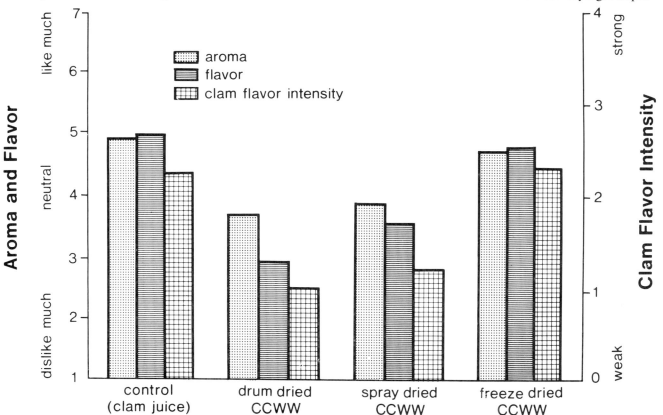

Fig 4   Effect of dehydration methods on the organoleptic quality of rehydrated dried clam flavour

357

ducts co-dried with the dextrin were non-hygroscopic and free-flowing. Freeze-drying produced a product that, when rehydrated, had an equivalent aroma, flavour and clam flavour intensity to clam juice (*Fig 4*). The drum-dried powder had a burned, caramelized flavour. In addition, it had poor solubility and dispersibility in water. Spray-drying yielded a more soluble and dispersible powder with a slightly better flavour than the drum-dried powder. Both spray-drying and drum-drying required larger concentrations of dextrin than freeze-drying in order to produce a powder with acceptable colour and physical properties. The higher amount of dextrin diluted the clam flavour and a grain-like flavour resulted.

The freeze-dried clam washwater had good dispersibility and solubility. The addition of the dextrin as a co-drying agent did not appear to affect the functional properties. Nevertheless, it was essential to include it in order to reduce the hygroscopicity of the dried product. The whiteness of the product was directly proportional to the dextrin concentration. The dehydrated clam flavour was packaged in the vacuum containers and in jars and stored for 90 days at 4°C, 24°C and 40°C. All products were judged to be equivalent after the 90-day storage period.

The dehydrated clam flavour can serve as an effective flavouring agent in seafood chowders. When 0·5% was added to chowder, the aroma, flavour, clam flavour intensity, and overall acceptability of the chowder were slightly improved (*Table I*). When it was added at the 1% level, organoleptic quality of the chowders was better than the chowder containing 0·5%. The taste panel judged the chowder containing 1% clam flavour and no clam meat to be slightly better than chowder made with 3·5% clam meat. Thus it is apparent that the dried clam juice can be used as either a replacement for clam meat or a flavour enhancer.

Table I
ORGANOLEPTIC QUALITY OF SEAFOOD CHOWDER MADE WITH DEHYDRATED CLAM FLAVOUR[1] INGREDIENT

Chowder	Aroma	Flavour	Clam flavour intensity	Overall acceptability
A[2]	5·0 ± 0·7	5·6 ± 0·7	2·5 ± 0·8	5·2 ± 1·3
A[3]	5·4 ± 1·0	5·8 ± 1·0	2·5 ± 0·9	5·4 ± 1·5
B[4]	5·0 ± 1·1	5·0 ± 1·0	2·2 ± 1·0	4·4 ± 1·5
B[5]	4·9 ± 0·8	5·2 ± 1·2	2·4 ± 0·9	4·6 ± 1·2
B[6]	5·6 ± 0·7	5·7 ± 0·6	2·8 ± 0·4	5·6 ± 1·3

[1] Mean ± SD, n = 10
[2] Cornell seafood chowder with 3·5% clam meat
[3] Cornell seafood chowder with 3·5% clam meat + 0·5% CFI
[4] Cornell seafood chowder without clam meat
[5] Cornell seafood chowder without clam meat + 0·5% CFI
[6] Cornell seafood chowder without clam meat + 1·0% CFI
Source: Joh and Hood, 1979

4 Recovery of meat from clam shells

In addition to the liquid effluents emanating from clam processing facilities, there are solid wastes generated such as shells, bellies, mantles, and parts of the adductor muscles. The meat represents about 30% of the total weight of the clam (*Table II*). We have examined various methods for removing and recovering the clam meat that adheres to the discarded shells (Cho, 1977; Zall and Cho, 1977). There is a substantial amount of meat in this

category, representing about one-half of the total edible clam meat.

In our studies, shells with adhering meat were collected before and after shell chopping and methods were devised for releasing the meat from the shell. Various techniques to accomplish this purpose were examined. They included immersing whole shells and shell parts with adhered meat fragments into boiling water with and without agitation, heating shell fragments in a muffle furnace at 200–600°C for up to 5min, and heating shell fragments with attached meat in a pressure cooker at 121°C at 15psi for up to 15min. Heating in boiling water for 2min caused the mantle to be released from the shell. However, the two adductor muscles, which control bivalve action, failed to come off the shell after a 15min cooking period. These muscles could be removed from the shell after cooking for 12min if minor scraping was used to free the meat. Anterior adductor muscles were always more difficult to remove from the shell than the posterior ones. Dry heat (muffle furnace) resulted in burned meat that could not be removed from the shell.

The preferred method for removing meat from shell fragments was by heating in a pressure cooker at 121°C at 15psi. This procedure produced a juice product in addition to the meat. Again, as in the case of other heating techniques, the anterior muscles were more difficult to remove than the posterior muscles. While pressure heating did free meat from shells, it also cooked the meat. Thus, the salvaged product could not be considered raw clam meat. In fact, the salvaged adductor muscle was much tougher (as measured by the Warner-Bratzler Shear Press method) than the unheated meat.

The amount of meat, juice and shells available for recovery by pressure heating is significant. Adductor muscles alone represent about 6% by weight of the total clam and about 20% of the total edible clam meat. The value of the slightly cooked muscle might be less than that of the foot or neck parts of the clam. Nevertheless, the recovered muscle could be utilized as an ingredient in cooked clam products such as chowder.

Table II
WEIGHTS OF PARTS OF THE SURF CLAM (*Spisula solidissima*)

Body part	Weight (% of total)
Shell	50–55
Juice	10–15
Meat	
Foot	11
Adductor muscle	6
Neck	2–3
Mantle	7–9
Belly	3–8

Source: Zall and Cho, 1977

5 Liquefied clam bellies

Not all fishery wastes can be readily converted to human food. Often the volume of the waste is not sufficient to economically justify processing, or the nature of the waste does not lend itself easily to handling and processing. Clam bellies are an example of these types of materials. They are the intestines and other visceral wastes and represent 3–8% of the total clam (*Table II*). The volumes of bellies generated are usually too low to justify any processing. In addition, they are slimy and full of intestinal contents and are difficult to handle and process. In a matter of hours after shucking, clam bellies will reach high microbiological counts, become odoriferous and must be quickly discarded. The challenge is to inexpensively stabilize clam belly waste and thereby produce useful by-products.

We have applied some of the procedures developed at the Torry Research Station for the preparation of fish silage to the preservation and liquefaction of clam belly waste. Clam bellies were preserved from microbial spoilage by treatment with formic acid (1·3%) or sodium chloride (19·5%). Specimens were stored in sealed glass jars at 4°C, 35°C and 55°C for up to 155 days. Microbial populations rapidly decreased, particularly at the elevated temperatures. Thermophilic plate counts, and yeast and mould counts, were negligible. In general, the brine-formic acid treatments arrested or reduced the microbial populations, thus preventing spoilage.

The use of either formic acid or brine would allow the processor to store the small daily volumes of waste in drums until they have enough by-product to economically ship to another area, to reprocess it into other products, or simply to dispose of it in a landfill site. In short, it provides a method for the short- or long-term storage of fishery waste for potential utilization in animal feed or as fish bait.

The concept has been extended to the development of a commercial bait for the crab and lobster fisheries. The product is blended with a gelling agent, canned and stored for an indefinite period. At the time of use, a hole is punched in the can, the can is placed in a trap, and the contents ooze out to attract lobsters or crabs. In addition, those animals that enter the trap cannot eat the remaining bait. The latter point is important because with many baits presently in use, the first animal that enters the trap eats up all the bait and the trap can no longer attract additional animals. We ran a 'taste panel' on lobsters, using the canned liquefied clam bellies. The canned product was equally as effective as redfish, the traditional lobster bait, in attracting lobsters into the traps. While we were unable to interview the lobsters that were attracted into the traps, they appeared content and satiated by the psuedo-redfish.

6 Utilization of clam shells

Clam shells are an underutilized resource. They represent about 50% of the total clam (*Table II*). After the clam is shucked and the foot removed, the shells with the adhering meat and bellies are either returned to the water or are taken to an isolated location where the meat 'disappears' over time. The remaining shells are pulverized and used for driveway coverings. In any case, the objective is to get rid of the shells in order to eliminate a pollution and environmental problem. Little emphasis has been placed on finding uses for the shells that would be of significant economic benefit to the shellfish processor.

In some cases, shellfish processors are located near agriculture croplands. It seemed logical to us that the shells might be useful as liming agents for agricultural lands. They are high in calcium ($CaCO_3$; carbonate equivalent is about 95%). Unfortunately, they contain very low levels of magnesium. This would suggest that the ground surf clam shells could serve as a liming material, but that additional magnesium would have to be included in some form.

The idea of using shellfish shells is not a new one. Oyster shells have been ground and used as chicken feed supplements. They are about 85–90% calcium carbonate. The feasibility of using clam, scallop or oyster shells will clearly depend upon the relative economic factors in the particular location being considered. Conversion of these shells to marketable products for the agricultural industry would result not only in economic benefits to the shellfish processor, but the correction of an environmental pollution problem. This problem is particulary significant in the summer months when the shells are stockpiled and the odour from the decaying meat becomes very objectionable.

7 Uses for other shellfish parts

In addition to those already described, there are parts of shellfish that are excluded during processing and that have heretofore been discarded. One example is the mantle of the bay or sea scallop. In scallop shucking, the adductor muscle is retained as the marketable scallop meat and the mantles and bellies are discarded. Like the clam situation, these discarded parts represent a wasted resource and a pollution problem. We have been evaluating these mantles as potential ingredients in processed seafood products such as seafood chowder.

Scallop mantles are flat, muscle-like membranes about $2 \times 8 \times 0.3$mm. They are often discarded with the shells in the waste stream. They are relatively easy to remove from the shell and can be chopped, washed, canned and marketed as a chowder ingredient through the same marketing channels as minced clams. A seafood chowder that has been developed by other Sea Grant researchers at Cornell University contains 10% of these scallop mantles (Baker *et al*, 1976). They impart a rich shellfish flavour to the product as well as contributing to the level of meat in the product. Since they are a low value product, they can compete economically with other meat ingredients (*ie* minced clams) as a chowder ingredient. The seafood chowder developed at Cornell also contains the clam juice that is manufactured from minced clam washwater. As stated earlier, it is possible to formulate this chowder without using minced clams by substituting the dried clam flavour derived from the washwater for clams (*Table I*). Thus, a marketable product has been developed from several heretofore unused waste materials from shellfish processing.

Scallop mantles can also be readily converted to a puree. The purée can be canned, retorted at 121°C (15psi) for 15min and kept for extended periods of time. Modified food starch and pyrophosphate-hexametaphosphate are included in the formulation. This purée could serve as a flavouring ingredient for dips, sauces and croquettes.

8 Marrying fin and shellfish wastes

Often shellfish and finfish processing plants are located near to each other. The concept of marrying fin- and shellfish waste to produce marketable products is one that has not been explored in any great detail. Obviously, it would require the co-operation of the plant operators. Such co-operation would undoubtedly be facilitated by the economic benefits to be derived. For example, on Long Island, New York, there is a shellfish processor and a flounder filleting plant located one mile from each other. The filleting operation generates large quantities of frames or racks after the fillets are removed. Commonly these frames are converted to fish meal. We have been exploring opportunities for converting these racks or extracts of them into food ingredients. Obviously, food ingredients would command a higher market value than fish meal. There has been some interest in mincing the cleaned racks and using them as a pet food ingredient. The concept that we have been investigating is to extract the flavour from the racks and to prepare a fish broth. This broth could be combined with the scallop mantles or salvaged clam meats from the clam processing to produce a seafood chowder base. Flavour problems have been encountered after retorting the fish broth. This is the subject of a thesis problem currently being carried out by a Sea Grant trainee at Cornell.

9 Fish scales as a coagulant

Fish scales represent an enormous resource for which no practical uses have been developed. They constitute about 1% of the total weight of the fish. The potential of the fish scales as a coagulant has been investigated by another Sea Grant trainee at Cornell (Welsh, 1978; Welsh and Zall, 1979). We see these materials as valuable aids in the pretreatment of food plant processing wastes. Chitosan has gained substantial notoriety in recent years as a flocculating agent. The results of our work suggest that dried and ground fish scales can function as effectively as chitosan as a flocculating agent (*Fig 5*). Preparation of the scales was as follows. Carp and porgy scales were dehydrated at 46°C for 24h and subsequently milled to less than 500μm. Dispersions (0·1%) were prepared and their coagulating capabilities compared to chitosan, alum and ferric chloride. Coagulating effectiveness was evaluated on egg washing and scallop shucking wastewater and on fruit juice processing effluents.

10 Brine recovery from fish processing

Salt (NaCl) brines are used in many fish processing or 'curing' operations. In many fresh fish filleting operations 2·5% to 5% (w/v) salt brines are often used to handle and store fresh fish fillets overnight. Some fillets are even packed with brine. Fish filleters are under increasing pressure from municipal sewage plant operations to reduce the salt content of the fishery's waste discharge. Consequently, we have been studying methods for the reclamation of brine. Simultaneously, we have explored methods for recovering the protein from the brine solution. The process is summarized in *Fig 6*

11 Separation of sand from press liquor

There is increasing interest among fish filleters in processing the frames or racks into fish meal because of the

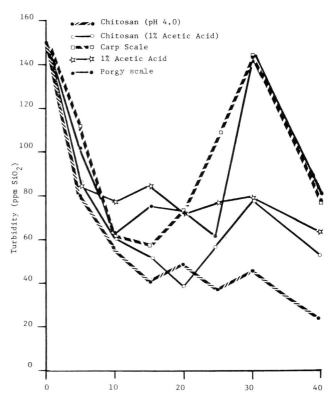

Fig 5 Effect of polymer concentration on the turbidity of scallop shucking wastewater. Determinations were made at the optimum pH for each coagulant

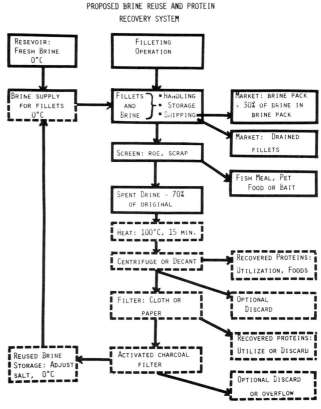

Fig 6 Proposed brine re-use and protein recovery system

increasing emphasis on pollution control. In the commonly used wet-rendering fish meal process, a press liquor by-product is produced which could be further processed by centrifugation and evaporation into fish oil and solubles. A processor who is already centrifuging press liquor has noted erosion of his centrifuge parts due to sand associated with the fish. The sand can potentially cause more economic damage to centrifuge parts than the returns from sale of fish oil and solubles. We found that the sand can effectively be removed by placing the fresh press liquor in Imhoff cones and allowing the sand to settle. Separation can be accomplished within 4h. After that time, the concentration of total solids in the bottom of the cone is about four times greater than the concentration in the top oily layer (0·5% *versus* 2% of the total solids on a dry weight basis). However, the bottom portion (settled sludge) represents only 5% of the total volume, whereas the top oily layer represents about 20–30%. The middle layer contains an average of about 1% sand (on a dry weight basis of the total solids). By allowing the bottom and most of the middle layer to bypass the centrifuge and go directly to the condenser, more than half of the sand can be excluded from the centrifuge. This procedure saves wear and tear on the centrifuge and reduces manpower and energy requirements.

12 Characterization and treatment of clam processing wastes

Not all waste material can be converted to marketable products. For a variety of reasons, some must be discharged from the plant as waste. In some cases, it is necessary to treat the waste streams coming from plants. Municipal sewage systems may not exist, or if they do, are not capable of handling the plant effluents. Therefore, knowledge of the treatability of effluents can be a valuable assistance to the processor.

We have measured the water flow rates and characterized the waste stream from a typical surf clam processing plant and have evaluated several methods for treatment of the waste material. The washwater from the first two washing stations and after the mincing operation were analysed for proximate composition. The following parameters were evaluated: BOD, COD, total solids, suspended solids, volatile suspended solids, TKN, NH_3, pH and PO_4. Effluent quality was determined by measuring turbidity, suspended solids and volatile suspended solids. The composite clam processing waste was amenable to aerobic biological treatment. After a retention time of two and a half days or longer, better than 90% of the COD and BOD was removed (*Table III*).

Results of these studies indicate that clam processing wastes are amenable to aerobic biological treatment. However, it would not appear to be possible to recover usable food products by the coagulation-sedimentation

Table III
REDUCTION IN CHEMICAL OXYGEN DEMAND (COD) AFTER AEROBIC BIOLOGICAL TREATMENT

Date	Solids retention time (SRT)			
	2·5 days	5 days	10 days	15 days
	%	%	%	%
8/1		94·5	96·4	98·9
8/4		94·5	97·8	98·9
8/6		95·8	97·9	98·9
8/7		95·8	96·9	98·9
8/13	96·6			
8/14		91·9	91·2	95·1
8/15	89·7	93·4	91·2	94·9
8/18	91·3			
8/20		93·2	89·9	94·0
8/21	94·4			
Average	93·0	94·1	94·4	97·0

process using chitosan, alum or ferric chloride. Less than 10% of the total solids were recovered by chemical coagulation and sedimentation. These waste streams are quite dilute (less than 1% solids) and therefore may in part account for the ineffectiveness of the coagulating-sedimentation treatments (Zall *et al*, 1976).

13 Acknowledgements

This research was sponsored by the New York Sea Grant Institute under a grant from the Office of Sea Grant, National Oceanic and Atmospheric Administration (NOAA), US Department of Commerce. Special thanks are due to our co-workers: D Brown, I Cho, R Conway (Taft), J Green, W Jewell, Y Joh, W Kim, S Pettigrew, J Plock, G O'Shea, M Switzenbaum and F Welsh.

14 References

BAKER, R C, REGENSTEIN, J M and DARFLER, J M. Seafood chowders. 1976 *Development of products from minced fish*. Booklet 1, New York Sea Grant Institute

CHO, I J. Cleaning surf clam meat and recovering edible food from the 1977 processing waste. MS thesis, Cornell University, Ithaca, NY

HOOD, L F, ZALL, R R, and CONWAY, R L. Conversion of minced 1976 clam washwater into clam juice: waste handling or product development? *Fd. Prod. Dev.*, 10(10), 86

JOH, Y. Preparation of clam flavoring ingredient from clam washwater. 1978 MS thesis, Cornell University, Ithaca, NY

JOH, Y and HOOD, L F. Preparation and properties of dehydrated clam 1979 flavor from clam processing wash water. *J. Fd. Sci.*, 44, 1612–1614, 1624

WELSH, F W. Fish scales: a coagulating aid to recover colloidal solids 1978 in food processing wastewater. MS thesis, Cornell University, Ithaca, NY

WELSH, F W and ZALL, R R. Fish scales: a coagulating aid for the 1979 recovery of food processing wastewater colloids. *Process Biochem.*, 14(8), 23–27

ZALL, R R and CHO, I L. Production of edible foods from surf clam 1977 wastes. *Trans. Amer. Soc. Agricult. Eng.*, 20(6), 1170

ZALL, R R, HOOD, L F, JEWELL, W J, CONWAY, R L and SWITZEN- 1976 BAUM, M S. Reclamation and treatment of clam washwater. *Proc: National symp. on food processing wastes, USEPA, Cincinatti, Ohio*

9 Fish technology and its transfer

The context of fish technology

A E Malaret

1 Introduction and background

Fishing technology and fish technology comprise very different aspects of primary production, industry and commerce of fish and shellfish.

Fishing technology covers devices and naval equipment that improve fishing performance, such as design and shipbuilding, electronic navigational aids, radar, propulsion systems, echo sounders, sonar and other underwater equipment for tactical information, nets, lines, traps, accessories and appliances to handle them. Catches depend on systems for handling fish, processing and its preservation on board.

Fish technology covers pre-treatment, processing, reprocessing, packaging and quality control.

In a restricted sense, fish technology is present during the services and trade stage in landing operations, handling and transport of raw materials and products, storage, distribution and exhibition of final products. In a wider sense, marketing and promotion of consumption is technology, too. Usually, features of frozen fish products are closely associated with selling and promotion methods.

The purpose of this paper is to examine institutional and socio-economic influences that affect fish technology; many innovations are directly taken from scientific and technological applications in other fields, such as food in general, packaging, transport, textiles, electronics.

Briefly, fish technology is developed within a double context: science and technology in general and the socio-economic circumstances of the world and each country.

Many authors on fish technology have pointed out the influences of this double context. Some make incidental references to aspects in which they are specifically interested; others give a more detailed analysis and show in an erudite way all the connections they find.

Among the former, Burgess (1974) reflects that 'the scientist or technologist, knowing the problems and the work necessary to find their solution, may see his contribution as the key to the difficulties of the manufacturer. The manufacturer, grappling also with questions of finance and management, may feel that the technology contribution is not necessarily the only key'. At the same time, attitudes towards technology of state-owned enterprises in the USSR differ from those of the free market world. According to Mathisen and Bevan (1968), to understand the differences it is necessary to know how investments and interests are treated in each area.

Trade and occupation policies, maritime jurisdiction and national strategy can have decisive effects on fish production (Chapman, 1968) and consumption can be affected by religious beliefs (Bell, 1968).

Kreuzer (1972), after extensively describing the role played by fish in culture since prehistoric times, refers to technological change at present and points out that 'more important than technology is the human element which is often overlooked in assistance programmes although it is the most complex factor, involving, as it does, customs and habits inherited through untold generations, religious and other beliefs, philosophy of life, ways of thinking, dressing, eating and patterns of behaviour, summed up by the broad concept of a people's culture. It is essential that in such programmes the considerable thought and care given to the economic and technical aspects of assistance are given in equal measure to consideration of the human element.'

Christy and Scott (1965) think that 'the rate of growth in innovation is far from uniform throughout the world. It varies considerably between low- and high-income countries, among major fishing nations and even within nations. In the low-income countries that are dependent upon coastal waters the adoption of established techniques can have significant effects upon their industries. The problems they face are not those of developing new devices but adopting and adapting old ones. They face considerable difficulties in revising marketing institutions, in the construction of transport facilities, and in adopting handling and processing techniques, as well as in developing their efficiency in locating and catching fish. The problems facing the major fishing nations, however, are quite different. The development in the fishing process, together with the growing demand, are leading to increasing competition for the fishery resources owned in common by the world. Technological innovations that tend to reduce the cost of catching fish place greater strain upon the resource.'

Using a historical approach, Coull (1972) explains interactions of culture, economy, politics and technology of fisheries in Europe. He thinks that present patterns of production, transport, storage and consumption are in large measure the creation of the greatly accelerating pace of development during the twentieth century; they have been built on foundations laid over a long historical period, and considerable sectors of the patterns are the legacy from previous centuries. Coull points out that acceleration of the tempo of economic activity in many fields leads to a more intensive exploitation of fishery resources, together with a generally increased range of fishing craft from their bases. Associated with this there have been outstanding technical innovations together with developments in marketing and commercial organization.

Coull abounds in special observations for different circumstances and levels of development. 'The spread of the railway network, which allowed a much wider market to be reached with iced fish from the coast, greatly stimulated trawling, as did the build-up of industrial city populations, which meant that consumption became concentrated more at particular points. It was, however, the application of steam-power to the vessels themselves which really made the superiority of trawling over other methods of white fishing decisive.' He says that political and customs unification between 1828 and 1871, in Germany, and the rapid growth of industrial population concentrations along with the extension of the railway system, helped the transition to a more technological and capital intensive fishery during the nineteenth century.

Coull's verifications are useful for forecasting: 'another notable trend has been the decline in importance of seasonal fisheries, which formerly accounted for the main efforts: big capital investment entails that the fleets and marketing system remain in operation all the year'.

From the point of view of the engineer, Eddie (1971a) carefully describes every achievement in fish technology. He speaks about advanced industrial countries and complains 'we have been slower to develop the fisheries than other food industries, as a result of the mis-match between the large range of scientific and technical expertise that needs to be deployed and the relatively unsophisticated organization and structure of much of the world fishing industries'. In another paper Eddie (1971b) insists on his preoccupation about the structure of the industry and its need for technical assistance: 'the evolutionary process is, however, too slow to provide an adequate response to rapid changes in the economic environment such as occurred to the British fishing industry in the 1950s'. He repeats 'technical development tends to benefit the user, not the manufacturer'. To be complete, Eddie's thinking should say that the bill for research should be paid by society, because it states the whole problem. The rapid changes in economic environment and the relatively unsophisticated organization and structure of fishing industries are part of the whole problem. If this is true for highly developed countries it is much more true for the developing world.

2 Technology and economic development

Technology makes it possible to incorporate new natural resources in economic activity. It reveals the virtues which are hidden or limited because of locality, concentration or intrinsic characteristics. It also increases work efficiency.

Since the Industrial Revolution the effects of technology on the economy have become increasingly more noticeable. Each innovation has broadened the economic horizon, while the existing open commercial system has allowed mankind a widespread participation in the benefits of technical progesss. There was relative harmony between the generation of savings to finance the corresponding investments and demand, resulting from demographic growth, habits of consumption and purchasing power.

Technology and the economic development of the world have gone hand in hand. The principal countries led the process, but the rest of the world followed at a reasonable distance. The innovations they generated could be applied universally. They were neither restricted in practice nor did they seem incompatible with the general usefulness for the peripheral countries.

But for the last 50 years technology has developed more systematically. The rate of increase in innovations has risen inexorably and each of the mentioned indicators which define the economic structure of each country has a different rhythm. The creation of technology has promoted the economic growth of industrial countries which in turn has given more opportunity for the development of technology, by increasing savings for investments and the power to purchase finished goods. But these effects do not spread with equal or similar dynamics to the peripheral countries.

Cultural and political reasons and restrictions on international trade prevent the wave generated by the creation of technology in leading countries from reaching those outside the sphere with the same strength and do not allow for the multiplication of the phenomenon of interaction on a world-wide scale.

Therefore, two interacting circuits are created. One is restricted but extraordinarily dynamic, encircling the most advanced countries which creates advanced technology suitable to their sophisticated standard of development. The second is universal but less active. Basically it comprises part of the technology created in the industrial countries which is used to prepare exports from the countries outside their sphere or to meet the needs of consumption of those with more purchasing power. A very high percentage of new technology is not within the grasp of, and does not resolve the needs of the greater part of the world population (between 60% and 80%).

On comparing the economic growth for the last few decades, as calculated by the World Bank (1978), it is noted that rates of the group of most advanced countries were three and a half times higher than those of the most backward countries (industrialized countries: 3.4%; countries of medium income: 2.8%; low income countries: 0.9%). As a consequence the gap in the respective gross national products has become wider. For the last 40 years, the ratio between the GNP of the high and medium income countries was 2:1, and that of the low income countries 10:1. Ten years ago the same ratios were 3:1, and 20:1, respectively. Now (1976 data) the ratios are 8:1 and 40:1.

To a great extent the distance now existing between the GNP of the industrialized countries (US$6 200 *per capita, per annum*) and those that follow (medium income: US$750, and low income: US$150) is the result of more efficient technology. The difference in income is spent on technology: new equipment is short-lived since rapid innovations soon make it obsolete; to meet basic needs consumer goods are more sophisticated; moreover, new needs calling for goods and services requiring a high percentage of technology have arisen.

The gap between incomes at the same time created and sustained by technology is growing wider since, in the last analysis, technology develops for the needs of the industrial world, and not for those of the rest of the world.

In like manner, the phenomenon of polarization appears on smaller scales. The group of medium income countries is quite large and among them there are marked differences in the rates of growth and income level which frequently reach 50% of average. Within each country and markedly more within the less developed ones, minorities representing 10% to 30% of the total population have average incomes 10–20 times higher than the mean income of the rest of the population.

Just as in countries of different levels of income as in the higher internal socio-economic strata of the peripheral countries, the degree of wealth is to a certain extent linked with the use of technology produced in industrial countries.

3 Appropriate technology

The adoption of technology created in industrial countries is the result of economic motivations of demand, commercial structures and imitation of consumer patterns considered convenient, comfortable or fashionable.

Frequently the techniques are efficient but they do not always suit local conditions perfectly, nor are they expected to solve all situations nor to develop all the opportunities arising from a social and physical environment which is usually different from the model for which they were designed.

Sometimes certain techniques are also very costly if the ratio of manpower and financing costs is considered, or they are too complex for the personnel to handle them. In like manner there are efficient techniques which, however, devastate resources and the environment.

All this has given fuel to a long debate to justify the need for the creation of techniques considering the needs of the peripheral countries. Different points of view consider 'an appropriate or intermediate technology', or 'technology with a mild or low effect' (according to the Sierra Club), 'an alternative technology' (in California), 'a village or rustic technology' (Swadeschi movement in India), 'ecotechnology, technology of the people or radical technology' (in more political versions), and also 'traditional technologies', or labour intensive or simple technology (Rybczynski, 1978). Schumacher's ideas also imply a different technology (1973).

At bottom, all the points of view converge in the need for a systematic development of techniques which consider different situations from those existing in the industrial countries (Herrera, 1978).

For the purpose of this paper it is necessary to add that techniques have to be of use to exploit fishing resources to the best advantage, taking into account the environmental and social situations of countries of various degrees of development and their continuing development in the future.

This does not mean that the growth and improvement of fishing production in the periphery will depend exclusively on its own local technology. What is intended is that apart from the technology imported from industrial countries, which frequently can be applied with no changes, a new technology will have to be created which is better suited to the aim of development in keeping with the needs of peripheral countries. The creation could be local and should be, for it can better be adapted to particular situations and because it would decidedly contribute to general economic development. But the creation of suitable technology can be originated in industrial countries too, if they expressly decide to do so.

Besides, fishing technology necessary for developing countries does not have to be generalized as simple or intensive work because conditions in the periphery are quite diverse.

4 Economic development and fisheries

The creation of technology is closely linked with development. Industrial countries are mainly 'manufacturers of technology' (Sabato, 1975), of any type, including fishing. But fish production – catching and processing – is a universal activity. Of the 16 most important countries whose catches amount to over 1m tonnes, only five can qualify as industrialized (non-socialist) and they contribute only 37% of the total of the elite which represents 77% of world production. If this is extended to those who can catch more than 100 000 tonnes, the number of industrialized countries barely reaches 13 out of a total of 52 with identical characteristics.

To better evaluate the various countries' performance in fishing production we have considered all those with more than a million inhabitants and whose catches are over 100 000 tonnes. Their economic structure and capacity to create technology have been taken into account. The production growth rate for each one for the period between 1970–1976 was calculated. The results have been grouped in similar intervals of growth for each group of countries. For this the criterion accepted by the World Bank was adopted.

Table I shows that 50% of the countries with medium income have had growth rates above 3%. For this reason they are the major users of technology. If the socialist countries, which have different mechanisms and criteria for commercialization, are excluded, the low income countries follow in size of their rate of growth in fishing production.

Forty-six per cent of the industrial countries have had negative rates in their fishing production and 15% have had positive rates under 1%. However, it is there where the greater part of technology is produced, thanks to their industrial capacity and fishing tradition.

Table I
DISTRIBUTION OF COUNTRIES BY GROWTH RATE IN EACH INCOME GROUP

Group of countries	Number of countries	Over 5%	Growth rate 3% to 5%	1% to 3%	Negative or under 1%
High income	13	7%	—	31%	62%
Medium income	22	41%	9%	23%	27%
Low income	11	18%	9%	37%	36%
Socialist	6	50%	—	17%	34%
	52				

Behind these booms and slumps is to be seen the effect of extending the 200-mile jurisdiction, which has allowed many developing countries to take advantage of the resources close to their coastlines and prevent others to continue exploiting them.

The rate of growth of fish production has been projected. The evolution that countries will have to have according to their respective rates of economic growth and other structural data have been taken into account. It is probable that high income countries are also those most qualified to create any type of technology, although it is also likely that these functions can also be carried out by lower income countries to a lesser degree. Apart from the claim in favour of creating local technology, fish technology in general cannot be very complex and can be open to developing countries although some sophisticated elements imported from leading countries are employed. Kahn and Wiener's classification (1976) which establishes six categories of countries according to their income *per capita* for the year 2000, was adopted.

Table II also presents the future medium groups as those with more growth in fish production and with a high number of countries which make them up. The socialist countries have been included. The most developed countries still have more negative results for the same reasons put forward in *Table I*. The negative results which are the consequence of the readjustment caused by the 200-mile jurisdiction will not be likely to recur in the future but production in developed countries is not likely to go on growing because additional natural resources are lacking. Developing countries in the southern hemisphere and in the intertropical belt can, however, depend on marine reserves and the possibility of increasing their aquaculture production.

The following conclusion is obvious on analysing both *Tables I* and *II*. The more developed countries will be more successful as 'manufacturers of technology' than as fish producers, and the developing countries will be the major buyers of technology because they will have the biggest catches. Without affecting the development of appropriate technology which will enable sectors to cope with situations not anticipated by industrialized countries, the validity of the principle of international division of labour is brought up to date. It is basically the task of the leading countries to realize the practicability of eliminating obstacles that restrict its widespread application.

5 Demand and technology

Demand is another conditioning factor of fishing technology and production. It depends on the population growth and, especially for international trade, on the purchasing power.

The 1960s were the scene of the explosive growth of production for reduction, mainly in Peru and Chile. The breeding boom in poultry and hogs to cope with increased purchasing power in Europe, Japan and United States brought about increased demand for fish meal and other components of balanced meal. The development of technology involved in that fishery was at the same time a consequence and an accelerating factor of that demand.

In the 1970s it seems that the predominant feature is processed or frozen pelagic fish. The dominant factor has been political, the universal extension of extended economic zones forced to re-orientate demand that was previously channelled to factory ships belonging to the older fishing countries. Argentina and Uruguay are the countries which have most benefited in South America. As a consequence there is an eager search for local technical solutions to solve problems. That is why, for example, a locally designed heading and gutting machine to be used on board has displaced prestigious European apparatus because it trebles performance, a most important detail for the voluminous hauls in the southwestern Atlantic Ocean. In like manner, freezer ships in this area try to use the thawing-processing-refrosting method so as to be able to compete with the wet fish ships in the US fish block market and thereby avoid exclusive dependence on European buyers.

These are historical and contemporary examples that show the influence of demand—in turn the result of other institutional or economic factors—over production that urge on and stimulate technology.

It is of interest to know what social or economic events are foreseeable so as to anticipate the requirements in the future.

On a world-wide scale population growth is a factor which will force fish production up. In South America, and especially in Argentina and Uruguay, it does not seem to be sufficient reason since a variety of food is abundant and demographic growth is insignificant.

Table II

DISTRIBUTION OF COUNTRIES BY GROWTH RATE ACCORDING TO PROJECTED *per capita* INCOME

Group of countries	Population, year 2000 (millions)	Number of countries selected	Over 5%	3% to 5%	Growth rate 1% to 3%	Negative or under 1%
A Post-industrial 2nd stage: ov. $20 000 *per capita*	665	9	11%	—	33%	55%
B Post-industrial 1st stage: $4 000/20 000 *per capita*	540	5	20%	—	20%	60%
C Mass consumption US$1 500 to 4 000 *per capita*	400	9	44%	—	22%	33%
D Advanced industrial US$600 to 1 500 *per capita*	700	9	44%	11%	11%	33%
E Partially industrial US$200 to 600 *per capita*	3 180	6	33%	17%	50%	—
F Pre-industrial US$50 to 200 *per capita*	850	13	31%	—	31%	38%
World population	6 335	51	Selected countries			

Demand will be channelled through the export trade which will be active provided that protectionism is eliminated and progress is accelerated so that the purchasing power of the underdeveloped countries will be increased.

In South America there are abundant non-exploited resources, generally of little-known species, which can be the object of a massive international effort. It will be necessary to redesign classifying and processing machines which take into account the dimension and shape of regional fishes. Some species, like 'polaca' (*Micromesistius australis*), and krill (*Euphausia superba*) can be exclusively treated on board factory ships but their catches are so large that they do not offer satisfactory solutions under present technical conditions.

It is advisable to anticipate a highly mechanized continuous process, equivalent to reduction but for human consumption. It would seem that creating new textures can be a viable way out to achieve acceptable and low-priced products. The objective would be to obtain intermediate goods for export, which would be finished by their respective food industries in the purchasing countries.

International co-operation for the marketing, industrialization and the transference of technology are apparently the means for significant and permanent results in the increase of production and exports of fish products in South America by the year 2000.

6 Other requirements

Underdeveloped countries are frequently poorly equipped with cold chainstores and demand low price products. New packaging, more economical than the metallic one which allows the sterilization of its contents, would be of great interest in developing markets for pelagic small species.

Studies on salting fish in Argentina have searched for substances from fish able to decrease the water activity both in fish products and other perishable foods. If results reach their objective a new utilization can be found for some abundant species and food production in general can have new opportunities (Lupin, personal communication).

No forward look can ignore the ever-increasing

energy costs and the shortage and pollution of waters. All the processes, including those that are working satisfactorily, are to be re-examined in order to ensure a major fuel and water economy.

Together with the concern in reducing costs it will be necessary to go on looking for alternatives which permit overseas transport with less volume and more flexibility to use general non-refrigerated cargo ships. Even at present, exports from South America to important markets of the periphery are limited because of lack of space in ships.

Finally, the outlook from the peripheral countries points forward problems that partially are different from those of industrialized countries. An appropriated technology must contribute the best performance for fish production and increase the purchasing power of more people in more countries. In such a way technology will serve for the general welfare of mankind.

7 References

BELL, F V. Economic and institutional factors affecting the demand
1968   for fish and shellfish. In: *The future of the fishing industry of the US*. University of Washington, Seattle. 345pp
BURGESS, G H O. The contribution of fish technology to the utilization
1974   of fish resources. In: *Fishery Products*, Ed R Kreuzer. Fishing News (Books) Ltd, London. 459pp
COULL, J R. *Fisheries of Europe*. G Bell and Sons Ltd, London. 235pp
1972
CHAPMAN, W M. Social and political factors in the development of fish
1968   production. In: *The future of the fishing industry of the US*. University of Washington, Seattle. 345pp
CHRISTY, F T and SCOTT, A. *The common wealth of ocean fisheries*. The
1965   Johns Hopkins Press, Baltimore. 280pp
EDDIE, G C. *Technical development in the fishing industry*. The Royal
1971a   Institution of Naval Architects, London. Paper No. 7
—— *The expansion of the fisheries: new applications of mechanical
1971b   power*. The Institution of Mechanical Engineers, London. Vol 185, 42/71
HERRERA, A. Tecnologías científicas y tradicionales en los paises en
1978   desarrollo. In: *Comercio Exterior*. Vol 28, No. 12
KAHN, H and WIENER. *The next 200 years*. Morrow, NY. 226pp
1976
KREUZER, R. Fish and its place in culture. In: *Fishery products*, Ed R
1972   Kreuzer. Fishing News Books Ltd., Farnham, Surrey
MATHISEN, O A and BEVAN, D E. *Some international aspects of Soviet
1968   fisheries*. Ohio State University Press, Ohio. 55pp
RYBCZYNSKI, W. Más allá de la tecnología adecuada. In: *Comercio
1978   Exterior*. Vol 28, No. 12
SABATO, J A. Empresas y fábricas de tecnología. In: *El pensamiento
1975   latinoamericano en la problemática ciencia-tecnología-desarrollo-dependencia*. Paidos, Buenos Aires. 345pp
SHUMACHER, J. *Small is beautiful*. Blond and Briggs, London
1973
WORLD BANK. World Development Report. Washington. 131pp
1978

Technology transfer to the small business sector in the fisheries and related industries

J Jaffray

1 Introduction

Any discussion of technology transfer is dependent on the definition of the word technology. Whereas science is principally concerned with increasing the boundaries of knowledge and understanding, technology is directed towards use. The output of technology, then, is a process, product, technique or material developed for a specific purpose (Creighton *et al*, 1970; Ettlie, 1973).

Technology transfer is an ambiguous phrase. The activity mainly involves an increase in the utilization of a

proven, existing technology rather than its expansion through further research. It is the application of technology to a new use or user. This may be a direct application or necessitate adaptation of the technology to match the needs of the new user. Technology transfer, then, may be regarded as either the process by which research is directed to a new use, or the secondary application of existing technology (Bieber, 1969; Sagal, 1978).

Economically, technology transfer affords us the opportunity to obtain a greater return on past investments in research and development (Hammond, 1974). It permits us to increase our use of the existing technological base and enhances the opportunities for bringing about technological change and progression which are essential elements of the economic growth of any country (Neale, 1977).

The fisheries industries in Canada have for a long time been well served by federal and provincial agencies involved in research and development and the subsequent transfer of technology. The Fisheries Research Board of Canada has long since been recognized internationally as a world leader in the fields of fisheries' biology and technology. The Inspection and Technology Branch of the Fisheries and Marine Service in Canada consistently acts as an important source of aid to the fisheries within the country. Educational institutions, such as the College of Fisheries in Newfoundland, through vocational programmes and province-wide extension services, offer up-to-date technological information on a wide range of fisheries skills including fishing, marine engineering, marine electronics and fish processing. At the present time centres of technological excellence in ocean sciences are being established and strengthened in Canada. These will provide resource centres in naval architecture and shipbuilding, cold water and ice-related engineering, fishing technology, marine resource management, and gear and vessel testing. The Ministry of State for Science and Technology in Canada is co-ordinating and supporting this programme. Despite serious cutbacks in current federal expenditures in support of fisheries technological development programmes and a centring of activities towards supporting the primary industry, it would appear that, on the whole, Canada will continue to be well served.

At present, technological information from research and development can be applied only by large organizations or companies who already possess a high degree of technical sophistication or employ technically trained staff capable of immediately analysing, utilizing and adapting the technologies generated (Roessner, 1975).

Many small or medium sized companies in the fisheries however have neither the time nor the technical capabilities to scan, select, analyse and adapt technological information material which in many instances could be of considerable economic value and could help them to be innovative and progressive. These small companies are frequently overlooked with respect to their special needs for technological information and assistance. In addition to companies directly involved with fishing or processing, there exist many small companies concerned with the manufacture of products which contribute technologically to operations or processes directly associated with the fisheries, for example refrigeration, materials handling, instrumentation, waste disposal, *etc*. Such firms produce everything from bearings, boxes and boats to trays, transformers and trucks. These are the so-called service industries, whose very existence is crucial to the efficiency and performance of the fisheries industries.

In Canada there are over 30 000 of these small companies contributing to the economy of the country to the extent of over half the industrial production and employment. They are a vitally important segment of the economy of the nation, and it is apparent that the well-being of these small enterprises is absolutely essential to the overall health of the industrial sector, as well as the fisheries. Typically the firms which comprise this industrial support system do not employ engineers or scientists, are not able to identify or describe clearly their technological problems, and have difficulty in using written technical information. Frequently, their productivity is 20% to 25% below that of comparable firms in the US and they suffer from one of the lowest rates of productivity improvement of any of the developed countries (Statistics Canada, 1978).

The implications are clear. While the results of research and development continue on the whole to favour the larger companies, efforts must be maintained to develop appropriate and effective mechanisms for the transfer of technology to meet the special needs of these small to medium sized companies (Wilson, 1973).

2 The Technical Information Service

The Technical Information Service (TIS) of the National Research Council is the only Canadian federal organization which provides assistance to Canadian industry, particularly the small industries, to improve production, operations, productivity and profitability by the better utilization of existing technology. The TIS provides industry with the most direct access to and application of current technology in the solution of industrial problems.

The service is administered by the National Research Council of Canada and delivered by Council staff and by personnel from the provincial research organizations in Canada. The TIS has a direct interface with industry and other federal and provincial industrial programmes, usually at the local level. They also work with the Federal Business Development Bank, the Department of Industry, Trade and Commerce, Canada Manpower and Immigration and with those provincial government services, concerned with industrial development.

2.1 *Field operations*
We are committed to the principle that technology transfer to the small business sector can best be done by competent technical people making personal contact with the firms they serve. The key element of TIS then is a field service with 16 offices in various parts of the country staffed by 40 professional engineers and scientists who make face-to-face contact with industry. Over 80% of Canada's small or medium sized firms are within 50 miles of such field offices.

TIS field engineers and scientists visit industries in their area, making clients aware of the availability and value of scientific and technical information. We assist our clients to identify and define their problems, then appropriate TIS personnel gather, analyse and interpret relevant technical information, propose solutions in terms which can be understood by the client, and assist in the implementation of the proposals. The service thereby helps manufacturing or processing industries transform or adapt existing scientific and technical information into practical use.

2.2 *Ottawa-based TIS operations*
The Scientific and Technical Advisory Section responds

to inquiries on a wide range of technical matters related to industrial processes or engineering. These requests for information or assistance originate directly from individuals within companies who know of the existence of the service, or through experienced field advisers following a visit to a company. Inquiries are handled in Ottawa by inquiry officers each of whom represents one or more speciality areas of technology, which include food technology, fisheries technology, electronics, mechanical engineering, electrical engineering, metallurgy and metal mechanics, packaging, plastics, hydraulics, chemical engineering, many of which are directly relevant to the fishing, fish processing and allied industries. Inquiries are handled as quickly as possible, usually within three weeks, drawing upon a wide range of resources including the individual and collective knowledge and skills of TIS scientific and engineering staff and the expertise of the scientists and engineers within the laboratories of the National Research Council. Scientists and engineers of federal, provincial, university and other research organizations in Canada are frequently contacted for help. In addition TIS makes extensive use of scientific and technological information available from various departments at all levels of provincial and federal governments in Canada, and has ready access to the scientific and technical literature of the world through the facilities of the Canada Institute for Scientific and Technical Information. TIS is, in fact, physically located within this institution.

The service works closely with the National Technical Information Services of a large number of countries world-wide, and with international agencies involved in the process of technology transfer such as the United Nations Industrial Development Organization (UNIDO) and the Organization of American States (OAS). Thus the Scientific and Technical Advisory Section is at the centre of a network of sources and resources for the acquisition, analysis and matching of scientific and technical information (*Fig 1*).

The technical information is conveyed to the client by the field officers who help the company with the interpretation and subsequent implementation of the technologies. Sometimes it is necessary for the Ottawa specialists to visit clients and assist in the adaptation of technologies. However, the service answers some 24 000 inquiries each year, and this number inhibits field operations by Ottawa-based staff in most instances.

The Manufacturing Science and Technology Section (Industrial Engineering) meets the demand from companies for information and assistance in providing skills and practical advice related to production operations, particularly in the elements of plant layout, materials handling, production planning, methods improvement, work measurement and quality and cost control. Industrial engineering is offered by TIS as a service designed principally for small enterprises. Suitable individuals within the client company receive industrial engineering assistance and TIS engineers guide, direct and advise them, thereby making self-help practicable. In addition, the Manufacturing Science and Technology Section of TIS has developed a comprehensive diagnostic tool as an aid in delivering their service. It provides an accurate profile of a company by analysing the elements of management, production, marketing, finance and work en-

Fig 1 Sources of scientific and technical information

Canada Institute for Scientific and Technical Information:

eg
Abstract and Indexing Services
External Data Bases
Monographs, Series and Journals
Inter-library Loan
Computerized Information
    Retrieval Services, *etc.*

Federal agencies:

eg
Department of Fisheries and
    Environment
Department of Industry, Trade
    and Commerce
Federal Business Development
    Bank
Ministry of State for Science and
    Technology
Agriculture Canada, *etc.*

National Technical Information
    Services:

eg
INDOTECH in Dominican
    Republic
CENDES in Ecuador
CONACYT in Mexico
ISSI in the Philippines
NTIS in the United States

Provincial departments,
    ministries or agencies:

eg
Departments of Fisheries
Departments of Agriculture and
    Food
Provincial Research Councils
Market Development Boards,
    etc.

Special information sources:

eg
Torry Research Station
SIDN in Georgia, USA
Fisheries and Marine Service
Tropical Products Institute,
    London, *etc.*

Suppliers:

eg
Canadian manufacturers
Importers
Brokers
Foreign embassies
Federal Department of Industry,
    Trade and Commerce, *etc.*

NRC Laboratories:

eg
Food technology
Atlantic regional
Mechanical engineering
Hydraulics
Electrical engineering, *etc.*

Universities and colleges:

eg
University of British Columbia
Memorial University,
    Newfoundland
Fisheries College,
    Newfoundland
University of Guelph, *etc.*

International agencies:

eg
Food and Agriculture
    Organization of the United
    Nations
United Nations Industrial
    Development Organization
International Development
    Research Centre
Canadian International
    Development Agency
TECHNONET, South-East
    Asian Information Network
Organization of American
    States, *etc.*

vironment within the organization. Recommendations may then be made towards improving any or all of these operations. Efforts are focused most often, however, on problems of production, *eg* plant layout and inventory control during periods of expansion, or cost control when profit margins tighten (Lapp, 1977).

TIS industrial engineers are able to observe and measure the impact of their work in client companies and quantify the benefits or savings brought about when recommended changes are implemented. Last year our clients reported that their combined identifiable dollar savings far exceeded the cost of operating the whole of TIS.

Another facet of our service is the Technical Awareness Section which keeps individuals and companies aware of development in technology by selecting literature relevant to their needs.

The section supplies information on such topics as new materials, products, manufacturing processes, and improvements in production management. The section reviews on a regular basis some 800 journals, reports, monographs, *etc* and locates, screens and evaluates new ideas and technologies. Industrial clients submit an interest profile and receive computer-matched lists of the titles of documents, copies of which may be sent upon request for a small nominal charge. Several variations of this selective dissemination of information system exist and materials are available in various forms – hard copy, microfiche, cassette tape or other – whichever is most appropriate. Nearly 34 000 items are analysed each year by the Technical Awareness Section, generating about 2 200 actual articles or Tech Briefs. Over 20 000 Canadian companies make use of this information.

To allow more time to be spent on the solution of problems with promising practical results, TIS has recently received approval to subsidize the hiring of 160 science and engineering students from universities and colleges of applied arts and technology in Canada. The students work in firms, preselected by TIS advisers, for periods up to three months, under the guidance of TIS personnel and university or college professors, helping companies to identify problems and assisting in the implementation of practical solutions. The programme extends the TIS capability in problem-solving and provides immediate and direct assistance to firms who are devoid of scientific and technical staff.

3 The small business sector

Many small companies recognize their need for technical assistance, but have difficulty locating sources of useful information. They usually wish to obtain 'capsule' advice which they can apply immediately to correct their difficulty (Bass, 1976).

Offering aid to the small industrial firms can be difficult in that managers frequently ask for the wrong kind of assistance.

In a recent study of the TIS (Lapp, 1977), visits to companies revealed certain common features among a number of Canadian small manufacturing or processing firms:

—Owner-managers frequently distrust people from the government who offer free services and advice.
—They resent being pressed for fast action by government agencies responsible for industrial development.
—They accept technical people more readily than financial advisers because the entrepreneur is more willing to concede his lack of appropriate technological background than his ability as a competent manager.
—They believe that an outsider can provide a fresh look and often see things differently.
—Owner-managers or senior company employees have little time to leave the plant to visit a government office unless their situation is desperate.
—They believe they have little clout with their suppliers because of the small size of their orders.
—The successful owner-manager is quite frugal by nature, which very likely accounts for his success in running and maintaining a small business.
—The majority of firms visited could neither afford nor would be prepared to pay for industrial engineering services from private consulting firms.

These agree strikingly, but not surprisingly, with the findings of other observers in Japan, Holland, Latin America and Asia (Jervis, 1975; Bass, 1976).

Small companies usually are so deeply involved with the everyday problems of production, marketing, financing and labour that they have relatively little time for technical reading and depend a great deal upon suppliers, competitors, trade associations, conferences and professional contacts for technical information. They may subscribe to a few technical journals dealing specifically with their own field but generally are unaware of other technical information sources and channels which could prove useful to them.

Marketing data, sources of financial assistance, data on internal systems costing, controls, *etc*, and scientific and technical information, while available from a wide variety of sources, is of little use if there is no one in the firm who can interpret it (Bourgault, 1972; Gee, 1974). A firm that cannot interpret marketing and financial information would not be in business long! The interpretation of technological information, however, normally requires staff with some form of engineering or technical capability but many small firms cannot afford to retain such experts.

Within this sector, approximately 20 000 Canadian firms, many of which are concerned with fisheries, have no scientific or engineering staff at all! (Statistics Canada, 1978). Such firms are usually run by an owner-manager who, though suspicious of government agencies offering free services and advice, frequently will accept a technical person from a non-regulatory agency, such as NRC, who is willing to come and work alongside the people in the plant, and whose stated purpose is to help the company use scientific and technical information (Jervis, 1975; Lapp, 1977).

Several studies have revealed that companies are likely to ask for technical aid before they seek any other form of help, apparently because technical problems are more easily delimited and rest mainly on demonstrable facts rather than opinions (Rubenstein, 1974; Roessner, 1975; Maninger, 1976). The initial route towards helping a company may often lie through the provision of technical information. Assistance services, however, should not disregard the possibilities of deeper, more complex, underlying problems existing within firms seeking help. Opportunities to improve managerial and production practices should be noted (Rubenstein, 1974) and the service agency should ensure that a complete diagnosis of the problem is made prior to offering advice to the client (Roessner, 1975). Recommendations, when they are made, must be clear, comprehensive and practical. With this in mind, then, the Technical Information Service of Canada employs what it terms the total approach to serving the needs of industrial clients; that is, an attempt is made to offer as complete and thorough assistance as is possible.

4 The total approach of TIS

In order to accomplish this, TIS endeavours to apply the following ideal system for each client.

4.1 Correctly identify and define the problem or need
Bearing in mind that the client may be unaware of under-

lying problems or needs, problem identification is made by visiting the company, talking with managers and operators and viewing the physical operations. TIS staff aid initially in identifying the industrial problem as clearly as possible, *eg* is it an innovative process which is required, a new product to be developed, a marketing or distribution problem, or a combination of these?

4.2 *Identify and locate any technologies which are appropriate*

TIS engineers and scientists examine the problem and bring their expertise to bear upon it, frequently calling upon NRC laboratories, or other specialized laboratories such as those of the Fisheries and Marine Service. When and if required, further scientific and technological information is acquired from other sources including provincial, federal or international agencies already mentioned.

4.3 *Adaptation to match the users' needs*

The information is collected and analysed by service scientists and engineers and is carefully adapted and adjusted to match the specific needs of the industrial client with regard to the technical and economic capabilities of the company, and the complexity of the client's problem. The appropriate technology is then transferred by TIS field officers, usually on site, and the client is assisted in the interpretation and comprehension of the technology from technical and socio-economic viewpoints.

4.4 *Assistance in the implementation of the technology*

In-plant assistance by TIS officers is often necessary at this phase in the transfer of technology and is accomplished by providing practical on-the-job suggestions and knowhow gained from previous experiences in industry. The industrial client receives ongoing aid and encouragement to develop further and to innovate technologically, where this is deemed advisable.

4.5 *Follow-up and assessment*

Even after providing assistance, problems may arise resulting in system failures. Proper follow-up helps to minimize these. The adequacy of the transferred technology is assessed and is later evaluated, usually after about six months, from the viewpoints of technology, industrial development, economic growth, modifications required, spin-offs, or technological innovations derived. Benefits are evaluated, *eg* profits, jobs saved or created, new products, *etc*. Unsatisfactory results will normally bring about an immediate review and TIS and the client initiate corrective measures.

5 Meeting the needs of the industries

Often small companies ask for the wrong kinds of assistance. They may try to deal with the symptoms of a problem rather than with the difficulty itself. Sometimes industrial problems are a mixture of technical and managerial difficulties so that diagnosis ranges from the plant floor to the front office. In such cases it is evident that field officers should not merely be technically competent, but also skilled in business and management practices, *ie* they should be industrial engineers. The ability to compare the cost of changing equipment, processes or products with the prospective savings or benefits derived from such changes is of invaluable assistance to the client.

As small businesses begin to grow their technological horizons expand. They need further scientific and technical information and knowledge of new technological developments which may affect the operations of the company. Hence the services offered by TIS, in-plant inquiry, industrial engineering and technological awareness, match very closely with the ongoing identifiable needs of companies.

How, then, does TIS differ from other agencies that help the fisheries industries?

First, it clearly should be established that the service does not compete in any way with other agencies involved with fisheries. Rather it is designed to complement and strengthen the activities of existing centres of technological excellence. It is, however, the only federal Canadian organization created to deal specifically with the problems of the small and medium sized manufacturers and processors. The skills of the service lie not in the development of new technologies but in the location and acquisition of existing technologies and in the application of these to the betterment of such small or medium sized companies. An understanding of the needs of small industry cannot be reached by simply scaling down the experiences and requirements of large companies. Their organization, resources, viewpoints, methods and problems are not comparable with large companies and vary greatly amongst themselves.

In addition to assisting with technical problems, field officers, by virtue of their numerous and continuous contacts with all kinds of industry, provide a cross-fertilization of ideas, techniques and procedures between companies, subject always to the maintenance of commercial confidentiality and security. They often put companies in touch with each other to build up local sources of supply and profitable business co-operation or, in some instances, mergers.

Studies have found that these strong personal relationships between industrial small-business clients and field officers have a considerable bearing on the degree of effectiveness of the successful transfer of technology (Jervis, 1975; Neale, 1977).

Industry's response to TIS activities (Lapp, 1977) is:

—Industrial clients appear to accept and trust the TIS field officer, particularly after he has demonstrated his sincerity and interest (often by literally rolling up his sleeves and working alongside others in the plant).

—They do not readily associate the National Research Council of Canada directly with "government", and thus feel such an agency is less "political" and more "objective" than other government agencies.

—Recipient companies have found the inquiry service and industrial engineering valuable, backing up their statement with specific benefits and results from applying the information or advice received.

—Cases were cited where TIS prevented the necessity for a firm to leave Canada, or saved the company from "going under".'

These findings attest to the value of the TIS philosophy that technology transfer to the small-business sector can best be carried out by means of competent technical people making personal contact with the firms they serve.

Equally important is the influence that the TIS field officer has concerning the type of information or assistance that the firm requires. Because TIS embodies the total approach system in its attempts to resolve clients' problems, the needs of industry for services are tempered by the very presence of the service, and the service itself has been sensitized to respond more perceptively to the needs of industry.

6 International activities

TIS/NRC is internationally recognized as a unique organization in the field of technology transfer to industry. A number of countries have already modelled their own services on Canada's TIS, with great success, while several other countries consult with TIS regularly. TIS has been providing technical assistance to developing nations for many years by replying to inquiries received, or, on a limited scale, by sending specialists on short-term missions. Our international activities have been provided in co-operation with or financed by various government departments and other organizations such as the International Development Research Centre (IDRC), the Canadian International Development Agency (CIDA), the United Nations Development Program (UNDP), the Organization of American States (OAS) and have benefited many countries including Mexico, Brazil, Colombia, Chile, Venezuela, Peru, Greece, Turkey, Egypt, India and others. Of particular interest is the assistance provided by TIS to IDRC in establishing TECHNONET, a technical information network in South East Asia serving six nations.

The programme strengthens TIS capabilities in generation of employment, small-scale industry development primarily in rural areas, in the application of foreign-generated technologies to Canadian manufacturing industries and in the establishment of new markets for Canadian manufactured goods.

7 References

BASS, L W. Technical managerial help for small enterprises. *World*
1976   *Develop.,* Vol 4, No. 4, 339–347
BIEBER, H. Technology transfer in practice. IEEE *Trans. Eng. Mgmt.,*
1969   V EM–16, N 4, Nov, 144–147
BOURGAULT, P L. Innovation and the structure of Canadian industry.
1972   Background Study for the Science Council of Canada, October
        Special Study No. 23, 100–106
CREIGHTON, J W, *et al. Enhancement of Research Development Output*
1970   *Utilization Efficiencies; Linker Concept Methodology in the
        Technology Transfer Process.* Naval Postgraduate School,
        NPS–55CF 72061A, June 30, 2–5.
ETTLIE, J. Technology transfer – from innovators to users. *Ind. Eng.,*
1973   V 5, N 6, Jun, 16–23
GEE, S. The role of technology transfer in innovation. *Res. Man.,* V 17,
1974   No. 6, Nov., 31–36
HAMMOND, R W. Is the cost of small industry assistance too high?
1974   *Small Ind. Develop. Network Newsletter.* 4th Quarter Office of
        International Programs, Engineering Experiment Station,
        Georgia Inst. Tech., Atlanta, Georgia, USA
JERVIS, P. Innovation and technology transfer – the roles and charac-
1975   ters of individuals. IEEE *Trans. Eng. Mgmt.,* V EM–22, N 1,
        Feb., 19–27
LAPP, P A. A study on the Tech. Inform. Serv.; a study conducted
1977   jointly for the Nat. Res. Council of Can. and the Assoc. of the
        Prov. Res. Org. Philip A. Lapp Ltd.
MANINGER, R C. Some commercial innovations from technology
1976   transfers from Federal Research and Development. *ASME
        Pap.* N76–WA/AERO-2, for Meet 5 Dec., 5 pp
NEALE, M J. Technology – the application in industry of the results of
1977   research. *Inst. Mech. Eng.* (Lond), Proc V 191, N 30, 333–378
NICHOLSON, W. *Microeconomic Theory.* Dryden Press Inc, Ill.
1972   216–217
ROESSNER, J D. Federal Technology Transfer: results of a survey of
1975   formal programs. *ASME Pap.* N75–WA/AERO-4, for Meet
        30 Nov.–4 Dec. 17pp
RUBENSTEIN, A H. Basic research in technology transfer. In: *Tech-
1974   nology Transfer,* Eds H F Davidson, M J Cetron and J D R
        Goldha. Proc: Nata Adv. Study Inst. on Techn. Trans., Ser E
        Appl. Sci., V 6, Noordhoff Int. Publ. 247–268
SAGAL, M W. Effective technology transfer – from laboratory to pro-
1978   duction line. *Mech. Eng.,* V 100, N 4, 32–35
STATISTICS CANADA. Manufacturing industries in Canada by type of
1978   organization and size of establishments: 1975. *Stat. Can.,* Cat.
        No. 31–210, February.
STEFFENS, J. Development and technology transfer. *ASME Pap.*
1976   N76–WA/TS–15, for Meet 5 Dec. 7pp
WILSON, A H. Governments and innovation. Background study for
1973   the Sci. Coun. of Can. Special Study No. 26, April, 20–21, 247

10 Seasonal changes

A seasonal study of the storage characteristics of mackerel stored at chill and ambient temperatures

J G M Smith, R Hardy and K W Young

1 Introduction

Over the past decade landings of mackerel in the UK have increased greatly and it has now become the major species landed in this country, even exceeding that of cod. Although a popular fish in some parts of the world mackerel were looked upon as a somewhat inferior fish in the UK. Today, not only are the whole fish and fillets sold throughout the country but smoked fish, especially hot-smoked, have become very popular and command good prices. One of the main over-wintering areas for mackerel, during October to March, is close inshore to the southwest coast off Cornwall and Devon in the English Channel, and landings in this Cornish fishery have increased from 2 000t in 1970 to 114 000t in 1977. It was decided therefore to carry out a bimonthly study of the storage characteristics of mackerel caught in this area, the whole and gutted mackerel being stored at ambient and chill temperatures, *ie* in ice and in chilled sea water (csw).

2 Experimental

2.1 *CSW tank*

The tank used was an insulated aluminium container of $2 \cdot 1 m^3$ capacity and the ratio of ice:water:fish placed in it was 1:1:1. The usual practice was to first add the ice and sea water to the tank before adding the fish. Immediately after loading the fish to the tank nitrogen was introduced to ensure circulation and uniform temperature distribution, the temperature being monitored at various levels in the tank by thermocouples. Once the temperature of the fish and water was at 0°C the flow of nitrogen could be reduced.

2.2 *Fish*

Storage trials were carried out on fish caught bimonthly from January onwards using a bottom trawl off the southwest coast of Devon. The fish were transferred immediately to the csw tank and also well iced in 8 stone aluminium boxes in the trawler's hold at +1°C to +3°C (all six trials). Mackerel gutted immediately after capture and after being held 24h at ambient temperature were stored well iced in boxes (May, July, September and November trials). Ungutted and gutted fish were also held at ambient temperature in the aluminium boxes on board the deck of the trawler for up to four days (March, May, July and September trials).

At set intervals samples of the fish were removed from each of the above storage experiments for analysis. Some analyses could not be carried out on board and, to overcome this, samples of the fish were frozen in an air-blast freezer, glazed and stored in polythene bags at −30°C.

2.3 *Methods*

At each sampling interval three fish were filleted, skinned (care being taken to remove all the flesh from the skin), minced, and portions of the well-mixed mince used to prepare extracts for the following analysis 2.3.1–2.3.4.

2.3.1 *Hypoxanthine (HX)*
Of the minced flesh 40g was homogenized with 200ml $0 \cdot 6 N$ perchloric acid and filtered. Filtrate of 10ml was then brought to pH 6–8 by adding buffered KOH and the solution was stored in a polythene bottle at −30°C to be analysed later by the method of Burt *et al* (1968). Results are expressed as μmoles HX/g fish.

2.3.2 *Trimethylamine (TMA)*
The remainder of the perchloric acid filtrate from the HX extract was used to estimate TMA by the automated method of Murray and Burt (1964) substituting sodium hydroxide for potassium hydroxide to prevent blockage of the system by the formation of insoluble potassium perchlorate. Results are expressed as mg TMA-N/100g fish.

2.3.3 *Histamine (HA)*
Of the minced flesh 20g was blended with 200ml $2 \cdot 5 N$ trichloroacetic acid and the HA content determined by the method of Hardy and Smith (1976). Results are expressed as mg HA/100g fish.

2.3.4 *pH methods*
This was carried out on a 2:1 water:fish homogenate using a glass electrode.

The following methods 2.3.5–2.3.7 were carried out on shore using the frozen mackerel. The three fish were thawed, filleted and the unskinned fillets minced and thoroughly mixed as before.

2.3.5 *Total lipid content*
This was determined on a 50g sample of the minced fillets using the extraction procedure of Bligh and Dyer (1959) but adding 0.001% of the antioxidant 2.6 ditertiary-butyl-*p*-cresol (BHT) to the chloroform.

2.3.6 *Peroxide value (PV)*
The PV of the lipid extracted in 2.3.5 above was determined by the method of Banks (1937). Results are expressed as ml $0 \cdot 002 N$ sodium thiosulphate/g lipid.

2.3.7 *Salt content*
This was determined as percentage sodium chloride on duplicate 5g samples of the mince by the method (18·014) of the Association of Official Analytical Chemists (Horwitz, 1970).

2.3.8 *Sensory assessment (SA)*
Fillets from the

thawed fish were steam casseroled and the panel was asked to score the fillets as follows and also to comment on the presence of any rancid flavours. None of the ambient stored fish was tasted.

5 fresh mackerel flavour
4 slight loss of fresh mackerel flavour
3 absence of fresh mackerel flavour
2 slight 'off' flavours and odours, slight rancidity present
1 definite 'off' flavours and odours, rancid

3 Results and discussion

The percentage lipid for the six trials are shown in *Table I*, where 18 to 24 samples were analysed each month. The lipid content is similar to that found by Hardy and Keay (1972) in their study on Cornish mackerel where the lowest lipid content is around May and highest during the winter months. *Table I* also shows, however, a wide spread of values during each month, especially during July, indicating that the fish in any one shoal were obviously at different nutritional and maturation levels.

Table I
LIPID CONTENT OF MACKEREL

Month caught	Average lipid %	Range of values
Jan	15·2	10·0–18·2
Mar	7·1	3·3–14·1
May	3·3	1·3– 8·0
July	12·0	3·7–20·0
Sep	17·2	8·3–23·8
Nov	19·0	15·3–25·0

The peroxide values are given in *Table II*. In general the increase in PV during storage is greater in the iced fish than in the csw fish. This difference could be due to the fact that the csw fish were protected from atmospheric oxygen, the csw being outgassed by nitrogen during the initial lowering of the sea water temperature. The

highest PVs are found in the iced fish during March and May when the lipid content is low. Banks (1952), on the other hand, in his detailed study on the cold storage of herring, found that the fattier summer fish oxidized more rapidly during frozen storage than the leaner winter fish.

The HX values shown in *Table III* indicate that the rate of increase in csw is in general slightly higher than that in the iced fish. With the exception of the March iced fish the rate of increase in HX is fairly constant, *ie* concentrations are all between 0·53 and 0·73 µmoles HX/g fish after 10 days' storage. The rate of increase in the csw fish appears more variable, the concentrations ranging between 0·5 and 1·68 µmoles HX/g fish after 10 days. These concentrations are lower when compared to November herring stored in ice and refrigerated sea water (rsw) where concentrations of 2·5 and 3·4 µmoles HX/g fish, respectively, were found after 10 days (Smith *et al*, 1979) but it must be pointed out that these herring had been held in a commercial trawler's hold for almost 8h before being transferred to ice and rsw on board this station's research trawler. Scad (*Trachurus trachurus*) show a rate of increase of 0·059 and 0·047 µmoles HX/g fish/day when stored in ice and csw, respectively (Smith *et al*, 1980a), the former figure being similar to that of mackerel stored in ice.

Greater differences in the TMA concentrations are found in mackerel stored in ice and csw, the higher concentrations in the csw fish being more apparent after six days' storage (*Table IV*). A similar result is found in the chilled storage of herring (Smith *et al*, 1979) and scad (Smith *et al*, 1980a). The rate of increase of TMA concentration in both iced and csw stored mackerel appears to change over the year, being low in January, increasing to a maximum in July and decreasing again towards the end of the year (*Fig 1*).

The pattern of HA production in the chilled fish (*Table V*) is somewhat similar to that of the TMA in that very little HA was detected at the beginning and end of the year after 10 days' storage but was determined as

Table II
PEROXIDE VALUE OF MACKEREL STORED IN ICE AND CSW
(ml 0·002 N sodium thiosulphate/g lipid)

Days storage	Jan		Mar		May		July		Sep		Nov	
	ice	csw	ice	csw	ice	csw	ice	csw	ice	csw	ice	csw
0	0·6	0·6	1·0	1·0	0·9	0·9	0·5	0·5	0·5	0·5	0·2	0·2
2	2·4	1·6	1·1	1·2	4·9	1·8	1·1	2·8	0·5	0·3	0·6	0·5
4	1·7	1·3	2·7	1·4	2·3	3·0	2·1	1·9	1·4	0·7	1·3	1·3
6	3·5	0·7	4·7	2·7	6·5	8·5	4·1	2·7	1·0	1·8	3·1	0·8
8	3·5	1·2	9·9	2·2	7·1	7·8	5·1	2·1	2·2	1·3	2·7	0·9
10	5·7	1·0	8·5	2·2	7·2	3·3	4·5	2·1	2·3	0·9	5·0	0·5

Table III
HYPOXANTHINE VALUE OF MACKEREL STORED IN ICE AND CSW
(µmoles HX/g fish)

Days storage	Jan		Mar		May		July		Sep		Nov	
	ice	csw	ice	csw	ice	csw	ice	csw	ice	csw	ice	csw
0	0·10	0·10	0·16	0·16	0·00	0·00	0·11	0·11	0·10	0·10	0·01	0·01
2	0·16	0·16	0·19	0·23	0·00	0·00	0·06	0·24	0·22	0·32	0·23	0·29
4	0·29	0·26	0·66	0·45	0·03	0·20	0·24	0·32	0·18	0·28	0·23	0·19
6	0·47	0·47	0·67	0·63	0·30	0·16	0·36	0·58	0·22	0·52	0·31	0·31
8	0·69	0·48	0·99	1·10	0·53	0·62	0·56	0·86	0·44	0·82	0·39	0·55
10	0·73	0·80	1·36	1·68	0·53	1·15	0·70	1·42	0·72	0·70	0·63	0·50

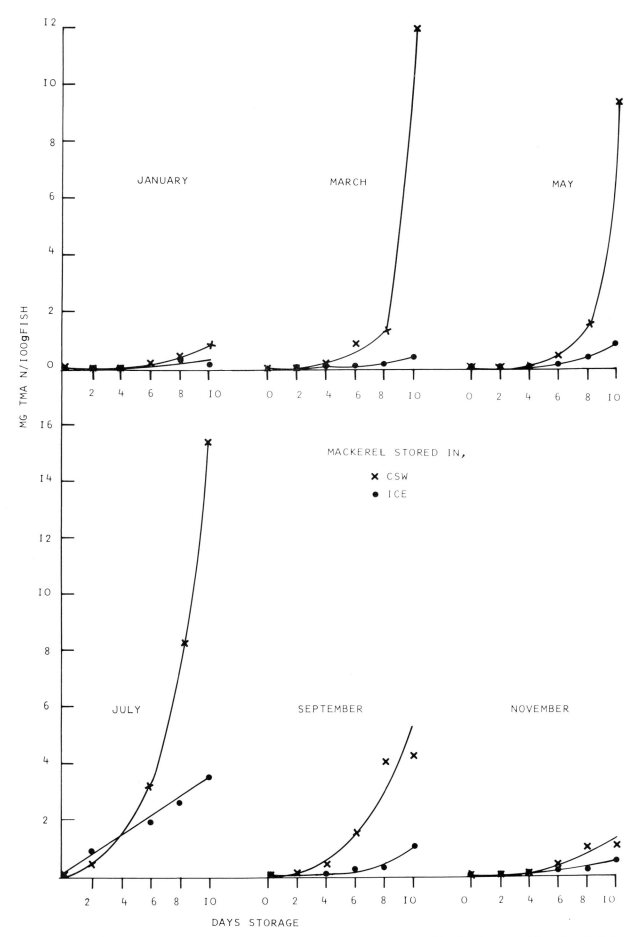

Fig 1 Production of trimethylamine in mackerel stored in ice and csw during the six bimonthly trips, January to November, inclusive

374

early as after four days in csw stored fish in July and in fish stored in both media in September after six days, the concentrations being higher in the csw stored fish. The highest concentrations were 4·8 and 7·0mg% in the eight- and 10-day fish caught in July and stored in csw, by which time they were considered unacceptable by the taste panel. In previous storage work on mackerel caught during November off Aberdeen, Scotland, no histamine could be found in ungutted fish stored in ice for up to 10 days (Hardy and Smith, 1976). Ungutted November herring, however, stored in ice and rsw, had concentrations of 2·8 and 8·9mg HA/100g fish after 10 days' storage (Smith et al, 1979) but, as pointed out earlier, these fish had been delayed some 8h before being transferred to the chilled media.

As can be seen in Table VI the pH values remain relatively low and constant during the 10 days' storage when compared to freshly caught white fish (Cutting, 1953). These results are similar when compared to those of herring stored in ice and rsw (Smith et al, 1979).

Sensory assessment scores (Table VII) show that mackerel stored in ice develop off flavours one to two days earlier than fish stored in csw. A similar result is found when ungutted scad (Smith et al, 1980a) and ungutted blue whiting (Micromesistius poutassou) (Smith et al, 1980b) are held in ice and csw. This finding, however, is a reversal to that found in herring (Smith et al, 1979) where fish stored in rsw developed off flavours approximately one day earlier than in iced fish. The use of nitrogen to circulate the water in the csw tank may have effected a change in the type of bacteria present in the water. In the case of herring stored in rsw the water could be considered to be aerated, whereas in these trials oxygen in the water was probably replaced by nitrogen

Table IV
TRIMETHYLAMINE VALUE OF MACKEREL STORED IN ICE AND CSW
(mg TMA-N/100g fish)

Days storage	Jan		Mar		May		July		Sep		Nov	
	ice	csw	ice	csw	ice	csw	ice	csw	ice	csw	ice	csw
0	0·00	0·00	0·00	0·00	0·00	0·00	0·00	0·00	0·00	0·00	0·00	0·00
2	0·00	0·01	0·02	0·02	0·21	0·01	0·91	1·48	0·15	0·15	0·04	0·04
4	0·01	0·00	0·15	0·11	0·14	0·16	1·77	1·82	0·15	0·49	0·14	0·14
6	0·07	0·24	0·08	0·89	0·18	0·53	2·00	3·82	0·30	1·72	0·28	0·28
8	0·27	0·31	0·18	1·39	0·49	1·72	2·68	8·23	0·30	4·08	0·28	1·02
10	0·14	0·91	0·45	12·05	0·95	9·56	3·53	15·39	1·16	4·34	0·66	1·16

Table V
HISTAMINE VALUE OF MACKEREL STORED IN ICE AND CSW
(mg HA/100g fish)

Days storage	Jan		Mar		May		July		Sep		Nov	
	ice	csw	ice	csw	ice	csw	ice	csw	ice	csw	ice	csw
0	—	—	—	—	—	—	—	—	—	—	—	—
2	—	—	—	—	—	—	—	—	—	—	—	—
4	—	—	—	—	—	—	—	0·23	—	—	—	—
6	—	—	—	—	—	—	—	0·51	0·23	1·13	—	—
8	—	—	—	—	—	0·14	—	4·79	0·57	2·44	—	—
10	—	—	0·11	0·11	0·17	2·59	0·03	7·01	0·37	0·99	0·11	0·11

Table VI
pH VALUE OF MACKEREL STORED IN ICE AND CSW

Days storage	Jan		Mar		May		July		Sep		Nov	
	ice	csw	ice	csw	ice	csw	ice	csw	ice	csw	ice	csw
0	6·45	6·45	6·5	6·5	6·4	6·35	6·4	6·4	6·65	6·65	6·2	6·2
2	6·5	6·4	6·8	6·5	6·7	6·6	6·3	6·5	6·4	6·4	6·3	6·25
4	6·4	6·5	6·6	6·5	6·5	6·8	6·35	6·6	6·5	6·3	6·35	6·05
6	6·35	6·5	6·8	6·6	6·3	6·8	6·6	6·5	6·4	6·2	6·2	6·3
8	6·5	6·4	6·8	6·3	6·5	6·6	6·6	6·4	6·3	6·45	6·4	6·25
10	6·6	6·5	6·85	6·5	6·6	6·8	6·7	6·4	6·5	6·3	6·3	6·35

Table VII
SENSORY ASSESSMENT OF MACKEREL STORED IN ICE AND CSW
(for scoring see 2.3.8)

Days storage	Jan		Mar		May		July		Sep		Nov	
	ice	csw	ice	csw	ice	csw	ice	csw	ice	csw	ice	csw
0	4·0	4·0	4·5	4·5	3·5	3·5	4·5	4·5	4·0	4·0	4·2	4·2
2	2·8	3·8	3·8	4·0	3·0	2·8	4·4	5·0	3·1	3·7	3·1	4·0
4	2·9	3·4	3·3	3·8	2·4	3·4	3·8	4·8	3·7	3·7	3·1	3·7
6	3·0	3·9	2·8	3·6	2·6	3·6	2·2	4·4	2·1	2·7	3·1	3·3
8	1·8	3·1	2·0	2·4	2·6	2·7	2·6	2·6	3·5	3·1	2·6	2·8
10	1·9	1·7	1·0	1·0	1·4	1·0	1·0	1·0	2·1	1·3	1·6	1·6

changing the ratio of aerobic to anaerobic bacteria. It was noticeable that after 10 days the water in the csw tank had developed strong off odours and similar odours and flavours could be detected in the csw fish. Periodic changing of the sea water could possibly prevent this. There was no evidence of rancidity in either the iced or csw fish during the 10 days' storage.

In the csw fish there is a gradual increase in salt content over 10 days' storage (*Table VIII*) with the lowest concentrations being found when the fish have the highest oil contents, *ie* September and November. The concentration of the salt in the mackerel after 10 days is lower than that found in herring stored for a similar period in rsw where the salt content increased at approximately 0·1g NaCl/100g fish/day (Smith *et al*, 1979).

Table VIII
SALT CONTENT OF MACKEREL STORED IN CSW (% NaCl)

Days in csw	Jan	Mar	May	July	Sep	Nov
0	0·21	0·29	0·29	0·26	0·21	0·22
2	0·31	0·40	0·43	0·46	0·36	0·33
4	0·40	0·62	0·51	0·40	0·46	0·42
6	0·55	0·71	0·76	0·53	0·62	0·47
8	0·76	0·64	0·76	0·42	0·44	0·53
10	0·82	0·71	0·76	0·55	0·46	0·49

Storing the fish for 24h at ambient temperature before gutting and icing tends to accelerate deterioration as measured by the various chemical parameters. The results are shown in *Tables IX to XVI* and it can be seen that in all four trials fish held at ambient temperature before gutting and icing have higher peroxide values than fish gutted and iced immediately. The ambient temperature also seems to affect the rate of increase of PV, as can be seen in the July mackerel where high values are found after only three days in the fish held at ambient temperature (12–23°C) for one day before gutting and icing (*Table XII*). In *Tables IX to XVI* the sensory assessment scores marked with an asterisk or dagger represent samples where slightly rancid or rancid flavours were detected by some of the taste panel members. It is interesting to note that slight rancid flavours were detected by panel members once the PV reached a level of 10 or over and definite rancid flavours when the PV was approximately 20 or over.

Comparing ungutted iced fish (*Table II*) with fish gutted and iced immediately (*Tables IX, XI, XIII and XV*) it can be seen that gutting appears to accelerate oxidation, probably due to greater exposure of the lipid to atmospheric oxygen. Hansen (1963) found that the development of rancidity in the fat of iced trout, as estimated by the PV and tasting tests, was more rapid in the gutted than in the ungutted fish. In fact, rancidity was found to limit the storage life of gutted, iced trout to between seven and 11 days. In these trials rancid flavours were not the limiting factor because other off flavours developed before rancidity could be detected.

Similarly HX, TMA and HA concentrations all increase at a faster rate when the fish are held at ambient temperature before gutting than in fish gutted immediately (*Tables IX to XVI*). HX and TMA concentrations in fish gutted and iced immediately (*Tables IX, XI, XIII and XV*) are similar to ungutted fish stored in ice (*Table*

III). Gutting and icing immediately, however, retards the formation of histamine, a result found in previous work carried out by Hardy and Smith (1976) on the chilled storage of mackerel.

Comparing the pH of mackerel in *Tables IX to XVI* with those in *Table VI* it can be seen that over 10 days there is little difference between ungutted and gutted fish iced immediately and fish held at ambient for 24h before gutting and icing.

Examining the two sets of scores for each month in *Tables IX to XVI* it can be seen that a delay of 24h at ambient temperature before gutting and icing lowers the SA scores, however on comparing *Tables IX, XI, XIII and XV* with *Table VII* it can also be seen that gutting and icing the fish immediately slightly improves the SA scores.

Storage of ungutted and gutted fish at ambient temperature was carried out during the four bimonthly trips, March to September inclusive, where the maximum temperature varied from 16°C in March to 23°C in July. The results are given in *Tables XVII to XX*. Both HX and TMA increase exponentially and lower concentrations due to gutting are really only evident after 96h, by which time the mackerel were obviously unfit to eat due to extreme autolysis and faecal odour. Over the 96h storage period both the pH and PV increase but little difference could be detected between gutted and ungutted fish. Histamine production appeared to be dependent on the temperature; for example in the July trial, which had the highest ambient temperature, 23°C, HA could be detected as early as after 4h storage and had reached concentrations of between 42mg and 52mg HA/100g fish after 96h. Compared to this, when the maximum temperature was 10°C in March the HA concentrations after 96h were only between 3mg and 5mg HA/100g. These figures compare well with previous experiments on the storage of mackerel at ambient temperature (6–12°C) by Hardy and Smith (1976) where after 76h and 81h the concentrations of HA in gutted and ungutted fish were 1·35 and 2·62mg HA/100g fish, respectively. As stated previously, none of the fish stored at ambient temperature was tasted by the panel.

From these trials results indicate that storage of mackerel in csw, which was found to be a simple and rapid method of cooling fish, is as good as storage of the fish when boxed in ice. Holding the fish at ambient temperature for 24h before gutting and icing tended to accelerate deterioration as measured by the various chemical methods used for monitoring spoilage. Of these chemical tests measurement of hypoxanthine possibly gives the best indication of spoilage in mackerel held for up to 10 days in ice. Increase of TMA concentrations in iced and csw fish varied throughout the year, the fastest rate being observed during the summer months. Rancidity as measured by the peroxide value is markedly reduced when the fish are stored in csw.

4 References

BANKS, A. Rancidity in fats. I. The effect of low temperature, sodium
1937   chloride and fish muscle on the oxidation of herring oil. *J. Soc. Chem. Ind.* London, 56, 13T–15T
——— *The freezing and cold storage of herrings.* DSIR Food Investiga-
1952   tion, Special Report No. 55, HMSO, London. 40pp
BLIGH, E G and DYER, W J. A rapid method of total lipid extraction
1959   and purification. *Can. J. Biochem. Physiol.*, 37, 911–917

BURT, J R, MURRAY, J and STROUD, G D. An improved automated
1968 analysis of hypoxanthine. *J. Fd. Technol.*, 3, 165–170
CUTTING, C L. Changes in the pH and buffering capacity of fish during
1953 spoilage. *J. Sci. Fd. Agric.*, 4, 597
HANSEN, P. Fat oxidation and storage life of iced trout. I. Influence of
1963 gutting. *J. Sci. Fd. Agric.*, 11, 781–786
HARDY, R and KEAY, J N. Seasonal variations in the chemical compos-
1972 ition of Cornish mackerel, *Scomber scombrus* (L), with
detailed reference to the lipids. *J. Fd. Technol.*, 7, 125–137
HARDY, R and SMITH, J G M. The storage of mackerel (*Scomber*
1976 *scombrus*). Development of histamine and rancidity. *J. Sci. Fd.
Agric.*, 27, 595–599
HORWITZ, W. *Official methods of analysis of the Association of Official
1970 Analytical Chemists.* ACAC, Washington. 11th edn, 296

MURRAY, C K and BURT, J R. An automated technique for determin-
1964 ing the concentration of trimethylamine in acid extracts of fish
muscle. *Proc: Technicon International Symposium, London,
England.* 9pp
SMITH, J G M, HARDY, R, McDONALD, I and TEMPLETON, J. The stor-
1979 age of herring (*Clupea harengus*) in ice, refrigerated sea water
and at ambient temperature. Chemical and sensory assessment. (*In press*)
SMITH, J G M, McGILL, A S, THOMSON, A and HARDY, R. Preliminary
1980a investigation into the chill and frozen storage characteristics of
scad (*Trachurus trachurus*) and its acceptability for human
consumption. *This volume*
SMITH, J G M, HARDY, R, THOMSON, A B, YOUNG, K W and PARSONS,
1980b E. Some observations on the ambient and chill storage of blue
whiting (*Micromesistius poutassou*). *This volume*

Table IX
CHEMICAL ANALYSIS OF MAY MACKEREL GUTTED AND ICED IMMEDIATELY

Days in ice	PV	HX	TMA	HA	pH	SA
0	0·9	0	0	0	6·5	3·5
4	2·4	0·46	0·89	0	6·6	3·2
7	4·7	0·44	0·61	0	6·5	2·6
10	7·2	0·55	0·68	0	6·45	2·8
14	14·9	0·90	4·42	0	6·45	2·6*
18	12·6	1·38	14·72	0·09	6·85	1·6*

* = slight rancid flavour

Table XIII
CHEMICAL ANALYSES OF SEPTEMBER MACKEREL GUTTED AND ICED IMMEDIATELY

Days in ice	PV	HX	TMA	HA	pH	SA
0	0·5	0·10	0	0	6·65	4·0
4	0·7	0·20	0·13	0	6·5	3·4
8	2·9	0·32	0·21	0	6·55	2·7
11	6·2	0·60	0·70	0·17	6·7	3·1
14	4·3	0·66	1·38	0·37	6·8	1·6
17	12·5	1·66	7·61	0·08	7·3	1·8*

* = slight rancid flavour

Table X
CHEMICAL ANALYSES OF MAY MACKEREL GUTTED AND ICED AFTER STORAGE FOR 24 HOURS AT AMBIENT (12–15°C)

Days in ice	PV	HX	TMA	HA	pH	SA
0	3·4	0·25	0·50	0	6·6	3·2
3	5·8	0·49	1·22	0	6·4	3·1
6	9·6	0·53	1·95	0·17	6·6	2·8
9	8·1	1·15	10·53	1·20	6·5	2·6
13	12·5	1·64	14·72	0	6·8	2·2*
17	32·4	2·05	23·93	23·46	7·1	1·3†

* = slight rancid flavour
† = rancid flavour

Table XIV
CHEMICAL ANALYSES OF SEPTEMBER MACKEREL GUTTED AND ICED AFTER STORAGE FOR 24 HOURS AT AMBIENT (12–20°C)

Days in ice	PV	HX	TMA	HA	pH	SA
0	0·9	0·23	0·64	0	6·55	3·3
3	4·3	0·32	1·31	0	6·45	3·1
7	5·2	0·54	2·08	0	6·25	3·2
10	10·1	0·82	2·21	0·40	6·65	2·8*
13	28·3	1·96	6·02	0·60	6·9	2·2†
16	21·9	3·16	12·46	4·42	7·1	1·9†

* = slight rancid flavour
† = rancid flavour

Table XI
CHEMICAL ANALYSES OF JULY MACKEREL GUTTED AND ICED IMMEDIATELY

Days in ice	PV	HX	TMA	HA	pH	SA
0	0·5	0·11	0	0	6·4	4·5
4	4·3	0·28	1·71	0	6·2	3·7
7	9·3	0·22	2·05	0	6·35	3·2
11	15·3	0·70	2·96	0·17	6·2	2·1*
14	19·0	1·21	3·98	0·80	6·25	1·4*

* = slight rancid flavour

Table XV
CHEMICAL ANALYSES OF NOVEMBER MACKEREL GUTTED AND ICED IMMEDIATELY

Days in ice	PV	HX	TMA	HA	pH	SA
0	0·2	0·01	0	0	6·2	4·2
4	0·9	0·10	0·14	0	6·25	3·9
7	3·8	0·52	0·28	0	6·1	3·6
10	10·2	0·36	0·41	0	6·4	1·9
13	12·9	0·71	0·56	0·06	6·7	2·2*
16	9·8	0·99	0·84	0·23	6·4	1·6*

* = slight rancid flavour

Table XII
CHEMICAL ANALYSES OF JULY MACKEREL GUTTED AND ICED AFTER STORAGE FOR 24 HOURS AT AMBIENT (12–23°C)

Days in ice	PV	HX	TMA	HA	pH	SA
0	0·9	0·25	0·93	0	6·15	4·1
3	13·5	0·58	2·96	0	6·15	3·4
6	14·2	0·98	6·61	0·34	6·2	3·1*
10	48·0	1·50	17·10	3·99	6·2	2·7†
13	38·0	1·60	14·95	34·55	6·4	1·1†

* = slight rancid flavour
† = rancid flavour

Table XVI
CHEMICAL ANALYSES OF NOVEMBER MACKEREL GUTTED AND ICED AFTER STORAGE FOR 24 HOURS AT AMBIENT (12–18°C)

Days in ice	PV	HX	TMA	HA	pH	SA
0	1·2	0·15	0·09	0	6·25	3·6
3	4·0	0·29	0·18	0	6·2	3·1
6	5·3	0·53	0·42	0	6·3	3·0
9	8·5	0·67	0·32	0	6·25	1·8
12	9·4	0·69	0·98	0·23	6·35	1·8*
15	16·7	1·02	2·94	0·23	6·55	1·4*

* = slight rancid flavour

Table XVII
CHEMICAL ANALYSES OF GUTTED (G) AND UNGUTTED (U) MARCH MACKEREL STORED AT AMBIENT TEMPERATURE (9–16°C)

Hours Storage	PV		HX		TMA		HA		pH	
	G	U	G	U	G	U	G	U	G	U
0	0·1	0·1	0·15	0·15	0·02	0·02	0	0	6·5	6·6
3	1·1	0·9	0·20	0·16	0·02	0·02	0	0	6·6	6·6
10	1·6	1·5	0·25	0·25	0·06	0·17	0	0	6·5	6·45
30	1·2	1·2	0·46	0·58	0·15	0·23	0·14	0·14	6·6	6·5
96	3·8	3·6	4·43	6·74	28·61	36·14	4·86	3·60	6·85	6·75

Table XVIII
CHEMICAL ANALYSES OF GUTTED (G) AND UNGUTTED (U) MAY MACKEREL STORED AT AMBIENT TEMPERATURE (10–18·3°C)

Hours Storage	PV		HX		TMA		HA		pH	
	G	U	G	U	G	U	G	U	G	U
0	0·9	0·9	0	0	0	0	0	0	6·4	6·4
3	3·4	2·8	0	0	0·28	0·21	0	0	6·35	6·6
10	2·7	4·0	0	0	0·25	0·35	0	0	6·5	6·3
30	2·0	3·5	0·33	0·43	0·49	1·90	0	0	6·5	6·45
96	3·3	3·2	2·93	2·96	21·10	47·82	5·95	8·82	7·2	6·9

Table XIX
CHEMICAL ANALYSES OF GUTTED (G) AND UNGUTTED (U) JULY MACKEREL STORED AT AMBIENT TEMPERATURE (12–23°C)

Hours Storage	PV		HX		TMA		HA		pH	
	G	U	G	U	G	U	G	U	G	U
0	0·5	0·5	0·11	0·11	0	0	0	0	6·4	6·4
4	—	—	0·30	0·30	1·37	1·14	0	0·37	6·2	6·2
10	2·6	3·5	0·20	0·26	1·25	1·60	0·40	0·68	6·3	6·3
36	4·2	5·2	0·80	0·90	6·84	6·50	0·71	0·57	6·7	6·75
96	8·9	10·3	5·78	6·50	39·87	54·67	42·8	51·3	7·15	7·2

Table XX
CHEMICAL ANALYSES OF GUTTED (G) AND UNGUTTED (U) SEPTEMBER MACKEREL STORED AT AMBIENT TEMPERATURE (12–20°C)

Hours Storage	PV		HX		TMA		HA		pH	
	G	U	G	U	G	U	G	U	G	U
0	0·5	0·5	0·10	0·10	0	0	0	0	6·65	6·65
12	1·0	1·0	0·28	0·20	0·14	0·14	0	0	6·35	6·35
27	1·7	1·5	0·52	0·40	0·54	0·29	0·28	0·40	6·40	6·30
48	1·8	1·3	0·96	0·82	3·40	18·44	4·90	7·00	7·00	6·5
100	4·6	4·1	5·30	6·12	25·13	31·12	27·80	28·30	7·7	7·4

Seasonal variability in blue whiting (*Micromesistius poutassou*) and its influence on processing

K J Whittle, I Robertson and I McDonald

1 Introduction

The effects of seasonal factors, fishing ground variations and environmental influences on the quality attributes of fish flesh, on the shelf-life of chilled or frozen fish and on the implications for processing were reviewed a few years ago for cod (*Gadus morhua*) by Love (1975) and again by the same author in one of the plenary lectures to this meeting. In recent years the blue whiting (*Micromesistius poutassou*), a gadoid abundant in the north-east Atlantic, has been of considerable interest to a number of research groups in Europe and Scandinavia attempting to assess its suitability for consumer pro-

ducts. Although its gross chemical composition is similar to cod (Dagbjartsson, 1975), research workers here at Torry and elsewhere (Ström, 1978) have been surprised by the much greater degree of seasonal variability it shows compared with cod or other gadoids.

Blue whiting can be caught in the EEC Economic Zone from about late February to early June and they spawn in waters to the west of Scotland about mid-April. At spawning and for some time afterwards, the fish are thin, watery and soft. An extreme case is shown in *Fig 1*. Our preliminary work up to 1978 showed that at this time the fish appeared to be emaciated or depleted with

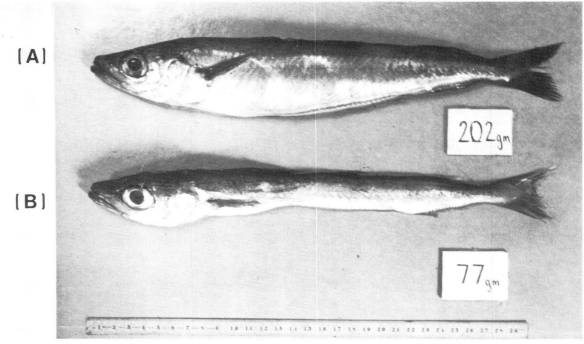

Fig 1   A comparison of blue whiting of similar length in (*a*) good condition, 202g, and (*b*) extremely poor condition, 77g

small livers, muscle water contents as high as 92% to 93% and the stomachs were usually empty or only contained what appeared to be fish scales. However, products formed from these fish, such as fish fingers, were still acceptable although they were not rated as highly as similar products made from cod.

The peak catching period, late March to May, in waters west of Scotland coincides with a high proportion of depleted fish but there is also the possibility of catching fish in better condition in late February and March. Accordingly, in 1978 we began to systematically examine and establish the changes in basic biological and chemical properties of the fish with season and to correlate these with possible changes in suitability for machine processing and acceptability of products. The fishery is relevant to EEC waters only from late February to early June, and this paper is concerned specifically with the preliminary analysis of the data from that period. In order to complete the seasonal cycle and follow the recovery of the fish in the post-spawning period, we intended to examine similar parameters in fish from other areas during the remainder of the season.

2 Sampling

Between February and June 1978, the FRV *G A Reay* carried out five cruises, designated A, B, C, D and F, respectively, in the test, to various fishing grounds within the area between northeast Faroe and St Kilda as shown in *Fig 2*. Blue whiting were caught during daylight hours by small pelagic trawl usually at depths between 280m and 450m. During each cruise, a random sample of about 200 whole fish from one haul in each area fished was stowed in ice for 24h and then air-blast frozen and stored at −30°C for basic biological and chemical measurements later in the laboratory ashore. From the same haul where possible on cruises A, B, D and F, much

larger samples of whole fish (1·5 tonnes to 2·5 tonnes) were subjected to delays of 4h, 8h and 12h either at ambient temperatures (10°C to 14°C) or boxed in ice before freezing in a vertical plate freezer on board in 50mm or 100mm thick blocks (25kg and 50kg, respectively) with and without added water. The treatments

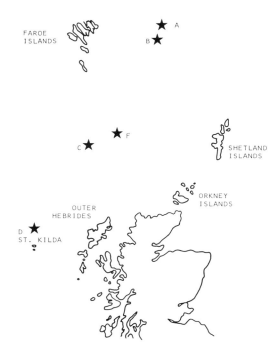

Fig 2   Fishing positions for blue whiting during cruises A to F of the FRV *G A Reay,* February to June 1978

Cruise A 30 January–6 February near 62°16′N 03°52′W
      B 15–24 February   62°02′N 03°56′W
      C 7–22 March   60°00′N 06°50′W
      D 6–19 April   57°50′N 09°30′W
      F 1–7 June   60°32′N 06°05′W

are summarized in *Table I*. A minimum of six blocks of each were produced and then stored at −30°C ashore until required for processing using the Baader 121 filleting machine.

Table I
PRE-FREEZING AND FREEZING TREATMENTS OF BLUE WHITING CAUGHT BY *G A Reay* CRUISES A, B, D AND F, FEBRUARY TO JUNE 1978

Pre-freezing delays	Pre-freezing stowage conditions		Treatments in the vertical plate freezer			
H	Boxed in ice	Process deck ambient 10–14°C	Dry blocks	Wet (added water) blocks	50 mm[1]	100 mm[1]
0			✓		✓	
		✓		✓		✓
		✓	✓	✓		
4	✓		✓		✓	
				✓		✓
		✓		✓		✓
			✓		✓	
			✓		✓	
		✓		✓		✓
8	Variables					
12	As for 4h delay above					

[1] Subsequently, it was possible only to complete machine processing of 50mm and 100mm variables for sample D

3 Biological/chemical assessment

Initially, length measurements and ungutted, gutted, liver and gonad weights as well as hand-cut, single fillet, skin-on weights were recorded. The fish were sexed and stomach contents and gonad development noted. Fillet 'softness' was assessed subjectively on a 10-point scale and the water content and pH of the muscle were determined. The softness value was probably influenced to some extent by the appearance of the fillet as it changed from bluish translucence to white and opaque, and these differences were more difficult to determine in thawed frozen fish.

Table II summarizes the range, mean and standard deviations of the values obtained for each seasonal sample and *Table III* shows the change in gross composition of the fish over the five-month period. Fortunately, mean length varied only within 1cm over the whole season, so that although the length/weight relationships were established it was not necessary to correct for differences in length in order to make a preliminary comparison of the sample means.

Ungutted weight remained relatively constant during February (A, B) and March (C), since decreasing liver size was largely compensated by increasing gonad size. The ungutted weight decreased by 21% in April (D) due largely to a decrease in total gut content from 12% to 15% (A, B, C) to 5% to 6% of body weight in April. Ungutted weight had increased slightly by June (F). The decrease in total gut content in April was due mainly to loss of gonad weight on spawning, reduction of liver size to a minimum and lack of food in the stomachs. The fish in February and March were characterized by having large stomachs full of euphausids.

Table III
VARIATION IN GROSS PERCENTAGE COMPOSITION DURING FEBRUARY TO JUNE

	A	B	C	D	F
1 As % ungutted weight					
Gutted weight	87·9	86·7	84·7	94·5	93·4
Guts total	12·1	13·3	15·3	5·5	6·6
Liver	6·9	6·2	5·0	2·1	2·5
Gonad (male)	3·2	2·9	8·4	1·0	1·0
Gonad (female)	2·0	1·8	5·5	1·5	0·8
Skin-on fillet	44·1	42·8	41·8	43·9	42·6
Skin-off fillet[1]	33·1	32·1	31·3	32·9	32·0
Residual[2]	43·8	43·9	42·9	50·6	50·6
2 As % gutted weight					
Skin-on fillet	50·2	49·3	49·3	46·4	45·7
Skin-off fillet[1]	37·7	37·0	37·0	34·8	34·3
Residual[3]	49·8	50·7	50·7	53·6	54·3

[1] Assuming yield on skinning is 75%
[2] Remainder after guts and fillets removed
[3] Remainder after fillets removed

Gutted weight tended to fall slowly to a minimum in April and had increased slightly by June, matched closely by the skin-on fillet weight. It is interesting to note that throughout this five-month period of changing body status the mean hand-fillet yield expressed as a percentage of body weight varied only within the range 41·5% to 44·0%, which compares well with a value of 40% for small fish noted under commercial conditions.

Mean water content in the muscle rose slowly at first from a 'normal' value of about 80% in February, and then more rapidly, to reach 83% in April immediately after spawning. Some individual values were as high as 91%. Water content had begun to fall again by June. The relationships between pH and water content in each seasonal sample were poor (correlation coefficients less than 0·4) except in June when pH was highest (7·0) and the correlation coefficient was 0·62 (*Fig 3*).

Although the subjective assessment of fillet softness cannot be reduced to a single uninfluenced parameter, it is interesting to note that the changes shown by this assessment appear to match the state of gonad development almost completely. The results suggest that the fish are firmest just before spawning in March when gonad development is maximal. The relative changes in the gross composition with time are compared graphically in *Fig 4*.

4 Machine processing

In October/November 1978, the frozen blocks from cruises A, B, D and F which had been subjected to various pre-freezing and freezing treatments were thawed overnight, separately weighed and kept chilled. They were fed treatment by treatment to a Baader 121 filleting machine (which had been developed for filleting blue whiting) and single, skinned fillets were prepared. Fish which were too thin or damaged, for example, decapitated, badly bent or with broken backbones were discarded by the machine operators and the total reject weight recorded for each treatment. Apart from relatively minor running adjustments, the machine blade settings remained the same throughout so that comparisons could be made on yields from the respective treatments. The water wash and spray flows were kept as low

380

		February A	February B	March C	April D	June F
Length cm	X	30·8	30·8	31·3	31·7	31·2
	SD	2·1	2·2	1·9	2·5	2·1
	range	27·1–35·5	25·4–36·7	27·4–35·3	25·6–35·7	27·3–37·3
Weight ungutted g	X	181	176	182	144	153
	SD	31	41	32	36	39
	range	115–273	82–287	137–281	82–294	96–342
Weight gutted g	X	159	153	154	136	142
	SD	28	35	26	34	34
	range	107–245	54–228	93–239	79–280	81–274
Liver g	X	12·6	11·0	9·1	3·1	3·9
	SD	4·7	6·2	4·4	1·9	3·9
	range	1·4–25·3	1·5–42·6	0·5–22·7	0·4–9·0	0·5–33·8
Gonad (male) g	X	5·9	5·2	15·4	1·4	1·5
	SD	4·3	3·7	6·6	0·7	1·4
	range	0·4–16·4	0·5–11·6	1·6–30·4	0·4–2·7	0·2–6·7
Gonad (female) g	X	3·7	3·2	10·0	2·1	1·1
	SD	1·4	2·4	3·7	1·1	0·7
	range	1·9–8·6	0·3–8·5	4·3–19·2	0·6–5·3	0·1–3·3
Skin-on[1] fillet L + R g	X	80·1	75·3	76·1	63·1	65·0
	SD	15·3	19·3	15·6	17·2	19·5
	range	44–127	25–148	28–122	37–134	25–147
Muscle water %	X	80·0	79·8	80·9	83·0	82·1
	SD	0·9	1·1	1·1	1·4	1·5
	range	77·4–83·0	77·0–83·9	79·2–86·7	80·0–91·4	79·7–88·0
pH	X	6·7	6·9	6·8	7·0	7·0
	SD	0·1	0·1	0·1	0·1	0·1
	range	6·4–7·0	6·7–7·1	6·6–7·0	6·7–7·2	6·7–7·4
Subjective fillet 'softness'	X	7·4	7·0	4·6	7·8	7·5
	SD	1·7	1·3	1·4	1·7	1·9
	range	3–10	4–10	2–9	2–10	3–10
Number samples		118	115	122	119	120

X = Mean
SD = Standard deviation
L + R = Left hand plus right hand
[1] Hand-cut fillets

as possible compatible with reliable functioning of the moving parts of the machine. The fillets leaving the skinning section were inspected continuously by three or four personnel and, where possible, trimmed by hand to remove adhering membranes, skin, fins, and tails. The trimmed fillets were collected in open-mesh baskets and designated 'good' fillets whilst those it was not possible to trim were collected separately and designated 'bad' fillets. Each was weighed according to a fairly strict time routine to overcome, as far as possible, variations due to differences in drainage time since it was noted that the water pick-up of the fillets immediately on leaving the machine could be as high as 25% of their original weight. However, this subsequently drained off almost completely as discussed later in the text.

The percentage of fish rejected before machine processing and the overall yields of good and total fillets (*ie* good plus bad) based on the fish actually processed by machine from the various treatments, are summarized in *Table IV*.

The proportion of fish rejected at the filleting machine was less than 1·3% in February, increased substantially to nearly 8·5% with poor condition fish in April and had decreased a little to about 6·5% by June.

The overwhelming proportion of the bad fillet weight

was due to single fillets which were not skinned properly. The problem was related to a fault on the skinner belt which can be rectified and would be expected to reduce this loss considerably. Thus, for practical purposes the total fillet yield values may be regarded as the more appropriate figures.

Figure 5 compares the good fillet yield for the various treatments at each seasonal sampling time and shows that sampling time appears to have the major effect on yield. In February, good fillet yield was 30% to 31%, but only 23·5% in April, rising to 25% by June. The rejects due to skinning were relatively constant throughout (1·5% to 2·5%) as might be expected from the continuing cause of the problem. It is interesting to compare these yields with the data on hand-cut skin-on yield from the biological assessment in the previous section. Earlier work on skinning problems had shown that the yield of skin-off fillets on skinning by Trio Skinning Machine was about 75%. Using this factor, coversion of the hand-cut, skin-on values to skin-off (31·5% to 33%) gives results remarkably similar to the total yield from the Baader 121 on good condition February fish (32·5% to 33·5%). As noted earlier in the text, the yield obtained by hand filleting remained fairly constant throughout the season and this serves to emphasize the large discrepancy be-

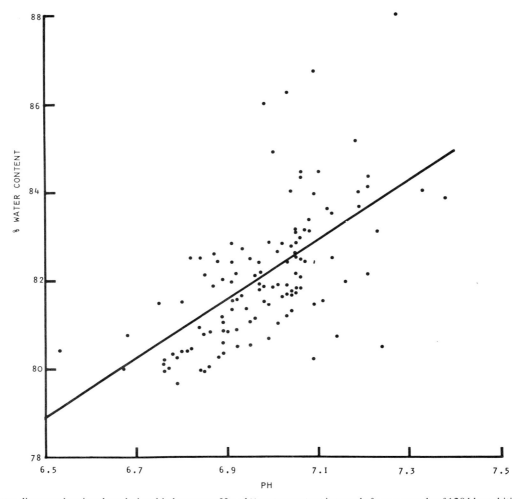

Fig 3 A scatter diagram showing the relationship between pH and % water content in muscle from a sample of 120 blue whiting caught in June. The correlation coefficient is 0·62 and the relationship is expressed by the regression line drawn: water content = 35·5 + 6·7 (pH)

tween the possible yield and the actual Baader yield on the poor-condition fish caught in April and June. Although the fish were apparently softer during this period and had a higher water content in the flesh, the reduced yield was probably influenced to a greater extent by the change in cross-sectional shape of the fish as it became thinner and belly depth reduced on spawning (*Fig 6*) such that filleting was less efficient. The result was that the Baader 121 removed only about 60% of the potential fillet.

As *Fig 5* shows, the various treatments had smaller effects on yield than season. Yield varied only between 1% to 1·5% with delays up to 8h before freezing and the effect was not consistent, whereas performance was consistently poorer with delays greater than 8h and the fillets were noticeably softer. In samples D and F delays of 12h at ambient consistently resulted in soft fillets with yields below 22%. In most cases iced was better than ambient storage and wet blocks gave 0·5% to 1% higher yields than dry blocks. It was only possible in D to test the effect on yield of 100mm and 50mm blocks. The latter gave 3% higher yields.

5 Product assessment

The good fillets from machine processing from each treatment were made into random-packed, 7·4kg frozen

fillet blocks (laminated blocks) in the traditional manner but without additives such as polyphosphate. Considerable accumulation of drip was noticed on standing from all batches of fillets but no problems were experienced in making the blocks unless considerable delays occurred between weighing the 7·4kg batches, packing the moulds and loading the horizontal plate freezer. Such delays resulted in excessive incidence of voids in the frozen block. Examination of the intact frozen blocks for various properties such as regularity, presence of voids, blemishes, colour, appearance and water content so far has not revealed much in the way of consistent differences between treatments but some seasonal effects were noted.

The appearance of the blocks did show an apparently greater degree of diffuseness of the individual fillet outline as delays before freezing increased, which probably relates to the tendency of the soft fillets to break up. The change in percentage water content of the frozen fillet blocks followed the changes in water content of the original fillet material (*Table II*) quite closely. It increased from 80·8% to 81·0% in February (about 1% higher than the original fillet) to 82·9% in April (*cf* 83·0% in the original fillet) and had decreased slightly to 82·7% by June (about 0·6% higher than the original fillet). This suggests that most of the water picked up during machine filleting subsequently drains off such

382

Table IV

MEAN YIELDS FROM BAADER 121 MACHINE PROCESSING IN SKIN-OFF, SINGLE FILLET MODE OF THE VARIOUS PRE-FREEZING AND FREEZING TREATMENTS OF FISH CAUGHT BETWEEN FEBRUARY AND JUNE 1978

	A (February)	B (February)	D (April)	F (June)
Percentage by weight of fish from all treatments rejected as unsuitable for machining				
negligible	1·3	8·3	6·4	
Total percentage yield fillets[1] (all treatments) from fish actually machined				
	33·6	32·6	25·2	27·3
Yield of good fillets[2] (all treatments) from fish actually machined				
	31·2	30·0	23·6	25·1
Effect of treatments on yield of good fillets:				
1 Pre-freezing delays h				
0	31·6	30·6	24·8	26·9
4	31·0	31·2	24·7	26·0
8	32·1	31·5	23·4	25·4
12	30·6	27·8	22·1	22·0
2 Temperature during delays				
ice	32·1	29·3	23·6	25·4
ambient (10–14°C)	30·3	30·3	23·1	23·7
3 Freezing with and without water added to the block				
wet	31·7	30·3	24·2	25·3
dry	30·8	29·8	23·0	24·9
4 Freezing blocks of different thickness				
50mm			24·9	
100mm			21·9	

[1] Total yield included fillets which were not properly skinned
[2] Good fillets represent those adequately trimmed and skinned and accepted for further processing

that the water content of the frozen block is likely to be enhanced by no more than 1%. The overall colour of the treatments from sample D (April) as a whole appeared lighter and whiter than the remaining seasonal samples which could be ranked in order of increasing darkness: F, A, B. This whiter appearance of samples D corresponds with the most depleted fish and the highest water content. Possible changes in pigmentation of the dark muscle which may contribute to the overall colour appearance were not measured but the whitening effect could also have been due to a change in the reflectance properties of the fillet as a result of the increase in muscle water content.

In April/May 1979, portions of the frozen laminated blocks from the various treatments were cut by handsaw into nominal, 20g fish finger blanks and coated with batter and crumb. These fish fingers were coded according to treatment, deep-fat fried and presented to two assessment panels for comparison with a well-known commercial brand of cod fish fingers. The expert panel of eight trained assessors compared the samples on a seven-point hedonic scale for texture, flavour, colour and overall acceptability. A consumer-type panel of 24 tasters drawn from all the staff of the research station compared the samples for acceptability on the same hedonic scale. Each panel was told that the cod fingers

were a reference sample and they were rated more highly by the expert panel (overall mean score, 6·5) than the consumer-type panel (overall mean score, 6·2). Both panels returned more variable results with blue whiting than cod. From the results of the consumer panel, no systematic differences between treatments or between seasons could be discerned and the overall mean scores for the four seasonal samples only varied between 4·7 and 5·1 on the hedonic scale, ie close to the 'like slightly' score (*Fig 7*). Thus, the consumer panel did not consistently discriminate against any of the treatments of the blue whiting fish fingers or note any seasonal differences but they did rate these products as a whole lower than cod. They scored the cod close to 'like' on the scale compared with 'like slightly' for blue whiting. The expert panel, however, was more discerning. Although no easily established systematic differences were noticed with treatment, inspection of the pooled data means for each seasonal sample showed quite obvious differences with time of catching among the attributes examined (*Fig 7*). Each of the four attributes was marked down by 0·5 to 0·8 score units for the poor-condition fish caught in April and June compared with good-condition fish caught in February and the relative ranking of the attributes was maintained within each seasonal sample. A slight improvement was noted in June. The expert panel scored the February samples 5·4, ie higher than the consumer panel, whilst the April and June samples were scored 4·7 to 4·8, ie lower than the consumer panel. Scores for overall acceptability seemed most closely related to flavour.

6 Conclusions

The changes in biological condition of blue whiting on spawning, which are also maintained for some time afterwards, result in markedly less efficient processing by the Baader 121 filleting machine. This seems most likely to be related to the marked change in shape of the fish during April to June when the total gut content is much reduced. There is also some decrease in the overall amount of the flesh on the fish. The result is that the fish becomes much thinner and belly depth reduces. Maintenance of good practices in handling and freezing the fish helps to keep the reduction in yield as small as possible but the seasonal difference remains the major effect.

In contrast, the potential percentage fillet yield by hand filleting, based on whole fish weight, appears to remain relatively constant during the season. Nevertheless, even if hand filleting was practicable, in order to produce the same quantity of fillets 80 fish would have to be filleted on average in April to June compared with only 65 in February and March, using the data in *Table II*. In the case of machine filleting, a much larger weight of fish has to be processed in April to June to produce the same amount of fillet as in February because of poorer yield and this increases even more the numbers of fish which need to be processed. Thus processing time is much prolonged because the numbers of fish which can be processed per minute per machine is a limiting factor.

However, the seasonal changes in blue whiting may not be so important from the point of view of consumer

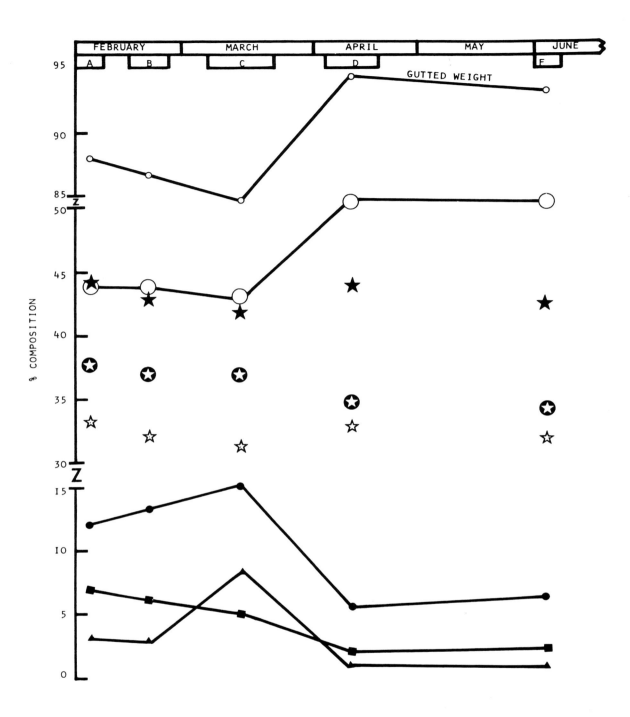

Fig 4 Change in gross composition of blue whiting with season
  ○ Gutted weight
  ● Total gut content
  ■ Liver
  ▲ Male gonad
  ★ Skin-on fillets cut by hand
  ☆ Skin-off fillets estimated as 75% of skin on-yield
  ○ The remainder of the fish after guts and fillet are removed. These are all expressed as percentages of the whole or ungutted weight of the fish caught in cruises A, B, C, D and F

  ✪ For comparison, skin-off fillet yield is also expressed as a percentage of the gutted weight

384

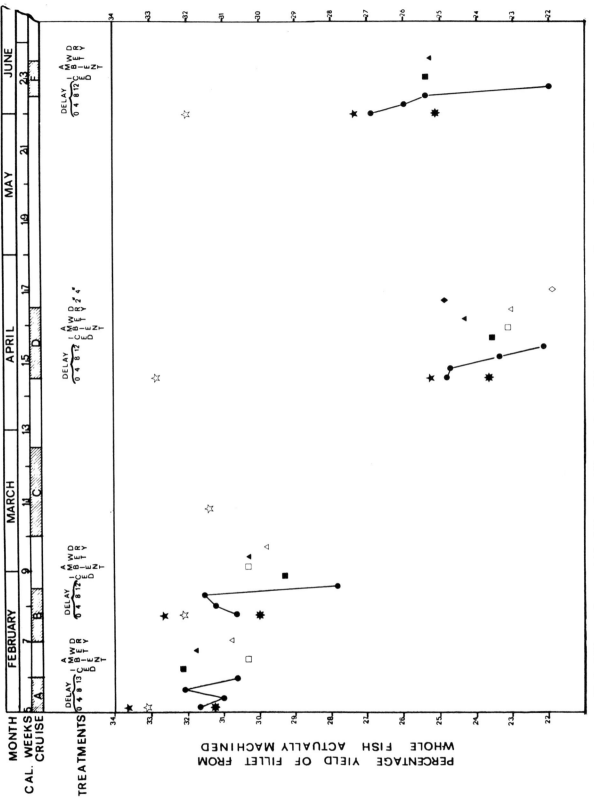

Fig 5   Differences in mean yields of skin-off good fillets from Baader 121 machine processing according to pre-freezing and freezing treatments for a series of seasonal samples from February to June, cruises A, B, D and F. (The good fillets represent the trimmed fillets accepted for processing to laminated blocks.)

●——● Delays 0, 4, 8 and 12h before freezing
▲ Pooled mean for blocks frozen with added water, wet
■ Pooled mean for delays in ice
△ Pooled mean for blocks frozen without added water, dry
□ Pooled mean for delays at ambient, 10°C to 14°C
● Overall pooled mean for good fillets
✱ These yields are compared with the total fillet yield from machine processing which mostly included those not skinned due to a skinning section fault
☆ They are also compared with the skin-off fillet yield estimated from the data obtained during the biological sampling (Table III)

385

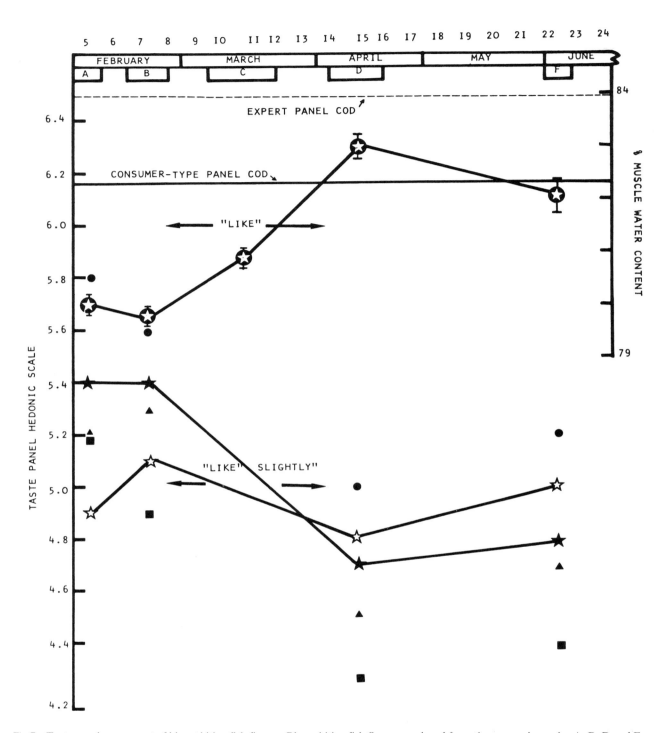

Fig 7 Taste panel assessment of blue whiting fish fingers. Blue whiting fish fingers produced from the seasonal samples A, B, D and F were compared with a commercial brand of cod fish fingers by a consumer-type and an expert panel. The horizontal lines at 6·16 and 6·49 respectively indicate the scores for cod. The consumer-type panel scored using a seven-point hedonic scale for acceptability ☆ and the expert panel using the same scale scored for texture ●, flavour ▲, colour ■ and overall acceptability ★. The points plotted are the overall mean values from the pooled data for all treatments in each seasonal sample. For comparison the water contents of the fillets from the comparable seasonal samples for biological and chemical measurements (*Table II*) are also shown and the bar lines indicate the 95% confidence limits about the mean

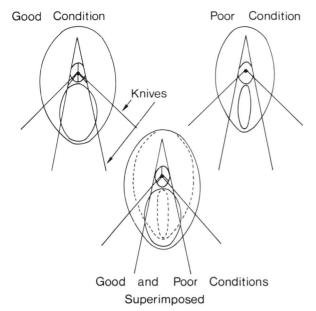

Good Condition   Poor Condition

Knives

Good and Poor Conditions
Superimposed

Fig 6 The change in sectional shape of good and poor condition blue whiting. A diagrammatic cross-sectional area representation of the gross composition by weight of fillet and total gut from blue whiting in good and poor condition. The representation is built up from actual data taken from *Table II* and also shows the superimposition of both conditions for comparison. The indicated position of the setting of the cutting knives of the Baader 121 shows how the efficiency of machine filleting is affected by shape

acceptability. All the blue whiting fish fingers were rated lower than the cod products, which confirms previous findings. However, the consumer-type panel did not differentiate between the fish fingers made from good-condition and poor-condition blue whiting, although the expert panel clearly considered that the products from poor-condition fish were less acceptable. The expert panel always gave the attribute of colour the lowest score. It is interesting to note that the blue whiting fish fingers were made and tasted some 11 to 15 months after catching. In the intervening period, the whole fish blocks

or the fillet blocks were stored at $-30°C$ and it is felt that this period more than covers the storage times likely to be encountered commercially. The results of a more systematic examination of cold storage stability are not available yet.

The water content of the fillet blocks remained substantially the same as the fish before processing so that the real seasonal differences were not distorted in processing and the high water pickup on the fillets during filleting and skinning apparently drained off under the conditions used.

The picture of variability shown by the samples from 1978 fits in very well with the results from May and June 1977, which are extremely similar to the April and June results in 1978. Further work from October 1977 and September/October 1978 suggests that the fish had completely recovered to good condition status by autumn and could be expected to process in much the same way as February-caught fish.

7 Acknowledgements

The Baader 121 filleting machine used in this work was on loan from the White Fish Authority and maintained by W Denton and P Wilson of the WFA.

G L Smith and J Lavety assisted by preparing programmes for use in analysing the data.

G P Hill and R A Anderson carried out some of the handling, freezing and preparative work at sea.

Miss A Cheyne arranged the taste panel assessments.

8 References

DAGBJARTSSON, B. Utilization of blue whiting *(Micromesistius poutas-*
1975 *sou)* for human consumption. *J. Fish. Res. Bd Can.,* 32, 747–751

LOVE, R M. Variability in Atlantic cod *(Gadus morhua)* from the
1975 northeast Atlantic: a review of seasonal and environmental influences on various attributes of the flesh. *J. Fish. Res. Bd Can.,* 32, 2333–2342

STRÖM, T. Processing of blue whiting. Survey of raw material handling,
1978 production and products. Paper No. 5, Nor Fishing 78, 16th Nordic Fishery Conference, August 1978

The effects of season and processing on the lipids of mandi *(Pimelodus clarias,* Bloch), a Brazilian freshwater fish

M Oetterer de Andrade and U de Almeida Lima

1 Introduction

The objectives of this work were to detect and to verify sexual and seasonal changes of some components of the lipid fraction (total lipids, total and free fatty acids, phospholipids, unsaponifiable fraction, cholesterol), to identify the fatty acids of fresh fishes and to compare the results with those of canned and stored preserves.

2 Material and methods

(*1*) Fish
Mandi, a Brazilian freshwater fish of the Piracicaba river, State of São Paulo; 20cm long and weighing 100g; freshly caught; analysed monthly over a whole year.
(*2*) Preserves

Five kinds of preserve were prepared with fishes caught in June and July, as recommended by Andrade (1975) and Andrade and Lima (1974; 1975): mandi in oil, mandi in tomato sauce, smoked fillets in oil, smoked fillets in tomato sauce and pastes.
(*3*) Sampling for analysis
Fresh fish, 10 males and 10 females, were each analysed 12 months consecutively. The preserves were analysed after 30, 60, 90, 120 and 210 days of storage.
(*4*) Chemical analysis
Total lipids after Folch *et al* (1957); total fatty acids, free fatty acids, unsaponifiable fraction and phospholipids as described by Elovson (1964); cholesterol after Kim and Golberg (1969), fatty acids of the total lipids by gas chromatography, methylation after Luddy (1960) and

the identification and quantification by equivalent chain length and internal normalization of the peak areas.

3 Results

Tables I to VI show the data for chemical analysis of fresh males and females of mandi.

Figure 1 illustrates chromatograms of the fatty acids of fresh males and females caught in September, fresh males caught in July and the same sample after processing and storage for 60 days.

Figures 2 and *3* record the lipid components of fresh and processed fish for the five kinds of preserves and periods of storage.

Figure 4 illustrates the six most representative fatty acids of the total lipids of fresh and processed fish.

4 Discussion

4.1 *Fresh fish*
Total lipids shown in *Table I* showed maxima in March

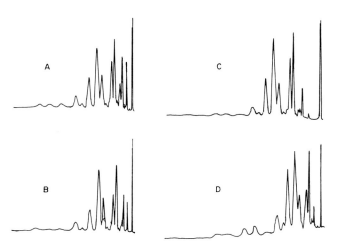

Fig 1   Chromatograms of the fatty acids of mandi

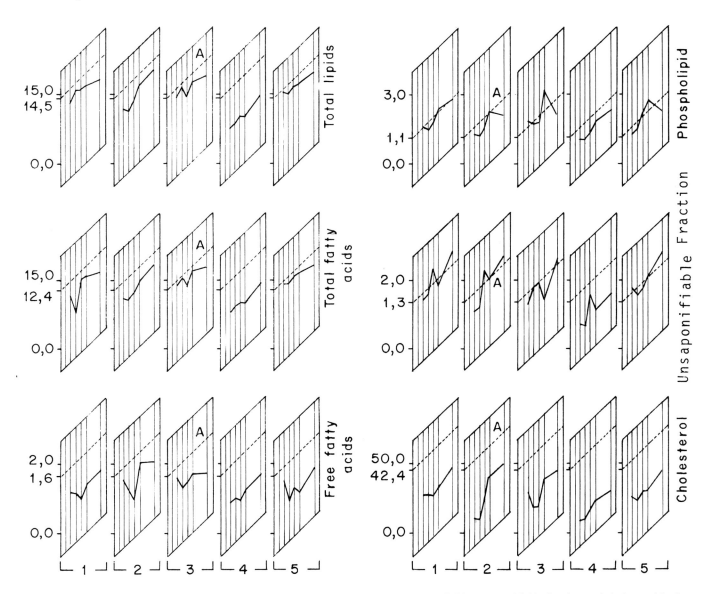

Fig 2   Total lipids, total and free fatty acids in the muscle of mandi processed and stored

Fig 3   Phospholipids, unsaponifiable fraction and cholesterol in the muscle of fresh mandi, processed and stored

388

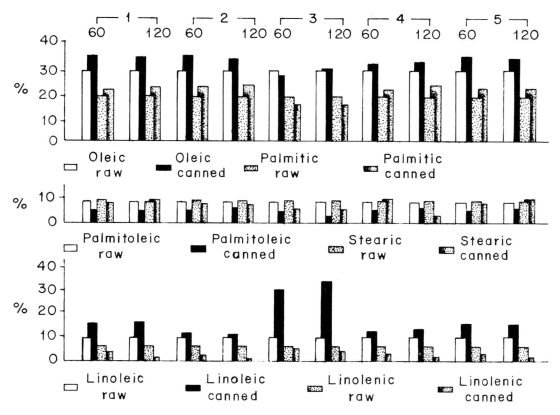

Fig 4 The fatty acids of fresh and processed fish stored during 60 and 120 days

and August, with little difference between the sexes.

The female annual mean was 13·66g/100g, near the male content of 13·85g/100g; the 2·53 maximum to minimum ratio is high but is the same for males and females. McCallum *et al* (1969) found similar amounts of total lipid in male and female migrating capelin.

Total fatty acids showed a significant monthly variation which was smaller than in the case of total lipids. Between sexes the fatty acid amounts are similar.

Sex and seasonal variations also occurred in free fatty acids (*Table II*). Shatunowski and Koslov (1973) observed a similar variation in *Notothemia rossimar-*

morata in which the young and adult females had free fatty acid contents similar to those found for mandi females. Fraser *et al* (1961) on the other hand found higher amounts of free fatty acids in white sucker than in mandi. McCallum *et al* (1969) found that the free fatty acids of capelin varied at different locations and during the migration period. These authors found that the minimum amount of total lipids corresponded to the maximum of free fatty acids. In female mandi the inverse was observed.

Total lipids in mandi varied throughout the year from 8·28g to 20·49g/100g muscle and free fatty acids from

Table I
TOTAL LIPIDS AND TOTAL FATTY ACIDS OF FRESH MANDI

	Total lipids g/100g of muscle[1]		Total fatty acids g/100g of lipids		g/100g of muscle[1]	
	Males	Females	Males	Females	Males	Females
Means for each month						
January	12·96	13·91	84·45	86·03	10·95	11·96
February	9·29	9·23	83·94	79·78	7·80	7·36
March	19·48	19·04	83·92	84·64	16·35	16·12
April	11·23	12·79	87·01	88·06	9·77	11·26
May	12·83	8·28	87·36	83·50	11·20	6·91
June	12·51	12·07	84·39	86·02	10·56	10·38
July	13·21	15·95	85·45	86·34	11·29	13·77
August	20·49	20·07	81·73	84·95	16·75	17·05
September	11·87	14·79	84·87	83·90	10·08	12·41
October	12·57	11·48	78·94	79·30	9·92	9·10
November	18·16	14·64	83·17	80·62	15·11	11·80
December	11·67	11·75	88·24	88·04	10·30	10·34
Year means	13·85	13·66	84·46	84·27	11·67	11·54
Maximum (M)	21·96	20·73	90·53	90·02		
Minimum (m)	8·68	8·18	77·47	75·84		
M/m	2·53	2·53	1·17	1·19		
Mean for males and females	13·75		84·36		11·61	

[1] In the moist material

0·55g to 3·04g. Toyomizu *et al* (1976) found similar wide variations in horse mackerel.

Significant monthly variations in phospholipid content but not between the sexes were observed (*Table II*). The average value was a little higher than that quoted by Stansby (1969) for fish muscle in general. Since Stansby's review the phospholipid contents of a number of species have been examined as follows: cod (Tarr, 1969), Indian freshwater fish (Gopakuman and Nair, 1971), mackerel (Hardy and Keay, 1972), *Notothemia* (Shatunowski and Koslov, 1973), veral (Gopakuman, 1975) and horse mackerel (Toyomizu *et al*, 1976). The variations in mandi fall within the values found for these species.

The unsaponifiable fraction (*Table III*) showed a significant seasonal variation but not much difference between the sexes. Jacquot (1961) records 0·5g and 15·0g/100g as the minimum and the maximum values in a number of species. Tsuchyia (1961) asserts that this fraction is extremely small in the lipids of sardines and herrings, but McMichael and Bailey (1951) found 0·5g and 1·5g of unsaponifiables/100g of lipids in these two species.

Cholesterol (*Table III*) varied widely with season but not between the sexes. Hardy and Keay (1972) found 900 to 1 050mg of cholesterol in the neutral lipids of mackerel. Stansby (1969) states that the amount of cholesterol is low in fish muscle being 90mg/100g in haddock, 75mg/100g in pollack and 95mg/100g in salmon. Similar values are recorded by Robinson (1965) for unspecified fish fillets.

Gas chromatography showed 28 fatty acid peaks of which 17 were identified. Oleic, palmitoleic, palmitic, stearic were present in highest amounts followed by linoleic and linolenic acids. In all the samples the total amount of unsaturated acids was approximately twice that of the saturated acids (*Tables IV, V* and *VI*). Myristic, heptadecenoic and linolenic acid varied significantly with season and myristic, myristoleic, pentadecanoic, palmitic, stearic, linoleic, linolenic and heptadecanoic

Table II
FREE FATTY ACIDS AND PHOSPHOLIPIDS OF FRESH MANDI

| | Free fatty acids | | | | Phospholipids | | | |
| | (g/100g of lipids) | | (g/100g of muscle[1]) | | (g/100g of lipids) | | (g/100g of muscle[1]) | |
	Males	Females	Males	Females	Males	Females	Males	Females
Means for each month								
January	7·26	6·86	0·92	0·78	12·29	15·45	1·61	2·15
February	6·21	6·66	0·58	0·61	11·99	8·38	1·11	0·76
March	15·60	16·86	3·04	3·21	7·02	5·48	1·37	0·99
April	17·67	14·23	1·99	1·81	6·84	8·78	0·76	0·99
May	6·96	13·10	0·89	1·08	9·05	8·35	1·16	0·69
June	10·70	10·04	1·34	1·21	7·57	9·15	0·95	1·09
July	7·40	15·53	0·96	2·48	9·58	7·16	1·25	1·14
August	5·91	5·88	1·21	1·18	9·00	10·06	1·86	2·01
September	4·82	7·88	0·55	1·18	13·11	13·04	1·56	1·91
October	5·01	7·06	0·62	0·85	6·55	8·94	0·81	1·03
November	9·62	5·89	1·45	0·87	7·48	7·11	0·69	1·04
December	13·97	14·03	1·64	1·65	11·94	12·60	1·39	1·47
Year means	9·26	10·33	1·27	1·39	9·37	9·54	1·21	1·27
Maximum (M)	20·06	17·54			15·25	17·10		
Minimum (m)	4·25	4·14			3·57	5·11		
M/m	4·72	4·23			4·27	3·34		
Means for males and females	9·80		1·29		9·45		1·24	

[1] In the moist material

Table III
UNSAPONIFIABLE FRACTION AND CHOLESTEROL OF FRESH MANDI

| | Unsaponifiable fraction | | | | Cholesterol | | | |
| | (g/100g of lipids) | | (g/100g of muscle[1]) | | (mg/100g of lipids) | | (mg/100g of muscle[1]) | |
	Males	Females	Males	Females	Males	Females	Males	Females
Means for each month								
January	10·75	9·35	1·39	1·30	377·80	350·59	48·85	48·22
February	11·23	15·64	1·04	1·44	518·20	314·67	48·50	28·97
March	11·79	12·55	2·30	2·39	590·93	627·50	115·08	113·25
April	9·44	5·98	1·06	0·76	433·31	640·06	72·13	81·80
May	7·60	12·09	0·97	1·00	466·78	465·63	59·87	38·54
June	10·99	9·17	1·37	1·11	486·06	470·27	61·16	56·71
July	9·65	9·21	1·27	1·47	410·79	255·52	53·86	40·92
August	13·53	10·09	2·77	2·03	344·63	272·16	70·33	54·22
September	10·15	11·38	1·20	1·68	386·68	302·41	46·39	44·76
October	16·26	16·19	2·04	1·86	287·15	284·63	35·99	32·64
November	16·57	14·92	3·01	2·18	166·44	148·80	30·07	21·88
December	7·14	7·37	0·38	0·86	439·28	554·58	73·04	65·29
Year means	11·26	11·16	1·61	1·51	409·00	390·57	59·61	52·27
Maximum (M)	18·47	19·85			690·19	688·20		
Minimum (m)	5·54	5·22			120·47	121·19		
M/m	3·33	3·80			5·73	5·68		
Means for males and females	11·21		1·56		399·78		55·94	

[1] In the moist material

differed between the sexes. *Figure 1* shows chromatograms of males and females in the same and in two different months.

Seasonal variations in the fatty acids of freshwater fish are mentioned by Gruger *et al* (1964).

In the work reported here the amounts of lauric, pentadecanoic, heptadecanoic, heptadecenoic, arachidic, gadoleic and Cf to Cj acids in particular showed great variations.

Notevarp (1961) and Kietzmann *et al* (1974) found that the relative proportions of the unsaturated fatty acids increase as the lipid content increases. In mandi only the heptadecenoic and linoleic acids showed this pattern.

4.2 *Preserves*

In the mandi in oil pack, there were general decreases in total lipid, total fatty acids and free fatty acids, little changes occurring in the other values measured. Canning caused a decrease in myristic, myristoleic, pentadecanoic, palmitoleic, heptadecanoic and heptadecenoic acids and an increase in palmitic, oleic and linoleic acids.

In the mandi in tomato sauce pack, the total lipid fell at 30, 60 and 90 days whilst total fatty acids decreased slightly at 90 and 120 days and cholesterol decreased at 30, 60 and 90 days. Myristic, pentadecanoic, palmitoleic, heptadecanoic, heptadecenoic and stearic acids decreased and palmitic, oleic and linoleic increased during the same period.

In smoked mandi fillets canned in oil the total lipid

decreased at 90 and 210 days, cholesterol decreased after 60 and 90 days and small variations occurred in the other values.

In smoked mandi in tomato sauce, the total lipid decreased at all periods of storage whilst total fatty acids increased at 30 and 60 days. Cholesterol decreased at 30, 60 and 210 days. Lauric, pentadecanoic, palmitoleic, heptadecanoic, heptadecenoic, and stearic acids decreased and palmitic, oleic and linoleic acids increased and others did not change.

In mandi pastes, total lipids and total fatty acids decreased at 210 days, free fatty acids at 60 days and cholesterol at 30, 60, 90 and 120 days. The majority of fatty acids decreased on canning and storage but palmitic, oleic and linoleic acids increased in amount.

In general for all five preserves total lipids and fatty acids, free fatty acids and cholesterol tended to be smaller than in the fresh fish and to decrease during storage.

In the stored preserves minor changes occurred in the concentration of individual fatty acids which were unrelated to the method of processing (*Fig 4*).

Little work has been reported which can be compared to the results obtained here. The effect of heat processing on the amount of phospholipid was studied by Bastavizi and Smirnova (1972) in carp and pike, but only small changes were observed. Shevchenko and Lapshin (1975) showed that after traditional smoking fish fillets a relative increase in the saturated fatty acids occurred after 60 days' storage. Meizies and Reichwold (1973) observed that little or no changes occurred in the com-

Table IV

FATTY ACIDS IN THE TOTAL LIPIDS OF FRESH MANDI ANALYSED IN JANUARY, FEBRUARY, MARCH AND APRIL

Fatty acids	Quantity (% of total fatty acids)							
	January		February		March		April	
	Males	Females	Males	Females	Males	Females	Males	Females
Octanoic	tr	—	tr	tr	tr	tr	tr	tr
Decanoic	tr	—	tr	tr	tr	tr	tr	tr
C(a)[1]	tr	—	tr	tr	tr	tr	tr	tr
C(b)[1]	tr	—	tr	tr	tr	tr	tr	tr
Lauric	tr	tr	tr	tr	0·50	3·30	1·50	1·35
C(c)[1]	tr	tr	tr	tr	tr	tr	tr	tr
C(d)[1]	tr	tr	tr	tr	tr	tr	tr	tr
C(e)[1]	tr	tr	tr	tr	tr	tr	tr	tr
Myristic	2·90	2·40	2·70	2·50	5·80	5·90	4·10	4·00
Myristoleic	1·00	2·00	0·60	tr	2·00	3·50	1·00	0·85
Pentadecanoic	2·10	1·20	tr	tr	2·90	3·50	1·10	3·00
Pentadecenoic	0·40	0·60	tr	—	—	tr	0·50	—
Palmitic	15·40	24·00	17·70	21·90	11·80	14·90	13·30	16·40
Palmitoleic	8·30	9·00	16·60	14·00	18·20	11·90	14·00	10·10
Heptadecanoic	6·00	3·10	—	—	0·60	0·50	2·50	3·00
Heptadecenoic	1·50	0·70	—	tr	—	—	—	—
Stearic	10·30	10·10	13·10	12·10	13·30	14·20	10·10	11·20
Oleic	34·30	37·00	34·70	36·80	31·00	30·40	32·10	30·50
Linoleic	3·80	3·00	4·80	4·80	11·70	3·50	7·50	3·50
Arachidic	0·50	0·40	0·50	tr	tr	1·50	1·20	1·50
Linolenic	5·80	5·00	7·50	7·60	2·00	3·34	4·00	5·00
Gadoleic	1·20	tr	—	—	tr	—	1·15	1·20
C(f)[1]	2·10	tr	1·50	—	tr	1·50	1·30	1·50
C(g)[1]	2·30	tr	tr	—	tr	tr	1·00	1·10
C(h)[1]	1·50	tr	tr	—	tr	tr	1·50	2·00
C(i)[1]	—	—	tr	—	—	tr	1·25	1·00
C(j)[1]	—	—	—	—	—	tr	—	—
C(l)[1]	—	—	—	—	—	—	—	—
Saturated	29·10	36·90	34·00	36·50	31·40	39·80	30·10	34·45
Unsaturated	55·90	56·70	64·20	63·20	64·90	52·64	59·75	51·15

[1] Peak not identified
tr = traces

Table V
FATTY ACIDS IN THE TOTAL LIPIDS OF FRESH MANDI ANALYSED IN MAY, JUNE, JULY AND AUGUST

Fatty acids	Quantity (% of total fatty acids)							
	May		June		July		August	
	Males	Females	Males	Females	Males	Females	Males	Females
Octanoic	tr	tr	—	—	—	—	—	—
Decanoic	tr	tr	tr	tr	tr	tr	—	—
C(a)[1]	tr	tr	tr	tr	tr	tr	—	—
C(b)[1]	tr	tr	tr	tr	tr	tr	—	—
Lauric	1·60	1·75	2·60	2·40	tr	tr	tr	tr
C(c)[1]	tr	tr	tr	tr	tr	tr	tr	tr
C(d)[1]	tr	tr	tr	tr	tr	tr	tr	tr
C(e)[1]	tr	tr	tr	tr	tr	tr	tr	—
Myristic	3·60	2·90	5·80	3·70	2·90	2·52	3·15	2·42
Myristoleic	2·60	2·50	0·52	0·50	1·30	1·48	1·50	tr
Pentadecanoic	1·70	2·72	1·26	1·20	1·24	1·00	1·25	tr
Pentadecenoic	—	tr	tr	tr	tr	tr	1·65	1·61
Palmitic	14·95	18·40	16·62	19·60	18·20	19·94	20·10	24·08
Palmitoleic	11·95	9·10	7·60	7·80	9·73	7·84	9·50	10·75
Heptadecanoic	2·10	2·30	2·52	2·30	2·16	2·36	4·50	2·65
Heptadecenoic	—	tr	1·50	1·78	0·84	1·08	tr	1·07
Stearic	10·10	10·60	13·80	13·16	9·21	9·79	7·65	8·32
Oleic	33·20	33·80	33·30	33·42	30·28	31·36	38·00	37·80
Linoleic	6·00	3·20	10·40	10·00	10·18	11·00	5·25	3·41
Arachidic	0·12	0·23	1·30	0·90	0·30	0·46	tr	tr
Linolenic	5·00	5·90	1·50	1·62	6·15	6·55	6·60	6·28
Gadoleic	1·00	1·10	1·24	1·60	0·64	0·20	tr	tr
C(f)[1]	1·25	1·50	tr	tr	0·10	0·64	tr	tr
C(g)[1]	1·17	1·00	tr	tr	1·44	0·60	tr	tr
C(h)[1]	1·85	1·90	tr	tr	1·60	1·58	tr	tr
C(i)[1]	tr	1·25	—	—	0·43	0·14	—	—
C(j)[1]	—	tr	—	—	1·19	1·23	—	—
C(l)[1]	—	—	—	—	—	—	—	—
Saturated	30·37	32·98	39·70	39·76	30·61	32·71	30·90	34·82
Unsaturated	59·15	55·60	56·06	56·72	59·12	59·51	60·85	59·31

[1] Peak not identified
tr = traces

Table VI
FATTY ACIDS IN THE TOTAL LIPIDS OF FRESH MANDI ANALYSED IN SEPTEMBER, OCTOBER, NOVEMBER AND DECEMBER

Fatty acids	Quantity (% of total fatty acids)							
	September		October		November		December	
	Males	Females	Males	Females	Males	Females	Males	Females
Octanoic	—	—	tr	tr	tr	tr	tr	—
Decanoic	tr	tr	tr	tr	tr	tr	tr	—
C(a)[1]	tr	tr	tr	tr	tr	tr	tr	—
C(b)[1]	tr	tr	tr	tr	tr	tr	tr	—
Lauric	1·62	1·01	0·60	tr	1·50	1·00	tr	tr
C(c)[1]	tr	tr	tr	tr	tr	tr	tr	tr
C(d)[1]	tr	tr	tr	tr	tr	tr	tr	tr
C(e)[1]	tr	tr	tr	tr	tr	tr	tr	tr
Myristic	3·65	3·22	3·40	2·79	2·80	2·44	2·20	2·10
Myristoleic	0·60	1·52	0·85	2·19	1·70	2·79	1·30	3·40
Pentadecanoic	tr	0·26	0·61	1·79	1·50	1·20	1·40	1·40
Pentadecenoic	tr	tr	tr	1·05	1·90	2·00	0·29	—
Palmitic	18·05	21·45	18·24	13·07	16·57	16·70	14·90	20·00
Palmitoleic	8·36	10·02	14·76	8·98	14·78	9·69	14·80	10·40
Heptadecanoic	1·34	2·44	0·76	3·59	3·50	4·24	1·00	4·90
Heptadecenoic	tr	0·22	tr	1·75	2·10	2·00	—	1·10
Stearic	9·83	11·02	8·56	9·45	7·63	10·37	6·70	11·30
Oleic	35·00	37·06	32·19	26·83	31·70	30·76	32·20	34·70
Linoleic	9·91	8·74	10·35	11·75	8·42	7·87	16·50	4·00
Arachidic	0·27	1·08	tr	0·60	0·90	tr	0·65	tr
Linolenic	3·45	4·17	5·14	9·65	6·87	7·87	8·60	6·10
Gadoleic	0·62	tr	tr	tr	tr	tr	1·50	tr
C(f)[1]	1·31	0·70	0·63	0·64	—	—	1·30	tr
C(g)[1]	1·30	0·25	tr	0·81	—	—	1·70	tr
C(h)[1]	0·50	0·84	1·23	3·93	—	—	2·90	—
C(i)[1]	—	tr	tr	tr	—	—	1·30	—
C(j)[1]	—	—	tr	tr	—	—	—	—
C(l)[1]	—	—	tr	tr	—	—	—	—
Saturated	33·42	37·78	30·80	25·91	29·40	30·51	24·45	33·40
Unsaturated	57·94	61·73	63·29	61·15	65·57	60·98	74·90	59·70

[1] Peak not identified
tr = traces

392

ponents of the lipid fraction or the fatty acids during smoking. Lapshin and Shevchenko (1973) observed that the free fatty acid content increased after smoking.

The changes observed during processing and storage cannot be explained at present. In some cases exchange of lipid components between the fish and the oil of the packing medium could be occurring; oxidation is also a likely possibility.

5 Acknowledgements

This work was carried out with the financial support of FAPESP (Foundation for the Support of Research of the State of São Paulo).

6 References

ANDRADE, M O DE and LIMA, U DE A. Aproveitamento tecnológico
1974   do mandi – Defumação. Anais IV. Jorn. Ci. FCMBB, 48
—— Aproveitamento tecnológico do mandi: sopas e 'corned fish'.
1975   Simp. Bras. Alim. Nutr., 19
—— Preparo, seleção, armazenamento e estudos químicos e sensoriais
1975   de conservas de mandi, Pimelodus clarias. Bloch, São Paulo.
      127pp. (Tse Mestrado FCF-USP)
BASTAVIZI, M A and SMIRNOVA, G A. Effect of heat treatment on fish
1972   muscle phospholipid composition. Vopr. Pitan, 31(3), 90–92
ELOVSON, J. Metabolism of some monohidroxystearic acid in the intact
1964   rat. Biochim. Biophys. Acta, 84, 275–293
FOLCH, J et al. A simple method for the isolation and purification of
1957   total lipids from animal tissues. J. Biol. Chem., 226, 497–509
FRASER, D I, et al. III. Sectional differences in the flesh of a species of
1961   chondrostei, one of chimarae and of some miscellaneous tele-
      osts. J. Fish. Res. Bd. Can., 18(6), 893–905
GOPAKUMAR, K and NAIR, M R. Phospholipids of five Indian food
1971   fishes. Fish. Technol., 8(2), 171–173
GOPAKUMAR, K. Fatty acid composition of three species of freshwater
1975   fishes. Fish. Technol., 12(1), 21–24

GRUGER, E H, et al. Fatty acid composition of oils from 21 species of
1964   marine fish, freshwater fish and shellfish. J. Am. Oil Chem.
      Soc., 41, 662–667
HARDY, R and KEAY, J N. Seasonal variations in the chemical compos-
1972   ition of cornish mackerel, Scomber scombrus L., with detailed
      reference to the lipids. J. Fd. Technol., 7, 125–137
JACQUOT, R. In: Fish as food. Academic Press, New York. Vol 1,
1961   725pp
KIETZMANN, U et al. In: Inspección veterinaria de pescados. Acribia,
1974   Zaragoza, 362pp
KIM, E and GOLBERG, M. Serum cholesterol assay using a stable
1969   Liebermann-Burchard reagent. Clim. Chem., 15(12),
      1171–1179
LAPSHIN, I I and SHEVCHENKO, M G. Change in lipid in marbled
1973   notothemia during the production of cured fish fillets. Ryb.
      Khoz., 6, 7373
LUDDY, F E. Direct conversion of lipids components to their fatty acid
1960   methyl esters. J. Am. Oil. Chem. Soc., Champaign, 37,
      447–451
MACCALLUM, W A et al. Newfoundland capelin: proximate composi-
1969   tion. J. Fish. Res. Bd. Can., 26(8), 2027–2035.
MCMICHAEL, C E and BAILEY, A E. In: The chemistry and technology
1951   of food and food products. Interscience, New York. Vol 2,
      834pp
MEIZIES, A and REICHWALD, I. Lipids in the muscle and roe of fresh
1973   and smoked fish. Z. Ernaehrungswiss, 12(4), 248–251
NOTEVARP, O. In: Fish as food. Academic Press, New York. Vol 1,
1961   725pp
ROBINSON, C H. In: Basic nutrition and diet therapy, MacMillan, New
1965   York. 308pp
SHATUNOVSKI, M I and KOSLOV, A N. Composition of fats of
1973   Notothemia rossimarmorata. Biol. Nauki., 16(4), 59–63
SHEVCHENKO, M G and LAPSHIN, I I. Effect of the method of smoking
1975   of large fish fillet products on the resistance of their lipids to
      oxidative decomposition, Rbyn. Khoz., 11, 74–76
STANSBY, M E. Nutritional properties of fish oils. World Rev. Nutr.
1969   Diet, 11, 47–96
TARR, H L A. Nutritional value of fish muscle and problems associated
1969   with its preservation. Can. Inst. Fd. Tech. J. 2(1), 42–45
TOYOMIZU, M, et al. Fatty acid composition of lipid from horse mack-
1976   erel muscle. Discussion of fatty acid composition of fish lipid.
      Bull. Jap. Soc. Sci. Fish, 42(1), 101–108
TSUCHIYA, T. In: Fish as food. Academic Press, New York. Vol 1,
1961   725pp.

11 Quality assessment

A comparison of different methods of freshness assessment of herring *A P Damoglou*

1 Introduction

This paper describes some work that was carried out using different methods to assess the freshness of herring caught in Northern Ireland waters and held at both ideal and non-ideal temperatures. The term freshness is used rather than spoilage because a measurement of freshness implies that the product may still be just edible when it has lost freshness whereas spoilage implies that the product is no longer edible once it has started to spoil.

Herring were chosen for the study because at that time very little appeared to have been published on the freshness assessment of herring as opposed to fish such as cod where many studies had been made over a number of years (Shewan *et al*, 1953; Shewan and Jones, 1957; Burt *et al*, 1975). At the time this study was carried out (January–February 1977) herring was of major economic importance to the local industry comprising 37% of the catch in weight terms and 22% of the monetary value of the total fish catch including shellfish (Department of Agriculture for Northern Ireland, 1978).

Traditionally, freshness of fish has been assessed solely by sensory methods which by their very nature are subjective rather than objective. It is for this reason that many workers have been looking for a rapid direct reading of freshness capable of being used either in the market place or not very far removed from it. The methods to assess freshness can be looked at under four headings:

(*1*) Sensory
(*2*) Chemical
(*3*) Instrumental
(*4*) Bacteriological

The sensory method for freshness assessment is the traditional and really the ultimate method but is expensive in terms of labour because specially trained personnel are required and for statistically meaningful results a panel of trained personnel is needed. The method has been put on a numerical basis with the method of Shewan *et al* (1953) and more recently further work at Torry (Burt *et al*, 1975; Burt *et al*, 1976b; Connell *et al*, 1976; Sanders and Smith, 1976) has shown the raw odour score to be a useful index of freshness.

Over the years a variety of chemical methods has been proposed as useful indicators of freshness: total volatile reducing substances, trimethylamine, dimethylamine, hypoxanthine and dephosphorylation of IMP. These methods have been the subject of a brief review by Martin *et al* (1978). These chemical methods depend for their assessment on the measurement of one of a large number of complex changes that take place when fish lose their freshness and begin to spoil. These

methods are all destructive and, with the possible exception of some newly developed strip tests for the measurement of hypoxanthine (Jahns *et al*, 1976), require expensive laboratory equipment.

Of the instrumental methods the Torrymeter is available commercially and appears to give a reliable estimate of freshness in a non-destructive manner using unskilled personnel (Burt *et al*, 1976a; 1976b; Connell *et al*, 1976). However, the results still need to be interpreted with care and there are several drawbacks to the use of the instrument. The Torrymeter cannot be used for fish that have been frozen because the freezing process changes the structure of the skin (Cheyne, 1975) giving an invalid reading. To obtain a reliable estimate of the reading for a batch of fish the average from 16 different fish must be taken.

Bacteriological assessment is often included as a measure of freshness although the usefulness of the result has been questioned on many occasions. However, the results can be of value in a study such as this where different methods are being compared, especially if the proportions of different genera are measured as well as the total count. The main drawback to its use is that the results are not available for perhaps five days, by which time the freshness of the fish under investigation can have changed quite dramatically.

2 Methods

A selection of the above methods was made, based on what was available and practicable.

2.1 Sensory methods

A simplified version of the Shewan *et al* (1953) raw odour scheme was used (*Fig 1*) because the expertise to utilize the full scheme was not available. The simplified scheme was operated by a single observer; the mean figure was based on observations of 16 fish assessed individually.

Raw odour	Score
Fresh seaweedy smell	5
Sweet sickly smell	4
Stale fishy smell	3
Ammonia smell	2
Putrid smell (hydrogen sulphide)	1

Fig 1 Simple raw odour scheme

2.2 Chemical methods

From the methods available trimethylamine and hypoxanthine appeared to be the most valuable.

2.2.1 *Trimethylamine* Trimethylamine was estimated by the method of Chang *et al* (1976) using a modified

specific ion electrode. This method appeared to be simpler than the picrate method of Murray and Gibson (1972). The fish was homogenized with formaldehyde and then made alkaline to convert the trimethylamine into the free base form for measurement with a modified ammonia-specific ion electrode. The response to ammonia was reduced by the use of formaldehyde and by replacing the internal NH_4Cl solution of the electrode with TMA.HCl.

2.2.2 *Hypoxanthine*
Hypoxanthine was measured using the traditional method of Jones *et al*, (1964) rather than the modern 'indicator type' method of Beuchat (1973). The fish was homogenized with perchloric acid and filtered. The extract was neutralized and centrifuged to remove the precipitate of potassium perchlorate. The extract was incubated at 35°C with xanthine oxidase and the uric acid released measured spectrophotometrically at 290nm.

2.3 *Instrumental method*
The Torrymeter (GR International Electronics Ltd) was used as in the instructions in the individual mode rather than in the averaging mode.

2.4 *Bacteriological method*
A portion (16cm²) was excised free of muscle using aseptic techniques and shaken with sharp sand suspended in diluent. Preparations were made from four fish and bulked. Serial decimal dilutions were made and

0·1ml spread on nutrient agar plates which were incubated at 22°C for five days. Colonies were picked at random from the plates and identified according to a diagnostic scheme (Shewan, personal communication).

2.5 *Fish*
The fish used for the study were herring (*Clupea harengus*) caught off the coast of Northern Ireland at Portavogie and delivered to the laboratory as soon as possible. Zero time was taken as the time of arrival in the laboratory 13h after catching.

The non-destructive tests were made every day up to day 11 and thereafter at day 15, 16 and 20. The destructive tests were carried out at 5, 10, 15 and 20 days. A total of 208 fish were used for the study. On arrival they were placed in numbered sterile bags (permeable to oxygen and open both ends). The fish were sampled at each stage using a table of random numbers. Sixteen fish were selected for sampling at time zero. The remainder were divided into three groups of 64 and stored at 0°C, 5°C and 10°C corresponding to melting ice, domestic refrigerator and low ambient temperatures.

3 Results

3.1 *Raw odour*
The results for the raw odour evaluation are plotted on *Fig 2*. Each point represents the mean of a single evaluation of 16 different fish. It can be seen that at 0°C a score of 5 which corresponds to 10 on the Shewan *et al* (1953)

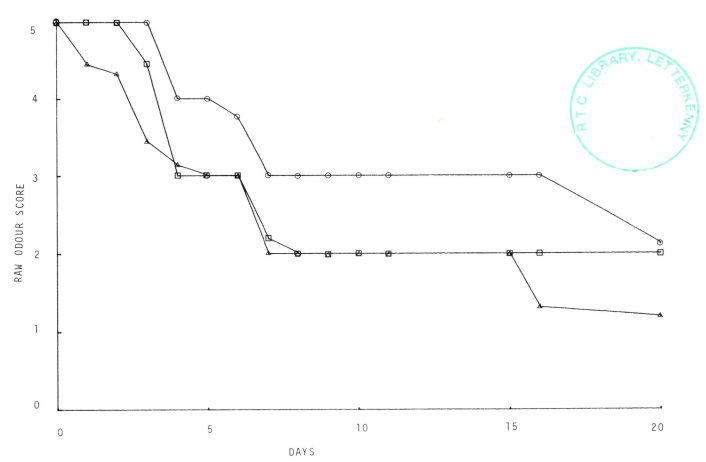

Fig 2   Relationship between raw odour score and time of storage at ○ 0°C, □ 5°C and △ 10°C

raw odour scheme was maintained for three days before falling off rapidly at first and then levelling off to a stale fishy smell. At 5°C the score of 5 was only maintained for two days followed by a rapid fall. At 10°C the fish lost freshness immediately and the average raw odour score fell even after one day and kept falling throughout the experiment.

Since there was only one observer it was not possible to statistically analyse the results. However, the results do show that it is possible for one untrained observer to determine when the fish lost their freshness and started to spoil. During the initial stages it was not possible to distinguish between the fish kept at 0°C and those at 5°C. The latter part of the study showed a difference between those at 0°C and those at 5°C and 10°C but not between those held at 5°C and those held at 10°C.

3.2 *Trimethylamine*

The readings from the specific ion electrode were converted to concentration of trimethylamine by means of a calibration curve. The results are shown in *Table I* and in graphical form in *Fig 3*. Because the original data was in logarithmic form the statistical analysis was performed on the results in the logarithmic form.

The analysis of the data using a multiple range test showed that at 0°C there was no significant difference between the initial reading and that at five days, but thereafter there were significant differences between each reading. At 5°C and 10°C there were differences between each reading that were significant at the 0·1% level. At each sampling period there were significant

differences between the three temperatures except at five days and 10 days; the difference between 5°C and 10°C was not significant. However, there was no dramatic increase in trimethylamine concentration at any of the three temperatures that could correlate with a loss of freshness.

Table I
TRIMETHYLAMINE PRODUCTION BY HERRING STORED AT 0°C, 5°C AND 10°C

Storage temperature	Trimethylamine in µmoles/g Mean of 4 determinations		
	0°C	*5°C*	*10°C*
Days			
0	11	—	—
5	15	26	36
10	27	62	103
15	49	142	284
20	85	258	1 417

3.3 *Hypoxanthine*

The results shown in *Table II* and graphically in *Fig 4* show that hypoxanthine increased with time at all three temperatures and a statistical analysis of the results using a multiple range test showed that although the differences were small they were all significant with exception of the readings at five and 10 days at 0°C and between the 5°C and 10°C readings at 10 days. The differences were all small except for the readings at 15 and 20 days where the fish have already spoiled.

During the first 10 days, by which time the fish were showing signs of spoilage at all temperatures, there were

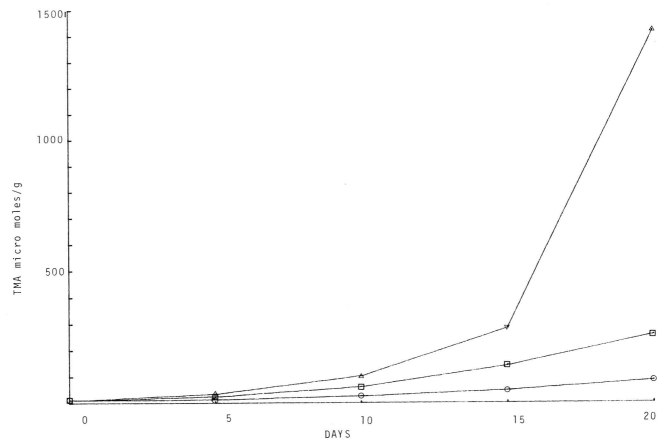

Fig 3 Relationship between trimethylamine in the fish flesh and time of storage at ○ 0°C, □ 5°C and △ 10°C

396

no dramatic changes in hypoxanthine concentration where a line could be drawn indicating when the fish had lost their freshness.

Table II
HYPOXANTHINE PRODUCTION BY HERRING STORED AT 0°C, 5°C AND 10°C

Storage temperature	Hypoxanthine in µmoles/g Mean of 4 determinations		
	0°C	5°C	10°C
Days			
0	0·178	—	—
5	0·307	0·422	0·513
10	0·474	0·635	0·658
15	0·474	1·540	1·969
20	0·566	1·836	2·889

Least significant difference where $P < 0.05 = 0.0289$, $P < 0.01 = 0.0387$, $P < 0.001 = 0.0508$

3.4 *Torrymeter*

An initial analysis of the Torrymeter readings indicated that the standard deviation of the mean of the 16 readings at each sampling varied over the course of the experiment so that a proper statistical analysis could not be made of the data as read on the meter. A log transform was carried out and the results are shown in *Table III* and graphically in *Fig 5* for the three storage temperatures up to 20 days of storage. An examination of the statistical analysis of the data showed that up to and including four days' storage there was no significant difference between the fish held at 0°C and those held at 5°C whilst the fish at 10°C showed a difference that was significant at the 0·1% level after three days. From then

until the tenth day there was a significant difference at the 0·1% level between all three curves. From the tenth day onwards there was no marked difference between fish held at 5°C and 10°C.

From these results it appeared that the Torrymeter could just distinguish the fish held at 10°C for more than two days and that it took up to the fifth day before it could distinguish between fish held at 0°C and those held at 5°C.

Considering the data at 0°C it was the fourth day before there was a statistically different reading even though the raw data appeared to show an apparent drop each day. This lag of four days before there was a statistically significant drop continued until the tenth day.

3.5 *Bacteriological results*

The results (*Fig 6*) showed quite clearly that the total number of colony-forming units (CFU) per cm² increased with time and those at 5°C and 10°C increased faster than those at 0°C with one major exception. Unfortunately the fish held at 10°C for 20 days dried out and this had a marked effect on the total CFU on the fish skin.

During the time that the fish are losing their freshness the CFU appeared to rise rapidly and showed its major change during this time.

The analysis of the flora obtained showed that while the initial flora was comprised almost entirely of Gram + ve cocci and rods, during storage the Gram + ve disappeared immediately and the flora was largely comprised of *Pseudomonas* spp and *Morax-*

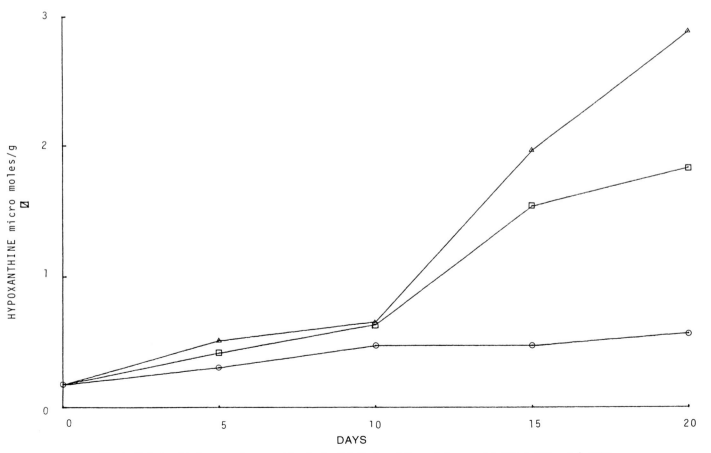

Fig 4 Relationship between hypoxanthine in the fish flesh and time of storage at ○ 0°C, □ 5°C and △ 10°C

Table III
ANALYSIS OF TORRYMETER RESULTS

Storage temperature	0°C		5°C		10°C	
Day	Log Tx[1]	Tx[2]	Log Tx[1]	Tx[2]	Log Tx[1]	Tx[2]
0	1·109	11·8				
1	1·061	10·5	1·120	12·2	1·024	9·6
2	1·043	10·0	1·096	11·5	0·959	8·1
3	1·017	9·4	0·996	8·9	0·581	2·8
4	0·980	8·5	0·909	7·1	0·363	1·3
5	0·936	7·6	0·715	4·2	0·254	0·8
6	0·931	7·5	0·704	4·1	0·445	1·8
7	0·919	7·3	0·649	3·5	0·320	1·1
8	0·860	6·2	0·637	3·4	0·171	0·5
9	0·782	5·1	0·540	2·5	0·143	0·4
10	0·495	2·1	0·127	0·3	0·127	0·3
11	0·229	0·7	0·108	0·3	0·108	0·3
15	0·232	0·7	0·038	0·1	0·075	0·2
16	0·094	0·2	0·105	0·3	0·075	0·2
20	0·094	0·2	0·116	0·3	0·000	0

[1] Log Tx = mean log (Torrymeter reading +1)
[2] Tx = antilog (log Tx) −1
Least significant difference of log means where $P < 0.05 = 0.110$, $P < 0.01 = 0.148$, $P < 0.001 = 0.190$

ella – Acinitobacter – like organisms in varying proportions.

4 Discussion

All of the measurements made on herring in this study have been shown to be useful in the determination of freshness for some other species of fish (Connell *et al*, 1976; Hiltz *et al*, 1971). It appears that no two species of fish spoil in exactly the same way, so that a test that is a

useful indicator for loss of freshness for one species, may be useless for another species.

In this study we have shown that even with one untrained observer an organoleptic assessment of herring corresponds to storage time and loss of freshness. The data from the Torrymeter correlates well with the sensory assessment where there appears to be a difference even after the first day at 10°C and at the third day between 0°C and 5°C.

The bacteriological results showed no lag phase even

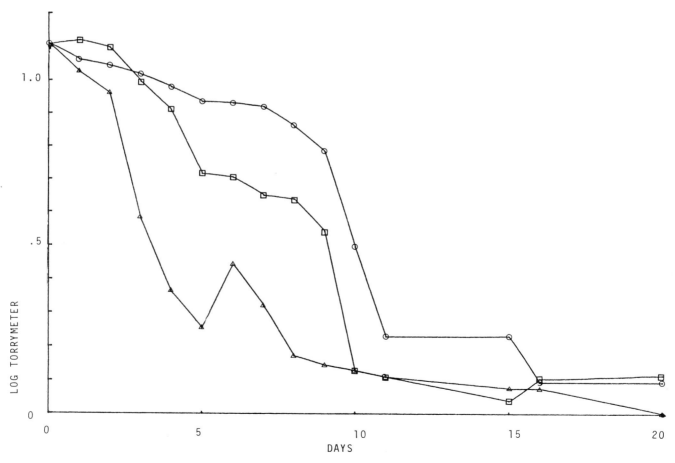

Fig 5   Relationship between log (Torrymeter reading + 1) and time of storage at ○ 0°C, □ 5°C and △ 10°C

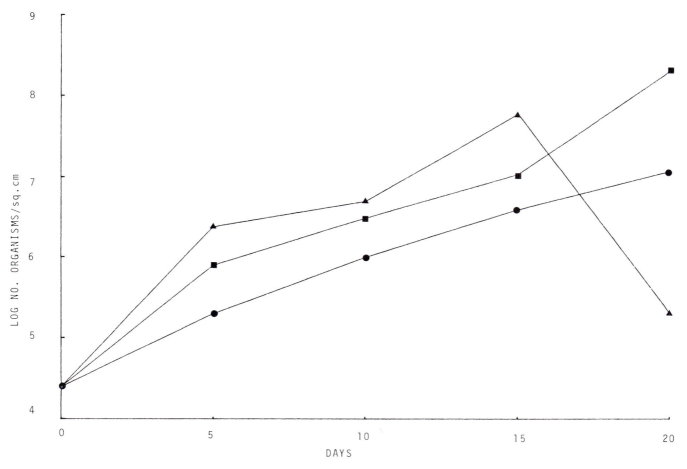

Fig 6 Relationship between log (number of organisms/cm² on the fish skin) and time of storage at ● 0°C, ■ 5°C and ▲ 10°C

at 0°C and whilst the results correlate well with the sensory data the initial count would obviously vary, depending on the handling of the fish before being tested and the waters from which they were taken.

The results for the chemical determinations both showed that while there were differences in the results during loss of freshness period these were not large enough to be used as a reliable indicator of loss of freshness.

From this it appears that the Torrymeter would be a useful instrument to determine the freshness of herring. Unfortunately, the initial Torrymeter reading for herring when it is caught varies with the fat content of the herring which shows a seasonal variation. The herring in this study had a fat content of 5–10% and the results correlated reasonably with those in the Torrymeter Handbook for herring of similar fat content. Since Irish Sea herring show a regular seasonal variation of fat content (Johnston, personal communication) it should be possible to set up a table for herring comparing Torrymeter reading with season to give an index of freshness.

5 References

BEUCHAT, L R. Hypoxanthine measurement in assessing the freshness
1973   of chilled channel catfish. *J. Agric. Fd. Chem.*, 21, 453–455
BURT, J R, GIBSON, D M, JASON, A C and SANDERS, H R. Comparison
1975   of methods of freshness assessment of wet fish. I. Sensory assessments of boxed experimental fish. *J. Fd. Technol.*, 10, 645–656

—— Comparison of methods of freshness assessment of wet fish. II.
1976a Instrumental and chemical assessments of boxed experimental fish. *J. Fd. Technol.*, 11, 73–89
—— Comparison of methods of freshness assessment of wet fish. III.
1976b Laboratory assessments of commercial fish. *J. Fd. Technol.*, 11, 117–128
CHANG, G W, CHANG, W L and LEW, K B K. Trimethylamine-specific
1976   electrode for fish quality control. *J. Fd. Sci.*, 41, 723–724
CHEYNE, A. How the GR Torrymeter aids quality control in the fishing
1975   industry. *Fish. News Inter.*, 14(12), 71–72, 75–76
CONNELL, J J, HOWGATE, P F, MACKIE, I M, SANDERS, H R and
1976   SMITH, G L. Comparison of methods of freshness assessment of wet fish. IV. Assessment of commercial fish at port markets. *J. Fd. Technol.*, 11, 297–308
DEPARTMENT OF AGRICULTURE FOR NORTHERN IRELAND. *Statistical*
1978   *Review of Northern Ireland Agriculture*
HILTZ, D J, DYER, W J, NOROLAN, S and DINGLE, J R. Variation of
1971   biochemical quality indices by biological and technological factors. In: *Fish Inspection and Quality Control*, Ed R Kreuzer. Fishing News (Books) Ltd, London. 191–195
JAHNS, F D, HOWE, J L, CODURI, R J and RAND, A G. A rapid visual
1976   enzyme test to assess fish freshness. *Fd. Technol.*, 30, 27–30
JONES, N R, MURRAY, J, LIVINGSTON, E I and MURRAY, C K. Rapid
1964   estimations of hypoxanthine concentrations as indices of the freshness of chill-stored fish. *J. Sci. Fd. Agric.*, 15, 763–773
MARTIN, R E, GRAY, R J H and PIERSON, M D. Quality assessment of
1978   fresh fish and the role of the naturally occurring microflora. *Fd. Tech.*, May, 188–192
MURRAY, C K and GIBSON, D M. An investigation of the method of
1972   determining trimethylamine in fish muscle extracts by the formation of the picrate salt. *J. Fd. Technol.*, 7, 35–51
SANDERS, H R and SMITH, G L. The construction of grading schemes
1976   based on freshness assessment of fish. *J. Fd. Technol.*, 11, 365–378
SHEWAN, J M and JONES, N R. Chemical changes occurring in cod
1957   muscle during chill storage and their possible use as objective indices of quality. *J. Sci. Food Agric.*, 8, 491–498
SHEWAN, J M. MACINTOSH, G, TUCKER C G and EHRENBERG, A S C.
1953   The development of a numerical scoring system for the sensory assessment of the spoilage of wet white fish stored in ice. *J. Sci. Fd. Agric.*, 4, 283–297

Assessment of the freshness of fillets by the GR Torrymeter   *A Lees and G L Smith*

1 Introduction

The Torry fish freshness meter and the commercial version, the GR Torrymeter (GR International Electronics Ltd, Almondbank, Perth, Scotland) were primarily intended to measure the freshness of intact wet fish (Jason and Richards, 1975). A typical application envisaged was quality assessment on the fish market by factory quality control personnel and the instrument is now established as a means of determining freshness in such situations. Nevertheless, as an aid to the exploitation of the instrument it is desirable to explore all possible applications for its use. One such application about which there have been frequent inquiries is the measurement of the quality of fillets. This is particularly relevant to retail fish outlets where a large proportion of the fish supplies is received from the wholesale merchants in the form of fillets rather than whole fish.

Even during early development of the meter it was recognized that readings could be obtained from fillets, especially from the skin side of skin-on fillets (Jason and Lees, 1971). These readings, though generally less than those obtained from the intact fish, declined with age-in-ice in a similar fashion to those from the intact fish. Readings taken from the flesh side of skin-on or from either side of skin-off fillets appeared to decrease rapidly during storage in ice to a minimum value over the first few days even when the fish were filleted immediately after death. Since little experimental work had been carried out expressly to investigate the performance of the meter on fillets the available data were limited and sometimes conflicting, particularly during the initial storage period.

Currently a series of experiments is being undertaken to establish the relationship between meter reading and days-in-ice for both skin-on and skin-off fillets soon after filleting and for fish stored as fillets. This paper presents the results from the first four experiments.

2 Experimental considerations

Since the small amount of available data seemed to indicate that the Torrymeter might provide an accurate relationship between fillet freshness and meter reading at least over the first two or three days' storage after death, it seemed appropriate to concentrate on establishing this part of the relationship.

A major constraint in planning the experiment was the availability at intervals over, say, a six-month period of supplies of codling in the size range 45cm to 60cm and of accurately known history. The quantities likely to be available determine the sample size, frequency of sampling and hence the duration of each run. In using the Torrymeter on intact fish, biological variation and general mixing during landing demand that a sample of 16 fish is required from a batch to give an accuracy in the estimation of freshness of $\pm\frac{1}{2}$ taste panel score unit (Connell *et al*, 1976). In the light of the existing data it seemed a reasonable assumption that the same accuracy could be obtained if necessary by using the fillets from both sides of only eight fish. With such a sample size and a sampling frequency of, say, once per day sufficient fish

is then required for each run to last a minimum time of between three and five days.

3 Experimental procedure

The fish used for the first run were cod (*Gadus morhua* L.) obtained from the RV *G A Reay* fishing near-water grounds and were caught in three hauls. Fish from each of the hauls were immediately gutted and thoroughly iced in plastic fish trunks labelled with the date and time of catching. Sufficient fish for the first run were procured to allow five sets of data to be obtained with a sample size of 13 fish. At the laboratory the trunks of fish were stored at $+1°C$ until required.

For the second and subsequent runs the fish were obtained from line vessels operating from Gourdon, a small fishing port some 50km south of Aberdeen. The fish were stowed ungutted at ambient temperature on the vessel and landed on each occasion at approximately 1400 hours. After initial icing at the port and transport to the laboratory all the fish were gutted and cleaned in cold fresh water before storage. From the batch of fish the first sample of eight was extracted, each fish being laid out individually in a shallow sampling box with ice and allowed to cool. The rest were carefully packed with an excess of ice in plastic fish trunks which were then placed in the bulk chillroom at $+1°C$. In all runs the fish were gutted and iced in the pre-rigor condition.

After the required sets of data had been obtained for the first batch of fish sampled during each run the resulting skin-off fillets were stored in ice and meter readings obtained at intervals over a storage period lasting up to nine days. As a protection against the effects of melt water the fillets were stored in a single layer between polythene sheets.

The instrument used throughout these experiments was a prototype meter equipped with an analogue display. For experimental purposes this display allows measurements to be made with a greater resolution than is possible with the digital presentation of the GR Torrymeter. The analogue instrument was calibrated so that:

$$\text{Meter reading} = 3\,CR$$

where C and R are the parallel capacitance and resistance of the sample of fish under test. In the present work, however, these readings were converted to freshness index (F) where:

$$F = 10\,\log_{10} CR$$

(The GR Torrymeter provides a direct reading of F.) Although the instrument used was equipped with the averaging facility it was used exclusively in the 'single fish' mode. All readings were converted to freshness index before any calculations were performed.

4 Results and discussion

The results from the first experiment are plotted in *Fig 1*. In this and subsequent figures the plotting symbol indicates the mean value of F, averaged over both sides of all

400

fish in a sample, and the vertical bars represent the 95% confidence interval of the mean, a measure of precision which reflects the fish-to-fish variation.

Within the range of storage times employed (up to 13 days) the values for skin side of skin-on fillets decreased linearly with time by almost 1 unit per day, while the rate of decrease of readings on the intact fish was about 0·5 unit per day. Readings obtained from flesh decreased rapidly within the first three days to about 4 for the bone side of skin-on fillets and about 2 for either side of skinned fillets. The results for intact fish appear to differ slightly from those previously published (Cheyne, 1975) and the rate of decrease is slightly more than would be expected. This may be due to factors such as season, condition of the fish or method of catching.

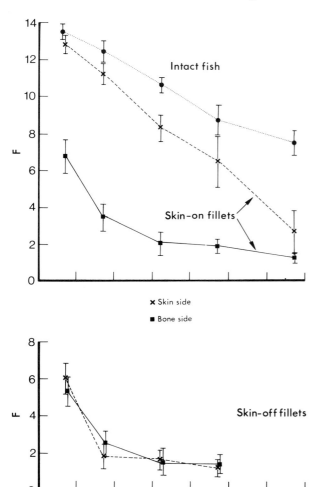

Fig 1

The results for the subsequent experiments are shown in *Figs 2–5*. The same general trend for skin side of skin-on fillets is evident in *Fig 2*. In *Figs 3–5* the rapid decrease in the first three days in the readings from flesh can be seen in more detail, but there are clear differences between the three experiments, which were conducted in February, March and May. In earlier work on physical assessment of freshness of intact fish (Jason and Lees, 1971) it was found that readings showed seasonal variation and fell during spring and summer by up to the

equivalent of 1·5 units. The results here suggest that the effect of season may be even more serious on fillets than on intact fish.

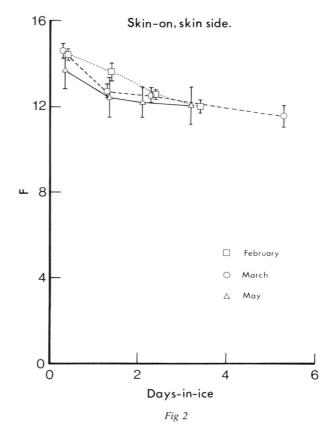

Fig 2

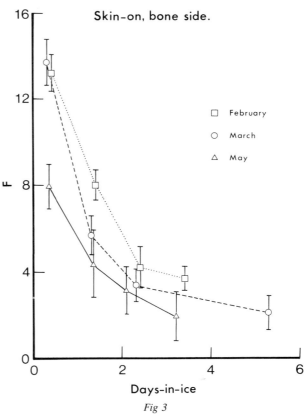

Fig 3

401

Skin-off fillets, skin side.

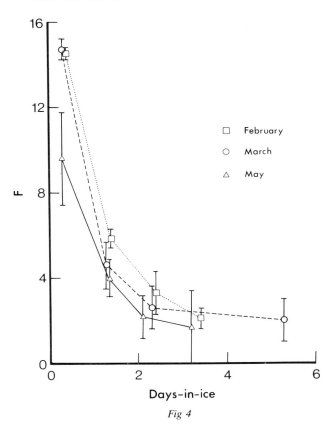

Fig 4

Skin-off fillets, bone side.

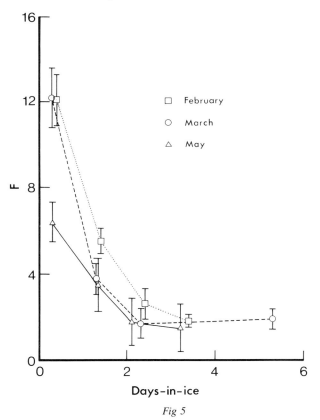

Fig 5

The discussion above on fillets relates to fish which had been stored whole for the given number of days and then filleted immediately before measurement. The relationship obtained is that between meter reading on the fillet and time of storage of the intact fish before filleting. *Figures* 6 and 7 show the relationship between meter reading of skin-off fillets and time of storage since catching for fish which had been filleted on the first day after catching. This was done in only the last two experiments. Variation between the experiments is again evident, but within each experiment the meter readings appear to decrease more slowly for fish which were filleted and then stored than for those which were stored intact and then filleted. For example, in the March run, readings do not reach a value of 2 until about six or seven days' storage as fillets, whereas it only takes three days' storage as whole fish for the same score to be obtained for skin-off fillets.

A vast amount of data has been accumulated from experimental runs relating meter reading to days-in-ice for different species of intact fish, including cod, obtained on a regular seasonal basis for several near- and middle-water grounds (Jason and Lees, 1971). Those fish were caught using standard trawling techniques, usually by Torry Research Station's research vessel. In a significant number of those runs the spoilage characteristic exhibited an initial though sometimes slight increase to a peak value over the first three or four days of spoilage before decreasing normally over the remaining storage period. This phenomenon which appeared to be random in its occurrence has been ascribed to some effect of the onset of the *rigor mortis* condition and could clearly influence meter readings obtained from fillets during the initial spoilage period.

None of the runs so far completed during the current series of experiments has provided a spoilage curve of this type for intact fish. This may be due to the random nature of its occurrence but it may also be attributable to the fish being caught by a different method and/or on fishing grounds where the available feeding could result in fish of quite different physical condition from those caught over adjacent fishing grounds suitable for trawling.

5 Conclusions

In each experiment the same forms of relationships were found. For intact fish and skin side of skin-on fillets relationships were linear within the range considered, while for measurements on flesh the initial decrease in readings was fast.

In the runs which have been completed there was a shift to the curves which may be a seasonal effect. This shift was more severe in flesh measurements than on skin.

Where the conditions under which the fish are caught and landed are known the Torrymeter is a sensitive indicator of freshness within the first two or three days, but because of the variation noted due to different conditions, *eg* season, prediction of freshness without prior information would not be possible.

Further data are required to investigate the change in the relationships due to season over a whole year cycle and also due to ground or method of catching.

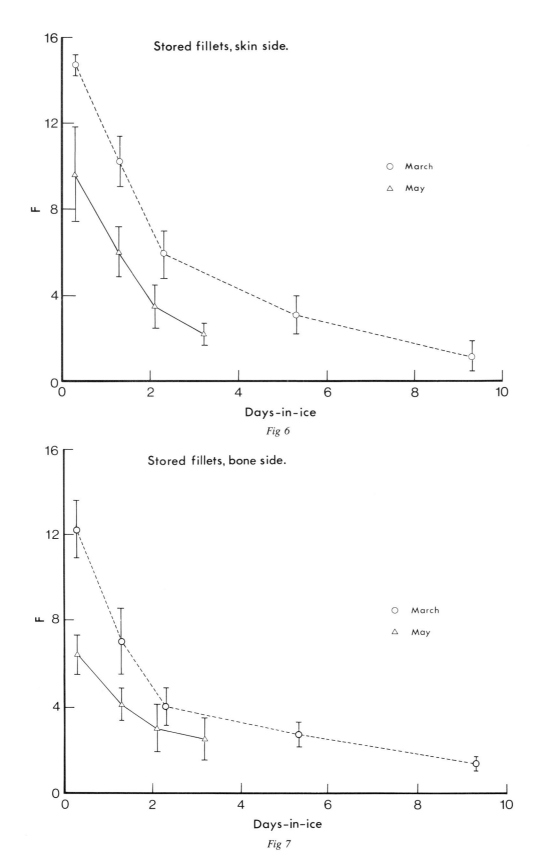

Fig 6

Fig 7

6 References

CHEYNE, A. How the GR Torrymeter aids quality control in the fishing
1975   industry. *Fish. News Int.,* 14(12), 71–76
CONNELL, J J, HOWGATE, P F, MACKIE, I M, SANDERS, H R and SMITH, G
1976   L. Comparison of methods of freshness assessment of wet fish.
        IV. Assessment of commercial fish at port markets. *J. Fd.
        Technol.,* 11, 297–308

JASON, A C and LEES, A. Estimation of fish freshness by dielectric
1971   measurement. Department of Trade and Industry Report No.
        71/7, Torry Research Station, Aberdeen
JASON, A C and RICHARDS, J C S. The development of an electronic
1975   fish freshness meter. *J. Phys. E.,* 8, 826–830

403

Consumer and instrumental edibility measures for grouping of fish species

*F J King, J G Kapsalis,
A V Cardello, and J R Brooker*

1 Introduction

The National Marine Fisheries Service, with support from consumers, industry and regulatory agencies, has undertaken a comprehensive project to develop and implement a new system for establishing market names for fishery products.

The four major problems which led to this project are as follows:

(1) There is an increasing use of fishery resources that have not been marketed previously and for which no common names exist that are familiar to consumers. In contrast, taxonomic or scientific names are used by scientists on a world-wide basis to recognize a given species. However, consumers use common names, not taxonomic ones, to recognize fish offered for sale. The present situation in names for marketing fish creates consumer confusion and often thwarts the orderly marketing and market development of fishery products.

(2) Many species of fish are known by different common names in different areas of a country or between different countries. In the USA there are approximately 300 different species of fish that are harvested commercially. This number includes several freshwater species as well as species of marine origin. World-wide, the number of species harvested may exceed 1 000. This number is increasing year to year as more and more countries establish 200-mile management zones in adjacent coastal waters. Many of these species are, or will be, exported to the USA where they are often processed into other products. When products of various forms of presentation and preservation are included, the number of combinations of products and species may become too large for recognition or identification. If the present system, or lack of a system, is used to market all of these various forms of products, one might expect consumers to be bewildered or confused.

(3) The marketability of many nutritious species is severely limited due to aesthetically objectionable common or local names. Often these names have arisen from an external feature (such as body shape or colour, appearance of head, tail or scales) which has no relation to the sensory characteristics of a fish's edible flesh. In recent years, interest has grown in development of markets for various underutilized species and consumer acceptance of nutritious products made from these fish. When a species has an aesthetically objectionable name, consumers, the industry and a country's regulatory agencies encounter problems in achieving a suitable product name which does not hamper development.

(4) New food-processing techniques present opportunities for developing new forms of seafood or alternative choices. This emphasizes the need for grouping of species on the basis of similar edibility characteristics. Most of the time, suitable common names do not exist.

For these reasons, we believe that a new nomenclature of standardized market names for labelling fish and fishery products should be developed and implemented. Such a system should meet the legitimate needs of consumers, the industry and regulatory agencies. To develop a comprehensive seafood product iden-

tification system, NMFS has initiated a series of projects with the following principal steps:

(i) identify the most significant edibility and related criteria about fishery products that are important to consumers;
(ii) apply these criteria to individual species by developing an edibility profile of its edible flesh;
(iii) group species that are similar on the basis of having similar edibility characteristics;
(iv) construct and implement an overall system or framework for naming and labelling seafood products.

We are currently at the stage of having completed a model retail identification plan. It provides a system for identifying species of fish with their forms of presentation and preservation. We believe that this retail identification plan has a potential for resolving many of the current labelling problems of common or usual name and ingredients for fishery products.

2 Previous work

The Brand Group, Inc (Chicago, Illinois) has proposed a model retail identification plan. It is based on three nomenclature components; 'Fish, Forms and Modifiers'. The components 'Forms' and 'Modifiers' cover those parts of the nomenclature that explain the form of presentation and form of preservation that are appropriate for each specific retail product. The component 'Fish' is a framework which includes all possible aquatic species. Given the hundreds of species that are presently available, there is a need to sort fish into a manageable, recognizable number of groups so that members of each group have similar edibility characteristics.

Scientists use systematic zoology to group species of fish into categories where the members of each category (*eg* genus, family) have similar characteristics. In systematic zoology, anatomical or physiological features are used most frequently and edibility characteristics of that part of a fish which is eaten are hardly ever described. In contrast, consumers are sensitive in perceiving species differences in terms of sensory characteristics of edible portions. They tend to be aware that some species from different zoological groupings have similar edibility characteristics but, individually, consumers are familiar with only a few of the anatomical or physiological characteristics used to classify species in systematic zoology.

An organization of finfish into groups on the basis of their edibility characteristics enables consumers to become familiar with a great many more species. The term 'comparative edibility' was selected to describe this consumer-oriented approach for establishing groups of similar species of finfish. After an extensive survey with seafood specialists and consumers, the Brand Group, Inc identified eight edibility factors as being generally the most important in grouping of species:

(1) Intensity of the *flavour*

(2) *Flakiness* of the meat (after cooking)

(3) *Fat* content

(4) *Firmness* of the meat (after cooking)

(5) Natural *odour* of the meat (when raw and fresh)

(6) *Coarseness* of the meat (after cooking)

(7) *Colour* of the meat (after cooking)

(8) *Moisture* of the meat (after cooking)

Flavour. Some fish are very mild-tasting. Others are, by nature, more robust. It is a matter of individual taste whether strong or mild flavour is preferred. It is wrong to asume that strong flavour in fish means poor quality.

Flakiness. It is a characteristic of certain species that the meat flakes readily when cooked, and of other species that the meat shows little or no flakiness. Stringiness and other textural characteristics may occur but flakiness appears to be the most important consideration.

Fat. All fish have some fat. Fish are often recommended for people on low-fat diets since the fat is lower in cholesterol than the fat of land animals. Fish that are high in fat are usually prepared differently to fish that have a low-fat content. In certain species, the kind of fat is important. Usually, the relative amount of fat is of most importance.

Firmness. When cooked the same way, the meat from various species of fish can range from very firm, almost resilient, to very soft or mushy.

Odour. Fresh fish has a mild, sweet odour like fresh meat. A strong odour means that the fish is not fresh. The tendency of a fish to give off a characteristic odour is closely related to the kind and amount of fat present in the meat. In a few species, the fat oxidizes so rapidly once the fish is removed from water that the fish is almost never encountered without a noticeable odour.

Coarseness. Certain species have a noticeable granular character in the meat, while others are smooth, almost creamy.

Colour. The colour of the fish is an aesthetic consideration. The American taste seems to prefer delicately flavoured, light or white-fleshed fish. But there are excellent species of fish that have very dark meat.

Moisture. Certain fish are characterized by flesh that remains moderately moist when it is properly cooked. Some fish tend to dry out more rapidly while others tend to remain more moist when cooked.

In its latest report, the Brand Group, Inc recommended three immediate and concurrent actions:

(1) Develop comprehensive and consistent information on edibility profiles and related information for all commercial aquatic species.

(2) Develop an interim programme for marketing seafood products on the basis of communicating edibility information.

(3) Develop guidelines and procedures for making interim decisions on seafood product labelling and nomenclature.

3 Present work – experimental plan

To develop the data bank of edibility profiles recommended in the retail identification plan, the US Army Natick Laboratories (NARADCOM, Natick, MA) has initiated a study whose objectives are to: *(1)* test the validity of the edibility characteristics, scaling procedure and grouping procedure suggested by Brand Group; *(2)* develop and evaluate appropriate standardized sensory and instrumental methods for assessing edibility characteristics of fish, as a prerequisite to *(3)* evaluating the correspondence between instrumental and organoleptic indices of edibility and *(4)* group species of fish into categories according to their similarities in edibility characteristics. In the present study, analyses are based on the fresh state (ice-chilled for 48 hours after harvest) of each species selected. The effects of storage (chilled or frozen) on each of the parameters being studied will be considered in subsequent studies. The requirement for very fresh fish limits the selection of species to those available in the New England area or nearby for the present study. In subsequent studies, it is planned to select species from other geographical areas of the USA and possibly species that are imported to the USA from other countries.

The boil-in-bag procedure for cooking fish samples is based on procedure 18.B01 of the Association of Official Analytical Chemists. The fish is cooked by immersing the bag in 160°F water for a time varying with the thickness of each sample. The times for the various thicknesses were established from heat-penetration measurements in order to cook all samples uniformly and to provide reproducibility from batch to batch.

Sensory evaluation of flavour is based on the flavour profile panel technique. Selection of panellists is based on each person's sensitivity and repeatability of making judgements. The panel first develops a vocabulary of terms related to the aroma, flavour and aftertaste of each species. Subsequent evaluations determine the sequence of perception and the intensity of each character note in this vocabulary. An overall sensory impression of aroma and another one for flavour are expressed numerically as amplitudes. *Figure 1* contains a sample flavour profile.

Sensory evaluation of texture is based on the General Foods' texture profile method. The panellists have been extensively trained in using this methodology which utilizes a well-defined master-set of sensory texture attributes, reflecting mechanical, geometrical and fat or moisture-related properties of food. In each of these categories, a specific attribute, such as firmness, chewiness, flakiness, moistness, can be evaluated as to its appropriateness for a given product, and each can be rated in a dimension of intensity using any one of a variety of scaling methodologies (see *Fig 2*).

Consumer panels are also used for sensory evaluations. These panellists represent a cross-section of sophisticated and unsophisticated consumers of fish. Initial research is focused on identifying sensory attributes of fish which are important and meaningful in discriminating among fish species. The consumers are given a variety of samples of fish to taste and are then asked to identify, from a comprehensive checklist of edibility attributes, those attributes of the samples which are most important or useful in distinguishing among them. The checklist of terms includes those sensory attributes of fish identified as being important by the flavour and texture profile panels of NARADCOM and Brand Group, Inc. The checklist also provides for alternative

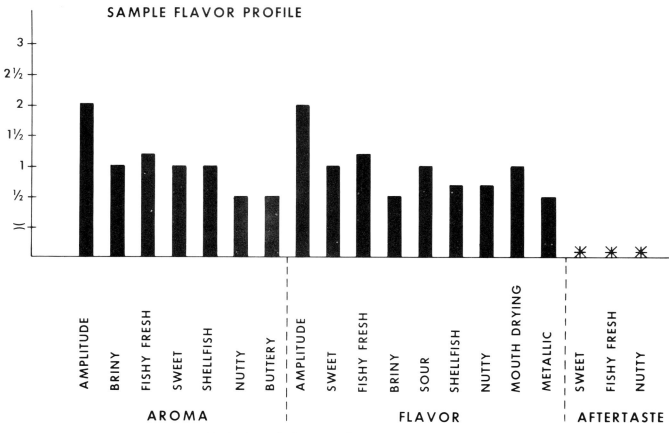

Fig 1   A sample flavour profile for one species of fillet

TEXTURE PROFILE BALLOT FOR FINFISH

DATE:_____

PANELIST:_____

		SAMPLES				
		A	B	C	D	E
FIRST BITE	Hardness					
	Flakiness (tongue against palate)					
MASTICATION	Chewiness					
	Fibrousness					
	Moistness					
	Cohesiveness of mass (at 10 chews)					
	Adhesiveness					
RESIDUALS	Oily mouthcoating					
	Astringent-like mouthcoating					
COLOR	Lightness: Skin side					
	Skeleton side					
	Uniformity of lightness: Skin side					
	Skeleton side					

Fig 2   Texture profile ballot for finfish

attributes to be suggested by the consumer at the time of testing. All tests are conducted in conventional taste-testing booths under controlled sensory testing conditions.

Instrumental texture measurements are based on an Instron universal testing system using uniaxial compression (*Fig 3*). Samples are compressed to 60% of their original thickness at which time the compressing force is removed. The stress (force per unit cross-section area) at 40% strain and the apparent modulus of elasticity (change in stress per unit change in strain over the linear portion of the curve) are used to quantify 'texture'. Other parameters such as strain energy required to compress the sample, recovered energy and curve fitting may be calculated from these curves.

Considerable difficulty has been experienced with sample preparation. One source of this difficulty arises from the requirement of mechanical compression testing for samples having flat, parallel surfaces of a uniform cross-section. A second difficulty is that the standard boil-in-bag cooking procedure distorts fillets to the point where no suitable flat surfaces can be obtained.

To compensate for lack of parallelism and surface flatness, a swivel-head compression plate was mounted on the moving cross-head of the Instron. Samples that could be easily cut into uniform cylinders (using a sharp edge cylindrical cutter) were preloaded on to this compression plate.

Fillets from the more flaky species, especially cooked fillets, are more difficult to handle. Cylindrical samples could not be obtained when the flakes separated from one another with much less force than is required to cut into a single flake.

To determine flakiness, a second instrumental test is being developed. This test measures the force required to pull apart a strip of fillet. Preliminary results indicate that some of the variation between samples is related to the depth of cutting when the fillets were skinned. If a membrane on the skin-side of a fillet is part of the sample, it greatly increases the force required to pull the strip apart.

To cook fillets with a minimum of distortion, an electric broiler is used. Instron compression data obtained on fillets cooked by this method appear to reflect sensory panel evaluation of texture (*Fig 4*).

Present data show that the evaluation of firmness and flakiness of fish requires two instrumental tests. Data obtained by uniaxial compression between flat plates are nearly linear in true stress-strain co-ordinates. The slope of the stress-strain plots is reasonably consistent within a species of fish and reflects the firmness of the fish. However, the test will not provide a measure of flakiness due to the lack of consistent yield points in the curve. Data so far show that a test which involves the pulling of the fish muscle to the rupture point is promising as a measure of flakiness.

Instrumental measurement of odour is based on combined gas chromatography, mass spectrometry and computer methodologies for analyses of volatile components (*Fig 5*). Experiences in using these methodologies in storage studies of fish fillets indicate that careful evaluation is needed to distinguish components that are characteristic of a species in a fresh state from components which arise during storage as a result

Fig 3   An Instron universal testing system equipped with a device to measure flakiness. A variety of other devices may be used depending on the material being examined and the attribute of texture to be measured

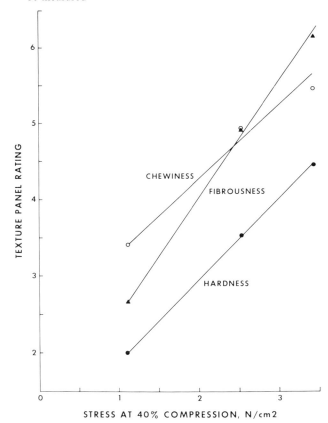

Fig 4   Preliminary correlations for three attributes of texture; chewiness, fibrousness and hardness. Sensory evaluations are based on the General Foods texture profile method. Instrumental measurements are based on an Instron universal testing system using uniaxial compression. The symbols ●, ▲ and ○ represent three different species of fillets

407

of microbial or autolytic activity.

Appearance characteristics of selected species of fish are being evaluated. They include colour (on a lightness-darkness scale) and uniformity of colour, both on each side of a cooked fillet and on minced samples. Colour is one of the eight edibility factors recommended by Brand Group.

Other analyses include moisture and fat content which are two other edibility factors recommended by Brand Group. Fat content is determined on raw flesh and moisture content is determined on raw and cooked flesh by methods of the Association of Official Analytical Chemists along with the sensory and instrumental data. The pH of raw and cooked flesh is also determined.

4 Present work – evaluation of data

Sensory scaling data are obtained using both classical category scaling methods as well as the method of magnitude estimation. These data are also being submitted to multi-dimensional scaling analysis as an alternative grouping procedure to that suggested by Brand Group, Inc. Advantages of the magnitude estimation technique over classical category scaling are: *(1)* it does not require specific intensity reference points and *(2)* it provides ratio data for comparison with objective measures. Advantages of multidimensional scaling analysis are: *(1)* it results in data that can be summarized in graphical form and *(2)* the magnitude of similarity among groups of species can be statistically tested.

Magnitude estimation is a scaling procedure that is rapidly replacing category scaling in sensory evaluation research. This scaling procedure is based on the free use of numbers by panellists to represent ratios of intensity of a sensory attribute (*eg* flavour). Thus, a panellist would be presented with one species of fish, asked to taste it and then assign a number to represent the flakiness, moistness, flavour, *etc* of that fish sample. A second species would then be presented for tasting and if the flakiness, flavour, *etc* of that species was twice that of the first, the panellist would assign a number twice as large as he assigned to the first species. If the second sample were one-half as flaky, flavourful, *etc*, he would assign a number one-half as large and so on. This direct scaling procedure produces ratio data from which profiles can be constructed that will enable the consumer to choose a product that is twice, one-half, one-fourth as flaky, flavourful, *etc* as another species. This advantage of the magnitude estimation technique adds useful and informative data to the edibility profiles that is precluded by category scaling procedures. Also, the ratio nature of the subjective data is necessary for valid subjective-objective correlations, since instrumental measures of flavour and/or texture are usually ratio in nature.

Multidimensional unfolding uses mathematical techniques by which edibility profiles of different fish species can be compared and grouped. It considers the matrix of fish species sensory attributes as a matrix of similarities. For example, if 20 species are rated on each of 8 attributes, then the judgements produce a 20×8 matrix of 160 numbers (scale values). The larger the scale value for an entry in the matrix, the 'closer' is that species to the attribute. Thus, the profile matrix may be interpreted as a matrix of the closeness or proximity of various fish species to the various sensory attributes, in the same way that one interprets the entries of a mileage chart as being the distance between two cities or other geographic locations. Through computer-based mathe-

A Nose in GC/MS/Computer Loop

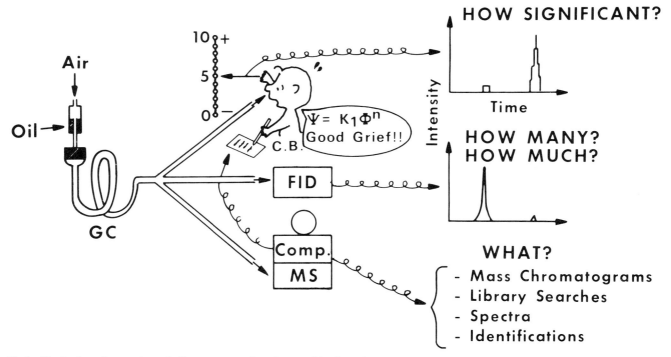

Fig 5   Illustration of measuring volatile components based on combined gas chromatography, mass spectrometry and computer methodologies (from *Food Engineering*, February 1979, 74)

matical techniques, a geometrical map of fish species and their sensory relationship to other species can be derived, just as one could work backwards from a mileage chart to reconstruct the geographical map of the United States. In such a 'fish map', species with similar edibility characteristics will form geometrical clusters, and will provide a useful tool for future consumer educa-

tion programmes. In addition, the groupings of similar species can be compared to the groupings arrived at by the grouping procedure suggested in the Brand Group, Inc report. Validity testing of both grouping procedures will be conducted by direct consumer panel ratings of overall similarity among the species.

Texture profile panelling: a systematic subjective method for describing and comparing the textures of fish materials, particularly partial comminutes

R B Weddle

1 Introduction

Minced fish blocks are becoming an increasingly important commodity. Some fish fingers are made wholly from fish mince, others may contain a proportion of mince. This mince, whether obtained from whole fish, off-cuts, or frames, has a texture which is different from and deteriorates more quickly than that of intact fillet. Products made from mince may therefore become unacceptable to the consumer after a relatively short time in cold storage.

The present study is concerned with describing the differences between intact fish and fish which had been comminuted to varying degrees, and describing how the texture of both deteriorates during frozen storage. The system is based on a method previously described by Horsfield and Taylor (1976) which was concerned with the texture and flavour of meat products; 18 attributes of texture, flavour and appearance of the meat product were identified and defined; panellists were trained to allocate scores between 1 and 10 on each attribute scale for each product sample. The Unilever system differs from that of Horsfield and Taylor in including attributes of texture and appearance only; Horsfield and Taylor also included flavour attributes as they wished to relate their data to consumers' acceptability scores, but within their limited product context only a few flavour attributes were relevant, whereas the Unilever fish scheme is more wide ranging and would consequently require a much larger number of flavour attribute scales.

2 Experimental procedure

2.1 *Raw materials*

Fresh cod, coley, plaice and salmon fillets were purchased from Aberdeen fish suppliers and frozen in rectangular blocks. Further quantities of the same batch of cod fillets were processed as follows:

(a) sliced by hand into strips about 1cm wide;
(b) chopped by hand into pieces roughly 1cm cubes;
(c) minced through 3mm die plate;
(d) some of the strips were agitated in a Hobart mixer at full speed for 2min using a 'paddle' mixer;
(e) some of the mince was also agitated in a Hobart mixer under the same conditions as *(d)* above.

These materials are referred in this report as 'strips', 'chopped', 'mince', 'bowl mixed strips' and 'bowl mixed mince', respectively. These five materials were subsequently frozen in rectangular blocks.

All blocks were stored at $-29°C$ except part of the mince batch, which was kept at $-10°C$ to allow denaturation to occur more rapidly. This batch is referred to as 'cold-stored mince' and had been subjected to this temperature regime for 24 days before being tasted.

Finally, a batch of mince which had been commercially produced and kept in cold storage for more than six months was obtained. Although this was of course derived from a different raw material from the above treatments and is therefore not strictly comparable, it was hoped that it would serve as an indicator of the gross effects of long-term storage.

2.2 *Development of scoring system*

The basic attribute scales were drawn up by a small group of people who tasted a wide range of fish and marine crustaceans and molluscs. Each person wrote a description of appearance, breakdown on the palate and in the mouth, and flavour. In subsequent discussion the descriptive words were grouped and scales were evolved where a significant variation was felt to exist within a particular grouping.

The resulting scoring system is shown in *Table I*, and definitions of the various terms are shown in *Table II*. The twelfth scale, overall liking, was included to give an indication of trends in acceptability; obviously a technique of this type is unsuitable for determining absolute acceptability. Flavour as well as texture was to be taken into account in scoring overall liking. Individual flavour scales were not incorporated in this general scheme because to do so would have made the system unwieldy; however, in particular applications, for example with fish fingers, the relatively few relevant flavour scales could easily be incorporated.

About 20 of the scientific staff from the laboratory were trained in the use of the profile form. This was done by presenting them with the end of scale and intermediate standards (see *Table I*) to familiarize them with the method of assessing each attribute. In later sessions they were given 'unknown' samples to score, followed by discussion to resolve any differences of opinion.

2.3 *Sample presentation*

The samples were presented in randomized groups of four as breaded steaks 76mm × 38mm × 12mm which had been fried at $180°C \pm 2°C$ in 2mm vegetable oil for

Table 1
FISH TEXTURE PROFILE FORM

1 Overall appearance of sample

	1	2	3	4	5	6	7	8	9	10	
unlike											identical
squid											fillet, fish, *eg* coley, herring, cod, *etc*

2 Apparent flake thickness

	1	2	3	4	5	6	7	8	9	10	
no flakes											v. large flakes
squid		plaice				cod				coley	

3 Resilience

	1	2	3	4	5	6	7	8	9	10	
no resilience											v. resilient
fish cake							spam		squid		

4 Brittleness

	1	2	3	4	5	6	7	8	9	10	
											brittle
non-brittle – squid pasty – fishcake					plaice				cod		

5 Resistance

	1	2	3	4	5	6	7	8	9	10	
no resistance											v. resistant
mashed potato			cod						squid		

6 Broken surface mouthfeel

	1	2	3	4	5	6	7	8	9	10	
smooth											random surface
squid				cod				threshings block			

7 Juiciness 1

	1	2	3	4	5	6	7	8	9	10	
dry											wet
minced pressure cooked dry cod				cod					kesp		

8 Fibrosity

	1	2	3	4	5	6	7	8	9	10	
no fibres											v. fibrous
squid					plaice				large cod		

9 Chewiness

	1	2	3	4	5	6	7	8	9	10	
not chewy											v. chewy
mashed potato		plaice			cod				squid		

10 Juiciness 2

	1	2	3	4	5	6	7	8	9	10	
dry											wet
FD cod fibres				cod				smoked salmon			

11 Ease of swallowing

	1	2	3	4	5	6	7	8	9	10	
difficult											v. easy
minced pressure cooked dry cod				cod				smoked salmon			

12 Overall liking

	1	2	3	4	5	6	7	8	9	10	
disliked					neither				liked		

4min/side. The panellists were to ignore the bread-crumb, which was only to facilitate cooking.

3 Results and discussion

3.1 Statistical treatment of results

The panel scores were processed statistically by multi-factor analysis of variance, taking into account day-to-day variation and individual panellist bias. The resulting adjusted mean scores for each sample on each attribute scale are given in *Tables III–V*. Comparison of texture profiles is facilitated by drawing lines connecting the 12 attribute scores for the samples to be compared. Examples of such a procedure are shown in *Figs 1–3*. However with more than two or three profiles, comparisons and trends are most clearly seen using principal components analysis.

Principal components analysis (PCA) is a statistical technique which consists basically of forming new vari-ables that are linear combinations of the old ones. The new variables are fewer in number than the old ones and are defined in such a way as to take account of the maximum amount of observed variance. In the present case, 11 variables (overall liking is excluded) are com-bined ('mapped') into three complex variables which together account for about 90% of the original variance.

The correlations between the 11 original variables and the three principal components are given in *Table VI*. The first principal component, that which explains the most variance, is obviously concerned with the struc-tural aspects of the material and is therefore referred to as the 'structure component', similarly the second and third principal components are the 'juiciness compo-nent' and the 'toughness component', respectively. Inci-dentally, overall liking, which was not part of the PCA procedure, correlates with the structure component.

Figures 4–8 give principal component 'maps' for a range of fish and fish comminutes.

Table II
TEXTURE PROFILE DEFINITIONS

1 *Overall appearance of sample*	How like is the sample to fillet fish when broken open?
2 *Apparent flake thickness*	Apparent thickness of flakes if any
3 *Resilience*	Force required to break structure on pressing the sample between the tongue and the roof of the mouth
4 *Brittleness*	The degree of brittleness/pastiness of the sample on pressing the sample between the tongue and the roof of the mouth
5 *Resistance*	The amount of effort required in attempting to bite through the sample using the front teeth only. (1, 2 or 3 bites)
6 *Broken surface mouthfeel*	Roughness of the surface of the broken pieces after 2 or 3 chews
7 *Juiciness 1*	The initial impression of moistness (free H_2O, flavour juices, liquid fat, oil and saliva) of the sample on initial chewing. (1, 2 or 3 chews)
8 *Fibrosity*	Impression of fibres in the sample on chewing
9 *Chewiness*	The total amount of chewing or effort required to convert the fish into a swallowable state
10 *Juiciness 2*	The total impression of succulence in the mouth just prior to swallowing
11 *Ease of swallowing*	The ease/difficulty in swallowing the sample
12 *Overall liking*	Taking everything into consideration your degree of liking for the sample

Table III
ADJUSTED MEAN SCORES FOR TEXTURE ATTRIBUTES OF FOUR SPECIES OF FISH

	Cod	Coley	Salmon	Plaice
Overall appearance	9·9	10·0	9·7	9·5
Apparent flake thickness	8·8	9·5	9·1	5·4
Resilience	4·4	4·7	6·4	3·9
Brittleness	8·3	9·3	8·0	5·9
Resistance	4·0	3·8	6·1	3·2
Broken surface mouthfeel	4·9	4·6	5·1	4·3
Juiciness I	5·6	6·1	4·4	6·8
Fibrosity	8·3	7·7	8·4	6·1
Chewiness	5·9	5·5	6·8	4·0
Juiciness II	5·4	5·6	4·0	5·7
Ease of swallowing	5·3	5·3	5·1	5·7
Overall liking	6·2	6·9	7·6	6·6

Table IV
ADJUSTED MEAN SCORES FOR TEXTURE ATTRIBUTES OF COD MUSCLE: INTACT, AND AFTER FIVE STAGES OF COMMINUTION

	Intact muscle	1cm strips	½—1cm pieces	Mince	Strips bowl mixed[1]	Mince bowl mixed[1]
Overall appearance	9·9	9·9	9·4	5·5	4·3	4·4
Apparent flake thickness	8·8	8·2	8·5	3·4	2·2	2·4
Resilience	4·4	4·7	5·3	4·6	6·1	6·6
Brittleness	8·3	8·3	7·9	6·1	4·9	5·1
Resistance	4·0	4·1	4·5	4·2	5·4	5·4
Broken surface mouthfeel	4·9	5·1	5·3	7·0	7·5	7·5
Juiciness I	5·6	5·7	5·9	5·8	5·7	5·7
Fibrosity	8·3	8·7	7·9	7·0	7·7	7·4
Chewiness	5·9	5·8	6·2	5·7	6·5	6·3
Juiciness II	5·4	5·3	4·9	5·0	4·7	4·5
Ease of swallowing	5·3	5·2	5·2	4·8	4·3	4·2
Overall liking	6·2	7·1	6·3	5·7	5·4	5·0

[1] The 1cm strips and mince were stirred vigorously in a bowl mixer to destroy the fibre alignment

Table V
ADJUSTED MEAN SCORES FOR TEXTURE ATTRIBUTES OF PARTIAL COMMINUTES OF COD, AFTER COLD STORAGE[1]

	1cm strips	½—1cm pieces	Mince	Commercial V-cut mince after 1 year
Overall appearance	9·5	9·0	5·8	4·7
Apparent flake thickness	8·5	7·2	4·0	3·1
Resilience	5·3	5·1	5·7	6·6
Brittleness	8·8	9·0	6·7	8·3
Resistance	4·8	4·9	5·3	6·4
Broken surface mouthfeel	5·6	5·6	7·9	9·2
Juiciness I	5·4	5·0	5·2	4·1
Fibrosity	8·2	7·9	8·2	8·3
Chewiness	6·7	6·8	7·5	7·7
Juiciness II	5·0	4·7	4·6	3·6
Ease of swallowing	5·4	5·0	4·2	2·6
Overall liking	6·9	6·7	4·6	2·6

[1] The strips, pieces and mince were derived from the same material as that in *Table IV*, but had been kept at $-10°C$ for about one month

Table VI
FACTOR-LOADING MATRIX SHOWING HOW EACH OF THE 11 TEXTURE ATTRIBUTES RELATES TO THE THREE PRINCIPAL COMPONENTS

	Principal component		
	1 Structure	*2 Juiciness*	*3 Toughness*
Overall appearance	0·93	−0·23	−0·17
Apparent flake thickness	0·93	−0·21	−0·07
Resilience	−0·36	−0·02	0·89
Brittleness	0·92	0·19	−0·23
Resistance	−0·21	−0·01	0·97
Broken surface mouthfeel	−0·30	0·85	−0·25
Juiciness I	−0·38	−0·74	−0·11
Fibrosity	0·77	0·33	−0·35
Chewiness	−0·33	0·04	0·95
Juiciness II	−0·07	−0·95	0·02
Ease of swallowing	0·46	−0·71	−0·37

3.2 Comparison of fish types and treatments

3.2.1 *Comparison of fish types* The textural relationships between cod, coley, salmon and plaice are shown in *Fig 4*. Cod and coley are very similar in all respects, plaice is softer and juicier and has a less obvious structure, whereas salmon is very much tougher and drier and has a rather more obvious structure.

3.2.2 *Effect of comminution* Figure 5 shows how the perceived texture of cod changes on progressive comminution. Only slight effects on juiciness and toughness are seen and the structure component is not affected until the fish is minced. A similar effect on structure can be achieved by stirring cod strips in a bowl mixer (*Fig 6*), however there is a concurrent change in toughness and dryness. Stirring cod mince produces a similar end product.

These results suggest that the effect on the structure component is due to the loss of flake structure, and this does not become fully apparent until the fish is minced. Within the mince particles, however, fibre structure and alignment are preserved. The fact that vigorous stirring does not significantly affect the structure component suggests that the residual structure component after mincing and/or stirring is due mainly to the presence of fibres and is not affected by their orientation. A further

411

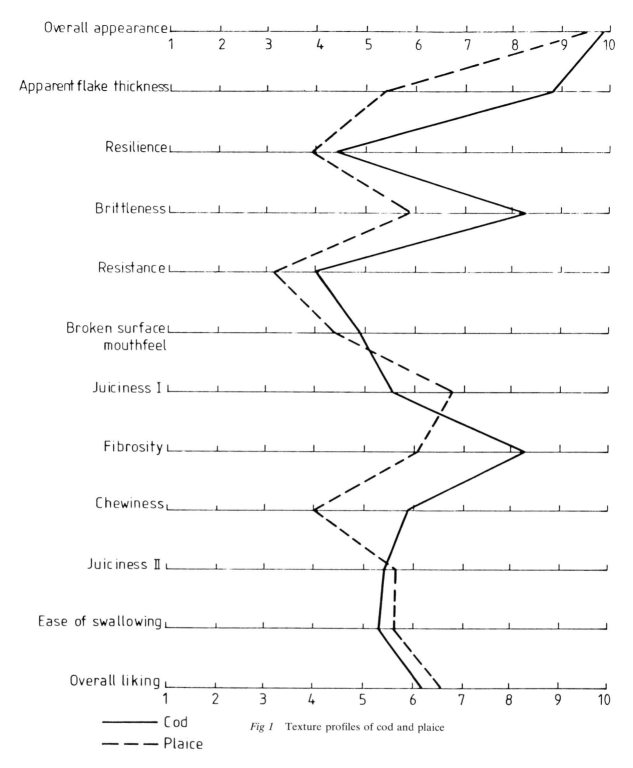

Overall appearance
Apparent flake thickness
Resilience
Brittleness
Resistance
Broken surface mouthfeel
Juiciness I
Fibrosity
Chewiness
Juiciness II
Ease of swallowing
Overall liking

——— Cod

— — — Plaice

Fig 1 Texture profiles of cod and plaice

reduction in the structure component would therefore only be achieved by a comminution method which would break up the fibres themselves – a bowl chopper, for example. This suggestion has been substantiated in subsequent work.

3.2.3 *Effect of cold storage*   *Figure 7* gives the three-dimensional texture data for cold-stored comminutes. Compared with the data for freshly-frozen raw material there is a significant increase in toughness, dryness and structure. These effects are more marked in the commi-

nuted material than in the intact fish.

The increase in structure component on cold storage is perhaps unexpected; it appears to be a result of the fibrous structure becoming more obvious after cold storage. *Figure 3* shows how this arises; there is little difference in the appearance attributes (overall appearance, apparent flake thickness) of the fresh mince and the cold-stored mince, but the other two attributes (brittleness, fibrosity) which map into the structure component are significantly increased by cold storage; that is, the structure breaks down more easily and the structural

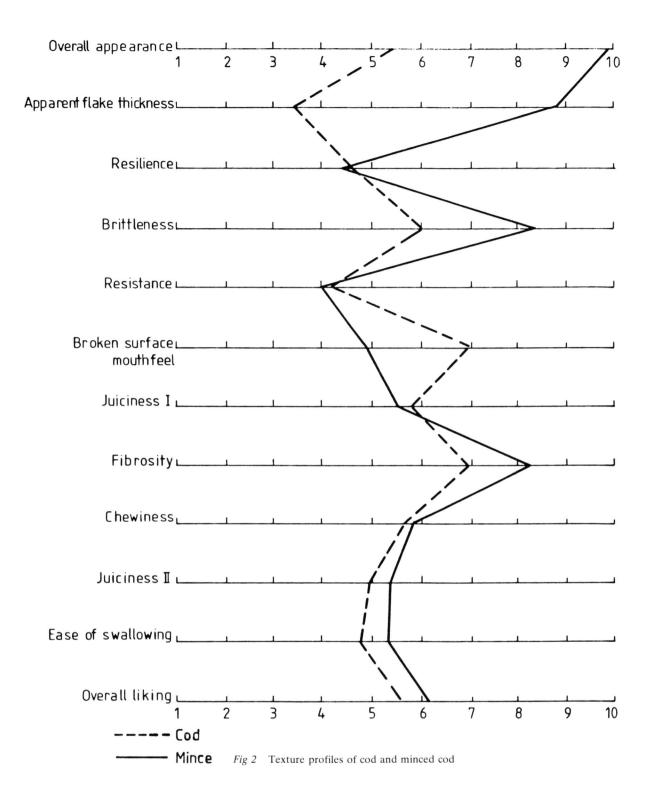

	1	2	3	4	5	6	7	8	9	10
Overall appearance										
Apparent flake thickness										
Resilience										
Brittleness										
Resistance										
Broken surface mouthfeel										
Juiciness I										
Fibrosity										
Chewiness										
Juiciness II										
Ease of swallowing										
Overall liking										

- - - - - Cod

——— Mince   *Fig 2*   Texture profiles of cod and minced cod

subunits are more clearly defined.

The fourth material shown in *Fig 7* is a sample of commercially produced cod V-cut mince which had been in cold storage at −30°C for about a year. Despite its different origin, it fits the general trend of increase in toughness, dryness, and structure with increase in storage time.

This concludes the description of the texture-profile method as applied to fish in general. Having characterized the material in this way we can go on to relate these data to consumers' judgements of acceptability, and within the more limited product contexts of such investigations the inclusion of flavour scales would be feasible. Obviously the consumer would be less analytical in his approach but would no doubt be able to show how his assessments of appearance, flavour and texture contributed to his final judgement. What we have in the texture profile is a method for interpreting and perhaps predicting the consumer's response to fish texture.

4 Reference
HORSFIELD, S and TAYLOR, L J. *J. Sci. Fd. Agric.*, 27, 1044–1056 1976

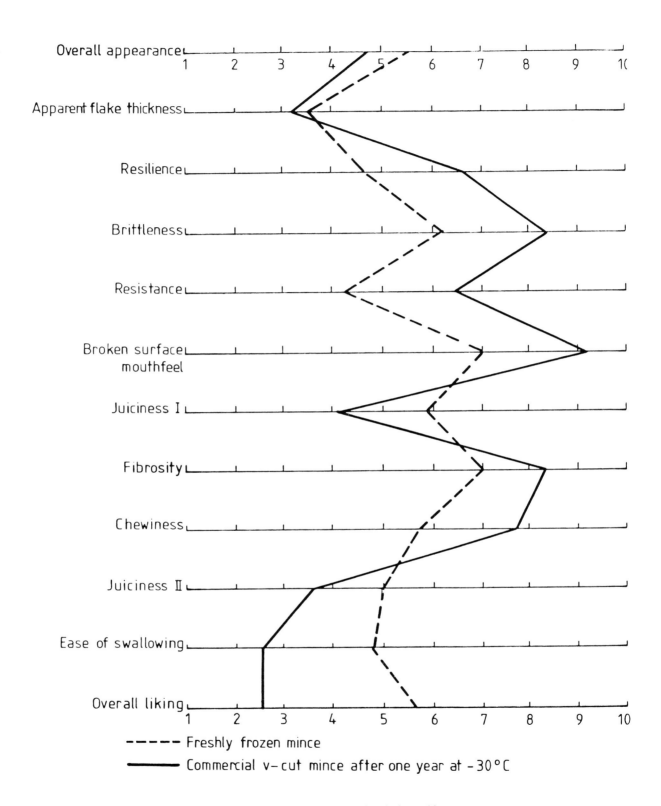

Fig 3 Texture profiles of minced cod after cold storage

414

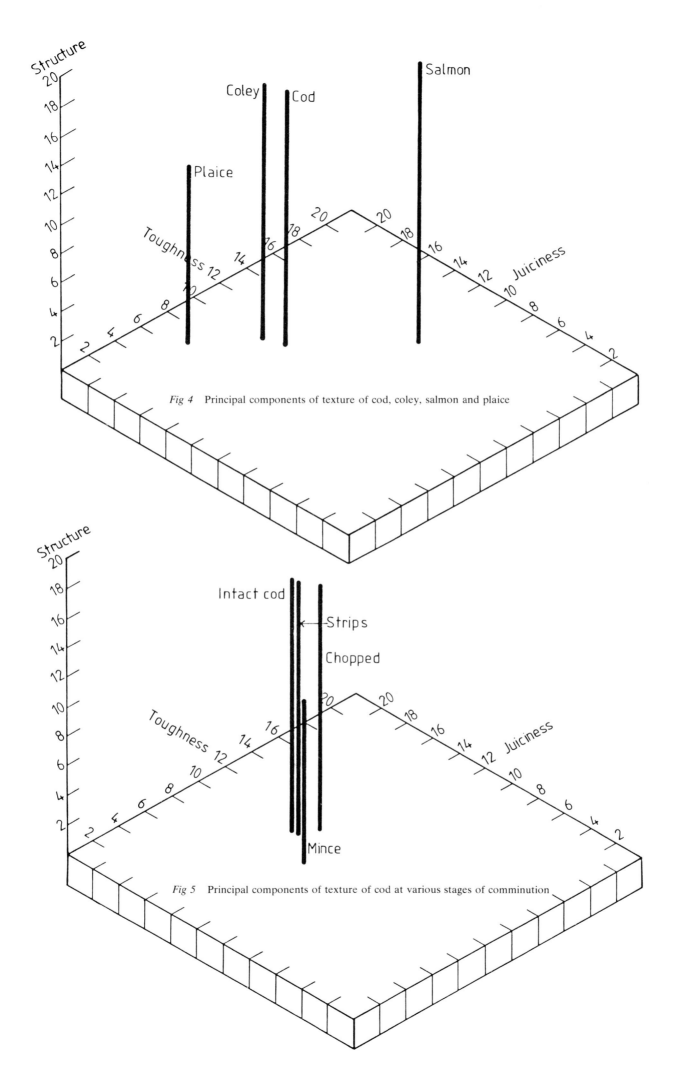

Fig 4   Principal components of texture of cod, coley, salmon and plaice

Fig 5   Principal components of texture of cod at various stages of comminution

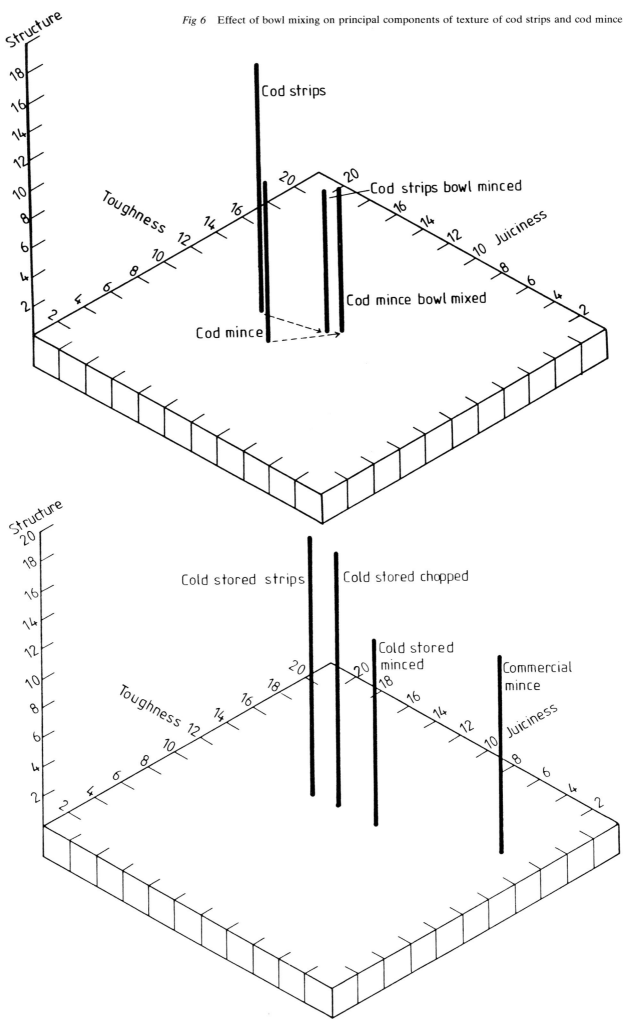

Fig 6 Effect of bowl mixing on principal components of texture of cod strips and cod mince

Fig 7 Effect of cold storage on principal components of texture of cod mince

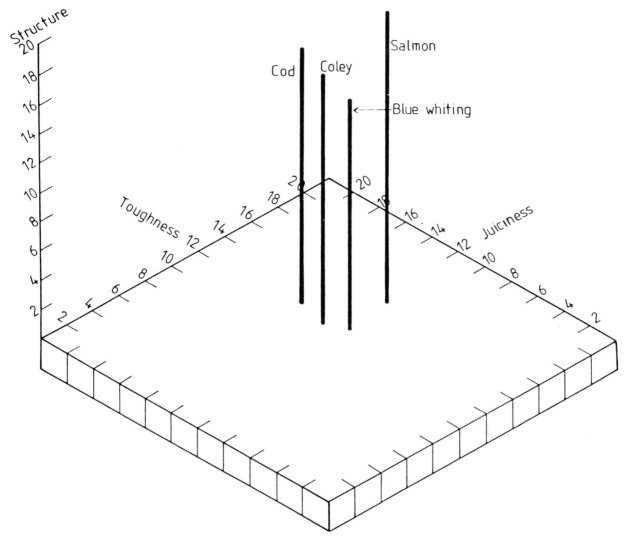

Fig 8 Effect of cold storage on principal components of texture of cod, coley, salmon and blue whiting

---

Spoilage patterns of wet and thawed frozen Cape hake during chilled storage

C K Simmonds and E C Lamprecht

1 Introduction

During the past decade there has been an increasing trend in South Africa, as in other countries, towards freezing fish at sea. A significant proportion of sea-frozen fish, particularly Cape hake, has been supplied to retail outlets, where it has subsequently been thawed and offered for sale in competition with wet fish. It was thus considered to be essential to determine the relative rates of spoilage of wet and thawed frozen hake.

2 Materials and methods

2.1 *Handling*
Headed and gutted Cape hake were obtained from the last catches of commercial wet fish trawlers, or from commercial freezer trawlers, at the time of off-loading. Wet fish were transported to the laboratory in ice, while frozen hake, in the form of 25kg cartons packed in polythene-lined cardboard cartons, were carried in a closed, unrefrigerated van, the transport time being of the order of 15–20min. The handling conditions and exact time elapsed since catching were not known, and the storage times for wet fish chill-stored in the laboratory were reckoned from the time of arrival there. Chilled fish were stored in containers of melting ice in a room maintained at 1°C, with the exception of three tests in which replicate samples were also stored in polythene bags at 5°C. Fish subjected to frozen storage were packed in polythene bags and placed overnight on a shelf in a freezer-room at the desired temperature and in such

a position as to be exposed to direct draught from the recirculating fans. They were subsequently packed in cartons for storage. Thawing was carried out in an air stream at ambient temperature, with a velocity of about 250m/min.

2.2 *Storage*
Four series of tests were carried out as follows:

Series 1 (two tests): Wet fish were chill-stored and compared with others from the same catch which were frozen immediately on arrival, thawed after one or two weeks' storage at −29°C and then chill-stored for the same lengths of time as the wet controls.

Series 2 (three tests): Wet fish were chill-stored for varying periods. At the same time that samples were removed from the ice for examination, others were frozen and stored for varying times at −29°C. In two of the tests samples were further chill-stored after thawing.

Series 3 (three tests): Wet and sea-frozen hake were obtained on the same day. The frozen fish were thawed, and half of each type of fish were iced, and the other half chill-stored at 5°C. The fish were thus caught at the same time of year, but were not necessarily equivalent in catch date or fishing ground.

Series 4 (two tests): Wet fish were frozen initially and after chill-storage at −29°C, −18°C and −7°C and stored at these temperatures for periods up to 20 weeks, except for those at the last named storage temperature which were stored for a maximum of eight weeks.

2.3 *Examination of samples*
In Series 1, 2 and 4, three to four fish per treatment were examined at a time, while in Series 3, three fish were examined initially and one and two at a time after each storage period.

The fish were placed on a flamed tray, and samples taken aseptically for microbiological examination. The fish were thereafter inspected by a trained panel and the freshness assessed using a raw odour scale on which 10 points represents sea-freshness, 0 points putridity, and 5 points the borderline at which the fish is just no longer acceptable (Rowan, 1956).

Separate counts were performed on skin and flesh. Skin samples (9·625cm²) were cut on the lateral line near the middle of the fish with a sterile metal cork borer and the adhering flesh removed with a scalpel. Flesh samples (10g) were removed from below the area from which the skin had been removed. Roll tube counts were performed at 20°C using a 75% sea water-based medium and incubating for four to five days.

Suitable dilutions from each fish sample were also set up as plates, poured with the same medium as used for the roll tubes. From these plates colonies were picked off for identification and were streaked on the same medium. Isolates obtained in this way were plated on appropriate media to obtain pure cultures. Ten skin and 10 flesh isolates were streaked per fish. Single colonies were also cultured in liquid medium, eg peptone water. Identification was based on morphology, motility, pig-

mentation, Gram reaction, type of growth in liquid culture, as well as biochemical tests such as reaction in litmus milk, liquefaction of gelatin, etc (Buchanan and Gibbons, 1974).

3 Results and discussion

3.1 *General*
Table I gives the bacterial counts and *Table II* the distribution of the genera for tests of Series 1, comparing the growth of micro-organisms and loss of quality during chill-storage of wet and thawed laboratory frozen hake stored for one to two weeks at −29°C. Mean figures for headed and gutted fish in two tests are given. Results obtained for fillets in the second test were not significantly different from those for the equivalent headed and gutted fish and are hence not recorded here.

Tables III and *IV* give figures for two tests of Series 2, for fish chill-stored before and after storage at −29°C for one month.

Tables V and *VI* show the counts and distribution of genera on iced wet fish compared with iced thawed sea-frozen samples, while *Tables VII* and *VIII* compare equivalent fish stored at 5°C (means for three tests of Series 3).

Tables IX and *X* show the die-off of bacteria over four months' storage at −29°C for three tests of Series 2, while *Table XI* gives the bacterial counts obtained during storage at −29°C, −18°C and −7°C in two tests of Series 4. The distribution of genera in the latter tests did not differ significantly at different storage temperatures, and were similar to those shown for −29°C storage in *Table X*.

All bacterial counts are expressed as logarithms to the base 10 of the numbers per square centimetre or gram, for skin and flesh samples, respectively.

3.2 *Growth rates during chilled storage*
Bacterial counts on wet fish started to rise virtually immediately, being significantly increased after even two days' storage, whereas those for thawed frozen fish were static for up to four days. In both cases the maximum rate of growth was between four and seven days (*Tables I* and *V*). From the same results it can be seen that after seven days, the growth on wet fish started to tail off, while that on sea-frozen fish continued to increase at least up to 13 days. By 11 to 13 days the bacterial counts on fish frozen in the laboratory, after a presumed one day in ice on board trawler, had caught up to those of the equivalent wet fish. Sea-frozen fish showed lower counts even at this stage.

The counts on fish after iced storage for nine to 10 days were similar whether the fish was stored in ice from catching, frozen fresh and subsequently thawed and chill-stored, or chill-stored partly before and partly after frozen storage (*Table III*). This effect was also noted for storage times at −29°C both shorter and longer than the one month shown in the table.

A similar pattern was observed for fish stored at 5°C (*Table VII*). Here the counts on wet fish had started to show an increase after one day, with accelerating growth up to six days, at which stage the fish was putrid. For sea-frozen fish, on the other hand, counts were static up to at least three days and by six days were still well below those for wet fish. Sea-frozen fish at this stage were still acceptable, although barely so.

3.3 Death rate during frozen storage

Initially, after one day frozen storage, there was little if any decrease in the bacterial count. There was a significant drop in numbers, on the other hand, by one week (*Table IX*), which was not however maintained thereafter, or at least not to an extent sufficient to show up against the variations from sample to sample. There was no significant difference in die-off at $-29°C$ or $-18°C$, but storage at $-7°C$ appeared to cause a slightly more rapid decrease than at the lower temperatures.

3.4 Odour ratings

The freshness of hake as assessed by odour ratings decreased more slowly for thawed frozen than for wet fish when stored in ice (*Tables I* and *V*). By the time that the wet fish had reached the borderline of acceptability, the bacterial counts of the thawed fish, owing to the long lag phase, had usually only just reached their maximum rate of increase. This was particularly so for sea-frozen hake, which was not merely acceptable but still of good quality at the end of the storage life of the wet fish (*Table V*). This effect was even more pronounced in the case of fish stored at 5°C (*Table VII*) where the thawed sea-frozen fish was still well above the borderline when the wet fish was putrid. There was a slight tendency for fresh frozen fish to exhibit a loss in apparent quality immediately after thawing, owing to development of cold storage odours. By contrast stale fish sometimes gave slightly better odour ratings after thawing.

3.5 Distribution of genera

The bacteria recovered from fresh wet hake consisted predominantly of *Achromobacter* with lower percentages of *Pseudomonas* and *Micrococcus*, and small numbers of other genera, mainly *Brevibacterium*, *Corynebacterium* and *Flavobacterium*. During chill storage the proportion of *Achromobacter* tended to increase, although not invariably so, while *Pseudomonas* generally increased markedly, while *Micrococcus* decreased, so that the first two genera dominated virtually completely by the time the fish was stale. During frozen storage, *Pseudomonas* died off most rapidly so that the percentage of *Pseudomonas* in freshly thawed frozen hake was extremely low. This was accompanied by a corresponding increase in *Achromobacter* which appeared to be most resistant to freezing. A slightly higher percentage of *Micrococcus* was recovered from the frozen fish, while there was no significant change in the proportions of other genera.

During chilled storage of thawed frozen fish, *Pseudomonas* increased markedly, but did not attain the levels found in chill-stored wet fish.

4 Conclusions

Due to a lowering in numbers of bacteria during frozen storage, and the time lag taken for them to recommence growth after thawing, thawed frozen, and especially sea-frozen, Cape hake has a longer shelf-life at chill temperatures than the equivalent wet fish. This effect may be enhanced by the relative susceptibility of *Pseudomonas* to these effects. The improvement in storage life is even more marked at 5°C than in iced storage. The fact that chill cabinets in retail stores, and domestic refrigerators, are likely to operate at temperatures somewhat above 0°C, makes this extension of shelf-life even more significant.

5 References

BUCHANAN, R E and GIBBONS, N E (Eds). *Bergey's Manual of Deter-*
1974 *minative Bacteriology*. Williams and Wilkins Company, Baltimore. 8th edn
ROWAN, A N. The assessment of the freshness of fish by odour. FIRI
1956 Annual Report 10, 8–11

Table I
GROWTH RATES OF BACTERIA ON ICED WET AND THAWED
LABORATORY-FROZEN HAKE (1–2 WEEKS AT $-29°C$)

Days in ice	Wet hake Skin log_{10} No./cm²	Wet hake Flesh log_{10} No./g	Odour rating	Frozen hake Skin log_{10} No./cm²	Frozen hake Flesh log_{10} No./g	Odour rating
0	4·688	4·337	8·0	4·364	4·012	8·0
4	6·390	5·311	6·7	4·229	4·081	7·3
7	7·096	6·707	5·0	6·270	6·176	5·4
11	7·736	6·973	4·9	7·298	6·545	5·2
14	7·640	7·303	3·4	7·873	7·569	4·2

Table II
DISTRIBUTION OF GENERA ON ICED WET AND THAWED LABORATORY-FROZEN HAKE (1–2 WEEKS AT $-29°C$)

Days in ice	Wet hake Achromobacter	Wet hake Pseudomonas	Wet hake Micrococcus	Wet hake Other	Frozen hake Achromobacter	Frozen hake Pseudomonas	Frozen hake Micrococcus	Frozen hake Other
0	70·9	16·5	4·0	8·6	81·6	1·4	10·2	6·8
4	79·5	19·9	—	0·6	81·6	8·1	4·4	5·9
7	67·1	32·2	—	0·7	88·2	11·2	0·6	—
11	65·0	35·0	—	—	87·5	12·5	—	—
14	85·0	12·5	2·5	—	90·0	10·0	—	—

Table III
GROWTH OF BACTERIA ON ICED WET AND THAWED FROZEN HAKE (1 MONTH AT −29°C)

Days in ice wet	Days in ice after freezing and thawing	Total time in ice	Skin $\log_{10}$ No. /cm²	Flesh $\log_{10}$ No./g	Odour rating	Treatment
0	0	0	4·778	4·426	8·0	Wet
6–7	0	6–7	6·659	5·821	6·3	Wet
9–11	0	9–11	7·149	6·895	4·3	Wet
0	0	0	3·795	3·633	8·0	Frozen fresh and thawed
0	6–7	6–7	6·673	5·790	6·0	Frozen fresh and thawed
0	9–11	9–11	7·422	6·674	5·5	Frozen fresh and thawed
6–7	0	6–7	5·859	5·417	6·0	Frozen after iced storage
6–7	4	10–11	7·097	6·765	5·3	Frozen after iced storage
9–11	0	9–11	6·596	6·290	4·7	Frozen after iced storage

Table IV
DISTRIBUTION OF GENERA ON ICED WET AND THAWED FROZEN HAKE (1 MONTH AT −29°C)

Days in ice wet	Days in ice after freezing and thawing	Total time in ice	% Genera Achromobacter	Pseudomonas	Micrococcus	Other	Treatment
0	0	0	48·1	31·3	11·3	9·3	Wet
6–7	0	6–7	63·0	35·8	0·7	0·5	Wet
9–11	0	9–11	52·5	46·9	0·6	—	Wet
0	0	0	85·0	3·8	6·9	4·3	Frozen fresh and thawed
0	6–7	6–7	85·1	13·1	0·6	1·2	Frozen fresh and thawed
0	9–11	9–11	86·3	13·1	—	0·6	Frozen fresh and thawed
6–7	0	6–7	90·0	8·2	0·6	1·2	Frozen after iced storage
6–7	4	10–11	80·0	18·8	1·2	—	Frozen after iced storage
9–11	0	9–11	76·3	23·1	0·6	—	Frozen after iced storage

Table V
GROWTH RATES OF BACTERIA ON ICED WET AND THAWED SEA-FROZEN HAKE

Days in ice	Wet hake Skin $\log_{10}$ No./cm²	Flesh $\log_{10}$ No./g	Odour rating	Frozen hake Skin $\log_{10}$ No./cm²	Flesh $\log_{10}$ No./g	Odour rating
0	4·555	4·325	8·0	3·730	3·554	8·0
2–3	5·346	4·621	7·7	4·105	3·246	8·0
7	7·166	6·157	5·7	4·879	4·323	7·3
10	7·677	6·841	4·0	5·353	5·408	5·5
13	8·344	7·437	3·3	6·603	6·301	4·3

Table VI
DISTRIBUTION OF GENERA ON ICED WET AND THAWED SEA-FROZEN HAKE

Days in ice	Wet hake Achromobacter	Pseudomonas	Micrococcus	Other	Frozen hake Achromobacter	Pseudomonas	Micrococcus	Other
0	53·3	35·0	7·8	3·9	57·8	6·1	20·0	16·1
2–3	74·0	22·0	2·0	2·0	76·0	15·0	6·0	3·0
7	65·8	33·3	0·8	—	74·2	20·0	4·2	1·6
10	33·3	66·7	—	—	58·3	40·8	0·8	—
13	32·5	67·5	—	—	54·2	45·0	0·8	—

Table VII
GROWTH RATES OF BACTERIA ON CHILLED (5°C) WET AND THAWED SEA-FROZEN HAKE

Days in ice	Wet hake Skin log_{10} No./cm²	Wet hake Flesh log_{10} No./g	Odour rating	Frozen hake Skin log_{10} No./cm²	Frozen hake Flesh log_{10} No./g	Odour rating
0	4·555	4·325	8·0	3·730	3·554	8·0
1	5·040	4·784	7·5	3·553	3·260	8·0
3	6·286	5·462	4·7	3·703	3·155	6·5
6	7·817	7·068	2·0	6·484	5·733	5·7

Table VIII
DISTRIBUTION OF GENERA ON CHILLED (5°C) WET AND THAWED SEA-FROZEN HAKE

Days in ice	Wet hake Achromobacter	Wet hake Pseudomonas	Micrococcus	Other	Frozen hake Achromobacter	Frozen hake Pseudomonas	Micrococcus	Other
0	53·3	35·0	7·8	3·9	57·8	6·1	20·0	16·1
1	50·0	42·0	4·0	4·0	62·0	2·0	31·0	5·0
3	50·8	49·2	—	—	66·7	18·4	14·0	0·9
6	35·9	64·1	—	—	73·0	24·3	2·7	2·7

Table IX
DEATH RATES OF BACTERIA AT −29°C

Storage time	0 days in ice Skin log_{10} No./cm²	0 days in ice Flesh log_{10} No./g	3–4 days in ice Skin log_{10} No./cm²	3–4 days in ice Flesh log_{10} No./g	6–7 days in ice Skin log_{10} No./cm²	6–7 days in ice Flesh log_{10} No./g	9–10 days in ice Skin log_{10} No./cm²	9–10 days in ice Flesh log_{10} No./g	13–14 days in ice Skin log_{10} No./cm²	13–14 days in ice Flesh log_{10} No./g
0	4·678	4·294	5·020	4·560	6·477	5·693	7·135	6·699	7·923	6·935
1 day	4·598	4·232	—	—	6·250	5·633	6·314	5·754	7·654	6·831
1 week	4·308	3·862	—	—	5·201	4·788	6·661	5·925	7·083	6·505
1 month	3·842	3·623	4·578	3·748	5·613	5·177	6·457	6·159	7·234	6·692
4 months	3·499	3·044	4·307	4·528	—	—	6·215	6·418	—	—

Table X
DISTRIBUTION OF GENERA DURING STORAGE OF HAKE AT −29°C

Storage time	0 days in ice Achromobacter	0 days in ice Pseudomonas	Micrococcus	Other	9–10 days in ice Achromobacter	9–10 days in ice Pseudomonas	Micrococcus	Other
0	44·2	27·8	17·1	10·9	55·3	43·8	2·5	0·4
1 day	63·3	15·5	14·1	7·1	66·3	30·0	2·5	1·2
1 week	56·1	6·0	19·2	18·7	74·4	20·7	3·8	1·1
1 month	63·8	3·8	17·5	14·9	81·3	16·2	1·7	0·8
4 months	81·0	5·0	12·7	1·3	100·0	—	—	—

Table XI
DEATH RATES OF BACTERIA AT THREE DIFFERENT STORAGE TEMPERATURES

Storage temperature °C	Storage time (days)	0 days in ice Skin log_{10} No./cm²	0 days in ice Flesh log_{10} No./g	6 days in ice Skin log_{10} No./cm²	6 days in ice Flesh log_{10} No./g	13 days in ice Skin log_{10} No./cm²	13 days in ice Flesh log_{10} No./g
—	0	5·095	4·490	6·609	5·838	7·932	6·999
−7	1	5·088	4·505	6·701	5·979	7·433	6·505
−7	28	3·726	3·136	5·286	4·841	5·792	5·301
−7	56	3·799	4·207	5·449	5·170	5·682	5·313
−18	1	4·753	4·140	6·473	5·612	7·214	6·606
−18	28	4·194	3·647	5·975	5·197	6·217	5·771
−18	56	3·461	4·538	6·390	5·189	7·056	6·547
−18	84	3·140	2·438	5·824	5·266	6·736	5·874
−18	140	3·473	3·810	5·337	4·895	5·719	5·483
−29	1	5·027	4·332	6·257	5·607	7·068	6·252
−29	28	4·298	3·657	5·606	5·112	6·918	6·066
−29	56	3·283	4·590	5·245	4·839	7·386	6·672
−29	84	3·260	2·695	4·734	4·592	6·325	5·625
−29	140	3·727	3·496	5·436	4·801	7·565	6·791

12 Protein studies

Analysis of the salt-soluble protein fraction of cod muscle by gel filtration

M Ohnishi and G W Rodger

1 Introduction

It is a well-known fact that fish muscle suffers deteriorative quality changes during frozen storage. However, there is a need to define differences between fresh and frozen muscle for a meaningful discussion of fish muscle preservation. Many studies have been carried out to detect changes in the chemical and physical properties of proteins – for example, measurement of solubility (Dyer, 1951; Connell, 1960), ATPase activity (Arai and Takashi, 1973), and the number of active SH groups (Connell, 1959). They all found that these properties were affected by freezing. Other studies have been performed to detect changes in the composition or the structure of muscle proteins. For this, gel filtration (Umemoto and Kanna, 1970; Seki and Arai, 1974), viscosity (Seagram, 1958), ultracentrifugation (Connell, 1963), electron microscopy (Tanaka, 1965; Tsuchiya *et al*, 1975) were used.

In the present study proteins were examined in an attempt to define any changes in their properties as a result of freezing, and to use this information as a monitor of changes occurring during frozen storage. For this reason, salt-soluble proteins (SSP) were extracted from fresh and frozen muscle in order to compare their gel filtration patterns. In addition, the eluted proteins were subjected to SDS-polyacrylamide gel electrophoresis (SDS-PAGE) to identify any changes occurring in electrophoretic patterns, and also compared by electron microscopy in an attempt to visualize any aggregates formed.

The method whereby 'denatured' proteins may be dissociated in their sub-units by SDS is well known (Connell, 1965), and by using this method, the sub-unit structure of even insoluble proteins may be found. Therefore, the salt-soluble proteins from fresh and frozen muscle were prepared and their gel-filtration patterns were compared with those obtained from SDS-solubilized fractions of the same muscle samples.

2 Materials and methods

A locally caught cod (*Gadus morhua*) 40cm long was killed by decapitation and its dorsal lateral muscle was cut out immediately as 1cm cubes, some of which were immediately frozen and stored at −29°C.

After six and twelve weeks, they were thawed at room temperature, and 20g of the frozen stored muscle homogenized for 30s in 180ml of 5% NaCl 0·02M NaHCO$_3$ solution. The homogenates were centrifuged for 30min at 18 000rpm on an MSE 18 centrifuge and the supernatants collected as the salt extracts.

2.1 *Gel filtration*
LKB ultragel AcA-22 (acrylamide 2% agarose 2%) was packed in a column (2·6cm × 100cm) and used as the gel-filtration medium. 5% NaCl-NaHCO$_3$ or 0·1% SDS-NaHCO$_3$ solution was used as the elution buffer. The protein concentration of the salt extract which was measured by the Biuret method (Umemoto, 1966) was adjusted to 6mg/ml, and 5ml was loaded on the gel using a peristaltic pump. When 0·1% SDS buffer was used, the salt extract was dialysed against water to remove salt, SDS added to 0·1% and the solution dialysed *v* 0·1% SDS for 3h at room temperature. The sample was then loaded on to the gel medium which had been pre-equilibrated with 0·1% SDS solution. The UV absorbance of the eluted proteins was measured at 280nm (LKB 2089 UVICORD III), and 4·5ml fractions were collected (Camlab Fracpac.). SDS-PAGE and electron microscopy were performed on each peak eluted from the column.

2.2 *SDS-PAGE analyses*
For the analysis of eluted proteins, 10% acrylamide gels in 0·025M tris-HCl buffer (pH 8·4), with 0·375M tris-glycine-SDS (pH 8·4) as the running buffer were used. The gel tubes measured 5cm with an id of 5mm. The system was pre-run for 1½h to stabilize the gels and to remove monomers at an applied voltage of 90V, at 3mA/tube (Pharmacia Fine Chemicals, power supply 500/400).

40 μl of the eluted samples were loaded on the gels after being made up to 4% SDS 1% mercaptoethanol, and electrophoresis was carried out at 100V, at 4mA/tube until a bromophenol blue marker dye had run 4·5cm.

2.3 *Electron microscopy of eluted proteins*
Samples gathered from the fraction collector were negatively stained using 3% uranyl acetate on carbon-coated formval mesh. They were observed using a JEM 100C electron microscope at an accelerating voltage of 80kV.

3 Results

Figures 1–3 show the gel-filtration patterns of the salt extracts of fresh, six and 12 weeks' frozen muscle, when using 5% NaCl-NaHCO$_3$ as the eluting buffer.

These figures show that the elution pattern of the salt extract of fresh muscle has three main peaks and SDS-PAGE showed that the first peak contained myosin heavy chain (MHC) and actinin (Act), and the second peak and shoulders contained actin (A), tropomyosin (TM), troponin (TN) and myosin light chain (MLC).

Tropomyosin was eluted faster than actin, and this tendency was shown in the salt extracts of both fresh and

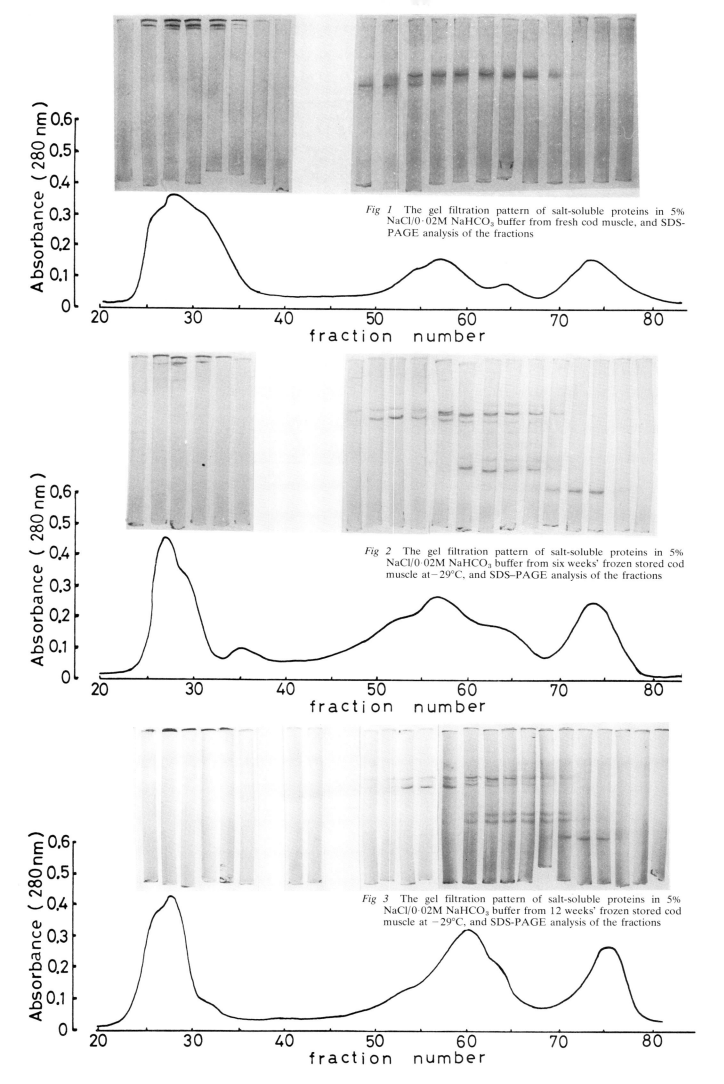

Fig 1 The gel filtration pattern of salt-soluble proteins in 5% NaCl/0·02M NaHCO₃ buffer from fresh cod muscle, and SDS-PAGE analysis of the fractions

Fig 2 The gel filtration pattern of salt-soluble proteins in 5% NaCl/0·02M NaHCO₃ buffer from six weeks' frozen stored cod muscle at −29°C, and SDS–PAGE analysis of the fractions

Fig 3 The gel filtration pattern of salt-soluble proteins in 5% NaCl/0·02M NaHCO₃ buffer from 12 weeks' frozen stored cod muscle at −29°C, and SDS-PAGE analysis of the fractions

frozen muscle. SDS-PAGE also indicated that the last peak comprised sarcoplasmic proteins. Although the absorption at 254nm is not shown on these figures, a simultaneous scan at this wavelength was performed This showed a fair degree of contamination of this peak, probably with nucleotides.

With the salt extract of frozen muscle, the first peak appeared clearer, though a shoulder appeared on it which may comprise Act. The second main eluted area comprised one main peak and two shoulders which SDS-PAGE suggested were TM, A, and a TN-MLC mixture.

For the 12 weeks' frozen stored sample (*Fig 3*) there was no MHC band on SDS gel electrophoresis performed on the first peak from gel filtration, though there was a protein band on the gel top. This is a result of the protein forming aggregates which are too large to enter the gel during electrophoresis.

The appearance of three major peaks and two shoulders on the second peak occurred in the elution profiles of both the six and 12 weeks' frozen stored muscle.

Figures 4–6 show the elution patterns when gel filtration was carried out using the SDS-NaHCO$_3$ buffer. The SDS-soluble protein of fresh muscle is separated into five peaks. The fact that the elution profile of the salt extract of frozen muscle is also separated into five definite peaks indicates that the proteins which aggregate during frozen storage are dissociable in SDS. A peak suspected as Act by SDS-PAGE appeared as a discrete entity using the SDS-buffer system, but occurred only as the shoulder of the first peak using the NaCl-buffer system.

The second main peak (around the fraction number 45–50) is suspected to be A and TM. Under this SDS-buffer system, A was eluted first and after that the A and TM mixture was eluted.

4 Electron microscopy

Electron microscopy studies, though performed only on proteins eluted using the NACl-buffer system, were carried out to detect if any 'visual' differences existed between the fresh and frozen muscle proteins.

For the fresh sample the first peak (fraction 24–26) contained some small particles which were thought to be aggregated myosin heavy chains.

In fraction numbers 30–32 some long filaments were observed. In fraction 60, small particles were again observed, but this time they were thought to be aggregated filaments of A and TM (*Plate 1*).

In the case of six weeks' frozen muscle, there were small particles, thought to be MHC, in fraction 26. In fractions 50–60 aggregated proteins which consisted of short filaments were observed, and were thought to be A. Small particles were again observed in fraction 62 (*Plate 2*).

For the 12 weeks' frozen muscle, all the samples from fractions showed only small particles and their aggregates (*Plate 3*).

5 Discussion

Up to the present, many experiments have been attempted to determine the extent of freeze denaturation of fish muscle. Methods which measured the changes in the chemical or physical properties of proteins have been used for a long time. Techniques which observe the change in components or the structure of proteins are currently being developed. Very few attempts have been made to compare fresh and frozen muscle using gel filtration although Umemoto (1970) and Seki and Arai (1974) have studied these differences to a certain extent.

Umemoto used Sepharose 2B and separated a salt extract into three components (myosin, actomyosin, sarcoplasmic proteins). He showed that the AM peak, which was large for fresh muscle, decreased after freezing, which indicated that the amount of extractable AM decreased as a result of freezing.

Seki and Arai showed, using Sephadex G–200, that myofibrillar proteins separated into MHC, A, TM, TN and MLC in an SDS-buffer system although only MHC, A-TM mixture, TN and MLC could be separated using the normal NaCl buffer.

In this study, a gel bed of AcA-22 was used, and this allowed salt extracts of fresh and frozen muscle to be separated clearly into their components. The elution profiles were then compared to define the differences between the two raw materials. NaCl and SDS-buffer systems were used to compare their relative dissociative effects on the fresh and frozen muscle. The number of peaks from gel filtration using the SDS buffer system was greater than from the NaCl-buffer system, as a result of the stronger dissociating effect of SDS on inter and intra molecular bonding (Connell, 1965). Therefore, in an NaCl-buffer system it is possible that A, TM, TN and MLC are all in the form of aggregates stabilized by these types of cross-linkage.

In the NaCl-buffer system the elution sequence of proteins was MHC, Act, TM and A-TN-MLC mixture.

For the SDS-buffer system, however, it was MHC, Act, A-TM and TN-MLC mixtures. In this study, pure A was eluted in the leading edge of the peak, but the majority of the peak comprised an A-TM mixture.

The order of A and TM is reversed between the NaCl and SDS buffer. Two suggestions arise from this observation. One is that in the NaCl-buffer system, TM retains its native dimeric structure which is not dissociated by NaCl. In the SDS-buffer system, the bonds in the TM dimers are broken by SDS, resulting in TM monomeric sub-units being eluted later than A. Another suggestion is that TN and MLC are complexed with A in the NaCl-buffer system, possibly because of intermolecular cross-linkages. In the SDS-buffer system separation is effected (although a little overlap of the peaks occurs) by SDS breaking the cross-linkage which stabilizes the aggregates. To make sure of this point, experiments with other dissociating agents, urea, guanidinium HCl, *etc*, should be carried out.

From the EM observations, the first peak of gel filtration (= MHC) comprised small particles, and may be slightly contaminated by sarcoplasmic reticulum fragments. In the second peak (= A, TM, TN, MLC) of fresh muscle, long filaments were observed by electron microscopy but with frozen muscle the filaments were much shorter. Why filaments shorten in this way during frozen storage is not clear but could be a result of the inter- and intramolecular bonds which stabilize the native struc-

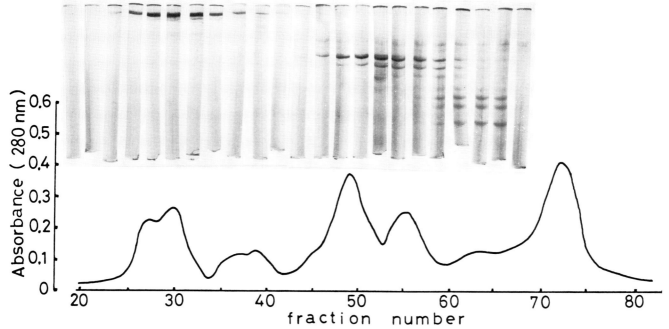

Fig 4　The gel filtration pattern of salt-soluble proteins in 1% SDS buffer from fresh cod muscle, and SDS-PAGE analysis of the fractions

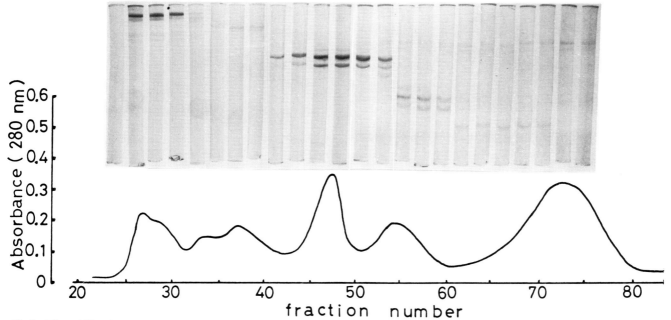

Fig 5　The gel filtration pattern of salt-soluble proteins in 1% SDS buffer from six weeks' frozen stored cod muscle at −29°C and SDS-PAGE analysis of the fractions

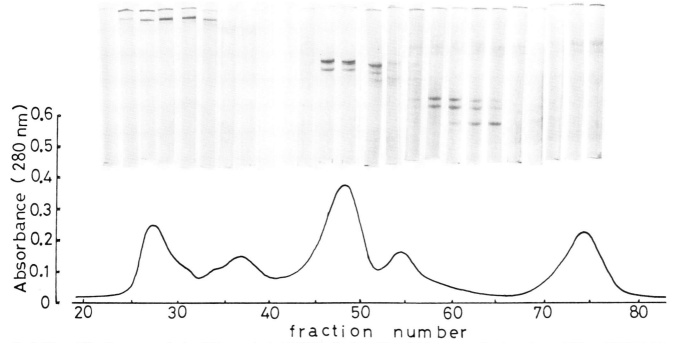

Fig 6　The gel filtration pattern of salt-soluble proteins in 1% SDS buffer from 12 weeks' frozen stored cod muscle at −29°C, and SDS-PAGE analysis of the fractions

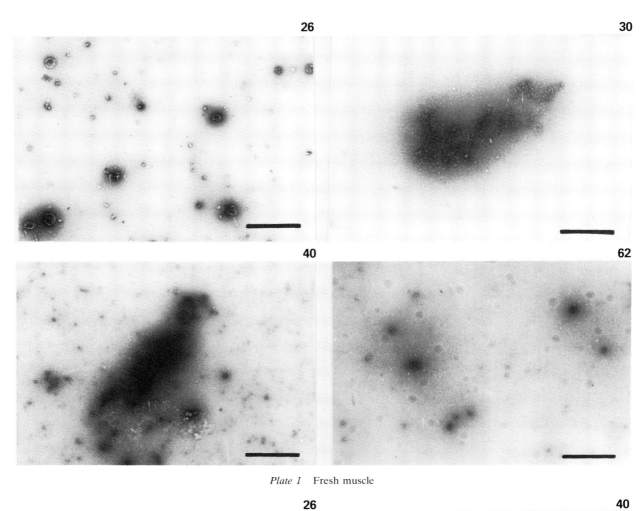

Plate 1 Fresh muscle

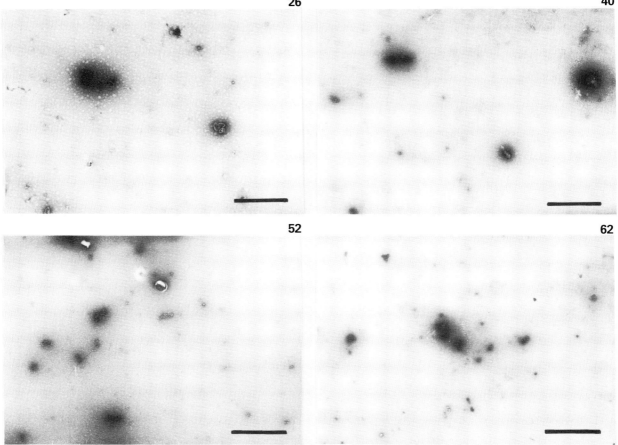

Plate 2 Six weeks' frozen stored muscle at −29°C

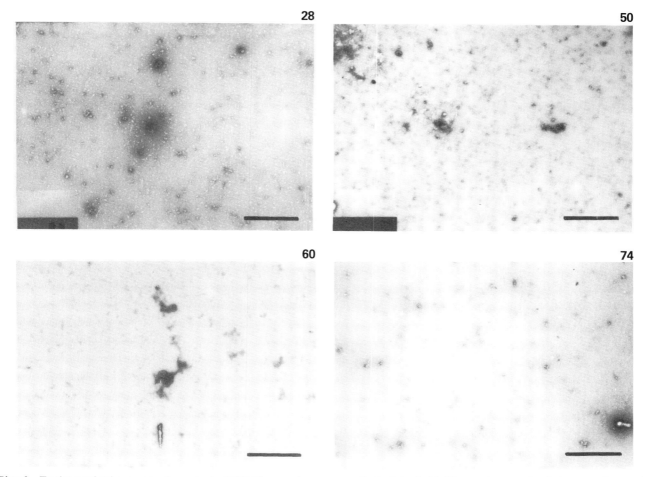

28　　　　　　　　　　　　　**50**

60　　　　　　　　　　　　　**74**

Plate 3　Twelve weeks' frozen stored muscle at −29°C. The samples were negatively stained with 3% uranylacetate. In all electron micrographs the bars represent 1 000nm

ture becoming weaker during frozen storage. Hydrogen, ionic, hydrophobic, disulphide and covalent bonds are all possible stabilizing influences in protein aggregates formed during frozen storage. Stabilization of native muscle proteins is generally by disulphide and other covalent bonds, but, to break covalent bonds, at least 80Kcal of energy are needed. If the assumption that filament shortening is caused by destabilization of these native covalent bonds is correct, it is unreasonable to expect extraction in an NaCl buffer to be capable of breaking these bonds. This suggests therefore that filament shortening is a direct result of the freezing process.

However, the most important thing arising from this study is further evidence that denaturation is caused by molecular aggregation. This supports the mechanism proposed by Buttkus (1971) and Johnson (1961) *ie* the formation of aggregates from monomers. It should be pointed out, however, that these authors worked with model myosin systems. Subsequently, many studies have also supported this, even when actomyosin and salt-soluble protein systems were used.

The loss of protein solubility is used as one of the many indicators of protein denaturation, and solubility loss is a well-known result of protein aggregation. When EM is used, shortening of filaments is also taken as a sign of protein denaturation. The 'denaturation', as shown up by solubility loss, will follow the mechanism of Johnson or Buttkus, after forming aggregates of shortened

filaments. When denatured proteins are studied by ultracentrifugal analysis, protein aggregates are visible (Connell, 1963). However, these proteins exhibit much shorter filament lengths than native AM, even though they have higher sedimentation coefficient values, which supports the view that denatured muscle proteins exist as aggregates of shortened filaments.

Therefore, as shown in this and a previous study (Ohnishi *et al*, 1978) by EM there is a need to determine whether denaturation which results in filament shortening is the result of some intramolecular bonding in the native filaments being broken by the freezing process. This would then allow the molecules to adopt conformations which are not only shorter than the native state, but also more amenable to aggregate formation.

It is worth pointing out that gel filtration may offer a means of preparing muscle proteins in purified form, albeit in small quantities, since methods are available for renaturing proteins from SDS solutions (Weber and Kuter, 1971).

6 Acknowledgement

The authors would like to thank Mr Brian Perry, the Institute of Marine Biochemistry, Aberdeen, for supplying us with carbon-coated formval meshes for electron microscopy.

427

7 References

ARAI, K and TAKASHI, R. Studies on muscular proteins of fish. XI.
1973   Effect of freezing on denaturation of actomyosin ATPase from carp muscle. *Bull. Jap. Soc. Sci. Fish.,* 39, 533–541

BUTTKUS, H. The sulphydryl content of rabbit and trout myosin in
1971   relation to protein stability. *Can. J. Biochem.,* 49, 97–107

CONNELL, J J. Aggregation of cod myosin during frozen storage.
1959   *Nature,* 183, 664–665

——— Changes in the actin of cod flesh during storage at −14°. *J. Sci. Fd.*
1960   *Agric.,* 11, 515–519

——— Sedimentation and aggregation of cod myosin. *Biochim. Biophys.*
1963   *Acta,* 74, 374–385

——— The use of sodium dodecyl sulphate in the study of protein
1965   interactions during the storage of cod flesh at −14°C. *J. Sci. Fd. Agric.,* 16, 769–783

DYER, W J. Protein denaturation in frozen and stored fish. *Food Res.,*
1951   16, 522–527

JOHNSON, P and ROWE, A J. The spontaneous transformation re-
1961   actions of myosin. *Biochim. Biophys. Acta,* 53, 343–360

OHNISHI, M, TSUCHIYA, T and MATSUMOTO, J J. Kinetic study on the

1978   denaturation mechanism of carp actomyosin during frozen storage. *Bull. Jap. Soc. Sci. Fish.,* 44, 27–37

SEAGRAM, H L. Analysis of the protein constituents of drip from
1958   thawed fish muscle. *Food Res.,* 23, 143–149

SEKI, N and ARAI, K. Gel filtration and electrophoresis of fish myo-
1974   fibrillar proteins in the presence of sodium dodecyl sulphate. *Bull. Jap. Soc. Sci. Fish.,* 40, 1187–1194

TANAKA, T. In: *The Technology of Fish Utilization,* Ed R Kreuzer,
1965   Fishing News (Books) Ltd, London

TSUCHIYA, T, TSUCHIYA, Y, NONOMURA, Y and MATSUMOTO, J J. Pre-
1975   vention of freezing denaturation of carp actomyosin by sodium glutamate. *J. Biochem.,* 77, 853–862

UMEMOTO, S. A modified method for estimation of fish muscle protein
1966   by biuret method. *Bull. Jap. Soc. Sci. Fish.,* 32, 427–435

UMEMOTO, S and KANNA, K. Studies on gel filtration on fish muscle
1970   protein. IV. Changes in elution patterns by gel filtration of extractable proteins of fish and rabbit muscle during cold storage at 20°C. *Bull. Jap. Soc. Sci. Fish.,* 36, 798–805

WEBER, K and KUTER, D J. Reversible denaturation of enzymes by
1971   sodium dodecyl sulphate. *J. Biol. Chem.,* 246(14), 4504

Studies of the changes in the proteins of cod-frame minces during frozen storage at −15°C

W M Laird, I M Mackie
and T Hattula

1 Introduction

In common with most other foods, when fish is held in the frozen state it undergoes deteriorative changes in both texture and flavour. The rates of deterioration are generally faster than for mammalian flesh but they vary markedly from one species to another. It is known, for example, that gadoids develop toughness more rapidly than flatfish (Sikorski *et al*, 1976) and that fatty species such as herring and mackerel produce rancid flavours more rapidly than white fish species (Ackman and Hardy, 1980).

Although a great deal of research has been carried out, there is as yet no completely satisfactory explanation for the development of toughness in fish flesh during frozen storage. It is known, however, that changes in the properties of the myofibrillar or contractile proteins are reflected in changes in texture and that as toughness develops the solubility of this group of proteins in strong solution progressively decreases. Indeed, this solubility in salt solution is used as an index of frozen storage deterioration. It has been postulated that this loss of solubility is due to cross-linking of the proteins through —S—S— bonds or through methylenic bridges formed between primary amino groups as in lysine and formaldehyde, which is known to accumulate in gadoid species. The information presently available, however, would suggest that non-covalent rather than covalent bonds are more likely to be formed (Connell *et al* 1978). The limited amount of evidence, based mostly on electrophoretic and ultracentrifugal analyses of sodium dodecylsulphate (SDS) extracts of proteins of unfrozen and frozen-stored muscle shows no change in the distribution of protein sub-units during frozen storage.

When recovered flesh from deboning machines is stored under similar conditions, the deteriorative changes in texture, particularly, are markedly accelerated due, it is believed, to variable contamination with parts of the kidney and other visceral organs. As formaldehyde and dimethylamine are produced in these tissues in relatively large amounts (Mackie and Thomson, 1974), the accelerated denaturation of the proteins has been attributed directly to the increased concentration of formaldehyde.

Recent work in our laboratory on the solution of proteins from such material with SDS has led us to re-examine the evidence for covalent bond formation, particularly in severely denatured material, as our preliminary results showed that the extent of solution and the distribution of the protein sub-units varied depending upon the conditions of extraction in 1% SDS.

This paper is concerned with the significance of these findings to present ideas on protein denaturation during frozen storage of fish.

2 Experimental

2.1 Materials

Cod (*Gadus morhua*) was purchased at Aberdeen fish market and filleted by hand at Torry Research Station. The resulting frames were put through a Baader 694 deboning machine and the recovered minced flesh collected in large beakers surrounded by ice. These frame minces were thoroughly mixed and aliquots were taken for analysis. The remainder, in approximately 100g portions, was sealed in polythene bags blast-frozen for 2–3h and stored at −15°C. Fillets for comparison were treated in the same manner.

2.2 Methods

2.2.1 *Sampling*   Frozen stored material was allowed to thaw and was thoroughly mixed by kneading the sealed polythene bag. The bags were then opened and samples taken for study.

2.2.2 *SDS — polyacrylamide gel electrophoresis (PAGE)*   SDS-PAGE was performed in 0·01M phosphate buffer containing 0·1% SDS in accordance with Weber and Osborn (1969) on a slab with a gel gradient of 2–16% (Pharmacia, Uppsala, Sweden). In some cases samples were subjected to electrophoresis without prior dialysis.

2.2.3 *Salt extraction of minces* Minces were either studied whole or after a salt extraction procedure similar to that of Groninger (1973). About 25g of mince were weighed out from the bag into a jar and enough (approx 200ml) 0·1M NaCl was added so that no air remained when the homogenizer shaft (fitted with a bung) was introduced. The jar contents were homogenized for 30s with an Ultra-Turrax Colloid Mill (Janke and Kunkel Kg., Staufen i. Br., West Germany) and the mixture spun for 20min at 13 000 × g on an MSE 18S centrifuge. The precipitate was returned to the jar which was topped up with 0·1M NaCl and the contents homogenized again for 30s before being spun down. This precipitate was treated likewise a third time and all three supernatants were combined and referred to as Wash. The resulting precipitate was extracted thrice in the same way with 0·6M NaCl. These supernatants were combined and referred to as Extract. The final precipitate was referred to as Residue.

2.2.4 *Extraction with SDS-containing buffers* Four extractants in 0·01M sodium phosphate buffer (pH 7) were used namely:
(A) 1% SDS
(B) 1% SDS and 1% 2-mercaptoethanol
(C) 1% SDS and 8M urea
(D) 1% SDS, 8M urea and 1% 2-mercaptoethanol

The minces (1·0g) were either homogenized for up to 30s by an Ultra-Turrax Colloid Mill or heated in a boiling water bath for 5–10min in 15ml of the SDS-containing solvents and allowed to stand at room temperature overnight. They were then centrifuged at approximately 10 000 × g for 15min and the supernatant solutions removed for immediate analysis. In the earlier studies, the material stood in solvent D for up to four weeks before the supernatant was subjected to electrophoresis.

2.2.5 *Protein content* Protein content was determined by Biuret reagent using Benedict's solution according to Bailey (1967). Crystalline egg albumin (Sigma Chemical Company, Grade V No. A–5503) was used as a standard.

2.2.6 *Solubility in 1% SDS-containing solvents* After centrifugation at 10 000 × g the residues from the above extractions were twice suspended in water and centrifuged. They were then boiled in aqueous ethanol, washed thoroughly with ethanol and air dried until the bulk of the solvent had evaporated. This air-dried matter was then dried further *in vacuo* overnight and weighed. Total dry matter was determined on 5·0g mince which was dried in alcohol as described above.

2.2.7 *Ultracentrifugal analyses* Sedimentation velocity runs were made in a Spinco model E ultracentrifuge equipped with Schlieren optics. The runs were carried out at 59 780rpm at ambient temperature on extracts obtained by homogenizing 25g of mince in 250ml of the SDS-containing solvent. The extracts were diluted × 2 with a 0·1% SDS solution prior to analysis.

3 Results and discussion

A direct comparison of the values for the extractability of the proteins of frame minces during storage unfrozen at 0°C and frozen stored at −15°C (*Fig 1*) shows that the myofibrillar fraction (the Extract) is largely inextractable after only four days. This period of storage compares with over 30 weeks for whole fish (Cowie and Mackie, 1968) and seven weeks for minced fillets (Thomson and Mackie, unpublished).

There is as yet no satisfactory explanation for either the toughening phenomenon of the flesh of frozen-

DISTRIBUTION OF PROTEIN IN FRACTIONS FROM COD FRAME MINCES.

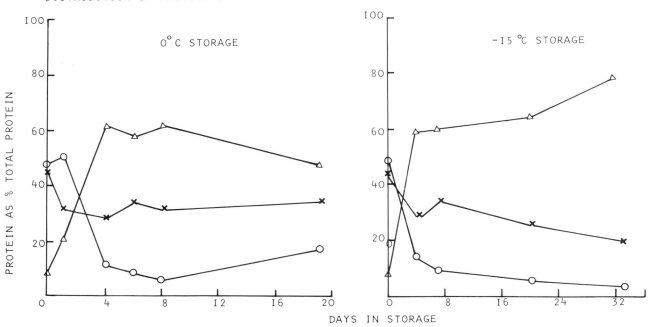

Fig 1 Changes in extractable proteins of frame mince during iced and frozen storage at −15°C: x 0·1M NaCl wash. ○ 0·6M NaCl extract △ Salt-insoluble residue

stored fish or the accelerated changes in comminuted flesh, particularly that contaminated with tissue of visceral organs. In the latter case, the greater possibility of localized heating during extrusion from the deboning machines and the greater opportunity for enzymes to react with substrates must be important contributory factors.

Although it has been generally found that proteins of the flesh of fish, as in frozen-stored fillets or whole fish, are completely soluble in 1% SDS, there is some evidence in our experience for incomplete solution after long-term cold storage of whole fish (Connell, 1975). A preliminary examination of frame mince showed that incomplete solution was obtained under conditions normally used for solution of proteins such as heating at 100°C for 10min or homogenization with an Ultra-Turrax Colloid Mill followed by storage overnight at room temperature. A range of 1% SDS-containing solvents (Table I) was used and after storage overnight at room temperature any undissolved material was separated by centrifugation, washed, dried and weighed. In all the SDS-containing solvents including (C) and (D), both of which contained 8M urea, the long-term frozen-stored frame material was only partially dissolved. Indeed there was incomplete solution of the unfrozen frame material after the heat treatment but not after homogenization. Further experiments would be necessary before attempting to draw any firm conclusions from these figures but they do none the less indicate that a relatively large proportion of the proteins of frame material even in the unfrozen state are inextractable in SDS under those particular conditions.

It is of interest that Dingle et al (1977) stated that 78.4% of the total protein of the mince of red hake (Urophycis chuss) remained inextractable in 1% SDS after frozen storage for 65 days at −10°C. Connell's (1975) value for frozen whole fish stored for one year at −15°C indicated that less than 5% of the protein had become inextractable in 1% SDS. The results presented here are in line with the view that severe denaturation of proteins takes place in frame material soon after preparation and some even prior to frozen storage.

The SDS electrophoretic analyses of the various SDS extracts of unfrozen frame mince are summarized in Fig 2. For each of the four solvents used, viz: (A) 1% SDS; (B) 1% SDS-containing 1% 2-mercaptoethanol (ME); (C) 1% SDS, 8M urea; (D) 1% SDS, 8M urea containing 1% ME, the distribution of the protein subunits after heating at 100°C for 10min and after homogenization with an Ultra-Turrax homogenizer respectively are compared, the extracts being examined by electrophoresis after storage overnight. The main conclusions are that the presence of 2-mercaptoethanol

made no difference to the distribution of the various peptides and that heating at 100°C for 10min produces substantially more of the higher molecular weight subunits particularly in extracts (A) and (B) which do not contain urea. The presence of 8M urea in itself increased the solution of these sub-units at room temperature to make them virtually indistinguishable from the heated extracts. As the extracts were applied directly to the polyacrylamide gel without prior dialysis against the SDS-containing electrolyte used for the electrophoresis the apparent distorted mobility of the 200 000 sub-units in solvent (C) could be due simply to a difference in ionic strength. Further investigations have not been made however.

A similar picture emerges with regard to the effectiveness of these four solvents when the corresponding extracts of frozen-stored frame material are examined (Fig 3a and 3b). After 35 weeks at −15°C the 200 000 molecular weight sub-unit of myosin is still extractable although, on comparing the patterns for the 18- and 35-week storage periods, respectively, with that for the unfrozen mince (Fig 2), there appears to be an overall decrease in this and the other larger sub-units. These changes are more easily detected in the heated extract using solvents (A) or (B); the strong background staining of the patterns of the Ultra-Turrax extracts essentially rules out any attempt to compare the relative distribution of the sub-units. More consistent agreement among the patterns appear to be obtained however for the heated extracts, regardless of the nature of the SDS solvent used.

The most striking difference between the patterns for flesh and frame material show up in the SDS solvents which have not been subjected to heating during the extraction. In the main, many of the higher molecular weight sub-units are present in relatively low concentrations in all the unheated extracts from frame material whereas comparable extracts from flesh show a preponderance of the higher molecular weight sub-units (Connell et al, 1978). As there is relatively less myofibrillar protein in frame tissue this distribution would be expected, but the proteolytic enzymes known to be present may also lead to significant breakdown during mincing and storage prior to freezing. As heated extracts, however, show relatively more of these higher molecular weight sub-units in the frame material other factors must also be considered. This difference between fillet and frame mince is confirmed by ultracentrifugal analysis (Fig 4) which compares the distribution of the proteins in sedimentation velocity runs. There is no obvious difference between the pictures for unfrozen and frozen-stored fillet mince or between unfrozen and frozen-stored frame mince, which is in agreement with

Table I

SOLUBILITY OF PROTEIN OF FRAME MINCE IN SDS-CONTAINING SOLVENTS. INSOLUBLE PROTEIN IS EXPRESSED AS A PERCENTAGE OF TOTAL PROTEIN

| | Heated at 100°C for 10min | | | | | Homogenized for 30s at room temperature | | | | | |
| | | Weeks in storage at −15°C | | | | | Weeks in storage at −15°C | | | |
Solvent	Unfrozen material	16	18	35	116	Unfrozen material	16	18	35	116
(A) 1% SDS	16	24	24	28	42	42	0	0	0	39	50	27
(B) 1% SDS and 1% ME	26	12	12	2	14	29	0	0	0	45	50	30
(C) 1% SDS, 8M urea	1	13	13	34	58	40	0	0	14	16	36	51
(D) 1% SDS, 8M urea, 1% ME	0	1	1	24	43	38	0	21	20	2	17	45

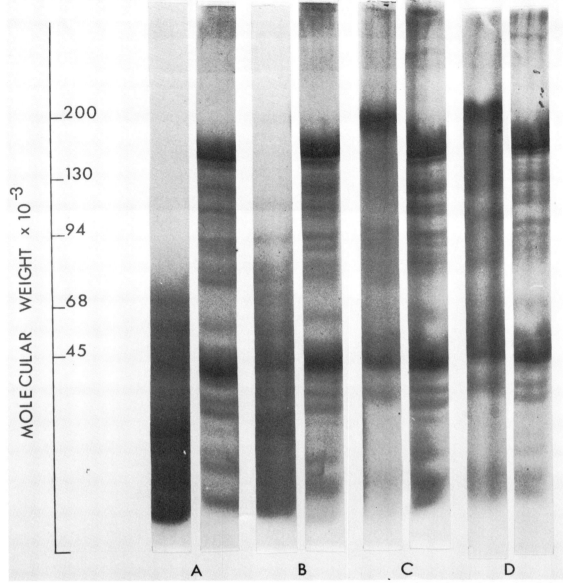

MOLECULAR WEIGHT × 10⁻³

Fig 2 SDS-PAGE (on gradient gels) of soluble fractions obtained by heating or triturating unfrozen frame mince in four different solvent systems. The heated sample is on the right hand side of each pair

previous observations (Connell, 1965). Fillet mince differs from frame mince, however, in having some active myosin which has not been broken down to its sub-units.

Further evidence for change in the electrophoretic patterns of frame tissue is given in _Fig 5_ which compares extracts in solvent (C) of the salt-inextractable residues of both fillet and frame mince (_Table I_). For frame material there appears to be a marked decrease in the myosin sub-unit during frozen storage. The difference for fillet mince is not so marked however but there is possibly some evidence for higher molecular weight units of 500 000 and upwards in the frozen material.

No firm conclusions can be drawn from these experiments which are presented to illustrate the problems of interpreting the results of sodium dodecylsulphate electrophoretic analyses when different SDS-containing solvents and different conditions of solution have been employed. For the frozen-stored frame material there is some evidence for covalent bond formation but whether it is —S—S— type or other has not been established. Further comparative investigations are planned in which particular attention will be paid to the nature of the minced flesh and the conditions of extraction of the protein for SDS electrophoresis.

4 References

ACKMAN, R G and HARDY, R. Lipids. _This volume_
1980
BAILEY, J L. _Techniques in Protein Chemistry_. Elsevier Publishing Co
1967   Ltd, Amsterdam. 2nd edn
CONNELL, J J. The use of sodium dodecyl sulphate in the study of
1965   protein interactions during the storage of cod flesh at −14°. _J. Sci. Fd. Agric._, 16, 769–783
—— The role of formaldehyde as a protein crosslinking agent during
1975   the frozen storage of cod. _J. Sci. Fd. Agric._, 26, 1925–1929
CONNELL, J J, LAIRD, W M, MACKIE, I M and RITCHIE, A H. Changes
1978   in proteins during frozen storage of cod as detected by SDS-electrophoresis. _Proc:_ V Int. Congress Food Sci. Technol.
COWIE, W P and MACKIE, I M. Examination of the protein extractabil-
1968   ity method for determining cold-storage protein denaturation in cod. _J. Sci. Fd. Agric._, 19, 696–700
DINGLE, J R, KEITH, R A and LALL, B. Protein instability in frozen
1977   storage induced in minced muscle of flatfishes by mixture with muscle of red hake. _J. Can. Inst. Fd. Sci. and Technol._ 10 (3), 143–146
GRONINGER, H S. Preparation and properties of succinylated fish
1973   myofibrillar protein. _J. Agr. Fd. Chem._, 21(6), 978–981
MACKIE, I M and THOMSON, B W. Decomposition of trimethylamine
1974   oxide during iced and frozen storage of whole and comminuted tissue. _Proc:_ IV Int. Congress Food Sci. Technol. Vol. 1, 243–250
SIKORSKI, Z, OLLEY, J and KOSTUCH, S. Protein changes in frozen fish
1976   CRC Crit. Rev. _Food Sci. and Nutr._, 8(1), 97–129
WEBER, K and OSBORN, M. The reliability of molecular weight deter-
1969   minations by dodecyl sulfate-polyacrylamide gel electrophoresis. _J. Biol. Chem._, 244(16), 4406–4412

431

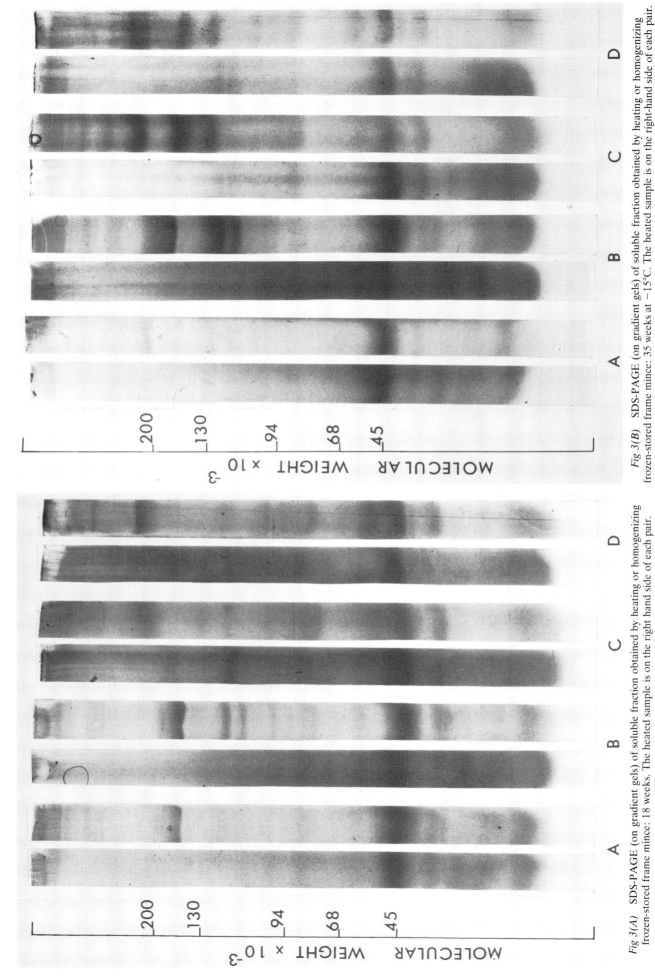

Fig 3(A) SDS-PAGE (on gradient gels) of soluble fraction obtained by heating or homogenizing frozen-stored frame mince: 18 weeks. The heated sample is on the right hand side of each pair.

Fig 3(B) SDS-PAGE (on gradient gels) of soluble fraction obtained by heating or homogenizing frozen-stored frame mince: 35 weeks at −15°C. The heated sample is on the right-hand side of each pair.

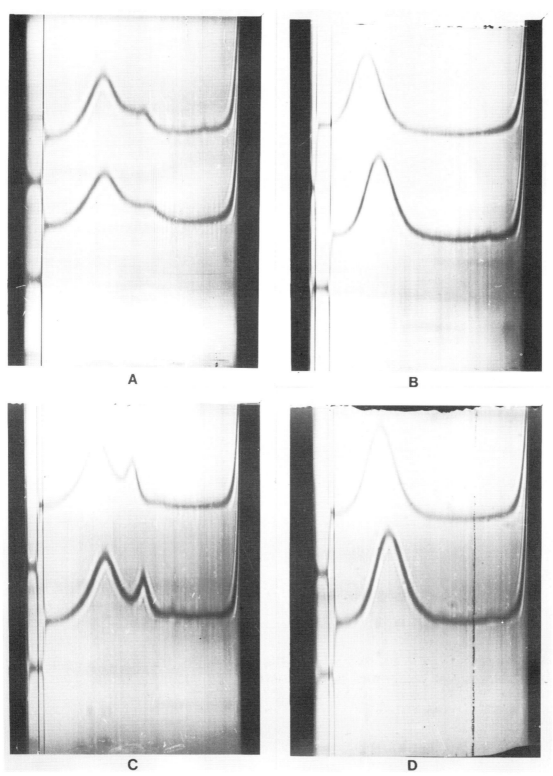

Fig 4 Ultracentrifugal examination at ambient temperature of protein extracts in 1% SDS solvents with and without 2-mercaptoethanol: (A) fillet unfrozen, (B) frame unfrozen, (C) fillet frozen stored eight weeks, (D) frame frozen stored five days. All photographs taken 96min after reaching full speed of 59 780rpm. (*Note:* upper traces are samples with 2-mercaptoethanol)

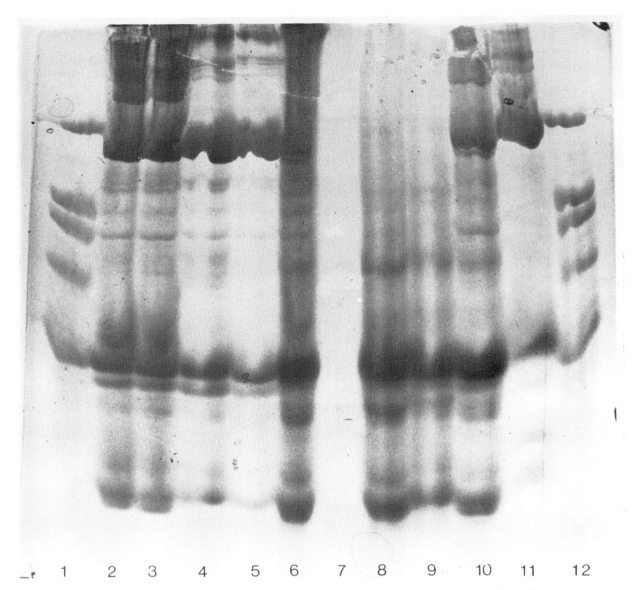

Figure 5 SDS-PAGE (on gradient gels) of extracts [in solvent (C)] from salt-inextractable residues

4 and 5:   unfrozen fillet
2 and 3:   frozen-stored fillet (19 weeks)
6 and 10: frozen stored frame (two weeks)
8 and 9:   frozen-stored frame (19 weeks)
11:        0·6M extract of unfrozen fillet for comparison
1 and 12: standard proteins (M wts = 200, 130, 94, 68 and 45 × 10³)

The nature of the cross-bridges constituting aggregates of frozen stored carp myosin and actomyosin

Y Tsuchiya, T Tsuchiya and J J Matsumoto

1 Introduction

1.1 *Frozen storage deterioration of fish muscle proteins*

The prevention of the deterioration of the eating and processing qualities of fish flesh during frozen storage is an important objective yet to be realized. Since the pioneering works of Reay *et al* (1937) and Dyer (1951), the deterioration has been ascribed to the denaturation of muscle proteins, especially the myofibrillar proteins.

In the early investigations, 'denaturation' of the muscle proteins was first defined on the basis of a loss of solubility. Connell (1959) proposed that the molecules of cod myosin form aggregates side-to-side during frozen storage. This view is now widely accepted.

1.2 *Cross-bridges in protein aggregates*

When protein molecules form aggregates, there must be some form of cross-bridging which keeps them together. Elucidation of the chemical nature of the cross-bridges is of importance in clarifying the mechanism of the denaturation and in controlling the mode of denaturation occurring during the storage and the processing of fish meat.

In his study on the frozen storage denaturation of cod myosin, Connell (1959) suggested that the cross-bridges are attributable to bonds other than disulphide bonds. Later, Connell (1965) was successful in dissolving nearly all the proteins of frozen-stored cod muscle with 1% sodium dodecylsulphate (SDS) and suggested that there were possible contributions from hydrogen bonds, non-polar bonds and others.

In his studies on the frozen storage denaturation of myosins of trout and rabbit, Buttkus (1970; 1971; 1974) was able to observe the side-to-side aggregates of myosin molecules under electron microscope. His interests were focused on the cross-bridges by disulphide bonds, though he was unable to detect any significant changes in the number of free SH groups before and after the frozen storage. The aggregated myosin formed on frozen storage was solubilized by several agents namely, 6M guanidine HCl containing 0·5M β-mercaptoethanol (ME), 0·3M sulphite, cyanide or Na-borohydride, and the cross-bridges were attributed to disulphide bonds, hydrophobic bonds and, possibly, hydrogen bonds.

None of the workers has definitely claimed that covalent bonds form the cross-bridges in the aggregates of myofibrillar proteins during the frozen storage.

During the frozen storage of cod muscle, no significant changes were found in the number of free SH-groups of actomyosin (Connell, 1960) and in the number of the titratable acidic and basic groups of myofibrils (Connell et al, 1964).

The following observations must be noted with respect to the aggregation of muscle proteins: (1) By the optical microscopy, fusion of the cod myofibrils was found to occur by freezing at liquid air temperature (Love, 1967). (2) Electron microscopy of actomyosins of carp (Tsuchiya et al, 1975; Oguni, 1975) and of cod (Jarenbäck et al, 1975) illustrated the aggregation of the actomyosin filaments after frozen storage.

The interaction of lipids with the proteins was claimed as a possible source of denaturation on frozen storage (King et al, 1962; Anderson et al, 1965; King, 1966). This was proved electron microscopically (Jarenbäck et al, 1975a). Nevertheless, Connell (1968) and Matsumoto (1979) refused to accept this interaction as the essential factor if, indeed, there was one in causing denaturation during frozen storage.

Another possibility is the reaction with formaldehyde (HCHO) which develops in frozen-stored fish muscle (Castell, 1971; Childs, 1973). However, Connell (1975) demonstrated that no methylene cross-bridge formation due to HCHO took place.

1.3 Programme of the present study

In the present study, differential solubilization of the carp myosin or actomyosin preparations has been determined following frozen storage at −20°C. The differential solubilization was conducted using various agents which differ from each other with respect to their ability to cleave intermolecular cross-bridges (Table I). The function of each agent was defined hypothetically, but it is based on accumulated knowledge from books and papers (Kauzmann, 1959; Lapanje, 1978). The

concentration of urea for hydrogen bond cleavage, 1·5M, has been adopted rather hypothetically because of the lack of information on the precise concentration required. Potassium hydroxide (KOH) was also used, though its function except for ionic bond cleaving is not clear.

Table I
HYPOTHETICAL CROSS-BRIDGE CLEAVING PROPERTIES OF VARIOUS SOLUBILIZING AGENTS

Water	Non-specific association forces
0·6M KCl	Ionic (electrostatic) bonds
1·5M urea	Hydrogen bonds
8M urea	Hydrogen bonds and non-polar (hydrophobic) bonds
0·5M β-mercapto-ethanol	S—S bonds
0·2M KOH	Ionic (electrostatic) bonds and other unspecified bonds involving (?) cleavage into sub-units and partial hydrolysis

2 Materials and methods

2.1 Carp
The dorsal lateral muscle of live carp, *Cyprinus carpio*, was employed.

2.2 Myosin
Myosin of carp was prepared as reported previously (Tsuchiya et al, 1975a). Two-thirds of the final myosin solution in 0·6M KCl was brought to 0·05M KCl by dialysis and the resultant myosin precipitate was suspended in 0·05M KCl. Both the suspension as well as the solution in 0·6M KCl were employed for the frozen storage experiments.

2.3 Actomyosin
Actomyosin of carp was prepared as reported previously (Tsuchiya et al, 1975). A part of the final solution was brought to suspension in 0·05M KCl and the suspension was used in the frozen storage experiments in parallel with the solution in 0·6M KCl. In this paper, the word actomyosin refers to the natural actomyosin which contains tropomyosin, troponin and other myofibrillar proteins.

Both the myosin and actomyosin preparations used here were homogeneous by ultracentrifugal analysis and showed, on SDS polyacrylamide gel electrophoresis, patterns typical for each protein.

2.4 Frozen storage
Each 10ml of the freshly prepared suspension or solution of myosin or actomyosin was divided in tubes, stoppered and frozen stored at −20°C. The concentration of protein in the system was adjusted to 10mg/ml. At intervals during frozen storage, the tubes were taken out in turn and thawed at 4°C.

2.5 Solubilization
The thawed solutions or homogenized suspensions were mixed with the solutions of solubilizing agents, the concentrations and volumes of which were calculated to give the desired levels of solubilizing agents in the final solutions and at the same time equal concentrations of protein under the different solubilizing conditions. The final concentrations of the solubilizing agents are illustrated in *Table II*. The mixtures of the proteins and the sol-

435

ubilizing agents were rehomogenized, held for 2h at 4°C, and centrifuged for 30min at 3 000rpm. Finally, the concentrations of the proteins in the supernatants were determined and the solubility expressed relative to the amount of the protein contained in the original unfrozen suspension or solution.

Table II
SOLUBILIZING CONDITIONS FOR FROZEN STORED MUSCLE PROTEINS

| Protein | Group | Storage condition (M KCl) | Water | Solubilizing agent | | | |
				KCl (M)	Urea (M)	ME[1] (M)	KOH (M)
Myosin or Actomyosin	A–a	0·05	added[2]	—	—	—	—
	–b	0·05	added	—	1·5	—	—
	–c	0·05	added	—	8	—	—
	–d	0·05	added	—	—	0·5	—
	–e	0·05	added	—	1·5	0·5	—
	–f	0·05	added	—	8	0·5	—
	–g	0·05	added	—	—	—	0·2
Myosin or Actomyosin	B–a	0·05	—	0·6	—	—	—
	–b	0·05	—	0·6	1·5	—	—
	–c	0·05	—	0·6	8	—	—
	–d	0·05	—	0·6	—	0·5	—
	–e	0·05	—	0·6	1·5	0·5	—
	–f	0·05	—	0·6	8	0·5	—
	–g	0·05	—	0·6	—	—	0·2
Myosin or Actomyosin	C–a	0·6	—	0·6	—	—	—
	–b	0·6	—	0·6	1·5	—	—
	–c	0·6	—	0·6	8	—	—
	–d	0·6	—	0·6	—	0·5	—
	–e	0·6	—	0·6	1·5	0·5	—
	–f	0·6	—	0·6	8	0·5	—
	–g	0·6	—	0·6	—	—	0·2

[1] β-Mercaptoethanol
[2] Concentration of KCl, less than 0·0025M

2.6 Determination of proteins

From the absorbance at 275nm, the protein concentration was calculated. Since the oxidized form of ME interfered with the absorbance at 275nm, only freshly prepared solutions were used and due attention was given to carrying out the treatments and analyses as quickly as possible.

3 Results

3.1 Solubilization of frozen-stored myosin

3.1.1 *Structure of the experiments* The following experiments were carried out in parallel:

(*1*) A suspension of myosin in 0·05M KCl was frozen-stored at −20°C and, after thawing, it was solubilized with the solubilizing agents given in *Table II* Group A a–g.

(*2*) A suspension of myosin was frozen-stored as in (*1*) and solubilized with the same solubilizing agents but containing 0·6M KCl (*Table II* Group B a–g) as the common factor.

(*3*) A myosin solution was frozen-stored in 0·6M KCl and solubilized with the same solubilizing agents containing 0·6M KCl as in (*2*) (*Table II* Group C a–g). The results are illustrated in *Fig 1*.

3.1.2 *Analysis of the results* The solubility in 0·6M KCl decreased with increasing time of frozen storage (B–a and C–a). After eight weeks frozen storage, about 10% myosin remained soluble in B–a, while 16% did so in C–a. These curves correspond to the conventionally defined 'frozen-storage denaturation' of myosin. The decrease in solubility must reflect the increase in the number of the cross-bridges other than ionic bonds being formed during frozen storage.

Addition of 1·5M or 8M urea, or 0·5M ME to 0·6M KCl improved the solubility to some extent (B–b, –c and –d, and C–b, –c and –d). The same was true with the combined use of 0·6M KCl, 1·5M urea and 0·5M ME (B–e and C–e). These results show that cleavage of the cross-bridges formed by hydrogen bonds, non-polar bonds and/or disulphide bonds and additionally ionic bonds, has taken place to give improved solubilization of the aggregates formed on frozen storage.

Nearly complete solubilization was obtained with the combined use of 0·6M KCl, 8M urea and ME, where all four kinds of cross-bridges are cleaved.

In Group A, the effects of cleavage of the hydrogen bonds, non-polar bonds and/or disulphide bonds are revealed to some extent (A–a, –c and –d). When 8M urea and ME were combined, fairly good solubility (*c* 90%) was obtained even in the absence of KCl (A–f).

If the results of Group A are compared with those of Group B, the effects of ionic bond cleavage are obvious (A–a *versus* B–a, A–b *versus* B–b, *etc*). However, the differences in solubility within each pair of solubilizing

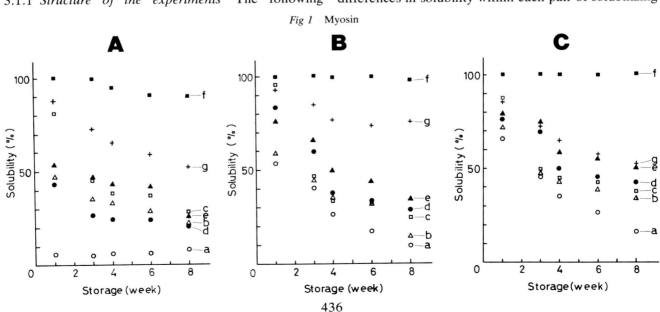

Fig 1 Myosin

conditions as shown in parenthesis above do not indicate a marked increase in ionic bond formation with the increasing duration of frozen storage. This suggests that most of the ionic cross-bridges had existed prior to freezing and storage and that they had been cleaved by solubilization with 0·6M KCl at any time before and during the frozen storage. Such solubilization corresponds to the usual treatments in extraction and determination of the native myosin.

The differences in the solubility values between those of Group A–f and Group B–f must reflect the increase in number, though not large, of the cross-bridges formed by ionic bonds during frozen storage.

The solubilities in 0·2M KOH turned out to be fairly good but much poorer than expected. Because of the uncertainty over the nature of the cross-bridges other than ionic bonds which are cleaved by KOH, it is not possible to discuss these results in detail. The ultracentrifugal data suggests, however, that myosin molecules have been broken down into their sub-units.

3.2 Solubilization of frozen-stored actomyosin
3.2.1 Structure of the experiments
The experiments with actomyosin were carried out in the same manner as described above for myosin and the results are illustrated in *Fig 2*.
3.2.2 Analysis of the results
In general, the solubilities in the respective solubilizing agents decreased during frozen storage. The curves of B–a and C–a correspond to the conventionally defined 'frozen-storage denaturation' of actomyosin. After eight weeks' frozen storage, about 30% actomyosin remained soluble in B–a, while the figure was 58% in C–a.

The effects of urea and ME on the solubilizing effect of KCl are obvious (B–a~e and C–a~e). Nearly complete solubilization was obtained but only by cleaving all the cross-bridges involving ionic, hydrogen, non-polar and disulphide bonds, respectively (B–f and C–f).

Comparison of the data of Group A with those of other groups leads one to believe, as was stated previously for myosin, that ionic bonds are of great significance. Solubilizing agents other than KCl produced similar effects as with myosin (A–a~d *versus* B–a~d).

Cleavage of the cross-bridges other than ionic bonds

by the combined use of 8M urea and ME led to 88% solubilization (A–f). The differences in the values between Group A–f and Group B–f reflect an increase in the formation of cross-bridges due to ionic bonds during the frozen storage.

4 Discussion
4.1 Comparison of myosin and actomyosin
When the data for the same solubilizing conditions are compared (*Figs 1* and *2*), it is clear that the rate of the decrease in solubility during frozen storage is quicker for myosin than for actomyosin. The storage in 0·05M KCl was more favourable for aggregation than storage in 0·6M KCl for both myosin and actomyosin (B's *versus* C's in *Figs 1* and *2*). The same is true with regard to the solubilization in the absence of KCl (*Fig 1A versus Fig 2A*).

4.2 The meaning of solubilization
In the present study, the concept of 'solubilization' and, thus, of 'denaturation' have been defined as the amount, whether great or small, of the proteins remaining in the supernatant after centrifugation at 3000rpm for 30min. As has been revealed by a microscopic study (Oguni *et al*, 1975), the amount of the proteins in the supernatant varies depending upon the centrifuging conditions. It is usual to find aggregated proteins in the supernatants obtained at moderate centrifugal speeds.

Therefore, the present results do not necessarily mean that the 'solubilized' proteins have been dispersed into a state in which all the protein molecules are present in their monomeric forms in the solution. What can be deduced from the present data is the extent of cleavage of the cross-bridges in the aggregates by each solubilizing agent, and hence the significance of cross-bridges in forming the aggregates.

5 Conclusion

The present study leads to the following conclusion:

When myosin and actomyosin are aggregated and insolubilized during frozen storage, the number of the cross-bridges between the protein molecules increases.

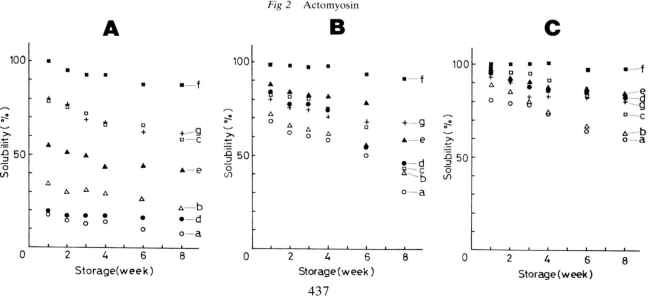

Fig 2 Actomyosin

In this cross-bridge formation, ionic (electrostatic) bonds, hydrogen bonds, non-polar (hydrophobic) bonds and disulphide (S–S) bonds all take part.

6 References

ANDERSON, M L, STEINBERG, M A and KING, F J. Some physical
1965   effects of freezing fish muscle and their relation to protein-fatty acid interaction. In: *The Technology of Fish Utilization*, Ed R Kreuzer. Fishing News (Books) Ltd, London. 105–109

BUTTKUS, H. Accelerated denaturation of myosin in frozen solution. *J.*
1970   *Fd. Sci.*, 35, 558–562

——— The sulfhydryl content of rabbit and trout myosins in relation to
1971   protein stability. *Can. J. Biochem.*, 49, 97–107

——— On the nature of the chemical and physical bonds which contri-
1974   bute to some structural properties of protein foods: A hypothesis. *J. Fd. Sci.*, 39, 484–489

CASTELL, C H. Metal catalyzed lipid oxidation and changes of proteins
1971   in fish. *J. Am. Oil Chem. Soc.*, 48, 645–649

CHILDS, E A. Interaction of formaldehyde with fish muscle *in vitro*. *J.*
1973   *Fd. Sci.*, 38, 1009–1011

CONNELL, J J. Aggregation of cod myosin during frozen storage.
1959   *Nature*, 183, 664–665

——— Changes in the adenosinetriphosphatase activity and sulphydryl
1960   groups of cod flesh during frozen storage. *J. Sci. Fd. Agric.*, 11, 245–249

——— The use of sodium dodecyl sulphate in the study of protein
1965   interactions during the storage of cod flesh at −14°. *J. Sci. Fd. Agric.*, 16, 769–783

——— The effect of freezing and frozen storage on the proteins of fish
1968   muscle, In: *Low Temperature Preservation of Foodstuffs*, Eds J Hawthorn and E J Rolfe. Pergamon Press, Oxford. 333–358

——— The role of formaldehyde as a protein cross-linking agent acting
1975   during the frozen storage of cod. *J. Fd. Sci.*, 29, 717–722

CONNELL, J J and HOWGATE, P F. The hydrogen ion titration curves of
1964   native, heat coagulated and frozen-stored myofibrils of cod and beef. *J. Fd. Sci.*, 29, 717–722

DYER, W J. Protein denaturation in frozen and stored fish. *Food Res.*,
1951   16, 522–527

JARENBÄCK, L and LILJEMARK, A. Ultrastructural changes during fro-
1975   zen storage of cod (*Gadus morhua*). I. Structure of myofibrils as revealed by freeze etching preparation. *J. Fd. Technol.*, 10, 229–239

——— Ultrastructural changes during frozen storage of cod (*Gadus*
1975a   *morhua*). II. Effects of linoleic acid and linoleic acid hydro-peroxides on myofibrillar proteins, *J. Fd. Technol.*, 10, 437–452

KAUZMANN, W. Some factors in the interpretation of protein denat-
1959   uration. In: *Advances in Protein Chemistry*. Academic Press, New York. Vol. 14, 1–63

KING, F J. Ultracentrifugal analysis of changes in the composition of
1966   myofibrillar protein extracts obtained from fresh and frozen cod muscle. *J. Fd. Sci.*, 31, 649–663

KING, F J., ANDERSON, M L and STEINBERG, M A. Reaction of cod
1962   actomyosin with linoleic and linolenic acid. *J. Fd. Sci.*, 27, 363–366

LAPANGE, S. *Physicochemical Aspects of Protein Denaturation*. John
1978   Wiley and Sons, New York. 331pp

LOVE, R M. The effect of initial freezing temperature on the behaviour
1967   of cod muscle proteins during subsequent storage: Histological study of homogenate. *Bull. Japan. Soc. Sci. Fish.*, 33, 746–752

MATSUMOTO, J J. Denaturation of fish muscle proteins during frozen
1979   storage, In: *Behaviour of Proteins at Low Temperatures*, Ed O Fennema. Advances in Chemistry Series, American Chemical Society, Washington DC (In press)

OGUNI, M, KUBO, T and MATSUMOTO, J J. Studies on the denaturation
of fish muscle proteins. I. Physico-chemical and electron microscopical studies of freeze-denaturation of carp actomyosin. *Bull. Japan. Soc. Sci. Fish.*, 41, 1113–1123

REAY, G A and KUCHEL, C C. The proteins of fish. GB Dept. Sci. Ind.
1937   Research Rept. Food Invest. Board 1936. 93–96

TSUCHIYA, T, TSUCHIYA, Y, NONOMURA, Y and MATSUMOTO, J J. Pre-
1975   vention of freeze denaturation of carp actomyosin by sodium glutamate. *J. Biochem.*, 77, 853–862

TSUCHIYA, T and MATSUMOTO, J J. Isolation, purification and structure
1975a   of carp myosin, HMM and LMM. *Bull. Japan. Soc. Sci. Fish.*, 41, 1319–1326

Evidence for the formation of covalent cross-linked myosin in frozen-stored cod minces

*A D Matthews, G R Park
and E M Anderson*

1 Introduction

Cod muscle mince recovered by mechanical deboning of V-cut and frame materials deteriorates rapidly during cold storage, particularly at temperatures above −20°C. This deterioration involves the toughening and loss of water-binding capacity of the muscle and a decline in the protein salt solubility and fat emulsifying capacity. The many factors implicated in these changes have been extensively reviewed by Sikorski *et al* (1976) and although conditions of high ionic strength in the un-frozen water and the accumulation of chemically reac-tive substances (formaldehyde and free fatty acids) which may interact with the myofibrillar protein (es-pecially myosin) are believed responsible; there is also evidence of aggregation and inextractability of actin, actomyosin, tropomyosin and whole myofibrils. The increasing difficulty of disrupting fibrils as cold storage progresses also suggests the possibility that changes to the sarcoplasmic reticulum may be important.

As Connell (1965) has pointed out this gradual toughening and loss of water-holding capacity are characteristic of what would be expected if progressive cross-linking of the myofibrillar proteins was occurring, but no direct proof of any such interactions *in situ* during frozen storage has been obtained although added for-maldehyde (Castell *et al*, 1973), free fatty acid and free fatty acid hydroperoxide have been shown to reduce the

salt solubility of the myofibrillar proteins (Jarenbäck and Liljemark, 1975).

Connell has concluded from ultracentrifugal experi-ments (Connell, 1965) and solubility experiments on sodium dodecylsulphate(SDS)-solubilized cold-stored cod fillets (Connell, 1975) that 'the bulk of the protein in cod muscle stored until it is extremely tough is not exten-sively cross-linked'. SDS solubility is not in itself proof of the absence of covalent bonding, however, since small aggregates may be soluble in this detergent. In addition, reactions may well occur in minces which are absent in intact fillet. The exact nature of the changes occurring during the frozen storage of mince are still, therefore, largely unknown.

Sodium dodecylsulphate and β-mercaptoethanol (ME) solubilization of mince proteins followed by gel electrophoresis of the resultant solutions should provide information as to the presence of primary bonds. SDS solubilization/electrophoresis will show if covalent bonds are present and SDS/ME electrophoresis whether these bonds are disulphide or not. An absolute require-ment, however, is that both the fresh and frozen ma-terials are totally solubilized (*ie* no differential solubil-ization) and that the fresh materials can be disrupted into single polypeptide chains in both solvents. As cold storage of the minces progresses the disappearance of bands from the gels and the appearance of new bands will indicate the presence of covalent cross-linking.

The present research was primarily undertaken, therefore, to determine if any covalent bonding could be detected in fish mince proteins after frozen storage. Measurement of protein solubility and the increase in free fatty acid and formaldehyde content was also undertaken, however, in an attempt to characterize the various minces used and possibly elucidate reaction mechanisms.

2 Materials and methods

2.1 *Preparation of minces*

Cod were obtained from an inshore Aberdeen fishing boat. They were all known to have been taken in the same shot and had been stored on ice for five days. After filleting, the kidney and backbone were band-sawn from half the frames leaving 'trimmed frames'.

Muscle material was recovered from the fillet V-cuts, frames and trimmed frames using a Baader 694 fitted with a drum having 5mm holes and operated at near maximum belt tension. Part of each mince obtained was washed by stirring for 5min with four parts of water at 2–3°C and dewatered by filtration through a fine nylon mesh. The six prepared materials and a number of whole fillets were packed into 2cm deep plate freezer trays and quick-frozen at −40°C. The frozen samples were packed in polythene bags and stored in a cabinet controlled at −7° ±0·25°C.

2.2 *Solubilization of minces for electrophoresis*

Total solubilization of both the fresh and cold-stored mince proteins was necessary in order that any changes which were occurring to the proteins could be detected on the gels and not remain undetected due to non-solubilization of part of the sample. According to Connell (1965) cold-stored fillet cod proteins will dissolve easily in 1% SDS, but his technique of room temperature solubilization overnight was found unsuitable for mince samples (particularly frame mince) due to bacterial growth. The possibility of S—S bond formation or other interactions during this prolonged solubilization also militated against the use of this procedure.

Homogenizers of the ultra-Turrax variety were also considered unsuitable due to the entrainment of substantial amounts of air and frothing during the solubilization, with again the possibility of the formation of bonds during the solubilization step.

Solubilization in a short period of time without air entrainment or heating was desired and this was eventually achieved by the use of an ultrasonic disintegrator. In all cases 6·25g of V-cut and trimmed frames and 7·0g of frame mince were added to 100ml of either 4% SDS or 4% SDS + 1% ME. A Rapidis 180 ultrasonic disintegrator with a 9mm tip was inserted into the mixture and disintegration was carried out for 35min at constant power. Cooling of the sample was achieved with a constant temperature bath (15°C). The temperature of the sample after 35min disintegration was 20–24°C.

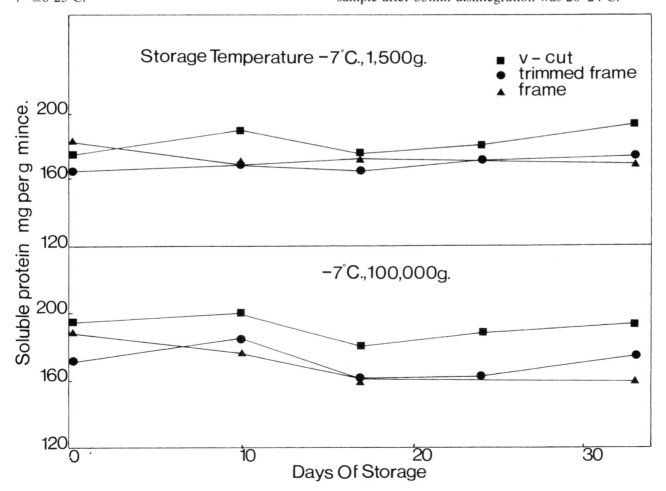

Fig 1   Sodium dodecyl sulphate (4%) solubilities of cold stored (−7°C) cod minces. Solutions centrifuged at 1 500g and 100 000g

Normally, samples for electrophoresis were spun at 1 500*g* for 10min to sediment insoluble collagen. In order to check that total solubilization of the protein was being achieved even after prolonged cold storage, protein solubilities were measured on 1 500*g* (10min) and 100 000*g* (20min) spun SDS solutions of −7°C cold-stored minces (*Fig 1*). Only small differences in the protein content of the 1 500*g* and 100 000*g* spun solutions were found and little if any decline in protein solubility occurred. It could be assumed, therefore, that the proteins were in true solution and were not present as miscellar aggregates.

2.3 *Gel electrophoresis*
The method employed for untreated blocks was similar to that described by Weber and Osborne (1969). 8·75% acrylamide gels were set in 5mm (internal diameter) × 75mm tubes. The gels and tank buffer contained 0·1% SDS. Tank buffer was 0·025M tris/HCl, pH 8·1 and gel buffer 0·375M tris/HCl, pH 8·3. 25% glycerol solution was added in the ratio 1 : 1 to increase the density of the protein solution and facilitate loading on to the gel. Generally 20μl of solution was applied. Electrophoresis was performed at 20°C at a constant current of 3mA/tube for 2h. The gels were stained overnight in a methanol : acetic acid : water solution (5 : 1 : 5) of 0·5% Naphthalene Black 2B and destained in 7% acetic acid.

The treated block samples were run on Pharmacia

polyacrylamide gradient slab gels PAA4/30 under conditions recommended in the Pharmacia Gel Electrophoresis System handbook.

2.4 *Free fatty acid*
Following chloroform/methanol extraction of the fat, free fatty acids were determined by the procedure of Olley and Lovern (1960).

2.5 *Dimethylamine determination (formaldehyde)*
The method of Dyer and Mounsey (1945) was used.

2.6 *Salt-soluble protein*
Five grams of sample and 195ml of 5% sodium chloride, 0·02M sodium bicarbonate solution (pH 7·6) were homogenized 30s at full speed with a Silverson sealed unit homogenizer. The homogenate was centrifuged at 38 000*g* for 30min at 3°C and protein in the supernatant determined by Biuret.

3 Protein-formaldehyde and protein-free fatty acid interactions during sample solubilization

The possibility of reactions between the myofibrillar protein and free fatty acids or formaldehyde (formed in the minces during cold storage) during solubilization, with the formation of cross-linked components not present in the original sample, required investigation.

To 6·25g samples of fresh V-cut mince were added

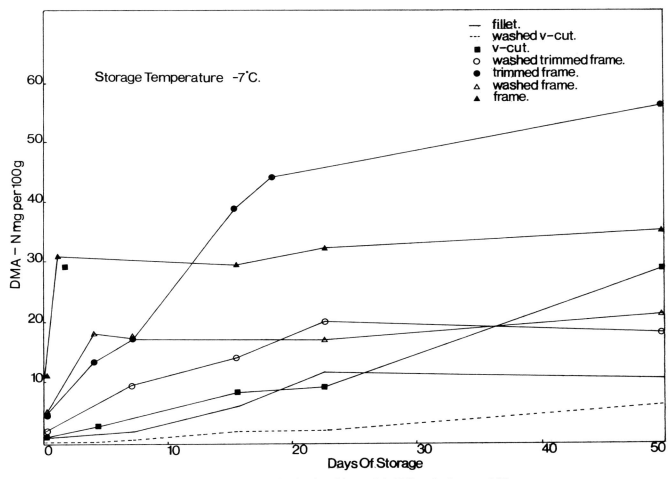

Fig 2   Development of dimethylamine in cold stored (−7°C) cod minces and fillets

440

100ml of 4% SDS or 100ml of 4% SDS + 1% ME. Individual mixtures were treated with one of the following:

(a) 5·4mg formaldehyde
(b) 54mg formaldehyde
(c) 25mg oleic acid
(d) 25mg of linolenic acid

5·4mg formaldehyde (0·86mg HCHO/g muscle) corresponding to the maximum amount detected in any of the minces (see *Fig 2*), 54mg formaldehyde, 10× the maximum amount and 25mg of the free fatty acids (4mg/g muscle) corresponding approximately to the maximum amount of free fatty acid detected. Sample disintegration and electrophoresis were performed as previously described.

The SDS and SDS/ME electrophoretograms are shown in *Plate 1*. No reactions between the proteins and formaldehyde or free fatty acids could be detected, the gels for the spiked samples being identical to the control. Extensive cross-linking could therefore not have occurred during the solubilization process.

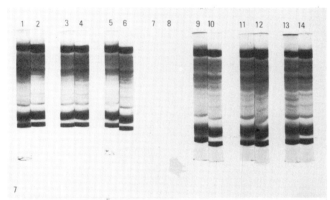

Plate 1 Polyacrylamide gel electrophoresis patterns of untreated, free fatty acid treated and formaldehyde treated V-cut cod mince. Samples 1, 3, 5, 9, *etc* are 4% SDS solubilized; samples 2, 4, 6, 8, *etc* are 4% SDS, 1% β-mercaptoethanol solubilized. This system of numbering applies throughout. Gels 1, 2, 9, 10 are untreated V-cut mince; gels 3, 4—25mg of added oleic acid; gels 5, 6—25mg of added linolenic acid; gels 11, 12—5·4mg of added formaldehyde; gels 13, 14—54mg of added formaldehyde

4 Results

4.1 *Untreated blocks*

Gels from a single storage experiment (with the occasional sample from a duplicate trial) are shown in *Plates 2* to *7*.

The initial effect for all samples is a decrease in the amount of myosin single polypeptide main chain on the SDS gels and the appearance of a band of material at the top of these gels as storage progresses. With β-mercaptoethanol in the solubilizing solution this band at the top of the gel is not present and the intensity of the myosin band is the same as at the beginning. As time of storage increases, however, a band also begins to appear on the top of the SDS/ME gels and the myosin band gradually decreases in intensity until finally very little, if any, myosin remains on the gel as the single 200 000 MW main chain. A major amount of protein is present

on the top of the gels and can be seen to be increasing as the myosin band decreases. The effect occurs most rapidly with unwashed materials and in the order frame mince, trimmed frame mince, V-cut mince. No other proteins appear to be affected. The conclusion to be drawn from these results is that covalent bonds are being formed during the cold storage of mince at −7°C and it is myosin which is involved. The initial step is one of disulphide bridging, followed at a later stage by the formation of other non-reducible covalent bonds until eventually myosin is totally cross-linked by these latter bonds.

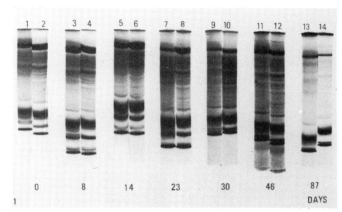

Plate 2 Polyacrylamide gel electrophoresis patterns of cold stored (−7°C) V-cut mince. Numbers below gels refer to number of days of storage

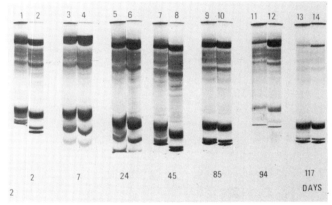

Plate 3 Polyacrylamide gel electrophoresis patterns of cold stored (−7°C) washed V-cut mince. 117-day sample from duplicate set of blocks

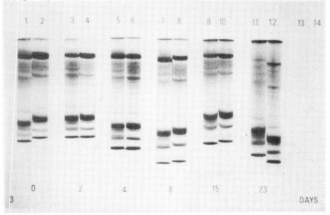

Plate 4 Polyacrylamide gel electrophoresis patterns of cold stored (−7°C) trimmed frame mince

441

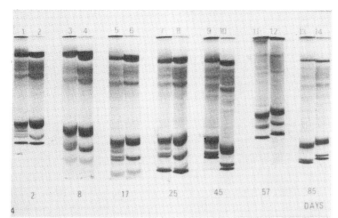

Plate 5 Polyacrylamide gel electrophoresis patterns of cold stored (−7°C) washed trimmed frame mince. 57-day sample from duplicate set of blocks

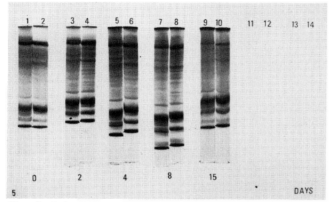

Plate 6 Polyacrylamide gel electrophoresis patterns of cold stored (−7°C) frame mince

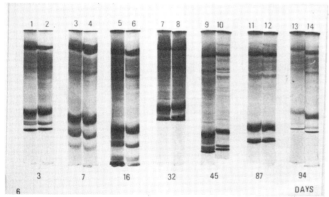

Plate 7 Polyacrylamide gel electrophoresis patterns of cold stored (−7°C) washed frame mince

4.2 *Sulphite and β-mercaptoethanol-treated blocks*

To gain confirmatory evidence for the involvement of disulphide links in the denaturation of cod minces, cod frame mince blocks were prepared with and without added sulphite (0.5% SO_2 as Na_2SO_3) or β-mercaptoethanol (1%) and cold stored at −7°C.

The level of dimethylamine in the blocks was determined after five days to check if the formation of formaldehyde was affected in any way by the presence of sulphite or β-mercaptoethanol. It can be seen that the formation of formaldehyde was not suppressed by these additives (*Table 1*).

Table 1
DIMETHYLAMINE LEVELS IN UNTREATED, SULPHITE AND β-MERCAPTOETHANOL TREATED COD FRAME MINCE BLOCKS AFTER FIVE DAYS' STORAGE AT −7°C

Sample	Level of dimethylamine mg DMA–N/100g
Frame mince	44
Frame mince + 0.5% SO_2	41
Frame mince + 1% β-mercaptoethanol	40.4

The gel electrophoresis patterns after 0, 11 and 18 days of storage are shown in *Plates 8, 9* and *10*. The 200 000 MW main chain myosin band of the untreated sample decreases rapidly until after 18 days only a very fine band can be seen on the gel. A substantial amount of protein remains in the slot at the origin. The samples treated with sulphite or β-mercaptoethanol remain unaffected.

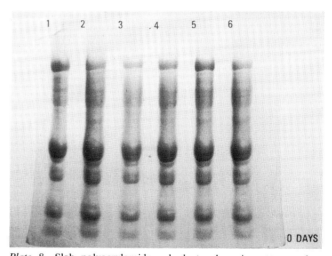

Plate 8 Slab polyacrylamide gel electrophoresis patterns of untreated, sulphite treated and β-mercaptoethanol treated frame minces. 0 day storage. Samples 1 and 2 are untreated; samples 3 and 4 0.5% sulphite treated; samples 5 and 6 1% β-mercaptoethanol treated

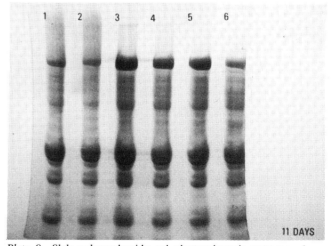

Plate 9 Slab polyacrylamide gel electrophoresis patterns of untreated, sulphite treated and β-mercaptoethanol treated frame minces. 11 days' storage −7°C. Sample numbering as for *Plate 8*

442

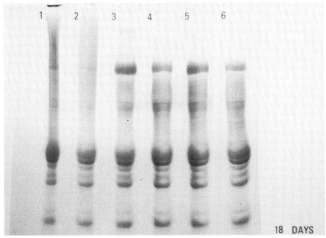

Plate 10 Slab polyacrylamide gel electrophoresis patterns of untreated, sulphite treated and β-mercaptoethanol treated frame mince. 18 days' storage −7°C. Sample numbering as for *Plate 8*. Distortion of pattern for sample 1 probably caused by large protein aggregates blocking gel at origin

It appears, therefore, that myosin is being cross-linked by disulphide bonds and other non-reducible covalent links but these reactions are being stopped by sulphite or mercaptoethanol. The non-S—S covalent links, therefore, will apparently not form if disulphide bridging is prevented. Sulphite of course may form an additional compound with formaldehyde and so stop formaldehyde-mediated bonding; this is not the case for mercaptoethanol, however, and yet the secondary covalent cross-links do not appear.

5 Origin of non-reducible covalent bonds

The increase of free fatty acid and formaldehyde (dimethylamine) in the untreated samples is shown in *Figs 2* and *3*. The formation of free fatty acid is rapid and almost reaches a peak value for all the samples within 10 days of storage. The formaldehyde content of frame mince rises very rapidly and trimmed frame mince also shows a rapid increase. In all other cases the rise is more gradual. These increases parallel in general (in storage time) the appearance of the non-reducible covalent bonds as detected by electrophoresis. The free fatty acid content on the other hand cannot be correlated with the appearance of non-reducible covalent bonds (*ie* for V-cut mince and trimmed frame mince the appearance of non-reducible covalent bonds does not take place until well beyond 20 days of storage) and although the free fatty acid may eventually take part in reactions with the proteins the evidence points to formaldehyde being the causative agent.

6 Discussion

The rapid fall in salt solubility of all samples apart from V-cut mince (*Fig 4*) indicates that covalent bonding will not completely explain myosin denaturation in cold-store minces and some other change must take place during the first days of storage. This change is possibly a deconformation process (as suggested also by Sikorski *et al*, 1976) which may allow and be necessary for disulphide bridging of myosin. The formation of further non-reducible covalent links apparently requires disul-

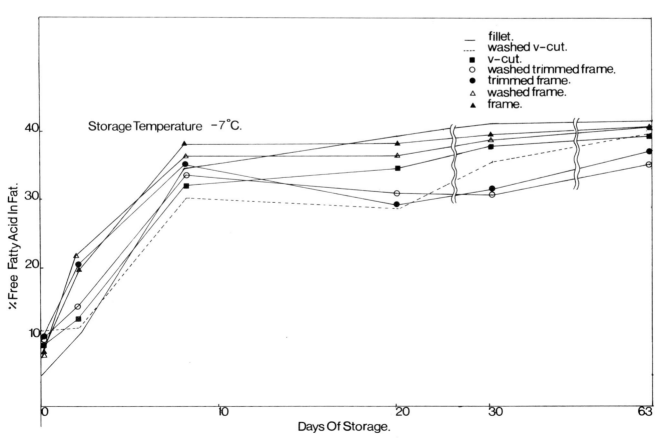

Fig 3 Development of free fatty acid in the fat of cold stored (−7°C) cod minces and fillets

443

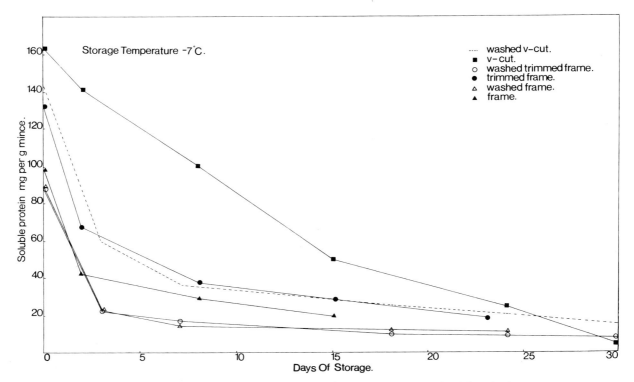

Fig 4 Effect of storage time on the protein solubility of cold stored (−7°C) cod minces

phide bridging to have occurred and circumstantial evidence suggests that these links are formaldehyde mediated.

Further work (ultracentrifuge and viscosity) is planned to verify these findings and to obtain an indication of the size of the aggregates being formed.

7 References

CONNELL, J J. The use of SDS in the study of protein interactions
1965 during the storage of cod flesh at −14°. *J. Sci. Fd. Agric.,* 16, 769–783

—— The role of formaldehyde as a protein cross-linking agent acting
1975 during the frozen storage of cod. *J. Sci. Fd. Agric.,* 26, 1925–1929

CASTELL, C H, SMITH, B and DYER, W J. Effects of formaldehyde on
1973 salt extractable proteins of gadoid muscle. *J. Fish Res. Bd. Can.,* 30, 1205–1213

DYER, W J and MOUNSEY, Y A. Amines in fish muscle. II. Develop-
1945 ment of trimethylamine and other amines.

JARENBÄCK, L and LILJEMARK, A. Ultrastructural changes during fro-
1975 zen storage of cod. III. Effects of linolenic acid and linolenic hydroperoxides on myofibrillar proteins. *J. Fd. Technol.,* 10, 437–452

OLLEY, J and LOVERN, J A. Phospholipid hydrolysis in cod flesh stored
1960 at various temperatures. *J. Sci. Fd. Agric.,* 11, 644–652

SIKORSKI, Z, OLLEY, J and KOSTUCH, S. Protein changes in frozen fish.
1976 Crit. Rev. in *Food Sci and Nutrit.,* 8(1), 97–129

WEBER, K and OSBORNE, M. The reliability of molecular weight de-
1969 terminations by SDS/polyacrylamide gel electrophoresis. *J. Biol. Chem.,* 244, 4406

A review of some recent applications of electrophoresis and iso-electric focusing in the identification of species of fish in fish and fish products

I Mackie

1 Introduction

It is the intention of this review to compare the various procedures that have been developed over the last few years to distinguish species of fish by electrophoretic analysis of extracts of the flesh proteins. It is reasonable to claim that the various electrophoretic techniques now available (Tsuyuki *et al*, 1966; Mackie, 1969; Morel, 1977) are the most reliable methods for identifying the species when the usual differences based on morphology are uncertain as, for example, in the closely related species or sub-species of hake *Merluccius* sp. (Mackie and Jones, 1978). In other situations, and these apply particularly to the commercially important species, it may be that only fillets or minces are available. In the interests of fair dealing, either in international trade or for the protection of the consumer, it is now widely accepted that reliable analytical methods are needed to check that specifications of fish and shellfish comply with labelling regulations. In the United Kingdom, under the Trades Description Act 1968, it is an offence to misrepresent any article offered for retail sale. More specifically, the permitted names for species of fish are set out in detail in two Statutory Instruments:

(*1*) Labelling of Food Regulations 1970 (Statutory Instrument 1970 No. 400 and amendments) which apply to England and Wales.

(*2*) Labelling of Food (Scotland) Regulations 1970 (Statutory Instrument 1970 No. 1127 and amendments).

Similar regulations apply in other countries.

It has been the case, however, that legislation has preceded the development of adequate methods of identifying species of fish. Until 1968 it was possible only to identify raw fish by electrophoresis. Papers demonstrating the application of this system to the identification of raw fish have been published by Cowie (1968), Tsuyuki (1966) and the author (Mackie, 1969). The water-soluble proteins which have a range of molecular weight from 20 000–60 000 are of an ideal size for electrophoretic separation by either starch or polyacrylamide gel electrophoresis. The electrophoretic patterns obtained were shown to be independent, by and large, of age, sex or physiological state of the fish and to be suitable for the unequivocal identification of an unknown sample of fish flesh (Tsuyuki, 1966; Mackie, 1969). Because of the loss of solubility of these proteins on heating, the method applied only to raw fish.

In 1969 the author extended the electrophoretic method to cooked fish (Mackie, 1969) by demonstrating that the protein fraction which dissolved in 6M urea could be separated into species-specific patterns and more recently Mackie and Taylor (1972) extended the method further to include canned fish which are known to be more severely heat-denatured than normally cooked fish. In the latter system, the muscle was allowed to react with cyanogen bromide, a reagent which splits proteins at the methionine residues (Gross, 1966). The protein fragments obtained could then be dissolved in 6M urea and separated into species-characteristic patterns by electrophoresis.

The purpose of this paper is to compare the suitability of zone electrophoresis on polyacrylamide gel with the newer technique of iso-electric focusing for the identification of raw, cooked and canned fish and fish products.

2 Experiments and methods

2.1 Samples of fish
The fish, apart from the commercial canned fish and salmon, used in this exercise were caught by the Station's research trawler and stored frozen at −30°C until required.

Raw samples of the following species of salmon were kindly made available to us by Dr Tsuyuki and Dr White of the Vancouver Laboratory of the Fisheries and Marine Service of Canada: chum (*Oncorhynchus keta*); coho (*O. kisutch*); pink (*O. gorbuscha*); sockeye (*O. nerka*); spring (*O. tschawytscha*). Altantic salmon (*Salmo salar*) was obtained locally. Authentic samples of herring (*Clupea harengus*), pilchard (*Sardina pilchardus*) and of bonito (*Sarda sarda*) together with the six species of salmon were canned on the research station.

Samples of cooked fish flesh (20–50g) were heated in a closed casserole for 30min on a steam bath.

2.2 Preparation of aqueous extracts
Raw fish (3–25g) was homogenized in an equal volume of water. After centrifugation of the homogenate at 3 000 × g for 20min the supernatant solution was stored at 0°C until required. Extracts for electrophoresis were diluted with an equal volume of 40% solution of sucrose but for iso-electric focusing they were used directly.

An alternative extraction procedure was to grind 50–150mg of the fish flesh in a Dryer tube (60 × 7mm)

in an equal volume of water or of 40% sucrose (depending on the type of analysis) with a stainless steel rod. After centrifugation in a bench-type centrifuge at 1000 × g for 30min, the supernatant solution was taken up on a 10μl Kirk-type ultra-micropipette and applied directly to the electrophoretic or iso-electric focusing systems.

2.3 Preparation of 6M urea extracts
The fish flesh (1·0g) was suspended in 6·0ml of 8M urea containing β-mercaptoethanol (0·02M). After standing for at least 24h at room temperature the suspension was centrifuged at 80 000 × g and the supernatant solution obtained was used directly for electrophoresis on iso-electric focusing.

2.4 Preparation of cyanogen bromide peptides
The fish flesh (1·0g) was suspended in 20·0ml 70% v/v formic acid solution to which was added 0·5g of cyanogen bromide to give a 200-fold excess. After allowing the flask to stand for 24h with occasional shaking, the cyanogen bromide and formic acid were removed under reduced pressure in a rotary evaporator at 40°C and the residue suspended in distilled water. After exhaustive dialysis against distilled water the extract was re-evaporated to dryness and then taken up in 6·0ml of 6M urea containing 0·02M β-mercaptoethanol or in 6·0ml of water. After centrifugation at 80 000 × g for 30min the aqueous solution was used directly for iso-electric focusing. (For electrophoresis it was diluted with an equal volume of 40% sucrose). The extracts could be stored frozen at −30°C for at least one year without any loss in the definition of the separation patterns.

2.5 Electrophoresis
The procedure previously described for disc electrophoresis (Mackie, 1969; Mackie, 1972) was used with the exception that the stock solution of TRIS-glycine was diluted × 10 and not × 14. 7·5% gels were used routinely and were prepared by dissolving acrylamide (2·85g) and bis-acrylamide (0·15g) in 20·0ml diluted buffer. The steps involving the addition of 10ml each of solutions of β-dimethylamino propionitrile (1·81%) and of ammonium persulphate (0·16% instead of 0·20%) were as previously described (Mackie, 1969).

2.6 Iso-electric focusing
Iso-electric focusing was carried out on thin layer polyacrylamide gel using the LKB 2117 Multophor (LKB Produkter AB).

The procedure for iso-electric focusing was as described in the LKB Application Note (Karlsson et al, 1973) and the pH range used was 3·5–9·5. Both prepared gels supplied by LKB and gels made in the laboratory were used. The 6M urea containing gels were prepared as in the LKB Application Note with the 7·5g of sucrose being replaced by 21·6g of urea. The temperature of the cooling plate was maintained at near 10°C either by running tap-water or by circulation of water through a refrigerated cooling bath.

The samples for iso-electric focusing were applied on strips of filter paper (10 × 5mm) previously dipped in the protein extracts. They were placed in a row in the

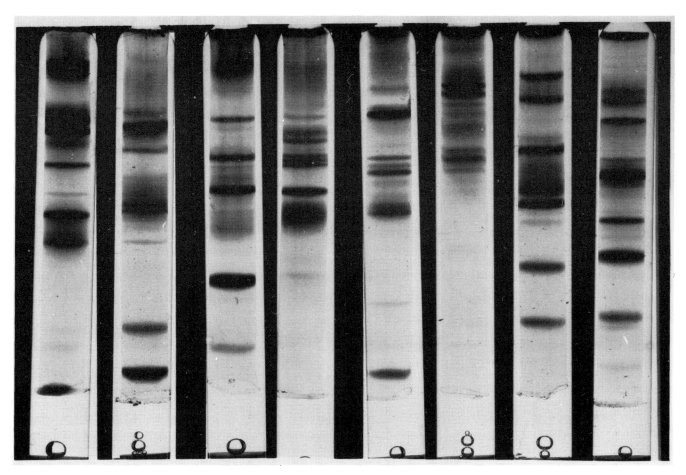

Plate 1 Disc electropherograms of aqueous extracts of (L–R) megrim (*Lepidorhombus whiffiagonis*); witch (*Glyptocephalus cynoglossus*); lemon sole (*Microstomus kitt*); plaice (*Pleuronectes platessa*); turbot (*Scophthalmus maximus*); salmon (*Salmo salar*); haddock (*Melanogrammus aeglefinus*); and cod (*Gadus morhua*)

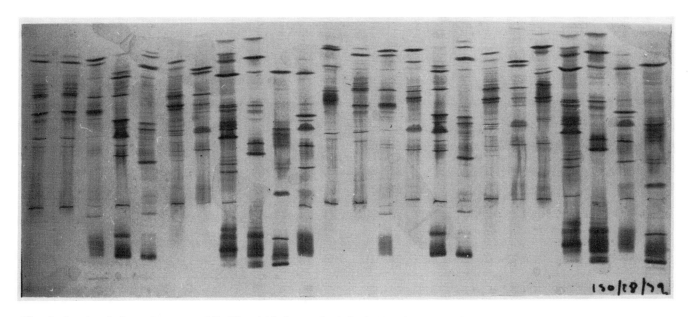

Plate 2 Iso-electric focused patterns of (L–R) cod (*Gadus morhua*) (in duplicate); haddock (*Melanogrammus aeglefinus*); saithe (*Pollacius virens*); hake (*Merluccius merluccius*); lythe (*Pollachius pollachius*); ling (*Molva molva*); lemon sole (*Microstomus kitt*); megrim (*Lepidorhombus whiffiagonis*); dogfish (*Squalus acanthias*); catfish (*Anarhichas lupus*); monkfish (*Lophius piscatorius*); cod, haddock, whiting (*Merlangus merlangus*); saithe; hake; lythe; ling; Norway pout (*Trisopterus esmarkii*); lemon sole; megrim; catfish; and dogfish

446

cathodal region of the gel surface. During the first hour of the run the current was maintained at 45mA as the voltage was increased to 1 000V. The strips of paper were then removed and the iso-electric focusing was allowed to proceed for a further hour. The gel was then removed from the apparatus and transferred to a fixing solution made up from 100ml methanol, 350ml water and containing 37·5g of trichloroacetic acid and 17·25g of sulphosalicylic acid. After standing for 1h to wash out the ampholytes and to precipitate the proteins, the gel was transferred to a bath containing a solution of Coomassie Blue R (0·115g in 100ml of destaining solvent) at 60°C. After 30min the gel was destained by washing in the destaining solution (500ml ethanol, 160ml acetic acid diluted to 2 litres with distilled water) and finally transferred to the same solution containing 10% glycerol. The gels were then covered with a plastic sheet, photographed and stored at room temperature.

3 Results and discussion

In comparison with normal zone electrophoresis, iso-electric focusing, in the main, gives even greater resolution of proteins. Its extreme resolving power is achieved through the pH gradient which forms between the electrodes. In this system the proteins in a mixture move to and then focus into very sharp zones corresponding to their iso-electric point on the pH gradient (Righetti and Drysdale, 1976). Very often small differences in the iso-electric point of proteins, which would be insufficient for resolution in conventional electrophoresis, lead to clearly separated zones on iso-electric focusing. In many cases, however, resolution of this order is not required for the differentiation of most of the commercially important species. There are, however, particular cases where the use of iso-electric focusing is preferred as for smoked salmon (Morel, 1977) and some where it may be essential; these points will be illustrated under the general headings of identification of raw, cooked and canned fish, respectively.

3.1 Raw fish

Typical separation patterns for disc electrophoresis and for iso-electric focusing of aqueous extracts of raw flesh of a range of commercially important species are given in Plates 1 and 2, respectively. To take the patterns for cod as typical it is evident that no more than 20 zones, some of which are very faint, can be seen on disc electrophoresis. However, on iso-electric focusing at least 40 zones are present. This does not in any way give the latter system an advantage as there is adequate differentiation of all species on Plate 1. It has been shown elsewhere (Mackie and Jones, 1978) that even at the subspecies level, normal electrophoresis can be used effectively. However, for closely related species such as cod, haddock, whiting, saithe, hake, lythe, ling and Norway pout, patterns for all of which are in Plate 2, the iso-electric focusing system gives readily distinguishable patterns.

As far as the differentiation of raw fish is concerned, iso-electric focusing could possibly be used with advantage to detect genetic variation within a species to the extent that it affects mobilities of the proteins (Markert et al, 1975). A preliminary observation on two stocks of

herring from the Irish Sea shows (Plate 3) that minor differences in the overall pattern do exist. Their significance has not yet been established and further analysis will be required before any conclusion can be drawn. An alternative to staining for total protein is staining for a specific enzyme such as lactate dehydrogenase (Chua et al, 1978; Mackie and Jones, 1978) where the frequency distribution of the iso-enzymes can be used to measure gene frequencies within a population. Extensive work by Markert et al (1978), Tsuyuki et al (1966) and others on starch gel has demonstrated the value of the technique. It is possible that a combination of staining for total protein and for specific enzymes could lead to improvements in the usual electrophoretic systems.

3.2 Cooked fish

The normal system for the identification of cooked fish is essentially electrophoresis in 6M urea-containing gels, of protein residues extracted in 6M urea (Mackie, 1969; 1972). Since this procedure was introduced by the author in 1968, it has been used routinely by public analysts in the country. A close examination of the patterns in Plate 4 shows that only a few zones can be used for differentiation of species. The actual differentiation is based on difference in mobility of usually not more than three or four zones which move in front of the intensely staining tropomyosin sub-unit which is readily extractable in the solvent. There can be some difficulty in differentiating the closely related gadoid species– saithe, cod, haddock and whiting– particularly if the fish has been heated or reheated for periods longer than 30min at 100°C.

An alternative procedure of iso-electric focusing of the 6M urea extracts in gels containing 6M urea can be used (Plate 5) but although much greater resolution of the protein residues is obtained, the number of species-specific zones is not correspondingly increased. In my experience there is little if any advantage in using this system as the same difficulty in differentiating gadoids, particularly cod and saithe, still exists.

The preferred system for identifying such closely related species and fish that has been heated for prolonged periods is iso-electric focusing of the cyanogen bromide peptides obtained as for canned fish (Mackie and Taylor, 1972). Typical separations of these water-soluble peptides are given in Plate 6. The patterns for raw and cooked cod, whiting and saithe, respectively, and of cooked hake and haddock are given for comparison; they are more distinguishable than in the other systems already discussed, usually by the presence or absence of one or two intense 'focused' zones along the pH gradient. An examination of all of the species on Plate 6 shows that each has a characteristic distribution of such zones.

3.3 Canned fish

When proteins are subjected to prolonged heating, and particularly to the high temperatures and pressure of canning, further covalent bonds are formed. The chemistry of the process is not clear but there is evidence for the formation of —S—S— bonds for example. Attempts to extract proteins or protein fragments from canned products with the usual solvents such as 6M urea or with acid or alkali have not been successful. In normal

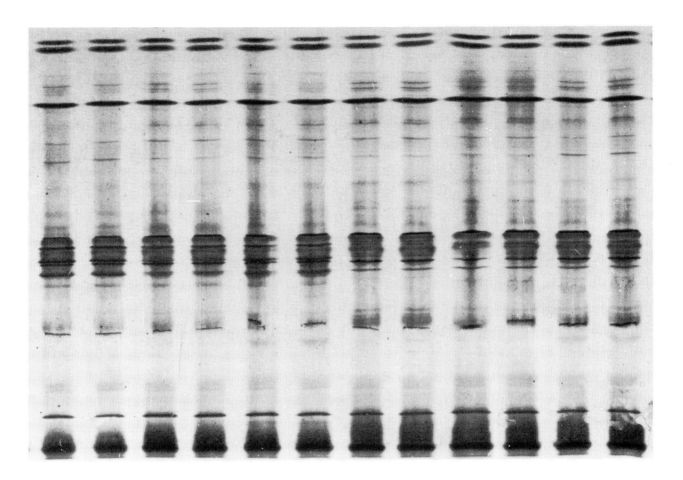

Plate 3 Iso-electric focused patterns of aqueous extracts of herring (in duplicate) (L–R) *stock 1 (1, 2 and 3) and stock 2 (4, 5 and 6)*

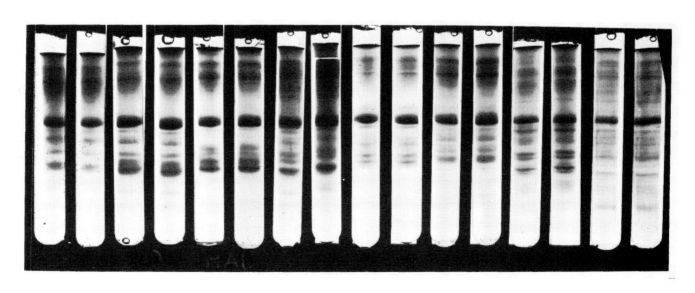

Plate 4 Disc electropherograms of 6M urea extracts of cooked fish (L–R) (in duplicate) saithe; cod; haddock; whiting; dogfish; herring; turbot; and plaice

zone electrophoresis of such extracts only a few zones are obtained and these are generally common to all species. In addition, the gels usually have strong background staining which adds to the problem of differentiation.

However, by allowing fish flesh to react with excess of cyanogen bromide in 80% formic acid, water-soluble peptides are released by specific cleavage at each methionine residue in the proteins (Gross, 1966). In my first paper on the problem (Mackie and Taylor, 1972) of identifying canned fish 6M urea was used as the extractant, but it has since been found that aqueous extraction followed by dialysis against distilled water gives a suitable extract for electrophoresis or iso-electric focusing. The patterns obtained vary only slightly between raw and cooked fish (*Plate 6*) or canned fish and are thus the least affected of all the electrophoretic patterns used for species identification. Typical patterns obtained after iso-electric focusing of aqueous extracts of cyanogen bromide peptides of a range of species of canned fish are given in *Plate 7*. What is of particular interest is the extent of differentiation of the six species of salmon. Two groupings have been obtained, namely chum, sockeye and pink (Group 1) and spring, coho and *Salmo salar* (Group 2). This grouping is in general agreement with one made by Utter *et al* (1974), who drew up a relationship based on biochemical genetic data. In their grouping pink and sockeye were shown to be closely related as were coho and spring. Chum was shown as intermediate but more closely related to the latter two.

Most of the commercially important canned species sold in the United Kingdom have been examined by this procedure and in all cases clear species-specific patterns have been obtained as, for example, for bonito and tuna and for pilchard and herring (*Plate 7*). Similarly, canned Japanese mackerel can be differentiated from canned Atlantic mackerel (*Scomber scombrus*). It is likely therefore that all species of fish will release cyanogen bromide fragments which will be as species-specific as the water-soluble proteins or myogens of raw fish.

While iso-electric focusing is the preferred method for the separation of the cyanogen bromide peptides of canned fish, it is not always necessary. In *Plate 8* typical disc electrophoretic patterns of bonito, herring, pilchard and *Salmo salar* are compared. In general, although differences are detectable in the latter system, there is usually a variable amount of background staining which adds to the problems of interpretation of the patterns. The expense of the reagents is, of course, a major consideration as far as iso-electric focusing is concerned. It is not required, in my opinion, for the identification of the species of raw fish but for some cooked fish it will clearly be advantageous to use it (Morel, 1977). For canned products it is the method of choice.

4 References

Cowie, W. Identification of fish species by thin slab polyacrylamide gel
1968   electrophoresis. *J. Sci. Fd. Agric.*, 19, 226–229
Chua, K E, Crossman, E J and Gilmour, C A. Lactate dehydro-
1978   genase (LDH) isoenzymes in muscle of freshwater fish by
      iso-electric focusing in thin layer polyacrylamide gel. *Science
      Tools*, 25(1), 9–11
Gross, E. The cyanogen bromide reaction. *Methods in Enzymology*.
1966   Vol. XI, 238–242

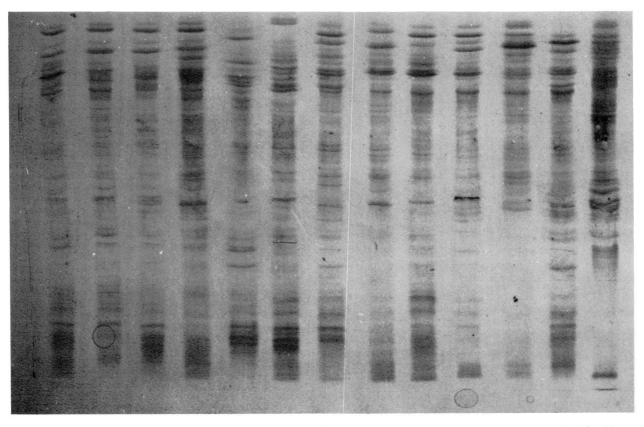

Plate 5   Iso-electric focused patterns of 6M urea extracts of cooked fish in polyacrylamide gel also containing 6M urea. (L–R) saithe; cod; haddock; whiting; dogfish; herring; turbot; plaice; lemon sole; witch; megrim; monkfish; and scampi

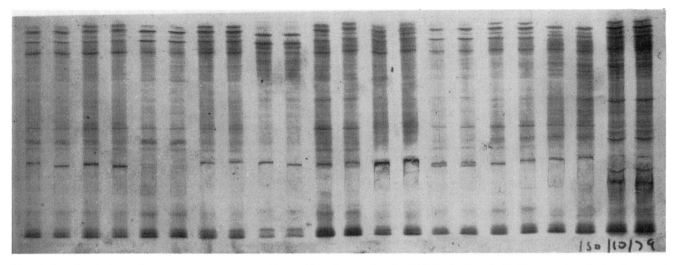

Plate 6 Iso-electric focused patterns of cyanogen bromide peptides of raw and cooked fish (in duplicate). (L–R) raw and cooked cod; raw and cooked whiting; raw and cooked saithe; cooked hake; scampi; haddock; monkfish; mackerel

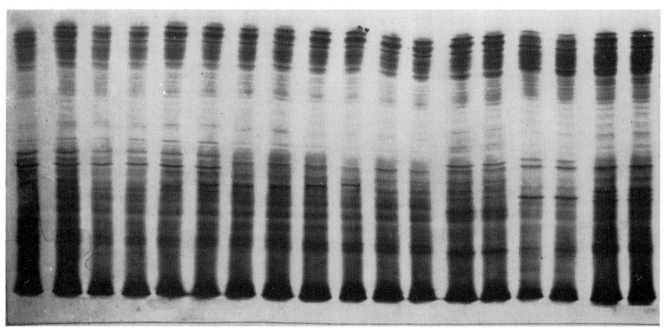

Plate 7 Iso-electric focused patterns of cyanogen bromide peptides of canned fish. (L–R) chum salmon; sockeye salmon; pink salmon; spring salmon; coho salmon; *Salmo salar*; pilchard; herring; and bonito

KARLSSON, C, DAVIES, H, ÖHMAN, J and ANDERSSON, U. LKB 2117
1973   Multiphor. I. Analytical thin layer gel electrofocusing in polyacrylamide gel. 13pp
MACKIE, I M. Identification of fish species by a modified polyac-
1969   rylamide disc electrophoresis technique. *J. Assoc. Publ. Anal.*, 83–87
—— Some improvements in the polyacrylamide disc electrophoretic
1972   method of identifying species of cooked fish. *J. Assoc. Publ. Anal.*, 18–20
MACKIE, I M and TAYLOR, T. Identification of species of heat-sterilized
1972   canned fish by polyacrylamide disc electrophoresis. *Analyst*, 97, 609–611
MACKIE, I M and JONES, B W. The use of electrophoresis of the
1978   water-soluble (sarcoplasmic) proteins of fish muscle to differentiate the closely related species of hake (*Merluccius* sp.). *Comp. Biochem. Physiol.*, 59B, 95–98

MARKERT, C L, SHAKLEE, J M and WHITT, G S. Evolution of a gene.
1975   *Science*, 189, 102–114
MOREL, M. Identification des espèces de poissons par electrofocalisa-
1977   tion en gel de polyacrylamide. *Sci. Peche*, 275, 1–8
RIGHETTI, P G and DRYSDALE, J W. Iso-electric focusing in laboratory
1976   techniques. In: *Biochemistry and Molecular Biology*, Eds T S Work and E Work. North Holland Publishing Co. 590pp
TSUYUKI, H, UTHE, J F, ROBERTS, E and CLARKE, L W. Comparative
1966   electropherograms of *Coregonis clupeoformis*, *Salvelinus namaycush*, *S. alpinus*, *S. malma* and *S. fontinalis* from the family Salmonidae. *J. Fish. Res. Bd. Can.*, 23(10), 1599–1606
UTTER, F M, HODGINS, H O and ALENDORF, F W. Biochemical genetic
1974   studies of fishes: potentialities and limitations. In: *Biochemical and Biophysical Perspectives*, Eds D C Malins and J R Sargent. Academic Press. Vol. 1, 213–238

450

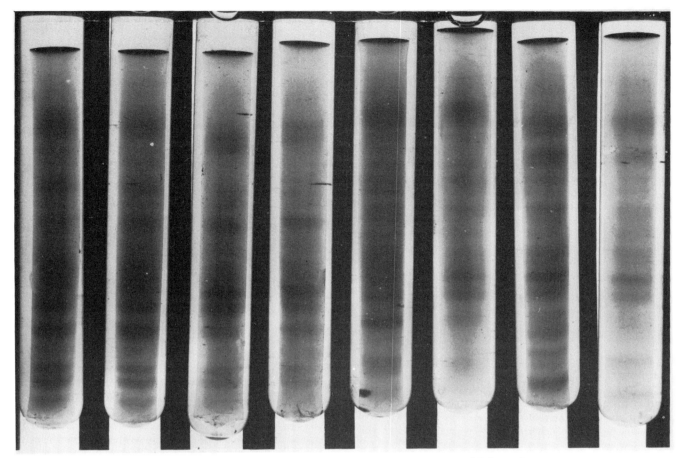

Plate 8 Disc electropherograms of cyanogen bromide peptides of bonito; herring; pilchard; and *Salmo salar*

The use of electrophoresis of the water-soluble muscle proteins in the quantitative analysis of the species components of a fish mince mixture

A Hume and I M Mackie

1 Introduction

Electrophoresis has been used for many years as a chemical means of identifying the species of fish either in the raw or cooked (Mackie, 1969) also canned states (Mackie and Taylor, 1972). As a qualitative method of identifying single species, the technique has proved successful and reproducible.

Attempts have been made to apply it quantitatively, particularly to the detection and estimation of soya bean proteins in cooked meat products (Penny and Hofmann, 1971; Llewellyn and Flaherty, 1976). All of the methods depend on clear separation of protein zones of interest as, for example, soya, from meat proteins or constituents of whey proteins (Darling and Butcher, 1975) so that densitometric measurements can be made. Varying degrees of success have been reported for such quantitative studies mainly because of the 'background staining' which is usually associated with heat denaturation of proteins, and which increases the error of measurement of the areas of the protein zones.

With the increasing amount of minced fish flesh being produced today, there is a need for a reliable method not only of identifying the species but also of detecting mixtures of species and of quantifying the components. For raw fish the problems of background staining are insignificant and provided that the respective separation patterns of the species under examination are sufficiently different from one another to allow densitometric measurement of at least one characteristic zone per species the method should have a useful application. In industrial practice mixtures of species are likely to be of closely related fish such as cod (*Gadus morhua*), saithe (*Pollachius virens*), whiting (*Merlangus merlangus*) or hake (*Merluccius merluccius*), all of which give similar separation patterns. Cod and saithe, however, have one or two zones in their respective patterns which are not present in the other and which, in a mixture of the two species, are sufficiently separated from other zones in the pattern to allow densitometric measurement to be made. Similar small differences are found for the other species.

This paper is an assessment of the value of densitometric measurement of electrophoretic separation patterns in the quantitative determination of the composition of mixed extracts and minces of two species. Cod and saithe were selected because of their commercial importance.

2 Experimental

2.1 Materials
Cod (*Gadus morhua*) and saithe (*Pollachius virens*) caught on different grounds were brought to Torry Research Station as gutted whole fish and stored in ice until required for analysis.

2.2 Preparation of protein extracts
Samples of flesh (100g) were removed from the fish and homogenized with two volumes of water and centrifuged for 1h at 10 000 × g. The supernatant solution obtained was diluted with an equal volume of 40% sucrose and kept at 0°C until required for electrophoresis. (To aid the application of the protein extract to the polyacrylamide gels, and to allow a visual check on the electrophoretic run, a few crystals of bromophenol blue were added to the 40% sucrose solution.) The blue solutions of protein were then applied as $10 \mu l$ aliquots to the gels and subjected to electrophoresis. Extracts from cod and saithe were combined to give a range of mixture compositions of the two species.

Mixtures of minces with a similar range in composition of cod and saithe were extracted as described above to give a minced species extract.

2.3 Methods
Various factors associated with the electrophoresis technique were examined including the acrylamide/bis ratio, gel setting, optimum run time and temperatures, staining and destaining techniques. The following procedure was adopted finally: gels of 7% concentration with an acrylamide/bis ratio of 97·2:2·8 were prepared and allowed to stand for at least 2h after polymerization. The tubes were inserted in a standard Shandon Electrophoresis Unit and a prerun of 20min duration at a potential difference of 200V was carried out at 1°C. The samples were applied and the separation done at 1°C for 40min using a potential difference of 280V. The gels were then stained at room temperature for 18h using a 1% Amido Black solution in 7% acetic acid followed by an electrolytic destaining technique. Densitometry was carried out at 540nm in a Pye Unicam spectrophotometer (SP 1700) with a densitometer attachment and the intensities of the bands were obtained as peaks, the areas of which were measured manually.

2.4 Data handling
Two methods of determining the percentage of cod/saithe in the extracts were used.

Method 1
(*a*) The percentage composition of cod or saithe in the extract was determined from a calibration plot of the area of cod- or saithe-specific peaks against the percentage of cod or saithe in standard mixtures of known protein concentrations.
(*b*) To reduce the error due to variation in intensity of staining from one run to another, standards were included in each electrophoretic run and a correction was made for any difference in protein concentration between the standard and unknown solutions. Normally six gels in a run were used for the calibration and the remaining two were used for the unknown samples.

Method 2
In this method the percentage composition of cod or saithe was obtained directly from the areas of the cod-specific and saithe-specific zones in a densitometric trace of the same gel. Using a previously prepared graph of the percentage composition of cod or saithe in known instances against the ratio of the area of one of the species-specific peaks to the sum of the area of the two species-specific peaks the percentage composition of the unknown mixture was obtained directly. This method is independent of variations in intensity of staining between gels but it does depend on the determination of the areas of the species-specific zones of both species. Corrections for variation in the concentration of protein between the unknown extract and standards are not required.

3 Results and discussion

An essential requirement of any attempt to quantify components of a mixture is that the area of the species-specific peak is linearly related to the concentration. For both cod- and saithe-specific zones (*Figs 1* and *2*) a linear relationship was found up to a concentration of protein in the extract of at least 0·3mg/ml (*Figs 3* and *4*). A statistical assessment of the value of the leading band 3c of cod in measuring the composition of cod in the mixtures was made. Seven extracts of cod and of saithe were mixed in pairs in varying amounts by volume and in all cases comparable amounts of protein were applied to the gels.

Gel No.	1	2	3	4	5	6	7	8
% cod	100	86	72	57	43	28	14	0
% saithe	0	14	28	43	57	72	86	100

For each mixture two gels were run and after destaining they were scanned and the area under the peak plotted against the gel number. A typical plot for one run

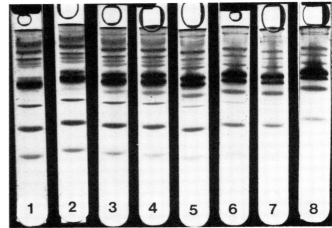

Fig 1 Electropherograms of mixtures of cod and saithe

L–R gel No.	1	2	3	4	5	6	7	8
% cod	100	86	72	57	43	28	14	0
% saithe	0	14	28	43	57	72	86	100

is given in *Fig 5*. However, the actual slopes and absolute values for the areas varied from run to run.

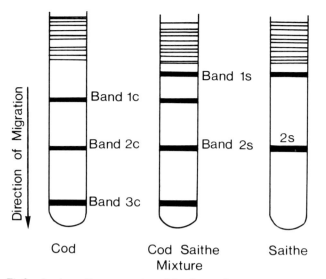

Fig 2   A schematic presentation of the electropherogram of pure cod, mixture of cod and saithe and pure saithe

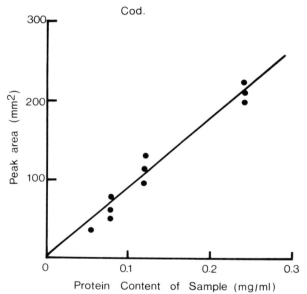

Fig 3   Plot of area of peak (1c) against concentration of protein in extract (mg/ml)

The data in *Table I* were analysed statistically and a regression line was drawn (*Fig 6*). This showed that an estimate of the cod content of a random cod and saithe mixture could be made at a 95% level of confidence within a range of ±12·0%.

In an attempt to reduce the error in estimating the content of a mixture which was due to run-to-run variations, method 1b was used. From the data presented in *Table II* which was obtained from calibration plots of the 1c and 3c for cod, and 1s for saithe, it is evident that all of these specific-zones are adequately separated from the other to give satisfactory densitometric estimates of the percentage composition of the mixture of cod and saithe.

In method 2, the ratio of the area of a species-specific zone to that of the sum of the areas of the species-specific zones of both cod and saithe (1c and 1s) is plotted

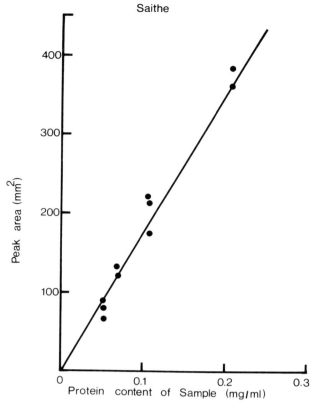

Fig 4   Plot of area of peak (1s) against concentration of protein in extract (mg/ml)

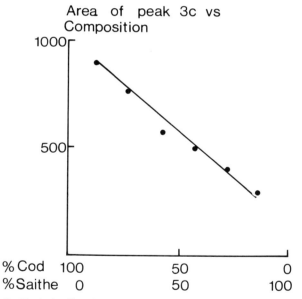

Fig 5   Typical calibration plot of area of peak (3c) against percentage composition for a single run (method 1a)

against the percentage of cod (or saithe) in the mixture. By taking 100% cod and 100% saithe samples, peak areas of the single species were established. This could also be done by taking standard mixtures and extrapolating to the 100% values for each species using the 1c and 1s peaks. When the ratio of the peaks was written

$$\frac{qs}{qs + pc}$$

453

Table I
AREAS OF LEADING PEAKS FOR SEVEN RUNS

Run	A	B	C	D	E	F	G
Gel No.							
1	1 300	1 360	995	1 128	1 150	1 241	1 179
	1 213	1 329	1 020	942	1 113	1 163	1 097
	(1 256·5)	(1 344·5)	(1 007·5)	(1 035·0)	(1 131·5)	(1 202·0)	(1 138·0)
2	1 015	1 177	868	930	992	962	1 000
	1 245	1 267	912	947	1 177	947	872
	(1 130·0)	(1 222·0)	(890·0)	(938·5)	(1 084·5)	(954·5)	(936·0)
3	934	1 012	739	823	—	894	747
	944	—	780	709	905	—	898
	(939·0)	(1 012·0)	(759·5)	(766·0)	(905·0)	(894·0)	(822·5)
4	673	752	605	665	893	694	708
	747	893	719	589	776	—	—
	(710·0)	(822·5)	(662·0)	(627·0)	(834·5)	(694·0)	(708·0)
5	583	640	540	514	524	538	560
	528	709	489	451	—	524	562
	(555·5)	(674·5)	(514·5)	(482·5)	(524·0)	(531·0)	(561·0)
6	454	517	323	339	331	333	367
	427	497	316	368	378	300	399
	(440·5)	(507·0)	(319·5)	(353·5)	(354·5)	(316·5)	(383·0)
7	112	302	176	132	248	200	217
	148	290	177	155	250	185	198
	(130·0)	(296·0)	(176·5)	(143·5)	(249·0)	(192·5)	(207·5)

Gel No.	1	2	3	4	5	6	7	8
Parts of cod	7	6	5	4	3	2	1	0
Parts of saithe	0	1	2	3	4	5	6	7

Table II
ANALYSES OF EXTRACTS BY METHOD 1

Sample	Actual composition (given as % cod)	Peak used	Reading of sample	Initial estimation of cod content (%)	Correction factor	Final estimation of cod content (%)
A	70	3c	912	84	0·91	76
A	70	1c	945	87	0·91	79
A	70	1s	439	79	1·00	79
B	30	3c	571	56	0·91	42
B	30	1c	544	49	0·91	44
B	30	1s	896	22	1·00	22
B	30	3c	470	38	0·91	35
B	30	1c	500	48	0·91	44
B	30	1s	889	23	1·00	23
C	70	3c	582	52	1·10	57
C	70	1c	804	68	1·10	75
C	70	1s	472	69	1·04	72
D	50	3c	504	41	1·10	45
D	50	1c	547	45	1·10	50
D	50	1s	589	42	1·04	44
C	70	3c	551	60	1·10	66
C	70	1c	662	65	1·10	72
C	70	1s	369	71	1·04	74
D	50	3c	469	49	1·10	54
D	50	1c	477	43	1·10	47
D	50	1s	562	49	1·04	51
E	90	3c	906	116	0·95	110
E	90	1c	1 193	116	0·95	110
E	90	1s	140	105	0·96	101
F	60	3c	519	66	0·95	63
F	60	1c	773	75	0·95	71
F	60	1s	457	62	0·96	59
E	90	3c	924	118	0·95	110
E	90	1c	968	108	0·95	103
E	90	1s	136	84	0·96	81
F	60	3c	521	64	0·95	61
F	60	1c	495	54	0·95	52
F	60	1s	336	38	0·96	37

(c is the value for 100% cod, s is the value for 100% saithe, $p = 0.1, 0.2 \ldots 0.9$ and $q = 1 - p$) and this formula plotted against the percentage of cod present, a curve as shown in *Fig 7* was produced. The curve is a theoretical one and is based only on the 100% values of each species as measured. By determining the ratio of peaks 1c and 1s from the mixture of unknown composition in the above form a direct reading of percentage of cod was obtained. The advantages of the method were that no correction factors were required (a ratio of peaks is independent of total protein concentration) and the system was less susceptible to variations in staining. *Table III* shows the analyses of mince mixtures using the ratio method. Most of the samples quoted were from minced mixtures made from market fish, *ie* the icing history and fishing grounds were not known.

It can be concluded that the major limitation in this work is not the quantitative aspect but the extreme

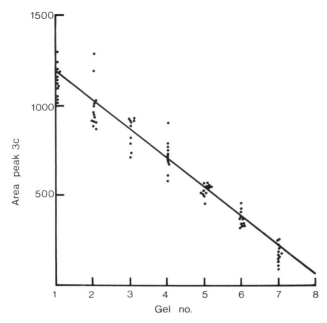

Fig 6 Regression line of area of peak (3c) against gel number for varying percentage mixtures analysed for cod and saithe, respectively

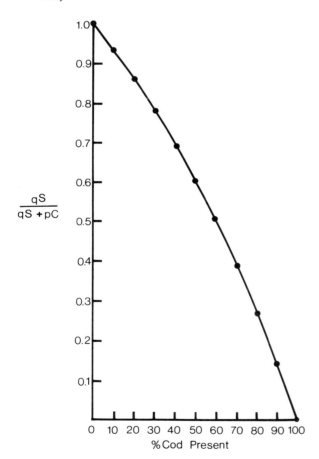

Fig 7 Plot of $\frac{qs}{qs + pc}$ against percentage of cod present

difficulty in identifying the species actually present in the mixture.

An estimate of the error within a run and between runs was made by running aliquots of an extract under different conditions. *Table IV* shows the area of band 1c

in four sets of eight gels, each group of eight having been prepared separately using the same stock solutions.

Table III
ANALYSES OF MINCE MIXTURES BY METHOD 2

Measurement method	Percentage of cod present		Observed
	Actual	*Observed*	*Actual*
Ratio-direct	75	69	92·0
Ratio-direct	75	76	98·7
Ratio-direct	40	41	102·5
Ratio-direct	80	79	98·8
Ratio-direct	60	44	73·3
Ratio-direct	40	32	80·0
Ratio-direct	20	14	70·0
		Mean	87·9 SD 13·3
Ratio-extrapolated	80	79	98·8
Ratio-extrapolated	60	56	93·3
Ratio-extrapolated	40	39	97·5
Ratio-extrapolated	20	20	100·0
Ratio-extrapolated	70	65	92·9
Ratio-extrapolated	60	62	103·3
Ratio-extrapolated	40	38	95·0
Ratio-extrapolated	30	35	116·7
Ratio-extrapolated	75	67	89·3
Ratio-extrapolated	45	40	88·9
Ratio-extrapolated	80	82	102·5
Ratio-extrapolated	60	56	93·3
Ratio-extrapolated	40	42	105·0
Ratio-extrapolated	20	25	125·0
		Mean	100·1 SD 10·2

Mean of all values 96·0
SD of all values 12·5
Standard error of the mean = 2·72

Table IV
AREA OF PEAK 1C FOR FOUR RUNS OF EIGHT GELS

Gel	Run			
	1	*2*	*3*	*4*
1	364·0	406·6	517·8	522·0
2	423·5	504·0	473·1	523·5
3	424·5	465·5	487·5	476·7
4	400·0	406·0	435·2	487·6
5	331·8	388·8	450·0	491·5
6	372·6	439·0	499·1	480·0
7	384·7	498·7	421·8	534·0
8	354·2	456·8	455·0	539·8
Mean	381·9	445·7	467·4	506·9
SD	32·8	43·3	32·9	25·5
SD/mean × 100	8·6	9·7	7·0	5·0

Variation in the results obtained was due to the fluctuation in both internal and external conditions. The gel-to-gel variation was due to variation in the internal parameters and operator error and is measured as

$$\frac{\Sigma \text{error of each run}}{\text{number of runs}} \times 100$$

in this case $\frac{8·6+9·7+7·0+5·0}{4} = 7·6\%$

The run-to-run variation was due to external factors including temperature fluctuation, variation in gel viscosity between runs, dye uptake, *etc* and is measured as

$$\frac{\text{standard deviation of the means}}{\text{mean of the run means}}$$

in this case $\frac{52·3}{450·9} = 11·6\%$

The technique error was measured by dividing the

Table V
DETERMINED AREAS FOR PEAK 1C FOR A SINGLE COD

Days in ice	0	1	2	6	7	8	13	14	15	Band	Mean (x)	SD (σ)	SD/x ×100
Peak	334·0	328·8	374·5	326·5	437·2	257·9	292·5	369·2	462·5				
	285·2	369·0	564·8		301·5								
	358·4	402·8											
1c	404·6	484·3								1c	353·3	85·8	24·3
	200·5	314·3											
	246·2	305·1											
	514·5	303·0	195·7	0	323·6	0	0	0	0				
	469·7	291·6	182·7		134·5								
	492·2	278·5											
2c	629·6	395·8								2c	245·7	185·9	75·6
	232·6	270·4											
	260·0	185·1											
	1 097·3	513·1	587·9	545·4	684·3	722·8	735·9	485·7	922·6				
	612·8	601·5	549·8		386·8								
	625·9	601·1											
3c	915·2	705·7								3c	629·9	175·8	27·9
	457·5	556·8											
	474·7	444·4											

standard deviation by the mean for all 32 peak areas and the percentage error was 12·5%.

To determine the effect of ice storage on peak area, extracts from single fish were taken over 15 days in ice storage and areas of the species-specific peaks measured. The figures shown in *Table V* are from a cod sample taken live from the tanks at Torry but the cod samples from different grounds showed similar variations.

The saithe pattern was also examined, but using fewer samples, and it was found that the 1s band was consistent and similar to the 1c band and that the 2s band was similar to the 2c. In consequence the 1c and 1s bands were used for all the analyses.

4 Conclusions

Optimal methods for quantitative electrophoresis of extracts of mixtures of species were established and the variation in the results measured.

Mixtures of cod and saithe were analysed and the percentage of each species was determined reasonably accurately (error of the means 2·7).

The methods have limited application for unknown mixtures in that it is difficult to identify unequivocally all species present and that the methods rely on single characteristic bands from the species to be analysed well separated from other zones.

5 Acknowledgements

The authors wish to thank S Graham and K Ridley, students from Trent Polytechnic, Nottingham for their contribution to this work.

6 References

DARLING, D F and BUTCHER, D W. Quantification of polyacrylamide
1975   gel electrophoresis for analyses of whey proteins. *J. Dairy Sci.*, 59(5), 863–867

LLEWELLYN, J W and FLAHERTY, B. The detection and estimation of
1976   soya protein in food products by iso-electric processing. *J. Fd. Technol.*, 11, 555–563

MACKIE, I M. Identification of fish species by a modified polyac-
1969   rylamide disc electrophoresis technique. *J. Assoc. Publ. Analyst*, 83–87

MACKIE, I M and TAYLOR, T. Identification of species of heat-sterilised
1972   canned fish by polyacrylamide disc electrophoresis. *Analyst*, 97, 609–611

PENNY, I F and HOFMANN, K. The detection of soya bean protein in
1971   meat products. 17th Meeting European Meat Research Workers, Bristol. 809–812

Fish tissue degradation by trypsin type enzymes

K Hjelmeland and J Raa

1 Introduction

Belly bursting of capelin results when digestive enzymes leak out and hydrolyse the surrounding belly tissues (Gildberg, 1978; Hjelmeland, 1978; Gildberg and Raa, 1979). Several different chemical linkages are broken during this process, but it is probably mainly the result of proteolysis.

The rate of the tissue degradation is determined not merely by the level of proteolytic enzymes in the fish (Gildberg and Raa, 1980). Additional factors which may affect the process are the physical/chemical condition of the connective tissue, the pH, and enzyme inhibitors.

Earlier studies (Hjelmeland, 1978; Gildberg and Raa, 1979) have shown that the tissues solubilize quickly at pH 6·5. This was unexpected since the proteolytic activity with a soluble protein substrate was at its minimum in the pH range 5·5–7. However, it may be that proteases have other pH optima when acting on intact tissue structures, or minced whole fish, than on soluble proteins. Moreover, different tissues are so different in their susceptibilities to digestive enzymes (Gildberg and Raa, 1979), that the relatively rapid solubilization at neutral pH of a whole fish, or a fish mince, may be the maximum sum effect of a co-operative action of a variety of factors. To examine these possibilities it has been necessary to study the process with purified enzymes and separate cellular components. The role of trypsin is dealt with in particular in this paper.

2 Materials and methods

Capelin (*Mallotus villosus*) was caught by trawl in the Barents Sea at the end of September 1978. The fish were wrapped individually in aluminium foil and stored at −20°C in sealed plastic bags immediately after catching.

Crude enzyme extract. The digestive tract, without stomach, was homogenized in an equal weight of distilled water, and 10vol% CCl$_4$ was added to remove lipids. The supernatant after centrifugation for 20min 40 000 × g was diluted 1:25 in distilled water before each experiment.

Purified trypsin. The crude enzyme extract was fractionated with (NH$_4$)$_2$SO$_4$, and the 40–70% fraction separated on a Sephadex G–100 column, followed by affinity chromatography on glycyl-glycin-*p*-aminobenzamidine substituted Sepharose 4 B (Grant *et al*, 1978). The trypsin-like enzymes were released from the column by benzamidine, which was subsequently removed by molecular sieving on Sephadex G 25. Two protein bands were detected by disc electrophoresis in 7% polyacryl amide gel, and both had specific trypsin activity. These tryptic enzymes differ from bovine trypsin in being acid labile and having acid pH, as will be described separately (Hjelmeland, in preparation). The trypsin preparation was used in the experiments at an activity which corresponds to total protease activity at pH 9 of the crude preparation.

Haemoglobin substrate. Bovine haemoglobin (Sigma) was dialysed for 24h against distilled water and the concentration adjusted to 8% (w/v).

Minced muscle substrate. Muscle tissue was homogenized at 4°C in three volumes of Johnson/Lindsay buffer at pH 8 (Johnson and Lindsay, 1939). The homogenate was divided in portions of 12g, and each portion diluted with 10ml distilled water before the pH was adjusted with 2N HCl or NaOH, and the weight finally adjusted to 24g with distilled water.

Sarcoplasmic and myofibril fractions. Cod muscle (50g) was homogenized at 4°C in 200ml 0·45M KCl-phosphate buffer at pH 7·5. After standing for 1h, 200ml of the same buffer was added and the mixture homogenized and centrifuged at 10 000 × g. The supernatant was dialysed for 24h against distilled water (5 litres) and centrifuged for 30min at 10 000 × g. The supernatant was defined as the *Sarcoplasmic fraction*, the pellet as the *Myofibril fraction*. Both fractions were lyophilized and prepared as substrates by dissolving in water at concentration of 8% (w/v).

Incubations. Substrate (0·25ml), enzyme (0·25ml) and Johnson/Lindsay buffer (0·5ml) were incubated for 1h at 30°C and the reaction terminated by adding 5ml 3% trichloroacetic acid (TCA). The references were incubated without substrate, which was added after TCA.

When minced muscle was the substrate, 2ml of the pH adjusted suspension was mixed with 1ml enzyme, and the reaction terminated after 12h by cooling in ice and centrifugation at 40 000 × g for 20min. One ml 10% TCA was added to 1ml of the supernatant and Folin-positive material in the TCA supernatant determined. The references were treated identically, except that the minced muscle substrates were incubated for 12h before TCA and enzyme were added. Folin-positive substances (Lowry *et al*, 1951) in the TCA extracts were determined according to Barret (1972), and the results are expressed as the difference between sample and the reference in colour reaction, recorded as A$_{700}$ with a Spectronic 20.

3 Results

Muscle tissue of capelin was more resistant to degradation by a crude extract of digestive enzymes than was haemoglobin (*Fig 1A*). The difference in degradation rates of these substrates varied with the pH, being about 15 times faster with haemoglobin than with minced muscle at pH 6, about 20 times at pH 7, and about 25 times higher at pH 8. Similar pH curves of degradation of muscle and haemoglobin were obtained with purified trypsin from capelin (*Fig 1B*). However, the rate of degradation of muscle tissue did not always increase with pH in the range from 6 to 9 if a crude enzyme preparation was used. With certain catches of capelin, autolysis of minced whole fish was even faster at pH 6·5 than at pH 8 to 9 (Hjelmeland, 1978).

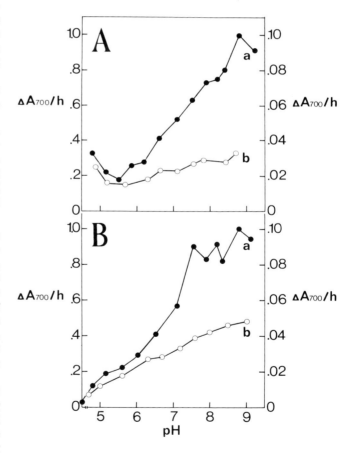

Fig 1 The effect of the pH on the rate of degradation of haemoglobin *(a)* and of minced muscle tissue of capelin *(b)*, in the presence of: A, a crude enzyme extract from the digestive system of capelin, and B, purified trypsin from capelin.
Abscissae: pH at start of incubation
Ordinates: rate of formation of TCA-soluble Folin-positive substances from haemoglobin (left) and from muscle mince (right)

457

Sarcoplasmic proteins of fish muscle were, like minced muscle, relatively resistant to degradation by a crude mixture of digestive enzymes and by trypsin, whereas the myofibrillar fraction was degraded twice as fast as was haemoglobin in the pH range 6 to 9 (*Fig 2*). The myofibrils were, however, relatively resistant to the proteases active at pH 4·5 present in the crude enzyme preparation.

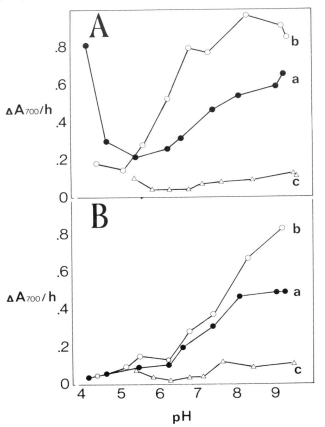

Fig 2 The effect of pH on the rate of degradation of haemoglobin *(a)*, of the myofibrillar fraction *(b)* and of the sarcoplasma proteins *(c)*, in the presence of: *A*, a crude enzyme extract from the digestive system of capelin, and *B*, purified trypsin from capelin.
Abscissae: pH at start of incubation
Ordinates: rate of formation of TCA-soluble Folin positive substances

The enzymatic degradation of the myofibrillar fraction was efficiently inhibited by the sarcoplasmic extract (*Fig 3*), as was the degradation of haemoglobin (*Fig 4*).

4 Discussion

Attempts to produce simple practical methods to be used for predicting the tendency of belly bursting in capelin, sardines or sprats, have not yet been successful. Although belly bursting is the result of proteolysis, the rate of the process is not determined merely by the activity of proteolytic enzymes in the fish, neither in capelin (Gildberg and Raa, 1980) nor in sardines (Marvik, 1974). Tissue pH, physical/chemical condition of the connective tissue, and the presence of enzyme inhibitors are additional factors which may affect the rate of the process.

The sarcoplasmic fraction of fish muscle was relatively resistant to degradation by the digestive enzymes in

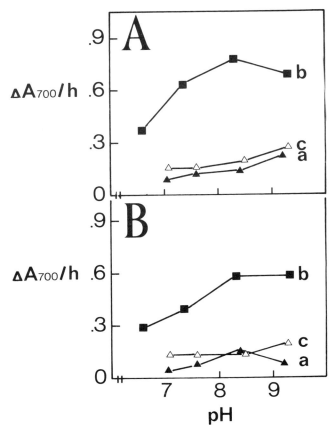

Fig 3 The rate of degradation by a crude enzyme preparation *(A)* and by pure trypsin *(B)* of a mixture of the myofibrilic and sarcoplasma fraction (curves *a*), of the myofibrill fraction alone (curves *b*) and of the sarcoplasmic fraction alone (curves *c*). The substrate concentrations were in *b* and *c* half of that described in Methods. In *a*, the concentrations of each two components were the same as in *b* and *c*.
Abscissae: pH at start of incubation
Ordinates: rate of formation of TCA-soluble Folin-positive substances

capelin, and by purified trypsin. This was apparently due to the presence of enzyme inhibitors, because the sarcoplasmic extract inhibited the enzymatic degradation of haemoglobin, both by the crude enzyme preparation and by purified trypsin. The myofibrillar fraction was a better substrate for the two enzyme preparations than was haemoglobin, but the sarcoplasmic fraction efficiently inhibited also the enzymatic attack on this substrate. This protective action of the sarcoplasmic components on the enzymatic degradation of the myofibrils may explain the relatively slow degradation of minced muscle by the digestive enzymes at alkaline pH. The presence of protease inhibitors further obscures the relationship between the tendency of belly bursting and the activity of proteolytic enzymes.

Inhibitors of trypsin and chymotrypsin were recently shown to be present also in bovine cardiac muscle (Waxman and Krebs, 1978).

It has earlier been observed (Gildberg and Raa, 1979; 1980) that the enzymatic degradation of fish muscle to water-soluble components may occur at a faster rate in the neutral/slightly acid pH range, than at the alkaline pH optimum of the proteases. This has occasionally been found also for the formation of TCA-soluble peptides/amino acids (Hjelmeland, 1978). The observation

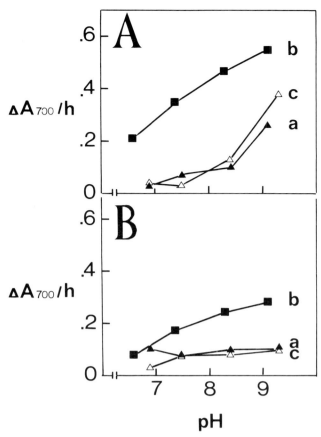

$\Delta A_{700}/h$

Fig 4 Corresponds to *Fig 3*, but with a mixture of haemoglobin and the sarcoplasma fraction (curves *a*), haemoglobin (curves *b*) and sarcoplasma (curves *c*)

than at the alkaline pH optimum of the proteases may be due to a combined action of inhibitors, which suppress proteolytic activity in the alkaline pH range, and of specific solubilizing enzymes, which are most active in the neutral pH range, but these possibilities have not yet been examined. However, with pure trypsin the rate of degradation of myofibrils at slightly acid pH was relatively lower than with the crude enzyme preparation, indicating that other enzymes than trypsin contribute essentially to this proteolysis.

5 Acknowledgements

This work was supported by the Norwegian Council of Fisheries Research (NFFR). The technical assistance by Torunn Arvesen and Elisabeth Grahl-Madsen is acknowledged.

6 References

BARRET, A J. Lysosomal enzymes. In: *Lysosomes – a Laboratory*
1972   *Handbook*, Ed J T Dingle. North-Holland, Amsterdam. 46–135
GILDBERG, A. Proteolytic activity and the frequency of burst bellies in
1978   capelin. *J. Fd. Technol.*, 13, 409–416
GILDBERG, A and RAA, J. Solubility and enzymatic solubilization of
1979   muscle and skin of capelin (*Mallotus villosus*) at different pH and temperature. *Comp. Biochem. Physiol.*, 63B, 309–314
1980   Tissue degradation and belly bursting in Capelin *This volume*
GRANT, D A W, MAGEE, A I and HERMON-TAYLOR, J. Optimisation of
1978   conditions for the affinity chromatography of human enterokinase on immobilised p-aminobenzamidine. Improvement of the preparative procedure by inclusion of negative affinity chromatography with glycylglycyl-aniline. *Eur. J. Biochem.*, 88, 183–189
HJELMELAND, K. Buksprenging hos lodde-Undersøkelse av
1978   fordøyelsesproteasenes rolle i vevsdegraderingen. Thesis, University of Tromsø, Norway. 115pp
JOHNSON, W C and LINDSAY, A J. An improved universal buffer.
1939   *Analyst. Lond.*, 64, 490–492
LOWRY, O H, ROSEBROUGH, N J, FARR, A L and RANDALL, R J. Pro-
1951   tein measurement with the Folin phenol reagent. *J. Biol. Chem.*, 193, 265–275
MARVIK, S. Report to: The Research Laboratory of the Norwegian
1974   Canning Industry, Stavanger, Norway
WAXMAN, L and KREBS, E G. Identification of two protease inhibitors
1978   from bovine cardiac muscle. *J. Biol. Chem.*, 253, 5888–5891

is noteworthy since Nicolaysen (unpublished) has found that sardines which were rejected for canning by the sorters, due to tendency of belly bursting, autolysed faster at pH 6·5 than at pH 8, to TCA-soluble peptides/amino acids.

A faster rate of autolysis at slightly acid/neutral pH

Effect of formaldehyde addition at different ionic strengths on the salt-soluble proteins of fish muscle

M Ohnishi and G W Rodger

1 Introduction

Many studies have been carried out to elucidate the mechanism of fish muscle protein denaturation during frozen storage. It has been shown that trimethylamine oxide (TMAO) in the muscle is converted into dimethylamine (DMA) and formaldehyde (FA) when fish muscle is stored, even after heating (Hughes, 1959) or freezing (Amano and Yamada, 1964; Tokunaga, 1964).

Recently, as a particular need has arisen to freeze and store fish muscle, many studies have been carried out on the formation of FA during storage. Connell (1975) investigated the process of denaturation of frozen muscle to which FA had been added before freezing. After extracting many proteins from frozen muscle with FA-containing buffer, Childs (1973) showed that

increasing FA concentration was associated with decreasing protein solubility. Castell *et al* (1973) added FA to a highly concentrated homogenate which they used as a muscle tissue model and, after measuring the percentage extractable protein nitrogen, suggested that actomyosin (AM) in a fish homogenate reacted with added FA. Recently, Poulter and Lawrie (1979) carried out experiments under the same conditions as Castell *et al*, and suggested that not only myofibrillar proteins but also sarcoplasmic proteins might be rendered insoluble by FA during frozen storage.

Although these studies were carried out with muscle homogenates at low ionic strength it is also important to determine effects at high ionic strength. In the present study, the relationship between protein solubility, FA concentration and ionic strength were examined to give more information about reaction mechanisms of protein denaturation in the presence of FA during frozen storage.

2 Experimental

A salt-soluble protein fraction (SSP–1) was obtained as previously described (Ohnishi and Rodger, unpublished), and the ionic strength and protein concentration were adjusted when necessary by adding concentrated NaCl solution or cold distilled water. FA solution was added to SSP–1 to obtain concentrations listed in *Table I* (studies 1–3). Although the FA solution contained 11% methanol to prevent polymerization, this is thought not to affect protein solubility (Castell and Smith, 1973).

Each sample was left in a chill-room (1–2°C) overnight (15–20h) after adding FA, to allow sufficient reaction time. Aliquots of SSP–1 were dispensed into test-tubes, frozen and stored at −29°C. After storage for the requisite period, each tube was taken out, thawed at room temperature and homogenized with a glass homogenizer. For studies 2 and 3, the ionic strength was adjusted to 0·86 and left 3h for the system to equilibrate. The solutions were then centrifuged for 15min at 2 500rpm, the supernatants collected and their protein concentration recorded.

2.1 Protein concentration measurement
The protein concentrations of the supernatants were measured by the Biuret method.

2.2 SDS-polyacrylamide gel electrophoresis (SDS-PAGE)
As in Ohnishi and Rodger, unpublished SDS-PAGE with 10% acrylamide gels was used for protein analyses of the supernatants.

2.3 Electron microscopy (EM) studies
Each sample was stained by the method outlined previously using 3% uranyl acetate. The stained samples were examined using the JEM 100C and Philips EM 400 electron microscopes.

3 Results

3.1 Effect of high ionic strength on the reaction of formaldehyde with muscle proteins (Study 1)
The solubility change with various FA concentrations (0–0·6% w/w) at high ionic strength ($\mu = 0·86$) during chill and frozen storage is shown in *Fig 1*. In the unfrozen state, samples which contain up to 0·1% FA hardly showed any changes relative to the control but, after frozen storage, they showed a decrease in solubility, the extent of which was related to the concentration of FA. In samples containing more than 0·2% FA, loss of solubility is apparent even in the unfrozen state, but after frozen storage they all reached the same low solubility levels.

Table I

| Study 1 | Ionic strength | 0·86 |
| | Initial protein concentration | 9·7mg/ml |

	Formaldehyde concentration	After chilling protein concentration
	0 (control)	9·7mg/ml
	0·005%	9·3
	0·01	9·0
	0·02	9·2
	0·04	9·3
	0·07	9·3
	0·1	9·3
	0·2	7·1
	0·4	4·7
	0·6	3·9

| Study 2 | Ionic strength | 0·30 |
| | Initial protein concentration | 6·2mg/ml |

	Formaldehyde concentration	After chilling protein concentration
	0 (control)	6·2mg/ml
	0·005%	5·5
	0·01	5·2
	0·02	2·8
	0·04	2·4
	0·07	2·3
	0·1	2·3
	0·2	2·3
	0·4	2·2
	0·6	2·1
	1·0	2·0

| Study 3 | Formaldehyde concentration | 0·01% |
| | Initial protein concentration | 6·2mg/ml |

	Ionic strength	After chilling protein concentration
	0·06	5·7mg/ml
	0·15	5·1
	0·30	5·3
	0·35	5·3
	0·45	6·0
	0·60	6·0
	0·80	5·8
	1·20	5·9

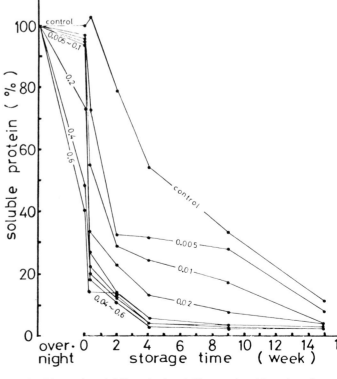

Fig 1 Change in solubility of salt-soluble protein with various formaldehyde concentrations during frozen storage at −29°C at high ionic strength ($\mu = 0·86$)

460

The fact that, after freezing, concentrations <0·1% insolubilize a greater amount of protein than the unfrozen control, suggests that the FA is freeze concentrated to an effective level >0·1%.

When the results in *Fig 1* are rearranged by plotting the relative solubility as the ordinates and the concentration of FA as the abscissae this pattern of solubility loss is shown more clearly (*Fig 2*). From this figure it can be seen that during chill storage, the solubility curves show a very slight inflection point at 0·1% FA, and that during frozen storage solubility is rapidly lost at all FA concentrations.

Why there should be a point of minimum solubility at 0·07–0·1% FA is not clear.

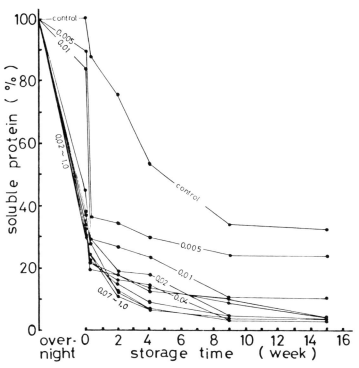

Fig 3 Changes in solubility of salt-soluble protein with various formaldehyde concentrations during frozen storage at −29°C at low ionic strength ($\mu = 0·3$)

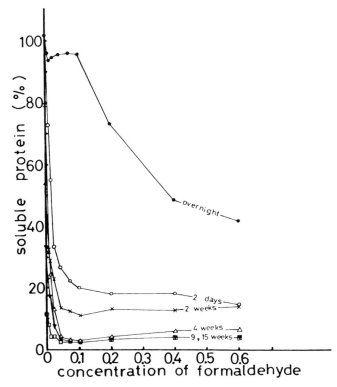

Fig 2 Change in solubility of salt-soluble protein as a function of formaldehyde concentration during frozen storage

3.2 Effect of low ionic strength on the reaction of FA with muscle proteins (Study 2)

The solubility change of SSP–1 containing 0–1·0% w/w FA at low ionic strength ($\mu = 0·30$) during chill and frozen storage is shown in *Fig 3*. The ionic strength was readjusted to 0·86 after thawing. These results therefore relate to the solubilities at $\mu = 0·86$ at ambient temperature. In the unfrozen state, the samples containing up to 0·02% FA showed little change, but the samples containing >0·04% FA showed dramatic changes which were related to the storage time. A big difference is shown when the abscissae are changed as before (*Fig 4*) and this figure compared with the one for high ionic strength (*Fig 2*). At low ionic strength much more protein in SSP–1 is rendered insoluble by a given FA concentration up to 0·1% than at high ionic strength, in the unfrozen state. The difference between 0 and 0·04% FA was dramatic but at concentrations greater than 0·04% there was no further dependence on the FA concentration,

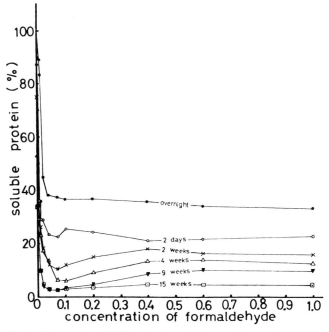

Fig 4 Change in solubility of salt-soluble protein as a function of formaldehyde concentration during frozen storage

indicating that FA at this concentration is in effect a saturated solution as far as its effect on protein solubility is concerned. However, during frozen storage, a further solubility loss occurs, which depends on the storage period. This suggests that there is a fraction of the salt-soluble protein which is not susceptible to FA 'denaturation' in solution but which can be 'denatured' during frozen storage. It may be that this could result from a further freeze concentration of the FA to >0·1%; on the other hand it is unlikely that >0·1% FA would be

461

needed to insolubilize all the myofibrillar proteins if it were the only causative agent of denaturation.

3.3 Effect of various ionic strengths on the reaction of FA with muscle proteins (Study 3)

Since differences exist in the results obtained from high and low ionic strength solutions (*Figs 1–4*), an effect of ionic strength on the mechanism of FA action is probable. *Figure 5* shows the solubility of SSP containing 0.01% w/w FA at ionic strengths $\mu = 0.06–1.2$, during chill and frozen storage. The ionic strengths indicated in *Fig 5* were the values before freezing, and they were all readjusted to 0.86 after thawing. These results therefore relate to the solubilities at $\mu = 0.86$ at ambient temperature. Differences in the apparent effect of ionic strength on the reaction of SSP–1 with 0.01% FA existed up till four weeks but after this period no differences were evident.

The solubility change as a function of the ionic strength in the presence of 0.01% FA is shown in *Fig 6*. A very slight difference was noticed for the chilled sample, the solubility being slightly higher at $\mu = 0.3$. After frozen storage for only two days, however, there was a very noticeable effect. Below $\mu = 0.3$, the solubility loss was considerable but little change is evident at $\mu = 0.6$. This suggests that under conditions of low ionic strength, the act of 'freezing' may be the major cause of initial solubility loss. Under conditions of high ionic strength with 0.01% FA, however, solubility is not affected by the freezing process but is a function of storage period, as evidenced by the 2-, 4-, 9- and 15-week curves.

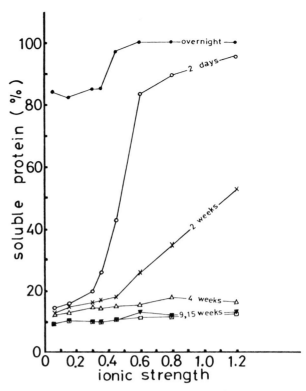

Fig 6   Change in solubility of salt-soluble protein as a function of ionic strength during frozen storage

4 SDS-PAGE analyses

Each supernatant derived from studies 1–3 was used for SDS-PAGE, and the results are shown in *Plates 1–3*.

With the unfrozen samples from studies 1 and 2 (*Plates 1* and *2*) protein bands originating from myofibrillar and sarcoplasmic proteins appeared very clearly, but they disappeared gradually at higher FA concentration ($>0.07\%$), especially the myosin heavy chain (MHC) and low molecular components such as myosin light chains (MLC) and troponins (TN). The Actin (A) and tropomyosin (TM) bands became progressively fainter when the FA content exceeded 0.02%. Correspondingly, the amount of protein at the origin became greater, the proteins having formed aggregates which were too large to enter the gel during electrophoresis, even though they were in a 4% SDS solution which can dissolve certain aggregated proteins (Connell, 1965; 1975). In study 3 (*Plate 3*) the SDS-PAGE patterns were typical of normal salt extracts at high ionic strengths. They also showed that the high molecular components such as MHC and actinin became fainter at low ionic strength.

For the frozen sample, after two weeks' frozen storage, in studies 1 and 2 (*Plates 1* and *2*) the protein bands became fainter at $0–0.1\%$ FA. At 0.04% FA, however, the proteins which actually penetrated the gel became more pronounced though they comprised only six broad bands, and the amount of protein aggregates left at the origin increased significantly. In study 3 the MHC band became progressively fainter with decreasing ionic strength though there was no similar decrease in the band intensity of the low molecular weight proteins such as A, TM, TN and MLC.

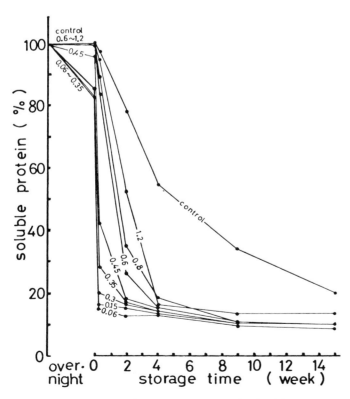

Fig 5   Change in solubility of salt-soluble protein with 0.01% formaldehyde during frozen storage at $-29°C$ at various ionic strengths

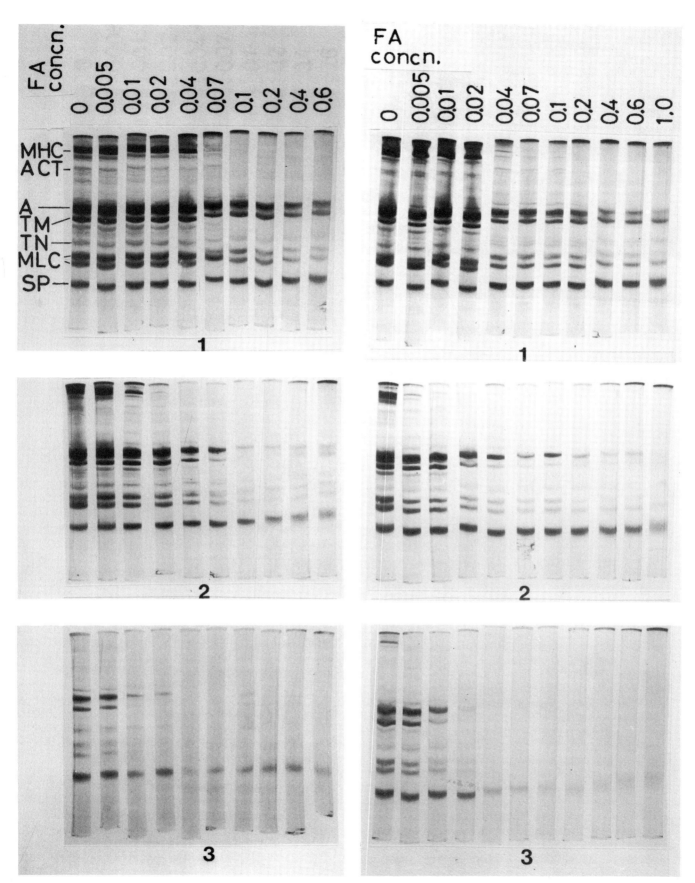

Plate 1 SDS-polyacrylamide gel electrophoresis patterns of salt-soluble proteins in various formaldehyde concentrations in high ionic strength ($\mu = 0 \cdot 86$). (1) unfrozen, (2) two weeks frozen stored samples, (3) 15 weeks frozen stored samples

Plate 2 SDS-polyacrylamide gel electrophoresis patterns of salt-soluble proteins in various formaldehyde concentrations in low ionic strength ($\mu = 0 \cdot 30$). (1) unfrozen, (2) two weeks frozen stored samples, (3) 15 weeks frozen stored samples

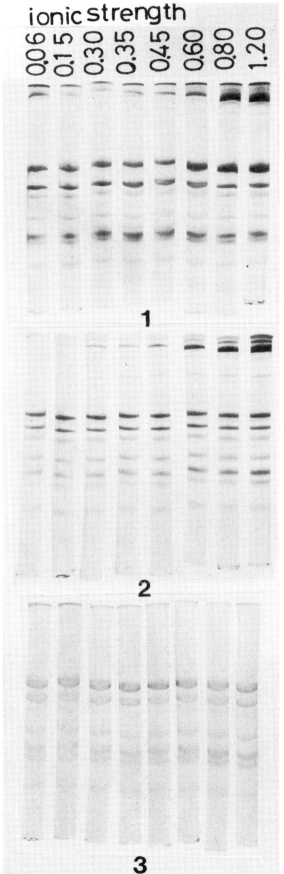

ionic strength

0.06 0.15 0.30 0.35 0.45 0.60 0.80 1.20

1

2

3

Plate 3 SDS-polyacrylamide gel electrophoresis patterns of salt-soluble proteins in various ionic strength with 0.01% formaldehyde. (1) unfrozen, (2) two weeks frozen stored samples, (3) 15 weeks frozen stored amples

5 Electron microscopy studies

EM observations were carried out only on unfrozen samples. In studies 1 and 2, AM filaments were observed in the control (*Plates 4* and *5*). At low FA concentration, 'short' and aggregated filaments were observed, comprising 'short' and 'granular' proteins. At high FA concentrations, there were no AM filaments and only granular particles were observed.

In study 3, aggregated and shortened AM was observed in all the samples (*Plate 6*) indicating that denaturation had occurred.

6 Discussion

It is well known that the muscle of many gadoid fish during frozen storage forms FA by conversion of TMAO to DMA and FA (Amano and Yamada, 1964; Tokunaga, 1964). Although many studies have been carried out to define the effect of FA on the muscle, only descriptions of the effect have been published. In this study, denaturation of a muscle homogenate during frozen storage was studied to elucidate the denaturation mechanism caused by the co-operative actions of FA, ionic strength and storage period. It is shown that the higher the FA concentration in the homogenate, the more denaturation is observed as evidenced by the loss of solubility, the disappearance of protein bands on SDS-PAGE, and the disappearance of AM filaments in EM images. All this supports previous work in this area. These observations also depended in part on the ionic strength of the system. Castell *et al* (1973) showed that cod muscle formed 134ppm FA after 60 days' frozen storage at $-5°C$. When ionic strengths similar to those occurring *in vivo* ($\mu = 0.15$) were used in the present studies, almost all the proteins lost their solubility after the addition of FA to 0.02%, supporting the hypothesis that enough FA can be formed *in vivo* to render all the proteins insoluble.

Several explanations are possible to explain the dependence of solubility of SSP-1 on the ionic strength as shown in *Fig 6*. One is that freezing by itself has a big effect, as shown by Love and Elerian (1964). Another is that FA can easily make intermolecular cross-linkages at low ionic strength because the proteins are already in 'aggregate' form, whereas a dissociating effect of salt exists at high ionic strength. At low ionic strength, ionic and other intermolecular bonds are present. Thus, when FA is present some cross-linking and hence protein aggregation due to covalent bonding is thought to occur easily. At low ionic strengths, salt cannot dissociate ionic bonds because of the number of inter-protein bonds, but when salt is present at high concentrations dissociation is possible. Therefore it seems that if the proteins are soluble at high ionic strength solution the effect of FA is moderated. In 1975, Connell added FA to fresh muscle and found a dependence of protein solubility in SDS on FA concentration. He therefore suggested cross-linkages formed by FA as being one of the causes of denaturation under the conditions used, but expressed doubts as to whether this actually occurred during frozen storage of muscle. As shown in *Plates 1-3*, the higher the FA concentration the more protein there is which does not penetrate the gels. Even although these proteins were soluble in SDS they were still too highly aggregated

464

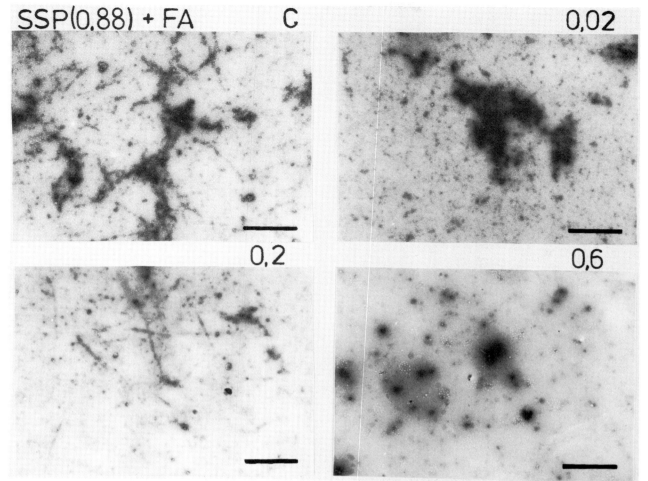

Plate 4 Electron micrographs of salt-soluble protein from fresh cod muscle in various formaldehyde concentration in high ionic strength ($\mu = 0.86$). The samples were negatively stained with 3% uranylacetate. In all electron micrographs the bars represent 1 000nm

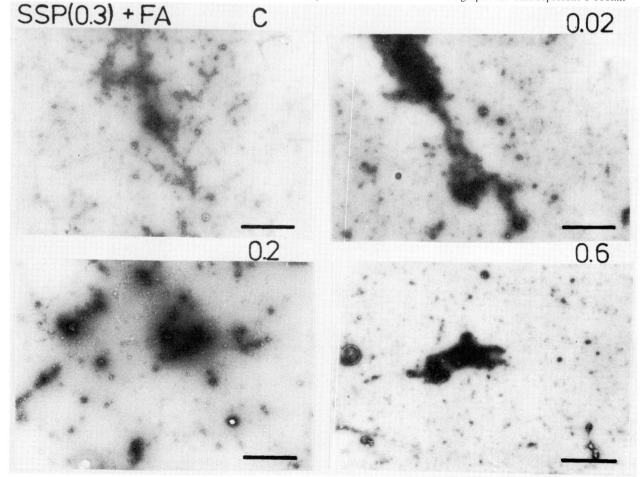

Plate 5 Electron micrographs of salt-soluble protein from fresh cod muscle in various formaldehyde concentration in low ionic strength ($\mu = 0.3$)

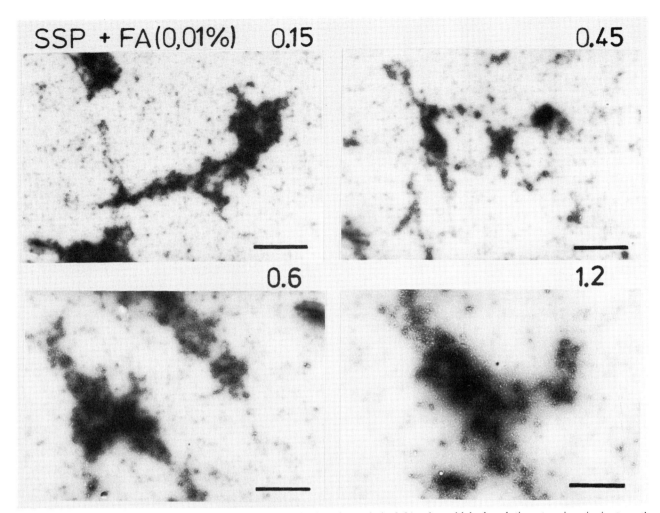

Plate 6   Electron micrographs of salt-soluble protein from fresh cod muscle in 0·01% formaldehyde solution at various ionic strength

to enter the 10% acrylamide gel. At what degree of cross-linking these proteins become totally inextractable even in SDS is not known.

In *Figs 2* and *4* it is shown that each frozen sample exhibits a slight solubility minimum at an FA concentration of 0·07–0·1%. At 0·1% FA, SDS-PAGE showed a disappearance of myofibrillar protein bands after two weeks' frozen storage (*Plate 1*). There are only six protein 'bands' which are not denatured even by high FA concentration; these are considered to be A, TM, TN, MLC, and sarcoplasmic protein. In addition, the intensity of these bands increases with increasing FA concentrations. The reason for this can be explained by the slight increase in protein concentration as shown in *Figs 2* and *4* at an FA concentration of 0·1%. Each gel was loaded with the same volume, therefore each band intensity must be proportional to the concentration of protein. From these results, myofibrillar proteins suffer most denaturation at 0·1% FA after two weeks' frozen storage, whereas sarcoplasmic proteins suffer denaturation at higher FA concentration. Poulter and Lawrie (1979) showed that sarcoplasmic protein which did not denature in NaCl buffer (Dyer, 1951) may be rendered insoluble by adding FA to a muscle homogenate. This study supports that finding, although the reason why the intensity of protein bands increased at 0·1% FA after two weeks' frozen storage on SDS-PAGE gel is not understood. *Figures 2* and *4* show that even after reac-

tion with 0·6 and 1·0% FA in solution overnight, there is still a further solubility loss in the samples during frozen storage. As pointed out previously, this could possibly be a result of an effective increase in FA concentration brought about by freezing, but it was suggested that this was unlikely since the resulting concentrations of FA would be very much in excess of those required for complete protein insolubilization. In addition, such an effect should lead to differences in the rates at which this further solubility loss occurs, depending on the original FA concentration. *Figures 2* and *4* show this is not the case.

It is therefore possible that only some of the proteins in SSP-1 are readily reactive towards FA (in this case 50% at high ionic strength and 60% at low), and that the remainder need to undergo a deconformation process, possibly brought about by freezing, before becoming so, or forming aggregates by another mechanism. This lends support to a similar suggestion advanced by Connell (1975) that the development of inextractability as a result of reaction with FA is greater for thawed frozen-stored cod than for fresh.

7 Acknowledgements

The authors would like to thank B Perry, The Institute of Marine Biochemistry, Aberdeen, for supplying carbon-coated formval meshes, and H H Eichelberger for the use of the Philips EM400 microscope at Aberdeen University.

8 References

AMANO, K and YAMADA, K. A biological formation of formaldehyde in
1964   the muscle tissue of gadoid fish. *Bull. Japan Soc. Sci. Fish.*, 30, 430–435

CASTELL, C H and SMITH, B. Measurement of formaldehyde in fish
1973   muscle using TCA extraction and the Nash reagent. *J. Fish. Res. Bd. Can.*, 30, 91–98

CASTELL, C H, SMITH, B and DYER, W J. Effects of formaldehyde on
1973   salt extractable proteins of gadoid muscle. *J. Fish. Res. Bd. Can.*, 30, 1205–1213

CHILDS, E A. Interaction of formaldehyde with fish muscle *in vitro. J.*
1973   *Fd. Sci.*, 38, 1009–1011

CONNELL, J J. The use of sodium dodecyl sulphate in the study of
1965   protein interactions during the storage of cod flesh at −14°C. *J. Sci. Fd. Agric.*, 16, 769–783

—— The role of formaldehyde as a protein crosslinking agent acting
1975   during the frozen storage of cod. *J. Sci. Fd. Agric.*, 26, 1925–1929

DYER, W J. Protein denaturation in frozen and stored fish. *Food Res.*,
1951   16, 522–527

HUGHES, R B. Chemical studies on the herring (*Clupea harengus*). I.
1959   Trimethylamine oxide and volatile amines in fresh, spoiling and cooked herring flesh. *J. Sci. Fd. Agric.*, 10, 431–436

LOVE, R M and ELERIAN, M K. Protein denaturation in frozen fish.
1964   VIII. The temperature of maximum denaturation in cod. *J. Sci. Fd. Agric.*, 15, 805–809

POULTER, R G and LAWRIE, R A. Studies on fish muscle protein. Nu-
1979   tritional consequences of adding low concentrations of formaldehyde and/or linoleic acid to cod muscle. *Lebensm. Wiss. U. Technol.* 12, 47–51

TOKUNAGA, T. Studies on the development of dimethylamine and for-
1964   maldehyde in Alaska pollack muscle during frozen storage. *Bull. Hokkaido Reg. Fish Res. Lab.*, 29, 108–122

13 Microbiology

Biochemical and fatty acids analyses of coliform bacteria isolated from fish fillets in violet red bile agar

J Gjerde and B Böe

1 Introduction

The coliform bacteria are generally considered to include all aerobic and facultative anaerobic Gram negative non-spore-forming rods which ferment lactose with acid and gas formation within 48h at 37°C.

Violet red bile agar (VRB) is commonly used for the direct enumeration of coliforms in milk, poultry, meat and cooked seafood. Official methods recommend to count as coliforms dark red colonies with a diameter of 0·5mm or greater, with a zone of precipitate (Nordic Committee on Food Analyses, 1975; Fishbein *et al*, 1976). Using this method only the acid production is observed. In a confirmatory procedure pure strains from each colony must be Gram stained and tested for gas production in a lactose medium.

Tennant and Reid (1961) observed that 26·7% of the coliform bacteria isolated in VRB from shellfish and sea water were incapable of rapid lactose fermentation at 35·5°C on subsequent isolation. It is also reported that besides bacteria in the family Enterobacteriaceae members of the family Vibrionaceae can grow in VRB (Rosen and Levin, 1970). As the genera *Vibrio* and *Aeromonas* are normally found on raw fish (Shewan, 1977), detection of such bacteria from fish fillets has little hygienic value.

This study was performed to obtain more information on which types of bacteria grow on VRB when raw fish or fish fillets are investigated.

2 Materials and methods

2.1 Biochemical tests

Samples of fresh, iced fish and commercially produced frozen fish fillets were obtained from different fish plants. From each sample 10g meat or 10cm² from the fish surface was aseptically removed and homogenized with a Stomacher Lab-Blender 400 in 90ml sterile physiological saline.

From the diluted sample, 1ml aliquots were transferred to petri dishes and about 12ml of liquefied VRB (Difco) at 45°C were poured into each dish and the contents were thoroughly mixed before the medium solidified. After solidification a thin layer of VRB was poured on to the surface. The plates were incubated in an incubator at 37°C.

After incubation for 18–24h dark red colonies with a diameter of 0·5mm or greater and with precipitate were collected and transferred to Eosin methylene blue agar (EMB Difco) and incubated at 37°C for about 24h. The strains were checked for purity and reinoculated into MacConkey (Difco) broth and incubated at 37°C for 48h. Strains which produced acid and gas were considered to be members of the Enterobacteriaceae and were not further examined.

Strains which failed to produce acid or gas were Gram stained and tested by a series of biochemical reactions according to the Analytab Products Inc. (API) 20E system for identification of Enterobacteriaceae and other Gram negative organisms (Moussa, 1975).

The biochemical tests include the following reactions:

Detection of beta-galactosidase, arginine dihydrolase, lysine decarboxylase, ornithine decarboxylase, urease and tryptophan deaminase, utilization of citrate as the sole carbon source, H_2S production from thiosulphate, indol formation from tryptophan, detection of acetoin as an intermediary glucose metabolite, liquefaction of gelatin, utilization of the following carbohydrates with acid formation: glucose, mannose, inositol, sorbitol, rhamnose, saccarose, melobiose, amylose, and arabinose, detection of cytochrome oxidase, catalase and reduction of nitrate.

Supplementary test: Anaerobic fermentation of glucose.

Strains which could not be clearly identified by API's procedure were analysed for fatty acid composition of whole cells and compared to standard strains from the families Enterobacteriaceae and Vibrionaceae.

2.2 Standard bacterial cultures

Vibrio parahaemolyticus ATCC 17802 was obtained from the American Type Culture Collection, *Citrobacter freundii* NCIB 11490, *Enterobacter aerogenes* NCIB 10102, *Enterobacter cloacae* NCIB 10101, *Klebsiella pneumoniae* NCIB 48, *Vibrio anguillarum* NCMB 6, *Vibrio alginolyticus* NCMB 1903, *Aeromonas hydrophila* NCMB 86, *Aeromonas salmonicida* NCMB 1102, from the National Collections of Marine and Industrial Bacteria, Torry Research Station, Aberdeen, Scotland, *Proteus morganii* NCTC 2815, *Proteus morganii* NCTC 2818, *Escherichia coli* NCTC 10082 from the National Collection of Type Cultures, Central Public Health Laboratory, England, and *Serratia marscescens* NVH 1026 from the Culture Collection, Veterinary College of Norway.

2.3 Preparation of bacterial cells and derivative formation

The cultures were grown on plate count agar (Difco) and incubated for 18–24h. *Vibrio anguillarum* and *Aeromonas salmonicida* were grown at 25°C, and the other strains were grown at 37°C. For growing the Vibrio cultures the medium was prepared with 2% NaCl.

The cells were carefully removed from the plate surface and transferred to a test tube containing 5ml of 5% NaOH in 50% aqueous methanol.

The cells were saponified, and liberated fatty acids were methylated with BCl_3-methanol, according to the literature (Moss *et al*, 1974).

468

2.4 Gas liquid chromatography

Gas liquid chromatography was performed using a Perkin-Elmer F22 gas chromatograph equipped with flame ionization detector and 60m × 0·25mm id glass capillary column coated with OV–101 (Perkin-Elmer). Operating conditions were: injector 230°C, detector 270°C, oven at 150°C for 2min, then programmed 4°/min to 225°C. One microlitre was injected on the column by the splitless technique (Grob and Grob, 1972; 1974). Quantitation was performed by electronic integration of peak areas using a Spectra Physics SP 4000 data system for chromatography. Tentative identifications were established by retention times relative to 16:0, compared to a standard mixture containing 23 fatty acid methyl esters (FAMEs). The standard was obtained from Supelco Inc. (No. 4–5436).

2.5 Numerical analysis

Raw data from *Table I* were logarithmically transformed according to $x = \ln(x+1)$. Similarities between pairs of bacteria were measured by correlation coefficients and also by the coefficient $1 - D/D_{max}$ based on Euclidean distances D. Clustering was achieved by the unweighted average linkage method (Sneath and Sokal, 1973), and results were presented as phenograms. Further details of the method will be published elsewhere (Böe and Gjerde, 1979).

3 Results and discussion

A total of 350 coliform bacteria were collected from VRB and pure cultures were obtained on EMB. *Figure 1* shows the schematic diagram of the classification procedure of the isolates.

When reinoculated in MacConkey broth only 180 strains (51%) produced acid and gas after incubation at 37°C for 48h. These strains were considered as Enterobacteriaceae and were not further examined.

The other group, 170 strains (49%), grew in MacConkey broth, but failed to produce detectable acid and gas. All these strains were found to be Gram negative rods which utilized glucose fermentatively with formation of acid.

In the identifying procedure according to API's system, the strains could be placed in four groups.

Group one consisted of 100 strains which clearly could be identified as species within the genera *Citrobacter*, *Enterobacter*, *Klebsiella* and *Serratia*. *Table II* shows the distribution of the species within the family Enterobacteriaceae.

Group two consisted of 38 strains which showed close similarity to genera within the family Enterobacteriaceae. The main distinction between these isolates and the species listed in the API manual was different utilization of carbohydrate. However, based on data from Edwards and Ewing (1972) the strains could be identified as members of the family Enterobacteriaceae.

The third group of 10 strains could not be identified according to the system and were listed as unidentified.

The fourth group of 22 strains displayed a weak oxidase reaction and could possibly belong to the family Vibrionaceae.

API's system arose from comparative studies of biochemical reactions given by large numbers of cultures isolated from clinical specimens and strains from culture collections. The system takes into consideration the variability which can exist among species within the family Enterobacteriaceae. Together with additional tests this system can also identify Gram negative rods which are not members of the family Enterobacteriaceae.

In this investigation a relatively large percentage of coliform bacteria isolated in VRB from raw fish (groups 2–4) could not be further identified according to the API system. Selected strains which showed close similarity to species within the family Enterobacteriaceae (group 2), all strains which remained unidentified (group 3), and all strains which could possibly belong to the family Vibrionaceae (group 4) were therefore analysed for the fatty acid composition of whole cells and compared to the standard strains.

Methyl esters of fatty acids ranging from C12 to C20 were identified in the extracts of bacterial whole cells (see *Table I*). Separation of the two families Enterobacteriaceae and Vibrionaceae was performed by utilizing the complete raw data matrix from *Table I*. The correlation matrix obtained for the 57 bacteria revealed that the only representatives from the family Vibrionaceae were the actual five standard strains, Nos. 51–55. Pairwise correlation coefficients within this family were above 0·90 while none of the other bacteria had coefficients above 0·85 with members of this family. With one exception the highest pairwise correlation coefficient for any given bacterium was well above 0·90. Bacterium No. 56 correlated best with No. 30 at a low level of 0·88.

Results comparable to the above were obtained when similarities between bacteria were measured by means of Euclidean distances. None of the five standard strains from the family Vibrionaceae was closely similar to any of the other 52 bacteria. The highest values of similarity coefficients involving Vibrionaceae were obtained with bacteria Nos. 24 and 37. These two bacteria were, however, closely related with others at much higher levels. Thus it may be concluded that there are no strains belonging to the family Vibrionaceae outside the set of standard strains. Of the remaining bacteria, No. 56 had a low similarity index of 0·83 with Nos. 22, 23 and 30, while all other bacteria had maximum values above 0·91. It was therefore decided to remove bacterium No. 56 from further analysis, as it probably did not belong to either of the families Enterobacteriaceae or Vibrionaceae. The conclusion based on fatty acid composition of whole cells is that the strains isolated as coliform bacteria from raw fish in VRB with one exception are related to the standard strains from the family Enterobacteriaceae.

The structure within the family Enterobacteriaceae was studied by basing the analysis on FAMEs >10% as recommended (Böe and Gjerde, 1979). Closely similar results were obtained using either correlation coefficients or the similarity index based on Euclidean distances. The phenogram based on correlation coefficients is shown in *Fig 2*. Bacterium No. 57 correlated best with No. 30 at a low level of 0·86, and was removed as unclassified. All the other 50 bacteria had maximum correlation coefficients above 0·90. As can be seen from *Fig 2* the two last groups joined at a level of 0·77. The smallest group contains four reference strains of a total of seven, while the greatest group includes four

Table I

FATTY ACID METHYL ESTERS ISOLATED FROM BACTERIA, RELATIVE AREAS IN PERCENTAGE

FAME		*12:0*	*13:0*	*14:0*	a– *15:0*	*15:0*	*16:1*	*16:0*	a– *17:0*	Δ *17:0*	*17:0*	*18:1*	*18:0*	Δ *19:0*	*19:0*	*20:0*	br– *16:0*	br– *18:0*
No.		*1*	*2*	*3*	*4*	*5*	*6*	*7*	*8*	*9*	*10*	*11*	*12*	*13*	*14*	*15*	*16*	*17*
	1	11	1	24	1	2	37	100	1	38	3	16	8	2	1	1	0	0
	2	8	0	21	0	1	41	100	0	38	2	11	3	1	0	0	0	0
	3	9	1	18	1	1	34	100	0	67	1	13	2	1	1	0	0	0
	4	10	1	20	1	2	28	100	1	71	1	15	3	2	1	0	0	0
	5	12	1	19	1	1	39	100	0	61	0	15	2	1	1	0	0	0
	6	12	1	19	1	1	22	100	0	82	1	10	2	2	1	1	0	0
	7	14	1	21	1	4	44	100	0	65	3	28	5	2	1	1	0	0
	8	13	1	18	0	3	47	100	0	63	4	31	7	2	1	0	0	0
	9	8	1	18	1	2	51	100	0	58	3	32	7	1	1	1	0	0
Escherichia coli NCTC 10082	10	17	1	29	0	4	43	100	0	33	4	29	4	3	0	1	0	0
	11	8	0	18	0	2	20	100	0	52	5	12	5	2	0	0	0	0
	12	7	0	13	0	2	15	100	0	62	6	7	5	2	3	0	0	0
	13	7	0	19	0	3	11	100	1	55	5	10	5	1	0	0	0	0
	14	7	0	19	0	5	20	100	1	48	27	8	5	3	0	0	0	0
	15	7	1	19	1	3	29	100	0	64	2	29	1	5	0	0	0	0
	16	10	1	19	1	3	39	100	1	57	3	31	3	5	0	0	0	0
	17	10	0	14	1	2	19	100	0	63	2	8	5	4	0	0	0	0
	18	7	0	16	1	3	26	100	0	66	4	13	4	4	0	0	0	0
	19	5	0	20	0	2	20	100	1	66	3	7	4	4	0	0	0	0
	20	25	2	32	1	1	15	100	1	51	2	14	4	4	0	0	0	1
	21	13	0	17	3	2	9	100	0	58	2	5	5	2	0	0	0	0
	22	13	1	21	1	8	68	100	1	26	4	21	2	0	0	0	0	0
	23	10	1	21	1	7	65	100	1	30	3	17	4	0	0	0	0	0
	24	23	2	34	0	4	77	100	1	40	3	20	3	0	0	0	0	0
	25	19	2	37	0	6	48	100	0	45	2	15	3	0	0	0	0	0
	26	14	1	22	1	4	62	100	1	52	3	25	4	1	1	0	0	0
	27	8	1	24	1	5	45	100	1	36	3	7	4	0	0	0	0	0
	28	9	0	21	1	5	43	100	5	37	3	11	8	1	0	0	0	0
	29	7	0	14	1	2	27	100	1	34	4	13	5	0	0	0	0	0
Citrobacter freundii NCIB 11490	30	12	0	28	1	1	41	100	5	10	1	17	3	0	0	0	0	0
Proteus morganii NCTC 2818	31	16	2	24	0	5	56	100	1	34	2	12	1	1	0	1	0	0
	32	9	0	21	1	5	40	100	1	34	3	11	5	1	0	0	0	0
	33	9	0	31	0	2	38	100	0	31	1	5	3	0	0	0	0	0
Proteus morganii NCTC 2815	34	12	1	26	1	4	36	100	0	40	2	8	1	1	1	1	0	0
	35	10	1	27	0	7	33	100	0	30	2	4	3	1	0	0	0	0
	36	9	0	24	1	3	36	100	1	20	1	5	3	2	0	0	0	0
	37	1	0	17	1	1	59	100	1	24	0	46	3	3	0	0	0	0
	38	1	0	14	0	0	48	100	1	34	0	37	3	3	0	0	0	0
	39	3	2	19	1	2	41	100	0	51	0	42	4	7	0	0	0	0
	40	4	0	17	1	2	50	100	1	40	0	38	4	5	0	0	0	0
	41	8	1	18	1	1	31	100	1	38	1	15	5	6	1	1	0	0
	42	6	2	26	0	2	33	100	1	55	0	21	2	5	0	0	0	0
	43	15	1	24	4	3	23	100	8	58	0	13	6	6	0	1	1	1
	44	6	0	13	1	3	7	100	0	75	4	7	6	6	0	0	0	0
	45	7	0	14	1	4	9	100	0	74	5	6	5	7	0	0	0	0
	46	8	1	17	1	6	8	100	0	65	7	5	5	6	0	0	0	0
Klebsiella pneumoniae NCIB 48	47	3	1	39	0	12	12	100	0	69	3	12	4	14	0	0	0	1
Enterobacter aerogenes NCIB 10102	48	3	1	30	0	9	12	100	0	48	5	14	2	14	0	0	0	2
Serratia marscescens NVH 1026	49	2	0	15	0	3	7	100	0	55	3	7	3	10	0	0	0	0
Enterobacter cloacae NCIB 10101	50	3	1	23	0	15	11	100	0	55	6	16	2	7	0	0	0	1
Vibrio parahaemolyticus ATCC 17802	51	13	1	26	3	10	122	100	10	7	22	127	16	0	1	0	20	7
Aeromonas hydrophila NCMB 86	52	31	1	16	1	4	210	100	2	4	4	86	3	0	0	0	2	1
Aeromonas salmonicida NCMB 1102	53	41	3	16	1	19	331	100	19	6	12	59	6	0	0	0	1	3
Vibrio alginolyticus NCMB 1903	54	21	1	36	2	10	169	100	8	13	13	88	7	0	1	0	22	5
Vibrio anguillarum NCMB 6	55	15	0	22	1	2	198	100	4	2	2	73	5	0	0	0	18	3
	56	8	0	23	49	1	59	100	14	12	1	10	17	6	0	15	1	1
	57	21	2	34	6	2	13	100	31	49	3	12	3	3	0	0	1	2

reference strains and 39 strains of bacteria isolated in this laboratory. As a consequence of the small number of standard strains in the latter group, many of the tight clusters do not contain any standards, and this makes it difficult to assign these bacteria to any particular genus. Standard strains representing the genera *Klebsiella*, *Enterobacter* and *Serratia* are not closely related to any of the bacteria isolated by us, and this result was obtained also by using Euclidean distances in the cluster analysis.

Examination of coliform bacteria is commonly used in bacteriological investigation of fish in international trade. To know the limits of VRB when raw fish are investigated, the bacteria which grow in this medium should be identified.

The results from this investigation show that fatty acid

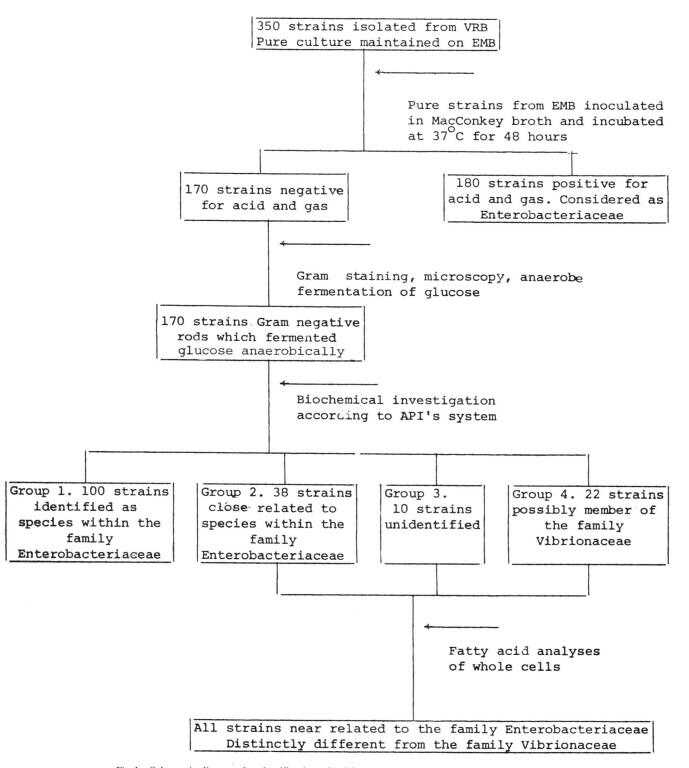

Fig 1 Schematic diagram for classification of coliform bacteria isolated in VRB from raw fish

analyses can prove useful in classification procedures where biochemical investigations do not give fully satisfying results.

R x 100

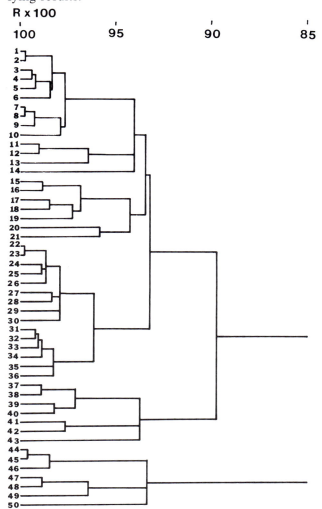

Fig 2 Phenogram of Enterobacteriaceae strains based on 10 fatty acid methyl esters. Numbers as in *Table I*

Table II
DISTRIBUTION OF SPECIES FROM GROUP 1 IDENTIFIED BY API'S SYSTEM

Species	Number of isolates
Citrobacter freundii	24
Enterobacter aerogenes	16
Enterobacter agglomerans	18
Enterobacter cloacae	20
Enterobacter hafnia	4
Klebsiella pneumoniae	5
Klebsiella rhinoscleromatis	5
Serratia marscescens	8

4 References

BÖE, B and GJERDE, J. Fatty acid patterns in the classification of
1979 some representatives of the families Enterobacteriaceae and Vibrionaceae. *J. Gen. Microbiol.* (In press)

EWARDS, P R and EWING, W H. *Identification of Enterobacteriaceae.*
1972 Burgess Publishing Company, Minneapolis, Minn. 3rd edn, 362pp

FISHBEIN, M, MEHLMAN, I J, CHUGG, L and OLSON, I C Jr. *Compen-*
1976 *dium of Methods for the Microbiological Examination of Foods.* American Public Health Association, Washington. Chap 24, 701pp

GROB, K and GROB, K Jr. Methodik der Kapillargas-chromatographie,
1972 I.teil: Die direkteinspritzung. *Chromatographia*, 5, 3–12

GROB, K and GROB, G. Isothermal analysis on capillary columns with-
1974 out stream splitting. *J. Chromatogr.*, 94, 53–64

MOSS, C W, LAMBERT, M A and MERWIN, W H. Comparison of rapid
1974 methods for analysis of bacterial fatty acids. *Appl. Microbiol*, 28, 80–85

MOUSSA, R S. Evaluation of the API, the pathotec and the improved
1975 enterotube system for the identification of Enterobacteriaceae. *New Approaches to the Identification of Microorganisms*, Eds C-G Hedén and T Illeni. John Wiley and Sons, New York. Chap 25, 466pp

Nordic Committee on Food Analyses, No. 44. Statens Livsmedels-
1975 verk, Uppsala, Sweden

ROSEN, A and LEVIN, R E. Vibriosis from fish pen slime which mimic
1970 *Escherichia coli* on violet red bile agar. *Appl. Microbiol.*, 20, 107–112

SHEWAN, I M. The bacteriology of fresh and spoiling fish and the
1977 biochemical changes induced by bacterial action. *Proc: Trop. Inst. Conf. Handl. Process. Mark. Trop. Fish.*, London. 51–66, 511pp

SNEATH, P H A and SOKAL, R P. *Numerical Taxonomy.* W H Freeman
1973 and Company, San Francisco. 573pp

TENNANT, A D and REID, J E. Coliform bacteria in sea water and
1961 shellfish. I. Lactose fermentation at 35·5°C and 44°C. *Can. J. Microbiol.*, 7, 725–731

Induced synthesis of membrane-bound c-type cytochromes and trimethylamine oxide reductase in *Escherichia coli*

M Sakaguchi, K Kan and A Kawai

1 Introduction

Trimethylamine (TMA) is an important component of the smell of spoiled fish and shellfish. TMA formation mainly results from the reduction of trimethylamine oxide (TMO), which is distributed in large quantities in the muscle of marine species by the action of microorganisms during spoilage.

With regard to mechanism of reduction, Neilands (1945) predicted the existence of a redox carrier or series of carriers between oxidizable substrates (electron donors) and TMO in the bacterial cells isolated from cod muscle. Recently, the participation of cytochromes as the carriers in TMO reduction has been reported in *Escherichia coli* (Ishimoto and Shimokawa, 1978;

Sakaguchi and Kawai, 1978a). Low temperature difference spectra of the membrane fragments from *E. coli* grown in the presence of TMO have demonstrated the multiplicity of the cytochrome components (Sakaguchi and Kawai, 1978a). Pyridine haemochrome spectra of the membrane fragments treated with HCl-acetone show the presence of protohaeme and haeme c, suggesting that the cytochromes correspond to at least b- and c-types, respectively. In addition, *E. coli* possesses TMO reductase, which is linked to reduced viologen dyes and able to catalyse the reduction of TMO (Sagai and Ishimoto, 1973; Sakaguchi and Kawai, 1975a). NAD (P) H and formate are effective physiological electron donors in the anaerobic respiration of *Vibrio para-*

472

haemolyticus and *E. coli* (Unemoto *et al*, 1965; Sakaguchi and Kawai, 1977a; Ishimoto and Shimokawa, 1978). In view of this evidence, we have proposed the following scheme as a possible mechanism for the reduction of TMO in *E. coli*.

NAD (P) H

$$\searrow \quad \text{b-Type} \longrightarrow \text{c-Type} \longrightarrow \text{TMO reductase} \rightarrow$$
$$\nearrow \quad \text{cytochromes} \quad \text{cytochromes} \quad \uparrow$$

Formate

Reduced
methyl viologen

$$\longrightarrow \begin{cases} \text{TMO} \\ \downarrow \text{TMA} \end{cases}$$

So far using the whole cells of *E. coli* we have also suggested that TMO reductase is inducible (Sakaguchi and Kawai, 1975a). Both the cytochromes and the enzyme are known to be largely of a membrane-bound nature as well (Sakaguchi and Kawai, 1975b; 1978a,b). This communication, therefore, describes the possibility of induced synthesis of the cytochromes and enzyme bound to the membrane in *E. coli*. The results show clearly that c-type cytochromes and TMO reductase respond to induced synthesis but that b-type cytochromes are hardly induced by TMO.

2 Experimental

2.1 *Growth condition and measurement of cell density*
E. coli (IFO 3301) was grown anaerobically at 37°C on a medium described previously (basal medium) (Sakaguchi and Kawai, 1976) and subsequently TMO was added to cultures at a concentration of 5×10^{-2}M when the cells were grown up to cell density of 0·13 to 0·95, or, when indicated, on a medium supplemented with varied concentrations of TMO instead of only a definite concentration of about 5×10^{-2}M (Sakaguchi and Kawai, 1977a). Cell density was measured at 660nm, expressed as OD660, in a cuvette with 1cm light-path length.

2.2 *Preparation of membrane fragments*
The cells were collected by centrifugation after choloramphenicol was added to cultures at a concentration of $144 \mu g/ml$ to stop further induction of proteins. The harvested cells were washed once with 0·05M potassium phosphate buffer (pH 7·2) and changed to sphaeroplasts by the method of Miura and Mizushima (1969). For determination of haemes, the sphaeroplasts were sonicated at 30s intervals for 1·5min in an ice bath with a sonicator (Kaijo Denki T–A–4201, 20kHz) in 0·1M tris–HCl buffer (pH 7·2) containing 5mM MgCl₂, 0·05M potassium phosphate (pH 7·2), and 0·3mM dithiothreitol. Undisrupted cells were removed by centrifugation at $6\,000 \times g$ for 10min, then the supernatant was centrifuged at $48\,000 \times g$ for 30min to collect membrane fragments. For assay of TMO reductase, the sphaeroplasts were broken osmotically, as described in the previous paper (Sakaguchi and Kawai, 1977a). The subsequent procedures for preparation of the membrane fragments were the same as stated above.

2.3 *Determination of protohaeme and haeme c*
Protohaeme and haeme c in the membrane fragments

were determined after fractionation by HCl acetone, as reported previously (Sakaguchi and Kawai, 1978b).

2.4 *Assay of TMO reductase*
The reaction mixture and the procedures for assay of TMO reductase were virtually the same as reported previously (Sakaguchi and Kawai, 1975a); instead of the cell suspension the membrane fragments (61–202μg protein) were used after suspending in 0·1M sodium phosphate buffer (pH 7·2) containing 0·3mM dithiothreitol. Incubation was performed at 30°C instead of 37°C.

2.5 *Determination of protein*
Contents of protein in the membrane fragments were estimated by the method of Lowry *et al* (1951) with bovine albumin as a standard.

3 Results

3.1 *Enhancement of cell growth by TMO*
It has been reported that *E. coli* can grow anaerobically under the same conditions as adopted in the present experiment and that the cell growth can be enhanced by TMO when it is given to the cells grown up to cell density of 0·43 (later exponential growth phase) (Sakaguchi and Kawai, 1975a). The effect of TMO on cell growth was again tested with the cells in other growth phases. When TMO was fed to the cells growing on the basal medium up to cell density of 0·13 (middle exponential growth phase), enhanced growth was observed about 45min after the addition of TMO (*Fig 1*). The cells fed with TMO at the density of 0·45 (later exponential growth phase) also showed the enhanced growth after about 25min, which is in accord with the period reported previously. In addition, the cells even at the density of 0·90 (earlier stationary growth phase) were subjected to promoted growth by TMO. The promotion appeared more rapidly after about 4min, indicating that the aged cells respond more quickly than the younger cells growing in the middle and later exponential phases.

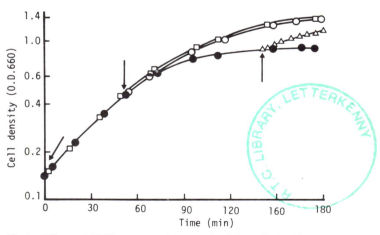

Fig 1 Effects of TMO on anaerobic growth of the cells. TMO was added at a concentration of 5×10^{-2}M to cultures containing the cells growing on the basal medium, when cell growth reached to cell density of 0·13, 0·45 and 0·90, as indicated by arrows. The cell growth after addition of TMO at cell density of 0·13 was represented by the symbol □—□, 0·45 by ○—○, 0·90 by △—△, and no addition of TMO by ●—●

473

3.2 Effects of TMO on the levels of membrane-bound protohaeme, haeme c and TMO reductase

In *E. coli*, b– and o-type cytochromes are the only known cytochromes whose prosthetic group is protohaeme (Castor and Chance, 1959; Jones, 1977). Both types of cytochromes have been found not only in the cells grown aerobically but also in those grown anaerobically or under lower oxygen tensions (Jones, 1977; Rice and Hempfling, 1978). Accordingly, the protohaeme detected in the membrane fragments from the cells grown in the presence of TMO (Sakaguchi and Kawai, 1978b) could be contained in b– as well as o-type cytochromes. The latter pigment, however, functions as a terminal redox carrier linked to molecular oxygen (Haddock and Jones, 1977), most probably having little functional activity in anaerobic respiration such as the reduction of TMO.

Increase of TMO concentration from 0 to 5×10^{-2}M in culture media failed to change protohaeme levels (*Table 1*). On the other hand, the increase resulted in increases in haeme c and TMO reductase. The evidence therefore suggests that TMO is able to induce both membrane-bound c-type cytochromes and TMO reductase but not b-type cytochromes in the cells.

Table 1
LEVELS OF PROTOHAEME, HAEME C AND TMO REDUCTASE IN THE MEMBRANE OF THE CELLS GROWN IN THE PRESENCE OF VARIOUS CONCENTRATIONS OF TMO

TMO added (M)	Protohaeme	Haeme c	TMO reductase
0	0·42	0·03	2·1
2×10^{-4}	0·20	0·08	3·5
2×10^{-3}	0·33	0·22	5·0
1×10^{-2}	0·14	0·32	9·0
5×10^{-2}	0·21	0·39	11·6

After cultivation for 8h in the medium containing various concentrations of TMO, the cells were harvested and submitted to preparation of the membrane fragments. Levels of haemes (protohaeme and haeme c) and TMO reductase were expressed as nmol/mg protein and TMA nmol formed/min/μg protein, respectively.

TMO, at 5×10^{-2}M in the medium gave the greatest values of haeme c and the enzyme as shown in *Table 1*. This was added to cultures grown on the basal medium up to cell densities of 0·48 (0min) and 0·95 (90min) to determine the induction patterns of the two cytochromes and TMO reductase. No significant increase in protohaeme levels was detected in either case (*Fig 2A*), again indicating that TMO is unable to induce b-type cytochromes in the cells. In contrast, addition of TMO at both times resulted in remarkable increases of haeme c and TMO reductase, as compared with the cells growing on the basal medium (*Fig 2B* and *2C*). Haeme c and TMO reductase showed the maximum levels at 100–110min, when addition of TMO was performed at 0min, then both the levels decreased. The apparent similarity of the induction patterns between haeme c and the enzyme suggests the co-ordinated induction of c-type cytochromes and TMO reductase in the cells. From the data in *Figs 2A–C* the levels of haeme c and protohaeme were plotted against the level of TMO reductase in the cells fed with TMO at 0min (*Fig 3*). The increase of

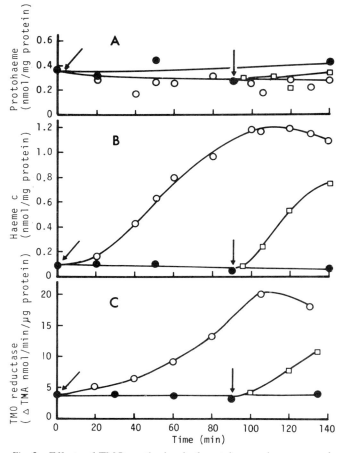

Fig 2   Effects of TMO on the level of protohaeme, haeme c, and TMO reductase in the membrane of the cells. TMO was added to cultures containing the cells growing on the basal medium at 5×10^{-2}M, when the growth reached to the density of 0·48 (0min) and 0·95 (90min), as indicated by arrows. Change in the level of the haemes and enzyme after addition of TMO was represented by the symbol ○—○ for the addition at 0min, □—□ for 90min, and no addition ●—●

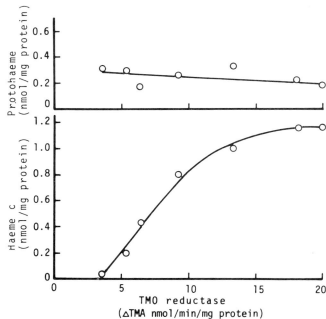

Fig 3   Relation of levels of protohaeme and haeme c to those of TMO reductase in the membrane of the cells. Each value was taken from levels of the haemes and enzyme in the cells fed with TMO at 0min, shown in *Fig 2A, 2B,* and *2C*

haeme was co-ordinated with that of enzyme until it reached a level of about 10nmol TMA formed/min/μg protein but not beyond this level. This observation suggests that synthesis of c-type cytochromes and the enzyme are only partly co-ordinated. Protohaeme, of course, never corresponded to either the enzyme or haeme c over the whole range examined.

3.3 Inhibition by choloramphenicol of the induced synthesis of c-type cytochromes and TMO reductase

Chloramphenicol (CP), a known inhibitor of protein synthesis, prevented the induced synthesis of TMO reductase in whole cells, as reported previously (Sakaguchi and Kawai, 1975a). The effect of CP on syntheses of membrane-bound protohaeme, haeme c and TMO reductase was tested by use of the cells growing up to cell density of 0·43. CP added to cultures 50min after the addition of TMO (0min) failed to affect the change in protohaeme levels in the membrane of the cells (*Fig 4A*). On the other hand, CP promptly prevented further increase in haeme c levels which was produced by TMO (*Fig 4B*). This marked depression made by CP possibly includes the inhibition of *de novo* synthesis of c-type cytochrome protein. Similarly, the inhibition of the enzyme level increase indicates the cessation of *de novo* synthesis of the enzyme protein (*Fig 4C*). This inhibition pattern of the enzyme synthesis is in accordance with that observed in the whole cells (Sakaguchi and Kawai, 1975a).

4 Discussion

The enhancement of growth of the cells by TMO demonstrates an energy gain from the anaerobic respiration coupled to the electron transfer system. There may be a possibility that the enhanced growth results from utilization of TMO and also TMA as a carbon or a nitrogen source, or both. However, this has been excluded by the fact that amounts of TMO plus TMA remained entirely unchanged during anaerobic growth of the cells (Sakaguchi and Kawai, 1975a). Similar enhancement of anaerobic growth by nitrate is well known in *E. coli* (Yamamoto and Ishimoto, 1977; Ishimoto and Shimokawa, 1978) and the role of nitrate as an electron acceptor has been established (Stouthamer, 1976). The associated electron transfer system differs considerably from the system suggested in TMO reduction. In nitrate reduction, a b-type cytochrome (b_1) plays a key role as a redox carrier, being more or less induced by nitrate (Wimpenny and Cole, 1967; Cole and Wimpenny, 1968; Ruiz-Herrela and DeMoss, 1969). In TMO reduction, other b-type cytochromes although in relatively low levels actually function as the carrier (Sakaguchi and Kawai, 1978a,b). These cytochromes are constitutive, since protohaeme in the membrane showed no response to TMO or to the inhibitor of protein synthesis (*Table I, Fig 2A* and *Fig 4A*).

The membrane-bound c-type cytochromes, unlike the b-type pigment, are induced by TMO, as supported by the findings that haeme c markedly responded to TMO (*Table I* and *Fig 2B*) and that the induced synthesis of the cytochrome protein was prevented by the inhibitor (*Fig 4B*). This type of pigment is in fact implicated in the reduction of TMO although other general properties are

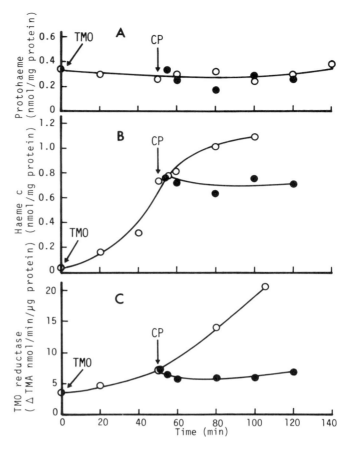

Fig 4 Effects of chloramphenicol on the levels of protohaeme, haeme c and TMO reductase in the membrane of the cells growing in cultures with TMO. TMO (5×10^{-2}M) was added at cell density of 0·04 (0min) and subsequently chloramphenicol (CP) was given at 0·95 (90min), as indicated by arrows. Change in the level of the haemes and enzyme after addition of TMO was represented by the symbol ●—● with CP, and ○—○ without CP

not yet well investigated. Only the low temperature (77K) difference spectrum is known, providing the pigment with an absorption maximum of α peak at 549·5nm and/or 551nm (Sakaguchi and Kawai, 1978a).

The pigment is truly of a membrane-bound nature (Sakaguchi and Kawai, 1978a,b). Data which suggest the occurrence of c-type cytochromes in the membrane are rather scanty in aerobically and anaerobically grown *E. coli*. A well-known c-type cytochrome (C_{552}), which takes part in nitrite reduction, is soluble but not tightly bound to the membrane of *E. coli* (Fujita and Sato, 1966).

The bulk of TMO reductase is also membrane-bound in the cells and catalyses the reduction of TMO to TMA in the presence of reduced viologen dyes (Sagai and Ishimoto, 1973; Sakaguchi and Kawai, 1978a,b), being analogous to nitrate reductase in nitrate reduction. The present results (*Table I, Fig 2C* and *Fig 4C*) thus demonstrate that the membrane-bound enzyme has an additional property of the inducibility. The induction pattern of the enzyme bears an apparent similarity to that of the c-type pigment (*Fig 3*). The induced synthesis of the two proteins was not fully co-ordinated (Wiseman, 1975); the two were simultaneously synthesized for about 1h during which time they might work together to reduce TMO for anaerobic growth of the cells.

475

Concerning other redox carriers for TMO reduction in *E. coli* such substances as flavoprotein, ion-sulphur protein and quinones (ubiquinone and menaquinone) seem to be involved, all of which operate as redox carriers in aerobic respiration as well as in anaerobic respiration on nitrate and fumarate (Haddock and Jones, 1977). Studies along these lines are in progress in our laboratory.

TMO functions as an inducer of membrane-bound c-type cytochromes and TMO reductase, both constituting a part of the electron transfer system in *E. coli*, as described above. Another function of TMO is to accept electrons from oxidizable substrates NAD (P) H and formate via the above redox carriers. This bifunctional role of TMO in cell respiration would practically be reproduced in the course of marine fish spoilage.

First, there should be complete consumption of molecular oxygen dissolved in fish tissues by the bacterial cells and the tissues themselves since the synthesis of TMO reductase and activity of the TMO-reducing system are known to be potently repressed by molecular oxygen (Sakaguchi and Kawai, 1976). Secondly, a specific factor such as redox potential inside or outside the bacterial cells seems to be required for commencing the synthesis of TMO reductase (Sakaguchi and Kawai, 1977b) and presumably that of membrane-bound c-type cytochromes as well. It has been suggested that the redox potential regulates nitrate reductase synthesis (Wimpenny and Cole, 1967; Showe and DeMoss, 1968).

It is under these conditions that the induced synthesis of TMO reductase and probably c-type cytochromes is achieved in the cells. This would in turn facilitate electron transfer from the substrates to the acceptor TMO on the cell membrane and concomitantly produce TMA. This process then provides the cells with energy to grow in anaerobic environments such as spoiling fish tissues, consequently forming more TMA.

5 References

Castor, L N and Chance, B. Photochemical determination of the
1959 oxidases of bacteria. *J. Biol. Chem.*, 234, 1587–1592
Cole, J A and Wimpenny, J W T. Metabolic pathways for nitrate
1968 reduction in *Escherichia coli. Biochim. Biophys. Acta*, 162, 39–48
Fujita, T and Sato, R. Studies on soluble cytochromes in Enterobac-
teriaceae. III. Localization of cytochrome C–552 in the surface layer of cells. *J. Biochem.*, 60, 568–577
Haddock, B A and Jones, C W. Bacterial respiration. *Bacteriol. Rev.*,
1977 41, 47–99
Ishimoto, M and Shimokawa, O. Reduction of trimethylamine
1978 N-oxide by *Escherichia coli* as anaerobic respiration. *Z. Allg. Mikrobiol.*, 18, 173–181
Jones, C W. Aerobic respiratory systems in bacteria. In: *Microbial*
1977 *energetics*, Eds B A Haddock and W A Hamilton. Cambridge University Press, Cambridge. 23–59
Lowry, O H, Rosebrough, N H, Farr, A L and Randall, R J. Pro-
1951 tein measurement with the Folin phenol reagent. *J. Biol. Chem.*, 193, 265–275
Miura, T and Mizushima, S. Separation and properties of outer and
1969 cytoplasmic membranes in *Escherichia coli. Biochim. Biophys. Acta*, 193, 268–276
Neilands, J B. Factors affecting triamineoxidease. I. Inhibition of the
1945 enzyme. *J. Fish. Res. Bd. Can.*, 6, 368–379
Rice, C W and Hempfling, W P. Oxygen-limited continuous culture
1978 and respiratory energy conservation in *Escherichia coli. J. Bacteriol.*, 134, 115–124
Ruiz-Herrera, J and DeMoss, J A. Nitrate reductase complex of
1969 *Escherichia coli* K–12: Participation of specific formate de-hydrogenase and cytochrome b_1 components in nitrate reduction. *J. Bacteriol.*, 99, 720–729
Sagai, M and Ishimoto, M. An enzyme reducing adenosine ^{1}N-oxide
1973 in *Escherichia coli*, amine N-oxide reductase. *J. Biochem.*, 73, 843–859
Sakaguchi, M and Kawai, A. Induction of trimethylamine N-oxide
1975a reductase in *Escherichia coli. Bull. Japan Soc. Sci. Fish.*, 41, 661–665
—— Trimethylamine N-oxide reductase: A membrane-bound enzyme
1975b in *Escherichia coli. Bull. Japan Soc. Sci. Fish.*, 41, 707
—— Effect of oxygen on formation and activity of trimethylamine
1976 N-oxide reductase in *Escherichia coli. Bull. Japan Soc. Sci. Fish.*, 42, 563–569
—— Electron donors and carriers for the reduction of trimethylamine
1977a N-oxide in *Escherichia coli. Bull. Japan Soc. Sci. Fish.*, 43, 437–442
—— Confirmation of a sequence of the events associated with
1977b trimethylamine formation by *Escherichia coli. Bull. Japan Soc. Sci. Fish.*, 43, 611
—— The participation of cytochromes in the reduction of
1978a trimethylamine N-oxide by *Escherichia coli. Bull. Japan Soc. Sci. Fish.*, 44, 511–516
—— Presence of b- and c-type cytochromes in the membrane of
1978b *Escherichia coli* induced by trimethylamine N-oxide. *Bull. Japan Soc. Sci. Fish.*, 44, 999–1002
Showe, M K and DeMoss, J. Location and regulation of synthesis of
1968 nitrate reductase in *Escherichia coli. J. Bacteriol.*, 95, 1305–1313
Stouthamer, A H. Biochemistry and genetics of nitrate reductase in
1976 bacteria. *Adv. Microbial Physiol.*, 14, 315–375
Unemoto, T, Hayashi, M and Miyaki, K. Intracellular localization and
1965 properties of trimethylamine-N-oxide reductase in *Vibrio parahaemolyticus. Biochim. Biophys. Acta*, 110, 319–328
Wimpenny, J W T and Cole, J A. The regulation of metabolism in
1967 facultative bacteria. III. The effect of nitrate. *Biochim. Biophys. Acta*, 148, 233–242
Wiseman, A. Enzyme induction in microbial organisms. In: *Enzyme*
1975 *Induction*, Ed D V Parke. Plenum Press, London. 1–26
Yamamoto, I and Ishimoto, M. Anaerobic growth of *Escherichia coli*
1977 on formate by reduction of nitrate, fumarate, and trimethylamine N-oxide. *Z. Allg. Mikrobiol.*, 17, 235–242

Toxin production by *Clostridium botulinum* type E in smoked fish in relation to the measured oxidation reduction potential (Eh), packaging method and the associated microflora

H H Huss, I Schœffer, A Pedersen and A Jepsen

1 Introduction

Clostridium botulinum type E grows well and produces toxin in lightly salted smoked fish (Kautter, 1964; Cann *et al*, 1965; Johannsen, 1965). However, not all factors governing growth and toxin production are fully understood. Important and well-established factors are storage time/temperature (Kautter, 1964; Cann *et al*, 1965), salt concentration in the aqueous phase (Christiansen *et al*, 1968; Cann and Taylor, 1979), inoculum size and fish species (Cann *et al*, 1965; 1967).

There are no reports relating the Eh of smoked fish to the botulinogenic property, and the effect of growth of other bacteria has only been assessed on raw fish (Abrahamsen *et al*, 1965; Valenzeula *et al*, 1967). As far as the packaging method is concerned it is generally believed that vacuum packing *per se* has little or no influence on toxin production in smoked fish (Thatcher *et al*, 1962; Kautter, 1964; Johannsen, 1965) but contributes to the botulism hazard by extending keeping time and thus providing more time and a better opportunity for toxin formation.

We have recently studied the Eh of fresh and smoked fish (Huss and Larsen 1977; 1979), and the effect of vacuum-packaging and the associated spoilage flora on Eh and toxin production in raw fish has also been evaluated (Huss *et al*, 1979). These studies have now been continued to include smoked fish. The effect on toxin production in smoked fish by packaging them in high concentrations of carbon dioxide and/or oxygen is also presented.

2 Materials and methods

2.1 *Fish*
Hot-smoked herring were prepared and collected at a commercial smoke-house, avoiding uncontrolled contamination as previously described (Huss and Larsen, 1979). The fish were either vacuum-packed singly in polyamide-laminated bags (Rylothene–S) or air-packed in polyethylene bags, or packed in atmospheres containing various amounts of carbon dioxide in Rylothene-S using a Multivac R′7000/S packaging machine.

2.2 *Organisms used in inoculated packs*
A specific spoilage organism isolated from spoiling fish was used. The organism was an aerobic, Gram negative, oxidase and catalase positive motile rod, which produced black colonies (H_2S) in iron-peptone-agar, fermented glucose oxidatively but weakly in Hugh and Leifson's medium, reduced nitrate, liquefied gelatin, contained the ornithine decarboxylase enzyme, and was able to reduce trimethylamine oxide (TMAO) in veal infusion broth (Difco) buffered to pH 6.8. Toxin production was tested using a strain of *C. botulinum* type E, National Collection of Industrial Bacteria (NCIB) 4207, provided by Torry Research Station, Aberdeen, Scotland. A spore suspension was prepared as described by Huss *et al* (1979). Spore counts on the suspensions used were carried out using the five-tube MPN technique. For safety reasons, all physical and chemical analyses were carried out on fish samples which had not been inoculated with *C. botulinum*, but otherwise handled exactly like the inoculated samples.

2.3 *Analysis*
The oxidation-reduction potential (Eh) was measured using a platinum electrode, as previously reported (Huss and Larsen, 1979). In the experiment related to *Fig 2*, the partial pressure of oxygen inside the bags was measured by use of an oxygen electrode (Huss, 1972). In the experiment related to *Fig 3*, however, the gas composition in the packs was measured on a Hewlet-Packard chromatograph, model 5700 A.

Toxin assays were carried out as previously reported (Huss *et al*, 1979).

All tests were carried out in duplicate.

3 Results

In a first experiment freshly smoked herring were inoculated intramuscularly in the loin with $10^2 g^{-1}$ type E spores. Some of the fish were further inoculated in the same place with $10^3 g^{-1}$ of a specific spoilage organism. All fish including controls, which were uninoculated fish handled aseptically after smoking, were vacuum-packed and stored at +15°C. The inoculation of a spoilage organism caused a rapid drop in Eh of the smoked herring (*Fig 1*). However, all fish inoculated with *C. botulinum* were toxic three days after inoculation and packaging. Thus, used in this way, the spoilage organism had little or no effect on toxin production neither *per se* nor via its effect on the Eh. Growth of *C. botulinum* had presumably been initiated at Eh levels of +100 to +250mV in both groups.

In a second experiment, vacuum-packed and air-packed smoked herring were surface-contaminated with a specific spoilage organism and/or type E spores. The results of this experiment are shown in *Fig 2*. The Eh was still positive in fish from all groups at the time when similarly handled fish inoculated with type E were toxic. A reduction in partial pressure of O_2 was found in vacuum-packed fish while in vacuum-packed fish also contaminated with a specific spoilage organism oxygen could not be detected.

Under these conditions toxin production took place as shown in *Fig 2*. In smoked herring surface-inoculated with *C. botulinum* alone and packed in air, toxin production was inconsistent, but six days after packaging toxin could be demonstrated in one sample. In contrast, toxin could be consistently demonstrated in duplicate samples of fish five days after vacuum-packaging. Surface inocu-

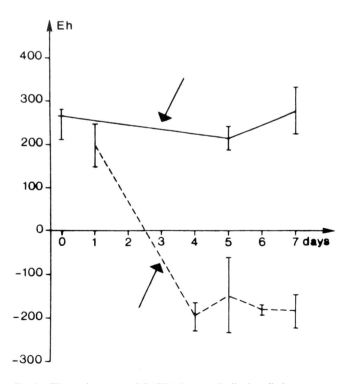

Fig 1 The redox potential (Eh) in aseptically handled, vacuum-packed smoked herring (———) and vacuum-packed smoked herring inoculated in the loin with $10^3 g^{-1}$ of a 'specific spoilage bacterium' (– – – –) and stored at 15°C. Similarly handled fish further inoculated in the loin with $10^2 g^{-1}$ of *Cl. botulinum* type E were toxic three days after packaging (arrow).

477

Code	O_2 tension in bags and and E_h in fish 0-2 days before similarly handled type E inoculated fish were toxic.		Days of storage at $+15^\circ$C																
	E_h – mV	O_2 – mm Hg	1	2	3	4	5	6	7	8	9	10	11	12	13	14	15	16
L	+250 – 0	120 – 80	NT	NT	NT														
V	+200 – 0	30 – 20	NT									←			NT				→
PL	+200 – +150	130 – 100	NT	NT								←			NT				→
PY	+200 – 0	0	NT	NT									←				NT		→

Fig 2   Toxin production in smoked herring which has been surface-inoculated with 10^3g^{-1} *Cl. botulinum* type E spores. Some fish were also surface-inoculated with 10^3g^{-1} of 'a specific spoilage organism' (code mark P). The fish were vacuum-packed (code mark V) or packed with air (code mark L). All samples were stored at $+15^\circ$C. Open area = no toxic sample; hatched area = one sample toxic; black area = duplicate sample toxic; NT = not tested

	Gas composition in pack (%)			Days of storage at $+15^\circ$C															
	N_2	CO_2	O_2	1	2	3	4	5	6	7	8	9	10	11	12	13	14	15
A	99,92	–	0,10	NT								←			NT			→
B	76,73	23,07	0,18	NT								←			NT			→
C	53,77	46,11	0,10	NT								←			NT			→
D	0,30	99,58	0,09	NT								←			NT			→
E	0,47	47,93	51,56	NT														
F	0,29	0,03	99,70	NT														
G	Vacuum packed (98% vac)			NT								←			NT			→
H	Air packed			NT														

Fig 3   Toxin production in smoked herring surface-inoculated with $5 \times 10^1g^{-1}$ type E spores, packed in controlled atmosphere and stored at $+15^\circ$C. Same symbols as in *Fig 2*

lation of the smoked fish with a specific spoilage organism also stabilized toxin production five to six days after packaging in both air-packed and vacuum-packed smoked herring.

In a third experiment, aseptically handled smoked herring were surface-inoculated with $5 \times 10^1g^{-1}$ type E spores, packed in carbon dioxide atmospheres and stored at $+15^\circ$C. Vacuum-packed smoked fish (92% vacuum) and all fish packed in the absence of oxygen were toxic after four days at $+15^\circ$C (*Fig 3*). Some fish packed in air were toxic after six days as before, but nearly two weeks of storage at $+15^\circ$C were required before duplicate samples of air-packed smoked fish became toxic. Increased amounts of oxygen in the atmosphere further repressed toxin production but even in fish stored in pure oxygen some toxin could be demonstrated after nine days. Greatest inhibition of toxin production was noted in fish stored in approximately 50% oxygen and 50% carbon dioxide.

4 Discussion

The effect of Eh and gaseous O_2 on the growth of anaerobes has been variously assessed as reviewed by Morris and O'Brien (1971). They concluded that some clostridia are able to tolerate limited exposure to air and also to create reducing conditions in that medium. Ando and Iida (1970) found germination of type E spores in a complex medium at any level of Eh but outgrowth into vegetative cells only at strictly anaerobic conditions and Eh levels below +198 mV. Barnes and Ingram (1956) found vegetative growth of *C. welchii* at Eh levels up to 231mV, while Pearson and Walker (1976) could grow *C. perfringens* at Eh levels up to +350mV. At this Eh the culture slowly died.

The Eh in smoked herring as measured by our technique (up to +250mV) is not inhibitory to the initiation of growth and toxin production by *C. botulinum* type E. Whether *C. botulinum* is able to create reducing con-

478

ditions is not known from our experiments. However, as Ando and Inoue (1957) have shown that type E is able to reduce TMAO, this organism would also be expected to influence the Eh of the fish (Huss and Larsen, 1979). It has recently been shown by Smoot and Pierson (1979) that *C. botulinum* type A grown at high Eh has increased sensitivity to salt. The high Eh measured in smoked fish may therefore be significant if these fish are also brined. Indeed, the lower salt tolerance of *C. botulinum* type E in smoked trout and mackerel reported by Cann and Taylor (1979) may be partly explained in this way.

While the effect of the Eh prevailing in fish flesh appears to be insignificant our results are in agreement with earlier reports (Abrahamsen *et al*, 1965; Huss *et al*, 1979), which emphasized the significance of available gaseous oxygen. Thus the vacuum-packing *per se* and surface contamination with an aerobic organism—each factor alone or in combination—enhance and stabilize the toxin production in fish surface contaminated with type E. This effect of vacuum-packing is not seen when the fish is inoculated with *C. botulinum* deep into the flesh (Johannsen, 1961; Cann and Taylor, personal communication).

It is probable that vacuum-packing and surface contamination with an aerobic organism exercise their influence by reducing the level of gaseous oxygen in the immediate environment on the fish surface. Also, the reduced toxin production in fish packed in equal amounts of O_2 and CO_2, as compared with pure O_2, may be explained in this way.

Storing perishable foodstuffs in carbon dioxide has long been known as a means of prolonging the storage life, but recently this practice has received new interest (Partman *et al*, 1978; Sander and Soo, 1978). However, Enfors and Molin (1978) found a strongly enhanced germination rate of clostridia in carbon dioxide at atmospheric pressure and therefore pointed to a possible hazard with regard to toxin production of *C. botulinum*. In our work we have not shown any enhanced toxin production in smoked herring packed in an atmosphere of high concentrations of carbon dioxide compared with other forms of packaging where gaseous oxygen is removed. In contrast, we found the safest packaging method in regard to toxin production to be a mixture of equal parts of oxygen and carbon dioxide. However, this may be due to a secondary effect of carbon dioxide, which is known to suppress the growth of aerobic fish organisms (Valley, 1928; Cann, personal communication) and thereby reduces the oxygen consumption of these organisms.

Although our results are in some agreement with those of Ando and Iida (1970) who stressed the necessity of strict anaerobic conditions for growth into vegetative cells of *C. botulinum*, it still remains an alarming fact that aseptically handled smoked herring, packed with air in open bags or even in an atmosphere of pure oxygen, and contaminated on the surface with only approximately $10^2 g^{-1}$ type E spores may be toxic in six to nine days at $+15°C$. Under these circumstances reduced oxygen tension or even anaerobic conditions may exist as minute foci or a thin film on the surface of the fish and below, dependent on the degree of contamination with oxygen-consuming micro-organisms.

5 References

ABRAHAMSEN, K, DE SILVA, N N and MOLIN, N. Toxin production by
1965 *Clostridium botulinum* type E in vacuum-packed, irradiated fresh fish in relation to changes in the associated microflora. *Can. J. Microbiol.*, 11, 523–529

ANDO, Y and INOUE, K. Studies on growth and toxin production of
1957 *Clostridium botulinum* type E in fish products. I. On the growth in relation to the oxidation-reduction potential in fish flesh. *Bull. Jap. Soc. Sci. Res.*, 23, 458–462

ANDO, Y and IIDA, H. Factors affecting the germination of spores of
1970 *Clostridium botulinum* type E. *Jap. J. Microbiol.*, 14(5), 361–370

BARNES, E M and INGRAM, M. The effect of redox potential on the
1956 growth of *Clostridium welchii* strains isolated from horse muscle. *J. Appl. Bact.*, 19, 117–128

CANN, D C, WILSON, B B, HOBBS, G and SHEWAN, J M. The growth
1965 and toxin production of *Clostridium botulinum* type E in certain vacuum-packed fish. *J. Appl. Bact.*, 28(3), 431–436

—— Toxin production by *Clostridium botulinum* type E in vacuum-
1967 packed fish. In: *Botulism 1966*, 202.207 Eds M Ingram and T A Roberts. Chapman and Hall Ltd, London

CANN, D C and TAYLOR, L Y. The control of the botulism hazard in
1979 hot-smoked trout and mackerel. *J. Fd. Technol.*, 14, 123–129

CHRISTIANSEN, L N, DEFFNER, J, FOSTER, E M and SUGIYAMA, H. Sur-
1968 vival and outgrowth of *Clostridium botulinum* type E spores in smoked fish. *Appl. Microbiol.*, 16, 133–137

ENFORS, S-O and MOLIN, G. The influence of high concentrations of
1978 carbon dioxide on the germination of bacterial spores. *J. Appl. Bact.*, 45, 279–285

HUSS, H H. Storage life of prepacked wet fish at 0°C. I. Plaice and
1972 haddock. *J. Fd. Technol.*, 7, 13–19

HUSS, H H and LARSEN, A. The post mortem changes in the
1977 oxidation-reduction potentials on fish muscle and internal organs. In: *Food as an Ecological Environment for Pathogenic and Index Microorganisms*, Eds K Sebolenska-Ceronik, E Ceronik and S Zaleski. Ars Polona, Poland. 265–279

—— Changes in the oxidation-reduction potential (Eh) of smoked fish
1979 during storage. *Lebensmittel-Wissen. und -Tech.* (In press)

HUSS, H H, SCHÆFFER, I, RYE PETERSEN, E and CANN, D C. Toxin
1979 production by *Clostridium botulinum* type E in fresh herring in relation to the measured oxidation-reduction potential. *Nord. Vet. Med.*, 31, 81–86

JOHANNSEN, A. Miljöbetingelser för växt och toxinbildning av *Clos-*
1961 *tridium botulinum* med särskild hänsyn til forhållandene i vacuumförpackede livsmedel. *SIK rapport Nr. 100*. Svenska Livsmedelinstitutet, Göteborg

—— *Clostridium botulinum* type E i rökta fiskvaror. *Svensk Veterinär-*
1965 *tidning*, 5

KAUTTER, D A. *Clostridium botulinum* type E in smoked fish. *J. Fd.*
1964 *Sci.*, 29, 843–849

MORRIS, J G and O'BRIEN, R W. Oxygen and clostridia. A review. In:
1971 *Spore Research*, Eds A N Barker, G W Gould and J Wolf. Academic Press, London, New York. 1–37

PARTMAN, W, BOMAR, M T, HAJEK, M, BOHLING, H and SCHLASZUS,
1978 H. Zur Haltbarkeit von gekühlten Schlachthühnern in kontrollierten Gasatmosphären. *Die Fleischwirtschaft*, 58, 837–843

PEARSON, C B and WALKER, H W. Effect of oxidation-reduction
1976 potential upon growth and sporulation of *Clostridium per-fringens*. *J. Milk Fd. Technol.*, 39(6), 421–425

SANDER, E H and SOO, HONG-MING. Increasing shelf life by carbon
1978 dioxide treatment and low temperature storage of bulk pack fresh chickens packed in nylon/surlyn film. *J. Fd. Sci.*, 43, 1519–1527

SMOOT, L A and PIERSON, M D. Effect of oxidation reduction potential
1979 on the outgrowth and chemical inhibition of *C. botulinum* 10755 A spores. *J. Fd. Sci.*, 44, 700–704

THATCHER, F S, ROBINSON, J and ERDMAN, J. The 'Vacuum Pack'
1962 method of packaging foods in relation to the formation of the botulinum and staphylococcal toxins. *J. Appl. Bact.*, 25(1), 120–124

VALENZEULA, S and NICKERSON, J T R. The effect of growth of other
1967 bacteria in radiation-sterilized haddock tissues on outgrowth and toxin production of *C. botulinum* type E. In: *Botulism 1966*, Eds M Ingram and T A Roberts. Chapman and Hall, London. 224–235

VALLEY, G. The effect of carbon dioxide on bacteria. *Quarterly Review*
1928 *of Biology*, 3, 209–224

Psychrotrophic *Lactobacillus plantarum* from fish and its ability to produce antibiotic substances

K Schrøder, E Clausen, A M Sandberg, and J Raa

1 Introduction

Lactobacilli have not usually been considered to be indigenous to the marine environment. However, Kraus (1961) found lactobacilli in the intestine of herring caught far from populated areas, and our own experience supports this. We have found that addition of glucose to freshly caught capelin, or to fish viscera, always selected lactobacilli, even at very low temperatures. Moreover, lactobacilli usually grew up after the initial phase of anaerobic spoilage of fish and krill at low temperature. It thus seemed evident that fish and krill in the arctic environment harboured cold-adapted lactobacilli. The possible use of such lactobacilli for preservation purposes, not only of fish raw materials, prompted the studies presented here, which describe some biochemical characteristics of one particular isolate from saithe (*Gadus virens*) resembling *Lactobacillus plantarum*, and its ability to produce antibacterial substances.

2 Materials and methods

2.1 *Isolation of bacteria*
Saithe (*Gadus virens*) from the fjord near Tromsø (Norway) was brought live to the laboratory. The gut was dissected aseptically and closed with clamps before samples were withdrawn by syringe from the mid-section of the gut. The samples were diluted in 0·9% aqueous sodium chloride and spread on 4% tryptone soya agar (TSA) plates (Oxoid). The small colourless and catalase negative colonies visible after three to four days at 20°C were maintained on TSA. The isolate described in this paper was chosen among a series with similar properties.

Lactobacilli were alternatively isolated from the gut of fish which were stored whole at 5–15°C for two days after injection of sterile glucose into the peritoneum. The isolate used in the present study has the code name *Lactobacillus plantarum*, UTC 101–11, in our culture collection.

2.2 *Growth media*
The following media were used for growth studies in liquid cultures:

Medium I: 5g glucose, 5g yeast extract (Difco), 5g bactotryptone (Difco) and 2g K_2HPO_4 in 1 litre tap water (pH 6·5).

Medium II (modified after Ford *et al*, 1958): glucose (10g), NaH_2PO_4 (6g), $MgCl_2·6H_2O$ (0·2g), $CaCl_2$ (50mg), $FeCl_3·6H_2O$ (5mg), $ZnSO_4·7H_2O$ (5mg), $MnSO_4·H_2O$ (5mg), $CoCl_2·6H_2O$ (2·5mg), $CuSO_4·5H_2O$ (2·5mg), $Na_2MoO_4·2H_2O$ (2·5mg), pyridoxine·HCl (2mg), nicotinic acid (1mg), Ca-pantothenate (1mg), thiamine·HCl (1mg), *p*-aminobenzoic acid (0·5mg), riboflavine (0·5mg), folic acid (0·25mg), biotine (0·01mg), cobalamine (0·001mg) and bactotryptone (10g) in 1 litre tap water (pH 6·0–6·5).

Medium III: 2g muscle of saithe and 0·1g glucose were homogenized in 100ml tap water (pH 6·5).

The media were sterilized by autoclaving for 20min at 120°C.

2.3 *Lactic acid*
The bacteria in 1ml suspension (Medium II) were precipitated with 9ml 5% perchloric acid, and the concentration of L-lactic acid in the supernatant determined enzymatically, using Lactate Test Combination (Boehringer Mannheim GmbH, Diagnostica). Total acid was determined titrimetrically.

2.4 *Diamino pimelic acid*
Bacterial cells (Medium II) were collected by centrifugation and washed twice in 0·9% NaCl. The cells were disrupted by freeze pressing (X-press–Domkraft AB Nike, Sweden). Undisrupted cells were removed by centrifugation at $3\,000 \times g$ and the cell wall fragments in the supernatant collected at $48\,000 \times g$. The crude cell wall pellet was digested with papain (Powell and Scholefield Ltd) for 24h and dialysed against distilled water for 24h. After concentration at reduced pressure, the cell walls were hydrolyzed in 6N HCl for 24h at 110°C, and analysed with a Jeol, JLC–6AH automatic amino acid analyser. The sample was oxidized with H_2O_2 in formic acid (Moore, 1963) to remove methionine, which otherwise co-chromatographs with diamino pimelic acid.

2.5 *Assays of antibiotics*
(*a*) *Growth kinetics* *Vibrio* sp. isolated from the gut of saithe was grown in liquid shake culture (100ml) in Medium II. A concentrated antibiotic culture filtrate of the lactobacillus isolate was added to the medium after autoclaving prior to inoculation, and growth measured turbidimetrically at 600nm, using a Spectronic 20.

(*b*) *Plate diffusion assay* TSA plates, 8·5cm, adjusted to pH 6·0 with HCl, were surface inoculated with *c* 10^4 of the *Vibrio* sp. The concentrated filtrate of *L. plantarum*, adjusted to pH 6·0 with NaOH, was filled in wells (7mm), and the zone of no growth after 25h at 20°C was measured (mm).

2.6 *Production of antibiotic culture filtrate of L. plantarum*
The *L. plantarum* was grown at 20°C in 250ml of Medium II in 1 litre conical flasks on a reciprocal shaker (100 strokes/min). After eight days the bacteria were removed by centrifugation and the filtrate concentrated 10 times by evaporation at low pressure at room temperature. Before assaying for antibiotics, the pH was adjusted to 6 with NaOH.

Culture filtrates of other bacteria were also added to the growth medium of *L. plantarum* to trigger the production of the antibiotics which were detected by the methods above. The results presented were obtained with *Bacillus thuringiensis* grown for nine days on Medium II (100ml) in 250ml conical shake culture flasks at 20°C. The cells were removed by centrifugation, and

5ml of the centrifugate added per 100ml Medium II, prior to incubation with the *L. plantarum*.

3 Results

The gut of cod, saithe and capelin contained bacteria which formed catalase negative and very small surface colonies on tryptone soya agar (TSA). These bacteria became predominant in the gut of dead whole fish which was stored for a few days at 5–15°C, if sterile glucose was injected into the peritoneum immediately after catch. Microscopical and standard biochemical identification revealed that these bacteria were lactobacilli.

The lactobacillus described in this paper was isolated from the gut content of saithe. It resembled *Lactobacillus plantarum* (Hansen, 1968; Buchanan and Gibbons, 1974) in most characteristics; it was a Gram positive, non-motile short rod which occurred singly, but formed long rods in chains at pH below 5·5. About 20mol% of the amino acids in the cell wall was diamino pimelic acid. It was homofermentative at high glucose levels in the absence of air. When aerated in Medium II at low glucose levels, it formed acetic acid and CO_2, in addition to lactic acid (Bøe, 1979). It was unable to degrade amino acids, except for arginine, which was converted to ornithine and ammonia when the glucose level was low (Jónsson, 1979). Unlike the archetype of *L. plantarum*, the isolate from saithe formed only L-lactic acid, and it was unable to ferment sorbitol. It grew moreover efficiently at low temperatures (*Fig 1*). The cell yield was

highest at 10°C, but the amount of acid produced increased with temperature up to 30°C. The growth yield was very low at 35°C.

The *L. plantarum* from saithe acted bactericidally on a mixed catalase positive bacterial population from the gut of saithe (*Fig 2*). The catalase positive bacteria initially grew fast, but they died rapidly when the lactobacilli started to grow and the pH had dropped to about 5·5. This pH in itself had no bactericidal action, because the variable number of the catalase positive bacteria remained unchanged after 24h in Medium II, even when the pH was adjusted to 4·5 with L-lactic acid.

The number of spoilage bacteria in homogenated fish decreased when lactic acid bacteria grew up as a result of glucose addition, even though the pH of the mince remained above 5·8. This is further support for the assumption that the lactobacilli can inhibit growth of other bacteria by mechanisms other than acid production.

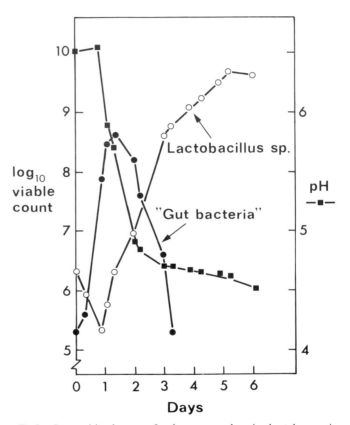

Fig 2 Competition between *L. plantarum* and a mixed catalase positive microflora from saithe gut in Medium III at 20°C. The medium was inoculated with the liquid gut content from saithe and with *L. plantarum* which had been grown on Medium II. The pH shown on the right ordinate

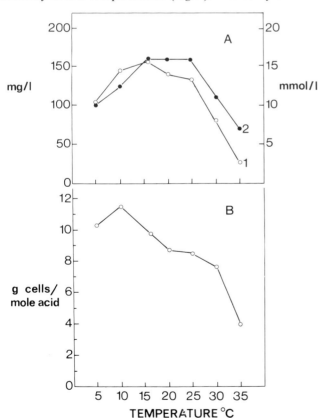

Fig 1 A: Growth yield (1:mg dry cells/litre) of *L. plantarum* and acid produced (2: mmoles/litre) after five days at different temperatures in liquid stand culture on Medium I. The cell density was calculated from a standard curve which related it to A_{600}.
B: Growth yield (g dry cells/mole lactic acid produced)

Nevertheless, filtrates of liquid cultures of *L. plantarum* (Medium II) did not usually contain antibacterial substances which could be detected by the plate diffusion assay or by the growth kinetics assay. However, other bacteria could induce the production of such substances by *L. plantarum* (*Fig 3*). *L. plantarum* produced substances which delayed growth of a *Vibrio* sp., when grown in the presence of a culture filtrate of *Bacillus thuringiensis*. The maximum growth rate of *Vibrio* sp. was not affected by the antibacterial culture filtrate of

L. plantarum, but the duration of the lag phase increased with increasing concentrations. At high concentrations (10% of medium volume) growth of *Vibrio* sp. (and of other bacteria) could be delayed for weeks. The number of viable bacteria was unchanged during the lag phase, even when this lasted for days.

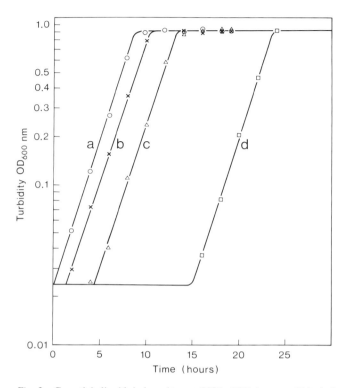

Fig 3   Growth in liquid shake culture at 20°C of *Vibrio* sp. on 100ml of Medium II (a) and the effect on growth of 0·5ml (b) 2·5ml (c) and 5ml (d) of a concentrated (10×) sterile culture filtrate from *L. plantarum* which had been triggered to produce antibiotics by *Bacillus thuringiensis*. The inoculum was *Vibrio* sp. in the maximum stationary phase after growth in Medium II.
Abscissae: incubation time
Ordinates: turbidity (A_{600}) of the culture

The antibacterial culture filtrate *L. plantarum* caused a rapid decrease of the respiration rate of the *Vibrio* sp. (*Fig 4*).

4 Discussion

Fish in the North Atlantic and the Barents Sea contained psychrotrophic lactobacilli. After injection of sterile glucose into the peritoneum of freshly caught fish, and storage for a few days at 5–15°C, the lactobacilli became the predominant bacteria in the gut. This was demonstrated with saithe (*Gadus virens*) and capelin (*Mallotus villosus*). Lactobacilli were selected also in Norwegian fjord krill (*Thysanoessa* sp.) when glucose was added. It thus seems evident that lactobacilli are indigenous in some marine animals, and are not merely contaminants from the terrestrial environment. Unlike typical marine bacteria, however the lactobacilli were not halophilic.

The lactobacillus described here was selected among a series of closely related isolates. Standard identification procedures revealed that it resembled *L. plantarum* so closely that we decided to use this species name. It differed, however, from the archetype by forming only

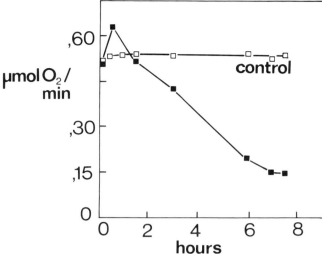

Fig 4   Respiration rate at 20°C of *Vibrio* sp. after increasing time of exposure to the antibiotic culture filtrate of *L. plantarum*, triggered to produce antibiotics by *B. thuringiensis*. Five ml of the 10× concentrated and neutralized culture filtrate of *L. plantarum* was added to 200ml of the *Vibrio* shake culture (5 conical flasks) in the maximum stationary phase (A_{600} 1·3) and 5ml samples withdrawn for measuring of the respiration rate. These samples were diluted 1:4 with water before the O_2 uptake was recorded, using a Clark O_2 electrode. The control shows the respiration rate in the absence of the culture filtrate of *L. plantarum*

L-lactic acid from glucose and by its inability to ferment sorbitol. The *L. plantarum* from saithe grew moreover, most efficiently at low temperature, as expected in view of its habitat. This is essentially different from the archetype of *L. plantarum* (Siegler, 1971).

Lactic acid bacteria can inhibit growth of other bacteria by lowering the pH (Tramer, 1966), by producing hydrogen peroxide (Dahiya and Speck, 1968; Gilliland and Speck, 1975) and by antibiotics (de Klerk and Coetzee, 1961; Reeves, 1972; Upreti and Hinsdill, 1973; Lindgren and Clevström, 1978a,b). Stimulation of the growth of lactic acid bacteria is therefore a method of preserving food or feed.

Addition of glucose to homogenized fish viscera, or to whole capelin, caused a significant preservation, concomitant with the selection and growth of indigenous lactobacilli. In such cases the number of spoilage bacteria declined, and the pH remained above 5·8. Factors other than the acid produced by the lactobacilli were therefore most likely involved in the suppression of spoilage.

The *L. plantarum* isolate was able to kill a mixed culture of catalase positive bacteria, after an initial phase of rapid growth of the catalase positives. The catalase positive bacteria died before the pH had dropped to a bactericidal level, suggesting that the lactobacillus produces antibacterial principles other than acid. However, filtrates of *L. plantarum* grown in shake culture had no antibiotic activity which could be detected by the methods used in this paper, unless extracts of other bacteria were added to the medium prior to growth of the lactobacillus. The transition from the active growth phase to a phase of rapid death of the catalase positives may accordingly be the result of synthesis of antibiotics in the lactobacilli, induced by the catalase positive bacteria themselves.

B. thuringiensis was used in model experiments to induce the production of antibiotics in *L. plantarum*. Other bacteria had qualitatively the same effect, but the quantitative response varied considerably.

The mode of action of the antibacterial factor(s) produced by *L. plantarum* is not yet known. Growth delay without an effect on the growth rate, as shown in *Fig 3*, might indicate that the antibacterial factor(s) were bactericidal to the majority of the vibrios in the inoculum and that a certain fraction of non-sensitive bacteria retained their normal growth potential. However, direct plate counts have not yet given conclusive results. The antibiotic culture filtrate seems to cause a repairable injury which affects cell division of the *Vibrio* sp., if the pH is 6 to 6·5. At pH 5·5 the inhibition was considerably higher, and at this pH also the growth rate was reduced by the antibiotic culture filtrate, and the bacteria lost viability. This may explain why the catalase positive bacteria in the mixed culture with *L. plantarum* lost their viability rapidly when pH decreased below 6.

The antibiotic culture filtrate caused a rapid decrease of the respiration rate of the *Vibrio* sp. This fact is inconsistent with the observation that the antibiotics delayed growth without affecting the maximum growth rate, unless it was bacteria which were resistant to the antibiotics that grew up. This has not yet been fully established. However, since the effects described in this paper are very sensitive to slight pH changes in the range 5·5 to 6·5, there may be other explanations. Moreover, bacteria in the maximum stationary phase are more sensitive to the antibiotic culture filtrate than bacteria in the logarithmic growth phase. Such phenomena were also observed with the chemically pure antibacterial compound, generated when nitrite was heated in nutrient media (Jónsson and Raa, 1979).

The chemical nature of antibiotics produced by other lactobacilli is still not fully known, but Lactocin 27 produced by *L. helveticus* is a glycoprotein (Upreti and Hinsdill, 1973) and the antibiotic activity of *L. fermenti* is associated with a lipo-glyco-protein (de Klerk and Smit, 1967). The antibiotic activity of several fermented foods and feeds resides in high molecular fractions (Lindgren and Clevström, 1978a,b) and is therefore difficult to detect by the conventional agar plate diffusion assay. The chemical nature of the antibiotic principles generated by *L. plantarum*, after exposure to an extract of *B. thuringiensis*, is in the molecular weight range 700–1 500.

5 Acknowledgement

This work was supported by the Norwegian Research Council of Fisheries (NFFR grant III 403.05).

6 References

BUCHANAN, R E and GIBBONS, N E. *Bergey's Manual of Determinative Bacteriology*. Williams and Wilkins Co, Baltimore, USA. 8th edn. 1974

BØE, U B. Vekst fysiologiske studier over *L. plantarum*, isolert frassei. Thesis, University of Tromsø, Norway. 1979

DAHIYA, R S and SPECK, M L. Hydrogen peroxide formation by lactobacilli and its effect on *Staphylococcus aureus*. *J. Dairy Science*, 51(10), 1568–1972. 1968

FORD, J E, PERRY, K D and BRIGGS, C A E. Nutrition of lactic acid bacteria isolated from the rumen. *J. Gen. Microbiol.*, 18, 273–284. 1958

GILLILAND, S E and SPECK, M L. Inhibition of psychrotrophic bacteria by lactobacilli and pediococci in nonfermented refrigerated foods. *J. Fd. Sci.*, 40, 903–905. 1975

HANSEN, P A. Type strains of lactobacillus species. A report by the taxonomic subcommittee on lactobacilli and closely related organisms. American Type Culture Collection, Rockville, Maryland, USA. 1968

JÓNSSON, S. Thesis, University of Tromsø, Norway. 1979

JÓNSSON, S and RAA, J. Conversion of cystine or cystein to the bacteriostatic compound bis- (2-amino-2-carboxyethyl) trisulfide (bactine), enhanced by ferric chloride and sodium nitrite. *J. Appl. Bact.* (submitted). 1979

DEKLERK, H C and COETZEE, J N. Antibiotics among lactobacilli. *Nature*, 192, 340. 1961

DEKLERK, H C and SMIT, J A. Properties of a *Lactobacillus fermenti* bacteriocin. *J. Gen. Microbiol.*, 48, 309–316. 1967

KRAUS, H. Kurze Mitteilung über das Vorkommen von Lactobazillen auf frischen Heringen. *Arch. Lebensmittelhygiene*, 12, 101–102. 1961

LINDGREN, S and CLEVSTRÖM, G. Antibacterial activity of lactic acid bacteria. 1. Activity of fish silage, a cereal starter and isolated organisms. *Swedish J. Agric. Res.*, 8, 61–66. 1978a

—— Antibacterial activity of lactic acid bacteria. 2. Activity in vegetable silages, Indonesian fermented foods and starter cultures. *Swedish J. Agric. Res.*, 8, 67–73. 1978b

MOORE, S. On the determination of cystine as cysteic acid. *J. Bio. Chem.*, 238, 235–237. 1963

REEVES, P. The bacteriocins. *Molecular Biology, Biochemistry and Biophysics*. Springer, New York, NY. Vol. 11. 1972

SIEGLER, E. Effect of temperature and pH on growth, yield, and macromolecular composition of *Lactobacillus plantarum*. University Microfilms International, London. Cat. No. 7202542, 158pp. 1971

TRAMER, J. Inhibitory effect of *Lactobacillus acidophilus*. *Nature*, 211, 204–205. 1966

UPRETI, G C and HINSDILL, R D. Isolation and characterization of a bacteriocin from a homofermentative lactobacillus. *Antimicrobial Agents and Chemotherapy*, 4(4), 487–494. 1973

14 Low molecular weight compounds

The analysis of a range of non-volatile constituents of cooked haddock (*Gadus aeglefinus*) and the influence of these on flavour

A B Thomson, A S McGill,
J Murray, R Hardy
and P F Howgate

1 Introduction

Over the years many compounds have been isolated from fish that were known or subsequently shown to have characteristics that contribute to flavour. Thus a range of volatile amines, aldehydes, ketones and sulphur compounds has been identified in fish, all of which could contribute to the smell (Laycock and Regier, 1971; Miller III *et al*, 1972a,b and c; McGill *et al*, 1974; 1977; Nishihari, 1976). Similarly, inorganic salts, purine base derivatives, amino acids, peptides, sugars and fats have been isolated which in turn could contribute to the taste and, of course, the overall flavour (Jones, 1967; Konosu, 1973; Partmann, 1973). In view of this it is surprising to observe how little has been done to find out just how much these components actually contribute to the flavour, be it fresh or spoiled.

The object of our studies in fish flavour is to try to rectify this situation by first identifying those components responsible for the flavour of cooked fresh fish and then characterizing the important components of the spoilage flavour.

In this report a description is given of the initial studies concerning the contribution non-volatile components make to the overall flavour of fresh cooked haddock.

2 Experimental

2.1 *Materials*

All fish used were supplied commercially and judged to be two to four days on ice with regard to freshness.

Amino acids, nucleotides, nucleosides, free bases, sugars and sugar phosphates were supplied by Sigma Chemical Co, Kingston upon Thames, Surrey.

Anserine was isolated from cod muscle (Jones, 1955).

Salts and solvents were of Analar grade and were purchased from BDH Ltd, Poole, Dorset.

2.2 *Methods*

2.2.1 *Fish matching experiment* Skinned fillet samples (100g) of cod, haddock, whiting, lemon sole and plaice were cooked in casseroles over boiling water for 30min. Each assessor was presented with a portion of each of the five species and a reference which was a portion of one of these five. The assessors were asked to taste the reference sample then the test samples and decide which, if any, were the same as the reference. They were told that none, one or more than one could be the same.

The assessors were 60 people from the staff of Torry Research Station who attended tasting sessions in groups of six.

2.2.2 *Chemical analyses* Sulphosalicylic acid extracts were used for the determination of amino acids (Mackie and Ritchie, 1974) and perchloric acid extracts for analyses of nucleotides, nucleosides, free bases (Jones and Murray, 1962), sugars and sugar phosphates (Meijbaum, 1939; Bergmyer, 1963). The cooked fish was prepared as described for tasting experiments.

Samples for the analysis of sodium, potassium, magnesium and calcium were ashed at 550°C prior to determination using a Unicam SP 90B atomic absorption spectrophotometer.

Acid digests were prepared for the determination of chloride and phosphorus according to the recommended Torry methods Nos. 5 and 15.

2.2.3 *Column chromatography of flavour isolates* Concentrated, neutralized perchloric acid extracts were eluted from a Sephadex G–15 column (100 × 2·5cm) and 11ml fractions collected for tasting.

2.2.4 *Flavour profile panel* The panel was trained according to the principles outlined originally by Cairncross and Sjostrom (1950) then extended and developed for beer and cider by Clapperton (1973) and Williams (1975), respectively, and will be reported in greater detail elsewhere. The profile consists of a list of terms used for describing the flavour of fish or extracts which are scored on an intensity scale from 0 (absent) to 5 (extremely strong). Panellists were presented with up to six different samples per tasting session and each sample consisted of approximately 7ml solution or 30g cooked fish muscle at room temperature.

3 Results and discussion

At the outset it was decided that there would be technical advantages in choosing to work on a fish that had a strong recognizable flavour. To do this a cross-section of the staff at the laboratories was asked whether they could match samples of white fish. A summary of the results is shown in *Table I*. Of the 60 people involved 24 made a correct identification on a single match. Other tasters thought that several of the fish samples had a flavour identical to the one that they were asked to match and of these 16 included the correct one in their choice. With the exception of plaice, where poor matching was obtained, no single species was matched in a statistically more significant manner than the others. Over all, the results suggest that the people involved, although used to eating fish but untrained in species differentiation, could not identify readily the species they were eating, indicating perhaps that to most people these fish have very similar flavours.

On the basis of these results we chose to work on

haddock because it had the highest number of correct matches even though it had proportionally more false matches than either lemon sole or whiting. Reinforcing this decision was the knowledge that the species is commercially more important than the other two and also its low lipid content, when compared to lemon sole, makes it somewhat easier to handle in flavour studies.

The work was carried out along two complementary lines. In the first of these a highly trained taste panel was used to evaluate flavour isolates prepared from cooked haddock which were then extensively analysed for non-volatile components. Secondly, simulated mixtures of the components based upon the analytical results were assessed by the panel.

Table I
ABILITY OF JUDGES TO MATCH WHITE FISH SPECIES

Species	Number of judges	Number of Correct Responses			Total Matches
		Single match	Multiple matches	Total No. correct matches	
Cod	12	3	6	9	20
Haddock	12	6	4	10	19
Whiting	12	7	2	9	14
Lemon sole	12	6	3	9	15
Plaice	12	2	1	3	14
Total	60	24	16	40	82

Preliminary high vacuum distillation experiments on steamed haddock indicated that the major flavour notes were not volatile. It is appreciated that complete distillation of volatile components under these conditions is not attained (Judson King, 1970) but the flavour of the distillate was slight and divorced from any recognizable haddock flavour. Subsequent to this, the residue from the distillation was fractionated on a gel permeation column and the flavours tasted. This experiment indicated that the major flavour components were of relatively low molecular weight eluting predominantly with the amino acid fraction. With these results in mind the decision was taken to analyse only for the low molecular weight non-volatile compounds and, furthermore, to restrict the work to the major constituents and to those components with known flavours, ie the amino acids, the sugars, the various purine derivatives, the major inorganic ions and the peptide anserine.

For some time considerable emphasis has been laid upon the fact that the relative proportions of non-volatile components and the interactions between them are important to the resultant overall flavour of marine foodstuffs (Jones, 1969; Konosu, 1973). As far as we can ascertain only the taste responses for one or two individual amino acids, nucleotides and a limited mixture of amino acids have been evaluated using concentrations determined for raw fish (Jones, 1961). In view of the possible changes in composition of these components during cooking, for example, due to reactions of amino acids with sugars (Hodge, 1967) and enzymatic breakdown, it was felt that the true relationship between fresh flavour and the composition of the non-volatiles could only be understood by designing our experiments on the amounts present in the cooked product.

The analytical results for raw and cooked haddock are presented in *Tables II–VI*.

The amino acid and anserine analysis (*Table II*) indicates that steam cooking causes little change in the anserine content and with the exception of glutamine, 1-methyl histidine and histidine which decrease in concentration, all the other amino acids increase. The changes however are small and well within the fish to fish variation observed in uncooked fish (Mackie and Ritchie, 1974). This variation is further exemplified by the results of the cooked muscle and liquor analyses (*Table II*) carried out on a single fish caught at a different time of the year from those used for the 'raw/cooked' determinations.

Table II
AMINO ACID ANALYSIS OF RAW AND COOKED HADDOCK (MG AMINO ACID/100G FLESH)

	Raw total	Cooked total	Cooked muscle[1]	Cooked liquor[1]
Taurine	54·1	60·4	31·2	24·9
Aspartic acid	1·3	2·0	0·4	—
Threonine	4·9	5·6	7·8	7·0
Serine	5·3	6·2	3·3	3·3
Glutamic acid	7·6	8·7	2·8	3·2
Glutamine	2·4	1·8	—	—
Proline	3·2	3·8	4·1	4·1
Glycine	21·8	23·2	4·0	3·6
Alanine	18·8	21·4	11·7	10·5
Valine	3·6	4·1	3·4	2·7
½ Cystine	—	—	—	1·1
Cystathionine	0·7	—	3·4	2·4
Methionine	2·4	3·3	2·0	1·8
Isoleucine	2·1	2·8	1·7	1·6
Leucine	2·8	4·9	3·4	3·1
Tyrosine	2·3	3·3	1·6	1·4
Phenylalanine	1·8	2·7	1·4	1·3
β-Alanine	13·4	14·0	1·2	1·2
Ornithine	0·9	0·9	0·2	0·3
Ethanolamine	0·4	0·5	0·2	0·2
Ammonia	5·2	5·9	1·6	1·0
Lysine	8·9	10·1	4·2	3·7
1-Methyl histidine	16·5	15·7	1·6	1·4
Histidine	6·2	5·5	1·2	1·3
Anserine	288·0	288·0	190·0	155·0
Arginine	2·7	3·8	1·0	1·2

Pooled extracts from 5 paired fillets

[1] Residue and liquor (38ml) from cooked haddock (100g)

Levels of NAD, ATP, ADP and AMP are such (see *Table III*) that we are unable to be certain of any changes that might be due to the cooking process. IMP shows a fall on cooking of 20% of the level present in raw muscle. Despite the fall in concentration IMP is by far the largest single purine component present in the cooked sample while inosine shows the largest increase. The amounts of the free bases are fairly similar in the raw and cooked fish. Denaturation of enzymes responsible for the autolysis process normally active during the chilled storage of white fish (Jones, 1955; 1962; Kassemsarn et al, 1963) is clearly demonstrated by the amount of the peptide anserine and the purine base hypoxanthine present in the cooked fish. It is perhaps slightly surprising that the rise in hypoxanthine level is so small considering the large change in the concentration of inosine, but inosine hydrolase activity is known to be the rate determining step in nucleotide breakdown during chilled storage (Jones, 1962) and it would appear that a similar situation exists during heat processing. This marginal change noted for haddock is in agreement with the observation of others with canned herring (Hughes and Jones, 1966).

485

Table III
NUCLEOTIDES, NUCLEOSIDES AND FREE BASES OF RAW AND COOKED HADDOCK

[μmol/g]	Raw	Cooked total	Cooked muscle[1]	Cooked liquor[1]
NAD	0·07	0·03	trace	trace
ATP	trace	0·06	0·07	0·02
ADP	0·29	0·18	0·09	0·02
AMP	0·02	0·09	0·09	0·03
IMP	3·81	2·89	1·93	0·80
Inosine	0·98	1·97	1·18	0·63
Hypoxanthine	0·19	0·25	0·17	0·08
Uracil	0·05	0·05	0·02	0·01
Guanine	0·35	0·39	0·34	trace

[1] Residue and liquor (29ml) from cooked haddock (100g)

The results of the analysis for chloride, metals and sugars are shown in *Tables IV, V* and *VI*. As expected, the values for the metals show no marked difference pre- and post-cooking and those that are found could be explained by biological variation between samples or, in the case of the calcium ion, the inclusion of a little bone in the sample.

Sugar levels (*Table VI*) indicate that while the total ribose value remains constant, there is an increase in non-phosphorylated ribose with a corresponding decrease in the phosphorylated ribose on cooking.

This is not unexpected and is in line with the decrease in inosine monophosphate and associated increase in inosine noted above. The total glucose shows a drop in cooking, almost all the fall being accounted for by the loss of glucose-6-phosphate, whereas only a slight decrease in free glucose is noted. It is conceivable that sugar/amino acid reactions occur and that the free glucose levels recorded represent an equilibrium with removal being matched by production from glucose-6-phosphate.

Steam cooking in casseroles results in the release of liquors but in general people taste the muscle only. It is interesting therefore to consider the differences in distribution of the various components between the muscle and free liquor after cooking. On the whole the amino acids, nucleosides, free bases and chloride composition of the muscle and liquor indicate that these components are homogeneously distributed throughout the aqueous phase whether associated with the free liquor or the residual muscle (*Tables II–VI*). This holds true for chloride irrespective of the amount of liquor obtained after cooking (23–43%). However, nucleotides, sugars and metals are distributed in favour of the muscle so that it was necessary in the preparation of the simulated mixtures to base the compositions on the analytical data of muscle plus liquor.

Table IV
DISTRIBUTION OF CHLORIDE IN FOUR SAMPLES OF COOKED HADDOCK

Weight of fish (g)	136·7	193	91·6	98·2
Free liquor collected (ml)	27	36	20	26
As % total water in fish	24·7	23·3	27·3	33·1
Titre (ml)	12·8	18·0	8·5	11·1
∴ titre/ml liquor	0·47	0·50	0·43	0·43
Water in residual muscle	82·4	118·4	53·3	52·6
Titre (ml)	36·2	55·2	25·7	25·6
∴ titre/ml residual water	0·44	0·47	0·48	0·49
Titre/ml liquor ⎯⎯⎯⎯⎯⎯⎯⎯ Titre/ml residual water	1·07	1·06	0·90	0·88
mg chloride/100g fish	127	135	132	132

Table V
METALS AND CATIONS OF RAW AND COOKED HADDOCK

[μmols/g]	Raw	Cooked total	Cooked muscle[1]	Cooked liquor[1]
Sodium	94·3	86·5	75·2	12·0
Potassium	56·0	63·7	52·9	11·8
Magnesium	8·6	7·9	7·6	0·3
Calcium	0·3	1·2	0·6	0·1
Phosphorus		70·3	59·4	10·7

[1] Residue and liquor (28ml) from cooked haddock (100g)

Table VI
SUGARS AND RELATED COMPOUNDS OF RAW AND COOKED HADDOCK

[μmols/g]	Raw	Cooked total	Cooked muscle[1]	Cooked liquor[1]
Total ribose	7·3	7·0	4·7	1·6
Phosphorylated ribose	5·6	4·7	2·9	0·9
Non-phosphorylated ribose	1·7	2·3	1·8	0·7
Total glucose	3·3	2·5	1·7	0·4
Glucose-6-phosphate	1·9	1·0	0·6	0·1
Free glucose	1·3	1·5	1·0	0·2

[1] Residue and liquor (29ml) from cooked haddock (100g)

Haddock muscle was described as possessing five major flavour notes by the panel, these being salty, sweet, meaty (boiled), chicken-like and boiled cabbage. These five notes were purported to be the major notes in the free liquors also but not in the same proportions, and certainly the flavour was different. How large an influence the texture of the muscle exerts is difficult to determine and an attempt was made to extract the flavour from the cooked fish using perchloric acid. The profile of this extract was not unlike the original muscle but more closely resembled that of the liquors (*Table VII*), suggesting that the protein substrate dramatically influences the response of the taster.

Tasting of the individual groups of compounds when compared to a distilled water blank (*Table VII*) shows the inorganics to be very salty, anserine to be bitter, and the amino acids to be slightly sweet and boiled cabbage in character.

The successive addition of the synthetic mixtures to the amino acids results in a gradual build-up of the major flavour notes characterized for haddock (*Table VII*) and changes the character of the meaty note.

Although anserine depresses the major notes and changes the character of the meaty note the total mixture shows a great deal of similarity with the cooked fish and the PCA extract (*Fig 1*). The build-up of flavour is not a simple additive process, confirming that synergistic effects play an active role in the overall integrated fish flavour. It is not possible to say which components specifically produce this synergistic effect and to do so would require a more detailed study using single compounds.

The analytical studies in this work and those carried out by others show the content of the individual components to vary to a marked extent so that the integrated effect of these may well account, at least in part, for the known flavour variations within the species.

To summarize, these experiments confirm that the non-volatile components are responsible for the major flavour notes associated with cooked haddock.

486

Table VII
FLAVOUR PROFILE RESPONSES

	Water blank	Inorganics	Sugars	Nucleotides, etc	Amino acids	Anserine	Amino acids +nucleotides, etc	Amino acids +nucleotides, etc +sugars	Amino acids +nucleotides, etc +sugars +inorganics	Amino acids +nucleotides, etc +sugars +inorganics +anserine	PCA extract	Liquor	Muscle
Overall intensity	1·18	2·73	1·26	1·75	1·79	1·59	2·58	2·67	3·42	3·27	3·08	3·35	2·17
Salty	0·02	2·07	0·09	0·29	0·30	0·09	0·75	0·67	1·50	1·36	1·15	1·53	0·57
Sweet	0·34	0·21	0·11	0·29	0·56	0·36	1·17	1·00	1·25	1·27	0·75	1·02	1·05
Acid	0·10	0·31	0·17	—	0·17	0·09	0·08	0·25	0·09	—	0·08	0·22	0·08
Bitter	0·55	0·54	0·67	0·82	0·54	1·00	0·67	0·67	—	0·18	0·39	0·38	0·10
Metallic	0·50	0·15	0·31	0·19	0·31	0·55	0·17	0·25	0·09	—	0·22	0·26	0·22
Astringent	0·02	0·03	0·03	0·12	0·03	0·05	—	0·08	—	—	0·03	0·06	—
Meaty (boiled)	—	0·07	0·03	0·04	0·14	—	0·50	0·25	1·24	0·36	1·04	0·96	1·08
Meaty (roast)	0·02	0·03	—	—	0·17	—	0·08	0·25	0·09	0·73	0·14	0·32	0·08
Creamy	0·02	0·18	—	0·04	0·07	0·18	0·17	0·17	0·42	0·18	0·20	0·28	0·27
Milky	0·05	—	—	—	—	—	—	—	0·18	0·27	0·03	0·04	0·10
Curry	—	—	—	—	—	—	—	—	—	—	—	0·02	0·02
Chicken-like	—	0·16	—	0·04	0·11	0·09	0·25	0·33	0·93	0·64	0·62	0·68	0·65
Musty	0·32	0·17	0·09	0·21	0·17	0·09	0·33	0·17	—	0·09	0·42	0·31	0·20
Cardboard	0·18	0·19	0·12	0·16	0·22	0·18	0·25	0·08	0·04	—	0·38	0·29	0·25
Turnip	0·14	0·21	—	0·19	0·22	0·09	0·17	0·33	0·27	0·27	0·37	0·37	0·24
Boiled cabbage	0·09	0·22	0·03	0·12	0·41	—	0·92	0·92	1·11	0·55	0·84	0·80	0·30
Oily	0·04	0·21	—	—	0·06	0·27	0·08	0·08	0·23	0·27	0·30	0·45	0·08
Sour milk	—	—	—	—	0·03	—	0·08	—	0·05	0·18	0·06	0·09	0·02
Cheese	—	—	—	—	0·05	—	—	—	—	—	0·01	0·05	0·02
Rubber	0·04	—	—	—	—	—	—	—	—	—	0·06	0·10	0·06
Sulphide	0·05	—	—	—	0·03	—	0·08	—	—	—	0·06	0·07	0·04
Soapy	0·02	—	—	—	—	0·09	0·08	—	—	—	0·12	0·15	0·14
Rancid oil	—	—	—	—	0·03	—	—	—	0·05	—	0·19	0·08	0·02
Painty	0·02	—	—	—	—	—	—	—	—	—	0·04	0·01	—

Although we have not yet comprehensively analysed other cooked white fish species it is not unreasonable to expect that the majority of the components will be present in approximately similar amounts. Since they obviously contribute to a large part of the overall flavour of the species this would explain the difficulty that the judges experience in distinguishing between species.

In future work it is hoped to overcome the difficulties arising from the tasting of solutions by presenting synthetic mixtures of a protein matrix.

4 Acknowledgements

We are grateful to Miss A Cheyne who organized the taste panel experiments, to A Ritchie for the amino acid analysis and I Robertson for the metal analysis.

5 References

BERGMYER, H U. *Methods of enzymatic analysis.* III, 1196, 1238
1963
CAIRNCROSS, S E and SJOSTROM, L B. Flavour profiles—A new
1950 approach to flavour problems. *Fd. Technol.*, 4, 308–311
CLAPPERTON, J F. Derivation of a profile method for sensory analysis
1973 of beer flavour. *J. Inst. Brew.*, 79, 495–508
HODGE, J E. In: *Chemistry and Physiology of Flavours.* Avi Publishing
1967 Co, Westport, Conn. 465–491
HUGHES, R B and JONES, N R. Measurement of hypoxanthine concen-
1966 tration in canned herring as an index of the freshness of the raw
materials, with a comment on flavour relations. *J. Sci. Fd.
Agric.*, 17, 434–436
JONES, N R. The free amino acids of fish. *Biochem. J.*, 60, 81–87
1955
—— Fish flavours. *Proc: Flavour. Chem. Symp.*, Campbell Soup Co,
1961 Camden, N J. 61–81
—— *Chemistry and Physiology of Flavours.* Avi Publishing Co, West-
1967 port, Conn. 267–295

—— Meat and fish flavours. *J. Agric. Fd. Chem.*, 17, 712–716
1969
JONES, N R and MURRAY, J. Degradation of adenine and hypoxanthine
1962 nucleotide in the muscle of chill-stored trawled cod (*Gadus
callarius*). *J. Sci. Fd. Agric.*, 13, 475–480
JUDSON KING, C. Freeze drying of foodstuffs. CRC critical reviews in
1970 *Fd. Techn.*, 1, 379–451
KASSEMSARN, B, SANPEREZ, B, MURRAY, J and JONES, N R. Nucleo-
1963 tide degradation in the muscle of iced haddock (*Gadus aeglefinus*), lemon sole (*Pleuronectes microcephalus*), and plaice
(*Pleuronectes platessa*). *J. Fd. Sci.*, 28, 28–37
KONOSU, S. The taste of seafood. *Shokuhin Kogyo*, 16, 65–73
1973
LAYCOCK, R A and REGIER, L W. Trimethylamine producing bacteria
1971 on haddock (*Melanogrammus aeglefinus*) fillets during refrigerated storage. *J. Fish. Res. Bd. Can.*, 28, 305–309
MACKIE, I M and RITCHIE, A H. Free amino acids of fish flesh. *Proc:
1974 IV Int. Congress. Food Sci. and Technol.*, 1, 29–38
MCGILL, A S, HARDY, R, BURT, J R and GUNSTONE, F D. Hept-cis-4-
1974 enal and its contribution to the off flavour in cold stored cod.
J. Sci. Fd. Agric., 25, 1477–1489
MCGILL, A S, HARDY, R and GUNSTONE, F D. Further analysis of the
1977 volatile components of frozen cold stored cod and the influence of these on flavour. *J. Sci. Fd. Agric.*, 28, 200–205
MEIJBAUM, W. Uber die Bestimmung Kleiner Pentosemengen,
1939 insbesondere in Derivaten der Adenylasaure. *Z. Physiol.
Chemice.*, 258, 117–120
MILLER, III, A, SCANLAN, R A, LEE, J S and LIBBEY, L M. Volatile
1972a compounds produced in ground muscle tissue of canary rockfish. *J. Fish. Res. Bd. Can.*, 29, 1125–1129
—— Volatile compounds produced in sterile fish muscle (*Sebastes
1972b mellanops*) by pseudomonas perolens. *Appl. Microbiol.*, 25,
257–261
—— Quantitative and selective gas chromatographic analysis of
1972c dimethyl and trimethylamine in fish. *J. Agr. Fd. Chem.*, 20,
709–711
NISHIHARI, K. The components of fish smells. *J. Fish Sausage*, 205,
1976 65–91
PARTMANN, W and SCHLASZUS, H. Uber die muster ninhydrinpositiver
1973 substanzen in muskelgewebe von knochenfischen. *Z. Lebensm.
u. Forsch.*, 152, 8–17
WILLIAMS, A A. The development of a vocabulary and profile assess-
1975 ment method for evaluating the flavour contribution of cider
and perry aroma constituents. *J. Sci. Fd. Agric.*, 26, 567–582

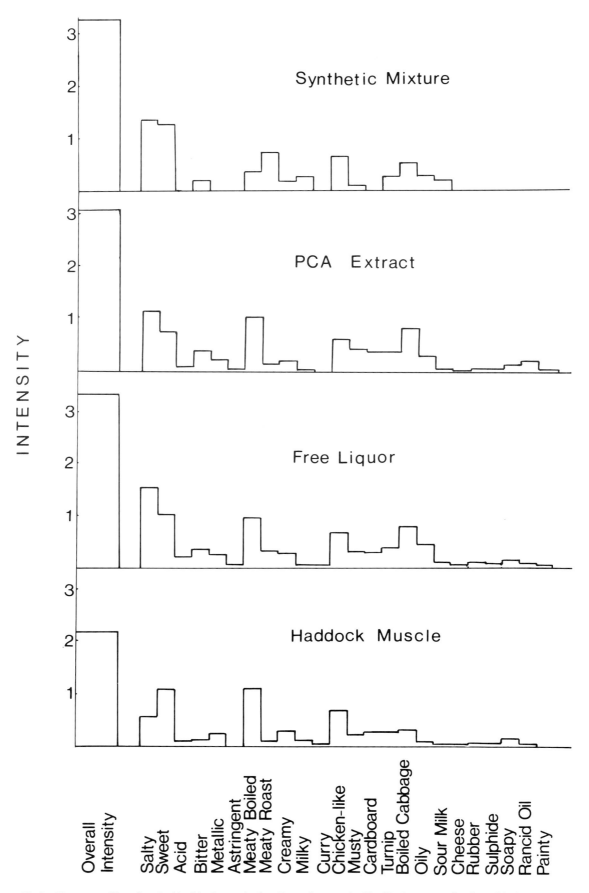

Fig 1 Flavour profiles of cooked haddock muscle, free liquor from cooked haddock, a neutralized perchloric acid extract of cooked haddock muscle plus free liquor and a synthetic mixture based on the analysis of cooked haddock muscle plus liquor

488

The formation of diamines and polyamines during storage of mackerel (*Scomber scombrus*)

A H Ritchie and I M Mackie

1 Introduction

It is well established that amines are formed during the spoilage of fish mainly by the action of bacterial enzymes on substrates such as trimethylamine oxide or free-amino acids. Much early work on amines was concerned with the determination of total volatile bases which, although non-specific, have been shown to be useful indices of the spoilage of fish during chilling. Tri-methylamine has been determined specifically, mostly by a procedure involving the formation of the picrate salt, and its concentration in the flesh has been shown to be equally useful in quality assessment, particularly of white fish (Shewan *et al*, 1971). Although it is known that trimethylamine is also produced in pelagic species (Hughes, 1959) it has been less often used as an index of quality; one example is in the spoilage of mackerel (Smith *et al*, 1980).

Developments in chromatographic procedures have made it possible to determine both volatile and non-volatile amines specifically. Concentrations of di-methylamine and trimethylamine, for example, can be determined routinely by GLC (Mackie and Thomson, 1974) and the higher non-volatile amines can be separated satisfactorily by GLC/HPLC and conventional liquid chromatography on ion exchange columns (Gehrke *et al*, 1977). Relatively little information is available on the formation of amines such as putrescine, cadaverine, spermidine and spermine during bacterial spoilage of fish flesh. These amines are produced by enzymes of spoilage bacteria acting on free-amino acids (*Fig 1*).

A considerable amount of information is available on the production of histamine in fish such as tuna and mackerel because of its implication with scombroid poisoning, an allergic-type condition often accompanied by nausea and headaches, which some individuals experience after eating these types of fish (Arnold and Brown, 1978). There is controversy over the role of histamine in scombroid poisoning since other species of fish such as herring (Hughes, 1959) which do not cause the condition also produce histamine during spoilage. It is possible that other amines present may act as syner-gists but this has not been clearly established (Arnold and Brown, 1978; Sinell, 1978). Histamine has been determined specifically by a fluorometric reaction with *o*-phthalaldehyde (OPT) (Shore, 1971) but analytical systems now available (Tabor and Tabor, 1973; Gehrke *et al*, 1977) allow for its determination routinely together with the other non-volatile amines of interest.

In this paper the production of non-volatile diamines and polyamines has been followed during the storage of mackerel and herring as part of a programme aimed at obtaining more information on their production in vari-ous species of fish. It is hoped to identify differences which may be important in explaining the cause of scombroid poisoning.

2 Materials and methods

2.1 *Fish samples*

Freshly caught mackerel (*Scomber scombrus*) and her-ring (*Clupea harengus*) were stored under three con-ditions. One batch of each species was held in ice at +1°C, one in an incubator at +10°C and one in an insulated box at an ambient temperature of about 25°C.

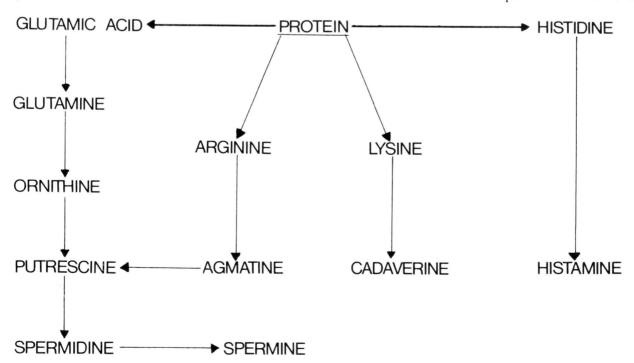

Fig 1 Biological pathways for the formation of non-volatile amines putrescine, histamine, cadaverine, spermidine and spermine from amino acids

The fish were sampled twice weekly at three- and four-day intervals for the samples at +1°C and at 24-hourly intervals for those at +10°C and +25°C. All the fish were held ungutted during storage and at least five fish were sampled at each storage time.

2.2 Reagents

2.2.1 Standards
Histamine, putrescine, cadaverine, spermidine and spermine were obtained as their hydrochloride salts from Sigma London Chemical Company. Solutions of the amines in 0·2M hydrochloric acid were prepared to give 1mg amine free base per ml of solution. These standards were kept refrigerated when not in use. A working standard solution of required concentration was made up weekly by dilution of the stock solutions. Standards in the range 0–20 µg of putrescine and 0–40 µg of the other amines were used to calibrate the analytical system.

2.2.2 Buffers
Ninhydrin buffer (4M sodium acetate pH 5·5)
Sodium acetate trihydrate (2 720g) was dissolved in 2l of hot distilled water, transferred to a 5l volumetric flask and 500ml glacial acetic acid added. The solution was cooled, the pH adjusted to 5·5 and made to the mark with distilled water.

Buffer 1
Trisodium citrate dihydrate (171·55g) and sodium chloride (584·4g) were dissolved in distilled water. Octanoic acid (0·5ml) 30% w/v Brij 35 (25ml) were added, the solution brought to pH 5·8 and made up to 5l.

Buffer 2
Trisodium citrate dihydrate (171·55g) and sodium chloride (628·3g) were dissolved in distilled water. Octanoic acid (0·5ml), 30% w/v Brij 35 (25ml) were added, the solution brought to pH 10·25 and made up to 5l.

0·2M sodium acetate buffer pH 4·6
Anhydrous sodium acetate (40·5g) and glacial acetic acid 29·3ml were dissolved in 5l distilled water and the pH adjusted to 4·6.

2.2.3 Ninhydrin
2-methoxy-ethanol (3 125ml), 4M sodium acetate buffer pH 5·5 (1 250ml) and distilled water (625ml) were mixed together and thoroughly purged with nitrogen. Ninhydrin (60g) was added to the solution which was stirred until all the ninhydrin had dissolved. Finally, stannous chloride dihydrate (1·667g) was added with continued stirring and maintaining a stream of nitrogen through the solution throughout the procedure.

2.3 Methods

2.3.1 Preparation of the extracts of muscle
Portions of the muscle (50–100g) from fillets of the fish were homogenized in three volumes of 5% w/v trichloroacetic acid (TCA) using an Atomix homogenizer. The slurry was filtered through Whatman 2V filter paper and aliquots of the filtrate from each treatment pooled. At least five fish from each treatment were sampled. The TCA solutions were extracted twice with 100ml diethyl ether to remove the TCA. The aqueous phases were taken to

dryness on a rotary evaporator and then dissolved in 25ml 0·2M sodium acetate buffer pH 4·6 prior to analysis.

2.3.2 Pre-column fractionation of extract
Prior to analysis the extracts (1–10ml) were applied to the top of the column (50 × 12mm id) of Amberlite CG 50 (Type 1 100–200 mesh) previously equilibrated with 0·2M sodium acetate buffer pH 4·6. The free amino acids which would otherwise interfere with the elution profile of the amines during the analysis on the amino acid analyser were eluted from the resin with 50ml 0·2M acetate buffer pH 4·6.

The adsorbed amines were released from the resin by washing the resin column with 50ml 0·2M hydrochloric acid.

2.3.3 Separation and determination of amines
The amines were separated on a Locarte amino acid analyser using a column of LA/48 resin (100 × 9mm) nominally 8% cross-linked and with a bead size of 8–12 µ. Flow rates of the buffers and ninhydrin were 50 and 35ml/h, respectively. A column temperature of 65°C and a reaction bath temperature of 100°C were maintained. A two-buffer system was employed according to the following programme:

Eluant	Time (min)
Buffer 1	80
Buffer 2	120
Regeneration (0·2M NaOH)	10
Equilibration Buffer 1	30
Total run time	240

The eluted amines were determined after reaction with ninhydrin reagent from areas of the peaks calculated by computer from a paper tape record of the mV output from the colorimeters.

The volume of amine extract normally applied to the column varied between 0·1ml and 0·8ml depending on the concentration of amines present.

3 Results and discussion

In our experience cation exchange chromatography lends itself most readily to the routine analysis of a large number of samples of amine extracts. The recorder chart of the standards (*Fig 2*) shows that complete resolution of these five amines can be obtained under our conditions. Histamine is well resolved from cadaverine but in some systems (Gehrke *et al*, 1977) it coelutes with cadaverine. Although the run time is undoubtedly longer than that obtained by Gehrke *et al* (1977) who used narrow-bore columns (2·8mm) the chromatogram compares very favourably with theirs.

Other methods of resolving and determining non-volatile amines have also been used with success. Karmas and Mietz (1978) have demonstrated that the dansyl derivatives can be separated satisfactorily by HPLC with even greater sensitivity that that obtained by Gehrke *et al* (1977). In our system preliminary separation of the amines from the amino acids is preferred to the procedure used by Tabor and Tabor (1973) as in our experience the relatively high concentration of amino

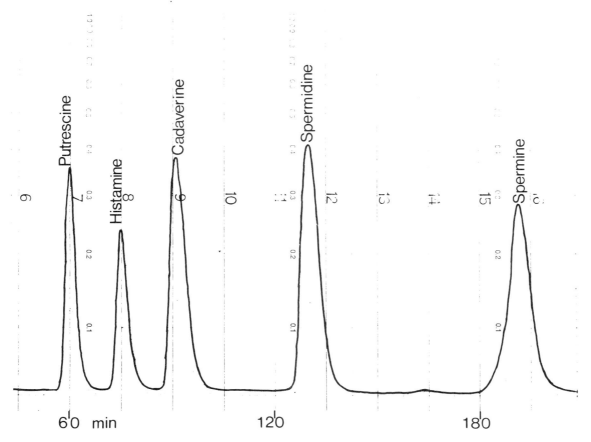

Fig 2 Chromatogram for calibration mixture of putrescine, histamine, cadaverine, spermine and spermidine

acids in extracts of fish muscle tends to raise the baseline prior to the elution of amines and thereby reduces the accuracy of the area determinations. The alternative procedure used by Gehrke *et al* (1977) of diverting the ninhydrin reagent until the amines are eluted has been tried but the problem of 'tailing' due to the relatively large amounts of free amino acids remained. By selecting a system which gave well separated peaks for these known amines it was hoped that if other unidentified amines developed during storage of the fish they would show up on the chromatogram.

Typical chromatograms for spoiling herring and mackerel are given in *Figs 3* and *4*. In addition to these five known amines other unknown peaks were found as the spoilage of fish progressed. They have not been studied further.

Analytical methods such as this which separate the components of a mixture have an obvious advantage over alternative procedures which depend on specific reactions as for example the OPT reaction for the determination of histamine (Shore, 1971). The success of the latter reaction depends on the absence of interfering substances such as histidine or other unidentified amines. A comparison of both procedures demonstrated that for extracts of mackerel and herring the OPT reaction did in fact measure only histamine and that the other amines present did not contribute to any significant extent to the fluorometric reaction (Ritchie *et al*, unpublished).

4 Storage experiments

Tables I and *II* show the concentrations of histamine, putrescine, cadaverine, spermidine and spermine during iced storage of mackerel and herring, respectively. The storage periods were extended well beyond the accepted times for edibility in order to give a full picture of the production of these amines from the fresh to the putrid state. It is of particular interest that the concentrations of histamine even in putrid mackerel are well below 100mg/100g of flesh, a concentration which has been accepted as high enough to cause poisoning (Arnold and Brown, 1978). Surprisingly, the concentrations of histamine in herring are significantly higher than in mackerel during storage beyond seven days; in both species the concentration remains relatively low. At the higher temperatures of storage (*Tables III* and *IV*) the difference in the rate of production of histamine in the two species is accentuated but only in the herring does the level exceed 100mg/100g flesh and then only after 72h at 25°C. These differences in the rates of production of histamine between the two species are unlikely to be due to seasonal variation because the mackerel for the +1°C storage experiment were caught in November 1978 and the herring and mackerel for the other trials were caught in June 1979. While it would appear that the concentration of histamine does not reach the level tentatively said to cause poisoning when the fish are stored in ice it is possible that at higher temperatures the concentration

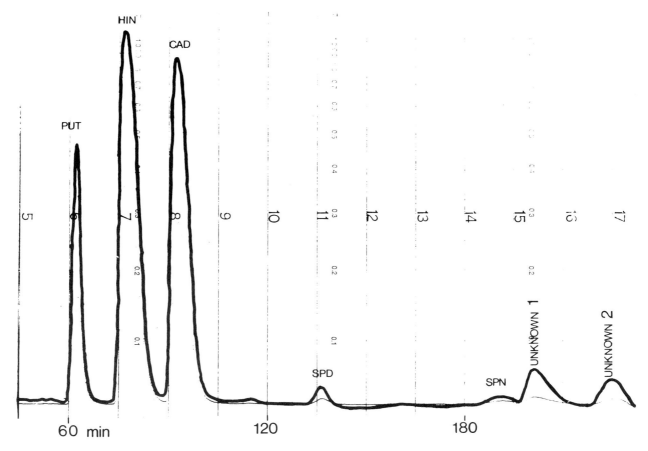

Fig 3 Chromatogram for herring after storage for 48h at 25°C

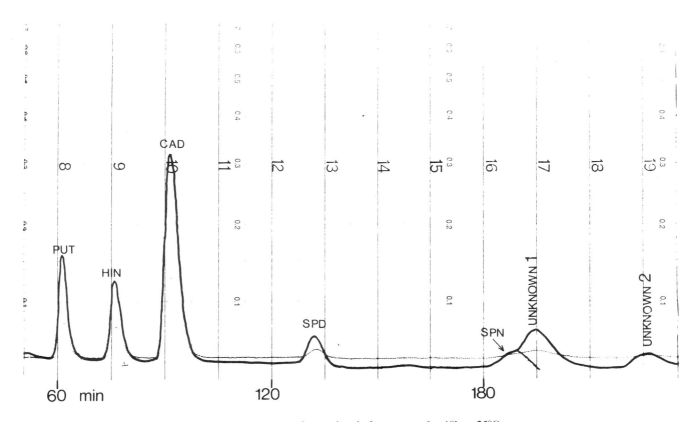

Fig 4 Chromatogram for mackerel after storage for 48h at 25°C

492

Table I
MACKEREL +1°C (MG AMINE/100G MUSCLE)

	Days								
	0	*3*	*7*	*10*	*14*	*17*	*21*	*24*	*28*
Putrescine	0·05	0·08	0·26	0·43	1·13	2·20	2·91	7·06	8·92
Histamine	0·01	0·41	2·16	2·37	5·25	12·04	32·44	52·14	57·94
Cadaverine	0·01	0·31	1·22	2·06	4·73	10·29	15·79	35·50	43·08
Spermidine	0·30	0·07	0·08	0·39	0·47	0·35	0·31	0·47	0·31
Spermine	0·37	0·12	0·09	0·56	0·57	0·43	0·41	0·40	0·55

Table II
HERRING +1°C (MG AMINE/100G MUSCLE)

	Days					
	0	*3*	*7*	*10*	*14*	*17*
Putrescine	0·32	1·09	0·98	1·94	3·02	5·88
Histamine	0·08	0·12	5·82	11·37	25·37	48·83
Cadaverine	0·53	2·32	5·81	11·57	14·78	34·78
Spermidine	0·49	0·46	0·24	0·46	0·16	2·48
Spermine	1·18	1·26	—	—	0·57	1·41

of histamine could rise above it. The danger of poisoning is therefore possibly more likely if the temperature of the fish after catching is not reduced quickly to 0°C. There is evidence that the concentration of histamine can be relatively high in viscera (Salguero and Mackie, 1979) and in the region of the belly wall (Kizevetter and Nasodkera, 1972). Thus removal of the viscera is likely to reduce the chances of high concentrations being produced in localized parts of the fish. It should be noted that Inukai (1976) showed that histamine is produced in thawed mackerel at a rate of about twice that in the unfrozen fish.

These results on mackerel and herring are similar to those described by Karmas and Mietz (1978) in tuna. The concentrations of the polyamines spermine and spermidine appear to decrease during the early stages of storage in ice, but beyond seven days they tend to increase.

A comparison of the formation of the three major amines putrescine, histamine and cadaverine during storage of mackerel and herring at +1°C is given in *Figs 5* and *6*. Both species behave similarly in that cadaverine

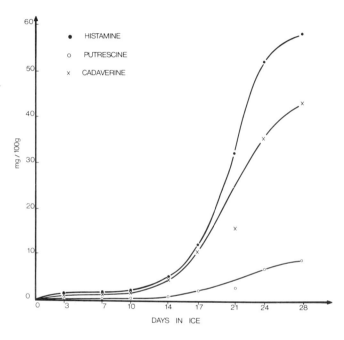

MACKEREL

(Scomber scombrus) at +1°c

- • HISTAMINE
- ○ PUTRESCINE
- × CADAVERINE

Fig 5 Formation of putrescine, histamine and cadaverine during storage of mackerel at 1°C

Table III
PRODUCTION OF AMINES (MG/100G MUSCLE) IN MACKEREL

	Hours at 10°C			Hours at 25°C		
	24	*48*	*72*	*24*	*48*	*72*
Putrescine	0·20	0·32	1·84	0·63	2·46	—
Histamine	0	0	0	0	5·90	—
Cadaverine	0·06	0·14	1·51	0·87	8·56	—
Spermidine	0·58	0·63	0·26	0·57	1·34	—
Spermine	0·92	0·56	0·43	0·68	0	—

Table IV
PRODUCTION OF AMINES (MG/100G MUSCLE) IN HERRING

	Hours at 10°C				Hours at 25°C			
	24	*48*	*72*	*96*	*24*	*48*	*72*	*96*
Putrescine	0·39	0·80	1·90	3·56	0·76	5·28	17·29	—
Histamine	0·61	2·19	15·32	47·18	3·14	59·29	107·36	—
Cadaverine	1·34	4·06	12·58	23·49	4·9	22·72	49·33	—
Spermidine	0·24	0·33	0·42	0·39	0·16	0·47	0·75	—
Spermine	0·25	0·35	0·72	0·81	0·16	0·75	1·57	—

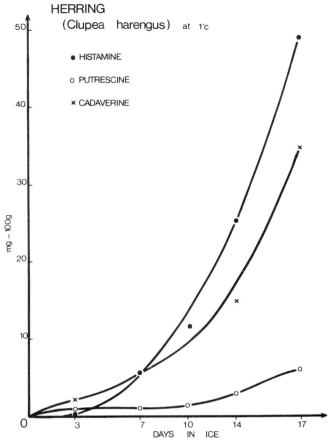

HERRING
(Clupea harengus) at 1°c

• HISTAMINE

o PUTRESCINE

× CADAVERINE

mg – 100g

DAYS   IN   ICE

Fig 6   Formation of putrescine, histamine and cadaverine during storage of herring at 1°C

and histidine are the major amines. In herring it appears that bacterial spoilage is greater than in mackerel stored under the same conditions. Histamine is formed later than putrescine or cadaverine and eventually its concentration reaches a higher level than any of the other amines.

The main conclusion to be drawn from the work is that during iced storage the production of the non-volatile amines is not significant even up to the putrid stage. At

higher temperatures however, the amines are produced in relatively large amounts and it would appear that at least for the formation of histamine, mesophilic rather than psychrophilic organisms are responsible. It is also clear that significant amounts of histamine and other amines can be produced in the non-scombroid species. Further work is necessary on the conditions required for the formation of these amines under other forms of processing and storage.

5 References

ARNOLD, S H and BROWN, W D. Histamine (?) toxicity from fish pro-
1978   ducts. *Adv. Fd. Res.*, 114–147

GEHRKE, C W, KUO, K C and ELLIS, R L. Polyamines–an improved
1977   automated ion-exchange method. *J. Chromatog*, 143, 345–361

HUGHES, R B. Chemical studies on the herring (*Clupea harengus*). II.
1959   The free amino acids of herring flesh and their behaviour during post-mortem storage. *J. Sci. Fd. Agric.*, 10, 558–564

INUKAI, K. Rate of spoilage of frozen mackerel after thawing. *Mie*
1976   *Igaka*, 20, 19–24

KARMAS, E and MIETZ, J L. Polyamine and histamine content of tuna
1978   fish and the relationship to decomposition. *Lebensm. Wiss. u Technol.*, 11, 333–337

KIZEVETTER, I V and NASODKERA, E A. The problems of histamine
1972   accumulation in the body tissues of Pacific Ocean Scomber. *Tikhookeanskii Nauchnoissledovatel'skii*, Institut Rybnogo Khozyaistva i Okeanografii, 83, 27–34

MACKIE, I M and THOMSON, B W. Decomposition of trimethylamine
1974   oxide during iced and frozen storage of whole and comminuted tissue of fish. *Proc*: IV Int. Congress Fd. Sci. and Technol., 1, 243–250

SALGUERO, J F and MACKIE, I M. Histidine metabolism in mackerel
1979   (*Scomber scombrus*). Studies on histidine decarboxylase activity and histamine formation during storage of flesh and liver under sterile and non-sterile conditions. *J. Fd. Technol.*, 14, 131–139

SHEWAN, J M, GIBSON, D M and MURRAY, C K. The estimation of
1971   trimethylamine. In: *Fish Inspection and Quality Control*, Ed R Kreuzer. Fishing News (Books) Ltd, London. 183–186

SHORE, P A. *Methods of Biochemical Analysis.* Analyses of biogenic
1971   amines and their related enzymes, Ed D Glick. Wiley-Interscience. (Supplemental Volume) 89–97

SINELL, H J. Biogene Amine ols RisiKofaktoren in der Fischhygiene.
1978   *Archiv. für Lebensmittelhygiene*, 29, 201–240

SMITH, J G M, HARDY, R and YOUNG, K W. A seasonal study of
1980   storage characteristics of mackerel. Storage at chill and ambient temperatures. *This volume*

TABOR, H and TABOR, C W. Quantitative determination of aliphatic
1973   diamines and polyamines by an automated liquid chromatography procedure. *Anal. Biochem.*, 55, 457–467

15 Water in fish

Water in fish tissue—a proton relaxation study of *post rigor* minced cod

P J Lillford, D V Jones and G W Rodger

1 Introduction

During frozen storage, fish minces undergo a series of deteriorative changes. Protein extractability decreases, textural changes occur, and water-holding capacity is progressively lost. Since the acceptability of many products incorporating fish mince depends in part on their succulence, this latter property is an important factor. For this reason we carried out a study on the state of the water in several fish minces (raw and cooked) considered to span a range in quality from 'good' to 'bad'; the technique used was pulse relaxation nuclear magnetic resonance.

In heterogeneous foodstuffs, water can be described as existing in a number of discrete domains, represented by discrete relaxation processes. The rate of phase or energy loss exhibited by the nuclei when the exciting pulse of rf energy is switched off, can give information about the interaction of the nucleus' parent molecule with other parts of the matrix. In biological tissues and foodstuffs, the matrix is seldom uniform, and contains pores, channels and even impermeable barriers. The water molecules experience all of these environments and 'sense' them via their relaxation rates which appear as an overlapping set of decay curves.

The development of mathematical treatments to analyse the observed complex decay curves into the components has allowed us to classify the water distribution in terms of the relaxation times for the fish minces studied.

2 Experimental

2.1 *Raw materials*
Samples of three V-cut minces considered to be of good, medium, and poor quality by virtue of their salt solubility levels ($> 80mg\ g^{-1}$, 50–$80mg\ g^{-1}$, and $< 50mg\ g^{-1}$) were chosen for this study. The minces were by necessity from different batches since the solubility difference arose from the minces having been stored at $-20°C$ for various periods.

2.2 *Methods*
2.2.1 *Preparation of samples*   The frozen samples were allowed to reach ambient in sealed NMR tubes and then analysed at $15°C$. The cooking regime entailed immersion of the sealed NMR tubes for 60min in a water bath set at $85°C$.

2.2.2 *NMR measurements and treatment of data*   A Bruker Spectrospin Pulse Spectrometer type SXP operating at 60 MHz was used to measure the spin–spin relaxation time (also referred to as transverse relaxation times and denoted by T_2) by the Carr–Purcell–Meiboom–Gill (CPMG) sequence.

The NMR probe temperature was maintained at $14°C$ with a Bruker temperature control device (dial setting $290°K$) for all the samples. Samples were packed to a height of 0.5–$0.7mm$ in the bottom of 8mm (od) NMR tubes, and left to equilibrate in the probe for at least 20min before recording of data. In all experiments (T_2 measurements) 100 scans were accumulated on an Intertechnique Dedac 800 signal averager to improve signal to noise, and the data displayed on a teletype recorder. For some of the samples the sampling rate from the signal averager was altered with a switching box, so that the whole decay curve including base line could be recorded over a broad line window from $80\mu s$ to 10s.

All the samples studied exhibited complex transverse relaxation behaviour, and the data were analysed by the following methods.

2.2.2.1 *Multiexponential least square analysis*   Using a weighted least squares programme, the various decay curves were fitted to 2, 3, 4 or in some cases even 5 exponential functions. Each exponential is described by two parameters, an amplitude (A) describing its percentage of the total magnetization decay, and a relaxation time (T) describing its rate.

2.2.2.2 *Deconvolution analysis*   Using a method described previously for deconvolution of exponential decay processes (Roessler, 1955), a distribution of relaxation times for each sample was obtained (*Figs 2* and *3*).

3 Results

In all of the samples studied, the decay of proton transverse magnetization could not be described by a single exponential process (a typical decay curve is shown in *Fig 1*). Using a weighted least squares programme, it was found that relaxation processes were capable of providing an adequate fit to the data. The results are given in *Table I*.

By deconvolution analysis, a binodal distribution of relaxation times was observed for all samples (*Figs 2* and *3*).

4 Discussion

Complex water proton relaxation has been reported previously in *post rigor* tissue (Pearson *et al*, 1972; Haslewood *et al*, 1974) but some disagreement remains as to the number of discrete processes describing the relaxation. In this experiment discrete processes were necessary to fit the experimental observations. Deconvolution analysis removes the necessity of assuming a number of discrete exponentials by deriving a relaxation time distribution directly from the experimental observations. This analysis results in the identification of a binodal distribution of relaxation times in all the samples examined (*Figs 2* and *3*).

Comparison of *Table 1* and *Figs 2* and *3* shows that the discrete exponentials lie within the envelope of the distribution function obtained by deconvolution.

The existence of a binodal distribution suggests that the assignment of water with different relaxation times might coincide with its location in intra- or extracellular space, *ie* compartmentalization by the membrane causing a non-time averaged relaxation behaviour. The effect of deterioration then appears to result in the transfer of water across the cell membrane, into extracellular space, resulting in an increased intracellular protein concentration.

Alternatively, aggregation of the myofibrillar component during storage may create sufficiently large spaces in the structure so that exchange between bound and free waters cannot be averaged over the entire sample. Such a mechanism does not require any compartmental-

ization of solutes by intact membranes. Deterioration on storage is then explained simply in terms of increasing aggregation, resulting in increasing number or size of interstitial spaces.

After heating the samples to temperatures well above those required to rupture membranes (85°), the binodal nature of the distribution became more marked, suggesting that the latter mechanism is more appropriate. Furthermore, the three samples were still distinguishable, indicating that the structural reorganization which occurs during frozen storage persists into the cooked state and may account for the noticeable textural differences in stored versus fresh minces.

5 References

HASLEWOOD, C F, *et al*. NMR transverse relaxation time of water 1974 protons in skeletal muscle. *Biophysical J.*, 14, 583
PEARSON, R T, *et al*. An NMR investigation of rigor in porcine muscle. 1972 *Biochem. Biophys. Res. Commun.*, 48, 873
ROESSLER, F C. Some applications of Fourrier Series in numerical 1955 treatment of linear behaviour. *Proc: Physical Society*, 68, 89–96

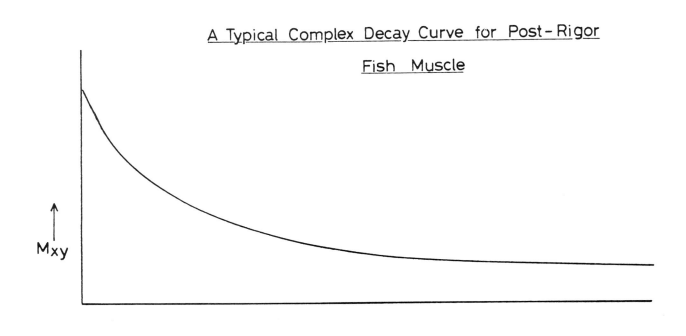

Fig 1 A typical complex decay curve for *post-rigor* fish muscle

Table 1

Sample type	Amplitude %	T_2 ms	Amplitude %	T_2 ms	Amplitude %	T_2 ms
Raw						
(a) Good quality	20·17	248·8	77·38	61·35	2·45	1·59
(b) Medium quality	9·88	373·6	26·82	115·0	63·28	42·26
(c) Poor quality	17·3	394·3	31·5	122·9	51·2	36·29
Cooked						
(a) Good quality	30·68	1·118	21·23	158·9	48·08	38·93
(b) Medium quality	28·77	1·085	18·88	132·0	52·35	33·4
(c) Poor quality	25·34	1·035	24·52	148·8	50·14	37·76

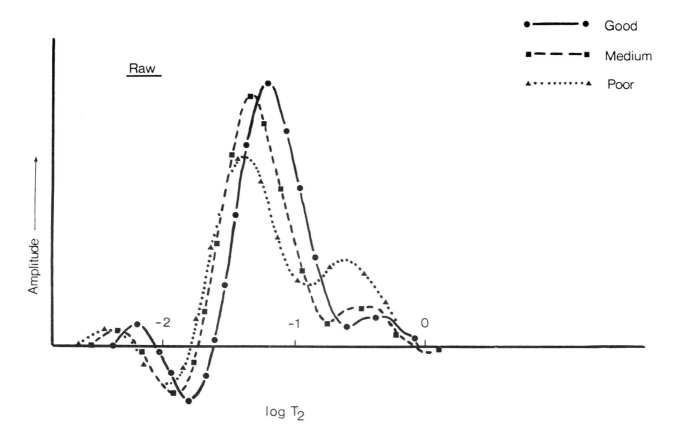

Fig 2 Deconvolution analysis of minces

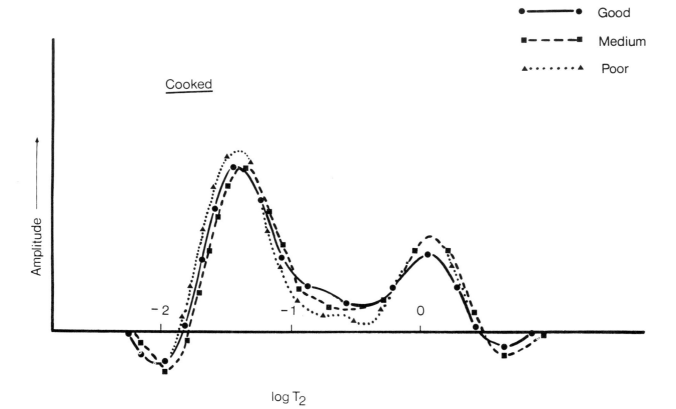

Fig 3 Deconvolution analysis of minces

497

Modes of dehydration of frozen fish flesh

R M Storey

1 Introduction

This paper is principally concerned with the more technical aspects of the dehydration of fish which occurs at subzero temperatures. It is not a review but is based on data obtained during studies of the migration of water within frozen cod flesh, which were undertaken as part of a wider study of the cold storage dehydration of fish and the resultant loss of eating quality.

The water relations of foodstuffs, including fish, are complex and are outside the scope of this paper. It is sufficient to comment that water exists in several states in frozen fish, each state possessing specific physical characteristics (Fennema, 1976). When high water content biological material, such as fish muscle, is cooled below its freezing point, water separates out as ice crystals. The proportion of water converted to ice is dependent on the temperature to which the fish is cooled and on its osmolality. The size and distribution of ice crystals depend upon the rate of freezing and on the *post mortem* state of the fish (Love, 1968). This paper is concerned with the contribution to dehydration of the various states of water on a macro rather than on a micro scale.

After freezing under normal commercial conditions, *ie* cooling to about $-20°C$ in under 2h, frozen fish muscle contains water in three major states: ice, unfrozen water which is freezable and unfrozen water which is not freezable. Storey and Stainsby (1970) showed that even though the liquid phase in frozen fish is ionic and not dilute, the relative lowering of its vapour pressure is directly proportional to the concentration of the solutes present. The unfrozen but freezable water may then be considered as a pseudo-ideal solvent. This work confirmed the observations of Duckworth and Smith (1963a,b) that some of the water present in frozen fish is available for the diffusion of radioactively labelled solutes such as glucose.

Much has been published on the proportion of water converted to ice and on the purity of the ice formed. Riedel (1956) probably published the earliest precise figures, obtained calorimetrically, for the proportion of water converted to ice in frozen fish. He also showed that 8% of the total water present was unfreezable.

Storey and Stainsby also noted that the equilibrium water vapour pressure of cod flesh was the same as that of ice over the range of temperature investigated ($-5°C$ to $-23°C$), which is at variance with Dyer *et al* (1966), Hill and Sunderland (1967) and Karel (1975), who claim that the vapour pressure of a number of flesh foods is up to 20% lower than that of ice at the same temperature, which would indicate that pure ice is not formed on freezing. However, Fennema and Berny (1974) who made very careful measurements of the equilibrium vapour pressure confirmed Storey and Stainsby's observations. In all probability, the ice which is formed when biological material is cooled is not pure, but the concentration of impurities is so low that the vapour pressure of the ice is for all practical purposes equal to that of pure ice (Schneeberger, 1977; Steinbach, 1977).

The fact that some water in biological tissue is unfreezable and therefore does not have the properties of ordinary water has been demonstrated many times from nuclear magnetic resonance studies, for example, by Sussman and Chinn (1966), Cope (1969) and others. Duckworth (1971), using differential thermal analysis, showed that when the water content of cod muscle is reduced below 24% (wet basis) no freezing takes place. This figure agrees closely with Riedel's observation that only 92% of the total water present can be frozen.

Dyer *et al* (1957) found that the initial solute concentration in cod was equivalent to a 1·4% solution of sodium chloride, whereas Jason and Long (1955) found a slightly higher value (1·6%) from the freezing point of cod flesh. Kelly and Dunnett (1969) measured osmolalities in cod which are equivalent to a 1·27–1·43% solution of sodium chloride. Storey and Stainsby (1970), applying Raoult's law to Riedel's data, showed that for the sample of cod from which Riedel obtained his data the initial concentration of solutes is equivalent to 1·34% sodium chloride. Thus when cooled below ice temperature cod behaves similarly to a solution of salt containing about 1·4g of sodium chloride per 100g of water. Storey (1975) developed the simple application of Raoult's law further and showed that the proportion of ice formed at temperatures within the range of interest ($-5°C$ to $-30°C$) could be calculated to be within 1% of Riedel's data, assuming that the tissue fluid contains solutes equivalent to 1·34% of sodium chloride.

Assuming that 8% of the total water of cod flesh cannot be frozen and assuming that the concentration of tissue fluid solutes is equivalent to 1·34% sodium chloride, the values of the proportion of water present in each of the three major water phases given in *Table I* can be calculated from simple Raoult law theory.

Table I
ESTIMATED PROPORTIONS OF WATER IN THE THREE MAJOR PHASES IN FROZEN COD

Temperature °C	Grams of water per 100g of fish		
	Ice	Freezable	Unfreezable
−5	61·4	12·2	6·4
−15	69·9	3·7	6·4
−30	72·2	1·7	6·4

It must be remembered that *Table I* is probably an over-simplification of the real situation. Water is a compound which can assume a variety of structures, particularly in the presence of solutes and protein, the solvent properties of which can change with structural change. The data are presented only to indicate the general pattern of the distribution of water in the three major phases.

No reference has yet been made to the possible freezing out, by precipitation, of tissue fluid solutes due to the concentrating effects of removing water from the system as ice. Van den Berg (1968) suggests that the observed changes in the pH of foods after freezing may be due to the precipitation of buffer salts during freezing. Storey

(1975) suggested that the numerically small differences between the amount of ice found to be present by Riedel and that calculated from simple theory could be explained by the precipitation of tissue fluid solutes. There is very good agreement between the two sets of data down to $-13°C$, below which temperature there is some deviation between the calculated and observed values. The deviation corresponds to about a 10% change in the effective solute concentration at $-25°C$ for example.

2 Dehydration at subzero temperatures

There are three distinct phases of water loss when frozen fish is allowed to dehydrate. They are most clearly demonstrated if the weight loss is plotted against the square root of the drying time, as illustrated in *Fig 1*. The weight loss occurring between A and B is the constant rate phase, B to C represents the first falling rate phase and C to D the second falling rate phase. All three phases appear to occur irrespective of the temperature, of the rate of drying and of the dimensions of the sample. The period of each phase is however dependent on temperature, humidity and air velocity.

2.1 *The constant rate phase*
This phase is typical of constant rate drying in that the rate of drying is determined by the rate at which water vapour can be removed from the surface of the fish. For as long as water can reach the surface of the fish at least as fast as it is removed, the constant rate will be maintained, provided that the external conditions which determine the rate of removal remain constant.

The actual rate of removal of water vapour depends on the difference between the water vapour pressure of the frozen fish and the partial pressure of water vapour in the surrounding air and also on the velocity of the air passing over the fish.

Powell (1939) demonstrated that the laws governing the rate of evaporation of water are applicable to the sublimation of ice. He found that the rate of sublimation is proportional to the difference between saturation water vapour pressure at the temperature of the air and the actual vapour pressure and is also proportional to the air velocity raised to the power 0.62.

Current work in this laboratory has shown that for small sections ($2.2cm$ diameter, $0.1cm$ thick) of cod flesh, the rate of loss of weight in the constant rate phase could be predicted from the following relation:

$$\log R = 0.536 + 0.615 \log U$$

where R is the rate of loss of weight per unit area per unit difference in water vapour pressure (g m^{-2} torr^{-1} h^{-1}) and U is the air velocity in cm s^{-1}, for air velocities above about $13cm$ s^{-1}. The velocity exponent is thus very close to that for ice found by Powell (1939). At air velocities below about $13cm$ s^{-1} the rate of loss of weight was practically independent of air velocity and was dependent solely on vapour pressure difference; it was approximately equal to $17g$ m^{-2} torr^{-1} h^{-1}. Discs of ice of the same dimensions as the fish samples behaved identically to fish during the constant rate phase, although, as expected, they did not exhibit a decreasing rate of sublimation until the surface area had been materially

reduced. Preliminary experiments with samples of fish with skin on indicated that the skin also exhibited a constant rate period, but the rate of drying was only about 60% that of exposed flesh.

The duration of the constant rate period depended on the rate of drying and, after the loss of about $150g$ m^{-2} of exposed surface, the rate of drying began to decrease. This value was not dependent on the rate of drying, or on temperature or on the dimensions of the sample. The loss of $150g$ m^{-2} is equivalent to the loss in weight of a layer approximately $0.02cm$ thick which dehydrated to a water content in equilibrium with the temperature and humidity of the surroundings.

It seems likely that this thickness of dried outer layer limits the removal of water vapour from the subliming ice surface as it recedes from the surface of the fish. Dehydration during constant rate drying may be considered as a form of freeze drying at atmospheric pressure. It is however much slower than freeze-drying; whereas a $0.1cm$ thick sample of biological material might be expected to freeze-dry in under half an hour (Stephenson, 1953), in experimental investigations at $-30°C$ the total time to reach equilibrium was at least a thousand times longer and was some 200 times longer at $-10°C$.

2.2 *The first falling rate phase*
After the loss of about $150g$ m^{-2} of surface the rate of loss of weight began to decrease and subsequently the loss in weight was proportional to the square root of time, as can be seen from *Fig 1*. During this phase of drying the loss of moisture was controlled by the rate at which it would migrate to the surface and was independent of the humidity and air velocity. The approximate relation between weight loss and time is given by the following relation (Storey, 1975):

$$W_0 - W_t = 4(W_0 - W_e)\frac{1}{l}\left[\frac{D_e}{\pi}\right]^{\frac{1}{2}} t^{\frac{1}{2}}$$

where W_0 = initial weight
W_t = weight at time t
W_e = equilibrium weight
l = thickness
t = time
D_e = coefficient of diffusion

Thus a plot of weight loss against the square root of time is a straight line, from the slope of which D_e can be calculated. It has been shown (Storey, 1975) that, to a first approximation, the value of D_e is halved for each 5°C decrease in temperature (*Table II*). The slope given by the above equation is therefore decreased by a factor of about $\sqrt{2}$ for each 5°C decrease in temperature.

The mechanism by which water is transferred to the surface during this phase of drying is thought to be one of molecular diffusion as described by Jason (1958). Evidence for this is suggested by the similarity between Jason's diffusion coefficients obtained for fish drying at above zero temperatures with those obtained at subzero temperatures, after correction for temperature and the amount of water remaining available for diffusion. Instead of the ice subliming and the resultant water vapour diffusing through the dry layer to the surface it is

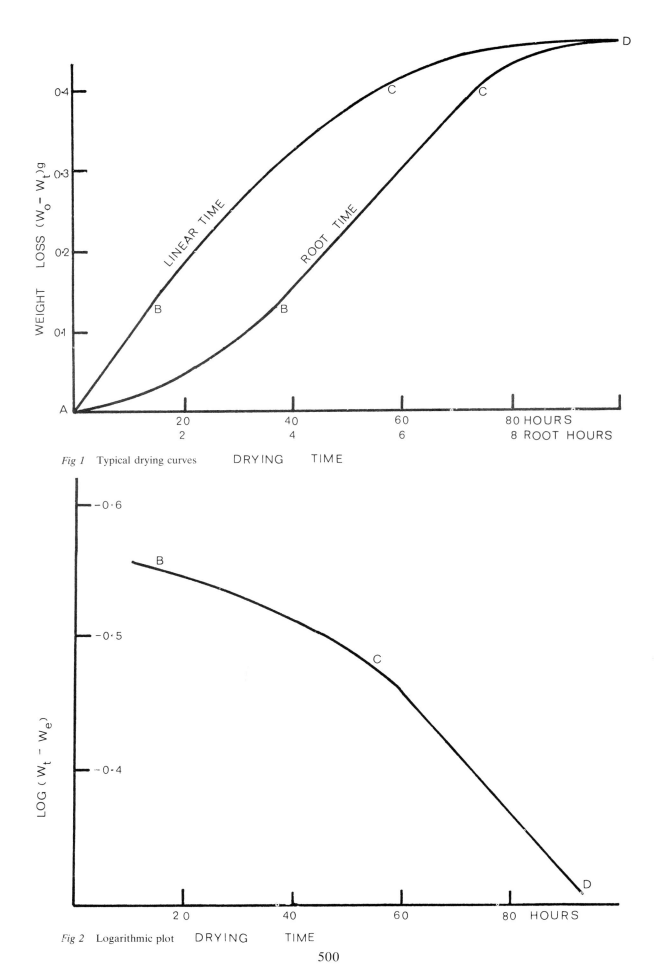

Fig 1 Typical drying curves   DRYING   TIME

Fig 2 Logarithmic plot   DRYING   TIME

500

Table II
EFFECTIVE DIFFUSION COEFFICIENTS

	$D_e \times 10^8\ cm^2\ s^{-1}$			
Temperature °C	From tracer measurements	From first falling rate period	From second falling rate period	After Jason[1]
−5	10·4	7·2	7·7	15·9
−10	5·2	3·6	3·8	6·5
−15	2·6	1·8	1·9	3·0
−20	1·2	0·86	0·89	1·6
−25	0·60	0·40	0·41	0·9

[1] Jason's 1958 data extrapolated to subzero temperatures and multiplied by the fraction of freezable water remaining at each temperature

probable that the ice melts and the water migrates to the surface via a liquid diffusion process. As water is removed from the system the remaining unfrozen tissue fluid will tend to become more concentrated and therefore its activity will tend to fall below that of the ice which, in consequence, will melt to maintain equilibrium. If freeze-drying is carried out at high pressures, Bralsford (1967) suggests that the distinct ice face present in freeze-drying becomes diffuse and that some of the migration of water to the surface is not by vapour diffusion. Further evidence for the proposed hypothesis is obtained from the magnitude of the weight loss achieved at the end of this drying phase. This weight loss is temperature dependent and the practical values are very close to the expected weight loss which would be obtained if all the ice was removed. This ranges from 61·4% at −5°C to 72·2% at −30°C (*Table I*).

2.3 *The second falling rate phase*

When all the ice had melted there was no 'reservoir' of water to maintain the higher, but decreasing, rates of drying obtained in the first falling rate phase. The rate of drying then decreased until equilibrium was attained. The equilibrium water content was dependent on the temperature and relative humidity and ranged from about 15% (wet basis) to over 25%, for rh in the range 40–85%. However this portion of the study is not yet complete.

For thin discs of fish the rate of drying in the second falling rate can be predicted from the following relation (Jason, 1958).

$$\text{Log}\ (W_t - W_e) = A - 0.4343\ \frac{\pi^2}{4} \cdot \frac{D_e}{C^2} t$$

where W_t = weight at time t
W_e = weight at equilibrium
A = constant
D_e = coefficient of diffusion
C = half thickness
t = time

As for the first falling rate phase, the values of D_e calculated from the linear plots of log $(W_t - W_e)$ and t (*Fig 2*) showed similarity to those extrapolated from Jason's data. This suggests that the mechanism of diffusion in this phase is also similar to that described by Jason.

Storey (1975) found similar values for D_e using deuterated water as a tracer in self-diffusion experiments with frozen cod muscle. The three observed effective diffusion coefficients are compared with Jason's extrapolated data in *Table II*. From *Table II* it can be

seen that the effective diffusion coefficients calculated from Jason's data are about twice as high as those observed. This may mean that only about half of the freezable water is available for diffusion.

3 Practical implications

3.1 *Constant rate drying*

Although freeze drying at atmospheric pressure is much slower than under reduced pressure, it is sufficiently rapid to have warranted consideration as a practical method of dehydration (Lewin and Mateles, 1962; Guoigo *et al*, 1974). This fact illustrates the speed at which dehydration can occur during cold storage if precautions are not taken to protect frozen stored products.

During constant rate drying the humidity and the velocity of air movement over the frozen product determine the rate of dehydration up to a loss of about 150g m^{-2} of exposed surface. Even at −29°C, Graham and Roger (1978) found that weight losses from exposed fish flesh in cold stores could be as high as 13·5g m^{-2} day^{-1}. Under these conditions constant rate drying would be completed in about 10 days. Generally, rates of drying were found to be higher in forced air cooled stores than in grid cooled stores. To obtain these high rates of dehydration it is not necessary for the humidity to be particularly low. The calculated relative humidities for a weight loss of 13·5g m^{-2} day^{-1} for a range of air velocities are given in *Table III*.

In contrast, if the air velocity had been of the order of 13cm s^{-1} at 97% rh the rate of drying would have been about 3·8g m^{-2} day^{-1}, or less than one-third of the rate at the higher velocity.

Table III
RELATION BETWEEN rh AND AIR VELOCITY FOR A WEIGHT LOSS OF 13·5g m^{-2} DAY^{-1}

Air velocity cm s^{-1}	Relative humidity %
100	97·0
50	95·4
25	92·8
13	89·3

Thus air velocities over the product should be kept to a minimum and the rh at a maximum. Both of these parameters will be dependent on cold store design and management. These comments apply to unprotected products and those protected by glaze, since the glaze will sublime at the same rate as ice from the fish under the same conditions. A simple protective cover to reduce the local air velocity over the product should be effective, and should also result in the generation within the cover of a microclimate of higher rh which should further reduce rates of drying.

The maximum loss in weight which can occur during constant rate drying, expressed as a percentage loss corresponding to the loss of 150g m^{-2} from the surface, is dependent on the surface area to volume ratio of the product and ranges from about a 10% loss for a 1cm cube to a 0·1% loss for a 1m cube.

3.2 *Falling rate drying*

The storage temperature and thickness of the product

become of greater importance than either the humidity or the air velocity in determining the time to lose a given amount of weight in falling rate drying. For example, during both the first and second falling rate period the diffusion of water to the surface of the product will be about four times faster at $-20°C$ than at $-30°C$, and about 20 times faster at $-10°C$. The degree of dehydration will be severe by the time the product has reached the end of the first falling rate period, having lost 60–70% in weight depending on the storage temperature, and therefore drying in the second falling rate period is probably academic from a practical point of view.

3.3 Packaged products

It can be assumed that the relative humidity within the voids of a sealed packaged product will be 100% and if such a product is stored at a constant temperature no dehydration can occur. If, however, the temperature of the product fluctuates, dehydration will take place. During each cycle of fluctuation there is a period when the wrapping is lower in temperature than the product, and water sublimed from the surface of the product at the higher temperature will condense on the inner surface of the wrapping. When the reverse is the case frost sublimes from the wrapping and forms on the surface of the product. The degree of dehydration depends not only on the storage temperature but also on the frequency and amplitude of the temperature fluctuation. A $1°C$ fluctuation at $-10°C$ has about a seven times greater effect on dehydration than $1°C$ at $-30°C$. The relative volume of voids is also important, for example small (60g) plaice fillets when packed loosely in a home freezer pack of 2kg in a polyethylene bag and stored in a chest freezer fluctuating between $-10°C$ and $-20°C$ once every 24h can lose as much as 6% in weight in four months (unpublished data).

4 References

BRALSFORD, R. Freeze drying of beef. 2. Measurement of freeze drying rates of beef. *J. Fd. Technol.*, 2, 353–363
1967

COPE, F W. Nuclear magnetic resonance evidence using D_2O for structured water in muscle and brain. *Biophysical J*, 9, 303–319
1969

DUCKWORTH, R B and SMITH, G M. *Diffusion of solutes at low*
1963a *Biochemistry and Biophysics in Food Research*, Eds J M Leitch and D N Rhodes. Butterworths, London. 230–238

—— The environment for chemical change in dried and frozen foods.
1963b *Proc. Nutr. Soc.*, 22, 182–189

—— Differential thermal analysis of frozen systems. I. The determi-
1971 nation of unfreezable water. *J. Fd. Technol*, 6, 317–327

DYER, W J, FRAZER, D I, ELLIS, D G and MACCALLUM, W A. Influ-
1957 ence of intermittent short storage periods at 15°F on quality of frozen cod stored at 0°F. *J. Fish. Res. Bd. Can.*, 14, 627–635

DYER, D F, CARPENTER, D K and SUNDERLAND, J E. Equilibrium vap-
1966 our pressure of frozen bovine muscle. *J. Fd. Science*, 31, 196–201

FENNEMA, O R and BERNY, L A. Equilibrium vapour pressure and
1974 water activity of food at sub-freezing temperatures. *Proc:* Int. Inst. of Food Sci. and Technol. Symp., Madrid

FENNEMA, O R. Water and Ice. Principles of Food Science Part 1.
1976 *Food Chemistry*. Marcel Dekker, New York. 792pp

GRAHAM, J AND ROGER, A. Measured weight loss in cold stores *Proc:*
1978 IIR Symposium on cooling freezing storage and transport—biological and technical aspects, Budapest. 95

GUOIGO, E I, KAOUKTCHECHVILI, E I, SAVCHENKO, A F and
1974 SIDOROVA, N D. Lyophilisation sous pression atmosphérique des matières premières d'origine animale. *Proc:* IIR symp. on the thermophysical properties of foodstuffs, Bressanone, Italy. 137

HILL, J E and SUNDERLAND, J E. Equilibrium vapour pressure and
1967 latent heat of sublimation for frozen meats. *Fd. Technol.*, 21, 112–114

JASON, A C and LONG, R A K. The specific heat and thermal conduc-
1955 tivity of fish muscle. *Proc:* Ninth IIR Int. Conf. of Refrig. 160–169

JASON, A C. A study of evaporation and diffusion processes in the
1958 drying of fish muscle. Fundamental aspects of the dehydration of foodstuffs. *Soc. of Chem. Industry*, London. 103–135

KAREL, M. Heat and mass transfer in drying. *Freeze Drying and*
1975 *Advanced Food Technology*, Eds S A Goldblith, L Rey and W W Rothmayr. Academic Press, New York

KELLY, T R and DUNNETT, J S. The effect of low temperature freezing
1969 on quality changes in cold stored cod. *J. Fd. Technol.*, 4, 105–115

LEWIN, L M and MATELES, R. Freeze drying without vacuum: a pre-
1962 liminary investigation. *Fd. Technol.*, 16, 94–96

LOVE, R M. Ice formation in frozen muscle. *Low Temperature Biology*
1968 *of Foodstuffs*, Ed J Hawthorne. Pergamon Press, Oxford. 105–124

POWELL, R W. On the rate of sublimation of ice. *Proc:* Brit. Assoc.
1939 Refrig., 36, 61–65

RIEDEL, L. Calorimetric investigations into the freezing of fish. *Kal-*
1956 *tetechnik*, 8, 374–377

SCHNEEBERGER, R, VOILLEY, A and WEISSER, H. Activity of water
1977 below 0°C. *Proc:* IIR Symp. on freezing, frozen storage and freeze drying of biological materials and foodstuffs, Karlsruhe. 73–85

STEINBACH, G. Phase equilibria in frozen solutions from refractometric
1977 measurements of freezing curves. *Proc:* IIR Symp. on freezing, frozen storage and freeze drying of biological materials and foodstuffs, Karlsruhe, 53–66

STEPHENSON, J L. Theory of the vacuum drying of frozen tissues. *Bull*
1953 *of Math. Biophysics*, 15, 411–429

STOREY, R M. The migration of water in frozen cod muscle. M Phil.
1975 Thesis, Procter Department, University of Leeds

STOREY, R M and STAINSBY, G. The equilibrium water vapour pressure
1970 of frozen cod. *J. Fd. Technol.*, 5, 157–163

SUSSMAN, M V and CHINN, L. Liquid water in frozen tissue. Study by
1966 nuclear magnetic resonance. *Science*, 151, 324–325

VAN DEN BERG, L. Physicochemical changes in foods during freezing
1968 and subsequent storage. *Recent Advances in Food Science*. Pergamon Press, London. Vol 4, 205–219

16 Histology

The histology of blue whiting

C K Murray and D M Gibson

1 Introduction

Blue whiting, a gadoid species, caught in deep water of 300–400m, occurs in such abundance that a fishery could be developed and would help compensate for the shortfall in landings of species traditionally used as food in the United Kingdom. It has not been used in the past for human consumption to any significant extent because of the availability of easier supplies and of species easier to process.

The catching season is short, February–April, when the fish are in shoals and spawning March–May; thus, in common with other species, at spawning they can exhibit wide variation in intrinsic condition. The fish are generally small and can be extremely soft.

It was evident from earlier catching operations that two main factors had to be contended with.

First, best catching rates were obtained when fish were in a poor nutritional condition, edible muscle yields also being low. Secondly, the texture of the muscle was very poor and damage during handling and processing could be excessive. Redesigning of filleting equipment resulted in improved yields and new netting techniques are being sought to improve catching rates, particularly during the season when the fish are in their prime but their numbers are more sporadic.

This investigation is concerned with the microscopic examination of the edible parts of blue whiting caught at different times, with the object of gaining an insight into the difficulties facing the processor. A detailed understanding of the mechanical properties of the material does require a knowledge of the micro-structure of the tissue and in particular the arrangements of its structural components, both within themselves and their relationships with the other components present. It is worth noting, of course, that the structure of the fish has evolved to be the most appropriate for the fish during its life and not to be the most suitable as food for another species. When compared with other food fishes, the environmental pressure on the specimens is high, over 40 atmospheres, and must contribute to the structure in some way. Previous histological work in this laboratory has been done on cod and plaice and highlights the structural differences between species probably resulting from the environmental effect. For example, although both cod and plaice exhibit well developed skin components, epidermis, dermis and subcutis, the epidermis of plaice, a bottom feeder, excretes copious quantities of mucus containing a number of complex compounds such as lysozyme (Murray and Fletcher, 1976), an antibacterial agent, whilst in cod, mainly a mid-water feeder, the epidermis has very little evidence of mucus.

The muscle of both of these species however is similar in composition although different in myomere architecture.

Apart from the skeleton, collagen is the main structural element in the soft tissue. Generally, collagen is synthesized by the fibroblasts and is excreted into the extracellular space where it is polymerized into a durable long-lived material. There are different types of intra- and intermolecular cross-links. Mohr (1971) has studied the collagen in cod in some detail. He has shown that intermolecular cross-links determine the mechanical properties of the tissue and relate to the tensile strength but reduce the extensibility of collagenous tissues. The skin of cod requires mechanical strength and is thus more highly cross-linked than the muscle.

Textural changes in blue whiting, like in most other fish, are generally the result of the biological or physiological state of the fish prior to catching and, of course, the effect of handling and storage procedures subsequent to catching. While nothing can be done about the prior to catching condition, it is considered that monitoring the tissue structure histologically during controlled storage experiments will explain and help reduce some of the unwanted changes by suggesting improved handling procedures.

2 Experimental

2.1 *Fish*

The fish examined during these experiments were caught by the RV *G A Reay* off the Faroe Islands at various times within the period June 1977 and June 1978. The experiments were designed to catch and examine fish in various stages of development.

The general appearance of the blue whiting was as follows:

June 1977   North Faroes (62°10′N, 04°15′W)
The fish caught in this period were mainly post-spawned, their thickness varied from moderately thin to extremely thin, 33cm to 29cm in length and around 2.5cm thick, when measured between the front of dorsal fin and directly below. One or two fish caught appeared to be in much better condition with deep conformation of the belly. The viscera and the organs of these fish appeared healthy, normal in size and infestation due to nematodes was considerably less than in the viscera of the thin spent fish. The spent fish had extremely wasted livers which, on histological examination, showed very poor cellular detail with a complete absence of fat when stained with Sudan III. The livers of the well-formed fish contained large quantities of fat.

February 1978   Northeast Faroes (62°16′N, 3°52′W)
All the fish examined were from one haul which yielded 4 tonnes. They were uniform in size, 31cm to 34cm long and 6cm to 7cm thick and appeared healthy, in good condition with no wasting of the muscle, and their livers were rich with oil; infestation with nematodes was considerably less than in the spawned fish described above.

Most of the specimens were females with their ovaries in various stages of maturation. Some fish showed bruising, probably as a result of handling after landing the net, a number having obvious net marks.

June 1978   Northwest Faroes (60°32′N, 06°05′W)
The fish examined during this sea voyage were predominantly males (61%), post-spawned, spent and in poor condition. There were occasionally a few large well-formed fish; autopsy revealed them to be non-spawned. In the spent fish the flesh was soft, the liver poor, in some almost non-existent, and the level of nematode infestation generally low. The state of the gonads indicated stages of development from recently spawned to just recovering from spawning. There was very little food in the stomachs.

2.2 *Sampling*
Blocks of skin and muscle approximately 20mm × 10mm × 10mm were removed from the same region (*Fig 1*) immediately after catching and rapidly frozen to a metal specimen chuck holder of a Pearse-Slee Cryostat type HR, by immersion in liquid nitrogen. Tissue Tek II (R A Lamb) was used to mount the tissue block on to the holder. Frozen sections, 15–20μ were cut from these blocks, mounted on coverglasses coated with an adhesive mixture of albumin and glycerol, stained, dehydrated in graded ethanols and permanently mounted in synthetic resin.

Blocks and sections were also prepared from fish stored for 4h, 8h, 12h, 24h and 264h under the following conditions:

(*a*) without ice in ambient temperatures of 8·5°C and 15°C
(*b*) in ice in a chill room at 4°C ambient
(*c*) in iced sea water

The above techniques were performed at sea on board the RV *G A Reay*.

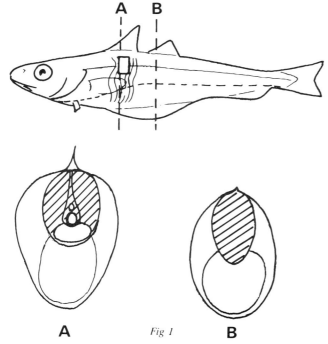

A   *Fig 1*   B

2.3 *Staining procedures*
Sudan III (BDH): 1% in propylene glycol, filter before use. Staining must be carried out on coverglass mounted thick frozen section (50μ).
   (*1*) Stain in Sudan III for 30s to 1min
   (*2*) Wash very carefully in water to remove excess stain
   (*3*) Mount on a microscope slide using Aquamount (Searle) Fat−orange/red.

Weigert Haematoxylin−Van Gieson supplied by Searle and R A Lamb: Mix equal volumes of Weigert Haematoxylin A and Weigert iron solution B immediately before use:

   (*1*) Stain in the mixture for 1min
   (*2*) Wash carefully in distilled water 15–30s
   (*3*) Counterstain with Van Gieson stain 30s
   (*4*) Wash rapidly in distilled water
   (*5*) Dehydrate in graded alcohols, clear in xylene and mount.

Nuclei−blue/black, collagen fibres−red, muscle−yellow/brown. Toluidine blue − borate supplied by BDH: 0·5% Toluidine blue in 0·5% sodium tetraborate (aqueous).

   (*1*) Stain sections in Toluidine blue−borate 30s
   (*2*) Differentiate in 70% ethanol 30s–1min
   (*3*) Dehydrate in alcohol, clear in xylene and mount.

Giemsa Stain (extra rapid) supplied by R A Lamb: Working solution freshly prepared from stock by diluting 1 part of stain with 20 parts water.

   (*1*) Stain section in working solution 10min
   (*2*) Carefully rinse with distilled water 5s
   (*3*) Differentiate carefully and rapidly in graded ethanols, clear in xylene and mount.

3 Results and discussion

3.1 *General condition*
Fish caught in February 1978 can be regarded as being in the best available condition, especially compared to those caught in June 1977. They also were less infested. The difference in the level of nematode infestation, considerably lower in February 1977 than in June 1977, could be due to two things. Perhaps healthy fish in good condition can resist infestation better than very soft spawned fish or, some shoals of blue whiting may simply be less parasitized than others. Fillets taken from well-formed fish had a normal translucent appearance as in cod or haddock with little formation of drip initially. Fillets from thin spent fish were extremely soft with an opaque milky appearance and formed a considerable amount of drip.

3.2 *Histology of skin*
There seems to be little difference in the skin of pre- and post-spawned fish. Although there is evidence of an epidermis in blue whiting it was impossible with the techniques used to demonstrate it in its entire form. It is extremely fragile, being composed mainly of spindle-

shaped atypical epithelia when stained with Giemsa. The dermis, as with gadoids, is made up of several layers of parallel collagen fibres containing poorly developed scales. The basement membrane is scant and the hypodermis, thin but well developed, is composed of loose connective tissue containing many chromatophores and refractile plates.

3.3 Histology of muscle

Figure 1 is a sketch of blue whiting showing the area from which the muscle block was sampled. The hatched area shows the reduction in muscle volume found in spent fish when compared with well-formed fish.

Sections from muscle blocks taken from spent fish demonstrate marked loss of myofibril components. The perimysium had separated and in some cases been replaced by ground substance, the endomysium is almost non-existent (*Plate 1*). These observations are even more pronounced in muscle sections from spent fish which had been stored in ice and iced sea water. The water-binding or holding capacity of the muscle is very poor in such fish.

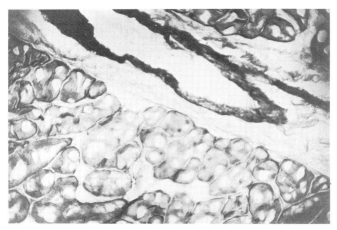

Plate 1 Newly caught blue whiting – poor nutritional condition. (Mag × 100)

Plate 2 Blue whiting – pre-spawned. (Mag × 200)

In the muscle sections of newly caught fish of good intrinsic condition the myofibril bundles are regular, closely packed, the cellular architecture is complete with few extracellular spaces (*Plate 2*), and the water-binding capacity of the tissue appears to be high. However, in a few of these fish, which had been stored in iced sea water (temperature approximately −1°C) grittiness could be felt at the point of the knife when cutting out tissue blocks. Thin sections from these blocks showed large tissue voids, probably the remains of large ice crystals (*Plate 3*). Such voids can be induced by allowing tissue to freeze very slowly at a few degrees below 0°C thus facilitating the formation of large ice crystals, which on melting leave behind their impression. However, the iced sea water temperature is a little too high for this to occur in normal fish tissue.

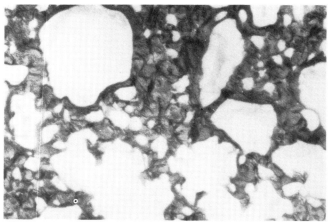

Plate 3 Pre-spawned blue whiting – ovaries well developed. Fish stored whole ungutted for 8h in iced sea water. Demonstrates the remains of large extracellular ice crystals. (Mag × 150)

These findings therefore prompted the histological examination of muscle from fish which had been stored in iced sea water for various lengths of time. While some muscle samples contained large voids, a few had smaller spaces and less cellular damage (*Plate 4*). It was noted that the gonads, particularly the ovaries of the females, of the fish exhibiting large ice crystals were well developed while those with poorly developed gonads exhibited no large void formation. An explanation may be that, because of the poor nutritional status in these fish and the depletion of muscle protein, a result of gonad development, there exists a low water-binding capacity which more readily allows the release of water

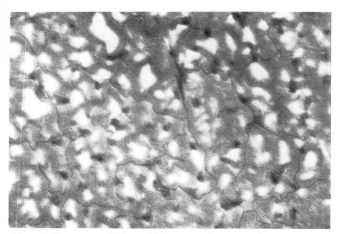

Plate 4 Pre-spawned blue whiting – ovaries poorly developed. Fish stored in iced sea water – whole ungutted for 8h. Demonstrates the remains of small ice crystals. (Mag × 150)

into the surrounding tissue where it accumulates. For water to freeze under these conditions it is inferred that there is free hypotonic water in the tissue, perhaps water of hydration released from the protein matrix on the onset of *rigor mortis*, and it freezes before equilibration with other tissue fluids. These fish were found to be very soft after thawing when compared to unfrozen fish of a similar storage period in ice. Even allowing for artefacts caused by the techniques applied and extrapolating from thin section to whole fish, it is possible that a large proportion of the edible part of the fish has been disrupted by the release of water from the tissue, the energy source being the contractile process of *rigor mortis*. After the resolution of *rigor mortis* some water may be taken up by the proteins more exposed to their environment (*Plates 5* and *6*) but the presence of large amounts of free water would certainly contribute to the softness of blue whiting tissue. The flesh of immediately *post rigor* fish was easily marked on palpation and of course at this point no bacterial spoilage had occurred.

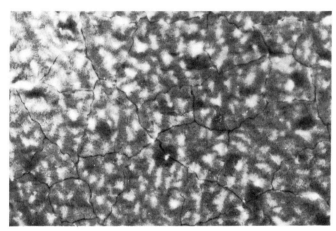

Plate 5 Pre-spawned blue whiting – whole ungutted fish. Stored in ice for 11 days (264h). Cell architecture almost returned to normal. (Mag × 150)

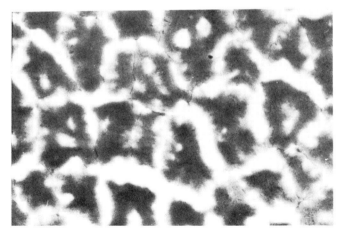

Plate 6 Pre-spawned blue whiting – ovaries moderately developed. Fish stored whole ungutted in iced sea water for 11 days. Demonstrates the remains of extracellular water. (Mag × 150)

3.4 *Water-binding capacity*
Wateriness in the white muscle of cod (*Gadus morhua*) and jelly cat (*Lycichthys denticulatus*) has already been investigated histologically (Love and Lavéty, 1977). They compared starved cod with normal jelly cat both having similar water content. Their findings suggested that it was less important for jelly cat to be tightly packed with contractile fibres than for the muscle of cod, the former probably being an adaptation to a permanent state of unusually high hydration.

Tanaka (1969) demonstrated the formation of intra- and extracellular ice crystals in the muscle of stored frozen Alaska pollack prepared from pre- and post-rigor fresh fish and pre- and post-spawned fish. The histological techniques in both these papers involved the use of fixatives while those on blue whiting in this work were on cryostat sections. Although both methods are subject to artefacts, the findings are remarkably similar. The histology of muscle blocks taken from pre- and post-spawned blue whiting demonstrated a similar intracellular ice formation to that found in pre- and post-spawned Alaska pollack, which Tanaka showed, in pollack, was related to the water content of the muscle. It would have been useful to have measured the water content of the blue whiting. However, as these results were not anticipated and the work was being carried out at sea, this was not possible on the samples taken.

3.5 *Collagen and increased 'macrophage' activity in spent fish*
In addition to being part of the 'gel' of water and muscle protein, collagen plays an important role in maintaining the structure of the fish. In blue whiting which had not spawned but which were in poor intrinsic condition, the principal collagen bands were relatively intact (*Plate 7*), whereas in fish of very poor condition after spawning there appeared to be marked changes in the collagen distribution. It is possible that with the dissolution of myofibril proteins and associated substances as a source of energy and nutrient, particularly during gonad development, the remaining non-utilized collagen may alter in structure. For example, changing solubility and structure *in situ* could help explain the concentration of collagen observed in the perimysium regions where it could remain in readiness for utilization during tissue regeneration. Evidence from other investigators (Traub and Piez, 1971) suggests that water plays an extremely important role in collagen structure and that in certain circumstances can destabilize the molecular fold formation. The stability of collagen seems dependent on water

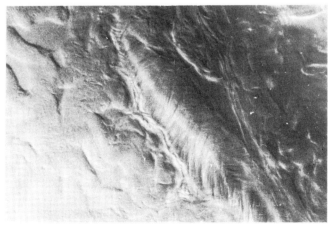

Plate 7 Blue whiting – pre-spawned showing regular striations occurring in the collagen of the perimysium. (Mag × 200)

molecules through intramolecular hydrophobic bonds, 'possibly by providing links between polar groups. Water appears to both stabilize and destabilize the collagen structure through interactions with specific groups, the amino acid sequence determining the contribution to stability of the water interactions, as well as those due to restrictions on rotational freedom' (Traub and Piez, 1971). This evidence, however, is from the study of *in vitro* experiments and makes interpretation as to how collagen functions *in vivo* under changing physiological conditions difficult. Nevertheless, in the light of these observations, alterations to the intimate involvement of water and collagen in the muscle of spent blue whiting, with its dramatic changes in water-binding capacity, could explain the apparent reorganization of collagen seen in cryostat sections of muscle blocks (*Plate 8*). Associated with this there is considerable cellular activity in tissue and collagenous regions surrounding depleted myofibres and these cells could be phagocytosing waste debris. In other species, such as cod, energy for locomotion and gonad development is produced at the expense of the musculature during starvation. *Plate 9* shows a ghost-like structure, the result of depletion of myofibril components in the muscle of post-spawned blue whiting. The structures are quite large at times, indicating that probably more than one muscle segment has been utilized. However, only a faint single collagenous membrane is visible. Again this suggests that the collagen may have reorganized to encompass the remaining ground substance. Changes in the structure of the muscle due to spoilage are minor compared with nutritional status. It is therefore concluded that, prior to and after spawning, the requirements for ovarian maturation at the expense of body protein greatly affect the texture of the edible muscle of blue whiting and thus the ease of processing and the desirability of the fish as food.

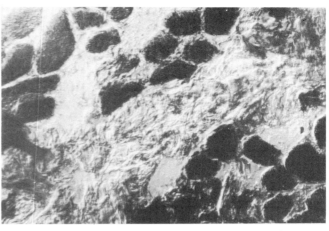

Plate 8 Blue whiting – post-spawned. Showing apparent alterations to structure and organization of collagen in the perimysium region. (Mag × 150)

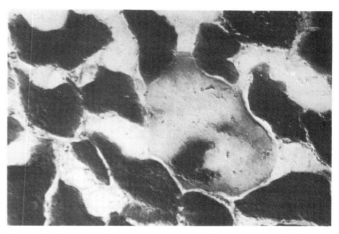

Plate 9 Blue whiting – post-spawned. Showing 'ghost' area of depleted myofibre with small area of remaining protein. Toluidine blue (Mag × 150)

4 References

LOVE, R M and LAVÉTY, J. Wateriness of white muscle: a comparison
1977   between cod (*Gadus morhua*) and jelly cat (*Lycichthys denticulatus*). *Marine Biol.*, 43, 117–121
MOHR, V. On the constitution and physical-chemical properties of the
1971   connective tissue of mammalian and fish skeletal muscle. PhD Thesis, University of Aberdeen
MURRAY, C K and FLETCHER, T C. The immunohistochemical localisa-
1976   tion of lysozyme in plaice (*Pleuronectes platessa* L.) tissues. *J. Fish. Biol.*, 9, 329–334

TANAKA, T. Relationship between freshness before freezing and cold
1969   storage deterioration in North Pacific Alaska pollack. *Bull. Tokai Reg. Fish. Res. Lab.*, No. 66, 143–168
TRAUB, W and PIEZ, K A. Chemistry and structure of collagen.
1971   *Advances in Protein Chemistry*, 25, 243–352

507

Index

Other books published by Fishing News Books Limited, Farnham, Surrey, England.

Free catalogue available on request

Advances in aquaculture
Aquaculture practices in Taiwan
Atlantic salmon: its future
Better angling with simple science
British freshwater fishes
Commercial fishing methods
Control of fish quality
Culture of bivalve molluscs
Echo sounding and sonar for fishing
The edible crab and its fishery in British waters
Eel capture, culture, processing and marketing
Eel culture
European inland water fish: a multilingual catalogue
FAO catalogue of fishing gear designs
FAO catalogue of small scale fishing gear
FAO investigates ferro-cement fishing craft
Farming the edge of the sea
Fish and shellfish farming in coastal waters
Fish catching methods of the world
Fish inspection and quality control
Fisheries of Australia
Fisheries oceanography
Fishermen's handbook
Fishery products
Fishing boats and their equipment
Fishing boats of the world 1
Fishing boats of the world 2
Fishing boats of the world 3
The fishing cadet's handbook
Fishing ports and markets
Fishing with electricity
Fishing with light
Freezing and irradiation of fish
Handbook of trout and salmon diseases

Handy medical guide for seafarers
How to make and set nets
Inshore fishing: its skills, risks, rewards
The lemon sole
A living from lobsters
Marine pollution and sea life
The marketing of shellfish
Mending of fishing nets
Modern deep sea trawling gear
Modern fishing gear of the world 1
Modern fishing gear of the world 2
Modern fishing gear of the world 3
More Scottish fishing craft and their work
Multilingual dictionary of fish and fish products
Navigation primer for fishermen
Netting materials for fishing gear
Pair trawling and pair seining – the technology of two boat fishing
Pelagic and semi-pelagic trawling gear
Planning of aquaculture development – an introductory guide
Power transmission and automation for ships and submersibles
Refrigeration on fishing vessels
Salmon and trout farming in Norway
Salmon fisheries of Scotland
Seafood fishing for amateur and professional
Stability and trim of fishing vessels
The stern trawler
Textbook of fish culture: breeding and cultivation of fish
Training fishermen at sea
Trout farming manual
Tuna: distribution and migration
Tuna fishing with pole and line